RS**Means**®

Light Commercial Cost Data

17th Annual Edition

1998

Senior Editor

John H. Ferguson, PE

Contributing Editors

Thomas J. Akins
Barbara Balboni
Howard M. Chandler
John H. Chiang, PE
Paul C. Crosscup
Jennifer L. Fordyce
Robert W. Mewis
Melville J. Mossman, PE
John J. Moylan
Jeannene D. Murphy
Peter T. Nightingale
Jesse R. Page
Stephen C. Plotner
Michael J. Regan
Kornelis Smit
William R. Tennyson, II
Phillip R. Waier, PE
James N. Wills
Rory Woolsey

Manager, Engineering Operations

John H. Ferguson, PE

President

Durwood S. Snead

Vice President and General Manager

Roger J. Grant

Vice President, Sales and Marketing

John M. Shea

Vice President, Operations

Andrew J. Centauro

Production Manager

Michael Kokernak

Production Coordinator

Marion E. Schofield

Technical Support

Michele S. Able
Wayne D. Anderson
Thomas J. Dion
Michael H. Donelan
Paul C. Hebert
Gary L. Hoitt
Mark H. Kaplan, Jr.
Marla A. Marek
Paula Reale-Camelio
Kathryn S. Rodriguez
Ali Vaghar

Art Director

Helen A. Marcella

Book & Cover Design

Norman R. Forgit

Editorial Advisory Board

James E. Armstrong
Energy Consultant
EnergyVision

William R. Barry
Chief Estimator
Mitchell Construction Company

Robert F. Cox, PhD
Assistant Professor
ME Rinker Sr. School of Bldg. Constr.
University of Florida

Roy F. Gilley, AIA
Principal
Gilley-Hinkel Architects

Kenneth K. Humphreys, PhD, PE, CCE
Secretary-Treasurer
International Cost
Engineering Council

Patricia L. Jackson, PE
Vice President, Project Management
Aguirre Corporation

Martin F. Joyce
Vice President, Utility Division
Bond Brothers, Inc.

First Printing

Foreword

R.S. Means Co., Inc. is a member of the **Construction Market Data Group,** a leading provider of construction information in North America. The group's other companies, besides **R.S. Means,** include **Architects' First Source,** the most extensive and widely circulated product selection resource for specifiers; **Construction Market Data (CMD),** the source for construction project information for sales leads, market analysis and marketing; **Manufacturers' Survey Associates,** a leading estimating and quantity survey company in the U.S.; and **BIMSA/Mexico,** the dominant distributor of information on building projects and construction costs throughout Mexico.

Our Mission

Since 1942, R.S. Means Company, Inc. has been actively engaged in construction cost publishing and consulting throughout North America.

Today, over fifty years after the company began, our primary objective remains the same: to provide you, the construction and facilities professional, with the most current and comprehensive construction cost data possible.

Whether you are a contractor, an owner, an architect, an engineer, a facilities manager, or anyone else who needs a quick construction cost estimate, you'll find this publication to be a highly useful and necessary tool.

Today, with the constant flow of new construction methods and materials, it's difficult to find the time to look at and evaluate all the different construction cost possibilities. In addition, because labor and material costs keep changing, last year's cost information is not a reliable basis for today's estimate or budget.

That's why so many construction professionals turn to R.S. Means. We keep track of the costs for you, along with a wide range of other key information, from city cost indexes . . . to productivity rates . . . to crew composition . . . to contractor's overhead and profit rates.

R.S. Means performs these functions by collecting data from all facets of the industry, and organizing it in a format that is instantly accessible to you. From the preliminary budget to the detailed unit price estimate, you'll find the data in this book useful for all phases of construction cost determination.

The Staff, the Organization, and Our Services

When you purchase one of R.S. Means' publications, you are in effect hiring the services of a full-time staff of construction and engineering professionals.

Our thoroughly experienced and highly qualified staff works daily at collecting, analyzing, and disseminating comprehensive cost information for your needs. These staff members have years of practical construction experience and engineering training prior to joining the firm. As a result, you can count on them not only for the cost figures, but also for additional background reference information that will help you create a realistic estimate.

The Means organization is always prepared to help you solve construction problems through its five major divisions: Construction and Cost Data Publishing, Electronic Products and Services, Consulting Services, Insurance Division, and Educational Services.

Besides a full array of construction cost estimating books, Means also publishes a number of other reference works for the construction industry. Subjects include construction estimating and project and business management; special topics such as HVAC, roofing, plumbing, and hazardous waste remediation; and a library of facility management references.

In addition, you can access all of our construction cost data through your computer with Means CostWorks '98 CD-ROM, an electronic tool that offers over 50,000 lines of Means construction cost data.

What's more, you can increase your knowledge and improve your construction estimating and management performance with a Means Construction Seminar or In-House Training Program. These two-day seminar programs offer unparalleled opportunities for everyone in your organization to get updated on a wide variety of construction-related issues.

Means also is a worldwide provider of construction cost management and analysis services for commercial and government owners and of claims and valuation services for insurers.

In short, R.S. Means can provide you with the tools and expertise for constructing accurate and dependable construction estimates and budgets in a variety of ways.

Robert Snow Means Established a Tradition of Quality That Continues Today

Robert Snow Means spent years building his company, making certain he always delivered a quality product.

Today, at R.S. Means, we do more than talk about the quality of our data and the usefulness of our books. We stand behind all of our data, from historical cost indexes... to construction materials and techniques... to current costs.

If you have any questions about our products or services, please call us toll-free at 1-800-334-3509. Our customer service representatives will be happy to assist you.

Table of Contents

COMMERCIAL/INDUSTRIAL/INSTITUTIONAL

BUILDING TYPES

UNIT PRICES

- GENERAL REQUIREMENTS — 1
- SITE WORK — 2
- CONCRETE — 3
- MASONRY — 4
- METALS — 5
- WOOD & PLASTICS — 6
- THERMAL & MOISTURE PROTECTION — 7
- DOORS & WINDOWS — 8
- FINISHES — 9
- SPECIALTIES — 10
- EQUIPMENT — 11
- FURNISHINGS — 12
- SPECIAL CONSTRUCTION — 13
- CONVEYING SYSTEMS — 14
- MECHANICAL — 15
- ELECTRICAL — 16

ASSEMBLIES

- FOUNDATIONS — 1
- SUBSTRUCTURES — 2
- SUPERSTRUCTURES — 3
- EXTERIOR CLOSURE — 4
- ROOFING — 5
- INTERIOR CONSTRUCTION — 6
- CONVEYING SYSTEMS — 7
- MECHANICAL — 8
- ELECTRICAL — 9
- SPECIAL CONSTRUCTION — 11
- SITE WORK — 12

REFERENCE INFORMATION

- REFERENCE NUMBERS
- CREWS
- COST INDEXES
- INDEX

How the Book Is Built: An Overview

A Powerful Construction Tool

You have in your hands one of the most powerful construction tools available today. A successful project is built on the foundation of an accurate and dependable estimate. This book will enable you to construct just such an estimate.

For the casual user the book is designed to be:

- quickly and easily understood so you can get right to your estimate
- filled with valuable information so you can understand the necessary factors that go into the cost estimate

For the regular user, the book is designed to be:

- a handy desk reference that can be quickly referred to for key costs
- a comprehensive, fully reliable source of current construction costs and productivity rates, so you'll be prepared to estimate any project
- a source book for preliminary project cost, product selections, and alternate materials and methods

To meet all of these requirements we have organized the book into the following clearly defined sections.

Quick Start

This one-page section (see following page) can quickly get you started on your assemblies or unit price estimate.

How To Use the Book: The Details

This section contains an in-depth explanation of how the book is arranged . . . and how you can use it to determine a reliable construction cost estimate. It includes information about how we develop our cost figures and how to completely prepare your estimate.

Commercial/Industrial/Institutional Section

This section lists square foot costs for typical light commercial construction projects. The individual projects are divided into ten common components of construction. An outline of a typical page layout, an explanation of square foot prices, and a Table of Contents are located at the beginning of the section.

Assemblies Section

The cost data in this section has been organized in an "Assemblies" format. These assemblies are the functional elements of a building and are arranged according to the 12 divisions of the UniFormat classification system. For a complete explanation of a typical "Assemblies" page, see "How To Use the Assemblies Cost Tables."

Unit Price Section

All cost data has been divided into the 16 divisions according to the MasterFormat system of classification and numbering as developed by the Construction Specifications Institute (CSI) and Construction Specifications Canada (CSC). For a listing of these divisions and an outline of their subdivisions, see the Unit Price Section Table of Contents.

Estimating tips are included at the beginning of each division.

Reference Section

This section includes information on Reference Numbers, Crew Listings, Historical Cost Indexes, Location Factors, and a listing of Abbreviations. It is visually identified by a vertical gray bar on the edge of pages.

Reference Numbers: Following selected major classifications throughout the book are "reference numbers" shown in bold squares. These numbers refer you to related information in other sections.

In these other sections, you'll find related tables, explanations, and estimating information. Also included are alternate pricing methods and technical data, along with information on design and economy in construction. You'll also find helpful tips on estimating and construction.

It is recommended that you refer to the Reference Section if a "reference number" appears within the major classification you are estimating.

Crew Listings: This section lists all the crews referenced. For the purposes of this book, a crew is composed of more than one trade classification and/or the addition of equipment to any trade classification. Power equipment is included in the cost of the crew. Costs are shown both with bare labor rates and with the installing contractor's overhead and profit added. For each, the total crew cost per eight-hour day and the composite cost per labor-hour are listed.

Historical Cost Indexes: These indexes provide you with data to adjust construction costs over time. If you know costs for a past project, you can use these indexes to estimate the cost to construct the same project today.

Location Factors: Obviously, costs vary depending on the regional economy. You can adjust the "national average" costs in this book to 930 major cities throughout the U.S. and Canada by using the data in this section.

Abbreviations: A listing of the abbreviations and the terms they represent, is included.

Index

A comprehensive listing of all terms and subjects in this book to help you find what you need quickly.

The Scope of This Book

This book is designed to be comprehensive and easy to use. To that end we have made certain assumptions:

1. We have established material prices based on a "national average."
2. We have computed labor costs based on a 30-city "national average" of open shop wage rates.
3. We have targeted the data for projects of a certain size range.

For a more detailed explanation of how the cost data is developed, see "How To Use the Book: The Details."

Project Size

This book is intended for use by those involved primarily in light commercial construction costing less than $1,000,000. This includes the construction of retail stores, offices, warehouses, apartments and motels.

With reasonable exercise of judgment the figures can be used for any building work. *For other types of projects, such as repair and remodeling or residential buildings, consult the appropriate MEANS publication for more information.*

Quick Start

If you feel you are ready to use this book and don't think you need the detailed instructions that begin on the following page, this Quick Start section is for you.

These steps will allow you to get started estimating in a matter of minutes.

1 First, decide whether you require a Unit Price or Assemblies type estimate. Unit price estimating requires a breakdown of the work to individual items. Assemblies estimates combine individual items or components into building systems.

If you need to estimate each line item separately, follow the instructions for **Unit Prices.**

If you can use an estimate for the entire assembly or system, follow the instructions for **Assemblies.**

2 Find each cost data section you need in the Table of Contents (either for Unit Prices or Assemblies).

Unit Prices: The cost data for Unit Prices has been divided into 16 divisions according to the CSI MasterFormat.

Assemblies: The cost data for Assemblies has been divided into 12 divisions according to the UniFormat.

3 Turn to the indicated section and locate the line item or assemblies table you need for your estimate. Portions of a sample page layout from both the Unit Price Listings and the Assemblies Cost Tables appear below.

Unit Prices: If there is a reference number listed at the beginning of the section, it refers to additional information you may find useful. See the referenced section for additional information.

- Note the crew code designation. You'll find full descriptions of crews in the Crew Listings including labor-hour and equipment costs.

Assemblies: The Assemblies (*not* shown in full here) are generally separated into three parts: 1) an illustration of the system to be estimated; 2) the components and related costs of a typical system; and 3) the costs for similar systems with dimensional and/or size variations. The Assemblies Section also contains reference numbers for additional useful information.

4 Determine the total number of units your job will require.

Unit Prices: Note the unit of measure for the material you're using is listed under "Unit."

- Bare Costs: These figures show unit costs for materials and installation. Labor and equipment costs are calculated according to crew costs and average daily output. Bare costs do not contain allowances for overhead, profit or taxes.
- "Labor-hours" allows you to calculate the total labor-hours to complete that task. Just multiply the quantity of work by this figure for an estimate of activity duration.

Assemblies: Note the unit of measure for the assembly or system you're estimating is listed in the Assemblies Table.

5 Then multiply the total units by . . .

Unit Prices: "Total Incl. O&P" which stands for the total cost including the installing contractor's overhead and profit. (See the "How To Use the Unit Price Pages" for a complete explanation.)

Assemblies: The "Total" in the right-hand column, which is the total cost **including the installing contractor's overhead and profit.** (See the "How To Use the Assemblies Cost Tables" section for a complete explanation.)

Material and equipment cost figures include a 10% markup. For labor markups, see the inside back cover of this book. If the work is to be subcontracted, add the general contractor's markup, approximately 10%.

6 The price you calculate will be an estimate for either an individual item of work or a completed assembly or *system.*

7 Compile a list of all items or assemblies included in the total project. Summarize cost information, and add project overhead.

For a more complete explanation of the way costs are derived, please see the following sections.

Editors' Note: We urge you to spend time reading and understanding all of the supporting material and to take into consideration the reference material such as Location Factors and the "reference numbers."

Unit Price Pages

061 | Rough Carpentry

061 100 | Wood Framing			Crew	Daily Output	Labor Hours	Unit	1998 Bare Costs Mat	Labor	Equip	Total	Total Incl O&P
128 0010	FRAMING, WALLS	R061-010									
2000	Headers over openings, 2" x 6"		F-2	360	.044	L.F.	.55	.72		1.27	1.83
2005	2" x 6", pneumatic nailed	R061-030	F-2A	432	.037		.55	.60		1.15	1.63
2050	2" x 8"		F-2	340	.047		.78	.76		1.54	2.16
2055	2" x 8", pneumatic nailed		F-2A	408	.039		.78	.63		1.41	1.95
2100	2" x 10"		F-2	320	.050		1.09	.81		1.90	2.59
2105	2" x 10", pneumatic nailed		F-2A	384	.042		1.09	.67		1.76	2.35
2150	2" x 12"		F-2	300	.053		1.41	.86		2.27	3.03
2155	2" x 12", pneumatic nailed		F-2A	360	.044		1.41	.72		2.13	2.78
2190	4" x 10"		F-2	240	.067		2.85	1.08		3.93	4.99
	4" x 10", pneumatic nailed									3.75	4.68

Assemblies Pages

INTERIOR CONSTR. | A6.1 | Partitions

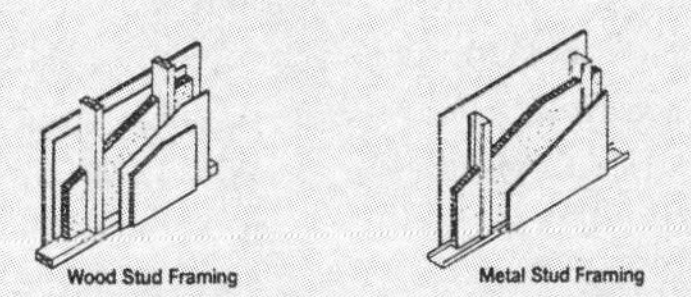

Wood Stud Framing

Metal Stud Framing

The Drywall Partitions/Stud Framing Systems are defined by type of drywall and number of layers, type and spacing of stud framing, and treatment on the opposite face. Components include taping and finishing.

Cost differences between regular and fire resistant drywall are negligible, and terminology is interchangeable. In some cases fiberglass insulation is included for additional sound deadening.

6.1-510 Drywall Partitions/Wood Stud Framing

	Face Layer	Base Layer	Framing	Opposite Face	Insulation	Cost per S.F. Mat.	Inst.	Total
1200	5/8" FR drywall	none	2 x 4, @ 16" O.C.	same	0	1.10	1.38	2.48
1250				5/8" reg. drywall	0	1.09	1.38	2.47
1300				nothing	0	.77	.90	1.67
1800		5/8" FR drywall	2 x 4 @ 24" O.C.	same	0	1.46	1.78	3.24
1850				5/8" FR drywall	0	1.23	1.54	2.77
1900				nothing	0	.90	1.06	1.96
2200		5/8" FR drywall	2 rows-2 x 4 16"O.C.	same	2" fiberglass	2.40	2.53	4.93
2250				5/8" FR drywall	2" fiberglass	2.17	2.29	4.46
2300				nothing	2" fiberglass	1.84	1.81	3.65
2400	5/8" WR drywall	none	2 x 4, @ 16" O.C.	same	0	1.20	1.38	2.58
2450				5/8" FR drywall	0	1.15	1.38	2.53
2500				nothing	0	.82	.90	1.72
2600		5/8" FR drywall	2 x 4, @ 24" O.C.	same	0	1.56	1.78	3.34
2650				5/8" FR drywall	0	1.28	1.54	2.82
2700				nothing	0	.95	1.06	2.01
2800	5/8 VF drywall	none	2 x 4, @ 16" O.C.	same	0	1.80	1.50	3.30
2850				5/8" FR drywall	0	1.45	1.44	2.89
2900				nothing	0	1.12		
				same	0			

How to Use the Book: The Details

What's Behind the Numbers? The Development of Cost Data

The staff at R.S. Means continuously monitors developments in the construction industry in order to ensure reliable, thorough and up-to-date cost information.

While *overall* construction costs may vary relative to general economic conditions, price fluctuations within the industry are dependent upon many factors. Individual price variations may, in fact, be opposite to overall economic trends. Therefore, costs are continually monitored and complete updates are published yearly. Also, new items are frequently added in response to changes in materials and methods.

Costs—$ (U.S.)

All costs represent U.S. national averages and are given in U.S. dollars. The Means Location Factors can be used to adjust costs to a particular location. The Location Factors for Canada can be used to adjust U.S. national averages to local costs in Canadian dollars.

Material Costs

The R.S. Means staff contacts manufacturers, dealers, distributors, and contractors all across the U.S. and Canada to determine national average material costs. If you have access to current material costs for your specific location, you may wish to make adjustments to reflect differences from the national average. Included within material costs are fasteners for a normal installation. R.S. Means engineers use manufacturers' recommendations, written specifications and/or standard construction practice for size and spacing of fasteners. Adjustments to material costs may be required for your specific application or location. Material costs do not include sales tax.

Labor Costs

Labor costs are based on the average of open shop wage rates from across the U.S. for construction trades for the current year. Rates along with overhead and profit markups are listed on the inside back cover of this book.

- If wage rates in your area vary from those used in this book, or if rate increases are expected within a given year, labor costs should be adjusted accordingly.

Labor costs reflect productivity based on actual working conditions. These figures include time spent during a normal workday on tasks other than actual installation, such as material receiving and handling, mobilization at site, site movement, breaks, and cleanup.

Productivity data is developed over an extended period so as not to be influenced by abnormal variations and reflects a typical average.

Equipment Costs

Equipment costs include not only rental, but also operating costs for equipment under normal use. The operating costs include parts and labor for routine servicing such as repair and replacement of pumps, filters and worn lines. Normal operating expendables such as fuel, lubricants, tires and electricity (where applicable) are also included. Extraordinary operating expendables with highly variable wear patterns such as diamond bits and blades are excluded. These costs are included under materials. Equipment rental rates are obtained from industry sources throughout North America—contractors, suppliers, dealers, manufacturers, and distributors.

Crew Equipment Cost/Day — The power equipment required for each crew is included in the crew cost. The daily cost for crew equipment is based on dividing the weekly bare rental rate by 5 (number of working days per week), and then adding the hourly operating cost times 8 (hours per day). This "Crew Equipment Cost/Day" is listed in Subdivision 016.

Factors Affecting Costs

Costs can vary depending upon a number of variables. Here's how we have handled the main factors affecting costs.

Quality — The prices for materials and the workmanship upon which productivity is based represent sound construction work. They are also in line with U.S. government specifications.

Overtime — We have made no allowance for overtime. If you anticipate premium time or work beyond normal working hours, be sure to make an appropriate adjustment to your labor costs.

Productivity — The productivity, daily output, and labor-hour figures for each line item are based on working an eight-hour day in daylight hours in moderate temperatures. For work that extends beyond normal work hours or is performed under adverse conditions, productivity may decrease. (See the section in "How To Use the Unit Price Pages" for more on productivity.)

Size of Project — The size, scope of work, and type of construction project will have a significant impact on cost. Economies of scale can reduce costs for large projects. Unit costs can often run higher for small projects. Costs in this book are intended for the size and type of project as previously described in "How the Book Is Built: An Overview." Costs for projects of a significantly different size or type should be adjusted accordingly.

Location — Material prices in this book are for metropolitan areas. However, in dense urban areas, traffic and site storage limitations may increase costs. Beyond a 20-mile radius of large cities, extra trucking or transportation charges may also increase the material costs slightly. On the other hand, lower wage rates may be in effect. Be sure to consider both these factors when preparing an estimate, particularly if the job site is located in a central city or remote rural location.

In addition, highly specialized subcontract items may require travel and per diem expenses for mechanics.

Other factors—

- season of year
- contractor management
- weather conditions
- local union restrictions
- building code requirements
- availability of:
  - adequate energy
  - skilled labor
  - building materials
- owner's special requirements/restrictions
- safety requirements
- environmental considerations

General Conditions—The "Assemblies" and "Commercial/Industrial/Institutional" sections of this book use costs that include the installing contractor's overhead and profit (O&P). The Unit Price section of this book presents cost data in two ways: Bare Costs and Total Cost including O&P. General Conditions, when applicable, should also be added to the Total Cost including O&P. The costs for General Conditions are listed in Division 1 of the Unit Price Section and the Reference Section of this book. General Conditions for the *Installing Contractor* may range from 0% to 10% of the Total Cost including O&P. For the *General* or *Prime Contractor*, costs for General Conditions may range from 5% to 15% of the Total Cost including O&P, with a figure of 10% as the most typical allowance.

Overhead & Profit—Total Cost including O&P for the *Installing Contractor* is shown in the last column on both the Unit Price and the Assemblies pages of this book. This figure is the sum of the bare material cost plus 10% for profit, the base labor cost plus total overhead and profit, and the bare equipment cost plus 10% for profit. Details for the calculation of Overhead and Profit on labor are shown on the inside back cover and in the Reference Section of this book. (See the "How to Use the Unit Price Pages" for an example of this calculation.).

Unpredictable Factors—General business conditions influence "in-place" costs of all items. Substitute materials and construction methods may have to be employed. These may affect the installed cost and/or life cycle costs. Such factors may be difficult to evaluate and cannot necessarily be predicted on the basis of the job's location in a particular section of the country. Thus, where these factors apply, you may find significant, but unavoidable cost variations for which you will have to apply a measure of judgment to your estimate.

Rounding of Costs

In general, all unit prices in excess of $5.00 have been rounded to make them easier to use and still maintain adequate precision of the results. The rounding rules we have chosen are in the following table.

Prices from . . .	Rounded to the nearest . . .
$.01 to $5.00	$.01
$5.01 to $20.00	$.05
$20.01 to $100.00	$.50
$100.01 to $300.00	$1.00
$300.01 to $1,000.00	$5.00
$1,000.01 to $10,000.00	$25.00
$10,000.01 to $50,000.00	$100.00
$50,000.01 and above	$500.00

Final Checklist

Estimating can be a straightforward process provided you remember the basics. Here's a checklist of some of the items you should remember to do before completing your estimate.

Did you remember to . . .

- utilize the appropriate Location Factor
- take into consideration which items have been marked up and by how much
- mark up the entire estimate sufficiently for your purposes
- read the background information on techniques and technical matters that could impact your project time span and cost
- include all components of your project in the final estimate
- double check your figures to be sure of your accuracy
- call R.S. Means if you have any questions about your estimate or the data you've found in our publications

Remember, R.S. Means stands behind its publications. If you have any questions about your estimate . . . about the costs you've used from our books . . . or even about the technical aspects of the job that may affect your estimate, feel free to call the R.S. Means editors at 1-781-585-7898 or 1-800-448-8182.

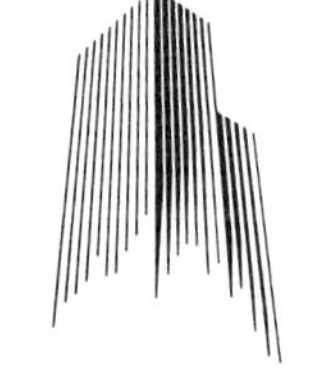

Commercial/Industrial/ Institutional Section

Table of Contents

How to Use the Commercial/Industrial/Institutional Section

The following is a detailed explanation of a sample entry in the Commercial/Industrial/Institutional Square Foot Cost Section. Each bold number below corresponds to the item being described on the following page with the appropriate component or cost of the sample entry following in parenthesis.

Prices listed are costs that include overhead and profit of the installing contractor and additional mark-ups for General Conditions and Architects' Fees.

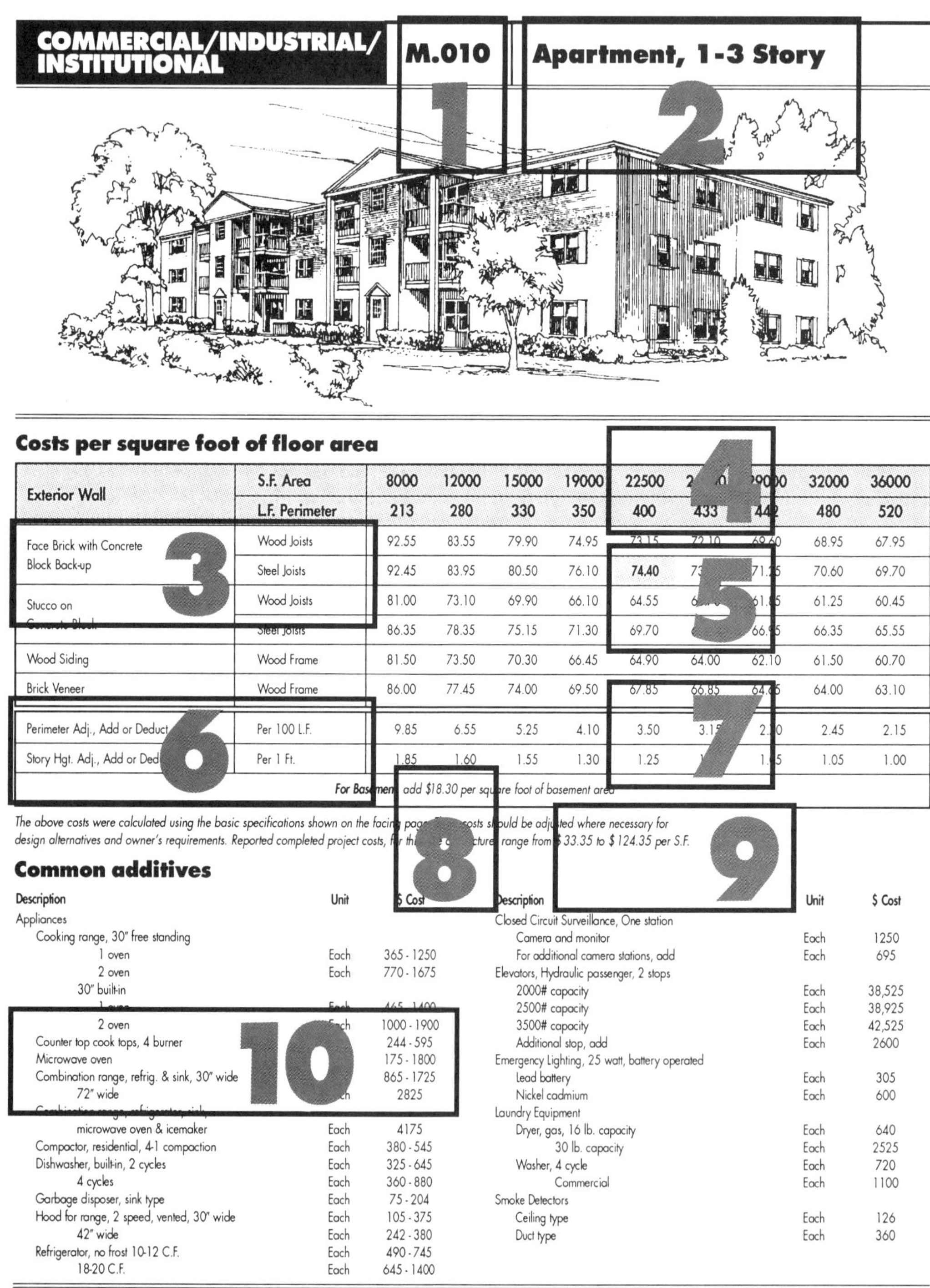

Costs per square foot of floor area

Exterior Wall	S.F. Area	8000	12000	15000	19000	22500	26000	29000	32000	36000
	L.F. Perimeter	213	280	330	350	400	433	442	480	520
Face Brick with Concrete Block Back-up	Wood Joists	92.55	83.55	79.90	74.95	73.15	72.10	69.60	68.95	67.95
	Steel Joists	92.45	83.95	80.50	76.10	74.40	[illegible]	71.25	70.60	69.70
Stucco on Concrete Block	Wood Joists	81.00	73.10	69.90	66.10	64.55	[illegible]	61.85	61.25	60.45
	Steel Joists	86.35	78.35	75.15	71.30	69.70	[illegible]	66.95	66.35	65.55
Wood Siding	Wood Frame	81.50	73.50	70.30	66.45	64.90	64.00	62.10	61.50	60.70
Brick Veneer	Wood Frame	86.00	77.45	74.00	69.50	67.85	66.85	64.65	64.00	63.10
Perimeter Adj., Add or Deduct	Per 100 L.F.	9.85	6.55	5.25	4.10	3.50	3.15	2.70	2.45	2.15
Story Hgt. Adj., Add or Deduct	Per 1 Ft.	1.85	1.60	1.55	1.30	1.25	[illegible]	1.15	1.05	1.00

For Basement, add $18.30 per square foot of basement area

The above costs were calculated using the basic specifications shown on the facing page. These costs should be adjusted where necessary for design alternatives and owner's requirements. Reported completed project costs, for this type of structure, range from $33.35 to $124.35 per S.F.

Common additives

Description	Unit	$ Cost
Appliances		
Cooking range, 30" free standing		
1 oven	Each	365 - 1250
2 oven	Each	770 - 1675
30" built-in		
1 oven	Each	465 - 1400
2 oven	Each	1000 - 1900
Counter top cook tops, 4 burner		244 - 595
Microwave oven		175 - 1800
Combination range, refrig. & sink, 30" wide		865 - 1725
72" wide	Each	2825
Combination range, refrigerator, sink, microwave oven & icemaker	Each	4175
Compactor, residential, 4-1 compaction	Each	380 - 545
Dishwasher, built-in, 2 cycles	Each	325 - 645
4 cycles	Each	360 - 880
Garbage disposer, sink type	Each	75 - 204
Hood for range, 2 speed, vented, 30" wide	Each	105 - 375
42" wide	Each	242 - 380
Refrigerator, no frost 10-12 C.F.	Each	490 - 745
18-20 C.F.	Each	645 - 1400

Description	Unit	$ Cost
Closed Circuit Surveillance, One station		
Camera and monitor	Each	1250
For additional camera stations, add	Each	695
Elevators, Hydraulic passenger, 2 stops		
2000# capacity	Each	38,525
2500# capacity	Each	38,925
3500# capacity	Each	42,525
Additional stop, add	Each	2600
Emergency Lighting, 25 watt, battery operated		
Lead battery	Each	305
Nickel cadmium	Each	600
Laundry Equipment		
Dryer, gas, 16 lb. capacity	Each	640
30 lb. capacity	Each	2525
Washer, 4 cycle	Each	720
Commercial	Each	1100
Smoke Detectors		
Ceiling type	Each	126
Duct type	Each	360

Important: See the Reference Section for Location Factors

1 Model Number (M.010)

"M" distinguishes this section of the book and stands for model. The number designation is a sequential number.

2 Type of Building (Apartment, 1-3 Story)

There are 43 different types of commercial/industrial/institutional buildings highlighted in this section.

3 Exterior Wall Construction and Building Framing Options (Face Brick with Concrete Block Back-up and Open Web Steel Bar Joists)

Three or more commonly used exterior walls, and in most cases, two typical building framing systems are presented for each type of building. The model selected should be based on the actual characteristics of the building being estimated.

4 Total Square Foot of Floor Area and Base Perimeter Used to Compute Base Costs (22,500 Square Feet and 400 Linear Feet)

Square foot of floor area is the total gross area of all floors at grade, and above, and does not include a basement. The perimeter in linear feet used for the base cost is for a generally rectangular economical building shape.

5 Cost per Square Foot of Floor Area ($74.40)

The highlighted cost is for a building of the selected exterior wall and framing system and floor area. Costs for buildings with floor areas other than those calculated may be interpolated between the costs shown.

6 Building Perimeter and Story Height Adjustments

Square foot costs for a building with a perimeter or floor to floor story height significantly different from the model used to calculate the base cost may be adjusted, add or deduct, to reflect the actual building geometry.

7 Cost per Square Foot of Floor Area for the Perimeter and/or Height Adjustment ($3.50 for Perimeter Difference and $1.25 for Story Height Difference)

Add (or deduct) $3.50 to the base square foot cost for each 100 feet of perimeter difference between the model and the actual building. Add (or deduct) $1.25 to the base square foot cost for each 1 foot of story height difference between the model and the actual building.

8 Optional Cost per Square Foot of Basement Floor Area ($18.30)

The cost of an unfinished basement for the building being estimated is $18.30 times the gross floor area of the basement.

9 Range of Cost per Square Foot of Floor Area for Similar Buildings ($33.35 to $124.35)

Many different buildings of the same type have been built using similar materials and systems. Means historical cost data of actual construction projects indicates a range of $33.35 to $124.35 for this type of building.

10 Common Additives

Common components and/or systems used in this type of building are listed. These costs should be added to the total building cost. Additional selections may be found in the Assemblies Section.

How to Use the Commercial/ Industrial/Institutional Section *(Continued)*

The following is a detailed explanation of a specification and costs for a model building in the Commercial/Industrial/Institutional Square Foot Cost Section. Each bold number below corresponds to the item being described on the following page with the appropriate component of the sample entry following in parenthesis.

Prices listed are costs that include overhead and profit of the installing contractor.

Model costs calculated for a 3 story building with 10' story height and 22,500 square feet of floor area   **1**

2   **Apartment, 1-3 Story**

				Unit	Unit Cost	Cost Per S.F.	% Of Sub-Total
1.0 Foundations							
.1	Footings & Foundations	Poured concrete; strip and spread footings and 4' foundation wall		S.F. Ground	5.13	1.71	
.4	Piles & Caissons	N/A		—	—	—	3.4%
.9	Excavation & Backfill	Site preparation for slab and trench for foundation wall and footing		S.F. Ground	.91	.30	
2.0 Substructure							
.1	Slab on Grade	4" reinforced concrete with vapor barrier and granular base		S.F. Slab	2.56	.85	1.4%
.2	Special Substructures	N/A		—			
3.0 Superstructure							
.1	Columns & Beams	Gypsum board fireproofing on columns; steel columns in 3.5 and 3.7		S.F. Floor	.63	.63	
.4	Structural Walls	N/A		—			
.5	Elevated Floors	Open web steel joists, slab form, concrete, interior steel columns		S.F. Floor	8.48	5.65	14.5%
.7	Roof	Open web steel joists with rib metal deck, interior steel columns		S.F. Roof	4.04	1.35	
.9	Stairs	Concrete filled metal pan		Flight	4725	1.05	
4.0 Exterior Closure							
.1	Walls	Face brick with concrete block backup	88% of wall	S.F. Wall	12.19	5.72	
.5	Exterior Wall Finishes	N/A		—	—	—	11.7%
.6	Doors	Aluminum and glass		Each	1017	.18	
.7	Windows & Glazed Walls	Aluminum horizontal sliding	12% of wall	Each	254	1.08	
5.0 Roofing							
.1	Roof Coverings	Built-up tar and gravel with flashing		S.F. Roof	2.16	.72	
.7	Insulation	Perlite/EPS composite		S.F. Roof	1.13	[illegible]	.8%
.8	Openings & Specialties	N/A		—			
6.0 Interior Construction							
.1	Partitions	Gypsum board and sound deadening board on metal studs	10 S.F. of Floor/L.F. Partition	S.F. Partition	2.62	2.33	
.4	Interior Doors	15% solid core wood, 85% hollow core wood	80 S.F. Floor/Door	Each	330	4.13	
.5	Wall Finishes	70% paint, 25% vinyl wall covering, 5% ceramic tile		S.F. Surface	.80	1.42	24.7%
.6	Floor Finishes	60% carpet, 30% vinyl composition tile, 10% ceramic tile		S.F. Floor	[illegible]	3.78	
.7	Ceiling Finishes	Painted gypsum board on resilient channels		S.F. Ceiling	[illegible]	2.20	
.9	Interior Surface/Exterior Wall	Painted gypsum board on furring	80% of wall	S.F. Wall	[illegible]	.96	
7.0 Conveying							
.1	Elevators	One hydraulic passenger elevator		Each	56,475	2.51	
.2	Special Conveyors	N/A		—	—	—	
8.0 Mechanical							
.1	Plumbing	Kitchen, bathroom and service fixtures, supply and drainage	1 Fixture/200 S.F. Floor	Each	1476	7.38	
.2	Fire Protection	Wet pipe sprinkler system		S.F. Floor	1.27	1.27	
.3	Heating	Oil fired hot water, baseboard radiation		S.F. Floor	3.70	3.70	29.0%
.4	Cooling	Chilled water, air cooled condenser system		S.F. Floor	5.05	5.05	
.5	Special Systems	N/A		—	—	—	
9.0 Electrical							
.1	Service & Distribution	400 ampere service, panel board and feeders		S.F. Floor	.68	.68	
.2	Lighting & Power	Incandescent fixtures, receptacles, switches, A.C. and misc. power		S.F. Floor	3.23	3.23	7.4%
.4	Special Electrical	Alarm systems and emergency lighting		S.F. Floor	.51	.51	
11.0 Special Construction							
.1	Specialties	Kitchen cabinets		S.F. Floor	1.14	1.14	1.9%
12.0 Site Work							
.1	Earthwork	N/A					
.3	Utilities	N/A					0.0%
.5	Roads & Parking	N/A					
.7	Site Improvements	N/A					
				Sub-Total		59.91	100%
	GENERAL CONDITIONS (Overhead & Profit)				15%	8.99	
	ARCHITECT FEES				8%	5.50	
				Total Building Cost		**74.40**	

1 *Building Description (Model costs are calculated for a three-story apartment building with 10' story height and 22,500 square feet of floor area)*

The model highlighted is described in terms of building type, number of stories, typical story height and square footage.

2 *Type of Building (Apartment, 1-3 Story)*

3 *Division 6.0 Interior Construction (6.4 Interior Doors)*

System costs are presented in divisions according to the 12-division UniFormat classifications. Each of the component systems are listed.

4 *Specification Highlights (15% solid core wood; 85% hollow core wood)*

All systems in each subdivision are described with the material and proportions used.

5 *Quality Criteria (80 S.F. Floor/Door)*

The criteria used in determining quantities for the calculations are shown.

6 *Unit (Each)*

The unit of measure shown in this column is the unit of measure of the particular system shown that corresponds to the unit cost.

7 *Unit Cost ($330)*

The cost per unit of measure of each system subdivision.

8 *Cost per Square Foot ($4.13)*

The cost per square foot for each system is the unit cost of the system times the total number of units divided by the total square feet of building area.

9 *% of Sub-Total (24.7%)*

The percent of sub-total is the total cost per square foot of all systems in the division divided by the sub-total cost per square foot of the building.

10 *Sub-Total ($59.91)*

The sub-total is the total of all the system costs per square foot.

11 *General Conditions (Overhead & Profit) (15%) (Architects' Fees) (8%)*

General Conditions to cover the overhead and profit of the General Contractor are added as a percentage of the sub-total. An Architect's Fee, also as a percentage of the sub-total, is also added. These values vary with the building type.

12 *Total Building Cost ($74.40)*

The total building cost per square foot of building area is the sum of the square foot costs of all the systems plus the General Contractor's overhead and profit and the Architect's fee. The total building cost is the amount which appears shaded in the Cost per Square Foot of Floor Area table shown previously.

BUILDING TYPES

Costs per square foot of floor area

Exterior Wall	S.F. Area	8000	12000	15000	19000	22500	25000	29000	32000	36000
	L.F. Perimeter	213	280	330	350	400	433	442	480	520
Face Brick with Concrete Block Back-up	Wood Joists	92.55	83.55	79.90	74.95	73.15	72.10	69.60	68.95	67.95
	Steel Joists	92.45	83.95	80.50	76.10	**74.40**	73.45	71.25	70.60	69.70
Stucco on Concrete Block	Wood Joists	81.00	73.10	69.90	66.10	64.55	63.70	61.85	61.25	60.45
	Steel Joists	86.35	78.35	75.15	71.30	69.70	68.85	66.95	66.35	65.55
Wood Siding	Wood Frame	81.50	73.50	70.30	66.45	64.90	64.00	62.10	61.50	60.70
Brick Veneer	Wood Frame	86.00	77.45	74.00	69.50	67.85	66.85	64.65	64.00	63.10
Perimeter Adj., Add or Deduct	Per 100 L.F.	9.85	6.55	5.25	4.10	3.50	3.15	2.70	2.45	2.15
Story Hgt. Adj., Add or Deduct	Per 1 Ft.	1.85	1.60	1.55	1.30	1.25	1.20	1.05	1.05	1.00

For Basement, add $18.30 per square foot of basement area

The above costs were calculated using the basic specifications shown on the facing page. These costs should be adjusted where necessary for design alternatives and owner's requirements. Reported completed project costs, for this type of structure, range from $33.35 to $124.35 per S.F.

Common additives

Description	Unit	$ Cost
Appliances		
Cooking range, 30" free standing		
1 oven	Each	365 - 1250
2 oven	Each	770 - 1675
30" built-in		
1 oven	Each	465 - 1400
2 oven	Each	1000 - 1900
Counter top cook tops, 4 burner	Each	244 - 595
Microwave oven	Each	175 - 1800
Combination range, refrig. & sink, 30" wide	Each	865 - 1725
72" wide	Each	2825
Combination range, refrigerator, sink,		
microwave oven & icemaker	Each	4175
Compactor, residential, 4-1 compaction	Each	380 - 545
Dishwasher, built-in, 2 cycles	Each	325 - 645
4 cycles	Each	360 - 880
Garbage disposer, sink type	Each	75 - 204
Hood for range, 2 speed, vented, 30" wide	Each	105 - 375
42" wide	Each	242 - 380
Refrigerator, no frost 10-12 C.F.	Each	490 - 745
18-20 C.F.	Each	645 - 1400

Description	Unit	$ Cost
Closed Circuit Surveillance, One station		
Camera and monitor	Each	1250
For additional camera stations, add	Each	695
Elevators, Hydraulic passenger, 2 stops		
2000# capacity	Each	38,525
2500# capacity	Each	38,925
3500# capacity	Each	42,525
Additional stop, add	Each	2600
Emergency Lighting, 25 watt, battery operated		
Lead battery	Each	305
Nickel cadmium	Each	600
Laundry Equipment		
Dryer, gas, 16 lb. capacity	Each	640
30 lb. capacity	Each	2525
Washer, 4 cycle	Each	720
Commercial	Each	1100
Smoke Detectors		
Ceiling type	Each	126
Duct type	Each	360

Important: See the Reference Section for Location Factors

Model costs calculated for a 3 story building with 10' story height and 22,500 square feet of floor area

				Unit	Unit Cost	Cost Per S.F.	% Of Sub-Total
1.0 Foundations							
	.1	Footings & Foundations	Poured concrete; strip and spread footings and 4' foundation wall	S.F. Ground	5.13	1.71	
	.4	Piles & Caissons	N/A	—	—	—	3.4%
	.9	Excavation & Backfill	Site preparation for slab and trench for foundation wall and footing	S.F. Ground	.91	.30	
2.0 Substructure							
	.1	Slab on Grade	4" reinforced concrete with vapor barrier and granular base	S.F. Slab	2.56	.85	1.4%
	.2	Special Substructures	N/A	—	—	—	
3.0 Superstructure							
	.1	Columns & Beams	Gypsum board fireproofing on columns; steel columns in 3.5 and 3.7	S.F. Floor	.63	.63	
	.4	Structural Walls	N/A	—	—	—	
	.5	Elevated Floors	Open web steel joists, slab form, concrete, interior steel columns	S.F. Floor	8.48	5.65	14.5%
	.7	Roof	Open web steel joists with rib metal deck, interior steel columns	S.F. Roof	4.04	1.35	
	.9	Stairs	Concrete filled metal pan	Flight	4725	1.05	
4.0 Exterior Closure							
	.1	Walls	Face brick with concrete block backup — 88% of wall	S.F. Wall	12.19	5.72	
	.5	Exterior Wall Finishes	N/A	—	—	—	
	.6	Doors	Aluminum and glass	Each	1017	.18	11.7%
	.7	Windows & Glazed Walls	Aluminum horizontal sliding — 12% of wall	Each	254	1.08	
5.0 Roofing							
	.1	Roof Coverings	Built-up tar and gravel with flashing	S.F. Roof	2.16	.72	
	.7	Insulation	Perlite/EPS composite	S.F. Roof	1.13	.38	1.8%
	.8	Openings & Specialties	N/A	—	—	—	
6.0 Interior Construction							
	.1	Partitions	Gypsum board and sound deadening board on metal studs — 10 S.F. of Floor/L.F. Partition	S.F. Partition	2.62	2.33	
	.4	Interior Doors	15% solid core wood, 85% hollow core wood — 80 S.F. Floor/Door	Each	330	4.13	
	.5	Wall Finishes	70% paint, 25% vinyl wall covering, 5% ceramic tile	S.F. Surface	.80	1.42	
	.6	Floor Finishes	60% carpet, 30% vinyl composition tile, 10% ceramic tile	S.F. Floor	3.78	3.78	24.7%
	.7	Ceiling Finishes	Painted gypsum board on resilient channels	S.F. Ceiling	2.20	2.20	
	.9	Interior Surface/Exterior Wall	Painted gypsum board on furring — 80% of wall	S.F. Wall	2.26	.96	
7.0 Conveying							
	.1	Elevators	One hydraulic passenger elevator	Each	56,475	2.51	4.2%
	.2	Special Conveyors	N/A	—	—	—	
8.0 Mechanical							
	.1	Plumbing	Kitchen, bathroom and service fixtures, supply and drainage — 1 Fixture/200 S.F. Floor	Each	1476	7.38	
	.2	Fire Protection	Wet pipe sprinkler system	S.F. Floor	1.27	1.27	
	.3	Heating	Oil fired hot water, baseboard radiation	S.F. Floor	3.70	3.70	29.0%
	.4	Cooling	Chilled water, air cooled condenser system	S.F. Floor	5.05	5.05	
	.5	Special Systems	N/A	—	—	—	
9.0 Electrical							
	.1	Service & Distribution	400 ampere service, panel board and feeders	S.F. Floor	.68	.68	
	.2	Lighting & Power	Incandescent fixtures, receptacles, switches, A.C. and misc. power	S.F. Floor	3.23	3.23	7.4%
	.4	Special Electrical	Alarm systems and emergency lighting	S.F. Floor	.51	.51	
11.0 Special Construction							
	.1	Specialties	Kitchen cabinets	S.F. Floor	1.14	1.14	1.9%
12.0 Site Work							
	.1	Earthwork	N/A	—	—	—	
	.3	Utilities	N/A	—	—	—	
	.5	Roads & Parking	N/A	—	—	—	0.0%
	.7	Site Improvements	N/A	—	—	—	
				Sub-Total		59.91	100%
		GENERAL CONDITIONS (Overhead & Profit)			15%	8.99	
		ARCHITECT FEES			8%	5.50	
				Total Building Cost		74.40	

Costs per square foot of floor area

Exterior Wall		2000	2700	3400	4100	4800	5500	6200	6900	7600
S.F. Area		2000	2700	3400	4100	4800	5500	6200	6900	7600
L.F. Perimeter		180	208	236	256	280	303	317	337	357
Face Brick with Concrete Block Back-up	Steel Frame	124.80	117.50	113.25	109.55	107.30	105.55	103.55	102.35	101.35
	R/Conc. Frame	132.60	125.30	121.05	**117.35**	115.10	113.35	111.35	110.15	109.15
Precast Concrete Panel	Steel Frame	120.65	114.00	110.05	106.70	104.65	103.05	101.25	100.10	99.20
	R/Conc. Frame	128.45	121.80	117.85	114.45	112.45	110.85	109.00	107.90	107.00
Limestone with Concrete Block Back-up	Steel Frame	134.40	125.70	120.65	116.20	113.50	111.45	109.00	107.55	106.35
	R/Conc. Frame	142.20	133.50	128.45	124.00	121.30	119.25	116.80	115.35	114.15
Perimeter Adj., Add or Deduct	Per 100 L.F.	22.30	16.55	13.15	10.90	9.30	8.10	7.20	6.45	5.90
Story Hgt. Adj., Add or Deduct	Per 1 Ft.	2.15	1.85	1.65	1.55	1.40	1.30	1.20	1.15	1.15

For Basement, add $16.90 per square foot of basement area

The above costs were calculated using the basic specifications shown on the facing page. These costs should be adjusted where necessary for design alternatives and owner's requirements. Reported completed project costs, for this type of structure, range from $77.10 to $189.80 per S.F.

Common additives

Description	Unit	$ Cost	Description	Unit	$ Cost
Bulletproof Teller Window, 44" x 60"	Each	3875	Service Windows, Pass thru, steel		
60" x 48"	Each	4850	24" x 36"	Each	2000
Closed Circuit Surveillance, One station			48" x 48"	Each	2675
Camera and monitor	Each	1250	72" x 40"	Each	3200
For additional camera stations, add	Each	695	For stainless steel frames, add	Each	
Counters, Complete	Station	3900	Smoke Detectors		
Door & Frame, 3' x 6'-8", bullet resistant steel			Ceiling type	Each	126
with vision panel	Each	3000 - 5200	Duct type	Each	360
Drive-up Window, Drawer & micr., not incl. glass	Each	5875 - 8875	Twenty-four Hour Teller		
Emergency Lighting, 25 watt, battery operated			Automatic deposit cash & memo	Each	40,500
Lead battery	Each	305	Vault Front, Door & frame		
Nickel cadmium	Each	600	1 hour test, 32"x 78"	Opening	3275
Night Depository	Each	7425 - 12,100	2 hour test, 32" door	Opening	3325
Package Receiver, painted	Each	1300	40" door	Opening	3800
stainless steel	Each	2075	4 hour test, 32" door	Opening	3475
Partitions, Bullet resistant to 8' high	L.F.	211 - 355	40" door	Opening	4100
Pneumatic Tube Systems, 2 station	Each	22,100	Time lock, two movement, add	Each	1525
With TV viewer	Each	41,900			

Important: See the Reference Section for Location Factors

Model costs calculated for a 1 story building with 14' story height and 4,100 square feet of floor area

				Unit	Unit Cost	Cost Per S.F.	% Of Sub-Total
1.0 Foundations							
.1	Footings & Foundations	Poured concrete; strip and spread footings and 4' foundation wall		S.F. Ground	4.69	4.69	
.4	Piles & Caissons	N/A		–	–	–	7.1%
.9	Excavation & Backfill	Site preparation for slab and trench for foundation wall and footing		S.F. Ground	1.81	1.81	
2.0 Substructure							
.1	Slab on Grade	4" reinforced concrete with vapor barrier and granular base		S.F. Slab	2.56	2.56	2.8%
.2	Special Substructures	N/A		–	–	–	
3.0 Superstructure							
.1	Columns & Beams	Concrete columns		L.F. Columns	51	2.51	
.4	Structural Walls	N/A		–	–	–	
.5	Elevated Floors	N/A		–	–	–	11.0%
.7	Roof	Cast-in-place concrete slab		S.F. Roof	7.64	7.64	
.9	Stairs	N/A		–	–	–	
4.0 Exterior Closure							
.1	Walls	Face brick with concrete block backup	80% of wall	S.F. Wall	15.80	11.05	
.5	Exterior Wall Finishes	N/A		–	–	–	
.6	Doors	Double aluminum and glass and hollow metal		Each	1682	.82	17.1%
.7	Windows & Glazed Walls	Horizontal aluminum sliding	20% of wall	Each	329	3.83	
5.0 Roofing							
.1	Roof Coverings	Built-up tar and gravel with flashing		S.F. Roof	2.29	2.29	
.7	Insulation	Perlite/EPS composite		S.F. Roof	1.13	1.13	4.0%
.8	Openings & Specialties	Gravel stop		L.F. Perimeter	4.78	.30	
6.0 Interior Construction							
.1	Partitions	Gypsum board on metal studs	20 S.F. of Floor/L.F. Partition	S.F. Partition	2.44	1.22	
.4	Interior Doors	Single leaf hollow metal	200 S.F. Floor/Door	Each	484	2.42	
.5	Wall Finishes	50% vinyl wall covering, 50% paint		S.F. Surface	.79	.79	
.6	Floor Finishes	50% carpet, 50% vinyl composition tile		S.F. Floor	3.68	3.68	13.4%
.7	Ceiling Finishes	Mineral fiber tile on concealed zee bars		S.F. Ceiling	2.62	2.62	
.9	Interior Surface/Exterior Wall	Painted gypsum board on furring	80% of wall	S.F. Wall	2.25	1.57	
7.0 Conveying							
.1	Elevators	N/A		–	–	–	0.0%
.2	Special Conveyors	N/A		–	–	–	
8.0 Mechanical							
.1	Plumbing	Toilet and service fixtures, supply and drainage	1 Fixture/580 S.F. Floor	Each	2847	4.91	
.2	Fire Protection	Wet pipe sprinkler system		S.F. Floor	1.66	1.66	
.3	Heating	Included in 8.4		–	–	–	16.2%
.4	Cooling	Single zone rooftop unit, gas heating, electric cooling		S.F. Floor	8.24	8.24	
.5	Special Systems	N/A		–	–	–	
9.0 Electrical							
.1	Service & Distribution	200 ampere service, panel board and feeders		S.F. Floor	1.28	1.28	
.2	Lighting & Power	Fluorescent fixtures, receptacles, switches, A.C. and misc. power		S.F. Floor	5.05	5.05	8.4%
.4	Special Electrical	Alarm systems, emergency lighting, and security television		S.F. Floor	1.43	1.43	
11.0 Special Construction							
.1	Specialties	Vault door, automatic teller, drive up window, closed circuit T.V., night depository		S.F. Floor	18.42	18.42	20.0%
12.0 Site Work							
.1	Earthwork	N/A		–	–	–	
.3	Utilities	N/A		–	–	–	
.5	Roads & Parking	N/A		–	–	–	0.0%
.7	Site Improvements	N/A		–	–	–	
			Sub-Total			91.92	100%
	GENERAL CONDITIONS (Overhead & Profit)				15%	13.79	
	ARCHITECT FEES				11%	11.64	
			Total Building Cost			**117.35**	

Costs per square foot of floor area

Exterior Wall	S.F. Area	12000	14000	16000	18000	20000	22000	24000	26000	28000
	L.F. Perimeter	460	491	520	540	566	593	620	645	670
Concrete Block	Steel Roof Deck	53.75	52.30	51.20	50.15	**49.40**	48.80	48.35	47.90	47.50
Decorative Concrete Block	Steel Roof Deck	56.05	54.40	53.15	51.95	51.10	50.40	49.90	49.40	48.95
Face Brick on Concrete Block	Steel Roof Deck	59.20	57.30	55.80	54.40	53.45	52.65	52.00	51.45	50.95
Jumbo Brick on Concrete Block	Steel Roof Deck	57.85	56.05	54.70	53.35	52.45	51.70	51.10	50.60	50.10
Stucco on Concrete Block	Steel Roof Deck	54.45	52.90	51.80	50.70	49.95	49.30	48.80	48.35	47.95
Precast Concrete Panel	Steel Roof Deck	55.00	53.45	52.25	51.10	50.35	49.70	49.20	48.70	48.30
Perimeter Adj., Add or Deduct	Per 100 L.F.	2.30	1.95	1.70	1.55	1.40	1.25	1.15	1.10	1.00
Story Hgt. Adj., Add or Deduct	Per 1 Ft.	.55	.45	.45	.40	.40	.35	.35	.35	.30

For Basement, add $16.60 per square foot of basement area

The above costs were calculated using the basic specifications shown on the facing page. These costs should be adjusted where necessary for design alternatives and owner's requirements. Reported completed project costs, for this type of structure, range from $34.50 to $84.50 per S.F.

Common additives

Description	Unit	$ Cost
Bowling Alleys, Incl. alley, pinsetter,		
scorer, counter & misc. supplies, average	Lane	43,400
For automatic scorer, add (maximum)	Lane	7475
Emergency Lighting, 25 watt, battery operated		
Lead battery	Each	305
Nickel cadmium	Each	600
Lockers, Steel, Single tier, 60" or 72"	Opening	128 - 207
2 tier, 60" or 72" total	Opening	79 - 99
5 tier, box lockers	Opening	41 - 53
Locker bench, lam., maple top only	L.F.	17.60
Pedestals, steel pipe	Each	39
Seating		
Auditorium chair, all veneer	Each	123
Veneer back, padded seat	Each	148
Upholstered, spring seat	Each	174
Sound System		
Amplifier, 250 watts	Each	1425
Speaker, ceiling or wall	Each	118
Trumpet	Each	222

Model costs calculated for a 1 story building with 14' story height and 20,000 square feet of floor area

				Unit	Unit Cost	Cost Per S.F.	% Of Sub-Total
1.0 Foundations							
	.1	Footings & Foundations	Poured concrete; strip and spread footings and 4' foundation wall	S.F. Ground	1.60	1.60	
	.4	Piles & Caissons	N/A	—	—	—	6.7%
	.9	Excavation & Backfill	Site preparation for slab and trench for foundation wall and footing	S.F. Ground	1.08	1.08	
2.0 Substructure							
	.1	Slab on Grade	4" reinforced concrete with vapor barrier and granular base	S.F. Slab	2.56	2.56	6.4%
	.2	Special Substructures	N/A	—	—	—	
3.0 Superstructure							
	.1	Columns & Beams	Columns included in 3.7	—	—	—	
	.4	Structural Walls	N/A	—	—	—	
	.5	Elevated Floors	N/A	—	—	—	11.8%
	.7	Roof	Metal deck on open web steel joists with columns and beams	S.F. Roof	4.75	4.75	
	.9	Stairs	N/A	—	—	—	
4.0 Exterior Closure							
	1	Walls	Concrete block — 90% of wall	S.F. Wall	6.59	2.35	
	.5	Exterior Wall Finishes	N/A	—	—	—	10.3%
	.6	Doors	Double aluminum and glass and hollow metal	Each	1234	.31	
	.7	Windows & Glazed Walls	Horizontal pivoted — 10% of wall	Each	905	1.49	
5.0 Roofing							
	.1	Roof Coverings	Built-up tar and gravel with flashing	S.F. Roof	1.74	1.74	
	.7	Insulation	Perlite/EPS composite	S.F. Roof	1.13	1.13	7.3%
	.8	Openings & Specialties	Gravel stop and hatches	S.F. Roof	.08	.08	
6.0 Interior Construction							
	.1	Partitions	Concrete block — 50 S.F. Floor/L.F. Partition	S.F. Partition	4.55	.91	
	.4	Interior Doors	Hollow metal single leaf — 1000 S.F. Floor/Door	Each	484	.48	
	.5	Wall Finishes	Paint and block filler	S.F. Surface	1.52	.61	
	.6	Floor Finishes	Vinyl tile — 25% of floor	S.F. Floor	1.84	.46	8.8%
	.7	Ceiling Finishes	Suspended fiberglass board — 25% of area	S.F. Ceiling	2.09	.52	
	.9	Interior Surface/Exterior Wall	Paint and block filler — 90% of wall	S.F. Wall	1.53	.55	
7.0 Conveying							
	.1	Elevators	N/A	—	—	—	0.0%
	.2	Special Conveyors	N/A	—	—	—	
8.0 Mechanical							
	.1	Plumbing	Toilet and service fixtures, supply and drainage — 1 Fixture/2200 S.F. Floor	Each	2354	1.07	
	.2	Fire Protection	Sprinklers, light hazard	S.F. Floor	1.09	1.09	
	.3	Heating	Included in 8.4	—	—	—	31.6%
	.4	Cooling	Single zone rooftop unit, gas heating, electric cooling	S.F. Floor	10.53	10.53	
	.5	Special Systems	N/A	—	—	—	
9.0 Electrical							
	1	Service & Distribution	800 ampere service, panel board and feeders	S.F. Floor	1.43	1.43	
	.2	Lighting & Power	Fluorescent fixtures, receptacles, switches, A.C. and misc. power	S.F. Floor	5.19	5.19	17.1%
	.4	Special Electrical	Alarm systems and emergency lighting	S.F. Floor	.23	.23	
11.0 Special Construction							
	.1	Specialties	N/A	—	—	—	0.0%
12.0 Site Work							
	.1	Earthwork	N/A	—	—	—	
	.3	Utilities	N/A	—	—	—	
	.5	Roads & Parking	N/A	—	—	—	0.0%
	.7	Site Improvements	N/A	—	—	—	
				Sub-Total		40.16	**100%**
		GENERAL CONDITIONS (Overhead & Profit)			15%	6.02	
		ARCHITECT FEES			7%	3.22	
		Total Building Cost				**49.40**	

Costs per square foot of floor area

Exterior Wall	S.F. Area	6000	8000	10000	12000	14000	16000	18000	20000	22000
	L.F. Perimeter	320	386	453	520	540	597	610	660	710
Face Brick with Concrete Block Back-up	Bearing Walls	73.70	70.80	69.05	67.95	65.85	65.15	63.70	63.25	62.85
	Steel Frame	74.35	71.50	69.75	68.65	66.60	65.95	64.50	64.05	63.65
Decorative Concrete Block	Bearing Walls	70.25	67.70	66.15	65.15	63.35	62.75	61.50	61.10	60.80
	Steel Frame	70.90	68.35	66.85	65.85	64.10	63.50	62.30	61.90	61.60
Precast Concrete Panels	Bearing Walls	70.20	67.65	66.10	65.10	63.30	62.70	61.45	61.10	60.75
	Steel Frame	70.85	68.30	66.80	65.80	64.05	63.50	62.25	61.85	61.55
Perimeter Adj., Add or Deduct	Per 100 L.F.	6.30	4.70	3.75	3.10	2.70	2.40	2.05	1.85	1.75
Story Hgt. Adj., Add or Deduct	Per 1 Ft.	1.05	.95	.90	.85	.75	.75	.65	.65	.60
Basement—Not Applicable										

The above costs were calculated using the basic specifications shown on the facing page. These costs should be adjusted where necessary for design alternatives and owner's requirements. Reported completed project costs, for this type of structure, range from $41.50 to $100.00 per S.F.

Common additives

Description	Unit	$ Cost
Directory Boards, Plastic, glass covered		
30" x 20"	Each	460
36" x 48"	Each	850
Aluminum, 24" x 18"	Each	410
36" x 24"	Each	495
48" x 32"	Each	590
48" x 60"	Each	1300
Emergency Lighting, 25 watt, battery operated		
Lead battery	Each	305
Nickel cadmium	Each	600
Benches, Hardwood	L.F.	61 - 96
Ticket Printer	Each	5175
Turnstiles, One way		
4 arm, 46" dia., manual	Each	370
Electric	Each	1375
High security, 3 arm		
65" dia., manual	Each	2975
Electric	Each	3925
Gate with horizontal bars		
65" dia., 7' high transit type	Each	3425

Important: See the Reference Section for Location Factors

Model costs calculated for a 1 story building with 14' story height and 12,000 square feet of floor area

			Unit	Unit Cost	Cost Per S.F.	% Of Sub-Total
1.0 Foundations						
.1	Footings & Foundations	Poured concrete; strip and spread footings and 4' foundation wall	S.F. Ground	2.85	2.85	6.9%
.4	Piles & Caissons	N/A	—	—	—	
.9	Excavation & Backfill	Site preparation for slab and trench for foundation wall and footing	S.F. Ground	.96	.96	
2.0 Substructure						
.1	Slab on Grade	6" reinforced concrete with vapor barrier and granular base	S.F. Slab	3.14	3.14	5.7%
.2	Special Substructures	N/A	—	—	—	
3.0 Superstructure						
.1	Columns & Beams	Columns included in 3.7	—	—	—	
.4	Structural Walls	N/A	—	—	—	
.5	Elevated Floors	N/A	—	—	—	5.7%
.7	Roof	Metal deck on open web steel joists with columns and beams	S.F. Roof	3.15	3.15	
.9	Stairs	N/A	—	—	—	
4.0 Exterior Closure						
.1	Walls	Face brick with concrete block backup — 70% of wall	S.F. Wall	15.00	6.71	
.5	Exterior Wall Finishes	N/A	—	—	—	18.7%
.6	Doors	Double aluminum and glass	Each	3160	1.05	
.7	Windows & Glazed Walls	Store front — 30% of wall	S.F. Window	14.15	2.58	
5.0 Roofing						
.1	Roof Coverings	Built-up tar and gravel with flashing	S.F. Roof	1.98	1.98	
.7	Insulation	Perlite/EPS composite	S.F. Roof	1.13	1.13	6.0%
.8	Openings & Specialties	Gravel stop	L.F. Perimeter	4.78	.21	
6.0 Interior Construction						
.1	Partitions	Lightweight concrete block — 15 S.F. Floor/L.F. Partition	S.F. Partition	4.57	3.05	
.4	Interior Doors	Hollow metal — 150 S.F. Floor/Door	Each	484	3.23	
.5	Wall Finishes	Glazed coating	S.F. Surface	.95	1.27	27.5%
.6	Floor Finishes	Quarry tile and vinyl composition tile	S.F. Floor	4.62	4.62	
.7	Ceiling Finishes	Mineral fiber tile on concealed zee bars	S.F. Ceiling	2.62	2.62	
.9	Interior Surface/Exterior Wall	Glazed coating — 70% of wall	S.F. Wall	.95	.40	
7.0 Conveying						
.1	Elevators	N/A	—	—	—	0.0%
.2	Special Conveyors	N/A	—	—	—	
8.0 Mechanical						
.1	Plumbing	Toilet and service fixtures, supply and drainage — 1 Fixture/850 S.F. Floor	Each	1989	2.34	
.2	Fire Protection	Wet pipe sprinkler system	S.F. Floor	1.26	1.26	
.3	Heating	Included in 8.4	—	—	—	20.9%
.4	Cooling	Single zone rooftop unit, gas heating, electric cooling	S.F. Floor	7.89	7.89	
.5	Special Systems	N/A	—	—	—	
9.0 Electrical						
.1	Service & Distribution	400 ampere service, panel board and feeders	S.F. Floor	.89	.89	
.2	Lighting & Power	Fluorescent fixtures, receptacles, switches, A.C. and misc. power	S.F. Floor	3.45	3.45	8.6%
.4	Special Electrical	Alarm systems and emergency lighting	S.F. Floor	.43	.43	
11.0 Special Construction						
.1	Specialties	N/A	—	—	—	0.0%
12.0 Site Work						
.1	Earthwork	N/A	—	—	—	
.3	Utilities	N/A	—	—	—	0.0%
.5	Roads & Parking	N/A	—	—	—	
.7	Site Improvements	N/A	—	—	—	
			Sub-Total		55.21	100%
	GENERAL CONDITIONS (Overhead & Profit)			15%	8.28	
	ARCHITECT FEES			7%	4.46	
			Total Building Cost		67.95	

Costs per square foot of floor area

Exterior Wall	S.F. Area	600	800	1000	1200	1600	2000	2400	3000	4000
	L.F. Perimeter	100	114	128	139	164	189	214	250	314
Brick with Concrete Block Back-up	Steel Frame	160.95	154.65	150.85	147.70	144.25	142.15	140.75	139.25	138.00
	Bearing Walls	160.00	153.65	149.90	146.75	143.25	141.20	139.80	138.30	137.00
Concrete Block	Steel Frame	141.80	138.25	136.15	134.40	132.45	131.30	130.50	129.65	128.95
	Bearing Walls	140.85	137.30	135.20	133.45	131.50	130.35	129.55	128.70	128.00
Galvanized Steel Siding	Steel Frame	140.70	137.65	135.85	134.30	132.65	131.65	131.00	130.25	129.65
Metal Sandwich Panel	Steel Frame	141.20	137.30	135.00	133.10	130.95	129.70	128.85	127.90	127.15
Perimeter Adj., Add or Deduct	Per 100 L.F.	43.35	32.55	26.05	21.70	16.30	13.00	10.85	8.70	6.50
Story Hgt. Adj., Add or Deduct	Per 1 Ft.	2.65	2.30	2.05	1.85	1.65	1.50	1.45	1.35	1.25
Basement—Not Applicable										

The above costs were calculated using the basic specifications shown on the facing page. These costs should be adjusted where necessary for design alternatives and owner's requirements. Reported completed project costs, for this type of structure, range from $65.75 to $195.00 per S.F.

Common additives

Description	Unit	$ Cost
Air Compressors, Electric		
1-1/2 H.P., standard controls	Each	3000
Dual controls	Each	3225
5 H.P., 115/230 volt, standard control	Each	4025
Dual controls	Each	4250
Emergency Lighting, 25 watt, battery operated		
Lead battery	Each	305
Nickel cadmium	Each	600
Fence, Chain link, 6' high		
9 ga. wire, galvanized	L.F.	11.45
6 ga. wire	L.F.	15.60
Gate	Each	192
Product Dispenser with vapor recovery		
for 6 nozzles	Each	16,500
Laundry Equipment		
Dryer, gas, 16 lb. capacity	Each	640
30 lb. capacity	Each	2525
Washer, 4 cycle	Each	720
Commercial	Each	1100

Description	Unit	$ Cost
Lockers, Steel, single tier, 60" or 72"	Opening	128 - 207
2 tier, 60" or 72" total	Opening	79 - 99
5 tier, box lockers	Opening	41 - 53
Locker bench, lam. maple top only	L.F.	17.60
Pedestals, steel pipe	Each	39
Paving, Bituminous		
Wearing course plus base course	Sq. Yard	9.65
Safe, Office type, 4 hour rating		
30" x 18" x 18"	Each	3475
62" x 33" x 20"	Each	7575
Sidewalks, Concrete 4" thick	S.F.	2.18
Yard Lighting,		
20' aluminum pole with 400 watt high pressure sodium fixture	Each	1855

Important: See the Reference Section for Location Factors

Model costs calculated for a 1 story building with 12′ story height and 800 square feet of floor area

			Unit	Unit Cost	Cost Per S.F.	% Of Sub-Total
1.0 Foundations						
.1	Footings & Foundations	Poured concrete; strip and spread footings and 4′ foundation wall	S.F. Ground	5.87	5.87	
.4	Piles & Caissons	N/A	—	—	—	5.8%
.9	Excavation & Backfill	Site preparation for slab and trench for foundation wall and footing	S.F. Ground	1.38	1.38	
2.0 Substructure						
.1	Slab on Grade	5″ reinforced concrete with vapor barrier and granular base	S.F. Slab	2.79	2.79	2.2%
.2	Special Substructures	N/A	—			
3.0 Superstructure						
.1	Columns & Beams	Steel columns included in 3.7	—	—	—	
.4	Structural Walls	N/A	—	—	—	
.5	Elevated Floors	N/A	—	—	—	2.4%
.7	Roof	Metal deck, open web steel joists, beams, columns	S.F. Roof	3.00	3.00	
.9	Stairs	N/A	—	—	—	
4.0 Exterior Closure						
.1	Walls	Face brick with concrete block backup *70% of wall*	S.F. Wall	15.80	18.91	
.5	Exterior Wall Finishes	N/A	—	—	—	24.1%
.6	Doors	Steel overhead hollow metal	Each	1581	7.90	
.7	Windows & Glazed Walls	Horizontal pivoted steel *5% of wall*	Each	361	3.43	
5.0 Roofing						
.1	Roof Coverings	Built-up tar and gravel with flashing	S.F. Roof	3.20	3.20	
.7	Insulation	Perlite/EPS composite	S.F. Roof	1.13	1.13	3.4%
.8	Openings & Specialties	N/A	—	—	—	
6.0 Interior Construction						
.1	Partitions	Concrete block *20 S.F. Floor/S.F. Partition*	S.F. Partition	4.70	2.35	
.4	Interior Doors	Hollow metal *600 S.F. Floor/Door*	Each	484	.81	
.5	Wall Finishes	Paint	S.F. Surface	.62	.62	
.6	Floor Finishes	N/A	—	—	—	3.0%
.7	Ceiling Finishes	N/A	—	—	—	
.9	Interior Surface/Exterior Wall	N/A	—	—	—	
7.0 Conveying						
.1	Elevators	N/A	—	—	—	0.0%
.2	Special Conveyors	N/A	—	—	—	
8.0 Mechanical						
.1	Plumbing	Toilet and service fixtures, supply and drainage *1 Fixture/160 S.F. Floor*	Each	4536	28.35	
.2	Fire Protection	N/A	—	—	—	
.3	Heating	Unit heaters	S.F. Floor	10.42	10.42	30.8%
.4	Cooling	N/A	—	—	—	
.5	Special Systems	N/A	—	—	—	
9.0 Electrical						
.1	Service & Distribution	400 ampere service, panel board and feeders	S.F. Floor	12.01	12.01	
.2	Lighting & Power	Fluorescent fixtures, receptacles, switches and misc. power	S.F. Floor	22	22.23	28.3%
.4	Special Electrical	Emergency power	S.F. Floor	1.26	1.26	
11.0 Special Construction						
.1	Specialties	N/A	—	—	—	0.0%
12.0 Site Work						
.1	Earthwork	N/A	—	—	—	
.3	Utilities	N/A	—	—	—	0.0%
.5	Roads & Parking	N/A	—	—	—	
.7	Site Improvements	N/A	—	—	—	
				Sub-Total	125.66	**100%**

GENERAL CONDITIONS (Overhead & Profit)		15%	18.85
ARCHITECT FEES		7%	10.14
	Total Building Cost		**154.65**

Costs per square foot of floor area

Exterior Wall	S.F. Area	2000	7000	12000	17000	22000	27000	32000	37000	42000
	L.F. Perimeter	180	340	470	540	640	740	762	793	860
Decorative Concrete Brick	Wood Arch	120.30	91.80	85.50	80.75	79.00	77.85	75.60	74.10	73.45
	Steel Truss	117.25	88.80	82.50	77.75	76.00	74.85	72.60	71.10	70.45
Stone with Concrete Block Back-up	Wood Arch	136.40	100.50	92.50	86.45	84.20	82.80	79.85	77.95	77.10
	Steel Truss	130.30	94.40	86.40	80.35	78.10	76.70	73.75	71.85	71.05
Face Brick with Concrete Block Back-up	Wood Arch	131.95	98.10	90.55	**84.85**	82.75	81.40	78.70	76.90	76.10
	Steel Truss	125.85	92.00	84.50	78.80	76.65	75.30	72.60	70.80	70.05
Perimeter Adj., Add or Deduct	Per 100 L.F.	30.90	8.80	5.15	3.60	2.80	2.25	1.95	1.65	1.50
Story Hgt. Adj., Add or Deduct	Per 1 Ft.	1.85	1.00	.80	.65	.60	.55	.50	.45	.40

For Basement, add $17.20 per square foot of basement area

The above costs were calculated using the basic specifications shown on the facing page. These costs should be adjusted where necessary for design alternatives and owner's requirements. Reported completed project costs, for this type of structure, range from $45.70 to $170.00 per S.F.

Common additives

Description	Unit	$ Cost	Description	Unit	$ Cost
Altar, Wood, custom design, plain	Each	1550	Pews/Benches, Hardwood	L.F.	61 - 96
Deluxe	Each	8800	Pulpits, Prefabricated, hardwood	Each	1225 - 7400
Granite or marble, average	Each	6150	Railing, Hardwood	L.F.	148
Deluxe	Each	14,000	Steeples, translucent fiberglass		
Ark, Prefabricated, plain	Each	2900	30" square, 15' high	Each	2700
Deluxe	Each	106,000	25' high	Each	4050
Baptistry, Fiberglass, incl. plumbing	Each	2125 - 4350	Painted fiberglass, 24" square, 14' high	Each	2750
Bells & Carillons, 48 bells	Each	440,000	28' high	Each	3850
24 bells	Each	165,000	Aluminum		
Confessional, Prefabricated wood			20' high, 3'- 6" base	Each	4275
Single, plain	Each	2175	35' high, 8'- 0" base	Each	16,200
Deluxe	Each	7775	60' high, 14'- 0" base	Each	36,400
Double, plain	Each	7400			
Deluxe	Each	17,700			
Emergency Lighting, 25 watt, battery operated					
Lead battery	Each	305			
Nickel cadmium	Each	600			
Lecterns, Wood, plain	Each	480			
Deluxe	Each	1500			

Important: See the Reference Section for Location Factors

Model costs calculated for a 1 story building with 24' story height and 17,000 square feet of floor area

				Unit	Unit Cost	Cost Per S.F.	% Of Sub-Total
1.0 Foundations							
.1	Footings & Foundations	Poured concrete; strip and spread footings and 4' foundation wall		S.F. Ground	3.64	3.64	
.4	Piles & Caissons	N/A		—	—	—	6.8%
.9	Excavation & Backfill	Site preparation for slab and trench for foundation wall and footing		S.F. Ground	.91	.91	
2.0 Substructure							
.1	Slab on Grade	4" reinforced concrete with vapor barrier and granular base		S.F. Slab	2.56	2.56	3.9%
.2	Special Substructures	N/A		—	—	—	
3.0 Superstructure							
.1	Columns & Beams	N/A		—	—	—	
.4	Structural Walls	N/A		—	—	—	
.5	Elevated Floors	N/A		—	—	—	18.1%
.7	Roof	Wood deck on laminated wood arches		S.F. Roof	11.36	12.06	
.9	Stairs	N/A		—	—	—	
4.0 Exterior Closure							
.1	Walls	Face brick with concrete block backup	80% of wall (adjusted for end walls)	S.F. Wall	18.76	11.44	
.5	Exterior Wall Finishes	N/A		—	—	—	21.6%
.6	Doors	Double hollow metal swinging, single hollow metal		Each	873	.31	
.7	Windows & Glazed Walls	Aluminum, top hinged, in-swinging and curtain wall panels	20% of wall	S.F. Window	17.05	2.60	
5.0 Roofing							
.1	Roof Coverings	Asphalt shingles with flashing		S.F. Roof	1.24	1.24	
.7	Insulation	Polystyrene		S.F. Ground	1.03	1.24	4.0%
.8	Openings & Specialties	Gutters and downspouts		S.F. Ground	.18	.18	
6.0 Interior Construction							
.1	Partitions	Plaster on metal studs	40 S.F. Floor/L.F. Partitions	S.F. Partition	5.42	3.25	
.4	Interior Doors	Hollow metal	400 S.F. Floor/Door	Each	484	1.21	
.5	Wall Finishes	Paint		S.F. Surface	.58	.70	
.6	Floor Finishes	Carpet		S.F. Floor	4.86	4.86	17.3%
.7	Ceiling Finishes	N/A		—	—	—	
.9	Interior Surface/Exterior Wall	Painted plaster	80% of wall	S.F. Wall	2.47	1.51	
7.0 Conveying							
.1	Elevators	N/A		—	—	—	0.0%
.2	Special Conveyors	N/A		—	—	—	
8.0 Mechanical							
.1	Plumbing	Kitchen, toilet and service fixtures, supply and drainage	1 Fixture/2430 S.F. Floor	Each	2235	.92	
.2	Fire Protection	Wet pipe sprinkler system		S.F. Floor	1.26	1.26	
.3	Heating	Oil fired hot water, wall fin radiation		S.F. Floor	4.74	4.74	19.4%
.4	Cooling	Split systems with air cooled condensing units		S.F. Floor	5.91	5.91	
.5	Special Systems	N/A		—	—		
9.0 Electrical							
.1	Service & Distribution	200 ampere service, panel board and feeders		S.F. Floor	.28	.28	
.2	Lighting & Power	Fluorescent fixtures, receptacles, switches, A.C. and misc. power		S.F. Floor	4.15	4.15	8.9%
.4	Special Electrical	Alarm systems, sound system and emergency lighting		S.F. Floor	1.52	1.52	
11.0 Special Construction							
.1	Specialties	N/A		—	—	—	0.0%
12.0 Site Work							
.1	Earthwork	N/A		—	—	—	
.3	Utilities	N/A		—	—	—	
.5	Roads & Parking	N/A		—	—	—	0.0%
.7	Site Improvements	N/A		—	—	—	
				Sub-Total		66.49	**100%**
	GENERAL CONDITIONS (Overhead & Profit)				15%	9.97	
	ARCHITECT FEES				11%	8.39	
				Total Building Cost		84.85	

Costs per square foot of floor area

Exterior Wall	S.F. Area	2000	4000	6000	8000	12000	15000	18000	20000	22000
	L.F. Perimeter	180	280	340	386	460	535	560	600	640
Stone Ashlar Veneer On Concrete Block	Wood Truss	120.05	105.70	98.40	94.05	89.10	87.65	85.60	85.00	84.55
	Steel Joists	118.40	104.10	96.75	92.40	87.50	86.05	83.95	83.40	82.95
Stucco on Concrete Block	Wood Truss	109.75	97.70	91.90	88.50	84.75	83.60	82.05	81.60	81.25
	Steel Joists	108.30	96.25	90.45	87.05	83.30	82.15	80.60	80.15	79.80
Brick Veneer	Wood Frame	115.45	102.10	95.40	91.45	87.00	85.70	83.85	83.30	82.90
Wood Shingles	Wood Frame	110.40	98.15	92.20	88.75	84.85	83.70	82.10	81.65	81.30
Perimeter Adj., Add or Deduct	Per 100 L.F.	19.25	9.60	6.40	4.80	3.20	2.55	2.15	1.95	1.75
Story Hgt. Adj., Add or Deduct	Per 1 Ft.	2.15	1.65	1.35	1.15	.90	.85	.75	.75	.70
For Basement, add $15.95 per square foot of basement area										

The above costs were calculated using the basic specifications shown on the facing page. These costs should be adjusted where necessary for design alternatives and owner's requirements. Reported completed project costs, for this type of structure, range from $46.75 to $153.00 per S.F.

Common additives

Description	Unit	$ Cost	Description	Unit	$ Cost
Bar, Front bar	L.F.	246	Sauna, Prefabricated, complete, 6' x 4'	Each	3550
Back bar	L.F.	193	6' x 9'	Each	5600
Booth, Upholstered, custom, straight	L.F.	125 - 232	8' x 8'	Each	5925
"L" or "U" shaped	L.F.	129 - 220	10' x 12'	Each	8500
Fireplaces, Brick not incl. chimney			Smoke Detectors		
or foundation, 30" x 24" opening	Each	1550	Ceiling type	Each	126
Chimney, standard brick			Duct type	Each	360
Single flue 16" x 20"	V.L.F.	46	Sound System		
20" x 20"	V.L.F.	52	Amplifier, 250 watts	Each	1425
2 flue, 20" x 32"	V.L.F.	77	Speaker, ceiling or wall	Each	118
Lockers, Steel, single tier, 60" or 72"	Opening	128 - 207	Trumpet	Each	222
2 tier, 60" or 72" total	Opening	79 - 99	Steam Bath, Complete, to 140 C.F.	Each	890
5 tier, box lockers	Opening	41 - 53	To 300 C.F.	Each	1075
Locker bench, lam. maple top only	L.F.	17.60	To 800 C.F.	Each	2925
Pedestals, steel pipe	Each	39	To 2500 C.F.	Each	3275
Refrigerators, Prefabricated, walk-in			Swimming Pool Complete, gunite	S.F.	39 - 48
7'-6" High, 6' x 6'	S.F.	114	Tennis Court, Complete with fence		
10' x 10'	S.F.	89	Bituminous	Each	19,100 - 23,400
12' x 14'	S.F.	79	Clay	Each	21,000 - 26,400
12' x 20'	S.F.	70			

Model costs calculated for a 1 story building with 12' story height and 6,000 square feet of floor area

				Unit	Unit Cost	Cost Per S.F.	% Of Sub-Total
1.0 Foundations							
	.1	Footings & Foundations	Poured concrete; strip and spread footings and 4' foundation wall	S.F. Ground	4.01	4.01	6.4%
	.4	Piles & Caissons	N/A	—	—	—	
	.9	Excavation & Backfill	Site preparation for slab and trench for foundation wall and footing	S.F. Ground	.96	.96	
2.0 Substructure							
	.1	Slab on Grade	4" reinforced concrete with vapor barrier and granular base	S.F. Slab	2.56	2.56	3.3%
	.2	Special Substructures	N/A	—	—	—	
3.0 Superstructure							
	.1	Columns & Beams	N/A	—	—	—	
	.4	Structural Walls	Load bearing partition walls, see item 6.1	—	—	—	
	.5	Elevated Floors	N/A	—	—	—	5.0%
	.7	Roof	Wood truss with plywood sheathing	S.F. Ground	3.87	3.87	
	.9	Stairs	N/A	—	—	—	
4.0 Exterior Closure							
	.1	Walls	Stone ashlar veneer on concrete block 65% of wall	S.F. Wall	17.51	7.74	
	.5	Exterior Wall Finishes	N/A	—	—	—	17 1%
	.6	Doors	Double aluminum and glass, hollow metal	Each	1433	1.43	
	.7	Windows & Glazed Walls	Aluminum horizontal sliding 35% of wall	Each	254	4.03	
5.0 Roofing							
	.1	Roof Coverings	Asphalt shingles	S.F. Roof	.81	1.00	
	.7	Insulation	N/A	—	—	—	1.7%
	.8	Openings & Specialties	Gutters and downspouts	S.F. Roof	.31	.31	
6.0 Interior Construction							
	.1	Partitions	Gypsum board on metal studs, load bearing 14 S.F. Floor/L.F. Partition	S.F. Partition	2.44	1.74	
	.4	Interior Doors	Single leaf wood 140 S.F. Floor/Door	Each	367	2.62	
	.5	Wall Finishes	40% vinyl wall covering, 40% paint, 20% ceramic tile	S.F. Surface	1.50	2.15	19.4%
	.6	Floor Finishes	50% carpet, 30% hardwood tile, 20% ceramic tile	S.F. Floor	5.13	5.13	
	.7	Ceiling Finishes	Gypsum plaster on wood furring	S.F. Ceiling	2.32	2.32	
	.9	Interior Surface/Exterior Wall	Painted gypsum board on furring 65% of wall	S.F. Wall	2.25	.99	
7.0 Conveying							
	.1	Elevators	N/A	—	—	—	0.0%
	.2	Special Conveyors	N/A	—	—	—	
8.0 Mechanical							
	.1	Plumbing	Kitchen, toilet and service fixtures, supply and drainage 1 Fixture/125 S.F. Floor	Each	1666	13.33	
	.2	Fire Protection	Wet pipe sprinkler system	S.F. Floor	1.66	1.66	
	.3	Heating	Included in 8.4	—	—	—	41.9%
	.4	Cooling	Multizone rooftop unit, gas heating, electric cooling	S.F. Floor	17.20	17.20	
	.5	Special Systems	N/A	—	—	—	
9.0 Electrical							
	.1	Service & Distribution	200 ampere service, panel board and feeders	S.F. Floor	1.07	1.07	
	.2	Lighting & Power	Fluorescent fixtures, receptacles, switches, A.C. and misc. power	S.F. Floor	2.59	2.59	5.2%
	.4	Special Electrical	Alarm systems and emergency lighting	S.F. Floor	.37	.37	
11.0 Special Construction							
	.1	Specialties	N/A	—	—	—	0.0%
12.0 Site Work							
	.1	Earthwork	N/A	—	—	—	
	.3	Utilities	N/A	—	—	—	0.0%
	.5	Roads & Parking	N/A	—	—	—	
	.7	Site Improvements	N/A	—	—	—	
				Sub-Total		77.08	**100%**

GENERAL CONDITIONS (Overhead & Profit)		15%	11.56
ARCHITECT FEES		11%	9.76

Total Building Cost	98.40

Costs per square foot of floor area

Exterior Wall	S.F. Area	4000	6000	8000	10000	12000	14000	16000	18000	20000
	L.F. Perimeter	260	340	420	453	460	510	560	610	600
Face Brick with Concrete Block Back-up	Bearing Walls	79.60	75.65	73.65	70.60	67.60	66.80	66.15	65.65	64.05
	Steel Frame	77.65	74.10	72.35	69.70	67.10	66.40	65.80	65.35	63.95
Decorative Concrete Block	Bearing Walls	70.55	67.40	65.80	63.40	61.10	60.45	59.90	59.50	58.25
	Steel Frame	71.85	69.10	67.70	65.65	63.70	63.15	62.70	62.35	61.30
Tilt Up Concrete Wall Panels	Bearing Walls	70.95	68.10	66.65	64.55	62.50	61.95	61.45	61.10	60.05
	Steel Frame	68.95	66.55	65.35	63.65	62.00	61.50	61.15	60.85	59.95
Perimeter Adj., Add or Deduct	Per 100 L.F.	10.10	6.75	5.05	4.00	3.40	2.85	2.55	2.25	2.00
Story Hgt. Adj., Add or Deduct	Per 1 Ft.	1.30	1.15	1.10	.90	.80	.75	.70	.70	.60

For Basement, add $16.70 per square foot of basement area

The above costs were calculated using the basic specifications shown on the facing page. These costs should be adjusted where necessary for design alternatives and owner's requirements. Reported completed project costs, for this type of structure, range from $ 45.00 to $ 143.50 per S.F.

Common additives

Description	Unit	$ Cost
Bar, Front bar	L.F.	246
Back bar	L.F.	193
Booth, Upholstered, custom straight	L.F.	125 - 232
"L" or "U" shaped	L.F.	129 - 220
Bowling Alleys, incl. alley, pinsetter		
Scorer, counter & misc. supplies, average	Lane	43,400
For automatic scorer, add	Lane	7475
Emergency Lighting, 25 watt, battery operated		
Lead battery	Each	305
Nickel cadmium	Each	600
Kitchen Equipment		
Broiler	Each	3400
Coffee urn, twin 6 gallon	Each	5875
Cooler, 6 ft. long	Each	2675
Dishwasher, 10-12 racks per hr.	Each	2525
Food warmer	Each	745
Freezer, 44 C.F., reach-in	Each	6575
Ice cube maker, 50 lb. per day	Each	1700
Range with 1 oven	Each	2300

Description	Unit	$ Cost
Movie Equipment		
Projector, 35mm	Each	10,900 - 14,500
Screen, wall or ceiling hung	S.F.	6.85 - 10.85
Partitions, Folding leaf, wood		
Acoustic type	S.F.	46 - 75
Seating		
Auditorium chair, all veneer	Each	123
Veneer back, padded seat	Each	148
Upholstered, spring seat	Each	174
Classroom, movable chair & desk	Set	65 - 120
Lecture hall, pedestal type	Each	123 - 360
Sound System		
Amplifier, 250 watts	Each	1425
Speaker, ceiling or wall	Each	118
Trumpet	Each	222
Stage Curtains, Medium weight	S.F.	12.95 - 375
Curtain Track, Light duty	L.F.	43
Swimming Pools, Complete, gunite	S.F.	39 - 48

Important: See the Reference Section for Location Factors

Model costs calculated for a 1 story building with 12' story height and 10,000 square feet of floor area

				Unit	Unit Cost	Cost Per S.F.	% Of Sub-Total
1.0 Foundations							
.1	Footings & Foundations	Poured concrete; strip and spread footings and 4' foundation wall		S.F. Ground	4.60	4.60	9.9%
.4	Piles & Caissons	N/A		–	–	–	
.9	Excavation & Backfill	Site preparation for slab and trench for foundation wall and footing		S.F. Ground	.96	.96	
2.0 Substructure							
.1	Slab on Grade	4" reinforced concrete with vapor barrier and granular base		S.F. Slab	2.56	2.56	4.5%
.2	Special Substructures	N/A		–	–	–	
3.0 Superstructure							
.1	Columns & Beams	N/A		–	–	–	
.4	Structural Walls	Concrete block		S.F. Wall	4.56	.75	
.5	Elevated Floors	N/A		–	–	–	7.5%
.7	Roof	Metal deck on open web steel joists		S.F. Roof	3.48	3.48	
.9	Stairs	N/A		–	–	–	
4.0 Exterior Closure							
.1	Walls	Face brick with concrete block backup	80% of wall	S.F. Wall	15.80	6.87	
.5	Exterior Wall Finishes	N/A		–	–	–	15.4%
.6	Doors	Double aluminum and glass and hollow metal		Each	1308	.52	
.7	Windows & Glazed Walls	Aluminum sliding	20% of wall	Each	370	1.26	
5.0 Roofing							
.1	Roof Coverings	Built-up tar and gravel with flashing		S.F. Roof	2.01	2.01	
.7	Insulation	Perlite/EPS composite		S.F. Roof	1.13	1.13	6.5%
.8	Openings & Specialties	Gravel stop and hatches		S.F. Roof	.52	.52	
6.0 Interior Construction							
.1	Partitions	Gypsum board on metal studs, toilet partitions	14 S.F. Floor/L.F. Partition	S.F. Partition	4.77	3.41	
.4	Interior Doors	Single leaf hollow metal	140 S.F. Floor/Door	Each	484	3.46	
.5	Wall Finishes	Paint		S.F. Surface	.42	.60	26.1%
.6	Floor Finishes	50% carpet, 50% vinyl tile		S.F. Floor	3.96	3.96	
.7	Ceiling Finishes	Mineral fiber tile on concealed zee bars		S.F. Ceiling	2.62	2.62	
.9	Interior Surface/Exterior Wall	Paint	80% of wall	S.F. Wall	1.53	.67	
7.0 Conveying							
.1	Elevators	N/A		–	–	–	0.0%
.2	Special Conveyors	N/A		–	–	–	
8.0 Mechanical							
.1	Plumbing	Kitchen, toilet and service fixtures, supply and drainage	1 Fixture/910 S.F. Floor	Each	3185	3.50	
.2	Fire Protection	Wet pipe sprinkler system		S.F. Floor	1.26	1.26	
.3	Heating	Included in 8.4		–	–	–	21.3%
.4	Cooling	Single zone rooftop unit, gas heating, electric cooling		S.F. Floor	7.25	7.25	
.5	Special Systems	N/A		–	–	–	
9.0 Electrical							
.1	Service & Distribution	200 ampere service, panel board and feeders		S.F. Floor	.53	.53	
.2	Lighting & Power	Incandescent fixtures, receptacles, switches, A.C. and misc. power		S.F. Floor	2.30	2.30	5.7%
.4	Special Electrical	Alarm systems and emergency lighting		S.F. Floor	.36	.36	
11.0 Special Construction							
.1	Specialties	Built-in coat racks, fume hoods, freezer, kitchen equipment		S.F. Floor	1.74	1.74	3.1%
12.0 Site Work							
.1	Earthwork	N/A		–	–	–	
.3	Utilities	N/A		–	–	–	0.0%
.5	Roads & Parking	N/A		–	–	–	
.7	Site Improvements	N/A		–	–	–	
				Sub-Total		56.32	**100%**

GENERAL CONDITIONS (Overhead & Profit)		15%	8.45
ARCHITECT FEES		9%	5.83

Total Building Cost		**70.60**

Costs per square foot of floor area

Exterior Wall	S.F. Area	16000	23000	30000	37000	44000	51000	58000	65000	72000
	L.F. Perimeter	597	763	821	968	954	1066	1090	1132	1220
Limestone with Concrete Block Back-up	R/Conc. Frame	101.85	98.55	**94.75**	93.70	90.95	90.35	89.05	88.15	87.80
	Steel Frame	103.30	99.95	96.15	95.10	92.35	91.75	90.45	89.55	89.25
Face Brick with Concrete Block Back-up	R/Conc. Frame	96.10	93.40	90.50	89.65	87.60	87.15	86.15	85.45	85.20
	Steel Frame	97.55	94.80	91.90	91.10	89.00	88.55	87.55	86.85	86.60
Stone with Concrete Block Back-up	R/Conc. Frame	96.90	94.10	91.05	90.20	88.05	87.55	86.50	85.80	85.50
	Steel Frame	98.35	95.55	92.50	91.65	89.50	89.00	87.95	87.25	86.95
Perimeter Adj., Add or Deduct	Per 100 L.F.	3.55	2.40	1.90	1.55	1.25	1.10	.95	.85	.80
Story Hgt. Adj., Add or Deduct	Per 1 Ft.	1.25	1.10	.90	.90	.70	.70	.65	.60	.55

For Basement, add $16.40 per square foot of basement area

The above costs were calculated using the basic specifications shown on the facing page. These costs should be adjusted where necessary for design alternatives and owner's requirements. Reported completed project costs, for this type of structure, range from $79.30 to $147.75 per S.F.

Common additives

Description	Unit	$ Cost		Description	Unit	$ Cost
Benches, Hardwood	L.F.	61 - 96		Flagpoles, Complete		
Clock System				Aluminum, 20' high	Each	1025
20 room	Each	9875		40' high	Each	2375
50 room	Each	23,800		70' high	Each	6925
Closed Circuit Surveillance, One station				Fiberglass, 23' high	Each	1400
Camera and monitor	Each	1250		39'-5" high	Each	2425
For additional camera stations, add	Each	695		59' high	Each	6875
Directory Boards, Plastic, glass covered				Intercom System, 25 station capacity		
30" x 20"	Each	460		Master station	Each	1725
36" x 48"	Each	850		Intercom outlets	Each	101
Aluminum, 24" x 18"	Each	410		Handset	Each	277
36" x 24"	Each	495		Safe, Office type, 4 hour rating		
48" x 32"	Each	590		30" x 18" x 18"	Each	3475
48" x 60"	Each	1300		62" x 33" x 20"	Each	7575
Emergency Lighting, 25 watt, battery operated				Smoke Detectors		
Lead battery	Each	305		Ceiling type	Each	126
Nickel cadmium	Each	600		Duct type	Each	360

Important: See the Reference Section for Location Factors

Model costs calculated for a 1 story building with 14′ story height and 30,000 square feet of floor area

					Unit	Unit Cost	Cost Per S.F.	% Of Sub-Total
1.0 Foundations								
	.1	Footings & Foundations	Poured concrete; strip and spread footings and 4′ foundation wall		S.F. Ground	1.88	1.88	
	.4	Piles & Caissons	N/A		—	—	—	3.6%
	.9	Excavation & Backfill	Site preparation for slab and trench for foundation wall and footing		S.F. Ground	.91	.91	
2.0 Substructure								
	.1	Slab on Grade	4″ reinforced concrete with vapor barrier and granular base		S.F. Slab	2.56	2.56	3.3%
	.2	Special Substructures	N/A		—	—	—	
3.0 Superstructure								
	.1	Columns & Beams	Concrete		L.F. Column	32	.82	
	.4	Structural Walls	N/A		—	—	—	
	.5	Elevated Floors	N/A		—	—	—	12.5%
	.7	Roof	Cast-in-place concrete waffle slab		S.F. Roof	8.84	8.84	
	.9	Stairs	N/A		—	—	—	
4.0 Exterior Closure								
	.1	Walls	Limestone panels with concrete block backup	75% of wall	S.F. Wall	27	7.97	
	.5	Exterior Wall Finishes	N/A		—	—	—	13.0%
	.6	Doors	Double wood		Each	1199	.28	
	.7	Windows & Glazed Walls	Aluminum with insulated glass	25% of wall	Each	420	1.75	
5.0 Roofing								
	.1	Roof Coverings	Built-up tar and gravel with flashing		S.F. Roof	1.81	1.81	
	.7	Insulation	Perlite/EPS composite		S.F. Roof	1.13	1.13	3.9%
	.8	Openings & Specialties	Gravel stop, hatches		S.F. Roof	.05	.05	
6.0 Interior Construction								
	.1	Partitions	Plaster on metal studs, toilet partitions	10 S.F. Floor/L.F. Partition	S.F. Partition	6.50	7.80	
	.4	Interior Doors	Single leaf wood	100 S.F. Floor/Door	Each	367	3.67	
	.5	Wall Finishes	70% paint, 20% wood paneling, 10% vinyl wall covering		S.F. Surface	1.52	3.64	36.8%
	.6	Floor Finishes	60% hardwood, 20% carpet, 20% terrazzo		S.F. Floor	7.59	7.59	
	.7	Ceiling Finishes	Gypsum plaster on metal lath, suspended		S.F. Ceiling	4.92	4.92	
	.9	Interior Surface/Exterior Wall	Painted plaster	75% of wall	S.F. Wall	2.47	.71	
7.0 Conveying								
	.1	Elevators	N/A		—	—	—	0.0%
	.2	Special Conveyors	N/A		—	—	—	
8.0 Mechanical								
	.1	Plumbing	Toilet and service fixtures, supply and drainage	1 Fixture/1110 S.F. Floor	Each	2553	2.30	
	.2	Fire Protection	Wet pipe sprinkler system		S.F. Floor	1.09	1.09	
	.3	Heating	Included in 8.4		—	—	—	19.8%
	.4	Cooling	Multizone unit, gas heating, electric cooling		S.F. Floor	11.81	11.81	
	.5	Special Systems	N/A		—	—	—	
9.0 Electrical								
	.1	Service & Distribution	400 ampere service, panel board and feeders		S.F. Floor	.50	.50	
	.2	Lighting & Power	Fluorescent fixtures, receptacles, switches, A.C. and misc. power		S.F. Floor	4.63	4.63	7.1%
	.4	Special Electrical	Alarm systems and emergency lighting		S.F. Floor	.32	.32	
11.0 Special Construction								
	.1	Specialties	N/A		—	—	—	0.0%
12.0 Site Work								
	.1	Earthwork	N/A		—	—	—	
	.3	Utilities	N/A		—	—	—	0.0%
	.5	Roads & Parking	N/A		—	—	—	
	.7	Site Improvements	N/A		—	—	—	
					Sub-Total		76.98	**100%**
		GENERAL CONDITIONS (Overhead & Profit)				15%	11.55	
		ARCHITECT FEES				7%	6.22	
					Total Building Cost		**94.75**	

Costs per square foot of floor area

Exterior Wall	S.F. Area	12000	18000	24000	30000	36000	42000	48000	54000	60000
	L.F. Perimeter	460	580	713	730	826	880	965	1006	1045
Concrete Block	Steel Frame	57.10	54.05	52.70	50.70	50.05	49.25	48.85	48.30	47.85
	Bearing Walls	56.35	53.30	51.95	49.95	49.30	48.50	48.10	47.55	47.10
Precast Concrete Panels	Steel Frame	60.15	56.60	55.05	52.65	51.85	50.90	50.45	49.80	49.25
Insulated Metal Panels	Steel Frame	58.10	54.85	53.45	51.35	50.60	49.80	49.35	48.80	48.30
Face Brick on Common Brick	Steel Frame	66.10	61.60	59.65	56.40	55.40	54.15	53.55	52.70	51.95
Tilt-up Concrete Panel	Steel Frame	57.70	54.55	53.15	51.10	50.40	49.55	49.15	48.60	48.15
Perimeter Adj., Add or Deduct	Per 100 L.F.	2.55	1.75	1.30	1.00	.85	.75	.65	.55	.50
Story Hgt. Adj., Add or Deduct	Per 1 Ft.	.40	.30	.30	.25	.20	.20	.20	.20	.15

For Basement, add $17.35 per square foot of basement area

The above costs were calculated using the basic specifications shown on the facing page. These costs should be adjusted where necessary for design alternatives and owner's requirements. Reported completed project costs, for this type of structure, range from $25.50 to $98.60 per S.F.

Common additives

Description	Unit	$ Cost		Description	Unit	$ Cost
Clock System				Dock Levelers, Hinged 10 ton cap.		
20 room	Each	9875		6' x 8'	Each	3825
50 room	Each	23,800		7' x 8'	Each	4050
Dock Bumpers, Rubber blocks				Partitions, Woven wire, 10 ga., 1-1/2" mesh		
4-1/2" thick, 10" high, 14" long	Each	52		4' wide x 7' high	Each	111
24" long	Each	74		8' high	Each	117
36" long	Each	92		10' High	Each	142
12" high, 14" long	Each	76		Platform Lifter, Portable, 6' x 6'		
24" long	Each	85		3000# cap.	Each	6000
36" long	Each	99		4000# cap.	Each	8425
6" thick, 10" high, 14" long	Each	74		Fixed, 6' x 8', 5000# cap.	Each	8350
24" long	Each	93				
36" long	Each	118				
20" high, 11" long	Each	122				
Dock Boards, Heavy						
60" x 60" Aluminum, 5,000# cap.	Each	1125				
9000# cap.	Each	1400				
15,000# cap.	Each	1550				

Important: See the Reference Section for Location Factors

Model costs calculated for a 1 story building with 20' story height and 30,000 square feet of floor area

				Unit	Unit Cost	Cost Per S.F.	% Of Sub-Total
1.0 Foundations							
	.1	Footings & Foundations	Poured concrete; strip and spread footings and 4' foundation wall	S.F. Ground	1.73	1.73	
	.4	Piles & Caissons	N/A	—	—	—	6.4%
	.9	Excavation & Backfill	Site preparation for slab and trench for foundation wall and footing	S.F. Ground	.91	.91	
2.0 Substructure							
	.1	Slab on Grade	4" reinforced concrete with vapor barrier and granular base	S.F. Slab	2.56	2.56	6.2%
	.2	Special Substructures	N/A	—	—	—	
3.0 Superstructure							
	.1	Columns & Beams	Steel columns included in 3.7	—	—	—	
	.4	Structural Walls	N/A	—	—	—	
	.5	Elevated Floors	N/A	—	—	—	10.7%
	.7	Roof	Metal deck, open web steel joists, beams and columns	S.F. Roof	4.40	4.40	
	9	Stairs	N/A	—	—	—	
4.0 Exterior Closure							
	.1	Walls	Concrete block /5% of wall	S.F. Wall	3.86	1.41	
	.5	Exterior Wall Finishes	N/A	—	—	—	9.8%
	.6	Doors	Double aluminum and glass, hollow metal, steel overhead	Each	1174	.59	
	.7	Windows & Glazed Walls	Industrial horizontal pivoted steel 25% of wall	Each	535	2.03	
5.0 Roofing							
	.1	Roof Coverings	Built-up tar and gravel with flashing	S.F. Roof	1.67	1.67	
	.7	Insulation	Perlite/EPS composite	S.F. Roof	1.13	1.13	7.5%
	.8	Openings & Specialties	Gravel stop, hatches, gutters and downspouts	S.F. Roof	.30	.30	
6.0 Interior Construction							
	.1	Partitions	Concrete block, toilet partitions 60 S.F. Floor/L.F. Partition	S.F. Partition	7.55	1.51	
	.4	Interior Doors	Single leaf hollow metal and fire doors 600 S.F. Floor/Door	Each	484	.81	
	.5	Wall Finishes	Paint	S.F. Surface	.88	.35	9.1%
	.6	Floor Finishes	Vinyl composition tile 10%of floor	S.F. Floor	2.50	.25	
	.7	Ceiling Finishes	Fiberglass board on exposed grid system 10% of area	S.F. Ceiling	2.62	.26	
	.9	Interior Surface/Exterior Wall	Paint and block filler 75% of wall	S.F. Wall	1.53	.56	
7.0 Conveying							
	.1	Elevators	N/A	—	—	—	0.0%
	.2	Special Conveyors	N/A	—	—	—	
8.0 Mechanical							
	.1	Plumbing	Toilet and service fixtures, supply and drainage 1 Fixture/1000 S.F. Floor	Each	2430	2.43	
	.2	Fire Protection	Sprinklers, ordinary hazard	S.F. Floor	1.63	1.63	
	.3	Heating	Oil fired hot water, unit heaters	S.F. Floor	4.61	4.61	36.8%
	.4	Cooling	Chilled water, air cooled condenser system	S.F. Floor	6.49	6.49	
	.5	Special Systems	N/A	—	—	—	
9.0 Electrical							
	.1	Service & Distribution	600 ampere service, panel board and feeders	S.F. Floor	.68	.68	
	.2	Lighting & Power	High intensity discharge fixtures, receptacles, switches, A.C. and misc. power	S.F. Floor	4.67	4.67	13.5%
	.4	Special Electrical	Alarm systems and emergency lighting	S.F. Floor	.22	.22	
11.0 Special Construction							
	.1	Specialties	N/A	—	—	—	0.0%
12.0 Site Work							
	.1	Earthwork	N/A	—	—	—	
	.3	Utilities	N/A	—	—	—	0.0%
	.5	Roads & Parking	N/A	—	—	—	
	.7	Site Improvements	N/A	—	—	—	
				Sub-Total		41.20	100%
		GENERAL CONDITIONS (Overhead & Profit)			15%	6.18	
		ARCHITECT FEES			7%	3.32	
				Total Building Cost		50.70	

Costs per square foot of floor area

Exterior Wall	S.F. Area	4000	4500	5000	5500	6000	6500	7000	7500	8000
	L.F. Perimeter	260	280	300	320	320	336	353	370	386
Face Brick Concrete Block Back-up	Steel Joists	77.75	76.30	75.10	74.10	72.05	71.25	70.60	70.00	69.45
	Precast Conc.	78.40	76.95	75.75	74.75	72.70	71.90	71.25	70.65	70.10
Decorative Concrete Block	Steel Joists	71.50	70.30	69.30	68.50	66.90	66.25	65.70	65.25	64.80
	Precast Conc.	72.25	71.05	70.10	69.25	67.65	67.00	66.50	66.00	65.60
Limestone with Concrete Block Back-up	Steel Joists	84.05	82.35	80.90	79.75	77.20	76.25	75.50	74.80	74.15
	Precast Conc.	84.85	83.10	81.70	80.50	78.00	77.00	76.25	75.55	74.95
Perimeter Adj., Add or Deduct	Per 100 L.F.	9.70	8.65	7.75	7.10	6.50	5.95	5.55	5.20	4.90
Story Hgt. Adj., Add or Deduct	Per 1 Ft.	1.25	1.20	1.15	1.15	1.05	1.00	1.00	1.00	.95

For Basement, add $19.40 per square foot of basement area

The above costs were calculated using the basic specifications shown on the facing page. These costs should be adjusted where necessary for design alternatives and owner's requirements. Reported completed project costs, for this type of structure, range from $41.75 to $123.45 per S.F.

Common additives

Description	Unit	$ Cost
Appliances		
Cooking range, 30" free standing		
1 oven	Each	365 - 1250
2 oven	Each	770 - 1675
30" built-in		
1 oven	Each	465 - 1400
2 oven	Each	1000 - 1900
Counter top cook tops, 4 burner	Each	244 - 595
Microwave oven	Each	175 - 1800
Combination range, refrig. & sink, 30" wide	Each	865 - 1725
60" wide	Each	2500
72" wide	Each	2825
Combination range refrigerator, sink microwave oven & icemaker	Each	4175
Compactor, residential, 4-1 compaction	Each	380 - 545
Dishwasher, built-in, 2 cycles	Each	325 - 645
4 cycles	Each	360 - 880
Garbage disposer, sink type	Each	75 - 204
Hood for range, 2 speed, vented, 30" wide	Each	105 - 375
42" wide	Each	242 - 380

Description	Unit	$ Cost
Appliances, cont.		
Refrigerator, no frost 10-12 C.F.	Each	490 - 745
14-16 C.F.	Each	560 - 820
18-20 C.F.	Each	645 - 1400
Lockers, Steel, single tier, 60" or 72"	Opening	128 - 207
2 tier, 60" or 72" total	Opening	79 - 99
5 tier, box lockers	Opening	41 - 53
Locker bench, lam. maple top only	L.F.	17.60
Pedestals, steel pipe	Each	39
Sound System		
Amplifier, 250 watts	Each	1425
Speaker, ceiling or wall	Each	118
Trumpet	Each	222

Important: See the Reference Section for Location Factors

Model costs calculated for a 1 story building with 14' story height and 6,000 square feet of floor area

				Unit	Unit Cost	Cost Per S.F.	% Of Sub-Total
1.0 Foundations							
	.1	Footings & Foundations	Poured concrete; strip and spread footings and 4' foundation wall	S.F. Ground	3.50	3.50	
	.4	Piles & Caissons	N/A	–	–	–	7.9%
	.9	Excavation & Backfill	Site preparation for slab and trench for foundation wall and footing	S.F. Ground	1.08	1.08	
2.0 Substructure							
	.1	Slab on Grade	4" reinforced concrete with vapor barrier and granular base	S.F. Slab	3.14	3.14	5.4%
	.2	Special Substructures	N/A	–			
3.0 Superstructure							
	.1	Columns & Beams	N/A	–	–	–	
	.4	Structural Walls	N/A	–	–	–	
	.5	Elevated Floors	N/A	–	–	–	4.1%
	.7	Roof	Metal deck, open web steel joists, beams	S.F. Roof	2.37	2.37	
	.9	Stairs	N/A	–			
4.0 Exterior Closure							
	.1	Walls	Face brick with concrete block backup _75% of wall_	S.F. Wall	15.80	8.85	
	.5	Exterior Wall Finishes	N/A	–	–		
	.6	Doors	Single aluminum and glass, overhead, hollow metal _15% of wall_	S.F. Door	18.21	2.04	20.9%
	.7	Windows & Glazed Walls	Aluminum insulated glass _10% of wall_	Each	535	1.25	
5.0 Roofing							
	.1	Roof Coverings	Built-up tar and gravel with flashing	S.F. Roof	2.14	2.14	
	.7	Insulation	Perlite/EPS composite	S.F. Roof	1.13	1.13	6.7%
	.8	Openings & Specialties	Gravel stop	S.F. Roof	.62	.62	
6.0 Interior Construction							
	.1	Partitions	Concrete block, toilet partitions _17 S.F. Floor/L.F. Partition_	S.F. Partition	4.74	2.79	
	.4	Interior Doors	Single leaf hollow metal _500 S.F. Floor/Door_	Each	484	.97	
	.5	Wall Finishes	Paint	S.F. Surface	.87	1.02	
	.6	Floor Finishes	50% vinyl tile, 50% paint	S.F. Floor	1.69	1.69	14.3%
	.7	Ceiling Finishes	Fiberglass board on exposed grid, suspended _50% of area_	S.F. Ceiling	2.62	1.31	
	.9	Interior Surface/Exterior Wall	Acrylic glazed coating _75% of wall_	S.F. Wall	.95	.53	
7.0 Conveying							
	.1	Elevators	N/A	–	–	–	0.0%
	.2	Special Conveyors	N/A	–	–	–	
8.0 Mechanical							
	.1	Plumbing	Kitchen, toilet and service fixtures, supply and drainage _1 Fixture/375 S.F. Floor_	Each	2051	5.47	
	.2	Fire Protection	Wet pipe sprinkler system	S.F. Floor	1.66	1.66	
	.3	Heating	Included in 8.4	–	–	–	33.6%
	.4	Cooling	Rooftop multizone unit system	S.F. Floor	12.30	12.30	
	.5	Special Systems	N/A	–	–	–	
9.0 Electrical							
	.1	Service & Distribution	200 ampere service, panel board and feeders	S.F. Floor	.77	.77	
	.2	Lighting & Power	Fluorescent fixtures, receptacles, switches, A.C. and misc. power	S.F. Floor	3.15	3.15	7.1%
	.4	Special Electrical	Alarm systems	S.F. Floor	.22	.22	
11.0 Special Construction							
	.1	Specialties	N/A	–	–	–	0.0%
12.0 Site Work							
	.1	Earthwork	N/A	–	–	–	
	.3	Utilities	N/A	–	–	–	
	.5	Roads & Parking	N/A	–	–	–	0.0%
	.7	Site Improvements	N/A	–	–	–	

		Sub-Total	58.00	100%
	GENERAL CONDITIONS (Overhead & Profit)	15%	8.70	
	ARCHITECT FEES	8%	5.35	
	Total Building Cost		**72.05**	

Costs per square foot of floor area

Exterior Wall	S.F. Area	4000	5000	6000	8000	10000	12000	14000	16000	18000
	L.F. Perimeter	180	205	230	260	300	340	353	386	420
Cedar Beveled Siding	Wood Frame	81.90	77.60	74.65	70.30	**68.00**	66.45	64.80	63.90	63.25
Aluminum Siding	Wood Frame	79.80	75.60	72.75	68.60	66.45	64.90	63.35	62.50	61.90
Board and Batten	Wood Frame	79.90	76.25	73.75	69.95	68.00	66.65	65.15	64.40	63.85
Face Brick on Block	Wood Joists	90.65	85.60	82.10	76.65	73.90	71.95	69.70	68.65	67.80
Stucco on Block	Wood Joists	81.25	77.00	74.05	69.75	67.45	65.90	64.25	63.40	62.75
Decorative Block	Wood Joists	84.65	80.10	77.00	72.30	69.85	68.20	66.35	65.40	64.70
Perimeter Adj., Add or Deduct	Per 100 L.F.	7.20	5.75	4.85	3.65	2.90	2.40	2.10	1.80	1.60
Story Hgt. Adj., Add or Deduct	Per 1 Ft.	.95	.85	.80	.70	.60	.60	.55	.50	.50

For Basement, add $12.30 per square foot of basement area

The above costs were calculated using the basic specifications shown on the facing page. These costs should be adjusted where necessary for design alternatives and owner's requirements. Reported completed project costs, for this type of structure, range from $59.95 to $115.60 per S.F.

Common additives

Description	Unit	$ Cost
Appliances		
Cooking range, 30" free standing		
1 oven	Each	365 - 1250
2 oven	Each	770 - 1675
30" built-in		
1 oven	Each	465 - 1400
2 oven	Each	1000 - 1900
Counter top cook tops, 4 burner	Each	244 - 595
Microwave oven	Each	175 - 1800
Combination range, refrig. & sink, 30" wide	Each	865 - 1725
60" wide	Each	2500
72" wide	Each	2825
Combination range, refrigerator, sink,		
microwave oven & icemaker	Each	4175
Compactor, residential, 4-1 compaction	Each	380 - 545
Dishwasher, built-in, 2 cycles	Each	325 - 645
4 cycles	Each	360 - 880
Garbage disposer, sink type	Each	75 - 204
Hood for range, 2 speed, vented, 30" wide	Each	105 - 375
42" wide	Each	242 - 380

Description	Unit	$ Cost
Appliances, cont.		
Refrigerator, no frost 10-12 C.F.	Each	490 - 745
14-16 C.F.	Each	560 - 820
18-20 C.F.	Each	645 - 1400
Elevators, Hydraulic passenger, 2 stops		
1500# capacity	Each	37,625
2500# capacity	Each	38,925
3500# capacity	Each	42,525
Laundry Equipment		
Dryer, gas, 16 lb. capacity	Each	640
30 lb. capacity	Each	2525
Washer, 4 cycle	Each	720
Commercial	Each	1100
Sound System		
Amplifier, 250 watts	Each	1425
Speaker, ceiling or wall	Each	118
Trumpet	Each	222

Important: See the Reference Section for Location Factors

Model costs calculated for a 2 story building with 10′ story height and 10,000 square feet of floor area

Fraternity/Sorority House

				Unit	Unit Cost	Cost Per S.F.	% Of Sub-Total
1.0 Foundations							
	.1	Footings & Foundations	Poured concrete; strip and spread footings and 4′ foundation wall	S.F. Ground	3.32	1.66	
	.4	Piles & Caissons	N/A	—	—	—	4.1%
	.9	Excavation & Backfill	Site preparation for slab and trench for foundation wall and footing	S.F. Ground	1.08	.54	
2.0 Substructure							
	.1	Slab on Grade	4″ reinforced concrete with vapor barrier and granular base	S.F. Slab	2.56	1.28	2.4%
	.2	Special Substructures	N/A	—	—	—	
3.0 Superstructure							
	.1	Columns & Beams	N/A	—	—	—	
	.4	Structural Walls	Included in 6.1	—	—	—	
	.5	Elevated Floors	Plywood on wood joists	S.F. Floor	2.46	1.23	5.3%
	.7	Roof	Plywood on wood rafters (pitched)	S.F. Roof	2.04	1.14	
	.9	Stairs	Wood	Flight	1240	.50	
4.0 Exterior Closure							
	.1	Walls	Cedar bevel siding on wood studs, insulated — 80% of wall	S.F. Wall	6.25	3.00	
	.5	Exterior Wall Finishes	N/A	—	—	—	9.9%
	.6	Doors	Solid core wood	Each	1325	.93	
	.7	Windows & Glazed Walls	Double hung wood — 20% of wall	Each	297	1.43	
5.0 Roofing							
	.1	Roof Coverings	Asphalt shingles with flashing (pitched)	S.F. Ground	1.06	.53	
	.7	Insulation	Fiberglass sheets	S.F. Ground	.88	.44	2.1%
	.8	Openings & Specialties	Gutters and downspouts	S.F. Ground	.32	.16	
6.0 Interior Construction							
	.1	Partitions	Gypsum board on wood studs — 25 S.F. Floor/L.F. Partition	S.F. Partition	2.47	.79	
	.4	Interior Doors	Single leaf wood — 200 S.F. Floor/Door	Each	367	1.84	
	.5	Wall Finishes	Paint	S.F. Surface	.42	.27	
	.6	Floor Finishes	50% hardwood, 50% carpet	S.F. Floor	5.87	5.87	21.5%
	.7	Ceiling Finishes	Gypsum board on wood furring	S.F. Ceiling	2.32	2.32	
	.9	Interior Surface/Exterior Wall	Painted gypsum board — 80% of wall	S.F. Wall	.98	.59	
7.0 Conveying							
	.1	Elevators	One hydraulic passenger elevator	Each	45,900	4.59	8.5%
	.2	Special Conveyors	N/A	—	—	—	
8.0 Mechanical							
	.1	Plumbing	Kitchen toilet and service fixtures, supply and drainage — 1 Fixture/150 S.F. Floor	Each	666	4.44	
	.2	Fire Protection	Wet pipe sprinkler system	S.F. Floor	1.37	1.37	
	.3	Heating	Oil fired hot water, baseboard radiation	S.F. Floor	3.41	3.41	24.5%
	.4	Cooling	Split system with air cooled condensing unit	S.F. Floor	4.14	4.14	
	.5	Special Systems	N/A	—	—	—	
9.0 Electrical							
	.1	Service & Distribution	600 ampere service, panel board and feeders	S.F. Floor	2.44	2.44	
	.2	Lighting & Power	Fluorescent fixtures, receptacles, switches, A.C. and misc. power	S.F. Floor	4.70	4.70	21.7%
	.4	Special Electrical	Alarm, communication system and generator set	S.F. Floor	4.65	4.65	
11.0 Special Construction							
	.1	Specialties	N/A	—	—	—	0.0%
12.0 Site Work							
	.1	Earthwork	N/A	—	—	—	
	.3	Utilities	N/A	—	—	—	0.0%
	.5	Roads & Parking	N/A	—	—	—	
	.7	Site Improvements	N/A	—	—	—	
			Sub-Total			54.26	**100%**
		GENERAL CONDITIONS (Overhead & Profit)			15%	8.14	
		ARCHITECT FEES			9%	5.60	
			Total Building Cost			68.00	

Costs per square foot of floor area

Exterior Wall	S.F. Area	4000	5000	6000	7000	8000	9000	10000	11000	12000
	L.F. Perimeter	260	300	340	353	386	420	425	435	460
Vertical Redwood Siding	Wood Frame	70.95	68.35	66.70	64.75	63.75	63.00	61.80	60.95	60.55
Brick Veneer	Wood Frame	75.35	72.45	70.60	68.25	67.15	66.30	64.85	63.85	63.30
Aluminum Siding	Wood Frame	69.60	67.10	65.55	63.75	62.85	62.10	61.05	60.30	59.90
Brick on Block	Wood Truss	79.00	75.90	73.90	71.30	70.10	69.20	67.60	66.45	65.85
Limestone on Block	Wood Truss	87.00	83.30	80.85	77.45	76.00	74.85	72.75	71.25	70.50
Stucco on Block	Wood Truss	71.70	69.15	67.50	65.60	64.65	63.90	62.80	62.00	61.55
Perimeter Adj., Add or Deduct	Per 100 L.F.	4.80	3.85	3.20	2.70	2.40	2.10	1.95	1.75	1.60
Story Hgt. Adj., Add or Deduct	Per 1 Ft.	.75	.70	.65	.55	.55	.55	.50	.45	.45

For Basement, add $15.40 per square foot of basement area

The above costs were calculated using the basic specifications shown on the facing page. These costs should be adjusted where necessary for design alternatives and owner's requirements. Reported completed project costs, for this type of structure, range from $58.50 to $165.15 per S.F.

Common additives

Description	Unit	$ Cost	Description	Unit	$ Cost
Autopsy Table, Standard	Each	6875	Planters, Precast concrete		
Deluxe	Each	8700	48" diam., 24" high	Each	495
Directory Boards, Plastic, glass covered			7" diam., 36" high	Each	1950
30" x 20"	Each	460	Fiberglass, 36" diam., 24" high	Each	370
36" x 48"	Each	850	60" diam., 24" high	Each	955
Aluminum, 24" x 18"	Each	410	Smoke Detectors		
36" x 24"	Each	495	Ceiling type	Each	126
48" x 32"	Each	590	Duct type	Each	360
48" x 60"	Each	1300			
Emergency Lighting, 25 watt, battery operated					
Lead battery	Each	305			
Nickel cadmium	Each	600			
Mortuary Refrigerator, End operated					
Two capacity	Each	10,900			
Six capacity	Each	20,700			

Important: See the Reference Section for Location Factors

Model costs calculated for a 1 story building with 10′ story height and 10,000 square feet of floor area

				Unit	Unit Cost	Cost Per S.F.	% Of Sub-Total
1.0 Foundations							
	.1	Footings & Foundations	Poured concrete; strip and spread footings and 4′ foundation wall	S.F. Ground	2.31	2.31	
	.4	Piles & Caissons	N/A	—	—	—	6.6%
	.9	Excavation & Backfill	Site preparation for slab and trench for foundation wall and footing	S.F. Ground	.96	.96	
2.0 Substructure							
	.1	Slab on Grade	4″ reinforced concrete with vapor barrier and granular base	S.F. Slab	2.56	2.56	5.2%
	.2	Special Substructures	N/A	—	—	—	
3.0 Superstructure							
	.1	Columns & Beams	N/A	—	—	—	
	.4	Structural Walls	N/A	—	—	—	
	.5	Elevated Floors	N/A	—	—	—	5.8%
	.7	Roof	Plywood on wood rafters (pitched)	S.F. Ground	2.88	2.88	
	.9	Stairs	N/A	—	—	—	
4.0 Exterior Closure							
	.1	Walls	1″ x 4″ vertical T & G redwood siding on wood studs — 90% of wall	S.F. Wall	6.75	2.58	
	.5	Exterior Wall Finishes	N/A	—	—	—	8.6%
	.6	Doors	Wood swinging double doors, single leaf hollow metal	Each	1254	.75	
	.7	Windows & Glazed Walls	Double hung wood — 10% of wall	Each	256	.91	
5.0 Roofing							
	.1	Roof Coverings	Asphalt shingles with flashing	S.F. Roof	1.26	1.26	
	.7	Insulation	Fiberglass sheets	S.F. Ground	.78	.78	4.6%
	.8	Openings & Specialties	Gutters and downspouts	S.F. Ground	.23	.23	
6.0 Interior Construction							
	.1	Partitions	Gypsum board on wood studs with sound deadening board — 15 S.F. Floor/L.F. Partition	S.F. Partition	3.94	2.10	
	.4	Interior Doors	Single leaf wood — 150 S.F. Floor/Door	Each	367	2.45	
	.5	Wall Finishes	50% wallpaper, 25% wood paneling, 25% paint	S.F. Surface	1.63	1.74	33.3%
	.6	Floor Finishes	70% carpet, 30% terrazzo	S.F. Floor	7.98	7.98	
	.7	Ceiling Finishes	Mineral fiberboard on wood furring	S.F. Ceiling	1.80	1.80	
	.9	Interior Surface/Exterior Wall	Painted gypsum board on wood furring — 90% of wall	S.F. Wall	1.22	.36	
7.0 Conveying							
	.1	Elevators	N/A	—	—	—	0.0%
	.2	Special Conveyors	N/A	—	—	—	
8.0 Mechanical							
	.1	Plumbing	Toilet and service fixtures, supply and drainage — 1 Fixture/770 S.F. Floor	Each	2679	3.48	
	.2	Fire Protection	Wet pipe sprinkler system	S.F. Floor	1.26	1.26	
	.3	Heating	Included in 8.4	—	—	—	29.3%
	.4	Cooling	Multizone rooftop unit, gas heating, electric cooling	S.F. Floor	9.69	9.69	
	.5	Special Systems	N/A	—	—	—	
9.0 Electrical							
	.1	Service & Distribution	200 ampere service, panel board and feeders	S.F. Floor	.46	.46	
	.2	Lighting & Power	Fluorescent fixtures, receptacles, switches, A.C. and misc. power	S.F. Floor	2.58	2.58	6.6%
	.4	Special Electrical	Alarm systems and emergency lighting	S.F. Floor	.20	.20	
11.0 Special Construction							
	.1	Specialties	N/A	—	—	—	0.0%
12.0 Site Work							
	.1	Earthwork	N/A	—	—	—	
	.3	Utilities	N/A	—	—	—	
	.5	Roads & Parking	N/A	—	—	—	0.0%
	.7	Site Improvements	N/A	—	—	—	

			Sub-Total	49.32	**100%**
GENERAL CONDITIONS (Overhead & Profit)		15%	7.40		
ARCHITECT FEES		9%	5.08		

Total Building Cost	**61.80**

Costs per square foot of floor area

Exterior Wall	S.F. Area	12000	14000	16000	19000	21000	23000	26000	28000	30000
	L.F. Perimeter	440	474	510	556	583	607	648	670	695
Metal Panel Curtain Walls	Steel Frame	54.60	53.20	52.15	50.85	50.15	49.50	48.80	48.35	48.00
Tilt-up Concrete Wall	Steel Frame	53.20	51.90	50.95	49.75	49.10	48.50	47.90	47.40	47.10
Face Brick with Concrete Block Back-up	Bearing Walls	55.85	54.15	52.95	51.35	50.50	49.75	48.90	48.30	47.90
	Steel Frame	58.15	56.45	55.20	53.65	52.75	52.00	51.20	50.60	50.20
Stucco on Concrete Block	Bearing Walls	51.35	50.00	49.00	47.75	47.10	46.45	45.80	45.35	45.05
	Steel Frame	53.85	52.50	51.50	50.25	49.60	48.95	48.30	47.85	47.55
Perimeter Adj., Add or Deduct	Per 100 L.F.	3.05	2.60	2.30	1.95	1.75	1.60	1.40	1.35	1.20
Story Hgt. Adj., Add or Deduct	Per 1 Ft.	.70	.65	.65	.55	.55	.50	.50	.50	.45
For Basement, add $18.35 per square foot of basement area										

The above costs were calculated using the basic specifications shown on the facing page. These costs should be adjusted where necessary for design alternatives and owner's requirements. Reported completed project costs, for this type of structure, range from $31.45 to $90.50 per S.F.

Common additives

Description	Unit	$ Cost
Emergency Lighting, 25 watt, battery operated		
Lead battery	Each	305
Nickel cadmium	Each	600
Smoke Detectors		
Ceiling type	Each	126
Duct type	Each	360
Sound System		
Amplifier, 250 watts	Each	1425
Speaker, ceiling or wall	Each	118
Trumpet	Each	222

Important: See the Reference Section for Location Factors

Model costs calculated for a 1 story building with 14' story height and 21,000 square feet of floor area

				Unit	Unit Cost	Cost Per S.F.	% Of Sub-Total
1.0 Foundations							
.1	Footings & Foundations	Poured concrete; strip and spread footings and 4' foundation wall		S.F. Ground	1.69	1.69	
.4	Piles & Caissons	N/A		—	—	—	6.5%
.9	Excavation & Backfill	Site preparation for slab and trench for foundation wall and footing		S.F. Ground	.96	.96	
2.0 Substructure							
.1	Slab on Grade	4" reinforced concrete with vapor barrier and granular base		S.F. Slab	3.14	3.14	7.7%
.2	Special Substructures	N/A		—	—	—	
3.0 Superstructure							
.1	Columns & Beams	Steel columns included in 3.7		—	—	—	
.4	Structural Walls	Metal siding support		S.F. Wall	2.86	.78	
.5	Elevated Floors	N/A		—	—	—	14.1%
.7	Roof	Metal deck, open web steel joists, beams, columns		S.F. Floor	4.96	4.96	
.9	Stairs	N/A		—	—	—	
4.0 Exterior Closure							
.1	Walls	Metal panel	70% of wall	S.F. Wall	7.68	2.09	
.5	Exterior Wall Finishes	N/A		—	—	—	18.0%
.6	Doors	Double aluminum and glass, hollow metal, steel overhead		Each	2215	1.90	
.7	Windows & Glazed Walls	Window wall	30% of wall	S.F. Window	28	3.33	
5.0 Roofing							
.1	Roof Coverings	Built-up tar and gravel with flashing		S.F. Roof	1.65	1.65	
.7	Insulation	Perlite/EPS composite		S.F. Roof	1.13	1.13	7.1%
.8	Openings & Specialties	Gravel stop and skylight		L.F. Perimeter	4.78	.13	
6.0 Interior Construction							
.1	Partitions	Gypsum board on metal studs	28 S.F. Floor/L.F. Partition	S.F. Partition	2.43	1.04	
.4	Interior Doors	Hollow metal	280 S.F. Floor/Door	Each	484	1.73	
.5	Wall Finishes	Paint		S.F. Surface	.42	.36	14.4%
.6	Floor Finishes	50% vinyl tile, 50% paint		S.F. Floor	1.69	1.69	
.7	Ceiling Finishes	Fiberglass board on exposed grid, suspended	50% of area	S.F. Ceiling	2.09	1.05	
.9	Interior Surface/Exterior Wall	N/A		—	—	—	
7.0 Conveying							
.1	Elevators	N/A		—	—	—	0.0%
.2	Special Conveyors	N/A		—	—	—	
8.0 Mechanical							
.1	Plumbing	Toilet and service fixtures, supply and drainage	1 Fixture/1500 S.F. Floor	Each	2640	1.76	
.2	Fire Protection	Wet pipe sprinkler system		S.F. Floor	1.66	1.66	
.3	Heating	Gas fired hot water, unit heaters (service area)		S.F. Floor	2.10	2.10	21.4%
.4	Cooling	Single zone rooftop unit, gas heating, electric (office and showroom)		S.F. Floor	3.04	3.04	
.5	Special Systems	Underfloor garage exhaust system		S.F. Floor	.20	.19	
9.0 Electrical							
.1	Service & Distribution	200 ampere service, panel board and feeders		S.F. Floor	.23	.23	
.2	Lighting & Power	Fluorescent fixtures, receptacles, switches, A.C. and misc. power		S.F. Floor	3.17	3.17	8.8%
.4	Special Electrical	Alarm systems and emergency lighting		S.F. Floor	.17	.17	
11.0 Special Construction							
.1	Specialties	Hoists, compressor, fuel pump		S.F. Floor	.80	.80	2.0%
12.0 Site Work							
.1	Earthwork	N/A		—	—	—	
.3	Utilities	N/A		—	—	—	0.0%
.5	Roads & Parking	N/A		—	—	—	
.7	Site Improvements	N/A		—	—	—	
				Sub-Total		40.75	**100%**
	GENERAL CONDITIONS (Overhead & Profit)				15%	6.11	
	ARCHITECT FEES				7%	3.29	
				Total Building Cost		50.15	

Costs per square foot of floor area

Exterior Wall	S.F. Area	2000	4000	6000	8000	10000	12000	14000	16000	18000
	L.F. Perimeter	180	260	340	420	500	580	586	600	610
Concrete Block	Wood Joists	74.20	64.00	60.60	58.90	57.90	57.15	55.45	54.30	53.30
	Steel Joists	74.45	64.60	61.30	59.70	58.70	58.00	56.30	55.15	54.20
Poured Concrete	Wood Joists	77.70	66.65	63.00	61.20	60.05	59.30	57.30	55.95	54.80
	Steel Joists	78.60	67.55	63.85	62.10	60.95	60.20	58.20	56.85	55.70
Insulated Metal Panels	Wood Frame	70.30	61.30	58.30	56.85	55.95	55.35	53.85	52.85	52.00
	Steel Frame	72.60	63.60	60.65	59.15	58.25	57.65	56.15	55.15	54.35
Perimeter Adj., Add or Deduct	Per 100 L.F.	11.65	5.85	3.90	2.90	2.30	1.95	1.65	1.45	1.30
Story Hgt. Adj., Add or Deduct	Per 1 Ft.	.95	.70	.60	.55	.50	.50	.45	.40	.35

For Basement, add $16.50 per square foot of basement area

The above costs were calculated using the basic specifications shown on the facing page. These costs should be adjusted where necessary for design alternatives and owner's requirements. Reported completed project costs, for this type of structure, range from $39.80 to $119.55 per S.F.

Common additives

Description	Unit	$ Cost
Air Compressors		
Electric 1-1/2 H.P., standard controls	Each	3000
Dual controls	Each	3225
5 H.P. 115/230 Volt, standard controls	Each	4025
Dual controls	Each	4250
Product Dispenser		
with vapor recovery for 6 nozzles	Each	16,500
Hoists, Single post		
8000# cap., swivel arm	Each	4975
Two post, adjustable frames, 11,000# cap.	Each	6750
24,000# cap.	Each	9575
7500# Frame support	Each	6400
Four post, roll on ramp	Each	5850
Lockers, Steel, single tier, 60" or 72"	Opening	128 - 207
2 tier, 60" or 72" total	Opening	79 - 99
5 tier, box lockers	Opening	41 - 53
Locker bench, lam. maple top only	L.F.	17.60
Pedestals, steel pipe	Each	39
Lube Equipment		
3 reel type, with pumps, no piping	Each	7500
Spray Painting Booth, 26' long, complete	Each	14,300

Important: See the Reference Section for Location Factors

Model costs calculated for a 1 story building with 14' story height and 4,000 square feet of floor area

			Unit	Unit Cost	Cost Per S.F.	% Of Sub-Total
1.0 Foundations						
.1	Footings & Foundations	Poured concrete; strip and spread footings and 4' foundation wall	S.F. Ground	3.41	3.41	8.4%
.4	Piles & Caissons	N/A	—	—	—	
.9	Excavation & Backfill	Site preparation for slab and trench for foundation wall and footing	S.F. Ground	.96	.96	
2.0 Substructure						
.1	Slab on Grade	4" reinforced concrete with vapor barrier and granular base	S.F. Slab	2.56	2.56	4.9%
.2	Special Substructures	N/A	—			
3.0 Superstructure						
.1	Columns & Beams	N/A	—	—	—	5.4%
.4	Structural Walls	N/A	—	—	—	
.5	Elevated Floors	N/A	—	—	—	
.7	Roof	Metal deck on open web steel joists	S.F. Roof	2.81	2.81	
.9	Stairs	N/A	—	—	—	
4.0 Exterior Closure						
.1	Walls	Concrete block	80% of wall	L.F. Wall	6.59	4.80	13.8%
.5	Exterior Wall Finishes	N/A	—	—	—	
.6	Doors	Steel overhead and hollow metal	15% of wall	S.F. Door	11.87	1.62	
.7	Windows & Glazed Walls	Hopper type commercial steel	5% of wall	Each	254	.77	
5.0 Roofing						
.1	Roof Coverings	Built-up tar and gravel	S.F. Roof	2.15	2.15	7.0%
.7	Insulation	Perlite/EPS composite	S.F. Roof	1.13	1.13	
.8	Openings & Specialties	Gravel stop and skylight	S.F. Roof	.36	.36	
6.0 Interior Construction						
.1	Partitions	Concrete block, toilet partitions	50 S.F. Floor/L.F. Partition	S.F. Partition	4.85	.97	6.4%
.4	Interior Doors	Single leaf hollow metal	3000 S.F. Floor/Door	Each	484	.16	
.5	Wall Finishes	Paint	S.F. Surface	.73	.29	
.6	Floor Finishes	90% metallic floor hardener, 10% vinyl composition tile	S.F. Floor	.55	.55	
.7	Ceiling Finishes	Gypsum board on wood joists in office and washrooms	10% of area	S.F. Ceiling	2.32	.23	
.9	Interior Surface/Exterior Wall	Paint	80% of wall	S.F. Wall	1.53	1.11	
7.0 Conveying						
.1	Elevators	N/A	—	—	—	0.0%
.2	Special Conveyors	N/A	—	—	—	
8.0 Mechanical						
.1	Plumbing	Toilet and service fixtures, supply and drainage	1 Fixture/500 S.F. Floor	Each	2300	4.60	32.2%
.2	Fire Protection	Sprinklers, ordinary hazard	S.F. Floor	1.82	1.82	
.3	Heating	Oil fired hot water, unit heaters	S.F. Floor	4.19	4.19	
.4	Cooling	Split systems with air cooled condensing units	S.F. Floor	5.36	5.36	
.5	Special Systems	Garage exhaust system	S.F. Floor	.74	.74	
9.0 Electrical						
.1	Service & Distribution	100 ampere service, panel board and feeders	S.F. Floor	.40	.40	7.7%
.2	Lighting & Power	Fluorescent fixtures, receptacles, switches, A.C. and misc. power	S.F. Floor	3.22	3.22	
.4	Special Electrical	Alarm systems and emergency lighting	S.F. Floor	.41	.41	
11.0 Special Construction						
.1	Specialties	Hoists	S.F. Floor	7.39	7.39	14.2%
12.0 Site Work						
.1	Earthwork	N/A	—	—	—	0.0%
.3	Utilities	N/A	—	—	—	
.5	Roads & Parking	N/A	—	—	—	
.7	Site Improvements	N/A	—	—	—	
			Sub-Total		52.01	**100%**
	GENERAL CONDITIONS (Overhead & Profit)			15%	7.80	
	ARCHITECT FEES			8%	4.79	
			Total Building Cost		64.60	

Costs per square foot of floor area

| Exterior Wall | S.F. Area | 600 | 800 | 1000 | 1200 | 1400 | 1600 | 1800 | 2000 | 2200 |
	L.F. Perimeter	100	120	126	140	153	160	170	180	190
Face Brick with Concrete Block Back-up	Wood Truss	102.30	94.40	85.85	81.95	**78.95**	75.70	73.60	71.95	70.60
	Steel Joists	98.35	90.45	81.85	78.00	75.00	71.75	69.65	68.05	66.65
Enameled Sandwich Panel Tile on Concrete Block	Steel Frame	91.20	84.10	76.60	73.15	70.50	67.70	65.85	64.40	63.20
	Steel Joists	108.30	99.50	89.65	85.25	81.85	78.15	75.75	73.85	72.30
Aluminum Siding Wood Siding	Wood Frame	88.30	81.80	75.25	72.15	69.80	67.30	65.70	64.40	63.35
	Wood Frame	89.15	82.55	75.90	72.75	70.30	67.80	66.15	64.90	63.80
Perimeter Adj., Add or Deduct	Per 100 L.F.	45.25	33.90	27.10	22.60	19.40	16.95	15.10	13.60	12.35
Story Hgt. Adj., Add or Deduct	Per 1 Ft.	3.50	3.15	2.65	2.45	2.30	2.05	2.00	1.90	1.80
Basement—Not Applicable										

The above costs were calculated using the basic specifications shown on the facing page. These costs should be adjusted where necessary for design alternatives and owner's requirements. Reported completed project costs, for this type of structure, range from $25.80 to $119.65 per S.F.

Common additives

Description	Unit	$ Cost
Air Compressors		
Electric 1-1/2 H.P., standard controls	Each	3000
Dual controls	Each	3225
5 H.P. 115/230 volt, standard controls	Each	4025
Dual controls	Each	4250
Product Dispenser		
with vapor recovery for 6 nozzles	Each	16,500
Hoists, Single post		
8000# cap. swivel arm	Each	4975
Two post, adjustable frames, 11,000# cap.	Each	6750
24,000# cap.	Each	9575
7500# cap.	Each	6400
Four post, roll on ramp	Each	5850
Lockers, Steel, single tier, 60" or 72"	Opening	128 - 207
2 tier, 60" or 72" total	Opening	79 - 99
5 tier, box lockers	Each	41 - 53
Locker bench, lam. maple top only	L.F.	17.60
Pedestals, steel pipe	Each	39
Lube Equipment		
3 reel type, with pumps, no piping	Each	7500

Model costs calculated for a 1 story building with 10' story height and 1,400 square feet of floor area

				Unit	Unit Cost	Cost Per S.F.	% Of Sub-Total
1.0 Foundations							
.1	Footings & Foundations	Poured concrete; strip and spread footings and 4' foundation wall		S.F. Ground	5.42	5.42	
.4	Piles & Caissons	N/A		–	–	–	10.7%
.9	Excavation & Backfill	Site preparation for slab and trench for foundation wall and footing		S.F. Ground	1.38	1.38	
2.0 Substructure							
.1	Slab on Grade	4" reinforced concrete with vapor barrier and granular base		S.F. Slab	2.56	2.56	4.0%
.2	Special Substructures	N/A		–	–	–	
3.0 Superstructure							
.1	Columns & Beams	N/A		–	–	–	
.4	Structural Walls	N/A		–	–	–	
.5	Elevated Floors	N/A		–	–	–	6.1%
.7	Roof	Plywood on wood trusses		S.F. Ground	3.87	3.87	
.9	Stairs	N/A		–	–	–	
4.0 Exterior Closure							
.1	Walls	Face brick with concrete block backup	60% of wall	S.F. Wall	15.80	10.36	
.5	Exterior Wall Finishes	N/A		–	–	–	31.7%
.6	Doors	Steel overhead, aluminum & glass and hollow metal	20% of wall	S.F. Door	22	4.94	
.7	Windows & Glazed Walls	Store front and metal top hinged outswinging	20% of wall	S.F. Window	22	4.83	
5.0 Roofing							
.1	Roof Coverings	Asphalt shingles with flashing		S.F. Ground	.96	.96	
.7	Insulation	Perlite/EPS composite		S.F. Ground	.78	.78	2.7%
.8	Openings & Specialties	N/A		–	–	–	
6.0 Interior Construction							
.1	Partitions	Concrete block, toilet partitions	25 S.F. Floor/L.F. Partition	S.F. Partition	8.53	2.73	
.4	Interior Doors	Single leaf hollow metal	700 S.F. Floor/Door	Each	484	.69	
.5	Wall Finishes	Paint		S.F. Surface	.88	.56	
.6	Floor Finishes	Vinyl composition tile	35% of floor area	S.F. Floor	2.49	.87	10.5%
.7	Ceiling Finishes	Painted gypsum board on wood joists in sales area & washrooms	35% of floor area	S.F. Ceiling	2.32	.81	
.9	Interior Surface/Exterior Wall	Paint	60% of wall	S.F. Wall	1.53	1.00	
7.0 Conveying							
.1	Elevators	N/A		–	–	–	0.0%
.2	Special Conveyors	N/A		–	–	–	
8.0 Mechanical							
.1	Plumbing	Toilet and service fixtures, supply and drainage	1 Fixture/235 S.F. Floor	Each	1475	6.28	
.2	Fire Protection	N/A		–	–	–	
.3	Heating	Oil fired hot water, wall fin radiation		S.F. Floor	4.19	4.19	24.7%
.4	Cooling	Split systems with air cooled condensing units		S.F. Floor	5.20	5.20	
.5	Special Systems	N/A		–	–	–	
9.0 Electrical							
.1	Service & Distribution	100 ampere service, panel board and feeders		S.F. Floor	1.85	1.85	
.2	Lighting & Power	Fluorescent fixtures, receptacles, switches, A.C. and misc. power		S.F. Floor	3.48	3.48	9.6%
.4	Special Electrical	Alarm systems and emergency lighting		S.F. Floor	.80	.80	
11.0 Special Construction							
.1	Specialties	N/A		–	–	–	0.0%
12.0 Site Work							
.1	Earthwork	N/A		–	–	–	
.3	Utilities	N/A		–	–	–	
.5	Roads & Parking	N/A		–	–	–	0.0%
.7	Site Improvements	N/A		–	–	–	

						Sub-Total	63.56	**100%**
	GENERAL CONDITIONS (Overhead & Profit)					15%	9.53	
	ARCHITECT FEES					8%	5.86	
						Total Building Cost	78.95	

BUILDING TYPES

Costs per square foot of floor area

Exterior Wall	S.F. Area	12000	16000	20000	25000	30000	35000	40000	45000	50000
	L.F. Perimeter	440	520	600	700	708	780	841	910	979
Reinforced Concrete Block	Lam. Wood Arches	77.50	74.95	73.45	72.20	70.25	69.55	68.90	68.50	68.15
	Rigid Steel Frame	74.60	72.05	70.50	69.30	67.35	66.65	66.00	65.55	65.20
Face Brick with Concrete Block Back-up	Lam. Wood Arches	89.40	85.55	83.20	81.30	77.95	76.80	75.75	75.05	74.50
	Rigid Steel Frame	86.50	82.60	80.30	78.40	75.00	73.90	72.85	72.15	71.60
Metal Sandwich Panels	Lam. Wood Arches	75.35	73.05	71.70	70.55	68.90	68.25	67.70	67.30	66.95
	Rigid Steel Frame	72.40	70.10	68.75	67.65	65.95	65.30	64.75	64.35	64.05
Perimeter Adj., Add or Deduct	Per 100 L.F.	3.10	2.35	1.90	1.50	1.25	1.05	.90	.80	.75
Story Hgt. Adj., Add or Deduct	Per 1 Ft.	.45	.40	.40	.35	.30	.25	.25	.25	.25
Basement—Not Applicable										

The above costs were calculated using the basic specifications shown on the facing page. These costs should be adjusted where necessary for design alternatives and owner's requirements. Reported completed project costs, for this type of structure, range from $43.55 to $130.70 per S.F.

Common additives

Description	Unit	$ Cost
Bleachers, Telescoping, manual		
To 15 tier	Seat	68 - 97
16-20 tier	Seat	147 - 181
21-30 tier	Seat	156 - 187
For power operation, add	Seat	29 - 46
Gym Divider Curtain, Mesh top		
Manual roll-up	S.F.	7.50
Gym Mats		
2" naugahyde covered	S.F.	4.19
2" nylon	S.F.	7.90
1-1/2" wall pads	S.F.	8.50
1" wrestling mats	S.F.	4.76
Scoreboard		
Basketball, one side	Each	2925 - 20,300
Basketball Backstop		
Wall mtd., 6' extended, fixed	Each	985 - 1775
Swing up, wall mtd.	Each	1575 - 2375

Description	Unit	$ Cost
Lockers, Steel, single tier, 60" or 72"	Opening	128 - 207
2 tier, 60" or 72" total	Opening	79 - 99
5 tier, box lockers	Opening	41 - 53
Locker bench, lam. maple top only	L.F.	17.60
Pedestals, steel pipe	Each	39
Sound System		
Amplifier, 250 watts	Each	1425
Speaker, ceiling or wall	Each	118
Trumpet	Each	222
Emergency Lighting, 25 watt, battery operated		
Lead battery	Each	305
Nickel cadmium	Each	600

Important: See the Reference Section for Location Factors

Model costs calculated for a 1 story building with 25' story height and 20,000 square feet of floor area

				Unit	Unit Cost	Cost Per S.F.	% Of Sub-Total
1.0 Foundations							
	.1	Footings & Foundations	Poured concrete; strip and spread footings and 4' foundation wall	S.F. Ground	1.93	1.93	
	.4	Piles & Caissons	N/A	—	—	—	4.8%
	.9	Excavation & Backfill	Site preparation for slab and trench for foundation wall and footing	S.F. Ground	.91	.91	
2.0 Substructure							
	.1	Slab on Grade	4" reinforced concrete with vapor barrier and granular base	S.F. Slab	2.56	2.56	4.3%
	.2	Special Substructures	N/A	—	—	—	
3.0 Superstructure							
	.1	Columns & Beams	N/A	—	—	—	
	.4	Structural Walls	N/A	—	—	—	
	.5	Elevated Floors	N/A	—	—	—	19.0%
	.7	Roof	Wood deck on laminated wood arches	S.F. Ground	11.32	11.32	
	.9	Stairs	N/A	—	—	—	
4.0 Exterior Closure							
	.1	Walls	Reinforced concrete block (end walls included) 90% of wall	S.F. Wall	6.50	4.39	
	.5	Exterior Wall Finishes	N/A	—	—	—	11.0%
	.6	Doors	Aluminum and glass, hollow metal, steel overhead	Each	834	.25	
	.7	Windows & Glazed Walls	Metal horizontal pivoted 10% of wall	Each	255	1.91	
5.0 Roofing							
	.1	Roof Coverings	EPDM, 60 mils, fully adhered	S.F. Ground	1.52	1.52	
	.7	Insulation	Polyisocyanurate	—	1.03	1.06	4.3%
	.8	Openings & Specialties	N/A	—	—	—	
6.0 Interior Construction							
	.1	Partitions	Concrete block, toilet partitions 50 S.F. Floor/L.F. Partition	S.F. Partition	5.25	1.05	
	.4	Interior Doors	Single leaf hollow metal 500 S.F. Floor/Door	Each	484	.97	
	.5	Wall Finishes	50% paint, 50% ceramic tile	S.F. Surface	2.63	1.05	
	.6	Floor Finishes	90% hardwood, 10% ceramic tile	S.F. Floor	8.27	8.27	21.4%
	.7	Ceiling Finishes	Mineral fiber tile on concealed zee bars 15% of area	S.F. Ceiling	2.62	.39	
	.9	Interior Surface/Exterior Wall	Paint 90% of wall	S.F. Wall	1.53	1.03	
7.0 Conveying							
	.1	Elevators	N/A	—	—	—	0.0%
	.2	Special Conveyors	N/A	—	—	—	
8.0 Mechanical							
	.1	Plumbing	Toilet and service fixtures, supply and drainage 1 Fixture/515 S.F. Floor	Each	2616	5.08	
	.2	Fire Protection	Wet pipe sprinkler system	S.F. Floor	1.26	1.26	
	.3	Heating	Included in 8.4	—	—	—	23.7%
	.4	Cooling	Single zone rooftop unit, gas heating, electric cooling	S.F. Floor	7.89	7.89	
	.5	Special Systems	N/A	—	—	—	
9.0 Electrical							
	.1	Service & Distribution	400 ampere service, panel board and feeders	S.F. Floor	.54	.54	
	.2	Lighting & Power	Fluorescent fixtures, receptacles, switches, A.C. and misc. power	S.F. Floor	4.17	4.17	9.8%
	.4	Special Electrical	Alarm systems, sound system and emergency lighting	S.F. Floor	1.11	1.11	
11.0 Special Construction							
	.1	Specialties	Bleachers, sauna, weight room	S.F. Floor	1.02	1.02	1.7%
12.0 Site Work							
	.1	Earthwork	N/A	—	—	—	
	.3	Utilities	N/A	—	—	—	0.0%
	.5	Roads & Parking	N/A	—	—	—	
	.7	Site Improvements	N/A	—	—	—	

			Sub-Total	59.68	**100%**
	GENERAL CONDITIONS (Overhead & Profit)		15%	8.95	
	ARCHITECT FEES		7%	4.82	

Total Building Cost	**73.45**

BUILDING TYPES

Costs per square foot of floor area

Exterior Wall	S.F. Area	1000	2000	3000	4000	5000	10000	15000	20000	25000
	L.F. Perimeter	126	179	219	253	283	400	490	568	632
Decorative Concrete Block	Steel Frame	94.85	85.70	81.60	79.15	77.55	73.40	71.55	70.45	69.70
	Bearing Walls	99.65	89.95	85.60	83.00	81.25	76.85	74.90	73.80	72.95
Face Brick with Concrete Block Back-up	Steel Frame	111.75	98.70	92.85	89.30	86.95	81.05	78.40	76.90	75.75
	Bearing Walls	112.00	98.75	92.75	89.20	86.80	80.80	78.10	76.55	75.45
Metal Sandwich Panel Precast Concrete Panel	Steel Frame	98.45	89.35	85.25	82.80	81.15	77.05	75.15	74.15	73.35
	Bearing Walls	96.05	87.40	83.50	81.20	79.65	75.75	73.95	72.95	72.25
Perimeter Adj., Add or Deduct	Per 100 L.F.	24.95	12.50	8.35	6.25	5.00	2.45	1.65	1.25	1.00
Story Hgt. Adj., Add or Deduct	Per 1 Ft.	1.70	1.25	1.00	.85	.75	.55	.45	.40	.35

For Basement, add $16.60 per square foot of basement area

The above costs were calculated using the basic specifications shown on the facing page. These costs should be adjusted where necessary for design alternatives and owner's requirements. Reported completed project costs, for this type of structure, range from $58.25 to $141.95 per S.F.

Common additives

Description	Unit	$ Cost
Closed Circuit Surveillance, One station		
Camera and monitor	Each	1250
For additional camera stations, add	Each	695
Emergency Lighting, 25 watt, battery operated		
Lead battery	Each	305
Nickel cadmium	Each	600
Laundry Equipment		
Dryers, coin operated 30 lb.	Each	2525
Double stacked	Each	5525
50 lb.	Each	2875
Dry cleaner 20 lb.	Each	32,000
30 lb.	Each	43,800
Washers, coin operated	Each	1100
Washer/extractor 20 lb.	Each	3875
30 lb.	Each	9175
50 lb.	Each	9700
75 lb.	Each	18,100
Smoke Detectors		
Ceiling type	Each	126
Duct type	Each	360

Important: See the Reference Section for Location Factors

Model costs calculated for a 1 story building with 12' story height and 3,000 square feet of floor area

				Unit	Unit Cost	Cost Per S.F.	% Of Sub-Total
1.0 Foundations							
	.1	Footings & Foundations	Poured concrete; strip and spread footings and 4' foundation wall	S.F. Ground	3.86	3.86	
	.4	Piles & Caissons	N/A	—	—	—	7.4%
	.9	Excavation & Backfill	Site preparation for slab and trench for foundation wall and footing	S.F. Ground	1.08	1.08	
2.0 Substructure							
	.1	Slab on Grade	5" reinforced concrete with vapor barrier and granular base	S.F. Slab	2.79	2.79	4.2%
	.2	Special Substructures	N/A	—	—	—	
3.0 Superstructure							
	.1	Columns & Beams	Fireproofing; steel columns included in 3.7	—	—	—	
	.4	Structural Walls	N/A	—	—	—	
	.5	Elevated Floors	N/A	—	—	—	4.2%
	.7	Roof	Metal deck, open web steel joists, beams, columns	S.F. Roof	2.81	2.81	
	.9	Stairs	N/A	—	—	—	
4.0 Exterior Closure							
	.1	Walls	Decorative concrete block 90% of wall	S.F. Wall	7.47	5.89	
	.5	Exterior Wall Finishes	N/A	—	—	—	13.2%
	.6	Doors	Double aluminum and glass	Each	2430	.81	
	.7	Windows & Glazed Walls	Store front 10% of wall	S.F. Window	23	2.04	
5.0 Roofing							
	.1	Roof Coverings	Built-up tar and gravel with flashing	S.F. Roof	2.69	2.69	
	.7	Insulation	Perlite/EPS composite	S.F. Roof	1.13	1.13	5.9%
	.8	Openings & Specialties	Gravel stop, gutters and downspouts	S.F. Roof	.10	.10	
6.0 Interior Construction							
	.1	Partitions	Gypsum board on metal studs 60 S.F. Floor/L.F. Partition	S.F. Partition	2.46	.41	
	.4	Interior Doors	Single leaf wood 750 S.F. Floor/Door	Each	367	.49	
	.5	Wall Finishes	Paint	S.F. Surface	.42	.14	
	.6	Floor Finishes	Vinyl composition tile	S.F. Floor	1.20	1.20	8.4%
	.7	Ceiling Finishes	Fiberglass board on exposed grid system	S.F. Ceiling	1.54	1.54	
	.9	Interior Surface/Exterior Wall	Painted gypsum board on furring 90% of wall	S.F. Wall	2.26	1.78	
7.0 Conveying							
	.1	Elevators	N/A	—	—	—	0.0%
	.2	Special Conveyors	N/A	—	—	—	
8.0 Mechanical							
	.1	Plumbing	Toilet and service fixtures, supply and drainage 1 Fixture/600 S.F. Floor	Each	9588	15.98	
	.2	Fire Protection	Sprinkler, ordinary hazard	S.F. Floor	1.82	1.82	
	.3	Heating	Included in 8.4	—	—	—	36.4%
	.4	Cooling	Rooftop single zone unit systems	S.F. Floor	6.31	6.31	
	.5	Special Systems	N/A	—	—	—	
9.0 Electrical							
	.1	Service & Distribution	200 ampere service, panel board and feeders	S.F. Floor	1.92	1.92	
	.2	Lighting & Power	Fluorescent fixtures, receptacles, switches, A.C. and misc. power	S.F. Floor	11.20	11.20	20.3%
	.4	Special Electrical	Alarm systems and emergency lighting	S.F. Floor	.34	.34	
11.0 Special Construction							
	.1	Specialties	N/A	—	—	—	0.0%
12.0 Site Work							
	.1	Earthwork	N/A	—	—	—	
	.3	Utilities	N/A	—	—	—	
	.5	Roads & Parking	N/A	—	—	—	0.0%
	.7	Site Improvements	N/A	—	—	—	
				Sub-Total		66.33	**100%**
		GENERAL CONDITIONS (Overhead & Profit)			*15%*	9.95	
		ARCHITECT FEES			*7%*	5.32	
				Total Building Cost		**81.60**	

Costs per square foot of floor area

Exterior Wall	S.F. Area	7000	10000	13000	16000	19000	22000	25000	28000	31000
	L.F. Perimeter	240	300	336	386	411	435	472	510	524
Face Brick with Concrete Block Back-up	R/Conc. Frame	88.05	83.30	79.35	77.55	75.30	73.65	72.80	72.15	71.05
	Steel Frame	88.75	84.05	80.10	78.30	76.00	74.35	73.50	72.85	71.75
Limestone with Concrete Block	R/Conc. Frame	100.05	93.80	88.45	86.00	82.90	80.55	79.40	78.50	76.95
	Steel Frame	100.80	94.55	89.15	86.75	83.60	81.30	80.15	79.25	77.70
Precast Concrete Panels	R/Conc. Frame	85.55	81.10	77.50	75.80	73.70	72.20	71.40	70.80	69.80
	Steel Frame	86.25	81.85	78.20	76.55	74.45	72.95	72.15	71.55	70.50
Perimeter Adj., Add or Deduct	Per 100 L.F.	10.85	7.60	5.85	4.75	4.00	3.50	3.10	2.75	2.45
Story Hgt. Adj., Add or Deduct	Per 1 Ft.	1.65	1.45	1.20	1.15	1.05	.95	.90	.90	.80

For Basement, add $14.35 per square foot of basement area

The above costs were calculated using the basic specifications shown on the facing page. These costs should be adjusted where necessary for design alternatives and owner's requirements. Reported completed project costs, for this type of structure, range from $55.85 to $143.25 per S.F.

Common additives

Description	Unit	$ Cost		Description	Unit	$ Cost
Carrels Hardwood	Each	640 - 835		Library Furnishings		
Closed Circuit Surveillance, One station				Bookshelf, 90" high, 10" shelf double face	L.F.	116
Camera and monitor	Each	1250		single face	L.F.	89
For additional camera stations, add	Each	695		Charging desk, built-in with counter		
Elevators, Hydraulic passenger, 2 stops				Plastic laminated top	L.F.	450
1500# capacity	Each	37,625		Reading table, laminated		
2500# capacity	Each	38,925		top 60" x 36"	Each	625
3500# capacity	Each	42,525				
Emergency Lighting, 25 watt, battery operated						
Lead battery	Each	305				
Nickel cadmium	Each	600				
Flagpoles, Complete						
Aluminum, 20' high	Each	1025				
40' high	Each	2375				
70' high	Each	6925				
Fiberglass, 23' high	Each	1400				
39'-5" high	Each	2425				
59' high	Each	6875				

Model costs calculated for a 2 story building with 14' story height and 22,000 square feet of floor area

				Unit	Unit Cost	Cost Per S.F.	% Of Sub-Total
1.0 Foundations							
	.1	Footings & Foundations	Poured concrete; strip and spread footings and 4' foundation wall	S.F. Ground	3.30	1.65	
	.4	Piles & Caissons	N/A	—	—	—	3.6%
	.9	Excavation & Backfill	Site preparation for slab and trench for foundation wall and footing	S.F. Ground	.96	.48	
2.0 Substructure							
	.1	Slab on Grade	4" reinforced concrete with vapor barrier and granular base	S.F. Slab	2.56	1.28	2.2%
	.2	Special Substructures	N/A	—	—	—	
3.0 Superstructure							
	.1	Columns & Beams	Concrete columns	L.F. Column	42	1.30	
	.4	Structural Walls	N/A	—	—	—	
	.5	Elevated Floors	Concrete waffle slab	S.F. Floor	9.31	4.66	18.0%
	.7	Roof	Concrete waffle slab	S.F. Roof	8.59	4.30	
	.9	Stairs	Concrete filled metal pan	Flight	4525	.41	
4.0 Exterior Closure							
	.1	Walls	Face brick with concrete block backup — 90% of wall	S.F. Wall	16.00	7.97	
	.5	Exterior Wall Finishes	N/A	—	—	—	16.6%
	.6	Doors	Double aluminum and glass, single leaf hollow metal	Each	3160	.29	
	.7	Windows & Glazed Walls	Window wall — 10% of wall	S.F. Wall	28	1.57	
5.0 Roofing							
	.1	Roof Coverings	Built-up tar and gravel with flashing	S.F. Roof	1.82	.91	
	.7	Insulation	Perlite/EPS composite	S.F. Roof	1.13	.56	2.7%
	.8	Openings & Specialties	Gravel stop and hatches	L.F. Roof	.22	.11	
6.0 Interior Construction							
	.1	Partitions	Gypsum board on metal studs — 30 S.F. Floor/L.F. Partition	S.F. Partition	3.15	1.26	
	.4	Interior Doors	Single leaf wood — 300 S.F. Floor/Door	Each	367	1.22	
	.5	Wall Finishes	Paint	S.F. Surface	.42	.34	17.1%
	.6	Floor Finishes	50% carpet, 50% vinyl tile	S.F. Floor	3.55	3.55	
	.7	Ceiling Finishes	Mineral fiber on concealed zee bars	S.F. Ceiling	2.62	2.62	
	.9	Interior Surface/Exterior Wall	Painted gypsum board on furring — 90% of wall	S.F. Wall	2.25	1.12	
7.0 Conveying							
	.1	Elevators	One hydraulic passenger elevator	Each	47,960	2.18	3.7%
	.2	Special Conveyors	N/A	—	—	—	
8.0 Mechanical							
	.1	Plumbing	Toilet and service fixtures, supply and drainage — 1 Fixture/1835 S.F. Floor	Each	2348	1.28	
	.2	Fire Protection	Wet pipe sprinkler system	S.F. Floor	1.13	1.13	
	.3	Heating	Included in 8.4	—	—	—	25.5%
	.4	Cooling	Multizone unit, gas heating, electric cooling	S.F. Floor	12.82	12.82	
	.5	Special Systems	N/A	—	—	—	
9.0 Electrical							
	.1	Service & Distribution	400 ampere service, panel board and feeders	S.F. Floor	.57	.57	
	.2	Lighting & Power	Fluorescent fixtures, receptacles, switches, A.C. and misc. power	S.F. Floor	5.28	5.28	10.6%
	.4	Special Electrical	Alarm systems and emergency lighting	S.F. Floor	.43	.43	
11.0 Special Construction							
	.1	Specialties	N/A	—	—	—	0.0%
12.0 Site Work							
	.1	Earthwork	N/A	—	—	—	
	.3	Utilities	N/A	—	—	—	0.0%
	.5	Roads & Parking	N/A	—	—	—	
	.7	Site Improvements	N/A	—	—	—	
				Sub-Total		59.29	**100%**
		GENERAL CONDITIONS (Overhead & Profit)			15%	8.89	
		ARCHITECT FEES			8%	5.47	
				Total Building Cost		**73.65**	

Costs per square foot of floor area

Exterior Wall	S.F. Area	4000	5500	7000	8500	10000	11500	13000	14500	16000
	L.F. Perimeter	280	320	380	440	453	503	510	522	560
Face Brick with Concrete Block Back-up	Steel Joists	85.30	80.70	79.00	77.90	75.70	75.00	73.45	72.35	71.95
	Wood Joists	88.70	84.10	82.40	81.30	79.10	78.40	76.85	75.75	75.35
Stucco on Concrete Block	Steel Joists	79.15	75.60	74.25	73.35	71.70	71.20	70.00	69.20	68.90
	Wood Joists	82.55	79.00	77.65	76.80	75.10	74.60	73.40	72.60	72.30
Brick Veneer	Wood Frame	86.50	-82.30	80.75	79.70	77.70	77.05	75.65	74.65	74.30
Wood Siding	Wood Frame	81.80	78.35	77.05	76.20	74.60	74.10	73.00	72.20	71.95
Perimeter Adj., Add or Deduct	Per 100 L.F.	7.75	5.65	4.45	3.65	3.10	2.70	2.40	2.15	1.95
Story Hgt. Adj., Add or Deduct	Per 1 Ft.	1.65	1.40	1.30	1.25	1.05	1.05	.95	.90	.85

For Basement, add $15.80 per square foot of basement area

The above costs were calculated using the basic specifications shown on the facing page. These costs should be adjusted where necessary for design alternatives and owner's requirements. Reported completed project costs, for this type of structure, range from $47.15 to $121.15 per S.F.

Common additives

Description	Unit	$ Cost
Cabinets, Hospital, base		
Laminated plastic	L.F.	239
Stainless steel	L.F.	286
Counter top, laminated plastic	L.F.	45
Stainless steel	L.F.	103
For drop-in sink, add	Each	460
Nurses station, door type		
Laminated plastic	L.F.	269
Enameled steel	L.F.	240
Stainless steel	L.F.	300
Wall cabinets, laminated plastic	L.F.	174
Enameled steel	L.F.	191
Stainless steel	L.F.	278

Description	Unit	$ Cost
Directory Boards, Plastic, glass covered		
30" x 20"	Each	460
36" x 48"	Each	850
Aluminum, 24" x 18"	Each	410
36" x 24"	Each	495
48" x 32"	Each	590
48" x 60"	Each	1300
Heat Therapy Unit		
Humidified, 26" x 78" x 28"	Each	2525
Smoke Detectors		
Ceiling type	Each	126
Duct type	Each	360
Tables, Examining, vinyl top with base cabinets	Each	1150 - 3175
Utensil Washer, Sanitizer	Each	8400
X-Ray, Mobile	Each	10,100 - 57,000

Important: See the Reference Section for Location Factors

Model costs calculated for a 1 story building with 10′ story height and 7,000 square feet of floor area

				Unit	Unit Cost	Cost Per S.F.	% Of Sub-Total
1.0 Foundations							
.1	Footings & Foundations	Poured concrete; strip and spread footings and 4′ foundation wall		S.F. Ground	2.70	2.70	
.4	Piles & Caissons	N/A		—	—	—	5.6%
.9	Excavation & Backfill	Site preparation for slab and trench for foundation wall and footing		S.F. Ground	.96	.96	
2.0 Substructure							
.1	Slab on Grade	4″ reinforced concrete with vapor barrier and granular base		S.F. Slab	2.56	2.56	3.9%
.2	Special Substructures	N/A		—	—	—	
3.0 Superstructure							
.1	Columns & Beams	N/A		—	—	—	
.4	Structural Walls	N/A		—	—	—	
.5	Elevated Floors	N/A		—	—	—	5.9%
.7	Roof	Plywood on wood trusses		S.F. Ground	3.87	3.87	
.9	Stairs	N/A		—	—	—	
4.0 Exterior Closure							
.1	Walls	Face brick with concrete block backup	70% of wall	S.F. Wall	16.00	6.08	
.5	Exterior Wall Finishes	N/A		—	—	—	16.6%
.6	Doors	Aluminum and glass doors and entrance with transoms		Each	873	1.50	
.7	Windows & Glazed Walls	Wood double hung	30% of wall	Each	350	3.35	
5.0 Roofing							
.1	Roof Coverings	Asphalt shingles with flashing (Pitched)		S.F. Ground	.99	.99	
.7	Insulation	Fiber glass sheeets		S.F. Ground	.78	.78	3.1%
.8	Openings & Specialties	Gutters and downspouts		S.F. Ground	.30	.30	
6.0 Interior Construction							
.1	Partitions	Gypsum bd. & sound deadening bd. on wood studs w/insul.	6 S.F. Floor/L.F. Partition	S.F. Partition	3.93	5.24	
.4	Interior Doors	Single leaf wood	60 S.F. Floor/Door	Each	367	6.12	
.5	Wall Finishes	50% paint, 50% vinyl wall covering		S.F. Surface	.78	2.09	30.7%
.6	Floor Finishes	50% carpet, 50% vinyl composition tile		S.F. Floor	4.09	4.09	
.7	Ceiling Finishes	Mineral fiber tile on concealed zee bars		S.F. Ceiling	1.80	1.80	
.9	Interior Surface/Exterior Wall	Painted gypsum board on furring	70% of wall	S.F. Wall	2.26	.86	
7.0 Conveying							
.1	Elevators	N/A		—	—	—	0.0%
.2	Special Conveyors	N/A		—	—	—	
8.0 Mechanical							
.1	Plumbing	Toilet and service fixtures, supply and drainage	1 Fixture/195 S.F. Floor	Each	1429	7.33	
.2	Fire Protection	Wet pipe sprinkler system		S.F. Floor	1.66	1.66	
.3	Heating	Included in 8.4		—	—	—	26.5%
.4	Cooling	Multizone unit, gas heating, electric cooling		S.F. Floor	8.38	8.38	
.5	Special Systems	N/A		—	—	—	
9.0 Electrical							
.1	Service & Distribution	200 ampere service, panel board and feeders		S.F. Floor	.77	.77	
.2	Lighting & Power	Fluorescent fixtures, receptacles, switches, A.C. and misc. power		S.F. Floor	3.48	3.48	7.7%
.4	Special Electrical	Alarm systems and emergency lighting		S.F. Floor	.84	.84	
11.0 Special Construction							
.1	Specialties	N/A		—	—	—	0.0%
12.0 Site Work							
.1	Earthwork	N/A		—	—	—	
.3	Utilities	N/A		—	—	—	0.0%
.5	Roads & Parking	N/A		—	—	—	
.7	Site Improvements	N/A		—	—	—	
				Sub-Total		65.75	100%
	GENERAL CONDITIONS (Overhead & Profit)				15%	9.86	
	ARCHITECT FEES				9%	6.79	
				Total Building Cost		82.40	

Costs per square foot of floor area

Exterior Wall	S.F. Area	4000	5500	7000	8500	10000	11500	13000	14500	16000
	L.F. Perimeter	180	210	240	270	286	311	336	361	386
Face Brick with Concrete Block Back-up	Steel Joists	99.85	95.10	92.40	90.65	88.70	87.65	86.85	86.20	85.70
	Wood Joists	101.90	97.15	94.50	92.75	90.75	89.70	88.90	88.25	87.75
Stucco on Concrete Block	Steel Joists	97.20	92.80	90.40	88.75	87.00	86.00	85.25	84.70	84.25
	Wood Joists	95.70	91.90	89.75	88.35	86.80	86.00	85.35	84.80	84.45
Brick Veneer	Wood Frame	99.25	94.90	92.50	90.85	89.05	88.15	87.40	86.80	86.35
Wood Siding	Wood Frame	95.35	91.60	89.50	88.10	86.60	85.75	85.15	84.65	84.25
Perimeter Adj., Add or Deduct	Per 100 L.F.	13.55	9.90	7.75	6.35	5.40	4.70	4.20	3.75	3.40
Story Hgt. Adj., Add or Deduct	Per 1 Ft.	2.00	1.70	1.55	1.40	1.30	1.20	1.15	1.10	1.05
For Basement, add $17.10 per square foot of basement area										

The above costs were calculated using the basic specifications shown on the facing page. These costs should be adjusted where necessary for design alternatives and owner's requirements. Reported completed project costs, for this type of structure, range from $47.15 to $121.15 per S.F.

Common additives

Description	Unit	$ Cost
Cabinets, Hospital, base		
Laminated plastic	L.F.	239
Stainless steel	L.F.	286
Counter top, laminated plastic	L.F.	45
Stainless steel	L.F.	103
For drop-in sink, add	Each	460
Nurses station, door type		
Laminated plastic	L.F.	269
Enameled steel	L.F.	240
Stainless steel	L.F.	300
Wall cabinets, laminated plastic	L.F.	174
Enameled steel	L.F.	191
Stainless steel	L.F.	278
Elevators, Hydraulic passenger, 2 stops		
1500# capacity	Each	37,625
2500# capacity	Each	38,925
3500# capacity	Each	42,525

Description	Unit	$ Cost
Directory Boards, Plastic, glass covered		
30" x 20"	Each	460
36" x 48"	Each	850
Aluminum, 24" x 18"	Each	410
36" x 24"	Each	495
48" x 32"	Each	590
48" x 60"	Each	1300
Emergency Lighting, 25 watt, battery operated		
Lead battery	Each	305
Nickel cadmium	Each	600
Heat Therapy Unit		
Humidified, 26" x 78" x 28"	Each	2525
Smoke Detectors		
Ceiling type	Each	126
Duct type	Each	360
Tables, Examining, vinyl top		
with base cabinets	Each	1150 - 3175
Utensil Washer, Sanitizer	Each	8400
X-Ray, Mobile	Each	10,100 - 57,000

Medical Office, 2 Story

Model costs calculated for a 2 story building with 10' story height and 7,000 square feet of floor area

				Unit	Unit Cost	Cost Per S.F.	% Of Sub-Total
1.0 Foundations							
.1	Footings & Foundations	Poured concrete; strip and spread footings and 4' foundation wall		S.F. Ground	4.22	2.11	
.4	Piles & Caissons	N/A		—	—	—	3.6%
.9	Excavation & Backfill	Site preparation for slab and trench for foundation wall and footing		S.F. Ground	.91	.46	
2.0 Substructure							
.1	Slab on Grade	4" reinforced concrete with vapor barrier and granular base		S.F. Slab	2.56	1.28	1.8%
.2	Special Substructures	N/A		—			
3.0 Superstructure							
.1	Columns & Beams	Included in 3.5 and 3.7		—	—	—	
.4	Structural Walls	N/A		—	—	—	
.5	Elevated Floors	Open web steel joists, slab form, concrete, columns		S.F. Floor	6.93	3.46	8.5%
.7	Roof	Metal deck, open web steel joists, beams, columns		S.F. Roof	2.70	1.35	
.9	Stairs	Concrete filled metal pan		Flight	4725	1.35	
4.0 Exterior Closure							
.1	Walls	Concrete block, insulated	70% of wall	S.F. Floor	3.77	3.77	
.5	Exterior Wall Finishes	Stucco on concrete block	70% of wall	S.F. Wall	4.73	2.27	14.5%
.6	Doors	Aluminum and glass doors with transoms		Each	3160	.90	
.7	Windows & Glazed Walls	Outward projecting metal	30% of wall	Each	254	3.48	
5.0 Roofing							
.1	Roof Coverings	Built-up tar and gravel with flashing		S.F. Roof	2.20	1.10	
.7	Insulation	Perlite/EPS composite		S.F. Roof	1.13	.56	2.6%
.8	Openings & Specialties	Gravel stop and hatches		S.F. Roof	.46	.23	
6.0 Interior Construction							
.1	Partitions	Gypsum bd. & sound deadening bd. on wood studs w/insul.	6 S.F. Floor/L.F. Partition	S.F. Partition	3.93	5.24	
.4	Interior Doors	Single leaf wood	60 S.F. Floor/Door	Each	367	6.12	
.5	Wall Finishes	50% paint, 50% vinyl wall coating, 5% ceramic tile		S.F. Surface	.78	2.09	28.3%
.6	Floor Finishes	50% carpet, 50% vinyl asbestos tile		S.F. Floor	4.09	4.09	
.7	Ceiling Finishes	Mineral fiber tile on concealed zee bars		S.F. Ceiling	1.80	1.80	
.9	Interior Surface/Exterior Wall	Painted gypsum board on furring	70% of wall	S.F. Wall	2.26	1.08	
7.0 Conveying							
.1	Elevators	One hydraulic hospital elevator		Each	56,070	8.01	11.1%
.2	Special Conveyors	N/A		—	—	—	
8.0 Mechanical							
.1	Plumbing	Toilet and service fixtures, supply and drainage	1 Fixture/160 S.F. Floor	Each	1041	6.51	
.2	Fire Protection	Wet pipe sprinkler system		S.F. Floor	1.37	1.37	
.3	Heating	Included in 8.4		—	—	—	22.5%
.4	Cooling	Multizone unit, gas heating, electric cooling		S.F. Floor	8.38	8.38	
.5	Special Systems	N/A		—	—	—	
9.0 Electrical							
.1	Service & Distribution	200 ampere service, panel board and feeders		S.F. Floor	.77	.77	
.2	Lighting & Power	Fluorescent fixtures, receptacles, switches, A.C. and misc. power		S.F. Floor	3.48	3.48	7.1%
.4	Special Electrical	Alarm systems and emergency lighting		S.F. Floor	.84	.84	
11.0 Special Construction							
.1	Specialties	N/A		—	—	—	0.0%
12.0 Site Work							
.1	Earthwork	N/A		—	—	—	
.3	Utilities	N/A		—	—	—	0.0%
.5	Roads & Parking	N/A		—	—	—	
.7	Site Improvements	N/A		—	—	—	

Sub-Total		72.10	**100%**
GENERAL CONDITIONS (Overhead & Profit)	15%	10.82	
ARCHITECT FEES	9%	7.48	
Total Building Cost		90.40	

Costs per square foot of floor area

| Exterior Wall | S.F. Area | 2000 | 3000 | 4000 | 6000 | 8000 | 10000 | 12000 | 14000 | 16000 |
	L.F. Perimeter	240	260	280	380	480	560	580	660	740
Brick Veneer	Wood Frame	87.95	77.55	72.30	69.80	68.50	67.20	65.00	64.60	64.25
Aluminum Siding	Wood Frame	79.55	71.45	67.40	65.35	64.30	63.30	61.60	61.30	61.00
Wood Siding	Wood Frame	79.80	71.65	67.55	65.45	64.40	63.40	61.70	61.40	61.10
Wood Shingles	Wood Frame	81.95	73.20	68.80	66.60	65.50	64.40	62.60	62.25	61.95
Precast Concrete Block	Wood Truss	78.85	70.95	66.95	64.95	63.95	62.95	61.30	61.00	60.70
Brick on Concrete Block	Wood Truss	91.55	80.10	74.40	71.65	70.30	68.90	66.45	66.00	65.60
Perimeter Adj., Add or Deduct	Per 100 L.F.	13.40	8.95	6.70	4.45	3.35	2.65	2.25	1.90	1.70
Story Hgt. Adj., Add or Deduct	Per 1 Ft.	2.35	1.70	1.40	1.25	1.20	1.10	.95	.95	.90

For Basement, add $12.70 per square foot of basement area

The above costs were calculated using the basic specifications shown on the facing page. These costs should be adjusted where necessary for design alternatives and owner's requirements. Reported completed project costs, for this type of structure, range from $38.65 to $93.35 per S.F.

Common additives

Description	Unit	$ Cost
Closed Circuit Surveillance, One station		
Camera and monitor	Each	1250
For additional camera stations, add	Each	695
Emergency Lighting, 25 watt, battery operated		
Lead battery	Each	305
Nickel cadmium	Each	600
Laundry Equipment		
Dryer, gas, 16 lb. capacity	Each	640
30 lb. capacity	Each	2525
Washer, 4 cycle	Each	720
Commercial	Each	1100
Sauna, Prefabricated, complete		
6' x 4'	Each	3550
6' x 6'	Each	4175
6' x 9'	Each	5600
8' x 8'	Each	5925
8' x 10'	Each	6500
10' x 12'	Each	8500
Smoke Detectors		
Ceiling type	Each	126
Duct type	Each	360

Description	Unit	$ Cost
Swimming Pools, Complete, gunite	S.F.	39 - 48
TV Antenna, Master system, 12 outlet	Outlet	212
30 outlet	Outlet	137
100 outlet	Outlet	133

Important: See the Reference Section for Location Factors

Model costs calculated for a 1 story building with 9' story height and 8,000 square feet of floor area

				Unit	Unit Cost	Cost Per S.F.	% Of Sub-Total
1.0 Foundations							
.1	Footings & Foundations	Poured concrete; strip and spread footings and 4' foundation wall		S.F. Ground	4.08	4.08	
.4	Piles & Caissons	N/A		—	—	—	9.1%
.9	Excavation & Backfill	Site preparation for slab and trench for foundation wall and footing		S.F. Ground	.96	.96	
2.0 Substructure							
.1	Slab on Grade	4" reinforced concrete with vapor barrier and granular base		S.F. Slab	2.56	2.56	4.6%
.2	Special Substructures	N/A		—			
3.0 Superstructure							
.1	Columns & Beams	N/A		—	—	—	
.4	Structural Walls	N/A		—	—	—	
.5	Elevated Floors	N/A		—	—	—	7.0%
.7	Roof	Plywood on wood trusses		S.F. Ground	3.87	3.87	
.9	Stairs	N/A		—			
4.0 Exterior Closure							
.1	Walls	Face brick on wood studs with sheathing, insulation and paper	80% of wall	S.F. Wall	12.45	5.38	
.5	Exterior Wall Finishes	N/A		—	—	—	18.3%
.6	Doors	Wood solid core		Each	873	2.40	
.7	Windows & Glazed Walls	Wood double hung	20% of wall	Each	315	2.43	
5.0 Roofing							
.1	Roof Coverings	Asphalt shingles with flashing (pitched)		S.F. Ground	.85	.85	
.7	Insulation	Fiberglass sheets		S.F. Ground	.78	.78	3.5%
.8	Openings & Specialties	Gutters and downspouts		S.F. Ground	.33	.33	
6.0 Interior Construction							
.1	Partitions	Gypsum bd. and sound deadening bd. on wood studs	9 S.F. Floor/L.F. Partition	S.F. Partition	3.93	3.49	
.4	Interior Doors	Single leaf hollow metal	300 S.F. Floor/Door	Each	323	1.08	
.5	Wall Finishes	90% paint, 10% ceramic tile		S.F. Surface	.82	1.45	
.6	Floor Finishes	85% carpet, 15% vinyl composition tile		S.F. Floor	5.51	5.51	26.4%
.7	Ceiling Finishes	Painted gypsum board on furring		S.F. Ceiling	2.32	2.32	
.9	Interior Surface/Exterior Wall	Painted gypsum board on furring	80% of wall	S.F. Wall	2.26	.85	
7.0 Conveying							
.1	Elevators	N/A		—	—	—	0.0%
.2	Special Conveyors	N/A		—	—	—	
8.0 Mechanical							
.1	Plumbing	Toilet and service fixtures, supply and drainage	1 Fixture/90 S.F. Floor	Each	792	8.81	
.2	Fire Protection	Wet pipe sprinkler system		S.F. Floor	1.66	1.66	
.3	Heating	Included in 8.4		—	—	—	24.1%
.4	Cooling	Through the wall electric heating and cooling units		S.F. Floor	2.97	2.97	
.5	Special Systems	N/A		—	—	—	
9.0 Electrical							
.1	Service & Distribution	100 ampere service, panel board and feeders		S.F. Floor	.45	.45	
.2	Lighting & Power	Fluorescent fixtures, receptacles, switches and misc. power		S.F. Floor	3.23	3.23	7.0%
.4	Special Electrical	Alarm systems		S.F. Floor	.22	.22	
11.0 Special Construction							
.1	Specialties	N/A		—	—	—	0.0%
12.0 Site Work							
.1	Earthwork	N/A		—	—	—	
.3	Utilities	N/A		—	—	—	0.0%
.5	Roads & Parking	N/A		—	—	—	
.7	Site Improvements	N/A		—	—	—	
				Sub-Total		55.68	100%
	GENERAL CONDITIONS (Overhead & Profit)				15%	8.35	
	ARCHITECT FEES				7%	4.47	
				Total Building Cost		68.50	

Costs per square foot of floor area

Exterior Wall	S.F. Area	25000	37000	49000	61000	73000	81000	88000	96000	104000
	L.F. Perimeter	433	593	606	720	835	911	978	1054	1074
Decorative Concrete Block	Wood Joists	65.85	64.20	61.70	61.15	60.70	60.50	60.40	60.25	59.80
	Precast Conc.	71.05	69.45	66.95	66.35	65.90	65.70	65.60	65.45	65.05
Stucco on Concrete Block	Wood Joists	64.75	63.15	60.80	60.20	59.80	59.60	59.50	59.35	58.95
	Precast Conc.	70.40	68.80	66.45	65.90	65.45	65.25	65.15	65.05	64.65
Wood Siding	Wood Frame	64.50	62.90	60.60	60.05	59.65	59.45	59.35	59.20	58.80
Brick Veneer	Wood Frame	66.90	65.15	62.35	61.70	61.25	61.00	60.90	60.75	60.25
Perimeter Adj., Add or Deduct	Per 100 L.F.	2.25	1.50	1.15	.90	.75	.70	.65	.60	.55
Story Hgt. Adj., Add or Deduct	Per 1 Ft.	.85	.75	.60	.55	.55	.55	.55	.50	.50

For Basement, add $15.75 per square foot of basement area

The above costs were calculated using the basic specifications shown on the facing page. These costs should be adjusted where necessary for design alternatives and owner's requirements. Reported completed project costs, for this type of structure, range from $39.15 to $94.05 per S.F.

Common additives

Description	Unit	$ Cost	Description	Unit	$ Cost
Closed Circuit Surveillance, One station			Sauna, Prefabricated, complete		
Camera and monitor	Each	1250	6' x 4'	Each	3550
For additional camera station, add	Each	695	6' x 6'	Each	4175
Elevators, Hydraulic passenger, 2 stops			6' x 9'	Each	5600
1500# capacity	Each	37,625	8' x 8'	Each	5925
2500# capacity	Each	38,925	8' x 10'	Each	6500
3500# capacity	Each	42,525	10' x 12'	Each	8500
Additional stop, add	Each	2600	Smoke Detectors		
Emergency Lighting, 25 watt, battery operated			Ceiling type	Each	126
Lead battery	Each	305	Duct type	Each	360
Nickel cadmium	Each	600	Swimming Pools, Complete, gunite	S.F.	39 - 48
Laundry Equipment			TV Antenna, Master system, 12 outlet	Outlet	212
Dryer, gas, 16 lb. capacity	Each	640	30 outlet	Outlet	137
30 lb. capacity	Each	2525	100 outlet	Outlet	133
Washer, 4 cycle	Each	720			
Commercial	Each	1100			

Model costs calculated for a 3 story building with 9' story height and 73,000 square feet of floor area

			Unit	Unit Cost	Cost Per S.F.	% Of Sub-Total
1.0 Foundations						
.1	Footings & Foundations	Poured concrete; strip and spread footings and 4' foundation wall	S.F. Ground	2.76	.92	
.4	Piles & Caissons	N/A	—	—	—	2.3%
.9	Excavation & Backfill	Site preparation for slab and trench for foundation wall and footing	S.F. Ground	.91	.30	
2.0 Substructure						
.1	Slab on Grade	4" reinforced concrete with vapor barrier and granular base	S.F. Slab	2.56	.85	1.6%
.2	Special Substructures	N/A				
3.0 Superstructure						
.1	Columns & Beams	N/A	—	—	—	
.4	Structural Walls	Reinforced concrete block	S.F. Wall	7.76	1.28	
.5	Elevated Floors	Precast concrete plank	S.F. Floor	5.62	3.75	14.3%
.7	Roof	Precast concrete plank	S.F. Roof	5.82	1.94	
.9	Stairs	Concrete filled metal pan	Flight	4725	.78	
4.0 Exterior Closure						
.1	Walls	Face brick with concrete block backup _85% of wall_	S.F. Wall	9.33	2.45	
.5	Exterior Wall Finishes	N/A	—	—	—	10.6%
.6	Doors	Aluminum and glass doors and entrance with transom	Each	1017	2.26	
.7	Windows & Glazed Walls	Aluminum sliding _15% of wall_	Each	329	1.02	
5.0 Roofing						
.1	Roof Coverings	Built-up tar and gravel with flashing	S.F. Roof	1.74	.58	
.7	Insulation	Perlite/EPS composite	S.F. Roof	1.13	.38	2.0%
.8	Openings & Specialties	Gravel stop, hatches, gutters and downspouts	S.F. Roof	.36	.12	
6.0 Interior Construction						
.1	Partitions	Gypsum board and sound deadening board on wood studs _7 S.F. Floor/L.F. Partition_	S.F. Partition	3.93	4.49	
.4	Interior Doors	Wood hollow core _70 S.F. Floor/Door_	Each	323	4.62	
.5	Wall Finishes	90% paint, 10% ceramic tile	S.F. Surface	.81	1.86	35.9%
.6	Floor Finishes	85% carpet, 15% vinyl composition tile	S.F. Floor	5.51	5.51	
.7	Ceiling Finishes	Painted gypsum board on furring	S.F. Ceiling	2.32	2.32	
.9	Interior Surface/Exterior Wall	Painted gypsum board on furring _85% of wall_	S.F. Wall	2.26	.59	
7.0 Conveying						
.1	Elevators	Two hydraulic passenger elevators	Each	56,940	1.56	2.9%
.2	Special Conveyors	N/A	—	—	—	
8.0 Mechanical						
.1	Plumbing	Toilet and service fixtures, supply and drainage _1 Fixture/180 S.F. Floor_	Each	1506	8.37	
.2	Fire Protection	Sprinklers, light hazard	S.F. Floor	.91	.91	
.3	Heating	Included in 8.4	—	—	—	22.8%
.4	Cooling	Through the wall electric heating and cooling units	S.F. Floor	3.11	3.11	
.5	Special Systems	N/A	—	—	—	
9.0 Electrical						
.1	Service & Distribution	400 ampere service, panel board and feeders	S.F. Floor	.26	.26	
.2	Lighting & Power	Fluorescent fixtures, receptacles, switches and misc. power	S.F. Floor	3.55	3.55	7.6%
.4	Special Electrical	Alarm systems and emergency lighting	S.F. Floor	.28	.28	
11.0 Special Construction						
.1	Specialties	N/A	—	—	—	0.0%
12.0 Site Work						
.1	Earthwork	N/A	—	—	—	
.3	Utilities	N/A	—	—	—	0.0%
.5	Roads & Parking	N/A	—	—	—	
.7	Site Improvements	N/A	—	—	—	
			Sub-Total		54.06	100%
	GENERAL CONDITIONS (Overhead & Profit)			15%	8.11	
	ARCHITECT FEES			6%	3.73	
			Total Building Cost		65.90	

BUILDING TYPES

Costs per square foot of floor area

Exterior Wall	S.F. Area	9000	10000	12000	13000	14000	15000	16000	18000	20000
	L.F. Perimeter	385	410	460	460	480	500	510	547	583
Decorative Concrete Block	Steel Joists	79.05	76.85	**73.50**	71.45	70.25	69.25	68.10	66.60	65.30
Painted Concrete Block	Steel Joists	75.00	72.95	69.85	68.10	67.00	66.10	65.10	63.70	62.55
Face Brick on Conc. Block	Steel Joists	84.70	82.25	78.55	76.10	74.75	73.65	72.30	70.55	69.15
Precast Concrete Panels	Steel Joists	83.60	81.20	77.60	75.25	73.90	72.80	71.55	69.80	68.45
Tilt-up Panels	Steel Joists	77.80	75.65	72.35	70.40	69.25	68.25	67.15	65.70	64.45
Metal Sandwich Panels	Steel Joists	72.05	70.15	67.20	65.65	64.65	63.80	62.90	61.60	60.55
Perimeter Adj., Add or Deduct	Per 100 L.F.	4.30	3.85	3.20	2.95	2.75	2.60	2.40	2.10	1.90
Story Hgt. Adj., Add or Deduct	Per 1 Ft.	.60	.60	.55	.55	.50	.50	.45	.45	.45
Basement—Not Applicable										

The above costs were calculated using the basic specifications shown on the facing page. These costs should be adjusted where necessary for design alternatives and owner's requirements. Reported completed project costs, for this type of structure, range from $46.10 to $120.40 per S.F.

Common additives

Description	Unit	$ Cost
Emergency Lighting, 25 watt, battery operated		
Lead battery	Each	305
Nickel cadmium	Each	600
Seating		
Auditorium chair, all veneer	Each	123
Veneer back, padded seat	Each	148
Upholstered, spring seat	Each	174
Classroom, movable chair & desk	Set	65 - 120
Lecture hall, pedestal type	Each	123 - 360
Smoke Detectors		
Ceiling type	Each	126
Duct type	Each	360
Sound System		
Amplifier, 250 watts	Each	1425
Speaker, ceiling or wall	Each	118
Trumpet	Each	222

Important: See the Reference Section for Location Factors

Model costs calculated for a 1 story building with 20' story height and 12,000 square feet of floor area

				Unit	Unit Cost	Cost Per S.F.	% Of Sub-Total
1.0 Foundations							
.1	Footings & Foundations	Poured concrete; strip and spread footings and 4' foundation wall		S.F. Ground	1.90	1.90	
.4	Piles & Caissons	N/A		—	—	—	6.6%
.9	Excavation & Backfill	Site preparation for slab and trench for foundation wall and footing		S.F. Ground	2.05	2.05	
2.0 Substructure							
.1	Slab on Grade	4" reinforced concrete with vapor barrier and granular base		S.F. Slab	2.56	2.56	4.3%
.2	Special Substructures	N/A		—	—	—	
3.0 Superstructure							
.1	Columns & Beams	N/A		—	—	—	
.4	Structural Walls	N/A		—	—	—	
.5	Elevated Floors	Open web steel joists, slab form, concrete	*mezzanine 2250 S.F.*	S.F. Mezz	5.60	1.05	12.0%
.7	Roof	Metal deck on open web steel joists		S.F. Roof	5.17	5.17	
.9	Stairs	Concrete filled metal pan		Flight	5550	.93	
4.0 Exterior Closure							
.1	Walls	Decorative concrete block	*100% of wall*	S.F. Wall	10.45	8.01	
.5	Exterior Wall Finishes	N/A		—	—	—	
.6	Doors	Sliding mallfront aluminum and glass and hollow metal		Each	1305	.65	14.5%
.7	Windows & Glazed Walls	N/A		—	—	—	
5.0 Roofing							
.1	Roof Coverings	Built-up tar and gravel with flashing		S.F. Roof	1.80	1.80	
.7	Insulation	Perlite/EPS composite		S.F. Roof	1.13	1.13	5.6%
.8	Openings & Specialties	Gravel stop, hatches, gutters and downspouts		S.F. Roof	.39	.39	
6.0 Interior Construction							
.1	Partitions	Concrete block, toilet partitions	*40 S.F. Floor/L.F. Partition*	S.F. Partition	5.05	2.02	
.4	Interior Doors	Single leaf hollow metal	*705 S.F. Floor/Door*	Each	484	.69	
.5	Wall Finishes	Paint		S.F. Surface	.88	.70	
.6	Floor Finishes	Carpet	*50% of area*	S.F. Floor	5.42	2.71	16.6%
.7	Ceiling Finishes	Mineral fiber tile on concealed zee runners suspended		S.F. Ceiling	2.62	2.62	
.9	Interior Surface/Exterior Wall	Paint	*100% of wall*	S.F. Wall	1.53	1.17	
7.0 Conveying							
.1	Elevators	N/A		—	—	—	0.0%
.2	Special Conveyors	N/A		—	—	—	
8.0 Mechanical							
.1	Plumbing	Toilet and service fixtures, supply and drainage	*1 Fixture/500 S.F. Floor*	Each	1235	2.47	
.2	Fire Protection	Wet pipe sprinkler system		S.F. Floor	1.26	1.26	
.3	Heating	Included in 8.4		—	—	—	18.3%
.4	Cooling	Single zone rooftop unit, gas heating, electric cooling		S.F. Floor	7.25	7.25	
.5	Special Systems	N/A		—	—	—	
9.0 Electrical							
.1	Service & Distribution	400 ampere service, panel board and feeders		S.F. Floor	.53	.53	
.2	Lighting & Power	Fluorescent fixtures, receptacles, switches, A.C. and misc. power		S.F. Floor	2.21	2.21	5.3%
.4	Special Electrical	Alarm systems, sound system and emergency lighting		S.F. Floor	.40	.40	
11.0 Special Construction							
.1	Specialties	Projection equipment, screen, seating		S.F. Floor	10.06	10.06	16.8%
12.0 Site Work							
.1	Earthwork	N/A		—	—	—	
.3	Utilities	N/A		—	—	—	
.5	Roads & Parking	N/A		—	—	—	0.0%
.7	Site Improvements	N/A		—	—	—	
				Sub-Total		59.73	**100%**
	GENERAL CONDITIONS (Overhead & Profit)				15%	8.96	
	ARCHITECT FEES				7%	4.81	
				Total Building Cost		73.50	

Costs per square foot of floor area

Exterior Wall	S.F. Area	10000	15000	20000	25000	30000	35000	40000	45000	50000
	L.F. Perimeter	286	370	453	457	513	568	624	680	735
Redwood Siding	Wood Frame	74.70	71.90	70.55	68.60	67.95	67.40	67.05	66.70	66.50
Brick Veneer	Wood Frame	79.30	75.90	74.20	71.55	70.75	70.05	69.60	69.20	68.90
Face Brick with Concrete Block Back-up	Wood Joists	81.55	77.85	75.95	73.00	72.05	71.30	70.80	70.35	70.05
	Steel Joists	88.35	84.65	82.75	79.80	78.90	78.10	77.60	77.15	76.85
Stucco on Concrete Block	Wood Joists	76.55	73.50	72.00	69.80	69.10	68.45	68.05	67.70	67.50
	Steel Joists	83.35	80.35	78.80	76.60	75.90	75.30	74.90	74.55	74.30
Perimeter Adj., Add or Deduct	Per 100 L.F.	3.45	2.30	1.70	1.40	1.10	1.00	.90	.80	.70
Story Hgt. Adj., Add or Deduct	Per 1 Ft.	.70	.60	.55	.45	.40	.40	.35	.35	.35

For Basement, add $16.55 per square foot of basement area

The above costs were calculated using the basic specifications shown on the facing page. These costs should be adjusted where necessary for design alternatives and owner's requirements. Reported completed project costs, for this type of structure, range from $49.85 to $123.80 per S.F.

Common additives

Description	Unit	$ Cost	Description	Unit	$ Cost
Beds, Manual	Each	865 - 1500	Kitchen Equipment, cont.		
Doctors In-Out Register, 200 names	Each	12,100	Ice cube maker, 50 lb. per day	Each	1700
Elevators, Hydraulic passenger, 2 stops			Range with 1 oven	Each	2300
1500# capacity	Each	37,625	Laundry Equipment		
2500# capacity	Each	38,925	Dryer, gas, 16 lb. capacity	Each	640
3500# capacity	Each	42,525	30 lb. capacity	Each	2525
Emergency Lighting, 25 watt, battery operated			Washer, 4 cycle	Each	720
Lead battery	Each	305	Commercial	Each	1100
Nickel cadmium	Each	600	Nurses Call System		
Intercom System, 25 station capacity			Single bedside call station	Each	188
Master station	Each	1725	Pillow speaker	Each	198
Intercom outlets	Each	101	Refrigerator, Prefabricated, walk-in		
Handset	Each	277	7'-6" high, 6' x 6'	S.F.	114
Kitchen Equipment			10' x 10'	S.F.	89
Broiler	Each	3400	12' x 14'	S.F.	79
Coffee urn, twin 6 gallon	Each	5875	12' x 20'	S.F.	70
Cooler, 6 ft. long	Each	2675	TV Antenna, Master system, 12 outlet	Outlet	212
Dishwasher, 10-12 racks per hr.	Each	2525	30 outlet	Outlet	137
Food warmer	Each	745	100 outlet	Outlet	133
Freezer, 44 C.F., reach-in	Each	6575	Whirlpool Bath, Mobile, 18" x 24" x 60"	Each	3475
			X-Ray, Mobile	Each	10,100 - 57,000

Important: See the Reference Section for Location Factors

Model costs calculated for a 2 story building with 10' story height and 25,000 square feet of floor area

				Unit	Unit Cost	Cost Per S.F.	% Of Sub-Total
1.0 Foundations							
	.1	Footings & Foundations	Poured concrete; strip and spread footings and 4' foundation wall	S.F. Ground	2.14	1.07	
	.4	Piles & Caissons	N/A	—	—	—	2.9%
	.9	Excavation & Backfill	Site preparation for slab and trench for foundation wall and footing	S.F. Ground	.96	.48	
2.0 Substructure							
	.1	Slab on Grade	4" reinforced concrete with vapor barrier and granular base	S.F. Slab	2.56	1.28	2.4%
	.2	Special Substructures	N/A	—	—	—	
3.0 Superstructure							
	.1	Columns & Beams	N/A	—	—	—	
	.4	Structural Walls	N/A	—	—	—	
	.5	Elevated Floors	Plywood on wood joists	S.F. Floor	3.02	1.51	5.6%
	.7	Roof	Plywood on wood joists	S.F. Roof	2.34	1.17	
	.9	Stairs	Wood	Flight	1328	.32	
4.0 Exterior Closure							
	.1	Walls	Vertical T & G redwood siding _85% of wall_	S.F. Wall	4.96	1.54	
	.5	Exterior Wall Finishes	N/A	—	—	—	5.6%
	.6	Doors	Double aluminum & glass doors, single leaf hollow metal	Each	1330	.27	
	.7	Windows & Glazed Walls	Wood double hung _15% of wall_	Each	329	1.20	
5.0 Roofing							
	.1	Roof Coverings	Built-up tar and gravel with flashing	S.F. Roof	1.78	.89	
	.7	Insulation	Perlite/EPS composite	S.F. Roof	1.13	.56	3.1%
	.8	Openings & Specialties	Gravel stop, hatches, gutters and downspouts	S.F. Roof	.42	.21	
6.0 Interior Construction							
	.1	Partitions	Gypsum board on wood studs _8 S.F. Floor/L.F. Partition_	S.F. Partition	2.48	2.48	
	.4	Interior Doors	Single leaf wood _80 S.F. Floor/Door_	Each	367	4.59	
	.5	Wall Finishes	50% vinyl wall coverings, 45% paint, 5% ceramic tile	S.F. Surface	.99	1.97	
	.6	Floor Finishes	50% carpet, 45% vinyl tile, 5% ceramic tile	S.F. Floor	4.26	4.26	30.4%
	.7	Ceiling Finishes	Painted gypsum board on wood furring	S.F. Ceiling	2.32	2.32	
	.9	Interior Surface/Exterior Wall	Painted gypsum board on wood furring _85% of wall_	S.F. Wall	2.26	.70	
7.0 Conveying							
	.1	Elevators	One hydraulic hospital elevator	Each	56,000	2.24	4.2%
	.2	Special Conveyors	N/A	—	—	—	
8.0 Mechanical							
	.1	Plumbing	Kitchen, toilet and service fixtures, supply and drainage _1 Fixture/230 S.F. Floor_	Each	2024	8.80	
	.2	Fire Protection	Sprinkler, light hazard	S.F. Floor	2.25	2.25	
	.3	Heating	Oil fired hot water, wall fin radiation	S.F. Floor	3.53	3.53	34.5%
	.4	Cooling	Split systems with air cooled condensing units	S.F. Floor	4.02	4.02	
	.5	Special Systems	N/A	—	—	—	
9.0 Electrical							
	.1	Service & Distribution	600 ampere service, panel board and feeders	S.F. Floor	.81	.81	
	.2	Lighting & Power	Incandescent fixtures, receptacles, switches, A.C. and misc. power	S.F. Floor	4.41	4.41	11.3%
	.4	Special Electrical	Alarm systems and emergency lighting	S.F. Floor	.86	.86	
11.0 Special Construction							
	.1	Specialties	N/A	—	—	—	0.0%
12.0 Site Work							
	.1	Earthwork	N/A	—	—	—	
	.3	Utilities	N/A	—	—	—	
	.5	Roads & Parking	N/A	—	—	—	0.0%
	.7	Site Improvements	N/A	—	—	—	

		Unit	Unit Cost	Cost Per S.F.	% Of Sub-Total
	Sub-Total			53.74	100%
GENERAL CONDITIONS (Overhead & Profit)			15%	8.06	
ARCHITECT FEES			11%	6.80	
	Total Building Cost			**68.60**	

Costs per square foot of floor area

Exterior Wall	S.F. Area	10000	22000	34000	46000	58000	63000	68000	73000	78000
	L.F. Perimeter	246	393	443	543	562	590	603	624	645
Face Brick with Concrete Block Back-up	Wood Joists	78.25	67.05	61.25	59.45	57.15	56.70	56.20	55.75	55.45
	Steel Joists	79.55	68.40	62.55	60.75	**58.45**	58.00	57.50	57.10	56.75
Glass and Metal Curtain Wall	Steel Frame	78.35	68.00	62.75	61.10	59.05	58.65	58.20	57.80	57.50
	R/Conc. Frame	79.15	68.80	63.55	61.90	59.85	59.45	59.00	58.65	58.30
Wood Siding	Wood Frame	67.85	58.65	54.25	52.80	51.15	50.80	50.40	50.10	49.85
Brick Veneer	Wood Frame	72.15	61.80	56.55	54.90	52.85	52.45	51.95	51.60	51.30
Perimeter Adj., Add or Deduct	Per 100 L.F.	8.90	4.05	2.60	1.95	1.50	1.40	1.25	1.25	1.15
Story Hgt. Adj., Add or Deduct	Per 1 Ft.	1.65	1.20	.85	.80	.65	.65	.55	.55	.55

For Basement, add $18.35 per square foot of basement area

The above costs were calculated using the basic specifications shown on the facing page. These costs should be adjusted where necessary for design alternatives and owner's requirements. Reported completed project costs, for this type of structure, range from $39.10 to $110.75 per S.F.

Common additives

Description	Unit	$ Cost
Clock System		
20 room	Each	9875
50 room	Each	23,800
Closed Circuit Surveillance, One station		
Camera and monitor	Each	1250
For additional camera stations, add	Each	695
Directory Boards, Plastic, glass covered		
30" x 20"	Each	460
36" x 48"	Each	850
Aluminum, 24" x 18"	Each	410
36" x 24"	Each	495
48" x 32"	Each	590
48" x 60"	Each	1300
Elevators, Hydraulic passenger, 2 stops		
1500# capacity	Each	37,625
2500# capacity	Each	38,925
3500# capacity	Each	42,525
Additional stop, add	Each	2600
Emergency Lighting, 25 watt, battery operated		
Lead battery	Each	305
Nickel cadmium	Each	600

Description	Unit	$ Cost
Smoke Detectors		
Ceiling type	Each	126
Duct type	Each	360
Sound System		
Amplifier, 250 watts	Each	1425
Speaker, ceiling or wall	Each	118
Trumpet	Each	222
TV Antenna, Master system, 12 outlet	Outlet	212
30 outlet	Outlet	137
100 outlet	Outlet	133

Model costs calculated for a 3 story building with 12' story height and 58,000 square feet of floor area

				Unit	Unit Cost	Cost Per S.F.	% Of Sub-Total
1.0 Foundations							
	.1	Footings & Foundations	Poured concrete; strip and spread footings and 4' foundation wall	S.F. Ground	2.64	.88	
	.4	Piles & Caissons	N/A	—	—	—	2.5%
	.9	Excavation & Backfill	Site preparation for slab and trench for foundation wall and footing	S.F. Ground	.91	.30	
2.0 Substructure							
	.1	Slab on Grade	4" reinforced concrete with vapor barrier and granular base	S.F. Slab	2.56	.85	1.8%
	.2	Special Substructures	N/A	—	—	—	
3.0 Superstructure							
	.1	Columns & Beams	Fireproofing; interior columns included in 3.5 and 3.7	L.F. Columns	16.94	.46	
	.4	Structural Walls	N/A	—	—	—	
	.5	Elevated Floors	Open web steel joists, slab form, concrete, columns	S.F. Floor	6.86	4.57	14.0%
	.7	Roof	Metal deck, open web steel joists, columns	S.F. Roof	3.21	1.07	
	.9	Stairs	Concrete filled metal pan	Flight	4725	.57	
4.0 Exterior Closure							
	.1	Walls	Face brick with concrete block backup *80% of wall*	S.F. Wall	15.80	4.41	
	.5	Exterior Wall Finishes	N/A	—	—	—	12.4%
	.6	Doors	Aluminum and glass, hollow metal	Each	2046	.21	
	.7	Windows & Glazed Walls	Steel outward projecting *20% of wall*	Each	420	1.27	
5.0 Roofing							
	.1	Roof Coverings	Built-up tar and gravel with flashing	S.F. Roof	1.77	.59	
	.7	Insulation	Perlite/EPS composite	S.F. Roof	1.13	.38	2.0%
	.8	Openings & Specialties	N/A	—	—	—	
6.0 Interior Construction							
	.1	Partitions	Gypsum board on metal studs, toilet partitions *20 S.F. Floor/L.F. Partition*	S.F. Partition	2.43	1.25	
	.4	Interior Doors	Single leaf hollow metal *200 S.F. Floor/Door*	Each	484	2.42	
	.5	Wall Finishes	60% vinyl wall covering, 40% paint	S.F. Surface	.85	.68	25.7%
	.6	Floor Finishes	60% carpet, 30% vinyl composition tile, 10% ceramic tile	S.F. Floor	4.61	4.61	
	.7	Ceiling Finishes	Mineral fiber tile on concealed zee bars	S.F. Ceiling	2.62	2.62	
	.9	Interior Surface/Exterior Wall	Painted gypsum board on furring *80% of wall*	S.F. Wall	2.26	.63	
7.0 Conveying							
	.1	Elevators	Two hydraulic passenger elevators	Each	58,580	2.02	4.3%
	.2	Special Conveyors	N/A	—	—	—	
8.0 Mechanical							
	.1	Plumbing	Toilet and service fixtures, supply and drainage *1 Fixture/1320 S.F. Floor*	Each	1755	1.33	
	.2	Fire Protection	Standpipes and hose systems	S.F. Floor	.17	.17	
	.3	Heating	Included in 8.4	—	—	—	23.7%
	.4	Cooling	Multizone unit gas heating, electric cooling	S.F. Floor	9.73	9.73	
	.5	Special Systems	N/A	—	—	—	
9.0 Electrical							
	.1	Service & Distribution	1000 ampere service, panel board and feeders	S.F. Floor	.81	.81	
	.2	Lighting & Power	Fluorescent fixtures, receptacles, switches, A.C. and misc. power	S.F. Floor	5.51	5.51	13.6%
	.4	Special Electrical	Alarm systems and emergency lighting	S.F. Floor	.16	.16	
11.0 Special Construction							
	.1	Specialties	N/A	—	—	—	0.0%
12.0 Site Work							
	.1	Earthwork	N/A	—	—	—	
	.3	Utilities	N/A	—	—	—	0.0%
	.5	Roads & Parking	N/A	—	—	—	
	.7	Site Improvements	N/A	—	—	—	
				Sub-Total		47.50	**100%**
		GENERAL CONDITIONS (Overhead & Profit)			15%	7.13	
		ARCHITECT FEES			7%	3.82	
				Total Building Cost		**58.45**	

COMMERCIAL/INDUSTRIAL/ INSTITUTIONAL · M.490 · Police Station

Costs per square foot of floor area

Exterior Wall	S.F. Area	7000	9000	11000	13000	15000	17000	19000	21000	23000
	L.F. Perimeter	240	280	303	325	354	372	397	422	447
Limestone with Concrete Block Back-up	Bearing Walls	117.20	109.35	102.60	97.90	94.95	91.95	90.00	88.50	87.15
	R/Conc. Frame	121.00	113.55	107.30	102.95	100.15	97.45	95.60	94.20	92.95
Face Brick with Concrete Block Back-up	Bearing Walls	102.20	95.65	90.45	86.80	84.40	82.20	80.65	79.45	78.45
	R/Conc. Frame	111.15	104.55	99.40	95.75	93.35	91.15	89.60	88.40	87.40
Decorative Concrete Block	Bearing Walls	97.80	91.60	86.90	83.60	81.40	79.35	77.95	76.90	75.95
	R/Conc. Frame	106.75	100.55	95.85	92.55	90.35	88.30	86.90	85.85	84.90
Perimeter Adj., Add or Deduct	Per 100 L.F.	15.40	12.00	9.80	8.30	7.20	6.35	5.70	5.15	4.70
Story Hgt. Adj., Add or Deduct	Per 1 Ft.	2.75	2.50	2.20	2.00	1.90	1.75	1.70	1.60	1.55

For Basement, add $14.45 per square foot of basement area

The above costs were calculated using the basic specifications shown on the facing page. These costs should be adjusted where necessary for design alternatives and owner's requirements. Reported completed project costs, for this type of structure, range from $65.45 to $168.10 per S.F.

Common additives

Description	Unit	$ Cost
Cells Prefabricated, 5'-6' wide, 7'-8' high, 7'-8' deep	Each	7900
Elevators, Hydraulic passenger, 2 stops		
1500# capacity	Each	37,625
2500# capacity	Each	38,925
3500# capacity	Each	42,525
Emergency Lighting, 25 watt, battery operated		
Lead battery	Each	305
Nickel cadmium	Each	600
Flagpoles, Complete		
Aluminum, 20' high	Each	1025
40' high	Each	2375
70' high	Each	6925
Fiberglass, 23' high	Each	1400
39'-5" high	Each	2425
59' high	Each	6875

Description	Unit	$ Cost
Lockers, Steel, Single tier, 60" to 72"	Opening	128 - 207
2 tier, 60" or 72" total	Opening	79 - 99
5 tier, box lockers	Opening	41 - 53
Locker bench, lam. maple top only	L.F.	17.60
Pedestals, steel pipe	Each	39
Safe, Office type, 4 hour rating		
30" x 18" x 18"	Each	3475
62" x 33" x 20"	Each	7575
Shooting Range, Incl. bullet traps, target provisions, and contols, not incl. structural shell	Each	20,600
Smoke Detectors		
Ceiling type	Each	126
Duct type	Each	360
Sound System		
Amplifier, 250 watts	Each	1425
Speaker, ceiling or wall	Each	118
Trumpet	Each	222

Important: See the Reference Section for Location Factors

Model costs calculated for a 2 story building with 12' story height and 11,000 square feet of floor area

			Unit	Unit Cost	Cost Per S.F.	% Of Sub-Total
1.0 Foundations						
.1	Footings & Foundations	Poured concrete; strip and spread footings and 4' foundation wall	S.F. Ground	4.00	2.00	
.4	Piles & Caissons	N/A	—	—	—	3.0%
.9	Excavation & Backfill	Site preparation for slab and trench for foundation wall and footing	S.F. Ground	.96	.48	
2.0 Substructure						
.1	Slab on Grade	4" reinforced concrete with vapor barrier and granular base	S.F. Slab	2.56	1.28	1.6%
.2	Special Substructures	N/A	—			
3.0 Superstructure						
.1	Columns & Beams	N/A	—	—	—	
.4	Structural Walls	N/A	—	—	—	
.5	Elevated Floors	Open web steel joists, slab form, concrete	S.F. Floor	5.94	2.97	6.4%
.7	Roof	Metal deck on open web steel joists	S.F. Roof	2.43	1.22	
.9	Stairs	Concrete filled metal pan	Flight	5550	1.01	
4.0 Exterior Closure						
.1	Walls	Limestone with concrete block backup — 80% of wall	S.F. Wall	27	14.68	
.5	Exterior Wall Finishes	N/A	—	—	—	26.0%
.6	Doors	Hollow metal	Each	1290	.94	
.7	Windows & Glazed Walls	Metal horizontal sliding — 20% of wall	Each	640	5.64	
5.0 Roofing						
.1	Roof Coverings	Built-up tar and gravel with flashing	S.F. Roof	2.16	1.08	
.7	Insulation	Perlite/EPS composite	S.F. Roof	1.13	.56	2.2%
.8	Openings & Specialties	Gravel stop	L.F. Perimeter	4.78	.13	
6.0 Interior Construction						
.1	Partitions	Concrete block, toilet partitions — 20 S.F. Floor/L.F. Partition	S.F. Partition	4.10	2.47	
.4	Interior Doors	Single leaf kalamein fire door — 200 S.F. Floor/Door	Each	484	2.42	
.5	Wall Finishes	90% paint, 10% ceramic tile	S.F. Surface	1.09	1.09	15.5%
.6	Floor Finishes	70% vinyl asbestos tile, 20% carpet, 10% ceramic tile	S.F. Floor	3.32	3.32	
.7	Ceiling Finishes	Mineral fiber tile on concealed zee bars	S.F. Ceiling	2.62	2.62	
.9	Interior Surface/Exterior Wall	Paint — 80% of wall	S.F. Wall	1.53	.81	
7.0 Conveying						
.1	Elevators	One hydraulic passenger elevator	Each	45,870	4.17	5.1%
.2	Special Conveyors	N/A	—	—	—	
8.0 Mechanical						
.1	Plumbing	Toilet and service fixtures, supply and drainage — 1 Fixture/580 S.F. Floor	Each	2076	3.58	
.2	Fire Protection	Wet pipe sprinkler system	S.F. Floor	1.37	1.37	
.3	Heating	Oil fired hot water, wall fin radiation	S.F. Floor	5.21	5.21	19.2%
.4	Cooling	Split systems with air cooled condensing units	S.F. Floor	5.61	5.61	
.5	Special Systems	N/A	—	—	—	
9.0 Electrical						
.1	Service & Distribution	400 ampere service, panel board and feeders	S.F. Floor	.97	.97	
.2	Lighting & Power	Fluorescent fixtures, receptacles, switches, A.C. and misc. power	S.F. Floor	5.17	5.17	8.2%
.4	Special Electrical	Alarm systems and emergency lighting	S.F. Floor	.58	.58	
11.0 Special Construction						
.1	Specialties	Lockers, detention rooms, cells	S.F. Floor	10.49	10.49	12.8%
12.0 Site Work						
.1	Earthwork	N/A	—	—	—	
.3	Utilities	N/A	—	—	—	0.0%
.5	Roads & Parking	N/A	—	—	—	
.7	Site Improvements	N/A	—	—	—	
			Sub-Total		81.87	**100%**
	GENERAL CONDITIONS (Overhead & Profit)			15%	12.28	
	ARCHITECT FEES			9%	8.45	
			Total Building Cost		102.60	

Costs per square foot of floor area

Exterior Wall	S.F. Area	5000	7000	9000	11000	13000	15000	17000	19000	21000
	L.F. Perimeter	300	380	420	486	468	513	540	580	620
Face Brick with Concrete Block Back-up	Steel Frame	72.90	69.30	65.50	64.05	60.30	59.35	58.15	57.50	57.00
	Bearing Walls	71.95	68.35	64.55	63.10	59.35	58.40	57.20	56.55	56.05
Limestone with Concrete Block Back-up	Steel Frame	83.00	78.40	73.30	71.45	66.35	65.10	63.50	62.65	62.00
	Bearing Walls	81.65	77.05	71.95	70.10	65.00	63.75	62.10	61.30	60.65
Decorative Concrete Block	Steel Frame	68.40	65.25	62.00	60.70	57.60	56.75	55.75	55.20	54.80
	Bearing Walls	67.45	64.30	61.00	59.75	56.65	55.80	54.80	54.25	53.85
Perimeter Adj., Add or Deduct	Per 100 L.F.	8.30	5.90	4.60	3.75	3.20	2.75	2.45	2.20	1.95
Story Hgt. Adj., Add or Deduct	Per 1 Ft.	1.30	1.20	1.00	.95	.80	.75	.70	.70	.65

For Basement, add $15.10 per square foot of basement area

The above costs were calculated using the basic specifications shown on the facing page. These costs should be adjusted where necessary for design alternatives and owner's requirements. Reported completed project costs, for this type of structure, range from $53.40 to $137.85 per S.F.

Common additives

Description	Unit	$ Cost
Closed Circuit Surveillance, One station		
Camera and monitor	Each	1250
For additional camera stations, add	Each	695
Emergency Lighting, 25 watt, battery operated		
Lead battery	Each	305
Nickel cadmium	Each	600
Flagpoles, Complete		
Aluminum, 20' high	Each	1025
40' high	Each	2375
70' high	Each	6925
Fiberglass, 23' high	Each	1400
39'-5" high	Each	2425
59' high	Each	6875

Description	Unit	$ Cost
Mail Boxes, Horizontal, key lock, 15" x 6" x 5"	Each	44
Double 15" x 12" x 5"	Each	77
Quadruple 15" x 12" x 10"	Each	140
Vertical, 6" x 5" x 15", aluminum	Each	36
Bronze	Each	61
Steel, enameled	Each	36
Scales, Dial type, 5 ton cap.		
8' x 6' platform	Each	6575
9' x 7' platform	Each	10,300
Smoke Detectors		
Ceiling type	Each	126
Duct type	Each	360

Important: See the Reference Section for Location Factors

Model costs calculated for a 1 story building with 14' story height and 13,000 square feet of floor area

				Unit	Unit Cost	Cost Per S.F.	% Of Sub-Total
1.0 Foundations							
	.1	Footings & Foundations	Poured concrete; strip and spread footings and 4' foundation wall	S.F. Ground	2.61	2.61	
	.4	Piles & Caissons	N/A	—	—	—	7.4%
	.9	Excavation & Backfill	Site preparation for slab and trench for foundation wall and footing	S.F. Ground	.96	.96	
2.0 Substructure							
	.1	Slab on Grade	4" reinforced concrete with vapor barrier and granular base	S.F. Slab	2.56	2.56	5.3%
	.2	Special Substructures	N/A	—	—	—	
3.0 Superstructure							
	.1	Columns & Beams	Fireproofing, steel columns included in 3.7	L.F. Column	16.44	.05	
	.4	Structural Walls	N/A	—	—	—	
	.5	Elevated Floors	N/A	—	—	—	9.2%
	.7	Roof	Metal deck, open web steel joists, columns	S.F. Roof	4.39	4.39	
	.9	Stairs	N/A	—	—	—	
4.0 Exterior Closure							
	.1	Walls	Face brick with concrete block backup 80% of wall	S.F. Wall	15.00	6.37	
	.5	Exterior Wall Finishes	N/A	—	—	—	
	.6	Doors	Double aluminum & glass, single aluminum, hollow metal, steel overhead	Each	1490	.69	18.5%
	.7	Windows & Glazed Walls	Double strength window glass 20% of wall	Each	420	1.84	
5.0 Roofing							
	.1	Roof Coverings	Built-up tar and gravel with flashing	S.F. Roof	1.86	1.86	
	.7	Insulation	Perlite/EPS composite	S.F. Roof	1.13	1.13	6.6%
	.8	Openings & Specialties	Gravel stop	L.F. Perimeter	4.78	.17	
6.0 Interior Construction							
	.1	Partitions	Concrete block, toilet partitions 15 S.F. Floor/L.F. Partition	S.F. Partition	4.10	3.63	
	.4	Interior Doors	Single leaf hollow metal 150 S.F. Floor/Door	Each	484	3.23	
	.5	Wall Finishes	Paint	S.F. Surface	.72	1.15	
	.6	Floor Finishes	50% vinyl tile, 50% paint	S.F. Floor	1.69	1.69	22.8%
	.7	Ceiling Finishes	Mineral fiber tile on concealed zee bars 25% of area	S.F. Ceiling	2.62	.66	
	.9	Interior Surface/Exterior Wall	Paint 80% of wall	S.F. Wall	1.53	.62	
7.0 Conveying							
	.1	Elevators	N/A	—	—	—	0.0%
	.2	Special Conveyors	N/A	—	—	—	
8.0 Mechanical							
	.1	Plumbing	Toilet and service fixtures, supply and drainage 1 Fixture/1180 S.F. Floor	Each	1581	1.34	
	.2	Fire Protection	Wet pipe sprinkler system	S.F. Floor	1.26	1.26	
	.3	Heating	Included in 8.4	—	—	—	18.0%
	.4	Cooling	Single zone, gas heating, electric cooling	S.F. Floor	6.07	6.07	
	.5	Special Systems	N/A	—	—	—	
9.0 Electrical							
	.1	Service & Distribution	400 ampere service, panel board and feeders	S.F. Floor	.02	.82	
	.2	Lighting & Power	Fluorescent fixtures, receptacles, switches, A.C. and misc. power	S.F. Floor	4.20	4.20	11.2%
	.4	Special Electrical	Alarm systems and emergency lighting	S.F. Floor	.35	.35	
11.0 Special Construction							
	.1	Specialties	Cabinets, lockers, shelving	S.F. Floor	.46	.46	1.0%
12.0 Site Work							
	.1	Earthwork	N/A	—	—	—	
	.3	Utilities	N/A	—	—	—	
	.5	Roads & Parking	N/A	—	—	—	0.0%
	.7	Site Improvements	N/A	—	—	—	
				Sub-Total		48.11	100%
		GENERAL CONDITIONS (Overhead & Profit)			15%	7.22	
		ARCHITECT FEES			9%	4.97	
				Total Building Cost		60.30	

Costs per square foot of floor area

Exterior Wall	S.F. Area	5000	10000	15000	21000	25000	30000	40000	50000	60000
	L.F. Perimeter	287	400	500	600	700	700	834	900	1000
Face Brick with Concrete Block Back-up	Steel Frame	116.70	98.40	91.70	87.15	86.30	82.60	80.40	78.10	76.95
	Bearing Walls	112.60	95.50	89.20	84.90	84.10	80.50	78.40	76.15	75.05
Concrete Block	Steel Frame	102.50	88.30	83.20	79.80	79.10	76.60	75.00	73.40	72.60
Brick Veneer	Steel Frame	110.90	93.45	87.05	82.80	81.95	78.60	76.50	74.40	73.35
Galvanized Steel Siding	Steel Frame	96.05	83.70	79.30	76.45	75.75	73.75	72.40	71.10	70.45
Metal Sandwich Panel	Steel Frame	96.15	83.30	78.70	75.75	75.05	72.90	71.50	70.15	69.45
Perimeter Adj., Add or Deduct	Per 100 L.F.	14.00	7.00	4.65	3.35	2.80	2.35	1.70	1.40	1.20
Story Hgt. Adj., Add or Deduct	Per 1 Ft.	2.90	2.05	1.70	1.45	1.45	1.20	1.05	.90	.85

For Basement, add $14.45 per square foot of basement area

The above costs were calculated using the basic specifications shown on the facing page. These costs should be adjusted where necessary for design alternatives and owner's requirements. Reported completed project costs, for this type of structure, range from $50.40 to $127.15 per S.F.

Common additives

Description	Unit	$ Cost	Description	Unit	$ Cost
Bar, Front Bar	L.F.	246	Lockers, Steel, single tier, 60" or 72"	Opening	128 - 207
Back Bar	L.F.	193	2 tier, 60" or 72" total	Opening	79 - 99
Booth, Upholstered, custom straight	L.F.	125 - 232	5 tier, box lockers	Opening	41 - 53
"L" or "U" shaped	L.F.	129 - 220	Locker bench, lam. maple top only	L.F.	17.60
Bleachers, Telescoping, manual			Pedestals, steel pipe	Each	39
To 15 tier	Seat	68 - 97	Sauna, Prefabricated, complete		
21-30 tier	Seat	156 - 187	6' x 4'	Each	3550
Courts			6' x 9'	Each	5600
Ceiling	Court	4500	8' x 8'	Each	5925
Floor	Court	8325	8' x 10'	Each	6500
Walls	Court	16,500	10' x 12'	Each	8500
Emergency Lighting, 25 watt, battery operated			Sound System		
Lead battery	Each	305	Amplifier, 250 watts	Each	1425
Nickel cadmium	Each	600	Speaker, ceiling or wall	Each	118
Kitchen Equipment			Trumpet	Each	222
Broiler	Each	3400	Steam Bath, Complete, to 140 C.F.	Each	890
Cooler, 6 ft. long, reach-in	Each	2675	To 300 C.F.	Each	1075
Dishwasher, 10-12 racks per hr.	Each	2525	To 800 C.F.	Each	2925
Food warmer, counter 1.2 KW	Each	745	To 2500 C.F.	Each	3275
Freezer, reach-in, 44 C.F.	Each	6575			
Ice cube maker, 50 lb. per day	Each	1700			

Important: See the Reference Section for Location Factors

Model costs calculated for a 2 story building with 12' story height and 30,000 square feet of floor area

		Description		Unit	Unit Cost	Cost Per S.F.	% Of Sub-Total
1.0 Foundations							
.1	Footings & Foundations	Poured concrete; strip and spread footings and 4' foundation wall		S.F. Ground	3.12	1.56	3.3%
.4	Piles & Caissons	N/A		—	—	—	
.9	Excavation & Backfill	Site preparation for slab and trench for foundation wall and footing		S.F. Ground	.96	.62	
2.0 Substructure							
.1	Slab on Grade	6" reinforced concrete with vapor barrier and granular base		S.F. Slab	5.04	3.28	4.9%
.2	Special Substructures	N/A		—	—	—	
3.0 Superstructure							
.1	Columns & Beams	Steel columns included in 3.5 and 3.7		—	—	—	10.0%
.4	Structural Walls	N/A		—	—	—	
.5	Elevated Floors	Open web steel joists, slab form, concrete, columns	50% of area	S.F. Floor	11.83	4.14	
.7	Roof	Metal deck on open web steel joists, columns		S.F. Roof	4.07	2.03	
.9	Stairs	Concrete filled metal pan		Flight	4725	.47	
4.0 Exterior Closure							
.1	Walls	Face brick with concrete block backup	95% of wall	S.F. Wall	16.00	8.51	15.5%
.5	Exterior Wall Finishes	N/A		—	—	—	
.6	Doors	Aluminum and glass and hollow metal		Each	1712	.22	
.7	Windows & Glazed Walls	Storefront	5% of wall	S.F. Window	57	1.60	
5.0 Roofing							
.1	Roof Coverings	Built-up tar and gravel with flashing		S.F. Roof	2.28	1.14	3.0%
.7	Insulation	Perlite/EPS composite		S.F. Roof	1.13	.73	
.8	Openings & Specialties	Gravel stop and hatches		S.F. Roof	.30	.15	
6.0 Interior Construction							
.1	Partitions	Concrete block, gypsum board on metal studs	25 S.F. Floor/L.F. Partition	S.F. Partition	4.10	1.80	12.4%
.4	Interior Doors	Single leaf hollow metal	810 S.F. Floor/Door	Each	484	.60	
.5	Wall Finishes	Paint		S.F. Surface	.69	.55	
.6	Floor Finishes	80% carpet, 20% ceramic tile	50% of floor area	S.F. Floor	4.88	2.44	
.7	Ceiling Finishes	Mineral fiber tile on concealed zee bars	60% of area	S.F. Ceiling	2.62	1.57	
.9	Interior Surface/Exterior Wall	Painted gypsum board on furring	95% of wall	S.F. Wall	2.26	1.27	
7.0 Conveying							
.1	Elevators	N/A		—	—	—	0.0%
.2	Special Conveyors	N/A		—	—	—	
8.0 Mechanical							
.1	Plumbing	Kitchen, bathroom and service fixtures, supply and drainage	1 Fixture/1000 S.F. Floor	Each	2820	2.82	30.6%
.2	Fire Protection	Sprinklers, light hazard		S.F. Floor	.11	.11	
.3	Heating	Included in 8.4		—	—	—	
.4	Cooling	Multizone unit, gas heating, electric cooling		S.F. Floor	17.40	17.40	
.5	Special Systems	N/A		—	—	—	
9.0 Electrical							
.1	Service & Distribution	400 ampere service, panel board and feeders		S.F. Floor	.44	.44	5.7%
.2	Lighting & Power	Fluorescent and high intensity discharge fixtures, receptacles, switches, A.C. and misc. power		S.F. Floor	3.18	3.18	
.4	Special Electrical	Alarm systems and emergency lighting		S.F. Floor	.20	.20	
11.0 Special Construction							
.1	Specialties	Courts, sauna baths		S.F. Floor	9.69	9.69	14.6%
12.0 Site Work							
.1	Earthwork	N/A		—	—	—	0.0%
.3	Utilities	N/A		—	—	—	
.5	Roads & Parking	N/A		—	—	—	
.7	Site Improvements	N/A		—	—	—	
				Sub-Total		66.52	100%
	GENERAL CONDITIONS (Overhead & Profit)				15%	9.98	
	ARCHITECT FEES				8%	6.10	
				Total Building Cost		82.60	

BUILDING TYPES

Costs per square foot of floor area

Exterior Wall	S.F. Area	2000	2800	3500	4200	5000	5800	6500	7200	8000
	L.F. Perimeter	180	212	240	268	300	314	336	344	368
Wood Siding	Wood Frame	110.70	102.70	98.70	96.00	93.85	91.50	90.35	88.85	88.00
Brick Veneer	Wood Frame	116.30	107.40	102.95	99.95	97.55	94.85	93.50	91.80	90.80
Face Brick with Concrete Block Back-up	Wood Joists	119.55	110.15	105.45	102.30	99.80	96.85	95.45	93.60	92.55
	Steel Joists	121.60	111.30	106.15	102.70	99.95	96.65	95.05	92.95	91.80
Stucco on Concrete Block	Wood Joists	112.30	·104.05	99.90	97.10	94.90	92.45	91.25	89.70	88.80
	Steel Joists	108.65	100.40	96.25	93.45	91.30	88.85	87.60	86.05	85.15
Perimeter Adj., Add or Deduct	Per 100 L.F.	12.80	9.10	7.30	6.10	5.10	4.40	3.90	3.55	3.20
Story Hgt. Adj., Add or Deduct	Per 1 Ft.	1.40	1.20	1.05	1.00	.95	.85	.80	.75	.70
For Basement, add $17.45 per square foot of basement area										

The above costs were calculated using the basic specifications shown on the facing page. These costs should be adjusted where necessary for design alternatives and owner's requirements. Reported completed project costs, for this type of structure, range from $67.85 to $158.50 per S.F.

Common additives

Description	Unit	$ Cost	Description	Unit	$ Cost
Bar, Front Bar	L.F.	246	Fireplace, Brick, not incl. chimney or foundation		
Back bar	L.F.	193	30" x 29" opening	Each	1550
Booth, Upholstered, custom straight	L.F.	125 - 232	Chimney, standard brick		
"L" or "U" shaped	L.F.	129 - 220	Single flue, 16" x 20"	V.L.F.	46
Cupola, Stock unit, redwood			20" x 20"	V.L.F.	52
30" square, 37" high, aluminum roof	Each	445	2 Flue, 20" x 24"	V.L.F.	64
Copper roof	Each	214	20" x 32"	V.L.F.	77
Fiberglass, 5'-0" base, 63" high	Each	2725 - 3250	Kitchen Equipment		
6'-0" base, 63" high	Each	4125 - 4450	Broiler	Each	3400
Decorative Wood Beams, Non load bearing			Coffee urn, twin 6 gallon	Each	5875
Rough sawn, 4" x 6"	L.F.	6.60	Cooler, 6 ft. long	Each	2675
4" x 8"	L.F.	8.00	Dishwasher, 10-12 racks per hr.	Each	2525
4" x 10"	L.F.	9.40	Food warmer, counter, 1.2 KW	Each	745
4" x 12"	L.F.	10.85	Freezer, 44 C.F., reach-in	Each	6575
8" x 8"	L.F.	13.55	Ice cube maker, 50 lb. per day	Each	1700
Emergency Lighting, 25 watt, battery operated			Range with 1 oven	Each	2300
Lead battery	Each	305	Refrigerators, Prefabricated, walk-in		
Nickel cadmium	Each	600	7'-6" high, 6' x 6'	S.F.	114
			10' x 10'	S.F.	89
			12' x 14'	S.F.	79
			12' x 20'	S.F.	70

Important: See the Reference Section for Location Factors

Model costs calculated for a 1 story building with 12' story height and 5,000 square feet of floor area

				Unit	Unit Cost	Cost Per S.F.	% Of Sub-Total
1.0 Foundations							
	.1	Footings & Foundations	Poured concrete; strip and spread footings and 4' foundation wall	S.F. Ground	3.11	3.11	
	.4	Piles & Caissons	N/A	—	—	—	5.5%
	.9	Excavation & Backfill	Site preparation for slab and trench for foundation wall and footing	S.F. Ground	1.08	1.08	
2.0 Substructure							
	.1	Slab on Grade	4" reinforced concrete with vapor barrier and granular base	S.F. Slab	2.56	2.56	3.4%
	.2	Special Substructures	N/A	—	—	—	
3.0 Superstructure							
	.1	Columns & Beams	Wood columns	S.F. Ground	.30	.30	
	.4	Structural Walls	N/A	—	—	—	
	.5	Elevated Floors	N/A	—	—	—	2.9%
	.7	Roof	Plywood on wood rafters (pitched)	S.F. Roof	1.68	1.88	
	.9	Stairs	N/A	—	—	—	
4.0 Exterior Closure							
	.1	Walls	Cedar siding on wood studs with insulation *70% of wall*	S.F. Wall	6.23	3.14	
	.5	Exterior Wall Finishes	N/A	—	—	—	14.2%
	.6	Doors	Aluminum and glass doors and entrance with transom	Each	2846	2.85	
	.7	Windows & Glazed Walls	Storefront windows *30% of wall*	S.F. Window	22	4.87	
5.0 Roofing							
	.1	Roof Coverings	Cedar shingles with flashing (pitched)	S.F. Roof	2.87	3.21	
	.7	Insulation	Fiberglass sheets	S.F. Roof	.78	.87	5.8%
	.8	Openings & Specialties	Gutters and downspouts and skylight	S.F. Roof	.38	.38	
6.0 Interior Construction							
	.1	Partitions	Gypsum board on wood studs, toilet partition *25 S.F. Floor/L.F. Partition*	S.F. Partition	688	1.54	
	.4	Interior Doors	Hollow core wood *250 S.F. Floor/Door*	Each	323	1.29	
	.5	Wall Finishes	75% paint, 25% ceramic tile	S.F. Surface	1.41	1.13	
	.6	Floor Finishes	65% carpet, 35% quarry tile	S.F. Floor	6.11	6.11	18.1%
	.7	Ceiling Finishes	Mineral fiber tile on concealed zee bars	S.F. Ceiling	2.62	2.62	
	.9	Interior Surface/Exterior Wall	Painted gypsum board on furring *70% of wall*	S.F. Wall	2.25	1.13	
7.0 Conveying							
	.1	Elevators	N/A	—	—	—	0.0%
	.2	Special Conveyors	N/A	—	—	—	
8.0 Mechanical							
	.1	Plumbing	Kitchen, bathroom and service fixtures, supply and drainage *1 Fixture/355 S.F. Floor*	Each	2513	7.08	
	.2	Fire Protection	Sprinklers, light hazard	S.F. Floor	1.26	1.26	
	.3	Heating	Included in 8.4	—	—	—	40.1%
	.4	Cooling	Multizone unit, gas heating, electric cooling	S.F. Floor	22	22.20	
	.5	Special Systems	N/A	—	—	—	
9.0 Electrical							
	.1	Service & Distribution	400 ampere service, panel board and feeders	S.F. Floor	2.23	2.23	
	.2	Lighting & Power	Fluorescent fixtures, receptacles, switches, A.C. and misc. power	S.F. Floor	4.80	4.80	10.0%
	.4	Special Electrical	Alarm systems and emergency lighting	S.F. Floor	.63	.63	
11.0 Special Construction							
	.1	Specialties	N/A	—	—	—	0.0%
12.0 Site Work							
	.1	Earthwork	N/A	—	—	—	
	.3	Utilities	N/A	—	—	—	
	.5	Roads & Parking	N/A	—	—	—	0.0%
	.7	Site Improvements	N/A	—	—	—	

			Sub-Total	**76.27**	**100%**
	GENERAL CONDITIONS (Overhead & Profit)		15%	11.44	
	ARCHITECT FEES		7%	6.14	
		Total Building Cost		**93.85**	

Costs per square foot of floor area

Exterior Wall	S.F. Area	2000	2800	3500	4000	5000	5800	6500	7200	8000
	L.F. Perimeter	180	212	240	260	300	314	336	344	368
Face Brick with Concrete Block Back-up	Bearing Walls	99.70	93.00	89.60	87.95	85.60	83.25	82.20	80.65	79.90
	Steel Frame	102.50	95.75	92.40	90.70	88.35	86.00	84.95	83.45	82.65
Concrete Block With Stucco	Bearing Walls	93.15	87.45	84.60	83.20	81.20	79.25	78.40	77.15	76.50
	Steel Frame	95.40	89.80	86.95	85.60	83.60	81.75	80.90	79.65	79.00
Wood Siding	Wood Frame	93.70	88.10	85.35	83.95	82.00	80.15	79.30	78.05	77.45
Brick Veneer	Steel Frame	99.60	93.30	90.15	88.60	86.40	84.25	83.30	81.85	81.15
Perimeter Adj., Add or Deduct	Per 100 L.F.	16.75	11.95	9.60	8.35	6.70	5.75	5.15	4.65	4.20
Story Hgt. Adj., Add or Deduct	Per 1 Ft.	2.20	1.85	1.70	1.60	1.50	1.35	1.25	1.20	1.15
Basement—Not Applicable										

The above costs were calculated using the basic specifications shown on the facing page. These costs should be adjusted where necessary for design alternatives and owner's requirements. Reported completed project costs, for this type of structure, range from $67.75 to $131.90 per S.F.

Common additives

Description	Unit	$ Cost		Description	Unit	$ Cost
Bar, Front Bar	L.F.	246		Refrigerators, Prefabricated, walk-in		
Back bar	L.F.	193		7'-6" High, 6' x 6'	S.F.	114
Booth, Upholstered, custom straight	L.F.	125 - 232		10' x 10'	S.F.	89
"L" or "U" shaped	L.F.	129 - 220		12' x 14'	S.F.	79
Drive-up Window	Each	5875 - 8875		12' x 20'	S.F.	70
Emergency Lighting, 25 watt, battery operated				Serving		
Lead battery	Each	305		Counter top (Stainless steel)	L.F.	103
Nickel cadmium	Each	600		Base cabinets	L.F.	239 - 286
Kitchen Equipment				Sound System		
Broiler	Each	3400		Amplifier, 250 watts	Each	1425
Coffee urn, twin 6 gallon	Each	5875		Speaker, ceiling or wall	Each	118
Cooler, 6 ft. long	Each	2675		Trumpet	Each	222
Dishwasher, 10-12 racks per hr.	Each	2525		Storage		
Food warmer, counter, 1.2 KW	Each	745		Shelving	S.F.	11.60
Freezer, 44 C.F., reach-in	Each	6575		Washing		
Ice cube maker, 50 lb. per day	Each	1700		Stainless steel counter	L.F.	103
Range with 1 oven	Each	2300				

Model costs calculated for a 1 story building with 10' story height and 4,000 square feet of floor area

Restaurant, Fast Food

				Unit	Unit Cost	Cost Per S.F.	% Of Sub-Total
1.0 Foundations							
	.1	Footings & Foundations	Poured concrete; strip and spread footings and 4' foundation wall	S.F. Ground	3.54	3.54	
	.4	Piles & Caissons	N/A	—	—	—	6.5%
	.9	Excavation & Backfill	Site preparation for slab and trench for foundation wall and footing	S.F. Ground	1.08	1.08	
2.0 Substructure							
	.1	Slab on Grade	4" reinforced concrete with vapor barrier and granular base	S.F. Slab	2.56	2.56	3.6%
	.2	Special Substructures	N/A				
3.0 Superstructure							
	.1	Columns & Beams	N/A	—	—	—	
	.4	Structural Walls	N/A	—	—	—	
	.5	Elevated Floors	N/A	—	—	—	4.0%
	.7	Roof	Metal deck on open web steel joists	S.F. Roof	2.81	2.81	
	.9	Stairs	N/A	—	—	—	
4.0 Exterior Closure							
	.1	Walls	Face brick with concrete block backup 70% of wall	S.F. Wall	13.34	7.19	
	.5	Exterior Wall Finishes	N/A	—	—	—	23.7%
	.6	Doors	Aluminum and glass	Each	2252	4.51	
	.7	Windows & Glazed Walls	Window wall 30% of wall	S.F Window	26	5.07	
5.0 Roofing							
	.1	Roof Coverings	Built-up tar and gravel with flashing	S.F. Roof	2.32	2.32	
	.7	Insulation	Perlite/EPS composite	S.F. Roof	1.13	1.13	5.2%
	.8	Openings & Specialties	Gravel stop and hatches	S.F. Roof	.26	.26	
6.0 Interior Construction							
	.1	Partitions	Gypsum board on metal studs 25 S.F. Floor/L.F. Partition	S.F. Partition	3.17	1.14	
	.4	Interior Doors	Hollow core wood 1000 S.F. Floor/Door	Each	323	.32	
	.5	Wall Finishes	Paint	S.F. Surface	.42	.30	17.6%
	.6	Floor Finishes	Quarry tile	S.F. Floor	7.40	7.40	
	.7	Ceiling Finishes	Mineral fiber tile on concealed zee bars	S.F. Floor	2.62	2.62	
	.9	Interior Surface/Exterior Wall	Paint 70% of wall	S.F. Wall	1.53	.70	
7.0 Conveying							
	.1	Elevators	N/A	—	—	—	0.0%
	.2	Special Conveyors	N/A	—	—	—	
8.0 Mechanical							
	.1	Plumbing	Kitchen, bathroom and service fixtures, supply and drainage 1 Fixture/400 S.F. Floor	Each	2248	5.62	
	.2	Fire Protection	Sprinklers, light hazard	S.F. Floor	1.66	1.66	
	.3	Heating	Included in 8.4	—	—	—	22.0%
	.4	Cooling	Multizone unit, gas heating, electric cooling	S.F. Floor	8.24	8.24	
	.5	Special Systems	N/A	—	—	—	
9.0 Electrical							
	.1	Service & Distribution	400 ampere service, panel board and feeders	S.F. Floor	2.35	2.35	
	.2	Lighting & Power	Fluorescent fixtures, receptacles, switches, A.C. and misc. power	S.F. Floor	5.22	5.22	11.4%
	.4	Special Electrical	Alarm systems and emergency lighting	S.F. Floor	.53	.53	
11.0 Special Construction							
	.1	Specialties	Walk-in refrigerator	Each	16,903	4.23	6.0%
12.0 Site Work							
	.1	Earthwork	N/A	—	—	—	
	.3	Utilities	N/A	—	—	—	0.0%
	.5	Roads & Parking	N/A	—	—	—	
	.7	Site Improvements	N/A	—	—	—	
				Sub-Total		70.80	**100%**
		GENERAL CONDITIONS (Overhead & Profit)			15%	10.62	
		ARCHITECT FEES			8%	6.53	
				Total Building Cost		87.95	

Costs per square foot of floor area

Exterior Wall	S.F. Area	10000	15000	20000	25000	30000	35000	40000	45000	50000
	L.F. Perimeter	450	500	600	700	740	822	890	920	966
Face Brick with Concrete Block Back-up	Steel Frame	104.60	95.75	92.85	91.15	88.75	87.75	86.85	85.55	84.75
	Lam. Wood Truss	99.10	90.60	87.80	86.15	83.85	82.90	82.05	80.80	80.05
Concrete Block	Steel Frame	88.00	82.55	80.65	79.50	78.10	77.45	76.90	76.15	75.65
	Lam. Wood Truss	88.45	83.00	81.10	79.95	78.55	77.90	77.35	76.60	76.10
Galvanized Steel Siding	Steel Frame	80.75	76.55	74.95	74.05	73.00	72.45	72.05	71.50	71.10
Metal Sandwich Panel	Steel Joists	81.75	77.25	75.60	74.65	73.50	72.95	72.50	71.90	71.55
Perimeter Adj., Add or Deduct	Per 100 L.F.	6.20	4.15	3.10	2.50	2.05	1.80	1.55	1.40	1.25
Story Hgt. Adj., Add or Deduct	Per 1 Ft.	1.00	.75	.65	.60	.55	.50	.50	.45	.45
Basement—Not Applicable										

The above costs were calculated using the basic specifications shown on the facing page. These costs should be adjusted where necessary for design alternatives and owner's requirements. Reported completed project costs, for this type of structure, range from $40.55 to $118.75 per S.F.

Common additives

Description	Unit	$ Cost
Bar, Front Bar	L.F.	246
Back bar	L.F.	193
Booth, Upholstered, custom straight	L.F.	125 - 232
"L" or "U" shaped	L.F.	129 - 220
Bleachers, Telescoping, manual		
To 15 tier	Seat	68 - 97
16-20 tier	Seat	147 - 181
21-30 tier	Seat	156 - 187
For power operation, add	Seat	29 - 46
Emergency Lighting, 25 watt, battery operated		
Lead battery	Each	305
Nickel cadmium	Each	600
Lockers, Steel, single tier, 60" or 72"	Opening	128 - 207
2 tier, 60" or 72" total	Opening	79 - 99
5 tier, box lockers	Opening	41 - 53
Locker bench, lam. maple top only	L.F.	17.60
Pedestals, steel pipe	Each	39

Description	Unit	$ Cost
Rink		
Dasher boards & top guard	Each	141,500
Mats, rubber	S.F.	15.00
Score Board	Each	12,300 - 33,000

Model costs calculated for a 1 story building with 24' story height and 30,000 square feet of floor area

Rink, Hockey/Indoor Soccer

					Unit	Unit Cost	Cost Per S.F.	% Of Sub-Total
1.0 Foundations								
	.1	Footings & Foundations	Poured concrete; strip and spread footings and 4' foundation wall		S.F. Ground	1.99	1.99	
	.4	Piles & Caissons	N/A		—	—	—	4.6%
	.9	Excavation & Backfill	Site preparation for slab and trench for foundation wall and footing		S.F. Ground	.91	.91	
2.0 Substructure								
	.1	Slab on Grade	6" reinforced concrete with vapor barrier and granular base		S.F. Slab	3.14	3.14	4.9%
	.2	Special Substructures	N/A		—	—	—	
3.0 Superstructure								
	.1	Columns & Beams	Wide flange beams and columns		S.F. Ground	7.63	7.63	
	.4	Structural Walls	N/A		—	—	—	
	.5	Elevated Floors	N/A		—	—	—	19.0%
	.7	Roof	Metal deck on steel joist		S.F. Roof	4.44	4.44	
	.9	Stairs	N/A		—	—	—	
4.0 Exterior Closure								
	.1	Walls	Concrete block	95% of wall	S.F. Wall	6.44	3.62	
	.5	Exterior Wall Finishes	N/A		—	—	—	7.6%
	.6	Doors	Aluminum and glass, hollow metal, overhead		Each	1678	.45	
	.7	Windows & Glazed Walls	Store front	5% of wall	S.F. Window	25	.75	
5.0 Roofing								
	.1	Roof Coverings	Elastomeric neoprene membrane with flashing		S.F. Roof	1.52	1.52	
	.7	Insulation	Perlite/EPS composite		S.F. Roof	1.13	1.13	4.6%
	.8	Openings & Specialties	Hatches		S.F. Roof	.24	.24	
6.0 Interior Construction								
	.1	Partitions	Concrete block	140 S.F. Floor/L.F. Partition	S.F. Partition	4.08	.35	
	.4	Interior Doors	Hollow metal	2500 S.F. Floor/Door	Each	484	.19	
	.5	Wall Finishes	Paint		S.F. Surface	.88	.15	
	.6	Floor Finishes	80% rubber mat, 20% paint	50% of floor area	S.F. Floor	4.26	2.13	6.2%
	.7	Ceiling Finishes	Mineral fiber tile on concealed zee bar	10% of area	S.F. Ceiling	2.62	.26	
	.9	Interior Surface/Exterior Wall	Paint	95% of wall	S.F. Wall	1.53	.86	
7.0 Conveying								
	.1	Elevators	N/A		—	—	—	0.0%
	.2	Special Conveyors	N/A		—	—	—	
8.0 Mechanical								
	.1	Plumbing	Toilet and service fixtures, supply and drainage	1 Fixture/1070 S.F. Floor	Each	3359	3.14	
	.2	Fire Protection	Standpipes and hose systems		—	—	—	
	.3	Heating	Oil fired hot water, unit heaters	10% of area	S.F. Floor	.42	.42	20.5%
	.4	Cooling	Single zone, electric cooling	90% of area	S.F. Floor	9.48	9.48	
	.5	Special Systems	N/A		—	—	—	
9.0 Electrical								
	.1	Service & Distribution	400 ampere service, panel board and feeders		S.F. Floor	.50	.50	
	.2	Lighting & Power	High intensity discharge and fluorescent fixtures, receptacles, switches, A.C. and misc. power		S.F. Floor	3.95	3.95	8.1%
	.4	Special Electrical	Alarm systems, emergency lighting and public address		S.F. Floor	.69	.69	
11.0 Special Construction								
	.1	Specialties	Dasher boards and rink (including ice making system)		S.F. Floor	15.52	15.52	24.5%
12.0 Site Work								
	.1	Earthwork	N/A		—	—	—	
	.3	Utilities	N/A		—	—	—	
	.5	Roads & Parking	N/A		—	—	—	0.0%
	.7	Site Improvements	N/A		—	—	—	
					Sub-Total		63.46	100%
		GENERAL CONDITIONS (Overhead & Profit)				15%	9.52	
		ARCHITECT FEES				7%	5.12	
					Total Building Cost		78.10	

Costs per square foot of floor area

Exterior Wall	S.F. Area	25000	30000	35000	40000	45000	50000	55000	60000	65000
	L.F. Perimeter	700	740	823	906	922	994	1033	1100	1166
Face Brick with Concrete Block Back-up	Steel Frame	69.30	67.45	66.65	66.10	64.95	64.55	63.95	63.70	63.45
	Bearing Walls	67.10	65.20	64.40	63.85	**62.70**	62.35	61.70	61.45	61.20
Stucco on Concrete Block	Steel Frame	67.05	65.45	64.80	64.25	63.30	62.95	62.40	62.20	62.00
	Bearing Walls	64.80	63.20	62.50	62.00	61.05	60.70	60.20	59.95	59.75
Decorative Concrete Block	Steel Frame	67.75	66.10	65.40	64.85	63.85	63.50	62.90	62.70	62.45
	Bearing Walls	65.55	63.85	63.15	62.60	61.60	61.25	60.65	60.45	60.20
Perimeter Adj., Add or Deduct	Per 100 L.F.	1.85	1.55	1.30	1.15	1.00	.90	.85	.75	.70
Story Hgt. Adj., Add or Deduct	Per 1 Ft.	.65	.55	.50	.50	.45	.45	.45	.40	.40

For Basement, add $13.30 per square foot of basement area

The above costs were calculated using the basic specifications shown on the facing page. These costs should be adjusted where necessary for design alternatives and owner's requirements. Reported completed project costs, for this type of structure, range from $45.45 to $121.15 per S.F.

Common additives

Description	Unit	$ Cost
Bleachers, Telescoping, manual		
To 15 tier	Seat	68 - 97
16-20 tier	Seat	147 - 181
21-30 tier	Seat	156 - 187
For power operation, add	Seat	29 - 46
Carrels Hardwood	Each	640 - 835
Clock System		
20 room	Each	9875
50 room	Each	23,800
Emergency Lighting, 25 watt, battery operated		
Lead battery	Each	305
Nickel cadmium	Each	600
Flagpoles, Complete		
Aluminum, 20' high	Each	1025
40' high	Each	2375
Fiberglass, 23' high	Each	1400
39'-5" high	Each	2425
Kitchen Equipment		
Broiler	Each	3400
Cooler, 6 ft. long, reach-in	Each	2675

Description	Unit	$ Cost
Kitchen Equipment, cont.		
Dishwasher, 10-12 racks per hr.	Each	2525
Food warmer, counter, 1.2 KW	Each	745
Freezer, 44 C.F., reach-in	Each	6575
Ice cube maker, 50 lb. per day	Each	1700
Range with 1 oven	Each	2300
Lockers, Steel, single tier, 60" to 72"	Opening	128 - 207
2 tier, 60" to 72" total	Opening	79 - 99
5 tier, box lockers	Opening	41 - 53
Locker bench, lam. maple top only	L.F.	17.60
Pedestals, steel pipe	Each	39
Seating		
Auditorium chair, all veneer	Each	123
Veneer back, padded seat	Each	148
Upholstered, spring seat	Each	174
Classroom, movable chair & desk	Set	65 - 120
Lecture hall, pedestal type	Each	123 - 360
Sound System		
Amplifier, 250 watts	Each	1425
Speaker, ceiling or wall	Each	118
Trumpet	Each	222

Important: See the Reference Section for Location Factors

Model costs calculated for a 1 story building with 12' story height and 45,000 square feet of floor area

				Unit	Unit Cost	Cost Per S.F.	% Of Sub-Total
1.0 Foundations							
	.1	Footings & Foundations	Poured concrete; strip and spread footings and 4' foundation wall	S.F. Ground	2.70	2.70	
	.4	Piles & Caissons	N/A	—	—	—	7.1%
	.9	Excavation & Backfill	Site preparation for slab and trench for foundation wall and footing	S.F. Ground	.91	.91	
2.0 Substructure							
	.1	Slab on Grade	4" reinforced concrete with vapor barrier and granular base	S.F. Slab	2.56	2.56	5.0%
	.2	Special Substructures	N/A	—	—	—	
3.0 Superstructure							
	.1	Columns & Beams	N/A	—	—	—	
	.4	Structural Walls	N/A	—	—	—	
	.5	Elevated Floors	N/A	—	—	—	4.4%
	.7	Roof	Metal deck on open web steel joists	S.F. Roof	2.22	2.22	
	.9	Stairs	N/A	—	—	—	
4.0 Exterior Closure							
	.1	Walls	Face brick with concrete block backup 70% of wall	S.F. Wall	15.80	2.72	
	.5	Exterior Wall Finishes	N/A	—	—	—	8.7%
	.6	Doors	Metal and glass 5% of wall	Each	2047	.36	
	.7	Windows & Glazed Walls	Steel outward projecting 25% of wall	Each	420	1.35	
5.0 Roofing							
	.1	Roof Coverings	Built-up tar and gravel with flashing	S.F. Roof	1.61	1.61	
	.7	Insulation	Perlite/EPS composite	S.F. Roof	1.13	1.13	5.6%
	.8	Openings & Specialties	Gravel stop	L.F. Perimeter	4.78	.10	
6.0 Interior Construction							
	.1	Partitions	Concrete block, toilet partitions 20 S.F. Floor/L.F. Partition	S.F. Partition	4.10	2.66	
	.4	Interior Doors	Single leaf kalamein fire doors 700 S.F. Floor/Door	Each	484	.69	
	.5	Wall Finishes	75% paint, 15% glazed coating, 10% ceramic tile	S.F. Surface	1.17	1.17	
	.6	Floor Finishes	65% vinyl composition tile, 25% carpet, 10% terrazzo	S.F. Floor	4.32	4.32	23.3%
	.7	Ceiling Finishes	Mineral fiber tile on concealed zee bars	S.F. Ceiling	2.62	2.62	
	.9	Interior Surface/Exterior Wall	Painted gypsum board on furring 70% of wall	S.F. Wall	2.26	.39	
7.0 Conveying							
	.1	Elevators	N/A	—	—	—	0.0%
	.2	Special Conveyors	N/A	—	—	—	
8.0 Mechanical							
	.1	Plumbing	Kitchen, bathroom and service fixtures, supply and drainage 1 Fixture/625 S.F. Floor	Each	1981	3.17	
	.2	Fire Protection	Sprinklers, light hazard	S.F. Floor	1.09	1.09	
	.3	Heating	Oil fired hot water, wall fin radiation	S.F. Floor	4.74	4.74	31.6%
	.4	Cooling	Split systems with air cooled condensing units	S.F. Floor	7.16	7.16	
	.5	Special Systems	N/A	—	—	—	
9.0 Electrical							
	.1	Service & Distribution	600 ampere service, panel board and feeders	S.F. Floor	.54	.54	
	.2	Lighting & Power	Fluorescent fixtures, receptacles, switches, A.C. and misc. power	S.F. Floor	4.96	4.96	12.7%
	.4	Special Electrical	Alarm systems, communications systems and emergency lighting	S.F. Floor	.96	.96	
11.0 Special Construction							
	.1	Specialties	Chalkboards	S.F. Floor	.83	.83	1.6%
12.0 Site Work							
	.1	Earthwork	N/A	—	—	—	
	.3	Utilities	N/A	—	—	—	
	.5	Roads & Parking	N/A	—	—	—	0.0%
	.7	Site Improvements	N/A	—	—	—	
				Sub-Total		50.96	100%
		GENERAL CONDITIONS (Overhead & Profit)			15%	7.64	
		ARCHITECT FEES			7%	4.10	
				Total Building Cost		**62.70**	

BUILDING TYPES

Costs per square foot of floor area

Exterior Wall	S.F. Area	50000	65000	80000	95000	110000	125000	140000	155000	170000
	L.F. Perimeter	816	1016	1000	1150	1890	2116	2287	2438	2552
Face Brick with Concrete Block Back-up	Steel Frame	70.10	69.40	67.10	66.75	**70.15**	69.90	69.45	69.00	68.50
	Bearing Walls	67.60	66.90	64.60	64.20	67.60	67.40	66.95	66.45	66.00
Concrete Block Stucco Face	Steel Frame	67.20	66.60	64.90	64.60	67.10	66.90	66.55	66.20	65.80
	Bearing Walls	64.70	64.10	62.35	62.10	64.55	64.40	64.05	63.70	63.30
Decorative Concrete Block	Steel Frame	68.10	67.50	65.55	65.25	68.00	67.85	67.45	67.05	66.65
	Bearing Walls	65.20	64.55	62.65	62.35	65.10	64.90	64.55	64.15	63.75
Perimeter Adj., Add or Deduct	Per 100 L.F.	1.35	1.05	.85	.75	.60	.55	.50	.45	.40
Story Hgt. Adj., Add or Deduct	Per 1 Ft.	.80	.75	.60	.60	.85	.85	.80	.80	.75

For Basement, add $17.95 per square foot of basement area

The above costs were calculated using the basic specifications shown on the facing page. These costs should be adjusted where necessary for design alternatives and owner's requirements. Reported completed project costs, for this type of structure, range from $52.25 to $119.40 per S.F.

Common additives

Description	Unit	$ Cost
Bleachers, Telescoping, manual		
To 15 tier	Seat	68 - 97
16-20 tier	Seat	147 - 181
21-30 tier	Seat	156 - 187
For power operation, add	Seat	29 - 46
Carrels Hardwood	Each	640 - 835
Clock System		
20 room	Each	9875
50 room	Each	23,800
Elevators, Hydraulic passenger, 2 stops		
1500# capacity	Each	37,625
2500# capacity	Each	38,925
Emergency Lighting, 25 watt, battery operated		
Lead battery	Each	305
Nickel cadmium	Each	600
Flagpoles, Complete		
Aluminum, 20' high	Each	1025
40' high	Each	2375
Fiberglass, 23' high	Each	1400
39'-5" high	Each	2425

Description	Unit	$ Cost
Kitchen Equipment		
Broiler	Each	3400
Cooler, 6 ft. long, reach-in	Each	2675
Dishwasher, 10-12 racks per hr.	Each	2525
Food warmer, counter, 1.2 KW	Each	745
Freezer, 44 C.F., reach-in	Each	6575
Lockers, Steel, single tier, 60" to 72"	Opening	128 - 207
2 tier, 60" to 72" total	Opening	79 - 99
5 tier, box lockers	Opening	41 - 53
Locker bench, lam. maple top only	L.F.	17.60
Pedestals, steel pipe	Each	39
Seating		
Auditorium chair, all veneer	Each	123
Veneer back, padded seat	Each	148
Upholstered, spring seat	Each	174
Classroom, movable chair & desk	Set	65 - 120
Lecture hall, pedestal type	Each	123 - 360
Sound System		
Amplifier, 250 watts	Each	1425
Speaker, ceiling or wall	Each	118
Trumpet	Each	222

Important: See the Reference Section for Location Factors

Model costs calculated for a 2 story building with 12' story height and 110,000 square feet of floor area

School, Jr High, 2-3 Story

				Unit	Unit Cost	Cost Per S.F.	% Of Sub-Total
1.0 Foundations							
	.1	Footings & Foundations	Poured concrete; strip and spread footings and 4' foundation wall	S.F. Ground	2.42	1.21	
	.4	Piles & Caissons	N/A	—	—	—	2.9%
	.9	Excavation & Backfill	Site preparation for slab and trench for foundation wall and footing	S.F. Ground	.91	.46	
2.0 Substructure							
	.1	Slab on Grade	4" reinforced concrete with vapor barrier and granular base	S.F. Slab	2.56	1.28	2.2%
	.2	Special Substructures	N/A	—	—	—	
3.0 Superstructure							
	.1	Columns & Beams	Fireproofing, steel columns included in 3.5 and 3.7	L.F. Column	16.44	.23	
	.4	Structural Walls	N/A	—	—	—	
	.5	Elevated Floors	Open web steel joists, slab form, concrete, columns	S.F. Floor	12.24	6.12	16.0%
	.7	Roof	Metal deck, open web steel joists, columns	S.F. Roof	4.96	2.48	
	.9	Stairs	Concrete filled metal pan	Flight	4725	.26	
4.0 Exterior Closure							
	.1	Walls	Face brick with concrete block backup — 75% of wall	S.F. Wall	15.81	4.89	
	.5	Exterior Wall Finishes	N/A	—	—	—	13.8%
	.6	Doors	Double aluminum & glass	Each	1221	.32	
	.7	Windows & Glazed Walls	Window wall — 25% of wall	S.F. Window	26	2.68	
5.0 Roofing							
	.1	Roof Coverings	Built-up tar and gravel with flashing	S.F. Roof	1.84	.92	
	.7	Insulation	Perlite/EPS composite	S.F. Roof	1.13	.56	2.8%
	.8	Openings & Specialties	Gravel stop and hatches	S.F. Roof	.20	.10	
6.0 Interior Construction							
	.1	Partitions	Concrete block, toilet partitions — 20 S.F. Floor/L.F. Partition	S.F. Partition	4.10	2.47	
	.4	Interior Doors	Single leaf kalamein fire doors — 750 S.F. Floor/Door	Each	484	.65	
	.5	Wall Finishes	50% paint, 40% glazed coatings, 10% ceramic tile	S.F. Surface	1.30	1.30	23.3%
	.6	Floor Finishes	50% vinyl composition tile, 30% carpet, 20% terrrazzo	S.F. Floor	5.55	5.55	
	.7	Ceiling Finishes	Mineral fiberboard on concealed zee bars	S.F. Ceiling	2.62	2.62	
	.9	Interior Surface/Exterior Wall	Painted gypsum board on furring — 75% of wall	S.F. Wall	2.26	.70	
7.0 Conveying							
	.1	Elevators	One hydraulic passenger elevator	Each	46,200	.42	0.7%
	.2	Special Conveyors	N/A	—	—	—	
8.0 Mechanical							
	.1	Plumbing	Kitchen, toilet and service fixtures, supply and drainage — 1 Fixture/1170 S.F. Floor	Each	2398	2.05	
	.2	Fire Protection	Sprinklers, light hazard — 10% of area	S.F. Floor	.19	.19	
	.3	Heating	Included in 8.4	—	—	—	24.7%
	.4	Cooling	Multizone unit, gas heating, electric cooling	S.F. Floor	11.81	11.81	
	.5	Special Systems	N/A	—	—	—	
9.0 Electrical							
	.1	Service & Distribution	1000 ampere service, panel board and feeders	S.F. Floor	.43	.43	
	.2	Lighting & Power	Fluorescent fixtures, receptacles, switches, A.C. and misc. power	S.F. Floor	4.97	4.97	12.4%
	.4	Special Electrical	Alarm systems, communications systems and emergency lighting	S.F. Floor	1.64	1.64	
11.0 Special Construction							
	.1	Specialties	Chalkboards, laboratory counters	S.F. Floor	.68	.68	1.2%
12.0 Site Work							
	.1	Earthwork	N/A	—	—	—	
	.3	Utilities	N/A	—	—	—	0.0%
	.5	Roads & Parking	N/A	—	—	—	
	.7	Site Improvements	N/A	—	—	—	
				Sub-Total		56.99	**100%**
		GENERAL CONDITIONS (Overhead & Profit)			15%	8.55	
		ARCHITECT FEES			7%	4.61	
				Total Building Cost		70.15	

Costs per square foot of floor area

Exterior Wall	S.F. Area	1000	2000	3000	4000	6000	8000	10000	12000	15000
	L.F. Perimeter	126	179	219	253	310	358	400	438	490
Wood Siding	Wood Frame	62.00	55.75	52.90	51.20	49.20	48.00	47.20	46.60	45.95
Face Brick Veneer	Wood Frame	75.50	65.10	60.40	57.60	54.35	52.35	51.00	50.00	48.90
Stucco on Concrete Block	Steel Frame	70.60	62.25	58.45	56.25	53.55	52.00	50.90	50.10	49.25
	Bearing Walls	69.35	61.00	57.20	55.00	52.35	50.75	49.70	48.90	48.00
Metal Sandwich Panel	Steel Frame	73.40	64.30	60.20	57.80	54.90	53.20	52.00	51.15	50.20
Precast Concrete	Steel Frame	84.20	71.80	66.20	62.85	58.90	56.55	54.95	53.75	52.45
Perimeter Adj., Add or Deduct	Per 100 L.F.	17.15	8.55	5.75	4.30	2.85	2.15	1.75	1.45	1.15
Story Hgt. Adj., Add or Deduct	Per 1 Ft.	1.30	.95	.75	.65	.55	.45	.40	.35	.35

For Basement, add $12.95 per square foot of basement area

The above costs were calculated using the basic specifications shown on the facing page. These costs should be adjusted where necessary for design alternatives and owner's requirements. Reported completed project costs, for this type of structure, range from $38.25 to $105.50 per S.F.

Common additives

Description	Unit	$ Cost
Check Out Counter		
Single belt	Each	2150
Double belt	Each	3675
Emergency Lighting, 25 watt, battery operated		
Lead battery	Each	305
Nickel cadmium	Each	600
Refrigerators, Prefabricated, walk-in		
7'-6" high, 6' x 6'	S.F.	114
10' x 10'	S.F.	89
12' x 14'	S.F.	79
12' x 20'	S.F.	70
Refrigerated Food Cases		
Dairy, multi deck, 12' long	Each	6925
Delicatessen case, single deck, 12' long	Each	4875
Multi deck, 18 S.F. shelf display	Each	7375
Freezer, self-contained chest type, 30 C.F.	Each	2900
Glass door upright, 78 C.F.	Each	6475

Description	Unit	$ Cost
Refrigerated Food Cases, cont.		
Frozen food, chest type, 12' long	Each	4725
Glass door reach-in, 5 door	Each	8975
Island case 12' long, single deck	Each	5350
Multi deck	Each	11,100
Meat cases, 12' long, single deck	Each	3875
Multi deck	Each	6625
Produce, 12' long single deck	Each	5125
Multi deck	Each	5700
Safe, Office type, 4 hour rating		
30" x 18" x 18"	Each	3475
62" x 33" x 20"	Each	7575
Smoke Detectors		
Ceiling type	Each	126
Duct type	Each	360
Sound System		
Amplifier, 250 watts	Each	1425
Speaker, ceiling or wall	Each	118
Trumpet	Each	222

Important: See the Reference Section for Location Factors

Model costs calculated for a 1 story building with 12' story height and 4,000 square feet of floor area

			Unit	Unit Cost	Cost Per S.F.	% Of Sub-Total
1.0 Foundations						
.1	Footings & Foundations	Poured concrete; strip and spread footings and 4' foundation wall	S.F. Ground	2.57	2.57	
.4	Piles & Caissons	N/A	—	—	—	8.8%
.9	Excavation & Backfill	Site preparation for slab and trench for foundation wall and footing	S.F. Ground	1.08	1.08	
2.0 Substructure						
.1	Slab on Grade	4" reinforced concrete with vapor barrier and granular base	S.F. Slab	2.56	2.56	6.2%
.2	Special Substructures	N/A	—	—	—	
3.0 Superstructure						
.1	Columns & Beams	N/A	—	—	—	
.4	Structural Walls	N/A	—	—	—	
.5	Elevated Floors	N/A	—	—	—	9.3%
.7	Roof	Wood truss with plywood sheathing	S.F. Ground	3.87	3.87	
.9	Stairs	N/A	—	—	—	
4.0 Exterior Closure						
.1	Walls	Wood siding on wood studs, insulated *80% of wall*	S.F. Wall	5.11	3.10	
.5	Exterior Wall Finishes	N/A	—	—	—	16.8%
.6	Doors	Double aluminum and glass, solid core wood	Each	1772	1.33	
.7	Windows & Glazed Walls	Storefront *20% of wall*	S.F. Window	25	2.55	
5.0 Roofing						
.1	Roof Coverings	Asphalt shingles	S.F. Roof	1.35	1.35	
.7	Insulation	Fiberglass sheets	S.F. Roof	.78	.78	5.3%
.8	Openings & Specialties	Gutters and downspouts	S.F. Roof	.07	.07	
6.0 Interior Construction						
.1	Partitions	Gypsum board on wood studs *60 S.F. Floor/L.F. Partition*	S.F. Partition	2.46	.41	
.4	Interior Doors	Single leaf wood, hollow metal *1300 S.F. Floor/Door*	Each	567	.44	
.5	Wall Finishes	Paint	S.F. Surface	.57	.19	
.6	Floor Finishes	Asphalt tile	S.F. Floor	1.70	1.70	12.7%
.7	Ceiling Finishes	Mineral fiber tile on wood furring	S.F. Ceiling	1.80	1.80	
.9	Interior Surface/Exterior Wall	Painted gypsum board on furring *80% of wall*	S.F. Wall	1.22	.74	
7.0 Conveying						
.1	Elevators	N/A	—	—	—	0.0%
.2	Special Conveyors	N/A	—	—	—	
8.0 Mechanical						
.1	Plumbing	Toilet and service fixtures, supply and drainage *1 Fixture/1000 S.F. Floor*	Each	1570	1.57	
.2	Fire Protection	Sprinkler, ordinary hazard	S.F. Floor	1.66	1.66	
.3	Heating	Included in 8.4	—	—	—	19.7%
.4	Cooling	Single zone rooftop unit, gas heating, electric cooling	S.F. Floor	5.01	5.01	
.5	Special Systems	N/A	—	—	—	
9.0 Electrical						
.1	Service & Distribution	200 ampere service, panel board and feeders	S.F. Floor	1.45	1.45	
.2	Lighting & Power	Fluorescent fixtures, receptacles, switches, A.C. and misc. power	S.F. Floor	5.05	5.05	21.2%
.4	Special Electrical	Alarm systems and emergency lighting	S.F. Floor	2.34	2.34	
11.0 Special Construction						
.1	Specialties	N/A	—	—	—	0.0%
12.0 Site Work						
.1	Earthwork	N/A	—	—	—	
.3	Utilities	N/A	—	—	—	0.0%
.5	Roads & Parking	N/A	—	—	—	
.7	Site Improvements	N/A	—	—	—	

					Sub-Total	41.62	100%
GENERAL CONDITIONS (Overhead & Profit)					15%	6.24	
ARCHITECT FEES					7%	3.34	
Total Building Cost						**51.20**	

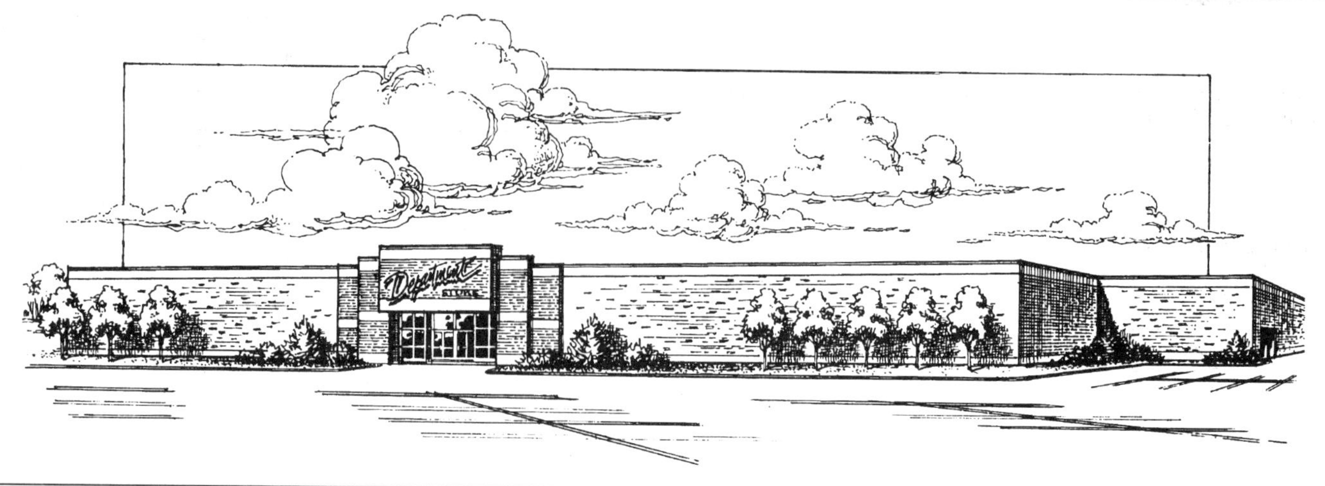

Costs per square foot of floor area

Exterior Wall	S.F. Area	50000	65000	80000	95000	110000	125000	140000	155000	170000
	L.F. Perimeter	920	1065	1167	1303	1333	1433	1533	1633	1733
Face Brick with Concrete Block Back-up	R/Conc. Frame	55.25	53.95	52.90	52.35	51.55	51.20	50.90	50.65	50.45
	Steel Frame	44.45	43.15	42.10	41.55	40.75	40.40	40.05	39.80	39.60
Decorative Concrete Block	R/Conc. Frame	52.95	51.90	51.10	50.65	50.05	49.75	49.50	49.30	49.15
	Steel Joists	42.60	41.50	40.65	40.15	39.50	39.20	38.95	38.75	38.55
Precast Concrete Panels	R/Conc. Frame	53.35	52.30	51.50	51.00	50.40	50.10	49.85	49.65	49.50
	Steel Joists	42.25	41.20	40.40	39.90	39.30	39.00	38.75	38.55	38.40
Perimeter Adj., Add or Deduct	Per 100 L.F.	.80	.65	.50	.45	.40	.35	.30	.25	.25
Story Hgt. Adj., Add or Deduct	Per 1 Ft.	.45	.40	.35	.30	.30	.25	.25	.25	.25

For Basement, add $12.60 per square foot of basement area

The above costs were calculated using the basic specifications shown on the facing page. These costs should be adjusted where necessary for design alternatives and owner's requirements. Reported completed project costs, for this type of structure, range from $33.05 to $73.00 per S.F.

Common additives

Description	Unit	$ Cost
Closed Circuit Surveillance, One station		
Camera and monitor	Each	1250
For additional camera stations, add	Each	695
Directory Boards, Plastic, glass covered		
30" x 20"	Each	460
36" x 48"	Each	850
Aluminum, 24" x 18"	Each	410
36" x 24"	Each	495
48" x 32"	Each	590
48" x 60"	Each	1300
Emergency Lighting, 25 watt, battery operated		
Lead battery	Each	305
Nickel cadmium	Each	600
Safe, Office type, 4 hour rating		
30" x 18" x 18"	Each	3475
62" x 33" x 20"	Each	7575
Sound System		
Amplifier, 250 watts	Each	1425
Speaker, ceiling or wall	Each	118
Trumpet	Each	222

Important: See the Reference Section for Location Factors

Model costs calculated for a 1 story building with 14' story height and 110,000 square feet of floor area

				Unit	Unit Cost	Cost Per S.F.	% Of Sub-Total
1.0 Foundations							
	.1	Footings & Foundations	Poured concrete; strip and spread footings and 4' foundation wall	S.F. Ground	.82	.82	
	.4	Piles & Caissons	N/A	—	—	—	4.0%
	.9	Excavation & Backfill	Site preparation for slab and trench for foundation wall and footing	S.F. Ground	.87	.87	
2.0 Substructure							
	.1	Slab on Grade	4" reinforced concrete with vapor barrier and granular base	S.F. Slab	2.56	2.56	6.1%
	.2	Special Substructures	N/A	—	—	—	
3.0 Superstructure							
	.1	Columns & Beams	Concrete columns	L.F. Column	32	.41	
	.4	Structural Walls	N/A	—	—	—	
	.5	Elevated Floors	N/A	—	—	—	31.4%
	.7	Roof	Precast concrete beam and plank	S.F. Roof	12.87	12.87	
	.9	Stairs	N/A	—	—	—	
4.0 Exterior Closure							
	.1	Walls	Face brick with concrete block backup	90% of wall	S.F. Wall	15.98	2.44	
	.5	Exterior Wall Finishes	N/A		—	—	—	7.7%
	.6	Doors	Sliding electric operated entrance, hollow metal		Each	4208	.23	
	.7	Windows & Glazed Walls	Storefront	10% of wall	S.F. Window	33	.57	
5.0 Roofing							
	.1	Roof Coverings	Built-up tar and gravel with flashing	S.F. Roof	1.44	1.44	
	.7	Insulation	Perlite/EPS composite	S.F. Roof	1.13	1.13	6.3%
	.8	Openings & Specialties	Gravel stop and hatches	S.F. Roof	.09	.09	
6.0 Interior Construction							
	.1	Partitions	Gypsum board on metal studs	60 S.F. Floor/L.F. Partition	S.F. Partition	3.18	.53	
	.4	Interior Doors	Single leaf hollow metal	600 S.F. Floor/Door	Each	484	.81	
	.5	Wall Finishes	Paint		S.F. Surface	.42	.14	
	.6	Floor Finishes	50% vinyl tile, 50% carpet		S.F. Floor	3.96	3.96	16.3%
	.7	Ceiling Finishes	Fiberglass board on exposed grid system, suspended		S.F. Ceiling	1.11	1.11	
	.9	Interior Surface/Exterior Wall	Painted gypsum board on furring	90% of wall	S.F. Wall	2.26	.35	
7.0 Conveying							
	.1	Elevators	N/A	—	—	—	0.0%
	.2	Special Conveyors	N/A	—	—	—	
8.0 Mechanical							
	.1	Plumbing	Toilet and service fixtures, supply and drainage	1 Fixture/4075 S.F. Floor	Each	2241	.55	
	.2	Fire Protection	Sprinklers, light hazard		S.F. Floor	1.09	1.09	
	.3	Heating	Included in 8.4		—	—	—	17.0%
	.4	Cooling	Single zone unit gas heating, electric cooling		S.F. Floor	5.57	5.57	
	.5	Special Systems	N/A		—	—	—	
9.0 Electrical							
	.1	Service & Distribution	1200 ampere service, panel board and feeders	S.F. Floor	.45	.45	
	.2	Lighting & Power	Fluorescent fixtures, receptacles, switches, A.C. and misc. power	S.F. Floor	4.04	4.04	11.2%
	.4	Special Electrical	Alarm systems and emergency lighting	S.F. Floor	.25	.25	
11.0 Special Construction							
	.1	Specialties	N/A	—	—	—	0.0%
12.0 Site Work							
	.1	Earthwork	N/A	—	—	—	
	.3	Utilities	N/A	—	—	—	
	.5	Roads & Parking	N/A	—	—	—	0.0%
	.7	Site Improvements	N/A	—	—	—	

				Sub-Total	42.28	100%
GENERAL CONDITIONS (Overhead & Profit)			15%		6.34	
ARCHITECT FEES			6%		2.93	
Total Building Cost					**51.55**	

BUILDING TYPES

Costs per square foot of floor area

Exterior Wall	S.F. Area	4000	6000	8000	10000	12000	15000	18000	20000	22000
	L.F. Perimeter	260	340	360	410	440	490	540	565	594
Split Face Concrete Block	Steel Joists	66.50	60.75	55.70	53.50	51.55	49.70	48.45	47.70	47.15
Stucco on Concrete Block	Steel Joists	63.20	57.80	53.35	51.35	49.65	48.00	46.85	46.20	45.75
Painted Concrete Block	Steel Joists	62.15	56.70	52.15	50.10	48.35	46.70	45.55	44.90	44.40
Face Brick on Concrete Block	Steel Joists	73.20	66.55	60.30	57.70	55.30	53.05	51.50	50.60	49.90
Painted Reinforced Concrete	Steel Joists	68.45	62.40	57.00	54.70	52.60	50.65	49.30	48.50	47.90
Tilt-up Concrete Panels	Steel Joists	63.00	57.70	53.25	51.25	49.55	47.90	46.80	46.15	45.65
Perimeter Adj., Add or Deduct	Per 100 L.F.	7.45	4.95	3.70	2.95	2.50	2.00	1.65	1.50	1.35
Story Hgt. Adj., Add or Deduct	Per 1 Ft.	1.00	.85	.65	.65	.55	.50	.45	.45	.40

For Basement, add $13.15 per square foot of basement area

The above costs were calculated using the basic specifications shown on the facing page. These costs should be adjusted where necessary for design alternatives and owner's requirements. Reported completed project costs, for this type of structure, range from $33.30 to $87.75 per S.F.

Common additives

Description	Unit	$ Cost
Emergency Lighting, 25 watt, battery operated		
Lead battery	Each	305
Nickel cadmium	Each	600
Safe, Office type, 4 hour rating		
30" x 18" x 18"	Each	3475
62" x 33" x 20"	Each	7575
Smoke Detectors		
Ceiling type	Each	126
Duct type	Each	360
Sound System		
Amplifier, 250 watts	Each	1425
Speaker, ceiling or wall	Each	118
Trumpet	Each	222

Important: See the Reference Section for Location Factors

Model costs calculated for a 1 story building with 14' story height and 8,000 square feet of floor area

				Unit	Unit Cost	Cost Per S.F.	% Of Sub-Total
1.0 Foundations							
.1	Footings & Foundations	Poured concrete; strip and spread footings and 4' foundation wall		S.F. Ground	2.29	2.29	
.4	Piles & Caissons	N/A		—	—	—	7.2%
.9	Excavation & Backfill	Site preparation for slab and trench for foundation wall and footing		S.F. Ground	.96	.96	
2.0 Substructure							
.1	Slab on Grade	4" reinforced concrete with vapor barrier and granular base		S.F. Slab	2.56	2.56	5.7%
.2	Special Substructures	N/A		—	—	—	
3.0 Superstructure							
.1	Columns & Beams	Interior columns included in 3.7		—	—	—	
.4	Structural Walls	N/A		—	—	—	
.5	Elevated Floors	N/A		—	—	—	7.3%
.7	Roof	Metal deck, open web steel joists, beams, interior columns		S.F. Roof	3.29	3.29	
.9	Stairs	N/A		—	—	—	
4.0 Exterior Closure							
.1	Walls	Decorative concrete block	90% of wall	S.F. Wall	9.35	5.30	
.5	Exterior Wall Finishes	N/A		—	—	—	16.1%
.6	Doors	Sliding entrance door and hollow metal service doors		Each	1460	.37	
.7	Windows & Glazed Walls	Storefront windows	10% of wall	S.F. Window	24	1.57	
5.0 Roofing							
.1	Roof Coverings	Built-up tar and gravel with flashing		S.F. Roof	1.89	1.89	
.7	Insulation	Perlite/EPS composite		S.F. Roof	1.13	1.13	7.4%
.8	Openings & Specialties	Gravel stop and hatches		S.F. Roof	.28	.28	
6.0 Interior Construction							
.1	Partitions	Gypsum board on metal studs	60 S.F. Floor/L.F. Partition	S.F. Partition	3.18	.53	
.4	Interior Doors	Single leaf hollow metal	600 S.F. Floor/Door	Each	484	.81	
.5	Wall Finishes	Paint		S.F. Surface	.42	.14	
.6	Floor Finishes	Vinyl tile		S.F. Floor	2.49	2.49	16.6%
.7	Ceiling Finishes	Mineral fiber tile on concealed zee bars		S.F. Ceiling	2.62	2.62	
.9	Interior Surface/Exterior Wall	Paint	90% of wall	S.F. Wall	1.53	.87	
7.0 Conveying							
.1	Elevators	N/A		—	—	—	0.0%
.2	Special Conveyors	N/A		—	—	—	
8.0 Mechanical							
.1	Plumbing	Toilet and service fixtures, supply and drainage	1 Fixture/890 S.F. Floor	Each	3382	3.80	
.2	Fire Protection	Wet pipe sprinkler system		S.F. Floor	1.66	1.66	
.3	Heating	Included in 8.4		—	—	—	24.7%
.4	Cooling	Single zone unit, gas heating, electric cooling		S.F. Floor	5.57	5.57	
.5	Special Systems	N/A		—	—	—	
9.0 Electrical							
.1	Service & Distribution	400 ampere service, panel board and feeders		S.F. Floor	1.35	1.35	
.2	Lighting & Power	Fluorescent fixtures, receptacles, switches, A.C. and misc. power		S.F. Floor	4.94	4.94	15.0%
.4	Special Electrical	Alarm systems and emergency lighting		S.F. Floor	.42	.42	
11.0 Special Construction							
.1	Specialties	N/A		—	—	—	0.0%
12.0 Site Work							
.1	Earthwork	N/A		—	—	—	
.3	Utilities	N/A		—	—	—	0.0%
.5	Roads & Parking	N/A		—	—	—	
.7	Site Improvements	N/A		—	—	—	
			Sub-Total			44.84	100%
	GENERAL CONDITIONS (Overhead & Profit)				15%	6.73	
	ARCHITECT FEES				8%	4.13	
			Total Building Cost			55.70	

Costs per square foot of floor area

Exterior Wall	S.F. Area	6000	8000	12000	16000	20000	25000	32000	38000	45000
	L.F. Perimeter	310	360	460	520	600	656	773	806	900
Face Brick with Concrete Block Back-up	Steel Frame	73.80	68.60	63.40	59.50	57.60	55.25	53.80	52.05	51.20
	Bearing Walls	72.60	67.40	62.25	58.30	56.45	54.05	52.60	50.85	50.00
Stucco on Concrete Block	Steel Frame	66.15	61.90	57.75	54.65	53.20	51.35	50.20	48.90	48.25
	Bearing Walls	66.70	62.45	58.25	55.20	53.75	51.90	50.75	49.45	48.80
Precast Concrete Panels	Steel Frame	72.80	67.70	62.65	58.85	57.05	54.75	53.35	51.65	50.85
Metal Sandwich Panels	Steel Frame	62.10	58.40	54.75	52.10	50.85	49.30	48.30	47.25	46.65
Perimeter Adj., Add or Deduct	Per 100 L.F.	8.95	6.70	4.45	3.35	2.70	2.15	1.65	1.40	1.20
Story Hgt. Adj., Add or Deduct	Per 1 Ft.	1.30	1.10	.95	.80	.75	.70	.60	.50	.50

For Basement, add $13.15 per square foot of basement area

The above costs were calculated using the basic specifications shown on the facing page. These costs should be adjusted where necessary for design alternatives and owner's requirements. Reported completed project costs, for this type of structure, range from $39.00 to $85.05 per S.F.

Common additives

Description	Unit	$ Cost
Check Out Counter		
Single belt	Each	2150
Double belt	Each	3675
Scanner, registers, guns & memory 2 lanes	Each	13,200
10 lanes	Each	125,500
Power take away	Each	4550
Emergency Lighting, 25 watt, battery operated		
Lead battery	Each	305
Nickel cadmium	Each	600
Refrigerators, Prefabricated, walk-in		
7'-6" High, 6' x 6'	S.F.	114
10' x 10'	S.F.	89
12' x 14'	S.F.	79
12' x 20'	S.F.	70
Refrigerated Food Cases		
Dairy, multi deck, 12' long	Each	6925
Delicatessen case, single deck, 12' long	Each	4875
Multi deck, 18 S.F. shelf display	Each	7375
Freezer, self-contained chest type, 30 C.F.	Each	2900
Glass door upright, 78 C.F.	Each	6475

Description	Unit	$ Cost
Refrigerated Food Cases, cont.		
Frozen food, chest type, 12' long	Each	4725
Glass door reach in, 5 door	Each	8975
Island case 12' long, single deck	Each	5350
Multi deck	Each	11,100
Meat case 12' long, single deck	Each	3875
Multi deck	Each	6625
Produce, 12' long single deck	Each	5125
Multi deck	Each	5700
Safe, Office type, 4 hour rating		
30" x 18" x 18"	Each	3475
62" x 33" x 20"	Each	7575
Smoke Detectors		
Ceiling type	Each	126
Duct type	Each	360
Sound System		
Amplifier, 250 watts	Each	1425
Speaker, ceiling or wall	Each	118
Trumpet	Each	222

Important: See the Reference Section for Location Factors

Model costs calculated for a 1 story building with 18' story height and 20,000 square feet of floor area

				Unit	Unit Cost	Cost Per S.F.	% Of Sub-Total
1.0 Foundations							
.1	Footings & Foundations	Poured concrete; strip and spread footings and 4' foundation wall		S.F. Ground	1.81	1.81	
.4	Piles & Caissons	N/A		—	—	—	5.9%
.9	Excavation & Backfill	Site preparation for slab and trench for foundation wall and footing		S.F. Ground	.91	.91	
2.0 Substructure							
.1	Slab on Grade	4" reinforced concrete with vapor barrier and granular base		S.F. Slab	2.56	2.56	5.6%
.2	Special Substructures	N/A		—	—	—	
3.0 Superstructure							
.1	Columns & Beams	Interior columns included in 3.7		—	—	—	
.4	Structural Walls	N/A		—	—	—	
.5	Elevated Floors	N/A		—	—	—	7.9%
.7	Roof	Metal deck, open web steel joists, beams, interior columns		S.F. Roof	3.63	3.63	
.9	Stairs	N/A		—	—	—	
4.0 Exterior Closure							
.1	Walls	Face brick with concrete block backup	85% of wall	S.F. Wall	15.80	7.25	
.5	Exterior Wall Finishes	N/A		—	—	—	
.6	Doors	Sliding entrance doors with electrical operator, hollow metal		Each	2929	1.61	25.7%
.7	Windows & Glazed Walls	Storefront windows	15% of wall	S.F. Window	36	2.94	
5.0 Roofing							
.1	Roof Coverings	Built-up tar and gravel with flashing		S.F. Roof	1.68	1.68	
.7	Insulation	Perlite/EPS composite		S.F. Roof	1.13	1.13	6.7%
.8	Openings & Specialties	Gravel stop and hatches		S.F. Roof	.27	.27	
6.0 Interior Construction							
.1	Partitions	50% concrete block, 50% gypsum board on metal studs	50 S.F. Floor/L.F. Partition	S.F. Partition	3.63	.87	
.4	Interior Doors	Single leaf hollow metal	2000 S.F. Floor/Door	Each	484	.24	
.5	Wall Finishes	Paint		S.F. Surface	.56	.27	
.6	Floor Finishes	Vinyl composition tile		S.F. Floor	2.49	2.49	15.7%
.7	Ceiling Finishes	Mineral fiber tile on concealed zee bars		S.F. Ceiling	2.62	2.62	
.9	Interior Surface/Exterior Wall	Paint	85% of wall	S.F. Wall	1.53	.70	
7.0 Conveying							
.1	Elevators	N/A		—	—	—	0.0%
.2	Special Conveyors	N/A		—	—	—	
8.0 Mechanical							
.1	Plumbing	Toilet and service fixtures, supply and drainage	1 Fixture/1820 S.F. Floor	Each	2730	1.50	
.2	Fire Protection	Sprinklers, light hazard		S.F. Floor	1.26	1.26	
.3	Heating	Included in 8.4		—	—	—	17.8%
.4	Cooling	Single zone rooftop unit, gas heating, electric cooling		S.F. Floor	5.41	5.41	
.5	Special Systems	N/A		—	—	—	
9.0 Electrical							
.1	Service & Distribution	400 ampere service, panel board and feeders		S.F. Floor	.74	.74	
.2	Lighting & Power	Fluorescent fixtures, receptacles, switches, A.C. and misc. power		S.F. Floor	5.68	5.68	14.7%
.4	Special Electrical	Alarm systems and emergency lighting		S.F. Floor	.30	.30	
11.0 Special Construction							
.1	Specialties	N/A		—	—	—	0.0%
12.0 Site Work							
.1	Earthwork	N/A		—	—	—	
.3	Utilities	N/A		—	—	—	
.5	Roads & Parking	N/A		—	—	—	0.0%
.7	Site Improvements	N/A		—	—	—	

	Sub-Total	45.87	100%
GENERAL CONDITIONS (Overhead & Profit)	15%	6.88	
ARCHITECT FEES	7%	3.70	

Total Building Cost	**56.45**

Costs per square foot of floor area

Exterior Wall	S.F. Area	2000	3000	4000	5000	6000	7000	8000	9000	10000
	L.F. Perimeter	180	220	260	286	320	353	368	397	425
Face Brick with Concrete Block Back-up	Steel Frame	96.95	86.95	81.95	77.70	75.45	73.80	71.55	70.45	69.55
	Bearing Walls	95.60	85.60	80.60	76.35	74.10	72.45	70.20	69.10	68.25
Limestone with Concrete Block Back-up	Steel Frame	110.10	97.70	91.45	86.05	83.25	81.15	78.25	76.90	75.80
	Bearing Walls	108.80	96.35	90.15	84.75	81.95	79.85	76.95	75.60	74.45
Decorative Concrete Block	Steel Frame	91.05	82.15	77.70	73.95	71.95	70.45	68.50	67.55	66.80
	Bearing Walls	89.70	80.80	76.35	72.60	70.60	69.15	67.20	66.25	65.45
Perimeter Adj., Add or Deduct	Per 100 L.F.	22.35	14.85	11.15	8.90	7.45	6.40	5.55	5.00	4.45
Story Hgt. Adj., Add or Deduct	Per 1 Ft.	2.65	2.15	1.90	1.70	1.55	1.50	1.35	1.30	1.25
For Basement, add $18.95 per square foot of basement area										

The above costs were calculated using the basic specifications shown on the facing page. These costs should be adjusted where necessary for design alternatives and owner's requirements. Reported completed project costs, for this type of structure, range from $53.85 to $190.15 per S.F.

Common additives

Description	Unit	$ Cost
Emergency Lighting, 25 watt, battery operated		
Lead battery	Each	305
Nickel cadmium	Each	600
Smoke Detectors		
Ceiling type	Each	126
Duct type	Each	360
Emergency Generators, complete system, gas		
15 kw	Each	12,300
85 kw	Each	27,300
170 kw	Each	70,500
Diesel, 50 kw	Each	23,600
150 kw	Each	41,700
350 kw	Each	63,000

Model costs calculated for a 1 story building with 12' story height and 5,000 square feet of floor area

Telephone Exchange

				Unit	Unit Cost	Cost Per S.F.	% Of Sub-Total
1.0 Foundations							
	.1	Footings & Foundations	Poured concrete; strip and spread footings and 4' foundation wall	S.F. Ground	3.37	3.37	7.3%
	.4	Piles & Caissons	N/A	—	—	—	
	.9	Excavation & Backfill	Site preparation for slab and trench for foundation wall and footing	S.F. Ground	1.08	1.08	
2.0 Substructure							
	.1	Slab on Grade	4" reinforced concrete with vapor barrier and granular base	S.F. Slab	2.56	2.56	4.2%
	.2	Special Substructures	N/A	—			
3.0 Superstructure							
	.1	Columns & Beams	Steel columns included in 3.7	—	—	—	
	.4	Structural Walls	N/A	—	—	—	
	.5	Elevated Floors	N/A	—	—	—	6.6%
	.7	Roof	Metal deck, open web steel joists, beams, columns	S.F. Roof	4.04	4.04	
	.9	Stairs	N/A				
4.0 Exterior Closure							
	.1	Walls	Face brick with concrete block backup — 80% of wall	S.F. Wall	15.81	8.68	
	.5	Exterior Wall Finishes	N/A	—	—	—	26.0%
	.6	Doors	Single aluminum glass with transom	Each	2010	.81	
	.7	Windows & Glazed Walls	Outward projecting steel — 20% of wall	Each	926	6.36	
5.0 Roofing							
	.1	Roof Coverings	Built-up tar and gravel with flashing	S.F. Roof	2.05	2.05	
	.7	Insulation	Perlite/EPS composite	S.F. Roof	1.13	1.13	5.7%
	.8	Openings & Specialties	Gravel stop and hatches	S.F. Roof	.27	.27	
6.0 Interior Construction							
	.1	Partitions	Double layer gypsum board on metal studs, toilet partitions — 15 S.F. Floor/L.F. Partition	S.F. Partition	4.55	3.03	
	.4	Interior Doors	Single leaf hollow metal — 150 S.F. Floor/Door	Each	484	3.23	
	.5	Wall Finishes	Paint	S.F. Surface	.42	.56	24.8%
	.6	Floor Finishes	90% carpet, 10% terrazzo	S.F. Floor	5.36	5.36	
	.7	Ceiling Finishes	Fiberglass board on exposed grid system, suspended	S.F. Ceiling	2.09	2.09	
	.9	Interior Surface/Exterior Wall	Paint — 80% of wall	S.F. Floor	1.53	.84	
7.0 Conveying							
	.1	Elevators	N/A	—	—	—	0.0%
	.2	Special Conveyors	N/A	—	—	—	
8.0 Mechanical							
	.1	Plumbing	Kitchen, toilet and service fixtures, supply and drainage — 1 Fixture/715 S.F. Floor	Each	2395	3.35	
	.2	Fire Protection	Wet pipe sprinkler system	S.F. Floor	1.82	1.82	
	.3	Heating	Included in 8.4	—	—	—	17.8%
	.4	Cooling	Single zone unit, gas heating, electric cooling	S.F. Floor	5.61	5.61	
	.5	Special Systems	N/A	—	—	—	
9.0 Electrical							
	.1	Service & Distribution	200 ampere service, panel board and feeders	S.F. Floor	.92	.92	
	.2	Lighting & Power	Fluorescent fixtures, receptacles, switches, A.C. and misc. power	S.F. Floor	2.84	2.84	7.6%
	.4	Special Electrical	Alarm systems and emergency lighting	S.F. Floor	.86	.86	
11.0 Special Construction							
	.1	Specialties	N/A	—	—	—	0.0%
12.0 Site Work							
	.1	Earthwork	N/A	—	—	—	
	.3	Utilities	N/A	—	—	—	0.0%
	.5	Roads & Parking	N/A	—	—	—	
	.7	Site Improvements	N/A	—	—	—	
				Sub-Total		60.86	100%
		GENERAL CONDITIONS (Overhead & Profit)		15%		9.13	
		ARCHITECT FEES		11%		7.71	
				Total Building Cost		77.70	

Costs per square foot of floor area

Exterior Wall	S.F. Area	5000	6500	8000	9500	11000	14000	17500	21000	24000
	L.F. Perimeter	300	360	386	396	435	510	550	620	680
Face Brick with Concrete Block Back-up	Steel Joists	76.65	73.80	70.45	67.50	66.35	64.70	62.50	61.55	61.00
	Wood Joists	75.50	72.55	69.15	66.15	64.95	63.25	61.00	60.05	59.45
Stone with Concrete Block Back-up	Steel Joists	77.75	74.75	71.30	68.25	67.05	65.35	63.10	62.10	61.50
	Wood Joists	76.55	73.55	70.00	66.90	65.65	63.90	61.60	60.55	59.95
Brick Veneer	Wood Frame	73.35	70.65	67.60	64.95	63.85	62.35	60.35	59.45	58.95
Wood Siding	Wood Frame	70.20	67.75	65.05	62.75	61.80	60.45	58.70	57.90	57.45
Perimeter Adj., Add or Deduct	Per 100 L.F.	7.35	5.65	4.60	3.85	3.30	2.65	2.10	1.75	1.55
Story Hgt. Adj., Add or Deduct	Per 1 Ft.	1.35	1.25	1.10	.95	.90	.85	.70	.65	.65
For Basement, add $15.70 per square foot of basement area										

The above costs were calculated using the basic specifications shown on the facing page. These costs should be adjusted where necessary for design alternatives and owner's requirements. Reported completed project costs, for this type of structure, range from $47.95 to $135.30 per S.F.

Common additives

Description	Unit	$ Cost	Description	Unit	$ Cost
Directory Boards, Plastic, glass covered			Smoke Detectors		
30" x 20"	Each	460	Ceiling type	Each	126
36" x 48"	Each	850	Duct type	Each	360
Aluminum, 24" x 18"	Each	410	Vault Front, Door & frame		
36" x 24"	Each	495	1 Hour test, 32" x 78"	Opening	3275
48" x 32"	Each	590	2 Hour test, 32" door	Opening	3325
48" x 60"	Each	1300	40" door	Opening	3800
Emergency Lighting, 25 watt, battery operated			4 Hour test, 32" door	Opening	3475
Lead battery	Each	305	40" door	Opening	4100
Nickel cadmium	Each	600	Time lock movement; two movement	Each	1525
Flagpoles, Complete					
Aluminum, 20' high	Each	1025			
40' high	Each	2375			
70' high	Each	6925			
Fiberglass, 23' high	Each	1400			
39'-5" high	Each	2425			
59' high	Each	6875			
Safe, Office type, 4 hour rating					
30" x 18" x 18"	Each	3475			
62" x 33" x 20"	Each	7575			

Important: See the Reference Section for Location Factors

Model costs calculated for a 1 story building with 12' story height and 11,000 square feet of floor area

				Unit	Unit Cost	Cost Per S.F.	% Of Sub-Total
1.0 Foundations							
.1	Footings & Foundations	Poured concrete; strip and spread footings and 4' foundation wall		S.F. Ground	2.22	2.22	
.4	Piles & Caissons	N/A		—	—	—	6.0%
.9	Excavation & Backfill	Site preparation for slab and trench for foundation wall and footing		S.F. Ground	.96	.96	
2.0 Substructure							
.1	Slab on Grade	4" reinforced concrete with vapor barrier and granular base		S.F. Slab	2.56	2.56	4.8%
.2	Special Substructures	N/A		—	—	—	
3.0 Superstructure							
.1	Columns & Beams	Steel columns included in 3.7		—	—	—	
.4	Structural Walls	N/A		—	—	—	
.5	Elevated Floors	N/A		—	—	—	6.5%
.7	Roof	Metal deck, open web steel joists, beams, interior columns		S.F. Roof	3.46	3.46	
.9	Stairs	N/A		—	—	—	
4.0 Exterior Closure							
.1	Walls	Face brick with concrete block backup	70% of wall	S.F. Wall	15.80	5.25	
.5	Exterior Wall Finishes	N/A		—	—	—	
.6	Doors	Metal and glass with transom		Each	1491	.55	15.9%
.7	Windows & Glazed Walls	Metal outward projecting	30% of wall	Each	420	2.60	
5.0 Roofing							
.1	Roof Coverings	Built-up tar and gravel with flashing		S.F. Roof	1.92	1.92	
.7	Insulation	Perlite/EPS composite		S.F. Roof	1.13	1.13	6.1%
.8	Openings & Specialties	Gravel stop and hatches		S.F. Roof	.19	.19	
6.0 Interior Construction							
.1	Partitions	Gypsum board on metal studs, toilet partition	20 S.F. Floor/L.F. Partition	S.F. Partition	3.16	1.95	
.4	Interior Doors	Wood solid core	200 S.F. Floor/Door	Each	367	1.84	
.5	Wall Finishes	90% paint, 10% ceramic tile		S.F. Surface	.82	.82	
.6	Floor Finishes	70% carpet, 15% terrazzo, 15% vinyl composition tile		S.F. Floor	6.34	6.34	27.1%
.7	Ceiling Finishes	Mineral fiber tile on concealed zee bars		S.F. Ceiling	2.62	2.62	
.9	Interior Surface/Exterior Wall	Painted gypsum board on furring	70% of wall	S.F. Wall	2.26	.75	
7.0 Conveying							
.1	Elevators	N/A		—	—	—	0.0%
.2	Special Conveyors	N/A		—	—	—	
8.0 Mechanical							
.1	Plumbing	Kitchen, toilet and service fixtures, supply and drainage	1 Fixture/500 S.F. Floor	Each	2080	4.16	
.2	Fire Protection	Wet pipe sprinkler system		S.F. Floor	1.26	1.26	
.3	Heating	Included in 8.4		—	—	—	21.7%
.4	Cooling	Multizone unit, gas heating, electric cooling		S.F. Floor	6.07	6.07	
.5	Special Systems	N/A		—	—	—	
9.0 Electrical							
.1	Service & Distribution	400 ampere service, panel board and feeders		S.F. Floor	.97	.97	
.2	Lighting & Power	Fluorescent fixtures, receptacles, switches, A.C. and misc. power		S.F. Floor	4.83	4.83	11.9%
.4	Special Electrical	Alarm systems and emergency lighting		S.F. Floor	.48	.48	
11.0 Special Construction							
.1	Specialties	N/A		—	—	—	0.0%
12.0 Site Work							
.1	Earthwork	N/A		—	—	—	
.3	Utilities	N/A		—	—	—	
.5	Roads & Parking	N/A		—	—	—	0.0%
.7	Site Improvements	N/A		—	—	—	
				Sub-Total		52.93	100%
	GENERAL CONDITIONS (Overhead & Profit)				15%	7.94	
	ARCHITECT FEES				9%	5.48	
				Total Building Cost		66.35	

Costs per square foot of floor area

Exterior Wall	S.F. Area	8000	10000	12000	15000	18000	24000	28000	35000	40000
	L.F. Perimeter	206	233	260	300	320	360	393	451	493
Face Brick with Concrete Block Back-up	Steel Frame	92.20	87.80	84.85	81.90	78.95	75.20	73.85	72.15	71.35
	R/Conc. Frame	91.80	87.40	84.50	81.50	78.55	74.80	73.45	71.75	70.95
Stone with Concrete Block Back-up	Steel Frame	87.85	84.05	81.50	78.95	76.35	73.10	71.90	70.40	69.70
	R/Conc. Frame	93.20	88.65	85.65	82.60	79.50	75.60	74.20	72.45	71.60
Limestone with Concrete Block Back-up	Steel Frame	101.95	96.60	93.05	89.45	85.65	80.85	79.15	77.00	75.95
	R/Conc. Frame	101.55	96.20	92.65	89.05	85.25	80.45	78.75	76.60	75.60
Perimeter Adj., Add or Deduct	Per 100 L.F.	11.35	9.10	7.60	6.10	5.05	3.80	3.25	2.60	2.25
Story Hgt. Adj., Add or Deduct	Per 1 Ft.	1.75	1.60	1.50	1.35	1.20	1.00	.95	.90	.85

For Basement, add $14.60 per square foot of basement area

The above costs were calculated using the basic specifications shown on the facing page. These costs should be adjusted where necessary for design alternatives and owner's requirements. Reported completed project costs, for this type of structure, range from $47.95 to $135.30 per S.F.

Common additives

Description	Unit	$ Cost	Description	Unit	$ Cost
Directory Boards, Plastic, glass covered			Flagpoles, Complete		
30" x 20"	Each	460	Aluminum, 20' high	Each	1025
36" x 48"	Each	850	40' high	Each	2375
Aluminum, 24" x 18"	Each	410	70' high	Each	6925
36" x 24"	Each	495	Fiberglass, 23' high	Each	1400
48" x 32"	Each	590	39'-5" high	Each	2425
48" x 60"	Each	1300	59' high	Each	6875
Elevators, Hydraulic passenger, 2 stops			Safe, Office type, 4 hour rating		
1500# capacity	Each	37,625	30" x 18" x 18"	Each	3475
2500# capacity	Each	38,925	62" x 33" x 20"	Each	7575
3500# capacity	Each	42,525	Smoke Detectors		
Additional stop, add	Each	2600	Ceiling type	Each	126
Emergency Lighting, 25 watt, battery operated			Duct type	Each	360
Lead battery	Each	305	Vault Front, Door & frame		
Nickel cadmium	Each	600	1 Hour test, 32" x 78"	Opening	3275
			2 Hour test, 32" door	Opening	3325
			40" door	Opening	3800
			4 Hour test, 32" door	Opening	3475
			40" door	Opening	4100
			Time lock movement; two movement	Each	1525

Important: See the Reference Section for Location Factors

Model costs calculated for a 3 story building with 12' story height and 18,000 square feet of floor area

Town Hall, 2-3 Story

			Unit	Unit Cost	Cost Per S.F.	% Of Sub-Total
1.0 Foundations						
.1	Footings & Foundations	Poured concrete; strip and spread footings and 4' foundation wall	S.F. Ground	4.08	1.36	
.4	Piles & Caissons	N/A	—	—	—	2.7%
.9	Excavation & Backfill	Site preparation for slab and trench for foundation wall and footing	S.F. Ground	.91	.30	
2.0 Substructure						
.1	Slab on Grade	4" reinforced concrete with vapor barrier and granular base	S.F. Slab	2.56	.85	1.4%
.2	Special Substructures	N/A	—	—	—	
3.0 Superstructure						
.1	Columns & Beams	Wide flange steel columns with gypsum board fireproofing	L.F. Column	42	1.27	
.4	Structural Walls	N/A	—	—	—	
.5	Elevated Floors	Open web steel joists, slab form, concrete	S.F. Floor	10.53	7.02	18.9%
.7	Roof	Metal deck, open web steel joists, beams, interior columns	S.F. Roof	4.05	1.35	
.9	Stairs	Concrete filled metal pan	Flight	5550	1.85	
4.0 Exterior Closure						
.1	Walls	Stone with concrete block backup 70% of wall	S.F. Wall	17.50	7.84	
.5	Exterior Wall Finishes	N/A	—	—	—	15.6%
.6	Doors	Metal and glass with transoms	Each	1380	.39	
.7	Windows & Glazed Walls	Metal outward projecting 10% of wall	Each	562	1.30	
5.0 Roofing						
.1	Roof Coverings	Built-up tar and gravel with flashing	S.F. Roof	2.16	.72	
.7	Insulation	Perlite/EPS composite	S.F. Roof	1.13	.38	1.8%
.8	Openings & Specialties	N/A	—	—	—	
6.0 Interior Construction						
.1	Partitions	Gypsum board on metal studs, toilet partitions 20 S.F. Floor/L.F. Partition	S.F. Partition	3.16	1.81	
.4	Interior Doors	Wood solid core 200 S.F. Floor/Door	Each	367	1.84	
.5	Wall Finishes	90% paint, 10% ceramic tile	S.F. Surface	.82	.82	23.7%
.6	Floor Finishes	70% carpet, 15% terrazzo, 15% vinyl composition tile	S.F. Floor	6.34	6.34	
.7	Ceiling Finishes	Mineral fiber tile on concealed zee bars	S.F. Ceiling	2.62	2.62	
.9	Interior Surface/Exterior Wall	Painted gypsum board on furring 70% of wall	S.F. Wall	2.26	1.01	
7.0 Conveying						
.1	Elevators	Two hydraulic elevators	Each	58,680	6.52	10.7%
.2	Special Conveyors	N/A	—	—	—	
8.0 Mechanical						
.1	Plumbing	Toilet and service fixtures, supply and drainage 1 Fixture/1385 S.F. Floor	Each	1828	1.32	
.2	Fire Protection	Sprinklers, light hazard	S.F. Floor	1.19	1.19	
.3	Heating	Included in 8.4	—	—	—	14.1%
.4	Cooling	Multizone unit, gas heating, electric cooling	S.F. Floor	6.07	6.07	
.5	Special Systems	N/A	—	—	—	
9.0 Electrical						
.1	Service & Distribution	400 ampere service, panel board and feeders	S.F. Floor	.82	.82	
.2	Lighting & Power	Fluorescent fixtures, receptacles, switches, A.C. and misc. power	S.F. Floor	5.41	5.41	11.1%
.4	Special Electrical	Alarm systems and emergency lighting	S.F. Floor	.50	.50	
11.0 Special Construction						
.1	Specialties	N/A	—	—	—	0.0%
12.0 Site Work						
.1	Earthwork	N/A	—	—	—	
.3	Utilities	N/A	—	—	—	0.0%
.5	Roads & Parking	N/A	—	—	—	
.7	Site Improvements	N/A	—	—	—	

		Sub-Total		60.90	**100%**
	GENERAL CONDITIONS (Overhead & Profit)		15%	9.14	
	ARCHITECT FEES		9%	6.31	
		Total Building Cost		**76.35**	

Costs per square foot of floor area

Exterior Wall	S.F. Area	10000	15000	20000	25000	30000	35000	40000	50000	60000
	L.F. Perimeter	410	500	600	700	700	766	833	966	1000
Brick with Concrete Block Back-up	Steel Frame	59.90	53.85	51.10	49.45	46.45	45.30	44.55	43.40	41.70
	Bearing Walls	59.35	53.10	50.25	48.60	45.50	44.35	43.60	42.45	40.70
Concrete Block	Steel Frame	49.65	45.45	43.50	42.35	40.50	39.75	39.25	38.50	37.45
	Bearing Walls	48.80	44.55	42.55	41.40	39.55	38.75	38.30	37.50	36.45
Galvanized Steel Siding	Steel Frame	49.85	45.90	44.10	43.05	41.35	40.65	40.20	39.45	38.55
Metal Sandwich Panels	Steel Frame	51.10	46.60	44.55	43.35	41.35	40.50	40.00	39.20	38.05
Perimeter Adj., Add or Deduct	Per 100 L.F.	5.85	3.95	2.95	2.35	1.95	1.70	1.45	1.20	1.00
Story Hgt. Adj., Add or Deduct	Per 1 Ft.	.85	.70	.60	.55	.50	.45	.40	.40	.35

For Basement, add $14.60 per square foot of basement area

The above costs were calculated using the basic specifications shown on the facing page. These costs should be adjusted where necessary for design alternatives and owner's requirements. Reported completed project costs, for this type of structure, range from $20.35 to $81.50 per S.F.

Common additives

Description	Unit	$ Cost
Dock Leveler, 10 ton cap.		
6' x 8'	Each	3825
7' x 8'	Each	4050
Emergency Lighting, 25 watt, battery operated		
Lead battery	Each	305
Nickel cadmium	Each	600
Fence, Chain link, 6' high		
9 ga. wire	L.F.	11.45
6 ga. wire	L.F.	15.60
Gate	Each	192
Flagpoles, Complete		
Aluminum, 20' high	Each	1025
40' high	Each	2375
70' high	Each	6925
Fiberglass, 23' high	Each	1400
39'-5" high	Each	2425
59' high	Each	6875
Paving, Bituminous		
Wearing course plus base course	S.Y.	9.65
Sidewalks, Concrete 4" thick	S.F.	2.18

Description	Unit	$ Cost
Sound System		
Amplifier, 250 watts	Each	1425
Speaker, ceiling or wall	Each	118
Trumpet	Each	222
Yard Lighting, 20' aluminum pole with 400 watt high pressure sodium fixture.	Each	1855

Important: See the Reference Section for Location Factors

Model costs calculated for a 1 story building with 24' story height and 30,000 square feet of floor area

Warehouse

					Unit	Unit Cost	Cost Per S.F.	% Of Sub-Total
1.0 Foundations								
	.1	Footings & Foundations	Poured concrete; strip and spread footings and 4' foundation wall		S.F. Ground	1.77	1.77	
	.4	Piles & Caissons	N/A		–	–	–	8.3%
	.9	Excavation & Backfill	Site preparation for slab and trench for foundation wall and footing		S.F. Ground	.91	.91	
2.0 Substructure								
	.1	Slab on Grade	5" reinforced concrete with vapor barrier and granular base		S.F. Slab	5.04	5.04	15.7%
	.2	Special Substructures	N/A		–	–	–	
3.0 Superstructure								
	.1	Columns & Beams	Steel columns included in 3.5 and 3.7		–	–	–	
	.4	Structural Walls	N/A		–	–	–	
	.5	Elevated Floors	Open web steel joists, slab form, concrete beams, columns, (above office)	10% of area	S.F. Floor	8.01	1.04	14.2%
	.7	Roof	Metal deck, open web steel joists, beams, columns		S.F. Roof	3.21	3.21	
	.9	Stairs	Steel gate with rails		Flight	4435	.30	
4.0 Exterior Closure								
	.1	Walls	Concrete block	95% of wall	S.F. Wall	6.60	3.51	
	.5	Exterior Wall Finishes	N/A		–	–	–	12.6%
	.6	Doors	Steel overhead, hollow metal	5% of wall	Each	1442	.53	
	.7	Windows & Glazed Walls	N/A		–	–	–	
5.0 Roofing								
	.1	Roof Coverings	Built-up tar and gravel with flashing		S.F. Roof	1.59	1.59	
	.7	Insulation	Perlite/EPS composite		S.F. Roof	1.13	1.13	9.3%
	.8	Openings & Specialties	Gravel stop, hatches and skylight		S.F. Roof	.28	.28	
6.0 Interior Construction								
	.1	Partitions	Concrete block (office and washrooms)	100 S.F. Floor/L.F. Partition	S.F. Partition	4.13	.33	
	.4	Interior Doors	Single leaf hollow metal	5000 S.F. Floor/Door	Each	484	.10	
	.5	Wall Finishes	Paint		S.F. Surface	.75	.12	
	.6	Floor Finishes	90% hardener, 10% vinyl composition tile		S.F. Floor	.87	.87	7.7%
	.7	Ceiling Finishes	Suspended mineral tile on zee channels in office area	10% of area	S.F. Ceiling	2.62	.26	
	.9	Interior Surface/Exterior Wall	Paint	95% of wall	S.F. Wall	1.53	.81	
7.0 Conveying								
	.1	Elevators	N/A		–	–	–	0.0%
	.2	Special Conveyors	N/A		–	–	–	
8.0 Mechanical								
	.1	Plumbing	Toilet and service fixtures, supply and drainage	1 Fixture/2500 S.F. Floor	Each	2975	1.19	
	.2	Fire Protection	Sprinklers, ordinary hazard		S.F. Floor	1.41	1.41	
	.3	Heating	Oil fired hot water, unit heaters	90% of area	S.F. Floor	2.75	2.75	18.6%
	.4	Cooling	Single zone unit gas, heating, electric cooling	10% of area	S.F. Floor	.60	.60	
	.5	Special Systems	N/A		–	–	–	
9.0 Electrical								
	.1	Service & Distribution	200 ampere service, panel board and feeders		S.F. Floor	.27	.27	
	.2	Lighting & Power	Fluorescent fixtures, receptacles, switches, A.C. and misc. power		S.F. Floor	2.30	2.30	8.7%
	.4	Special Electrical	Alarm systems		S.F. Floor	.22	.22	
11.0 Special Construction								
	.1	Specialties	Dock boards, dock levelers		S.F. Floor	1.59	1.59	4.9%
12.0 Site Work								
	.1	Earthwork	N/A		–	–	–	
	.3	Utilities	N/A		–	–	–	
	.5	Roads & Parking	N/A		–	–	–	0.0%
	.7	Site Improvements	N/A		–	–	–	

Sub-Total		32.13	**100%**
GENERAL CONDITIONS (Overhead & Profit)	15%	4.82	
ARCHITECT FEES	7%	2.60	
Total Building Cost		**39.55**	

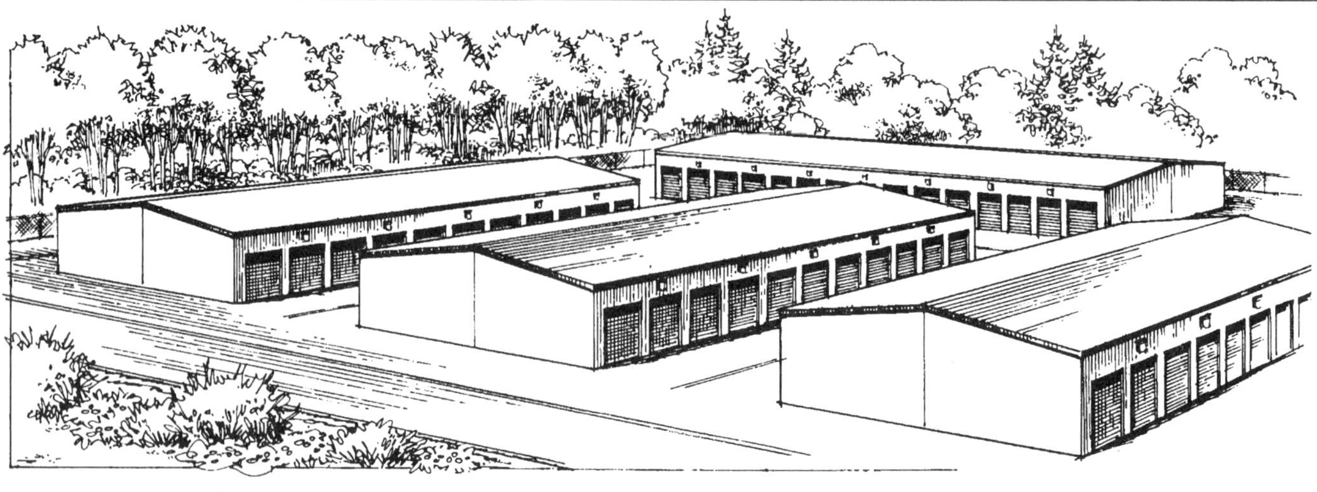

Costs per square foot of floor area

Exterior Wall	S.F. Area	2000	3000	5000	8000	12000	20000	30000	50000	100000
	L.F. Perimeter	**179**	**220**	**284**	**350**	**438**	**566**	**693**	**894**	**1266**
Concrete Block	Steel Frame	153.15	114.05	81.95	63.10	52.60	**43.65**	38.95	34.90	31.65
	R/Conc. Frame	74.30	68.20	62.55	58.50	56.25	53.95	52.55	51.20	49.85
Metal Sandwich Panel	Steel Frame	69.50	63.05	57.05	52.85	50.45	48.00	46.50	45.05	43.65
Tilt-up Concrete Panel	R/Conc. Frame	77.35	71.20	65.40	61.30	59.00	56.65	55.20	53.80	52.45
Precast Concrete Panel	Steel Frame	70.20	63.60	57.40	52.95	50.45	47.85	46.30	44.75	43.30
	Bearing Wall	82.35	75.15	68.40	63.55	60.80	57.95	56.20	54.45	52.80
Perimeter Adj., Add or Deduct	Per 100 L.F.	11.20	7.45	4.50	2.80	1.85	1.10	.75	.45	.20
Story Hgt. Adj., Add or Deduct	Per 1 Ft.	.60	.50	.40	.30	.25	.20	.15	.10	.10
Basement—Not Applicable										

The above costs were calculated using the basic specifications shown on the facing page. These costs should be adjusted where necessary for design alternatives and owner's requirements. Reported completed project costs, for this type of structure, range from $16.95 to $71.90 per S.F.

Common additives

Description	Unit	$ Cost
Dock Leveler, 10 ton cap.		
6' x 8'	Each	3825
7' x 8'	Each	4050
Emergency Lighting, 25 watt, battery operated		
Lead battery	Each	305
Nickel cadmium	Each	600
Fence, Chain link, 6' high		
9 ga. wire	L.F.	11.45
6 ga. wire	L.F.	15.60
Gate	Each	192
Flagpoles, Complete		
Aluminum, 20' high	Each	1025
40' high	Each	2375
70' high	Each	6925
Fiberglass, 23' high	Each	1400
39'-5" high	Each	2425
59' high	Each	6875
Paving, Bituminous		
Wearing course plus base course	S.Y.	9.65
Sidewalks, Concrete 4" thick	S.F.	2.18

Description	Unit	$ Cost
Sound System		
Amplifier, 250 watts	Each	1425
Speaker, ceiling or wall	Each	118
Trumpet	Each	222
Yard Lighting,	Each	1855
20' aluminum pole with		
400 watt high pressure		
sodium fixture		

Important: See the Reference Section for Location Factors

Model costs calculated for a 1 story building with 12' story height and 20,000 square feet of floor area

				Unit	Unit Cost	Cost Per S.F.	% Of Sub-Total
1.0 Foundations							
	.1	Footings & Foundations	Poured concrete; strip and spread footings and 4' foundation wall	S.F. Ground	.53	.53	4.2%
	.4	Piles & Caissons	N/A	—	—	—	
	.9	Excavation & Backfill	Site preparation for slab and trench for foundation wall and footing	S.F. Ground	.96	.96	
2.0 Substructure							
	.1	Slab on Grade	4" reinforced concrete with vapor barrier and granular base	S.F. Slab	4.23	4.23	11.9%
	.2	Special Substructures	N/A	—			
3.0 Superstructure							
	.1	Columns & Beams	Column fireproofing, steel columns included in 3.7	S.F. Floor	2.32	2.32	
	.4	Structural Walls	N/A	—	—	—	
	.5	Elevated Floors	N/A	—	—	—	20.5%
	.7	Roof	Metal deck, open web steel joists, beams, columns	S.F. Roof	4.96	4.96	
	.9	Stairs	N/A	—	—	—	
4.0 Exterior Closure							
	.1	Walls	Concrete block 70% of wall	S.F. Wall	7.99	1.90	
	.5	Exterior Wall Finishes	N/A	—	—	—	22.4%
	.6	Doors	Steel overhead, hollow metal 25% of wall	Each	720	5.83	
	.7	Windows & Glazed Walls	Aluminum projecting 5% of wall	Each	556	.22	
5.0 Roofing							
	.1	Roof Coverings	Built-up tar and gravel with flashing	S.F. Roof	1.61	1.61	
	.7	Insulation	Perlite/EPS composite	S.F. Roof	.67	.67	6.4%
	.8	Openings & Specialties	N/A	—	—	—	
6.0 Interior Construction							
	.1	Partitions	Concrete block, gypsum board on metal studs 10.65 S.F. Floor/L.F. Partition	S.F. Partition	6.34	4.48	
	.4	Interior Doors	Single leaf hollow metal 1835 S.F. Floor/Door	Each	57	.19	
	.5	Wall Finishes	Paint	S.F. Floor	.30	.30	14.7%
	.6	Floor Finishes	Carpet, asphalt tile, ceramic tile	S.F. Floor	.23	.23	
	.7	Ceiling Finishes	N/A	—	—	—	
	.9	Interior Surface/Exterior Wall	N/A	—	—	—	
7.0 Conveying							
	.1	Elevators	N/A	—	—	—	0.0%
	.2	Special Conveyors	N/A	—	—	—	
8.0 Mechanical							
	.1	Plumbing	Toilet and service fixtures, supply and drainage 1 Fixture/5000 S.F. Floor	Each	4050	.81	
	.2	Fire Protection	Wet pipe sprinkler system	S.F. Floor	1.66	1.66	
	.3	Heating	Oil fired hot water, unit heaters	S.F. Floor	9319	.47	8.3%
	.4	Cooling	N/A	—	—	—	
	.5	Special Systems	N/A	—	—	—	
9.0 Electrical							
	.1	Service & Distribution	200 ampere service, panel board and feeders	S.F. Floor	.82	.82	
	.2	Lighting & Power	Fluorescent fixtures, receptacles, switches and misc. power	S.F. Floor	2.54	2.54	10.1%
	.4	Special Electrical	Alarm systems	S.F. Floor	.21	.21	
11.0 Special Construction							
	.1	Specialties	Bathroom access, appliances, cabinets	S.F. Floor	.52	.52	1.5%
12.0 Site Work							
	.1	Earthwork	N/A	—	—	—	
	.3	Utilities	N/A	—	—	—	0.0%
	.5	Roads & Parking	N/A	—	—	—	
	.7	Site Improvements	N/A	—	—	—	
				Sub-Total		35.46	**100%**
		GENERAL CONDITIONS (Overhead & Profit)			15%	5.32	
		ARCHITECT FEES			7%	2.87	
				Total Building Cost		43.65	

Assemblies Section

Table of Contents

Table of Contents

How to Use the Assemblies Cost Tables

The following is a detailed explanation of a sample Assemblies Cost Table. Most Assembly Tables are separated into three parts: 1) an illustration of the system to be estimated; 2) the components and related costs of a typical system; and 3) the costs for similar systems with dimensional and/or size variations. For costs of the components that comprise these systems or "assemblies" refer to the Unit Price Section. Next to each bold number below is the item being described with the appropriate component of the sample entry following in parenthesis. In most cases, if the work is to be subcontracted, the general contractor will need to add an additional markup (R.S. Means suggests using 10%) to the "Total" figures.

1 System/Line Numbers (A6.1-510-2400)

Each Assemblies Cost Line has been assigned a unique identification number based on the UniFormat classification system.

UniFormat Division

6.1 510 2400

Means Subdivision
Means Major Classification
Means Individual Line Number

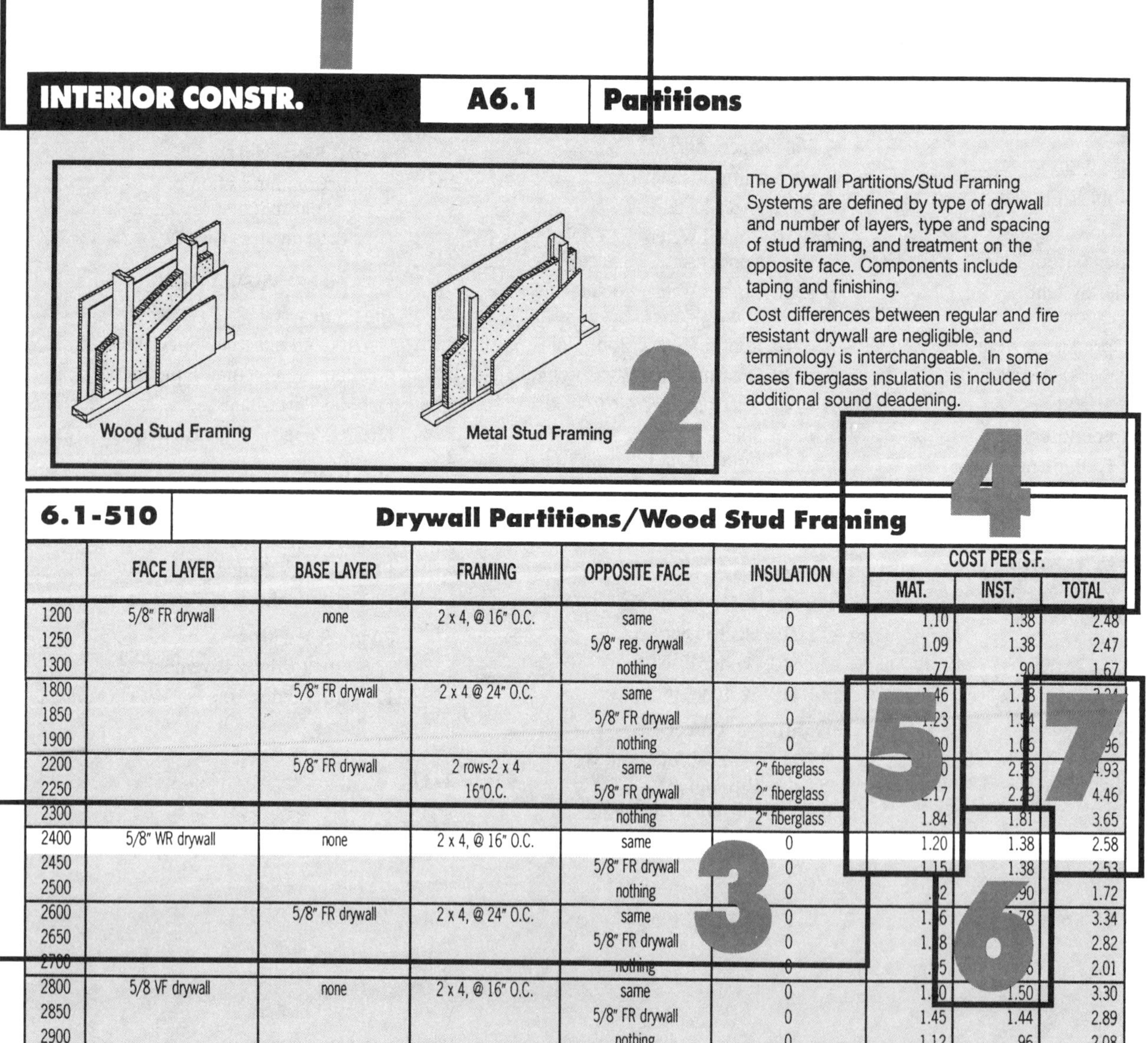

6.1-510	Drywall Partitions/Wood Stud Framing					COST PER S.F.		
	FACE LAYER	BASE LAYER	FRAMING	OPPOSITE FACE	INSULATION	MAT.	INST.	TOTAL
1200	5/8" FR drywall	none	2 x 4, @ 16" O.C.	same	0	1.10	1.38	2.48
1250				5/8" reg. drywall	0	1.09	1.38	2.47
1300				nothing	0	.77	.90	1.67
1800		5/8" FR drywall	2 x 4 @ 24" O.C.	same	0	1.46	1.78	3.24
1850				5/8" FR drywall	0	1.23	1.54	2.77
1900				nothing	0	.90	1.06	1.96
2200		5/8" FR drywall	2 rows-2 x 4	same	2" fiberglass	2.40	2.53	4.93
2250			16"O.C.	5/8" FR drywall	2" fiberglass	2.17	2.29	4.46
2300				nothing	2" fiberglass	1.84	1.81	3.65
2400	5/8" WR drywall	none	2 x 4, @ 16" O.C.	same	0	1.20	1.38	2.58
2450				5/8" FR drywall	0	1.15	1.38	2.53
2500				nothing	0	.82	.90	1.72
2600		5/8" FR drywall	2 x 4, @ 24" O.C.	same	0	1.56	1.78	3.34
2650				5/8" FR drywall	0	1.18	1.64	2.82
2700				nothing	0	.85	1.16	2.01
2800	5/8 VF drywall	none	2 x 4, @ 16" O.C.	same	0	1.80	1.50	3.30
2850				5/8" FR drywall	0	1.45	1.44	2.89
2900				nothing	0	1.12	.96	2.08
3000	5/8 VF drywall	5/8" FR drywall	2 x 4 , 24" O.C.	same	0	2.16	1.90	4.06
						1.58	1.60	

Illustration

At the top of most assembly pages is an illustration, a brief description, and the design criteria used to develop the cost.

System Description

The components of a typical system are listed in the description to show what has been included in the development of the total system price. The rest of the table contains prices for other similar systems with dimensional and/or size variations.

Unit of Measure for Each System (S.F.)

Costs shown in the three right hand columns have been adjusted by the component quantity and unit of measure for the entire system. In this example, "S.F. Cost" is the unit of measure for this system or "assembly."

Materials (1.20)

This column contains the Materials Cost of each component. These cost figures are bare costs plus 10% for profit.

Installation (1.38)

Installation includes labor and equipment plus the installing contractor's overhead and profit. Equipment costs are the bare rental costs plus 10% for profit. The labor overhead and profit is defined on the inside back cover of this book.

Total (2.58)

The figure in this column is the sum of the material and installation costs.

Material Cost	+	Installation Cost	=	Total
$1.20	+	$1.38	=	$2.58

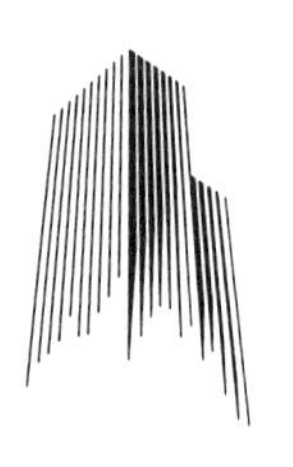

Division 1
Foundations

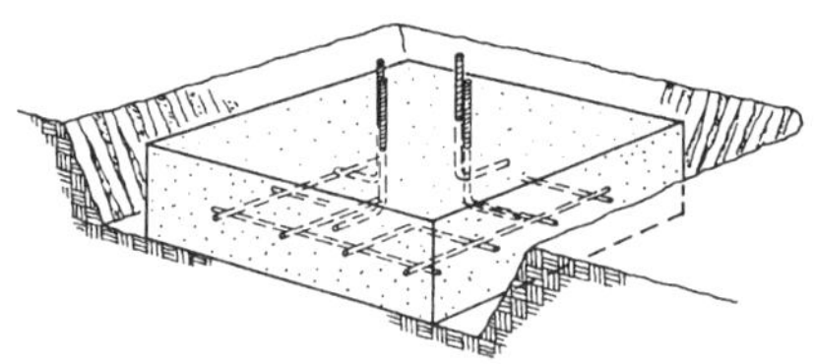

The Spread Footing System includes: excavation; backfill; forms (four uses); all reinforcement; 3,000 p.s.i. concrete (chute placed); and screed finish.

Footing systems are priced per individual unit. The Expanded System Listing at the bottom shows footings that range from 3′ square x 12″ deep, to 18′ square x 52″ deep. It is assumed that excavation is done by a truck mounted hydraulic excavator with an operator and oiler.

Backfill is with a dozer, and compaction by air tamp. The excavation and backfill equipment is assumed to operate at 30 C.Y. per hour.

Please see the reference section for further design and cost information.

1.1-120	Spread Footings	COST EACH		
		MAT.	INST.	TOTAL
7090	Spread footings, 3000 psi concrete, chute delivered			
7100	Load 25K, soil capacity 3 KSF, 3′-0″ sq. x 12″ deep	36	50	86
7150	Load 50K, soil capacity 3 KSF, 4′-6″ sq. x 12″ deep	74.50	87	161.50
7200	Load 50K, soil capacity 6 KSF, 3′-0″ sq. x 12″ deep	36	50	86
7250	Load 75K, soil capacity 3 KSF, 5′-6″ sq. x 13″ deep	117	124	241
7300	Load 75K, soil capacity 6 KSF, 4′-0″ sq. x 12″ deep	61	74.50	135.50
7350	Load 100K, soil capacity 3 KSF, 6′-0″ sq. x 14″ deep	148	149	297
7410	Load 100K, soil capacity 6 KSF, 4′-6″ sq. x 15″ deep	91.50	102	193.50
7450	Load 125K, soil capacity 3 KSF, 7′-0″ sq. x 17″ deep	233	213	446
7500	Load 125K, soil capacity 6 KSF, 5′-0″ sq. x 16″ deep	118	123	241
7550	Load 150K, soil capacity 3 KSF 7′-6″ sq. x 18″ deep	280	249	529
7610	Load 150K, soil capacity 6 KSF, 5′-6″ sq. x 18″ deep	156	155	311
7650	Load 200K, soil capacity 3 KSF, 8′-6″ sq. x 20″ deep	395	330	725
7700	Load 200K, soil capacity 6 KSF, 6′-0″ sq. x 20″ deep	203	192	395
7750	Load 300K, soil capacity 3 KSF, 10′-6″ sq. x 25″ deep	725	540	1,265
7810	Load 300K, soil capacity 6 KSF, 7′-6″ sq. x 25″ deep	380	325	705
7850	Load 400K, soil capacity 3 KSF, 12′-6″ sq. x 28″ deep	1,150	810	1,960
7900	Load 400K, soil capacity 6 KSF, 8′-6″ sq. x 27″ deep	530	420	950
8010	Load 500K, soil capacity 6 KSF, 9′-6″ sq. x 30″ deep	725	550	1,275
8100	Load 600K, soil capacity 6 KSF, 10′-6″ sq. x 33″ deep	975	710	1,685
8200	Load 700K, soil capacity 6 KSF, 11′-6″ sq. x 36″ deep	1,250	870	2,120
8300	Load 800K, soil capacity 6 KSF, 12′-0″ sq. x 37″ deep	1,400	955	2,355
8400	Load 900K, soil capacity 6 KSF, 13′-0″ sq. x 39″ deep	1,725	1,150	2,875
8500	Load 1000K, soil capacity 6 KSF, 13′-6″ sq. x 41″ deep	1,950	1,275	3,225

1.1-140	Strip Footings	COST PER L.F.		
		MAT.	INST.	TOTAL
2100	Strip footing, load 2.6KLF, soil capacity 3KSF, 16″wide x 8″deep plain	4.30	6.05	10.35
2300	Load 3.9 KLF, soil capacity, 3 KSF, 24″wide x 8″deep, plain	5.40	6.75	12.15
2500	Load 5.1KLF, soil capacity 3 KSF, 24″wide x 12″deep, reinf.	8.55	10.15	18.70
2700	Load 11.1KLF, soil capacity 6 KSF, 24″wide x 12″deep, reinf.	8.55	10.15	18.70
2900	Load 6.8 KLF, soil capacity 3 KSF, 32″wide x 12″deep, reinf.	10.45	11.10	21.55
3100	Load 14.8 KLF, soil capacity 6 KSF, 32″wide x 12″deep, reinf.	10.45	11.10	21.55
3300	Load 9.3 KLF, soil capacity 3 KSF, 40″wide x 12″deep, reinf.	12.35	12.10	24.45
3500	Load 18.4 KLF, soil capacity 6 KSF, 40″wide x 12″deep, reinf.	12.45	12.15	24.60
4500	Load 10KLF, soil capacity 3 KSF, 48″wide x 16″deep, reinf.	17.55	15.15	32.70
4700	Load 22KLF, soil capacity 6 KSF, 48″wide, 16″deep, reinf.	17.95	15.50	33.45
5700	Load 15KLF, soil capacity 3 KSF, 72″wide x 20″deep, reinf.	30.50	21.50	52
5900	Load 33KLF, soil capacity 6 KSF, 72″wide x 20″deep, reinf.	32.50	23.50	56

1.1-210 — Walls, Cast in Place

	WALL HEIGHT (FT.)	PLACING METHOD	CONCRETE (C.Y./L.F.)	REINFORCING (LBS./L.F.)	WALL THICKNESS (IN.)	COST PER L.F.		
						MAT.	INST.	TOTAL
1500	4'	direct chute	.074	3.3	6	9.35	18.90	28.25
1520			.099	4.8	8	11.35	19.55	30.90
1540			.123	6.0	10	13.20	19.90	33.10
1561			.148	7.2	12	15.15	20.50	35.65
1580			.173	8.1	14	17	20.50	37.50
1600			.197	9.44	16	18.90	21	39.90
3000	6'	direct chute	.111	4.95	6	14	28	42
3020			.149	7.20	8	17.05	29	46.05
3040			.184	9.00	10	19.80	29.50	49.30
3061			.222	10.8	12	22.50	30.50	53
5000	8'	direct chute	.148	6.6	6	18.65	37.50	56.15
5020			.199	9.6	8	23	39	62
5040			.250	12	10	26.50	40	66.50
5061			.296	14.39	12	30.50	41	71.50
6020	10'	direct chute	.248	12	8	28.50	49	77.50
6040			.307	14.99	10	33	50	83
6061			.370	17.99	12	38	51.50	89.50
7220	12'	pumped	.298	14.39	8	34	61.50	95.50
7240			.369	17.99	10	39.50	63	102.50
7262			.444	21.59	12	45.50	65	110.50
9220	16'	pumped	.397	19.19	8	45.50	82	127.50
9240			.492	23.99	10	53	84	137
9260			.593	28.79	12	60.50	87.50	148

1.1-292 — Foundation Dampproofing

	Foundation Dampproofing	COST PER L.F.		
		MAT.	INST.	TOTAL
1000	Foundation dampproofing, bituminous, 1 coat, 4' high	.28	1.93	2.21
1400	8' high	.56	3.86	4.42
1800	12' high	.84	5.95	6.79
2000	2 coats, 4' high	.44	2.37	2.81
2400	8' high	.88	4.74	5.62
2800	12' high	1.32	7.30	8.62
3000	Asphalt with fibers, 1/16" thick, 4' high	.68	2.37	3.05
3400	8' high	1.36	4.74	6.10
3800	12' high	2.04	7.30	9.34
4000	1/8" thick, 4' high	1.16	2.85	4.01
4400	8' high	2.32	5.70	8.02
4800	12' high	3.48	8.70	12.18
5000	Asphalt coated board and mastic, 1/4" thick, 4' high	2.60	2.57	5.17
5400	8' high	5.20	5.15	10.35
5800	12' high	7.80	7.90	15.70
6000	1/2" thick, 4' high	3.90	3.59	7.49
6400	8' high	7.80	7.20	15
6800	12' high	11.70	10.95	22.65
7000	Cementitious coating, on walls, 1/8" thick coating, 4' high	6.30	3.36	9.66
7400	8' high	12.65	6.70	19.35
7800	12' high	18.95	10.10	29.05
8000	Cementitious/metallic slurry, 2 coat, 1/4" thick, 2' high	37.50	89.50	127
8400	4' high	75	179	254
8800	6' high	113	269	382

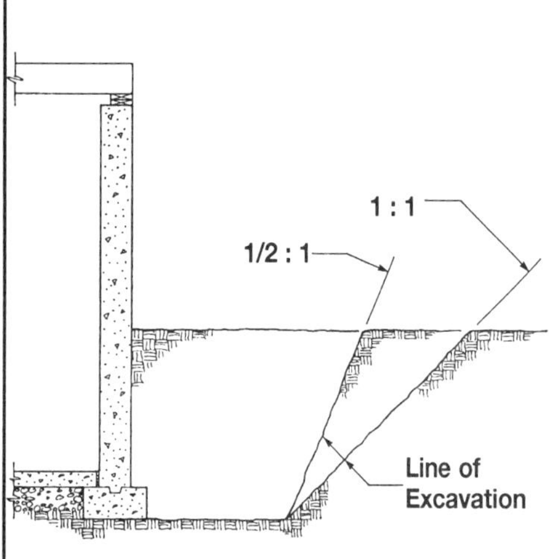

Pricing Assumptions: Two-thirds of excavation is by 2-1/2 C.Y. wheel mounted front end loader and one-third by 1-1/2 C.Y. hydraulic excavator.

Two-mile round trip haul by 12 C.Y. tandem trucks is included for excavation wasted and storage of suitable fill from excavated soil. For excavation in clay, all is wasted and the cost of suitable backfill with two-mile haul is included.

Sand and gravel assumes 15% swell and compaction; common earth assumes 25% swell and 15% compaction; clay assumes 40% swell and 15% compaction (non-clay).

In general, the following items are accounted for in the costs in the table below.

1. Excavation for building or other structure to depth and extent indicated.
2. Backfill compacted in place.
3. Haul of excavated waste.
4. Replacement of unsuitable material with bank run gravel.

Note: Additional excavation and fill beyond this line of general excavation for the building (as required for isolated spread footings, strip footings, etc.) are included in the cost of the appropriate component systems.

1.9-100	Building Excavation & Backfill	COST PER S.F.		
		MAT.	INST.	TOTAL
2220	Excav & fill, 1000 S.F. 4' sand, gravel, or common earth, on site storage		1.38	1.38
2240	Off site storage		2.99	2.99
2260	Clay excavation, bank run gravel borrow for backfill	.91	2.67	3.58
2280	8' deep, sand, gravel, or common earth, on site storage		3.19	3.19
2300	Off site storage		7.60	7.60
2320	Clay excavation, bank run gravel borrow for backfill	2.19	6.10	8.29
2340	16' deep, sand, gravel, or common earth, on site storage		8.60	8.60
2350	Off site storage		18.50	18.50
2360	Clay excavation, bank run gravel borrow for backfill	6.05	15.60	21.65
3380	4000 S.F., 4' deep, sand, gravel, or common earth, on site storage		1.08	1.08
3400	Off site storage		1.81	1.81
3420	Clay excavation, bank run gravel borrow for backfill	.42	1.68	2.10
3440	8' deep, sand, gravel, or common earth, on site storage		2.34	2.34
3460	Off site storage		4.30	4.30
3480	Clay excavation, bank run gravel borrow for backfill	1	3.68	4.68
3500	16' deep, sand, gravel, or common earth, on site storage		5.55	5.55
3520	Off site storage		11.65	11.65
3540	Clay, excavation, bank run gravel borrow for backfill	2.67	8.75	11.42
4560	10,000 S.F., 4' deep, sand gravel, or common earth, on site storage		.96	.96
4580	Off site storage		1.41	1.41
4600	Clay excavation, bank run gravel borrow for backfill	.26	1.35	1.61
4620	8' deep, sand, gravel, or common earth, on site storage		2.05	2.05
4640	Off site storage		3.22	3.22
4660	Clay excavation, bank run gravel borrow for backfill	.60	2.89	3.49
4680	16' deep, sand, gravel, or common earth, on site storage		4.62	4.62
4700	Off site storage		8.20	8.20
4720	Clay excavation, bank run gravel borrow for backfill	1.60	6.55	8.15
5740	30,000 S.F., 4' deep, sand, gravel, or common earth, on site storage		.91	.91
5760	Off site storage		1.16	1.16
5780	Clay excavation, bank run gravel borrow for backfill	.15	1.12	1.27
5860	16' deep, sand, gravel, or common earth, on site storage		4.04	4.04
5880	Off site storage		6	6
5900	Clay excavation, bank run gravel borrow for backfill	.89	5.10	5.99
6910	100,000 S.F., 4' deep, sand, gravel, or common earth, on site storage		.87	.87
6940	8' deep, sand, gravel, or common earth, on site storage		1.78	1.78
6970	16' deep, sand, gravel, or common earth, on site storage		3.72	3.72
6980	Off site storage		4.75	4.75
6990	Clay excavation, bank run gravel borrow for backfill	.48	4.27	4.75

For information about Means Estimating Seminars, see yellow pages 11 and 12 in back of book

Important: See the Reference Section for critical supporting data - Reference Nos., Crews & Location Factors

Division 2
Substructures

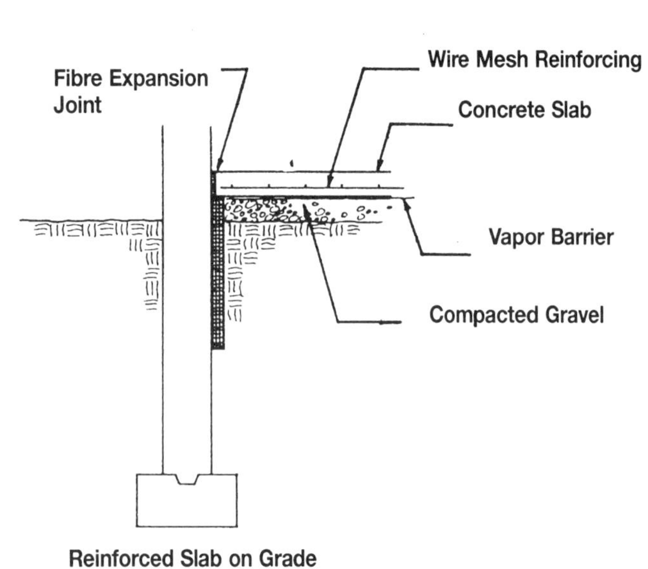

A Slab on Grade system includes fine grading; 6″ of compacted gravel; vapor barrier; 3500 p.s.i. concrete; bituminous fiber expansion joint; all necessary edge forms 4 uses; steel trowel finish; and sprayed on membrane curing compound. Wire mesh reinforcing used in all reinforced slabs.

Non-industrial slabs are for foot traffic only with negligible abrasion. Light industrial slabs are for pneumatic wheels and light abrasion. Industrial slabs are for solid rubber wheels and moderate abrasion. Heavy industrial slabs are for steel wheels and severe abrasion.

2.1-200	Plain & Reinforced	COST PER S.F.		
		MAT.	INST.	TOTAL
2220	Slab on grade, 4″ thick, non industrial, non reinforced	1.08	1.23	2.31
2240	Reinforced	1.17	1.39	2.56
2280	Reinforced	1.41	1.65	3.06
2300	Industrial, non reinforced	1.66	2.32	3.98
2320	Reinforced	1.75	2.48	4.23
3340	5″ thick, non industrial, non reinforced	1.28	1.26	2.54
3360	Reinforced	1.37	1.42	2.79
3380	Light industrial, non reinforced	1.52	1.52	3.04
3400	Reinforced	1.61	1.68	3.29
3420	Heavy industrial, non reinforced	2.11	2.68	4.79
3440	Reinforced	2.18	2.86	5.04
4460	6″ thick, non industrial, non reinforced	1.54	1.24	2.78
4480	Reinforced	1.68	1.46	3.14
4500	Light industrial, non reinforced	1.78	1.50	3.28
4520	Reinforced	2.01	1.80	3.81
4540	Heavy industrial, non reinforced	2.38	2.72	5.10
4560	Reinforced	2.52	2.94	5.46
5580	7″ thick, non industrial, non reinforced	1.75	1.28	3.03
5600	Reinforced	1.95	1.51	3.46
5620	Light industrial, non reinforced	2	1.54	3.54
5640	Reinforced	2.20	1.77	3.97
5660	Heavy industrial, non reinforced	2.59	2.68	5.27
5680	Reinforced	2.73	2.90	5.63
6700	8″ thick, non industrial, non reinforced	1.95	1.30	3.25
6720	Reinforced	2.12	1.50	3.62
6740	Light industrial, non reinforced	2.20	1.56	3.76
6760	Reinforced	2.37	1.76	4.13
6780	Heavy industrial, non reinforced	2.81	2.72	5.53
6800	Reinforced	3.03	2.93	5.96

For information about Means Estimating Seminars, see yellow pages 11 and 12 in back of book

Important: See the Reference Section for critical supporting data - Reference Nos., Crews & Location Factors

Division 3
Superstructures

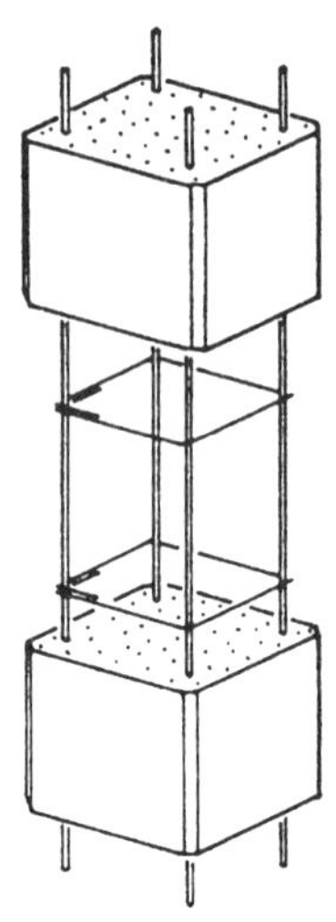

General: It is desirable for purposes of consistency and simplicity to maintain constant column sizes throughout building height. To do this, concrete strength may be varied (higher strength concrete at lower stories and lower strength concrete at upper stories), as well as varying the amount of reinforcing.

The table provides probably minimum column sizes with related costs and weight per lineal foot of story height.

3.1-114 — C.I.P. Column, Square Tied

	LOAD (KIPS)	STORY HEIGHT (FT.)	COLUMN SIZE (IN.)	COLUMN WEIGHT (P.L.F.)	CONCRETE STRENGTH (PSI)	COST PER V.L.F. MAT.	COST PER V.L.F. INST.	COST PER V.L.F. TOTAL
0640	100	10	10	96	4000	7	19.30	26.30
0680		12	10	97	4000	6.85	19.10	25.95
0720		14	12	142	4000	8.85	23.50	32.35
0840	200	10	12	140	4000	8.95	23.50	32.45
0860		12	12	142	4000	8.90	23.50	32.40
0900		14	14	196	4000	11.15	27	38.15
0920	300	10	14	192	4000	11.40	27	38.40
0960		12	14	194	4000	11.30	27	38.30
0980		14	16	253	4000	12.75	29.50	42.25
1020	400	10	16	248	4000	13.90	31	44.90
1060		12	16	251	4000	13.75	31	44.75
1080		14	16	253	4000	13.65	30.50	44.15
1200	500	10	18	315	4000	18.30	37	55.30
1250		12	20	394	4000	18.60	38.50	57.10
1300		14	20	397	4000	18.50	38.50	57
1350	600	10	20	388	4000	21.50	42.50	64
1400		12	20	394	4000	21.50	42	63.50
1600		14	20	397	4000	21.50	42	63.50
3400	900	10	24	560	4000	31	55.50	86.50
3800		12	24	567	4000	30.50	55	85.50
4000		14	24	571	4000	30	54.50	84.50
7300	300	10	14	192	6000	11.30	27	38.30
7500		12	14	194	6000	11.20	27	38.20
7600		14	14	196	6000	11.15	27	38.15
8000	500	10	16	248	6000	13.90	31	44.90
8050		12	16	251	6000	13.75	31	44.75
8100		14	16	253	6000	13.65	30.50	44.15
8200	600	10	18	315	6000	16.70	35.50	52.20
8300		12	18	319	6000	16.55	35.50	52.05
8400		14	18	321	6000	16.40	35	51.40
8800	800	10	20	388	6000	18.80	39	57.80
8900		12	20	394	6000	18.60	38.50	57.10
9000		14	20	397	6000	18.50	38.50	57

Important: See the Reference Section for critical supporting data - Reference Nos., Crews & Location Factors

3.1-114 — C.I.P. Column, Square Tied

	LOAD (KIPS)	STORY HEIGHT (FT.)	COLUMN SIZE (IN.)	COLUMN WEIGHT (P.L.F.)	CONCRETE STRENGTH (PSI)	COST PER V.L.F.		
						MAT.	INST.	TOTAL
9100	900	10	20	388	6000	27	49	76
9300		12	20	394	6000	26.50	48	74.50
9600		14	20	397	6000	26	48	74

3.1-114 — C.I.P. Column, Square Tied-Minimum Reinforcing

	LOAD (KIPS)	STORY HEIGHT (FT.)	COLUMN SIZE (IN.)	COLUMN WEIGHT (P.L.F.)	CONCRETE STRENGTH (PSI)	COST PER V.L.F.		
						MAT.	INST.	TOTAL
9912	150	10-14	12	135	4000	8.70	23	31.70
9918	300	10-14	16	240	4000	12.65	29.50	42.15
9924	500	10-14	20	375	4000	18.30	38.50	56.80
9930	700	10-14	24	540	4000	26	49.50	75.50
9936	1000	10-14	28	740	4000	32.50	58.50	91
9942	1400	10-14	32	965	4000	43	68	111
9948	1800	10-14	36	1220	4000	52	70.50	130.50

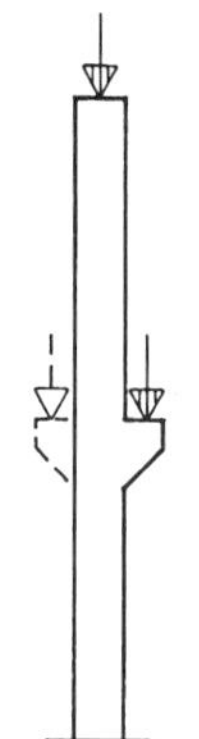

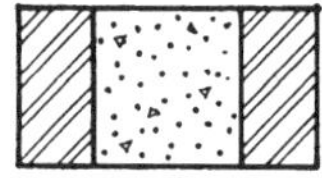

Concentric Load

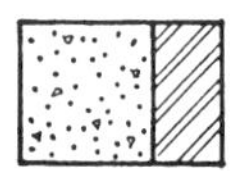

Eccentric Load

General: Data presented here is for plant produced members transported 50 miles to 100 miles to the site and erected.

Design and pricing assumptions:

Normal wt. concrete, f'c = 5 KSI

Main reinforcement, fy = 60 KSI
Ties, fy = 40 KSI

Minimum design eccentricity, 0.1t.

Concrete encased structural steel haunches are assumed where practical; otherwise galvanized rebar haunches are assumed.

Base plates are integral with columns.

Foundation anchor bolts, nuts and washers are included in price.

3.1-120   Tied, Concentric Loaded Precast Concrete Columns

	LOAD (KIPS)	STORY HEIGHT (FT.)	COLUMN SIZE (IN.)	COLUMN WEIGHT (P.L.F.)	LOAD LEVELS	COST PER V.L.F.		
						MAT.	INST.	TOTAL
0560	100	10	12x12	164	2	34.50	5.20	39.70
0570		12	12x12	162	2	34	4.34	38.34
0580		14	12x12	161	2	33	4.34	37.34
0590	150	10	12x12	166	3	33.50	4.34	37.84
0600		12	12x12	169	3	31.50	3.91	35.41
0610		14	12x12	162	3	32.50	3.91	36.41
0620	200	10	12x12	168	4	34.50	4.77	39.27
0630		12	12x12	170	4	35	4.34	39.34
0640		14	14x14	220	4	37.50	4.34	41.84

3.1-120   Tied, Eccentric Loaded Precast Concrete Columns

	LOAD (KIPS)	STORY HEIGHT (FT.)	COLUMN SIZE (IN.)	COLUMN WEIGHT (P.L.F.)	LOAD LEVELS	COST PER V.L.F.		
						MAT.	INST.	TOTAL
1130	100	10	12x12	161	2	31.50	5.20	36.70
1140		12	12x12	159	2	33.50	4.34	37.84
1150		14	12x12	159	2	32.50	4.34	36.84
1390	600	10	18x18	385	4	62	4.77	66.77
1400		12	18x18	380	4	60.50	4.34	64.84
1410		14	18x18	375	4	60	4.34	64.34
1480	800	10	20x20	490	4	79.50	4.77	84.27
1490		12	20x20	480	4	78.50	4.34	82.84
1500		14	20x20	475	4	78.50	4.34	82.84

Important: See the Reference Section for critical supporting data - Reference Nos., Crews & Location Factors

(A) Wide Flange

(B) Pipe

(C) Pipe, Concrete Filled

(D) Square Tube

(E) Square Tube Concrete Filled

(F) Rectangular Tube

(G) Rectangular Tube, Concrete Filled

General: The following pages provide data for seven types of steel columns: wide flange, round pipe, round pipe concrete filled, square tube, square tube concrete filled, rectangular tube and rectangular tube concrete filled.

Design Assumptions: Loads are concentric; wide flange and round pipe bearing capacity is for 36 KSI steel. Square and rectangular tubing bearing capacity is for 46 KSI steel.

The effective length factor K=1.1 is used for determining column values in the tables. K=1.1 is within a frequently used range for pinned connections with cross bracing.

How To Use Tables:
a. Steel columns usually extend through two or more stories to minimize splices. Determine floors with splices.
b. Enter Table No. below with load to column at the splice. Use the unsupported height.
c. Determine the column type desired by price or design.

Cost:
a. Multiply number of columns at the desired level by the total height of the column by the cost/VLF.
b. Repeat the above for all tiers.

Please see the reference section for further design and cost information.

3.1-130 — Steel Columns

	LOAD (KIPS)	UNSUPPORTED HEIGHT (FT.)	WEIGHT (P.L.F.)	SIZE (IN.)	TYPE	COST PER V.L.F.		
						MAT.	INST.	TOTAL
1000	25	10	13	4	A	10.20	4.84	15.04
1020			7.58	3	B	9.25	4.84	14.09
1040			15	3-1/2	C	11.25	4.84	16.09
1120			20	4x3	G	12.50	4.84	17.34
1200		16	16	5	A	11.65	3.63	15.28
1220			10.79	4	B	12.20	3.63	15.83
1240			36	5-1/2	C	16.90	3.63	20.53
1320			64	8x6	G	25	3.63	28.63
1600	50	10	16	5	A	12.60	4.84	17.44
1620			14.62	5	B	17.85	4.84	22.69
1640			24	4-1/2	C	13.40	4.84	18.24
1720			28	6x3	G	16.60	4.84	21.44
1800		16	24	8	A	17.45	3.63	21.08
1840			36	5-1/2	C	16.90	3.63	20.53
1920			64	8x6	G	25	3.63	28.63
2000	50	20	28	8	A	19.30	3.63	22.93
2040			49	6-5/8	C	21	3.63	24.63
2120			64	8x6	G	24	3.63	27.63
2200	75	10	20	6	A	15.75	4.84	20.59
2240			36	4-1/2	C	34	4.84	38.84
2320			35	6x4	G	18.70	4.84	23.54
2400		16	31	8	A	22.50	3.63	26.13
2440			49	6-5/8	C	22	3.63	25.63
2520			64	8x6	G	25	3.63	28.63
2600		20	31	8	A	21.50	3.63	25.13
2640			81	8-5/8	C	31.50	3.63	35.13
2720			64	8x6	G	24	3.63	27.63
2800	100	10	24	8	A	18.90	4.84	23.74
2840			35	4-1/2	C	34	4.84	38.84
2920			46	8x4	G	23	4.84	27.84
3000		16	31	8	A	22.50	3.63	26.13
3040			56	6-5/8	C	33	3.63	36.63
3120			64	8x6	G	25	3.63	28.63

	LOAD (KIPS)	UNSUPPORTED HEIGHT (FT.)	WEIGHT (P.L.F.)	SIZE (IN.)	TYPE	COST PER V.L.F.		
						MAT.	INST.	TOTAL
3200	100	20	40	8	A	27.50	3.63	31.13
3240			81	8-5/8	C	31.50	3.63	35.13
3320			70	8x6	G	34	3.63	37.63
3400	125	10	31	8	A	24.50	4.84	29.34
3440			81	8	C	36	4.84	40.84
3520			64	8x6	G	27	4.84	31.84
3600		16	40	8	A	29	3.63	32.63
3640			81	8	C	33	3.63	36.63
3720			64	8x6	G	25	3.63	28.63
3800		20	48	8	A	33	3.63	36.63
3840			81	8	C	31.50	3.63	35.13
3920			60	8x6	G	34	3.63	37.63
4000	150	10	35	8	A	27.50	4.84	32.34
4040			81	8-5/8	C	36	4.84	40.84
4120			64	8x6	G	27	4.84	31.84
4200		16	45	10	A	33	3.63	36.63
4240			81	8-5/8	C	33	3.63	36.63
4320			70	8x6	G	36	3.63	39.63
4400		20	49	10	A	34	3.63	37.63
4440			123	10-3/4	C	45	3.63	48.63
4520			86	10x6	G	33.50	3.63	37.13
4600	200	10	45	10	A	35.50	4.84	40.34
4640			81	8-5/8	C	36	4.84	40.84
4720			70	8x6	G	39	4.84	43.84
4800		16	49	10	A	35.50	3.63	39.13
4840			123	10-3/4	C	47	3.63	50.63
4920			85	10x6	G	42	3.63	45.63
5200	300	10	61	14	A	48	4.84	52.84
5240			169	12-3/4	C	62.50	4.84	67.34
5320			86	10x6	G	58	4.84	62.84
5400		16	72	12	A	52.50	3.63	56.13
5440			169	12-3/4	C	58	3.63	61.63
5600		20	79	12	A	54.50	3.63	58.13
5640			169	12-3/4	C	55	3.63	58.63
5800	400	10	79	12	A	62	4.84	66.84
5840			178	12-3/4	C	82	4.84	86.84
6000		16	87	12	A	63.50	3.63	67.13
6040			178	12-3/4	C	76	3.63	79.63
6400	500	10	99	14	A	78	4.84	82.84
6600		16	109	14	A	79.50	3.63	83.13
6800		20	120	12	A	82.50	3.63	86.13
7000	600	10	120	12	A	94.50	4.84	99.34
7200		16	132	14	A	96	3.63	99.63
7400		20	132	14	A	91	3.63	94.63
7600	700	10	136	12	A	107	4.84	111.84
7800		16	145	14	A	106	3.63	109.63
8000		20	145	14	A	100	3.63	103.63
8200	800	10	145	14	A	114	4.84	118.84
8300		16	159	14	A	116	3.63	119.63
8400		20	176	14	A	121	3.63	124.63
8800	900	10	159	14	A	125	4.84	129.84
8900		16	176	14	A	128	3.63	131.63
9000		20	193	14	A	133	3.63	136.63

Important: See the Reference Section for critical supporting data - Reference Nos., Crews & Location Factors

| 3.1-130 | Steel Columns |

	LOAD (KIPS)	UNSUPPORTED HEIGHT (FT.)	WEIGHT (P.L.F.)	SIZE (IN.)	TYPE	COST PER V.L.F.		
						MAT.	INST.	TOTAL
9100	1000	10	176	14	A	138	4.84	142.84
9200		16	193	14	A	141	3.63	144.63
9300		20	211	14	A	145	3.63	148.63

For expanded coverage of these items see *Means Assemblies Cost Data 1998*

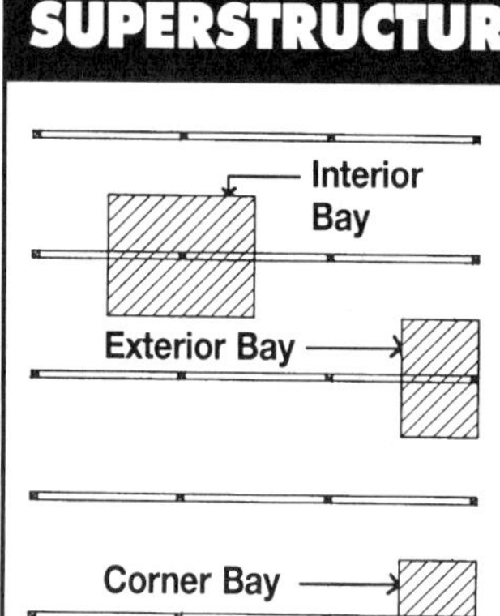

Description: Table below lists costs of columns per S.F. of bay for wood columns of various sizes and unsupported heights and the maximum allowable total load per S.F. by bay size.

Design Assumptions: Columns are concentrically loaded and are not subject to bending.

Fiber stress (f) is 1200 psi maximum.

Modulus of elasticity is 1,760,000. Use table to factor load capacity figures for modulus of elasticity other than 1,760,000.

The cost of columns per S.F. of exterior bay is proportional to the area supported. For exterior bays, multiply the costs below by two. For corner bays, multiply the cost by four.

Modulus of Elasticity	Factor
1,210,000 psi	0.69
1,320,000 psi	0.75
1,430,000 psi	0.81
1,540,000 psi	0.87
1,650,000 psi	0.94
1,760,000 psi	1.00

3.1-140 Wood Columns

	NOMINAL COLUMN SIZE (IN.)	BAY SIZE (FT.)	UNSUPPORTED HEIGHT (FT.)	MATERIAL (BF/M.S.F.)	TOTAL LOAD (P.S.F.)	COST PER S.F. MAT.	COST PER S.F. INST.	COST PER S.F. TOTAL
1000	4 x 4	10 x 8	8	133	100	.18	.09	.27
1050			10	167	60	.23	.12	.35
1200		10 x 10	8	106	80	.15	.07	.22
1250			10	133	50	.18	.09	.27
1400		10 x 15	8	71	50	.10	.05	.15
1450			10	88	30	.12	.06	.18
1600		15 x 15	8	47	30	.06	.03	.09
1650			10	59	15	.08	.04	.12
2000	6 x 6	10 x 15	8	160	230	.33	.10	.43
2050			10	200	210	.42	.13	.55
2200		15 x 15	8	107	150	.22	.07	.29
2250			10	133	140	.28	.09	.37
2400		15 x 20	8	80	110	.17	.05	.22
2450			10	100	100	.21	.07	.28
2600		20 x 20	8	60	80	.12	.04	.16
2650			10	75	70	.16	.05	.21
2800		20 x 25	8	48	60	.10	.03	.13
2850			10	60	50	.12	.04	.16
3400	8 x 8	20 x 20	8	107	160	.24	.06	.30
3450			10	133	160	.30	.08	.38
3600		20 x 25	8	85	130	.12	.06	.18
3650			10	107	130	.15	.08	.23
3800		25 x 25	8	68	100	.14	.04	.18
3850			10	85	100	.18	.05	.23
4200	10 x 10	20 x 25	8	133	210	.28	.08	.36
4250			10	167	210	.35	.09	.44
4400		25 x 25	8	107	160	.22	.06	.28
4450			10	133	160	.28	.08	.36
4700	12 x 12	20 x 25	8	192	310	.40	.10	.50
4750			10	240	310	.50	.13	.63
4900		25 x 25	8	154	240	.32	.08	.40
4950			10	192	240	.40	.10	.50

Important: See the Reference Section for critical supporting data - Reference Nos., Crews & Location Factors

Listed below are costs per V.L.F. for fireproofing by material, column size, thickness and fire rating. Weights listed are for the fireproofing material only.

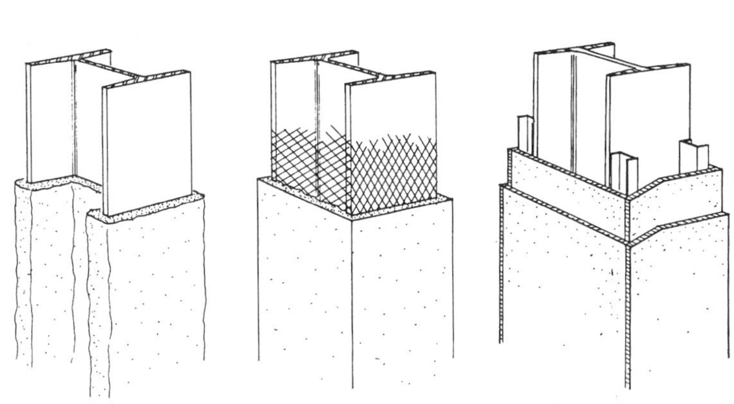

3.1-190 — Steel Column Fireproofing

	ENCASEMENT SYSTEM	COLUMN SIZE (IN.)	THICKNESS (IN.)	FIRE RATING (HRS.)	WEIGHT (P.L.F.)	MAT.	INST.	TOTAL
3000	Concrete	8	1	1	110	3.92	14.45	18.37
3300		14	1	1	258	6.40	21	27.40
3400			2	3	325	7.70	24	31.70
3450	Gypsum board	8	1/2	2	8	2.44	9.40	11.84
3550	1 layer	14	1/2	2	18	2.68	10.05	12.73
3600	Gypsum board	8	1	3	14	3.41	12	15.41
3650	1/2" fire rated	10	1	3	17	3.69	12.75	16.44
3700	2 layers	14	1	3	22	3.84	13.10	16.94
3750	Gypsum board	8	1-1/2	3	23	4.59	15.15	19.74
3800	1/2" fire rated	10	1-1/2	3	27	5.20	16.75	21.95
3850	3 layers	14	1-1/2	3	35	5.80	18.35	24.15
3900	Sprayed fiber	8	1-1/2	2	6.3	3.05	3.79	6.84
3950	Direct application		2	3	8.3	4.21	5.25	9.46
4050		10	1-1/2	2	7.9	3.69	4.58	8.27
4200		14	1-1/2	2	10.8	4.58	5.70	10.28

3.1-224 — "T" Shaped Precast Beams

	SPAN (FT.)	SUPERIMPOSED LOAD (K.L.F.)	SIZE W X D (IN.)	BEAM WEIGHT (P.L.F.)	TOTAL LOAD (K.L.F.)	MAT.	INST.	TOTAL
2300	15	2.8	12x16	260	3.06	96	9.45	105.45
2500		8.37	12x28	515	8.89	123	9.10	132.10
8900	45	3.34	12x60	1165	4.51	220	8.70	228.70
9900		6.5	24x60	1915	8.42	286	8.70	294.70

3.1-226 — "L" Shaped Precast Beams

	SPAN (FT.)	SUPERIMPOSED LOAD (K.L.F.)	SIZE W X D (IN.)	BEAM WEIGHT (P.L.F.)	TOTAL LOAD (K.L.F.)	MAT.	INST.	TOTAL
2250	15	2.58	12x16	230	2.81	72.50	9.45	81.95
2400		5.92	12x24	370	6.29	88.50	8.70	97.20
4000	25	2.64	12x28	435	3.08	98	5.20	103.20
4450		6.44	18x36	790	7.23	134	6.50	140.50
5300	30	2.80	12x36	565	3.37	117	5.20	122.20
6400		8.66	24x44	1245	9.90	187	9.55	196.55

SUPERSTRUCTURES

3

Description: The table below lists costs, $/S.F., for channel girts with sag rods and connector angles top and bottom for various column spacings, building heights and wind loads. Additive costs are shown for wind columns.

How to Use this Table: Add the cost of girts, sag rods, and angles to the framing costs of steel buildings clad in metal or composition siding. If the column spacing is in excess of the column spacing shown, use intermediate wind columns. Additive costs are shown under "Wind Columns".

Column Spacing 20′–0″ Girts	$1.89
Wind Columns	.69
Total Costs for Girts	$2.58/S.F.

Design and Pricing Assumptions: Structural steel is A36.

3.4-300 — Metal Siding Support

	BLDG. HEIGHT (FT.)	WIND LOAD (P.S.F.)	COL. SPACING (FT.)	INTERMEDIATE COLUMNS	MAT.	INST.	TOTAL
3000	18	20	20		1.31	.51	1.82
3100				wind cols.	.56	.13	.69
3200		20	25		1.41	.54	1.95
3300				wind cols.	.44	.10	.54
3400		20	30		1.53	.57	2.10
3500				wind cols.	.37	.09	.46
3600		20	35		1.67	.60	2.27
3700				wind cols.	.32	.07	.39
3800		30	20		1.43	.54	1.97
3900				wind cols.	.56	.13	.69
4000		30	25		1.53	.57	2.10
4100				wind cols.	.44	.10	.54
4200		30	30		1.67	.60	2.27
4300				wind cols.	.50	.12	.62
4600	30	20	20		1.20	.41	1.61
4700				wind cols.	.79	.18	.97
4800		20	25		1.33	.43	1.76
4900				wind cols.	.63	.15	.78
5000		20	30		1.48	.47	1.95
5100				wind cols.	.62	.15	.77
5200	30	20	35		1.63	.51	2.14
5300				wind cols.	.61	.14	.75
5400		30	20		1.35	.44	1.79
5500				wind cols.	.93	.21	1.14
5600		30	25		1.48	.47	1.95
5700				wind cols.	.89	.21	1.10
5800		30	30		1.64	.51	2.15
5900				wind cols.	.83	.19	1.02

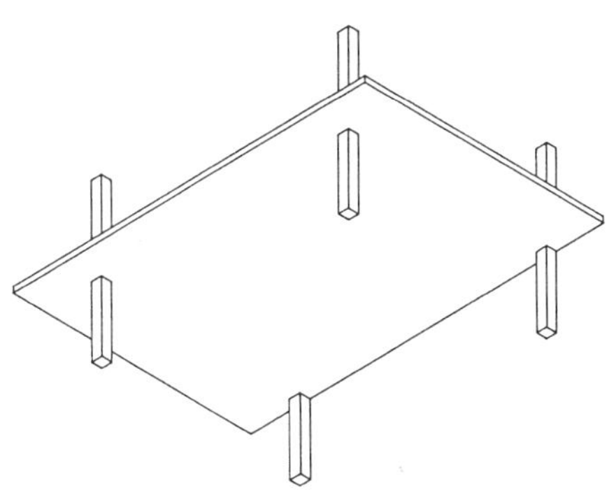

General: Flat Plates: Solid uniform depth concrete two way slab without drops or interior beams. Primary design limit is shear at columns.

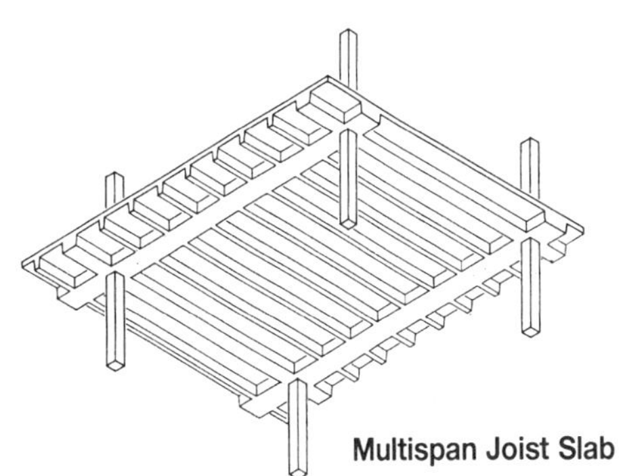

Multispan Joist Slab

General: Combination of thin concrete slab and monolithic ribs at uniform spacing to reduce dead weight and increase rigidity.

3.5-150 — Cast in Place Flat Plate

	BAY SIZE (FT.)	SUPERIMPOSED LOAD (P.S.F.)	MINIMUM COL. SIZE (IN.)	SLAB THICKNESS (IN.)	TOTAL LOAD (P.S.F.)	COST PER S.F.		
						MAT.	INST.	TOTAL
3000	15 x 20	40	14	7	127	2.92	3.59	6.51
3400		75	16	7-1/2	169	3.11	3.67	6.78
3600		125	22	8-1/2	231	3.41	3.79	7.20
3800		175	24	8-1/2	281	3.43	3.80	7.23
4200	20 x 20	40	16	7	127	2.91	3.59	6.50
4400		75	20	7-1/2	175	3.13	3.68	6.81
4600		125	24	8-1/2	231	3.41	3.77	7.18
5000		175	24	8-1/2	281	3.45	3.80	7.25
5600	20 x 25	40	18	8-1/2	146	3.39	3.78	7.17
6000		75	20	9	188	3.51	3.83	7.34
6400		125	26	9-1/2	244	3.80	3.95	7.75
6600		175	30	10	300	3.94	4.01	7.95
7000	25 x 25	40	20	9	152	3.50	3.82	7.32
7400		75	24	9-1/2	194	3.72	3.92	7.64
7600		125	30	10	250	3.96	4.02	7.98

3.5-160 — Cast in Place Multispan Joist Slab

	BAY SIZE (FT.)	SUPERIMPOSED LOAD (P.S.F.)	MINIMUM COL. SIZE (IN.)	RIB DEPTH (IN.)	TOTAL LOAD (P.S.F.)	COST PER S.F.		
						MAT.	INST.	TOTAL
2000	15 x 15	40	12	8	115	2.98	4.33	7.31
2100		75	12	8	150	2.99	4.34	7.33
2200		125	12	8	200	3.07	4.39	7.46
2300		200	14	8	275	3.19	4.57	7.76
2600	15 x 20	40	12	8	115	3.03	4.34	7.37
2800		75	12	8	150	3.11	4.43	7.54
3000		125	14	8	200	3.24	4.66	7.90
3300		200	16	8	275	3.42	4.73	8.15
3600	20 x 20	40	12	10	120	3.11	4.27	7.38
3900		75	14	10	155	3.27	4.52	7.79
4000		125	16	10	205	3.30	4.60	7.90
4100		200	18	10	280	3.49	4.81	8.30
6200	30 x 30	40	14	14	131	3.51	4.49	8
6400		75	18	14	166	3.64	4.64	8.28
6600		125	20	14	216	3.88	4.88	8.76
6700		200	24	16	297	4.19	5.10	9.29

For expanded coverage of these items see *Means Assemblies Cost Data 1998*

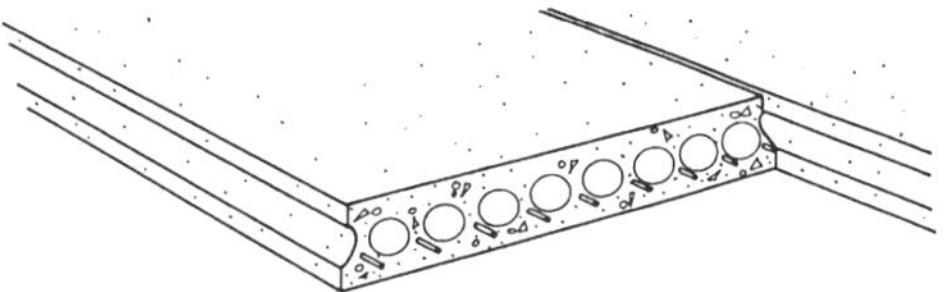

Precast Plank with No Topping

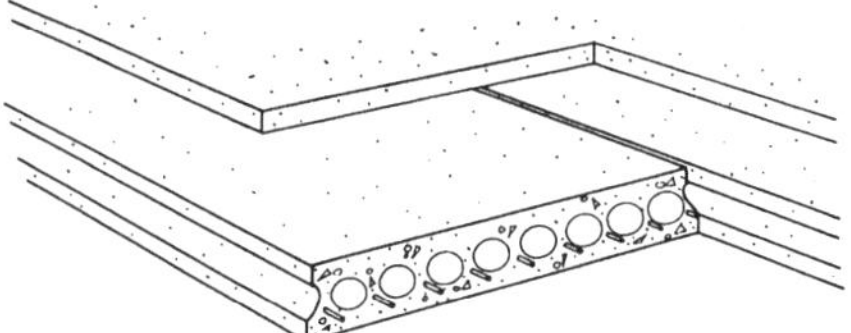

Precast Plank with 2″ Concrete Topping

3.5-210 — Precast Plank with No Topping

	SPAN (FT.)	SUPERIMPOSED LOAD (P.S.F.)	TOTAL DEPTH (IN.)	DEAD LOAD (P.S.F.)	TOTAL LOAD (P.S.F.)	COST PER S.F.		
						MAT.	INST.	TOTAL
0720	10	40	4	50	90	5.25	.97	6.22
0750		75	6	50	125	5.05	.77	5.82
0770		100	6	50	150	5.05	.77	5.82
0800	15	40	6	50	90	5.05	.77	5.82
0820		75	6	50	125	5.05	.77	5.82
0850		100	6	50	150	5.05	.77	5.82
0950	25	40	6	50	90	5.05	.77	5.82
0970		75	8	55	130	5	.62	5.62
1000		100	8	55	155	5	.62	5.62
1200	30	40	8	55	95	5	.62	5.62
1300		75	8	55	130	5	.62	5.62
1400		100	10	70	170	5.85	.39	6.24
1500	40	40	10	70	110	5.85	.39	6.24
1600		75	12	70	145	5.25	.43	5.68
1700	45	40	12	70	110	5.25	.43	5.68

3.5-210 — Precast Plank with 2″ Concrete Topping

	SPAN (FT.)	SUPERIMPOSED LOAD (P.S.F.)	TOTAL DEPTH (IN.)	DEAD LOAD (P.S.F.)	TOTAL LOAD (P.S.F.)	COST PER S.F.		
						MAT.	INST.	TOTAL
2000	10	40	6	75	115	5.80	1.83	7.63
2100		75	8	75	150	5.60	1.63	7.23
2200		100	8	75	175	5.60	1.63	7.23
2500	15	40	8	75	115	5.60	1.63	7.23
2600		75	8	75	150	5.60	1.63	7.23
2700		100	8	75	175	5.60	1.63	7.23
3100	25	40	8	75	115	5.60	1.63	7.23
3200		75	8	75	150	5.60	1.63	7.23
3300		100	10	80	180	5.55	1.48	7.03
3400	30	40	10	80	120	5.55	1.48	7.03
3500		75	10	80	155	5.55	1.48	7.03
3600		100	10	80	180	5.55	1.48	7.03
4000	40	40	12	95	135	6.40	1.25	7.65
4500		75	14	95	170	5.80	1.29	7.09
5000	45	40	14	95	135	5.80	1.29	7.09

Important: See the Reference Section for critical supporting data - Reference Nos., Crews & Location Factors

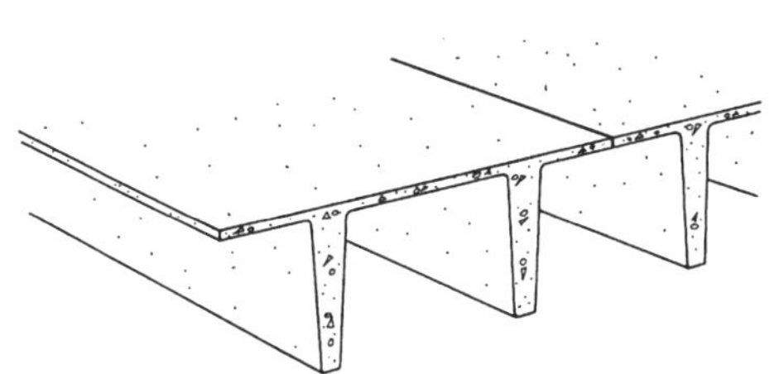

Most widely used for moderate span floors and roofs. At shorter spans, they tend to be competitive with hollow core slabs. They are also used as wall panels.

3.5-230 — Precast Double "T" Beams with No Topping

	SPAN (FT.)	SUPERIMPOSED LOAD (P.S.F.)	DBL. "T" SIZE D (IN.) W (FT.)	CONCRETE "T" TYPE	TOTAL LOAD (P.S.F.)	COST PER S.F.		
						MAT.	INST.	TOTAL
4300	50	30	20x8	Lt. Wt.	66	5.15	.66	5.81
4400		40	20x8	Lt. Wt.	76	5.50	.59	6.09
4500		50	20x8	Lt. Wt.	86	5.75	.72	6.47
4600		75	20x8	Lt. Wt.	111	6	.82	6.82
5600	70	30	32x10	Lt. Wt.	78	5.95	.45	6.40
5750		40	32x10	Lt. Wt.	88	6.15	.55	6.70
5900		50	32x10	Lt. Wt.	98	6.35	.63	6.98
6000		75	32x10	Lt. Wt.	123	6.70	.76	7.46
6100		100	32x10	Lt. Wt.	148	7.35	1.03	8.38
6200	80	30	32x10	Lt. Wt.	78	6.35	.63	6.98
6300		40	32x10	Lt. Wt.	88	6.70	.76	7.46
6400		50	32x10	Lt. Wt.	98	7	.90	7.90

3.5-230 — Precast Double "T" Beams With 2″ Topping

	SPAN (FT.)	SUPERIMPOSED LOAD (P.S.F.)	DBL. "T" SIZE D (IN.) W (FT.)	CONCRETE "T" TYPE	TOTAL LOAD (P.S.F.)	COST PER S.F.		
						MAT.	INST.	TOTAL
7100	40	30	18x8	Reg. Wt.	120	4.76	1.39	6.15
7200		40	20x8	Reg. Wt.	130	4.55	1.32	5.87
7300		50	20x8	Reg. Wt.	140	4.84	1.44	6.28
7400		75	20x8	Reg. Wt.	165	4.97	1.50	6.47
7500		100	20x8	Reg. Wt.	190	5.40	1.66	7.06
7550	50	30	24x8	Reg. Wt.	120	5.05	1.38	6.43
7600		40	24x8	Reg. Wt.	130	5.15	1.44	6.59
7750		50	24x8	Reg. Wt.	140	5.20	1.44	6.64
7800		75	24x8	Reg. Wt.	165	5.60	1.61	7.21
7900		100	32x10	Reg. Wt.	189	5.95	1.32	7.27

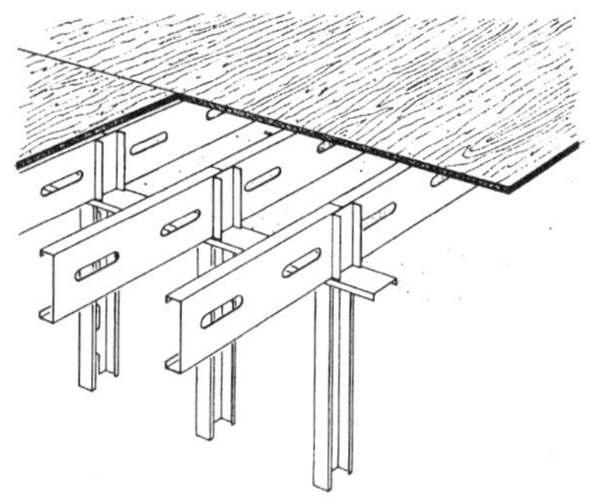

Description: Table below lists costs for light gauge CEE or PUNCHED DOUBLE joists to suit the span and loading with the minimum thickness subfloor required by the joist spacing.

Design Assumptions:

Maximum live load deflection is 1/360 of the clear span.

Maximum total load deflection is 1/240 of the clear span.

Bending strength is 20,000 psi.

8% allowance has been added to framing quantities for overlaps, double joists at openings under partitions, etc.; 5% added to glued & nailed subfloor for waste.

Maximum span is in feet and is the unsupported clear span.

3.5-360 Light Gauge Steel Floor Systems

	SPAN (FT.)	SUPERIMPOSED LOAD (P.S.F.)	FRAMING DEPTH (IN.)	FRAMING SPAC. (IN.)	TOTAL LOAD (P.S.F.)	COST PER S.F.		
						MAT.	INST.	TOTAL
1500	15	40	8	16	54	1.69	.76	2.45
1550			8	24	54	1.79	.79	2.58
1600		65	10	16	80	2.02	.89	2.91
1650			10	24	80	2.02	.88	2.90
1700		75	10	16	90	2.02	.89	2.91
1750			10	24	90	2.02	.88	2.90
1800		100	10	16	116	2.58	1.12	3.70
1850			10	24	116	2.02	.88	2.90
1900		125	10	16	141	2.58	1.12	3.70
1950			10	24	141	2.02	.88	2.90
2500	20	40	8	16	55	1.93	.85	2.78
2550			8	24	55	1.79	.79	2.58
2600		65	8	16	80	2.22	.96	3.18
2650			10	24	80	2.03	.89	2.92
2700		75	10	16	90	1.91	.85	2.76
2750			12	24	90	2.12	.80	2.92
2800		100	10	16	115	1.91	.85	2.76
2850			10	24	116	2.36	.98	3.34
2900		125	12	16	142	2.73	.99	3.72
2950			12	24	141	2.43	1.05	3.48
3500	25	40	10	16	55	2.02	.89	2.91
3550			10	24	55	2.02	.88	2.90
3600		65	10	16	81	2.58	1.12	3.70
3650			12	24	81	2.33	.86	3.19
3700		75	12	16	92	2.72	.99	3.71
3750			12	24	91	2.43	1.05	3.48
3800		100	12	16	117	3.04	1.08	4.12
3850		125	12	16	143	3.20	1.38	4.58
4500	30	40	12	16	57	2.72	.99	3.71
4550			12	24	56	2.32	.86	3.18
4600		65	12	16	82	3.03	1.08	4.11
4650		75	12	16	92	3.03	1.08	4.11

Important: See the Reference Section for critical supporting data - Reference Nos., Crews & Location Factors

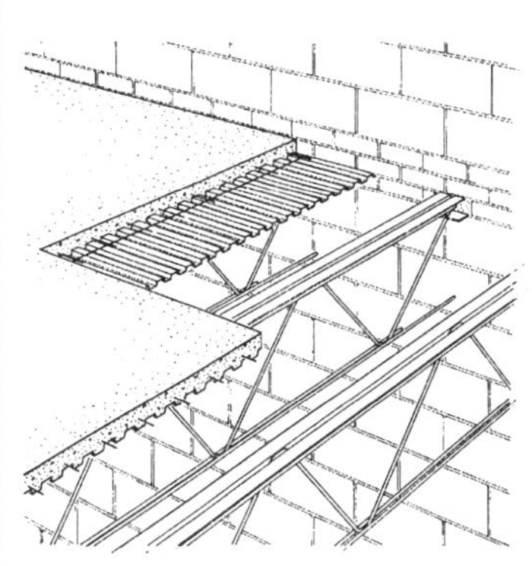

Description: Table below lists cost per S.F. for a floor system on bearing walls using open web steel joists, galvanized steel slab form and 2-1/2″ concrete slab reinforced with welded wire fabric.

Design and Pricing Assumptions:
Concrete f'c = 3 KSI placed by pump.
WWF 6 x 6 – W1.4 x W1.4 (10 x 10)
Joists are spaced as shown.
Slab form is 28 gauge galvanized.
Joists costs include appropriate bridging. Deflection is limited to 1/360 of the span. Screeds and steel trowel finish.

Design Loads	Min.	Max.
Joists	3.0 PSF	7.6 PSF
Slab Form	1.0	1.0
2-1/2″ Concrete	27.0	27.0
Ceiling	3.0	3.0
Misc.	9.0	9.4
	43.0 PSF	48.0 PSF

3.5-420 — Deck & Joists on Bearing Walls

	SPAN (FT.)	SUPERIMPOSED LOAD (P.S.F.)	JOIST SPACING FT./IN.	DEPTH (IN.)	TOTAL LOAD (P.S.F.)	COST PER S.F. MAT.	COST PER S.F. INST.	COST PER S.F. TOTAL
1050	20	40	2-0	14-1/2	83	2.93	1.71	4.64
1070		65	2-0	16-1/2	109	3.20	1.82	5.02
1100		75	2-0	16-1/2	119	3.20	1.82	5.02
1120		100	2-0	18-1/2	145	3.22	1.82	5.04
1150		125	1-9	18-1/2	170	4.12	2.16	6.28
1170	25	40	2-0	18-1/2	84	3.69	2.02	5.71
1200		65	2-0	20-1/2	109	3.84	2.07	5.91
1220		75	2-0	20-1/2	119	3.59	1.93	5.52
1250		100	2-0	22-1/2	145	3.65	1.95	5.60
1270		125	1-9	22-1/2	170	4.21	2.13	6.34
1300	30	40	2-0	22-1/2	84	3.56	1.79	5.35
1320		65	2-0	24-1/2	110	3.88	1.88	5.76
1350		75	2-0	26-1/2	121	4.03	1.91	5.94
1370		100	2-0	26-1/2	146	4.32	1.98	6.30
1400		125	2-0	24-1/2	172	4.84	2.54	7.38
1420	35	40	2-0	26-1/2	85	4.56	2.43	6.99
1450		65	2-0	28-1/2	111	4.72	2.49	7.21
1470		75	2-0	28-1/2	121	4.72	2.49	7.21
1500		100	1-11	28-1/2	147	5.15	2.65	7.80
1520		125	1-8	28-1/2	172	5.70	2.85	8.55
1550	5/8″ gyp. fireproof.							
1560	On metal furring, add					.53	1.33	1.86

Table below lists costs for a floor system on exterior bearing walls and interior columns and beams using open web steel joists, galvanized steel slab form, 2-1/2″ concrete slab reinforced with welded wire fabric.

Design and Pricing Assumptions:

Structural Steel is A36.
Concrete f'c = 3 KSI placed by pump.
WWF 6 x 6 – W1:4 x W1.4 (10 x 10)
Columns are 12′ high.
Building is 4 bays long by 4 bays wide.
Joists are 2′ O.C. ± and span the long direction of the bay.

Joists at columns have bottom chords extended and are connected to columns.

Slab form is 28 gauge galvanized. Column costs in table are for columns to support 1 floor plus roof loading in a 2-story building; however, column costs are from ground floor to 2nd floor only. Joist costs include appropriate bridging. Deflection is limited to 1/360 of the span. Screeds and steel trowel finish.

Design Loads	Min.	Max.
S.S & Joists	4.4 PSF	11.5 PSF
Slab Form	1.0	1.0
2-1/2″ Concrete	27.0	27.0
Ceiling	3.0	3.0
Misc.	7.6	5.5
	43.0 PSF	48.0 PSF

3.5-440 — Steel Joists on Beam & Wall

	BAY SIZE (FT.)	SUPERIMPOSED LOAD (P.S.F.)	DEPTH (IN.)	TOTAL LOAD (P.S.F.)	COLUMN ADD	COST PER S.F.		
						MAT.	INST.	TOTAL
1720	25x25	40	23	84		4.44	2.16	6.60
1730					columns	.19	.06	.25
1750	25x25	65	29	110		4.68	2.25	6.93
1760					columns	.19	.06	.25
1770	25x25	75	26	120		4.68	2.18	6.86
1780					columns	.22	.06	.28
1800	25x25	100	29	145		5.30	2.37	7.67
1810					columns	.22	.06	.28
1820	25x25	125	29	170		5.55	2.46	8.01
1830					columns	.24	.07	.31
2020	30x30	40	29	84		4.52	1.99	6.51
2030					columns	.17	.05	.22
2050	30x30	65	29	110		5.15	2.16	7.31
2060					columns	.17	.05	.22
2070	30x30	75	32	120		5.30	2.20	7.50
2080					columns	.19	.06	.25
2100	30x30	100	35	145		5.85	2.35	8.20
2110					columns	.22	.06	.28
2120	30x30	125	35	172		6.55	2.97	9.52
2130					columns	.27	.07	.34
2170	30x35	40	29	85		5.15	2.16	7.31
2180					columns	.16	.04	.20
2200	30x35	65	29	111		5.90	2.77	8.67
2210					columns	.18	.06	.24
2220	30x35	75	32	121		5.90	2.77	8.67
2230					columns	.19	.06	.25
2250	30x35	100	35	148		6.25	2.44	8.69
2260					columns	.23	.07	.30
2320	35x35	40	32	85		5.30	2.20	7.50
2330					columns	.16	.05	.21
2350	35x35	65	35	111		6.25	2.86	9.11
2360					columns	.20	.06	.26
2370	35x35	75	35	121		6.35	2.90	9.25
2380					columns	.20	.06	.26
2400	35x35	100	38	148		6.65	3.02	9.67
2410					columns	.24	.07	.31
2460	5/8 gyp. fireproof.							
2475	On metal furring, add					.53	1.33	1.86

Important: See the Reference Section for critical supporting data - Reference Nos., Crews & Location Factors

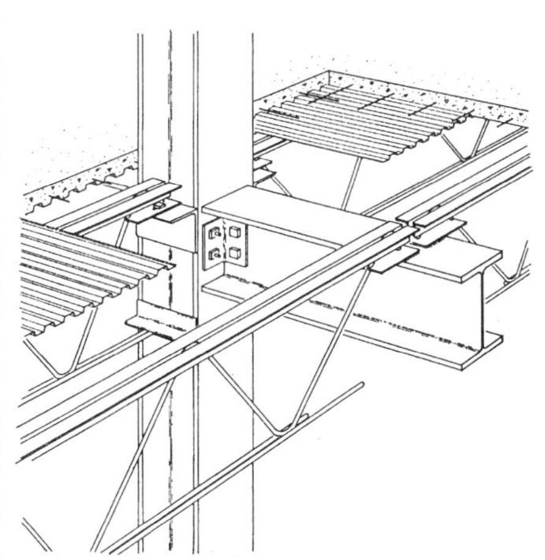

Table below lists costs for a floor system on steel columns and beams using open web steel joists, galvanized steel slab form, and 2-1/2″ concrete slab reinforced with welded wire fabric.

Design and Pricing Assumptions:
Structural Steel is A36.
Concrete f'c = 3 KSI placed by pump.
WWF 6 x 6 – W1.4 x W1.4 (10 x 10)
Columns are 12′ high.
Building is 4 bays long by 4 bays wide.
Joists are 2′ O.C. ± and span the long direction of the bay.

Joists at columns have bottom chords extended and are connected to columns.

Slab form is 28 gauge galvanized. Column costs in table are for columns to support 1 floor plus roof loading in a 2-story building; however, column costs are from ground floor to 2nd floor only. Joist costs include appropriate bridging. Deflection is limited to 1/360 of the span. Screeds and steel trowel finish.

Design Loads	Min.	Max.
S.S. & Joists	6.3 PSF	15.3 PSF
Slab Form	1.0	1.0
2-1/2″ Concrete	27.0	27.0
Ceiling	3.0	3.0
Misc.	5.7	1.7
	43.0 PSF	48.0 PSF

3.5-460   Steel Joists, Beams & Slab on Columns

	BAY SIZE (FT.)	SUPERIMPOSED LOAD (P.S.F.)	DEPTH (IN.)	TOTAL LOAD (P.S.F.)	COLUMN ADD	MAT.	INST.	TOTAL
2350	15x20	40	17	83		4.09	1.99	6.08
2400					column	.61	.18	.79
2450	15x20	65	19	108		4.51	2.10	6.61
2500					column	.61	.18	.79
2550	15x20	75	19	119		4.72	2.16	6.88
2600					column	.67	.19	.86
2650	15x20	100	19	144		5	2.25	7.25
2700					column	.67	.19	.86
2750	15x20	125	19	170		5.80	2.55	8.35
2800					column	.89	.26	1.15
2850	20x20	40	19	83		4.41	2.07	6.48
2900					column	.50	.15	.65
2950	20x20	65	23	109		4.87	2.21	7.08
3000					column	.67	.19	.86
3100	20x20	75	26	119		5.15	2.28	7.43
3200					column	.67	.19	.86
3400	20x20	100	23	144		5.30	2.34	7.64
3450					column	.67	.19	.86
3500	20x20	125	23	170		5.95	2.53	8.48
3600					column	.80	.23	1.03
3700	20x25	40	44	83		5.20	2.38	7.58
3800					column	.54	.16	.70
3900	20x25	65	26	110		5.70	2.52	8.22
4000					column	.54	.16	.70
4100	20x25	75	26	120		5.45	2.40	7.85
4200					column	.64	.19	.83
4300	20x25	100	26	145		5.75	2.49	8.24
4400					column	.64	.19	.83
4500	20x25	125	29	170		6.45	2.69	9.14
4600					column	.75	.22	.97
4700	25x25	40	23	84		5.60	2.47	8.07
4800					column	.51	.15	.66
4900	25x25	65	29	110		5.90	2.58	8.48
5000					column	.51	.15	.66
5100	25x25	75	26	120		6	2.55	8.55
5200					column	.60	.18	.78
5300	25x25	100	29	145		6.70	2.76	9.46
5400					column	.60	.18	.78

3.5-460 Steel Joists, Beams & Slab on Columns

	BAY SIZE (FT.)	SUPERIMPOSED LOAD (P.S.F.)	DEPTH (IN.)	TOTAL LOAD (P.S.F.)	COLUMN ADD	COST PER S.F. MAT.	INST.	TOTAL
5500	25x25	125	32	170		7.10	2.88	9.98
5600					column	.67	.19	.86
5700	25x30	40	29	84		5.65	2.59	8.24
5800					column	.50	.15	.65
5900	25x30	65	29	110		5.90	2.69	8.59
6000					column	.50	.15	.65
6050	25x30	75	29	120		6.25	2.47	8.72
6100					column	.55	.16	.71
6150	25x30	100	29	145		6.80	2.62	9.42
6200					column	.55	.16	.71
6250	25x30	125	32	170		7.40	3.18	10.58
6300					column	.64	.19	.83
6350	30x30	40	29	84		5.80	2.35	8.15
6400					column	.46	.13	.59
6500	30x30	65	29	110		6.60	2.58	9.18
6600					column	.46	.13	.59
6700	30x30	75	32	120		6.75	2.61	9.36
6800					column	.53	.16	.69
6900	30x30	100	35	145		7.50	2.82	10.32
7000					column	.62	.18	.80
7100	30x30	125	35	172		8.35	3.48	11.83
7200					column	.69	.20	.89
7300	30x35	40	29	85		6.60	2.55	9.15
7400					column	.40	.12	.52
7500	30x35	65	29	111		7.45	3.21	10.66
7600					column	.51	.15	.66
7700	30x35	75	32	121		7.45	3.21	10.66
7800					column	.52	.15	.67
7900	30x35	100	35	148		7.95	2.92	10.87
8000					column	.64	.19	.83
8100	30x35	125	38	173		8.85	3.16	12.01
8200					column	.65	.19	.84
8300	35x35	40	32	85		6.75	2.61	9.36
8400					column	.46	.13	.59
8500	35x35	65	35	111		7.85	3.31	11.16
8600					column	.55	.16	.71
9300	35x35	75	38	121		8.05	3.37	11.42
9400					column	.55	.16	.71
9500	35x35	100	38	148		8.65	3.59	12.24
9600					column	.67	.19	.86
9750	35x35	125	41	173		9.20	3.28	12.48
9800					column	.69	.20	.89
9810	5/8 gyp. fireproof.							
9815	On metal furring, add					.53	1.33	1.86

Important: See the Reference Section for critical supporting data - Reference Nos., Crews & Location Factors

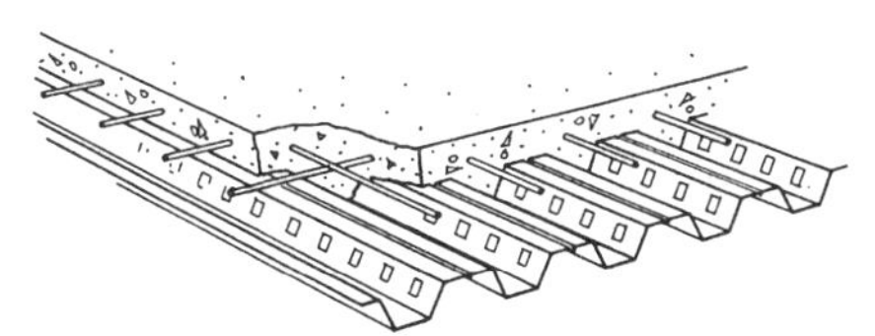

How to Use Tables: Enter any table at superimposed load and support spacing for deck.

Cellular decking tends to be stiffer and has the potential for utility integration with deck structure.

3.5-580 — Metal Deck/Concrete Fill

	SUPERIMPOSED LOAD (P.S.F.)	DECK SPAN (FT.)	DECK GAGE DEPTH	SLAB THICKNESS (IN.)	TOTAL LOAD (P.S.F.)	COST PER S.F. MAT.	COST PER S.F. INST.	COST PER S.F. TOTAL
0900	125	6	22 1-1/2	4	164	1.51	1.10	2.61
0920		7	20 1-1/2	4	164	1.60	1.17	2.77
0950		8	20 1-1/2	4	165	1.60	1.17	2.77
0970		9	18 1-1/2	4	165	1.88	1.17	3.05
1000		10	18 2	4	165	1.97	1.23	3.20
1020		11	18 3	5	169	2.45	1.33	3.78

3.5-710 — Wood Joist

		COST PER S.F. MAT.	COST PER S.F. INST.	COST PER S.F. TOTAL
2902	Wood joist, 2" x 8",12" O.C.	1.42	.71	2.13
2950	16" O.C.	1.19	.61	1.80
3000	24" O.C.	1.17	.58	1.75
3300	2"x10", 12" O.C.	1.81	.81	2.62
3350	16" O.C.	1.48	.68	2.16
3400	24" O.C.	1.36	.62	1.98
3700	2"x12", 12" O.C.	2.20	.82	3.02
3750	16" O.C.	1.77	.69	2.46
3800	24" O.C.	1.55	.63	2.18
4100	2"x14", 12" O.C.	2.58	.90	3.48
4150	16" O.C.	2.06	.75	2.81
4200	24" O.C.	1.75	.67	2.42
7101	Note: Subfloor cost is included in these prices			

3.5-720 — Wood Beam & Joist

	BAY SIZE (FT.)	SUPERIMPOSED LOAD (P.S.F.)	GIRDER BEAM (IN.)	JOISTS (IN.)	TOTAL LOAD (P.S.F.)	COST PER S.F. MAT.	COST PER S.F. INST.	COST PER S.F. TOTAL
2000	15x15	40	8 x 12 4 x 12	2 x 6 @ 16	53	4.58	1.79	6.37
2050		75	8 x 16 4 x 16	2 x 8 @ 16	90	6.35	1.95	8.30
2100		125	12 x 16 6 x 16	2 x 8 @ 12	144	9.60	2.52	12.12
2150		200	14 x 22 12 x 16	2 x 10 @ 12	227	17.70	3.90	21.60
3000	20x20	40	10 x 14 10 x 12	2 x 8 @ 16	63	6.70	1.79	8.49
3050		75	12 x 16 8 x 16	2 x 10 @ 16	102	8.70	2.02	10.72
3100		125	14 x 22 12 x 16	2 x 10 @ 12	163	14.30	3.26	17.56

SUPERSTRUCTURES

3

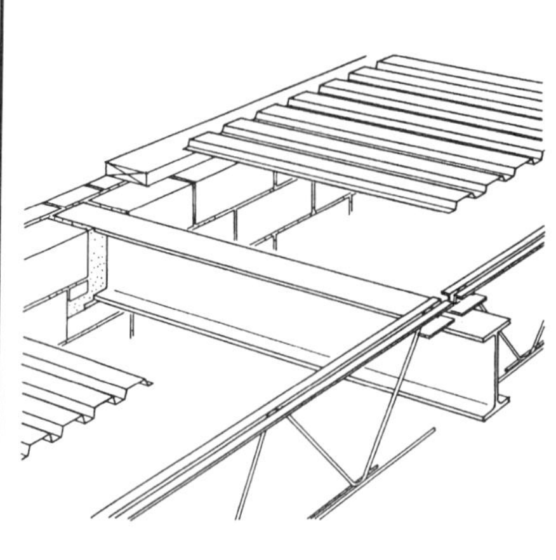

Table below lists the cost per S.F. for a roof system with steel columns, beams, and deck using open web steel joists and 1-1/2″ galvanized metal deck. Perimeter of system is supported on bearing walls.

Design and Pricing Assumptions:
Columns are 18′ high.
Joists are 5′-0″ O.C. and span the long direction of the bay.

Joists at columns have bottom chords extended and are connected to columns. Column costs are not included but are listed separately per S.F. of floor.

Roof deck is 1-1/2″, 22 gauge galvanized steel. Joist cost includes appropriate bridging. Deflection is limited to 1/240 of the span. Fireproofing is not included.

Costs/S.F. are based on a building 4 bays long and 4 bays wide.

Design Loads	Min.	Max.
Joists & Beams	3 PSF	5 PSF
Deck	2	2
Insulation	3	3
Roofing	6	6
Misc.	6	6
Total Dead Load	20 PSF	22 PSF

3.7-410 — Steel Joists, Beams & Deck on Columns & Walls

	BAY SIZE (FT.)	SUPERIMPOSED LOAD (P.S.F.)	DEPTH (IN.)	TOTAL LOAD (P.S.F.)	COLUMN ADD	COST PER S.F. MAT.	COST PER S.F. INST.	COST PER S.F. TOTAL
3000	25x25	20	18	40		2.08	.62	2.70
3100					columns	.17	.04	.21
3200		30	22	50		2.41	.74	3.15
3300					columns	.23	.06	.29
3400		40	20	60		2.48	.73	3.21
3500					columns	.23	.06	.29
3600	25x30	20	22	40		2.20	.61	2.81
3700					columns	.19	.04	.23
3800		30	20	50		2.42	.78	3.20
3900					columns	.19	.04	.23
4000		40	25	60		2.59	.70	3.29
4100					columns	.23	.06	.29
4200	30x30	20	25	42		2.40	.66	3.06
4300					columns	.16	.03	.19
4400		30	22	52		2.62	.72	3.34
4500					columns	.19	.04	.23
4600		40	28	62		2.72	.74	3.46
4700					columns	.19	.04	.23
4800	30x35	20	22	42		2.51	.69	3.20
4900					columns	.17	.04	.21
5000		30	28	52		2.64	.71	3.35
5100					columns	.17	.04	.21
5200		40	25	62		2.86	.77	3.63
5300					columns	.19	.04	.23
5400	35x35	20	28	42		2.52	.69	3.21
5500					columns	.14	.03	.17

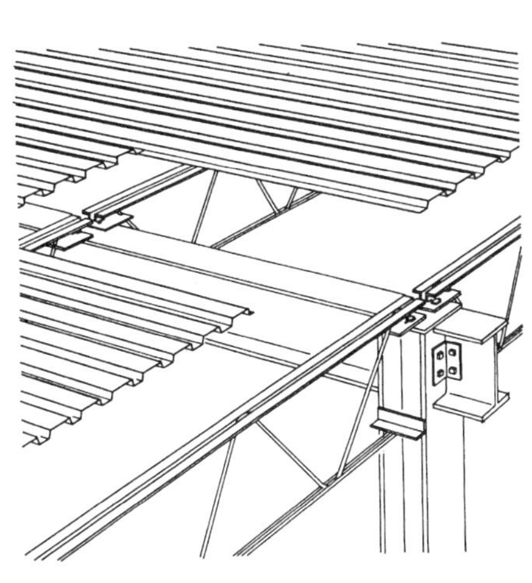

Description: Table below lists the cost per S.F. for a roof system with steel columns, beams, and deck, using open web steel joists and 1-1/2″ galvanized metal deck.

Design and Pricing Assumptions:
Columns are 18′ high.
Building is 4 bays long by 4 bays wide.
Joists are 5′-0″ O.C. and span the long direction of the bay.
Joists at columns have bottom chords extended and are connected to columns. Column costs are not included but are listed separately per S.F. of floor.

Roof deck is 1-1/2″, 22 gauge galvanized steel. Joist cost includes appropriate bridging. Deflection is limited to 1/240 of the span. Fireproofing is not included.

Design Loads	Min.	Max.
Joists & Beams	3 PSF	5 PSF
Deck	2	2
Insulation	3	3
Roofing	6	6
Misc.	6	6
Total Dead Load	20 PSF	22 PSF

3.7-420 — Steel Joists, Beams, & Deck on Columns

	BAY SIZE (FT.)	SUPERIMPOSED LOAD (P.S.F.)	DEPTH (IN.)	TOTAL LOAD (P.S.F.)	COLUMN ADD	COST PER S.F. MAT.	INST.	TOTAL
1100	15x20	20	16	40		2.06	.62	2.68
1200					columns	1	.23	1.23
1500		40	18	60		2.31	.69	3
1600					columns	1	.23	1.23
2300	20x25	20	18	40		2.35	.68	3.03
2400					columns	.60	.14	.74
2700		40	20	60		2.64	.77	3.41
2800					columns	.80	.18	.98
2900	25x25	20	18	40		2.72	.77	3.49
3000					columns	.48	.11	.59
3100		30	22	50		3.01	.88	3.89
3200					columns	.64	.15	.79
3300		40	20	60		3.15	.89	4.04
3400					columns	.64	.15	.79
3500	25x30	20	22	40		2.66	.72	3.38
3600					columns	.54	.12	.66
3900		40	25	60		3.21	.84	4.05
4000					columns	.64	.15	.79
4100	30x30	20	25	42		3	.79	3.79
4200					columns	.45	.10	.55
4300		30	22	52		3.30	.86	4.16
4400					columns	.54	.12	.66
4500		40	28	62		3.45	.90	4.35
4600					columns	.54	.12	.66
5300	35x35	20	28	42		3.28	.87	4.15
5400					columns	.39	.09	.48
5500		30	25	52		3.65	.95	4.60
5600					columns	.46	.11	.57
5700		40	28	62		3.94	1.02	4.96
5800					columns	.51	.12	.63

SUPERSTRUCTURES

3

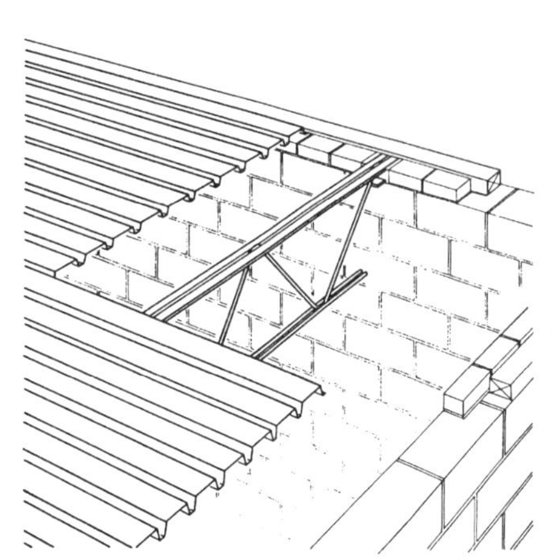

Description: Table below lists cost per S.F. for a roof system using open web steel joists and 1-1/2" galvanized metal deck. The system is assumed supported on bearing walls or other suitable support. Costs for the supports are not included.

Design and Pricing Assumptions:
Joists are 5'-0" O.C.
Roof deck is 1-1/2", 22 gauge galvanized.

3.7-430 Steel Joists & Deck on Bearing Walls

	BAY SIZE (FT.)	SUPERIMPOSED LOAD (P.S.F.)	DEPTH (IN.)	TOTAL LOAD (P.S.F.)		COST PER S.F. MAT.	INST.	TOTAL
1100	20	20	13-1/2	40		1.37	.45	1.82
1200		30	15-1/2	50		1.40	.46	1.86
1300		40	15-1/2	60		1.51	.49	2
1400	25	20	17-1/2	40		1.51	.49	2
1500		30	17-1/2	50		1.64	.53	2.17
1600		40	19-1/2	60		1.67	.55	2.22
1700	30	20	19-1/2	40		1.66	.54	2.20
1800		30	21-1/2	50		1.70	.55	2.25
1900		40	23-1/2	60		1.84	.59	2.43
2000	35	20	23-1/2	40		1.78	.51	2.29
2100		30	25-1/2	50		1.84	.53	2.37
2200		40	25-1/2	60		1.96	.56	2.52
2300	40	20	25-1/2	41		1.99	.57	2.56
2400		30	25-1/2	51		2.12	.60	2.72
2500		40	25-1/2	61		2.19	.62	2.81
2600	45	20	27-1/2	41		2.28	.83	3.11
2700		30	31-1/2	51		2.42	.88	3.30
2800		40	31-1/2	61		2.55	.93	3.48
2900	50	20	29-1/2	42		2.56	.93	3.49
3000		30	31-1/2	52		2.80	1.02	3.82
3100		40	31-1/2	62		2.98	1.09	4.07
3200	60	20	37-1/2	42		2.95	.94	3.89
3300		30	37-1/2	52		3.27	1.04	4.31
3400		40	37-1/2	62		3.27	1.04	4.31
3500	70	20	41-1/2	42		3.26	1.04	4.30
3600		30	41-1/2	52		3.48	1.11	4.59
3700		40	41-1/2	64		4.23	1.34	5.57
3800	80	20	45-1/2	44		4.46	1.51	5.97
3900		30	45-1/2	54		4.46	1.51	5.97
4000		40	45-1/2	64		4.93	1.68	6.61
4400	100	20	57-1/2	44		3.89	1.23	5.12
4500		30	57-1/2	54		4.64	1.47	6.11
4600		40	57-1/2	65		5.10	1.62	6.72
4700	125	20	69-1/2	44		5.50	1.71	7.21
4800		30	69-1/2	56		6.45	1.98	8.43
4900		40	69-1/2	67		7.35	2.26	9.61

Important: See the Reference Section for critical supporting data - Reference Nos., Crews & Location Factors

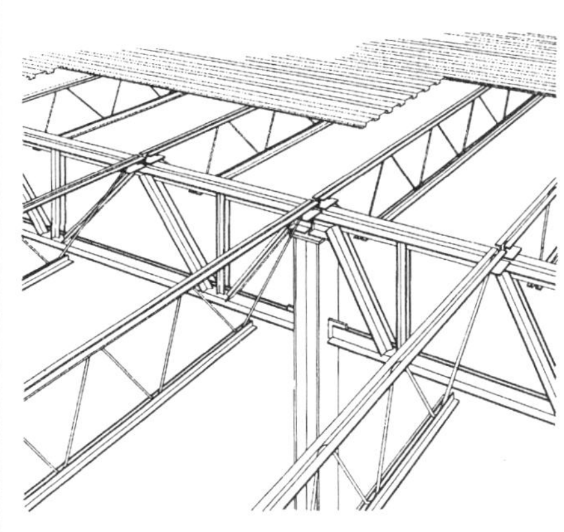

Description: Table below lists costs for a roof system supported on exterior bearing walls and interior columns. Costs include bracing, joist girders, open web steel joists and 1-1/2″ galvanized metal deck.

Design and Pricing Assumptions:
Columns are 18′ high.
Joists are 5′-0″ O.C.
Joist girders and joists have bottom chords connected to columns. Roof deck is 1-1/2″, 22 gauge galvanized steel. Costs include bridging and bracing. Deflection is limited to 1/240 of the span.

Fireproofing is not included.
Costs/S.F. are based on a building 4 bays long and 4 bays wide.
Costs for bearing walls are not included.

Column costs are not included but are listed separately per S.F. of floor.

3.7-440 — Steel Joists, Joist Girders & Deck on Columns & Walls

	BAY SIZE (FT.) GIRD X JOISTS	SUPERIMPOSED LOAD (P.S.F.)	DEPTH (IN.)	TOTAL LOAD (P.S.F.)	COLUMN ADD	COST PER S.F.		
						MAT.	INST.	TOTAL
2350	30x35	20	32-1/2	40		1.97	.68	2.65
2400					columns	.17	.04	.21
2550		40	36-1/2	60		2.21	.75	2.96
2600					columns	.19	.04	.23
3000	35x35	20	36-1/2	40		2.13	.87	3
3050					columns	.14	.03	.17
3200		40	36-1/2	60		2.42	.95	3.37
3250					columns	.18	.04	.22
3300	35x40	20	36-1/2	40		2.18	.77	2.95
3350					columns	.14	.03	.17
3500		40	36-1/2	60		2.46	.85	3.31
3550					columns	.16	.04	.20
3900	40x40	20	40-1/2	41		2.48	.98	3.46
3950					columns	.14	.03	.17
4100		40	40-1/2	61		2.78	1.06	3.84
4150					columns	.14	.03	.17
5100	45x50	20	52-1/2	41		2.80	1.12	3.92
5150					columns	.10	.03	.13
5300		40	52-1/2	61		3.41	1.34	4.75
5350					columns	.14	.03	.17
5400	50x45	20	56-1/2	41		2.71	1.08	3.79
5450					columns	.10	.03	.13
5600		40	56-1/2	61		3.49	1.36	4.85
5650					columns	.14	.03	.17
5700	50x50	20	56-1/2	42		3.06	1.23	4.29
5750					columns	.10	.03	.13
5900		40	59	64		3.66	1.37	5.03
5950					columns	.14	.03	.17
6300	60x50	20	62-1/2	43		3.08	1.25	4.33
6350					columns	.16	.04	.20
6500		40	71	65		3.75	1.42	5.17
6550					columns	.21	.05	.26

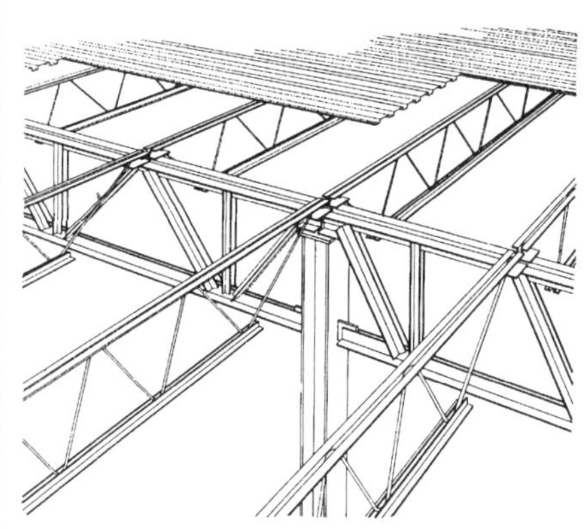

Description: Table below lists the cost per S.F. for a roof system supported on columns. Costs include joist girders, open web steel joists and 1-1/2″ galvanized metal deck.

Design and Pricing Assumptions:
Columns are 18′ high.
Joists are 5′-0″ O.C.
Joist girders and joists have bottom chords connected to columns. Roof deck is 1-1/2″, 22 gauge galvanized steel. Costs include bridging and bracing. Deflection is limited to 1/240 of the span. Fireproofing is not included.

Costs/S.F. are based on a building 4 bays long and 4 bays wide.

Costs for columns are not included, but are listed separately per S.F. of area.

3.7-450 — Steel Joists & Joist Girders on Columns

	BAY SIZE (FT.) GIRD X JOISTS	SUPERIMPOSED LOAD (P.S.F.)	DEPTH (IN.)	TOTAL LOAD (P.S.F.)	COLUMN ADD	COST PER S.F. MAT.	INST.	TOTAL
2000	30x30	20	17-1/2	40		2	.82	2.82
2050					columns	.54	.13	.67
2100		30	17-1/2	50		2.19	.86	3.05
2150					columns	.54	.13	.67
2200		40	21-1/2	60		2.24	.89	3.13
2250					columns	.54	.13	.67
2300	30x35	20	32-1/2	40		2.11	.76	2.87
2350					columns	.46	.10	.56
2500		40	36-1/2	60		2.39	.84	3.23
2550					columns	.54	.13	.67
3200	35x40	20	36-1/2	40		2.71	1.19	3.90
3250					columns	.40	.09	.49
3400		40	36-1/2	60		3.06	1.32	4.38
3450					columns	.45	.10	.55
3800	40x40	20	40-1/2	41		2.74	1.20	3.94
3850					columns	.39	.09	.48
4000		40	40-1/2	61		3.08	1.32	4.40
4050					columns	.39	.09	.48
4100	40x45	20	40-1/2	41		2.95	1.39	4.34
4150					columns	.35	.09	.44
4200		30	40-1/2	51		3.21	1.48	4.69
4250					columns	.35	.09	.44
4300		40	40-1/2	61		3.58	1.60	5.18
4350					columns	.39	.09	.48
5000	45x50	20	52-1/2	41		3.12	1.40	4.52
5050					columns	.28	.06	.34
5200		40	52-1/2	61		3.83	1.65	5.48
5250					columns	.36	.09	.45
5300	50x45	20	56-1/2	41		3.03	1.36	4.39
5350					columns	.28	.06	.34
5500		40	56-1/2	61		3.92	1.68	5.60
5550					columns	.36	.09	.45
5600	50x50	20	56-1/2	42		3.38	1.48	4.86
5650					columns	.28	.07	.35
5800		40	59	64		3.22	1.37	4.59
5850					columns	.39	.09	.48

Important: See the Reference Section for critical supporting data - Reference Nos., Crews & Location Factors

The table below lists prices per S.F. for roof rafters and sheathing by nominal size and spacing. Sheathing is 5/16″ CDX for 12″ and 16″ spacing and 3/8″ CDX for 24″ spacing.

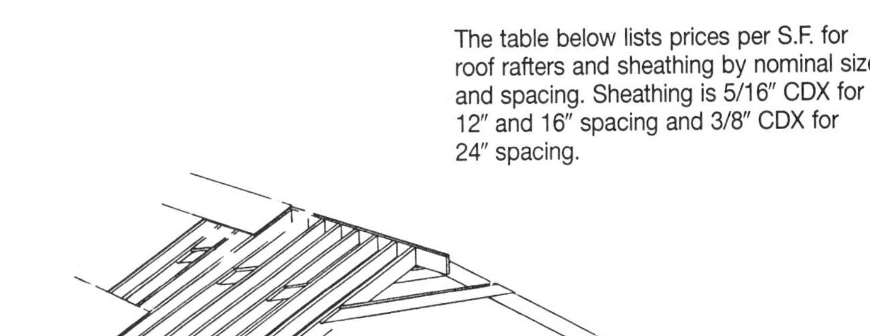

Factors for Converting Inclined to Horizontal

Roof Slope	Approx. Angle	Factor	Roof Slope	Approx. Angle	Factor
Flat	0°	1.000	12 in 12	45.0°	1.414
1 in 12	4.8°	1.003	13 in 12	47.3°	1.474
2 in 12	9.5°	1.014	14 in 12	49.4°	1.537
3 in 12	14.0°	1.031	15 in 12	51.3°	1.601
4 in 12	18.4°	1.054	16 in 12	53.1°	1.667
5 in 12	22.6°	1.083	17 in 12	54.8°	1.734
6 in 12	26.6°	1.118	18 in 12	56.3°	1.803
7 in 12	30.3°	1.158	19 in 12	57.7°	1.873
8 in 12	33.7°	1.202	20 in 12	59.0°	1.943
9 in 12	36.9°	1.250	21 in 12	60.3°	2.015
10 in 12	39.8°	1.302	22 in 12	61.4°	2.088
11 in 12	42.5°	1.357	23 in 12	62.4°	2.162

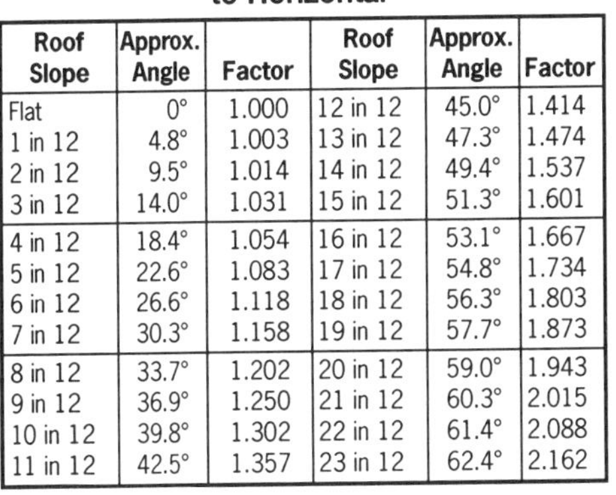

3.7-510	Wood/Flat or Pitched	COST PER S.F.		
		MAT.	INST.	TOTAL
2500	Flat rafter, 2″x4″, 12″ O.C.	.84	.65	1.49
2550	16″ O.C.	.73	.56	1.29
2600	24″ O.C.	.62	.48	1.10
2900	2″x6″, 12″ O.C.	1.04	.65	1.69
2950	16″ O.C.	.88	.56	1.44
3000	24″ O.C.	.72	.48	1.20
3300	2″x8″, 12″ O.C.	1.31	.70	2.01
3350	16″ O.C.	1.08	.60	1.68
3400	24″ O.C.	.86	.51	1.37
3700	2″x10″, 12″ O.C.	1.70	.80	2.50
3750	16″ O.C.	1.37	.67	2.04
3800	24″ O.C.	1.05	.55	1.60
4100	2″x12″, 12″ O.C.	2.09	.81	2.90
4150	16″ O.C.	1.66	.68	2.34
4200	24″ O.C.	1.24	.56	1.80
4500	2″x14″, 12″ O.C.	2.47	1.19	3.66
4550	16″ O.C.	1.95	.97	2.92
4600	24″ O.C.	1.44	.75	2.19
4900	3″x6″, 12″ O.C.	2.36	.69	3.05
4950	16″ O.C.	1.87	.59	2.46
5000	24″ O.C.	1.39	.50	1.89
5300	3″x8″, 12″ O.C.	3.02	.77	3.79
5350	16″ O.C.	2.36	.65	3.01
5400	24″ O.C.	1.71	.54	2.25
5700	3″x10″, 12″ O.C.	3.68	.88	4.56
5750	16″ O.C.	2.86	.73	3.59
5800	24″ O.C.	2.05	.59	2.64
6100	3″x12″, 12″ O.C.	4.34	1.06	5.40
6150	16″ O.C.	3.36	.87	4.23
6200	24″ O.C.	2.37	.68	3.05
7001	Wood truss, 4 in 12 slope, 24″ O.C., 24′ to 29′ span	2.61	1.09	3.70
7100	30′ to 43′ span	2.78	1.09	3.87
7200	44′ to 60′ span	2.94	1.09	4.03

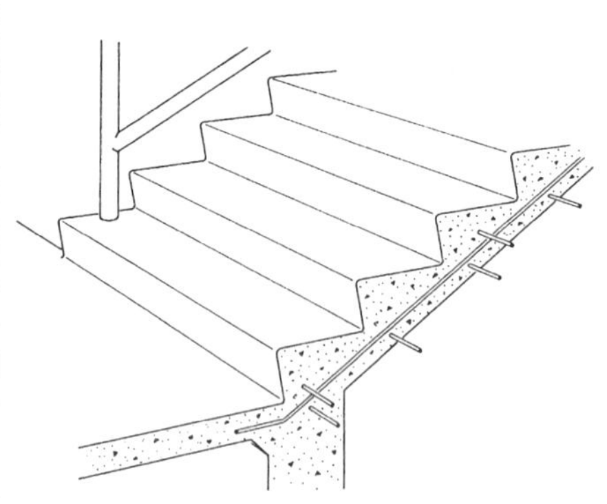

General Design: See reference section for code requirements. Maximum height between landings is 12′; usual stair angle is 20° to 50° with 30° to 35° best. Usual relation of riser to treads is:

    Riser + tread = 17.5.
    2x (Riser) + tread = 25.
    Riser x tread = 70 or 75.

Maximum riser height is 7″ for commercial, 8-1/4″ for residential.
Usual riser height is 6-1/2″ to 7-1/4″.

Minimum tread width is 11″ for commercial and 9″ for residential.

For additional information please see reference section.

Cost Per Flight: Table below lists the cost per flight for 4′-0″ wide stairs. Side walls are not included. Railings are included.

3.9-100	Stairs	MAT.	INST.	TOTAL
0470	Stairs, C.I.P. concrete, w/o landing, 12 risers, w/o nosing	655	1,025	1,680
0480	With nosing	1,000	1,175	2,175
0550	W/landing, 12 risers, w/o nosing	735	1,275	2,010
0560	With nosing	1,100	1,425	2,525
0570	16 risers, w/o nosing	910	1,600	2,510
0580	With nosing	1,375	1,800	3,175
0590	20 risers, w/o nosing	1,100	1,925	3,025
0600	With nosing	1,675	2,175	3,850
0610	24 risers, w/o nosing	1,275	2,250	3,525
0620	With nosing	1,975	2,550	4,525
0630	Steel, grate type w/nosing & rails, 12 risers, w/o landing	1,625	365	1,990
0640	With landing	2,525	565	3,090
0660	16 risers, with landing	3,075	685	3,760
0680	20 risers, with landing	3,625	810	4,435
0700	24 risers, with landing	4,175	935	5,110
0710	Cement fill metal pan & picket rail, 12 risers, w/o landing	2,100	365	2,465
0720	With landing	3,300	630	3,930
0740	16 risers, with landing	3,975	750	4,725
0760	20 risers, with landing	4,675	875	5,550
0780	24 risers, with landing	5,375	995	6,370
0790	Cast iron tread & pipe rail, 12 risers, w/o landing	2,125	365	2,490
0800	With landing	3,325	630	3,955
1120	Wood, prefab box type, oak treads, wood rails 3′-6″ wide, 14 risers	1,125	203	1,328
1150	Prefab basement type, oak treads, wood rails 3′-0″ wide, 14 risers	815	162	977

For information about Means Estimating Seminars, see yellow pages 11 and 12 in back of book

Important: See the Reference Section for critical supporting data - Reference Nos., Crews & Location Factors

Division 4
Exterior Closure

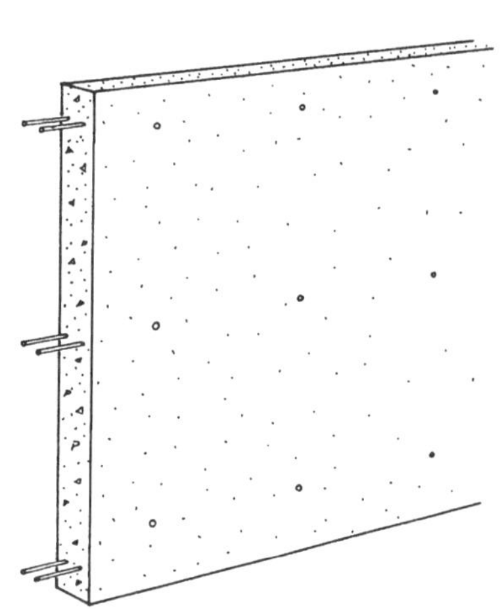

The table below describes a concrete wall system for exterior closure. There are several types of wall finishes priced from plain finish to a finish with 3/4″ rustication strip.

Design Assumptions:
Conc. f'c = 3000 to 5000 psi
Reinf. fy = 60,000 psi

4.1-110	Cast In Place Concrete	COST PER S.F.		
		MAT.	INST.	TOTAL
2100	Conc wall reinforced, 8′ high, 6″ thick, plain finish, 3000 PSI	2.77	6.50	9.27
2700	Aged wood liner, 3000 PSI	4.53	6.85	11.38
3000	Sand blast light 1 side, 3000 PSI	2.99	7.05	10.04
3700	3/4″ bevel rustication strip, 3000 PSI	2.93	7.95	10.88
4000	8″ thick, plain finish, 3000 PSI	3.25	6.70	9.95
4100	4000 PSI	3.33	6.70	10.03
4300	Rub concrete 1 side, 3000 PSI	3.33	8	11.33
4550	8″ thick, aged wood liner, 3000 PSI	5	7.05	12.05
4750	Sand blast light 1 side, 3000 PSI	3.47	7.25	10.72
5300	3/4″ bevel rustication strip, 3000 PSI	3.41	8.15	11.56
5600	10″ thick, plain finish, 3000 PSI	3.72	6.90	10.62
5900	Rub concrete 1 side, 3000 PSI	3.80	8.15	11.95
6200	Aged wood liner, 3000 PSI	5.50	7.20	12.70
6500	Sand blast light 1 side, 3000 PSI	3.94	7.45	11.39
7100	3/4″ bevel rustication strip, 3000 PSI	3.88	8.35	12.23
7400	12″ thick, plain finish, 3000 PSI	4.26	7.15	11.41
7700	Rub concrete 1 side, 3000 PSI	4.34	8.40	12.74
8000	Aged wood liner, 3000 PSI	6	7.45	13.45
8300	Sand blast light 1 side, 3000 PSI	4.48	7.70	12.18
8900	3/4″ bevel rustication strip, 3000 PSI	4.42	8.55	12.97

Important: See the Reference Section for critical supporting data - Reference Nos., Crews & Location Factors

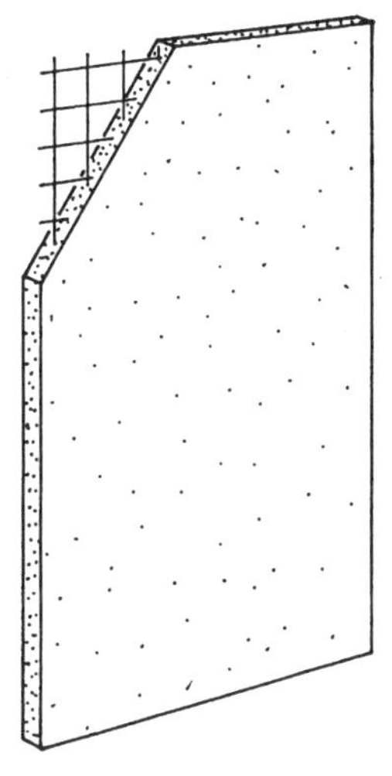

Precast concrete wall panels are either solid or insulated with plain, colored or textured finishes. Transportation is an important cost factor. Prices below are based on delivery within fifty miles of a plant. Engineering data is available from fabricators to assist with construction details. Usual minimum job size for economical use of panels is about 5000 S.F. Small jobs can double the prices below. For large, highly repetitive jobs, deduct up to 15% from the prices below.

4.1-140 — Flat Precast Concrete

	THICKNESS (IN.)	PANEL SIZE (FT.)	FINISHES	RIGID INSULATION (IN)	TYPE	COST PER S.F. MAT.	COST PER S.F. INST.	COST PER S.F. TOTAL
3000	4	5x18	smooth gray	none	low rise	5.20	3.31	8.51
3050		6x18				4.35	2.77	7.12
3100		8x20				6.35	1.04	7.39
3150		12x20				6	.98	6.98
3200	6	5x18	smooth gray	2	low rise	6	3.55	9.55
3250		6x18				5.15	3.01	8.16
3300		8x20				7.25	1.32	8.57
3350		12x20				6.60	1.22	7.82
3400	8	5x18	smooth gray	2	low rise	9.30	1.65	10.95
3450		6x18				8.80	1.57	10.37
3500		8x20				8.10	1.45	9.55
3550		12x20				7.40	1.34	8.74
3600	4	4x8	white face	none	low rise	15.90	1.91	17.81
3650		8x8				11.90	1.43	13.33
3700		10x10				10.45	1.26	11.71
3750		20x10				9.45	1.14	10.59
3800	5	4x8	white face	none	low rise	16.20	1.95	18.15
3850		8x8				12.25	1.47	13.72
3900		10x10				10.85	1.31	12.16
3950		20x20				9.95	1.20	11.15
4000	6	4x8	white face	none	low rise	16.80	2.02	18.82
4050		8x8				12.75	1.53	14.28
4100		10x10				11.25	1.36	12.61
4150		20x10				10.35	1.24	11.59
4200	6	4x8	white face	2	low rise	17.65	2.29	19.94
4250		8x8				13.60	1.80	15.40
4300		10x10				12.10	1.63	13.73
4350		20x10				10.35	1.24	11.59
4400	7	4x8	white face	none	low rise	17.20	2.08	19.28
4450		8x8				13.20	1.59	14.79
4500		10x10				11.85	1.43	13.28
4550		20x10				10.85	1.31	12.16
4600	7	4x8	white face	2	low rise	18.05	2.35	20.40
4650		8x8				14.05	1.86	15.91
4700		10x10				12.70	1.70	14.40
4750		20x10				11.70	1.58	13.28

4.1-140 — Flat Precast Concrete

	THICKNESS (IN.)	PANEL SIZE (FT.)	FINISHES	RIGID INSULATION (IN)	TYPE	COST PER S.F.		
						MAT.	INST.	TOTAL
4800	8	4x8	white face	none	low rise	17.60	2.12	19.72
4850		8x8				13.55	1.63	15.18
4900		10x10				12.20	1.47	13.67
4950		20x10				11.25	1.35	12.60
5000	8	4x8	white face	2	low rise	18.45	2.39	20.84
5050		8x8				14.40	1.90	16.30
5100		10x10				13.05	1.74	14.79
5150		20x10				12.05	1.62	13.67

4.1-140 — Fluted Window or Mullion Precast Concrete

	THICKNESS (IN.)	PANEL SIZE (FT.)	FINISHES	RIGID INSULATION (IN)	TYPE	COST PER S.F.		
						MAT.	INST.	TOTAL
5200	4	4x8	smooth gray	none	high rise	11.90	7.60	19.50
5250		8x8				8.50	5.40	13.90
5300		10x10				12.15	1.99	14.14
5350		20x10				10.65	1.75	12.40
5400	5	4x8	smooth gray	none	high rise	12.10	7.70	19.80
5450		8x8				8.80	5.60	14.40
5500		10x10				12.70	2.07	14.77
5550		20x10				11.20	1.83	13.03
5600	6	4x8	smooth gray	none	high rise	12.45	7.90	20.35
5650		8x8				9.10	5.80	14.90
5700		10x10				13.05	2.13	15.18
5750		20x10				11.60	1.89	13.49
5800	6	4x8	smooth gray	2	high rise	13.25	8.20	21.45
5850		8x8				9.95	6.05	16
5900		10x10				13.90	2.40	16.30
5950		20x10				12.45	2.16	14.61
6000	7	4x8	smooth gray	none	high rise	12.70	8.10	20.80
6050		8x8				9.35	5.95	15.30
6100		10x10				13.60	2.23	15.83
6150		20x10				11.95	1.95	13.90
6200	7	4x8	smooth gray	2	high rise	13.55	8.35	21.90
6250		8x8				10.20	6.25	16.45
6300		10x10				14.45	2.50	16.95
6350		20x10				12.80	2.22	15.02
6400	8	4x8	smooth gray	none	high rise	12.85	8.20	21.05
6450		8x8				9.60	6.10	15.70
6500		10x10				14.05	2.30	16.35
6550		20x10				12.55	2.05	14.60
6600	8	4x8	smooth gray	2	high rise	13.70	8.45	22.15
6650		8x8				10.45	6.40	16.85
6700		10x10				14.90	2.57	17.47
6750		20x10				13.40	2.32	15.72

4.1-140 — Precast Concrete Specialties

	TYPE	SIZE				COST PER L.F.		
						MAT.	INST.	TOTAL
6773	Coping, precast	6"wide				10.55	5.80	16.35
6774	Stock units	10"wide				11.10	6.20	17.30
6775		12"wide				10	6.70	16.70
6776		14" wide				15.30	7.25	22.55

Important: See the Reference Section for critical supporting data - Reference Nos., Crews & Location Factors

4.1-140	Precast Concrete Specialties

	TYPE	SIZE				COST PER L.F.		
						MAT.	INST.	TOTAL
6777	Window sills	6" wide				9.85	6.20	16.05
6778	Precast	10" wide				9.15	7.25	16.40
6779		14" wide				13.35	8.70	22.05
6780								

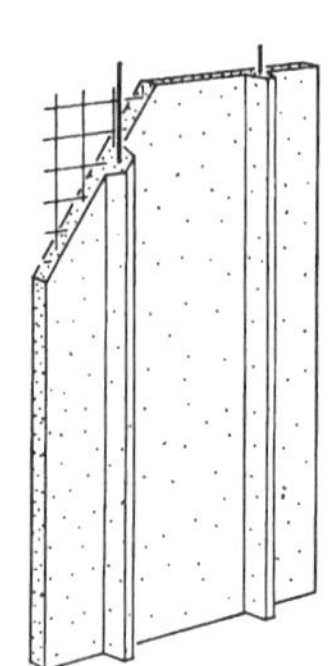

Ribbed Precast Panel

4.1-140 — Ribbed Precast Concrete

	THICKNESS (IN.)	PANEL SIZE (FT.)	FINISHES	RIGID INSULATION(IN.)	TYPE	COST PER S.F. MAT.	INST.	TOTAL
6800	4	4x8	aggregate	none	high rise	12.95	7.60	20.55
6850		8x8				9.45	5.55	15
6900		10x10				13.45	2.01	15.46
6950		20x10				11.90	1.77	13.67
7000	5	4x8	aggregate	none	high rise	13.20	7.70	20.90
7050		8x8				9.75	5.70	15.45
7100		10x10				14	2.09	16.09
7150		20x10				12.50	1.86	14.36
7200	6	4x8	aggregate	none	high rise	13.45	7.85	21.30
7250		8x8				10	5.85	15.85
7300		10x10				14.45	2.15	16.60
7350		20x10				12.95	1.93	14.88
7400	6	4x8	aggregate	2	high rise	14.30	8.15	22.45
7450		8x8				10.85	6.10	16.95
7500		10x10				15.30	2.42	17.72
7550		20x10				13.80	2.20	16
7600	7	4x8	aggregate	none	high rise	13.75	8.05	21.80
7650		8x8				10.25	6	16.25
7700		10x10				14.90	2.22	17.12
7750		20x10				13.30	1.99	15.29
7800	7	4x8	aggregate	2	high rise	14.60	8.30	22.90
7850		8x8				11.10	6.25	17.35
7900		10x10				15.75	2.49	18.24
7950		20x10				14.15	2.26	16.41
8000	8	4x8	aggregate	none	high rise	14	8.20	22.20
8050		8x8				10.55	6.15	16.70
8100		10x10				15.40	2.30	17.70
8150		20x10				13.90	2.07	15.97
8200	8	4x8	aggregate	2	high rise	14.85	8.45	23.30
8250		8x8				11.40	6.45	17.85
8300		10x10				16.25	2.57	18.82
8350		20x10				14.75	2.34	17.09

Important: See the Reference Section for critical supporting data - Reference Nos., Crews & Location Factors

The advantage of tilt up construction is in the low cost of forms and placing of concrete and reinforcing. Tilt up has been used for several types of buildings, including warehouses, stores, offices, and schools. The panels are cast in forms on the ground, or floor slab. Most jobs use 5-1/2″ thick solid reinforced concrete panels.

Design Assumptions:
Conc. f'c = 3000 psi
Reinf. fy = 60,000

4.1-160	Tilt-Up Concrete Panel	COST PER S.F.		
		MAT.	INST.	TOTAL
3200	Tilt up conc panels, broom finish, 5-1/2″ thick, 3000 PSI	2.40	2.01	4.41
3250	5000 PSI	2.47	1.97	4.44
3300	6″ thick, 3000 PSI	2.63	2.08	4.71
3350	5000 PSI	2.69	2.03	4.72
3400	7-1/2″ thick, 3000 PSI	3.27	2.19	5.46
3450	5000 PSI	3.36	2.14	5.50
3500	8″ thick, 3000 PSI	3.50	2.25	5.75
3550	5000 PSI	3.61	2.20	5.81
3700	Steel trowel finish, 5-1/2″ thick, 3000 PSI	2.41	2.09	4.50
3750	5000 PSI	2.47	2.04	4.51
3800	6″ thick, 3000 PSI	2.63	2.15	4.78
3850	5000 PSI	2.69	2.10	4.79
3900	7-1/2″ thick, 3000 PSI	3.27	2.26	5.53
3950	5000 PSI	3.36	2.21	5.57
4000	8″ thick, 3000 PSI	3.50	2.32	5.82
4050	5000 PSI	3.61	2.27	5.88
4200	Exp. aggregate finish, 5-1/2″ thick, 3000 PSI	2.87	2.10	4.97
4250	5000 PSI	2.92	2.05	4.97
4300	6″ thick, 3000 PSI	3.09	2.16	5.25
4350	5000 PSI	3.15	2.11	5.26
4400	7-1/2″ thick, 3000 PSI	3.73	2.27	6
4450	5000 PSI	3.82	2.22	6.04
4500	8″ thick, 3000 PSI	3.95	2.32	6.27
4550	5000 PSI	4.07	2.28	6.35
4600	Exposed aggregate & vert. rustication 5-1/2″ thick, 3000 PSI	4.22	3.72	7.94
4650	5000 PSI	4.27	3.67	7.94
4700	6″ thick, 3000 PSI	4.44	3.78	8.22
4750	5000 PSI	4.50	3.73	8.23
4800	7-1/2″ thick, 3000 PSI	5.10	3.89	8.99
4850	5000 PSI	5.15	3.84	8.99
4900	8″ thick, 3000 PSI	5.30	3.94	9.24
4950	5000 PSI	5.40	3.90	9.30
5000	Vertical rib & light sandblast, 5-1/2″ thick, 3000 PSI	4.51	2.60	7.11
5100	6″ thick, 3000 PSI	4.73	2.66	7.39
5200	7-1/2″ thick, 3000 PSI	5.35	2.77	8.12
5300	8″ thick, 3000 PSI	5.60	2.82	8.42
6000	Broom finish w/2″ urethane insulation, 6″ thick, 3000 PSI	2.39	2.62	5.01
6100	Broom finish 2″ fiberplank insulation, 6″ thick, 3000 PSI	2.89	2.59	5.48
6200	Exposed aggregate w/2″ urethane insulation, 6″ thick, 3000 PSI	2.79	2.73	5.52
6300	Exposed aggregate 2″ fiberplank insulation, 6″ thick, 3000 PSI	3.29	2.70	5.99

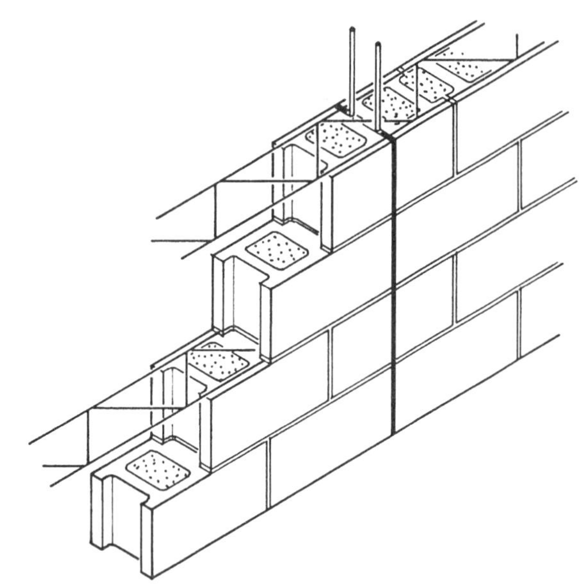

Exterior concrete block walls are defined in the following terms; structural reinforcement, weight, percent solid, size, strength and insulation. Within each of these categories, two to four variations are shown. No costs are included for brick shelf or relieving angles.

4.1-211 — Concrete Block Wall - Regular Weight

	TYPE	SIZE (IN.)	STRENGTH (P.S.I.)	CORE FILL		COST PER S.F. MAT.	INST.	TOTAL
1200	Hollow	4x8x16	2,000	none		.92	2.70	3.62
1250			4,500	none		1.37	2.70	4.07
1400		8x8x16	2,000	perlite		2	3.27	5.27
1410				styrofoam		2.41	3.09	5.50
1440				none		1.60	3.09	4.69
1450			4,500	perlite		2.83	3.27	6.10
1460				styrofoam		3.24	3.09	6.33
1490				none		2.43	3.09	5.52
2000	75% solid	4x8x16	2,000	none		1.08	2.73	3.81
2100		6x8x16	2,000	perlite		1.44	3.01	4.45
2500	Solid	4x8x16	2,000	none		1.35	2.79	4.14
2700		8x8x16	2,000	none		2.21	3.21	5.42

4.1-211 — Concrete Block Wall - Lightweight

	TYPE	SIZE (IN.)	WEIGHT (P.C.F.)	CORE FILL		COST PER S.F. MAT.	INST.	TOTAL
3100	Hollow	8x4x16	105	perlite		1.45	3.08	4.53
3110				styrofoam		1.86	2.90	4.76
3200		4x8x16	105	none		1.03	2.64	3.67
3250			85	none		1.27	2.58	3.85
3300		6x8x16	105	perlite		1.55	2.97	4.52
3310				styrofoam		2.09	2.82	4.91
3340				none		1.28	2.82	4.10
3400		8x8x16	105	perlite		1.96	3.19	5.15
3410				styrofoam		2.37	3.01	5.38
3440				none		1.56	3.01	4.57
3450			85	perlite		2.15	3.12	5.27
4000	75% solid	4x8x16	105	none		1.37	2.67	4.04
4050			85	none		2.13	2.61	4.74
4100		6x8x16	105	perlite		2.17	2.94	5.11
4500	Solid	4x8x16	105	none		1.39	2.76	4.15
4700		8x8x16	105	none		2.30	3.17	5.47

Important: See the Reference Section for critical supporting data - Reference Nos., Crews & Location Factors

4.1-211 — Reinforced Concrete Block Wall - Regular Weight

	TYPE	SIZE (IN.)	STRENGTH (P.S.I.)	VERT. REINF & GROUT SPACING		COST PER S.F.		
						MAT.	INST.	TOTAL
5200	Hollow	4x8x16	2,000	#4 @ 48"		1.01	3	4.01
5300		6x8x16	2,000	#4 @ 48"		1.28	3.23	4.51
5330				#5 @ 32"		1.41	3.37	4.78
5340				#5 @ 16"		1.70	3.86	5.56
5350			4,500	#4 @ 28"		1.92	3.23	5.15
5390				#5 @ 16"		2.34	3.86	6.20
5400		8x8x16	2,000	#4 @ 48"		1.83	3.45	5.28
5430				#5 @ 32"		1.94	3.69	5.63
5440				#5 @ 16"		2.29	4.27	6.56
5450			4,500	#4 @ 48"		2.62	3.50	6.12
5490				#5 @ 16"		3.12	4.27	7.39
5500		12x8x16	2,000	#4 @ 48"		2.43	4.41	6.84
5540				#5 @ 16"		3.11	5.20	8.31
6100	75% solid	6x8x16	2,000	#4 @ 48"		1.38	3.19	4.57
6140				#5 @ 16"		1.64	3.63	5.27
6150			4,500	#4 @ 48"		2.06	3.19	5.25
6190				#5 @ 16"		2.32	3.63	5.95
6200		8x8x16	2,000	#4 @ 48"		1.64	3.43	5.07
6230				#5 @ 32"		1.74	3.54	5.28
6240				#5 @ 16"		1.92	3.96	5.88
6250			4,500	#4 @ 48"		2.96	3.43	6.39
6280				#5 @ 32"		3.06	3.54	6.60
6290				#5 @ 16"		3.24	3.96	7.20
6500	Solid-double Wythe	2-4x8x16	2,000	#4 @ 48" E.W.		3.23	6.40	9.63
6530				#5 @ 16" E.W.		3.63	6.80	10.43
6550			4,500	#4 @ 48" E.W.		3.49	6.35	9.84
6580				#5 @ 16" E.W.		3.89	6.75	10.64

4.1-211 — Reinforced Concrete Block Wall - Lightweight

	TYPE	SIZE (IN.)	WEIGHT (P.C.F.)	VERT REINF. & GROUT SPACING		COST PER S.F.		
						MAT.	INST.	TOTAL
7100	Hollow	8x4x16	105	#4 @ 48"		1.24	3.31	4.55
7140				#5 @ 16"		1.74	4.08	5.82
7150			85	#4 @ 48"		3.02	3.24	6.26
7190				#5 @ 16"		3.52	4.01	7.53
7400		8x8x16	105	#4 @ 48"		1.75	3.42	5.17
7440				#5 @ 16"		2.25	4.19	6.44
7450		8x8x16	85	#4 @ 48"		1.94	3.35	5.29
7490				#5 @ 16"		2.44	4.12	6.56
7800		8x8x24	105	#4 @ 48"		2.26	3.67	5.93
7840				#5 @ 16"		2.76	4.44	7.20
7850			85	#4 @ 48"		3.30	3.15	6.45
7890				#5 @ 16"		3.80	3.92	7.72
8100	75% solid	6x8x16	105	#4 @ 48"		2.11	3.12	5.23
8130				#5 @ 32"		2.20	3.21	5.41
8150			85	#4 @ 48"		2.53	3.05	5.58
8180				#5 @ 32"		2.62	3.14	5.76
8200		8x8x16	105	#4 @ 48"		3.01	3.35	6.36
8230				#5 @ 32"		3.11	3.46	6.57
8250			85	#4 @ 48"		3.43	3.27	6.70
8280				#5 @ 32"		3.53	3.38	6.91

4.1-211 — Reinforced Concrete Block Wall - Lightweight

	TYPE	SIZE (IN.)	WEIGHT (P.C.F.)	VERT REINF. & GROUT SPACING		COST PER S.F.		
						MAT.	INST.	TOTAL
8500	Solid-double	2-4x8x16	105	#4 @ 48"		3.31	6.35	9.66
8530	Wythe		105	#5 @ 16"		3.71	6.75	10.46
8600		2-6x8x16	105	#4 @ 48"		4.54	6.80	11.34
8630				#5 @ 16"		4.94	7.20	12.14
8650			85	#4 @ 48"		5.85	6.50	12.35

Important: See the Reference Section for critical supporting data - Reference Nos., Crews & Location Factors

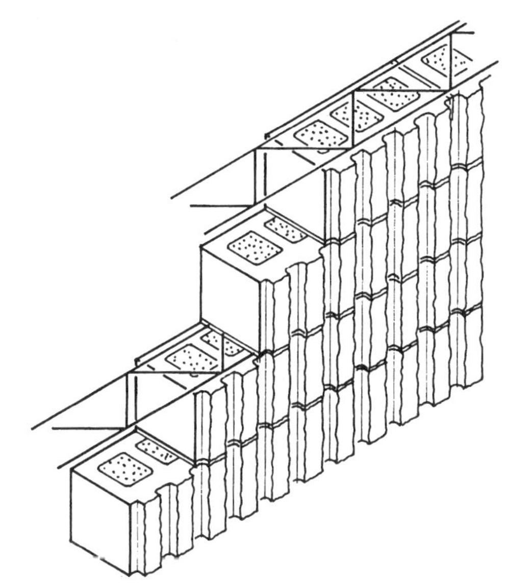

Exterior split ribbed block walls are defined in the following terms; structural reinforcement, weight, percent solid, size, number of ribs and insulation. Within each of these categories two to four variations are shown. No costs are included for brick shelf or relieving angles. Costs include control joints every 20′ and horizontal reinforcing.

4.1-212 — Split Ribbed Block Wall - Regular Weight

	TYPE	SIZE (IN.)	RIBS	CORE FILL		COST PER S.F. MAT.	COST PER S.F. INST.	COST PER S.F. TOTAL
1220	Hollow	4x8x16	4	none		2.05	3.34	5.39
1250			8	none		2.39	3.34	5.73
1280			16	none		1.94	3.38	5.32
1430		8x8x16	8	perlite		4.10	3.95	8.05
1440				styrofoam		4.51	3.77	8.28
1450				none		3.70	3.77	7.47
1530		12x8x16	8	perlite		5.15	5.25	10.40
1540				styrofoam		5.45	4.88	10.33
1550				none		4.47	4.88	9.35
2120	75% solid	4x8x16	4	none		2.57	3.38	5.95
2150			8	none		3.01	3.38	6.39
2180			16	none		2.46	3.43	5.89
2520	Solid	4x8x16	4	none		3.11	3.43	6.54
2550			8	none		3.65	3.43	7.08
2580			16	none		2.94	3.48	6.42

4.1-212 — Reinforced Split Ribbed Block Wall - Regular Weight

	TYPE	SIZE (IN.)	RIBS	VERT. REINF. & GROUT SPACING		COST PER S.F. MAT.	COST PER S.F. INST.	COST PER S.F. TOTAL
5200	Hollow	4x8x16	4	#4 @ 48″		2.14	3.64	5.78
5230			8	#4 @ 48″		2.48	3.64	6.12
5260			16	#4 @ 48″		2.03	3.68	5.71
5430		8x8x16	8	#4 @ 48″		3.89	4.18	8.07
5440				#5 @ 32″		4.04	4.37	8.41
5450				#5 @ 16″		4.39	4.95	9.34
5530		12x8x16	8	#4 @ 48″		4.75	5.30	10.05
5540				#5 @ 32″		4.95	5.50	10.45
5550				#5 @ 16″		5.45	6.15	11.60
6230	75% solid	6x8x16	8	#4 @ 48″		3.83	3.85	7.68
6240				#5 @ 32″		3.92	3.94	7.86
6250				#5 @ 16″		4.09	4.29	8.38
6330		8x8x16	8	#4 @ 48″		4.80	4.13	8.93
6340				#5 @ 32″		4.90	4.24	9.14
6350				#5 @ 16″		5.10	4.66	9.76

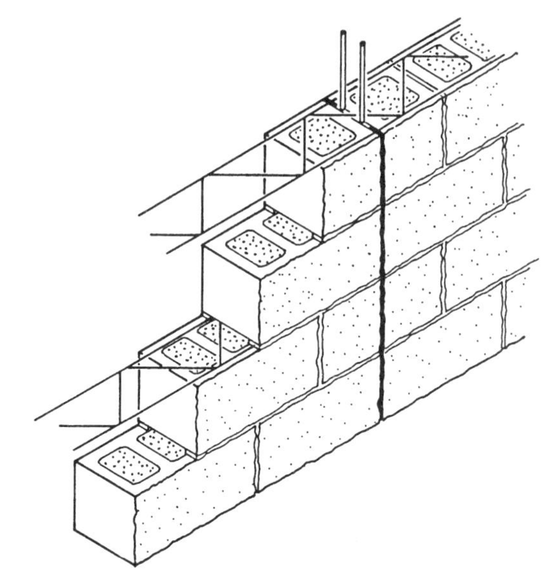

Exterior split face block walls are defined in the following terms; structural reinforcement, weight, percent solid, size, scores and insulation. Within each of these categories two to four variations are shown. No costs are included for brick shelf or relieving angles. Costs include control joints every 20′ and horizontal reinforcing.

4.1-213 — Split Face Block Wall - Regular Weight

	TYPE	SIZE (IN.)	SCORES	CORE FILL		MAT.	INST.	TOTAL
1200	Hollow	8x4x16	0	perlite		5.95	4.35	10.30
1210				styrofoam		6.40	4.17	10.57
1240				none		5.55	4.17	9.72
1250			1	perlite		6.35	4.35	10.70
1260				styrofoam		6.80	4.17	10.97
1290				none		5.95	4.17	10.12
1300		12x4x16	0	perlite		7.75	4.96	12.71
1310				styrofoam		8.10	4.60	12.70
1340				none		7.10	4.60	11.70
1350			1	perlite		8.20	4.96	13.16
1360				styrofoam		8.55	4.60	13.15
1390				none		7.55	4.60	12.15
1400		4x8x16	0	none		2.05	3.29	5.34
1450			1	none		2.31	3.34	5.65
1500		6x8x16	0	perlite		2.92	3.79	6.71
1510				styrofoam		3.46	3.64	7.10
1540				none		2.65	3.64	6.29
1550			1	perlite		3.17	3.85	7.02
1560				styrofoam		3.71	3.70	7.41
1590				none		2.90	3.70	6.60
1600		8x8x16	0	perlite		3.53	4.07	7.60
1610				styrofoam		3.94	3.89	7.83
1640				none		3.13	3.89	7.02
1650		8x8x16	1	perlite		3.74	4.14	7.88
1660				styrofoam		4.15	3.96	8.11
1690				none		3.34	3.96	7.30
1700		12x8x16	0	perlite		4.52	5.35	9.87
1710				styrofoam		4.86	4.97	9.83
1740				none		3.86	4.97	8.83
1750			1	perlite		4.73	5.40	10.13
1760				styrofoam		5.05	5.05	10.10
1790				none		4.07	5.05	9.12

Important: See the Reference Section for critical supporting data - Reference Nos., Crews & Location Factors

4.1-213   Split Face Block Wall - Regular Weight

	TYPE	SIZE (IN.)	SCORES	CORE FILL		MAT.	INST.	TOTAL
						\- COST PER S.F. \-		
1800	75% solid	8x4x16	0	perlite		7.20	4.33	11.53
1840				none		7	4.24	11.24
1852		8x4x16	1	perlite		7.65	4.33	11.98
1890				none		7.45	4.24	11.69
2000		4x8x16	0	none		2.57	3.34	5.91
2050			1	none		2.78	3.38	6.16
2400	Solid	8x4x16	0	none		8.55	4.32	12.87
2450			1	none		8.95	4.32	13.27
2900		12x8x16	0	none		5.90	5.20	11.10
2950			1	none		6.10	5.30	11.40

4.1-213   Reinforced Split Face Block Wall - Regular Weight

	TYPE	SIZE (IN.)	SCORES	VERT. REINF. & GROUT SPACING		MAT.	INST.	TOTAL
						\- COST PER S.F. \-		
5200	Hollow	8x4x16	0	#4 @ 48"		5.75	4.58	10.33
5210				#5 @ 32"		5.90	4.77	10.67
5240				#5 @ 16"		6.25	5.35	11.60
5250			1	#4 @ 48"		6.15	4.58	10.73
5260				#5 @ 32"		6.30	4.77	11.07
5290				#5 @ 16"		6.65	5.35	12
5700		12x8x16	0	#4 @ 48"		4.14	5.40	9.54
5710				#5 @ 32"		4.34	5.55	9.89
5740				#5 @ 16"		4.82	6.20	11.02
5750			1	#4 @ 48"		4.35	5.50	9.85
5760				#5 @ 32"		4.55	5.65	10.20
5790				#5 @ 16"		5.05	6.30	11.35
6000	75% solid	8x4x16	0	#4 @ 48"		7.10	4.54	11.64
6010				#5 @ 32"		7.20	4.65	11.85
6040				#5 @ 16"		7.40	5.05	12.45
6050			1	#4 @ 48"		7.55	4.54	12.09
6060				#5 @ 32"		7.65	4.65	12.30
6090				#5 @ 16"		7.85	5.05	12.90
6100		12x4x16	0	#4 @ 48"		9.15	5.05	14.20
6110				#5 @ 32"		9.25	5.15	14.40
6140				#5 @ 16"		9.50	5.60	15.10
6150			1	#4 @ 48"		9.50	5.05	14.55
6160				#5 @ 32"		9.60	5.15	14.75
6190				#5 @ 16"		9.90	5.65	15.55
6700	Solid-double Wythe	2-4x8x16	0	#4 @ 48" E.W.		6.75	7.60	14.35
6710				#5 @ 32" E.W.		6.90	7.65	14.55
6740				#5 @ 16" E.W.		7.15	8	15.15
6750			1	#4 @ 48" E.W.		7.15	7.70	14.85
6760				#5 @ 32" E.W.		7.30	7.75	15.05
6790				#5 @ 16" E.W.		7.55	8.10	15.65
6800		2-6x8x16	0	#4 @ 48" E.W.		8.75	8.40	17.15
6810				#5 @ 32" E.W.		8.90	8.45	17.35
6840				#5 @ 16" E.W.		9.15	8.80	17.95
6850			1	#4 @ 48" E.W.		9.20	8.50	17.70
6860				#5 @ 32" E.W.		9.35	8.55	17.90
6890				#5 @ 16" E.W.		9.60	8.90	18.50

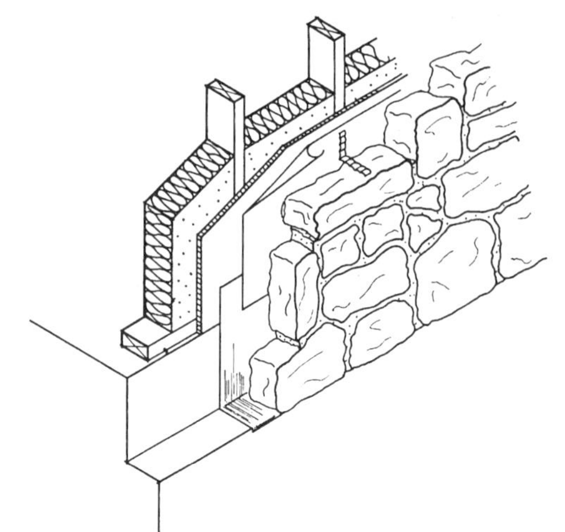

The table below lists costs per S.F. for stone veneer walls on various backup using different stone. Typical components for a system are shown in the component block below.

4.1-242	Stone Veneer	COST PER S.F.		
		MAT.	INST.	TOTAL
2000	Ashlar veneer,4", 2"x4" stud backup,16" O.C., 8' high, low priced stone	6.50	8.20	14.70
2100	Metal stud backup, 8' high, 16" O.C.	7	8.60	15.60
2150	24" O.C.	6.85	8.45	15.30
2200	Conc. block backup, 4" thick	6.75	10.20	16.95
2300	6" thick	6.90	10.35	17.25
2350	8" thick	7.05	10.45	17.50
2400	10" thick	7.50	10.55	18.05
2500	12" thick	7.55	11.20	18.75
3100	High priced stone, wood stud backup, 10' high, 16" O.C.	11.20	9.25	20.45
3200	Metal stud backup, 10' high, 16" O.C.	11.65	9.70	21.35
3250	24" O.C.	11.50	9.50	21
3300	Conc. block backup, 10' high, 4" thick	11.40	11.30	22.70
3350	6" thick	11.55	11.30	22.85
3400	8" thick	11.70	11.55	23.25
3450	10" thick	12.15	11.60	23.75
3500	12" thick	12.20	12.25	24.45
4000	Indiana limestone 2" thick,sawn finish,wood stud backup,10' high,16" O.C.	13.05	7.35	20.40
4100	Metal stud backup, 10' high, 16" O.C.	13.55	7.80	21.35
4150	24" O.C.	13.40	7.60	21
4200	Conc. block backup, 4" thick	13.30	9.40	22.70
4250	6" thick	13.45	9.50	22.95
4300	8" thick	13.60	9.65	23.25
4350	10" thick	14.05	9.70	23.75
4400	12" thick	14.10	10.35	24.45
4450	2" thick, smooth finish, wood stud backup, 8' high, 16" O.C.	13.05	7.35	20.40
4550	Metal stud backup, 8' high, 16" O.C.	13.55	7.70	21.25
4600	24" O.C.	13.40	7.55	20.95
4650	Conc. block backup, 4" thick	13.30	9.30	22.60
4700	6" thick	13.45	9.40	22.85
4750	8" thick	13.60	9.55	23.15
4800	10" thick	14.05	9.70	23.75
4850	12" thick	14.10	10.35	24.45
5350	4" thick, smooth finish, wood stud backup, 8' high, 16" O.C.	17.55	7.35	24.90
5450	Metal stud backup, 8' high, 16" O.C.	18.05	7.80	25.85
5500	24" O.C.	17.90	7.60	25.50
5550	Conc. block backup, 4" thick	17.80	9.40	27.20
5600	6" thick	17.95	9.50	27.45
5650	8" thick	18.10	9.65	27.75
5700	10" thick	18.55	9.70	28.25
5750	12" thick	18.60	10.35	28.95

Important: See the Reference Section for critical supporting data - Reference Nos., Crews & Location Factors

4.1-242	Stone Veneer	COST PER S.F.		
		MAT.	INST.	TOTAL
6000	Granite, grey or pink, 2″ thick, wood stud backup, 8′ high, 16″ O.C.	18.45	11.65	30.10
6100	Metal studs, 8′ high, 16″ O.C.	18.95	12.05	31
6150	24″ O.C.	18.80	11.90	30.70
6200	Conc. block backup, 4″ thick	18.70	13.65	32.35
6250	6″ thick	18.85	13.75	32.60
6300	8″ thick	19	13.90	32.90
6350	10″ thick	19.45	13.95	33.40
6400	12″ thick	19.50	14.65	34.15
6900	4″ thick, wood stud backup, 8′ high, 16″ O.C.	30	13.15	43.15
7000	Metal studs, 8′ high, 16″ O.C.	30.50	13.55	44.05
7050	24″ O.C.	30.50	13.40	43.90
7100	Conc. block backup, 4″ thick	30	15.15	45.15
7150	6″ thick	30.50	15.25	45.75
7200	8″ thick	30.50	15.40	45.90
7250	10″ thick	31	15.45	46.45
7300	12″ thick	31	16.15	47.15

Exterior brick veneer/stud backup walls are defined in the following terms: type of brick and studs, stud spacing and bond. All systems include a back-up wall, a control joint every 20′, a brick shelf every 12′ of height, ties to the backup and the necessary dampproofing, flashing and insulation.

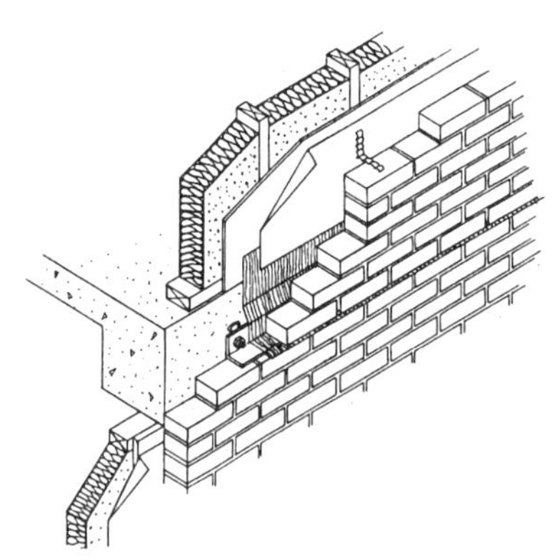

4.1-252   Brick Veneer/Wood Stud Backup

	FACE BRICK	STUD BACKUP	STUD SPACING (IN.)	BOND		COST PER S.F.		
						MAT.	INST.	TOTAL
1100	Standard	2x4-wood	16	running		4.91	7.30	12.21
1120				common		5.50	8.30	13.80
1140				Flemish		5.90	9.70	15.60
1160				English		6.40	10.20	16.60
1400		2x6-wood	16	running		5.10	7.35	12.45
1420				common		5.70	8.35	14.05
1440				Flemish		6.10	9.75	15.85
1460				English		6.60	10.25	16.85
1700	Glazed	2x4-wood	16	running		9.95	7.55	17.50
1720				common		11.50	8.65	20.15
1740				Flemish		12.60	10.20	22.80
1760				English		13.90	10.85	24.75
2300	Engineer	2x4-wood	16	running		5.20	6.55	11.75
2320				common		5.85	7.30	13.15
2340				Flemish		6.30	8.65	14.95
2360				English		6.85	9	15.85
2900	Roman	2x4-wood	16	running		7	6.75	13.75
2920				common		8.05	7.55	15.60
2940				Flemish		8.70	8.80	17.50
2960				English		9.60	9.45	19.05
4100	Norwegian	2x4-wood	16	running		5.35	5.25	10.60
4120				common		6	5.80	11.80
4140				Flemish		6.45	6.75	13.20
4160				English		7.05	7	14.05

Important: See the Reference Section for critical supporting data - Reference Nos., Crews & Location Factors

4.1-252 — Brick Veneer/Metal Stud Backup

	FACE BRICK	STUD BACKUP	STUD SPACING (IN.)	BOND		COST PER S.F.		
						MAT.	INST.	TOTAL
5100	Standard	25ga.x6"NLB	24	running		4.62	7.40	12.02
5120				common		5.20	8.40	13.60
5140				Flemish		5.35	9.30	14.65
5160				English		6.10	10.30	16.40
5200		20ga.x3-5/8"NLB	16	running		4.97	7.45	12.42
5220				common		5.55	8.45	14
5240				Flemish		5.95	9.85	15.80
5260				English		6.45	10.35	16.80
5400		16ga.x3-5/8"LB	16	running		5.35	7.80	13.15
5420				common		5.90	8.80	14.70
5440				Flemish		6.30	10.20	16.50
5460				English		6.85	10.70	17.55
5700	Glazed	25ga.x6"NLB	24	running		9.65	7.65	17.30
5720				common		11.20	8.75	19.95
5740				Flemish		12.30	10.30	22.60
5760				English		13.60	10.95	24.55
5800		20ga.x3-5/8"NLB	24	running		9.85	7.60	17.45
5820				common		11.40	8.70	20.10
5840				Flemish		12.50	10.25	22.75
5860				English		13.80	10.90	24.70
6000		16ga.x3-5/8"LB	16	running		10.35	8.05	18.40
6020				common		11.90	9.15	21.05
6040				Flemish		13	10.70	23.70
6060				English		14.30	11.35	25.65
6300	Engineer	25ga.x6"NLB	24	running		4.91	6.60	11.51
6320				common		5.55	7.40	12.95
6340				Flemish		6	8.75	14.75
6360				English		6.55	9.10	15.65
6400		20ga.x3-5/8"NLB	16	running		5.25	6.65	11.90
6420				common		5.90	7.45	13.35
6440				Flemish		6.35	8.80	15.15
6460				English		6.90	9.15	16.05
6900	Roman	25ga.x6"NLB	24	running		6.70	6.80	13.50
6920				common		7.75	7.65	15.40
6940				Flemish		8.40	8.90	17.30
6960				English		9.30	9.55	18.85
7000		20ga.x3-5/8"NLB	16	running		7.05	6.85	13.90
7020				common		8.10	7.70	15.80
7040				Flemish		8.75	8.95	17.70
7060				English		9.65	9.60	19.25
7500	Norman	25ga.x6"NLB	24	running		5.80	5.80	11.60
7520				common		6.60	6.45	13.05
7540				Flemish		10.50	7.55	18.05
7560				English		7.90	7.95	15.85
7600		20ga.x3-5/8"NLB	24	running		6	5.80	11.80
7620				common		6.85	6.45	13.30
7640				Flemish		10.70	7.50	18.20
7660				English		8.10	7.90	16
8100	Norwegian	25ga.x6"NLB	24	running		5.05	5.30	10.35
8120				common		5.70	5.85	11.55
8140				Flemish		6.15	6.80	12.95
8160				English		6.75	7.10	13.85

4.1-252 — Brick Veneer/Metal Stud Backup

	FACE BRICK	STUD BACKUP	STUD SPACING (IN.)	BOND		COST PER S.F.		
						MAT.	INST.	TOTAL
8400	Norwegian	16ga.x3-5/8"LB	16	running		5.75	5.75	11.50
8420				common		6.45	6.30	12.75
8440				Flemish		6.85	7.25	14.10
8460				English		7.45	7.50	14.95

Important: See the Reference Section for critical supporting data - Reference Nos., Crews & Location Factors

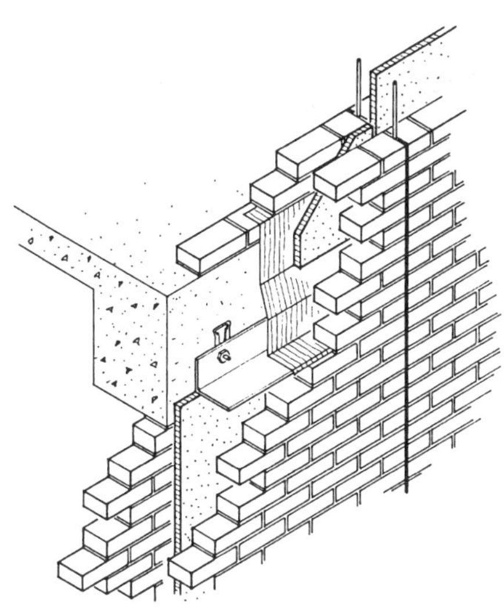

Exterior brick face cavity walls are defined in the following terms: cavity treatment, type of face brick, backup masonry, total thickness and insulation. Seven types of face brick are shown with fourteen types of backup. All systems include a brick shelf, ties to the backups and necessary dampproofing, flashing, and control joints every 20′.

4.1-273 — Brick Face Cavity Wall

	FACE BRICK	BACKUP MASONRY	TOTAL THICKNESS (IN.)	CAVITY INSULATION		COST PER S.F.		
						MAT.	INST.	TOTAL
1000	Standard	4″ common brick	10	polystyrene		6	11.55	17.55
1020				none		5.80	11.25	17.05
1040		6″ SCR brick	12	polystyrene		8.40	10.30	18.70
1060				none		8.20	10	18.20
1080		4″ conc. block	10	polystyrene		4.77	9.30	14.07
1100				none		4.57	9	13.57
1120		6″ conc. block	12	polystyrene		4.99	9.50	14.49
1140				none		4.79	9.20	13.99
1160		4″ L.W. block	10	polystyrene		4.88	9.25	14.13
1180				none		4.68	8.95	13.63
1200		6″ L.W. block	12	polystyrene		5.15	9.40	14.55
1220				none		4.95	9.10	14.05
1240		4″ glazed block	10	polystyrene		9.55	9.95	19.50
1260				none		9.35	9.65	19
1280		6″ glazed block	12	polystyrene		9.60	9.95	19.55
1300				none		9.40	9.65	19.05
1320		4″ clay tile	10	polystyrene		8.10	8.95	17.05
1340				none		7.90	8.65	16.55
1360		4″ glazed tile	10	polystyrene		11.55	11.75	23.30
1380				none		11.35	11.45	22.80
1500	Glazed	4″ common brick	10	polystyrene		11.05	11.80	22.85
1520				none		10.85	11.50	22.35
1580		4″ conc. block	10	polystyrene		9.80	9.55	19.35
1600				none		9.60	9.25	18.85
1660		4″ L.W. block	10	polystyrene		9.90	9.50	19.40
1680				none		9.70	9.20	18.90
1740		4″ glazed block	10	polystyrene		14.60	10.20	24.80
1760				none		14.40	9.90	24.30
1820		4″ clay tile	10	polystyrene		13.10	9.20	22.30
1840				none		12.90	8.90	21.80
1860		4″ glazed tile	10	polystyrene		13.25	9.45	22.70
1880				none		13.05	9.15	22.20
2000	Engineer	4″ common brick	10	polystyrene		6.30	10.80	17.10
2020				none		6.10	10.50	16.60
2080		4″ conc. block	10	polystyrene		5.05	8.55	13.60
2100				none		4.86	8.25	13.11

4.1-273 Brick Face Cavity Wall

	FACE BRICK	BACKUP MASONRY	TOTAL THICKNESS (IN.)	CAVITY INSULATION		COST PER S.F.		
						MAT.	INST.	TOTAL
2162	Engineer	4" L.W. block	10	polystyrene		5.15	8.50	13.65
2180				none		4.97	8.20	13.17
2240		4" glazed block	10	polystyrene		9.85	9.20	19.05
2260				none		9.65	8.90	18.55
2320		4" clay tile	10	polystyrene		8.35	8.15	16.50
2340				none		8.15	7.85	16
2360		4" glazed tile	10	polystyrene		11.85	11	22.85
2380				none		11.65	10.70	22.35
2500	Roman	4" common brick	10	polystyrene		8.10	10.95	19.05
2520				none		7.90	10.65	18.55
2580		4" conc. block	10	polystyrene		6.85	8.70	15.55
2600				none		6.65	8.40	15.05
2660		4" L.W. block	10	polystyrene		6.95	8.65	15.60
2680				none		6.75	8.35	15.10
2740		4" glazed block	10	polystyrene		11.65	9.35	21
2760				none		11.45	9.05	20.50
2820		4" clay tile	10	polystyrene		10.15	8.35	18.50
2840				none		9.95	8.05	18
2860		4" glazed tile	10	polystyrene		13.65	11.15	24.80
2880				none		13.45	10.85	24.30
3000	Norman	4" common brick	10	polystyrene		7.20	10	17.20
3020				none		7	9.70	16.70
3080		4" conc. block	10	polystyrene		5.95	7.75	13.70
3100				none		5.75	7.45	13.20
3160		4" L.W. block	10	polystyrene		6.05	7.70	13.75
3180				none		5.85	7.40	13.25
3240		4" glazed block	10	polystyrene		10.75	8.40	19.15
3260				none		10.55	8.10	18.65
3320		4" clay tile	10	polystyrene		9.25	7.35	16.60
3340				none		9.05	7.05	16.10
3360		4" glazed tile	10	polystyrene		12.75	10.20	22.95
3380				none		12.55	9.90	22.45
3500	Norwegian	4" common brick	10	polystyrene		6.45	9.50	15.95
3520				none		6.25	9.20	15.45
3580		4" conc. block	10	polystyrene		5.20	7.25	12.45
3600				none		5	6.95	11.95
3660		4" L.W. block	10	polystyrene		5.30	7.15	12.45
3680				none		5.10	6.85	11.95
3740		4" glazed block	10	polystyrene		10	7.85	17.85
3760				none		9.80	7.55	17.35
3820		4" clay tile	10	polystyrene		8.50	6.85	15.35
3840				none		8.30	6.55	14.85
3860		4" glazed tile	10	polystyrene		12	9.70	21.70
3880				none		11.80	9.40	21.20
4000	Utility	4" common brick	10	polystyrene		6.10	9	15.10
4020				none		5.90	8.70	14.60
4080		4" conc. block	10	polystyrene		4.86	6.75	11.61
4100				none		4.66	6.45	11.11
4160		4" L.W. block	10	polystyrene		4.97	6.65	11.62
4180				none		4.77	6.35	11.12
4240		4" glazed block	10	polystyrene		9.65	7.35	17
4260				none		9.45	7.05	16.50
4320		4" clay tile	10	polystyrene		8.15	6.35	14.50
4340				none		7.95	6.05	14

Important: See the Reference Section for critical supporting data - Reference Nos., Crews & Location Factors

4.1-273 — Brick Face Cavity Wall

	FACE BRICK	BACKUP MASONRY	TOTAL THICKNESS (IN.)	CAVITY INSULATION		COST PER S.F.		
						MAT.	INST.	TOTAL
4360	Utility	4″ glazed tile	10	polystyrene		11.65	9.20	20.85
4380				none		11.45	8.90	20.35

4.1-273 — Brick Face Cavity Wall - Insulated Backup

	FACE BRICK	BACKUP MASONRY	TOTAL THICKNESS (IN.)	BACKUP CORE FILL		COST PER S.F. MAT.	INST.	TOTAL
5100	Standard	6" conc. block	10	perlite		5.05	9.35	14.40
5120				styrofoam		5.60	9.20	14.80
5180		6" L.W. block	10	perlite		5.20	9.25	14.45
5200				styrofoam		5.75	9.10	14.85
5260		6" glazed block	10	perlite		10.35	9.95	20.30
5280				styrofoam		10.85	9.80	20.65
5340		6" clay tile	10	none		8.05	8.90	16.95
5360		8" clay tile	12	none		9	9.20	18.20
5600	Glazed	6" conc. block	10	perlite		10.05	9.60	19.65
5620				styrofoam		10.60	9.45	20.05
5680		6" L.W. block	10	perlite		10.20	9.50	19.70
5700				styrofoam		10.75	9.35	20.10
5760		6" glazed block	10	perlite		15.35	10.20	25.55
5780				styrofoam		15.90	10.05	25.95
5840		6" clay tile	10	none		13.05	9.15	22.20
5860		8"clay tile	8	none		14	9.45	23.45
6100	Engineer	6" conc. block	10	perlite		5.35	8.60	13.95
6120				styrofoam		5.85	8.45	14.30
6180		6" L.W. block	10	perlite		5.50	8.50	14
6200				styrofoam		6.05	8.35	14.40
6260		6" glazed block	10	perlite		10.60	9.15	19.75
6280				styrofoam		11.15	9	20.15
6340		6" clay tile	10	none		8.35	8.10	16.45
6360		8" clay tile	12	none		9.25	8.45	17.70
6600	Roman	6" conc. block	10	perlite		7.15	8.75	15.90
6620				styrofoam		7.65	8.60	16.25
6680		6" L.W. block	10	perlite		7.30	8.70	16
6700				styrofoam		7.85	8.55	16.40
6760		6" glazed block	10	perlite		12.40	9.35	21.75
6780				styrofoam		12.95	9.20	22.15
6840		6" clay tile	10	none		10.15	8.30	18.45
6860		8" clay tile	12	none		11.05	8.60	19.65
7100	Norman	6" conc. block	10	perlite		6.20	7.80	14
7120				styrofoam		6.75	7.65	14.40
7180		6" L.W. block	10	perlite		6.40	7.70	14.10
7200				styrofoam		6.90	7.55	14.45
7260		6" glazed block	10	perlite		11.50	8.35	19.85
7280				styrofoam		12.05	8.20	20.25
7340		6" clay tile	10	none		9.20	7.30	16.50
7360		8" clay tile	12	none		10.15	7.65	17.80
7600	Norwegian	6" conc. block	10	perlite		5.45	7.25	12.70
7620				styrofoam		6	7.10	13.10
7680		6" L.W. block	10	perlite		5.65	7.20	12.85
7700				styrofoam		6.15	7.05	13.20
7760		6" glazed block	10	perlite		10.75	7.85	18.60
7780				styrofoam		11.30	7.70	19
7840		6" clay tile	10	none		8.45	6.80	15.25
7860		8" clay tile	12	none		9.40	7.10	16.50
8100	Utility	6" conc. block	10	perlite		5.15	6.75	11.90
8120				styrofoam		5.65	6.60	12.25
8180		6" L.W. block	10	perlite		5.30	6.70	12
8200				styrofoam		5.85	6.55	12.40
8260		6" glazed block	10	perlite		10.40	7.35	17.75
8280				styrofoam		10.95	7.20	18.15

Important: See the Reference Section for critical supporting data - Reference Nos., Crews & Location Factors

4.1-273   Brick Face Cavity Wall - Insulated Backup

	FACE BRICK	BACKUP MASONRY	TOTAL THICKNESS (IN.)	BACKUP CORE FILL		COST PER S.F.		
						MAT.	INST.	TOTAL
8340	Utility	6″ clay tile	10	none		8.15	6.30	14.45
8360		8″ clay tile	12	none		9.05	6.60	15.65

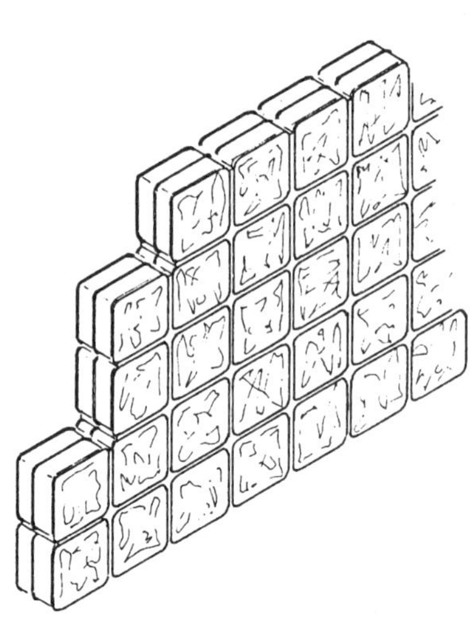

The table below lists costs per S.F. for glass block walls. Included in the costs are the following special accessories required for glass block walls.

Glass block accessories required for proper installation.

Wall ties: Galvanized double steel mesh full length of joint.

Fiberglass expansion joint at sides and top.

Silicone caulking: One gallon does 95 L.F.

Oakum: One lb. does 30 L.F.

Asphalt emulsion: One gallon does 600 L.F.

If block are not set in wall chase, use 2'-0" long wall anchors at 2'-0" O.C.

4.1-282	Glass Block	COST PER S.F.		
		MAT.	INST.	TOTAL
2300	Glass block 4" thick, 6"x6" plain, under 1,000 S.F.	17.70	10.80	28.50
2400	1,000 to 5,000 S.F.	16.60	9.35	25.95
2500	Over 5,000 S.F.	15.50	8.80	24.30
2600	Solar reflective, under 1,000 S.F.	25	14.70	39.70
2700	1,000 to 5,000 S.F.	23	12.65	35.65
2800	Over 5,000 S.F.	21.50	11.90	33.40
3500	8"x8" plain, under 1,000 S.F.	12.50	8.10	20.60
3600	1,000 to 5,000 S.F.	11.85	7	18.85
3700	Over 5,000 S.F.	11.25	6.30	17.55
3800	Solar reflective, under 1,000 S.F.	17.45	10.90	28.35
3900	1,000 to 5,000 S.F.	16.55	9.40	25.95
4000	Over 5,000 S.F.	15.70	8.40	24.10
5000	12"x12" plain, under 1,000 S.F.	14.95	7.50	22.45
5100	1,000 to 5,000 S.F.	14.15	6.30	20.45
5200	Over 5,000 S.F.	13.30	5.75	19.05
5300	Solar reflective, under 1,000 S.F.	21	10.10	31.10
5400	1,000 to 5,000 S.F.	19.80	8.40	28.20
5600	Over 5,000 S.F.	18.60	7.65	26.25
5800	3" thinline, 6"x6" plain, under 1,000 S.F.	14.85	10.80	25.65
5900	Over 5,000 S.F.	14.85	10.80	25.65
6000	Solar reflective, under 1,000 S.F.	21	14.70	35.70
6100	Over 5,000 S.F.	20	11.90	31.90
6200	8"x8" plain, under 1,000 S.F.	9.15	8.10	17.25
6300	Over 5,000 S.F.	8.65	6.30	14.95
6400	Solar reflective, under 1,000 S.F.	12.80	10.90	23.70
6500	Over 5,000 S.F.	12.10	8.40	20.50

Important: See the Reference Section for critical supporting data - Reference Nos., Crews & Location Factors

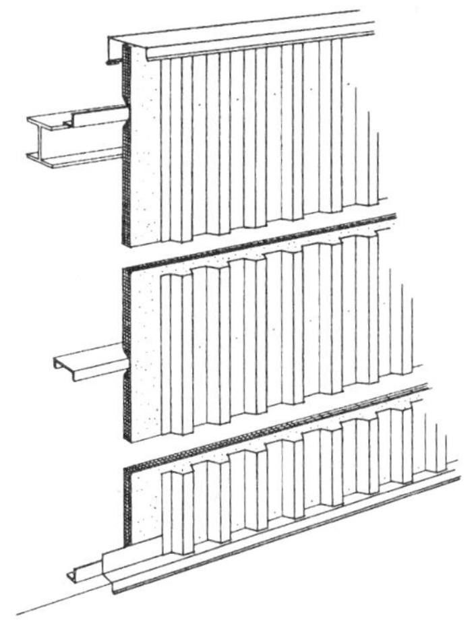

The table below lists costs for metal siding of various descriptions, not including the steel frame, or the structural steel, of a building. Costs are per S.F. including all accessories and insulation.

For steel frame support see System A3.4-300.

4.1-384	Metal Siding Panel	COST PER S.F.		
		MAT.	INST.	TOTAL
1400	Metal siding aluminum panel, corrugated, .024" thick, natural	1.43	1.65	3.08
1450	Painted	1.65	1.65	3.30
1500	.032" thick, natural	1.84	1.65	3.49
1550	Painted	2.19	1.65	3.84
1600	Ribbed 4" pitch, .032" thick, natural	1.86	1.65	3.51
1650	Painted	2.15	1.65	3.80
1700	.040" thick, natural	2.20	1.65	3.85
1750	Painted	2.52	1.65	4.17
1800	.050" thick, natural	2.52	1.65	4.17
1850	Painted	2.82	1.65	4.47
1900	8" pitch panel, .032" thick, natural	1.77	1.58	3.35
1950	Painted	2.06	1.59	3.65
2000	.040" thick, natural	2.11	1.59	3.70
2050	Painted	2.41	1.61	4.02
2100	.050" thick, natural	2.43	1.60	4.03
2150	Painted	2.75	1.61	4.36
3000	Steel, corrugated or ribbed, 29 Ga. .0135" thick, galvanized	1.22	1.48	2.70
3050	Colored	1.62	1.50	3.12
3100	26 Ga. .0179" thick, galvanized	1.26	1.49	2.75
3150	Colored	1.55	1.51	3.06
3200	24 Ga. .0239" thick, galvanized	1.34	1.49	2.83
3250	Colored	1.73	1.51	3.24
3300	22 Ga. .0299" thick, galvanized	1.52	1.50	3.02
3350	Colored	2.07	1.52	3.59
3400	20 Ga. .0359" thick, galvanized	1.52	1.50	3.02
3450	Colored	2.20	1.61	3.81
4100	Sandwich panels, factory fab., 1" polystyrene, steel core, 26 Ga., galv.	2.96	2.29	5.25
4200	Colored, 1 side	3.72	2.29	6.01
4300	2 sides	4.85	2.29	7.14
4400	2" polystyrene, steel core, 26 Ga., galvanized	3.53	2.29	5.82
4500	Colored, 1 side	4.29	2.29	6.58
4600	2 sides	5.40	2.29	7.69
4700	22 Ga., baked enamel exterior	7.20	2.42	9.62
4800	Polyvinyl chloride exterior	7.60	2.42	10.02
5100	Textured aluminum, 4' x 8' x 5/16" plywood backing, single face	2.40	1.39	3.79
5200	Double face	3.61	1.39	5

The table below lists costs per S.F. for exterior walls with wood siding. A variety of systems are presented using both wood and metal studs at 16″ and 24″ O.C.

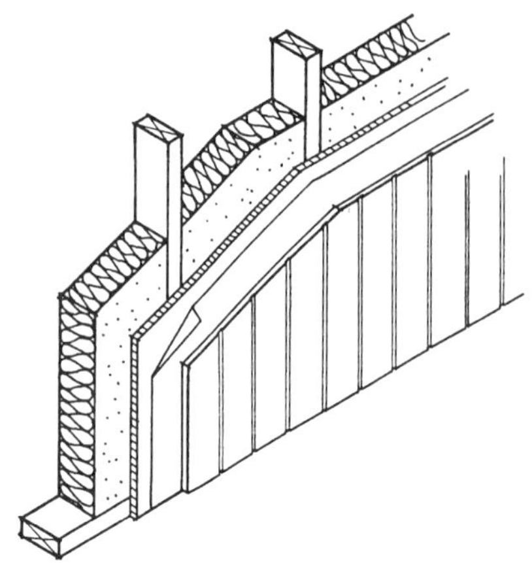

4.1-412	Wood & Other Siding	COST PER S.F.		
		MAT.	INST.	TOTAL
1400	Wood siding w/2″x4″studs, 16″O.C., insul. wall, 5/8″text 1-11 fir plywood	2.48	2.28	4.76
1450	5/8″ text 1-11 cedar plywood	3.02	2.20	5.22
1500	1″ x 4″ vert T.&G. redwood	4.38	3.06	7.44
1600	1″ x 8″ vert T.&G. redwood	3.97	2.78	6.75
1650	1″ x 5″ rabbetted cedar bev. siding	3.64	2.62	6.26
1700	1″ x 6″ cedar drop siding	3.69	2.64	6.33
1750	1″ x 12″ rough sawn cedar	2.66	2.30	4.96
1800	1″ x 12″ sawn cedar, 1″ x 4″ battens	3.66	2.58	6.24
1850	1″ x 10″ redwood shiplap siding	3.90	2.50	6.40
1900	18″ no. 1 red cedar shingles, 5-1/2″ exposed	3.65	3.16	6.81
1950	6″ exposed	3.49	3.05	6.54
2000	6-1/2″ exposed	3.33	2.95	6.28
2100	7″ exposed	3.16	2.84	6
2150	7-1/2″ exposed	3	2.73	5.73
3000	8″ wide aluminum siding	2.59	1.95	4.54
3150	8″ plain vinyl siding	1.92	2.08	4
3250	8″ insulated vinyl siding	2.08	2.08	4.16
3300				
3400	2″ x 6″ studs, 16″ O.C., insul. wall, w/ 5/8″ text 1-11 fir plywood	2.74	2.37	5.11
3500	5/8″ text 1-11 cedar plywood	3.28	2.37	5.65
3600	1″ x 4″ vert T.&G. redwood	4.64	3.15	7.79
3700	1″ x 8″ vert T.&G. redwood	4.23	2.87	7.10
3800	1″ x 5″ rabbetted cedar bev siding	3.90	2.71	6.61
3900	1″ x 6″ cedar drop siding	3.95	2.73	6.68
4000	1″ x 12″ rough sawn cedar	2.92	2.39	5.31
4200	1″ x 12″ sawn cedar, 1″ x 4″ battens	3.92	2.67	6.59
4500	1″ x 10″ redwood shiplap siding	4.16	2.59	6.75
4550	18″ no. 1 red cedar shingles, 5-1/2″ exposed	3.91	3.25	7.16
4600	6″ exposed	3.75	3.14	6.89
4650	6-1/2″ exposed	3.59	3.04	6.63
4700	7″ exposed	3.42	2.93	6.35
4750	7-1/2″ exposed	3.26	2.82	6.08
4800	8″ wide aluminum siding	2.85	2.04	4.89
4850	8″ plain vinyl siding	2.18	2.17	4.35
4900	8″ insulated vinyl siding	2.34	2.17	4.51
4910				
5000	2″ x 6″ studs, 24″ O.C., insul. wall, 5/8″ text 1-11, fir plywood	2.59	2.24	4.83
5050	5/8″ text 1-11 cedar plywood	3.13	2.24	5.37
5100	1″ x 4″ vert T.&G. redwood	4.49	3.02	7.51
5150	1″ x 8″ vert T.&G. redwood	4.08	2.74	6.82
5200	1″ x 5″ rabbetted cedar bev siding	3.75	2.58	6.33
5250	1″ x 6″ cedar drop siding	3.80	2.60	6.40

Important: See the Reference Section for critical supporting data - Reference Nos., Crews & Location Factors

4.1-412	Wood & Other Siding	COST PER S.F.		
		MAT.	**INST.**	**TOTAL**
5300	1" x 12" rough sawn cedar	2.77	2.26	5.03
5400	1" x 12" sawn cedar, 1" x 4" battens	3.77	2.54	6.31
5450	1" x 10" redwood shiplap siding	4.01	2.46	6.47
5500	18" no. 1 red cedar shingles, 5-1/2" exposed	3.76	3.12	6.88
5550	6" exposed	3.60	3.01	6.61
5650	7" exposed	3.27	2.80	6.07
5700	7-1/2" exposed	3.11	2.69	5.80
5750	8" wide aluminum siding	2.70	1.91	4.61
5800	8" plain vinyl siding	2.03	2.04	4.07
5850	8" insulated vinyl siding	2.19	2.04	4.23
5900	3-5/8" metal studs, 16 Ga., 16" OC insul.wall, 5/8"text 1-11 fir plywood	3.01	2.70	5.71
5950	5/8" text 1-11 cedar plywood	3.55	2.70	6.25
6000	1" x 4" vert T.&G. redwood	4.91	3.48	8.39
6050	1" x 8" vert T.&G. redwood	4.50	3.20	7.70
6100	1" x 5" rabbetted cedar bev siding	4.17	3.04	7.21
6150	1" x 6" cedar drop siding	4.22	3.06	7.28
6200	1" x 12" rough sawn cedar	3.19	2.72	5.91
6250	1" x 12" sawn cedar, 1" x 4" battens	4.19	3	7.19
6300	1" x 10" redwood shiplap siding	4.43	2.92	7.35
6350	18" no. 1 red cedar shingles, 5-1/2" exposed	4.18	3.58	7.76
6500	6" exposed	4.02	3.47	7.49
6550	6-1/2" exposed	3.86	3.37	7.23
6600	7" exposed	3.69	3.26	6.95
6650	7-1/2" exposed	3.69	3.18	6.87
6700	8" wide aluminum siding	3.12	2.37	5.49
6750	8" plain vinyl siding	2.45	2.42	4.87
6800	8" insulated vinyl siding	2.61	2.42	5.03
7000	3-5/8" metal studs, 16 Ga. 24" OC insul wall, 5/8" text 1-11 fir plywood	2.86	2.44	5.30
7050	5/8" text 1-11 cedar plywood	3.40	2.44	5.84
7100	1" x 4" vert T.&G. redwood	4.76	3.30	8.06
7150	1" x 8" vert T.&G. redwood	4.35	3.02	7.37
7200	1" x 5" rabbetted cedar bev siding	4.02	2.86	6.88
7250	1" x 6" cedar drop siding	4.06	2.88	6.94
7300	1" x 12" rough sawn cedar	3.04	2.54	5.58
7350	1" x 12" sawn cedar 1" x 4" battens	4.04	2.82	6.86
7400	1" x 10" redwood shiplap siding	4.28	2.74	7.02
7450	18" no. 1 red cedar shingles, 5-1/2" exposed	4.20	3.51	7.71
7500	6" exposed	4.03	3.40	7.43
7550	6-1/2" exposed	3.87	3.29	7.16
7600	7" exposed	3.71	3.19	6.90
7650	7-1/2" exposed	3.38	2.97	6.35
7700	8" wide aluminum siding	2.97	2.19	5.16
7750	8" plain vinyl siding	2.30	2.32	4.62
7800	8" insul. vinyl siding	2.46	2.32	4.78

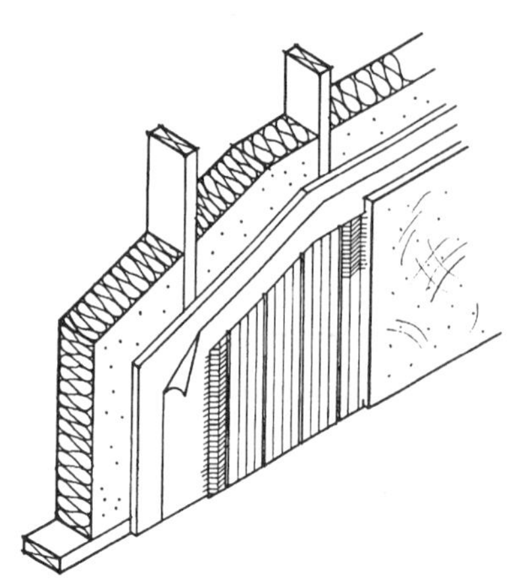

The table below lists costs for some typical stucco walls including all the components as demonstrated in the component block below. Prices are presented for backup walls using wood studs, metal studs and CMU.

4.5-110	Stucco Wall	COST PER S.F.		
		MAT.	INST.	TOTAL
2100	Cement stucco, 7/8" th., plywood sheathing, stud wall, 2" x 4", 16" O.C.	2.19	2.68	4.87
2200	24" O.C.	2.09	2.58	4.67
2300	2" x 6", 16" O.C.	2.45	2.77	5.22
2400	24" O.C.	2.30	2.64	4.94
2500	No sheathing, metal lath on stud wall, 2" x 4", 16" O.C.	1.60	2.21	3.81
2600	24" O.C.	1.50	2.11	3.61
2700	2" x 6", 16" O.C.	1.86	2.30	4.16
2800	24" O.C.	1.71	2.17	3.88
2900	1/2" gypsum sheathing, 3-5/8" metal studs, 16" O.C.	2.16	2.57	4.73
2950	24" O.C.	2.01	2.39	4.40
3000	Cement stucco, 5/8" th., 2 coats on std. CMU block, 8"x 16", 8" thick	1.93	4.08	6.01
3100	10" Thick	2.74	4.24	6.98
3200	12" Thick	2.81	4.98	7.79
3300	Std. light Wt. block 8" x 16", 8" Thick	1.75	3.89	5.64
3400	10" Thick	2.27	4.01	6.28
3500	12" Thick	2.52	4.74	7.26
3600	3 coat stucco, self furring metal lath 3.4 Lb/SY, on 8" x 16", 8" thick	1.97	4.18	6.15
3700	10" Thick	2.78	4.34	7.12
3800	12" Thick	2.85	5.10	7.95
3900	Lt. Wt. block, 8" Thick	1.79	3.99	5.78
4000	10" Thick	2.31	4.11	6.42
4100	12" Thick	2.56	4.84	7.40

Important: See the Reference Section for critical supporting data - Reference Nos., Crews & Location Factors

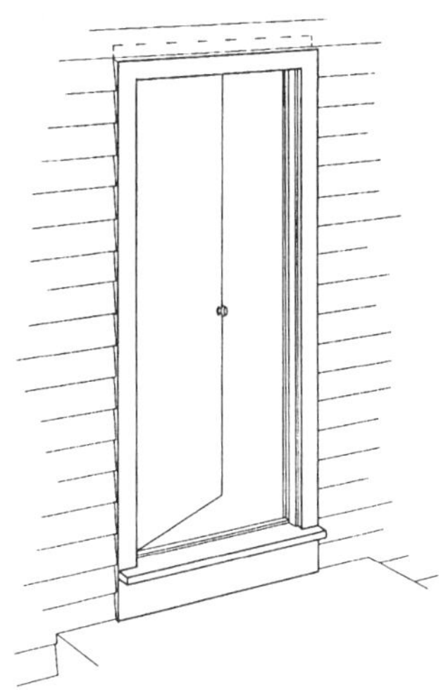

Costs are listed for exterior door systems by material, type and size. Prices between sizes listed can be interpolated with reasonable accuracy. Prices are per opening for a complete door system including frame as illustrated in the component block.

4.6-100 — Wood, Steel & Aluminum

	MATERIAL	TYPE	DOORS	SPECIFICATION	OPENING	COST PER OPNG. MAT.	INST.	TOTAL
2350	Birch	solid core	single door	hinged	2'-6" x 6'-8"	735	125	860
2400					2'-6" x 7'-0"	740	126	866
2450					2'-8" x 7'-0"	745	126	871
2500					3'-0" x 7'-0"	745	128	873
2550			double door	hinged	2'-6" x 6'-8"	1,375	228	1,603
2600					2'-6" x 7'-0"	1,400	231	1,631
2650					2'-8" x 7'-0"	1,400	231	1,631
2700					3'-0" x 7'-0"	1,400	235	1,635
2750	Wood	combination	storm & screen	hinged	3'-0" x 6'-8"	229	28.50	257.50
2800					3'-0" x 7'-0"	252	31.50	283.50
2850		overhead	panels, H.D.	manual oper.	8'-0" x 8'-0"	415	243	658
2900					10'-0" x 10'-0"	635	270	905
2950					12'-0" x 12'-0"	885	325	1,210
3000					14'-0" x 14'-0"	1,525	375	1,900
3050					20'-0" x 16'-0"	3,175	745	3,920
3100				electric oper.	8'-0" x 8'-0"	820	365	1,185
3150					10'-0" x 10'-0"	1,050	390	1,440
3200					12'-0" x 12'-0"	1,300	445	1,745
3250					14'-0" x 14'-0"	1,925	495	2,420
3300					20'-0" x 16'-0"	3,600	990	4,590
3350	Steel 18 Ga.	hollow metal	1 door w/frame	no label	2'-6" x 7'-0"	805	130	935
3400					2'-8" x 7'-0"	805	130	935
3450					3'-0" x 7'-0"	805	130	935
3500					3'-6" x 7'-0"	880	137	1,017
3550					4'-0" x 8'-0"	1,100	137	1,237
3600			2 doors w/frame	no label	5'-0" x 7'-0"	1,550	243	1,793
3650					5'-4" x 7'-0"	1,550	243	1,793
3700					6'-0" x 7'-0"	1,550	243	1,793
3750					7'-0" x 7'-0"	1,700	256	1,956
3800					8'-0" x 8'-0"	2,125	257	2,382
3850			1 door w/frame	"A" label	2'-6" x 7'-0"	970	158	1,128
3900					2'-8" x 7'-0"	975	160	1,135
3950					3'-0" x 7'-0"	975	160	1,135
4000					3'-6" x 7'-0"	1,050	163	1,213
4050					4'-0" x 8'-0"	1,050	171	1,221
4100			2 doors w/frame	"A" label	5'-0" x 7'-0"	1,875	295	2,170

	MATERIAL	TYPE	DOORS	SPECIFICATION	OPENING	COST PER OPNG.		
						MAT.	INST.	TOTAL
4150	Steel 18 ga.	hollow metal	2 doors w/frame	"A" label	5'-4" x 7'-0"	1,850	299	2,149
4200					6'-0" x 7'-0"	1,850	299	2,149
4250					7'-0" x 7'-0"	2,000	305	2,305
4300					8'-0" x 8'-0"	2,000	305	2,305
4350	Steel 24 Ga.	overhead	sectional	manual oper.	8'-0" x 8'-0"	370	243	613
4400					10'-0" x 10'-0"	480	270	750
4450					12'-0" x 12'-0"	650	325	975
4500					20'-0" x 14'-0"	1,625	695	2,320
4550				electric oper.	8'-0" x 8'-0"	775	365	1,140
4600					10'-0" x 10'-0"	885	390	1,275
4650					12'-0" x 12'-0"	1,050	445	1,495
4700					20'-0" x 14'-0"	2,050	940	2,990
4750	Steel	overhead	rolling	manual oper.	8'-0" x 8'-0"	710	375	1,085
4800					10'-0" x 10'-0"	950	430	1,380
4850					12'-0" x 12'-0"	1,250	500	1,750
4900					14'-0" x 14'-0"	1,700	750	2,450
4950					20'-0" x 12'-0"	2,525	670	3,195
5000					20'-0" x 16'-0"	3,575	1,000	4,575
5050				electric oper.	8'-0" x 8'-0"	1,650	495	2,145
5100					10'-0" x 10'-0"	1,900	550	2,450
5150					12'-0" x 12'-0"	2,200	620	2,820
5200					14'-0" x 14'-0"	2,650	870	3,520
5250					20'-0" x 12'-0"	3,500	790	4,290
5300					20'-0" x 16'-0"	4,550	1,125	5,675
5350				fire rated	10'-0" x 10'-0"	1,675	545	2,220
5400			rolling grill	manual oper.	10'-0" x 10'-0"	1,600	600	2,200
5450					15'-0" x 8'-0"	1,875	750	2,625
5500		vertical lift	1 door w/frame	motor operator	16'-0" x 16"-0"	16,100	2,750	18,850
5550					32'-0" x 24'-0"	29,900	1,825	31,725
5600	St. Stl. & glass	revolving	stock unit	manual oper.	6'-0" x 7'-0"	26,400	4,000	30,400
5650				auto Cntrls.	6'-10" x 7'-0"	37,400	4,250	41,650
5700	Bronze	revolving	stock unit	manual oper.	6'-10" x 7'-0"	30,800	8,000	38,800
5750				auto Cntrls.	6'-10" x 7'-0"	41,800	8,250	50,050
5800	St. Stl. & glass	balanced	standard	economy	3'-0" x 7'-0"	5,700	675	6,375
5850				premium	3'-0" x 7'-0"	9,875	880	10,755
6000	Aluminum	combination	storm & screen	hinged	3'-0" x 6'-8"	188	30.50	218.50
6050					3'-0" x 7'-0"	207	33.50	240.50
6100		overhead	rolling grill	manual oper.	12'-0" x 12'-0"	2,575	1,050	3,625
6150				motor oper.	12'-0" x 12'-0"	3,375	1,175	4,550
6200	Alum. & Fbrgls.	overhead	heavy duty	manual oper.	12'-0" x 12'-0"	1,300	325	1,625
6250				electric oper.	12'-0" x 12'-0"	1,700	445	2,145
6300	Alum. & glass	w/o transom	narrow stile	w/panic Hrdwre.	3'-0" x 7'-0"	945	445	1,390
6350				dbl. door, Hrdwre.	6'-0" x 7'-0"	1,700	730	2,430
6400			wide stile	hdwre.	3'-0" x 7'-0"	1,350	435	1,785
6450				dbl. door, Hdwre.	6'-0" x 7'-0"	2,700	870	3,570
6500			full vision	hdwre.	3'-0" x 7'-0"	1,400	695	2,095
6550				dbl. door, Hdwre.	6'-0" x 7'-0"	2,100	985	3,085
6600			non-standard	hdwre.	3'-0" x 7'-0"	1,525	435	1,960
6650				dbl. door, Hdwre.	6'-0" x 7'-0"	3,025	870	3,895
6700			bronze fin.	hdwre.	3'-0" x 7'-0"	1,550	435	1,985
6750				dbl. door, Hrdwre.	6'-0" x 7'-0"	3,075	870	3,945
6800			black fin.	hdwre.	3'-0" x 7'-0"	1,600	435	2,035
6850				dbl. door, Hdwre.	6'-0" x 7'-0"	3,175	870	4,045
6900		w/transom	narrow stile	hdwre.	3'-0" x 10'-0"	1,350	505	1,855

Important: See the Reference Section for critical supporting data - Reference Nos., Crews & Location Factors

4.6-100 Wood, Steel & Aluminum

	MATERIAL	TYPE	DOORS	SPECIFICATION	OPENING	COST PER OPNG. MAT.	INST.	TOTAL
6950	Aluminum & glass	w/transom		dbl. door, Hdwre.	6'-0" x 10'-0"	2,275	885	3,160
7000			wide stile	hdwre.	3'-0" x 10'-0"	1,575	605	2,180
7050				dbl. door, Hdwre.	6'-0" x 10'-0"	2,625	1,050	3,675
7100			full vision	hdwre.	3'-0" x 10'-0"	1,725	675	2,400
7150				dbl. door, Hdwre.	6'-0" x 10'-0"	2,875	1,150	4,025
7200			non-standard	hdwre.	3'-0" x 10'-0"	1,550	470	2,020
7250				dbl. door, Hdwre.	6'-0" x 10'-0"	3,100	940	4,040
7300			bronze fin.	hdwre.	3'-0" x 10'-0"	1,575	470	2,045
7350				dbl. door, Hdwre.	6'-0" x 10'-0"	3,150	940	4,090
7400			black fin.	hdwre.	3'-0" x 10'-0"	1,625	470	2,095
7450				dbl. door, Hdwre.	6'-0" x 10'-0"	3,250	940	4,190
7500		revolving	stock design	minimum	6'-10" x 7'-0"	14,100	1,600	15,700
7550				average	6'-0" x 7'-0"	17,600	2,000	19,600
7600				maximum	6'-10" x 7'-0"	23,000	2,675	25,675
7650				min., automatic	6'-10" x 7'-0"	25,100	1,850	26,950
7700				avg., automatic	6'-10" x 7'-0"	28,600	2,250	30,850
7750				max., automatic	6'-10" x 7'-0"	34,000	2,925	36,925
7800		balanced	standard	economy	3'-0" x 7'-0"	3,600	670	4,270
7850				premium	3'-0" x 7'-0"	4,975	860	5,835
7900		mall front	sliding panels	alum. fin.	16'-0" x 9'-0"	2,200	360	2,560
7950					24'-0" x 9'-0"	3,200	665	3,865
8000				bronze fin.	16'-0" x 9'-0"	2,575	420	2,995
8050					24'-0" x 9'-0"	3,725	775	4,500
8100			fixed panels	alum. fin.	48'-0" x 9'-0"	5,950	520	6,470
8150				bronze fin.	48'-0" x 9'-0"	6,925	605	7,530
8200		sliding entrance	5' x 7' door	electric oper.	12'-0" x 7'-6"	5,600	665	6,265
8250		sliding patio	temp. glass	economy	6'-0" x 7'-0"	695	121	816
8300			temp. glass	economy	12'-0" x 7'-0"	2,300	162	2,462
8350				premium	6'-0" x 7'-0"	1,050	182	1,232
8400					12'-0" x 7'-0"	3,450	243	3,693

For expanded coverage of these items see *Means Assemblies Cost Data 1998*

The table below lists window systems by material, type and size. Prices between sizes listed can be interpolated with reasonable accuracy. Prices include frame, hardware, and casing as illustrated in the component block below.

4.7-110   Wood, Steel & Aluminum

	MATERIAL	TYPE	GLAZING	SIZE	DETAIL	MAT.	INST.	TOTAL
						COST PER UNIT		
3000	Wood	double hung	std. glass	2'-8" x 4'-6"		152	104	256
3050				3'-0" x 5'-6"		177	120	297
3100			insul. glass	2'-8" x 4'-6"		184	104	288
3150				3'-0" x 5'-6"		230	120	350
3200		sliding	std. glass	3'-4" x 2'-7"		157	86.50	243.50
3250				4'-4" x 3'-3"		220	95	315
3300				5'-4" x 6'-0"		288	114	402
3350			insul. glass	3'-4" x 2'-7"		190	102	292
3400				4'-4" x 3'-3"		267	110	377
3450				5'-4" x 6'-0"		350	130	480
3500		awning	std. glass	2'-10" x 1'-9"		190	52.50	242.50
3600				4'-4" x 2'-8"		227	50	277
3700			insul. glass	2'-10" x 1'-9"		233	60.50	293.50
3800				4'-4" x 2'-8"		277	55	332
3900		casement	std. glass	1'-10" x 3'-2"	1 lite	234	68.50	302.50
3950				4'-2" x 4'-2"	2 lite	365	91	456
4000				5'-11" x 5'-2"	3 lite	615	117	732
4050				7'-11" x 6'-3"	4 lite	780	141	921
4100				9'-11" x 6'-3"	5 lite	980	164	1,144
4150			insul. glass	1'-10" x 3'-2"	1 lite	264	68.50	332.50
4200				4'-2" x 4'-2"	2 lite	465	91	556
4250				5'-11" x 5'-2"	3 lite	810	117	927
4300				7'-11" x 6'-3"	4 lite	1,025	141	1,166
4350				9'-11" x 6'-3"	5 lite	1,300	164	1,464
4400		picture	std. glass	4'-6" x 4'-6"		223	118	341
4450				5'-8" x 4'-6"		273	131	404
4500	Wood	picture	insul. glass	4'-6" x 4'-6"		270	137	407
4550				5'-8" x 4'-6"		330	153	483
4600		fixed bay	std. glass	8' x 5'		640	217	857
4650				9'-9" x 5'-4"		750	305	1,055
4700			insul. glass	8' x 5'		795	217	1,012
4750				9'-9" x 5'-4"		960	305	1,265
4800		casement bay	std. glass	8' x 5'		675	249	924
4850			insul. glass	8' x 5'		950	249	1,199
4900		vert. bay	std. glass	8' x 5'		1,200	249	1,449
4950			insul. glass	8' x 5'		1,450	249	1,699
5000	Steel	double hung	1/4" tempered	2'-8" x 4'-6"		520	82.50	602.50
5050				3'-4" x 5'-6"		795	126	921

Important: See the Reference Section for critical supporting data - Reference Nos., Crews & Location Factors

	MATERIAL	TYPE	GLAZING	SIZE	DETAIL	COST PER UNIT		
						MAT.	INST.	TOTAL
5100	Steel	double hung	insul. glass	2'-8" x 4'-6"		535	95	630
5150				3'-4" x 5'-6"		820	145	965
5202		horiz. pivoted	std. glass	2' x 2'		124	27.50	151.50
5250				3' x 3'		278	62	340
5300				4' x 4'		495	110	605
5350				6' x 4'		740	165	905
5400			insul. glass	2' x 2'		129	31.50	160.50
5450				3' x 3'		290	71.50	361.50
5500				4' x 4'		515	127	642
5550				6' x 4'		775	190	965
5600		picture window	std. glass	3' x 3'		171	62	233
5650				6' x 4'		455	165	620
5700			insul. glass	3' x 3'		183	71.50	254.50
5750				6' x 4'		490	190	680
5800		industrial security	std. glass	2'-9" x 4'-1"		505	77.50	582.50
5850				4'-1" x 5'-5"		995	152	1,147
5900			insul. glass	2'-9" x 4'-1"		520	89	609
5950				4'-1" x 5'-5"		1,025	175	1,200
6000		comm. projected	std. glass	3'-9" x 5'-5"		740	140	880
6050				6'-9" x 4'-1"		1,000	190	1,190
6100			insul. glass	3'-9" x 5'-5"		765	161	926
6150				6'-9" x 4'-1"		1,050	218	1,268
6200		casement	std. glass	4'-2" x 4'-2"	2 lite	605	120	725
6250			insul. glass	4'-2" x 4'-2"		630	137	767
6300			std. glass	5'-11" x 5'-2"	3 lite	1,225	211	1,436
6350			insul. glass	5'-11" x 5'-2"		1,275	242	1,517
6400	Aluminum	projecting	std. glass	3'-1" x 3'-2"		195	60	255
6450				4'-5" x 5'-3"		275	75	350
6500			insul. glass	3'-1" x 3'-2"		234	72	306
6550				4'-5" x 5'-3"		330	90	420
6600		sliding	std. glass	3' x 2'		147	60	207
6650				5' x 3'		187	67	254
6700				8' x 4'		270	100	370
6750				9' x 5'		410	150	560
6800			insul. glass	3' x 2'		163	60	223
6850				5' x 3'		262	67	329
6900				8' x 4'		435	100	535
6950				9' x 5'		650	150	800
7000		single hung	std. glass	2' x 3'		131	60	191
7050				2'-8" x 6'-8"		278	75	353
7100				3'-4" x 5'0"		179	67	246
7150			insul. glass	2' x 3'		159	60	219
7200				2'-8" x 6'-8"		360	75	435
7250				3'-4" x 5'		252	67	319
7300		double hung	std. glass	2' x 3'		188	41.50	229.50
7350				2'-8" x 6'-8"		560	123	683
7400				3'-4" x 5'		525	115	640
7450			insul. glass	2' x 3'		196	47.50	243.50
7500				2'-8" x 6'-8"		580	141	721
7550				3'-4" x 5'-0"		545	132	677
7600		casements	std. glass	3'-1" x 3'-2"		143	67	210
7650				4'-5" x 5'-3"		345	163	508
7700			insul. glass	3'-1" x 3'-2"		156	77	233
7750				4'-5" x 5'-3"		375	187	562

	MATERIAL	TYPE	GLAZING	SIZE	DETAIL	COST PER UNIT		
						MAT.	INST.	TOTAL
7800	Aluminum	hinged swing	std. glass	3' x 4'		335	82.50	417.50
7850				4' x 5'		560	138	698
7900			insul. glass	3' x 4'		350	95	445
7950				4' x 5'		585	158	743
8002		folding type	std. glass	3'-0" x 4'-0"		197	43	240
8050				4'-0" x 5'-0"		275	43	318
8100			insul. glass	3'-0" x 4'-0"		247	43	290
8150				4'-0" x 5'-0"		360	43	403
8200		picture unit	std. glass	2'-0" x 3'-0"		97	41.50	138.50
8250				2'-8" x 6'-8"		287	123	410
8300				3'-4" x 5'-0"		269	115	384
8350			insul. glass	2'-0" x 3'-0"		105	47.50	152.50
8400				2'-8" x 6'-8"		310	141	451
8450				3'-4" x 5'-0"		291	132	423
8500		awning type	std. glass	3'-0" x 3'-0"	2 lite	320	43	363
8550				3'-0" x 4'-0"	3 lite	375	60	435
8600				3'-0" x 5'-4"	4 lite	455	60	515
8650				4'-0" x 5'-4"	4 lite	500	67	567
8700			insul. glass	3'-0" x 3'-0"	2 lite	345	43	388
8750				3'-0" x 4'-0"	3 lite	435	60	495
8800				3'-0" x 5'-4"	4 lite	535	60	595
8850				4'-0" x 5'-4"	4 lite	595	67	662
8900		jalousie type	std. glass	1'-7" x 3'-2"		118	60	178
8950				2'-3" x 4'-0"		169	60	229
9000				3'-1" x 2'-0"		116	60	176
9050				3'-1" x 5'-3"		241	60	301

4.7-582	**Tubular Aluminum Framing**	COST/S.F. OPNG.		
		MAT.	INST.	TOTAL
1100	Alum flush tube frame, for 1/4"glass, 1-3/4"x4", 5'x6'opng, no inter horizontals	8.50	5.75	14.25
1150	One intermediate horizontal	11.55	6.75	18.30
1200	Two intermediate horizontals	14.60	7.70	22.30
1250	5' x 20' opening, three intermediate horizontals	7.65	4.77	12.42
1400	1-3/4" x 4-1/2", 5' x 6' opening, no intermediate horizontals	9.65	5.75	15.40
1450	One intermediate horizontal	12.90	6.75	19.65
1500	Two intermediate horizontals	16.05	7.70	23.75
1550	5' x 20' opening, three intermediate horizontals	8.60	4.77	13.37
1700	For insulating glass, 2"x4-1/2", 5'x6' opening, no intermediate horizontals	10.85	6	16.85
1750	One intermediate horizontal	14.20	7.10	21.30
1800	Two intermediate horizontals	17.50	8.10	25.60
1850	5' x 20' opening, three intermediate horizontals	9.50	5	14.50
2000	Thermal break frame, 2-1/4"x4-1/2", 5'x6'opng, no intermediate horizontals	11.65	6.10	17.75
2050	One intermediate horizontal	15.65	7.35	23
2100	Two intermediate horizontals	19.60	8.55	28.15
2150	5' x 20' opening, three intermediate horizontals	10.65	5.20	15.85

Important: See the Reference Section for critical supporting data - Reference Nos., Crews & Location Factors

The table below lists costs of curtain wall and spandrel panels per S.F. Costs do not include structural framing used to hang the panels from.

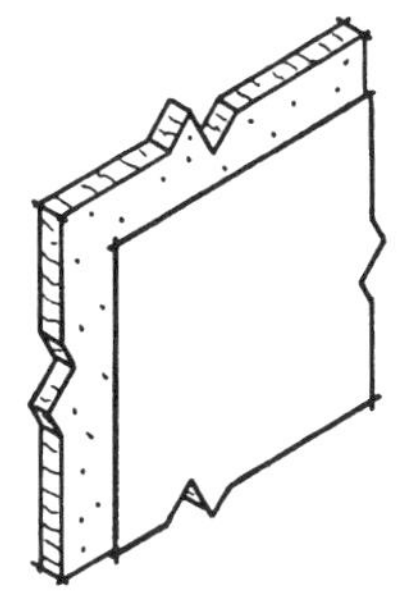

Spandrel Glass Panel

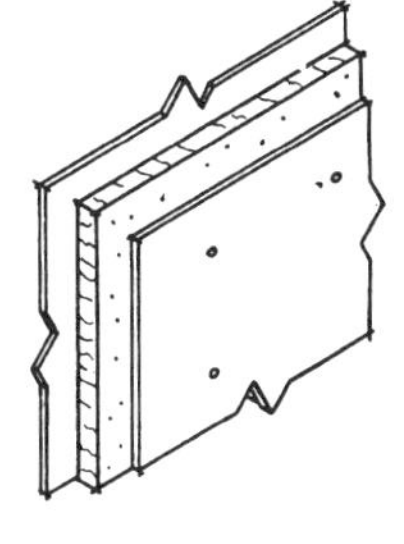

Sandwich Panel

4.7-584	Curtain Wall Panels	COST PER S.F.		
		MAT.	INST.	TOTAL
1000	Glazing panel, insulating, 1/2" thick, 2 lites 1/8" float, clear	5.70	4.92	10.62
1100	Tinted	8.50	4.92	13.42
1200	5/8" thick units, 2 lites 3/16" float, clear	6.45	5.20	11.65
1300	Tinted	7.20	5.20	12.40
1400	1" thick units, 2 lites, 1/4" float, clear	9.90	6.25	16.15
1500	Tinted	9.15	6.25	15.40
1600	Heat reflective film inside	13.90	5.50	19.40
1700	Light and heat reflective glass, tinted	15.20	5.50	20.70
2000	Plate glass, 1/4" thick, clear	3.99	3.89	7.88
2050	Tempered	4.39	3.89	8.28
2100	Tinted	4.08	3.89	7.97
2200	3/8" thick, clear	6.70	6.25	12.95
2250	Tempered	9.15	6.25	15.40
2300	Tinted	7.90	6.25	14.15
2400	1/2" thick, clear	12.70	8.50	21.20
2450	Tempered	14.65	8.50	23.15
2500	Tinted	13.55	8.50	22.05
2600	3/4" thick, clear	17.55	13.35	30.90
2650	Tempered	20.50	13.35	33.85
3000	Spandrel glass, panels, 1/4" plate glass insul w/fiberglass, 1" thick	8.45	3.89	12.34
3100	2" thick	9.85	3.89	13.74
3200	Galvanized steel backing, add	2.92		2.92
3300	3/8" plate glass, 1" thick	14.15	3.89	18.04
3400	2" thick	15.55	3.89	19.44
4000	Polycarbonate, masked, clear or colored, 1/8" thick	5.20	2.75	7.95
4100	3/16" thick	6.05	2.83	8.88
4200	1/4" thick	6.90	3.01	9.91
4300	3/8" thick	11.80	3.11	14.91
5000	Facing panel, textured al, 4' x 8' x 5/16" plywood backing, sgl face	2.40	1.30	3.79
5100	Double face	3.61	1.39	5
5200	4' x 10' x 5/16" plywood backing, single face	2.29	1.39	3.68
5300	Double face	3.47	1.39	4.86
5400	4' x 12' x 5/16" plywood backing, single face	3.29	1.39	4.68
5500	Sandwich panel, 22 Ga. galv., both sides 2" insulation, enamel exterior	7.20	2.42	9.62
5600	Polyvinylidene floride exterior finish	7.60	2.42	10.02
5700	26 Ga., galv. both sides, 1" insulation, colored 1 side	3.72	2.29	6.01
5800	Colored 2 sides	4.85	2.29	7.14

For information about Means Estimating Seminars, see yellow pages 11 and 12 in back of book

EXTERIOR CLOSURE

4

For expanded coverage of these items see *Means Assemblies Cost Data 1998*

Division 5
Roofing

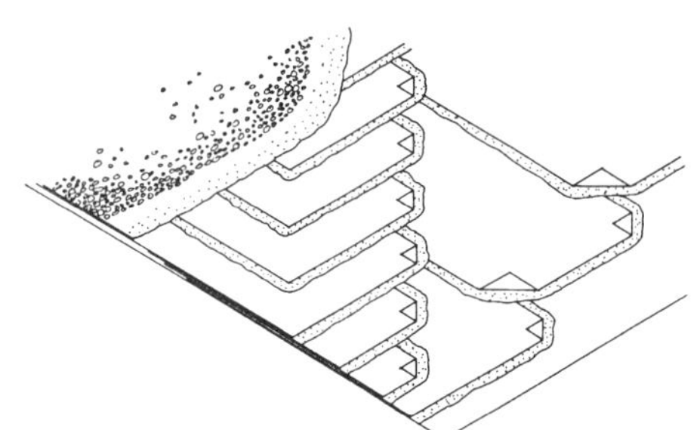

Multiple ply roofing is the most popular covering for minimum pitch roofs. Lines 1200 through 6300 list the costs of the various types, plies and weights per S.F.

5.1-103	Built-Up	MAT.	INST.	TOTAL
1200	Asphalt flood coat w/gravel; not incl. insul, flash., nailers			
1300				
1400	Asphalt base sheets & 3 plies #15 asphalt felt, mopped	.38	.90	1.28
1500	On nailable deck	.43	.94	1.37
1600	4 plies #15 asphalt felt, mopped	.54	.99	1.53
1700	On nailable deck	.49	1.04	1.53
1800	Coated glass base sheet, 2 plies glass (type IV), mopped	.41	.90	1.31
1900	For 3 plies	.49	.99	1.48
2000	On nailable deck	.47	1.04	1.51
2300	4 plies glass fiber felt (type IV), mopped	.59	.99	1.58
2400	On nailable deck	.55	1.04	1.59
2500	Organic base sheet & 3 plies #15 organic felt, mopped	.41	.96	1.37
2600	On nailable deck	.43	1.04	1.47
2700	4 plies #15 organic felt, mopped	.52	.90	1.42
2750				
2800	Asphalt flood coat, smooth surface			
2900	Asphalt base sheet & 3 plies #15 asphalt felt, mopped	.39	.82	1.21
3000	On nailable deck	.37	.86	1.23
3100	Coated glass fiber base sheet & 2 plies glass fiber felt, mopped	.36	.79	1.15
3200	On nailable deck	.34	.82	1.16
3300	For 3 plies, mopped	.43	.86	1.29
3400	On nailable deck	.41	.90	1.31
3700	4 plies glass fiber felt (type IV), mopped	.51	.86	1.37
3800	On nailable deck	.49	.90	1.39
3900	Organic base sheet & 3 plies #15 organic felt, mopped	.39	.82	1.21
4000	On nailable decks	.37	.86	1.23
4100	4 plies #15 organic felt, mopped	.46	.90	1.36
4200	Coal tar pitch with gravel surfacing			
4300	4 plies #15 tarred felt, mopped	1.07	.94	2.01
4400	3 plies glass fiber felt (type IV), mopped	.88	1.04	1.92
4500	Coated glass fiber base sheets 2 plies glass fiber felt, mopped	.88	1.04	1.92
4600	On nailable decks	.78	1.10	1.88
4800	3 plies glass fiber felt (type IV), mopped	1.20	.94	2.14
4900	On nailable decks	1.10	.99	2.09

Important: See the Reference Section for critical supporting data - Reference Nos., Crews & Location Factors

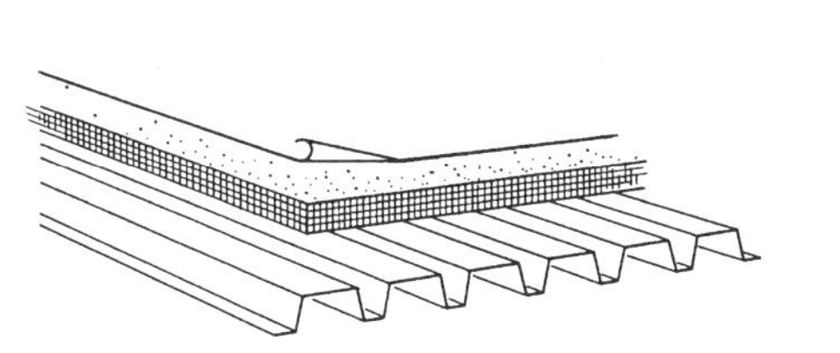

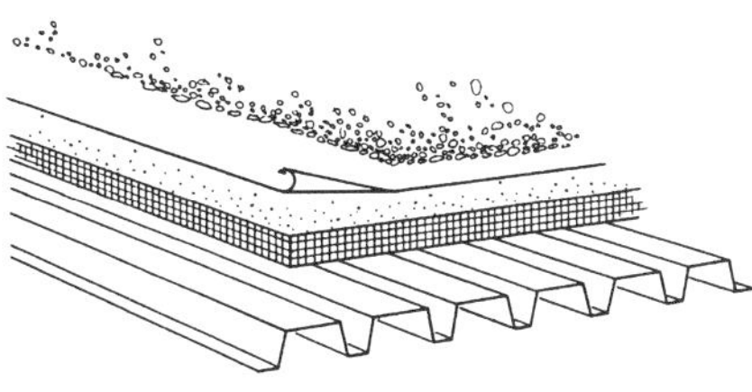

The systems listed below reflect only the cost for the single ply membrane.

For additional components see:

Insulation	A5.7-101
Base Flashing	A5.1-510
Roof Edge	A5.1-520
Roof Openings	A5.8-100

5.1-220	Single Ply Membrane	COST PER S.F.		
		MAT.	INST.	TOTAL
1000	CSPE (Chlorosulfonated polyethylene), 35 mils, fully adhered	1.34	.48	1.82
2000	EPDM (Ethylene propylene diene monomer), 45 mils, fully adhered	.83	.48	1.31
3000	55 mils, fully adhered	.97	.48	1.45
4000	Modified bit., SBS modified, granule surface cap sheet, mopped, 150 mils	.42	.98	1.40
4500	APP modified, granule surface cap sheet, torched, 180 mils	.42	.62	1.04
6000	Reinforced PVC, 48 mils, loose laid and ballasted with stone	.95	.25	1.20
6200	Fully adhered with adhesive	1.22	.48	1.70
6202				

5.1-310	Preformed Metal Roofing	COST PER S.F.		
		MAT.	INST.	TOTAL
0200	Corrugated roofing, aluminum, mill finish, .0175" thick, .272 P.S.F.	.73	.73	1.46
0250	.0215 thick, .334 P.S.F.	.92	.73	1.65

5.1-330	Formed Metal	COST PER S.F.		
		MAT.	INST.	TOTAL
1000	Batten seam, formed copper roofing, 3"min slope, 16 oz., 1.2 P.S.F.	4.86	2.40	7.26
1100	18 oz., 1.35 P.S.F.	5.40	2.64	8.04
2000	Zinc copper alloy, 3" min slope, .020" thick, .88 P.S.F.	5.85	2.21	8.06
3000	Flat seam, copper, 1/4" min. slope, 16 oz., 1.2 P.S.F.	4.31	2.21	6.52
3100	18 oz., 1.35 P.S.F.	4.86	2.30	7.16
5000	Standing seam, copper, 2-1/2" min. slope, 16 oz., 1.25 P.S.F.	4.66	2.04	6.70
5100	18 oz., 1.40 P.S.F.	5.20	2.21	7.41
6000	Zinc copper alloy, 2-1/2" min. slope, .020" thick, .87 P.S.F.	5.70	2.21	7.91
6100	.032" thick, 1.39 P.S.F.	7.75	2.40	10.15

5 ROOFING

Shingles and tiles are practical in applications where the roof slope is more than 3-1/2″ per foot of rise. Table below lists the various materials and the weight per S.F.

5.1-410 — Shingle & Tile

		COST PER S.F.		
		MAT.	INST.	TOTAL
1095	Asphalt roofing			
1100	Strip shingles, 4″ slope, inorganic class A 210-235 lb/Sq.	.32	.49	.81
1150	Organic, class C, 235-240 lb./sq.	.42	.54	.96
1200	Premium laminated multi-layered, class A, 260-300 lb./Sq.	.55	.73	1.28
1545	Metal roofing			
1550	Alum., shingles, colors, 3″min slope, .019″thick, 0.4 PSF	1.88	.53	2.41
1850	Steel, colors, 3″ min slope, 26 gauge, 1.0 PSF	2.15	1.10	3.25
2795	Slate roofing			
2800	4″ min. slope, shingles, 3/16″ thick, 8.0 PSF	6.10	1.37	7.47
3495	Wood roofing			
3500	4″ min slope, cedar shingles, 16″ x 5″, 5″ exposure 1.6 PSF	1.82	1.05	2.87
4000	Shakes, 18″, 8-1/2″ exposure, 2.8 PSF	1.64	1.29	2.93
5095	Tile roofing			
5100	Aluminum, mission, 3″ min slope, .019″ thick, 0.65 PSF	3.81	1.01	4.82
6000	Clay, Americana, 3″ minimum slope, 8 PSF	5.45	1.45	6.90
6002				

5.1-520 — Roof Edges

	EDGE TYPE	DESCRIPTION	SPECIFICATION	FACE HEIGHT		COST PER L.F.		
						MAT.	INST.	TOTAL
1000	Aluminum	mill finish	.050″ thick	4″		7.05	4.46	11.51
1100				6″		7.60	4.60	12.20
1300		duranodic	.050″ thick	4″		8.10	4.46	12.56
1400				6″		8.85	4.60	13.45
1600		painted	.050″ thick	4″		8.75	4.46	13.21
1700				6″		9.60	4.60	14.20
2000	Copper	plain	16 oz.	4″		5.70	4.46	10.16
2100				6″		6.60	4.60	11.20
2300			20 oz.	4″		6.25	5.05	11.30
2400				6″		6.95	4.90	11.85
2700	Sheet Metal	galvanized	20 Ga.	4″		6.40	5.25	11.65
2800				6″		7.65	5.25	12.90
3000			24 Ga.	4″		5.70	4.46	10.16
3100				6″		6.55	4.46	11.01

5.1-620	Flashing							
	MATERIAL	BACKING	SIDES	SPECIFICATION	QUANTITY	COST PER S.F.		
						MAT.	INST.	TOTAL
0040	Aluminum	none		.019″		.87	1.79	2.66
0050				.032″		1.05	1.79	2.84
0300		fabric	2	.004″		.94	.79	1.73
0400		mastic		.004″		.89	.79	1.68
0700	Copper	none		16 oz.	<500 lbs.	3.68	2.26	5.94
0800				24 oz.	<500 lbs.	5.50	2.48	7.98
2000	Copper lead	fabric	1	2 oz.		1.54	.79	2.33
3500	PVC black	none		.010″		.14	.81	.95
3700				.030″		.33	.81	1.14
4200	Neoprene			1/16″		1.62	.81	2.43
4500	Stainless steel	none		.015″	<500 lbs.	3.30	2.26	5.56
4600	Copper clad				>2000 lbs.	3.28	1.68	4.96
5000	Plain			32 ga.		2.37	1.68	4.05
5009								

5.7-101	Roof Deck Rigid Insulation	COST PER S.F.		
		MAT.	INST.	TOTAL
0100	Fiberboard low density, 1/2" thick, R1.39			
0150	1" thick R2.78	.35	.29	.64
0300	1 1/2" thick R4.17	.53	.29	.82
0350	2" thick R5.56	.70	.29	.99
0370	Fiberboard high density, 1/2" thick R1.3	.20	.23	.43
0380	1" thick R2.5	.37	.29	.66
0390	1 1/2" thick R3.8	.61	.29	.90
0410	Fiberglass, 3/4" thick R2.78	.43	.23	.66
0450	15/16" thick R3.70	.55	.23	.78
0550	1-5/16" thick R5.26	.96	.23	1.19
0650	2 7/16" thick R10	1.15	.29	1.44
1510	Polyisocyanurate 2#/CF density, 1" thick R7.14	.34	.17	.51
1550	1 1/2" thick R10.87	.37	.19	.56
1600	2" thick R14.29	.48	.21	.69
1650	2 1/2" thick R16.67	.55	.22	.77
1700	3" thick R21.74	.69	.23	.92
1750	3 1/2" thick R25	.80	.23	1.03
1800	Tapered for drainage	.41	.17	.58
1810	Expanded polystyrene, 1#/CF density, 3/4" thick R2.89	.22	.15	.37
1820	2" thick R7.69	.35	.19	.54
1830	Extruded polystyrene, 15 PSI compressive strength, 1" thick R5	.26	.15	.41
1835	2" thick R10	.35	.19	.54
1840	3" thick R15	.80	.23	1.03
2550	40 PSI compressive strength, 1" thick R5	.37	.15	.52
2600	2" thick R10	.73	.19	.92
2650	3" thick R15	1.09	.23	1.32
2700	4" thick R20	1.45	.23	1.68
2750	Tapered for drainage	.52	.17	.69
2810	60 PSI compressive strength, 1" thick R5	.44	.16	.60
2850	2" thick R10	.81	.19	1
2900	Tapered for drainage	.63	.17	.80
2910	115 PSI compressive strength, 1" thick R5	.95	.17	1.12
2950	2" thick R10	1.89	.20	2.09
3000	Tapered for drainage	1.06	.17	1.23

Important: See the Reference Section for critical supporting data - Reference Nos., Crews & Location Factors

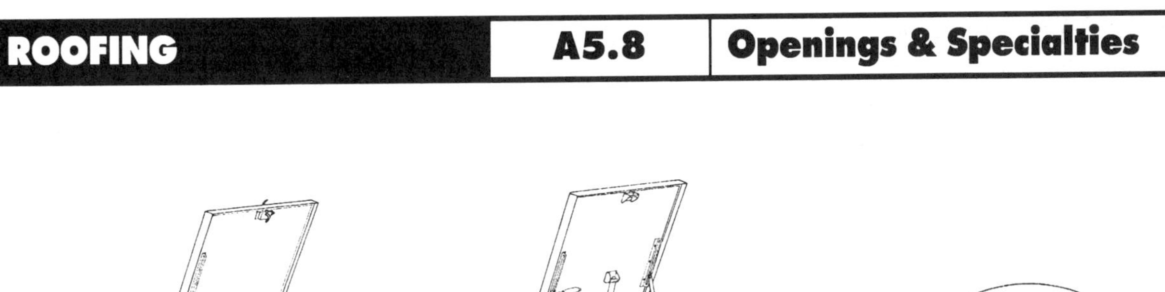

5.8-100	Hatches	COST PER OPNG.		
		MAT.	INST.	TOTAL
0200	Roof hatches, with curb, and 1" fiberglass insulation, 2'-6"x3'-0", aluminum	445	87	532
0300	Galvanized steel 165 lbs.	375	87	462
0400	Primed steel 164 lbs.	330	87	417
0500	2'-6"x4' 6" aluminum curb and cover, 150 lbs.	615	97	712
0600	Galvanized steel 220 lbs.	525	97	622
0650	Primed steel 218 lbs.	505	97	602
0800	2'x6"x8'-0" aluminum curb and cover, 260 lbs.	1,100	132	1,232
0900	Galvanized steel, 360 lbs.	1,000	132	1,132
0950	Primed steel 358 lbs.	995	132	1,127
1200	For plexiglass panels, add to the above	380		380
2100	Smoke hatches, unlabeled not incl. hand winch operator, 2'-6"x3', galv.	455	105	560
2200	Plain steel, 160 lbs.	400	105	505
2400	2'-6"x8'-0", galvanized steel, 360 lbs.	1,100	144	1,244
2500	Plain steel, 350 lbs.	1,100	144	1,244
3000	4'-0"x8'-0", double leaf low profile, aluminum cover, 359 lb.	1,575	109	1,684
3100	Galvanized steel 475 lbs.	1,375	109	1,484
3200	High profile, aluminum cover, galvanized curb, 361 lbs.	1,425	109	1,534

5.8-100	Skylights	COST PER S.F.		
		MAT.	INST.	TOTAL
5100	Skylights, plastic domes, insul curbs, nom. size to 10 S.F., single glaze	19.55	7.40	26.95
5200	Double glazing	22	8.15	30.15
5300	10 S.F. to 20 S.F., single glazing	24.50	9.10	33.60
5400	Double glazing	17.95	3.38	21.33
5500	20 S.F. to 30 S.F., single glazing	17.30	2.54	19.84
5600	Double glazing	18.35	2.70	21.05
5700	30 S.F. to 65 S.F., single glazing	17.75	1.94	19.69
5800	Double glazing	19.10	2.28	21.38
6000	Sandwich panels fiberglass, 9-1/16" thick, 2 S.F. to 10 S.F.	14.90	4.36	19.26
6100	10 S.F. to 18 S.F.	13.35	3.29	16.64
6200	2-3/4" thick, 25 S.F. to 40 S.F.	21.50	2.95	24.45
6300	40 S.F. to 70 S.F.	17.60	2.64	20.24
6301				

5 ROOFING

5.8-400 Gutters

	SECTION	MATERIAL	THICKNESS	SIZE	FINISH	COST PER L.F. MAT.	INST.	TOTAL
0050	Box	aluminum	.027"	5"	enameled	1.15	2.17	3.32
0100					mill	.95	2.17	3.12
0200			.032"	5"	enameled	1.44	2.17	3.61
0500		copper	16 Oz.	4"	lead coated	7.65	2.17	9.82
0600					mill	3.91	2.17	6.08
1000		steel galv.	28 Ga.	5"	enameled	.92	2.17	3.09
1200			26 Ga.	5"	mill	.90	2.17	3.07
1800		vinyl		4"	colors	.94	2.21	3.15
1900				5"	colors	1.10	2.21	3.31
2300		hemlock or fir		4"x5"	treated	8	2.43	10.43
3000	Half round	copper	16 Oz.	4"	lead coated	5.35	2.17	7.52
3100					mill	3.55	2.17	5.72
3600		steel galv.	28 Ga.	5"	enameled	.92	2.17	3.09
4102		stainless steel		5"	mill	5.50	2.17	7.67
5000		vinyl		4"	white	.75	2.21	2.96
5002								

5.8-500 Downspouts

	MATERIALS	SECTION	SIZE	FINISH	THICKNESS	COST PER V.L.F. MAT.	INST.	TOTAL
0100	Aluminum	rectangular	2"x3"	embossed mill	.020"	.77	1.37	2.14
0150				enameled	.020"	.71	1.37	2.08
0250			3"x4"	enameled	.024"	1.49	1.86	3.35
0300		round corrugated	3"	enameled	.020"	.94	1.37	2.31
0350			4"	enameled	.025"	1.42	1.86	3.28
0500	Copper	rectangular corr.	2"x3"	mill	16 Oz.	4.42	1.37	5.79
0600		smooth		mill	16 Oz.	5.20	1.37	6.57
0700		rectangular corr.	3"x4"	mill	16 Oz.	5.90	1.79	7.69
1300	Steel	rectangular corr.	2"x3"	galvanized	28 Ga.	.58	1.37	1.95
1350				epoxy coated	24 Ga.	1.11	1.37	2.48
1400		smooth		galvanized	28 Ga.	.87	1.37	2.24
1450		rectangular corr.	3"x4"	galvanized	28 Ga.	1.63	1.79	3.42
1500				epoxy coated	24 Ga.	1.88	1.79	3.67
1550		smooth		galvanized	28 Ga.	1.33	1.79	3.12
1652		round corrugated	3"	galvanized	28 Ga.	.78	1.37	2.15
1700			4"	galvanized	28 Ga.	1.01	1.79	2.80
1750			5"	galvanized	28 Ga.	1.35	2	3.35
2000	Steel pipe	round	4"	black	X.H.	6.70	13	19.70
2552	S.S. tubing sch.5	rectangular	3"x4"	mill		17.95	1.79	19.74
2702								

5.8-500 Gravel Stop

	MATERIALS	SECTION	SIZE	FINISH	THICKNESS	COST PER L.F. MAT.	INST.	TOTAL
5100	Aluminum	extruded	4"	mill	.050"	2.99	1.79	4.78
5200			4"	duranodic	.050"	4.06	1.79	5.85
5300			8"	mill	.050"	4.37	2.08	6.45
5400			8"	duranodic	.050"	5.50	2.08	7.58
6000			12"-2 pc.	duranodic	.050"	6.90	2.60	9.50
6100	Stainless	formed	6"	mill	24 Ga.	7.85	1.93	9.78

For information about Means Estimating Seminars, see yellow pages 11 and 12 in back of book

Important: See the Reference Section for critical supporting data - Reference Nos., Crews & Location Factors

Division 6
Interior Construction

The Concrete Block Partition Systems are defined by weight and type of block, thickness, type of finish and number of sides finished. System components include joint reinforcing on alternate courses and vertical control joints.

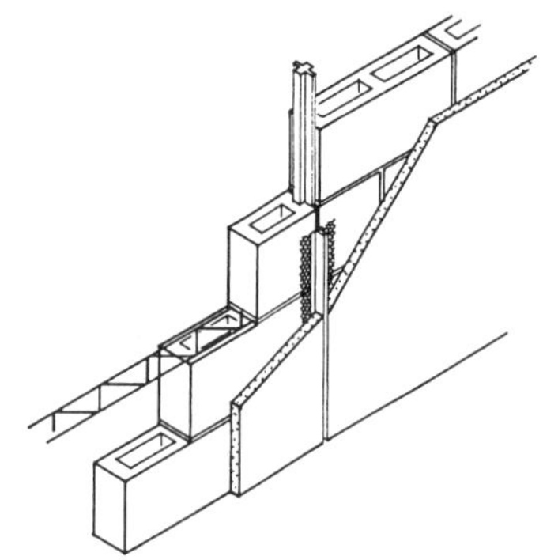

6.1-210 — Concrete Block Partitions - Regular Weight

	TYPE	THICKNESS (IN.)	TYPE FINISH	SIDES FINISHED		COST PER S.F. MAT.	INST.	TOTAL
1000	Hollow	4	none	0		.92	2.70	3.62
1010			gyp. plaster 2 coat	1		1.23	3.92	5.15
1020				2		1.53	5.15	6.68
1200			portland - 3 coat	1		1.13	4.10	5.23
1400			5/8" drywall	1		1.27	3.59	4.86
1500		6	none	0		1.12	2.89	4.01
1510			gyp. plaster 2 coat	1		1.43	4.11	5.54
1520				2		1.73	5.30	7.03
1700			portland - 3 coat	1		1.33	4.29	5.62
1900			5/8" drywall	1		1.47	3.78	5.25
1910				2		1.82	4.67	6.49
2000		8	none	0		1.60	3.09	4.69
2010			gyp. plaster 2 coat	1		1.91	4.31	6.22
2020			gyp. plaster 2 coat	2		2.21	5.50	7.71
2200			portland - 3 coat	1		1.81	4.49	6.30
2400			5/8" drywall	1		1.95	3.98	5.93
2410				2		2.30	4.87	7.17
2500		10	none	0		1.94	3.23	5.17
2510			gyp. plaster 2 coat	1		2.25	4.45	6.70
2520				2		2.55	5.65	8.20
2700			portland - 3 coat	1		2.15	4.63	6.78
2900			5/8" drywall	1		2.29	4.12	6.41
2910				2		2.64	5	7.64
3000	Solid	2	none	0		1.16	2.67	3.83
3010			gyp. plaster	1		1.47	3.89	5.36
3020				2		1.77	5.10	6.87
3200			portland - 3 coat	1		1.37	4.07	5.44
3400			5/8" drywall	1		1.51	3.56	5.07
3410				2		1.86	4.45	6.31
3500		4	none	0		1.35	2.79	4.14
3510			gyp. plaster	1		1.73	4.02	5.75
3520				2		1.96	5.20	7.16
3700			portland - 3 coat	1		1.56	4.19	5.75
3900			5/8" drywall	1		1.70	3.68	5.38
3910				2		2.05	4.57	6.62

Important: See the Reference Section for critical supporting data - Reference Nos., Crews & Location Factors

6.1-210 — Concrete Block Partitions - Regular Weight

	TYPE	THICKNESS (IN.)	TYPE FINISH	SIDES FINISHED		COST PER S.F. MAT.	INST.	TOTAL
4002	Solid	6	none	0		1.82	3	4.82
4010			gyp. plaster	1		2.13	4.22	6.35
4020				2		2.43	5.45	7.88
4200			portland - 3 coat	1		2.03	4.40	6.43
4400			5/8" drywall	1		2.17	3.89	6.06
4410				2		2.52	4.78	7.30

6.1-210 — Concrete Block Partitions - Lightweight

	TYPE	THICKNESS (IN.)	TYPE FINISH	SIDES FINISHED		COST PER S.F. MAT.	INST.	TOTAL
5000	Hollow	4	none	0		1.03	2.64	3.67
5010			gyp. plaster	1		1.34	3.86	5.20
5020				2		1.64	5.05	6.69
5200			portland - 3 coat	1		1.24	4.04	5.28
5400			5/8" drywall	1		1.38	3.53	4.91
5410				2		1.73	4.42	6.15
5500		6	none	0		1.28	2.82	4.10
5510			gyp. plaster	1		1.59	4.04	5.63
5520			gyp. plaster	2		1.89	5.25	7.14
5700			portland - 3 coat	1		1.49	4.22	5.71
5900			5/8" drywall	1		1.63	3.71	5.34
5910				2		1.98	4.60	6.58
6000		8	none	0		1.56	3.01	4.57
6010			gyp. plaster	1		1.87	4.23	6.10
6020				2		2.17	5.45	7.62
6200			portland - 3 coat	1		1.77	4.41	6.18
6400			5/8" drywall	1		1.91	3.90	5.81
6410				2		2.26	4.79	7.05
6500		10	none	0		2.08	3.15	5.23
6510			gyp. plaster	1		2.39	4.37	6.76
6520				2		2.69	5.60	8.29
6700			portland - 3 coat	1		2.29	4.55	6.84
6900			5/8" drywall	1		2.43	4.04	6.47
6910				2		2.78	4.93	7.71
7000	Solid	4	none	0		1.39	2.76	4.15
7010			gyp. plaster	1		1.70	3.98	5.68
7020				2		2	5.20	7.20
7200			portland - 3 coat	1		1.60	4.16	5.76
7400			5/8" drywall	1		1.74	3.65	5.39
7410				2		2.09	4.54	6.63
7500		6	none	0		2	2.96	4.96
7510			gyp. plaster	1		2.38	4.21	6.59
7520				2		2.61	5.40	8.01
7700			portland - 3 coat	1		2.21	4.36	6.57
7900			5/8" drywall	1		2.35	3.85	6.20
7910				2		2.70	4.74	7.44
8000		8	none	0		2.30	3.17	5.47
8010			gyp. plaster	1		2.61	4.39	7
8020				2		2.91	5.60	8.51
8200			portland - 3 coat	1		2.51	4.57	7.08
8400			5/8" drywall	1		2.65	4.06	6.71
8410				2		3	4.95	7.95

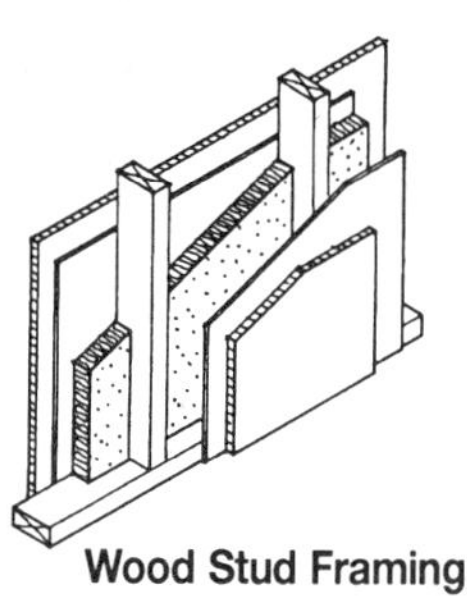

Wood Stud Framing

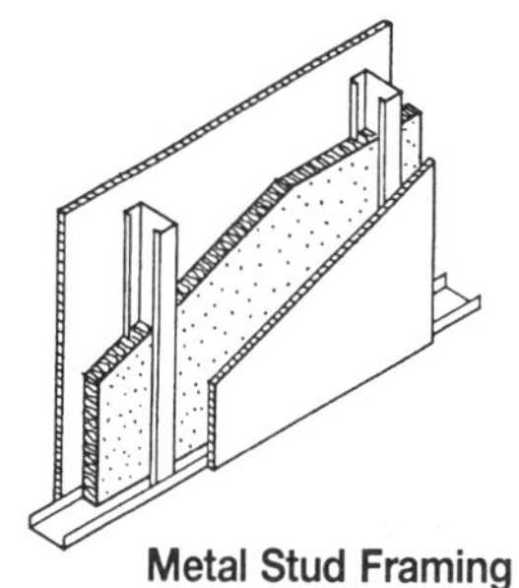

Metal Stud Framing

The Drywall Partitions/Stud Framing Systems are defined by type of drywall and number of layers, type and spacing of stud framing, and treatment on the opposite face. Components include taping and finishing.

Cost differences between regular and fire resistant drywall are negligible, and terminology is interchangeable. In some cases fiberglass insulation is included for additional sound deadening.

6.1-510 — Drywall Partitions/Wood Stud Framing

	FACE LAYER	BASE LAYER	FRAMING	OPPOSITE FACE	INSULATION	MAT.	INST.	TOTAL
1200	5/8" FR drywall	none	2 x 4, @ 16" O.C.	same	0	1.10	1.38	2.48
1250				5/8" reg. drywall	0	1.09	1.38	2.47
1300				nothing	0	.77	.90	1.67
1400		1/4" SD gypsum	2 x 4 @ 16" O.C.	same	1-1/2" fiberglass	1.77	2.16	3.93
1450				5/8" FR drywall	1-1/2" fiberglass	1.60	1.89	3.49
1500				nothing	1-1/2" fiberglass	1.27	1.41	2.68
1600		resil. channels	2 x 4 @ 16", O.C.	same	1-1/2" fiberglass	1.68	2.76	4.44
1650				5/8" FR drywall	1-1/2" fiberglass	1.55	2.19	3.74
1700				nothing	1-1/2" fiberglass	1.22	1.71	2.93
1800		5/8" FR drywall	2 x 4 @ 24" O.C.	same	0	1.46	1.78	3.24
1850				5/8" FR drywall	0	1.23	1.54	2.77
1900				nothing	0	.90	1.06	1.96
2200		5/8" FR drywall	2 rows-2 x 4	same	2" fiberglass	2.40	2.53	4.93
2250			16"O.C.	5/8" FR drywall	2" fiberglass	2.17	2.29	4.46
2300				nothing	2" fiberglass	1.84	1.81	3.65
2400	5/8" WR drywall	none	2 x 4, @ 16" O.C.	same	0	1.20	1.38	2.58
2450				5/8" FR drywall	0	1.15	1.38	2.53
2500				nothing	0	.82	.90	1.72
2600		5/8" FR drywall	2 x 4, @ 24" O.C.	same	0	1.56	1.78	3.34
2650				5/8" FR drywall	0	1.28	1.54	2.82
2700				nothing	0	.95	1.06	2.01
2800	5/8 VF drywall	none	2 x 4, @ 16" O.C.	same	0	1.80	1.50	3.30
2850				5/8" FR drywall	0	1.45	1.44	2.89
2900				nothing	0	1.12	.96	2.08
3000	5/8 VF drywall	5/8" FR drywall	2 x 4 , 24" O.C.	same	0	2.16	1.90	4.06
3050				5/8" FR drywall	0	1.58	1.60	3.18
3100				nothing	0	1.25	1.12	2.37
3200	1/2" reg drywall	3/8" reg drywall	2 x 4, @ 16" O.C.	same	0	1.30	1.86	3.16
3252	1/2" reg. drywall	3/8" reg. drywall	2 x 4, @ 16"O.C.	5/8" FR drywall	0	1.20	1.62	2.82
3300				nothing	0	.87	1.14	2.01

6.1-510 — Drywall Partitions/Metal Stud Framing

	FACE LAYER	BASE LAYER	FRAMING	OPPOSITE FACE	INSULATION	MAT.	INST.	TOTAL
5200	5/8" FR drywall	none	1-5/8" @ 24" O.C.	same	0	.82	1.43	2.25
5250				5/8" reg. drywall	0	.81	1.43	2.24
5300				nothing	0	.49	.95	1.44
5400			3-5/8" @ 24" O.C.	same	0	.88	1.45	2.33
5450				5/8" reg. drywall	0	.87	1.45	2.32
5500				nothing	0	.55	.97	1.52

Important: See the Reference Section for critical supporting data - Reference Nos., Crews & Location Factors

6.1-510 Drywall Partitions/Metal Stud Framing

	FACE LAYER	BASE LAYER	FRAMING	OPPOSITE FACE	INSULATION	COST PER S.F.		
						MAT.	INST.	TOTAL
5600	5/8" FR drywall	1/4" SD gypsum	1-5/8" @ 24" O.C.	same	0	1.16	1.97	3.13
5650				5/8" FR drywall	0	.99	1.70	2.69
5700				nothing	0	.66	1.22	1.88
5800			2-1/2" @ 24" O.C.	same	0	1.18	1.98	3.16
5850				5/8" FR drywall	0	1.01	1.71	2.72
5900				nothing	0	.68	1.23	1.91
6000		5/8" FR drywall	2-1/2" @ 16" O.C.	same	0	1.65	1.99	3.64
6050				5/8" FR drywall	0	1.42	1.75	3.17
6100				nothing	0	1.09	1.27	2.36
6200			3-5/8" @ 24" O.C.	same	0	1.34	1.93	3.27
6250				5/8"FR drywall	3-1/2" fiberglass	1.39	1.84	3.23
6300				nothing	0	.78	1.21	1.99
6400	5/8" WR drywall	none	1-5/8" @ 24" O.C.	same	0	.92	1.43	2.35
6450				5/8" FR drywall	0	.87	1.43	2.30
6500				nothing	0	.54	.95	1.49
6600			3-5/8" @ 24" O.C.	same	0	.98	1.45	2.43
6650				5/8" FR drywall	0	.93	1.45	2.38
6700				nothing	0	.60	.97	1.57
6800		5/8" FR drywall	2-1/2" @ 16" O.C.	same	0	1.75	1.99	3.74
6850				5/8" FR drywall	0	1.47	1.75	3.22
6900				nothing	0	1.14	1.27	2.41
7000			3-5/8" @ 24" O.C.	same	0	1.44	1.93	3.37
7050				5/8"FR drywall	3-1/2" fiberglass	1.44	1.84	3.28
7100				nothing	0	.83	1.21	2.04
7200	5/8" VF drywall	none	1-5/8" @ 24" O.C.	same	0	1.52	1.55	3.07
7250				5/8" FR drywall	0	1.17	1.49	2.66
7300				nothing	0	.84	1.01	1.85
7400			3-5/8" @ 24" O.C.	same	0	1.58	1.57	3.15
7450				5/8" FR drywall	0	1.23	1.51	2.74
7500				nothing	0	.90	1.03	1.93
7600		5/8" FR drywall	2-1/2" @ 16" O.C.	same	0	2.35	2.11	4.46
7650				5/8" FR drywall	0	1.77	1.81	3.58
7700				nothing	0	1.44	1.33	2.77
7800			3-5/8" @ 24" O.C.	same	0	2.04	2.05	4.09
7850				5/8"FR drywall	3-1/2" fiberglass	1.74	1.90	3.64
7900				nothing	0	1.13	1.27	2.40

INTERIOR CONSTRUCTION

6

6.1-580	Drywall Components	COST PER S.F.		
		MAT.	INST.	TOTAL
0060	Metal studs, 24″ O.C. including track, load bearing, 20 gage, 2-1/2″	.52	.67	1.19
0080	3-5/8″	.60	.70	1.30
0100	4″	.63	.74	1.37
0120	6″	.77	.77	1.54
0140	Metal studs, 24″ O.C. including track, load bearing, 18 gage, 2-1/2″	.52	.67	1.19
0160	3-5/8″	.60	.70	1.30
0180	4″	.63	.74	1.37
0200	6″	.77	.77	1.54
0220	16 gage, 2-1/2″	.62	.74	1.36
0240	3-5/8″	.85	.77	1.62
0260	4″	.86	.81	1.67
0280	6″	1.05	1.08	2.13
0300	Non load bearing, 25 gage, 1-5/8″	.16	.47	.63
0340	3-5/8″	.22	.49	.71
0360	4″	.23	.50	.73
0380	6″	.29	.51	.80
0400	20 gage, 2-1/2″	.42	.48	.90
0420	3-5/8″	.51	.49	1
0440	4″	.54	.50	1.04
0460	6″	.69	.51	1.20
0540	Wood studs including blocking, shoe and double top plate, 2″x4″, 12″O.C.	.55	.53	1.08
0560	16″ O.C.	.44	.42	.86
0580	24″ O.C.	.34	.34	.68
0600	2″x6″, 12″ O.C.	.78	.61	1.39
0620	16″ O.C.	.63	.47	1.10
0640	24″ O.C.	.48	.37	.85
0642	Furring one side only, steel channels, 3/4″, 12″ O.C.	.19	.98	1.17
0644	16″ O.C.	.17	.86	1.03
0646	24″ O.C.	.12	.65	.77
0647	1-1/2″ , 12″ O.C.	.27	1.09	1.36
0648	16″ O.C.	.25	.95	1.20
0649	24″ O.C.	.16	.75	.91
0650	Wood strips, 1″ x 3″, on wood, 12″ O.C.	.31	.44	.75
0651	16″ O.C.	.23	.33	.56
0652	On masonry, 12″ O.C.	.31	.49	.80
0653	16″ O.C.	.23	.37	.60
0654	On concrete, 12″ O.C.	.31	.93	1.24
0655	16″ O.C.	.23	.70	.93
0665	Gypsum board, one face only, exterior sheathing, 1/2″	.19	.38	.57
0680	Fire resistant, 1/2″	.23	.24	.47
0700	5/8″	.23	.24	.47
0720	Sound deadening board 1/4″	.17	.27	.44
0740	Standard drywall 3/8″	.17	.24	.41
0760	1/2″	.16	.24	.40
0780	5/8″	.22	.24	.46
0800	Tongue & groove coreboard 1″	.57	1.01	1.58
0820	Water resistant, 1/2″	.25	.24	.49
0840	5/8″	.28	.24	.52
0860	Add for the following:, foil backing	.08		.08
0880	Fiberglass insulation, 3-1/2″	.28	.15	.43
0900	6″	.33	.18	.51
0920	Rigid insulation 1″	.35	.24	.59
0940	Resilient furring @ 16″ O.C.	.16	.76	.92
0960	Taping and finishing	.10	.24	.34
0980	Texture spray	.13	.29	.42
1000	Thin coat plaster	.13	.30	.43
1040	2″x4″ staggered studs 2″x6″ plates & blocking	.63	.46	1.09
1050				

Important: See the Reference Section for critical supporting data - Reference Nos., Crews & Location Factors

6.1-680	Plaster Partition Components	COST PER S.F.		
		MAT.	INST.	TOTAL
0060	Metal studs, 16″ O.C., including track, non load bearing, 25 gage, 1-5/8″	.23	.55	.78
0080	2-1/2″	.23	.55	.78
0100	3-1/4″	.26	.56	.82
0120	3-5/8″	.26	.56	.82
0140	4″	.29	.58	.87
0160	6″	.36	.59	.95
0180	Load bearing, 20 gage, 2-1/2″	.65	.81	1.46
0200	3-5/8″	.74	.85	1.59
0220	4″	.79	.90	1.69
0240	6″	.96	.95	1.91
0260	16 gage 2-1/2″	.77	.90	1.67
0280	3-5/8″	1	.95	1.95
0300	4″	1.08	1.01	2.09
0320	6″	1.32	1.28	2.60
0340	Wood studs, including blocking, shoe and double plate, 2″x4″ , 12″ O.C.	.55	.53	1.08
0360	16″ O.C.	.44	.42	.86
0380	24″ O.C.	.34	.34	.68
0400	2″x6″ , 12″ O.C.	.78	.61	1.39
0420	16″ O.C.	.63	.47	1.10
0440	24″ O.C.	.48	.37	.85
0460	Furring one face only, steel channels, 3/4″, 12″ O.C.	.19	.98	1.17
0480	16″ O.C.	.17	.86	1.03
0500	24″ O.C.	.12	.65	.77
0520	1-1/2″ , 12″ O.C.	.27	1.09	1.36
0540	16″ O.C.	.25	.95	1.20
0560	24″O.C.	.16	.75	.91
0580	Wood strips 1″x3″, on wood., 12″ O.C.	.31	.44	.75
0600	16″O.C.	.23	.33	.56
0620	On masonry, 12″ O.C.	.31	.49	.80
0640	16″ O.C.	.23	.37	.60
0660	On concrete, 12″ O.C.	.31	.93	1.24
0680	16″ O.C.	.23	.70	.93
0700	Gypsum lath. plain or perforated, nailed to studs, 3/8″ thick	.40	.30	.70
0720	1/2″ thick	.47	.32	.79
0740	Clipped to studs, 3/8″ thick	.45	.34	.79
0760	1/2″ thick	.47	.36	.83
0780	Metal lath, diamond painted, nailed to wood studs, 2.5 lb.	.22	.30	.52
0800	3.4 lb.	.25	.32	.57
0820	Screwed to steel studs, 2.5 lb.	.22	.32	.54
0840	3.4 lb.	.26	.34	.60
0860	Rib painted, wired to steel, 2.75 lb	.21	.34	.55
0880	3.4 lb	.39	.36	.75
0900	4.0 lb	.34	.39	.73
0910				
0920	Gypsum plaster, 2 coats	.35	1.16	1.51
0940	3 coats	.49	1.40	1.89
0960	Perlite or vermiculite plaster, 2 coats	.43	1.32	1.75
0980	3 coats	.68	1.65	2.33
1000	Stucco, 3 coats, 1″ thick, on wood framing	.69	1.09	1.78
1020	On masonry	.24	.61	.85
1100	Metal base galvanized and painted 2-1/2″ high	.47	.96	1.43

6 INTERIOR CONSTRUCTION

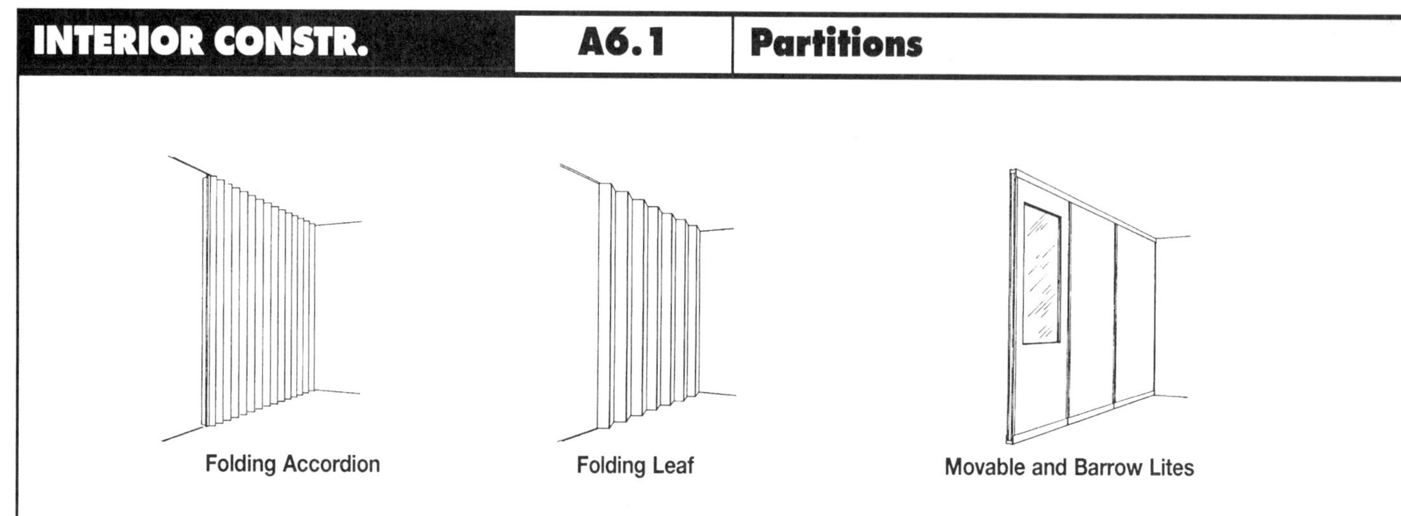

6.1-820	Partitions	COST PER S.F.		
		MAT.	INST.	TOTAL
0360	Folding accordion, vinyl covered, acoustical, 3 lb. S.F., 17 ft max. hgt.	19.25	4.86	24.11
0380	5 lb. per S.F. 27 ft max height	28.50	5.10	33.60
0400	5.5 lb. per S.F., 17 ft. max height	32	5.40	37.40
0420	Commercial, 1.75 lb per S.F., 8 ft. max height	12.75	2.16	14.91
0440	2.0 Lb per S.F., 17 ft. max height	15.50	3.24	18.74
0460	Industrial, 4.0 lb. per S.F. 27 ft max height	26	6.45	32.45
0480	Vinyl clad wood or steel, electric operation 6 psf	32	3.04	35.04
0500	Wood, non acoustic, birch or mahogany	17.10	1.62	18.72
0560	Folding leaf, aluminum framed acoustical 12 ft.high.,5.5 lb per S.F.,min.	27	8.10	35.10
0580	Maximum	32.50	16.20	48.70
0600	6.5 lb. per S.F., minimum	28.50	8.10	36.60
0620	Maximum	35	16.20	51.20
0640	Steel acoustical, 7.5 per S.F., vinyl faced, minimum	40	8.10	48.10
0660	Maximum	49	16.20	65.20
0680	Wood acoustic type, vinyl faced to 18' high 6 psf, minimum	38	8.10	46.10
0700	Average	45	10.80	55.80
0720	Maximum	58.50	16.20	74.70
0740	Formica or hardwood faced, minimum	39	8.10	47.10
0760	Maximum	41.50	16.20	57.70
0780	Wood, low acoustical type to 12 ft. high 4.5 psf	28.50	9.70	38.20
0840	Demountable, trackless wall, cork finish, semi acous, 1-5/8"th, min.	19.70	1.49	21.19
0860	Maximum	24	2.56	26.56
0880	Acoustic, 2" thick, minimum	20.50	1.59	22.09
0900	Maximum	30	2.16	32.16
0920	In-plant modular office system, w/prehung steel door			
0940	3" thick honeycomb core panels			
0960	12' x 12', 2 wall	10.30	.28	10.58
0970	4 wall	10.60	.37	10.97
0980	16' x 16', 2 wall	10.25	.20	10.45
0990	4 wall	7.35	.20	7.55
1000	Gypsum, demountable, 3" to 3-3/4" thick x 9' high, vinyl clad	2.72	1.12	3.84
1020	Fabric clad	8.45	1.23	9.68
1040	1.75 system, vinyl clad hardboard, paper honeycomb core panel			
1060	1-3/4" to 2-1/2" thick x 9' high	5.70	1.12	6.82
1080	Unitized gypsum panel system, 2" to 2-1/2" thick x 9' high			
1100	Vinyl clad gypsum	7.70	1.12	8.82
1120	Fabric clad gypsum	13.90	1.23	15.13
1140	Movable steel walls, modular system			
1160	Unitized panels, 48" wide x 9' high			
1180	Baked enamel, pre-finished	9	.90	9.90
1200	Fabric clad	13.75	.96	14.71
1203				

Important: See the Reference Section for critical supporting data - Reference Nos., Crews & Location Factors

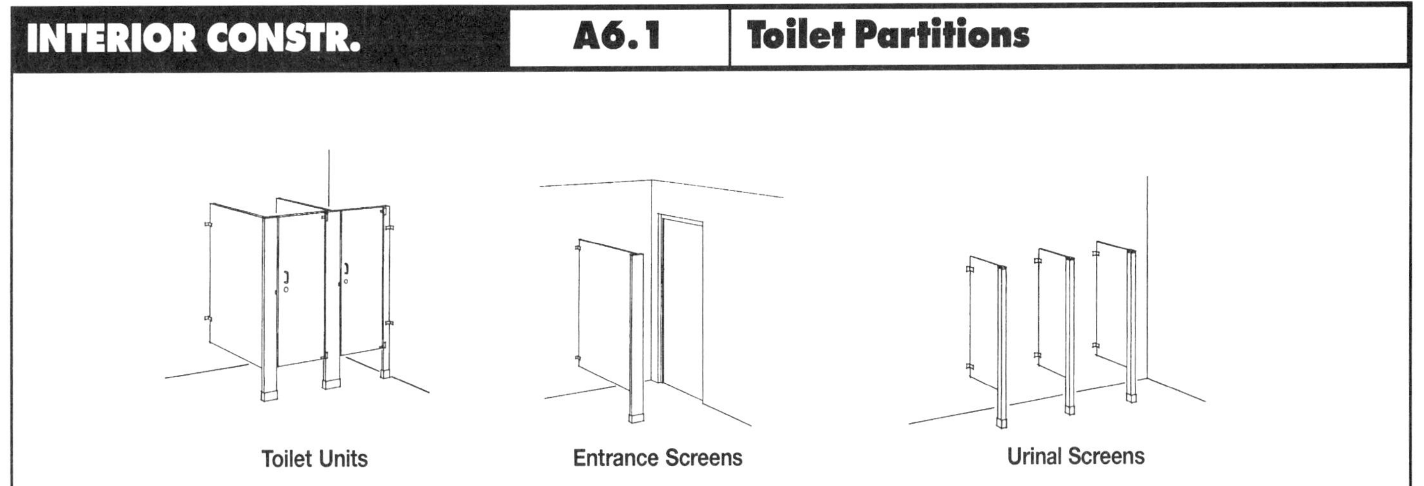

6.1-870	Toilet Partitions	COST PER UNIT		
		MAT.	INST.	TOTAL
0380	Toilet partitions, cubicles, ceiling hung, marble	1,375	242	1,617
0400	Painted metal	420	121	541
0420	Plastic laminate	565	121	686
0460	Stainless steel	1,025	121	1,146
0480	Handicap addition	268		268
0520	Floor and ceiling anchored, marble	1,500	194	1,694
0540	Painted metal	425	97	522
0560	Plastic laminate	565	97	662
0600	Stainless steel	1,200	97	1,297
0620	Handicap addition	268		268
0660	Floor mounted marble	890	162	1,052
0680	Painted metal	385	69.50	454.50
0700	Plastic laminate	565	69.50	634.50
0740	Stainless steel	1,150	69.50	1,219.50
0760	Handicap addition	268		268
0780	Juvenile deduction	38.50		38.50
0820	Floor mounted with handrail marble	1,025	162	1,187
0840	Painted metal	395	81	476
0860	Plastic laminate	550	81	631
0900	Stainless steel	1,175	81	1,256
0920	Handicap addition	268		268
0960	Wall hung, painted metal	485	69.50	554.50
1020	Stainless steel	1,150	69.50	1,219.50
1040	Handicap addition	268		268
1080	Entrance screens, floor mounted, 54" high, marble	565	54	619
1100	Painted metal	177	32.50	209.50
1140	Stainless steel	640	32.50	672.50
1300	Urinal screens, floor mounted, 24" wide, laminated plastic	256	60.50	316.50
1320	Marble	530	72.50	602.50
1340	Painted metal	183	60.50	243.50
1380	Stainless steel	530	60.50	590.50
1428	Wall mounted wedge type, painted metal	207	48.50	255.50
1460	Stainless steel	425	48.50	473.50

INTERIOR CONSTRUCTION

6

Sliding Panel-Mall Front

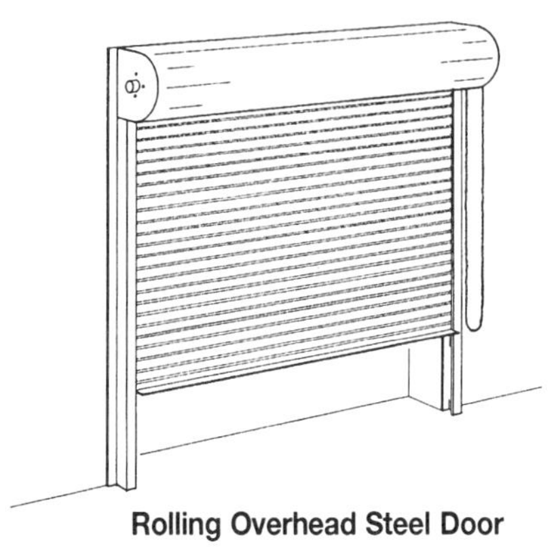

Rolling Overhead Steel Door

6.4-100	Special Doors	COST PER OPNG.		
		MAT.	INST.	TOTAL
2500	Single leaf, wood, 3'-0"x7'-0"x1 3/8", birch, solid core	282	85	367
2510	Hollow core	242	81.50	323.50
2530	Hollow core, lauan	221	81.50	302.50
2540	Louvered pine	335	81.50	416.50
2550	Paneled pine	330	81.50	411.50
2600	Hollow metal, comm. quality, flush, 3'-0"x7'-0"x1-3/8"	395	89	484
2650	3'-0"x10'-0" openings with panel	630	114	744
2700	Metal fire, comm. quality, 3'-0"x7'-0"x1-3/8"	445	92.50	537.50
2800	Kalamein fire, comm. quality, 3'-0"x7'-0"x1-3/4"	380	179	559
3200	Double leaf, wood, hollow core, 2 - 3'-0"x7'-0"x1-3/8"	380	158	538
3300	Hollow metal, comm. quality, B label, 2'-3'-0"x7'-0"x1-3/8"	840	195	1,035
3400	6'-0"x10'-0" opening, with panel	1,200	237	1,437
3500	Double swing door system, 12'-0"x7'-0", mill finish	5,250	1,350	6,600
3700	Black finish	5,750	1,375	7,125
3800	Sliding entrance door and system mill finish	6,250	1,175	7,425
3900	Bronze finish	6,875	1,275	8,150
4000	Black finish	7,175	1,325	8,500
4100	Sliding panel mall front, 16'x9' opening, mill finish	2,200	360	2,560
4200	Bronze finish	2,850	470	3,320
4300	Black finish	3,525	575	4,100
4400	24'x9' opening mill finish	3,200	665	3,865
4500	Bronze finish	4,150	865	5,015
4600	Black finish	5,125	1,075	6,200
4700	48'x9' opening mill finish	5,950	520	6,470
4800	Bronze finish	7,725	675	8,400
4900	Black finish	9,525	830	10,355
5000	Rolling overhead steel door, manual, 8' x 8' high	710	375	1,085
5100	10' x 10' high	950	430	1,380
5200	20' x 10' high	2,425	600	3,025
5300	12' x 12' high	1,250	500	1,750
5400	Motor operated, 8' x 8' high	1,650	495	2,145
5500	10' x 10' high	1,900	550	2,450
5600	20' x 10' high	3,375	720	4,095
5700	12' x 12' high	2,200	620	2,820
5800	Roll up grille, aluminum, manual, 10' x 10' high, mill finish	1,800	735	2,535
5900	Bronze anodized	2,850	735	3,585
6000	Motor operated, 10' x 10' high, mill finish	2,600	855	3,455
6100	Bronze anodized	3,650	855	4,505
6200	Steel, manual, 10' x 10' high	1,600	600	2,200
6300	15' x 8' high	1,875	750	2,625
6400	Motor operated, 10' x 10' high	2,400	720	3,120
6500	15' x 8' high	2,675	870	3,545

Important: See the Reference Section for critical supporting data - Reference Nos., Crews & Location Factors

6.5-100	Paint & Covering	COST PER S.F.		
		MAT.	INST.	TOTAL
0060	Painting, interior on plaster and drywall, brushwork, primer & 1 coat	.09	.34	.43
0080	Primer & 2 coats	.13	.45	.58
0100	Primer & 3 coats	.17	.55	.72
0120	Walls & ceilings, roller work, primer & 1 coat	.09	.23	.32
0140	Primer & 2 coats	.13	.29	.42
0160	Woodwork incl. puttying, brushwork, primer & 1 coat	.09	.49	.58
0180	Primer & 2 coats	.13	.65	.78
0200	Primer & 3 coats	.17	.88	1.05
0260	Cabinets and casework, enamel, primer & 1 coat	.09	.55	.64
0280	Primer & 2 coats	.13	.68	.81
0300	Masonry or concrete, latex, brushwork, primer & 1 coat	.16	.46	.62
0320	Primer & 2 coats	.21	.66	.87
0340	Addition for block filler	.10	.56	.66
0380	Fireproof paints, intumescent, 1/8" thick 3/4 hour	1.67	.45	2.12
0400	3/16" thick 1 hour	3.72	.68	4.40
0420	7/16" thick 2 hour	4.79	1.57	6.36
0440	1-1/16" thick 3 hour	7.80	3.14	10.94
0480	Miscellaneous metal brushwork, exposed metal, primer & 1 coat	.06	.55	.61
0500	Gratings, primer & 1 coat	.15	.69	.84
0600	Pipes over 12" diameter	.40	2.20	2.60
0700	Structural steel, brushwork, light framing 300-500 S.F./Ton	.05	.88	.93
0720	Heavy framing 50-100 S.F./Ton	.05	.44	.49
0740	Spraywork, light framing 300-500 S.F./Ton	.06	.20	.26
0760	Heavy framing 50-100 S.F./Ton	.06	.22	.28
0800	Varnish, interior wood trim, no sanding sealer & 1 coat	.07	.55	.62
0820	Hardwood floor, no sanding 2 coats	.12	.12	.24
0840	Wall coatings, acrylic glazed coatings, minimum	.24	.42	.66
0860	Maximum	.52	.72	1.24
0880	Epoxy coatings, minimum	.32	.42	.74
0900	Maximum	.98	1.29	2.27
0940	Exposed epoxy aggregate, troweled on, 1/16" to 1/4" aggregate, minimum	.50	.94	1.44
0960	Maximum	1.06	1.69	2.75
0980	1/2" to 5/8" aggregate, minimum	.97	1.69	2.66
1000	Maximum	1.66	2.75	4.41
1020	1" aggregate, minimum	1.69	2.44	4.13
1040	Maximum	2.60	4	6.60
1060	Sprayed on, minimum	.46	.75	1.21
1080	Maximum	.85	1.52	2.37
1100	High build epoxy 50 mil, minimum	.53	.56	1.09
1120	Maximum	.92	2.32	3.24
1140	Laminated epoxy with fiberglass minimum	.58	.75	1.33
1160	Maximum	1.05	1.52	2.57
1180	Sprayed perlite or vermiculite 1/16" thick, minimum	.20	.07	.27
1200	Maximum	.59	.34	.93
1260	Wall coatings, vinyl plastic, minimum	.26	.30	.56
1280	Maximum	.66	.92	1.58
1300	Urethane on smooth surface, 2 coats, minimum	.19	.19	.38
1320	Maximum	.45	.33	.78
1340	3 coats, minimum	.26	.26	.52
1360	Maximum	.59	.47	1.06
1380	Ceramic-like glazed coating, cementitious, minimum	.40	.50	.90
1400	Maximum	.65	.64	1.29
1420	Resin base, minimum	.26	.34	.60
1440	Maximum	.44	.67	1.11
1460	Wall coverings, aluminum foil	.83	.81	1.64
1480	Copper sheets, .025" thick, phenolic backing	5.65	.93	6.58
1500	Vinyl backing	4.37	.93	5.30
1520	Cork tiles, 12"x12", light or dark, 3/16" thick	2.80	.93	3.73
1540	5/16" thick	2.92	.95	3.87
1560	Basketweave, 1/4" thick	4.48	.93	5.41

For expanded coverage of these items see *Means Assemblies Cost Data 1998*

6.5-100	Paint & Covering	COST PER S.F.		
		MAT.	INST.	TOTAL
1580	Natural, non-directional, 1/2" thick	6.45	.93	7.38
1600	12"x36", granular, 3/16" thick	.97	.58	1.55
1620	1" thick	1.25	.60	1.85
1640	12"x12", polyurethane coated, 3/16" thick	3.03	.93	3.96
1660	5/16" thick	4.29	.95	5.24
1661	Paneling, prefinished plywood, birch	1.13	1.01	2.14
1662	Mahogany, African	2.10	1.06	3.16
1663	Philippine (lauan)	.90	.85	1.75
1664	Oak or cherry	2.95	1.06	4.01
1665	Rosewood	4.19	1.33	5.52
1666	Teak	2.95	1.06	4.01
1667	Chestnut	4.37	1.13	5.50
1668	Pecan	1.87	1.06	2.93
1669	Walnut	4.76	1.06	5.82
1670	Wood board, knotty pine, finished	1.60	1.83	3.43
1671	Rough sawn cedar	2	1.83	3.83
1672	Redwood	4.37	1.83	6.20
1673	Aromatic cedar	3.35	1.96	5.31
1680	Cork wallpaper, paper backed, natural	1.72	.46	2.18
1700	Color	2.12	.46	2.58
1720	Gypsum based, fabric backed, minimum	.68	.28	.96
1740	Average	.95	.31	1.26
1760	Maximum	1.06	.35	1.41
1780	Vinyl wall covering, fabric back, light weight	.55	.35	.90
1800	Medium weight	.69	.46	1.15
1820	Heavy weight	1.24	.51	1.75
1840	Wall paper, double roll, solid pattern, avg. workmanship	.29	.35	.64
1860	Basic pattern, avg. workmanship	.51	.42	.93
1880	Basic pattern, quality workmanship	.97	.51	1.48
1900	Grass cloths with lining paper, minimum	.62	.56	1.18
1920	Maximum	1.98	.64	2.62
1940	Ceramic tile, thin set, 4-1/4" x 4-1/4"	2.23	2.15	4.38

6.5-100	Paint Trim	COST PER L.F.		
		MAT.	INST.	TOTAL
2040	Painting, wood trim, to 6" wide, enamel, primer & 1 coat	.09	.28	.37
2060	Primer & 2 coats	.13	.35	.48
2080	Misc. metal brushwork, ladders	.29	2.75	3.04
2100	Pipes, to 4" dia.	.05	.58	.63
2120	6" to 8" dia.	.11	1.16	1.27
2140	10" to 12" dia.	.32	1.73	2.05
2160	Railings, 2" pipe	.10	1.38	1.48
2180	Handrail, single	.08	.55	.63

6.6-100 — Tile & Covering

		COST PER S.F.		
		MAT.	INST.	TOTAL
0060	Carpet tile, nylon, fusion bonded, 18" x 18" or 24" x 24", 24 oz.	3	.31	3.31
0080	35 oz.	3.72	.31	4.03
0100	42 oz.	4.33	.31	4.64
0140	Carpet, tufted, nylon, roll goods, 12' wide, 26 oz.	2.83	.36	3.19
0160	36 oz.	4.50	.36	4.86
0180	Woven, wool, 36 oz.	7.10	.36	7.46
0200	42 oz.	7.25	.36	7.61
0220	Padding, add to above, minimum	.38	.17	.55
0240	Maximum	.65	.17	.82
0260	Composition flooring, acrylic, 1/4" thick	1.21	2.33	3.54
0280	3/8" thick	1.60	2.68	4.28
0300	Epoxy, minimum	2.15	1.79	3.94
0320	Maximum	2.58	2.47	5.05
0340	Epoxy terrazzo, minimum	4.46	2.66	7.12
0360	Maximum	7.45	3.54	10.99
0380	Mastic, hot laid, 1-1/2" thick, minimum	3.14	1.75	4.89
0400	Maximum	4.02	2.33	6.35
0420	Neoprene 1/4" thick, minimum	3.08	2.22	5.30
0440	Maximum	4.23	2.81	7.04
0460	Polyacrylate with ground granite 1/4", minimum	2.37	1.64	4.01
0480	Maximum	4.84	2.52	7.36
0500	Polyester with colored quart 2 chips 1/16", minimum	2.15	1.14	3.29
0520	Maximum	3.85	1.79	5.64
0540	Polyurethane with vinyl chips, minimum	6	1.14	7.14
0560	Maximum	8.65	1.41	10.06
0600	Concrete topping, granolithic concrete, 1/2" thick	.17	1.05	1.22
0620	1" thick	.35	1.08	1.43
0640	2" thick	.69	1.24	1.93
0660	Heavy duty 3/4" thick, minimum	.23	1.63	1.86
0680	Maximum	.23	1.94	2.17
0700	For colors, add to above, minimum	.46	.38	.84
0720	Maximum	2.05	.41	2.46
0740	Exposed aggregate finish, minimum	.40	.36	.76
0760	Maximum	1.25	.48	1.73
0780	Abrasives, .25 P.S.F. add to above, minimum	.16	.26	.42
0800		.19	.26	.45
0820	Dust on coloring, add, minimum	.62	.17	.79
0840	Maximum	2.16	.36	2.52
0860	1/2" integral, minimum	.19	1.05	1.24
0880	Maximum	.19	1.05	1.24
0900	Dustproofing, add, minimum	.07	.12	.19
0920	Maximum	.12	.17	.29
0930	Paint	.21	.66	.87
0940	Hardeners, metallic add, minimum	.22	.26	.48
0960	Maximum	.49	.39	.88
0980	Non-metallic, minimum	.13	.26	.39
1000	Maximum	.37	.39	.76
1020	Integral topping, 3/16" thick	.06	.62	.68
1040	1/2" thick	.16	.65	.81
1060	3/4" thick	.25	.73	.98
1080	1" thick	.32	.83	1.15
1100	Terrazzo, minimum	2.42	4.09	6.51
1120	Maximum	4.51	8.85	13.36
1280	Resilient, asphalt tile, 1/8" thick on concrete, minimum	1.03	.57	1.60
1300	Maximum	1.13	.57	1.70
1320	On wood, add for felt underlay	.20		.20
1340	Cork tile, minimum	2.66	.72	3.38
1360	Maximum	10.10	.72	10.82
1380	Polyethylene, in rolls, minimum	2.41	.82	3.23
1400	Maximum	4.68	.82	5.50

INTERIOR CONSTRUCTION

6.6-100	Tile & Covering	COST PER S.F.		
		MAT.	INST.	TOTAL
1420	Polyurethane, thermoset, minimum	3.71	2.26	5.97
1440	Maximum	4.40	4.53	8.93
1460	Rubber, sheet goods, minimum	3.31	1.89	5.20
1480	Maximum	5.45	2.52	7.97
1500	Tile, minimum	3.33	.57	3.90
1520	Maximum	6.45	.82	7.27
1580	Vinyl, composition tile, minimum	.75	.45	1.20
1600	Maximum	2.04	.45	2.49
1620	Tile, minimum	1.69	.45	2.14
1640	Maximum	4.92	.45	5.37
1660	Sheet goods, minimum	1.38	.91	2.29
1680	Maximum	3.93	1.18	5.11
1720	Tile, ceramic natural clay	3.85	2.23	6.08
1730	Marble, synthetic 12"x12"x5/8"	7.40	6.80	14.20
1740	Porcelain type, minimum	4.15	2.23	6.38
1760	Maximum	4.38	2.15	6.53
1800	Quarry tile, mud set, minimum	3.20	2.92	6.12
1820	Maximum	4.97	3.71	8.68
1840	Thin set, deduct		.58	.58
1860	Terrazzo precast, minimum	7.20	3.02	10.22
1880	Maximum	8.05	3.02	11.07
1900	Non-slip, minimum	16.35	15	31.35
1920	Maximum	17.60	21	38.60
1960	Wood, block, end grain factory type, creosoted, 2" thick	2.92	.82	3.74
1980	2-1/2" thick	3.03	1.94	4.97
2000	3" thick	2.97	1.94	4.91
2020	Natural finish, 2" thick	3.03	1.94	4.97
2040	Fir, vertical grain, 1"x4", no finish, minimum	2.38	.95	3.33
2060	Maximum	2.53	.95	3.48
2080	Prefinished white oak, prime grade, 2-1/4" wide	6.75	1.43	8.18
2100	3-1/4" wide	8.35	1.31	9.66
2120	Maple strip, sanded and finished, minimum	3.62	2.03	5.65
2140	Maximum	4.28	2.03	6.31
2160	Oak strip, sanded and finished, minimum	4.38	2.03	6.41
2180	Maximum	3.96	2.03	5.99
2200	Parquetry, sanded and finished, minimum	2.40	2.12	4.52
2220	Maximum	6.40	3.03	9.43
2260	Add for sleepers on concrete, treated, 24" O.C., 1"x2"	1.10	1.10	2.20
2280	1"x3"	1.10	.85	1.95
2300	2"x4"	.87	.43	1.30
2340	Underlayment, plywood, 3/8" thick	.50	.28	.78
2350	1/2" thick	.58	.29	.87
2360	5/8" thick	.69	.30	.99
2370	3/4" thick	.83	.33	1.16
2380	Particle board, 3/8" thick	.34	.28	.62
2390	1/2" thick	.36	.29	.65
2400	5/8" thick	.39	.30	.69
2410	3/4" thick	.45	.33	.78
2420	Hardboard, 4' x 4', .215" thick	.45	.28	.73

Important: See the Reference Section for critical supporting data - Reference Nos., Crews & Location Factors

Plaster Partitions are defined as follows: type of plaster, type and spacing of framing, type of lath and treatment of opposite face.

Included in the system components are expansion joints. Metal studs are assumed to be nonload bearing.

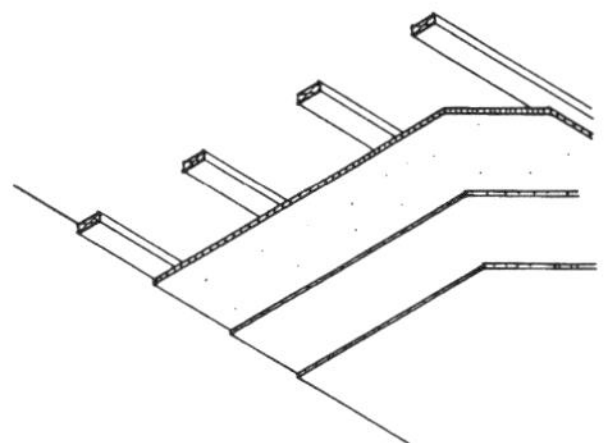

2 Coats of Plaster on Gypsum Lath on Wood Furring

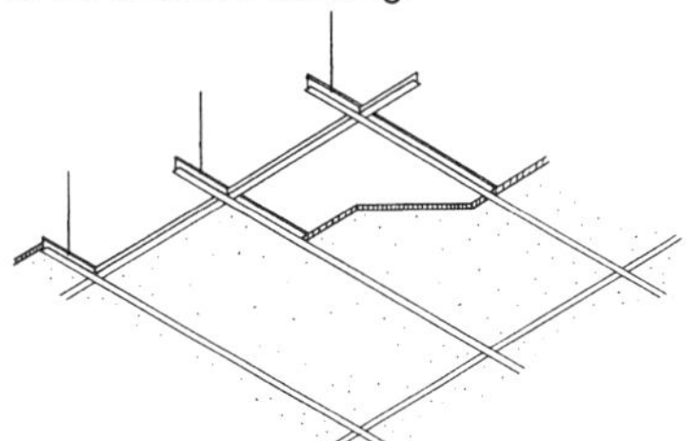

Fiberglass Board on Exposed Suspended Grid System

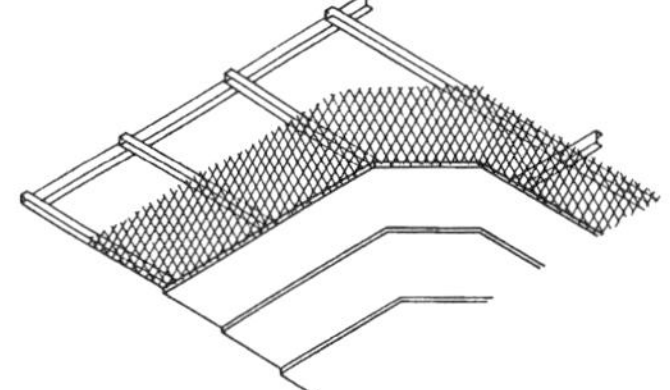

Plaster and Metal Lath on Metal Furring

6.7-100 — Plaster Ceilings

	TYPE	LATH	FURRING	SUPPORT		MAT.	INST.	TOTAL
2400	2 coat gypsum	3/8" gypsum	1"x3" wood, 16" O.C.	wood		1.07	2.49	3.56
2500	Painted			masonry		1.07	2.54	3.61
2600				concrete		1.07	2.84	3.91
2700	3 coat gypsum	3.4# metal	1"x3" wood, 16" O.C.	wood		1.06	2.66	3.72
2800	Painted			masonry		1.06	2.71	3.77
2900				concrete		1.06	3.01	4.07
3000	2 coat perlite	3/8" gypsum	1"x3" wood, 16" O.C.	wood		1.15	2.59	3.74
3100	Painted			masonry		1.15	2.64	3.79
3200				concrete		1.15	2.94	4.09
3300	3 coat perlite	3.4# metal	1"x3" wood, 16" O.C.	wood		1	2.65	3.65
3400	Painted			masonry		1	2.70	3.70
3500				concrete		1	3	4
3600	2 coat gypsum	3/8" gypsum	3/4" CRC, 12" O.C.	1-1/2" CRC, 48"O.C.		1.03	3.06	4.09
3700	Painted		3/4" CRC, 16" O.C.	1-1/2" CRC, 48"O.C.		1.01	2.76	3.77
3800			3/4" CRC, 24" O.C.	1-1/2" CRC, 48"O.C.		.96	2.52	3.48
3900	2 coat perlite	3/8" gypsum	3/4" CRC, 12" O.C.	1-1/2" CRC, 48"O.C		1.11	3.28	4.39
4000	Painted		3/4" CRC, 16" O.C.	1-1/2" CRC, 48"O.C.		1.09	2.98	4.07
4100			3/4" CRC, 24" O.C.	1-1/2" CRC, 48"O.C.		1.04	2.74	3.78
4200	3 coat gypsum	3.4# metal	3/4" CRC, 12" O.C.	1-1/2" CRC, 36" O.C.		1.28	3.96	5.24
4300	Painted		3/4" CRC, 16" O.C.	1-1/2" CRC, 36" O.C.		1.26	3.66	4.92
4400			3/4" CRC, 24" O.C.	1-1/2" CRC, 36" O.C.		1.21	3.42	4.63
4500	3 coat perlite	3.4# metal	3/4" CRC, 12" O.C.	1-1/2" CRC, 36" O.C.		1.47	4.34	5.81
4600	Painted		3/4" CRC, 16" O.C.	1-1/2" CRC, 36" O.C.		1.45	4.04	5.49
4700			3/4" CRC, 24" O.C.	1-1/2" CRC, 36" O.C.		1.40	3.80	5.20

6.7-100 — Drywall Ceilings

	TYPE	FINISH	FURRING	SUPPORT		MAT.	INST.	TOTAL
4800	1/2" F.R. drywall	painted and textured	1"x3" wood, 16" O.C.	wood		.78	1.54	2.32
4900				masonry		.78	1.59	2.37
5000				concrete		.78	1.89	2.67
5100	5/8" F.R. drywall	painted and textured	1"x3" wood, 16" O.C.	wood		.78	1.54	2.32
5200				masonry		.78	1.59	2.37
5300				concrete		.78	1.89	2.67
5400	1/2" F.R. drywall	painted and textured	7/8"resil. channels	24" O.C.		.65	1.50	2.15
5500			1"x2" wood	stud clips		.78	1.45	2.23
5602	1/2" F.R. drywall	painted	1-5/8" metal studs	24" O.C.		.71	1.49	2.20

INTERIOR CONSTRUCTION

6

6.7-100 — Drywall Ceilings

	TYPE	FINISH	FURRING	SUPPORT		COST PER S.F.		
						MAT.	INST.	TOTAL
5700 5702	5/8" F.R. drywall	painted and textured	1-5/8"metal studs	24" O.C.		.71	1.49	2.20

6.7-100 — Acoustical Ceilings

	TYPE	TILE	GRID	SUPPORT		COST PER S.F.		
						MAT.	INST.	TOTAL
5800	5/8" fiberglass board	24" x 48"	tee	suspended		.84	.70	1.54
5900		24" x 24"	tee	suspended		.92	.77	1.69
6000	3/4" fiberglass board	24" x 48"	tee	suspended		1.38	.71	2.09
6100		24" x 24"	tee	suspended		1.46	.78	2.24
6500	5/8" mineral fiber	12" x 12"	1"x3" wood, 12" O.C.	wood		.87	.93	1.80
6600				masonry		.87	1	1.87
6700				concrete		.87	1.40	2.27
6800	3/4" mineral fiber	12" x 12"	1"x3" wood, 12" O.C.	wood		1.54	.93	2.47
6900				masonry		1.54	.93	2.47
7000				concrete		1.54	.93	2.47
7100 7102	3/4"mineral fiber on 5/8" F.R. drywall	12" x 12"	25 ga. channels	runners		1.71	1.29	3
7200 7201 7202	5/8" plastic coated Mineral fiber	12" x 12"		adhesive backed		.87	.24	1.11
7300 7301 7302	3/4" plastic coated Mineral fiber	12" x 12"		adhesive backed		1.54	.24	1.78
7400 7401 7402	3/4" mineral fiber	12" x 12"	conceal 2" bar & channels	suspended		1.38	1.24	2.62

6.7-810 — Acoustical Ceilings

		COST PER S.F.		
		MAT.	INST.	TOTAL
2480	Ceiling boards, eggcrate, acrylic, 1/2" x 1/2" x 1/2" cubes	1.33	.49	1.82
2500	Polystyrene, 3/8" x 3/8" x 1/2" cubes	1.11	.48	1.59
2520	1/2" x 1/2" x 1/2" cubes	1.50	.49	1.99
2540	Fiberglass boards, plain, 5/8" thick	.50	.39	.89
2560	3/4" thick	1.04	.40	1.44
2580	Grass cloth faced, 3/4" thick	1.62	.49	2.11
2600	1" thick	1.83	.50	2.33
2620	Luminous panels, prismatic, acrylic	1.61	.61	2.22
2640	Polystyrene	.83	.61	1.44
2660	Flat or ribbed, acrylic	2.82	.61	3.43
2680	Polystyrene	1.91	.61	2.52
2700	Drop pan, white, acrylic	4.11	.61	4.72
2720	Polystyrene	3.44	.61	4.05
2740	Mineral fiber boards, 5/8" thick, standard	.57	.36	.93
2760	Plastic faced	.89	.61	1.50
2780	2 hour rating	.79	.36	1.15
2800	Perforated aluminum sheets, .024 thick, corrugated painted	1.64	.50	2.14
2820	Plain			
3080	Mineral fiber, plastic coated, 12" x 12" or 12" x 24", 5/8" thick	.56	.24	.80
3100	3/4" thick	1.23	.24	1.47
3120	Fire rated, 3/4" thick, plain faced	.83	.24	1.07
3140	Mylar faced	.96	.24	1.20
3160	Add for ceiling primer	.12		.12
3180	Add for ceiling cement	.31		.31
3240	Suspension system, furring, 1" x 3" wood 12" O.C.	.31	.69	1
3260	T bar suspension system, 2' x 4' grid	.33	.30	.63

Important: See the Reference Section for critical supporting data - Reference Nos., Crews & Location Factors

Acoustical Ceilings

		COST PER S.F.		
		MAT.	INST.	TOTAL
	2' grid	.41	.37	.78
3300	Concealed Z bar suspension system 12" module	.36	.47	.83
3320	Add to above for 1-1/2" carrier channels 4' O.C.	.18	.52	.70
3340	Add to above for carrier channels for recessed lighting	.35	.53	.88

6.7-820 — Plaster Ceilings

		COST PER S.F.		
		MAT.	INST.	TOTAL
0060	Plaster, gypsum incl. finish, 2 coats			
0080	3 coats	.49	1.55	2.04
0100	Perlite, incl. finish, 2 coats	.43	1.54	1.97
0120	3 coats	.68	1.93	2.61
0140	Thin coat on drywall	.13	.30	.43
0200	Lath, gypsum, 3/8" thick	.40	.42	.82
0220	1/2" thick	.47	.44	.91
0240	5/8" thick	.48	.51	.99
0260	Metal, diamond, 2.5 lb.	.22	.34	.56
0280	3.4 lb.	.25	.36	.61
0300	Flat rlb, 2.75 lb.	.21	.34	.55
0320	3.4 lb.	.39	.36	.75
0440	Furring, steel channels, 3/4" galvanized , 12" O.C.	.19	1.09	1.28
0460	16" O.C.	.17	.79	.96
0480	24" O.C.	.12	.55	.67
0500	1-1/2" galvanized , 12" O.C.	.27	1.21	1.48
0520	16" O.C.	.25	.88	1.13
0540	24" O.C.	.16	.59	.75
0560	Wood strips, 1"x3", on wood, 12" O.C.	.31	.69	1
0580	16" O.C.	.23	.52	.75
0600	24" O.C.	.16	.35	.51
0620	On masonry, 12" O.C.	.31	.76	1.07
0640	16" O.C.	.23	.57	.80
0660	24" O.C.	.16	.38	.54
0680	On concrete, 12" O.C.	.31	1.16	1.47
0700	16" O.C.	.23	.87	1.10
0720	24" O.C.	.16	.58	.74
0722				
0740				
0940	Paint on plaster or drywall, roller work, primer + 1 coat	.09	.23	.32
0960	Primer + 2 coats	.13	.29	.42

For information about Means Estimating Seminars, see yellow pages 11 and 12 in back of book

INTERIOR CONSTRUCTION

6

For expanded coverage of these items see *Means Assemblies Cost Data 1998*

Division 7
Conveying Systems

The hydraulic elevator obtains its motion from the movement of liquid under pressure in the piston connected to the car bottom. These pistons can provide travel to a maximum rise of 70' and are sized for the intended load. As the rise reaches the upper limits the cost tends to exceed that of a geared electric unit.

7.1-100	Hydraulic	COST EACH		
		MAT.	INST.	TOTAL
1300	Pass. elev., 1500 lb., 2 Floors, 100 FPM	32,100	5,525	37,625
1400	5 Floors, 100 FPM	43,500	14,700	58,200
1600	2000 lb., 2 Floors, 100 FPM	33,000	5,525	38,525
1700	5 floors, 100 FPM	44,300	14,700	59,000
1900	2500 lb., 2 Floors, 100 FPM	33,400	5,525	38,925
2000	5 floors, 100 FPM	44,800	14,700	59,500
2200	3000 lb., 2 Floors, 100 FPM	34,900	5,525	40,425
2300	5 floors, 100 FPM	46,200	14,700	60,900
2500	3500 lb., 2 Floors, 100 FPM	37,000	5,525	42,525
2600	5 floors, 100 FPM	48,400	14,700	63,100
2800	4000 lb., 2 Floors, 100 FPM	38,000	5,525	43,525
2900	5 floors, 100 FPM	49,400	14,700	64,100
3100	4500 lb., 2 Floors, 100 FPM	38,900	5,525	44,425
3200	5 floors, 100 FPM	50,500	14,700	65,200
4000	Hospital elevators, 3500 lb., 2 Floors, 100 FPM	46,400	5,525	51,925
4100	5 floors, 100 FPM	70,000	14,700	84,700
4300	4000 lb., 2 Floors, 100 FPM	46,400	5,525	51,925
4400	5 floors, 100 FPM	70,000	14,700	84,700
4600	4500 lb., 2 Floors, 100 FPM	52,000	5,525	57,525
4800	5 floors, 100 FPM	75,500	14,700	90,200
4900	5000 lb., 2 Floors, 100 FPM	54,500	5,525	60,025
5000	5 floors, 100 FPM	78,000	14,700	92,700
6700	Freight elevators (Class "B"), 3000 lb., 2 Floors, 50 FPM	44,600	7,050	51,650
6800	5 floors, 100 FPM	64,000	18,500	82,500
7000	4000 lb., 2 Floors, 50 FPM	47,500	7,050	54,550
7100	5 floors, 100 FPM	67,000	18,500	85,500
7500	10,000 lb., 2 Floors, 50 FPM	59,000	7,050	66,050
7600	5 floors, 100 FPM	78,500	18,500	97,000
8100	20,000 lb., 2 Floors, 50 FPM	71,500	7,050	78,550
8200	5 Floors, 100 FPM	90,500	18,500	109,000

For information about Means Estimating Seminars, see yellow pages 11 and 12 in back of book

Important: See the Reference Section for critical supporting data - Reference Nos., Crews & Location Factors

Division 8
Mechanical

Installation includes piping and fittings within 10′ of heater. Gas heaters require vent piping (not included with these units).

8.1-120	Gas Fired Water Heaters - Residential Systems	COST EACH		
		MAT.	INST.	TOTAL
2200	Gas fired water heater, residential, 100° F rise			
2220	20 gallon tank, 25 GPH	540	535	1,075
2260	30 gallon tank, 32 GPH	545	545	1,090
2300	40 gallon tank, 32 GPH	620	610	1,230
2340	50 gallon tank, 63 GPH	700	610	1,310
2380	75 gallon tank, 63 GPH	1,075	685	1,760
2420	100 gallon tank, 63 GPH	1,525	720	2,245
2422				

8.1-130	Oil Fired Water Heaters - Residential Systems	COST EACH		
		MAT.	INST.	TOTAL
2200	Oil fired water heater, residential, 100° F rise			
2220	30 gallon tank, 103 GPH	1,025	505	1,530
2260	50 gallon tank, 145 GPH	1,425	565	1,990
2300	70 gallon tank, 164 GPH	1,625	635	2,260
2340	85 gallon tank, 181 GPH	2,200	655	2,855
2344				

Important: See the Reference Section for critical supporting data - Reference Nos., Crews & Location Factors

Systems below include piping and fittings within 10' of heater. Electric water heaters do not require venting.

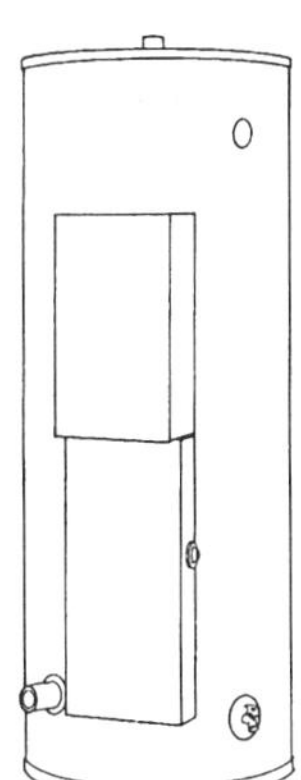

8.1-160	Electric Water Heaters - Commercial Systems	COST EACH		
		MAT.	INST.	TOTAL
1800	Electric water heater, commercial, 100° F rise			
1820	50 gallon tank, 9 KW 37 GPH	2,925	455	3,380
1860	80 gal, 12 KW 49 GPH	3,725	565	4,290
1900	36 KW 147 GPH	5,075	610	5,685
1940	120 gal, 36 KW 147 GPH	5,475	660	6,135
1980	150 gal, 120 KW 490 GPH	18,700	705	19,405
2020	200 gal, 120 KW 490 GPH	19,600	725	20,325
2060	250 gal, 150 KW 615 GPH	21,800	845	22,645
2100	300 gal, 180 KW 738 GPH	33,800	890	34,690
2140	350 gal, 30 KW 123 GPH	16,700	960	17,660
2180	180 KW 738 GPH	24,400	960	25,360
2220	500 gal, 30 KW 123 GPH	21,500	1,125	22,625
2260	240 KW 984 GPH	32,100	1,125	33,225
2300	700 gal, 30 KW 123 GPH	26,100	1,250	27,350
2340	300 KW 1230 GPH	41,500	1,250	42,750
2380	1000 gal, 60 KW 245 GPH	31,700	1,725	33,425
2420	480 KW 1970 GPH	56,000	1,725	57,725
2460	1500 gal, 60 KW 245 GPH	46,900	2,125	49,025
2500	480 KW 1970 GPH	66,500	2,125	68,625

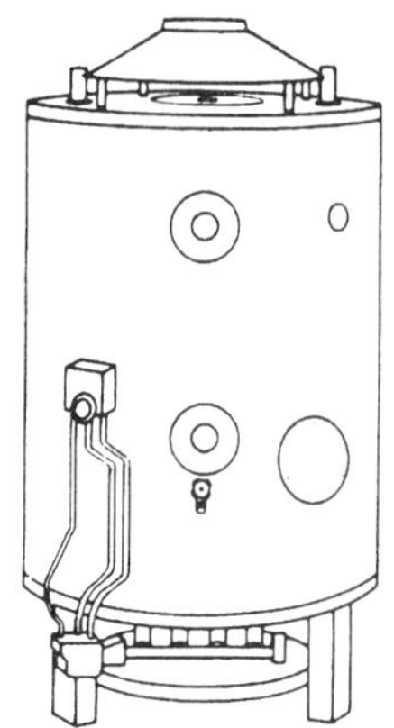

Units may be installed in multiples for increased capacity.

Included below is the heater with self-energizing gas controls, safety pilots, insulated jacket, hi-limit aquastat and pressure relief valve.

Installation includes piping and fittings within 10′ of heater. Gas heaters require vent piping (not included in these prices).

8.1-170 — Gas Fired Water Heaters - Commercial Systems

		COST EACH		
		MAT.	INST.	TOTAL
1760	Gas fired water heater, commercial, 100° F rise			
1780	75.5 MBH input, 63 GPH	1,600	705	2,305
1820	95 MBH input, 86 GPH	2,475	705	3,180
1860	100 MBH input, 91 GPH	2,675	735	3,410
1900	115 MBH input, 110 GPH	2,075	760	2,835
1980	155 MBH input, 150 GPH	2,400	850	3,250
2020	175 MBH input, 168 GPH	2,725	910	3,635
2060	200 MBH input, 192 GPH	3,200	1,025	4,225
2100	240 MBH input, 230 GPH	3,775	1,125	4,900
2140	300 MBH input, 278 GPH	4,825	1,275	6,100
2180	390 MBH input, 374 GPH	5,425	1,300	6,725
2220	500 MBH input, 480 GPH	7,150	1,400	8,550
2260	600 MBH input, 576 GPH	8,100	1,500	9,600
2300	800 MBH input, 768 GPH	9,775	1,675	11,450
2340	1000 MBH input, 960 GPH	11,500	1,700	13,200
2380	1200 MBH input, 1150 GPH	14,400	2,100	16,500
2420	1500 MBH input, 1440 GPH	17,500	2,075	19,575
2460	1800 MBH input, 1730 GPH	19,600	2,300	21,900
2500	2450 MBH input, 2350 GPH	25,700	2,675	28,375
2540	3000 MBH input, 2880 GPH	31,000	3,225	34,225
2580	3750 MBH input, 3600 GPH	38,700	3,325	42,025

Important: See the Reference Section for critical supporting data - Reference Nos., Crews & Location Factors

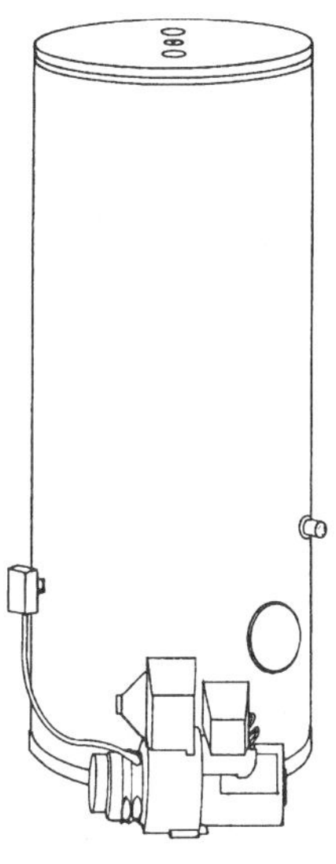

Units may be installed in multiples for increased capacity.

Included below is the heater, wired-in flame retention burners, cadmium cell primary controls, hi-limit controls, ASME pressure relief valves, draft controls, and insulated jacket.

Oil-fired water heater systems include piping and fittings within 10′ of heater. Oil-fired heaters require vent piping (not included in these systems).

8.1-180	Oil Fired Water Heaters - Commercial Systems	COST EACH		
		MAT.	INST.	TOTAL
1800	Oil fired water heater, commercial, 100° F rise			
1820	103 MBH output, 116 GPH	2,100	615	2,715
1900	134 MBH output, 161 GPH	2,400	785	3,185
1940	161 MBH output, 192 GPH	2,850	900	3,750
1980	187 MBH output, 224 GPH	3,100	1,050	4,150
2060	262 MBH output, 315 GPH	3,850	1,200	5,050
2100	341 MBH output, 409 GPH	5,100	1,275	6,375
2140	420 MBH output, 504 GPH	5,775	1,375	7,150
2180	525 MBH output, 630 GPH	7,325	1,550	8,875
2220	630 MBH output, 756 GPH	8,100	1,550	9,650
2260	735 MBH output, 880 GPH	9,975	1,800	11,775
2300	840 MBH output, 1000 GPH	11,300	1,800	13,100
2340	1050 MBH output, 1260 GPH	13,600	1,775	15,375
2380	1365 MBH output, 1640 GPH	17,100	2,025	19,125
2420	1680 MBH output, 2000 GPH	20,400	2,550	22,950
2460	2310 MBH output, 2780 GPH	25,900	3,100	29,000
2500	2835 MBH output, 3400 GPH	31,900	3,100	35,000
2540	3150 MBH output, 3780 GPH	35,700	4,250	39,950

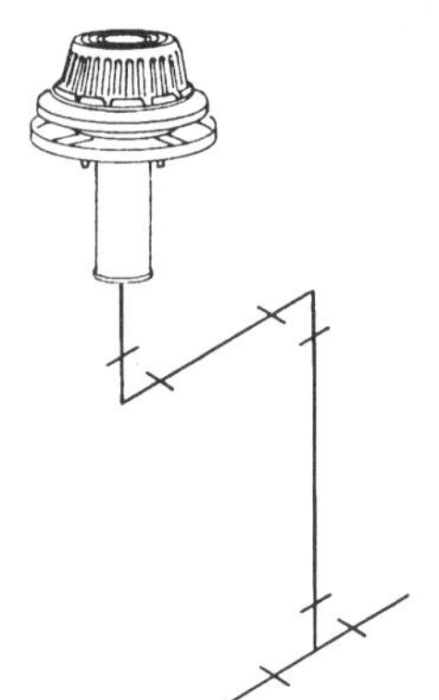

Design Assumptions: Vertical conductor size is based on a maximum rate of rainfall of 4" per hour. To convert roof area to other rates multiply "Max. S.F. Roof Area" shown by four and divide the result by desired local rate. The answer is the local roof area that may be handled by the indicated pipe diameter.

Basic cost is for roof drain, 10' of vertical leader and 10' of horizontal, plus connection to the main.

Pipe Dia.	Max. S.F. Roof Area	Gallons per Min.
2"	544	23
3"	1610	67
4"	3460	144
5"	6280	261
6"	10,200	424
8"	22,000	913

8.1-310	Roof Drain Systems	COST PER SYSTEM		
		MAT.	INST.	TOTAL
1880	Roof drain, DWV PVC, 2" diam., piping, 10' high	110	277	387
1920	For each additional foot add	2.44	8.05	10.49
1960	3" diam., 10' high	147	330	477
2000	For each additional foot add	3.95	8.95	12.90
2040	4" diam., 10' high	202	370	572
2080	For each additional foot add	6	9.90	15.90
2120	5" diam., 10' high	375	395	770
2160	For each additional foot add	7.65	11.05	18.70
2200	6" diam., 10' high	510	470	980
2240	For each additional foot add	9.20	12.15	21.35
2280	8" diam., 10' high	995	715	1,710
2320	For each additional foot add	14.15	14.25	28.40
3940	C.I., soil, single hub, service wt., 2" diam. piping, 10' high	190	289	479
3980	For each additional foot add	4.72	7.55	12.27
4120	3" diam., 10' high	246	315	561
4160	For each additional foot add	6.50	7.90	14.40
4200	4" diam., 10' high	320	340	660
4240	For each additional foot add	9.10	8.65	17.75
4280	5" diam., 10' high	440	355	795
4320	For each additional foot add	11.15	9	20.15
4360	6" diam., 10' high	520	380	900
4400	For each additional foot add	13.60	9.40	23
4440	8" diam., 10' high	1,075	790	1,865
4480	For each additional foot add	21.50	16.10	37.60
6040	Steel galv. sch 40 threaded, 2" diam. piping, 10' high	197	280	477
6080	For each additional foot add	3.61	7.40	11.01
6120	3" diam., 10' high	360	405	765
6160	For each additional foot add	6.55	11.05	17.60
6200	4" diam., 10' high	530	520	1,050
6240	For each additional foot add	9.10	13.20	22.30
6280	5" diam., 10' high	755	650	1,405
6320	For each additional foot add	16.90	18.25	35.15
6360	6" diam, 10' high	1,175	745	1,920
6400	For each additional foot add	22	22	44
6440	8" diam., 10' high	2,225	1,025	3,250
6480	For each additional foot add	32	25.50	57.50

Important: See the Reference Section for critical supporting data - Reference Nos., Crews & Location Factors

Table 8.1-401 Minimum Plumbing Fixture Requirements

Type of Building/Use	Water Closets		Urinals		Lavatories		Bathtubs or Showers		Drinking Fountain	Other
	Persons	Fixtures	Persons	Fixtures	Persons	Fixtures	Persons	Fixtures	Fixtures	Fixtures
Assembly Halls Auditoriums Theater Public assembly	1-100 101-200 201-400	1 2 3	1-200 201-400 401-600	1 2 3	1-200 201-400 401-750	1 2 3			1 for each 1000 persons	1 service sink
	Over 400 add 1 fixt. for ea. 500 men; 1 fixt. for ea. 300 women		Over 600 add 1 fixture for each 300 men		Over 750 add 1 fixture for each 500 persons					
Assembly Public Worship	300 men 150 women	1 1	300 men	1	men women	1 1			1	
Dormitories	Men: 1 for each 10 persons Women: 1 for each 8 persons		1 for each 25 men; over 150 add 1 fixture for each 50 men		1 for ea. 12 persons 1 separate dental lav. for each 50 persons recom.		1 for ea. 8 persons For women add 1 additional for each 30. Over 150 persons add 1 for each 20.		1 for each 75 persons	Laundry trays 1 for each 50 serv. sink 1 for ea. 100
Dwellings Apartments and homes	1 fixture for each unit				1 fixture for each unit		1 fixture for each unit			
Hospitals Indiv. Room Ward Waiting room	8 persons	1 1 1			10 persons	1 1 1	20 persons	1 1	1 for 100 patients	1 service sink per floor
Industrial Mfg. plants Warehouses	1-10 11-25 26-50 51-75 76-100	1 2 3 4 5	0-30 31-80 81-160 161-240	1 2 3 4	1-100 over 100	1 for ea. 10 1 for ea. 15	1 Shower for each 15 persons subject to excessive heat or occupational hazard		1 for each 75 persons	
	1 fixture for each additional 30 persons									
Public Buildings Businesses Offices	1-15 16-35 36-55 56-80 81-110 111-150	1 2 3 4 5 6	Urinals may be provided in place of water closets but may not replace more than 1/3 required number of men's water closets		1-15 16-35 36-60 61-90 91-125	1 2 3 4 5			1 for each 75 persons	1 service sink per floor
	1 fixture for ea. additional 40 persons				1 fixture for ea. additional 45 persons					
Schools Elementary	1 for ea. 30 boys 1 for ea. 25 girls		1 for ea. 25 boys		1 for ea. 35 boys 1 for ea. 35 girls		For gym or pool shower room 1/5 of a class		1 for each 40 pupils	
Schools Secondary	1 for ea. 40 boys 1 for ea. 30 girls		1 for ea. 25 boys		1 for ea. 40 boys 1 for ea. 40 girls		For gym or pool shower room 1/5 of a class		1 for each 50 pupils	

8 MECHANICAL

Systems are complete with trim and rough-in (supply, waste and vent) to connect to supply branches and waste mains.

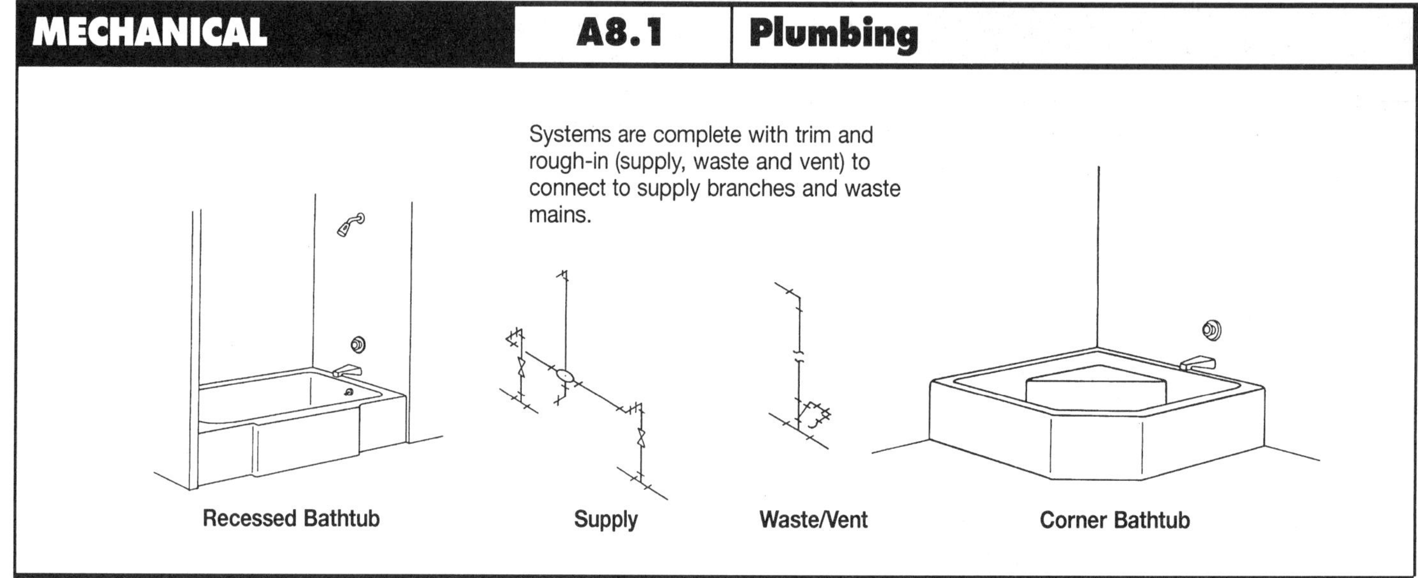

8.1-410	Bathtub Systems	COST EACH		
		MAT.	INST.	TOTAL
1960	Bathtub, recessed, P.E. on Cl., 42" x 37"	970	325	1,295
2000	48" x 42"	1,425	350	1,775
2040	72" x 36"	1,500	390	1,890
2080	Mat bottom, 5' long	565	340	905
2120	5'-6" long	1,175	350	1,525
2160	Corner, 48" x 42"	1,600	340	1,940
2200	Formed steel, enameled, 4'-6" long	445	310	755
2240	5' long	425	315	740

8.1-420	Drinking Fountain Systems	COST EACH		
		MAT.	INST.	TOTAL
1740	Drinking fountain, one bubbler, wall mounted			
1760	Non recessed			
1800	Bronze, no back	1,025	185	1,210
1840	Cast iron, enameled, low back	495	185	680
1880	Fiberglass, 12" back	565	185	750
1920	Stainless steel, no back	905	185	1,090
1960	Semi-recessed, poly marble	670	185	855
2040	Stainless steel	705	185	890
2080	Vitreous china	520	185	705
2120	Full recessed, poly marble	755	185	940
2200	Stainless steel	650	185	835
2240	Floor mounted, pedestal type, aluminum	485	251	736
2320	Bronze	1,125	251	1,376
2360	Stainless steel	1,125	251	1,376

Important: See the Reference Section for critical supporting data - Reference Nos., Crews & Location Factors

Systems are complete with trim and rough-in (supply, waste and vent) to connect to supply branches and waste mains.

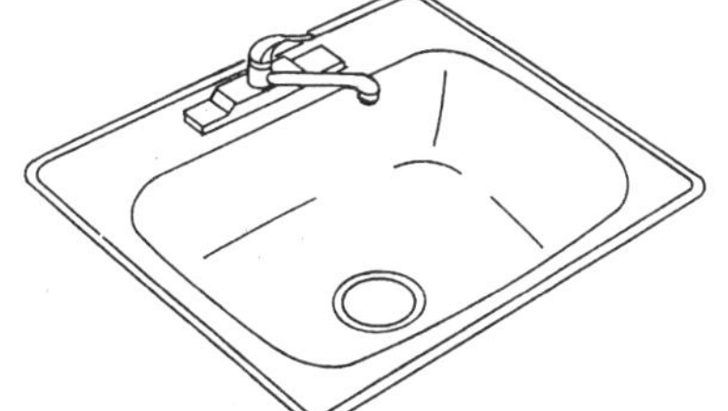

Countertop
Single Bowl

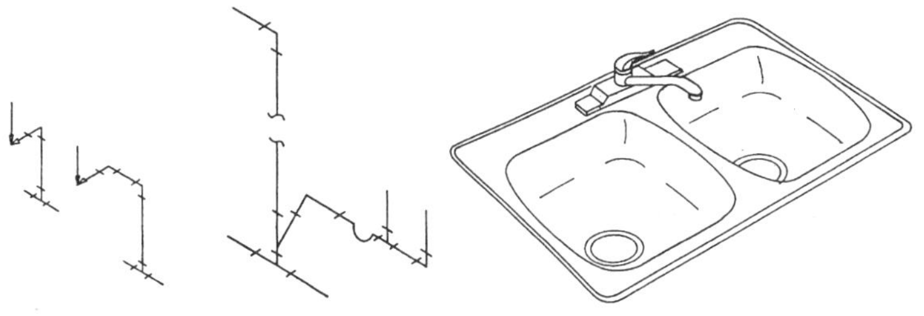

Supply   Waste/Vent   Countertop Double Bowl

8.1-431	Kitchen Sink Systems	COST EACH		
		MAT.	INST.	TOTAL
1720	Kitchen sink w/trim, countertop, PE on CI, 24"x21", single bowl	350	310	660
1760	30" x 21" single bowl	425	310	735
1800	32" x 21" double bowl	470	330	800
1840	42" x 21" double bowl	630	335	965
1880	Stainless steel, 19" x 18" single bowl	390	310	700
1920	25" x 22" single bowl	415	310	725
1960	33" x 22" double bowl	560	330	890
2000	43" x 22" double bowl	630	335	965
2040	44" x 22" triple bowl	810	350	1,160
2080	44" x 24" corner double bowl	575	335	910
2120	Steel, enameled, 24" x 21" single bowl	229	310	539
2160	32" x 21" double bowl	255	330	585
2240	Raised deck, PE on CI, 32" x 21", dual level, double bowl	470	420	890
2280	42" x 21" dual level, triple bowl	740	460	1,200
2281				
2282				

8.1-432	Laundry Sink Systems	COST EACH		
		MAT.	INST.	TOTAL
1740	Laundry sink w/trim, PE on CI, black iron frame			
1760	24" x 20", single compartment	440	300	740
1840	48" x 21" double compartment	920	330	1,250
1920	Molded stone, on wall, 22" x 21" single compartment	242	300	542
1960	45"x 21" double compartment	350	330	680
2040	Plastic, on wall or legs, 18" x 23" single compartment	203	296	499
2080	20" x 24" single compartment	232	296	528
2120	36" x 23" double compartment	261	320	581
2160	40" x 24" double compartment	325	320	645
2200	Stainless steel, countertop, 22" x 17" single compartment	380	300	680
2240	19" x 22" single compartment	450	300	750
2280	33" x 22" double compartment	435	330	765

MECHANICAL

8

Systems are complete with trim and rough-in (supply, waste and vent) to connect to supply branches and waste mains.

Vanity Top

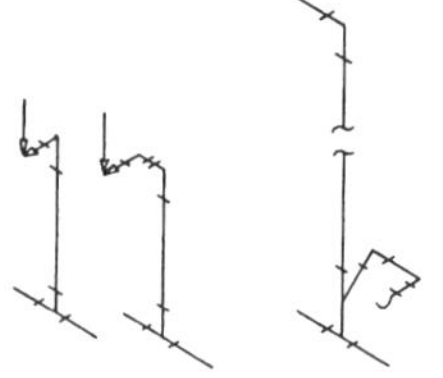

Supply   Waste/Vent

Wall Hung

8.1-433	Lavatory Systems	COST EACH		
		MAT.	INST.	TOTAL
1560	Lavatory w/trim, vanity top, PE on CI, 20" x 18", Vanity top by others.	296	281	577
1600	19" x 16" oval	289	281	570
1640	18" round	300	281	581
1680	Cultured marble, 19" x 17"	209	281	490
1720	25" x 19"	239	281	520
1760	Stainless, self-rimming, 25" x 22"	265	281	546
1800	17" x 22"	261	281	542
1840	Steel enameled, 20" x 17"	198	289	487
1880	19" round	196	289	485
1920	Vitreous china, 20" x 16"	300	295	595
1960	19" x 16"	300	295	595
2000	22" x 13"	310	295	605
2040	Wall hung, PE on CI, 18" x 15"	595	310	905
2080	19" x 17"	570	310	880
2120	20" x 18"	420	310	730
2160	Vitreous china, 18" x 15"	415	320	735
2200	19" x 17"	350	320	670
2240	24" x 20"	485	320	805

8.1-434	Laboratory Sink Systems	COST EACH		
		MAT.	INST.	TOTAL
1580	Laboratory sink w/trim, polyethylene, single bowl,			
1600	Double drainboard, 54" x 24" O.D.	910	395	1,305
1640	Single drainboard, 47" x 24" O.D.	945	395	1,340
1680	70" x 24" O.D.	965	395	1,360
1760	Flanged, 14-1/2" x 14-1/2" O.D.	350	355	705
1800	18-1/2" x 18-1/2" O.D.	315	355	670
1840	23-1/2" x 20-1/2" O.D.	365	355	720
1920	Polypropylene, cup sink, oval, 7" x 4" O.D.	163	315	478
1960	10" x 4-1/2" O.D.	174	315	489
1961				

8.1-434	Service Sink Systems	COST EACH		
		MAT.	INST.	TOTAL
4260	Service sink w/trim, PE on CI, corner floor, 28" x 28", w/rim guard	870	395	1,265
4300	Wall hung w/rim guard, 22" x 18"	780	460	1,240
4340	24" x 20"	820	460	1,280
4380	Vitreous china, wall hung 22" x 20"	835	460	1,295

Important: See the Reference Section for critical supporting data - Reference Nos., Crews & Location Factors

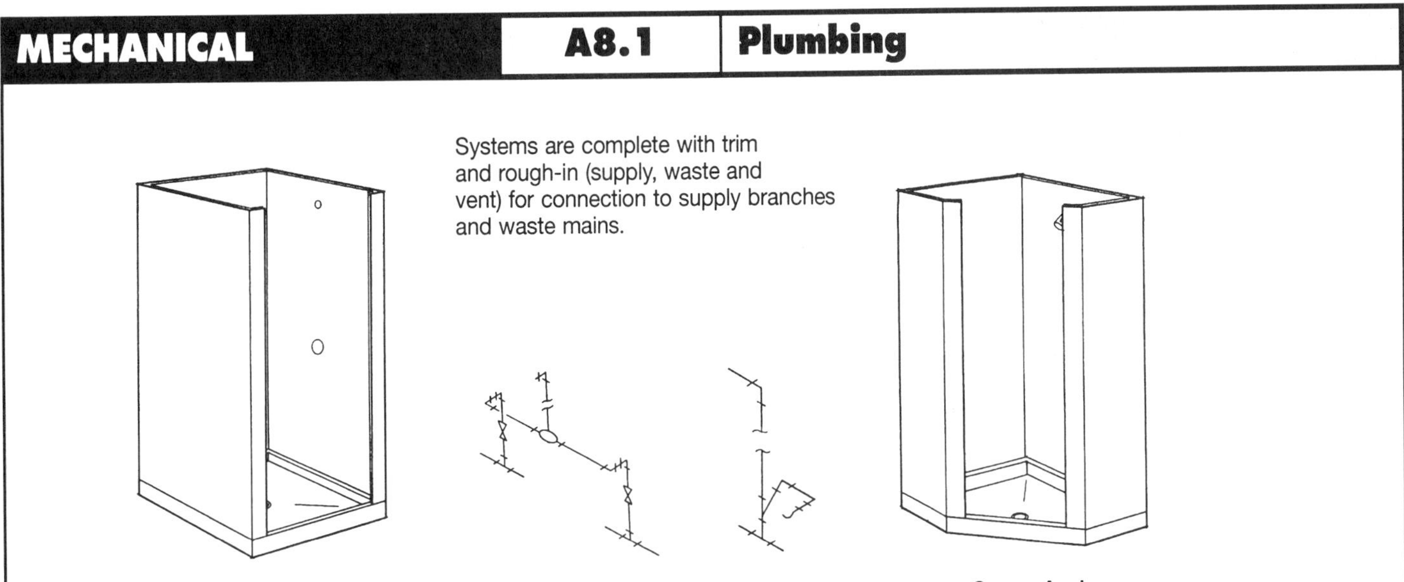

Systems are complete with trim and rough-in (supply, waste and vent) for connection to supply branches and waste mains.

8.1-440	Shower Systems	COST EACH		
		MAT.	INST.	TOTAL
1560	Shower, stall, baked enamel, molded stone receptor, 30" square	475	470	945
1600	32" square	495	470	965
1640	Terrazzo receptor, 32" square	930	470	1,400
1680	36" square	990	495	1,485
1720	36" corner angle	910	495	1,405
1800	Fiberglass one piece, three walls, 32" square	530	430	960
1840	36" square	585	430	1,015
1880	Polypropylene, molded stone receptor, 30" square	485	470	955
1920	32" square	495	470	965
1960	Built-in head, arm, bypass, stops and handles	96	122	218

8.1-450	Urinal Systems	COST EACH		
		MAT.	INST.	TOTAL
2000	Urinal, vitreous china, wall hung	460	340	800
2040	Stall type	780	375	1,155

8.1-460	Water Cooler Systems	COST EACH		
		MAT.	INST.	TOTAL
1840	Water cooler, electric, wall hung, 8.2 GPH	595	238	833
1880	Dual height, 14.3 GPH	840	244	1,084
1920	Wheelchair type, 7.5 G.P.H.	1,300	238	1,538
1960	Semi recessed, 8.1 G.P.H.	790	238	1,028
2000	Full recessed, 8 G.P.H.	1,200	255	1,455
2040	Floor mounted, 14.3 G.P.H.	625	207	832
2080	Dual height, 14.3 G.P.H.	860	251	1,111
2120	Refrigerated compartment type, 1.5 G.P.H.	1,100	207	1,307

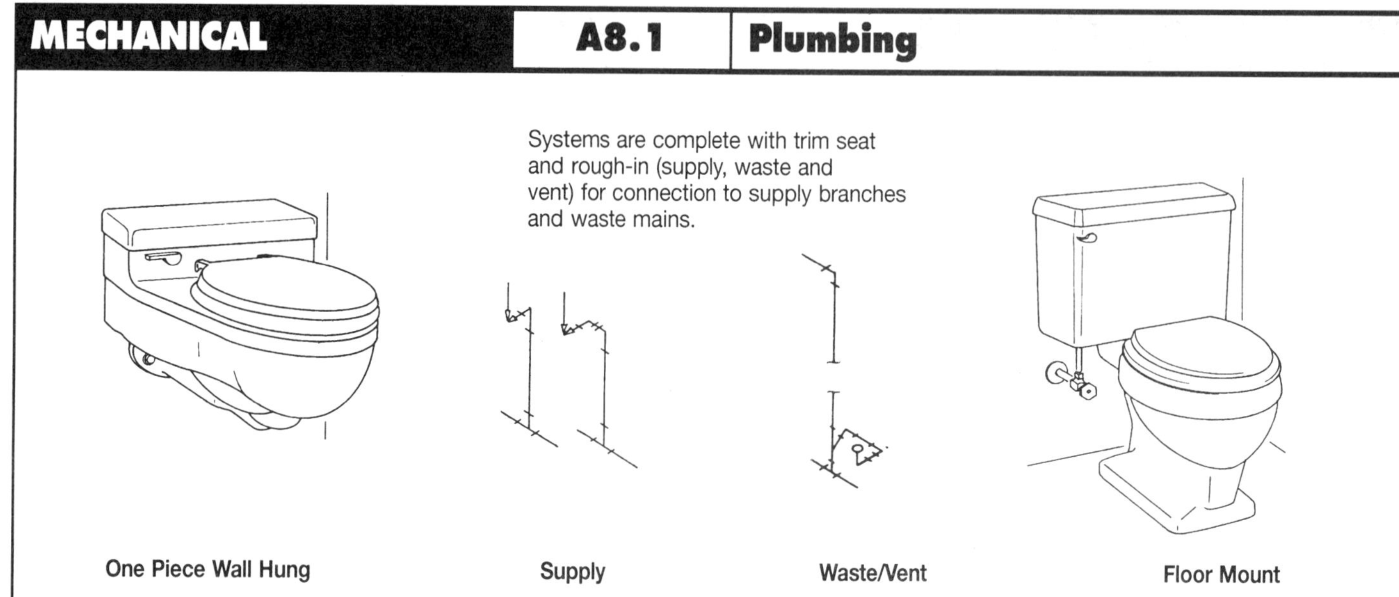

8.1-470	**Water Closet Systems**	COST EACH		
		MAT.	INST.	TOTAL
1800	Water closet, vitreous china, elongated			
1840	Tank type, wall hung, one piece	900	275	1,175
1880	Close coupled two piece	570	275	845
1920	Floor mount, one piece	630	298	928
1960	One piece low profile	1,025	298	1,323
2000	Two piece close coupled	310	298	608
2040	Bowl only with flush valve			
2080	Wall hung	640	315	955
2120	Floor mount	505	305	810
2122				

8.1-510	**Water Closets, Group**	COST EACH		
		MAT.	INST.	TOTAL
1760	Water closets, battery mount, wall hung, side by side, first closet	670	320	990
1800	Each additional water closet, add	640	305	945
3000	Back to back, first pair of closets	1,225	435	1,660
3100	Each additional pair of closets, back to back	1,200	425	1,625

8.1-560	**Group Wash Fountain Systems**	COST EACH		
		MAT.	INST.	TOTAL
1740	Group wash fountain, precast terrazzo			
1760	Circular, 36" diameter	2,150	490	2,640
1800	54" diameter	2,625	535	3,160
1840	Semi-circular, 36" diameter	1,975	490	2,465
1880	54" diameter	2,375	535	2,910
1960	Stainless steel, circular, 36" diameter	2,475	455	2,930
2000	54" diameter	3,150	505	3,655
2040	Semi-circular, 36" diameter	2,175	455	2,630
2080	54" diameter	2,750	505	3,255
2160	Thermoplastic, circular, 36" diameter	1,900	375	2,275
2200	54" diameter	2,175	430	2,605
2240	Semi-circular, 36" diameter	1,750	375	2,125
2280	54" diameter	2,125	430	2,555

Important: See the Reference Section for critical supporting data - Reference Nos., Crews & Location Factors

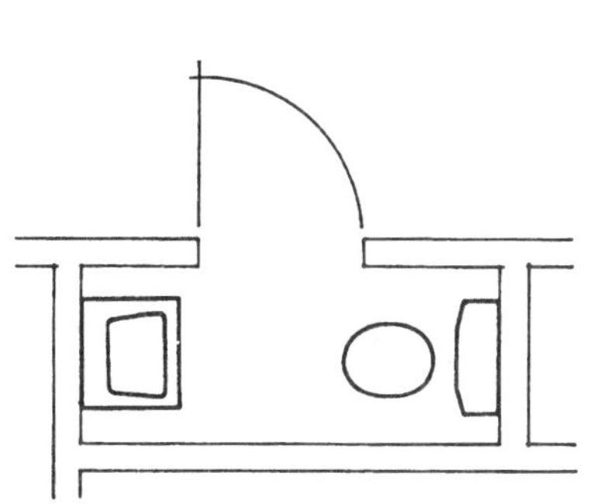

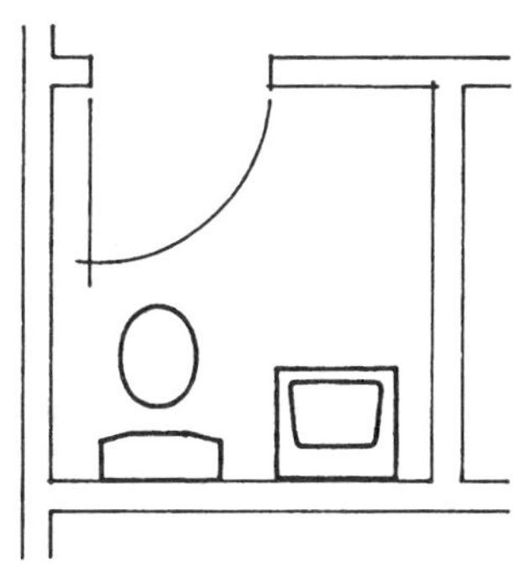

Two Fixture Bathroom Systems consisting of a lavatory, water closet, and rough-in service piping.

- Prices for plumbing and fixtures only.

*Common wall is with an adjacent bathroom

8.1-620	Two Fixture Bathroom, Two Wall Plumbing	COST EACH		
		MAT.	INST.	TOTAL
1180	Bathroom, lavatory & water closet, 2 wall plumbing, stand alone	840	775	1,615
1200	Share common plumbing wall*	785	665	1,450

8.1-620	Two Fixture Bathroom, One Wall Plumbing	COST EACH		
		MAT.	INST.	TOTAL
2220	Bathroom, lavatory & water closet, one wall plumbing, stand alone	800	695	1,495
2240	Share common plumbing wall*	715	590	1,305

8.1-630	Three Fixture Bathroom, One Wall Plumbing	COST EACH		
		MAT.	INST.	TOTAL
1150	Bathroom, three fixture, one wall plumbing			
1160	Lavatory, water closet & bathtub			
1170	Stand alone	1,250	905	2,155
1180	Share common plumbing wall *	1,075	655	1,730

8.1-630	Three Fixture Bathroom, Two Wall Plumbing	COST EACH		
		MAT.	INST.	TOTAL
2130	Bathroom, three fixture, two wall plumbing			
2140	Lavatory, water closet & bathtub			
2160	Stand alone	1,275	915	2,190
2180	Long plumbing wall common *	1,150	730	1,880
3610	Lavatory, bathtub & water closet			
3620	Stand alone	1,350	1,050	2,400
3640	Long plumbing wall common *	1,275	945	2,220
4660	Water closet, corner bathtub & lavatory			
4680	Stand alone	2,325	930	3,255
4700	Long plumbing wall common *	2,175	700	2,875
6100	Water closet, stall shower & lavatory			
6120	Stand alone	1,300	1,175	2,475
6140	Long plumbing wall common *	1,250	1,100	2,350
7060	Lavatory, corner stall shower & water closet			
7080	Stand alone	1,625	1,075	2,700
7100	Short plumbing wall common *	1,475	785	2,260

MECHANICAL

8

Table 8.2-101   Sprinkler Systems (Automatic)

Sprinkler systems may be classified by type as follows:

1. **Wet Pipe System.** A system employing automatic sprinklers attached to a piping system containing water and connected to a water supply so that water discharges immediately from sprinklers opened by a fire.

2. **Dry Pipe System.** A system employing automatic sprinklers attached to a piping system containing air under pressure, the release of which as from the opening of sprinklers permits the water pressure to open a valve known as a "dry pipe valve". The water then flows into the piping system and out the opened sprinklers.

3. **Pre-Action System.** A system employing automatic sprinklers attached to a piping system containing air that may or may not be under pressure, with a supplemental heat responsive system of generally more sensitive characteristics than the automatic sprinklers themselves, installed in the same areas as the sprinklers; actuation of the heat responsive system, as from a fire, opens a valve which permits water to flow into the sprinkler piping system and to be discharged from any sprinklers which may be open.

4. **Deluge System.** A system employing open sprinklers attached to a piping system connected to a water supply through a valve which is opened by the operation of a heat responsive system installed in the same areas as the sprinklers. When this valve opens, water flows into the piping system and discharges from all sprinklers attached thereto.

5. **Combined Dry Pipe and Pre-Action Sprinkler System.** A system employing automatic sprinklers attached to a piping system containing air under pressure with a supplemental heat responsive system of generally more sensitive characteristics than the automatic sprinklers themselves, installed in the same areas as the sprinklers; operation of the heat responsive system, as from a fire, actuates tripping devices which open dry pipe valves simultaneously and without loss of air pressure in the system. Operation of the heat responsive system also opens approved air exhaust valves at the end of the feed main which facilitates the filling of the system with water which usually precedes the opening of sprinklers. The heat responsive system also serves as an automatic fire alarm system.

6. **Limited Water Supply System.** A system employing automatic sprinklers and conforming to these standards but supplied by a pressure tank of limited capacity.

7. **Chemical Systems.** Systems using halon, carbon dioxide, dry chemical or high expansion foam as selected for special requirements. Agent may extinguish flames by chemically inhibiting flame propagation, suffocate flames by excluding oxygen, interrupting chemical action of oxygen uniting with fuel or sealing and cooling the combustion center.

8. **Firecycle System.** Firecycle is a fixed fire protection sprinkler system utilizing water as its extinguishing agent. It is a time delayed, recycling, preaction type which automatically shuts the water off when heat is reduced below the detector operating temperature and turns the water back on when that temperature is exceeded. The system senses a fire condition through a closed circuit electrical detector system which controls water flow to the fire automatically. Batteries supply up to 90 hour emergency power supply for system operation. The piping system is dry (until water is required) and is monitored with pressurized air. Should any leak in the system piping occur, an alarm will sound, but water will not enter the system until heat is sensed by a firecycle detector.

Area coverage sprinkler systems may be laid out and fed from the supply in any one of several patterns as shown below. It is desirable, if possible, to utilize a central feed and achieve a shorter flow path from the riser to the furthest sprinkler. This permits use of the smallest sizes of pipe possible with resulting savings.

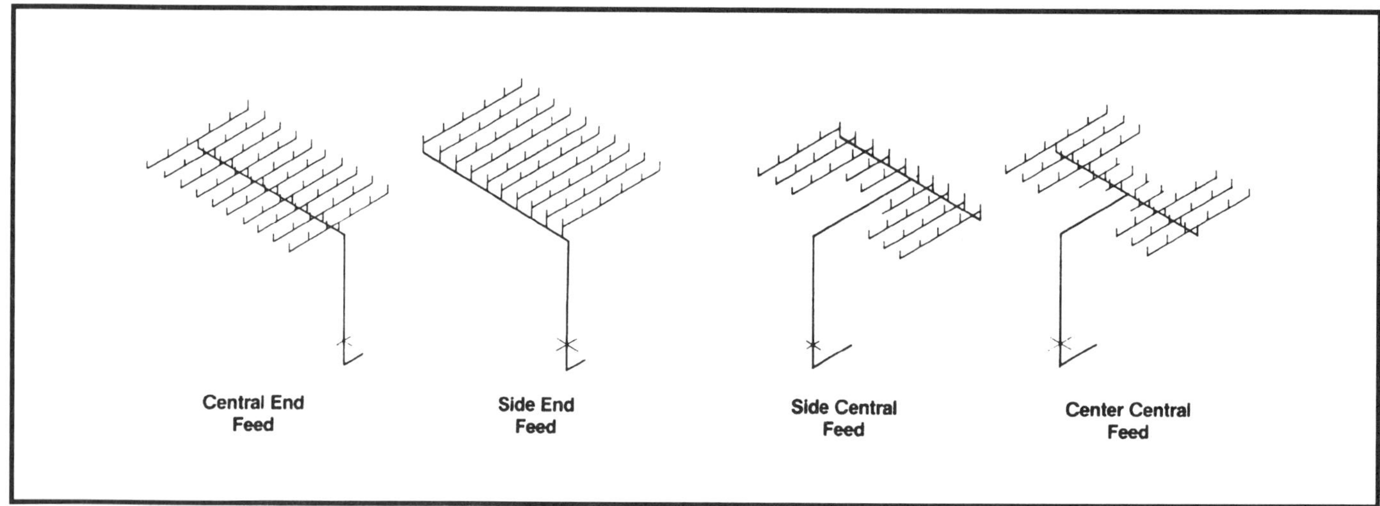

Table 8.2-102   System Classification

System Classification

Rules for installation of sprinkler systems vary depending on the classification of occupancy falling into one of three categories as follows:

Light Hazard Occupancy

The protection area allotted per sprinkler should not exceed 200 S.F. with the maximum distance between lines and sprinklers on lines being 15'. The sprinklers do not need to be staggered. Branch lines should not exceed eight sprinklers on either side of a cross main. Each large area requiring more than 100 sprinklers and without a sub-dividing partition should be supplied by feed mains or risers sized for ordinary hazard occupancy.

Included in this group are:

Auditoriums	Museums
Churches	Nursing Homes
Clubs	Offices
Educational	Residential
Hospitals	Restaurants
Institutional	Schools
Libraries	Theaters
(except large stack rooms)	

Ordinary Hazard Occupancy

The protection area allotted per sprinkler shall not exceed 130 S.F. of noncombustible ceiling and 120 S.F. of combustible ceiling. The maximum allowable distance between sprinkler lines and sprinklers on line is 15'. Sprinklers shall be staggered if the distance between heads exceeds 12'. Branch lines should not exceed eight sprinklers on either side of a cross main.

Included in this group are:

Automotive garages	Electric generating stations
Bakeries	Feed mills
Beverage manufacturing	Grain elevators
Bleacheries	Ice manufacturing
Boiler houses	Laundries
Canneries	Machine shops
Cement plants	Mercantiles
Clothing factories	Paper mills
Cold storage warehouses	Printing and Publishing
Dairy products manufacturing	Shoe factories
Distilleries	Warehouses
Dry cleaning	Wood product assembly

Extra Hazard Occupancy

The protection area allotted per sprinkler shall not exceed 90 S.F. of noncombustible ceiling and 80 S.F. of combustible ceiling. The maximum allowable distance between lines and between sprinklers on lines is 12'. Sprinklers on alternate lines shall be staggered if the distance between sprinklers on lines exceeds 8'. Branch lines should not exceed six sprinklers on either side of a cross main.

Included in this group are:

Aircraft hangars	Paint shops
Chemical works	Shade cloth manufacturing
Explosives manufacturing	Solvent extracting
Linoleum manufacturing	Varnish works
Linseed oil mills	Volatile flammable
Oil refineries	liquid manufacturing & use

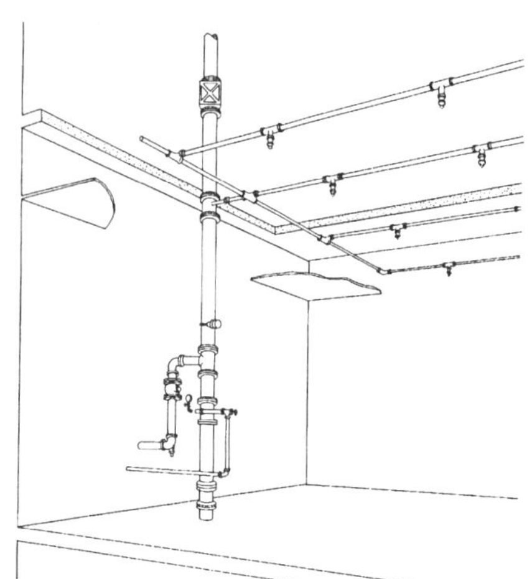

Wet Pipe System. A system employing automatic sprinklers attached to a piping system containing water and connected to a water supply so that water discharges immediately from sprinklers opened by heat from a fire.

All areas are assumed to be open.

8.2-110	Wet Pipe Sprinkler Systems	COST PER S.F.		
		MAT.	INST.	TOTAL
0520	Wet pipe sprinkler systems, steel, black, sch. 40 pipe			
0530	Light hazard, one floor, 500 S.F.	1.11	1.21	2.32
0560	1000 S.F.	1.53	1.26	2.79
0580	2000 S.F.	1.56	1.26	2.82
0600	5000 S.F.	.80	.86	1.66
0620	10,000 S.F.	.52	.74	1.26
0640	50,000 S.F.	.40	.69	1.09
0660	Each additional floor, 500 S.F.	.58	1.02	1.60
0680	1000 S.F.	.49	.95	1.44
0700	2000 S.F.	.45	.89	1.34
0720	5000 S.F.	.35	.73	1.08
0740	10,000 S.F.	.32	.67	.99
0760	50,000 S.F.	.28	.55	.83
1000	Ordinary hazard, one floor, 500 S.F.	1.23	1.30	2.53
1020	1000 S.F.	1.50	1.24	2.74
1040	2000 S.F.	1.63	1.34	2.97
1060	5000 S.F.	.89	.93	1.82
1080	10,000 S.F.	.68	.98	1.66
1100	50,000 S.F.	.62	1.01	1.63
1140	Each additional floor, 500 S.F.	.70	1.16	1.86
1160	1000 S.F.	.46	.95	1.41
1180	2000 S.F.	.53	.97	1.50
1200	5000 S.F.	.54	.91	1.45
1220	10,000 S.F.	.49	.92	1.41
1240	50,000 S.F.	.48	.89	1.37
1500	Extra hazard, one floor, 500 S.F.	2.86	1.99	4.85
1520	1000 S.F.	1.83	1.70	3.53
1540	2000 S.F.	1.74	1.79	3.53
1560	5000 S.F.	1.17	1.52	2.69
1580	10,000 S.F.	1.13	1.49	2.62
1600	50,000 S.F.	1.21	1.39	2.60
1660	Each additional floor, 500 S.F.	.78	1.41	2.19
1680	1000 S.F.	.74	1.33	2.07
1700	2000 S.F.	.67	1.38	2.05
1720	5000 S.F.	.60	1.21	1.81
1740	10,000 S.F.	.71	1.11	1.82
1760	50,000 S.F.	.72	1.06	1.78
2020	Grooved steel, black sch. 40 pipe, light hazard, one floor, 2000 S.F.	1.56	1.07	2.63
2060	10,000 S.F.	.60	.67	1.27
2100	Each additional floor, 2000 S.F.	.46	.70	1.16
2150	10,000 S.F.	.29	.58	.87
2200	Ordinary hazard, one floor, 2000 S.F.	1.57	1.16	2.73

Important: See the Reference Section for critical supporting data - Reference Nos., Crews & Location Factors

8.2-110   Wet Pipe Sprinkler Systems

		COST PER S.F.		
		MAT.	INST.	TOTAL
2250	10,000 S.F.	.59	.82	1.41
2300	Each additional floor, 2000 S.F.	.47	.79	1.26
2350	10,000 S.F.	.40	.76	1.16
2400	Extra hazard, one floor, 2000 S.F.	1.72	1.44	3.16
2450	10,000 S.F.	.84	1.07	1.91
2500	Each additional floor, 2000 S.F.	.67	1.10	1.77
2550	10,000 S.F.	.58	.94	1.52
3050	Grooved steel black sch. 10 pipe, light hazard, one floor, 2000 S.F.	1.53	1.07	2.60
3100	10,000 S.F.	.47	.64	1.11
3150	Each additional floor, 2000 S.F.	.43	.70	1.13
3200	10,000 S.F.	.27	.57	.84
3250	Ordinary hazard, one floor, 2000 S.F.	1.54	1.15	2.69
3300	10,000 S.F.	.56	.80	1.36
3350	Each additional floor, 2000 S.F.	.44	.78	1.22
3400	10,000 S.F.	.37	.74	1.11
3450	Extra hazard, one floor, 2000 S.F.	1.70	1.44	3.14
3500	10,000 S.F.	.79	1.05	1.84
3550	Each additional floor, 2000 S.F.	.65	1.10	1.75
3600	10,000 S.F.	.57	.93	1.50
4050	Copper tubing, type M, light hazard, one floor, 2000 S.F.	1.66	1.07	2.73
4100	10,000 S.F.	.66	.64	1.30
4150	Each additional floor, 2000 S.F.	.57	.71	1.28
4200	10,000 S.F.	.46	.57	1.03
4250	Ordinary hazard, one floor, 2000 S.F.	1.74	1.19	2.93
4300	10,000 S.F.	.77	.75	1.52
4350	Each additional floor, 2000 S.F.	.67	.78	1.45
4400	10,000 S.F.	.56	.67	1.23
4450	Extra hazard, one floor, 2000 S.F.	1.95	1.47	3.42
4500	10,000 S.F.	1.44	1.17	2.61
4550	Each additional floor, 2000 S.F.	.90	1.13	2.03
4600	10,000 S.F.	1	1.03	2.03
5050	Copper tubing, type M, T-drill system, light hazard, one floor			
5060	2000 S.F.	1.67	1	2.67
5100	10,000 S.F.	.61	.51	1.12
5150	Each additional floor, 2000 S.F.	.58	.64	1.22
5200	10,000 S.F.	.41	.44	.85
5250	Ordinary hazard, one floor, 2000 S.F.	1.67	1.01	2.68
5300	10,000 S.F.	.73	.67	1.40
5350	Each additional floor, 2000 S.F.	.57	.64	1.21
5400	10,000 S.F.	.54	.61	1.15
5450	Extra hazard, one floor, 2000 S.F.	1.75	1.20	2.95
5500	10,000 S.F.	1.12	.85	1.97
5550	Each additional floor, 2000 S.F.	.73	.88	1.61
5600	10,000 S.F.	.68	.71	1.39

MECHANICAL

8

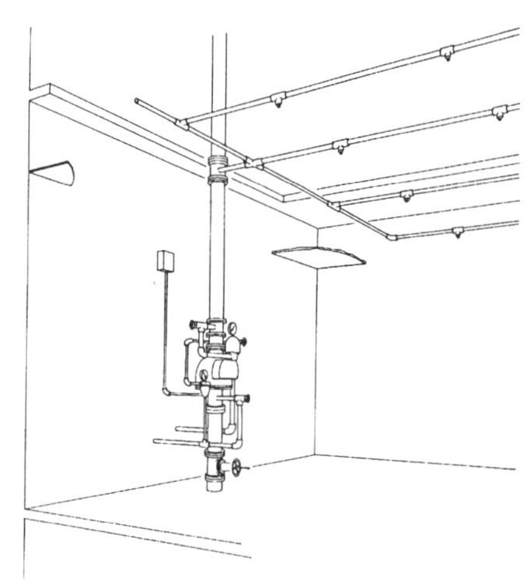

Dry Pipe System: A system employing automatic sprinklers attached to a piping system containing air under pressure, the release of which as from the opening of sprinklers permits the water pressure to open a valve known as a "dry pipe valve". The water then flows into the piping system and out the opened sprinklers.

All areas are assumed to be open.

8.2-120	Dry Pipe Sprinkler Systems	COST PER S.F.		
		MAT.	INST.	TOTAL
0520	Dry pipe sprinkler systems, steel, black, sch. 40 pipe			
0530	Light hazard, one floor, 500 S.F.	4.15	2.39	6.54
0560	1000 S.F.	2.23	1.38	3.61
0580	2000 S.F.	1.90	1.40	3.30
0600	5000 S.F.	.99	.92	1.91
0620	10,000 S.F.	.67	.78	1.45
0640	50,000 S.F.	.51	.69	1.20
0660	Each additional floor, 500 S.F.	.81	1.15	1.96
0680	1000 S.F.	.60	.94	1.54
0700	2000 S.F.	.57	.90	1.47
0720	5000 S.F.	.47	.74	1.21
0740	10,000 S.F.	.43	.68	1.11
0760	50,000 S.F.	.40	.62	1.02
1000	Ordinary hazard, one floor, 500 S.F.	4.20	2.41	6.61
1020	1000 S.F.	2.25	1.40	3.65
1040	2000 S.F.	1.97	1.48	3.45
1060	5000 S.F.	1.12	.99	2.11
1080	10,000 S.F.	.88	1.02	1.90
1100	50,000 S.F.	.80	1.04	1.84
1140	Each additional floor, 500 S.F.	.86	1.17	2.03
1160	1000 S.F.	.68	1.06	1.74
1180	2000 S.F.	.69	.98	1.67
1200	5000 S.F.	.63	.83	1.46
1220	10,000 S.F.	.57	.81	1.38
1240	50,000 S.F.	.56	.76	1.32
1500	Extra hazard, one floor, 500 S.F.	5.45	2.98	8.43
1520	1000 S.F.	3.12	2.11	5.23
1540	2000 S.F.	2.14	1.89	4.03
1560	5000 S.F.	1.28	1.39	2.67
1580	10,000 S.F.	1.31	1.36	2.67
1600	50,000 S.F.	1.38	1.28	2.66
1660	Each additional floor, 500 S.F.	1.01	1.43	2.44
1680	1000 S.F.	.97	1.35	2.32
1700	2000 S.F.	.90	1.40	2.30
1720	5000 S.F.	.79	1.21	2
1740	10,000 S.F.	.93	1.10	2.03
1760	50,000 S.F.	.95	1.06	2.01
2020	Grooved steel, black, sch. 40 pipe, light hazard, one floor, 2000 S.F.	1.86	1.21	3.07
2060	10,000 S.F.	.64	.69	1.33
2100	Each additional floor, 2000 S.F.	.58	.71	1.29
2150	10,000 S.F.	.40	.59	.99
2200	Ordinary hazard, one floor, 2000 S.F.	1.91	1.30	3.21

    Important: See the Reference Section for critical supporting data - Reference Nos., Crews & Location Factors

8.2-120 | Dry Pipe Sprinkler Systems

		MAT.	INST.	TOTAL
2250	10,000 S.F.	.79	.86	1.65
2300	Each additional floor, 2000 S.F.	.63	.80	1.43
2350	10,000 S.F.	.56	.77	1.33
2400	Extra hazard, one floor, 2000 S.F.	2.13	1.59	3.72
2450	10,000 S.F.	1.12	1.12	2.24
2500	Each additional floor, 2000 S.F.	.90	1.12	2.02
2550	10,000 S.F.	.82	.96	1.78
3050	Grooved steel black sch. 10 pipe, light hazard, one floor, 2000 S.F.	1.83	1.21	3.04
3100	10,000 S.F.	.62	.68	1.30
3150	Each additional floor, 2000 S.F.	.55	.71	1.26
3200	10,000 S.F.	.38	.58	.96
3250	Ordinary hazard, one floor, 2000 S.F.	1.88	1.29	3.17
3300	10,000 S.F.	.76	.84	1.60
3350	Each additional floor, 2000 S.F.	.60	.79	1.39
3400	10,000 S.F.	.53	.75	1.28
3450	Extra hazard, one floor, 2000 S.F.	2.11	1.59	3.70
3500	10,000 S.F.	1.07	1.10	2.17
3550	Each additional floor, 2000 S.F.	.88	1.12	2
3600	10,000 S.F.	.81	.95	1.76
4050	Copper tubing, type M, light hazard, one floor, 2000 S.F.	1.96	1.21	3.17
4100	10,000 S.F.	.81	.68	1.49
4150	Each additional floor, 2000 S.F.	.69	.72	1.41
4200	10,000 S.F.	.57	.58	1.15
4250	Ordinary hazard, one floor, 2000 S.F.	2.08	1.33	3.41
4300	10,000 S.F.	.97	.79	1.76
4350	Each additional floor, 2000 S.F.	.91	.82	1.73
4400	10,000 S.F.	.72	.68	1.40
4450	Extra hazard, one floor, 2000 S.F.	2.36	1.62	3.98
4500	10,000 S.F.	1.74	1.22	2.96
4550	Each additional floor, 2000 S.F.	1.13	1.15	2.28
4600	10,000 S.F.	1.24	1.05	2.29
5050	Copper tubing, type M, T-drill system, light hazard, one floor			
5060	2000 S.F.	1.97	1.14	3.11
5100	10,000 S.F.	.76	.55	1.31
5150	Each additional floor, 2000 S.F.	.70	.65	1.35
5200	10,000 S.F.	.52	.45	.97
5250	Ordinary hazard, one floor, 2000 S.F.	2.01	1.15	3.16
5300	10,000 S.F.	.93	.71	1.64
5350	Each additional floor, 2000 S.F.	.73	.65	1.38
5400	10,000 S.F.	.66	.59	1.25
5450	Extra hazard, one floor, 2000 S.F.	2.16	1.35	3.51
5500	10,000 S.F.	1.42	.90	2.32
5550	Each additional floor, 2000 S.F.	.93	.88	1.81
5600	10,000 S.F.	.92	.73	1.65

COST PER S.F.

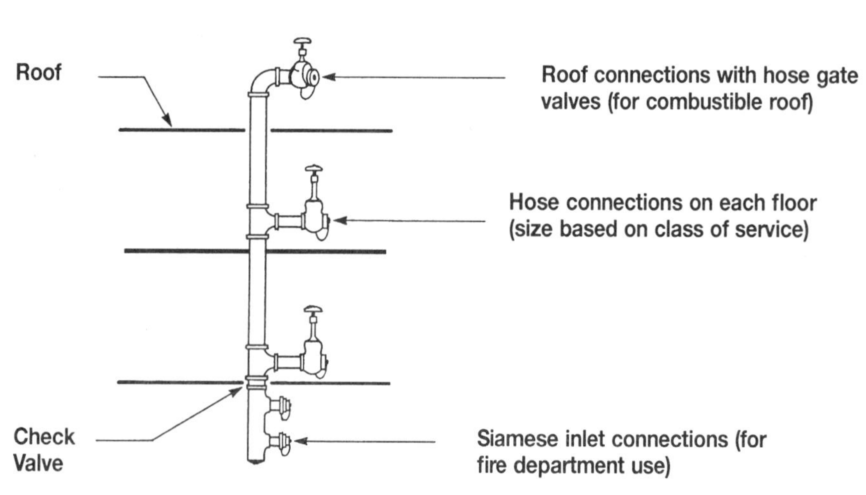

8.2-310	Wet Standpipe Risers, Class I	COST PER FLOOR		
		MAT.	INST.	TOTAL
0550	Wet standpipe risers, Class I, steel black sch. 40, 10′ height			
0560	4″ diameter pipe, one floor	1,900	1,350	3,250
0580	Additional floors	555	425	980
0600	6″ diameter pipe, one floor	3,350	2,300	5,650
0620	Additional floors	945	660	1,605
0640	8″ diameter pipe, one floor	4,875	2,750	7,625
0660	Additional floors	1,250	805	2,055

8.2-310	Wet Standpipe Risers, Class II	COST PER FLOOR		
		MAT.	INST.	TOTAL
1030	Wet standpipe risers, Class II, steel black sch. 40, 10′ height			
1040	2″ diameter pipe, one floor	800	475	1,275
1060	Additional floors	284	185	469
1080	2-1/2″ diameter pipe, one floor	1,025	700	1,725
1100	Additional floors	310	215	525

8.2-310	Wet Standpipe Risers, Class III	COST PER FLOOR		
		MAT.	INST.	TOTAL
1530	Wet standpipe risers, Class III, steel black sch. 40, 10′ height			
1540	4″ diameter pipe, one floor	1,975	1,350	3,325
1560	Additional floors	450	355	805
1580	6″ diameter pipe, one floor	3,425	2,300	5,725
1600	Additional floors	980	660	1,640
1620	8″ diameter pipe, one floor	4,950	2,750	7,700
1640	Additional floors	1,275	805	2,080

Important: See the Reference Section for critical supporting data - Reference Nos., Crews & Location Factors

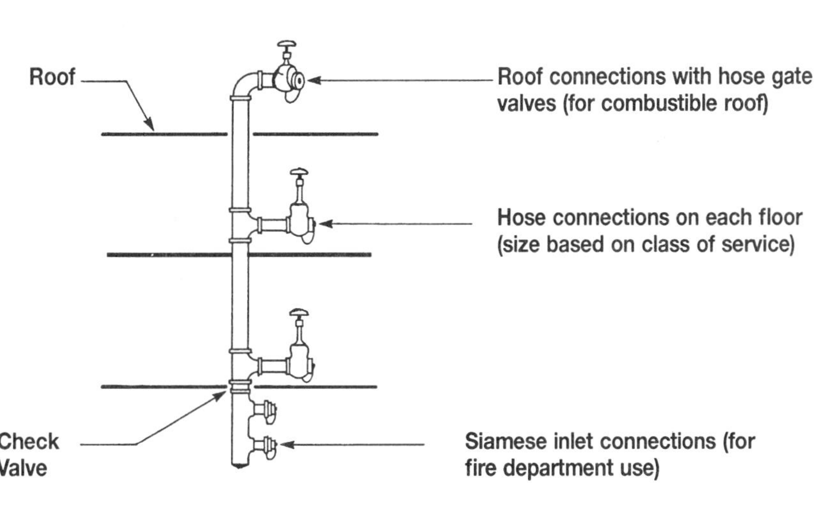

8.2-320	**Dry Standpipe Risers, Class I**	COST PER FLOOR		
		MAT.	INST.	TOTAL
0530	Dry standpipe riser, Class I, steel black sch. 40, 10' height			
0540	4" diameter pipe, one floor	1,275	1,075	2,350
0560	Additional floors	455	385	840
0580	6" diameter pipe, one floor	2,600	1,825	4,425
0600	Additional floors	850	625	1,475
0620	8" diameter pipe, one floor	3,800	2,200	6,000
0640	Additional floors	1,150	770	1,920

8.2-320	**Dry Standpipe Risers, Class II**	COST PER FLOOR		
		MAT.	INST.	TOTAL
1030	Dry standpipe risers, Class II, steel black sch. 40, 10' height			
1040	2" diameter pipe, one floor	760	495	1,255
1060	Additional floor	251	162	413
1080	2-1/2" diameter pipe, one floor	845	580	1,425
1100	Additional floors	278	193	471

8.2-320	**Dry Standpipe Risers, Class III**	COST PER FLOOR		
		MAT.	INST.	TOTAL
1530	Dry standpipe risers, Class III, steel black sch. 40, 10' height			
1540	4" diameter pipe, one floor	1,300	1,075	2,375
1560	Additional floors	365	350	715
1580	6" diameter pipe, one floor	2,625	1,825	4,450
1600	Additional floors	885	625	1,510
1620	8" diameter pipe, one floor	3,850	2,200	6,050
1640	Additional floor	1,200	770	1,970

MECHANICAL

8

8.2-390	Standpipe Equipment	COST EACH		
		MAT.	INST.	TOTAL
0100	Adapters, reducing, 1 piece, FxM, hexagon, cast brass, 2-1/2" x 1-1/2"	36.50		36.50
0200	Pin lug, 1-1/2" x 1"	11.55		11.55
0250	3" x 2-1/2"	45		45
0300	For polished chrome, add 75% mat.			
0400	Cabinets, D.S. glass in door, recessed, steel box, not equipped			
0500	Single extinguisher, steel door & frame	72.50	60	132.50
0550	Stainless steel door & frame	169	60	229
0600	Valve, 2-1/2" angle, steel door & frame	80	40	120
0650	Aluminum door & frame	123	40	163
0700	Stainless steel door & frame	172	40	212
0750	Hose rack assy, 2-1/2" x 1-1/2" valve & 100' hose, steel door & frame	162	80	242
0800	Aluminum door & frame	280	80	360
0850	Stainless steel door & frame	385	80	465
0900	Hose rack assy,& extinguisher,2-1/2"x1-1/2" valve & hose,steel door & frame	203	96	299
0950	Aluminum	355	96	451
1000	Stainless steel	450	96	546
1550	Compressor, air, dry pipe system, automatic, 200 gal., 1/3 H.P.	770	206	976
1600	520 gal., 1 H.P.	805	206	1,011
1650	Alarm, electric pressure switch (circuit closer)	120	10.30	130.30
2500	Couplings, hose, rocker lug, cast brass, 1-1/2"	28		28
2550	2-1/2"	58.50		58.50
3000	Escutcheon plate, for angle valves, polished brass, 1-1/2"	13.75		13.75
3050	2-1/2"	34		34
3500	Fire pump, electric, w/controller, fittings, relief valve			
3550	4" pump, 30 H.P., 500 G.P.M.	13,900	1,425	15,325
3600	5" pump, 40 H.P., 1000 G.P.M.	20,300	1,600	21,900
3650	5" pump, 100 H.P., 1000 G.P.M.	22,800	1,775	24,575
3700	For jockey pump system, add	2,500	240	2,740
5000	Hose, per linear foot, synthetic jacket, lined,			
5100	300 lb. test, 1-1/2" diameter	2.11		2.11
5150	2-1/2" diameter	3.41		3.41
5200	500 lb. test, 1-1/2" diameter	2.50		2.50
5250	2-1/2" diameter	4.30		4.30
5500	Nozzle, plain stream, polished brass, 1-1/2" x 10"	32		32
5550	2-1/2" x 15" x 13/16" or 1-1/2"	116		116
5600	Heavy duty combination adjustable fog and straight stream w/handle 1-1/2"	253		253
5650	2-1/2" direct connection	360		360
6000	Rack, for 1-1/2" diameter hose 100 ft. long, steel	32.50	24	56.50
6050	Brass	51.50	24	75.50
6500	Reel, steel, for 50 ft. long 1-1/2" diameter hose	87	34.50	121.50
6550	For 75 ft. long 2-1/2" diameter hose	141	34.50	175.50
7050	Siamese, w/plugs & chains, polished brass, sidewalk, 4" x 2-1/2" x 2-1/2"	299	192	491
7100	6" x 2-1/2" x 2-1/2"	455	240	695
7200	Wall type, flush, 4" x 2-1/2" x 2-1/2"	345	96	441
7250	6" x 2-1/2" x 2-1/2"	470	105	575
7300	Projecting, 4" x 2-1/2" x 2-1/2"	315	96	411
7350	6" x 2-1/2" x 2-1/2"	515	105	620
7400	For chrome plate, add 15% mat.			
8000	Valves, angle, wheel handle, 300 Lb., rough brass, 1-1/2"	35	22.50	57.50
8050	2-1/2"	65.50	38	103.50
8100	Combination pressure restricting, 1-1/2"	54.50	22.50	77
8150	2-1/2"	116	38	154
8200	Pressure restricting, adjustable, satin brass, 1-1/2"	63	22.50	85.50
8250	2-1/2"	105	38	143
8300	Hydrolator, vent and drain, rough brass, 1-1/2"	33.50	22.50	56
8350	2-1/2"	96.50	38	134.50
8400	Cabinet assy, incls. 2-1/2" valve, adapter, rack, hose, nozzle & hydrolator	670	180	850

Important: See the Reference Section for critical supporting data - Reference Nos., Crews & Location Factors

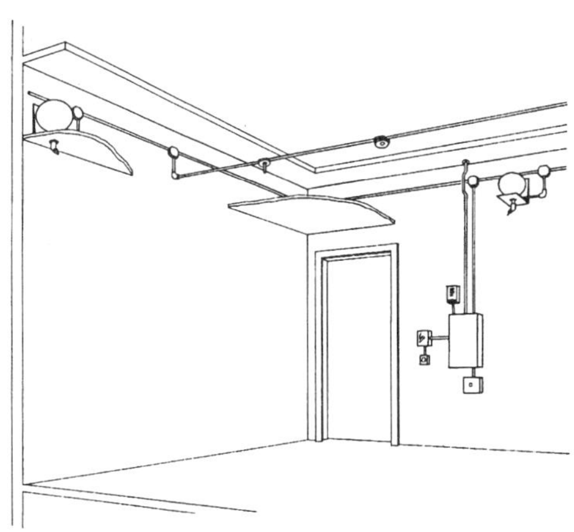

General: Automatic fire protection (suppression) systems other than water sprinklers may be desired for special environments, high risk areas, isolated locations or unusual hazards. Some typical applications would include:

Paint dip tanks
Securities vaults
Electronic data processing
Tape and data storage
Transformer rooms
Spray booths
Petroleum storage
High rack storage

Piping and wiring costs are dependent on the individual application and must be added to the component costs shown below.

All areas are assumed to be open.

8.2-810	FM200 Fire Suppression	COST EACH		
		MAT.	INST.	TOTAL
0020	Detectors with brackets			
0040	Fixed temperature heat detector	31	32.50	63.50
0060	Rate of temperature rise detector	37	32.50	69.50
0080	Ion detector (smoke) detector	82.50	41.50	124
0200	Extinguisher agent			
0240	200 lb FM200, container	6,375	119	6,494
0280	75 lb carbon dioxide cylinder	1,050	79	1,129
0320	Dispersion nozzle			
0340	FM200 1-1/2" dispersion nozzle	55	18.85	73.85
0380	Carbon dioxide 3" x 5" dispersion nozzle	55	14.65	69.65
0420	Control station			
0440	Single zone control station with batteries	1,525	258	1,783
0470	Multizone (4) control station with batteries	2,700	515	3,215
0500	Electric mechanical release	127	82	209
0550	Manual pull station	49.50	44	93.50
0640	Battery standby power 10" x 10" x 17"	760	64.50	824.50
0740	Bell signalling device	54.50	32.50	87

8.2-810	FM200 Systems	COST PER C.F.		
		MAT.	INST.	TOTAL
0820	Average FM200 system, minimum			1.38
0840	Maximum			2.75

Table 8.3-011 Factor for Determining Heat Loss for Various Types of Buildings

General: While the most accurate estimates of heating requirements would naturally be based on detailed information about the building being considered, it is possible to arrive at a reasonable approximation using the following procedure.

1. Calculate the cubic volume of the room or building.
2. Select the appropriate factor from Table 8.3-011. Note that the factors apply only to inside temperatures listed in the first column and to 0°F outside temperature.
3. If the building has bad north and west exposures, multiply the heat loss factor by 1.1.
4. If the outside design temperature is other than 0°F, multiply the factor from Table 8.3-011 by the factor from Table 8.3-012.
5. Multiply the cubic volume by the factor selected from Table 8.3-011. This will give the estimated BTUH heat loss which must be made up to maintain inside temperature.

Building Type	Conditions	Qualifications	Loss Factor*
Factories & Industrial Plants General Office Areas 70°F	One Story	Skylight in Roof	6.2
		No Skylight in Roof	5.7
	Multiple Story	Two Story	4.6
		Three Story	4.3
		Four Story	4.1
		Five Story	3.9
		Six Story	3.6
	All Walls Exposed	Flat Roof	6.9
		Heated Space Above	5.2
	One Long Warm Common Wall	Flat Roof	6.3
		Heated Space Above	4.7
	Warm Common Walls on Both Long Sides	Flat Roof	5.8
		Heated Space Above	4.1
Warehouses 60°F	All Walls Exposed	Skylights in Roof	5.5
		No Skylights in Roof	5.1
		Heated Space Above	4.0
	One Long Warm Common Wall	Skylight in Roof	5.0
		No Skylight in Roof	4.9
		Heated Space Above	3.4
	Warm Common Walls On Both Long Sides	Skylight In Roof	4.7
		No Skylight in Roof	4.4
		Heated Space Above	3.0

Note: This table tends to be conservative particularly for new buildings designed for minimum energy consumption.

Table 8.3-012 Outside Design Temperature Correction Factor (for Degrees Fahrenheit)

Outside Design Temperature	50	40	30	20	10	0	-10	-20	-30
Correction Factor	0.29	0.43	0.57	0.72	0.86	1.00	1.14	1.28	1.43

Table 8.3-013 and Table 8.3-014 provide a way to calculate heat transmission of various construction materials from their U values and the TD (Temperature Difference).

1. From the Exterior Enclosure Division or elsewhere, determine U values for the construction desired.

2. Determine the coldest design temperature. The difference between this temperature and the desired interior temperature is the TD (temperature difference).

3. Enter Table 8.3-013 or Table 8.3-014 at correct U Value. Cross horizontally to the intersection with appropriate TD. Read transmission per square foot from bottom of figure.

4. Multiply this value of BTU per hour transmission per square foot of area by the total surface area of that type of construction.

Important: See the Reference Section for critical supporting data - Reference Nos., Crews & Location Factors

Table 8.3-013   Transmission of Heat

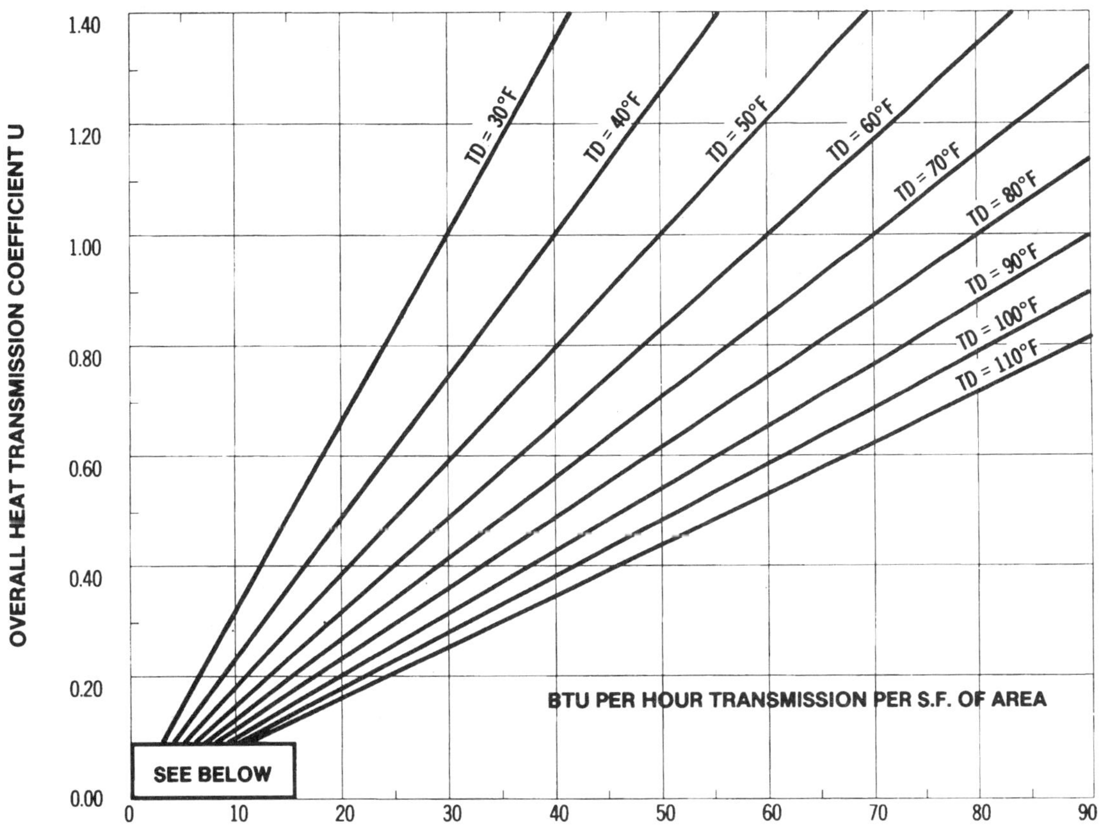

Table 8.3-014   Transmission of Heat (Low Rate)

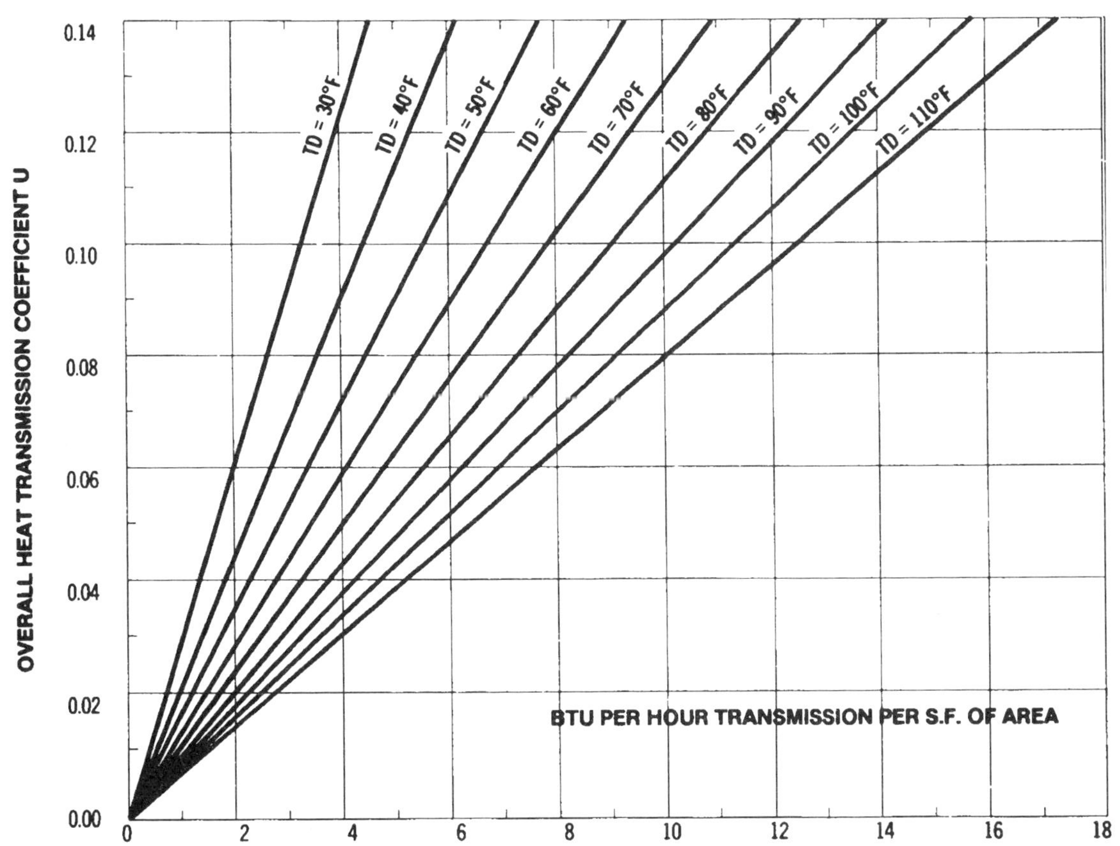

MECHANICAL

8

For expanded coverage of these items see *Means Mechanical or Plumbing Cost Data 1998*

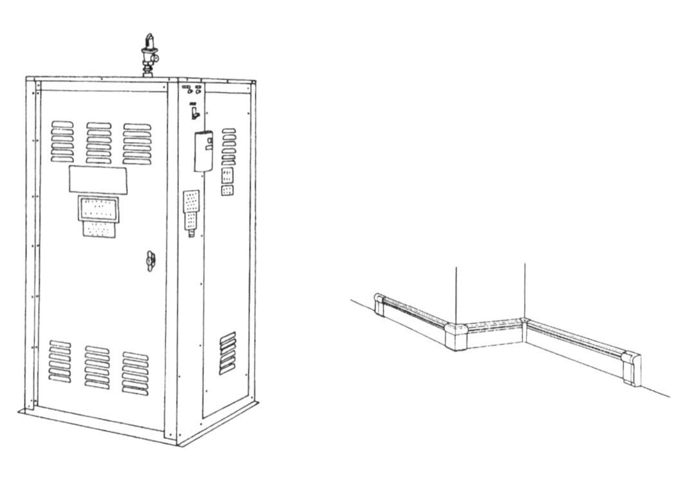

Boiler **Baseboard Radiation**

Small Electric Boiler
System Considerations:
1. Terminal units are fin tube baseboard radiation rated at 720 BTU/hr with 200° water temperature or 820 BTU/hr steam.
2. Primary use being for residential or smaller supplementary areas, the floor levels are based on 7-1/2′ ceiling heights.
3. All distribution piping is copper for boilers through 205 MBH. All piping for larger systems is steel pipe.

8.3-110 — Small Heating Systems, Hydronic, Electric Boilers

		COST PER S.F.		
		MAT.	INST.	TOTAL
1100	Small heating systems, hydronic, electric boilers			
1120	Steam, 1 floor, 1480 S.F., 61 M.B.H.	8.65	3.30	11.95
1160	3,000 S.F., 123 M.B.H.	5.20	2.89	8.09
1200	5,000 S.F., 205 M.B.H.	3.99	2.65	6.64
1240	2 floors, 12,400 S.F., 512 M.B.H.	2.90	2.64	5.54
1280	3 floors, 24,800 S.F., 1023 M.B.H.	2.56	2.61	5.17
1320	34,750 S.F., 1,433 M.B.H.	2.29	2.53	4.82
1360	Hot water, 1 floor, 1,000 S.F., 41 M.B.H.	7.80	1.84	9.64
1400	2,500 S.F., 103 M.B.H.	5	3.29	8.29
1440	2 floors, 4,850 S.F., 205 M.B.H.	4.69	3.95	8.64
1480	3 floors, 9,700 S.F., 410 M.B.H.	4.67	4.10	8.77

8.3-120 — Large Heating Systems, Hydronic, Electric Boilers

		COST PER S.F.		
		MAT.	INST.	TOTAL
1230	Large heating systems, hydronic, electric boilers			
1240	9,280 S.F., 150 K.W., 510 M.B.H., 1 floor	3.22	1.25	4.47
1280	14,900 S.F., 240 K.W., 820 M.B.H., 2 floors	3.36	2.06	5.42
1320	18,600 S.F., 300 K.W., 1,024 M.B.H., 3 floors	3.53	2.26	5.79
1360	26,100 S.F., 420 K.W., 1,432 M.B.H., 4 floors	3.38	2.18	5.56
1400	39,100 S.F., 630 K.W., 2,148 M.B.H., 4 floors	2.98	1.81	4.79
1440	57,700 S.F., 900 K.W., 3,071 M.B.H., 5 floors	2.84	1.79	4.63
1480	111,700 S.F., 1,800 K.W., 6,148 M.B.H., 6 floors	2.54	1.53	4.07
1520	149,000 S.F., 2,400 K.W., 8,191 M.B.H., 8 floors	2.48	1.53	4.01
1560	223,300 S.F., 3,600 K.W., 12,283 M.B.H., 14 floors	2.51	1.75	4.26

MECHANICAL

Important: See the Reference Section for critical supporting data - Reference Nos., Crews & Location Factors

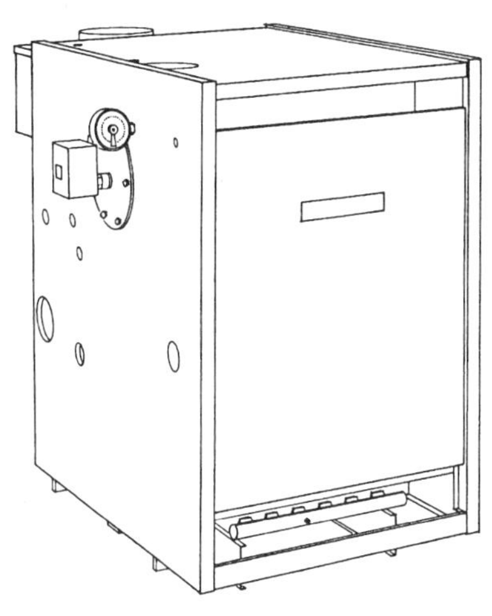

Boiler Selection: The maximum allowable working pressures are limited by ASME "Code for Heating Boilers" to 15 PSI for steam and 160 PSI for hot water heating boilers, with a maximum temperature limitation of 250°F. Hot water boilers are generally rated for a working pressure of 30 PSI. High pressure boilers are governed by the ASME "Code for Power Boilers" which is used almost universally for boilers operating over 15 PSIG. High pressure boilers used for a combination of heating/process loads are usually designed for 150 PSIG.

Boiler ratings are usually indicated as either Gross or Net Output. The Gross Load is equal to the Net Load plus a piping and pickup allowance. When this allowance cannot be determined, divide the gross output rating by 1.25 for a value equal to or greater than the next heat loss requirement of the building.

Table below lists installed cost per boiler and includes insulating jacket, standard controls, burner and safety controls. Costs do not include piping or boiler base pad. Outputs are Gross.

8.3-130	Boilers, Hot Water & Steam	COST EACH		
		MAT.	INST.	TOTAL
0600	Boiler, electric, steel, hot water, 12 K.W., 41 M.B.H.	3,650	565	4,215
0620	30 K.W., 103 M.B.H.	4,000	615	4,615
0640	60 K.W., 205 M.B.H.	5,100	670	5,770
0660	120 K.W., 410 M.B.H.	6,500	820	7,320
0680	210 K.W., 716 M.B.H.	10,800	1,225	12,025
0700	510 K.W., 1,739 M.B.H.	20,400	2,275	22,675
0720	720 K.W., 2,452 M.B.H.	25,600	2,575	28,175
0740	1,200 K.W., 4,095 M.B.H.	36,300	2,950	39,250
0760	2,100 K.W., 7,167 M.B.H.	57,000	3,700	60,700
0780	3,600 K.W., 12,283 M.B.H.	78,000	6,250	84,250
0820	Steam, 6 K.W., 20.5 M.B.H.	8,700	615	9,315
0840	24 K.W., 81.8 M.B.H.	9,075	670	9,745
0860	60 K.W., 205 M.B.H.	10,500	735	11,235
0880	150 K.W., 512 M.B.H.	13,600	1,125	14,725
0900	510 K.W., 1,740 M.B.H.	22,200	2,775	24,975
0920	1,080 K.W., 3,685 M.B.H.	34,000	4,000	38,000
0940	2,340 K.W., 7,984 M.B.H.	64,500	6,250	70,750
0980	Gas, cast iron, hot water, 80 M.B.H.	1,175	655	1,830
1000	100 M.B.H.	1,325	710	2,035
1020	163 M.B.H.	1,775	955	2,730
1040	280 M.B.H.	2,500	1,075	3,575
1060	544 M.B.H.	3,900	1,600	5,500
1080	1,088 M.B.H.	6,575	1,925	8,500
1100	2,000 M.B.H.	11,300	2,525	13,825
1120	2,856 M.B.H.	16,000	2,900	18,900
1140	4,720 M.B.H.	45,800	5,325	51,125
1160	6,970 M.B.H.	77,000	12,000	89,000
1180	For steam systems under 2,856 M.B.H., add 8%			
1240	Steel, hot water, 72 M.B.H.	1,925	345	2,270
1260	101 M.B.H.	2,200	385	2,585
1280	132 M.B.H.	2,500	405	2,905
1300	150 M.B.H.	2,900	460	3,360
1320	240 M.B.H.	4,450	530	4,980
1340	400 M.B.H.	6,350	865	7,215
1360	640 M.B.H.	8,625	1,150	9,775
1380	800 M.B.H.	10,100	1,375	11,475
1400	960 M.B.H.	12,500	1,525	14,025
1420	1,440 M.B.H.	17,400	1,975	19,375
1440	2,400 M.B.H.	28,100	3,450	31,550
1460	3,000 M.B.H.	34,700	4,600	39,300
1520	Oil, cast iron, hot water, 109 M.B.H.	1,300	795	2,095
1540	173 M.B.H.	2,200	955	3,155

MECHANICAL

8

8.3-130	Boilers, Hot Water & Steam	COST EACH		
		MAT.	INST.	TOTAL
1560	236 M.B.H.	2,550	1,125	3,675
1580	1,084 M.B.H.	9,125	2,275	11,400
1600	1,600 M.B.H.	12,100	3,075	15,175
1620	2,480 M.B.H.	17,100	3,825	20,925
1640	3,550 M.B.H.	16,900	4,350	21,250
1660	Steam systems same price as hot water			
1700	Steel, hot water, 103 M.B.H.	2,275	365	2,640
1720	137 M.B.H.	2,425	430	2,855
1740	225 M.B.H.	3,425	495	3,920
1760	315 M.B.H.	5,050	630	5,680
1780	420 M.B.H.	5,675	865	6,540
1800	630 M.B.H.	7,325	1,150	8,475
1820	735 M.B.H.	9,775	1,250	11,025
1840	1,050 M.B.H.	13,500	1,650	15,150
1860	1,365 M.B.H.	17,300	1,875	19,175
1880	1,680 M.B.H.	20,300	2,100	22,400
1900	2,310 M.B.H.	26,200	2,875	29,075
1920	2,835 M.B.H.	32,500	4,075	36,575
1940	3,150 M.B.H.	36,500	5,325	41,825

Important: See the Reference Section for critical supporting data - Reference Nos., Crews & Location Factors

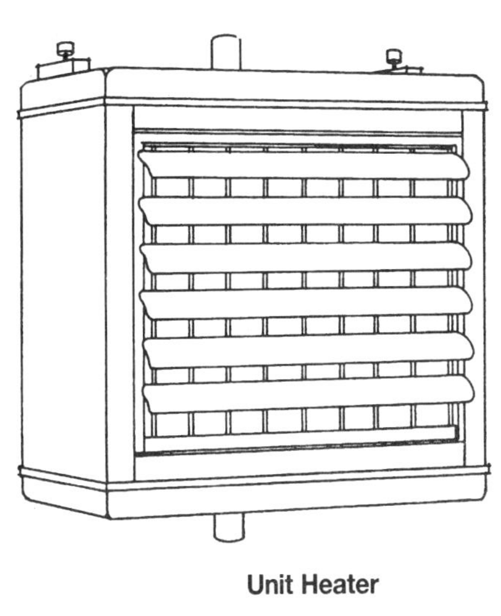

Unit Heater

Fossil Fuel Boiler System Considerations:

1. Terminal units are horizontal unit heaters. Quantities are varied to accommodate total heat loss per building.
2. Unit heater selection was determined by their capacity to circulate the building volume a minimum of three times per hour in addition to the BTU output.
3. Systems shown are forced hot water. Steam boilers cost slightly more than hot water boilers. However, this is compensated for by the smaller size or fewer terminal units required with steam.
4. Floor levels are based on 10' story heights.
5. MBH requirements are gross boiler output.

8.3-141	Heating Systems, Unit Heaters	COST PER S.F.		
		MAT.	INST.	TOTAL
1260	Heating systems, hydronic, fossil fuel, terminal unit heaters,			
1280	Cast iron boiler, gas, 80 M.B.H., 1,070 S.F. bldg.	6.55	3.87	10.42
1320	163 M.B.H., 2,140 S.F. bldg.	4.46	2.62	7.08
1360	544 M.B.H., 7,250 S.F. bldg.	2.93	1.86	4.79
1400	1,088 M.B.H., 14,500 S.F. bldg.	2.60	1.77	4.37
1440	3,264 M.B.H., 43,500 S.F. bldg.	2.19	1.31	3.50
1480	5,032 M.B.H., 67,100 S.F. bldg.	2.56	1.37	3.93
1520	Oil, 109 M.B.H., 1,420 S.F. bldg.	6.25	3.45	9.70
1560	235 M.B.H., 3,150 S.F. bldg.	4.30	2.52	6.82
1600	940 M.B.H., 12,500 S.F. bldg.	3.29	1.61	4.90
1640	1,600 M.B.H., 21,300 S.F. bldg.	3.15	1.52	4.67
1680	2,480 M.B.H., 33,100 S.F. bldg.	3.12	1.37	4.49
1720	3,350 M.B.H., 44,500 S.F. bldg.	2.61	1.42	4.03
1760	Coal, 148 M.B.H., 1,975 S.F. bldg.	5.05	2.25	7.30
1800	300 M.B.H., 4,000 S.F. bldg.	3.98	1.77	5.75
1840	2,360 M.B.H., 31,500 S.F. bldg.	2.81	1.46	4.27
1880	Steel boiler, gas, 72 M.B.H., 1,020 S.F. bldg.	6.15	2.87	9.02
1920	240 M.B.H., 3,200 S.F. bldg.	4.25	2.31	6.56
1960	480 M.B.H., 6,400 S.F. bldg.	3.52	1.69	5.21
2000	800 M.B.H., 10,700 S.F. bldg.	3.10	1.50	4.60
2040	1,960 M.B.H., 26,100 S.F. bldg.	2.77	1.38	4.15
2080	3,000 M.B.H., 40,000 S.F. bldg.	2.72	1.40	4.12
2120	Oil, 97 M.B.H., 1,300 S.F. bldg.	6.95	3.18	10.13
2160	315 M.B.H., 4,550 S.F. bldg.	4.13	1.63	5.76
2200	525 M.B.H., 7,000 S.F. bldg.	4.14	1.66	5.80
2240	1,050 M.B.H., 14,000 S.F. bldg.	3.62	1.69	5.31
2280	2,310 M.B.H. 30,800 S.F. bldg.	3.43	1.47	4.90
2320	3,150 M.B.H., 42,000 S.F. bldg.	3.26	1.51	4.77

MECHANICAL

8

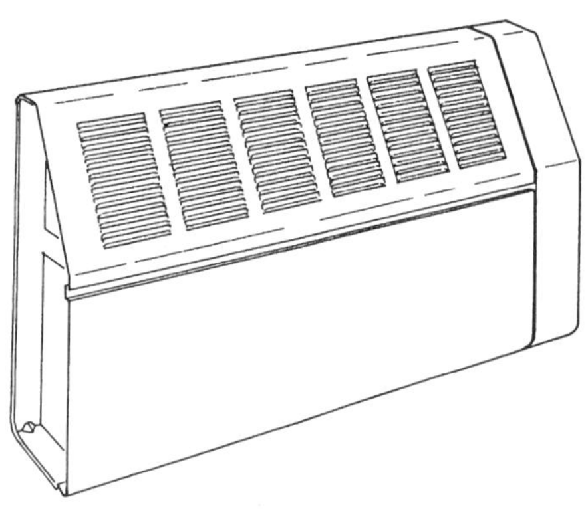

Fin Tube Radiator

Fossil Fuel Boiler System Considerations:

1. Terminal units are commercial steel fin tube radiation. Quantities are varied to accommodate total heat loss per building.
2. Systems shown are forced hot water. Steam boilers cost slightly more than hot water boilers. However, this is compensated for by the smaller size or fewer terminal units required with steam.
3. Floor levels are based on 10' story heights.
4. MBH requirements are gross boiler output.

8.3-142	Heating System, Fin Tube Radiation	COST PER S.F.		
		MAT.	INST.	TOTAL
3230	Heating systems, hydronic, fossil fuel, fin tube radiation			
3240	Cast iron boiler, gas, 80 MBH, 1,070 S.F. bldg.	7.60	6.15	13.75
3280	169 M.B.H., 2,140 S.F. bldg.	4.76	3.89	8.65
3320	544 M.B.H., 7,250 S.F. bldg.	3.79	3.31	7.10
3360	1,088 M.B.H., 14,500 S.F. bldg.	3.55	3.27	6.82
3400	3,264 M.B.H., 43,500 S.F. bldg.	3.27	2.85	6.12
3440	5,032 M.B.H., 67,100 S.F. bldg.	3.66	2.94	6.60
3480	Oil, 109 M.B.H., 1,420 S.F. bldg.	8.60	6.70	15.30
3520	235 M.B.H., 3,150 S.F. bldg.	5.15	3.98	9.13
3560	940 M.B.H., 12,500 S.F. bldg.	4.27	3.12	7.39
3600	1,600 M.B.H., 21,300 S.F. bldg.	4.23	3.10	7.33
3640	2,480 M.B.H., 33,100 S.F. bldg.	4.21	2.94	7.15
3680	3,350 M.B.H., 44,500 S.F. bldg.	3.68	2.98	6.66
3720	Coal, 148 M.B.H., 1,975 S.F. bldg.	5.85	3.72	9.57
3760	300 M.B.H., 4,000 S.F. bldg.	4.77	3.18	7.95
3800	2,360 M.B.H., 31,500 S.F. bldg.	3.84	2.98	6.82
3840	Steel boiler, gas, 72 M.B.H., 1,020 S.F. bldg.	8.55	5.85	14.40
3880	240 M.B.H., 3,200 S.F. bldg.	5.15	3.86	9.01
3920	480 M.B.H., 6,400 S.F. bldg.	4.42	3.17	7.59
3960	800 M.B.H., 10,700 S.F. bldg.	4.42	3.37	7.79
4000	1,960 M.B.H., 26,100 S.F. bldg.	3.80	2.89	6.69
4040	3,000 M.B.H., 40,000 S.F. bldg.	3.74	2.91	6.65
4080	Oil, 97 M.B.H., 1,300 S.F. bldg.	7.30	5.45	12.75
4120	315 M.B.H., 4,550 S.F. bldg.	4.94	2.98	7.92
4160	525 M.B.H., 7,000 S.F. bldg.	5.15	3.20	8.35
4200	1,050 M.B.H., 14,000 S.F. bldg.	4.61	3.22	7.83
4240	2,310 M.B.H., 30,800 S.F. bldg.	4.45	2.98	7.43
4280	3,150 M.B.H., 42,000 S.F. bldg.	4.30	3.05	7.35

Important: See the Reference Section for critical supporting data - Reference Nos., Crews & Location Factors

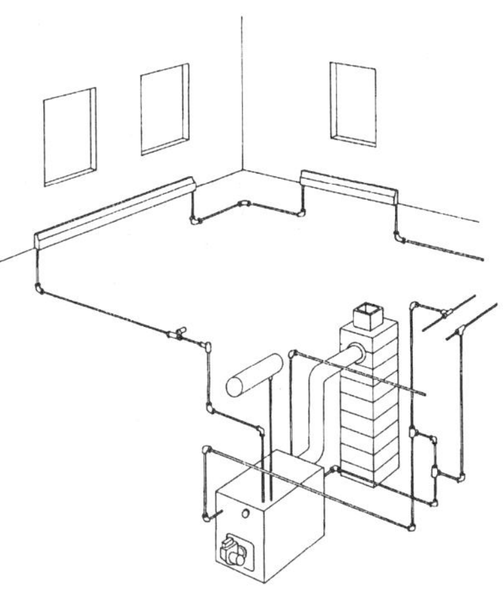

Basis for Heat Loss Estimate, Apartment Type Structures:

1. Masonry walls and flat roof are insulated. U factor is assumed at .08.
2. Window glass area taken as BOCA minimum, 1/10th of floor area. Double insulating glass with 1/4″ air space, U = .65.
3. Infiltration = 0.3 C.F. per hour per S.F. of net wall.
4. Concrete floor loss is 2 BTUH per S.F.
5. Temperature difference taken as 70° F.
6. Ventilating or makeup air has not been included and must be added if desired. Air shafts are not used.

8.3-151 Apartment Building Heating - Fin Tube Radiation

		COST PER S.F.		
		MAT.	INST.	TOTAL
1740	Heating systems, fin tube radiation, forced hot water			
1760	1,000 S.F. area, 10,000 C.F. volume	3.87	2.47	6.34
1800	10,000 S.F. area, 100,000 C.F. volume	1.56	1.54	3.10
1840	20,000 S.F. area, 200,000 C.F. volume	1.65	1.71	3.36
1880	30,000 S.F. area, 300,000 C.F. volume	1.56	1.68	3.24

8.3-161 Commercial Building Heating - Fin Tube Radiation

		COST PER S.F.		
		MAT.	INST.	TOTAL
1940	Heating systems, fin tube radiation, forced hot water			
1960	1,000 S.F. bldg, one floor	8.75	5.70	14.45
2000	10,000 S.F., 100,000 C.F., total two floors	2.53	2.21	4.74
2040	100,000 S.F., 1,000,000 C.F., total three floors	1.16	1	2.16
2080	1,000,000 S.F., 10,000,000 C.F., total five floors	.52	.54	1.06

8.3-162 Commercial Bldg. Heating - Terminal Unit Heaters

		COST PER S.F.		
		MAT.	INST.	TOTAL
1860	Heating systems, terminal unit heaters, forced hot water			
1880	1,000 S.F. bldg., one floor	9.15	5.20	14.35
1920	10,000 S.F. bldg., 100,000 C.F. total two floors	2.36	1.83	4.19
1960	100,000 S.F. bldg., 1,000,000 C.F. total three floors	1.29	.91	.2.20
2000	1,000,000 S.F. bldg., 10,000,000 C.F. total five floors	.82	.59	1.41

MECHANICAL

8

Table 8.4-001 Air Conditioning

General: The purpose of air conditioning is to control the environment of a space so that comfort is provided for the occupants and/or conditions are suitable for the processes or equipment contained therein. The several items which should be evaluated to define system objectives are:

Temperature Control
Humidity Control
Cleanliness
Odor, smoke and fumes
Ventilation

Efforts to control the above parameters must also include consideration of the degree or tolerance of variation, the noise level introduced, the velocity of air motion and the energy requirements to accomplish the desired results.

The variation in **temperature** and **humidity** is a function of the sensor and the controller. The controller reacts to a signal from the sensor and produces the appropriate suitable response in either the terminal unit, the conductor of the transporting medium (air, steam, chilled water, etc.), or the source (boiler, evaporating coils, etc.).

The **noise level** is a by-product of the energy supplied to moving components of the system. Those items which usually contribute the most noise are pumps, blowers, fans, compressors and diffusers. The level of noise can be partially controlled through use of vibration pads, isolators, proper sizing, shields, baffles and sound absorbing liners.

Some **air motion** is necessary to prevent stagnation and stratification. The maximum acceptable velocity varies with the degree of heating or cooling which is taking place. Most people feel air moving past them at velocities in excess of 25 FPM as an annoying draft. However, velocities up to 45 FPM may be acceptable in certain cases. Ventilation, expressed as air changes per hour and percentage of fresh air, is usually an item regulated by local codes.

Selection of the system to be used for a particular application is usually a trade-off. In some cases the building size, style, or room available for mechanical use limits the range of possibilities. Prime factors influencing the decision are first cost and total life (operating, maintenance and replacement costs). The accuracy with which each parameter is determined will be an important measure of the reliability of the decision and subsequent satisfactory operation of the installed system.

Heat delivery may be desired from an air conditioning system. Heating capability usually is added as follows: A gas fired burner or hot water/steam/electric coils may be added to the air handling unit directly and heat all air equally. For limited or localized heat requirements the water/steam/electric coils may be inserted into the duct branch supplying the cold areas. Gas fired duct furnaces are also available.

Note: When water or steam coils are used the cost of the piping and boiler must also be added. For a rough estimate use the cost per square foot of the appropriate sized hydronic system with unit heaters. This will provide a cost for the boiler and piping, and the unit heaters of the system would equate to the approximate cost of the heating coils. The installed cost of electric and gas heaters, boilers and other heat related items on a unit basis may be located in Section 155 of *Means Mechanical Cost Data.*

Table 8.4-002 Air Conditioning Requirements

BTU's per hour per S.F. of floor area and S.F. per ton of air conditioning.

Type of Building	BTU per S.F.	S.F. per Ton	Type of Building	BTU per S.F.	S.F. per Ton	Type of Building	BTU per S.F.	S.F. per Ton
Apartments, Individual	26	450	Dormitory, Rooms	40	300	Libraries	50	240
Corridors	22	550	Corridors	30	400	Low Rise Office, Exterior	38	320
Auditoriums & Theaters	40	300/18*	Dress Shops	43	280	Interior	33	360
Banks	50	240	Drug Stores	80	150	Medical Centers	28	425
Barber Shops	48	250	Factories	40	300	Motels	28	425
Bars & Taverns	133	90	High Rise Office—Ext. Rms.	46	263	Office (small suite)	43	280
Beauty Parlors	66	180	Interior Rooms	37	325	Post Office, Individual Office	42	285
Bowling Alleys	68	175	Hospitals, Core	43	280	Central Area	46	260
Churches	36	330/20*	Perimeter	46	260	Residences	20	600
Cocktail Lounges	68	175	Hotel, Guest Rooms	44	275	Restaurants	60	200
Computer Rooms	141	85	Corridors	30	400	Schools & Colleges	46	260
Dental Offices	52	230	Public Spaces	55	220	Shoe Stores	55	220
Dept. Stores, Basement	34	350	Industrial Plants, Offices	38	320	Shop'g. Ctrs., Supermarkets	34	350
Main Floor	40	300	General Offices	34	350	Retail Stores	48	250
Upper Floor	30	400	Plant Areas	40	300	Specialty	60	200

*Persons per ton
12,000 BTU = 1 ton of air conditioning

Important: See the Reference Section for critical supporting data - Reference Nos., Crews & Location Factors

Table 8.4-004 Recommended Ventilation Air Changes

Table below lists range of time in minutes per change for various types of facilities.

Assembly Halls	2-10	Dance Halls	2-10	Laundries	1-3
Auditoriums	2-10	Dining Rooms	3-10	Markets	2-10
Bakeries	2-3	Dry Cleaners	1-5	Offices	2-10
Banks	3-10	Factories	2-5	Pool Rooms	2-5
Bars	2-5	Garages	2-10	Recreation Rooms	2-10
Beauty Parlors	2-5	Generator Rooms	2-5	Sales Rooms	2-10
Boiler Rooms	1-5	Gymnasiums	2-10	Theaters	2-8
Bowling Alleys	2-10	Kitchens-Hospitals	2-5	Toilets	2-5
Churches	5-10	Kitchens-Restaurant	1-3	Transformer Rooms	1-5

CFM air required for changes = Volume of room in cubic feet ÷ Minutes per change.

Table 8.4-005 Ductwork

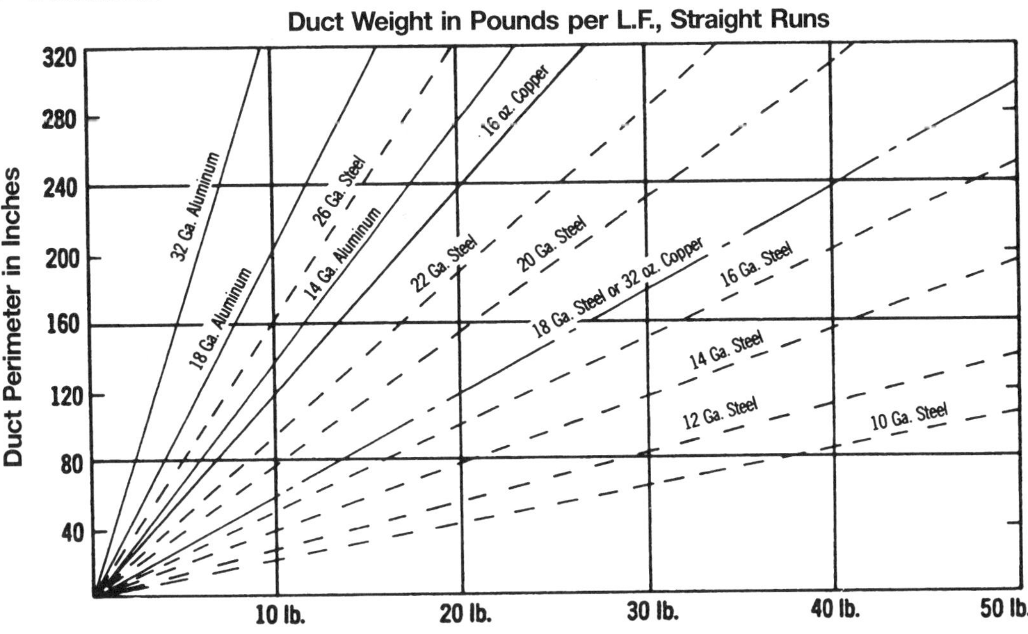

Add to the above for fittings; 90° elbow is 3 L.F.; 45° elbow is 2.5 L.F.; offset is 4 L.F.; transition offset is 6 L.F.; square-to-round transition is 4 L.F.; 90° reducing elbow is 5 L.F. For bracing and waste, add 20% to aluminum and copper, 15% to steel.

Table 8.4-006 Diffuser Evaluation

CFM = V × An × K where V = Outlet velocity in feet per minute. An = Neck area in square feet and K = Diffuser delivery factor. An undersized diffuser for a desired CFM will produce a high velocity and noise level. When air moves past people at a velocity in excess of 25 FPM, an annoying draft is felt. An oversized diffuser will result in low velocity with poor mixing. Consideration must be given to avoid vertical stratification or horizontal areas of stagnation.

MECHANICAL

8

R157-100   Sheet Metal Calculator (Weight in Lb./Ft. of Length)

Gauge	26	24	22	20	18	16	Gauge	26	24	22	20	18	16
Wt.-Lb./S.F.	.906	1.156	1.406	1.656	2.156	2.656	Wt.-Lb./S.F.	.906	1.156	1.406	1.656	2.156	2.656
SMACNA Max. Dimension – Long Side		30″	54″	84″	85″ Up		SMACNA Max. Dimension – Long Side		30″	54″	84″	85″ Up	
Sum-2 sides							Sum-2 Sides						
2	.3	.40	.50	.60	.80	.90	56	9.3	12.0	14.0	16.2	21.3	25.2
3	.5	.65	.80	.90	1.1	1.4	57	9.5	12.3	14.3	16.5	21.7	25.7
4	.7	.85	1.0	1.2	1.5	1.8	58	9.7	12.5	14.5	16.8	22.0	26.1
5	.8	1.1	1.3	1.5	1.9	2.3	59	9.8	12.7	14.8	17.1	22.4	26.6
6	1.0	1.3	1.5	1.7	2.3	2.7	60	10.0	12.9	15.0	17.4	22.8	27.0
7	1.2	1.5	1.8	2.0	2.7	3.2	61	10.2	13.1	15.3	17.7	23.2	27.5
8	1.3	1.7	2.0	2.3	3.0	3.6	62	10.3	13.3	15.5	18.0	23.6	27.9
9	1.5	1.9	2.3	2.6	3.4	4.1	63	10.5	13.5	15.8	18.3	24.0	28.4
10	1.7	2.2	2.5	2.9	3.8	4.5	64	10.7	13.7	16.0	18.6	24.3	28.8
11	1.8	2.4	2.8	3.2	4.2	5.0	65	10.8	13.9	16.3	18.9	24.7	29.3
12	2.0	2.6	3.0	3.5	4.6	5.4	66	11.0	14.1	16.5	19.1	25.1	29.7
13	2.2	2.8	3.3	3.8	4.9	5.9	67	11.2	14.3	16.8	19.4	25.5	30.2
14	2.3	3.0	3.5	4.1	5.3	6.3	68	11.3	14.6	17.0	19.7	25.8	30.6
15	2.5	3.2	3.8	4.4	5.7	6.8	69	11.5	14.8	17.3	20.0	26.2	31.1
16	2.7	3.4	4.0	4.6	6.1	7.2	70	11.7	15.0	17.5	20.3	26.6	31.5
17	2.8	3.7	4.3	4.9	6.5	7.7	71	11.8	15.2	17.8	20.6	27.0	32.0
18	3.0	3.9	4.5	5.2	6.8	8.1	72	12.0	15.4	18.0	20.9	27.4	32.4
19	3.2	4.1	4.8	5.5	7.2	8.6	73	12.2	15.6	18.3	21.2	27.7	32.9
20	3.3	4.3	5.0	5.8	7.6	9.0	74	12.3	15.8	18.5	21.5	28.1	33.3
21	3.5	4.5	5.3	6.1	8.0	9.5	75	12.5	16.1	18.8	21.8	28.5	33.8
22	3.7	4.7	5.5	6.4	8.4	9.9	76	12.7	16.3	19.0	22.0	28.9	34.2
23	3.8	5.0	5.8	6.7	8.7	10.4	77	12.8	16.5	19.3	22.3	29.3	34.7
24	4.0	5.2	6.0	7.0	9.1	10.8	78	13.0	16.7	19.5	22.6	29.6	35.1
25	4.2	5.4	6.3	7.3	9.5	11.3	79	13.2	16.9	19.8	22.9	30.0	35.6
26	4.3	5.6	6.5	7.5	9.9	11.7	80	13.3	17.1	20.0	23.2	30.4	36.0
27	4.5	5.8	6.8	7.8	10.3	12.2	81	13.5	17.3	20.3	23.5	30.8	36.5
28	4.7	6.0	7.0	8.1	10.6	12.6	82	13.7	17.5	20.5	23.8	31.2	36.9
29	4.8	6.2	7.3	8.4	11.0	13.1	83	13.8	17.8	20.8	24.1	31.5	37.4
30	5.0	6.5	7.5	8.7	11.4	13.5	84	14.0	18.0	21.0	24.4	31.9	37.8
31	5.2	6.7	7.8	9.0	11.8	14.0	85	14.2	18.2	21.3	24.7	32.3	38.3
32	5.3	6.9	8.0	9.3	12.2	14.4	86	14.3	18.4	21.5	24.9	32.7	38.7
33	5.5	7.1	8.3	9.6	12.5	14.9	87	14.5	18.6	21.8	25.2	33.1	39.2
34	5.7	7.3	8.5	9.9	12.9	15.3	88	14.7	18.8	22.0	25.5	33.4	39.6
35	5.8	7.5	8.8	10.2	13.3	15.8	89	14.8	19.0	22.3	25.8	33.8	40.1
36	6.0	7.8	9.0	10.4	13.7	16.2	90	15.0	19.3	22.5	26.1	34.2	40.5
37	6.2	8.0	9.3	10.7	14.1	16.7	91	15.2	19.5	22.8	26.4	34.6	41.0
38	6.3	8.2	9.5	11.0	14.4	17.1	92	15.3	19.7	23.0	26.7	35.0	41.4
39	6.5	8.4	9.8	11.3	14.8	17.6	93	15.5	19.9	23.3	27.0	35.3	41.9
40	6.7	8.6	10.0	11.6	15.2	18.0	94	15.7	20.1	23.5	27.3	35.7	42.3
41	6.8	8.8	10.3	11.9	15.6	18.5	95	15.8	20.3	23.8	27.6	36.1	42.8
42	7.0	9.0	10.5	12.2	16.0	18.9	96	16.0	20.5	24.0	27.8	36.5	43.2
43	7.2	9.2	10.8	12.5	16.3	19.4	97	16.2	20.8	24.3	28.1	36.9	43.7
44	7.3	9.5	11.0	12.8	16.7	19.8	98	16.3	21.0	24.5	28.4	37.2	44.1
45	7.5	9.7	11.3	13.1	17.1	20.3	99	16.5	21.2	24.8	28.7	37.6	44.6
46	7.7	9.9	11.5	13.3	17.5	20.7	100	16.7	21.4	25.0	29.0	38.0	45.0
47	7.8	10.1	11.8	13.6	17.9	21.2	101	16.8	21.6	25.3	29.3	38.4	45.5
48	8.0	10.3	12.0	13.9	18.2	21.6	102	17.0	21.8	25.5	29.6	38.8	45.9
49	8.2	10.5	12.3	14.2	18.6	22.1	103	17.2	22.0	25.8	29.9	39.1	46.4
50	8.3	10./7	12.5	14.5	19.0	22.5	104	17.3	22.3	26.0	30.2	39.5	46.8
51	8.5	11.0	12.8	14.8	19.4	23.0	105	17.5	22.5	26.3	30.5	39.9	47.3
52	8.7	11.2	13.0	15.1	19.8	23.4	106	17.7	22.7	26.5	30.7	40.3	47.7
53	8.8	11.4	13.3	15.4	20.1	23.9	107	17.8	22.9	26.8	31.0	40.7	48.2
54	9.0	11.6	13.5	15.7	20.5	24.3	108	18.0	23.1	27.0	31.3	41.0	48.6
55	9.2	11.8	13.8	16.0	20.9	24.8	109	18.2	23.3	27.3	31.6	41.4	49.1
							110	18.3	23.5	27.5	31.9	41.8	49.5

Example: If duct is 34″ x 20″ x 15′ long, 34″ is greater than 30″ maximum, for 24 ga. so must be 22 ga. 34″ + 20″ = 54″ going across from 54″ find 13.5 lb. per foot. 13.5 x 15′ = 202.5 lbs. For S.F. of surface area

202.5 ÷ 1.406 = 144 S.F.

Note: Figures include an allowance for scrap.

Important: See the Reference Section for critical supporting data - Reference Nos., Crews & Location Factors

Cooling

Chilled Water Supply & Return Piping

Air Cooled Water Chiller Unit

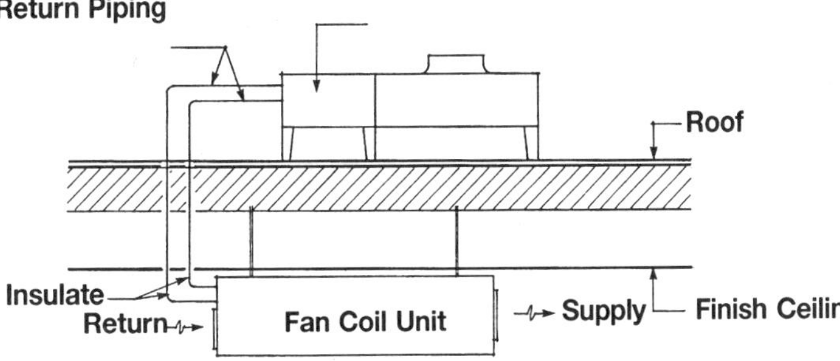

Design Assumptions: The chilled water, air cooled systems priced, utilize reciprocating hermetic compressors and propeller-type condenser fans. Piping with pumps and expansion tanks is included based on a two pipe system. No ducting is included and the fan-coil units are cooling only. Water treatment and balancing are not included. Chilled water piping is insulated. Area distribution is through the use of multiple fan coil units. Fewer but larger fan coil units with duct distribution would be approximately the same S.F. cost.

* Suggest using a water cooled unit.

8.4-110	Chilled Water, Air Cooled Condenser Systems	MAT.	INST.	TOTAL
1180	Packaged chiller, air cooled, with fan coil unit			
1200	Apartment corridors, 3,000 S.F., 5.50 ton	4.55	3.22	7.77
1240	6,000 S.F., 11.00 ton	3.57	2.58	6.15
1280	10,000 S.F., 18.33 ton	3.13	1.95	5.08
1320	20,000 S.F., 36.66 ton	2.53	1.45	3.98
1360	40,000 S.F., 73.33 ton	3	1.46	4.46
1440	Banks and libraries, 3,000 S.F., 12.50 ton	6.75	3.76	10.51
1480	6,000 S.F., 25.00 ton	6.25	3.05	9.30
1520	10,000 S.F., 41.66 ton	5.10	2.21	7.31
1560	20,000 S.F., 83.33 ton	5.50	2.07	7.57
1600	40,000 S.F., 167 ton*			
1680	Bars and taverns, 3,000 S.F., 33.25 ton	12.65	4.66	17.31
1720	6,000 S.F., 66.50 ton	11.90	4.01	15.91
1760	10,000 S.F., 110.83 ton	9.60	1.87	11.47
1800	20,000 S.F., 220 ton*			
1840	40,000 S.F., 440 ton*			
1920	Bowling alleys, 3,000 S.F., 17.00 ton	8.55	4.20	12.75
1960	6,000 S.F., 34.00 ton	6.90	3.08	9.98
2000	10,000 S.F., 56.66 ton	6.30	2.44	8.74
2040	20,000 S.F., 113.33 ton	6.10	2.16	8.26
2080	40,000 S.F., 227 ton*			
2160	Department stores, 3,000 S.F., 8.75 ton	6.45	3.58	10.03
2200	6,000 S.F., 17.50 ton	4.74	2.76	7.50
2240	10,000 S.F., 29.17 ton	3.96	2.05	6.01
2280	20,000 S.F., 58.33 ton	3.52	1.63	5.15
2320	40,000 S.F., 116.66 ton	3.82	1.64	5.46
2400	Drug stores, 3,000 S.F., 20.00 ton	9.80	4.33	14.13
2440	6,000 S.F., 40.00 ton	8.15	3.37	11.52
2480	10,000 S.F., 66.66 ton	8.45	3	11.45
2520	20,000 S.F., 133.33 ton	7.40	2.49	9.89
2560	40,000 S.F., 267 ton*			
2640	Factories, 2,000 S.F., 10.00 ton	5.85	3.61	9.46
2680	6,000 S.F., 20.00 ton	5.35	2.90	8.25
2720	10,000 S.F., 33.33 ton	4.34	2.10	6.44
2760	20,000 S.F., 66.66 ton	4.69	1.93	6.62
2800	40,000 S.F., 133.33 ton	4.20	1.70	5.90
2880	Food supermarkets, 3,000 S.F., 8.50 ton	6.30	3.56	9.86
2920	6,000 S.F., 17.00 ton	4.65	2.75	7.40
2960	10,000 S.F., 28.33 ton	3.78	2	5.78
3000	20,000 S.F., 56.66 ton	3.40	1.59	4.99
3040	40,000 S.F., 113.33 ton	3.69	1.62	5.31
3120	Medical centers, 3,000 S.F., 7.00 ton	5.60	3.47	9.07

Table header spanning columns: **COST PER S.F.** (MAT. / INST. / TOTAL)

8.4-110   Chilled Water, Air Cooled Condenser Systems

		COST PER S.F.		
		MAT.	INST.	TOTAL
3160	6,000 S.F., 14.00 ton	4.11	2.66	6.77
3200	10,000 S.F., 23.33 ton	3.72	2.02	5.74
3240	20,000 S.F., 46.66 ton	3.06	1.53	4.59
3280	40,000 S.F., 93.33 ton	3.30	1.51	4.81
3360	Offices, 3,000 S.F., 9.50 ton	5.45	3.53	8.98
3400	6,000 S.F., 19.00 ton	5.15	2.88	8.03
3440	10,000 S.F., 31.66 ton	4.19	2.08	6.27
3480	20,000 S.F., 63.33 ton	4.64	1.95	6.59
3520	40,000 S.F., 126.66 ton	4.11	1.71	5.82
3600	Restaurants, 3,000 S.F., 15.00 ton	7.65	3.89	11.54
3640	6,000 S.F., 30.00 ton	6.50	3.08	9.58
3680	10,000 S.F., 50.00 ton	5.95	2.44	8.39
3720	20,000 S.F., 100.00 ton	6.10	2.25	8.35
3760	40,000 S.F., 200 ton*			
3840	Schools and colleges, 3,000 S.F., 11.50 ton	6.40	3.70	10.10
3880	6,000 S.F., 23.00 ton	5.90	2.99	8.89
3920	10,000 S.F., 38.33 ton	4.80	2.17	6.97
3960	20,000 S.F., 76.66 ton	5.30	2.07	7.37
4000	40,000 S.F., 153 ton*			

Important: See the Reference Section for critical supporting data - Reference Nos., Crews & Location Factors

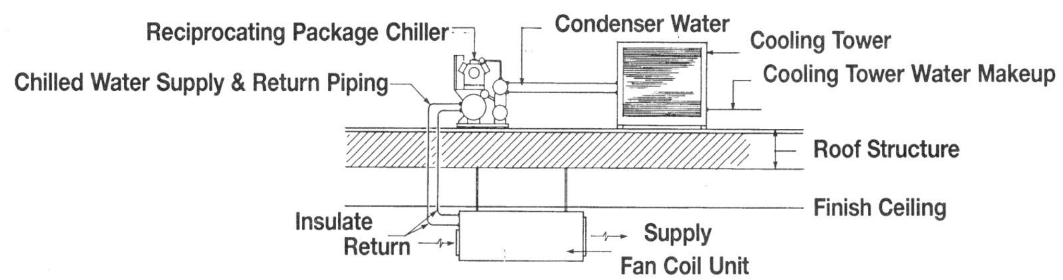

General: Water cooled chillers are available in the same sizes as air cooled units. They are also available in larger capacities.

Design Assumptions: The chilled water systems with water cooled condenser include reciprocating hermetic compressors, water cooling tower, pumps, piping and expansion tanks and are based on a two pipe system. Chilled water piping is insulated. No ducts are included and fan-coil units are cooling only. Area distribution is through use of multiple fan coil units. Fewer but larger fan coil units with duct distribution would be approximately the same S.F. cost. Water treatment and balancing are not included.
* Suggest using multiple chillers

8.4-120	Chilled Water, Cooling Tower Systems	COST PER S.F.		
		MAT.	INST.	TOTAL
1300	Packaged chiller, water cooled, with fan coil unit			
1320	Apartment corridors, 4,000 S.F., 7.33 ton	5.50	3.22	8.72
1360	6,000 S.F., 11.00 ton	4.30	2.70	7
1400	10,000 S.F., 18.33 ton	3.62	2.03	5.65
1440	20,000 S.F., 26.66 ton	2.85	1.52	4.37
1480	40,000 S.F., 73.33 ton	3.20	1.55	4.75
1520	60,000 S.F., 110.00 ton	3.14	1.59	4.73
1600	Banks and libraries, 4,000 S.F., 16.66 ton	8.20	3.54	11.74
1640	6,000 S.F., 25.00 ton	6.60	3.09	9.69
1680	10,000 S.F., 41.66 ton	5.25	2.32	7.57
1720	20,000 S.F., 83.33 ton	5.90	2.28	8.18
1760	40,000 S.F., 166.66 ton	5.70	2.91	8.61
1800	60,000 S.F., 250.00 ton	5.60	3.15	8.75
1880	Bars and taverns, 4,000 S.F., 44.33 ton	12.65	4.51	17.16
1920	6,000 S.F., 66.50 ton	12.90	4.53	17.43
1960	10,000 S.F., 110.83 ton	12.75	3.99	16.74
2000	20,000 S.F., 221.66 ton	11.90	4.49	16.39
2040	40,000 S.F., 440 ton*			
2080	60,000 S.F., 660 ton*			
2160	Bowling alleys, 4,000 S.F., 22.66 ton	9.55	3.84	13.39
2200	6,000 S.F., 34.00 ton	8	3.41	11.41
2240	10,000 S.F., 56.66 ton	6.50	2.55	9.05
2280	20,000 S.F., 113.33 ton	6.95	2.45	9.40
2320	40,000 S.F., 226.66 ton	6.70	3.10	9.80
2360	60,000 S.F., 340 ton			
2440	Department stores, 4,000 S.F., 11.66 ton	6.10	3.38	9.48
2480	6,000 S.F., 17.50 ton	6	2.88	8.88
2520	10,000 S.F., 29.17 ton	4.34	2.14	6.48
2560	20,000 S.F., 58.33 ton	3.34	1.60	4.94
2600	40,000 S.F., 116.66 ton	4.02	1.70	5.72
2640	60,000 S.F., 175.00 ton	4.59	2.82	7.41
2720	Drug stores, 4,000 S.F., 26.66 ton	9.95	3.95	13.90
2760	6,000 S.F., 40.00 ton	8.30	3.44	11.74
2800	10,000 S.F., 66.66 ton	8.15	3.03	11.18
2840	20,000 S.F., 133.33 ton	7.95	2.67	10.62
2880	40,000 S.F., 266.67 ton	7.75	3.52	11.27
2920	60,000 S.F., 400 ton*			
3000	Factories, 4,000 S.F., 13.33 ton	6.95	3.36	10.31
3040	6,000 S.F., 20.00 ton	6	2.88	8.88
3080	10,000 S.F., 33.33 ton	4.79	2.21	7

MECHANICAL

8

8.4-120	Chilled Water, Cooling Tower Systems	COST PER S.F.		
		MAT.	INST.	TOTAL
3120	20,000 S.F., 66.66 ton	4.48	1.92	6.40
3160	40,000 S.F., 133.33 ton	4.47	1.80	6.27
3200	60,000 S.F., 200.00 ton	4.94	2.96	7.90
3280	Food supermarkets, 4,000 S.F., 11.33 ton	6	3.37	9.37
3320	6,000 S.F., 17.00 ton	5.25	2.81	8.06
3360	10,000 S.F., 28.33 ton	4.24	2.13	6.37
3400	20,000 S.F., 56.66 ton	3.43	1.62	5.05
3440	40,000 S.F., 113.33 ton	3.97	1.70	5.67
3480	60,000 S.F., 170.00 ton	4.52	2.81	7.33
3560	Medical centers, 4.000 S.F., 9.33 ton	5.15	3.09	8.24
3600	6,000 S.F., 14.00 ton	5.10	2.76	7.86
3640	10,000 S.F., 23.33 ton	3.91	2.04	5.95
3680	20,000 S.F., 46.66 ton	3.03	1.57	4.60
3720	40,000 S.F., 93.33 ton	3.63	1.62	5.25
3760	60,000 S.F., 140.00 ton	4.16	2.76	6.92*
3840	Offices, 4,000 S.F., 12.66 ton	6.70	3.33	10.03
3880	6,000 S.F., 19.00 ton	5.85	2.95	8.80
3920	10,000 S.F., 31.66 ton	4.72	2.22	6.94
3960	20,000 S.F., 63.33 ton	4.40	1.93	6.33
4000	40,000 S.F., 126.66 ton	4.79	2.74	7.53
4040	60,000 S.F., 190.00 ton	4.78	2.92	7.70
4120	Restaurants, 4,000 S.F., 20.00 ton	8.55	3.57	12.12
4160	6,000 S.F., 30.00 ton	7.25	3.20	10.45
4200	10,000 S.F., 50.00 ton	5.95	2.45	8.40
4240	20,000 S.F., 100.00 ton	6.70	2.41	9.11
4280	40,000 S.F., 200.00 ton	6	2.95	8.95
4320	60,000 S.F., 300.00 ton	6.35	3.30	9.65
4400	Schools and colleges, 4,000 S.F., 15.33 ton	7.70	3.47	11.17
4440	6,000 S.F., 23.00 ton	6.25	3.03	9.28
4480	10,000 S.F., 38.33 ton	4.92	2.27	7.19
4520	20,000 S.F., 76.66 ton	5.60	2.23	7.83
4560	40,000 S.F., 153.33 ton	5.35	2.84	8.19
4600	60,000 S.F., 230.00 ton	5.20	3.03	8.23
4603				

Important: See the Reference Section for critical supporting data - Reference Nos., Crews & Location Factors

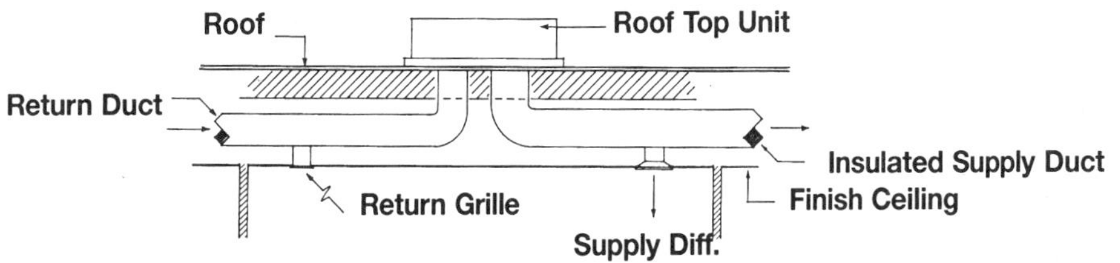

System Description: Rooftop single zone units are electric cooling and gas heat. Duct systems are low velocity, galvanized steel supply and return. Price variations between sizes are due to several factors. Jumps in the cost of the rooftop unit occur when the manufacturer shifts from the largest capacity unit on a small frame to the smallest capacity on the next larger frame, or changes from one compressor to two. As the unit capacity increases for larger areas the duct distribution grows in proportion. For most applications there is a tradeoff point where it is less expensive and more efficient to utilize smaller units with short simple distribution systems. Larger units also require larger initial supply and return ducts which can create a space problem. Supplemental heat may be desired in colder locations.

The table below is based on one unit supplying the area listed. The 10,000 S.F. unit for bars and taverns is not listed because a nominal 110 ton unit would be required and this is above the normal single zone rooftop capacity.

* Use multiple units.

8.4-210	Rooftop Single Zone Unit Systems	COST PER S.F.		
		MAT.	INST.	TOTAL
1260	Rooftop, single zone, air conditioner			
1280	Apartment corridors, 500 S.F., .92 ton	4.62	1.23	5.85
1320	1,000 S.F., 1.83 ton	4.59	1.23	5.82
1360	1500 S.F., 2.75 ton	2.66	1.09	3.75
1400	3,000 S.F., 5.50 ton	2.28	1.20	3.48
1440	5,000 S.F., 9.17 ton	2.30	1.17	3.47
1480	10,000 S.F., 18.33 ton	2.47	1.17	3.64
1560	Banks or libraries, 500 S.F., 2.08 ton	10.45	2.79	13.24
1600	1,000 S.F., 4.17 ton	6.05	2.47	8.52
1640	1,500 S.F., 6.25 ton	5.20	2.73	7.93
1680	3,000 S.F., 12.50 ton	5.25	2.65	7.90
1720	5,000 S.F., 20.80 ton	5.60	2.64	8.24
1760	10,000 S.F., 41.67 ton	5.25	2.64	7.89
1840	Bars and taverns, 500 S.F. 5.54 ton	12.50	4.58	17.08
1880	1,000 S.F., 11.08 ton	12.60	4.35	16.95
1920	1,500 S.F., 16.62 ton	11.30	4.33	15.63
1960	3,000 S.F., 33.25 ton	12.90	4.30	17.20
2000	5,000 S.F., 55.42 ton	12.30	4.32	16.62
2040	10,000 S.F., 110.83 ton*			
2080	Bowling alleys, 500 S.F., 2.83 ton	8.25	3.37	11.62
2120	1,000 S.F., 5.67 ton	7.05	3.72	10.77
2160	1,500 S.F., 8.50 ton	7.10	3.60	10.70
2200	3,000 S.F., 17.00 ton	6.45	3.59	10.04
2240	5,000 S.F., 28.33 ton	7.25	3.57	10.82
2280	10,000 S.F., 56.67 ton	6.95	3.58	10.53
2360	Department stores, 500 S.F., 1.46 ton	7.30	1.96	9.26
2400	1,000 S.F., 2.92 ton	4.24	1.74	5.98
2480	3,000 S.F., 8.75 ton	3.67	1.85	5.52
2520	5,000 S.F., 14.58 ton	3.31	1.84	5.15
2560	10,000 S.F., 29.17 ton	3.74	1.83	5.57
2640	Drug stores, 500 S.F., 3.33 ton	9.70	3.95	13.65
2680	1,000 S.F., 6.67 ton	8.30	4.37	12.67
2720	1,500 S.F., 10.00 ton	8.35	4.23	12.58
2760	3,000 S.F., 20.00 ton	8.95	4.23	13.18
2800	5,000 S.F., 33.33 ton	8.55	4.20	12.75
2840	10,000 S.F., 66.67 ton	8.20	4.21	12.41
2920	Factories, 500 S.F., 1.67 ton	8.35	2.23	10.58
3000	1,500 S.F., 5.00 ton	4.16	2.19	6.35
3040	3,000 S.F., 10.00 ton	4.19	2.12	6.31
3080	5,000 S.F., 16.67 ton	3.79	2.11	5.90

MECHANICAL

8

8.4-210	Rooftop Single Zone Unit Systems	COST PER S.F.		
		MAT.	INST.	TOTAL
3120	10,000 S.F., 33.33 ton	4.28	2.10	6.38
3200	Food supermarkets, 500 S.F., 1.42 ton	7.10	1.90	9
3240	1,000 S.F., 2.83 ton	4.08	1.67	5.75
3280	1,500 S.F., 4.25 ton	3.53	1.85	5.38
3320	3,000 S.F., 8.50 ton	3.56	1.80	5.36
3360	5,000 S.F., 14.17 ton	3.22	1.79	5.01
3400	10,000 S.F., 28.33 ton	3.63	1.78	5.41
3480	Medical centers, 500 S.F., 1.17 ton	5.85	1.56	7.41
3520	1,000 S.F., 2.33 ton	5.85	1.56	7.41
3560	1,500 S.F., 3.50 ton	3.39	1.38	4.77
3640	5,000 S.F., 11.67 ton	2.93	1.48	4.41
3680	10,000 S.F., 23.33 ton	3.14	1.48	4.62
3760	Offices, 500 S.F., 1.58 ton	7.95	2.12	10.07
3800	1,000 S.F., 3.17 ton	4.61	1.88	6.49
3840	1,500 S.F., 4.75 ton	3.95	2.08	6.03
3880	3,000 S.F., 9.50 ton	3.98	2.02	6
3920	5,000 S.F., 15.83 ton	3.60	2.01	5.61
3960	10,000 S.F., 31.67 ton	4.07	2	6.07
4000	Restaurants, 500 S.F., 2.50 ton	12.55	3.36	15.91
4040	1,000 S.F., 5.00 ton	6.25	3.29	9.54
4080	1,500 S.F., 7.50 ton	6.30	3.18	9.48
4120	3,000 S.F., 15.00 ton	5.70	3.17	8.87
4160	5,000 S.F., 25.00 ton	6.75	3.18	9.93
4200	10,000 S.F., 50.00 ton	6.15	3.17	9.32
4240	Schools and colleges, 500 S.F., 1.92 ton	9.65	2.57	12.22
4280	1,000 S.F., 3.83 ton	5.55	2.28	7.83
4360	3,000 S.F., 11.50 ton	4.81	2.44	7.25
4400	5,000 S.F., 19.17 ton	5.15	2.43	7.58
4441	10,000 S.F., 38.33 ton	4.82	2.43	7.25

Important: See the Reference Section for critical supporting data - Reference Nos., Crews & Location Factors

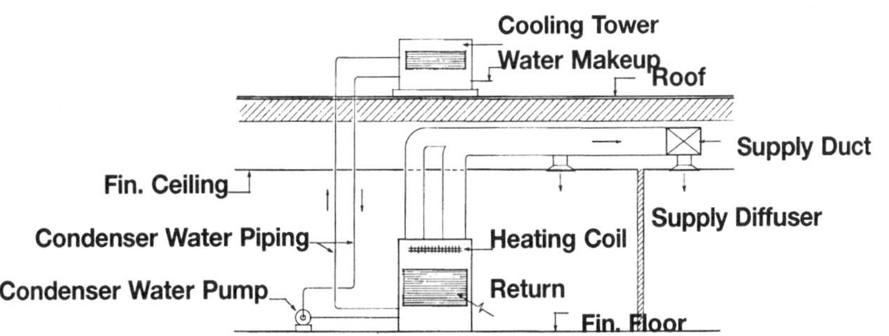

System Description: Self-contained, single package water cooled units include cooling tower, pump, piping allowance. Systems for 1000 S.F. and up include duct and diffusers to provide for even distribution of air. Smaller units distribute air through a supply air plenum, which is integral with the unit.

Returns are not ducted and supplies are not insulated.

Hot water or steam heating coils are included but piping to boiler and the boiler itself is not included.

Where local codes or conditions permit single pass cooling for the smaller units, deduct 10%.

8.4-230	Self-contained, Water Cooled Unit Systems	COST PER S.F.		
		MAT.	INST.	TOTAL
1280	Self-contained, water cooled unit	2.54	1.18	3.72
1300	Apartment corridors, 500 S.F., .92 ton	2.54	1.19	3.73
1320	1,000 S.F., 1.83 ton	2.53	1.18	3.71
1360	3,000 S.F., 5.50 ton	1.90	1	2.90
1400	5,000 S.F., 9.17 ton	1.93	.94	2.87
1440	10,000 S.F., 18.33 ton	1.76	.84	2.60
1520	Banks or libraries, 500 S.F., 2.08 ton	5.45	1.16	6.61
1560	1,000 S.F., 4.17 ton	4.34	2.28	6.62
1600	3,000 S.F., 12.50 ton	4.41	2.14	6.55
1640	5,000 S.F., 20.80 ton	4.01	1.92	5.93
1680	10,000 S.F., 41.66 ton	3.51	1.92	5.43
1760	Bars and taverns, 500 S.F., 5.54 ton	10.65	1.96	12.61
1800	1,000 S.F., 11.08 ton	11.25	3.64	14.89
1840	3,000 S.F., 33.25 ton	9.25	2.89	12.14
1880	5,000 S.F., 55.42 ton	8.70	3.07	11.77
1920	10,000 S.F., 110.00 ton	8.50	3.03	11.53
2000	Bowling alleys, 500 S.F., 2.83 ton	7.40	1.57	8.97
2040	1,000 S.F., 5.66 ton	5.90	3.10	9
2080	3,000 S.F., 17.00 ton	5.45	2.61	8.06
2120	5,000 S.F., 28.33 ton	4.97	2.53	7.50
2160	10,000 S.F., 56.66 ton	4.67	2.61	7.28
2200	Department stores, 500 S.F., 1.46 ton	3.83	.81	4.64
2240	1,000 S.F., 2.92 ton	3.04	1.59	4.63
2280	3,000 S.F., 8.75 ton	3.09	1.50	4.59
2320	5,000 S.F., 14.58 ton	2.82	1.34	4.16
2360	10,000 S.F., 29.17 ton	2.56	1.30	3.86
2440	Drug stores, 500 S.F., 3.33 ton	8.75	1.86	10.61
2480	1,000 S.F., 6.66 ton	6.95	3.65	10.60
2520	3,000 S.F., 20.00 ton	6.45	3.08	9.53
2560	5,000 S.F., 33.33 ton	6.75	3.04	9.79
2600	10,000 S.F., 66.66 ton	5.50	3.07	8.57
2680	Factories, 500 S.F., 1.66 ton	4.36	.93	5.29
2720	1,000 S.F. 3.37 ton	3.51	1.85	5.36
2760	3,000 S.F., 10.00 ton	3.53	1.71	5.24
2800	5,000 S.F., 16.66 ton	3.22	1.54	4.76
2840	10,000 S.F., 33.33 ton	2.92	1.49	4.41
2920	Food supermarkets, 500 S.F., 1.42 ton	3.70	.79	4.49
2960	1,000 S.F., 2.83 ton	3.92	1.84	5.76
3000	3,000 S.F., 8.50 ton	2.95	1.55	4.50
3040	5,000 S.F., 14.17 ton	3	1.45	4.45

8.4-230	Self-contained, Water Cooled Unit Systems	COST PER S.F.		
		MAT.	**INST.**	**TOTAL**
3080	10,000 S.F., 28.33 ton	2.48	1.26	3.74
3160	Medical centers, 500 S.F., 1.17 ton	3.05	.65	3.70
3200	1,000 S.F., 2.33 ton	3.23	1.51	4.74
3240	3,000 S.F., 7.00 ton	2.43	1.27	3.70
3280	5,000 S.F., 11.66 ton	2.46	1.20	3.66
3320	10,000 S.F., 23.33 ton	2.25	1.07	3.32
3400	Offices, 500 S.F., 1.58 ton	4.15	.88	5.03
3440	1,000 S.F., 3.17 ton	4.39	2.05	6.44
3480	3,000 S.F., 9.50 ton	3.35	1.62	4.97
3520	5,000 S.F., 15.83 ton	3.06	1.46	4.52
3560	10,000 S.F., 31.67 ton	2.78	1.41	4.19
3640	Restaurants, 500 S.F., 2.50 ton	6.55	1.39	7.94
3680	1,000 S.F., 5.00 ton	5.25	2.74	7.99
3720	3,000 S.F., 15.00 ton	5.30	2.57	7.87
3760	5,000 S.F., 25.00 ton	4.84	2.31	7.15
3800	10,000 S.F., 50.00 ton	3.04	2.13	5.17
3880	Schools and colleges, 500 S.F., 1.92 ton	5	1.07	6.07
3920	1,000 S.F., 3.83 ton	4	2.10	6.10
3960	3,000 S.F., 11.50 ton	4.06	1.97	6.03
4000	5,000 S.F., 19.17 ton	3.70	1.77	5.47
4040	10,000 S.F., 38.33 ton	3.24	1.77	5.01

Important: See the Reference Section for critical supporting data - Reference Nos., Crews & Location Factors

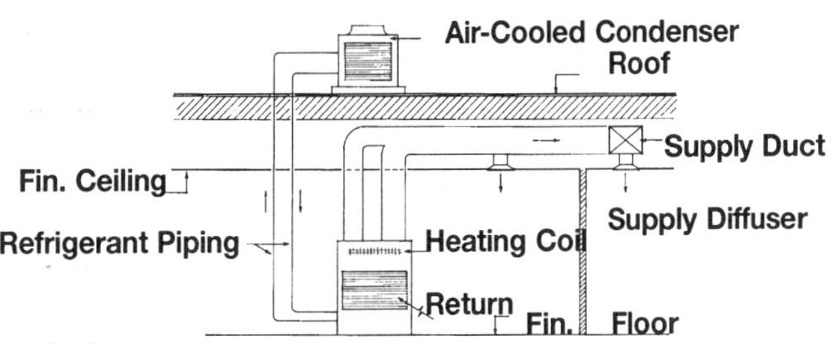

System Description: Self-contained air cooled units with remote air cooled condenser and interconnecting tubing. Systems for 1000 S.F. and up include duct and diffusers. Smaller units distribute air directly.

Returns are not ducted and supplies are not insulated.

Potential savings may be realized by using a single zone rooftop system or through-the-wall unit, especially in the smaller capacities, if the application permits.

Hot water or steam heating coils are included but piping to boiler and the boiler itself is not included.

Condenserless models are available for 15% less where remote refrigerant source is available.

8.4-240	Self-contained, Air Cooled Unit Systems	COST PER S.F.		
		MAT.	INST.	TOTAL
1300	Self-contained, air cooled unit			
1320	Apartment corridors, 500 S.F., .92 ton	4.12	1.57	5.69
1360	1,000 S.F., 1.83 ton	4.06	1.56	5.62
1400	3,000 S.F., 5.50 ton	3.27	1.44	4.71
1440	5,000 S.F., 9.17 ton	2.82	1.37	4.19
1480	10,000 S.F., 18.33 ton	2.40	1.27	3.67
1560	Banks or libraries, 500 S.F., 2.08 ton	8.90	1.99	10.89
1600	1,000 S.F., 4.17 ton	7.40	3.26	10.66
1640	3,000 S.F., 12.50 ton	6.40	3.12	9.52
1680	5,000 S.F., 20.80 ton	5.50	2.88	8.38
1720	10,000 S.F., 41.66 ton	5	2.85	7.85
1800	Bars and taverns, 500 S.F., 5.54 ton	16.05	4.35	20.40
1840	1,000 S.F., 11.08 ton	16.60	6.25	22.85
1880	3,000 S.F., 33.25 ton	13.10	5.50	18.60
1920	5,000 S.F., 55.42 ton	12.60	5.65	18.25
1960	10,000 S.F., 110.00 ton	12.75	5.55	18.30
2040	Bowling alleys, 500 S.F., 2.83 ton	12.15	2.73	14.88
2080	1,000 S.F., 5.66 ton	10.10	4.44	14.54
2120	3,000 S.F., 17.00 ton	7.45	3.92	11.37
2160	5,000 S.F., 28.33 ton	6.90	3.85	10.75
2200	10,000 S.F., 56.66 ton	6.65	3.91	10.56
2240	Department stores, 500 S.F., 1.46 ton	6.25	1.40	7.65
2280	1,000 S.F., 2.92 ton	5.20	2.29	7.49
2320	3,000 S.F., 8.75 ton	4.48	2.19	6.67
2360	5,000 S.F., 14.58 ton	4.48	2.19	6.67
2400	10,000 S.F., 29.17 ton	3.56	1.99	5.55
2480	Drug stores, 500 S.F., 3.33 ton	14.30	3.20	17.50
2520	1,000 S.F., 6.66 ton	11.85	5.20	17.05
2560	3,000 S.F., 20.00 ton	8.80	4.61	13.41
2600	5,000 S.F., 33.33 ton	8.15	4.53	12.68
2640	10,000 S.F., 66.66 ton	7.90	4.63	12.53
2720	Factories, 500 S.F., 1.66 ton	7.25	1.63	8.88
2760	1,000 S.F., 3.33 ton	6	2.62	8.62
2800	3,000 S.F., 10.00 ton	5.10	2.49	7.59
2840	5,000 S.F., 16.66 ton	4.34	2.31	6.65
2880	10,000 S.F., 33.33 ton	4.06	2.27	6.33
2960	Food supermarkets, 500 S.F., 1.42 ton	6.05	1.37	7.42
3000	1,000 S.F., 2.83 ton	6.25	2.41	8.66
3040	3,000 S.F., 8.50 ton	5.05	2.23	7.28
3080	5,000 S.F., 14.17 ton	4.35	2.13	6.48

MECHANICAL

8

8.4-240	Self-contained, Air Cooled Unit Systems	COST PER S.F.		
		MAT.	INST.	TOTAL
3120	10,000 S.F., 28.33 ton	3.46	1.93	5.39
3200	Medical centers, 500 S.F., 1.17 ton	5	1.13	6.13
3240	1,000 S.F., 2.33 ton	5.20	1.99	7.19
3280	3,000 S.F., 7.00 ton	4.16	1.83	5.99
3320	5,000 S.F., 16.66 ton	3.60	1.75	5.35
3360	10,000 S.F., 23.33 ton	3.06	1.61	4.67
3440	Offices, 500 S.F., 1.58 ton	6.80	1.53	8.33
3480	1,000 S.F., 3.16 ton	7.10	2.70	9.80
3520	3,000 S.F., 9.50 ton	4.88	2.37	7.25
3560	5,000 S.F., 15.83 ton	4.17	2.20	6.37
3600	10,000 S.F., 31.66 ton	3.87	2.15	6.02
3680	Restaurants, 500 S.F., 2.50 ton	10.65	2.40	13.05
3720	1,000 S.F., 5.00 ton	8.95	3.92	12.87
3760	3,000 S.F., 15.00 ton	7.70	3.75	11.45
3800	5,000 S.F., 25.00 ton	6.05	3.40	9.45
3840	10,000 S.F., 50.00 ton	6.10	3.43	9.53
3920	Schools and colleges, 500 S.F., 1.92 ton	8.20	1.84	10.04
3960	1,000 S.F., 3.83 ton	6.90	3.01	9.91
4000	3,000 S.F., 11.50 ton	5.90	2.87	8.77
4040	5,000 S.F., 19.17 ton	5.05	2.66	7.71
4080	10,000 S.F., 38.33 ton	4.62	2.63	7.25

Important: See the Reference Section for critical supporting data - Reference Nos., Crews & Location Factors

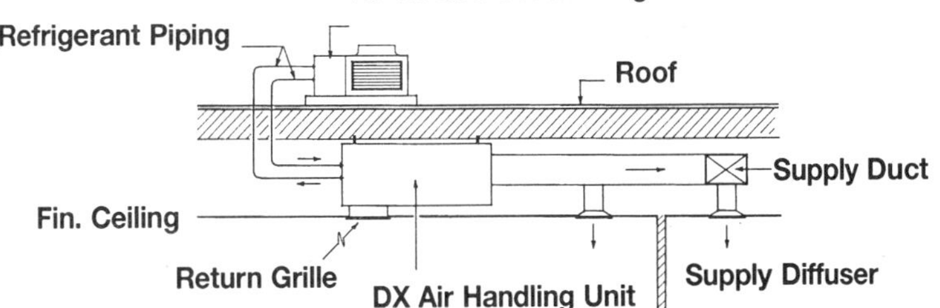

General: Split systems offer several important advantages which should be evaluated when a selection is to be made. They provide a greater degree of flexibility in component selection which permits an accurate match-up of the proper equipment size and type with the particular needs of the building. This allows for maximum use of modern energy saving concepts in heating and cooling. Outdoor installation of the air cooled condensing unit allows space savings in the building and also isolates the equipment operating sounds from building occupants.

Design Assumptions: The systems below are comprised of a direct expansion air handling unit and air cooled condensing unit with interconnecting copper tubing. Ducts and diffusers are also included for distribution of air. Systems are priced for cooling only. Heat can be added as desired either by putting hot water/steam coils into the air unit or into the duct supplying the particular area of need. Gas fired duct furnaces are also available. Refrigerant liquid line is insulated.
* Use multiple systems.

8.4-250	Split Systems With Air Cooled Condensing Units	COST PER S.F.		
		MAT.	INST.	TOTAL
1260	Split system, air cooled condensing unit			
1280	Apartment corridors, 1,000 S.F., 1.83 ton	1.65	.98	2.63
1320	2,000 S.F., 3.66 ton	1.45	.98	2.43
1360	5,000 S.F., 9.17 ton	1.58	1.24	2.82
1400	10,000 S.F., 18.33 ton	1.61	1.34	2.95
1440	20,000 S.F., 36.66 ton	1.64	1.36	3
1520	Banks and libraries, 1,000 S.F., 4.17 ton	3.31	2.25	5.56
1560	2,000 S.F., 8.33 ton	3.62	2.82	6.44
1600	5,000 S.F., 20.80 ton	3.67	3.03	6.70
1640	10,000 S.F., 41.66 ton	3.75	3.10	6.85
1680	20,000 S.F., 83.32 ton	4.55	3.14	7.69
1760	Bars and taverns, 1,000 S.F., 11.08 ton	8.80	4.42	13.22
1800	2,000 S.F., 22.16 ton	10.75	5.15	15.90
1840	5,000 S.F., 55.42 ton	8.80	4.88	13.68
1880	10,000 S.F., 110.84 ton	11.80	4.97	16.77
1920	20,000 S.F., 220 ton*			
2000	Bowling alleys, 1,000 S.F., 5.66 ton	4.62	4.22	8.84
2040	2,000 S.F., 11.33 ton	4.91	3.83	8.74
2080	5,000 S.F., 28.33 ton	4.99	4.13	9.12
2120	10,000 S.F., 56.66 ton	5.10	4.21	9.31
2160	20,000 S.F., 113.32 ton	7	4.43	11.43
2320	Department stores, 1,000 S.F., 2.92 ton	2.33	1.55	3.88
2360	2,000 S.F., 5.83 ton	2.38	2.17	4.55
2400	5,000 S.F., 14.58 ton	2.54	1.98	4.52
2440	10,000 S.F., 29.17 ton	2.57	2.13	4.70
2480	20,000 S.F., 58.33 ton	2.63	2.16	4.79
2560	Drug stores, 1,000 S.F., 6.66 ton	5.45	4.96	10.41
2600	2,000 S.F., 13.32 ton	5.80	4.51	10.31
2640	5,000 S.F., 33.33 ton	6	4.95	10.95
2680	10,000 S.F., 66.66 ton	7.40	5.05	12.45
2720	20,000 S.F., 133.32 ton*			
2800	Factories, 1,000 S.F., 3.33 ton	2.65	1.77	4.42
2840	2,000 S.F., 6.66 ton	2.72	2.48	5.20
2880	5,000 S.F., 16.66 ton	2.93	2.43	5.36
2920	10,000 S.F., 33.33 ton	2.99	2.47	5.46
2960	20,000 S.F., 66.66 ton	3.70	2.52	6.22
3040	Food supermarkets, 1,000 S.F., 2.83 ton	2.25	1.50	3.75
3080	2,000 S.F., 5.66 ton	2.31	2.11	4.42
3120	5,000 S.F., 14.66 ton	2.47	1.92	4.39
3160	10,000 S.F., 28.33 ton	2.50	2.07	4.57

8.4-250	Split Systems With Air Cooled Condensing Units	COST PER S.F.		
		MAT.	INST.	TOTAL
3200	20,000 S.F., 56.66 ton	2.56	2.10	4.66
3280	Medical centers, 1,000 S.F., 2.33 ton	1.88	1.20	3.08
3320	2,000 S.F., 4.66 ton	1.90	1.73	3.63
3360	5,000 S.F., 11.66 ton	2.03	1.57	3.60
3400	10,000 S.F., 23.33 ton	2.06	1.70	3.76
3440	20,000 S.F., 46.66 ton	2.10	1.73	3.83
3520	Offices, 1,000 S.F., 3.17 ton	2.52	1.68	4.20
3560	2,000 S.F., 6.33 ton	2.58	2.35	4.93
3600	5,000 S.F., 15.83 ton	2.75	2.13	4.88
3640	10,000 S.F., 31.66 ton	2.79	2.31	5.10
3680	20,000 S.F., 63.32 ton	3.52	2.39	5.91
3760	Restaurants, 1,000 S.F., 5.00 ton	4.08	3.74	7.82
3800	2,000 S.F., 10.00 ton	4.34	3.39	7.73
3840	5,000 S.F., 25.00 ton	4.41	3.66	8.07
3880	10,000 S.F., 50.00 ton	4.50	3.72	8.22
3920	20,000 S.F., 100.00 ton	6.15	3.92	10.07
4000	Schools and colleges, 1,000 S.F., 3.83 ton	3.03	2.07	5.10
4040	2,000 S.F., 7.66 ton	3.33	2.59	5.92
4080	5,000 S.F., 19.17 ton	3.37	2.80	6.17
4120	10,000 S.F., 38.33 ton	3.44	2.85	6.29
4161	20,000 S.F., 76.66 ton	4.28	2.88	7.16

Important: See the Reference Section for critical supporting data - Reference Nos., Crews & Location Factors

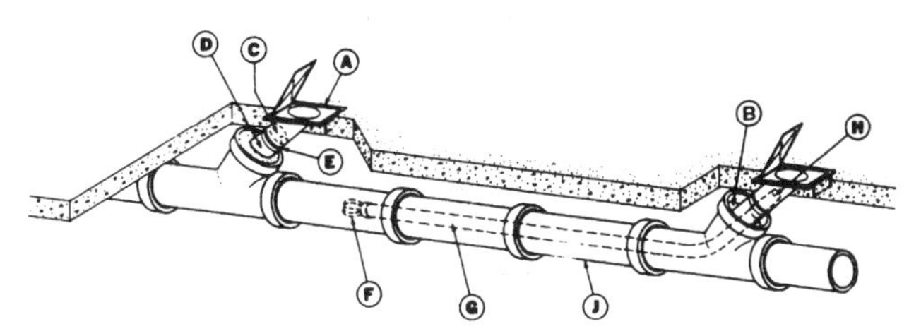

Vitrified Clay Garage Exhaust System

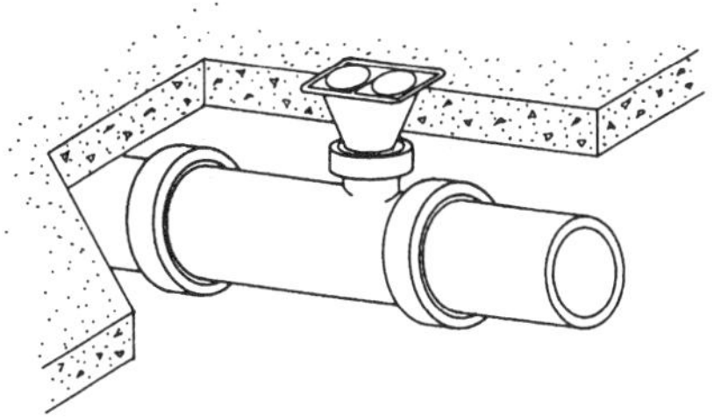

Dual Exhaust System

8.5-110	Garage Exhaust Systems	COST PER BAY		
		MAT.	INST.	TOTAL
1040	Garage, single exhaust, 3″ outlet, cars & light trucks, one bay	1,675	525	2,200
1060	Additional bays up to seven bays	305	78	383
1500	4″ outlet, trucks, one bay	1,675	525	2,200
1520	Additional bays up to six bays	315	78	393
1600	5″ outlet, diesel trucks, one bay	1,725	525	2,250
1650	Additional single bays up to six	375	90.50	465.50
1700	Two adjoining bays	1,725	525	2,250
2000	Dual exhaust, 3″ outlets, pair of adjoining bays	1,975	590	2,565
2100	Additional pairs of adjoining bays	570	90.50	660.50

For information about Means Estimating Seminars, see yellow pages 11 and 12 in back of book

Division 9
Electrical

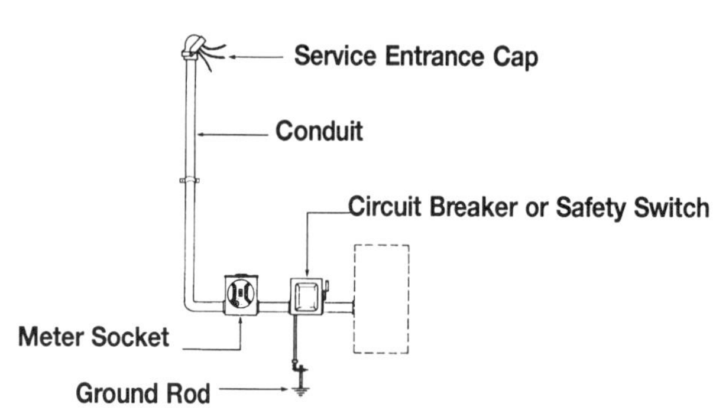

9.1-210	Electric Service, 3 Phase - 4 Wire	COST EACH		
		MAT.	INST.	TOTAL
0200	Service installation, includes breakers, metering, 20' conduit & wire			
0220	3 phase, 4 wire, 120/208 volts, 60 amp	500	400	900
0240	100 amps	620	480	1,100
0280	200 amps	920	740	1,660
0320	400 amps	2,150	1,350	3,500
0360	600 amps	4,100	1,825	5,925
0400	800 amps	5,375	2,200	7,575
0440	1000 amps	6,875	2,525	9,400
0480	1200 amps	8,200	2,600	10,800
0520	1600 amps	16,800	3,725	20,525
0560	2000 amps	18,600	4,250	22,850
0570	Add 25% for 277/480 volt			

9.1-310	Feeder Installation	COST PER L.F.		
		MAT.	INST.	TOTAL
0200	Feeder installation 600 volt, including conduit and wire, 60 amperes	3.14	4.81	7.95
0240	100 amperes	5.70	6.35	12.05
0280	200 amperes	12.30	9.85	22.15
0320	400 amperes	24.50	19.75	44.25
0360	600 amperes	51.50	32	83.50
0400	800 amperes	66	38.50	104.50
0440	1000 amperes	84.50	49	133.50
0480	1200 amperes	94	50	144
0520	1600 amperes	132	76.50	208.50
0560	2000 amperes	169	98.50	267.50

9.1-410	Switchgear	COST EACH		
		MAT.	INST.	TOTAL
0200	Switchgear inst., incl. swbd., panels & circ bkr, 400 amps, 120/208volt	3,375	1,625	5,000
0240	600 amperes	7,875	2,200	10,075
0280	800 amperes	10,100	3,125	13,225
0320	1200 amperes	13,600	4,800	18,400
0360	1600 amperes	18,900	6,725	25,625
0400	2000 amperes	23,600	8,550	32,150
0410	Add 20% for 277/480 volt			

Important: See the Reference Section for critical supporting data - Reference Nos., Crews & Location Factors

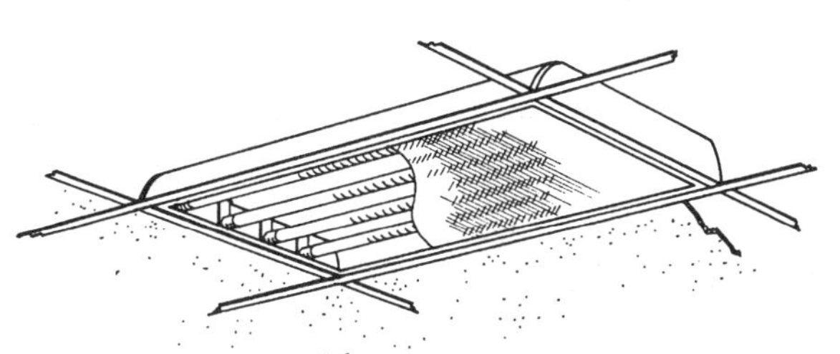

Type C. Recessed, mounted on grid ceiling suspension system, 2′ x 4′, four 40 watt lamps, acrylic prismatic diffusers.

5.3 watts per S.F. for 100 footcandles.

3 watts per S.F. for 57 footcandles.

9.2-213	Fluorescent Fixtures (by Wattage)	COST PER S.F.		
		MAT.	INST.	TOTAL
0190	Fluorescent fixtures recess mounted in ceiling			
0200	1 watt per S.F., 20 FC, 5 fixtures per 1000 S.F.	.53	.76	1.29
0240	2 watts per S.F., 40 FC, 10 fixtures per 1000 S.F.	1.06	1.49	2.55
0280	3 watts per S.F., 60 FC, 15 fixtures per 1000 S.F	1.59	2.25	3.84
0320	4 watts per S.F., 80 FC, 20 fixtures per 1000 S.F.	2.11	2.98	5.09
0400	5 watts per S.F., 100 FC, 25 fixtures per 1000 S.F.	2.64	3.74	6.38
0402				

9.2-223	Incandescent Fixture (by Wattage)	COST PER S.F.		
		MAT.	INST.	TOTAL
0190	Incandescent fixture recess mounted, type A			
0200	1 watt per S.F., 8 FC, 6 fixtures per 1000 S.F.	.51	.61	1.12
0240	2 watt per S.F., 16 FC, 12 fixtures per 1000 S.F.	1.05	1.21	2.26
0280	3 watt per S.F., 24 FC, 18 fixtures, per 1000 S.F.	1.57	1.80	3.37
0320	4 watt per S.F., 32 FC, 24 fixtures per 1000 S.F.	2.09	2.41	4.50
0400	5 watt per S.F., 40 FC, 30 fixtures per 1000 S.F.	2.60	3.03	5.63

9.2-235	H.I.D. Fixture, High Bay (by Wattage)	COST PER S.F.		
		MAT.	INST.	TOTAL
0190	High intensity discharge fixture, 16′ above work plane			
0200	1 watt/S.F., type D, 23 FC, 1 fixture/1000 S.F.	.65	.60	1.25
0240	Type E, 42 FC, 1 fixture/1000 S.F.	.77	.63	1.40
0280	Type G, 52 FC, 1 fixture/1000 S.F.	.77	.63	1.40
0320	Type C, 54 FC, 2 fixture/1000 S.F.	.90	.72	1.62
0400	2 watt/S.F., type D, 45 FC, 2 fixture/1000 S.F.	1.30	1.23	2.53
0440	Type E, 84 FC, 2 fixture/1000 S.F.	1.55	1.31	2.86
0480	Type G, 105 FC, 2 fixture/1000 S.F.	1.55	1.31	2.86
0520	Type C, 108 FC, 4 fixture/1000 S.F.	1.82	1.42	3.24
0600	3 watt/S.F., type D, 68 FC, 3 fixture/1000 S.F.	1.95	1.81	3.76
0640	Type E, 126 FC, 3 fixture/1000 S.F.	2.33	1.94	4.27
0680	Type G, 157 FC, 3 fixture/1000 S.F.	2.33	1.94	4.27
0720	Type C, 162 FC, 6 fixture/1000 S.F.	2.72	2.13	4.85
0800	4 watt/ S.F., type D, 91 FC, 4 fixture/1000 S.F.	2.80	2.85	5.65
0840	Type E, 168 FC, 4 fixture/1000 S.F.	3.10	2.59	5.69
0880	Type G, 210 FC, 4 fixture/1000 S.F.	3.10	2.59	5.69
0920	Type C, 243 FC, 9 fixture/1000 S.F.	3.98	2.98	6.96
1000	5 watt/S.F., type D, 113 FC, 5 fixture/1000 S.F.	3.26	3.03	6.29
1040	Type E, 210 FC, 5 fixture/1000 S.F.	3.89	3.23	7.12
1080	Type G, 262 FC, 5 fixture/1000 S.F.	3.89	3.23	7.12
1120	Type C, 297 FC, 11 fixture/1000 S.F.	4.89	3.68	8.57

ELECTRICAL

9

9.2-239	H.I.D. Fixture, High Bay (by Wattage)	COST PER S.F.		
		MAT.	INST.	TOTAL
0190	High intensity discharge fixture, 30' above work plane			
0200	1 watt/S.F., type D, 23 FC, 1 fixture/1000 S.F.	.74	.78	1.52
0240	Type E, 37 FC, 1 fixture/1000 S.F.	.86	.81	1.67
0280	Type G, 45 FC., 1 fixture/1000 S.F.	.86	.81	1.67
0320	Type F, 50 FC, 1 fixture/1000 S.F.	.72	.62	1.34
0400	2 watt/S.F., type D, 40 FC, 2 fixtures/1000 S.F.	1.44	1.55	2.99
0440	Type E, 74 FC, 2 fixtures/1000 S.F.	1.68	1.62	3.30
0480	Type G, 92 FC, 2 fixtures/1000 S.F.	1.68	1.62	3.30
0520	Type F, 100 FC, 2 fixtures/1000 S.F.	1.45	1.29	2.74
0600	3 watt/S.F., type D, 60 FC, 3 fixtures/1000 S.F.	2.19	2.32	4.51
0640	Type E, 110 FC, 3 fixtures/1000 S.F.	2.56	2.45	5.01
0680	Type G, 138FC, 3 fixtures/1000 S.F.	2.56	2.45	5.01
0720	Type F, 150 FC, 3 fixtures/1000 S.F.	2.18	1.93	4.11
0800	4 watt/ S.F., type D, 80 FC, 4 fixtures/1000 S.F.	2.90	3.07	5.97
0840	Type E, 148 FC, 4 fixtures/1000 S.F.	3.40	3.25	6.65.
0880	Type G, 185 FC, 4 fixtures/1000 S.F.	3.40	3.25	6.65
0920	Type F, 200 FC, 4 fixtures/1000 S.F.	2.92	2.58	5.50
1000	5 watt/ S.F., type D, 100 FC 5 fixtures/1000 S.F.	3.64	3.86	7.50
1040	Type E, 185 FC, 5 fixtures/1000 S.F.	4.26	4.07	8.33
1080	Type G, 230 FC, 5 fixtures/1000 S.F.	4.26	4.07	8.33
1120	Type F, 250 FC, 5 fixtures/1000 S.F.	3.64	3.21	6.85
1121				

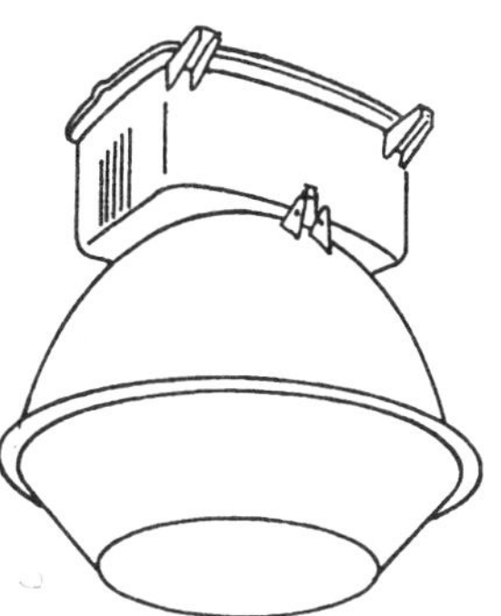

LOW BAY FIXTURES
H. Mercury vapor 250 watt
J. Metal halide 250 watt
K. High pressure sodium 150 watt

9.2-242	H.I.D. Fixture, Low Bay (by Wattage)	COST PER S.F.		
		MAT.	INST.	TOTAL
0190	High intensity discharge fixture, 8'-10' above work plane			
0200	1 watt/S.F., type H, 19 FC, 4 fixtures/1000 S.F.	1.70	1.30	3
0240	Type J, 30 FC, 4 fixtures/1000 S.F.	1.91	1.31	3.22
0280	Type K, 29 FC, 5 fixtures/1000 S.F.	1.83	1.18	3.01
0360	2 watt/S.F. type H, 33 FC, 7 fixtures/1000 S.F.	3.08	2.50	5.58
0400	Type J, 52 FC, 7 fixtures/1000 S.F.	3.39	2.41	5.80
0440	Type K, 63 FC, 11 fixtures/1000 S.F.	3.97	2.49	6.46
0520	3 watt/S.F., type H, 51 FC, 11 fixtures/1000 S.F.	4.78	3.84	8.62
0560	Type J, 81 FC, 11 fixtures/1000 S.F.	5.25	3.65	8.90
0600	Type K, 92 FC, 16 fixtures/1000 S.F.	5.80	3.67	9.47
0680	4 watt/S.F., type H, 65 FC, 14 fixtures/1000 S.F.	6.15	5.05	11.20
0720	Type J, 103 FC, 14 fixtures/1000 S.F.	6.75	4.81	11.56
0760	Type K, 127 FC, 22 fixtures/1000 S.F.	7.95	4.97	12.92
0840	5 watt/S.F., type H, 84 FC, 18 fixtures/1000 S.F.	7.85	6.35	14.20
0880	Type J, 133 FC, 18 fixtures/1000 S.F.	8.65	6.05	14.70
0920	Type K, 155 FC, 27 fixtures/1000 S.F.	9.75	6.15	15.90

9.2-244	H.I.D. Fixture, Low Bay (by Wattage)	COST PER S.F.		
		MAT.	INST.	TOTAL
0190	High intensity discharge fixture, mounted 16' above work plane			
0200	1 watt/S.F., type H, 19 FC, 4 fixtures/1000 S.F.	1.78	1.48	3.26
0240	Type J, 28 FC, 4 fixt./1000 S.F.	1.99	1.49	3.48
0280	Type K, 27 FC, 5 fixt./1000 S.F.	2.05	1.67	3.72
0360	2 watts/S.F., type H, 30 FC, 7 fixt/1000 S.F.	3.24	2.86	6.10
0400	Type J, 48 FC, 7 fixt/1000 S.F.	3.60	2.88	6.48
0440	Type K, 58 FC, 11 fixt/1000 S.F.	4.40	3.43	7.83
0520	3 watts/S.F., type H, 47 FC, 11 fixt/1000 S.F.	5.05	4.36	9.41
0560	Type J, 75 FC, 11 fixt/1000 S.F.	5.60	4.39	9.99
0600	Type K, 85 FC, 16 fixt/1000 S.F.	6.45	5.10	11.55
0680	4 watts/S.F., type H, 60 FC, 14 fixt/1000 S.F.	6.50	5.75	12.25
0720	Type J, 95 FC, 14 fixt/1000 S.F.	7.20	5.80	13
0760	Type K, 117 FC, 22 fixt/1000 S.F.	8.80	6.80	15.60
0840	5 watts/S.F., type H, 77 FC, 18 fixt/1000 S.F.	8.35	7.45	15.80
0880	Type J, 122 FC, 18 fixt/1000 S.F.	9.20	7.25	16.45
0920	Type K, 143 FC, 27 fixt/1000 S.F.	10.85	8.50	19.35

ELECTRICAL

9

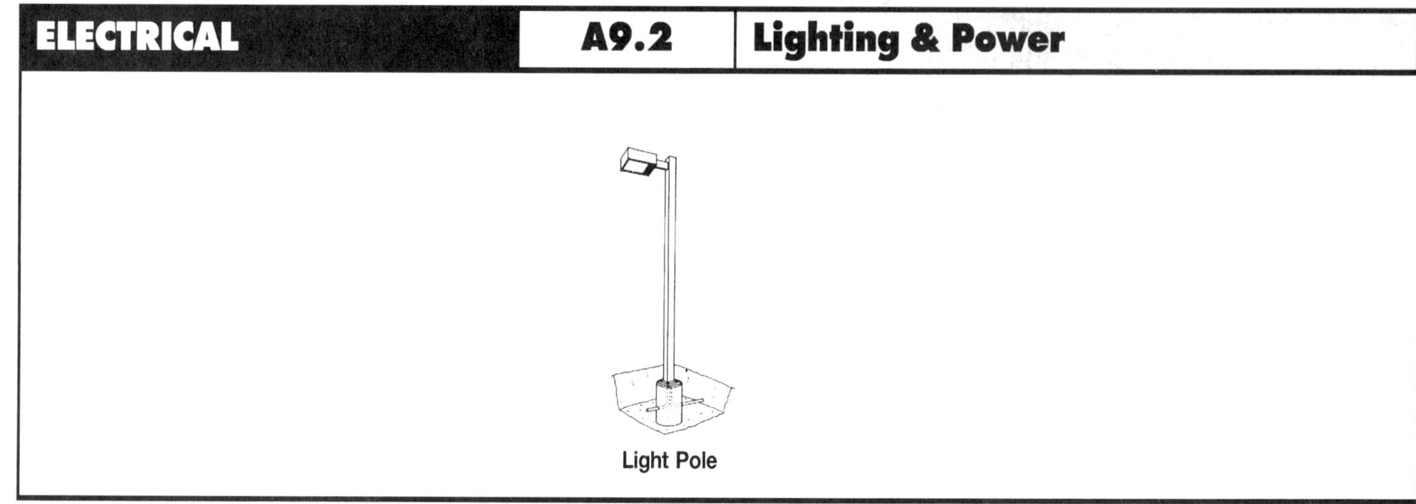

Light Pole

9.2-252	Light Pole (Installed)	COST EACH		
		MAT.	INST.	TOTAL
0200	Light pole, aluminum, 20' high, 1 arm bracket	780	485	1,265
0240	2 arm brackets	865	485	1,350
0280	3 arm brackets	945	500	1,445
0320	4 arm brackets	1,025	500	1,525
0360	30' high, 1 arm bracket	1,375	605	1,980
0400	2 arm brackets	1,450	605	2,055
0440	3 arm brackets	1,525	620	2,145
0480	4 arm brackets	1,600	620	2,220
0680	40' high, 1 arm bracket	1,650	810	2,460
0720	2 arm brackets	1,725	810	2,535
0760	3 arm brackets	1,825	825	2,650
0800	4 arm brackets	1,900	825	2,725
0840	Steel, 20' high, 1 arm bracket	980	515	1,495
0880	2 arm brackets	1,050	515	1,565
0920	3 arm brackets	1,075	535	1,610
0960	4 arm brackets	1,150	535	1,685
1000	30' high, 1 arm bracket	1,125	650	1,775
1040	2 arm brackets	1,200	650	1,850
1080	3 arm brackets	1,225	665	1,890
1120	4 arm brackets	1,300	665	1,965
1320	40' high, 1 arm bracket	1,500	875	2,375
1360	2 arm brackets	1,575	875	2,450
1400	3 arm brackets	1,575	890	2,465
1440	4 arm brackets	1,675	890	2,565

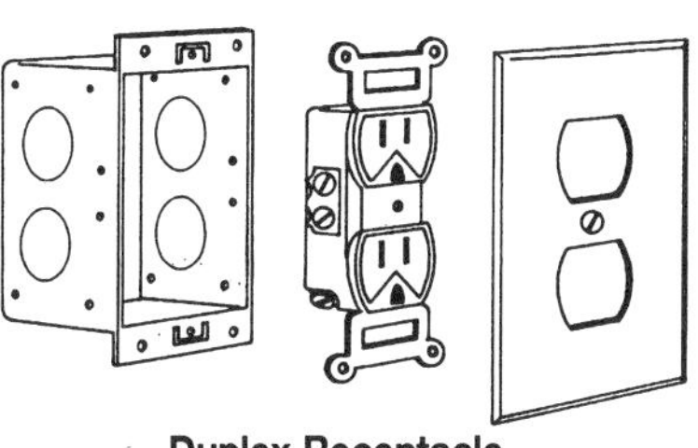

Duplex Receptacle

9.2-522	Receptacle (by Wattage)	COST PER S.F.		
		MAT.	INST.	TOTAL
0190	Receptacles include plate, box, conduit, wire & transformer when required			
0200	2.5 per 1000 S.F., .3 watts per S.F.	.24	.58	.82
0240	With transformer	.27	.61	.88
0280	4 per 1000 S.F., .5 watts per S.F.	.28	.68	.96
0320	With transformer	.33	.72	1.05
0360	5 per 1000 S.F., .6 watts per S.F.	.32	.79	1.11
0400	With transformer	.38	.85	1.23
0440	8 per 1000 S.F., .9 watts per S.F.	.35	.87	1.22
0480	With transformer	.44	.95	1.39
0520	10 per 1000 S.F., 1.2 watts per S.F.	.36	.95	1.31
0560	With transformer	.51	1.08	1.59
0600	16.5 per 1000 S.F., 2.0 watts per S.F.	.44	1.19	1.63
0640	With transformer	.69	1.41	2.10
0680	20 per 1000 S.F., 2.4 watts per S.F.	.47	1.30	1.77
0720	With transformer	.76	1.56	2.32

9.2-524	Receptacles	COST PER S.F.		
		MAT.	INST.	TOTAL
0200	Receptacle systems, underfloor duct, 5' on center, low density	3	1.27	4.27
0240	High density	3.24	1.63	4.87
0280	7' on center, low density	2.41	1.10	3.51
0320	High density	2.65	1.46	4.11
0400	Poke thru fittings, low density	.87	.59	1.46
0440	High density	1.74	1.15	2.89
0520	Telepoles, using Romex, low density	.64	.38	1.02
0560	High density	1.28	.77	2.05
0600	Using EMT, low density	.66	.51	1.17
0640	High density	1.35	1.01	2.36
0720	Conduit system with floor boxes, low density	.73	.43	1.16
0760	High density	1.47	.88	2.35
0840	Undercarpet power system, 3 conductor with 5 conductor feeder, low density	1.03	.17	1.20
0880	High density	2.04	.32	2.36

ELECTRICAL

9

Description: Table 9.2-542 includes the cost for switch, plate, box, conduit in slab or EMT exposed and copper wire. Add 20% for exposed conduit.

No power required for switches.

Federal energy guidelines recommend the maximum lighting area controlled per switch shall not exceed 1000 S.F. and that areas over 500 S.F. shall be so controlled that total illumination can be reduced by at least 50%.

9.2-542	Wall Switch by Sq. Ft.	COST PER S.F.		
		MAT.	INST.	TOTAL
0200	Wall switches, 1.0 per 1000 S.F.	.04	.09	.13
0240	1.2 per 1000 S.F.	.04	.10	.14
0280	2.0 per 1000 S.F.	.05	.16	.21
0320	2.5 per 1000 S.F.	.08	.19	.27
0360	5.0 per 1000 S.F.	.16	.40	.56
0400	10.0 per 1000 S.F.	.34	.80	1.14

9.2-582	Miscellaneous Power	COST PER S.F.		
		MAT.	INST.	TOTAL
0200	Miscellaneous power, to .5 watts	.02	.05	.07
0240	.8 watts	.03	.06	.09
0280	1 watt	.04	.08	.12
0320	1.2 watts	.05	.11	.16
0360	1.5 watts	.06	.12	.18
0400	1.8 watts	.07	.14	.21
0440	2 watts	.08	.16	.24
0480	2.5 watts	.09	.20	.29
0520	3 watts	.11	.24	.35

9.2-610	Central A. C. Power (by Wattage)	COST PER S.F.		
		MAT.	INST.	TOTAL
0200	Central air conditioning power, 1 watt	.05	.10	.15
0220	2 watts	.05	.11	.16
0240	3 watts	.07	.13	.20
0280	4 watts	.10	.17	.27
0320	6 watts	.18	.24	.42
0360	8 watts	.21	.25	.46
0400	10 watts	.28	.29	.57

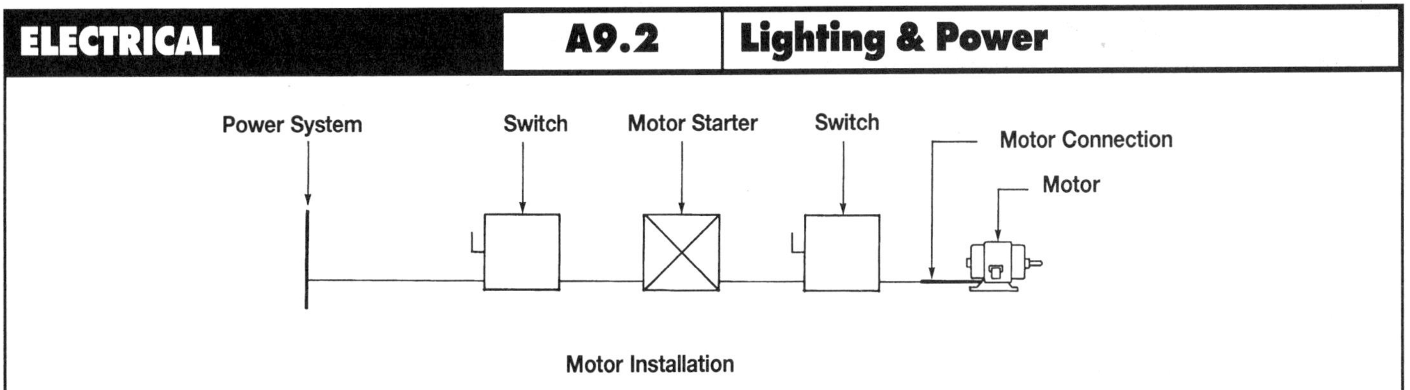

9.2-710	Motor Installation	COST EACH		
		MAT.	INST.	TOTAL
0200	Motor installation, single phase, 115V, to and including 1/3 HP motor size	385	390	775
0240	To and incl. 1 HP motor size	400	390	790
0280	To and incl. 2 HP motor size	435	415	850
0320	To and incl. 3 HP motor size	490	425	915
0360	230V, to and including 1 HP motor size	385	400	785
0400	To and incl. 2 HP motor size	405	400	805
0440	To and incl. 3 HP motor size	455	425	880
0520	Three phase, 200V, to and including 1-1/2 HP motor size	440	435	875
0560	To and incl. 3 HP motor size	475	475	950
0600	To and incl. 5 HP motor size	525	530	1,055
0640	To and incl. 7-1/2 HP motor size	540	540	1,080
0680	To and incl. 10 HP motor size	880	680	1,560
0720	To and incl. 15 HP motor size	1,200	755	1,955
0760	To and incl. 20 HP motor size	1,475	870	2,345
0800	To and incl. 25 HP motor size	1,500	875	2,375
0840	To and incl. 30 HP motor size	2,400	1,025	3,425
0880	To and incl. 40 HP motor size	2,850	1,225	4,075
0920	To and incl. 50 HP motor size	5,200	1,425	6,625
0960	To and incl. 60 HP motor size	5,325	1,500	6,825
1000	To and incl. 75 HP motor size	6,700	1,700	8,400
1040	To and incl. 100 HP motor size	13,700	2,000	15,700
1080	To and incl. 125 HP motor size	14,000	2,200	16,200
1120	To and incl. 150 HP motor size	16,600	2,600	19,200
1160	To and incl. 200 HP motor size	20,000	3,175	23,175
1240	230V, to and including 1-1/2 HP motor size	420	430	850
1280	To and incl. 3 HP motor size	455	470	925
1320	To and incl. 5 HP motor size	505	525	1,030
1360	To and incl. 7-1/2 HP motor size	505	525	1,030
1400	To and incl. 10 HP motor size	805	640	1,445
1440	To and incl. 15 HP motor size	905	700	1,605
1480	To and incl. 20 HP motor size	1,400	845	2,245
1520	To and incl. 25 HP motor size	1,475	870	2,345
1560	To and incl. 30 HP motor size	1,500	875	2,375
1600	To and incl. 40 HP motor size	2,800	1,175	3,975
1640	To and incl. 50 HP motor size	2,925	1,250	4,175
1680	To and incl. 60 HP motor size	5,200	1,425	6,625
1720	To and incl. 75 HP motor size	6,175	1,600	7,775
1760	To and incl. 100 HP motor size	6,950	1,800	8,750
1800	To and incl. 125 HP motor size	14,000	2,075	16,075
1840	To and incl. 150 HP motor size	14,900	2,350	17,250
1880	To and incl. 200 HP motor size	16,000	2,625	18,625
1960	460V, to and including 2 HP motor size	515	435	950
2000	To and incl. 5 HP motor size	555	475	1,030
2040	To and incl. 10 HP motor size	590	525	1,115
2080	To and incl. 15 HP motor size	805	600	1,405
2120	To and incl. 20 HP motor size	825	645	1,470
2160	To and incl. 25 HP motor size	890	675	1,565
2200	To and incl. 30 HP motor size	1,175	730	1,905

ELECTRICAL

9

9.2-710 — Motor Installation

		COST EACH		
		MAT.	INST.	TOTAL
2240	To and incl. 40 HP motor size	1,475	780	2,255
2280	To and incl. 50 HP motor size	1,625	870	2,495
2320	To and incl. 60 HP motor size	2,525	1,025	3,550
2360	To and incl. 75 HP motor size	2,850	1,125	3,975
2400	To and incl. 100 HP motor size	3,100	1,250	4,350
2440	To and incl. 125 HP motor size	5,375	1,425	6,800
2480	To and incl. 150 HP motor size	6,525	1,600	8,125
2520	To and incl. 200 HP motor size	7,375	1,800	9,175
2600	575V, to and including 2 HP motor size	515	435	950
2640	To and incl. 5 HP motor size	555	475	1,030
2680	To and incl. 10 HP motor size	590	525	1,115
2720	To and incl. 20 HP motor size	805	600	1,405
2760	To and incl. 25 HP motor size	825	645	1,470
2800	To and incl. 30 HP motor size	1,175	730	1,905
2840	To and incl. 50 HP motor size	1,250	755	2,005
2880	To and incl. 60 HP motor size	2,500	1,025	3,525
2920	To and incl. 75 HP motor size	2,525	1,025	3,550
2960	To and incl. 100 HP motor size	2,850	1,125	3,975
3000	To and incl. 125 HP motor size	5,300	1,400	6,700
3040	To and incl. 150 HP motor size	5,375	1,425	6,800
3080	To and incl. 200 HP motor size	6,625	1,625	8,250

9.2-720 — Motor Feeder

		COST PER L.F.		
		MAT.	INST.	TOTAL
0200	Motor feeder systems, single phase, feed up to 115V 1HP or 230V 2 HP	1.35	3.05	4.40
0240	115V 2HP, 230V 3HP	1.43	3.10	4.53
0280	115V 3HP	1.62	3.23	4.85
0360	Three phase, feed to 200V 3HP, 230V 5HP, 460V 10HP, 575V 10HP	1.42	3.29	4.71
0440	200V 5HP, 230V 7.5HP, 460V 15HP, 575V 20HP	1.54	3.36	4.90
0520	200V 10HP, 230V 10HP, 460V 30HP, 575V 30HP	1.83	3.56	5.39
0600	200V 15HP, 230V 15HP, 460V 40HP, 575V 50HP	2.36	4.06	6.42
0680	200V 20HP, 230V 25HP, 460V 50HP, 575V 60HP	3.39	5.15	8.54
0760	200V 25HP, 230V 30HP, 460V 60HP, 575V 75HP	3.78	5.25	9.03
0840	200V 30HP	4.20	5.40	9.60
0920	230V 40HP, 460V 75HP, 575V 100HP	5.30	5.90	11.20
1000	200V 40HP	6.10	6.30	12.40
1080	230V 50HP, 460V 100HP, 575V 125HP	7.35	6.95	14.30
1160	200V 50HP, 230V 60HP, 460V 125HP, 575V 150HP	8.60	7.40	16
1240	200V 60HP, 460V 150HP	10.45	8.65	19.10
1320	230V 75HP, 575V 200HP	12.10	9	21.10
1400	200V 75HP	13.80	9.25	23.05
1480	230V 100HP, 460V 200HP	17.75	10.75	28.50
1560	200V 100HP	24.50	13.45	37.95
1640	230V 125HP	28.50	14.55	43.05
1720	200V 125HP, 230V 150HP	29	16.75	45.75
1800	200V 150HP	34	18.15	52.15
1880	200V 200HP	49	27	76
1960	230V 200HP	45.50	19.95	65.45

Important: See the Reference Section for critical supporting data - Reference Nos., Crews & Location Factors

9.4-100	Communication & Alarm Systems	COST EACH		
		MAT.	INST.	TOTAL
0200	Communication & alarm systems, includes outlets, boxes, conduit & wire			
0210	Sound system, 6 outlets	4,050	3,275	7,325
0220	12 outlets	5,600	5,225	10,825
0240	30 outlets	9,675	9,875	19,550
0280	100 outlets	29,000	33,100	62,100
0320	Fire detection systems, 12 detectors	2,150	2,700	4,850
0360	25 detectors	3,775	4,575	8,350
0400	50 detectors	7,250	8,975	16,225
0440	100 detectors	13,200	16,200	29,400
0480	Intercom systems, 6 stations	2,350	2,225	4,575
0560	25 stations	7,975	8,575	16,550
0640	100 stations	30,500	31,400	61,900
0680	Master clock systems, 6 rooms	3,675	3,650	7,325
0720	12 rooms	5,450	6,200	11,650
0760	20 rooms	7,300	8,800	16,100
0800	30 rooms	11,600	16,200	27,800
0840	50 rooms	18,400	27,300	45,700
0920	Master TV antenna systems, 6 outlets	2,050	2,300	4,350
0960	12 outlets	3,825	4,300	8,125
1000	30 outlets	7,525	9,950	17,475
1040	100 outlets	24,900	32,400	57,300

Description: System below tabulates the installed cost for generators by KW. Included in costs are battery, charger, muffler, and transfer switch.

No conduit, wire, or terminations included.

9.4-310	Generators (by KW)	COST PER KW		
		MAT.	INST.	TOTAL
0190	Generator sets, include battery, charger, muffler & transfer switch			
0200	Gas/gasoline operated, 3 phase, 4 wire, 277/480 volt, 7.5 KW	880	128	1,008
0240	10 KW	905	113	1,018
0280	15 KW	735	84	819
0320	30 KW	440	48	488
0360	70 KW	325	28.50	353.50
0400	85 KW	292	28.50	320.50
0440	115 KW	390	24.50	414.50
0480	170 KW	395	18.75	413.75
0560	Diesel engine with fuel tank, 30 KW	585	48	633
0600	50 KW	430	38	468
0640	75 KW	375	30	405
0680	100 KW	315	26	341
0760	150 KW	257	20.50	277.50
0840	200 KW	209	16.60	225.60
0880	250 KW	183	13.90	196.90
0960	350 KW	169	11.40	180.40
1040	500 KW	167	8.85	175.85

For information about Means Estimating Seminars, see yellow pages 11 and 12 in back of book

Important: See the Reference Section for critical supporting data - Reference Nos., Crews & Location Factors

Division 11
Special Construction

11.1-100 — Architectural Specialties/Each

		COST EACH		
		MAT.	**INST.**	**TOTAL**
1000	Specialties, bathroom accessories, st. steel, curtain rod, 5' long, 1" diam	30	18.70	48.70
1100	Dispenser, towel, surface mounted	45	15.20	60.20
1200	Grab bar, 1-1/4" diam., 12" long	39	10.10	49.10
1300	Mirror, framed with shelf, 18" x 24"	99.50	12.15	111.65
1340	72" x 24"	680	40.50	720.50
1400	Toilet tissue dispenser, surface mounted, single roll	10.55	8.10	18.65
1500	Towel bar, 18" long	30	10.55	40.55
1600	Canopies, wall hung, prefinished aluminum, 8' x 10'	1,250	785	2,035
1700	Chutes, linen or refuse incl. sprinklers, galv. steel, 18" diam., per floor	800	149	949
1740	Aluminized steel, 36" diam., per floor	1,075	186	1,261
1800	Mail, 8-3/4" x 3-1/2", aluminum & glass, per floor	570	104	674
1840	Bronze or stainless, per floor	885	116	1,001
1900	Control boards, magnetic, porcelain finish, framed, 24" x 18"	106	60.50	166.50
1940	96" x 48"	480	97	577
2000	Directory boards, outdoor, black plastic, 36" x 24"	605	243	848
2040	36" x 36"	695	325	1,020
2100	Indoor, economy, open faced, 18" x 24"	86.50	69.50	156
2300	Disappearing stairways, folding, pine, 8'-6" ceiling	91.50	60.50	152
2400	Fire escape, galvanized steel, 8'-0" to 10'-6" ceiling	1,225	485	1,710
2500	Automatic electric, wood, 8' to 9' ceiling	6,075	485	6,560
2600	Display cases, freestanding, glass and aluminum, 3'-6" x 3' x 1'-0" deep	490	60.50	550.50
2700	Wall mounted, glass and aluminum, 3' x 4' x 1'-4" deep	940	97	1,037
2900	Fireplace prefabricated, freestanding or wall hung, economy	995	187	1,182
2940	Deluxe	2,875	270	3,145
3000	Woodburning stoves, cast iron, economy	730	325	1,055
3040	Deluxe	2,650	530	3,180
3100	Flagpoles, on grade, aluminum, tapered, 20' high	710	315	1,025
3140	70' high	6,125	785	6,910
3240	59' high	6,150	700	6,850
3300	Concrete, internal halyard, 20' high	1,325	252	1,577
3340	100' high	13,400	625	14,025
3400	Lockers, steel, single tier, 5' to 6' high, per opening, min.	108	18.55	126.55
3440	Maximum	184	21.50	205.50
3700	Mail boxes, horizontal, rear loaded, aluminum, 5" x 6" x 15" deep	37	7.15	44.15
3740	Front loaded, aluminum, 10" x 12" x 15" deep	141	12.15	153.15
3800	Vertical, front loaded, aluminum, 15" x 5" x 6" deep	28.50	7.15	35.65
3840	Bronze, duranodic finish	53.50	7.15	60.65
3900	Letter slot, post office	208	30.50	238.50
4000	Mail counter, window, post office, with grille	560	121	681
4100	Medicine cabinets, sliding mirror doors, 20" x 16" x 4-3/4", unlighted	88.50	34.50	123
4140	24" x 19" x 8-1/2", lighted	149	48.50	197.50
4400	Partitions, shower stall, single wall, painted steel, 2'-8" x 2'-8"	625	104	729
4440	Fiberglass, 2'-8" x 2'-8"	510	116	626
4500	Double wall, enameled steel, 2'-8" x 2'-8"	635	104	739
4540	Stainless steel, 2'-8" x 2'-8"	1,150	104	1,254
4700	Tub enclosure, sliding panels, tempered glass, aluminum frame	275	130	405
4740	Chrome/brass frame-deluxe	785	173	958
5400	Projection screens, wall hung, manual operation, 50 S.F., economy	293	48.50	341.50
5440	Electric operation, 100 S.F., deluxe	1,575	243	1,818
5500	Scales, dial type, built in floor, 5 ton capacity, 8' x 6' platform	5,050	1,450	6,500
5540	10 ton capacity, 9' x 7' platform	9,525	2,075	11,600
5600	Truck (including weigh bridge), 20 ton capacity, 24' x 10'	9,600	2,425	12,025
5900	Security gates-scissors type, painted steel, single, 6' high, 5-1/2' wide	125	150	275
5940	Double gate, 7-1/2' high, 14' wide	400	300	700
6000	Shelving, metal industrial, braced, 3' wide, 1' deep	19.95	5.15	25.10
6040	2' deep	28.50	5.45	33.95
6200	Signs, interior electric exit sign, wall mounted, 6"	41.50	32.50	74
6240	Street, reflective alum., dbl. face, 4 way, w/bracket	79.50	16.20	95.70

Important: See the Reference Section for critical supporting data - Reference Nos., Crews & Location Factors

11.1-100	Architectural Specialties/Each	COST EACH		
		MAT.	INST.	TOTAL
6300	Letters, cast aluminum, 1/2" deep, 4" high	15	13.50	28.50
6340	1" deep, 10" high	45.50	13.50	59
6400	Plaques, cast aluminum, 20" x 30"	890	121	1,011
6440	Cast bronze, 36" x 48"	3,500	243	3,743
6700	Turnstiles, one way, 4' arm, 46" diam., manual	268	97	365
6740	Electric	945	405	1,350
6800	3 arm, 5'-5' diam. & 7' high, manual	2,475	485	2,960
6840	Electric	3,075	810	3,885

11.1-100	Architectural Specialties/S.F.	COST PER S.F.		
		MAT.	INST.	TOTAL
7100	Bulletin board, cork sheets, no frame, 1/4" thick	2.68	1.67	4.35
7200	Aluminum frame, 1/4" thick, 3' x 5'	6.95	2.04	8.99
7500	Chalkboards, wall hung, alum, frame & chalktrough	9.60	1.07	10.67
7540	Wood frame & chalktrough	8.85	1.16	10.01
7600	Sliding board, one board with back panel	31	1.04	32.04
7640	Two boards with back panel	47	1.04	48.04
7700	Liquid chalk type, alum. frame & chalktrough	10.25	1.07	11.32
7740	Wood frame & chalktrough	17.50	1.07	18.57

11.1-100	Architectural Specialties/L.F.	COST PER L.F.		
		MAT.	INST.	TOTAL
8600	Partitions, hospital curtain, ceiling hung, polyester oxford cloth	12.05	2.37	14.42
8640	Designer oxford cloth	23.50	3	26.50

11.1-200	Architectural Equipment/Each	COST EACH		
		MAT.	INST.	TOTAL
0200	Arch. equip., appliances, range, cook top, 4 burner, economy	199	43	242
0240	Built in, single oven 30" wide, economy	455	8.80	463.80
0300	Standing, single oven-21" wide, economy	330	35	365
0340	Double oven-30" wide, deluxe	1,650	35	1,685
0400	Compactor, residential, economy	330	48.50	378.50
0440	Deluxe	460	81	541
0500	Dish washer, built-in, 2 cycles, economy	239	82	321
0540	4 or more cycles, deluxe	710	164	874
0600	Garbage disposer, sink type, economy	40.50	33	73.50
0640	Deluxe	169	33	202
0700	Refrigerator, no frost, 10 to 12 C.F., economy	455	35	490
0740	21 to 29 C.F., deluxe	2,500	117	2,617
0800	Automotive equipment, compressors, electric, 1-1/2 H.P., std. controls	2,700	285	2,985
0840	5 H.P., dual controls	3,800	430	4,230
0900	Hoists, single post, 4 ton capacity, swivel arms	3,850	1,075	4,925
0940	Dual post, 12 ton capacity, adjustable frame	6,600	2,850	9,450
1000	Lube equipment, 3 reel type, with pumps	6,600	855	7,455
1040	Product dispenser, 6 nozzles, w/vapor recovery, not incl. piping, installed	16,500		16,500
1100	Bank equipment, drive up window, drawer & mike, no glazing, economy	5,250	600	5,850
1140	Deluxe	7,625	1,200	8,825
1200	Night depository, economy	6,800	600	7,400
1240	Deluxe	10,800	1,200	12,000
1300	Pneumatic tube systems, 2 station, standard	20,800	1,225	22,025
1340	Teller, automated, 24 hour, single unit	39,200	1,225	40,425
1400	Teller window, bullet proof glazing, 44" x 60"	3,475	390	3,865
1440	Pass through, painted steel, 72" x 40"	2,425	750	3,175
1500	Barber equipment, chair, hydraulic, economy	440	10.10	450.10
1540	Deluxe	2,975	15.20	2,990.20
1700	Church equipment, altar, wood, custom, plain	1,375	173	1,548
1740	Granite, custom, deluxe	11,400	2,425	13,825
1800	Baptistry, fiberglass, economy	1,550	555	2,105
1840	Bells & carillons, keyboard operation	9,900	3,875	13,775

SPECIAL CONSTRUCTION

11

11.1-200 — Architectural Equipment/Each

Line	Description	MAT.	INST.	TOTAL
1900	Confessional, wood, single, economy	1,750	405	2,155
1940	Double, deluxe	16,400	1,225	17,625
2000	Steeples, translucent fiberglas, 30" square, 15' high	1,875	790	2,665
2100	Checkout counter, single belt	2,125	35	2,160
2140	Double belt, power take-away	3,625	39	3,664
2600	Dental equipment, central suction system, economy	2,425	220	2,645
2640	Compressor-air, deluxe	6,825	495	7,320
2700	Chair, hydraulic, economy	3,325	495	3,820
2740	Deluxe	7,675	985	8,660
2800	Drill console with accessories, economy	4,175	154	4,329
2840	Deluxe	10,900	154	11,054
2900	X-ray unit, portable	4,700	61.50	4,761.50
2940	Panoramic unit	12,700	410	13,110
3000	Detention equipment, cell front rolling door, 7/8" bars, 5' x 7' high	4,475	660	5,135
3040	Cells, prefab., including front, 5' x 7' x 7' deep	6,975	880	7,855
3100	Doors and frames, 3' x 7', single plate	3,325	330	3,655
3140	Double plate	4,100	330	4,430
3200	Toilet apparatus, incl wash basin	2,175	370	2,545
3240	Visitor cubicle, vision panel, no intercom	2,525	660	3,185
3300	Dock bumpers, rubber blocks, 4-1/2" thick, 10" high, 14" long	42.50	9.35	51.85
3340	6" thick, 20" high, 11" long	102	18.70	120.70
3400	Dock boards, H.D., 5' x 5', aluminum, 5000 lb. capacity	1,125		1,125
3440	16,000 lb. capacity	1,550		1,550
3500	Dock levelers, hydraulic, 7' x 8', 10 ton capacity	6,125	665	6,790
3540	Dock lifters, platform, 6' x 6', portable, 3000 lb. capacity	6,000		6,000
3600	Dock shelters, truck, scissor arms, economy	1,025	243	1,268
3640	Deluxe	1,350	485	1,835
3700	Kitchen equipment, bake oven, single deck	3,475	59.50	3,534.50
3740	Broiler, without oven	3,350	59.50	3,409.50
3800	Commercial dish washer, semiautomatic, 50 racks/hr.	5,000	365	5,365
3840	Automatic, 275 racks/hr.	20,600	1,375	21,975
3900	Cooler, beverage, reach-in, 6 ft. long	2,600	79	2,679
3940	Food warmer, counter, 1.65 kw	745		745
4000	Fryers, with submerger, single	3,525	68	3,593
4040	Double	4,525	95	4,620
4100	Kettles, steam jacketed, 20 gallons	4,575	96	4,671
4140	Range, restaurant type, burners, 2 ovens and 24" griddle	4,150	79	4,229
4200	Range hood, incl. carbon dioxide system, economy	2,700	158	2,858
4240	Deluxe	19,400	475	19,875
4300	Laboratory equipment, glassware washer, distilled water, economy	4,850	182	5,032
4340	Deluxe	18,800	330	19,130
4440	Radio isotope	12,300		12,300
4600	Laundry equipment, dryers, gas fired, residential, 16 lb. capacity	550	88	638
4640	Commercial, 30 lb. capacity, single	2,425	88	2,513
4700	Dry cleaners, electric, 20 lb. capacity	30,300	1,650	31,950
4740	30 lb. capacity	41,500	2,175	43,675
4800	Ironers, commercial, 120" with canopy, 8 roll	146,000	4,675	150,675
4840	Institutional, 110", single roll	25,800	1,300	27,100
4900	Washers, residential, 4 cycle	630	88	718
4940	Commercial, coin operated, deluxe	2,625	88	2,713
5000	Library equipment, carrels, metal, economy	195	48.50	243.50
5040	Hardwood, deluxe	770	60.50	830.50
5100	Medical equipment, autopsy table, standard	6,600	264	6,864
5140	Deluxe	8,250	440	8,690
5200	Incubators, economy	2,875		2,875
5240	Deluxe	15,000		15,000
5300	Station, scrub-surgical, single, economy	3,900		3,900
5340	Dietary, medium, with ice	11,800		11,800
5400	Sterilizers, general purpose, single door, 20" x 20" x 28"	11,600		11,600
5440	Floor loading, double door, 28" x 67" x 52"	154,000		154,000

Important: See the Reference Section for critical supporting data - Reference Nos., Crews & Location Factors

11.1-200	Architectural Equipment/Each	COST EACH		
		MAT.	INST.	TOTAL
5500	Surgery tables, standard	10,500	430	10,930
5540	Deluxe	23,800	600	24,400
5600	Tables, standard, with base cabinets, economy	985	162	1,147
5640	Deluxe	2,925	243	3,168
5700	X-ray, mobile, economy	10,100		10,100
5740	Stationary, deluxe	164,500		164,500
5800	Movie equipment, changeover, economy	385		385
5840	Film transport, incl. platters and autowind, economy	7,825		7,825
5900	Lamphouses, incl. rectifiers, xenon, 1000W	6,475	129	6,604
5940	4000W	10,600	172	10,772
6000	Projector mechanisms, 35 mm, economy	10,900		10,900
6040	Deluxe	14,500		14,500
6100	Sound systems, incl. amplifier, single, economy	2,025	287	2,312
6140	Dual, Dolby/super sound	13,900	645	14,545
6200	Seating, painted steel, upholstered, economy	101	13.85	114.85
6240	Deluxe	260	17.35	277.35
6300	Parking equipment, automatic gates, 8 ft. arm, one way	2,775	470	3,245
6340	Traffic detectors, single treadle	2,050	215	2,265
6400	Booth for attendant, economy	4,400		4,400
6440	Deluxe	22,000		22,000
6500	Ticket printer/dispenser, rate computing	6,425	370	6,795
6540	Key station on pedestal	360	126	486
6700	Frozen food, chest type, 12 ft. long	4,575	145	4,720
7000	Safe, office type, 1 hr. rating, 34" x 20" x 20"	1,900		1,900
7040	4 hr. rating, 62" x 33" x 20"	7,575		7,575
7100	Data storage, 4 hr. rating, 63" x 44" x 16"	12,200		12,200
7140	Jewelers, 63" x 44" x 16"	20,900		20,900
7200	Money, "B" label, 9" x 14" x 14"	480		480
7240	Tool and torch resistive, 24" x 24" x 20"	8,250		8,250
7300	Sauna, prefabricated, incl. heater and controls, 7' high, 6' x 4'	3,225	305	3,530
7340	10' x 12'	7,800	670	8,470
7400	Heaters, wall mounted, to 200 C.F.	550		550
7440	Floor standing, to 1000 C.F., 12500 W	1,725	86	1,811
7500	School equipment, basketball backstops, wall mounted, wood, fixed	540	425	965
7540	Suspended type, electrically operated	2,650	805	3,455
7600	Bleachers-telescoping, manual operation, 15 tier, economy (per seat)	55	13.05	68.05
7640	Power operation, 30 tier, deluxe (per seat)	209	23	232
7700	Weight lifting gym, universal, economy	3,675	350	4,025
7740	Deluxe	13,200	700	13,900
7800	Scoreboards, basketball, 1 side, economy	2,525	385	2,910
7840	4 sides, deluxe	27,500	5,300	32,800
8300	Vocational shop equipment, benches, metal	248	97	345
8340	Wood	425	97	522
8400	Dust collector, not incl. ductwork, 6' diam.	2,350	236	2,586
8440	Planer, 13" x 6"	1,475	121	1,596
8500	Waste handling, compactors, single bag, 250 lbs./hr., hand fed	6,250	178	6,428
8540	Heavy duty industrial, 5 C.Y. capacity	16,300	855	17,155
8600	Incinerator, electric, 100 lbs./hr., economy	9,900	1,350	11,250
8640	Gas, 2000 lbs./hr., deluxe	171,000	9,500	180,500
8700	Shredder, no baling, 35 tons/hr.	211,000		211,000
8740	Incl. baling, 50 tons/day	422,000		422,000

SPECIAL CONSTRUCTION

11

11.1-200 — Architectural Equipment/S.F.

		COST PER S.F.		
		MAT.	INST.	TOTAL
8910	Arch. equip., lab equip., counter tops, acid proof, economy	15.40	5.90	21.30
8940	Stainless steel	65.50	5.90	71.40
9000	Movie equipment, projection screens, rigid in wall, acrylic, 1/4" thick	35	2.40	37.40
9040	1/2" thick	40.50	3.59	44.09
9100	School equipment, gym mats, naugahyde cover, 2" thick	4.19		4.19
9140	Wrestling, 1" thick, heavy duty	4.76		4.76
9200	Stage equipment, curtains, velour, medium weight	12.10	.81	12.91
9240	Silica based yarn, fireproof	27.50	9.70	37.20
9300	Stages, portable with steps, folding legs, 8" high	7.65		7.65
9340	Telescoping platforms, aluminum, deluxe	44	12.60	56.60

11.1-200 — Architectural Equipment/L.F.

		COST PER L.F.		
		MAT.	INST.	TOTAL
9410	Arch. equip., church equip. pews, bench type, hardwood, economy	49	12.15	61.15
9440	Deluxe	79	16.20	95.20
9500	Laboratory equipment, cabinets, wall, open	70.50	24.50	95
9540	Base, drawer units	230	27	257
9600	Fume hoods, not incl. HVAC, economy	630	90	720
9640	Deluxe incl. fixtures	1,250	202	1,452
9700	Library equipment, book shelf, metal, single face, 90" high x 10" shelf	68	20	88
9740	Double face, 90" high x 10" shelf	194	43.50	237.50
9800	Charging desk, built-in, with counter, plastic laminate	415	34.50	449.50
9900	Stage equipment, curtain track, heavy duty	26	27	53
9940	Lights, border, quartz, colored	132	12.90	144.90

11.1-500 — Furnishings/Each

		COST EACH		
		MAT.	INST.	TOTAL
1000	Furnishings, blinds, exterior, aluminum, louvered, 1'-4" wide x 3'-0" long	32.50	24.50	57
1040	1'-4" wide x 6'-8" long	52.50	27	79.50
1100	Hemlock, solid raised, 1'-4" wide x 3'-0" long	56	24.50	80.50
1140	1'-4" wide x 6'-9" long	105	27	132
1200	Polystyrene, louvered, 1'-3" wide x 3'-3" long	43	24.50	67.50
1240	1'-3" wide x 6'-8" long	93.50	27	120.50
1300	Interior, wood folding panels, louvered, 7" x 20" (per pair)	25	14.30	39.30
1340	18" x 40" (per pair)	79	14.30	93.30
1800	Hospital furniture, beds, manual, economy	865		865
1840	Deluxe	1,500		1,500
1900	All electric, economy	1,450		1,450
1940	Deluxe	3,575		3,575
2000	Patient wall systems, no utilities, economy, per room	870		870
2040	Deluxe, per room	1,600		1,600
2200	Hotel furnishings, standard room set, economy, per room	1,625		1,625
2240	Deluxe, per room	7,850		7,850
2400	Office furniture, standard employee set, economy, per person	360		360
2440	Deluxe, per person	2,025		2,025
2800	Posts, portable, pedestrian traffic control, economy	90		90
2840	Deluxe	315		315
3000	Restaurant furniture, booth, molded plastic, stub wall and 2 seats, economy	370	121	491
3040	Deluxe	790	162	952
3100	Upholstered seats, foursome, single-economy	530	48.50	578.50
3140	Foursome, double-deluxe	2,850	81	2,931
3300	Seating, lecture hall, pedestal type, economy	108	13.85	121.85
3340	Deluxe	335	24.50	359.50
3400	Auditorium chair, veneer construction	108	13.85	121.85
3440	Fully upholstered, spring seat	159	13.85	172.85

Important: See the Reference Section for critical supporting data - Reference Nos., Crews & Location Factors

11.1-500 — Furnishings/S.F.

		COST PER S.F.		
		MAT.	INST.	TOTAL
4010	Furnishings, blinds-interior, venetian-aluminum, stock, 2" slats, economy	1.35	.41	1.76
4040	Custom, 1" slats, deluxe	8.10	.55	8.65
4100	Vertical, PVC or cloth, T&B track, economy	7.15	.53	7.68
4140	Deluxe	11.95	.61	12.56
4300	Draperies, unlined, economy	2.06		2.06
4440	Lightproof, deluxe	6.20		6.20
4700	Floor mats, recessed, inlaid black rubber, 3/8" thick, solid	21	1.13	22.13
4740	Colors, 1/2" thick, perforated	33	1.13	34.13
4800	Link-including nosings, steel-galvanized, 3/8" thick	5.70	1.13	6.83
4840	Vinyl, in colors	17.80	1.13	18.93
5000	Shades, mylar, wood roller, single layer, non-reflective	5.70	.35	6.05
5040	Metal roller, triple layer, heat reflective	10.05	.35	10.40
5100	Vinyl, light weight, 4 ga.	.41	.35	.76
5140	Heavyweight, 6 ga.	1.27	.35	1.62
5200	Vinyl coated cotton, lightproof decorator shades	1.27	.35	1.62
5300	Woven aluminum, 3/8" thick, light and fireproof	4.05	.69	4.74

11.1-500 — Furnishings/L.F.

		COST PER L.F.		
		MAT.	INST.	TOTAL
5510	Furnishings, cabinets, hospital, base, laminated plastic	188	48.50	236.50
5540	Stainless steel	235	48.50	283.50
5600	Counter top, laminated plastic, no backsplash	32.50	12.15	44.65
5640	Stainless steel	90	12.15	102.15
5700	Nurses station, door type, laminated plastic	218	48.50	266.50
5740	Stainless steel	251	48.50	299.50
5900	Household, base, hardwood, one top drawer & one door below x 12" wide	122	19.60	141.60
5940	Four drawer x 24" wide	190	22	212
6000	Wall, hardwood, 30" high with one door x 12" wide	91	22	113
6040	Two doors x 48" wide	188	26.50	214.50
6100	Counter top-laminated plastic, stock, economy	5.40	8.10	13.50
6140	Custom-square edge, 7/8" thick	11.35	18.20	29.55
6300	School, counter, wood, 32" high	149	24.50	173.50
6340	Metal, 84" high	230	32.50	262.50
6500	Dormitory furniture, desk top (built-in),laminated plastc, 24"deep, economy	24	9.70	33.70
6540	30" deep, deluxe	105	12.15	117.15
6600	Dressing unit, built-in, economy	177	40.50	217.50
6640	Deluxe	525	60.50	585.50
7000	Restaurant furniture, bars, built-in, back bar	142	48.50	190.50
7040	Front bar	195	48.50	243.50
7200	Wardrobes & coatracks, standing, steel, single pedestal, 30" x 18" x 63"	77.50		77.50
7300	Double face rack, 39" x 26" x 70"	131		131
7340	Wall mounted rack, steel frame & shelves, 12" x 15" x 26"	55	3.46	58.46
7400	12" x 15" x 50"	43.50	1.81	45.31

SPECIAL CONSTRUCTION

11

11.1-700	Special Construction/Each	COST EACH		
		MAT.	INST.	TOTAL
1000	Special construction, bowling alley incl. pinsetter, scorer etc., economy	35,500	4,850	40,350
1040	Deluxe	40,400	5,400	45,800
1100	For automatic scorer, economy, add	6,600		6,600
1140	Deluxe, add	7,475		7,475
1200	Chimney, metal, high temp. steel jacket, factory lining, 24" diameter	7,750	2,250	10,000
1240	60" diameter	58,000	9,750	67,750
1300	Poured concrete, brick lining, 10' diam., 200' high			1,120,000
1340	20' diam., 500' high			5,000,000
1400	Radial brick, 3'-6" I.D., 75' high			184,000
1440	7' I.D., 175' high			555,500
1500	Control tower, modular, 12' x 10', incl. instrumentation, economy			450,000
1540	Deluxe			600,000
1600	Garage costs, residential, prefab, wood, single car economy	2,800	425	3,225
1640	Two car deluxe	8,275	850	9,125
1700	Grandstands, permanent, closed deck, steel, economy (per seat)			25
1740	Deluxe (per seat)			80
1800	Composite design, economy (per seat)			30
1840	Deluxe (per seat)			90
1900	Hangars, prefab, galv. steel, bottom rolling doors, economy (per plane)	8,950	2,325	11,275
1940	Electrical bi-folding doors, deluxe (per plane)	12,500	3,200	15,700
2000	Ice skating rink, 85' x 200', 55° system, 5 mos., 100 ton			325,000
2040	90° system, 12 mos., 135 ton			350,000
2100	Dash boards, acrylic screens, polyethylene coated plywood	126,500	14,100	140,600
2140	Fiberglass and aluminum construction	137,500	14,100	151,600
2200	Integrated ceilings, radiant electric, 2' x 4' panel, manila finish	44.50	10.30	54.80
2240	ABS plastic finish	81.50	19.85	101.35
2300	Kiosks, round, 5' diam., 1/4" fiberglass wall, 8' high	6,050		6,050
2340	Rectangular, 5' x 9', 1" insulated dbl. wall fiberglass, 7'-6" high	10,500		10,500
2400	Portable booth, acoustical, 27 db 1000 hz., 15 S.F. floor	3,175		3,175
2440	55 S.F. flr.	6,500		6,500
2500	Radio towers, guyed, 40 lb. section, 50' high, 70 MPH basic wind speed	1,725	600	2,325
2540	90 lb. section, 400' high, wind load 70 MPH basic wind speed	22,400	7,275	29,675
2600	Self supporting, 60' high, 70 MPH basic wind speed	3,600	1,275	4,875
2640	190' high, wind load 90 MPH basic wind speed	21,400	5,075	26,475
2700	Shelters, aluminum frame, acrylic glazing, 8' high, 3' x 9'	2,800	525	3,325
2740	9' x 12'	4,300	825	5,125
2800	Shielding, lead x-ray protection, radiography room, 1/16" lead, economy	5,050	1,825	6,875
2840	Deluxe	6,600	3,050	9,650
2900	Deep therapy x-ray, 1/4" lead, economy	17,100	5,725	22,825
2940	Deluxe	23,100	7,650	30,750
3000	Shooting range incl. bullet traps, controls, separators, ceilings, economy	11,600	3,175	14,775
3040	Deluxe	22,100	5,325	27,425
3100	Silos, concrete stave, industrial, 12' diam., 35' high	16,500	10,100	26,600
3140	25' diam., 75' high	54,000	22,300	76,300
3200	Steel prefab, 30,000 gal., painted, economy	12,000	2,675	14,675
3240	Epoxy-lined, deluxe	24,000	5,350	29,350
3300	Sport court, squash, regulation, in existing building, economy			14,500
3340	Deluxe			27,100
3400	Racketball, regulation, in existing building, economy	1,925	4,400	6,325
3440	Deluxe	25,800	8,425	34,225
3500	Swimming pool equipment, diving stand, stainless steel, 1 meter	4,625	180	4,805
3540	3 meter	5,450	1,225	6,675
3600	Diving boards, 16 ft. long, aluminum	2,050	180	2,230
3640	Fiberglass	1,525	180	1,705
3700	Filter system, sand, incl. pump, 6000 gal./hr.	1,100	293	1,393
3800	Lights, underwater, 12 volt with transformer, 300W	190	645	835
3900	Slides, fiberglass with aluminum handrails & ladder, 6' high, straight	4,550	305	4,855
3940	12' high, straight with platform	14,100	405	14,505
4000	Tanks, steel, ground level, 100,000 gal.			96,000
4040	10,000,000 gal.			2,200,000

Important: See the Reference Section for critical supporting data - Reference Nos., Crews & Location Factors

11.1-700	Special Construction/Each	COST EACH		
		MAT.	INST.	TOTAL
4100	Elevated water, 50,000 gal.			163,000
4140	1,000,000 gal.			797,000
4200	Cypress wood, ground level, 3,000 gal.	5,500	4,425	9,925
4240	Redwood, ground level, 45,000 gal.	27,500	12,000	39,500

11.1-700	Special Construction/S.F.	COST PER S.F.		
		MAT.	INST.	TOTAL
4500	Special construction, acoustical, enclosure, 4" thick, 8 psf panels	26.50	10.10	36.60
4540	Reverb chamber, 4" thick, parallel walls	32	12.15	44.15
4600	Sound absorbing panels, 2'-6" x 8', painted metal	9.55	3.39	12.94
4640	Vinyl faced	7.75	3.04	10.79
4700	Flexible transparent curtain, clear	6	3.63	9.63
4740	With absorbing foam, 75% coverage	8.35	3.63	11.98
4800	Strip entrance, 2/3 overlap	3.91	5.80	9.71
4840	Full overlap	4.83	6.80	11.63
4900	Audio masking system, plenum mounted, over 10,000 S.F.	.57	.12	.69
4940	Ceiling mounted, under 5,000 S.F.	1.02	.22	1.24
5000	Air supported struc., polyester vinyl fabric, 24oz., warehouse, 5000 S.F.	16.95	.14	17.09
5040	50,000 S.F.	6.80	.11	6.91
5100	Tennis, 7,200 S.F.	14.30	.12	14.42
5140	24,000 S.F.	9.35	.12	9.47
5200	Woven polyethylene, 6 oz., shelter, 3,000 S.F.	6.50	.23	6.73
5240	24,000 S.F.	2.96	.12	3.08
5300	Teflon coated fiberglass, stadium cover, economy	38.50	.06	38.56
5340	Deluxe	44	.08	44.08
5400	Air supported storage tank covers, reinf. vinyl fabric, 12 oz., 400 S.F.	13.20	.20	13.40
5440	18,000 S.F.	3.91	.18	4.09
5500	Anechoic chambers, 7' high, 100 cps cutoff, 25 S.F.			1,925
5540	200 cps cutoff, 100 S.F.			1,000
5600	Audiometric rooms, under 500 S.F.	42	9.90	51.90
5640	Over 500 S.F.	39.50	8.10	47.60
5700	Comfort stations, prefab, mobile on steel frame, economy	41		41
5740	Permanent on concrete slab, deluxe	220	22	242
5800	Darkrooms, shell, not including door, 240 S.F., 8' high	35	4.05	39.05
5840	64 S.F., 12' high	82.50	7.60	90.10
5900	Domes, bulk storage, wood framing, wood decking, 50' diam.	27.50	.79	28.29
5940	116' diam.	16.50	.92	17.42
6000	Steel framing, metal decking, 150' diam.	29.50	5.30	34.80
6040	400' diam.	24	4.05	28.05
6100	Geodesic, wood framing, wood panels, 30' diam.	15.55	.76	16.31
6140	60' diam.	12.90	.56	13.46
6200	Aluminum framing, acrylic panels, 40' diam.			75
6240	Aluminum panels, 400' diam.			25
6300	Garden house, prefab, wood, shell only, 48 S.F.	14.80	2.12	16.92
6340	200 S.F.	27	8.85	35.85
6400	Greenhouse, shell-stock, residential, lean-to, 8'-6" long x 3'-10" wide	34	14.30	48.30
6440	Freestanding, 8'-6" long x 13'-6" wide	26.50	4.50	31
6500	Commercial-truss frame, under 2000 S.F., deluxe			38
6540	Over 5,000 S.F., economy			23
6600	Institutional-rigid frame, under 500 S.F., deluxe			99
6640	Over 2,000 S.F., economy			39
6700	Hangar, prefab, galv. roof and walls, bottom rolling doors, economy	8.55	2.11	10.66
6740	Electric bifolding doors, deluxe	10.10	2.75	12.85
6800	Integrated ceilings, Luminaire, suspended, 5' x 5' modules, 50% lighted	3.88	3.41	7.29
6840	100% lighted	5.80	6.15	11.95
6900	Dimensionaire, 2' x 4' module tile system, no air bar	2.35	.97	3.32
6940	With air bar, deluxe	3.40	.97	4.37
7000	Music practice room, modular, perforated steel, under 500 S.F.	23	6.95	29.95
7040	Over 500 S.F.	17.45	6.05	23.50

11.1-700	Special Construction/S.F.	COST PER S.F.		
		MAT.	INST.	TOTAL
7100	Pedestal access floors, stl pnls, no stringers, vinyl covering, >6000 S.F.	11.40	1.08	12.48
7140	With stringers, high pressure laminate covering, under 6000 S.F.	10.20	1.21	11.41
7200	Carpet covering, under 6000 S.F.	11.85	1.21	13.06
7240	Aluminum panels, no stringers, no covering	22.50	.97	23.47
7300	Refrigerators, prefab, walk-in, 7'-6" high, 6' x 6'	105	8.85	113.85
7340	12' x 20'	66	4.43	70.43
7400	Shielding, lead, gypsum board, 5/8" thick, 1/16" lead	5.45	2.92	8.37
7440	1/8" lead	11.55	3.34	14.89
7500	Lath, 1/16" thick	5.20	3.40	8.60
7540	1/8" thick	11	3.82	14.82
7600	Radio frequency, copper, prefab screen type, economy	26.50	2.70	29.20
7640	Deluxe	34	3.35	37.35
7700	Swimming pool enclosure, transluscent, freestanding, economy	11	2.43	13.43
7740	Deluxe	275	6.95	281.95
7800	Swimming pools, residential, vinyl liner, metal sides	9.25	3.34	12.59
7840	Concrete sides	11.05	6.25	17.30
7900	Gunite shell, plaster finish, 350 S.F.	18.30	12.95	31.25
7940	800 S.F.	14.75	7.50	22.25
8000	Motel, gunite shell, plaster finish	22.50	16.25	38.75
8040	Municipal, gunite shell, tile finish, formed gutters	94	28	122
8100	Tension structures, steel frame, polyester vinyl fabric, 12,000 S.F.	9.10	1.12	10.22
8140	20,800 S.F.	9.80	1.01	10.81

11.1-700	Special Construction/L.F.	COST PER L.F.		
		MAT.	INST.	TOTAL
8500	Spec. const., air curtains, shipping & receiving, 8'high x 5'wide, economy	141	104	245
8540	20' high x 8' wide, heated, deluxe	1,250	200	1,450
8600	Customer entrance, 10' high x 5' wide, economy	208	104	312
8640	12' high x 4' wide, heated, deluxe	370	186	556

For information about Means Estimating Seminars, see yellow pages 11 and 12 in back of book

Important: See the Reference Section for critical supporting data - Reference Nos., Crews & Location Factors

SPECIAL CONSTRUCTION 11

Division 12
Site Work

12.3-110 — Trenching

		COST PER L.F.		
		MAT.	INST.	TOTAL
1310	Trenching, backhoe, 0 to 1 slope, 2' wide, 2' deep, 3/8 C.Y. bucket		1.94	1.94
1330	4' deep, 3/8 C.Y. bucket		3.28	3.28
1360	10' deep, 1 C.Y. bucket		5.85	5.85
1400	4' wide, 2' deep, 3/8 C.Y. bucket		4.01	4.01
1420	4' deep, 1/2 C.Y. bucket		5.73	5.73
1450	10' deep, 1 C.Y. bucket		11.90	11.90
1480	18' deep, 2-1/2 C.Y. bucket		18.70	18.70
1520	6' wide, 6' deep, 5/8 C.Y. bucket		12.19	12.19
1540	10' deep, 1 C.Y. bucket		16.20	16.20
1570	20' deep, 3-1/2 C.Y. bucket		25.85	25.85
1640	8' wide, 12' deep, 1-1/4 C.Y. bucket		21.40	21.40
1680	24' deep, 3-1/2 C.Y. bucket		57.50	57.50
1730	10' wide, 20' deep, 3-1/2 C.Y. bucket		45.85	45.85
1740	24' deep, 3-1/2 C.Y. bucket		73	73
3500	1 to 1 slope, 2' wide, 2' deep, 3/8 C.Y. bucket		2.99	2.99
3540	4' deep, 3/8 C.Y. bucket		6.75	6.75
3600	10' deep, 1 C.Y. bucket		24.90	24.90
3800	4' wide, 2' deep, 3/8 C.Y. bucket		5.08	5.08
3840	4' deep, 1/2 C.Y. bucket		8.43	8.43
3900	10' deep, 1 C.Y. bucket		26.85	26.85
4030	6' wide, 6' deep, 5/8 C.Y. bucket		16.30	16.30
4050	10' deep, 1 C.Y. bucket		32.25	32.25
4080	20' deep, 3-1/2 C.Y. bucket		84	84
4500	8' wide, 12' deep, 1-1/4 C.Y. bucket		40.60	40.60
4650	24' deep, 3-1/2 C.Y. bucket		174	174
4800	10' wide, 20' deep, 3-1/2 C.Y. bucket		103	103
4850	24' deep, 3-1/2 C.Y. bucket		186.50	186.50

12.3-310 — Pipe Bedding

		COST PER L.F.		
		MAT.	INST.	TOTAL
1440	Pipe bedding, side slope 0 to 1, 1' wide, pipe size 6" diameter	.26	.50	.76
1460	2' wide, pipe size 8" diameter	.58	1.09	1.67
1500	Pipe size 12" diameter	.61	1.15	1.76
1600	4' wide, pipe size 20" diameter	1.53	2.85	4.38
1660	Pipe size 30" diameter	1.62	3.03	4.65
1680	6' wide, pipe size 32" diameter	2.83	5.30	8.13
1740	8' wide, pipe size 60" diameter	4.72	8.80	13.52
1780	12' wide, pipe size 84" diameter	9.25	17.30	26.55

12.3-710 — Manholes & Catch Basins

		COST PER EACH		
		MAT.	INST.	TOTAL
1920	Manhole/catch basin, brick, 4' I.D. riser, 4' deep	655	760	1,415
1980	10' deep	1,375	1,775	3,150
3200	Block, 4' I.D. riser, 4' deep	700	615	1,315
3260	10' deep	1,350	1,475	2,825
4620	Concrete, cast-in-place, 4' I.D. riser, 4' deep	725	1,075	1,800
4680	10' deep	1,550	2,525	4,075
5820	Concrete, precast, 4' I.D. riser, 4' deep	715	555	1,270
5880	10' deep	1,325	1,300	2,625
6200	6' I.D. riser, 4' deep	1,275	830	2,105
6260	10' deep	2,500	1,975	4,475

12.5-110	Roadway Pavement	COST PER L.F.		
		MAT.	INST.	TOTAL
1500	Roadways, bituminous conc. paving, 2-1/2" thick, 20' wide	34	28.50	62.50
1580	30' wide	40.50	32.50	73
1800	3" thick, 20' wide	36	27	63
1880	30' wide	43	33	76
2100	4" thick, 20' wide	39	27	66
2180	30' wide	47.50	32.50	80
2400	5" thick, 20' wide	45.50	27.50	73
2480	30' wide	57	33.50	90.50
3000	6" thick, 20' wide	50.50	26.50	77
3080	30' wide	65	31.50	96.50
3300	8" thick 20' wide	58	26.50	84.50
3380	30' wide	76	32	108
3600	12" thick 20' wide	75	27	102
3700	32' wide	106	34	140

12.5-510	Parking Lots	COST PER CAR		
		MAT.	INST.	TOTAL
1500	Parking lot, 90° angle parking, 3" bituminous paving, 6" gravel base	242	162	404
1540	10" gravel base	262	190	452
1560	4" bituminous paving, 6" gravel base	287	161	448
1600	10" gravel base	305	190	495
1620	6" bituminous paving, 6" gravel base	385	168	553
1660	10" gravel base	405	196	601
1800	60° angle parking, 3" bituminous paving, 6" gravel base	278	177	455
1840	10" gravel base	300	210	510
1860	4" bituminous paving, 6" gravel base	330	176	506
1900	10" gravel base	355	209	564
1920	6" bituminous paving, 6" gravel base	445	183	628
1960	10" gravel base	470	217	687
2200	45° angle parking, 3" bituminous paving, 6" gravel base	315	193	508
2240	10" gravel base	340	230	570
2260	4" bituminous paving, 6" gravel base	375	191	566
2300	10" gravel base	400	229	629
2320	6" bituminous paving, 6" gravel base	505	200	705
2360	10" gravel base	530	237	767

For information about Means Estimating Seminars, see yellow pages 11 and 12 in back of book

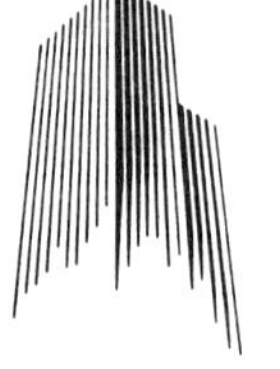

Unit Price Section

Table of Contents

How to Use the Unit Price Pages

The following is a detailed explanation of a sample entry in the Unit Price Section. Next to each bold number below is the item being described with appropriate component of the sample entry following in parenthesis. Some prices are listed as bare costs, others as costs that include overhead and profit of the installing contractor. In most cases, if the work is to be subcontracted, the general contractor will need to add an additional markup (R.S. Means suggests using 10%) to the figures in the column "Total Incl. O&P."

1 Division Number/Title (061/Rough Carpentry)

Use the Unit Price Section Table of Contents to locate specific items. The sections are classified according to the CSI MasterFormat.

2 Line Numbers (061 128 2050)

Each unit price line item has been assigned a unique 10-digit code based on the 5-digit CSI MasterFormat classification.

MasterFormat Mediumscope
MasterFormat Division

061 100
061 128 2050

Means Subdivision
Means Major Classification
Means Individual Line Number

3 Description (Headers Over Openings, Etc.)

Each line item is described in detail. Sub-items and additional sizes are indented beneath the appropriate line items. The first line or two after the main item (in boldface) may contain descriptive information that pertains to all line items beneath this boldface listing.

4 Reference Number Information

R061-010 You'll see reference numbers shown in bold squares at the beginning of some major classifications. These refer to related items in the Reference Section, visually identified by a vertical gray bar on the edge of pages.

The relation may be: (1) an estimating procedure that should be read before estimating, (2) an alternate pricing method, or (3) technical information.

The "R" designates the Reference Section. The numbers refer to the MasterFormat classification system.

It is strongly recommended that you review all reference numbers that appear within the major classification you are estimating.

Example: The square number above is directing you to refer to the reference number R061-010. This particular reference number shows how the unit price lines for blocking were formulated and costs derived.

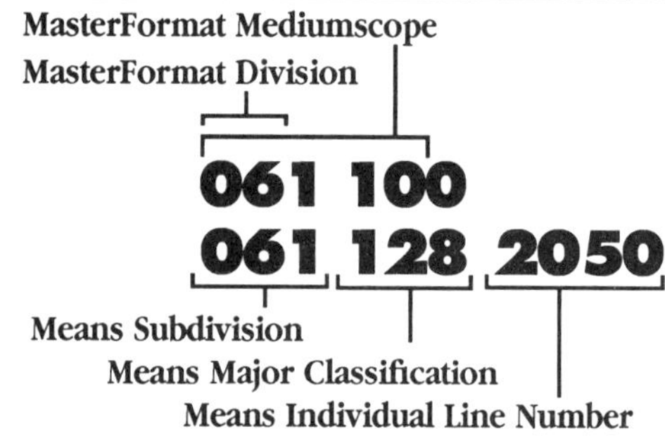

				CREW	DAILY OUTPUT	LABOR-HOURS	UNIT	MAT.	LABOR	EQUIP.	TOTAL	TOTAL INCL O&P
128	0010	**FRAMING, WALLS**											128
	2000	Headers over openings, 2" x 6"		F-2	300	.044	L.F.	.55	.72		1.27	1.83
	2000	2" x 6", pneumatic nailed		F-2A	432	.037		.55	.60		1.15	1.63
	2050	2" x 8"		F-2	340	.047		.78	.76		1.54	2.16
	2050	2" x 8", pneumatic nailed		F-2A	408	.039		.78	.63		1.41	1.95
	2100	2" x 10"		F-2	320	.050		1.09	.81		1.90	2.49
	2100	2" x 10", pneumatic nailed		F-2A	384	.042		1.09	.67		1.76	2.26
		2" x 12"		F-2	300	.053		1.41	.86		2.27	3.03
		2" x 12", pneumatic nailed		F-2A	360	.044		1.41	.72		2.13	2.78
	2190	4" x 10"		F-2	240	.067		2.85	1.08		3.93	4.99
		4" x 10", pneumatic nailed		F-2A	288	.056		2.85	.90		3.75	4.68
	2200	4" x 12"		F-2	190	.084		3.42	1.36		4.78	6.10
	2205	4" x 12", pneumatic nailed		F-2A	228	.070		3.42	1.14		4.56	5.70
	2240	6" x 10"		F-2	165	.097		7.05	1.57		8.62	10.45
	2245	6" x 10", pneumatic nailed			198	.081		7.05	1.31			

Crew (F-2)

The "Crew" column designates the typical trade or crew used to install the item. If an installation can be accomplished by one trade and requires no power equipment, that trade and the number of workers are listed (for example, "2 Carp"). If an installation requires a composite crew, a crew code designation is listed (for example, "F-2"). You'll find full details on all composite crews in the Crew Listings.

- For a complete list of all trades utilized in this book and their abbreviations, see the inside back cover.

Crews

Crew No.	Bare Costs		Incl. Subs O & P		Cost Per Labor-Hour	
Crew F-2	Hr.	Daily	Hr.	Daily	Bare Costs	Incl. O&P
1 Carpenters	$18.50	$148.00	$31.70	$253.60	$16.18	$27.70
1 Carpenter Helper	13.85	110.80	23.70	189.60		
16 L.H., Daily Totals		$258.80		$443.20	$16.18	$27.70

Productivity: Daily Output (340)/ Labor-Hours (.047)

The "Daily Output" represents the typical number of units the designated crew will install in a normal 8-hour day. To find out the number of days the given crew would require to complete the installation, divide your quantity by the daily output. For example:

Quantity	÷	Daily Output	=	Duration
1000 LF	÷	340 LF/ Crew Day	=	3 Crew Days

The "Labor-Hours" figure represents the number of labor-hours required to install one unit of work. To find out the number of labor-hours required for your particular task, multiply the quantity of the item times the number of labor-hours shown. For example:

Quantity	x	Productivity Rate	=	Duration
1000 LF	x	.047 Labor-Hours/ LF	=	47 Labor-Hours

Unit (LF)

The abbreviated designation indicates the unit of measure upon which the price, production, and crew are based (LF = Linear Foot). For a complete listing of abbreviations refer to the Abbreviations Listing in the Reference Section of this book.

Bare Costs:
Mat. (Bare Material Cost) (.78)

The unit material cost is the "bare" material cost with no overhead and profit included. *Costs shown reflect national average material prices for January of the current year and include delivery to the job site. No sales taxes are included.*

Labor (.76)

The unit labor cost is derived by multiplying bare labor-hour costs for Crew F-2 by labor-hour units. The bare labor-hour cost is found in the Crew Section under F-2. (If a trade is listed, the hourly labor cost—the wage rate—is found on the inside back cover.)

Labor-Hour Cost Crew F-1	x	Labor-Hour Units	=	Labor
$16.18	x	.047	=	$.76

Equip. (Equipment) (.00)

Equipment costs for each crew are listed in the description of each crew. Tools or equipment whose value justifies purchase or ownership by a contractor are considered overhead as shown on the inside back cover. The unit equipment cost is derived by multiplying the bare equipment hourly cost by the labor-hour units.

Equipment Cost Crew F-1	x	Labor-Hour Units	=	Equip.
$.00	x	.047	=	$.00

Total (1.54)

The total of the bare costs is the arithmetic total of the three previous columns: mat., labor, and equip.

Material	+	Labor	+	Equip.	=	Total
$.78	+	$.76	+	$.00	=	$1.54

Total Costs Including O&P

This figure is the sum of the bare material cost plus 10% for profit; the bare labor cost plus total overhead and profit (per the inside back cover or, if a crew is listed, from the crew listings); and the bare equipment cost plus 10% for profit.

Material is Bare Material Cost + 10% = $.78 + $.08	=	$ .86
Labor for Crew F-2 = Labor-Hour Cost ($27.70) x Labor-Hour Units (.047)	=	$1.30
Equip. is Bare Equip. Cost + 10% = $.00 + $.00	=	$ .00
Total	=	$2.16

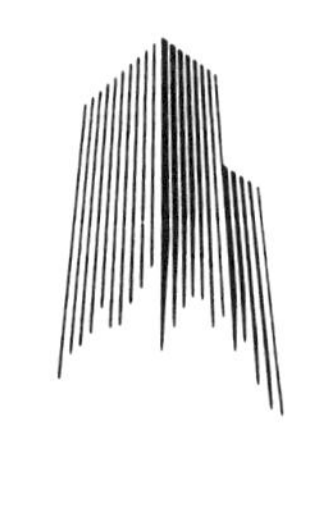

Division 1
General Requirements

Estimating Tips

The General Requirements of any contract are very important to both the bidder and the owner. These lay the ground rules under which the contract will be executed and have a significant influence on the cost of operations. Therefore, it is extremely important to thoroughly read and understand the General Requirements both before preparing an estimate and when the estimate is complete, to ascertain that nothing in the contract is overlooked. Caution should be exercised when applying items listed in Division 1 to an estimate. Many of the items are included in the unit prices listed in the other divisions such as mark-ups on labor and company overhead.

010 Overhead & Miscellaneous Data
- Before determining a final cost estimate, it is a good practice to review all the items listed in subdivision 010 to make final adjustments for items that may need customizing to specific job conditions.

013 Submittals
- Requirements for initial and periodic submittals can represent a significant cost to the General Requirements of a job. Thoroughly check the submittal specifications when estimating a project to determine any costs that should be included.

014 Quality Control
- All projects will require some degree of Quality Control. This cost is not included in the unit cost of construction listed in each division. Depending upon the terms of the contract, the various costs of inspection and testing can be the responsibility of either the owner or the contractor. Be sure to include the required costs in your estimate.

015 Construction Facilities & Temporary Controls
- Barricades, access roads, safety nets, scaffolding, security and many more requirements for the execution of a safe project are elements of direct cost. These costs can easily be overlooked when preparing an estimate. When looking through the major classifications of this subdivision, determine which items apply to each division in your estimate.

016 Material & Equipment
- This subdivision contains transportation, handling, storage, protection and product options and substitutions. Listed in this cost manual are average equipment rental rates for all types of equipment. This is useful information when estimating the time and materials requirement of any particular operation in order to establish a unit or total cost.
- A good rule of thumb is that weekly rental is 3 times daily rental and that monthly rental is 3 times weekly rental.

- The figures in the column for Crew Equipment Cost represent the rental rate used in determining the daily cost of equipment in a crew. It is calculated by dividing the weekly rate by 5 days and adding the hourly operating cost times 8 hours.

017 Contract Closeout
- When preparing an estimate, read the specifications to determine the requirements for Contract Closeout thoroughly. Final cleaning, record documentation, operation and maintenance data, warranties and bonds, and spare parts and maintenance materials can all be elements of cost for the completion of a contract. Do not overlook these in your estimate.

018 Maintenance
- If maintenance and repair are included in your contract, they require special attention. To estimate the cost to remove and replace any unit usually requires a site visit to determine the accessibility and the specific difficulty at that location. Obstructions, dust control, safety, and often overtime hours must be considered when preparing your estimate.

Reference Numbers
Reference numbers are shown in bold squares at the beginning of some major classifications. These numbers refer to related items in the Reference Section. The reference information may be an estimating procedure, an alternate pricing method or technical information.

Note: Not all subdivisions listed here necessarily appear in this publication.

GENERAL REQUIREMENTS — 1

010 000	Overhead		CREW	DAILY OUTPUT	LABOR-HOURS	UNIT	1998 BARE COSTS				TOTAL INCL O&P	
							MAT.	LABOR	EQUIP.	TOTAL		
004	0011	**ARCHITECTURAL FEES**										004
	0020	For new construction	R010 -010									
	0060	Minimum				Project					4.90%	
016	0011	**CONSTRUCTION MANAGEMENT FEES**										016
	0060	For work to $10,000				Project					10%	
	0070	To $25,000									9%	
	0090	To $100,000									6%	
020	0010	**CONTINGENCIES** Allowance to add at conceptual stage				Project					15%	020
	0150	Final working drawing stage				"					2%	
034	0010	**FIELD OFFICE EXPENSE**										034
	0100	Field office expense, office equipment rental average				Month	130			130	143	
	0120	Office supplies, average					84			84	92.50	
	0140	Telephone bill; avg. bill/month incl. long dist.					230			230	253	
	0160	Field office lights & HVAC					86			86	94.50	
040	0010	**INSURANCE** Builders risk, standard, minimum	R010 -040				Job					.22%	040
	0050	Maximum									.59%	
	0200	All-risk type, minimum	R010 -060									.25%	
	0250	Maximum									.62%	
	0400	Contractor's equipment floater, minimum	R010 -080				Value					.50%	
	0450	Maximum					"					1.50%	
	0600	Public liability, average				Job					1.55%	
	0800	Workers' compensation & employer's liability, average										
	0850	by trade, carpentry, general				Payroll		19.88%				
	0900	Clerical						.48%				
	0950	Concrete						19.18%				
	1000	Electrical						7.13%				
	1050	Excavation						11.63%				
	1100	Glazing						14.85%				
	1150	Insulation						18.65%				
	1200	Lathing						12.25%				
	1250	Masonry						17.87%				
	1300	Painting & decorating						15.60%				
	1350	Pile driving						30.47%				
	1400	Plastering						15.59%				
	1450	Plumbing						9.18%				
	1500	Roofing						35.21%				
	1550	Sheet metal work (HVAC)						12.73%				
	1600	Steel erection, structural						42.56%				
	1650	Tile work, interior ceramic						10.30%				
	1700	Waterproofing, brush or hand caulking						8.52%				
	1800	Wrecking						42.28%				
	2000	Range of 35 trades in 50 states, excl. wrecking, minimum						2.02%				
	2100	Average						18.50%				
	2200	Maximum						132.92%				
048	0010	**MAIN OFFICE EXPENSE** Average for General Contractors	R010 -050										048
	0020	As a percentage of their annual volume										
	0125	Annual volume under 1 million dollars				% Vol.				13.60%		
	0145	Up to 2.5 million dollars								8%		
	0150	Up to 4.0 million dollars								6.80%		
	0200	Up to 7.0 million dollars								5.60%		
062	0010	**OVERHEAD & PROFIT** Allowance to add to items in this	R010 -070										062
	0020	book that do not include Subs O&P, average				%				25%		
	0100	Allowance to add to items in this book that										
	0110	do include Subs O&P, minimum				%					5%	

Important: See the Reference Section for critical supporting data - Reference Nos., Crews, & Location Factors

010 | Overhead & Miscellaneous Data

010 000 | Overhead

			CREW	DAILY OUTPUT	LABOR-HOURS	UNIT	1998 BARE COSTS MAT.	LABOR	EQUIP.	TOTAL	TOTAL INCL O&P	
062	0150	Average				%					10%	062
	0200	Maximum	R010 -070								15%	
	0290	Typical, by size of project, under $50,000								40%		
	0310	$50,000 to $100,000								35%		
	0320	$100,000 to $500,000								25%		
	0330	$500,000 to $1,000,000								20%		
070	0010	**PERMITS** Rule of thumb, most cities, minimum				Job					.50%	070
	0100	Maximum				"					2%	
082	0010	**SMALL TOOLS** As % of contractor's work, minimum				Total					.50%	082
	0100	Maximum				"					2%	
086	0010	**TAXES** Sales tax, State, average	R010 -090			%	4.78%					086
	0050	Maximum					7.25%					
	0200	Social Security, on first $65,400 of wages	R010 -100					7.65%				
	0300	Unemployment, MA, combined Federal and State, minimum						3.40%				
	0350	Average						7%				
	0400	Maximum						9.30%				

013 | Submittals

013 300 | Survey Data

			CREW	DAILY OUTPUT	LABOR-HOURS	UNIT	1998 BARE COSTS MAT.	LABOR	EQUIP.	TOTAL	TOTAL INCL O&P	
306	0010	**SURVEYING** Conventional, topographical, minimum	A-7	3.30	7.273	Acre	16	117		133	214	306
	0100	Maximum	A-8	.60	53.333		48	825		873	1,450	
	0300	Lot location and lines, minimum, for large quantities	A-7	2	12		25	193		218	355	
	0320	Average	"	1.25	19.200		45	310		355	565	
	0400	Maximum, for small quantities	A-8	1	32		72	495		567	915	
	0600	Monuments, 3' long	A-7	10	2.400	Ea.	16	38.50		54.50	82	
	0800	Property lines, perimeter, cleared land	"	1,000	.024	L.F.	.03	.39		.42	.68	
	0900	Wooded land	A-8	875	.037	"	.05	.57		.62	1.01	
	1100	Crew for building layout, 2 person crew	A-6	1	16	Day		275		275	455	
	1200	3 person crew	A-7	1	24	"		385		385	645	

013 400 | Shop Drawings

			CREW	DAILY OUTPUT	LABOR-HOURS	UNIT	1998 BARE COSTS MAT.	LABOR	EQUIP.	TOTAL	TOTAL INCL O&P	
408	0010	**RENDERINGS** Water color, matted, 20" x 30", eye level,										408
	0020	1 building, minimum				Ea.	425			425	470	
	0050	Average					975			975	1,075	
	0100	Maximum					1,575			1,575	1,725	

014 | Quality Control

014 100 | Testing Services

			CREW	DAILY OUTPUT	LABOR-HOURS	UNIT	1998 BARE COSTS MAT.	LABOR	EQUIP.	TOTAL	TOTAL INCL O&P	
108	0010	**TESTING**										108
	0015	For concrete building costing $1,000,000, minimum				Project					5,000	
	0020	Maximum									40,000	
	0050	Steel building, minimum									5,000	

014 100	Testing Services	CREW	DAILY OUTPUT	LABOR-HOURS	UNIT	1998 BARE COSTS				TOTAL INCL O&P	
						MAT.	LABOR	EQUIP.	TOTAL		
108	0070	Maximum				Project					15,000	**108**
	0600	Concrete testing, aggregates, abrasion				Ea.					120	
	0650	Absorption									46	
	1800	Compressive strength, cylinders, delivered to lab									13	
	1900	Picked up by lab, minimum									15	
	1950	Average									20	
	2000	Maximum									30	
	2200	Compressive strength, cores (not incl. drilling)									40	
	2250	Core drilling, 4" diameter (plus technician)				Inch					25	
	2260	Technician for core drilling				Hr.					50	
	2300	Patching core holes				Ea.					24	

015 200	Temporary Construction	CREW	DAILY OUTPUT	LABOR-HOURS	UNIT	1998 BARE COSTS				TOTAL INCL O&P	
						MAT.	LABOR	EQUIP.	TOTAL		
251	0010	**STAGING AIDS** and fall protection equipment										**251**
	0110	Cost each per day, based on 250 days use				Day	.13			.13	.14	
	0200	Guard post, buy				Ea.	15			15	16.50	
	0210	Cost each per day, based on 250 days use				Day	.06			.06	.07	
	0300	End guard chains, buy per set				Ea.	25			25	27.50	
	0310	Cost per set per day, based on 250 days use				Day	.13			.13	.14	
	1010	Cost each per day, based on 250 days use				"	.03			.03	.03	
	1100	Wood bracket, buy				Ea.	11.50			11.50	12.60	
	1110	Cost each per day, based on 250 days use				Day	.05			.05	.05	
	2010	Cost per pair per day, based on 250 days use				"	.32			.32	.35	
	2100	Steel siderail jack, buy per pair				Pair	69			69	76	
	2110	Cost per pair per day, based on 250 days use				Day	.28			.28	.30	
	3010	Cost each per day, based on 250 days use				"	.17			.17	.19	
	3100	Aluminum scaffolding plank, 20" wide x 24' long, buy				Ea.	760			760	835	
	3110	Cost each per day, based on 250 days use				Day	3.04			3.04	3.35	
	4010	Cost each per day, based on 250 days use				"	1.04			1.04	1.15	
	4100	Rope for safety line, 5/8" x 100' nylon, buy				Ea.	35			35	38.50	
	4110	Cost each per day, based on 250 days use				Day	.14			.14	.15	
	4200	Permanent U-Bolt roof anchor, buy				Ea.	37.50			37.50	41	
	4300	Temporary (one use) roof ridge anchor, buy				"	17			17	18.70	
	5000	Installation (setup and removal) of staging aids										
	5010	Sidewall staging bracket	2 Carp	64	.250	Ea.		4.63		4.63	7.95	
	5020	Guard post with 2 wood rails	"	64	.250			4.63		4.63	7.95	
	5030	End guard chains, set	1 Carp	64	.125			2.31		2.31	3.96	
	5100	Roof shingling bracket		96	.083			1.54		1.54	2.64	
	5200	Ladder jack		64	.125			2.31		2.31	3.96	
	5300	Wood plank, 2x10x16'	2 Carp	80	.200			3.70		3.70	6.35	
	5310	Aluminum scaffold plank, 20" x 24'	"	40	.400			7.40		7.40	12.70	
	5410	Safety rope	1 Carp	40	.200			3.70		3.70	6.35	
	5420	Permanent U-Bolt roof anchor (install only)	2 Carp	40	.400			7.40		7.40	12.70	
	5430	Temporary roof ridge anchor (install only)	1 Carp	64	.125			2.31		2.31	3.96	
254	0014	**SCAFFOLDING, STEEL TUBULAR** Rent, 1 use per mo., no plank R015 -100										**254**
	0015	Set up and take down										

015 200 | Temporary Construction

		CREW	DAILY OUTPUT	LABOR-HOURS	UNIT	1998 BARE COSTS				TOTAL INCL O&P	
						MAT.	LABOR	EQUIP.	TOTAL		
254	0090	Building exterior, wall face, 1 to 5 stories	3 Carp	24	1	C.S.F.	19.05	18.50		37.55	52.50	254
	0090	Building exterior, wall face, 1 to 5 stories	3 Carp	24	1	C.S.F.	19.05	18.50		37.55	52.50	
	0200	6 to 12 stories	4 Carp	21.20	1.509		19.05	28		47.05	69	
	0090	Building exterior, wall face, 1 to 5 stories	3 Carp	24	1	C.S.F.	19.05	18.50		37.55	52.50	
	0310	13 to 20 stories	5 Carp	20	2		19.05	37		56.05	84.50	
	0460	Building interior, wall face area, up to 16' high	3 Carp	25	.960		19.05	17.75		36.80	51.50	
	0560	16' to 40' high		23	1.043		19.05	19.30		38.35	54	
	0800	Building interior floor area, up to 30' high		312	.077	C.C.F.	1.47	1.42		2.89	4.05	
	0900	Over 30' high	4 Carp	275	.116	"	1.47	2.15		3.62	5.30	
	2000	Steel tubular, regular, rent/mo.										
	2100	Frames 3' high 5' wide				Ea.	3.75			3.75	4.13	
	2150	5' high 5' wide					3.75			3.75	4.13	
	2200	6'-4" high 5' wide					3.75			3.75	4.13	
	2250	7'-6" high 6' wide					7			7	7.70	
	2500	Accessories, cross braces					.60			.60	.66	
	2550	Guardrail post					1			1	1.10	
	2600	Guardrail 7' section					.75			.75	.83	
	2650	Screw jacks & plates					1.50			1.50	1.65	
	2700	Sidearm brackets					1.50			1.50	1.65	
	2750	8" casters					6			6	6.60	
	2800	Outrigger for rolling tower										
	2850	Plank 2" x 10" x 16'-0"				Ea.	5			5	5.50	
	6820	Erect and dismantle frames, 1st tier	4 Clab	45	.711			9.55		9.55	16.30	
	6830	Erect and dismantle frames, 2nd tier		93	.344			4.61		4.61	7.90	
	6840	Erect and dismantle frames, 3rd tier		87	.368			4.93		4.93	8.45	
	6850	Erect and dismantle frames, 4th tier		75	.427			5.70		5.70	9.80	

015 600 | Temporary Controls

		CREW	DAILY OUTPUT	LABOR-HOURS	UNIT	1998 BARE COSTS				TOTAL INCL O&P	
						MAT.	LABOR	EQUIP.	TOTAL		
602	0010	**TARPAULINS** Cotton duck, 10 oz. to 13.13 oz. per S.Y., minimum				S.F.	.38			.38	.42	602
	0050	Maximum					.55			.55	.61	
	0200	Reinforced polyethylene 3 mils thick, white					.10			.10	.11	
	0300	4 mils thick, white, clear or black					.12			.12	.13	
	0730	Polyester reinforced w/ integral fastening system 11 mils thick					1			1	1.10	

For the cell at 0090 line 1: crew reference **R015 -100**.

016 400 | Equipment Rental

			UNIT	HOURLY OPER. COST	RENT PER DAY	RENT PER WEEK	RENT PER MONTH	CREW EQUIPMENT COST/DAY	
406	0010	**CONCRETE EQUIPMENT RENTAL**							406
	0100	without operators							
	0200	Bucket, concrete lightweight, 1/2 C.Y.	Ea.	.14	26.50	80	240	17.10	
	0300	1 C.Y.		.17	41.50	125	375	26.35	
	0400	1-1/2 C.Y.		.24	43.50	130	390	27.90	
	0500	2 C.Y.		.25	53.50	160	480	34	
	0600	Cart, concrete, self propelled, operator walking, 10 C.F.		1.20	68.50	205	615	50.60	
	0700	Operator riding, 18 C.F.		2.35	110	330	990	84.80	
	0800	Conveyer for concrete, portable, gas, 16″ wide, 26′ long		3.50	167	500	1,500	128	
	0900	46′ long		3.60	227	680	2,050	164.80	
	1000	56′ long		3.75	250	750	2,250	180	
	1100	Core drill, electric, 2-1/2 H.P., 1″ to 8″ bit diameter		.40	68.50	205	615	44.20	
	1150	11 H.P., 8″ to 18″ cores		.65	90	270	810	59.20	
	1200	Finisher, concrete floor, gas, riding trowel, 48″ diameter		2.60	107	320	960	84.80	
	1300	Gas, manual, 3 blade, 36″ trowel		2.20	58.50	175	525	52.60	
	1400	4 blade, 48″ trowel		2.30	53.50	160	480	50.40	
	1500	Float, hand-operated (Bull float) 48″ wide		.10	6.65	20	60	4.80	
	1570	Curb builder , 14 H.P., gas, single screw		1.55	52	156	470	43.60	
	1590	Double screw		2.25	55	165	495	51	
	1600	Grinder, concrete and terrazzo, electric, floor		1.30	56	168	505	44	
	1700	Wall grinder		.60	33.50	100	300	24.80	
	1800	Mixer, powered, mortar and concrete, gas, 6 C.F., 18 H.P.		.90	56.50	170	510	41.20	
	1900	10 C.F., 25 H.P.		1.30	66.50	200	600	50.40	
	2000	16 C.F.		1.55	83.50	250	750	62.40	
	2100	Concrete, stationary, tilt drum, 2 C.Y.		7.70	325	980	2,950	257.60	
	2120	Pump, concrete, truck mounted, 4″ line, 80′ boom		11.60	900	2,700	8,100	632.80	
	2140	5″ line, 110′ boom		12.90	1,075	3,200	9,600	743.20	
	2160	Mud jack, 47 C.F. per hr.		3.08	66.50	200	600	64.65	
	2180	225 C.F. per hr.		6.10	300	900	2,700	228.80	
	2600	Saw, concrete, manual, gas, 18 H.P.		2.75	75	225	675	67	
	2650	Self-propelled, gas, 30 H.P.		5.42	107	320	960	107.35	
	2700	Vibrators, concrete, electric, 60 cycle, 2 H.P.		.30	23.50	70	210	16.40	
	2800	3 H.P.		.35	33.50	100	300	22.80	
	2900	Gas engine, 5 H.P.		.65	33.50	100	300	25.20	
	3000	8 H.P.		.90	50	150	450	37.20	
408	0010	**EARTHWORK EQUIPMENT RENTAL** Without operators							408
	0040	Aggregate spreader, push type 8′ to 12′ wide	Ea.	1.10	100	300	900	68.80	
	0050	Augers for truck or trailer mounting, vertical drilling							
	0060	4″ to 36″ diam., 54 H.P., gas, 10′ spindle travel	Ea.	15.45	535	1,610	4,825	445.60	
	0070	14′ spindle travel		15.90	635	1,900	5,700	507.20	
	0080	Auger, horizontal boring machine, 12″ to 36″ diameter, 45 H.P.		7.10	315	950	2,850	246.80	
	0090	12″ to 48″ diameter, 65 H.P.		12.10	635	1,900	5,700	476.80	
	0100	Backhoe, diesel hydraulic, crawler mounted, 1/2 C.Y. cap.		11.10	410	1,225	3,675	333.80	
	0120	5/8 C.Y. capacity		15.08	460	1,375	4,125	395.65	
	0140	3/4 C.Y. capacity		16.88	500	1,500	4,500	435.05	
	0150	1 C.Y. capacity		21.55	600	1,800	5,400	532.40	
	0200	1-1/2 C.Y. capacity		25.82	835	2,500	7,500	706.55	
	0300	2 C.Y. capacity		41.60	1,125	3,400	10,200	1,013	
	0320	2-1/2 C.Y. capacity		57.65	2,025	6,100	18,300	1,681	
	0340	3-1/2 C.Y. capacity		77.13	2,500	7,500	22,500	2,117	
	0341	Attachments							
	0342	Bucket thumbs		.45	335	1,000	3,000	203.60	
	0345	Grapples		.30	227	680	2,050	138.40	
	0350	Gradall type, truck mounted, 3 ton @ 15′ radius, 5/8 C.Y.		25.70	700	2,100	6,300	625.60	
	0370	1 C.Y. capacity		27.40	950	2,850	8,550	789.20	
	0400	Backhoe-loader, wheel type, 40 to 45 H.P., 5/8 C.Y. capacity		7.10	233	700	2,100	196.80	
	0450	45 H.P. to 60 H.P., 3/4 C.Y. capacity		8.35	245	735	2,200	213.80	

Important: See the Reference Section for critical supporting data - Reference Nos., Crews, & Location Factors

016 400 | Equipment Rental

		UNIT	HOURLY OPER. COST	RENT PER DAY	RENT PER WEEK	RENT PER MONTH	CREW EQUIPMENT COST/DAY	
408	0460	80 H.P., 1-1/4 C.Y. capacity	Ea.	12.15	315	950	2,850	287.20	408
	0480	Attachments							
	0482	Compactor, 20,000 lb		1.25	183	550	1,650	120	
	0485	Hydraulic hammer, 750 ft-lbs		.40	297	890	2,675	181.20	
	0486	Hydraulic hammer, 1000 ft-lbs		.45	335	1,000	3,000	203.60	
	0500	Brush chipper, gas engine, 6″ cutter head, 35 H.P.		4	133	400	1,200	112	
	0550	12″ cutter head, 130 H.P.		12.15	185	555	1,675	208.20	
	0600	15″ cutter head, 165 H.P.		15.70	248	745	2,225	274.60	
	0750	Bucket, clamshell, general purpose, 3/8 C.Y.		.78	58.50	175	525	41.25	
	0800	1/2 C.Y.		1.08	66.50	200	600	48.65	
	0850	3/4 C.Y.		1.09	75	225	675	53.70	
	0900	1 C.Y.		1.35	90	270	810	64.80	
	0950	1-1/2 C.Y.		1.60	125	375	1,125	87.80	
	1000	2 C.Y.		1.80	140	420	1,250	98.40	
	1010	Bucket, dragline, medium duty, 1/2 C.Y.		.35	36.50	110	330	24.80	
	1020	3/4 C.Y.		.55	46.50	140	420	32.40	
	1030	1 C.Y.		.58	50	150	450	34.65	
	1040	1-1/2 C.Y.		.75	68.50	205	615	47	
	1050	2 C.Y.		.82	83.50	250	750	56.55	
	1070	3 C.Y.		1.30	117	350	1,050	80.40	
	1200	Compactor, roller, 2 drum, 2000 lb., operator walking		1.83	127	380	1,150	90.65	
	1250	Rammer compactor, gas, 1000 lb. blow		.55	43.50	130	390	30.40	
	1300	Vibratory plate, gas, 13″ plate, 1000 lb. blow		.69	40	120	360	29.50	
	1350	24″ plate, 5000 lb. blow		1.70	60	180	540	49.60	
	1750	Extractor, piling, see lines 2500 to 2750							
	1860	Grader, self-propelled, 25,000 lb.	Ea.	14.40	565	1,690	5,075	453.20	
	1910	30,000 lb.		17.30	700	2,100	6,300	558.40	
	1920	40,000 lb.		19.90	835	2,500	7,500	659.20	
	1930	55,000 lb.		32.50	1,250	3,750	11,300	1,010	
	1950	Hammer, pavement demo., hyd., gas, self-prop., 1000 to 1250 lb.		18.05	395	1,190	3,575	382.40	
	2000	Diesel 1300 to 1500 lb.		8.75	440	1,320	3,950	334	
	3000	Roller, tandem, gas, 3 to 5 ton		6.05	167	500	1,500	148.40	
	3050	Diesel, 8 to 12 ton		8.75	280	840	2,525	238	
	3100	Towed type, vibratory, gas 12.5 H.P., 2 ton		3.75	133	400	1,200	110	
	3150	Sheepsfoot, double 60″ x 60″		4.60	140	420	1,250	120.80	
	3200	Pneumatic tire diesel roller, 12 ton		8.40	283	850	2,550	237.20	
	3250	21 to 25 ton		15.50	335	1,000	3,000	324	
	3300	Sheepsfoot roller, self-propelled, 4 wheel, 130 H.P.		20.35	635	1,900	5,700	542.80	
	3320	300 H.P.		25.90	745	2,230	6,700	653.20	
	3350	Vibratory steel drum & pneumatic tire, diesel, 18,000 lb.		11.15	420	1,260	3,775	341.20	
	3400	29,000 lb.		12.35	515	1,550	4,650	408.80	
	3450	Scrapers, towed type, 9 to 12 C.Y. capacity		3.10	83.50	250	750	74.80	
	3500	12 to 17 C.Y. capacity		5.76	250	750	2,250	196.10	
	3550	Scrapers, self-propelled, 4 x 4 drive, 2 engine, 14 C.Y. capacity		59.10	1,825	5,500	16,500	1,573	
	3600	2 engine, 24 C.Y. capacity		71.80	2,125	6,400	19,200	1,854	
	3650	Self-loading, 11 C.Y. capacity		26.52	760	2,275	6,825	667.15	
	3700	22 C.Y. capacity		32.11	1,175	3,500	10,500	956.90	
	3710	Screening plant 110 hp. w / 5' x 10'screen		15.50	415	1,250	3,750	374	
	3720	5' x 16' screen		17.10	500	1,500	4,500	436.80	
	3850	Shovels, see Cranes division 016-460							
	3860	Shovel/backhoe bucket, 1/2 C.Y.	Ea.	1.10	75	225	675	53.80	
	3870	3/4 C.Y.		3.10	125	375	1,125	99.80	
	3880	1 C.Y.		3.40	167	500	1,500	127.20	
	3890	1-1/2 C.Y.		4.09	207	620	1,850	156.70	
	3910	3 C.Y.		7.30	385	1,150	3,450	288.40	
	4110	Tractor, crawler, with bulldozer, torque converter, diesel 75 H.P.		10.65	365	1,100	3,300	305.20	
	4150	105 H.P.		14.25	500	1,500	4,500	414	
	4200	140 H.P.		16.50	605	1,820	5,450	496	

016 400 | Equipment Rental

			UNIT	HOURLY OPER. COST	RENT PER DAY	RENT PER WEEK	RENT PER MONTH	CREW EQUIPMENT COST/DAY	
408	4260	200 H.P.	Ea.	29	1,000	3,000	9,000	832	408
	4310	300 H.P.		37.70	1,375	4,100	12,300	1,122	
	4360	410 H.P.		46.10	1,600	4,800	14,400	1,329	
	4380	700 H.P.		92.80	3,325	10,000	30,000	2,742	
	4400	Loader, crawler, torque conv., diesel, 1-1/2 C.Y., 80 H.P.		12.15	435	1,300	3,900	357.20	
	4450	1-1/2 to 1-3/4 C.Y., 95 H.P.		14.52	445	1,340	4,025	384.15	
	4510	1-3/4 to 2-1/4 C.Y., 130 H.P.		18.25	585	1,750	5,250	496	
	4530	2-1/2 to 3-1/4 C.Y., 190 H.P.		29.75	1,000	3,000	9,000	838	
	4560	4-1/2 to 5 C.Y., 275 H.P.		41.56	1,325	4,000	12,000	1,132	
	4610	Tractor loader, wheel, torque conv., 4 x 4, 1 to 1-1/4 C.Y., 65 H.P.		9.06	273	820	2,450	236.50	
	4620	1-1/2 to 1-3/4 C.Y., 80 H.P.		11.54	365	1,100	3,300	312.30	
	4650	1-3/4 to 2 C.Y., 100 H.P.		12.11	395	1,180	3,550	332.90	
	4710	2-1/2 to 3-1/2 C.Y., 130 H.P.		18.26	485	1,450	4,350	436.10	
	4730	3 to 4-1/2 C.Y., 170 H.P.		20.25	665	2,000	6,000	562	
	4760	5-1/4 to 5-3/4 C.Y., 270 H.P.		37.19	1,000	3,000	9,000	897.50	
	4810	7 to 8 C.Y., 375 H.P.		55.22	1,325	4,000	12,000	1,242	
	4870	12-1/2 C.Y., 690 H.P.		112.23	2,325	7,000	21,000	2,298	
	4880	Wheeled, skid steer, 10 C.F., 30 H.P. gas		4.40	117	350	1,050	105.20	
	4890	1 C.Y., 78 H.P., diesel		6.10	300	900	2,700	228.80	
	4891	Attachments for all skid steer loaders							
	4892	Auger	Ea.	.12	81.50	245	735	49.95	
	4893	Backhoe		.15	107	320	960	65.20	
	4894	Broom		.16	103	310	930	63.30	
	4895	Forks		.08	36.50	110	330	22.65	
	4896	Grapple		.12	83.50	250	750	50.95	
	4897	Concrete hammer		.24	175	525	1,575	106.90	
	4898	Tree spade		.35	125	375	1,125	77.80	
	4899	Trencher		.40	233	700	2,100	143.20	
	4900	Trencher, chain, boom type, gas, operator walking, 12 H.P.		2.10	125	375	1,125	91.80	
	4910	Operator riding, 40 H.P.		6.17	267	800	2,400	209.35	
	5000	Wheel type, diesel, 4' deep, 12" wide		12.06	500	1,500	4,500	396.50	
	5100	Diesel, 6' deep, 20" wide		15.19	735	2,200	6,600	561.50	
	5150	Ladder type, diesel, 5' deep, 8" wide		9.14	365	1,100	3,300	293.10	
	5200	Diesel, 8' deep, 16" wide		17.10	650	1,950	5,850	526.80	
	5250	Truck, dump, tandem, 12 ton payload		16.90	365	1,100	3,300	355.20	
	5300	Three axle dump, 16 ton payload		19.69	455	1,370	4,100	431.50	
	5350	Dump trailer only, rear dump, 16-1/2 C.Y.		3.35	170	510	1,525	128.80	
	5400	20 C.Y.		3.41	173	520	1,550	131.30	
	5450	Flatbed, single axle, 1-1/2 ton rating		10.55	138	415	1,250	167.40	
	5500	3 ton rating		11.10	143	430	1,300	174.80	
	5550	Off highway rear dump, 25 ton capacity		19.90	865	2,600	7,800	679.20	
	5600	35 ton capacity		31.33	1,275	3,800	11,400	1,011	
420	0010	**GENERAL EQUIPMENT RENTAL** Without operators							420
	0150	Aerial lift, scissor type, to 15' high, 1000 lb. cap., electric	Ea.	1.15	83.50	250	750	59.20	
	0160	To 25' high, 2000 lb. capacity		1.75	127	380	1,150	90	
	0170	Telescoping boom to 40' high, 750 lb. capacity, gas		6.23	365	1,100	3,300	269.85	
	0180	1000 lb. capacity		7.75	445	1,340	4,025	330	
	0195	Air compressor, portable, 6.5 CFM, electric		.12	33.50	100	300	20.95	
	0196	gasoline		.13	41.50	125	375	26.05	
	0200	Air compressor, portable, gas engine, 60 C.F.M.		5.25	53.50	160	480	74	
	0300	160 C.F.M.		6.60	70	210	630	94.80	
	0400	Diesel engine, rotary screw, 250 C.F.M.		6.30	117	350	1,050	120.40	
	0500	365 C.F.M.		9.02	165	495	1,475	171.15	
	0600	600 C.F.M.		15.52	240	720	2,150	268.15	
	0700	750 C.F.M.		17.06	243	730	2,200	282.50	
	0800	For silenced models, small sizes, add		3%	5%	5%	5%		
	0900	Large sizes, add		5%	7%	7%	7%		
	0920	Air tools and accessories							

Important: See the Reference Section for critical supporting data - Reference Nos., Crews, & Location Factors

016 400 | Equipment Rental

		UNIT	HOURLY OPER. COST	RENT PER DAY	RENT PER WEEK	RENT PER MONTH	CREW EQUIPMENT COST/DAY	
420	0930	Breaker, pavement, 60 lb.	Ea.	.18	28.50	85	255	18.45	**420**
	0940	80 lb.		.20	33.50	100	300	21.60	
	0950	Drills, hand (jackhammer) 65 lb.		.22	25	75	225	16.75	
	0960	Track or wagon, swing boom, 4" drifter		10.26	365	1,100	3,300	302.10	
	0970	5" drifter		11.05	635	1,900	5,700	468.40	
	0980	Dust control per drill		2.05	11.65	35	105	23.40	
	0990	Hammer, chipping, 12 lb.		.12	20	60	180	12.95	
	1000	Hose, air with couplings, 50' long, 3/4" diameter		.15	5	15	45	4.20	
	1100	1" diameter		.15	6	18	54	4.80	
	1200	1-1/2" diameter		.15	11.65	35	105	8.20	
	1300	2" diameter		.20	16.65	50	150	11.60	
	1400	2-1/2" diameter		.21	20	60	180	13.70	
	1410	3" diameter		.22	25	75	225	16.75	
	1450	Drill, steel, 7/8" x 2'			3.33	10	30	2	
	1460	7/8" x 6'			4	12	36	2.40	
	1520	Moil points		.80	2	6	18	7.60	
	1525	Pneumatic nailer w/accessories		.12	30	90	270	18.95	
	1530	Sheeting driver for 60 lb. breaker		.15	10	30	90	7.20	
	1540	For 90 lb. breaker		.15	16.65	50	150	11.20	
	1550	Spade, 25 lb.		.08	8.35	25	75	5.65	
	1560	Tamper, single, 35 lb.		.10	23.50	70	210	14.80	
	1570	Triple, 140 lb.		1.75	40	120	360	38	
	1580	Wrenches, impact, air powered, up to 3/4" bolt			21.50	65	195	13	
	1590	Up to 1-1/4" bolt		.34	41.50	125	375	27.70	
	1600	Barricades, barrels, reflectorized, 1 to 50 barrels			2	6	18	1.20	
	1610	100 to 200 barrels			1.33	4	12	.80	
	1620	Barrels with flashers, 1 to 50 barrels			3	9	27	1.80	
	1630	100 to 200 barrels			2	6	18	1.20	
	1640	Barrels with steady burn type C lights			3.33	10	30	2	
	1650	Illuminated board, trailer mounted, with generator		.76	83.50	250	750	56.10	
	1670	Portable, stock, with flashers, 1 to 6 units			.67	2	6	.40	
	1680	25 to 50 units			.67	2	6	.40	
	1700	Carts, brick, hand powered, 1000 lb. capacity		1.08	21.50	65	195	21.65	
	1800	Gas engine, 1500 lb., 7-1/2' lift		1.60	91.50	275	825	67.80	
	1830	Distributor, asphalt, trailer mtd, 2000 gal., 38 H.P. diesel		7.32	450	1,350	4,050	328.55	
	1840	3000 gal., 38 H.P. diesel		7.88	485	1,450	4,350	353.05	
	1850	Drill, rotary hammer, electric, 1-1/2" diameter		.15	23.50	70	210	15.20	
	1860	Carbide bit for above			5.35	16	48	3.20	
	1870	Emulsion sprayer, 65 gal., 5 H.P. gas engine		.61	53.50	160	480	36.90	
	1880	200 gal., 5 H.P. engine		.65	56.50	170	510	39.20	
	1920	Floodlight, mercury, vapor or quartz, on tripod							
	1930	1000 watt	Ea.	.12	16.65	50	150	10.95	
	1940	2000 watt	"	.12	38.50	115	345	23.95	
	1950	Floodlights, trailer mounted with generator, 1-300 watt light		4.82	16.65	50	150	48.55	
	2000	4-300 watt lights	Ea.	6.12	33.50	100	300	68.95	
	2020	Forklift, wheeled, for brick, 18', 3000 lb., 2 wheel drive, gas		9.01	170	510	1,525	174.10	
	2040	28', 4000 lb., 4 wheel drive, diesel		6.20	210	630	1,900	175.60	
	2100	Generator, electric, gas engine, 1.5 KW to 3 KW		1.10	36.50	110	330	30.80	
	2200	5 KW		1.45	55	165	495	44.60	
	2300	10 KW		2.32	133	400	1,200	98.55	
	2400	25 KW		6.61	147	440	1,325	140.90	
	2500	Diesel engine, 20 KW		4.22	107	320	960	97.75	
	2600	50 KW		6.66	120	360	1,075	125.30	
	2700	100 KW		11.01	183	550	1,650	198.10	
	2800	250 KW		28.15	315	950	2,850	415.20	
	2850	Hammer, hydraulic, for mounting on boom, to 500 ft.-lb.		.95	125	375	1,125	82.60	
	2860	500 to 1200 ft.-lb.		2.45	258	775	2,325	174.60	
	2900	Heaters, space, oil or electric, 50 MBH		.12	26.50	80	240	16.95	

016 400 | Equipment Rental

		UNIT	HOURLY OPER. COST	RENT PER DAY	RENT PER WEEK	RENT PER MONTH	CREW EQUIPMENT COST/DAY	
420	3000	100 MBH	Ea.	.12	30	90	270	18.95	**420**
	3100	300 MBH		.12	48.50	145	435	29.95	
	3150	500 MBH		.15	65	195	585	40.20	
	3200	Hose, water, suction with coupling, 20' long, 2" diameter		.06	10	30	90	6.50	
	3210	3" diameter		.06	15	45	135	9.50	
	3220	4" diameter		.06	20	60	180	12.50	
	3230	6" diameter		.06	33.50	100	300	20.50	
	3240	8" diameter		.07	43.50	130	390	26.55	
	3250	Discharge hose with coupling, 50' long, 2" diameter		.05	6.65	20	60	4.40	
	3260	3" diameter		.05	8.35	25	75	5.40	
	3270	4" diameter		.06	13.35	40	120	8.50	
	3280	6" diameter		.06	30	90	270	18.50	
	3290	8" diameter		.08	31.50	95	285	19.65	
	3300	Ladders, extension type, 16' to 36' long			15	45	135	9	
	3400	40' to 60' long			30	90	270	18	
	3410	Level, laser type, for pipe laying, self leveling			98.50	295	885	59	
	3430	Manual leveling			76.50	230	690	46	
	3440	Rotary beacon with rod and sensor			100	300	900	60	
	3460	Builders level with tripod and rod			28.50	85	255	17	
	3500	Light towers, towable, with diesel generator, 2000 watt		1.50	113	340	1,025	80	
	3600	4000 watt		1.91	133	400	1,200	95.30	
	3700	Mixer, powered, plaster and mortar, 6 C.F., 7 H.P.		.80	51.50	155	465	37.40	
	3800	10 C.F., 9 H.P.		1.18	75	225	675	54.45	
	3850	Nailer, pneumatic		.08	23.50	70	210	14.65	
	3900	Paint sprayers complete, 8 CFM		.08	40	120	360	24.65	
	4000	17 CFM		.08	58.50	175	525	35.65	
	4020	Pavers, bituminous, rubber tires, 8' wide, 52 H.P., gas		14.50	575	1,725	5,175	461	
	4030	8' wide, 64 H.P., diesel		15	1,000	3,000	9,000	720	
	4050	Crawler, 10' wide, 78 H.P., gas		21.55	1,400	4,200	12,600	1,012	
	4060	10' wide, 87 H.P., diesel		22.14	1,075	3,200	9,600	817.10	
	4070	Concrete paver, 12' to 24' wide, 250 H.P.		23.56	1,525	4,550	13,700	1,098	
	4080	Placer-spreader-trimmer, 24' wide, 300 H.P.		31.09	1,775	5,320	16,000	1,313	
	4100	Pump, centrifugal gas pump, 1-1/2", 4 MGPH		.45	30	90	270	21.60	
	4200	2", 8 MGPH		.50	33.50	100	300	24	
	4300	3", 15 MGPH		1.20	45	135	405	36.60	
	4400	6", 90 MGPH		9.40	133	400	1,200	155.20	
	4500	Submersible electric pump, 1-1/4", 55 GPM		.35	35	105	315	23.80	
	4600	1-1/2", 83 GPM		.38	38.50	115	345	26.05	
	4700	2", 120 GPM		.40	43.50	130	390	29.20	
	4800	3", 300 GPM		.70	53.50	160	480	37.60	
	4900	4", 560 GPM		1.15	66.50	200	600	49.20	
	5000	6", 1590 GPM		5.20	200	600	1,800	161.60	
	5100	Diaphragm pump, gas, single, 1-1/2" diameter		.48	25	75	225	18.85	
	5200	2" diameter		.50	40	120	360	28	
	5300	3" diameter		.75	43.50	130	390	32	
	5400	Double, 4" diameter		1.50	83.50	250	750	62	
	5500	Trash pump, self-priming, gas, 2" diameter		1.20	39.50	118	355	33.20	
	5600	Diesel, 4" diameter		1.90	93.50	280	840	71.20	
	5650	Diesel, 6" diameter		4.88	157	470	1,400	133.05	
	5660	Rollers, see division 016-408							
	5700	Salamanders, L.P. gas fired, 100,000 B.T.U.	Ea.	.68	20	60	180	17.45	
	5720	Sandblaster, portable, open top, 3 C.F. capacity		.18	48.50	145	435	30.45	
	5730	6 C.F. capacity		.32	58.50	175	525	37.55	
	5740	Accessories for above		.06	16.65	50	150	10.50	
	5750	Sander, floor		.12	38.50	115	345	23.95	
	5760	Edger		.12	26.50	80	240	16.95	
	5800	Saw, chain, gas engine, 18" long		.52	38.50	115	345	27.15	
	5900	36" long		1.10	73.50	220	660	52.80	

Important: See the Reference Section for critical supporting data - Reference Nos., Crews, & Location Factors

016 400 | Equipment Rental

		UNIT	HOURLY OPER. COST	RENT PER DAY	RENT PER WEEK	RENT PER MONTH	CREW EQUIPMENT COST/DAY	
420	6000	Masonry, table mounted, 14" diameter, 5 H.P.	Ea.	1.77	50	150	450	44.15	**420**
	6100	Circular, hand held, electric, 7-1/4" diameter		.16	18.35	55	165	12.30	
	6200	12" diameter		.26	33.50	100	300	22.10	
	6300	Steam cleaner, 100 gallons per hour		.42	51.50	155	465	34.35	
	6310	200 gallons per hour		.68	58.50	175	525	40.45	
	6340	Tar Kettle/Pot, 400 gallon		.48	66.50	200	600	43.85	
	6350	Torch, cutting, acetylene-oxygen, 150' hose		7.31	23.50	70	210	72.50	
	6360	Hourly operating cost includes tips and gas		7.40				59.20	
	6410	Toilet, portable chemical			13.35	40	120	8	
	6420	Recycle flush type			16.65	50	150	10	
	6450	Toilet, trailers, minimum			33.50	100	300	20	
	6460	Maximum			100	300	900	60	
	6500	Trailers, platform, flush deck, 2 axle, 25 ton capacity		1.35	120	360	1,075	82.80	
	6600	40 ton capacity		1.66	222	665	2,000	146.30	
	6700	3 axle, 50 ton capacity		2.77	237	710	2,125	164.15	
	6800	75 ton capacity		3.55	315	945	2,825	217.40	
	6900	Water tank, engine driven discharge, 5000 gallons		7.90	233	700	2,100	203.20	
	7000	10,000 gallons		10.75	335	1,000	3,000	286	
	7020	Transit with tripod			33.50	100	300	20	
	7030	Trench box, 3000 lbs. 6'x8'		.78	95	285	855	63.25	
	7040	7200 lbs. 6'x20'		1.42	167	500	1,500	111.35	
	7050	8000 lbs., 8' x 16'		1.56	177	530	1,600	118.50	
	7060	9500 lbs., 8'x20'		1.77	183	550	1,650	124.15	
	7065	11,000 lbs., 8'x24'		1.82	235	705	2,125	155.55	
	7070	12,000 lbs., 10' x 20'		1.92	243	730	2,200	161.35	
	7100	Truck, pickup, 3/4 ton, 2 wheel drive		10.31	71.50	215	645	125.50	
	7200	4 wheel drive		11.71	80	240	720	141.70	
	7300	Tractor, 4 x 2, 30 ton capacity, 195 H.P.		10.81	385	1,150	3,450	316.50	
	7410	250 H.P.		14.56	415	1,250	3,750	366.50	
	7500	6 x 2, 40 ton capacity, 240 H.P.		17.39	460	1,375	4,125	414.10	
	7600	6 x 4, 45 ton capacity, 240 H.P.		22.48	565	1,690	5,075	517.85	
	7620	Vacuum truck, hazardous material, 2500 gallon		13.49	297	890	2,675	285.90	
	7625	5,000 gallon		14.58	335	1,000	3,000	316.65	
	7650	Vacuum, H.E.P.A., 16 gal., wet/dry		.90	50	150	450	37.20	
	7655	55 gal, wet/dry		.78	50	150	450	36.25	
	7690	Large production vacuum loader, 3150 CFM		11.51	700	2,100	6,300	512.10	
	7700	Welder, electric, 200 amp		.79	30	90	270	24.30	
	7800	300 amp		1.06	66.50	200	600	.48.50	
	7900	Gas engine, 200 amp		4.20	50	150	450	63.60	
	8000	300 amp		5.02	66.50	200	600	80.15	
	8100	Wheelbarrow, any size			8.35	25	75	5	
	8200	Wrecking ball, 4000 lb.		.40	60	180	540	39.20	
460	0010	**LIFTING AND HOISTING EQUIPMENT RENTAL**							**460**
	0100	without operators							
	0600	Crawler, cable, 1/2 C.Y., 15 tons at 12' radius	Ea.	18.45	500	1,500	4,500	447.60	
	0700	3/4 C.Y., 20 tons at 12' radius		19.61	510	1,530	4,600	462.90	
	0800	1 C.Y., 25 tons at 12' radius		20.24	560	1,675	5,025	496.90	
	0900	Crawler, cable, 1-1/2 C.Y., 40 tons at 12' radius		29.65	865	2,600	7,800	757.20	
	1000	2 C.Y., 50 tons at 12' radius		34.55	1,000	3,000	9,000	876.40	
	1100	3 C.Y., 75 tons at 12' radius		42.39	1,025	3,100	9,300	959.10	
	1200	100 ton capacity, standard boom		40.22	1,400	4,200	12,600	1,162	
	1300	165 ton capacity, standard boom		62.88	2,325	7,000	21,000	1,903	
	1400	200 ton capacity, 150' boom		117.21	2,525	7,600	22,800	2,458	
	1500	450' boom		131.55	3,200	9,600	28,800	2,972	
	1600	Truck mounted, cable operated, 6 x 4, 20 tons at 10' radius		13.66	700	2,100	6,300	529.30	
	1700	25 tons at 10' radius		20.19	1,100	3,300	9,900	821.50	
	1800	8 x 4, 30 tons at 10' radius		27.78	635	1,900	5,700	602.25	
	1900	40 tons at 12' radius		28.49	765	2,300	6,900	687.90	

016 400 | Equipment Rental

460			UNIT	HOURLY OPER. COST	RENT PER DAY	RENT PER WEEK	RENT PER MONTH	CREW EQUIPMENT COST/DAY	460
	2000	8 x 4, 60 tons at 15' radius	Ea.	43.56	930	2,790	8,375	906.50	
	2100	90 tons at 15' radius		47.52	1,075	3,250	9,750	1,030	
	2200	115 tons at 15' radius		49.66	1,925	5,750	17,300	1,547	
	2300	150 tons at 18' radius		74.38	1,600	4,800	14,400	1,555	
	2400	Truck mounted, hydraulic, 12 ton capacity		22.26	435	1,300	3,900	438.10	
	2500	25 ton capacity		22.98	590	1,775	5,325	538.85	
	2550	33 ton capacity		23.66	835	2,500	7,500	689.30	
	2600	55 ton capacity		33.15	865	2,600	7,800	785.20	
	2700	80 ton capacity		36.20	1,325	4,000	12,000	1,090	
	2800	Self-propelled, 4 x 4, with telescoping boom, 5 ton		9.88	315	950	2,850	269.05	
	2900	12-1/2 ton capacity		15.94	565	1,700	5,100	467.50	
	3000	15 ton capacity		17.67	465	1,400	4,200	421.35	
	3100	25 ton capacity		20.38	665	2,000	6,000	563.05	
	3200	Derricks, guy, 20 ton capacity, 60' boom, 75' mast		8.45	283	850	2,550	237.60	
	3300	100' boom, 115' mast		15.91	500	1,500	4,500	427.30	
	3400	Stiffleg, 20 ton capacity, 70' boom, 37' mast		11.22	365	1,100	3,300	309.75	
	3500	100' boom, 47' mast		17.53	615	1,850	5,550	510.25	
	3600	Hoists, chain type, overhead, manual, 3/4 ton		.06	6	18	54	4.10	
	3900	10 ton		.24	25	75	225	16.90	
	4000	Hoist and tower, 5000 lb. cap., portable electric, 40' high		3.91	167	500	1,500	131.30	
	4100	For each added 10' section, add			10	30	90	6	
	5200	Jacks, hydraulic, 20 ton		.15	2.67	8	24	2.80	
	5500	100 ton		.18	20	60	180	13.45	
	6000	Jacks, hydraulic, climbing with 50' jackrods							
	6010	and control consoles, minimum 3 mo. rental							
	6100	30 ton capacity	Ea.	.06	100	300	900	60.50	
	6150	For each added 10' jackrod section, add			3	9	27	1.80	
	6300	50 ton capacity			160	480	1,450	96	
	6350	For each added 10' jackrod section, add			3.33	10	30	2	
	6500	125 ton capacity			435	1,300	3,900	260	
	6550	For each added 10' jackrod section, add			25	75	225	15	
	6600	Cable jack, 10 ton capacity with 200' cable			80	240	720	48	
	6650	For each added 50' of cable, add			8.35	25	75	5	

Important: See the Reference Section for critical supporting data - Reference Nos., Crews, & Location Factors

017 100	Final Cleaning	CREW	DAILY OUTPUT	LABOR-HOURS	UNIT	1998 BARE COSTS				TOTAL INCL O&P	
						MAT.	LABOR	EQUIP.	TOTAL		
104	0010 **CLEANING UP** After job completion, allow, minimum				Job					.30%	104
	0040　　　Maximum				"					1%	

For information about Means Estimating Seminars, see yellow pages 11 and 12 in back of book

GENERAL REQUIREMENTS

For expanded coverage of these items see *Means Building Construction Cost Data 1998*

	CREW	DAILY OUTPUT	LABOR-HOURS	UNIT	1998 BARE COSTS				TOTAL INCL O&P
					MAT.	LABOR	EQUIP.	TOTAL	

Division 2
Site Work

Estimating Tips

020 Subsurface Investigation & Demolition

In preparing estimates on structures involving earthwork or foundations, all information concerning soil characteristics should be obtained. Look particularly for hazardous waste, evidence of prior dumping of debris, and previous stream beds.

- The costs shown for selective demolition do not include rubbish handling or disposal. These items should be estimated separately using Means data or other sources.

021 Site Preparation & Excavation Support

- If possible visit the site and take an inventory of the type, quantity and size of the trees. Certain trees may have a landscape resale value or firewood value. Stump disposal can be very expensive, particularly if they cannot be buried at the site. Consider using a bulldozer in lieu of hand cutting trees.

- Estimators should visit the site to determine the need for haul road, access, storage of materials, and security considerations. When estimating for access roads on unstable soil, consider using a geotextile stabilization fabric. It can greatly reduce the quantity of crushed stone or gravel. Sites of limited size and access can cause cost overruns due to lost productivity. Theft and damage is another consideration if the location is isolated. A temporary fence or security guards may be required. Investigate the site thoroughly.

022 Earthwork

- Estimating the actual cost of performing earthwork requires careful consideration of the variables involved. This includes items such as type of soil, whether or not water will be encountered, dewatering, whether or not banks need bracing, disposal of excavated earth, length of haul to fill or spoil sites, etc. If the project has large quantities of cut or fill, consider raising or lowering the site to reduce costs while paying close attention to the effect on site drainage and utilities if doing this.

- If the project has large quantities of fill, creating a borrow pit on the site can significantly lower the costs. It is very important to consider what time of year the project is scheduled for completion. Bad weather can create large cost overruns from dewatering, site repair and lost productivity from cold weather.

025 Paving & Surfacing

- When estimating paving, keep in mind the project schedule. If an asphaltic paving project is in a colder climate and runs through to the spring, consider placing the base course in the autumn, then topping it in the spring just prior to completion. This could save considerable costs in spring repair. Keep in mind that prices for asphalt and concrete are generally higher in the cold seasons.

026 Piped Utilities
027 Sewerage & Drainage

- Never assume that the water, sewer and drainage lines will go in at the early stages of the project. Consider the site access needs before dividing the site in half with open trenches, loose pipe, and machinery obstructions. Always inspect the site to establish that the site drawings are complete. Check off all existing utilities on your drawing as you locate them. If you find any discrepancies, mark up the site plan for further research. Differing site conditions can be very costly if discovered later in the project.

029 Landscaping

- The timing of planting and guarantee specifications often dictate the costs for establishing tree and shrub growth and a stand of grass or ground cover. Establish the work performance schedule to coincide with the local planting season. Maintenance and growth guarantees can add from 20% to 100% to the total landscaping cost. The cost to replace trees and shrubs can be as high as 5% of the total cost depending on the planting zone, soil conditions and time of year.

Reference Numbers

Reference numbers are shown in bold squares at the beginning of some major classifications. These numbers refer to related items in the Reference Section. The reference information may be an estimating procedure, an alternate pricing method or technical information.

Note: Not all subdivisions listed here necessarily appear in this publication.

020 120 | Std Penetration Tests

		CREW	DAILY OUTPUT	LABOR-HOURS	UNIT	MAT.	LABOR	EQUIP.	TOTAL	TOTAL INCL O&P	
123	0010	**BORINGS** Initial field stake out and determination of elevations	A-6	1	16	Day		275		275	455	**123**
	0100	Drawings showing boring details				Total		170		170	245	
	0200	Report and recommendations from P.E.						375		375	540	
	0300	Mobilization and demobilization, minimum	B-55	4	4			57	155	212	268	
	0350	For over 100 miles, per added mile		450	.036	Mile		.51	1.38	1.89	2.38	
	0600	Auger holes in earth, no samples, 2-1/2" diameter		78.60	.204	L.F.		2.90	7.90	10.80	13.60	
	0800	Cased borings in earth, with samples, 2-1/2" diameter		55.50	.288	"		4.11	11.20	15.31	19.25	
	1400	Drill rig and crew with truck mounted auger		1	16	Day		228	620	848	1,075	
	1500	For inner city borings add, minimum									10%	
	1510	Maximum									20%	

020 550 | Site Demolition

		CREW	DAILY OUTPUT	LABOR-HOURS	UNIT	MAT.	LABOR	EQUIP.	TOTAL	TOTAL INCL O&P	
554	0010	**SITE DEMOLITION** No hauling, abandon catch basin or manhole	B-6	7	3.429	Ea.		51.50	30.50	82	121	**554**
	0020	Remove existing catch basin or manhole, masonry		4	6			90	53.50	143.50	211	
	0030	Catch basin or manhole frames and covers, stored		13	1.846			27.50	16.45	43.95	65	
	0040	Remove and reset		7	3.429			51.50	30.50	82	121	
	0600	Fencing, barbed wire, 3 strand	2 Clab	430	.037	L.F.		.50		.50	.85	
	0650	5 strand	"	280	.057			.77		.77	1.31	
	0700	Chain link, remove only, 8' to 10' high	B-6	445	.054			.81	.48	1.29	1.90	
	0750	Remove and reset	"	70	.343			5.15	3.05	8.20	12.05	
	1000	Masonry walls, block or tile, solid, remove	B-5	1,800	.022	C.F.		.33	.56	.89	1.18	
	1100	Cavity wall		2,200	.018			.27	.46	.73	.97	
	1200	Brick, solid		900	.044			.66	1.12	1.78	2.37	
	1300	With block back-up		1,130	.035			.53	.90	1.43	1.88	
	1400	Stone, with mortar		900	.044			.66	1.12	1.78	2.37	
	1500	Dry set		1,500	.027			.40	.67	1.07	1.42	
	1710	Pavement removal, bituminous roads, 3" thick	B-38	690	.035	S.Y.		.52	.56	1.08	1.50	
	1750	4" to 6" thick		420	.057			.86	.92	1.78	2.47	
	1800	Bituminous driveways		640	.038			.56	.61	1.17	1.62	
	1900	Concrete to 6" thick, hydraulic hammer, mesh reinforced		255	.094			1.41	1.52	2.93	4.07	
	2000	Rod reinforced		200	.120			1.80	1.94	3.74	5.20	
	2300	With hand held air equipment, bituminous, to 6" thick	B-39	1,900	.025	S.F.		.35	.09	.44	.69	
	2320	Concrete to 6" thick, no reinforcing		1,200	.040			.55	.14	.69	1.10	
	2340	Mesh reinforced		1,400	.034			.47	.12	.59	.95	
	2360	Rod reinforced		765	.063			.86	.23	1.09	1.73	
	2400	Curbs, concrete, plain	B-6	360	.067	L.F.		1	.59	1.59	2.34	
	2500	Reinforced		275	.087			1.31	.78	2.09	3.07	
	2600	Granite		360	.067			1	.59	1.59	2.34	
	2700	Bituminous		528	.045			.68	.40	1.08	1.60	
	2900	Pipe removal, sewer/water, no excavation, 12" diameter		175	.137			2.05	1.22	3.27	4.82	
	2960	24" diameter		120	.200			3	1.78	4.78	7	
	4000	Sidewalk removal, bituminous, 2-1/2" thick		325	.074	S.Y.		1.11	.66	1.77	2.59	
	4050	Brick, set in mortar		185	.130			1.94	1.16	3.10	4.56	
	4100	Concrete, plain, 4"		160	.150			2.25	1.34	3.59	5.25	
	4200	Mesh reinforced		150	.160			2.40	1.43	3.83	5.65	
	5000	Slab on grade removal, plain	B-5	45	.889	C.Y.		13.25	22.50	35.75	47	
	5100	Mesh reinforced		33	1.212			18.10	30.50	48.60	64	
	5200	Rod reinforced		25	1.600			24	40.50	64.50	85	
	5500	For congested sites or small quantities, add up to								200%	200%	
	5550	For disposal on site, add	B-11A	232	.069			1.12	3.59	4.71	5.80	
	5600	To 5 miles, add	B-34D	76	.105			1.62	7.20	8.82	10.60	

020 600 | Building Demolition

		CREW	DAILY OUTPUT	LABOR-HOURS	UNIT	MAT.	LABOR	EQUIP.	TOTAL	TOTAL INCL O&P	
604	0010	**BUILDING DEMOLITION** Large urban projects, incl. 20 Mi. haul										**604**
	0500	Small bldgs, or single bldgs, no salvage included, steel	B-3	14,800	.003	C.F.		.05	.11	.16	.21	

Important: See the Reference Section for critical supporting data - Reference Nos., Crews, & Location Factors

020 600 | Building Demolition

		CREW	DAILY OUTPUT	LABOR-HOURS	UNIT	MAT.	LABOR	EQUIP.	TOTAL	TOTAL INCL O&P	
604	0600	Concrete	B-3	11,300	.004	C.F.		.07	.15	.22	.28	**604**
	0605	Concrete, plain	B-5	33	1.212	C.Y.		18.10	30.50	48.60	64	
	0610	Reinforced		25	1.600			24	40.50	64.50	85	
	0615	Concrete walls		34	1.176			17.55	30	47.55	62.50	
	0620	Elevated slabs		26	1.538			23	39	62	82	
	0650	Masonry	B-3	14,800	.003	C.F.		.05	.11	.16	.21	
	0700	Wood	"	14,800	.003	"		.05	.11	.16	.21	
	1000	Single family, one story house, wood, minimum				Ea.				2,300	2,700	
	1020	Maximum								4,000	4,800	
	1200	Two family, two story house, wood, minimum								3,000	3,600	
	1220	Maximum								5,800	7,000	
	1300	Three family, three story house, wood, minimum								4,000	4,800	
	1320	Maximum								7,000	8,400	
	1400	Gutting building, see division 020-716										
612	0010	**DUMP CHARGES** Typical urban city, tipping fees only										**612**
	0100	Building construction materials				Ton					50	
	0200	Trees, brush, lumber									40	
	0300	Rubbish only									45	
	0500	Reclamation station, usual charge									80	
620	0010	**RUBBISH HANDLING** The following are to be added to the										**620**
	0020	demolition prices										
	0400	Chute, circular, prefabricated steel, 18" diameter	B-1	40	.600	L.F.	17.65	8.45		26.10	34	
	0440	30" diameter	"	30	.800	"	24	11.25		35.25	46	
	0600	Dumpster, weekly rental, 1 dump/week, 6 C.Y. capacity (2 Tons)				Ea.					275	
	0700	10 C.Y. capacity (4 Tons)									350	
	0800	30 C.Y. capacity (10 Tons)									575	
	0840	40 C.Y. capacity (13 Tons)									680	
	1000	Dust partition, 6 mil polyethylene, 4' x 8' panels, 1" x 3" frame	2 Carp	2,000	.008	S.F.	.40	.15		.55	.69	
	1080	2" x 4" frame	"	2,000	.008	"	.50	.15		.65	.80	
	2000	Load, haul to chute & dumping into chute, 50' haul	2 Clab	24	.667	C.Y.		8.95		8.95	15.30	
	2040	100' haul		16.50	.970			13		13	22.50	
	2080	Over 100' haul, add per 100 L.F.		35.50	.451			6.05		6.05	10.35	
	2120	In elevators, per 10 floors, add		140	.114			1.53		1.53	2.62	
	3000	Loading & trucking, including 2 mile haul, chute loaded	B-16	56	.571			8.25	7.70	15.95	22.50	
	3040	Hand loading truck, 50' haul	"	48	.667			9.60	9	18.60	26.50	
	3080	Machine loading truck	B-17	120	.267			4.02	4.74	8.76	12·	
	5000	Haul, per mile, up to 8 C.Y. truck	B-34B	1,165	.007			.11	.37	.48	.59	
	5100	Over 8 C.Y. truck	"	1,550	.005			.08	.28	.36	.44	

020 700 | Selective Demolition

		CREW	DAILY OUTPUT	LABOR-HOURS	UNIT	MAT.	LABOR	EQUIP.	TOTAL	TOTAL INCL O&P	
702	0010	**CEILING DEMOLITION**										**702**
	0200	Drywall, furred and nailed	2 Clab	800	.020	S.F.		.27		.27	.46	
	1000	Plaster, lime and horse hair, on wood lath, incl. lath		700	.023			.31		.31	.52	
	1200	Suspended ceiling, mineral fiber, 2'x2' or 2'x4'		1,500	.011			.14		.14	.24	
	1250	On suspension system, incl. system		1,200	.013			.18		.18	.31	
	1500	Tile, wood fiber, 12" x 12", glued		900	.018			.24		.24	.41	
	1540	Stapled		1,500	.011			.14		.14	.24	
	2000	Wood, tongue and groove, 1" x 4"		1,000	.016			.21		.21	.37	
	2040	1" x 8"		1,100	.015			.19		.19	.33	
	2400	Plywood or wood fiberboard, 4' x 8' sheets		1,200	.013			.18		.18	.31	
704	0010	**CUTOUT DEMOLITION** Conc., elev. slab, light reinf., under 6 C.F.	B-9C	65	.615	C.F.		8.50	2.67	11.17	17.50	**704**
	0050	Light reinforcing, over 6 C.F.	"	75	.533	"		7.35	2.31	9.66	15.15	
	0200	Slab on grade to 6" thick, not reinforced, under 8 S.F.	B-9	85	.471	S.F.		6.50	2.04	8.54	13.35	
	0250	Not reinforced, over 8 S.F.		175	.229	"		3.15	.99	4.14	6.50	

020 700 | Selective Demolition

		CREW	DAILY OUTPUT	LABOR-HOURS	UNIT	1998 BARE COSTS MAT.	LABOR	EQUIP.	TOTAL	TOTAL INCL O&P	
704	0600	Walls, not reinforced, under 6 C.F.	B-9	60	.667	C.F.		9.20	2.89	12.09	18.95	**704**
	0650	Not reinforced, over 6 C.F.	↓	65	.615			8.50	2.67	11.17	17.50	
	1000	Concrete, elevated slab, bar reinforced, under 6 C.F.	B-9C	45	.889			12.25	3.86	16.11	25.50	
	1050	Bar reinforced, over 6 C.F.	"	50	.800	↓		11.05	3.47	14.52	22.50	
	1200	Slab on grade to 6" thick, bar reinforced, under 8 S.F.	B-9	75	.533	S.F.		7.35	2.31	9.66	15.15	
	1250	Bar reinforced, over 8 S.F.	"	105	.381	"		5.25	1.65	6.90	10.80	
	1400	Walls, bar reinforced, under 6 C.F.	B-9C	50	.800	C.F.		11.05	3.47	14.52	22.50	
	1450	Bar reinforced, over 6 C.F.	"	55	.727	"		10.05	3.16	13.21	20.50	
	2000	Brick, to 4 S.F. opening, not including toothing										
	2040	4" thick	B-9C	30	1.333	Ea.		18.40	5.80	24.20	38	
	2060	8" thick		18	2.222			30.50	9.65	40.15	63	
	2080	12" thick		10	4			55	17.35	72.35	114	
	2400	Concrete block, to 4 S.F. opening, 2" thick		35	1.143			15.75	4.96	20.71	32.50	
	2420	4" thick		30	1.333			18.40	5.80	24.20	38	
	2440	8" thick		27	1.481			20.50	6.45	26.95	42	
	2460	12" thick	↓	24	1.667			23	7.25	30.25	47.50	
	2600	Gypsum block, to 4 S.F. opening, 2" thick	B-9	80	.500			6.90	2.17	9.07	14.20	
	2620	4" thick		70	.571			7.90	2.48	10.38	16.25	
	2640	8" thick		55	.727			10.05	3.16	13.21	20.50	
	2800	Terra cotta, to 4 S.F. opening, 4" thick		70	.571			7.90	2.48	10.38	16.25	
	2840	8" thick		65	.615			8.50	2.67	11.17	17.50	
	2880	12" thick	↓	50	.800	↓		11.05	3.47	14.52	22.50	
	3000	Toothing masonry cutouts, brick, soft old mortar	1 Brhe	40	.200	V.L.F.		2.93		2.93	4.96	
	3100	Hard mortar		30	.267			3.91		3.91	6.60	
	3200	Block, soft old mortar		70	.114			1.67		1.67	2.83	
	3400	Hard mortar	↓	50	.160	↓		2.34		2.34	3.97	
	4000	For toothing masonry, see Division 045-290										
	6000	Walls, interior, not including re-framing,										
	6010	openings to 5 S.F.										
	6100	Drywall to 5/8" thick	A-1	24	.333	Ea.		4.47	2.79	7.26	10.70	
	6200	Paneling to 3/4" thick		20	.400			5.35	3.35	8.70	12.90	
	6300	Plaster, on gypsum lath		20	.400			5.35	3.35	8.70	12.90	
	6340	On wire lath	↓	14	.571	↓		7.65	4.79	12.44	18.35	
	7000	Wood frame, not including re-framing, openings to 5 S.F.										
	7200	Floors, sheathing and flooring to 2" thick	A-1	5	1.600	Ea.		21.50	13.40	34.90	51.50	
	7310	Roofs, sheathing to 1" thick, not including roofing		6	1.333			17.85	11.15	29	43	
	7410	Walls, sheathing to 1" thick, not including siding	↓	7	1.143	↓		15.30	9.60	24.90	36.50	
706	0010	**DOOR DEMOLITION**										**706**
	0200	Doors, exterior, 1-3/4" thick, single, 3' x 7' high	1 Clab	16	.500	Ea.		6.70		6.70	11.50	
	0220	Double, 6' x 7' high		12	.667			8.95		8.95	15.30	
	0500	Interior, 1-3/8" thick, single, 3' x 7' high		20	.400			5.35		5.35	9.20	
	0520	Double, 6' x 7' high		16	.500			6.70		6.70	11.50	
	0700	Bi-folding, 3' x 6'-8" high		20	.400			5.35		5.35	9.20	
	0720	6' x 6'-8" high		18	.444			5.95		5.95	10.20	
	0900	Bi-passing, 3' x 6'-8" high		16	.500			6.70		6.70	11.50	
	0940	6' x 6'-8" high	↓	14	.571			7.65		7.65	13.10	
	1500	Remove and reset, minimum	1 Carp	8	1			18.50		18.50	31.50	
	1520	Maximum	"	6	1.333			24.50		24.50	42.50	
	2000	Frames, including trim, metal	A-1	8	1			13.40	8.40	21.80	32	
	2200	Wood	F-2	32	.500	↓		8.10		8.10	13.85	
	2201	Alternate pricing method	A-1	200	.040	L.F.		.54	.34	.88	1.29	
	3000	Special doors, counter doors	F-2	6	2.667	Ea.		43		43	74	
	3300	Glass, sliding, including frames		12	1.333			21.50		21.50	37	
	3400	Overhead, commercial, 12' x 12' high		4	4			64.50		64.50	111	
	3500	Residential, 9' x 7' high	↓	8	2	↓		32.50		32.50	55.50	

Important: See the Reference Section for critical supporting data - Reference Nos., Crews, & Location Factors

020 700 | Selective Demolition

		CREW	DAILY OUTPUT	LABOR-HOURS	UNIT	1998 BARE COSTS MAT.	LABOR	EQUIP.	TOTAL	TOTAL INCL O&P	
706	3540	16' x 7' high	F-2	7	2.286	Ea.		37		37	63.50	**706**
	3600	Remove and reset, minimum		4	4			64.50		64.50	111	
	3620	Maximum	↓	2.50	6.400			104		104	177	
	3660	Remove and reset elec. garage door opener	1 Carp	8	1			18.50		18.50	31.50	
	4000	Residential lockset, exterior		30	.267	↓		4.93		4.93	8.45	
	4200	Deadbolt lock	↓	32	.250	↓		4.63		4.63	7.95	
708	0010	**ELECTRICAL DEMOLITION**										**708**
	0020	Conduit to 15' high, including fittings & hangers										
	0100	Rigid galvanized steel, 1/2" to 1" diameter	1 Elec	242	.033	L.F.		.68		.68	1.11	
	0120	1-1/4" to 2"	"	200	.040	"		.82		.82	1.35	
	0270	Armored cable, (BX) avg. 50' runs										
	0280	#14, 2 wire	1 Elec	690	.012	L.F.		.24		.24	.39	
	0290	#14, 3 wire		571	.014			.29		.29	.47	
	0300	#12, 2 wire		605	.013			.27		.27	.45	
	0310	#12, 3 wire		514	.016			.32		.32	.52	
	0320	#10, 2 wire		514	.016			.32		.32	.52	
	0330	#10, 3 wire		425	.019			.39		.39	.63	
	0340	#8, 3 wire	↓	342	.023	↓		.48		.48	79	
	0350	Non metallic sheathed cable (Romex)										
	0360	#14, 2 wire	1 Elec	720	.011	L.F.		.23		.23	.37	
	0370	#14, 3 wire		657	.012			.25		.25	.41	
	0380	#12, 2 wire		629	.013			.26		.26	.43	
	0390	#10, 3 wire	↓	450	.018	↓		.37		.37	.60	
	0400	Wiremold raceway, including fittings & hangers										
	0440	No. 4000	1 Elec	217	.037	L.F.		.76		.76	1.24	
	0460	No. 6000	"	166	.048	"		.99		.99	1.62	
	0500	Channels, steel, including fittings & hangers										
	0520	3/4" x 1-1/2"	1 Elec	308	.026	L.F.		.54		.54	.88	
	0540	1-1/2" x 1-1/2"		269	.030	↓		.61		.61	1	
	0560	1-1/2" x 1-7/8"		229	.035	↓		.72		.72	1.18	
	1180	400 amp	↓	3.40	2.353	Ea.		48.50		48.50	79.50	
	1210	Panel boards, incl. removal of all breakers,										
	1220	pipe terminations & wire connections										
	1720	Junction boxes, 4" sq. & oct.	1 Elec	80	.100	Ea.		2.06		2.06	3.37	
	1760	Switch box		107	.075			1.54		1.54	2.52	
	1780	Receptacle & switch plates	↓	257	.031	↓		.64		.64	1.05	
	1800	Wire, THW-THWN-THHN, removed from										
	1810	in place conduit, to 15' high										
	1830	#14	1 Elec	65	.123	C.L.F.		2.54		2.54	4.15	
	1840	#12		55	.145			3		3	4.90	
	1850	#10	↓	45.50	.176	↓		3.62		3.62	5.95	
	2000	Interior fluorescent fixtures, incl. supports										
	2010	& whips, to 15' high										
	2100	Recessed drop-in 2' x 2', 2 lamp	2 Elec	35	.457	Ea.		9.40		9.40	15.40	
	2140	2' x 4', 4 lamp	"	30	.533	"		11		11	17.95	
	2180	Surface mount, acrylic lens & hinged frame										
	2220	2' x 2', 2 lamp	2 Elec	44	.364	Ea.		7.50		7.50	12.25	
	2260	2' x 4', 4 lamp	"	33	.485	"		10		10	16.35	
	2300	Strip fixtures, surface mount										
	2320	4' long, 1 lamp	2 Elec	53	.302	Ea.		6.20		6.20	10.15	
	2380	8' long, 2 lamp	"	40	.400	"		8.25		8.25	13.50	
	2600	Exterior fixtures, incandescent, wall mount										
	2620	100 Watt	2 Elec	50	.320	Ea.		6.60		6.60	10.80	
712	0010	**FLOORING DEMOLITION**										**712**
	0200	Brick with mortar	A-1	300	.027	S.F.		.36	.22	.58	.86	

020 700 | Selective Demolition

		CREW	DAILY OUTPUT	LABOR-HOURS	UNIT	MAT.	LABOR	EQUIP.	TOTAL	TOTAL INCL O&P	
712	0400	Carpet, bonded, including surface scraping	2 Clab	2,000	.008	S.F.		.11		.11	.18	**712**
	0480	Tackless		9,000	.002			.02		.02	.04	
	0800	Resilient, sheet goods		1,400	.011			.15		.15	.26	
	0900	Tile, 12″ x 12″		1,000	.016			.21		.21	.37	
	2000	Tile, ceramic, thin set	A-1	400	.020			.27	.17	.44	.64	
	2020	Mud set	″	350	.023			.31	.19	.50	.73	
	3000	Wood, block, on end	1 Carp	400	.020			.37		.37	.63	
	3200	Parquet		450	.018			.33		.33	.56	
	3400	Strip flooring, interior, 2-1/4″ x 25/32″ thick		325	.025			.46		.46	.78	
	3500	Exterior, porch flooring, 1″ x 4″		220	.036			.67		.67	1.15	
	3800	Subfloor, tongue and groove, 1″ x 6″		325	.025			.46		.46	.78	
	3820	1″ x 8″		430	.019			.34		.34	.59	
	3840	1″ x 10″		520	.015			.28		.28	.49	
	4000	Plywood, nailed		600	.013			.25		.25	.42	
	4100	Glued and nailed		400	.020			.37		.37	.63	
714	0010	**FRAMING DEMOLITION**										**714**
	3000	Wood framing, beams, 6″ x 8″	B-2	275	.145	L.F.		2.01		2.01	3.44	
	3040	6″ x 10″		220	.182			2.51		2.51	4.30	
	3080	6″ x 12″		185	.216			2.98		2.98	5.10	
	3120	8″ x 12″		140	.286			3.94		3.94	6.75	
	3160	10″ x 12″		110	.364			5		5	8.60	
	3400	Fascia boards, 1″ x 6″	1 Clab	500	.016			.21		.21	.37	
	3440	1″ x 8″		450	.018			.24		.24	.41	
	3480	1″ x 10″		400	.020			.27		.27	.46	
	3800	Headers over openings, 2 @ 2″ x 6″		110	.073			.97		.97	1.67	
	3840	2 @ 2″ x 8″		100	.080			1.07		1.07	1.84	
	3880	2 @ 2″ x 10″		90	.089			1.19		1.19	2.04	
	4230	Joists, 2″ x 6″	2 Clab	970	.016			.22		.22	.38	
	4240	2″ x 8″		940	.017			.23		.23	.39	
	4250	2″ x 10″		910	.018			.24		.24	.40	
	4280	2″ x 12″		880	.018			.24		.24	.42	
	5400	Posts, 4″ x 4″		800	.020			.27		.27	.46	
	5440	6″ x 6″		400	.040			.54		.54	.92	
	5480	8″ x 8″		300	.053			.71		.71	1.22	
	5500	10″ x 10″		240	.067			.89		.89	1.53	
	5800	Rafters, ordinary, 2″ x 6″		850	.019			.25		.25	.43	
	5840	2″ x 8″		720	.022			.30		.30	.51	
	6200	Stairs and stringers, minimum		40	.400	Riser		5.35		5.35	9.20	
	6240	Maximum		26	.615	″		8.25		8.25	14.10	
	6600	Studs, 2″ x 4″		2,000	.008	L.F.		.11		.11	.18	
	6640	2″ x 6″		1,600	.010	″		.13		.13	.23	
	7000	Trusses, 2″ x 4″ flat wood construction										
	7050	12′ span	2 Clab	74	.216	Ea.		2.90		2.90	4.96	
	7150	24′ span		66	.242			3.25		3.25	5.55	
	7200	26′ span		64	.250			3.35		3.35	5.75	
	7250	28′ span		62	.258			3.46		3.46	5.90	
	7300	30′ span		58	.276			3.70		3.70	6.35	
	7350	32′ span		56	.286			3.83		3.83	6.55	
	7400	34′ span		54	.296			3.97		3.97	6.80	
	7450	36′ span		52	.308			4.12		4.12	7.05	
	9500	See Div. 020-620 for rubbish handling										
716	0010	**GUTTING** Building interior, including disposal, dumpster fees not included										**716**
	0500	Residential building										
	0560	Minimum	B-16	400	.080	SF Flr.		1.15	1.08	2.23	3.15	
	0580	Maximum	″	360	.089	″		1.28	1.20	2.48	3.50	

Important: See the Reference Section for critical supporting data - Reference Nos., Crews, & Location Factors

020 700	Selective Demolition	CREW	DAILY OUTPUT	LABOR-HOURS	UNIT	1998 BARE COSTS				TOTAL INCL O&P	
						MAT.	LABOR	EQUIP.	TOTAL		
716	0900	Commercial building									**716**
	1000	Minimum	B-16	350	.091	SF Flr.		1.32	1.23	2.55	3.60
	1020	Maximum	"	250	.128	"		1.84	1.73	3.57	5.05
718	0010	**HVAC DEMOLITION**									**718**
	0100	Air conditioner, split unit, 3 ton	Q-5	2	8	Ea.		151		151	251
	0150	Package unit, 3 ton	Q-6	3	8	"		146		146	241
	0260	Baseboard, hydronic fin tube, 1/2"	Q-5	117	.137	L.F.		2.58		2.58	4.28
	0300	Boiler, electric	Q-19	2	12	Ea.		234		234	385
	0340	Gas or oil, steel, under 150 MBH	Q-6	3	8	"		146		146	241
	1000	Ductwork, 4" high, 8" wide	1 Clab	200	.040	L.F.		.54		.54	.92
	1100	6" high, 8" wide		165	.048			.65		.65	1.11
	1200	10" high, 12" wide		125	.064			.86		.86	1.47
	1300	12"-14" high, 16"-18" wide		85	.094			1.26		1.26	2.16
	1500	30" high, 36" wide		56	.143			1.91		1.91	3.28
	2800	Heat pump, package unit, 3 ton	Q-5	2.40	6.667	Ea.		126		126	209
	2840	Split unit, 3 ton		2	8			151		151	251
	2950	Tank, steel, oil, 275 gal., above ground		10	1.600			30		30	50
	2960	Remove and reset		3	5.333			101		101	167
720	0010	**MILLWORK AND TRIM DEMOLITION**									**720**
	1000	Cabinets, wood, base cabinets	2 Clab	80	.200	L.F.		2.68		2.68	4.59
	1020	Wall cabinets		80	.200			2.68		2.68	4.59
	1100	Steel, painted, base cabinets		60	.267			3.57		3.57	6.10
	1500	Counter top, minimum		200	.080			1.07		1.07	1.84
	1510	Maximum		120	.133			1.79		1.79	3.06
	2000	Paneling, 4' x 8' sheets, 1/4" thick		2,000	.008	S.F.		.11		.11	.18
	2100	Boards, 1" x 4"		700	.023			.31		.31	.52
	2120	1" x 6"		750	.021			.29		.29	.49
	2140	1" x 8"		800	.020			.27		.27	.46
	3000	Trim, baseboard, to 6" wide		1,200	.013	L.F.		.18		.18	.31
	3040	12" wide		1,000	.016			.21		.21	.37
	3100	Ceiling trim		1,000	.016			.21		.21	.37
	3120	Chair rail		1,200	.013			.18		.18	.31
	3140	Railings with balusters		240	.067			.89		.89	1.53
	3160	Wainscoting		700	.023	S.F.		.31		.31	.52
	4000	Curtain rod	1 Clab	80	.100	Ea.		1.34		1.34	2.30
724	0010	**PLUMBING DEMOLITION**									**724**
	1020	Fixtures, including 10' piping									
	1100	Bath tubs, cast iron	1 Plum	4	2	Ea.		41.50		41.50	69
	1120	Fiberglass		6	1.333			28		28	46
	1140	Steel		5	1.600			33.50		33.50	55.50
	1200	Lavatory, wall hung		10	.800			16.70		16.70	27.50
	1220	Counter top		8	1			21		21	34.50
	1300	Sink, steel or cast iron, single		8	1			21		21	34.50
	1320	Double		7	1.143			24		24	39.50
	1400	Water closet, floor mounted		8	1			21		21	34.50
	1420	Wall mounted		7	1.143			24		24	39.50
	2000	Piping, metal, to 2" diameter		200	.040	L.F.		.83		.83	1.38
	2050	2" to 4" diameter		150	.053			1.11		1.11	1.84
	2100	4" to 8" diameter	2 Plum	100	.160			3.34		3.34	5.55
	2250	Water heater, 40 gal.	1 Plum	6	1.333	Ea.		28		28	46
	3000	Submersible sump pump		24	.333			6.95		6.95	11.50
	6000	Remove and reset fixtures, minimum		6	1.333			28		28	46
	6100	Maximum		4	2			41.50		41.50	69

020 700 | Selective Demolition

		CREW	DAILY OUTPUT	LABOR-HOURS	UNIT	1998 BARE COSTS MAT.	LABOR	EQUIP.	TOTAL	TOTAL INCL O&P
726	0010	**ROOFING AND SIDING DEMOLITION**									**726**
	1200	Wood, boards, tongue and groove, 2" x 6"	2 Clab	960	.017	S.F.		.22		.22	.38
	1220	2" x 10"		1,040	.015			.21		.21	.35
	1280	Standard planks, 1" x 6"		1,080	.015			.20		.20	.34
	1320	1" x 8"		1,160	.014			.18		.18	.32
	1340	1" x 12"		1,200	.013			.18		.18	.31
	1360	Flashing, aluminum	1 Clab	290	.028			.37		.37	.63
	2000	Gutters, aluminum or wood, edge hung	"	240	.033	L.F.		.45		.45	.77
	2010	Remove and reset, aluminum	1 Shee	125	.064			1.29		1.29	2.18
	2020	Remove and reset, vinyl	1 Carp	125	.064			1.18		1.18	2.03
	2100	Built-in	1 Clab	100	.080			1.07		1.07	1.84
	2500	Roof accessories, plumbing vent flashing		14	.571	Ea.		7.65		7.65	13.10
	2600	Adjustable metal chimney flashing		9	.889	"		11.90		11.90	20.50
	2650	Coping, sheet metal, up to 12" wide		240	.033	L.F.		.45		.45	.77
	2660	Concrete, up to 12" wide	2 Clab	160	.100	"		1.34		1.34	2.30
	3000	Roofing, built-up, 5 ply roof, no gravel	B-2	1,600	.025	S.F.		.35		.35	.59
	3100	Gravel removal, minimum		5,000	.008			.11		.11	.19
	3120	Maximum		2,000	.020			.28		.28	.47
	3400	Roof insulation board, up to 2" thick		3,900	.010			.14		.14	.24
	3450	Roll roofing, cold adhesive	1 Clab	12	.667	Sq.		8.95		8.95	15.30
	4000	Shingles, asphalt strip, 1 layer	B-2	3,500	.011	S.F.		.16		.16	.27
	4100	Slate		2,500	.016			.22		.22	.38
	4300	Wood		2,200	.018			.25		.25	.43
	4500	Skylight to 10 S.F.	1 Clab	8	1	Ea.		13.40		13.40	23
	5000	Siding, metal, horizontal		444	.018	S.F.		.24		.24	.41
	5020	Vertical		400	.020			.27		.27	.46
	5200	Wood, boards, vertical		400	.020			.27		.27	.46
	5220	Clapboards, horizontal		380	.021			.28		.28	.48
	5240	Shingles		350	.023			.31		.31	.52
	5260	Textured plywood		725	.011			.15		.15	.25
732	0010	**WALLS AND PARTITIONS DEMOLITION**									**732**
	1000	Drywall, nailed	1 Clab	1,000	.008	S.F.		.11		.11	.18
	1500	Fiberboard, nailed	"	900	.009			.12		.12	.20
	2200	Metal or wood studs, finish 2 sides, fiberboard	B-1	520	.046			.65		.65	1.11
	2250	Lath and plaster		260	.092			1.30		1.30	2.22
	2300	Plasterboard (drywall)		520	.046			.65		.65	1.11
	2350	Plywood		450	.053			.75		.75	1.29
	3000	Plaster, lime and horsehair, on wood lath	1 Clab	400	.020			.27		.27	.46
	3020	On metal lath	"	335	.024			.32		.32	.55
734	0010	**WINDOW DEMOLITION**									**734**
	0200	Aluminum, including trim, to 12 S.F.	1 Clab	16	.500	Ea.		6.70		6.70	11.50
	0240	To 25 S.F.		11	.727			9.75		9.75	16.70
	0280	To 50 S.F.		5	1.600			21.50		21.50	36.50
	0320	Storm windows, to 12 S.F.		27	.296			3.97		3.97	6.80
	0360	To 25 S.F.		21	.381			5.10		5.10	8.75
	0400	To 50 S.F.		16	.500			6.70		6.70	11.50
	0600	Glass, minimum		200	.040	S.F.		.54		.54	.92
	0620	Maximum		150	.053	"		.71		.71	1.22
	2000	Wood, including trim, to 12 S.F.		22	.364	Ea.		4.87		4.87	8.35
	2020	To 25 S.F.		18	.444			5.95		5.95	10.20
	2060	To 50 S.F.		13	.615			8.25		8.25	14.10
	5020	Remove and reset window, minimum	1 Carp	6	1.333			24.50		24.50	42.50
	5040	Average		4	2			37		37	63.50
	5080	Maximum		2	4			74		74	127

SITE WORK 2

020 700	Selective Demolition	CREW	DAILY OUTPUT	LABOR-HOURS	UNIT	MAT.	LABOR	EQUIP.	TOTAL	TOTAL INCL O&P	
							1998 BARE COSTS				
754	**0010** **FOOTINGS AND FOUNDATIONS DEMOLITION**										**754**
0200	Floors, concrete slab on grade,										
0240	4" thick, plain concrete	B-9C	500	.080	S.F.		1.10	.35	1.45	2.27	
0280	Reinforced, wire mesh		470	.085			1.17	.37	1.54	2.42	
0300	Rods		400	.100			1.38	.43	1.81	2.84	
0400	6" thick, plain concrete		375	.107			1.47	.46	1.93	3.03	
0420	Reinforced, wire mesh		340	.118			1.62	.51	2.13	3.34	
0440	Rods		300	.133			1.84	.58	2.42	3.79	
1000	Footings, concrete, 1' thick, 2' wide	B-5	300	.133	L.F.		1.99	3.37	5.36	7.10	
1080	1'-6" thick, 2' wide		250	.160			2.39	4.05	6.44	8.50	
1120	3' wide		200	.200			2.98	5.05	8.03	10.60	
1200	Average reinforcing, add								10%	10%	
2000	Walls, block, 4" thick	A-1	200	.040	S.F.		.54	.34	.88	1.29	
2040	6" thick		190	.042			.56	.35	.91	1.36	
2080	8" thick		180	.044			.60	.37	.97	1.43	
2100	12" thick		175	.046			.61	.38	.99	1.47	
2400	Concrete, plain concrete, 6" thick	D-9	160	.250			3.45	1.08	4.53	7.10	
2420	8" thick		140	.286			3.94	1.24	5.18	8.10	
2440	10" thick		120	.333			4.60	1.45	6.05	9.50	
2500	12" thick		100	.400			5.50	1.74	7.24	11.35	
2600	For average reinforcing, add								10%	10%	
4000	For congested sites or small quantities, add up to								200%	200%	
4200	Add for disposal, on site	B-11A	232	.069	C.Y.		1.12	3.59	4.71	5.80	
4250	To five miles	B-30	220	.109	"		1.81	7.15	8.96	10.85	
758	**0010** **MASONRY DEMOLITION**										**758**
1000	Chimney, 16" x 16", soft old mortar	A-1	24	.333	V.L.F.		4.47	2.79	7.26	10.70	
1020	Hard mortar		18	.444			5.95	3.72	9.67	14.30	
1080	20" x 20", soft old mortar		12	.667			8.95	5.60	14.55	21.50	
1100	Hard mortar		10	.800			10.70	6.70	17.40	25.50	
1140	20" x 32", soft old mortar		10	.800			10.70	6.70	17.40	25.50	
1160	Hard mortar		8	1			13.40	8.40	21.80	32	
1200	48" x 48", soft old mortar		5	1.600			21.50	13.40	34.90	51.50	
1220	Hard mortar		4	2			27	16.75	43.75	64.50	
2000	Columns, 8" x 8", soft old mortar		48	.167			2.23	1.40	3.63	5.35	
2020	Hard mortar		40	.200			2.68	1.68	4.36	6.45	
2060	16" x 16", soft old mortar		16	.500			6.70	4.19	10.89	16.10	
2100	Hard mortar		14	.571			7.65	4.79	12.44	18.35	
2140	24" x 24", soft old mortar		8	1			13.40	8.40	21.80	32	
2160	Hard mortar		6	1.333			17.85	11.15	29	43	
2200	36" x 36", soft old mortar		4	2			27	16.75	43.75	64.50	
2220	Hard mortar		3	2.667			35.50	22.50	58	85.50	
3000	Copings, precast or masonry, to 8" wide										
3020	Soft old mortar	A-1	180	.044	L.F.		.60	.37	.97	1.43	
3040	Hard mortar	"	160	.050	"		.67	.42	1.09	1.61	
3100	To 12" wide										
3120	Soft old mortar	A-1	160	.050	L.F.		.67	.42	1.09	1.61	
3140	Hard mortar	"	140	.057	"		.77	.48	1.25	1.84	
4000	Fireplace, brick, 30" x 24" opening										
4020	Soft old mortar	A-1	2	4	Ea.		53.50	33.50	87	129	
4040	Hard mortar		1.25	6.400			86	53.50	139.50	206	
4100	Stone, soft old mortar		1.50	5.333			71.50	44.50	116	171	
4120	Hard mortar		1	8			107	67	174	258	
5000	Veneers, brick, soft old mortar		140	.057	S.F.		.77	.48	1.25	1.84	
5020	Hard mortar		125	.064			.86	.54	1.40	2.06	
5100	Granite and marble, 2" thick		180	.044			.60	.37	.97	1.43	
5120	4" thick		170	.047			.63	.39	1.02	1.51	

020 700 | Selective Demolition

		CREW	DAILY OUTPUT	LABOR-HOURS	UNIT	1998 BARE COSTS MAT.	LABOR	EQUIP.	TOTAL	TOTAL INCL O&P	
758	5140	Stone, 4″ thick	A-1	180	.044	S.F.		.60	.37	.97	1.43	**758**
	5160	8″ thick		175	.046			.61	.38	.99	1.47	
	5400	Alternate pricing method, stone, 4″ thick		60	.133	C.F.		1.79	1.12	2.91	4.29	
	5420	8″ thick		85	.094	″		1.26	.79	2.05	3.03	
810	0010	**ASBESTOS ABATEMENT EQUIPMENT** and supplies, buy [R020 -820]										**810**
	0200	Air filtration device, 2000 C.F.M.				Ea.	2,550			2,550	2,800	
	0250	Large volume air sampling pump, minimum					460			460	505	
	0260	Maximum					1,000			1,000	1,100	
	0300	Airless sprayer unit, 2 gun					2,075			2,075	2,275	
	0350	Light stand, 500 watt					225			225	248	
	0400	Personal respirators										
	0410	Negative pressure, 1/2 face, dual operation, min.				Ea.	18.50			18.50	20.50	
	0420	Maximum					22.50			22.50	25	
	0450	P.A.P.R., full face, minimum					450			450	495	
	0460	Maximum					700			700	770	
	0470	Supplied air, full face, incl. air line, minimum					380			380	420	
	0480	Maximum					550			550	605	
	0500	Personnel sampling pump, minimum					440			440	485	
	0510	Maximum					700			700	770	
	1500	Power panel, 20 unit, incl. G.F.I.					1,750			1,750	1,925	
	1600	Shower unit, including pump and filters					1,125			1,125	1,225	
	1700	Supplied air system (type C)					10,000			10,000	11,000	
	1750	Vacuum cleaner, HEPA, 16 gal., stainless steel, wet/dry					900			900	990	
	1760	55 gallon					2,250			2,250	2,475	
	1800	Vacuum loader, 9-18 ton/hr					85,000			85,000	93,500	
	1900	Water atomizer unit, including 55 gal. drum					225			225	248	
	2000	Worker protection, whole body, foot, head cover & gloves, plastic					30			30	33	
	2500	Respirator, single use					8.90			8.90	9.80	
	2550	Cartridge for respirator					37			37	40.50	
	2570	Glove bag, 7 mil, 50″ x 64″					8.75			8.75	9.65	
	2580	10 mil, 44″ x 60″					8.50			8.50	9.35	
	3000	HEPA vacuum for work area, minimun					1,100			1,100	1,200	
	3050	Maximum					4,400			4,400	4,850	
	6000	Disposable polyethelene bags, 6 mil, 3 C.F.					1.10			1.10	1.21	
	6300	Disposable fiber drums, 3 C.F.					6.30			6.30	6.95	
	6400	Pressure sensitive caution lables, 3″ x 5″					.10			.10	.11	
	6450	11″ x 17″					.20			.20	.22	
	6500	Negative air machine, 1800 C.F.M.					775			775	855	
820	0010	**ASBESTOS ABATEMENT WORK AREA** Containment and preparation.										**820**
	0100	Pre-cleaning, HEPA vacuum and wet wipe, flat surfaces	A-10	12,000	.005	S.F.		.10		.10	.18	
	0200	Protect carpeted area, 2 layers 6 mil poly on 3/4″ plywood	″	1,000	.064		1.40	1.26		2.66	3.74	
	0300	Separation barrier, 2″ x 4″ @ 16″, 1/2″ plywood ea. side, 8′ high	F-2	400	.040		1.20	.65		1.85	2.43	
	0310	12′ high		320	.050		1.35	.81		2.16	2.88	
	0320	16′ high		200	.080		1.60	1.29		2.89	3.98	
	0400	Personnel decontam. chamber, 2″ x 4″ @ 16″, 3/4″ ply ea. side		280	.057		2.50	.92		3.42	4.33	
	0450	Waste decontam. chamber, 2″ x 4″ studs @ 16″, 3/4″ ply ea. side		360	.044		3.15	.72		3.87	4.70	
	0500	Cover surfaces with polyethelene sheeting										
	0501	Including glue and tape										
	0550	Floors, each layer, 6 mil	A-10	8,000	.008	S.F.	.10	.16		.26	.38	
	0551	4 mil		9,000	.007		.05	.14		.19	.30	
	0560	Walls, each layer, 6 mil		6,000	.011		.09	.21		.30	.47	
	0561	4 mil		7,000	.009		.06	.18		.24	.38	
	0570	For heights above 12′, add						20%				
	0575	For heights above 20′, add						30%				
	0580	For fire retardant poly, add					100%					
	0590	For large open areas, deduct					10%	20%				

Important: See the Reference Section for critical supporting data - Reference Nos., Crews, & Location Factors

SITE WORK | 2

		020 700	Selective Demolition	CREW	DAILY OUTPUT	LABOR-HOURS	UNIT	1998 BARE COSTS MAT.	LABOR	EQUIP.	TOTAL	TOTAL INCL O&P	
820	0600	Seal floor penetrations with foam firestop to 36 Sq. In.	F-2	200	.080	Ea.	6.20	1.29		7.49	9	820
	0610	36 Sq. In. to 72 Sq. In.		125	.128		12.40	2.07		14.47	17.20	
	0615	72 Sq. In. to 144 Sq. In.		80	.200		25	3.24		28.24	33	
	0620	Wall penetrations, to 36 square inches		180	.089		6.20	1.44		7.64	9.25	
	0630	36 Sq. In. to 72 Sq. In.		100	.160		12.40	2.59		14.99	18.10	
	0640	72 Sq. In. to 144 Sq. In.		60	.267		25	4.31		29.31	35	
	0800	Caulk seams with latex	1 Carp	230	.035	L.F.	.16	.64		.80	1.28	
	0900	Set up neg. air machine, 1-2k C.F.M. /25 M.C.F. volume	1 Asbe	4.30	1.860	Ea.		36.50		36.50	63.50	
830	0010	**DEMOLITION IN ASBESTOS CONTAMINATED AREA**										830
	0200	Ceiling, including suspension system, plaster and lath	A-9	2,100	.030	S.F.		.60		.60	1.05	
	0210	Finished plaster, leaving wire lath		585	.109			2.15		2.15	3.76	
	0220	Suspended acoustical tile		3,500	.018			.36		.36	.63	
	0230	Concealed tile grid system		3,000	.021			.42		.42	.73	
	0240	Metal pan grid system		1,500	.043			.84		.84	1.47	
	0250	Gypsum board		2,500	.026			.50		.50	.88	
	0260	Lighting fixtures up to 2' x 4'		72	.889	Ea.		17.45		17.45	30.50	
	0400	Partitions, non load bearing										
	0410	Plaster, lath, and studs	A-9	690	.093	S.F.	.55	1.82		2.37	3.80	
	0450	Gypsum board and studs	"	1,390	.046	"		.90		.90	1.58	
	9000	For type C (supplied air) respirator equipment, add				%					10%	
840	0010	**BULK ASBESTOS REMOVAL**										840
	0020	Includes disposable tools and 4 suits and respirators/day/worker										
	0100	Beams, W 10 x 19	A-9	235	.272	L.F.		5.35		5.35	9.35	
	0110	W 12 x 22		210	.305			6		6	10.45	
	0120	W 14 x 26		180	.356			6.95		6.95	12.20	
	0130	W 16 x 31		160	.400			7.85		7.85	13.75	
	0140	W 18 x 40		140	.457			8.95		8.95	15.70	
	0150	W 24 x 55		110	.582			11.40		11.40	20	
	0160	W 30 x 108		85	.753			14.75		14.75	26	
	0170	W 36 x 150		72	.889			17.45		17.45	30.50	
	0200	Boiler insulation		480	.133	S.F.	.06	2.61		2.67	4.65	
	0210	With metal lath add				%				50%		
	0300	Boiler breeching or flue insulation	A-9	520	.123	S.F.		2.41		2.41	4.23	
	0310	For active boiler, add				%				100%		
	0400	Duct or AHU insulation	A-10B	440	.073	S.F.		1.43		1.43	2.51	
	0500	Duct vibration isolation joints, up to 24 Sq. In. duct	A-9	56	1.143	Ea.		22.50		22.50	39.50	
	0520	25 Sq. In. to 48 Sq. In. duct		48	1.333			26		26	46	
	0530	49 Sq. In. to 76 Sq. In. duct		40	1.600			31.50		31.50	55	
	0600	Pipe insulation, air cell type, up to 4" diameter pipe		900	.071	L.F.		1.39		1.39	2.44	
	0610	4" to 8" diameter pipe		800	.080			1.57		1.57	2.75	
	0620	10" to 12" diameter pipe		700	.091			1.79		1.79	3.14	
	0630	14" to 16" diameter pipe		550	.116			2.28		2.28	4	
	0650	Over 16" diameter pipe		650	.098	S.F.		1.93		1.93	3.38	
	0700	With glove bag up to 3" diameter pipe		100	.640	L.F.	3.15	12.55		15.70	25.50	
	1000	Pipe fitting insulation up to 4" diameter pipe		320	.200	Ea.		3.92		3.92	6.85	
	1100	6" to 8" diameter pipe		304	.211			4.13		4.13	7.25	
	1110	10" to 12" diameter pipe		192	.333			6.55		6.55	11.45	
	1120	14" to 16" diameter pipe		128	.500			9.80		9.80	17.20	
	1130	Over 16" diameter pipe		176	.364	S.F.		7.15		7.15	12.50	
	1200	With glove bag, up to 8" diameter pipe		40	1.600	Ea.	6.55	31.50		38.05	62	
	2000	Scrape foam fireproofing from flat surface		2,400	.027	S.F.		.52		.52	.92	
	2100	Irregular surfaces		1,200	.053			1.05		1.05	1.83	
	3000	Remove cementitious material from flat surface		800	.080			1.57		1.57	2.75	
	3100	Irregular surface		400	.160			3.14		3.14	5.50	
	4000	Scrape acoustical coating/fireproofing, from ceiling		3,200	.020			.39		.39	.69	
	5000	Remove VAT from floor by hand		2,400	.027			.52		.52	.92	

020 700 | Selective Demolition

		CREW	DAILY OUTPUT	LABOR-HOURS	UNIT	1998 BARE COSTS				TOTAL INCL O&P	
						MAT.	LABOR	EQUIP.	TOTAL		
840	5100	By machine	A-11	4,800	.013	S.F.		.26	.01	.27	.47	**840**
	5150	For 2 layers, add				%				50%	
	6000	Remove contaminated soil from crawl space by hand	A-9	400	.160	C.F.		3.14		3.14	5.50
	6100	With large production vacuum loader	A-12	700	.091	"		1.79	.73	2.52	3.94
	7000	Radiator backing, not including radiator removal	A-9	1,200	.053	S.F.		1.05		1.05	1.83
	8000	Cement-asbestos transite board	2 Asbe	2,000	.008		.08	.16		.24	.36
	8100	Transite shingle siding	A-10D	750	.043		.08	.80	.92	1.80	2.46
	8200	Shingle roofing	A-10B	2,000	.016		.07	.31		.38	.63
	8250	Built-up, no gravel, non-friable	B-2	1,400	.029		.07	.39		.46	.76
	8300	Asbestos millboard	2 Asbe	1,000	.016		.08	.31		.39	.63
	9000	For type C (supplied air) respirator equipment, add				%					10%
850	0010	**WASTE PACKAGING, HANDLING, & DISPOSAL**										**850**
	0100	Collect and bag bulk material, 3 C.F. bags, by hand	A-9	400	.160	Ea.	1.10	3.14		4.24	6.70
	0200	Large production vacuum loader	A-12	880	.073		.75	1.43	.58	2.76	3.97
	1000	Double bag and decontaminate	A-9	960	.067		17	1.31		18.31	21
	2000	Containerize bagged material in drums, per 3 C.F. drum	"	800	.080		6.30	1.57		7.87	9.70
	3000	Cart bags 50' to dumpster	2 Asbe	400	.040			.78		.78	1.37
	5000	Disposal charges, not including haul, minimum				C.Y.					42
	5020	Maximum				"					160
	5100	Remove refrigerant from system	1 Plum	40	.200	Lb.		4.17		4.17	6.90
	9000	For type C (supplied air) respirator equipment, add				%					10%
860	0010	**DECONTAMINATION CONTAINMENT AREA DEMOLITION** and clean-up										**860**
	0100	Spray exposed substrate with surfactant (bridging)									
	0200	Flat surfaces	A-9	6,000	.011	S.F.	.30	.21		.51	.70
	0250	Irregular surfaces		4,000	.016	"	.35	.31		.66	.94
	0300	Pipes, beams, and columns		2,000	.032	L.F.	.60	.63		1.23	1.76
	1000	Spray encapsulate polyethelene sheeting		8,000	.008	S.F.	.25	.16		.41	.55
	1100	Roll down polyethelene sheeting		8,000	.008	"		.16		.16	.27
	1500	Bag polyethelene sheeting		400	.160	Ea.	.70	3.14		3.84	6.25
	2000	Fine clean exposed substrate, with nylon brush		2,400	.027	S.F.		.52		.52	.92
	2500	Wet wipe substrate		4,800	.013			.26		.26	.46
	2600	Vacuum surfaces, fine brush		6,400	.010			.20		.20	.34
	3000	Structural demolition									
	3100	Wood stud walls	A-9	2,800	.023	S.F.		.45		.45	.79
	3500	Window manifolds, not incl. window replacement		4,200	.015			.30		.30	.52
	3600	Plywood carpet protection		2,000	.032			.63		.63	1.10
	4000	Remove custom decontamination facility	A-10A	8	3	Ea.	11	59		70	116
	4100	Remove portable decontamination facility	3 Asbe	12	2	"	11.50	39		50.50	81
	5000	HEPA vacuum, shampoo carpeting	A-9	4,800	.013	S.F.	.05	.26		.31	.52
	9000	Final cleaning of protected surfaces	A-10A	8,000	.003	"		.06		.06	.10
870	0010	**ENCAPSULATION WITH SEALANTS**										**870**
	0100	Ceilings and walls, minimum	A-9	21,000	.003	S.F.	.25	.06		.31	.38
	0110	Maximum		10,600	.006		.35	.12		.47	.60
	0200	Columns and beams, minimum		13,300	.005		.25	.09		.34	.45
	0210	Maximum		5,325	.012		.35	.24		.59	.80
	0300	Pipes to 12" diameter including minor repairs, minimum		800	.080	L.F.	.35	1.57		1.92	3.14
	0310	Maximum		400	.160	"	1	3.14		4.14	6.60
890	0010	**OSHA TESTING**										**890**
	0100	Certified technician, minimum				Day					300
	0110	Maximum				"					475
	0200	Personal sampling, PCM analysis, minimum	1 Asbe	8	1	Ea.	2.66	19.55		22.21	37.50
	0210	Maximum	"	4	2	"	2.58	39		41.58	71.50
	0300	Industrial hygenist, minimum				Day					375

Important: See the Reference Section for critical supporting data - Reference Nos., Crews, & Location Factors

020 700	Selective Demolition	CREW	DAILY OUTPUT	LABOR-HOURS	UNIT	1998 BARE COSTS				TOTAL INCL O&P	
						MAT.	LABOR	EQUIP.	TOTAL		
890 0310	Maximum				Day					500	**890**
1000	Cleaned area samples	1 Asbe	8	1	Ea.	2.58	19.55		22.13	37.50	
1100	PCM analysis		8	1		30	19.55		49.55	67.50	
1110	Maximum	↓	4	2		3.09	39		42.09	72	
1200	TEM analysis, minimum									400	
1210	Maximum				↓					1,000	

021 100	Site Clearing	CREW	DAILY OUTPUT	LABOR-HOURS	UNIT	1998 BARE COSTS				TOTAL INCL O&P	
						MAT.	LABOR	EQUIP.	TOTAL		
104 0010	**CLEAR AND GRUB** Cut & chip light, trees to 6" diam.	B-7	1	48	Acre		705	1,150	1,855	2,475	**104**
0150	Grub stumps and remove	B-30	2	12			199	785	984	1,200	
0200	Cut & chip medium, trees to 12" diam.	B-7	.70	68.571			1,000	1,650	2,650	3,500	
0250	Grub stumps and remove	B-30	1	24			400	1,575	1,975	2,400	
0300	Cut & chip heavy, trees to 24" diam.	B-7	.30	160			2,350	3,850	6,200	8,225	
0350	Grub stumps and remove	B-30	.50	48			795	3,150	3,945	4,775	
0400	If burning is allowed, reduce cut & chip				↓					40%	
108 0010	**CLEARING** Brush with brush saw	A-1	.25	32	Acre		430	268	698	1,025	**108**
0100	By hand	"	.12	66.667			895	560	1,455	2,150	
0300	With dozer, ball and chain, light clearing	B-11A	2	8			130	415	545	680	
0400	Medium clearing	"	1.50	10.667	↓		173	555	728	900	

022 100	Grading	CREW	DAILY OUTPUT	LABOR-HOURS	UNIT	1998 BARE COSTS				TOTAL INCL O&P	
						MAT.	LABOR	EQUIP.	TOTAL		
104 0010	**GRADING** Site excav. & fill, see div 022-200										**104**
0020	Fine grading, see div 025-122										

022 200	Excav./Backfill/Compact.	CREW	DAILY OUTPUT	LABOR-HOURS	UNIT	1998 BARE COSTS				TOTAL INCL O&P	
						MAT.	LABOR	EQUIP.	TOTAL		
204 0010	**BACKFILL** By hand, no compaction, light soil	1 Clab	14	.571	C.Y.		7.65		7.65	13.10	**204**
0100	Heavy soil		11	.727			9.75		9.75	16.70	
0300	Compaction in 6" layers, hand tamp, add to above	↓	20.60	.388			5.20		5.20	8.90	
0500	Air tamp, add	B-9C	190	.211			2.91	.91	3.82	6	
0600	Vibrating plate, add	A-1	60	.133			1.79	1.12	2.91	4.29	
0800	Compaction in 12" layers, hand tamp, add to above	1 Clab	34	.235			3.15		3.15	5.40	
1300	Dozer backfilling, bulk, up to 300' haul, no compaction	B-10B	1,200	.007			.13	.69	.82	.97	
1400	Air tamped	B-11B	240	.067	↓		1.08	4.31	5.39	6.55	
212 0010	**BORROW** Buy and load at pit, haul 2 miles round trip										**212**
0020	and spread, with 200 H.P. dozer, no compaction										
0100	Bank run gravel	B-15	600	.047	C.Y.	4.90	.75	2.83	8.48	9.75	
0200	Common borrow		600	.047		3.50	.75	2.83	7.08	8.20	
0300	Crushed stone, (1.40 tons per CY), 1-1/2"		600	.047		14.85	.75	2.83	18.43	20.50	
0320	3/4"	↓	600	.047	↓	15.40	.75	2.83	18.98	21.50	

022 200 | Excav./Backfill/Compact.

			CREW	DAILY OUTPUT	LABOR-HOURS	UNIT	1998 BARE COSTS MAT.	LABOR	EQUIP.	TOTAL	TOTAL INCL O&P	
212	0340	1/2"	B-15	600	.047	C.Y.	15.90	.75	2.83	19.48	22	212
	0360	3/8"		600	.047		16.50	.75	2.83	20.08	22.50	
	0400	Sand, washed, concrete		600	.047		10.45	.75	2.83	14.03	15.85	
	0500	Dead or bank sand		600	.047		3.58	.75	2.83	7.16	8.30	
222	0010	**COMPACTION, STRUCTURAL** Steel wheel tandem roller, 5 tons R022-220	B-10E	8	1	Hr.		19	18.55	37.55	52	222
	0050	Air tamp, 6" to 8" lifts, common fill	B-9	250	.160	C.Y.		2.21	.69	2.90	4.54	
	0060	Select fill	"	300	.133			1.84	.58	2.42	3.79	
	0600	Vibratory plate, 8" lifts, common fill	A-1	200	.040			.54	.34	.88	1.29	
	0700	Select fill	"	216	.037			.50	.31	.81	1.19	
238	0010	**EXCAVATING, BULK BANK MEASURE** Common earth piled R022-240										238
	0020	For loading onto trucks, add								15%	15%	
	0200	Backhoe, hydraulic, crawler mtd., 1 C.Y. cap. = 75 C.Y./hr.	B-12A	600	.013	C.Y.		.26	.89	1.15	1.42	
	0310	Wheel mounted, 1/2 C.Y. cap. = 30 C.Y./hr.	B-12E	240	.033			.66	1.39	2.05	2.62	
	1200	Front end loader, track mtd., 1-1/2 C.Y. cap. = 70 C.Y./hr.	B-10N	560	.014			.27	.64	.91	1.15	
	1500	Wheel mounted, 3/4 C.Y. cap. = 45 C.Y./hr.	B-10R	360	.022			.42	.66	1.08	1.42	
	8000	For hauling excavated material, see div. 022-266										
250	0010	**EXCAVATING, STRUCTURAL** Hand, pits to 6' deep, sandy soil	1 Clab	8	1	C.Y.		13.40		13.40	23	250
	0100	Heavy soil or clay		4	2			27		27	46	
	1100	Hand loading trucks from stock pile, sandy soil		12	.667			8.95		8.95	15.30	
	1300	Heavy soil or clay		8	1			13.40		13.40	23	
	1500	For wet or muck hand excavation, add to above				%				50%	50%	
254	0010	**EXCAVATING, TRENCH** or continuous footing, common earth										254
	0050	1' to 4' deep, 3/8 C.Y. tractor loader/backhoe	B-11C	150	.107	C.Y.		1.73	1.43	3.16	4.48	
	0060	1/2 C.Y. tractor loader/backhoe	B-11M	200	.080			1.30	1.44	2.74	3.76	
	0090	4' to 6' deep, 1/2 C.Y. tractor loader/backhoe	"	200	.080			1.30	1.44	2.74	3.76	
	0100	5/8 C.Y. hydraulic backhoe	B-12Q	250	.032			.63	1.58	2.21	2.79	
	0300	1/2 C.Y. hydraulic excavator, truck mounted	B-12J	200	.040			.79	3.13	3.92	4.75	
	1400	By hand with pick and shovel 2' to 6' deep, light soil	1 Clab	8	1			13.40		13.40	23	
	1500	Heavy soil	"	4	2			27		27	46	
258	0010	**EXCAVATING, UTILITY TRENCH** Common earth										258
	0050	Trenching with chain trencher, 12 H.P., operator walking										
	0100	4" wide trench, 12" deep	B-53	800	.010	L.F.		.13	.11	.24	.36	
	1000	Backfill by hand including compaction, add										
	1050	4" wide trench, 12" deep	A-1	800	.010	L.F.		.13	.08	.21	.32	
262	0010	**FILL** Spread dumped material, by dozer, no compaction	B-10B	1,000	.008	C.Y.		.15	.83	.98	1.17	262
	0100	By hand	1 Clab	12	.667	"		8.95		8.95	15.30	
	0500	Gravel fill, compacted, under floor slabs, 4" deep	B-37	10,000	.005	S.F.	.12	.07	.01	.20	.27	
	0600	6" deep		8,600	.006		.17	.08	.02	.27	.35	
	0700	9" deep		7,200	.007		.29	.10	.02	.41	.50	
	0800	12" deep		6,000	.008		.40	.12	.02	.54	.67	
	1000	Alternate pricing method, 4" deep		120	.400	C.Y.	8.65	5.80	1.24	15.69	21	
	1100	6" deep		160	.300		8.65	4.36	.93	13.94	17.95	
	1200	9" deep		200	.240		8.65	3.49	.74	12.88	16.30	
	1300	12" deep		220	.218		8.65	3.17	.67	12.49	15.70	
266	0011	**HAULING** Excavated or borrow material, loose cubic yards										266
	0015	no loading included, highway haulers										
	0020	6 C.Y. dump truck, 1/4 mile round trip, 5.0 loads/hr.	B-34A	195	.041	C.Y.		.63	1.82	2.45	3.06	
	0200	4 mile round trip, 1.8 loads/hr.	"	70	.114			1.76	5.05	6.81	8.55	
	0310	12 C.Y. dump truck, 1/4 mile round trip 3.7 loads/hr.	B-34B	288	.028			.43	1.50	1.93	2.37	
	0500	4 mile round trip, 1.6 loads/hr.	"	125	.064			.99	3.45	4.44	5.45	
274	0010	**MOBILIZATION OR DEMOBILIZATION** Up to 50 miles										274
	0020	Dozer, loader, backhoe or excavator, 70 H.P.- 250 H.P.	B-34K	6	1.333	Ea.		20.50	149	169.50	199	

Important: See the Reference Section for critical supporting data - Reference Nos., Crews, & Location Factors

022 200 | Excav./Backfill/Compact.

		Description	CREW	DAILY OUTPUT	LABOR-HOURS	UNIT	1998 BARE COSTS				TOTAL INCL O&P	
							MAT.	LABOR	EQUIP.	TOTAL		
274	0900	Shovel or dragline, 3/4 C.Y.	B-34K	3.60	2.222	Ea.		34	249	283	330	274
	1100	Delivery charge for small equipment on flatbed trailer, minimum									40	
	1150	Maximum									100	
286	0010	**LOAM OR TOPSOIL** Remove and stockpile on site										286
	0700	Furnish and place, truck dumped, screened, 4" deep	B-10S	1,300	.006	S.Y.	2.08	.12	.24	2.44	2.74	
	0800	6" deep	"	820	.010	"	3.32	.19	.38	3.89	4.39	
	0900	Fine grading and seeding, incl. lime, fertilizer & seed,										
	1000	With equipment	B-14	1,000	.048	S.Y.	.28	.70	.21	1.19	1.73	

022 300 | Pavement Base

		Description	CREW	DAILY OUTPUT	LABOR-HOURS	UNIT	1998 BARE COSTS				TOTAL INCL O&P	
							MAT.	LABOR	EQUIP.	TOTAL		
304	0010	**BASE** Prepare and roll sub-base, small areas to 2500 S.Y.	B-32A	1,500	.016	S.Y.		.27	.64	.91	1.17	304
308	0010	**BASE COURSE** For roadways and large paved areas										308
	0050	Crushed 3/4" stone base, compacted, 3" deep	B-36B	4,600	.014	S.Y.	5.40	.23	.69	6.32	7.10	
	0100	6" deep		4,500	.014		7.70	.24	.71	8.65	9.65	
	0200	9" deep		3,300	.019		11.55	.32	.97	12.84	14.30	
	0300	12" deep		2,700	.024		15.40	.40	1.18	16.98	18.90	
	0301	Crushed 1-1/2" stone base, compacted to 4" deep		6,000	.011		5.05	.18	.53	5.76	6.50	
	0302	6" deep		4,500	.014		7.70	.24	.71	8.65	9.65	
	0303	8" deep		3,500	.018		10.30	.31	.91	11.52	12.85	
	0304	12" deep		2,000	.032		15.35	.53	1.60	17.48	19.55	
	0350	Bank run gravel, spread and compacted										
	0370	6" deep	B-32	6,000	.005	S.Y.	2.45	.09	.27	2.81	3.16	
	0390	9" deep		4,900	.007		3.68	.11	.33	4.12	4.60	
	0400	12" deep		3,600	.009		4.90	.16	.45	5.51	6.15	
	8900	For small and irregular areas, add						50%	50%			

022 700 | Slope/Erosion Control

		Description	CREW	DAILY OUTPUT	LABOR-HOURS	UNIT	1998 BARE COSTS				TOTAL INCL O&P	
							MAT.	LABOR	EQUIP.	TOTAL		
704	0010	**EROSION CONTROL** Jute mesh, 100 S.Y. per roll, 4' wide, stapled	B-80A	2,400	.010	S.Y.	.78	.13	.07	.98	1.17	704
	0100	Plastic netting, stapled, 2" x 1" mesh, 20 mil	B-1	2,500	.010		.40	.14		.54	.67	
	0200	Polypropylene mesh, stapled, 6.5 oz./S.Y.		2,500	.010		1.75	.14		1.89	2.16	
	0300	Tobacco netting, or jute mesh #2, stapled		2,500	.010		.06	.14		.20	.30	
708	0010	**RETAINING WALLS** Aluminized steel bin, excavation										708
	0020	and backfill not included, 10' wide										
	0100	4' high, 5.5' deep	B-13	650	.074	S.F.	13.85	1.09	1.21	16.15	18.40	
	0200	8' high, 5.5' deep		615	.078		15.90	1.15	1.28	18.33	21	
	0300	10' high, 7.7' deep		580	.083		16.70	1.22	1.35	19.27	22	
	0400	12' high, 7.7' deep		530	.091		18.05	1.34	1.48	20.87	24	
	0500	16' high, 7.7' deep		515	.093		19.05	1.38	1.52	21.95	25	
	1800	Concrete gravity wall with vertical face including excavation & backfill										
	1850	No reinforcing										
	1900	6' high, level embankment	C-17C	36	2.306	L.F.	42	44	11.35	97.35	135	
	2000	33° slope embankment	"	32	2.594	"	43.50	49.50	12.75	105.75	148	
	2800	Reinforced concrete cantilever, incl. excavation, backfill & reinf.										
	2900	6' high, 33° slope embankment	C-17C	35	2.371	L.F.	39	45.50	11.65	96.15	134	
716	0010	**STONE WALL** Including excavation, concrete footing and										716
	0020	stone 3' below grade. Price is exposed face area.										
	0200	Decorative random stone, to 6' high, 1'-6" thick, dry set	D-1	35	.457	S.F.	7.50	7.70		15.20	21.50	
	0300	Mortar set		40	.400		9	6.70		15.70	21.50	
	0500	Cut stone, to 6' high, 1'-6" thick, dry set		35	.457		11	7.70		18.70	25	
	0600	Mortar set		40	.400		13	6.70		19.70	25.50	
	0800	Retaining wall, random stone, 6' to 10' high, 2' thick, dry set		45	.356		9.25	5.95		15.20	20.50	
	0900	Mortar set		50	.320		11.40	5.40		16.80	21.50	

022 | Earthwork

022 700 | Slope/Erosion Control

			CREW	DAILY OUTPUT	LABOR-HOURS	UNIT	1998 BARE COSTS				TOTAL INCL O&P	
							MAT.	LABOR	EQUIP.	TOTAL		
716	1100	Cut stone, 6' to 10' high, 2' thick, dry set	D-1	45	.356	S.F.	14.50	5.95		20.45	26	716
	1200	Mortar set	↓	50	.320	↓	15.50	5.40		20.90	26	

022 800 | Soil Treatment

			CREW	DAILY OUTPUT	LABOR-HOURS	UNIT	MAT.	LABOR	EQUIP.	TOTAL	TOTAL INCL O&P	
804	0010	**TERMITE PRETREATMENT**										804
	0020	Slab and walls, residential	1 Skwk	1,200	.007	SF Flr.	.16	.13		.29	.39	
	0400	Insecticides for termite control, minimum		14.20	.563	Gal.	10	10.55		20.55	29	
	0500	Maximum	↓	11	.727	"	17.10	13.65		30.75	42.50	

023 100 | Tunnel Construction

			CREW	DAILY OUTPUT	LABOR-HOURS	UNIT	MAT.	LABOR	EQUIP.	TOTAL	TOTAL INCL O&P	
150	0010	**MICROTUNNELING** Not including excavation, backfill, shoring,										150
	0020	or dewatering, average 50'/day, slurry method										
	0100	24" to 48" outside diameter, minimum				L.F.					600	
	0110	Adverse conditions, add				%					50%	
	1000	Rent microtunneling machine, average monthly lease				Month					100,000	
	1010	Operating technician				Day					600	
	1100	Mobilization and demobilization, minimum				Job					40,000	
	1110	Maximum				"					400,000	

025 | Paving & Surfacing

025 100 | Walk/Rd/Parkng Paving

			CREW	DAILY OUTPUT	LABOR-HOURS	UNIT	1998 BARE COSTS				TOTAL INCL O&P	
							MAT.	LABOR	EQUIP.	TOTAL		
120	0010	**CONCRETE PAVEMENT** Including joints, finishing, and curing										120
	0020	Fixed form, 12' pass, unreinforced, 6" thick	C-14H	3,000	.016	S.Y.	10.55	.29	.01	10.85	12.15	
	0100	8" thick	"	2,700	.018	↓	13.70	.32	.01	14.03	15.65	
	0700	Finishing, broom finish small areas	2 Cefi	120	.133			2.37		2.37	3.87	
124	0010	**PAVING** Asphaltic concrete	R025 -110									124
	0020	6" stone base, 2" binder course, 1" topping	B-25	10,125	.009	S.F.	1.14	.13	.17	1.44	1.67	
	0300	Binder course, 1-1/2" thick		41,400	.002		.29	.03	.04	.36	.41	
	0400	2" thick		20,700	.004		.36	.06	.08	.50	.59	
	0500	3" thick		14,850	.006		.50	.09	.12	.71	.83	
	0600	4" thick		10,800	.008		.66	.12	.16	.94	1.12	
	0800	Sand finish course, 3/4" thick		51,300	.002		.10	.03	.03	.16	.19	
	0900	1" thick	↓	34,650	.003		.14	.04	.05	.23	.28	
	1000	Fill pot holes, hot mix, 2" thick	B-16	4,200	.008		.10	.11	.10	.31	.41	
	1100	4" thick		3,500	.009		.15	.13	.12	.40	.52	
	1120	6" thick	↓	3,100	.010		.21	.15	.14	.50	.63	
	1140	Cold patch, 2" thick	B-51	3,000	.016		.38	.22	.06	.66	.85	
	1160	4" thick		2,700	.018		.47	.25	.06	.78	1.02	
	1180	6" thick	↓	1,900	.025	↓	.63	.35	.09	1.07	1.39	
128	0010	**SIDEWALKS, DRIVEWAYS, & PATIOS** No base										128
	0020	Asphaltic concrete, 2" thick	B-37	720	.067	S.Y.	3.51	.97	.21	4.69	5.75	
	0100	2-1/2" thick	"	660	.073	"	4.02	1.06	.22	5.30	6.45	
	0300	Concrete, 3000 psi, CIP, 6 x 6 - W1.4 x W1.4 mesh,										
	0310	broomed finish, no base, 4" thick	B-24	600	.040	S.F.	.96	.66		1.62	2.18	
	0350	5" thick	↓	545	.044		1.28	.73		2.01	2.64	
	0400	6" thick	↓	510	.047		1.49	.78		2.27	2.95	
	0450	For bank run gravel base, 4" thick, add	B-18	2,500	.010		.09	.14	.02	.25	.35	
	0520	8" thick, add	"	1,600	.015	↓	.24	.21	.03	.48	.66	
	1000	Crushed stone, 1" thick, white marble	2 Clab	1,700	.009	↓	.17	.13		.30	.40	

Important: See the Reference Section for critical supporting data - Reference Nos., Crews, & Location Factors

		025 100 | **Walk/Rd/Parkng Paving**	CREW	DAILY OUTPUT	LABOR-HOURS	UNIT	1998 BARE COSTS MAT.	LABOR	EQUIP.	TOTAL	TOTAL INCL O&P	
128	1050	Bluestone	2 Clab	1,700	.009	S.F.	.18	.13		.31	.41	128
	1700	Redwood, prefabricated, 4' x 4' sections	F-2	316	.051		4	.82		4.82	5.80	
	1750	Redwood planks, 1" thick, on sleepers	"	240	.067		3	1.08		4.08	5.15	

025 150 | Unit Pavers

			CREW	DAILY OUTPUT	LABOR-HOURS	UNIT	MAT.	LABOR	EQUIP.	TOTAL	TOTAL INCL O&P	
154	0010	**ASPHALT BLOCKS**, 6"x12"x1-1/4", w/bed & neopr. adhesive	D-1	135	.119	S.F.	2.90	1.99		4.89	6.55	154
	0100	3" thick		130	.123		4.06	2.07		6.13	7.95	
	0300	Hexagonal tile, 8" wide, 1-1/4" thick		135	.119		3	1.99		4.99	6.65	
	0400	2" thick		130	.123		4.20	2.07		6.27	8.10	
	0500	Square, 8" x 8", 1-1/4" thick		135	.119		2.95	1.99		4.94	6.60	
	0600	2" thick		130	.123		4.13	2.07		6.20	8.05	
158	0010	**BRICK PAVING** 4" x 8" x 1-1/2", without joints (4.5 brick/S.F.)	D-1	110	.145	S.F.	1.96	2.44		4.40	6.30	158
	0100	Grouted, 3/8" joint (3.9 brick/S.F.)		90	.178		2.35	2.99		5.34	7.65	
	0200	4" x 8" x 2-1/4", without joints (4.5 bricks/S.F.)		110	.145		2.60	2.44		5.04	7	
	0300	Grouted, 3/8" joint (3.9 brick/S.F.)		90	.178		2.40	2.99		5.39	7.70	
	1500	Brick on 1" thick sand bed laid flat, 4.5 per S.F.		100	.160		2.90	2.69		5.59	7.75	
	2000	Laid on edge, 7.2 per S.F.		70	.229		1.93	3.84		5.77	8.05	
166	0010	**STONE PAVERS**										166
	1100	Flagging, bluestone, irregular, 1" thick,	D-1	81	.198	S.F.	1.80	3.32		5.12	7.60	
	1150	Snapped random rectangular, 1" thick		92	.174		2.73	2.92		5.65	7.95	
	1200	1-1/2" thick		85	.188		3.28	3.16		6.44	8.95	
	1250	2" thick		83	.193		3.82	3.24		7.06	9.70	
	1300	Slate, natural cleft, irregular, 3/4" thick		92	.174		1.75	2.92		4.67	6.90	
	1350	Random rectangular, gauged, 1/2" thick		105	.152		3.79	2.56		6.35	8.50	
	1400	Random rectangular, butt joint, gauged, 1/4" thick		150	.107		4.08	1.79		5.87	7.50	
	1450	For sand rubbed finish, add					2.50			2.50	2.75	
	1550	Granite blocks, 3-1/2" x 3-1/2" x 3-1/2"	D-1	92	.174		5.25	2.92		8.17	10.70	

025 250 | Curbs

			CREW	DAILY OUTPUT	LABOR-HOURS	UNIT	MAT.	LABOR	EQUIP.	TOTAL	TOTAL INCL O&P	
254	0010	**CURBS** Asphaltic, machine formed, 8" wide, 6" high, 40 L.F./ton	B-27	1,000	.032	L.F.	.57	.44	.07	1.08	1.47	254
	0150	Asphaltic berm, 12" W, 3"-6" H, 35 L.F./ton, before pavement	"	700	.046		.77	.64	.10	1.51	2.05	
	0200	12" W, 1-1/2" to 4" H, 60 L.F. per ton, laid with pavement	B-2	1,050	.038		.47	.53		1	1.42	
	0300	Concrete, wood forms, 6" x 18", straight	C-2A	500	.096		1.77	1.72		3.49	4.87	
	0400	6" x 18", radius	"	200	.240		1.89	4.29		6.18	9.35	
	0550	Precast, 6" x 18", straight	B-29	700	.069		6.25	1.01	.89	8.15	9.60	
	0600	6" x 18", radius	"	325	.148		7.75	2.18	1.92	11.85	14.35	
	1000	Granite, split face, straight, 5" x 16"	D-13	500	.096		10	1.70	.94	12.64	14.90	
	1100	6" x 18"	"	450	.107		13.15	1.89	1.04	16.08	18.80	
	1300	Radius curbing, 6" x 18", over 10' radius	B-29	260	.185		16.10	2.73	2.41	21.24	25	
	1400	Corners, 2' radius		80	.600	Ea.	54	8.85	7.80	70.65	83	
	1600	Edging, 4-1/2" x 12", straight		300	.160	L.F.	5	2.37	2.08	9.45	11.80	
	1800	Curb inlets, (guttermouth) straight		41	1.171	Ea.	120	17.30	15.25	152.55	178	
258	0010	**EDGING**										258
	0050	Aluminum alloy, including stakes, 1/8" x 4", mill finish	B-1	390	.062	L.F.	1.70	.87		2.57	3.35	
	0051	Black paint		390	.062		1.97	.87		2.84	3.65	
	0052	Black anodized		390	.062		2.28	.87		3.15	3.99	
	0100	Brick, set horizontally, 1-1/2 bricks per L.F.	D-1	370	.043		.90	.73		1.63	2.22	
	0150	Set vertically, 3 bricks per L.F.	"	135	.119		1.82	1.99		3.81	5.35	
	0200	Corrugated aluminum, roll, 4" wide	1 Carp	650	.012		.25	.23		.48	.67	
	0250	6" wide	"	550	.015		.31	.27		.58	.80	
	0600	Railroad ties, 6" x 8"	F-2	170	.094		2.25	1.52		3.77	5.10	
	0650	7" x 9"	"	136	.118		2.50	1.90		4.40	6	
	0700	Redwood										
	0750	2" x 4"	F-2	330	.048	L.F.	2.67	.78		3.45	4.27	

025 250 | Curbs

		CREW	DAILY OUTPUT	LABOR-HOURS	UNIT	1998 BARE COSTS				TOTAL INCL O&P	
						MAT.	LABOR	EQUIP.	TOTAL		
258	0800	Steel edge strips, incl. stakes, 1/4" x 5"	B-1	390	.062	L.F.	2.65	.87		3.52	4.40	258
	0850	3/16" x 4"	"	390	.062	↓	2.09	.87		2.96	3.78	

025 450 | Surfacing

		CREW	DAILY OUTPUT	LABOR-HOURS	UNIT	MAT.	LABOR	EQUIP.	TOTAL	TOTAL INCL O&P	
458	0010	**SEALCOATING** 2 coat coal tar pitch emulsion over 10,000 S.Y.	B-45	5,000	.003	S.Y.	.38	.05	.07	.50	.57	458
	0030	1000 to 10,000 S.Y.	"	3,000	.005		.38	.08	.12	.58	.67	
	0100	Under 1000 S.Y.	B-1	1,050	.023		.38	.32		.70	.96	
	0300	Petroleum resistant, over 10,000 S.Y.	B-45	5,000	.003		.50	.05	.07	.62	.71	
	0320	1000 to 10,000 S.Y.	"	3,000	.005		.50	.08	.12	.70	.81	
	0400	Under 1000 S.Y.	B-1	1,050	.023	↓	.50	.32		.82	1.10	

026 010 | Piped Utilities

		CREW	DAILY OUTPUT	LABOR-HOURS	UNIT	1998 BARE COSTS				TOTAL INCL O&P	
						MAT.	LABOR	EQUIP.	TOTAL		
012	0010	**BEDDING** For pipe and conduit, not incl. compaction										012
	0050	Crushed or screened bank run gravel	B-6	150	.160	C.Y.	7.35	2.40	1.43	11.18	13.75	
	0100	Crushed stone 3/4" to 1/2"		150	.160		15.40	2.40	1.43	19.23	22.50	
	0200	Sand, dead or bank	↓	150	.160		3.58	2.40	1.43	7.41	9.55	
	0500	Compacting bedding in trench	A-1	90	.089	↓		1.19	.74	1.93	2.86	

026 050 | Manholes & Cleanouts

		CREW	DAILY OUTPUT	LABOR-HOURS	UNIT	MAT.	LABOR	EQUIP.	TOTAL	TOTAL INCL O&P	
054	0010	**UTILITY VAULTS** Precast concrete, 6" thick										054
	0050	5' x 10' x 6' high, I.D.	B-13	2	24	Ea.	1,850	355	395	2,600	3,075	
	0350	Hand hole, precast concrete, 1-1/2" thick										
	0400	1'-0" x 2'-0" x 1'-9", I.D., light duty	B-1	4	6	Ea.	325	84.50		409.50	505	
	0450	4'-6" x 3'-2" x 2'-0", O.D., heavy duty	B-6	3	8	"	600	120	71.50	791.50	940	

026 650 | Water Systems

		CREW	DAILY OUTPUT	LABOR-HOURS	UNIT	MAT.	LABOR	EQUIP.	TOTAL	TOTAL INCL O&P	
686	0010	**PIPING, WATER DISTRIBUTION SYSTEMS** Pipe laid in trench,										686
	0020	excavation and backfill not included										
	1400	Ductile Iron, cement lined, class 50 water pipe, 18' lengths										
	1410	Mechanical joint, 4" diameter	B-20	144	.167	L.F.	6.80	2.35		9.15	11.50	
	2650	Polyvinyl chloride pipe, class 160, S.D.R.-26, 1-1/2" diameter		300	.080		.26	1.13		1.39	2.22	
	2700	2" diameter		250	.096		.36	1.35		1.71	2.70	
	2750	2-1/2" diameter		250	.096		.51	1.35		1.86	2.87	
	2800	3" diameter		200	.120		.73	1.69		2.42	3.69	
	2850	4" diameter	↓	200	.120	↓	1.15	1.69		2.84	4.16	

026 700 | Water Wells

		CREW	DAILY OUTPUT	LABOR-HOURS	UNIT	MAT.	LABOR	EQUIP.	TOTAL	TOTAL INCL O&P	
704	0010	**WELLS** Domestic water										704
	0100	Drilled, 4" to 6" diameter	B-23	120	.333	L.F.		4.60	15.85	20.45	25.50	
	1500	Pumps, installed in wells to 100' deep, 4" submersible										
	1520	3/4 H.P.	Q-1	2.66	6.015	Ea.	390	113		503	610	
	1600	1 H.P.	"	2.29	6.987		400	131		531	655	
	1800	2 H.P.	Q-22	1.33	12.030	↓	535	226	330	1,091	1,325	

Important: See the Reference Section for critical supporting data - Reference Nos., Crews, & Location Factors

026 850 | Gas Distribution System

		CREW	DAILY OUTPUT	LABOR-HOURS	UNIT	1998 BARE COSTS MAT.	LABOR	EQUIP.	TOTAL	TOTAL INCL O&P	
852	0010	**GAS SERVICE & DISTRIBUTION** Not including excavation									852
	0050	or backfill									
	0100	Polyethylene, 60 psi, coils, 1/2" diameter, SDR 9.3	B-20	450	.053	L.F.	.28	.75		1.03	1.60
	0150	1-1/4" diameter, SDR 11		400	.060		.62	.84		1.46	2.13
	0200	2" diameter, SDR 11	↓	360	.067		.79	.94		1.73	2.48
	0300	40' joints with coupling, 3" diameter, SDR 11	B-21	300	.093		1.98	1.39	.45	3.82	5.05
	0350	4" diameter, SDR 11		260	.108	↓	2.84	1.60	.52	4.96	6.40
	0400	6" diameter, SDR 11	↓	240	.117		6.15	1.74	.56	8.45	10.35
	0500	Steel, schedule 40, plain end, tar coated & wrapped									
	0550	1" diameter	Q-4	300	.107	L.F.	3.16	2.11	.16	5.43	7.15
	0600	2" diameter		280	.114		3.60	2.26	.17	6.03	7.90
	0650	3" diameter	↓	260	.123		6.95	2.44	.19	9.58	11.90
	0700	4" diameter	B-35	255	.157		9.95	2.77	1.90	14.62	17.70
	0750	5" diameter		220	.182		14.35	3.21	2.20	19.76	23.50
	0800	6" diameter	↓	180	.222	↓	18.25	3.92	2.69	24.86	29.50

027 100 | Subdrainage Systems

		CREW	DAILY OUTPUT	LABOR-HOURS	UNIT	1998 BARE COSTS MAT.	LABOR	EQUIP.	TOTAL	TOTAL INCL O&P	
108	0010	**PIPING, SUBDRAINAGE, CONCRETE**									108
	0021	Not including excavation and backfill									
	3000	Porous wall concrete underdrain, std. strength, 4" diameter	B-20	335	.072	L.F.	1.75	1.01		2.76	3.66
	3020	6" diameter	"	315	.076		2.28	1.07		3.35	4.35
	3040	8" diameter	B-21	310	.090	↓	2.81	1.34	.43	4.58	5.85
110	0010	**PIPING, SUBDRAINAGE, CORRUGATED METAL**									110
	0021	Not including excavation and backfill									
	2010	Aluminum, perforated									
	2020	6" diameter, 18 ga.	B-14	380	.126	L.F.	2.57	1.84	.56	4.97	6.55
	2200	8" diameter, 16 ga.		370	.130		3.70	1.88	.58	6.16	7.90
	2220	10" diameter, 16 ga.	↓	360	.133	↓	4.63	1.94	.59	7.16	9.05
	3000	Uncoated galvanized, perforated									
	3020	6" diameter, 18 ga.	B-20	380	.063	L.F.	3.95	.89		4.84	5.85
	3200	8" diameter, 16 ga.	"	370	.065		5.45	.91		6.36	7.55
	3220	10" diameter, 16 ga.	B-21	360	.078	↓	8.20	1.16	.37	9.73	11.40
	4000	Steel, perforated, asphalt coated									
	4020	6" diameter 18 ga.	B-20	380	.063	L.F.	3	.89		3.89	4.82
	4030	8" diameter 18 ga	"	370	.065		4.60	.91		5.51	6.60
	4040	10" diameter 16 ga	B-21	360	.078		5.60	1.16	.37	7.13	8.55
	4050	12" diameter 16 ga		285	.098		6.50	1.46	.47	8.43	10.15
	4060	18" diameter 16 ga	↓	205	.137	↓	8.70	2.03	.66	11.39	13.75
112	0010	**PIPING, SUBDRAINAGE, VITRIFIED CLAY**									112
	4000	Channel pipe, 4" diameter	B-20	430	.056	L.F.	1.85	.79		2.64	3.39
	4060	8" diameter	"	295	.081	"	4.25	1.14		5.39	6.65

027 150 | Sewage Systems

		CREW	DAILY OUTPUT	LABOR-HOURS	UNIT	1998 BARE COSTS MAT.	LABOR	EQUIP.	TOTAL	TOTAL INCL O&P	
152	0010	**CATCH BASINS OR MANHOLES** not including footing, excavation,									152
	0020	backfill, frame and cover									

R027-110

2 SITE WORK

027 150 | Sewage Systems

			CREW	DAILY OUTPUT	LABOR-HOURS	UNIT	MAT.	LABOR	EQUIP.	TOTAL	TOTAL INCL O&P	
								1998 BARE COSTS				
152	0050	Brick, 4' inside diameter, 4' deep	D-1	1	16	Ea.	251	269		520	730	152
	1110	Precast, 4' I.D., 4' deep	B-22	4.10	7.317		305	111	49	465	580	
	1600	Frames & covers, C.I., 24" square, 500 lb.	B-6	7.80	3.077		210	46	27.50	283.50	340	
162	0010	PIPING, DRAINAGE & SEWAGE, CONCRETE										162
	0020	Not including excavation or backfill	R027 -110									
	1020	8" diameter	B-14	224	.214	L.F.	3.74	3.11	.95	7.80	10.45	
	1030	10" diameter	"	216	.222	"	4.15	3.23	.99	8.37	11.15	
	3780	Concrete slotted pipe, class 4 mortar joint										
	3800	12" diameter	B-21	168	.167	L.F.	10.05	2.48	.80	13.33	16.15	
	3840	18" diameter	"	152	.184	"	15.95	2.74	.88	19.57	23	
	3900	Class 4 O-ring										
	3940	12" diameter	B-21	168	.167	L.F.	11.10	2.48	.80	14.38	17.30	
	3960	18" diameter	"	152	.184	"	17.35	2.74	.88	20.97	24.50	
164	0010	PIPING, STORM DRAINAGE, CORRUGATED METAL										164
	0020	Not including excavation or backfill										
	2040	8" diameter, 16 ga.	B-14	330	.145	L.F.	7.40	2.11	.65	10.16	12.45	
168	0010	PIPING, DRAINAGE & SEWAGE, POLYVINYL CHLORIDE										168
	0020	Not including excavation or backfill										
	2000	10' lengths, S.D.R. 35, B&S, 4" diameter	B-20	375	.064	L.F.	2.16	.90		3.06	3.92	
	2040	6" diameter		350	.069		3.96	.96		4.92	6	
	2080	8" diameter		335	.072		3.50	1.01		4.51	5.60	
	2120	10" diameter	B-21	330	.085		4.53	1.26	.41	6.20	7.60	
172	0010	PIPING, DRAINAGE & SEWAGE, VITRIFIED CLAY C700										172
	0020	Not including excavation or backfill,										
	4030	Extra strength, compression joints, C425										
	5000	4" diameter x 4' long	B-20	265	.091	L.F.	1.62	1.27		2.89	3.96	
	5020	6" diameter x 5' long	"	200	.120		2.62	1.69		4.31	5.75	
	5040	8" diameter x 5' long	B-21	200	.140		3.77	2.08	.67	6.52	8.45	
	5060	10" diameter x 5' long	"	190	.147		6.15	2.19	.71	9.05	11.25	

027 400 | Septic Systems

			CREW	DAILY OUTPUT	LABOR-HOURS	UNIT	MAT.	LABOR	EQUIP.	TOTAL	TOTAL INCL O&P	
404	0010	SEPTIC TANKS Not incl. excav. or piping, precast, 1,000 gallon	B-21	8	3.500	Ea.	420	52	16.80	488.80	565	404
	0100	2,000 gallon		5	5.600		840	83.50	27	950.50	1,100	
	0600	High density polyethylene, 1,000 gallon		6	4.667		745	69.50	22.50	837	965	
	0700	1,500 gallon		4	7		980	104	33.50	1,117.50	1,300	
	1000	Distribution boxes, concrete, 7 outlets	2 Clab	16	1		70	13.40		83.40	100	
	1100	9 outlets	"	8	2		212	27		239	279	
	1150	Leaching field chambers, 13' x 3'-7" x 1'-4", standard	B-13	16	3		645	44.50	49	738.50	840	
	1420	Leaching pit, 6', dia, 3' deep complete					480			480	530	
	2200	Excavation for septic tank, 3/4 C.Y. backhoe	B-12F	145	.055	C.Y.		1.09	3	4.09	5.10	
	2400	4' trench for disposal field, 3/4 C.Y. backhoe	"	335	.024	L.F.		.47	1.30	1.77	2.21	
	2600	Gravel fill, run of bank	B-6	150	.160	C.Y.	11	2.40	1.43	14.83	17.75	
	2800	Crushed stone, 3/4"	"	150	.160	"	17.15	2.40	1.43	20.98	24.50	

Important: See the Reference Section for critical supporting data - Reference Nos., Crews, & Location Factors

		028 100	Irrigation Systems	CREW	DAILY OUTPUT	LABOR-HOURS	UNIT	1998 BARE COSTS MAT.	LABOR	EQUIP.	TOTAL	TOTAL INCL O&P	
104	0010		**SPRINKLER IRRIGATION SYSTEM** For lawns										**104**
	0800		Residential system, custom, 1" supply	B-20	2,619	.009	S.F.	.22	.13		.35	.46	
	0900		1-1/2" supply	"	2,311	.010	"	.20	.15		.35	.47	

		028 300	Fences & Gates	CREW	DAILY OUTPUT	LABOR-HOURS	UNIT	MAT.	LABOR	EQUIP.	TOTAL	TOTAL INCL O&P	
304	0011		**CHAIN LINK FENCE**										**304**
	0350		Aluminized steel, 9 ga. wire, 3' high	B-1	185	.130	L.F.	4.31	1.83		6.14	7.85	
	0400		6' high		115	.209	"	6.50	2.94		9.44	12.20	
	0450		Add for gate 3' wide, 1-3/8" frame 3' high		12	2	Ea.	43	28		71	95.50	
	0490		6' high		10	2.400		85	34		119	152	
	0500		Add for gate 4' wide, 1-3/8" frame 3' high		10	2.400		58	34		92	122	
	0540		6' high		8	3		115	42		157	200	
	0860		Tennis courts, 11 ga. wire, 2 1/2" post 10' O.C., 1-5/8" top rail										
	0900		2-1/2" corner post, 10' high	B-1	95	.253	L.F.	9.90	3.55		13.45	16.95	
	0920		12' high		80	.300	"	11.60	4.22		15.82	20	
	1000		Add for gate 3' wide, 1-5/8" frame 10' high		10	2.400	Ea.	122	34		156	192	
	1040		Aluminized, 11 ga. wire 10' high		95	.253	L.F.	12.05	3.55		15.60	19.35	
	1100		12' high		80	.300	"	14.05	4.22		18.27	23	
	1140		Add for gate 3' wide, 1-5/8" frame, 10' high		10	2.400	Ea.	157	34		191	231	
	1250		Vinyl covered 11 ga. wire, 10' high		95	.253	L.F.	11	3.55		14.55	18.20	
	1300		12' high		80	.300	"	13	4.22		17.22	21.50	
	1400		Add for gate 3' wide, 1-3/8" frame, 10' high		10	2.400	Ea.	174	34		208	250	
320	0010		**FENCE, MISC. METAL** Chicken wire, posts @ 4', 1" mesh, 4' high	B-80	410	.059	L.F.	1.09	.82	1.14	3.05	3.86	**320**
	0100		2" mesh, 6' high		350	.069		.99	.96	1.33	3.28	4.21	
	0200		Galv. steel, 12 ga., 2" x 4" mesh, posts 5' O.C., 3' high		300	.080		1.30	1.13	1.56	3.99	5.05	
	0300		5' high		300	.080		1.90	1.13	1.56	4.59	5.75	
	0400		14 ga., 1" x 2" mesh, 3' high		300	.080		1.31	1.13	1.56	4	5.10	
	0500		5' high		300	.080		1.90	1.13	1.56	4.59	5.75	
	1000		Kennel fencing, 1-1/2" mesh, 6' long, 3'-6" wide, 6'-2" high	2 Clab	4	4	Ea.	230	53.50		283.50	345	
	1050		12' long		4	4		280	53.50		333.50	400	
	1200		Top covers, 1-1/2" mesh, 6' long		15	1.067		43	14.30		57.30	72	
	1250		12' long		12	1.333		77	17.85		94.85	115	
	1300		For kennel doors, see division 083-729										
	5000		Snow fence on steel posts 10' O.C., 4' high	B-1	500	.048	L.F.	1.55	.68		2.23	2.87	
324	0010		**FENCE, RAIL** Picket, No. 2 cedar, Gothic, 2 rail, 3' high		160	.150	"	4.31	2.11		6.42	8.35	**324**
	0050		Gate, 3'-6" wide		9	2.667	Ea.	37	37.50		74.50	106	
	0400		3 rail, 4' high		150	.160	L.F.	4.95	2.25		7.20	9.30	
	0500		Gate, 3'-6" wide		9	2.667	Ea.	44.50	37.50		82	114	
	1200		Stockade, No. 2 cedar, treated wood rails, 6' high		160	.150	L.F.	5.40	2.11		7.51	9.50	
	1250		Gate, 3' wide		9	2.667	Ea.	44	37.50		81.50	113	
	1300		No. 1 cedar, 3-1/4" cedar rails, 6' high		160	.150	L.F.	13.25	2.11		15.36	18.15	
	1500		Gate, 3' wide		9	2.667	Ea.	104	37.50		141.50	179	
	2700		Prefabricated redwood or cedar, 4' high		160	.150	L.F.	10.25	2.11		12.36	14.85	
	2800		6' high		150	.160		13.55	2.25		15.80	18.75	
	3300		Board, shadow box, 1" x 6", treated pine, 6' high		160	.150		7.95	2.11		10.06	12.35	
	3400		No. 1 cedar, 6' high		150	.160		15.70	2.25		17.95	21	
	3900		Basket weave, No. 1 cedar, 6' high		160	.150		15.60	2.11		17.71	21	
	4200		Gate, 3'-6" wide		9	2.667	Ea.	46.50	37.50		84	116	
328	0010		**FENCE, WOOD** Basket weave, 3/8" x 4" boards, 2" x 4"										**328**
	0020		stringers on spreaders, 4" x 4" posts										
	0050		No. 1 cedar, 6' high	B-1	160	.150	L.F.	6.05	2.11		8.16	10.25	
	0860		Open rail fence, split rails, 2 rail 3' high, no. 1 cedar		160	.150		4.19	2.11		6.30	8.20	
	0870		No. 2 cedar		160	.150		3.36	2.11		5.47	7.30	
	0880		3 rail, 4' high, no. 1 cedar		150	.160		5.35	2.25		7.60	9.70	

028 300 | Fences & Gates

		CREW	DAILY OUTPUT	LABOR-HOURS	UNIT	1998 BARE COSTS				TOTAL INCL O&P	
						MAT.	LABOR	EQUIP.	TOTAL		
328	0890	No. 2 cedar	B-1	150	.160	L.F.	3.97	2.25		6.22	8.20	**328**
	1240	Stockade fence, no. 1 cedar, 3-1/4″ rails, 6′ high		160	.150		8.30	2.11		10.41	12.70
	1260	8′ high		155	.155		10.95	2.18		13.13	15.80
	1320	Gate, 3′-6″ wide		8	3	Ea.	45.50	42		87.50	123

028 400 | Walk/Road/Parkg Appurt

		CREW	DAILY OUTPUT	LABOR-HOURS	UNIT	MAT.	LABOR	EQUIP.	TOTAL	TOTAL INCL O&P	
416	0010	STEPS Incl. excav., borrow & concrete base, where applicable										**416**
	0100	Brick steps	B-24	35	.686	LF Riser	7.30	11.35		18.65	27
	0200	Railroad ties	2 Clab	25	.640		2.69	8.60		11.29	17.65
	0300	Bluestone treads, 12″ x 2″ or 12″ x 1-1/2″	B-24	30	.800		16	13.25		29.25	40

028 700 | Site/Street Furnishings

		CREW	DAILY OUTPUT	LABOR-HOURS	UNIT	MAT.	LABOR	EQUIP.	TOTAL	TOTAL INCL O&P	
704	0010	BENCHES Park, precast concrete, w/backs, wood rails, 4′ long	2 Clab	5	3.200	Ea.	293	43		336	395	**704**
	0100	8′ long		4	4		625	53.50		678.50	780
	0500	Steel barstock pedestals w/backs, 2″ x 3″ wood rails, 4′ long		10	1.600		630	21.50		651.50	725
	0510	8′ long		7	2.286		745	30.50		775.50	875
	0800	Cast iron pedestals, back & arms, wood slats, 4′ long		8	2		460	27		487	550
	0820	8′ long		5	3.200		765	43		808	915
	1700	Steel frame, fir seat, 10′ long		10	1.600		141	21.50		162.50	192
716	0010	PLANTERS Concrete, sandblasted, precast, 48″ diameter, 24″ high	2 Clab	15	1.067	Ea.	425	14.30		439.30	495	**716**
	0300	Fiberglass, circular, 36″ diameter, 24″ high		15	1.067		315	14.30		329.30	370
	1200	Wood, square, 48″ side, 24″ high		15	1.067		755	14.30		769.30	855
	1300	Circular, 48″ diameter, 30″ high		10	1.600		675	21.50		696.50	780
	1600	Planter/bench, 72″		5	3.200		2,100	43		2,143	2,400

029 100 | Shrub/Tree Transplanting

		CREW	DAILY OUTPUT	LABOR-HOURS	UNIT	1998 BARE COSTS				TOTAL INCL O&P	
						MAT.	LABOR	EQUIP.	TOTAL		
104	0010	TREE GUYING Including stakes, guy wire and wrap										**104**
	0100	Less than 3″ caliper, 2 stakes	2 Clab	35	.457	Ea.	15	6.15		21.15	27
	0200	3″ to 4″ caliper, 3 stakes	″	21	.762	″	14.50	10.20		24.70	33.50
	1000	Including arrowhead anchor, cable, turnbuckles and wrap									
	1100	Less than 3″ caliper, 3″ anchors	2 Clab	20	.800	Ea.	45	10.70		55.70	68
	1200	3″ to 6″ caliper, 4″ anchors		15	1.067		60	14.30		74.30	90.50
	1300	6″ caliper, 6″ anchors		12	1.333		78	17.85		95.85	117
	1400	8″ caliper, 8″ anchors		9	1.778		80	24		104	129

029 200 | Soil Preparation

		CREW	DAILY OUTPUT	LABOR-HOURS	UNIT	MAT.	LABOR	EQUIP.	TOTAL	TOTAL INCL O&P	
208	0010	PLANT BED PREPARATION										**208**
	0100	Backfill planting pit, by hand, on site topsoil	2 Clab	18	.889	C.Y.		11.90		11.90	20.50
	0200	Prepared planting mix	″	24	.667			8.95		8.95	15.30
	0300	Skid steer loader, on site topsoil	B-62	340	.071			1.06	.31	1.37	2.13
	0400	Prepared planting mix	″	410	.059			.88	.26	1.14	1.76
	1000	Excavate planting pit, by hand, sandy soil	2 Clab	16	1			13.40		13.40	23
	1100	Heavy soil or clay	″	8	2			27		27	46
	1200	1/2 C.Y. backhoe, sandy soil	B-11C	150	.107			1.73	1.43	3.16	4.48
	1300	Heavy soil or clay	″	115	.139			2.25	1.86	4.11	5.85
	2000	Mix planting soil, incl. loam, manure, peat, by hand	2 Clab	60	.267		24.50	3.57		28.07	33

Important: See the Reference Section for critical supporting data - Reference Nos., Crews, & Location Factors

		CREW	DAILY OUTPUT	LABOR-HOURS	UNIT	1998 BARE COSTS				TOTAL INCL O&P	
	029 200 | Soil Preparation					MAT.	LABOR	EQUIP.	TOTAL		
208 2100	Skid steer loader	B-62	150	.160	C.Y.	24.50	2.40	.70	27.60	32	**208**
3000	Pile sod, skid steer loader	"	2,800	.009	S.Y.		.13	.04	.17	.26	
4000	Remove sod, F.E. loader	B-10S	2,000	.004			.08	.16	.24	.30	
4100	Sod cutter	B-12K	3,200	.002	↓		.05	.25	.30	.35	
	029 300 | Lawns & Grasses										
304 0010	**SEEDING** Mechanical seeding, 215 lb./acre R029 -310	A-1	.31	25.806	Acre	480	345	216	1,041	1,350	**304**
0100	44 lb./M.S.Y.	"	1,550	.005	S.Y.	.12	.07	.04	.23	.30	
0300	Fine grading and seeding incl. lime, fertilizer & seed,										
0310	with equipment	B-14	1,000	.048	S.Y.	.15	.70	.21	1.06	1.60	
0600	Limestone hand push spreader, 50 lbs. per M.S.F.	A-1	200	.040	M.S.F.	3	.54	.34	3.88	4.59	
0800	Grass seed hand push spreader, 4.5 lbs. per M.S.F.	"	200	.040	"	10.10	.54	.34	10.98	12.40	
316 0010	**SODDING** 1" deep, bluegrass sod, on level ground, over 8 M.S.F.	B-63	8	5	M.S.F.	200	67	13.15	280.15	350	**316**
0200	4 M.S.F.		7	5.714	"	220	76.50	15.05	311.55	390	
0300	1000 S.F.		3.50	11.429	Ea.	250	153	30	433	570	
0500	Sloped ground, over 8 M.S.F.		6	6.667	M.S.F.	200	89.50	17.55	307.05	390	
0600	4 M.S.F.		5	8		220	107	21	348	450	
0700	1000 S.F.		4	10		250	134	26.50	410.50	535	
1000	Bent grass sod, on level ground, over 6 M.S.F.		20	2		460	27	5.25	492.25	555	
1100	3 M.S.F.		18	2.222		500	30	5.85	535.85	605	
1200	Sodding 1000 S.F. or less		14	2.857		540	38.50	7.50	586	670	
1500	Sloped ground, over 6 M.S.F.		15	2.667		460	35.50	7	502.50	575	
1600	3 M.S.F.		13.50	2.963		500	39.50	7.80	547.30	625	
1700	1000 S.F.	↓	12	3.333	↓	295	44.50	8.75	348.25	410	
	029 500 | Trees/Plants/Grnd Cover										
504 0010	**GROUND COVER** Plants, pachysandra, in prepared beds	B-1	15	1.600	C	32	22.50		54.50	73.50	**504**
0200	Vinca minor, 1 yr, bare root		12	2	"	38	28		66	90	
0600	Stone chips, in 50 lb. bags, Georgia marble		520	.046	Bag	3.20	.65		3.85	4.63	
0700	Onyx gemstone		260	.092		12.50	1.30		13.80	15.95	
0800	Quartz		260	.092	↓	5.25	1.30		6.55	8	
0900	Pea gravel, truckload lots	↓	28	.857	Ton	22	12.05		34.05	44.50	
516 0010	**MULCH**										**516**
0100	Aged barks, 3" deep, hand spread	1 Clab	100	.080	S.Y.	1.98	1.07		3.05	4.02	
0150	Skid steer loader	B-63	13.50	2.963	M.S.F.	220	39.50	7.80	267.30	320	
0200	Hay, 1" deep, hand spread	1 Clab	475	.017	S.Y.	.20	.23		.43	.61	
0250	Power mulcher, small	B-64	180	.089	M.S.F.	18.20	1.27	1.55	21.02	24	
0350	Large	B-65	530	.030	"	18.20	.43	.83	19.46	21.50	
0400	Humus peat, 1" deep, hand spread	1 Clab	700	.011	S.Y.	1.10	.15		1.25	1.47	
0450	Push spreader	A-1	2,500	.003	"	1.38	.04	.03	1.45	1.62	
0550	Tractor spreader	B-66	700	.011	M.S.F.	103	.21	.28	103.49	115	
0600	Oat straw, 1" deep, hand spread	1 Clab	475	.017	S.Y.	.26	.23		.49	.68	
0650	Power mulcher, small	B-64	180	.089	M.S.F.	24.50	1.27	1.55	27.32	31	
0700	Large	B-65	530	.030	"	24.50	.43	.83	25.76	28.50	
0750	Add for asphaltic emulsion	B-45	1,770	.009	Gal.	1.58	.13	.20	1.91	2.18	
0800	Peat moss, 1" deep, hand spread	1 Clab	900	.009	S.Y.	1.71	.12		1.83	2.08	
0850	Push spreader	A-1	2,500	.003	"	1.55	.04	.03	1.62	1.80	
0950	Tractor spreader	B-66	700	.011	M.S.F.	173	.21	.28	173.49	191	
1000	Polyethylene film, 6 mil.	2 Clab	2,000	.008	S.Y.	.18	.11		.29	.38	
1100	Redwood nuggets, 3" deep, hand spread	1 Clab	150	.053	"	5.90	.71		6.61	7.70	
1150	Skid steer loader	B-63	13.50	2.963	M.S.F.	595	39.50	7.80	642.30	730	
1200	Stone mulch, hand spread, ceramic chips, economy	1 Clab	125	.064	S.Y.	5.65	.86		6.51	7.65	
1250	Deluxe	"	95	.084	"	8.55	1.13		9.68	11.35	
1300	Granite chips	B-1	10	2.400	C.Y.	28	34		62	89	
1400	Marble chips		10	2.400		103	34		137	172	
1500	Onyx gemstone	↓	10	2.400	↓	305	34		339	395	

SITE WORK 2

029 500 | Trees/Plants/Grnd Cover		CREW	DAILY OUTPUT	LABOR-HOURS	UNIT	1998 BARE COSTS				TOTAL INCL O&P	
						MAT.	LABOR	EQUIP.	TOTAL		
516	1600	Pea gravel	B-1	28	.857	C.Y.	22	12.05		34.05	44.50	**516**
	1700	Quartz	↓	10	2.400	↓	132	34		166	203	
	1800	Tar paper, 15 Lb. felt	1 Clab	800	.010	S.Y.	.38	.13		.51	.65	
	1900	Wood chips, 2″ deep, hand spread	"	220	.036	"	2.03	.49		2.52	3.06	
	1950	Skid steer loader	B-63	20.30	1.970	M.S.F.	105	26.50	5.20	136.70	167	
520	0010	**PLANTING** Moving shrubs on site, 12″ ball	B-62	28	.857	Ea.		12.85	3.75	16.60	25.50	**520**
	0100	24″ ball	"	22	1.091			16.35	4.78	21.13	33	
	0300	Moving trees on site, 36″ ball	B-6	3.75	6.400			96	57	153	225	
	0400	60″ ball	"	1	24	↓		360	214	574	845	
524	0010	**SHRUBS** Broadleaf evergreen, planted in prepared beds										**524**
	0100	Andromeda, 15″-18″, container	B-1	96	.250	Ea.	16.50	3.52		20.02	24	
	0200	Azalea, 15″ - 18″, container		96	.250		18	3.52		21.52	26	
	0300	Barberry, 9″-12″, container		130	.185		7	2.60		9.60	12.15	
	0400	Boxwood, 15″-18″, B & B		96	.250		24.50	3.52		28.02	32.50	
	0500	Euonymus, emerald gaiety, 12″ to 15″, container		115	.209		12	2.94		14.94	18.25	
	0600	Holly, 15″-18″, B & B		96	.250		16.75	3.52		20.27	24.50	
	0900	Mount laurel, 18″ - 24″, B & B		80	.300		52	4.22		56.22	64.50	
	1000	Paxistema, 9 - 12″ high		130	.185		14	2.60		16.60	19.85	
	1100	Rhododendron, 18″-24″, container		48	.500		25.50	7.05		32.55	40	
	1200	Rosemary, 1 gal container		600	.040		50	.56		50.56	56	
	2000	Deciduous, amelanchier, 2′-3′, B & B		57	.421		65	5.90		70.90	81.50	
	2100	Azalea, 15″-18″, B & B		96	.250		25	3.52		28.52	33.50	
	2300	Bayberry, 2′-3′, B & B		57	.421		22.50	5.90		28.40	34.50	
	2600	Cotoneaster, 15″-18″, B & B	↓	80	.300		21	4.22		25.22	30.50	
	2800	Dogwood, 3′-4′, B & B	B-17	40	.800		21	12.05	14.20	47.25	59	
	2900	Euonymus, alatus compacta, 15″ to 18″, container	B-1	80	.300		20	4.22		24.22	29.50	
	3200	Forsythia, 2′-3′, container	"	60	.400		17.50	5.65		23.15	29	
	3300	Hibiscus, 3′-4′, B & B	B-17	75	.427		28	6.45	7.60	42.05	50	
	3400	Honeysuckle, 3′-4′, B & B	B-1	60	.400		26	5.65		31.65	38	
	3500	Hydrangea, 2′-3′, B & B	"	57	.421		13	5.90		18.90	24.50	
	3600	Lilac, 3′-4′, B & B	B-17	40	.800		34.50	12.05	14.20	60.75	74	
	3900	Privet, bare root, 18″-24″	B-1	80	.300		5	4.22		9.22	12.75	
	4100	Quince, 2′-3′, B & B	"	57	.421		27	5.90		32.90	39.50	
	4200	Russian olive, 3′-4′, B & B	B-17	75	.427		17	6.45	7.60	31.05	38	
	4400	Spirea, 3′-4′, B & B	B-1	70	.343		38	4.82		42.82	50.50	
	4500	Viburnum, 3′-4′, B & B	B-17	40	.800	↓	32	12.05	14.20	58.25	71	
528	0010	**SHRUBS AND TREES** Evergreen, in prepared beds, B & B										**528**
	0100	Arborvitae pyramidal, 4′-5′	B-17	30	1.067	Ea.	32	16.10	18.95	67.05	83	
	0150	Globe, 12″-15″	B-1	96	.250		10	3.52		13.52	17.05	
	0300	Cedar, blue, 8′-10′	B-17	18	1.778		112	27	31.50	170.50	204	
	0500	Hemlock, canadian, 2-1/2′-3′	B-1	36	.667		26	9.40		35.40	44.50	
	0550	Holly, Savannah, 8′ - 10′ H		9.68	2.479		450	35		485	555	
	0600	Juniper, andorra, 18″-24″		80	.300		13.30	4.22		17.52	22	
	0620	Wiltoni, 15″-18″	↓	80	.300		14.30	4.22		18.52	23	
	0640	Skyrocket, 4-1/2′-5′	B-17	55	.582		37.50	8.80	10.35	56.65	67	
	0660	Blue pfitzer, 2′-2-1/2′	B-1	44	.545		15.35	7.65		23	30	
	0680	Ketleerie, 2-1/2′-3′		50	.480		26.50	6.75		33.25	40.50	
	0700	Pine, black, 2-1/2′-3′		50	.480		30	6.75		36.75	44.50	
	0720	Mugo, 18″-24″	↓	60	.400		28	5.65		33.65	40.50	
	0740	White, 4′-5′	B-17	75	.427		41	6.45	7.60	55.05	64.50	
	0800	Spruce, blue, 18″-24″	B-1	60	.400		28	5.65		33.65	40.50	
	0840	Norway, 4′-5′	B-17	75	.427		45	6.45	7.60	59.05	68.50	
	0900	Yew, denisforma, 12″-15″	B-1	60	.400		18	5.65		23.65	29.50	
	1000	Capitata, 18″-24″	↓	30	.800	↓	25	11.25		36.25	47	

Important: See the Reference Section for critical supporting data - Reference Nos., Crews, & Location Factors

029 500		Trees/Plants/Grnd Cover	CREW	DAILY OUTPUT	LABOR-HOURS	UNIT	1998 BARE COSTS				TOTAL INCL O&P	
							MAT.	LABOR	EQUIP.	TOTAL		
528	1100	Hicksi, 2'-2-1/2'	B-1	30	.800	Ea.	25	11.25		36.25	47	**528**
536	0010	**TREES** Deciduous, in prep. beds, balled & burlapped (B&B) [R029-545]										**536**
	0100	Ash, 2" caliper	B-17	8	4	Ea.	90	60.50	71	221.50	279	
	0200	Beech, 5'-6'		50	.640		200	9.65	11.40	221.05	249	
	0300	Birch, 6'-8', 3 stems		20	1.600		80	24	28.50	132.50	160	
	0500	Crabapple, 6'-8'		20	1.600		75	24	28.50	127.50	155	
	0600	Dogwood, 4'-5'		40	.800		48	12.05	14.20	74.25	89	
	0700	Eastern redbud 4'-5'		40	.800		65	12.05	14.20	91.25	108	
	0800	Elm, 8'-10'		20	1.600		62	24	28.50	114.50	140	
	0900	Ginkgo, 6'-7'		24	1.333		75	20	23.50	118.50	143	
	1000	Hawthorn, 8'-10', 1" caliper		20	1.600		85	24	28.50	137.50	166	
	1100	Honeylocust, 10'-12', 1-1/2" caliper		10	3.200		90	48.50	57	195.50	243	
	1300	Larch, 8'		32	1		45	15.10	17.80	77.90	94.50	
	1400	Linden, 8'-10', 1" caliper		20	1.600		75	24	28.50	127.50	155	
	1500	Magnolia, 4'-5'		20	1.600		55	24	28.50	107.50	133	
	1600	Maple, red, 8'-10', 1-1/2" caliper		10	3.200		100	48.50	57	205.50	254	
	1700	Mountain ash, 8'-10', 1" caliper		16	2		75	30	35.50	140.50	173	
	1800	Oak, 2-1/2"-3" caliper		3	10.667		258	161	190	609	765	
	2100	Planetree, 9'-11', 1-1/4" caliper		10	3.200		90	48.50	57	195.50	243	
	2200	Plum, 6'-8', 1" caliper		20	1.600		75	24	28.50	127.50	155	
	2300	Poplar, 9'-11', 1-1/4" caliper		10	3.200		60	48.50	57	165.50	210	
	2500	Sumac, 2'-3'		75	.427		30	6.45	7.60	44.05	52	
	2700	Tulip, 5'-6'		40	.800		45	12.05	14.20	71.25	85.50	
	2800	Willow, 6'-8', 1" caliper		20	1.600		35.50	24	28.50	88	111	

For information about Means Estimating Seminars, see yellow pages 11 and 12 in back of book

For expanded coverage of these items see *Means Site Work and Landscape Cost Data 1998*

	CREW	DAILY OUTPUT	LABOR-HOURS	UNIT	1998 BARE COSTS				TOTAL INCL O&P
					MAT.	LABOR	EQUIP.	TOTAL	
	CREW	DAILY OUTPUT	LABOR-HOURS	UNIT	MAT.	LABOR	EQUIP.	TOTAL	TOTAL INCL O&P

Division 3
Concrete

Estimating Tips

General

- Carefully check all the plans and specifications. Concrete often appears on drawings other than structural drawings, including mechanical and electrical drawings for equipment pads. The cost of cutting and patching is often difficult to estimate. See Division 020 for demolition costs.
- Always obtain concrete prices from suppliers near the job site. A volume discount can often be negotiated depending upon competition in the area. Remember to add for waste, particularly for slabs and footings on grade.

031 Concrete Formwork

- A primary cost for concrete construction is forming. Most jobs today are constructed with prefabricated forms. The selection of the forms best suited for the job and the total square feet of forms required for efficient concrete forming and placing are key elements in estimating concrete construction. Enough forms must be available for erection to make efficient use of the concrete placing equipment and crew.
- Concrete accessories for forming and placing depend upon the systems used. Study the plans and specifications to assure that all special accessory requirements have been included in the cost estimate such as anchor bolts, inserts and hangers.

032 Concrete Reinforcement

- Ascertain that the reinforcing steel supplier has included all accessories, cutting, bending and an allowance for lapping, splicing and waste. A good rule of thumb is 10% for lapping, splicing and waste. Also, 10% waste should be allowed for welded wire fabric.

033 Cast-in-Place Concrete

- When estimating structural concrete, pay particular attention to requirements for concrete additives, curing methods and surface treatments. Special consideration for climate, hot or cold, must be included in your estimate. Be sure to include requirements for concrete placing equipment and concrete finishing.

034 Precast Concrete
035 Cementitious Decks & Toppings

- The cost of hauling precast concrete structural members is often an important factor. For this reason, it is important to get a quote from the nearest supplier. It may become economically feasible to set up precasting beds on the site if the hauling costs are prohibitive.

Reference Numbers

Reference numbers are shown in bold squares at the beginning of some major classifications. These numbers refer to related items in the Reference Section. The reference information may be an estimating procedure, an alternate pricing method or technical information.

Note: Not all subdivisions listed here necessarily appear in this publication.

		031 100	**Struct C.I.P. Formwork**	CREW	DAILY OUTPUT	LABOR-HOURS	UNIT	MAT.	LABOR	EQUIP.	TOTAL	TOTAL INCL O&P	
									1998 BARE COSTS				
125	0010	**SLAB TEXTURE STAMPING,** buy											**125**
	0020	Approx. 3 S.F.- 5 S.F. each, minimum					Ea.	44			44	48.50	
	0030	Average					"	218			218	240	
	0120	Per S.F. of tool, average					S.F.	55			55	60.50	
	0200	Commonly used chemicals for texture systems											
	0210	Hardeners w/colors average					S.F.	.38			.38	.42	
	0220	Release agents w/colors, average						.12			.12	.13	
	0225	Release agents, clear, average						.08			.08	.09	
	0230	Sealers, clear, average						.10			.10	.11	
	0240	Sealers, w/colors, average						.12			.12	.13	
158	0010	**FORMS IN PLACE, FOOTINGS** Continuous wall, plywood, 1 use	C-1	375	.085	SFCA	2.39	1.37		3.76	4.97	**158**
	0150	4 use	"	485	.066	"	.78	1.06		1.84	2.68	
	1500	Keyway, 4 use, tapered wood, 2" x 4"	1 Carp	530	.015	L.F.	.21	.28		.49	.71	
	1550	2" x 6"	"	500	.016	"	.27	.30		.57	.81	
	5000	Spread footings, plywood, 1 use	C-1	305	.105	SFCA	1.70	1.68		3.38	4.76	
	5150	4 use	"	414	.077	"	.55	1.24		1.79	2.74	
170	0010	**FORMS IN PLACE, SLAB ON GRADE**											**170**
	1000	Bulkhead forms with keyway, wood, 1 use, 2 piece	C-1	510	.063	L.F.	.65	1.01		1.66	2.44	
	1400	Bulkhead forms w/keyway, 1 piece expanded metal, left in place										
	1410	In lieu of 2 piece form	C-1	1,375	.023	L.F.	1.15	.37		1.52	1.91	
	1420	In lieu of 3 piece form		1,200	.027	"	1.15	.43		1.58	2	
	2000	Curb forms, wood, 6" to 12" high, on grade, 1 use		215	.149	SFCA	1.72	2.39		4.11	6	
	2150	4 use		275	.116	"	.56	1.87		2.43	3.82	
	3000	Edge forms, wood, 4 use, on grade, to 6" high		600	.053	L.F.	.43	.86		1.29	1.94	
	3050	7" to 12" high		435	.074	SFCA	.90	1.18		2.08	3.01	
	4000	For slab blockouts, to 12" high, 1 use		200	.160	L.F.	.67	2.57		3.24	5.15	
	4100	Plastic (extruded), to 6" high, multiple use, on grade		800	.040	"	.23	.64		.87	1.35	
174	0010	**FORMS IN PLACE, STAIRS** (Slant length x width), 1 use	C-2	165	.291	S.F.	3.53	4.78		8.31	12.10	**174**
	0150	4 use		190	.253		.99	4.15		5.14	8.20	
	2000	Stairs, cast on sloping ground (length x width), 1 use		220	.218		2.30	3.58		5.88	8.70	
	2100	4 use		240	.200		1.30	3.29		4.59	7.10	
182	0010	**FORMS IN PLACE, WALLS**											**182**
	0100	Box out for wall openings, to 16" thick, to 10 S.F.	C-2	24	2	Ea.	20.50	33		53.50	79	
	0150	Over 10 S.F. (use perimeter)	"	280	.171	L.F.	1.85	2.82		4.67	6.85	
	0250	Brick shelf, 4" w, add to wall forms, use wall area abv shelf										
	0260	1 use	C-2	240	.200	SFCA	2.09	3.29		5.38	7.95	
	0350	4 use		300	.160	"	.84	2.63		3.47	5.40	
	0500	Bulkhead forms, with keyway, 1 use, 2 piece		265	.181	L.F.	2.51	2.98		5.49	7.85	
	0550	3 piece		175	.274	"	3.15	4.51		7.66	11.15	
	0600	Bulkhead forms w/keyway, 1 piece expanded metal, left in place										
	0610	In lieu of 2 piece form	C-1	800	.040	L.F.	1.07	.64		1.71	2.28	
	0620	In lieu of 3 piece form	"	525	.061	"	1.08	.98		2.06	2.87	
	2000	Job built plyform wall forms, to 8' high, 1 use, below grade	C-2	370	.130	SFCA	1.87	2.13		4	5.70	
	2150	4 use		505	.095		.62	1.56		2.18	3.36	
	2400	Over 8' to 16' high, 1 use		280	.171		2.96	2.82		5.78	8.10	
	2550	4 use		395	.122		.67	2		2.67	4.16	
	3000	For architectural finish, add		1,820	.026		.60	.43		1.03	1.40	
	3500	Polystyrene (expanded) wall forms										
	3510	To 8' high, 1 use, left in place	1 Carp	295	.027	SFCA	1.60	.50		2.10	2.62	
	7800	Modular prefabricated plywood, to 8' high, 1 use per month	C-2	910	.053		.95	.87		1.82	2.53	
	7860	4 use per month		970	.049		.31	.81		1.12	1.73	
	8000	To 16' high, 1 use per month		550	.087		1.05	1.43		2.48	3.61	
	8060	4 use per month		610	.079		.34	1.29		1.63	2.60	
	8100	Over 16' high, 1 use per month		550	.087		1.26	1.43		2.69	3.85	
	8160	4 use per month		610	.079		.41	1.29		1.70	2.67	

Important: See the Reference Section for critical supporting data - Reference Nos., Crews, & Location Factors

031 | Concrete Formwork

031 100	Struct C.I.P. Formwork	CREW	DAILY OUTPUT	LABOR-HOURS	UNIT	1998 BARE COSTS				TOTAL INCL O&P	
						MAT.	LABOR	EQUIP.	TOTAL		
190	**0010** SCAFFOLDING See division 015-254										190
198	**0010** WATERSTOP PVC, ribbed 3/16" thick, 4" wide	1 Carp	155	.052	L.F.	1.17	.95		2.12	2.93	198
	0050 6" wide		145	.055		1.75	1.02		2.77	3.68	
	0500 Ribbed, PVC, with center bulb, 9" wide, 3/16" thick		135	.059		1.72	1.10		2.82	3.77	
	0550 3/8" thick		130	.062		2.88	1.14		4.02	5.10	

032 | Concrete Reinforcement

032 100	Reinforcing Steel	CREW	DAILY OUTPUT	LABOR-HOURS	UNIT	1998 BARE COSTS				TOTAL INCL O&P	
						MAT.	LABOR	EQUIP.	TOTAL		
107	**0010** REINFORCING IN PLACE A615 Grade 60										107
	0500 Footings, #4 to #7	4 Rodm	2.10	15.238	Ton	520	300		820	1,150	
	0550 #8 to #18		3.60	8.889		495	176		671	875	
	0700 Walls, #3 to #7		3	10.667		520	211		731	970	
	0750 #8 to #18		4	8		520	158		678	870	
	2400 Dowels, 2 feet long, deformed, #3	2 Rodm	520	.031	Ea.	.23	.61		.84	1.39	
	2410 #4		480	.033		.40	.66		1.06	1.67	
	2420 #5		435	.037		.64	.73		1.37	2.06	
	2430 #6		360	.044		.91	.88		1.79	2.65	

032 200	Welded Wire Fabric	CREW	DAILY OUTPUT	LABOR-HOURS	UNIT	1998 BARE COSTS				TOTAL INCL O&P	
						MAT.	LABOR	EQUIP.	TOTAL		
207	**0010** WELDED WIRE FABRIC ASTM A185										207
	0050 Sheets										
	0100 6 x 6 - W1.4 x W1.4 (10 x 10) 21 lb. per C.S.F.	2 Rodm	35	.457	C.S.F.	8.35	9.05		17.40	26	
	0300 6 x 6 - W2.9 x W2.9 (6 x 6) 42 lb. per C.S.F.		29	.552		15.25	10.90		26.15	37.50	
	0500 4 x 4 - W1.4 x W1.4 (10 x 10) 31 lb. per C.S.F.		31	.516		10.45	10.20		20.65	30.50	
	0750 Rolls										
	0900 2 x 2 - #12 galv. for gunite reinforcing	2 Rodm	6.50	2.462	C.S.F.	17.05	48.50		65.55	110	
	0950 Material prices for above include 10% lap										
240	**0010** FIBROUS REINFORCING										240
	0100 Synthetic fibers				Lb.	4.11			4.11	4.52	
	0110 1-1/2 lb. per C.Y., add to concrete				C.Y.	6.35			6.35	7	
	0150 Steel fibers				Lb.	.47			.47	.52	
	0160 50 lb. per C.Y., add to concrete				C.Y.	24			24	26	
	0170 75 lb. per C.Y., add to concrete					36.50			36.50	40	
	0180 100 lb. per C.Y., add to concrete					47.50			47.50	52.50	

033 | Cast-In-Place Concrete

033 100	Structural Concrete	CREW	DAILY OUTPUT	LABOR-HOURS	UNIT	1998 BARE COSTS				TOTAL INCL O&P	
						MAT.	LABOR	EQUIP.	TOTAL		
126	**0010** CONCRETE, READY MIX Regular weight	R033 -070									126
	0020 2000 psi				C.Y.	54.50			54.50	60	

For expanded coverage of these items see *Means Concrete & Masonry Cost Data 1998*

033 100 | Structural Concrete

		CREW	DAILY OUTPUT	LABOR-HOURS	UNIT	1998 BARE COSTS				TOTAL INCL O&P	
						MAT.	LABOR	EQUIP.	TOTAL		
126	0100	2500 psi (R033 -070)				C.Y.	56			56	61.50	**126**
	0150	3000 psi					57.50			57.50	63.50	
	0200	3500 psi					59			59	65	
	0250	3750 psi					59			59	65	
	0300	4000 psi					60.50			60.50	66.50	
	0350	4500 psi					62			62	68.50	
	0400	5000 psi					63.50			63.50	70	
	0411	6000 psi					72.50			72.50	79.50	
	0412	8000 psi					118			118	130	
	0413	10,000 psi					168			168	184	
	0414	12,000 psi					203			203	223	
	1000	For high early strength cement, add					10%					
	2000	For all lightweight aggregate, add					45%					
130	0010	**CONCRETE IN PLACE** Including forms (4 uses), reinforcing										**130**
	0050	steel, including finishing unless otherwise indicated										
	0500	Chimney foundations, industrial, minimum (R033 -130)	C-14C	32.22	3.476	C.Y.	120	60	1.15	181.15	237	
	0510	Maximum		23.71	4.724		140	82	1.56	223.56	298	
	3800	Footings, spread under 1 C.Y.		38.07	2.942		86.50	51	.97	138.47	185	
	3850	Over 5 C.Y.		81.04	1.382		79	24	.46	103.46	129	
	3900	Footings, strip, 18" x 9", plain		41.04	2.729		76.50	47	.90	124.40	168	
	3950	36" x 12", reinforced		61.55	1.820		79.50	31.50	.60	111.60	143	
	4000	Foundation mat, under 10 C.Y.		38.67	2.896		110	50	.96	160.96	209	
	4050	Over 20 C.Y.		56.40	1.986		97.50	34.50	.66	132.66	167	
	4520	Handicap access ramp, railing both sides, 3' wide	C-14H	14.58	3.292	L.F.	103	59.50	2.57	165.07	219	
	4525	5' wide		12.22	3.928		105	71	3.06	179.06	241	
	4530	With cheek walls and rails both sides, 3' wide		8.55	5.614		105	101	4.38	210.38	296	
	4535	5' wide		7.31	6.566		105	119	5.10	229.10	325	
	4760	or reinforcing, over 10,000 S.F., 4" thick slab	C-14F	3,425	.021	S.F.	.77	.35	.01	1.13	1.43	
	4820	6" thick slab	"	3,350	.021	"	1.12	.36	.01	1.49	1.83	
	5000	Slab on grade, incl. textured finish, not incl. forms										
	5001	or reinforcing, 4" thick slab	C-14G	2,873	.019	S.F.	.75	.32	.01	1.08	1.35	
	5010	6" thick		2,590	.022		1.17	.35	.01	1.53	1.89	
	5020	8" thick		2,320	.024		1.53	.39	.02	1.94	2.35	
	6203	Retaining walls, gravity, 4' high				C.Y.	73.50			73.50	80.50	
	6800	Stairs, not including safety treads, free standing, 3'-6" wide	C-14H	83	.578	LF Nose	5.50	10.45	.45	16.40	24.50	
	6850	Cast on ground		125	.384	"	3.86	6.95	.30	11.11	16.60	
	7000	Stair landings, free standing		200	.240	S.F.	2.19	4.34	.19	6.72	10.10	
	7050	Cast on ground		475	.101	"	1.24	1.83	.08	3.15	4.61	
134	0010	**CURING** Burlap, 4 uses assumed, 7.5 oz.	2 Clab	55	.291	C.S.F.	2.50	3.90		6.40	9.45	**134**
	0100	12 oz.		55	.291		3.86	3.90		7.76	10.95	
	0200	Waterproof curing paper, 2 ply, reinforced		70	.229		7	3.06		10.06	12.95	
	0300	Sprayed membrane curing compound		95	.168		2.53	2.26		4.79	6.65	
	0710	Electrically, heated pads, 15 watts/S.F., 20 uses, minimum				S.F.	.16			.16	.17	
	0800	Maximum				"	.26			.26	.29	
152	0012	**FLOOR PATCHING** 1/4" thick, small areas, regular	1 Cefi	170	.047	S.F.	.92	.84		1.76	2.38	**152**
	0100	Epoxy	"	100	.080	"	3.95	1.42		5.37	6.65	
172	0010	**PLACING CONCRETE** and vibrating, including labor & equipment										**172**
	1400	Elevated slabs, less than 6" thick, pumped	C-20	140	.457	C.Y.		6.80	5.05	11.85	17.10	
	1450	With crane and bucket	C-7	95	.758			11.40	9.30	20.70	29.50	
	1500	6" to 10" thick, pumped (R033 -090)	C-20	160	.400			5.95	4.42	10.37	14.95	
	1550	With crane and bucket	C-7	110	.655			9.85	8.05	17.90	25.50	
	1600	Slabs over 10" thick, pumped	C-20	180	.356			5.30	3.93	9.23	13.25	
	1650	With crane and bucket	C-7	130	.554			8.35	6.80	15.15	21.50	
	1900	Footings, continuous, shallow, direct chute	C-6	120	.400			5.80	.62	6.42	10.50	

Important: See the Reference Section for critical supporting data - Reference Nos., Crews, & Location Factors

033 100 | Structural Concrete

		CREW	DAILY OUTPUT	LABOR-HOURS	UNIT	1998 BARE COSTS				TOTAL INCL O&P	
						MAT.	LABOR	EQUIP.	TOTAL		
172	1950	Pumped	C-20	150	.427	C.Y.		6.35	4.71	11.06	15.95	172
	2000	With crane and bucket	C-7	90	.800			12.05	9.85	21.90	31.50
	2400	Footings, spread, under 1 C.Y., direct chute	C-6	55	.873			12.65	1.35	14	23
	2600	Footings, spread, over 5 C.Y., direct chute		120	.400			5.80	.62	6.42	10.50
	2900	Foundation mats, over 20 C.Y., direct chute		350	.137			1.98	.21	2.19	3.60
	4300	Slab on grade, 4" thick, direct chute		110	.436			6.30	.68	6.98	11.45
	4350	Pumped	C-20	130	.492			7.35	5.45	12.80	18.40
	4400	With crane and bucket	C-7	110	.655			9.85	8.05	17.90	25.50
	4900	Walls, 8" thick, direct chute	C-6	90	.533			7.70	.83	8.53	14
	4950	Pumped	C-20	100	.640			9.55	7.05	16.60	24
	5000	With crane and bucket	C-7	80	.900			13.55	11.05	24.60	35
	5050	12" thick, direct chute	C-6	100	.480			6.95	.74	7.69	12.60
	5100	Pumped	C-20	110	.582			8.65	6.45	15.10	21.50
	5200	With crane and bucket	C-7	90	.800			12.05	9.85	21.90	31.50
	5600	Wheeled concrete dumping, add to placing costs above				C.Y.					
	5610	Walking cart, 50' haul, add	C-18	32	.281			3.83	1.58	5.41	8.30
	5620	150' haul, add		24	.375			5.10	2.11	7.21	11.05
	5700	250' haul, add		18	.500			6.80	2.81	9.61	14.75
	5800	Riding cart, 50' haul, add	C-19	80	.112			1.53	1.06	2.59	3.79
	5810	150' haul, add		60	.150			2.04	1.41	3.45	5.05
	5900	250' haul, add		45	.200			2.72	1.88	4.60	6.75
196	0010	**WINTER PROTECTION** For heated ready mix, add, minimum				C.Y.	3.61			3.61	3.97	196
	0050	Maximum				"	4.64			4.64	5.10
	0100	Protecting concrete and temporary heat, add, minimum	2 Clab	6,000	.003	S.F.	.10	.04		.14	.17
	0200	Temporary shelter for slab on grade, wood frame and polyethylene									
	0201	sheeting, minimum	F-2	10	1.600	M.S.F.	242	26		268	310
	0210	Maximum	"	3	5.333	"	455	86.50		541.50	650

033 450 | Concrete Finishing

		CREW	DAILY OUTPUT	LABOR-HOURS	UNIT	1998 BARE COSTS				TOTAL INCL O&P	
						MAT.	LABOR	EQUIP.	TOTAL		
454	0010	**FINISHING FLOORS** Monolithic, screed finish	1 Cefi	900	.009	S.F.		.16		.16	.26	454
	0050	Darby finish		750	.011			.19		.19	.31
	0100	Float finish		725	.011			.20		.20	.32
	0150	Broom finish		675	.012			.21		.21	.34
	0200	Steel trowel finish, for resilient tile		625	.013			.23		.23	.37
	0250	For finish floor		550	.015			.26		.26	.42
	1600	Exposed local aggregate finish, minimum		625	.013		.36	.23		.59	.77
	1650	Maximum		465	.017		1.14	.31		1.45	1.75
458	0010	**FINISHING WALLS** Break ties and patch voids	1 Cefi	540	.015	S.F.	.02	.26		.28	.45	458
	0050	Burlap rub with grout	"	450	.018		.08	.32		.40	.61
	0300	Bush hammer, green concrete	B-39	1,000	.048		.02	.66	.17	.85	1.34
	0350	Cured concrete	"	650	.074		.03	1.01	.27	1.31	2.06

034 100 | Structural Precast

		CREW	DAILY OUTPUT	LABOR-HOURS	UNIT	1998 BARE COSTS				TOTAL INCL O&P	
						MAT.	LABOR	EQUIP.	TOTAL		
136	0010	**SLABS** Prestressed roof/floor members, solid, grouted, 4" thick	C-11	3,600	.016	S.F.	4.78	.30	.43	5.51	6.25	136
	0050	6" thick		4,500	.012		4.58	.24	.35	5.17	5.85
	0100	8" thick, hollow		5,600	.010		4.55	.19	.28	5.02	5.65
	0150	10" thick		8,800	.006		5.30	.12	.18	5.60	6.25

For expanded coverage of these items see *Means Concrete & Masonry Cost Data 1998*

034 100		Structural Precast	CREW	DAILY OUTPUT	LABOR-HOURS	UNIT	1998 BARE COSTS				TOTAL INCL O&P	
							MAT.	LABOR	EQUIP.	TOTAL		
136	0200	12" thick	C-11	8,000	.007	S.F.	4.78	.13	.19	5.10	5.70	136

034 800		Precast Specialties										
802	0010	LINTELS										802
	0800	Precast concrete, 4" wide, 8" high, stock units to 5' long	D-1	175	.091	L.F.	5.80	1.54		7.34	9	
	0850	5'-12' long	D-4	190	.211		5.75	3.21	.85	9.81	12.70	
	1000	6" wide, 8" high, solid, stock units to 5' long		185	.216		8.20	3.30	.87	12.37	15.55	
	1050	5'-12' long	↓	190	.211	↓	6.80	3.21	.85	10.86	13.90	
804	0010	STAIRS, Precast concrete treads on steel stringers, 3' wide	C-12	75	.640	Riser	53	11.65	5.85	70.50	85	804
	0300	Front entrance, 5' wide with 48" platform, 2 risers		16	3	Flight	273	54.50	27.50	355	425	
	0350	5 risers		12	4		310	73	36.50	419.50	505	
	0500	6' wide, 2 risers	↓	15	3.200		310	58	29	397	470	
	1200	Basement entrance stairs, steel bulkhead doors, minimum	B-51	22	2.182		455	30.50	7.60	493.10	560	
	1250	Maximum	"	11	4.364	↓	675	61	15.25	751.25	860	

For information about Means Estimating Seminars, see yellow pages 11 and 12 in back of book

Important: See the Reference Section for critical supporting data - Reference Nos., Crews, & Location Factors

Division 4
Masonry

Estimating Tips

041 Mortar & Masonry Accessories

- The terms *mortar* and *grout* are often used interchangeably, and incorrectly. Mortar is used to bed masonry units, seal the entry of air and moisture, provide architectural appearance, and allow for size variations in the units. Grout is used primarily in reinforced masonry construction and is used to bond the masonry to the reinforcing steel. Common mortar types are M(2500 psi), S(1800 psi), N(750 psi), and O(350 psi), and conform to ASTM C270. Grout is either fine or coarse, conforms to ASTM C476, and in-place strengths generally exceed 2500 psi. Mortar and grout are different components of masonry construction and are placed by entirely different methods. An estimator should be aware of their unique uses and costs.

042 Unit Masonry

- The most common types of unit masonry are brick and concrete masonry. The major classifications of brick are building brick (ASTM C62), facing brick (ASTM C216) and glazed brick, fire brick and pavers. Many varieties of texture and appearance can exist within these classifications, and the estimator would be wise to check local custom and availability within the project area. On repair and remodeling jobs, matching the existing brick may be the most important criteria.
- Brick and concrete block are priced by the piece and then converted into a price per square foot of wall. Openings less than two square feet are generally ignored by the estimator because any savings in units used is offset by the cutting and trimming required.

- All masonry walls, whether interior or exterior, require bracing. The cost of bracing walls during construction should be included by the estimator and this bracing must remain in place until permanent bracing is complete. Permanent bracing of masonry walls is accomplished by masonry itself, in the form of pilasters or abutting wall corners, or by anchoring the walls to the structural frame. Accessories in the form of anchors, anchor slots and ties are used, but their supply and installation can be by different trades. For instance, anchor slots on spandrel beams and columns are supplied and welded in place by the steel fabricator, but the ties from the slots into the masonry are installed by the bricklayer. Regardless of the installation method the estimator must be certain that these accessories are accounted for in pricing.

Reference Numbers

Reference numbers are shown in bold squares at the beginning of some major classifications. These numbers refer to related items in the Reference Section. The reference information may be an estimating procedure, an alternate pricing method or technical information.

Note: Not all subdivisions listed here necessarily appear in this publication.

041 000 | Mortar

		CREW	DAILY OUTPUT	LABOR-HOURS	UNIT	MAT.	LABOR	EQUIP.	TOTAL	TOTAL INCL O&P	
008 0010	**CEMENT** Gypsum 80 lb. bag, T.L. lots — R041-100				Bag	11.25			11.25	12.40	**008**
0050	L.T.L. lots					11.75			11.75	12.95	
0100	Masonry, 70 lb. bag, T.L. lots — R041-200					5.40			5.40	5.95	
0150	L.T.L. lots					5.40			5.40	5.95	
0200	White masonry cement, 70 lb. bag, T.L. lots					15.30			15.30	16.85	
0250	L.T.L. lots					15.85			15.85	17.45	
016 0010	**GROUTING** Bond bms. & lintels, 8" dp., pumped, not incl. block										**016**
0200	Concrete block cores, solid, 4" thk., by hand, 0.067 C.F./S.F.	D-8	1,100	.036	S.F.	.20	.63		.83	1.28	
0210	6" thick, pumped, 0.175 C.F. per S.F.	D-4	720	.056		.52	.85	.22	1.59	2.27	
0250	8" thick, pumped, 0.258 C.F. per S.F.		680	.059		.77	.90	.24	1.91	2.63	
0300	10" thick, pumped, 0.340 C.F. per S.F.		660	.061		1.02	.92	.24	2.18	2.96	
0350	12" thick, pumped, 0.422 C.F. per S.F.		640	.063		1.27	.95	.25	2.47	3.29	

041 500 | Masonry Accessories

		CREW	DAILY OUTPUT	LABOR-HOURS	UNIT	MAT.	LABOR	EQUIP.	TOTAL	TOTAL INCL O&P	
504 0010	**ANCHOR BOLTS** Hooked type with nut and washer, 1/2" diam., 8" long	1 Bric	200	.040	Ea.	.43	.76		1.19	1.76	**504**
0030	12" long		190	.042		.67	.80		1.47	2.09	
0060	3/4" diameter, 8" long		160	.050		1.76	.95		2.71	3.54	
0070	12" long		150	.053		2.01	1.01		3.02	3.92	
512 0010	**JOINT REINFORCING** Steel bars, placed horiz., #3 & #4 bars — R041-500	1 Bric	450	.018	Lb.	.27	.34		.61	.87	**512**
0050	Placed vertical, #3 & #4 bars		350	.023		.27	.43		.70	1.03	
0060	#5 & #6 bars		650	.012		.27	.23		.50	.70	
0200	Wire strips, regular truss, to 6" wide, galvanized		30	.267	C.L.F.	14	5.05		19.05	24	
0250	12" wide		20	.400		16	7.60		23.60	30.50	
0400	Cavity truss with drip section to 6" wide		30	.267		14	5.05		19.05	24	
0450	12" wide		20	.400		18	7.60		25.60	32.50	
520 0010	**WALL TIES** To brick veneer, galv., corrugated, 7/8" x 7", 22 gauge	1 Bric	10.50	.762	C	4.20	14.45		18.65	29	**520**
0150	16 gauge		10.50	.762		11.50	14.45		25.95	37	
0200	Buck anchors, galv., corrugated, 16 gauge, 2" bend. 8" x 2"		10.50	.762		51.50	14.45		65.95	81	
0250	8" x 3"		10.50	.762		67	14.45		81.45	98.50	
0600	Cavity wall, Z type, galvanized, 6" long, 1/4" diameter		10.50	.762		14.75	14.45		29.20	41	
0650	3/16" diameter		10.50	.762		7.15	14.45		21.60	32.50	
0800	8" long, 1/4" diameter		10.50	.762		21	14.45		35.45	47.50	
0850	3/16" diameter		10.50	.762		8.15	14.45		22.60	33.50	
1000	Rectangular type, galvanized, 1/4" diameter, 2" x 6"		10.50	.762		32.50	14.45		46.95	60.50	
1050	2" x 8" or 4" x 6"		10.50	.762		35.50	14.45		49.95	63.50	
1100	3/16" diameter, 2" x 6"		10.50	.762		14.05	14.45		28.50	40	
1150	2" x 8" or 4" x 6"		10.50	.762		18.80	14.45		33.25	45	
1500	Rigid partition anchors, plain, 8" long, 1" x 1/8"		10.50	.762		49.50	14.45		63.95	79	
1550	x 1/4"		10.50	.762		97.50	14.45		111.95	132	
1580	1-1/2" x 1/8"		10.50	.762		69	14.45		83.45	100	
1600	x 1/4"		10.50	.762		163	14.45		177.45	205	
1650	2" x 1/8"		10.50	.762		86	14.45		100.45	119	
1700	x 1/4"		10.50	.762		233	14.45		247.45	281	

Important: See the Reference Section for critical supporting data - Reference Nos., Crews, & Location Factors

042 050 | Chimneys

			DAILY OUTPUT	LABOR-HOURS	UNIT	1998 BARE COSTS				TOTAL INCL O&P	
		CREW				MAT.	LABOR	EQUIP.	TOTAL		
054	0010	**CHIMNEY** For foundation, add to prices below, see div. 033-130-0500									054
	0100	Brick, 16" x 16", 8" flue, scaff. not incl.	D-1	18.20	.879	V.L.F.	12.90	14.75		27.65	39
	0150	16" x 20" with one 8" x 12" flue R042 -050		16	1		15.95	16.80		32.75	46
	0200	16" x 24" with two 8" x 8" flues		14	1.143		20.50	19.20		39.70	55
	0250	20" x 20" with one 12" x 12" flue		13.70	1.168		17.85	19.60		37.45	52.50
	0300	20" x 24" with two 8" x 12" flues		12	1.333		24.50	22.50		47	64.50
	0350	20" x 32" with two 12" x 12" flues		10	1.600		29	27		56	77.50

042 100 | Brick Masonry

			DAILY OUTPUT	LABOR-HOURS	UNIT	1998 BARE COSTS				TOTAL INCL O&P	
		CREW				MAT.	LABOR	EQUIP.	TOTAL		
108	0010	**COLUMNS** Brick, scaffolding not included									108
	0050	8" x 8", 9 brick	D-1	56	.286	V.L.F.	3.22	4.80		8.02	11.70
	0100	12" x 8", 13.5 brick		37	.432		4.83	7.25		12.08	17.60
	0200	12" x 12", 20 brick		25	.640		7.15	10.75		17.90	26
	0300	16" x 12", 27 brick		19	.842		9.65	14.15		23.80	34.50
	0400	16" x 16", 36 brick		14	1.143		12.90	19.20		32.10	46.50
	0500	20" x 16", 45 brick		11	1.455		16.10	24.50		40.60	59.50
	0600	20" x 20", 56 brick		9	1.778		20	30		50	72.50
110	0010	**COMMON BUILDING BRICK** C62, material only R042 -100									110
	0020	Standard, minimum				M	240			240	264
	0050	Average (select)				"	300			300	330
116	0010	**COPING** Stock units									116
	0050	Precast concrete, 10" wide, 4" tapers to 3-1/2", 8" wall	D-1	75	.213	L.F.	9.60	3.58		13.18	16.60
	0100	12" wide, 3-1/2" tapers to 3", 10" wall		70	.229		10.10	3.84		13.94	17.60
	0150	16" wide, 4" tapers to 3-1/2", 14" wall		60	.267		13.90	4.48		18.38	23
	0300	Limestone for 12" wall, 4" thick		90	.178		12.90	2.99		15.89	19.20
	0350	6" thick		80	.200		15.15	3.36		18.51	22.50
	0500	Marble to 4" thick, no wash, 9" wide		90	.178		18.70	2.99		21.69	25.50
	0550	12" wide		80	.200		27.50	3.36		30.86	36
	0700	Terra cotta, 9" wide		90	.178		4.55	2.99		7.54	10.05
	0750	12" wide		80	.200		7.60	3.36		10.96	14.05
	0800	Aluminum, for 12" wall		80	.200		10.60	3.36		13.96	17.35
120	0010	**CORNICES** Brick cornice on existing building									120
	0020	Not including scaffolding									
	0110	Face bricks, 12 brick/S.F., minimum	D-1	30	.533	SF Face	4.60	8.95		13.55	20
	0150	15 brick/S.F., maximum	"	23	.696	"	5.50	11.70		17.20	26
124	0010	**FACE BRICK** C216, TL lots, material only R042 -120									124
	0300	Standard modular, 4" x 2-2/3" x 8", minimum				M	320			320	350
	0350	Maximum				"	450			450	495
	2160	Fire Brick, 2" x 2-2/3" x 9", minimum									
	2165	Maximum				M	885			885	975
	2170	For less than truck load lots, add					10				
	2180	For buff or gray brick, add					15				
134	0010	**LINTELS** See division 051-232									134
158	0010	**SCAFFOLDING & SWING STAGING** Div. 015-254, 255 & 256									158
184	0016	**WALLS** R042 -180									184
	0800	Common, 4" x 2-2/3" x 8", 4" wall, as face brick	D-8	215	.186	S.F.	2.43	3.21		5.64	8.15
	0850	4" thick, as back up, 6.75 bricks per S.F. R042 -185		240	.167		1.88	2.87		4.75	6.90
	0900	8" thick wall, 13.50 brick per S.F.		135	.296		3.84	5.10		8.94	12.90
	1000	12" thick wall, 20.25 bricks per S.F. R042 -550		95	.421		5.80	7.25		13.05	18.65
	1050	16" thick wall, 27.00 bricks per S.F.		75	.533		7.80	9.20		17	24

MASONRY   4

042 100 | Brick Masonry

			CREW	DAILY OUTPUT	LABOR-HOURS	UNIT	MAT.	LABOR	EQUIP.	TOTAL	TOTAL INCL O&P	
184	1200	Reinforced, 4" x 2-2/3" x 8" , 4" wall	D-8	205	.195	S.F.	1.89	3.36		5.25	7.80	**184**
	1250	8" thick wall, 13.50 brick per S.F. `R042 -180`		130	.308		3.85	5.30		9.15	13.25	
	1300	12" thick wall, 20.25 bricks per S.F.		90	.444		5.80	7.65		13.45	19.30	
	1350	16" thick wall, 27.00 bricks per S.F. `R042 -185`		70	.571		7.80	9.85		17.65	25.50	
194	0010	**WINDOW SILL** Bluestone, thermal top, 10" wide, 1-1/2" thick	D-1	85	.188	S.F.	12.25	3.16		15.41	18.85	**194**
	0050	2" thick		75	.213	"	14.15	3.58		17.73	21.50	
	0100	Cut stone, 5" x 8" plain		48	.333	L.F.	10	5.60		15.60	20.50	
	0200	Face brick on edge, brick, 8" wide		80	.200		1.80	3.36		5.16	7.70	
	0400	Marble, 9" wide, 1" thick		85	.188		6.75	3.16		9.91	12.80	
	0600	Precast conc., 9" wide, 4" tapers to 3", 8" wall		70	.229		9	3.84		12.84	16.35	
	0650	11" wide		60	.267		8.35	4.48		12.83	16.75	
	0700	13" wide, 3 1/2" tapers to 2 1/2", 12" wall		50	.320		12.15	5.40		17.55	22.50	
	0900	Slate, colored, unfading, honed, 12" wide, 1" thick		85	.188		14.65	3.16		17.81	21.50	
	0950	2" thick		70	.229		20.50	3.84		24.34	29	

042 200 | Concrete Unit Masonry

		CREW	DAILY OUTPUT	LABOR-HOURS	UNIT	MAT.	LABOR	EQUIP.	TOTAL	TOTAL INCL O&P	
216 0010	**CONCRETE BLOCK, BACK-UP** Scaffolding not included										**216**
0020	Sand aggregate, tooled joint 1 side										
1000	Reinforced, alternate courses, 8"x16" units 4" thick	D-8	435	.092	S.F.	.75	1.58		2.33	3.51	
1100	6" thick		415	.096		.83	1.66		2.49	3.72	
1150	8" thick		395	.101		.96	1.74		2.70	4.01	
1200	10" thick		385	.104		1.35	1.79		3.14	4.52	
1250	12" thick	D-9	365	.132		1.41	2.21		3.62	5.30	
220 0010	**CONCRETE BLOCK, DECORATIVE** Scaffolding and reinf not included										**220**
0020	Embossed, simulated brick face										
0100	8" x 16" units, 4" thick	D-8	400	.100	S.F.	1.96	1.72		3.68	5.10	
0200	8" thick		340	.118		2.68	2.03		4.71	6.40	
0250	12" thick		300	.133		3.42	2.30		5.72	7.65	
0400	Embossed both sides										
0500	8" thick	D-8	300	.133	S.F.	3	2.30		5.30	7.20	
0550	12" thick	"	275	.145	"	3.99	2.51		6.50	8.65	
1000	Fluted high strength										
1100	Flutes 1 side, 8" x 16" x 4" thick	D-8	345	.116	S.F.	2.47	2		4.47	6.10	
1150	Flutes 2 sides, 8" x 16" x 4" thick		335	.119		3	2.06		5.06	6.80	
1200	8" thick		300	.133		3.83	2.30		6.13	8.10	
1250	For special colors, add					.24			.24	.26	
1400	Deep grooved, smooth face										
1450	8" x 16" x 4" thick	D-8	345	.116	S.F.	1.47	2		3.47	5	
1500	8" thick	"	300	.133	"	2.47	2.30		4.77	6.60	
2000	Formblock, incl. inserts & reinforcing										
2100	8" x 16" x 8" thick	D-8	345	.116	S.F.	2.68	2		4.68	6.35	
2150	12" thick	"	310	.129	"	3.25	2.22		5.47	7.35	
2500	Ground face										
2600	8" x 16" x 4" thick	D-8	345	.116	S.F.	3.76	2		5.76	7.50	
2650	6" thick		310	.129		4.05	2.22		6.27	8.25	
2700	8" thick		290	.138		4.22	2.38		6.60	8.65	
2750	12" thick	D-9	265	.181		4.81	3.04		7.85	10.45	
2900	For special colors, add, minimum					15%					
2950	For special colors, add, maximum					45%					
4000	Slump block										
4100	4" face height x 16" x 4" thick	D-1	165	.097	S.F.	1.81	1.63		3.44	4.75	
4150	6" thick		160	.100		2.21	1.68		3.89	5.30	
4200	8" thick		155	.103		3.04	1.73		4.77	6.30	
4250	10" thick		140	.114		4.59	1.92		6.51	8.30	
4300	12" thick		130	.123		4.90	2.07		6.97	8.90	

Important: See the Reference Section for critical supporting data - Reference Nos., Crews, & Location Factors

042 200 | Concrete Unit Masonry

		CREW	DAILY OUTPUT	LABOR-HOURS	UNIT	1998 BARE COSTS				TOTAL INCL O&P
						MAT.	LABOR	EQUIP.	TOTAL	
220 4400	6" face height x 16" x 6" thick	D-1	155	.103	S.F.	1.81	1.73		3.54	4.93
4450	8" thick		150	.107		2.37	1.79		4.16	5.65
4500	10" thick		130	.123		3.66	2.07		5.73	7.55
4550	12" thick		120	.133		3.81	2.24		6.05	8
5000	Split rib profile units, 1" deep ribs, 8 ribs									
5100	8" x 16" x 4" thick	D-8	345	.116	S.F.	1.24	2		3.24	4.75
5150	6" thick		325	.123		1.41	2.12		3.53	5.15
5200	8" thick		305	.131		1.67	2.26		3.93	5.65
5250	12" thick	D-9	275	.175		2.08	2.93		5.01	7.25
5400	For special deeper colors, 4" thick, add					.15			.15	.17
5450	12" thick, add					.31			.31	.34
5600	For white, 4" thick, add					.68			.68	.75
5650	6" thick, add					.90			.90	.99
5700	8" thick, add					1.11			1.11	1.22
5750	12" thick, add					1.55			1.55	1.70
6000	Split face or scored split face									
6100	8" x 16" x 4" thick	D-8	350	.114	S.F.	1.65	1.97		3.62	5.15
6150	6" thick		315	.127		2.24	2.19		4.43	6.15
6200	8" thick		295	.136		1.68	2.34		4.02	5.80
6250	12" thick	D-9	270	.178		2.10	2.99		5.09	7.35
6400	For special deeper colors, 4" thick, add					.18			.18	.20
6450	6" thick, add					.20			.20	.22
6500	8" thick, add					.21			.21	.23
6550	12" thick, add					.39			.39	.43
6650	For white, 4" thick, add					.72			.72	.79
6700	6" thick, add					.94			.94	1.03
6750	8" thick, add					1.17			1.17	1.29
6800	12" thick, add					1.61			1.61	1.77
7000	Scored ground face, 2 to 5 scores									
7100	8" x 16" x 4" thick	D-8	345	.116	S.F.	3.45	2		5.45	7.20
7150	6" thick		310	.129		4.15	2.22		6.37	8.35
7200	8" thick		290	.138		4.65	2.38		7.03	9.10
7250	12" thick	D-9	265	.181		5.85	3.04		8.89	11.60
8000	Hexagonal face profile units, 8" x 16" units									
8100	4" thick, hollow	D-8	345	.116	S.F.	1.89	2		3.89	5.45
8200	Solid		345	.116		2.72	2		4.72	6.35
8300	6" thick, hollow		310	.129		2.13	2.22		4.35	6.10
8350	8" thick, hollow		290	.138		2.95	2.38		5.33	7.25
8500	For stacked bond, add						26%			
8550	For high rise construction, add per story	D-8	67.80	.590	M.S.F.		10.15		10.15	17.20
8600	For scored block, add					10%				
8650	For honed or ground face, per face, add				Ea.	.20			.20	.22
8700	For honed or ground end, per end, add				"	2.10			2.10	2.31
8750	For bullnose block, add					10%				
8800	For special color, add					13%				
232 0010	**CONCRETE BLOCK, PARTITIONS** Scaffolding not included									
1000	Lightweight block, tooled joints, 2 sides, hollow									
1100	Not reinforced, 8" x 16" x 4" thick	D-8	440	.091	S.F.	.79	1.57		2.36	3.52
1150	6" thick		410	.098		1.02	1.68		2.70	3.97
1200	8" thick		385	.104		1.27	1.79		3.06	4.42
1250	10" thick		370	.108		1.74	1.86		3.60	5.05
1300	12" thick	D-9	350	.137		1.96	2.30		4.26	6.05
4000	Regular block, tooled joints, 2 sides, hollow									
4100	Not reinforced, 8" x 16" x 4" thick	D-8	430	.093	S.F.	.69	1.60		2.29	3.47
4150	6" thick		400	.100		.88	1.72		2.60	3.88

042 200 | Concrete Unit Masonry

			CREW	DAILY OUTPUT	LABOR-HOURS	UNIT	1998 BARE COSTS MAT.	LABOR	EQUIP.	TOTAL	TOTAL INCL O&P	
232	4200	8" thick	D-8	375	.107	S.F.	1.30	1.84		3.14	4.54	232
	4250	10" thick	↓	360	.111		1.61	1.91		3.52	5	
	4300	12" thick	D-9	340	.141	↓	1.79	2.37		4.16	6	
252	0010	**INSULATION** See also division 072-108										252
	0100	Inserts, styrofoam, plant installed, add to block prices										
	0200	8" x 16" units, 6" thick				S.F.	.74			.74	.81	
	0250	8" thick					.74			.74	.81	
	0300	10" thick					.87			.87	.96	
	0350	12" thick				↓	.91			.91	1	

042 300 | Reinforced Unit Masonry

			CREW	DAILY OUTPUT	LABOR-HOURS	UNIT	1998 BARE COSTS MAT.	LABOR	EQUIP.	TOTAL	TOTAL INCL O&P	
304	0010	**CONCRETE BLOCK BOND BEAM** Scaffolding not included R042-200										304
	0020	Not including grout or reinforcing										
	0100	Regular block, 8" high, 8" thick	D-8	565	.071	L.F.	1.36	1.22		2.58	3.56	
	0150	12" thick	D-9	510	.094	"	1.85	1.58		3.43	4.72	
310	0010	**CONCRETE BLOCK, EXTERIOR** Not including scaffolding										310
	0020	Reinforced, tooled joints 2 sides, styrofoam inserts										
	0100	Regular, 8" x 16" x 6" thick	D-8	390	.103	S.F.	1.04	1.77		2.81	4.13	
	0200	8" thick	↓	365	.110		1.25	1.89		3.14	4.58	
	0250	10" thick	↓	355	.113		1.63	1.94		3.57	5.10	
	0300	12" thick	D-9	330	.145	↓	1.91	2.44		4.35	6.25	
320	0010	**CONCRETE BLOCK FOUNDATION WALL** Scaffolding not included										320
	0050	Sand aggregate, trowel cut joints, not reinf., parged 1/2" thick										
	0200	Regular, 8" x 16" x 6" thick	D-8	450	.089	S.F.	1.18	1.53		2.71	3.88	
	0250	8" thick	↓	430	.093		1.43	1.60		3.03	4.28	
	0300	10" thick	↓	420	.095		2.07	1.64		3.71	5.05	
	0350	12" thick	D-9	395	.122		2.22	2.04		4.26	5.90	
	0500	Solid, 8" x 16" block, 6" thick	D-8	440	.091		1.82	1.57		3.39	4.65	
	0550	8" thick	"	415	.096		2.22	1.66		3.88	5.25	
	0600	12" thick	D-9	380	.126	↓	3.33	2.12		5.45	7.25	
330	0010	**CONCRETE BLOCK, LINTELS** Scaffolding not included										330
	0100	Including grout and reinforcing										
	0200	8" x 8" x 8", 1 #4 bar	D-4	300	.133	L.F.	3.25	2.03	.54	5.82	7.60	
	0250	2 #4 bars		295	.136		3.37	2.07	.55	5.99	7.80	
	1000	12" x 8" x 8", 1 #4 bar		275	.145		4.55	2.22	.59	7.36	9.40	
	1150	2 #5 bars	↓	270	.148	↓	4.81	2.26	.60	7.67	9.80	

042 550 | Masonry Veneer

			CREW	DAILY OUTPUT	LABOR-HOURS	UNIT	1998 BARE COSTS MAT.	LABOR	EQUIP.	TOTAL	TOTAL INCL O&P	
554	0010	**BRICK VENEER** Scaffolding not included, truck load lots R042-550										554
	0015	Material costs incl. 3% brick and 25% mortar waste										
	2000	Standard, sel. common, 4" x 2-2/3" x 8", (6.75/S.F.)	D-8	230	.174	S.F.	2.46	3		5.46	7.75	
	2020	Standard, red, 4" x 2-2/3" x 8", running bond (6.75/SF)		220	.182		2.46	3.13		5.59	8	
	2050	Full header every 6th course (7.88/S.F.)		185	.216		2.87	3.73		6.60	9.45	
	2100	English, full header every 2nd course (10.13/S.F.)		140	.286		3.68	4.92		8.60	12.40	
	2150	Flemish, alternate header every course (9.00/S.F.)		150	.267		3.27	4.59		7.86	11.40	
	2200	Flemish, alt. header every 6th course (7.13/S.F.)		205	.195		2.60	3.36		5.96	8.55	
	2250	Full headers throughout (13.50/S.F.)		105	.381		4.90	6.55		11.45	16.50	
	2300	Rowlock course (13.50/S.F.)		100	.400		4.90	6.90		11.80	17.05	
	2350	Rowlock stretcher (4.50/S.F.)		310	.129		1.65	2.22		3.87	5.60	
	2400	Soldier course (6.75/S.F.)		200	.200		2.46	3.45		5.91	8.55	
	2450	Sailor course (4.50/S.F.)		290	.138		1.65	2.38		4.03	5.85	
	2600	Buff or gray face, running bond, (6.75/S.F.)		220	.182		2.61	3.13		5.74	8.15	
	2700	Glazed face brick, running bond		210	.190		7.10	3.28		10.38	13.35	
	2750	Full header every 6th course (7.88/S.F.)	↓	170	.235	↓	8.25	4.05		12.30	15.95	

042 550 | Masonry Veneer

		CREW	DAILY OUTPUT	LABOR-HOURS	UNIT	MAT.	LABOR	EQUIP.	TOTAL	TOTAL INCL O&P	
554	3000	Jumbo, 6" x 4" x 12" running bond (3.00/S.F.) [R042-550]	D-8	435	.092	S.F.	3.62	1.58		5.20	6.65	**554**
	3050	Norman, 4" x 2-2/3" x 12" running bond, (4.5/S.F.)		320	.125		2.35	2.15		4.50	6.25	
	3100	Norwegian, 4" x 3-1/5" x 12" (3.75/S.F.)		375	.107		2.10	1.84		3.94	5.40	
	3150	Economy, 4" x 4" x 8" (4.50/S.F.)		310	.129		1.98	2.22		4.20	5.95	
	3200	Engineer, 4" x 3-1/5" x 8" (5.63/S.F.)		260	.154		1.87	2.65		4.52	6.55	
	3250	Roman, 4" x 2" x 12" (6.00/S.F.)		250	.160		4.59	2.76		7.35	9.70	
	3300	SCR, 6" x 2-2/3" x 12" (4.50/S.F.)		310	.129		4.10	2.22		6.32	8.30	
	3350	Utility, 4" x 4" x 12" (3.00/S.F.)		450	.089		2.40	1.53		3.93	5.25	
	3400	For cavity wall construction, add					15%					
	3450	For stacked bond, add					10%					
	3500	For interior veneer construction, add					15%					
	3550	For curved walls, add					30%					

042 700 | Glass Unit Masonry

		CREW	DAILY OUTPUT	LABOR-HOURS	UNIT	MAT.	LABOR	EQUIP.	TOTAL	TOTAL INCL O&P	
704	0010	GLASS BLOCK Scaffolding not included [R042-700]										**704**
	0150	8" x 8" block	D-8	160	.250	S.F.	11.25	4.31		15.56	19.70	
	0200	12" x 12" block	"	175	.229	"	13.50	3.94		17.44	21.50	
	0700	For solar reflective blocks, add					100%					
	0800	Under 1,000 S.F., 4" x 8" blocks	D-8	145	.276	S.F.	19.15	4.75		23.90	29	
	1000	Thinline, plain, 3-1/8" thick, under 1,000 S.F., 6" x 6" block		115	.348		13.40	6		19.40	25	
	1050	8" x 8" block		160	.250		8.20	4.31		12.51	16.35	
	1400	For cleaning block after installation (both sides), add		1,000	.040		.10	.69		.79	1.28	

042 900 | Adobe Masonry

		CREW	DAILY OUTPUT	LABOR-HOURS	UNIT	MAT.	LABOR	EQUIP.	TOTAL	TOTAL INCL O&P	
904	0010	ADOBE BRICK Unstabilized, with adobe mortar (Southwestern States)										**904**
	0260	Brick, 4" x 3" x 8" (6.0 per S.F.)	D-8	266	.150	S.F.	1.80	2.59		4.39	6.35	
	0280	4" x 4" x 8" (4.5 per S.F.)		333	.120		1.83	2.07		3.90	5.50	
	0300	4" x 4" x 14" (2.5 per S.F.)		540	.074		1.25	1.28		2.53	3.54	
	0320	8" x 3" x 16" (3.0 per S.F.)		466	.086		1.52	1.48		3	4.17	
	0340	4" x 3" x 12" (4.0 per S.F.)		362	.110		1.86	1.90		3.76	5.25	
	0360	6" x 3" x 12" (4.0 per S.F.)		362	.110		1.92	1.90		3.82	5.35	
	0380	4" x 5" x 16" (1.80 per S.F.)		600	.067		.89	1.15		2.04	2.93	
	0400	8" x 4" x 16" (2.25 per S.F.)		577	.069		1.32	1.19		2.51	3.47	
	0420	10" x 4" x 14" (2.57 per S.F.)		555	.072		1.41	1.24		2.65	3.65	
	0440	Adobe, partially stabilized, add					10%					
	0480	Fully stabilized, add					25%					

044 100 | Rough Stone

		CREW	DAILY OUTPUT	LABOR-HOURS	UNIT	MAT.	LABOR	EQUIP.	TOTAL	TOTAL INCL O&P	
104	0011	ROUGH STONE WALL, Dry										**104**
	0100	Random fieldstone, under 18" thick	D-12	60	.533	C.F.	9.75	8.85		18.60	25.50	
	0150	Over 18" thick	"	63	.508	"	12.45	8.40		20.85	28	

044 550 | Marble

		CREW	DAILY OUTPUT	LABOR-HOURS	UNIT	MAT.	LABOR	EQUIP.	TOTAL	TOTAL INCL O&P	
554	0010	MARBLE Base, 3/4" thick, polished, group A, 4" thick	D-1	60	.267	L.F.	9.20	4.48		13.68	17.75	**554**
	1000	Facing, polished finish, cut to size, 3/4" to 7/8" thick										
	1050	Average	D-10	130	.308	S.F.	17.80	5.40	3.60	26.80	32.50	
	1100	Maximum	"	130	.308	"	41	5.40	3.60	50	58.50	

044 550 | Marble

		CREW	DAILY OUTPUT	LABOR-HOURS	UNIT	MAT.	LABOR	EQUIP.	TOTAL	TOTAL INCL O&P	
554	2200	Window sills, 6" x 3/4" thick	D-1	85	.188	L.F.	6.70	3.16		9.86	12.70	**554**
	2500	Flooring, polished tiles, 12" x 12" x 3/8" thick										
	2510	Thin set, average	D-11	90	.267	S.F.	8.25	4.58		12.83	16.80	
	2600	Maximum		90	.267		30	4.58		34.58	41	
	2700	Mortar bed, average		65	.369		8.40	6.35		14.75	20	
	2740	Maximum		65	.369		27.50	6.35		33.85	41	
	2780	Travertine, 3/8" thick, average	D-10	130	.308		11.40	5.40	3.60	20.40	25.50	
	2790	Maximum	"	130	.308		28	5.40	3.60	37	43.50	
	3500	Thresholds, 3' long, 7/8" thick, 4" to 5" wide, plain	D-12	24	1.333	Ea.	12.85	22		34.85	51.50	
	3550	Beveled		24	1.333	"	14.95	22		36.95	54	
	3700	Window stools, polished, 7/8" thick, 5" wide		85	.376	L.F.	12	6.25		18.25	24	

044 600 | Limestone

		CREW	DAILY OUTPUT	LABOR-HOURS	UNIT	MAT.	LABOR	EQUIP.	TOTAL	TOTAL INCL O&P	
604	0012	LIMESTONE, Cut to size										**604**
	0020	Veneer facing panels										
	0500	Texture finish, light stick, 4-1/2" thick, 5' x 12'	D-4	300	.133	S.F.	17.80	2.03	.54	20.37	23.50	
	0750	5" thick, 5' x 14' panels	D-10	275	.145		21.50	2.56	1.70	25.76	29.50	
	1000	Sugarcube finish, 2" Thick, 3' x 5' panels		275	.145		10.55	2.56	1.70	14.81	17.80	
	1050	3" Thick, 4' x 9' panels		275	.145		11	2.56	1.70	15.26	18.30	
	1200	4" Thick, 5' x 11' panels		275	.145		14.65	2.56	1.70	18.91	22.50	
	1400	Sugarcube, textured finish, 4-1/2" thick, 5' x 12'		275	.145		18.95	2.56	1.70	23.21	27	
	1450	5" thick, 5' x 14' panels		275	.145		22	2.56	1.70	26.26	30	
	2000	Coping, sugarcube finish, top & 2 sides		30	1.333	C.F.	49	23.50	15.60	88.10	110	
	2100	Sills, lintels, jambs, trim, stops, sugarcube finish, average		20	2		44	35	23.50	102.50	134	
	2150	Detailed		20	2		59	35	23.50	117.50	150	
	2300	Steps, extra hard, 14" wide, 6" rise		50	.800	L.F.	39	14.05	9.35	62.40	77	
	3000	Quoins, plain finish, 6"x12"x12"	D-12	25	1.280	Ea.	96	21		117	142	
	3050	6"x16"x24"	"	25	1.280	"	128	21		149	177	

044 650 | Granite

		CREW	DAILY OUTPUT	LABOR-HOURS	UNIT	MAT.	LABOR	EQUIP.	TOTAL	TOTAL INCL O&P	
651	0010	GRANITE Cut to size										**651**
	0050	Veneer, polished face, 3/4" to 1-1/2" thick										
	0150	Low price, gray, light gray, etc.	D-10	130	.308	S.F.	18.55	5.40	3.60	27.55	33.50	
	0220	High price, red, black, etc.	"	130	.308	"	31.50	5.40	3.60	40.50	48	
	0300	1-1/2" to 2-1/2" thick, veneer										
	0350	Low price, gray, light gray, etc.	D-10	130	.308	S.F.	20.50	5.40	3.60	29.50	35.50	
	0550	High price, red, black, etc.	"	130	.308	"	36	5.40	3.60	45	52.50	
	0700	2-1/2" to 4" thick, veneer										
	0750	Low price, gray, light gray, etc.	D-10	110	.364	S.F.	26	6.40	4.25	36.65	44	
	0950	High price, red, black, etc.	"	110	.364		41	6.40	4.25	51.65	61	
	1000	For bush hammered finish, deduct					5%					
	1050	Coarse rubbed finish, deduct					10%					
	1100	Honed finish, deduct					5%					
	1150	Thermal finish, deduct					18%					
	2450	For radius under 5', add				L.F.	100%					
	2500	Steps, copings, etc., finished on more than one surface										
	2550	Minimum	D-10	50	.800	C.F.	72	14.05	9.35	95.40	113	
	2600	Maximum	"	50	.800	"	108	14.05	9.35	131.40	153	
	2800	Pavers, 4" x 4" x 4" blocks, split face and joints										
	2850	Minimum	D-11	80	.300	S.F.	10.30	5.15		15.45	20	
	2900	Maximum	"	80	.300	"	20	5.15		25.15	31	
	3500	Curbing, city street type, See Division 025-254										
	4000	Soffits, 2" thick, minimum	D-13	35	1.371	S.F.	31	24.50	13.35	68.85	89.50	
	4100	Maximum		35	1.371		62	24.50	13.35	99.85	124	

Important: See the Reference Section for critical supporting data - Reference Nos., Crews, & Location Factors

4 · MASONRY

044 650 | Granite

			CREW	DAILY OUTPUT	LABOR-HOURS	UNIT	1998 BARE COSTS				TOTAL INCL O&P	
							MAT.	LABOR	EQUIP.	TOTAL		
651	4200	4" thick, minimum	D-13	35	1.371	S.F.	42	24.50	13.35	79.85	102	651
	4300	Maximum	↓	35	1.371	↓	79.50	24.50	13.35	117.35	143	

044 700 | Sandstone

			CREW	DAILY OUTPUT	LABOR-HOURS	UNIT	MAT.	LABOR	EQUIP.	TOTAL	TOTAL INCL O&P	
704	0011	**SANDSTONE OR BROWNSTONE**										704
	0100	Sawed face veneer, 2-1/2" thick, to 2' x 4' panels	D-10	130	.308	S.F.	12.45	5.40	3.60	21.45	27	
	0150	4' thick, to 3'-6" x 8'panels		100	.400		12.45	7.05	4.68	24.18	30.50	
	0300	Split face, random sizes	↓	100	.400	↓	8.95	7.05	4.68	20.68	27	
	0350	Cut stone trim (limestone)										
	0360	Ribbon stone, 4" thick, 5' pieces	D-8	120	.333	Ea.	110	5.75		115.75	131	
	0370	Cove stone, 4" thick, 5' pieces		105	.381		110	6.55		116.55	132	
	0380	Cornice stone, 10" to 12" wide		90	.444		136	7.65		143.65	162	
	0390	Band stone, 4" thick, 5' pieces		145	.276		70	4.75		74.75	85	
	0410	Window and door trim, 3" to 4" wide		160	.250		59.50	4.31		63.81	73	
	0420	Key stone, 18" long	↓	60	.667	↓	63	11.50		74.50	88.50	

044 750 | Slate

			CREW	DAILY OUTPUT	LABOR-HOURS	UNIT	MAT.	LABOR	EQUIP.	TOTAL	TOTAL INCL O&P	
754	0010	**SLATE** Pennsylvania, blue gray to gray black; Vermont,										754
	3500	Stair treads, sand finish, 1" thick x 12" wide										
	3600	3 L.F. to 6 L.F.	D-10	120	.333	L.F.	13.90	5.85	3.90	23.65	29.50	
	3700	Ribbon, sand finish, 1" thick x 12" wide										
	3750	To 6 L.F.	D-10	120	.333	L.F.	9.30	5.85	3.90	19.05	24.50	

045 100 | Masonry Cleaning

			CREW	DAILY OUTPUT	LABOR-HOURS	UNIT	1998 BARE COSTS				TOTAL INCL O&P	
							MAT.	LABOR	EQUIP.	TOTAL		
106	0010	**CLEAN AND POINT** Smooth brick	1 Bric	300	.027	S.F.	.24	.51		.75	1.12	106
	0100	Rough brick	"	265	.030	"	.30	.57		.87	1.30	
108	0010	**WASHING BRICK** Acid wash, smooth brick R045-100	1 Bric	560	.014	S.F.	.22	.27		.49	.70	108
	0050	Rough brick	↓	400	.020		.23	.38		.61	.89	
	0060	Stone, acid wash	↓	600	.013	↓	.22	.25		.47	.67	
	1000	Muriatic acid, price per gallon in 5 gallon lots				Gal.	7.50			7.50	8.25	

045 550 | Flue Liners

			CREW	DAILY OUTPUT	LABOR-HOURS	UNIT	MAT.	LABOR	EQUIP.	TOTAL	TOTAL INCL O&P	
554	0010	**FLUE LINING** Including mortar joints, 8" x 8"	D-1	125	.128	V.L.F.	2.65	2.15		4.80	6.55	554
	0100	8" x 12"		103	.155		4	2.61		6.61	8.80	
	0200	12" x 12"		93	.172		5	2.89		7.89	10.40	
	0300	12" x 18"		84	.190		8.50	3.20		11.70	14.75	
	0400	18" x 18"		75	.213		11.60	3.58		15.18	18.80	
	0500	20" x 20"		66	.242		17.15	4.07		21.22	26	
	0600	24" x 24"		56	.286		23.50	4.80		28.30	33.50	
	1000	Round, 18" diameter		66	.242		20.50	4.07		24.57	29.50	
	1100	24" diameter	↓	47	.340	↓	28.50	5.70		34.20	41	

045 650 | Fire Brick

			CREW	DAILY OUTPUT	LABOR-HOURS	UNIT	MAT.	LABOR	EQUIP.	TOTAL	TOTAL INCL O&P	
656	0010	**FIREPLACE** For prefabricated fireplace, see div. 103-054										656
	0100	Brick fireplace, not incl. foundations or chimneys										

MASONRY **4**

045 650	Fire Brick	CREW	DAILY OUTPUT	LABOR-HOURS	UNIT	1998 BARE COSTS				TOTAL INCL O&P	
						MAT.	LABOR	EQUIP.	TOTAL		
656 0110	30" x 29" opening, incl. chamber, plain brickwork	D-1	.40	40	Ea.	355	670		1,025	1,550	**656**
0200	Fireplace box only (110 brick)	"	2	8	"	115	134		249	355	
0300	For elaborate brickwork and details, add					35%	35%				
0400	For hearth, brick & stone, add	D-1	2	8	Ea.	130	134		264	370	
0410	For steel angle, damper, cleanouts, add		4	4		90	67		157	213	
0600	Plain brickwork, incl. metal circulator		.50	32		695	540		1,235	1,675	
0800	Face brick only, standard size, 8" x 2-2/3" x 4"		.30	53.333	M	360	895		1,255	1,925	
0900	Stone fireplace, fieldstone, add				SF Face	9.25			9.25	10.20	
1000	Cut stone, add				"	9.75			9.75	10.75	

For information about Means Estimating Seminars, see yellow pages 11 and 12 in back of book

Important: See the Reference Section for critical supporting data - Reference Nos., Crews, & Location Factors

Division 5
Metals

Estimating Tips

050 Materials, Coatings & Fastenings

- Nuts, bolts, washers, connection angles and plates can add a significant amount to both the tonnage of a structural steel job as well as the estimated cost. As a rule of thumb add 10% to the total weight to account for these accessories.

- Type 2 steel construction, commonly referred to as "simple construction," consists generally of field bolted connections with lateral bracing supplied by other elements of the building, such as masonry walls or x-bracing. The estimator should be aware, however, that shop connections may be accomplished by welding or bolting. The method may be particular to the fabrication shop and may have an impact on the estimated cost.

052 Steel Joists

- In any given project the total weight of open web steel joists is determined by the loads to be supported and the design. However, economies can be realized in minimizing the amount of labor used to place the joists. This is done by maximizing the joist spacing and therefore minimizing the number of joists required to be installed on the job. Certain spacings and locations may be required by the design, but in other cases maximizing the spacing and keeping it as uniform as possible will keep the costs down.

053 Metal Decking

- The takeoff and estimating of metal deck involves more than simply the area of the floor or roof and the type of deck specified or shown on the drawings. Many different sizes and types of openings may exist. Small openings for individual pipes or conduits may be drilled after the floor/roof is installed, but larger openings may require special deck lengths as well as reinforcing or structural support. The estimator should determine who will be supplying this reinforcing. Additionally, some deck terminations are part of the deck package, such as screed angles and pour stops, and others will be part of the steel contract, such as angles attached to structural members and cast-in-place angles and plates. The estimator must ensure that all pieces are accounted for in the complete estimate.

055 Metal Fabrications

- The most economical steel stairs are those that use common materials, standard details and most importantly, a uniform and relatively simple method of field assembly. Commonly available A36 channels and plates are very good choices for the main stringers of the stairs, as are angles and tees for the carrier members. Risers and treads are usually made by specialty shops, and it is most economical to use a typical detail in as many places as possible. The stairs should be pre-assembled and shipped directly to the site. The field connections should be simple and straightforward to be accomplished efficiently and with a minimum of equipment and labor.

Reference Numbers

Reference numbers are shown in bold squares at the beginning of some major classifications. These numbers refer to related items in the Reference Section. The reference information may be an estimating procedure, an alternate pricing method or technical information.

Note: Not all subdivisions listed here necessarily appear in this publication.

050 500 | Metal Fastening

		CREW	DAILY OUTPUT	LABOR-HOURS	UNIT	1998 BARE COSTS MAT.	LABOR	EQUIP.	TOTAL	TOTAL INCL O&P
508 0010	**BOLTS & HEX NUTS** Steel, A307									**508**
0100	1/4" diameter, 1/2" long				Ea.	.06			.06	.07
0200	1" long					.07			.07	.08
0300	2" long					.09			.09	.10
0400	3" long					.14			.14	.15
0500	4" long					.22			.22	.24
0600	3/8" diameter, 1" long					.13			.13	.14
0700	2" long					.17			.17	.19
0800	3" long					.22			.22	.24
0900	4" long					.29			.29	.32
1000	5" long					.39			.39	.43
1100	1/2" diameter, 1-1/2" long					.26			.26	.29
1200	2" long					.30			.30	.33
1300	4" long					.46			.46	.51
1400	6" long					.64			.64	.70
1500	8" long					1.07			1.07	1.18
1600	5/8" diameter, 1-1/2" long					.46			.46	.51
1700	2" long					.52			.52	.57
1800	4" long					.77			.77	.85
1900	6" long					1.03			1.03	1.13
2000	8" long					1.66			1.66	1.83
2100	10" long					1.91			1.91	2.10
2200	3/4" diameter, 2" long					.72			.72	.79
2300	4" long					1.14			1.14	1.25
2400	6" long					1.53			1.53	1.68
2500	8" long					2.01			2.01	2.21
2600	10" long					2.02			2.02	2.22
2700	12" long					2.31			2.31	2.54
2800	1" diameter, 3" long					2.01			2.01	2.21
2900	6" long					2.35			2.35	2.58
3000	12" long					5.45			5.45	6
3100	For galvanized, add					75%				
3200	For stainless, add					350%				
515 0010	**DRILLING** And layout for anchors, per									**515**
0050	inch of depth, concrete or brick walls and floors									
0100	1/4" diameter	1 Carp	75	.107	Ea.	.09	1.97		2.06	3.48
0200	3/8" diameter		63	.127		.11	2.35		2.46	4.15
0300	1/2" diameter		50	.160		.14	2.96		3.10	5.20
0400	5/8" diameter		48	.167		.20	3.08		3.28	5.50
0500	3/4" diameter		45	.178		.24	3.29		3.53	5.90
0600	7/8" diameter		43	.186		.34	3.44		3.78	6.25
0700	1" diameter		40	.200		.43	3.70		4.13	6.80
0800	1-1/4" diameter		38	.211		1.17	3.89		5.06	7.95
0900	1-1/2" diameter		35	.229		2.04	4.23		6.27	9.50
1000	For ceiling installations add						40%			
1100	Drilling & layout for drywall or plaster walls									
1200	Holes, 1/4" diameter	1 Carp	150	.053	Ea.	.04	.99		1.03	1.74
1300	3/8" diameter		140	.057		.06	1.06		1.12	1.87
1400	1/2" diameter		130	.062		.07	1.14		1.21	2.03
1500	3/4" diameter		120	.067		.12	1.23		1.35	2.24
1600	1" diameter		110	.073		.21	1.35		1.56	2.54
1700	1-1/4" diameter		100	.080		.59	1.48		2.07	3.19
1800	1-1/2" diameter		90	.089		1.02	1.64		2.66	3.94
1900	For ceiling installations add						40%			

Important: See the Reference Section for critical supporting data - Reference Nos., Crews, & Location Factors

050 500 | Metal Fastening

		CREW	DAILY OUTPUT	LABOR-HOURS	UNIT	1998 BARE COSTS				TOTAL INCL O&P
						MAT.	LABOR	EQUIP.	TOTAL	
520	0010	**EXPANSION ANCHORS** & shields									520
	0100	Bolt anchors for concrete, brick or stone, no layout and drilling									
	0200	Expansion shields, zinc, 1/4" diameter, 1" long, single	1 Carp	90	.089	Ea.	.85	1.64		2.49	3.76
	0300	1-3/8" long, double		85	.094		.93	1.74		2.67	4.01
	0500	2" long, double		80	.100		1.73	1.85		3.58	5.05
	0700	2-1/2" long, double		75	.107		2.24	1.97		4.21	5.85
	0900	3" long, double		70	.114		3.31	2.11		5.42	7.25
	1100	4" long, double		65	.123		6.55	2.28		8.83	11.15
	1300	1" diameter, 6" long, double		60	.133		23	2.47		25.47	29.50
	1410	Concrete anchor, w/rod & epoxy cartridge, 1-3/4 diameter x 15" long	E-22	20	1.200		31	23.50	.10	54.60	74
	1415	18" long		17	1.412		33	27.50	.11	60.61	83.50
	1420	2" diameter x 18" long		16	1.500		33.50	29	.12	62.62	87
	1425	24" long		15	1.600		36	31	.13	67.13	93
	1430	Chemical anchor, w/rod & epoxy cartridge, 3/4" diam. x 9-1/2" long		27	.889		7.80	17.25	.07	25.12	38
	1435	1" diameter x 11-3/4 long		24	1		15.05	19.40	.08	34.53	50
	1440	1-1/4" diameter x 14" long		21	1.143		28	22	.09	50.09	68.50
	2100	Hollow wall anchors for gypsum board,									
	2200	plaster, tile or wall board									
	2500	3/16" diameter, short				Ea.	.57			.57	.62
	3000	Toggle bolts, bright steel, 1/8" diameter, 2" long	1 Carp	85	.094		.28	1.74		2.02	3.29
	3100	4" long		80	.100		.37	1.85		2.22	3.57
	3200	3/16" diameter, 3" long		80	.100		.35	1.85		2.20	3.56
	3300	6" long		75	.107		.53	1.97		2.50	3.96
	3400	1/4" diameter, 3" long		75	.107		.39	1.97		2.36	3.81
	3500	6" long		70	.114		.58	2.11		2.69	4.26
	3600	3/8" diameter, 3" long		70	.114		.85	2.11		2.96	4.55
	3700	6" long		60	.133		1.26	2.47		3.73	5.60
	3800	1/2" diameter, 4" long		60	.133		2.69	2.47		5.16	7.20
	3900	6" long		50	.160		3.59	2.96		6.55	9
	4000	Nailing anchors									
	4100	Nylon anchor standard nail, 1/4" diameter, 1" long				C	14.95			14.95	16.40
	4200	1-1/2" long					19.20			19.20	21
	4300	2" long					32			32	35
	4400	Zamac anchor stainless nail, 1/4" diameter, 1" long					21			21	23
	4500	1-1/2" long					29			29	31.50
	4600	2" long					39			39	42.50
	5000	Screw anchors for concrete, masonry,									
	5100	stone & tile, no layout or drilling included									
	5200	Jute fiber, #6, #8, & #10, 1" long				Ea.	.18			.18	.20
	5300	#12, 1-1/2" long					.26			.26	.28
	5400	#14, 2" long					.40			.40	.44
	5500	#16, 2" long					.42			.42	.46
	5600	#20, 2" long					.67			.67	.74
	5700	Lag screw shields, 1/4" diameter, short					.41			.41	.45
	5800	Long					.48			.48	.53
	5900	3/8" diameter, short					.70			.70	.77
	6000	Long					.80			.80	.88
	6100	1/2" diameter, short					1.06			1.06	1.16
	6200	Long					1.24			1.24	1.37
	6300	3/4" diameter, short					2.27			2.27	2.50
	6400	Long					2.91			2.91	3.20
	6600	Lead, #6 & #8, 3/4" long					.17			.17	.18
	6700	#10 - #14, 1-1/2" long					.25			.25	.28
	6800	#16 & #18, 1-1/2" long					.35			.35	.38
	6900	Plastic, #6 & #8, 3/4" long					.03			.03	.04
	7000	#8 & #10, 7/8" long					.04			.04	.04

5

METALS

050 500	Metal Fastening	CREW	DAILY OUTPUT	LABOR-HOURS	UNIT	1998 BARE COSTS				TOTAL INCL O&P	
						MAT.	LABOR	EQUIP.	TOTAL		
520	7100	#10 & #12, 1″ long				Ea.	.05			.05	.05	**520**
	7200	#14 & #16, 1-1/2″ long					.06			.06	.07	
530	0010	**LAG SCREWS** Steel, 1/4″ diameter, 2″ long	1 Carp	200	.040	Ea.	.05	.74		.79	1.32	**530**
	0100	3/8″ diameter, 3″ long		150	.053		.15	.99		1.14	1.86	
	0200	1/2″ diameter, 3″ long		130	.062		.26	1.14		1.40	2.24	
	0300	5/8″ diameter, 3″ long		120	.067		.56	1.23		1.79	2.73	
545	0010	**RIVETS** 1/2″ grip length										**545**
	0100	Aluminum rivet & mandrel, 1/8″ diameter				C	4.53			4.53	4.98	
	0200	3/16″ diameter					7			7	7.70	
	0300	Aluminum rivet, steel mandrel, 1/8″ diameter					4.23			4.23	4.65	
	0400	3/16″ diameter					6.35			6.35	7	
	0500	Copper rivet, steel mandrel, 1/8″ diameter					6.25			6.25	6.85	
	0600	Monel rivet, steel mandrel, 1/8″ diameter					19.30			19.30	21	
	0700	3/16″ diameter					35.50			35.50	39.50	
	0800	Stainless rivet & mandrel, 1/8″ diameter					12			12	13.20	
	0900	3/16″ diameter					18.20			18.20	20	
	1000	Stainless rivet, steel mandrel, 1/8″ diameter					7.60			7.60	8.35	
	1100	3/16″ diameter					13.75			13.75	15.10	
	1200	Steel rivet and mandrel, 1/8″ diameter					4.75			4.75	5.20	
	1300	3/16″ diameter					7.15			7.15	7.85	
	1400	Hand riveting tool, minimum				Ea.	76.50			76.50	84	
	1500	Maximum					340			340	370	
	1600	Power riveting tool, minimum					660			660	730	
	1700	Maximum					1,750			1,750	1,925	
550	0010	**STUDS** .22 caliber stud driver, minimum				Ea.	330			330	360	**550**
	0100	Maximum				″	670			670	740	
	0300	Powder charges for above, low velocity				C	14.90			14.90	16.40	
	0400	Standard velocity					26			26	29	
	0600	Drive pins & studs, 1/4″ & 3/8″ diam., to 3″ long, minimum					25			25	27.50	
	0700	Maximum					74			74	81	

051 | Structural Metal Framing

051 100	Bracing	CREW	DAILY OUTPUT	LABOR-HOURS	UNIT	1998 BARE COSTS				TOTAL INCL O&P	
						MAT.	LABOR	EQUIP.	TOTAL		
108	0010	**BRACING**										**108**
	0300	Let-in, ″T″ shaped, 22 ga. galv. steel, studs at 16″ O.C. R051-220	1 Carp	5.80	1.379	C.L.F.	37	25.50		62.50	84	
	0400	Studs at 24″ O.C.		6	1.333		37	24.50		61.50	83	
	0500	16 ga. galv. steel straps, studs at 16″ O.C.		6	1.333		48	24.50		72.50	95	
	0600	Studs at 24″ O.C.		6.20	1.290		48	24		72	93.50	

051 200	Structural Steel	CREW	DAILY OUTPUT	LABOR-HOURS	UNIT	1998 BARE COSTS				TOTAL INCL O&P	
						MAT.	LABOR	EQUIP.	TOTAL		
215	0010	**CEILING SUPPORTS**										**215**
	1000	Entrance door/folding partition supports	E-4	60	.533	L.F.	12.95	10.90	1.33	25.18	37	
	1100	Linear accelerator door supports		14	2.286		59	46.50	5.70	111.20	163	
	1200	Lintels or shelf angles, hung, exterior hot dipped galv.		267	.120		9.55	2.45	.30	12.30	15.70	
	1250	Two coats primer paint instead of galv.		267	.120		7.65	2.45	.30	10.40	13.55	
	1400	Monitor support, ceiling hung, expansion bolted		4	8	Ea.	205	164	20	389	565	
	1450	Hung from pre-set inserts		6	5.333		215	109	13.35	337.35	465	
	1600	Motor supports for overhead doors		4	8		104	164	20	288	455	

Important: See the Reference Section for critical supporting data - Reference Nos., Crews, & Location Factors

051 200 | Structural Steel

			CREW	DAILY OUTPUT	LABOR-HOURS	UNIT	1998 BARE COSTS				TOTAL INCL O&P	
							MAT.	LABOR	EQUIP.	TOTAL		
215	1700	Partition support for heavy folding partitions, without pocket	E-4	24	1.333	L.F.	29.50	27.50	3.33	60.33	89.50	**215**
	1750	Supports at pocket only		12	2.667		59	54.50	6.65	120.15	179	
	2000	Rolling grilles & fire door supports		34	.941		25	19.25	2.35	46.60	68	
	2100	Spider-leg light supports, expansion bolted to ceiling slab		8	4	Ea.	84	82	10	176	265	
	2150	Hung from pre-set inserts		12	2.667	"	88	54.50	6.65	149.15	211	
	2400	Toilet partition support		36	.889	L.F.	29.50	18.20	2.22	49.92	71	
	2500	X-ray travel gantry support		12	2.667	"	101	54.50	6.65	162.15	225	
220	0005	**COLUMNS**										**220**
	0800	Steel, concrete filled, extra strong pipe, 3-1/2" diameter	E-2	660	.073	L.F.	11	1.47	1.56	14.03	16.65	
	0830	4" diameter		780	.062		13.70	1.25	1.32	16.27	18.95	
	0890	5" diameter		1,020	.047		18.40	.95	1.01	20.36	23.50	
	0930	6" diameter		1,200	.040		26.50	.81	.86	28.17	31.50	
	1000	Lightweight units, 3-1/2" diameter		780	.062		3.36	1.25	1.32	5.93	7.55	
	1050	4" diameter		900	.053		3.75	1.08	1.14	5.97	7.45	
	1100	For galvanizing, add				Lb.	.41			.41	.45	
	1300	For web ties, angles, etc., add per added lb.	1 Sswk	945	.008		.81	.17		.98	1.22	
	1500	Steel pipe, extra strong, no concrete, 3" to 5" diameter	E-2	16,000	.003		.96	.06	.06	1.08	1.25	
	1600	6" to 12" diameter		14,000	.003		.85	.07	.07	.99	1.15	
	2400	Structural tubing, rect, 5" to 6" wide, light section		11,200	.004		.87	.09	.09	1.05	1.23	
	2700	12" x 8" x 1/2" thk wall		24,000	.002		.88	.04	.04	.96	1.10	
	2800	Heavy section		32,000	.002		.96	.03	.03	1.02	1.15	
	8000	Lally columns, to 8', 3-1/2" diameter	F-2	24	.667	Ea.	25	10.80		35.80	46	
	8080	4" diameter	"	20	.800	"	28	12.95		40.95	53	
232	0010	**LINTELS** Plain steel angles, under 500 lb.	1 Bric	550	.015	Lb.	.59	.28		.87	1.12	**232**
	0100	500 to 1000 lb.		640	.013	"	.56	.24		.80	1.01	
	2000	Steel angles, 3-1/2" x 3", 1/4" thick, 2'-6" long		50	.160	Ea.	7.95	3.03		10.98	13.85	
	2100	4'-6" long		45	.178		14.30	3.37		17.67	21.50	
	2600	4" x 3-1/2", 1/4" thick, 5'-0" long		40	.200		18.20	3.79		21.99	26.50	
	2700	9'-0" long		35	.229		33	4.33		37.33	43.50	
	3500	For precast concrete lintels, see div. 034-802										
255	0010	**STRUCTURAL STEEL PROJECTS** Bolted, unless noted otherwise										**255**
	0200	Apartments, nursing homes, etc., 1 to 2 stories	E-5	10.30	6.990	Ton	1,175	141	108	1,424	1,700	
	0700	Offices, hospitals, etc., steel bearing, 1 to 2 stories		10.30	6.990		1,175	141	108	1,424	1,700	
	3100	Roof trusses, minimum		13	5.538		1,325	112	85.50	1,522.50	1,750	
	3200	Maximum		8.30	8.675		2,250	175	134	2,559	2,950	
260	0010	**STRUCTURAL STEEL** Bolted, incl. fabrication										**260**
	0050	Beams, W 6 x 9	E-2	720	.067	L.F.	5.80	1.35	1.43	8.58	10.55	
	0100	W 8 x 10		720	.067		6.45	1.35	1.43	9.23	11.25	
	0200	Columns, W 6 x 15		540	.089		9.70	1.80	1.91	13.41	16.20	
	0250	W 8 x 31		540	.089		20	1.80	1.91	23.71	27.50	

At item 0700 (255): R051 -220

---

052 100 | Steel Joists

			CREW	DAILY OUTPUT	LABOR-HOURS	UNIT	1998 BARE COSTS				TOTAL INCL O&P	
							MAT.	LABOR	EQUIP.	TOTAL		
115	0010	**OPEN WEB JOISTS**										**115**
	1000	Bar joists installed, no material included, 15' span	E-2	1,050	.046	L.F.		.93	.98	1.91	2.86	

At item 0010 (115): R054 -210

052 100 | Steel Joists

		CREW	DAILY OUTPUT	LABOR-HOURS	UNIT	MAT.	LABOR	EQUIP.	TOTAL	TOTAL INCL O&P	
115	1100	20' span	E-2	1,300	.037	L.F.		.75	.79	1.54	2.31	115
	1200	25' span		1,500	.032			.65	.69	1.34	2	
	1300	30' span		1,650	.029			.59	.62	1.21	1.82	
	1400	35' span		1,750	.027			.56	.59	1.15	1.72	
	1500	40' span		1,600	.030			.61	.64	1.25	1.88	
	1600	45' span		1,665	.029			.58	.62	1.20	1.80	
	1700	50' span		1,750	.027			.56	.59	1.15	1.72	
	3000	Add per pound for material, incl. bridging				Lb.	.46			.46	.51	

R054-210

053 100 | Steel Deck

		CREW	DAILY OUTPUT	LABOR-HOURS	UNIT	MAT.	LABOR	EQUIP.	TOTAL	TOTAL INCL O&P	
104	0010	METAL DECKING Steel decking										104
	2100	Open type, galv., 1-1/2" deep wide rib, 22 gauge, under 50 squares	E-4	4,500	.007	S.F.	.93	.15	.02	1.10	1.34	
	2600	20 gauge, under 50 squares		3,865	.008		.98	.17	.02	1.17	1.42	
	2900	18 gauge, under 50 squares		3,800	.008		1.35	.17	.02	1.54	1.84	
	3700	4-1/2" deep, long span roof, over 50 squares, 20 gauge		2,700	.012		2.14	.24	.03	2.41	2.86	
	6100	Slab form, steel, 28 gauge, 9/16" deep, uncoated		4,000	.008		.45	.16	.02	.63	.84	
	6200	Galvanized		4,000	.008		.54	.16	.02	.72	.93	
	6220	24 gauge, 1" deep, uncoated		3,900	.008		.62	.17	.02	.81	1.03	
	6240	Galvanized		3,900	.008		.72	.17	.02	.91	1.14	
	6300	24 gauge, 1-5/16" deep, uncoated		3,800	.008		.75	.17	.02	.94	1.19	
	6400	Galvanized		3,800	.008		.90	.17	.02	1.09	1.34	
	6500	22 gauge, 1-5/16" deep, uncoated		3,700	.009		.87	.18	.02	1.07	1.32	
	6600	Galvanized		3,700	.009		1.02	.18	.02	1.22	1.49	
	6700	22 gauge, 3" deep uncoated		3,600	.009		.80	.18	.02	1	1.26	
	6800	Galvanized		3,600	.009		.93	.18	.02	1.13	1.40	

055 100 | Metal Stairs

		CREW	DAILY OUTPUT	LABOR-HOURS	UNIT	MAT.	LABOR	EQUIP.	TOTAL	TOTAL INCL O&P	
104	0012	STAIR										104
	1700	Pre-erected, steel pan tread, 3'-6" wide, 2 line pipe rail	E-2	87	.552	Riser	218	11.15	11.85	241	275	
	1800	With flat bar picket rail	"	87	.552		305	11.15	11.85	328	370	
	1810	Spiral aluminum, 5'-0" diameter, stock units	E-4	45	.711		170	14.55	1.78	186.33	217	
	1820	Custom units		45	.711		325	14.55	1.78	341.33	385	
	1900	Spiral, cast iron, 4'-0" diameter, ornamental, minimum		45	.711		155	14.55	1.78	171.33	200	
	1920	Maximum		25	1.280		206	26	3.20	235.20	282	

055 150 | Ladders

		CREW	DAILY OUTPUT	LABOR-HOURS	UNIT	MAT.	LABOR	EQUIP.	TOTAL	TOTAL INCL O&P	
156	0010	FIRE ESCAPE STAIRS One story, disappearing, stainless steel	2 Sswk	20	.800	V.L.F.	145	15.95		160.95	192	156
	0100	Portable ladder				Ea.	170			170	187	

Important: See the Reference Section for critical supporting data - Reference Nos., Crews, & Location Factors

055 150 | Ladders

		CREW	DAILY OUTPUT	LABOR-HOURS	UNIT	1998 BARE COSTS MAT.	LABOR	EQUIP.	TOTAL	TOTAL INCL O&P	
158	0010	**LADDER** Steel, 20" wide, bolted to concrete, with cage	E-4	50	.640	V.L.F.	56.50	13.10	1.60	71.20	90	158
	0100	Without cage		85	.376	"	27	7.70	.94	35.64	46	
	0300	Aluminum, bolted to concrete, with cage		50	.640	V.L.F.	100	13.10	1.60	114.70	138	
	0400	Without cage		85	.376		58	7.70	.94	66.64	80	
	1350	Alternating tread stair, 56/68°, steel, standard paint color	2 Sswk	50	.320		127	6.40		133.40	152	
	1360	Non-standard paint color		50	.320		136	6.40		142.40	162	
	1370	Galvanized steel		50	.320		144	6.40		150.40	171	
	1380	Stainless steel		50	.320		217	6.40		223.40	252	
	1390	68°, aluminum		50	.320		157	6.40		163.40	185	

055 200 | Handrails & Railings

		CREW	DAILY OUTPUT	LABOR-HOURS	UNIT	1998 BARE COSTS MAT.	LABOR	EQUIP.	TOTAL	TOTAL INCL O&P	
203	0010	**RAILING, PIPE** Aluminum, 2 rail, satin finish, 1-1/4" diameter	E-4	160	.200	L.F.	12.35	4.09	.50	16.94	22	203
	0030	Clear anodized		160	.200		15.30	4.09	.50	19.89	25.50	
	0040	Dark anodized		160	.200		17.25	4.09	.50	21.84	27.50	
	0080	1-1/2" diameter, satin finish		160	.200		14.80	4.09	.50	19.39	25	
	0090	Clear anodized		160	.200		16.50	4.09	.50	21.09	27	
	0100	Dark anodized		160	.200		18.35	4.09	.50	22.94	28.50	
	0140	Aluminum, 3 rail, 1-1/4" diam., satin finish		137	.234		18.95	4.78	.58	24.31	31	
	0150	Clear anodized		137	.234		23.50	4.78	.58	28.86	36	
	0160	Dark anodized		137	.234		26.50	4.78	.58	31.86	39	
	0200	1-1/2" diameter, satin finish		137	.234		22.50	4.78	.58	27.86	35	
	0210	Clear anodized		137	.234		26	4.78	.58	31.36	38.50	
	0220	Dark anodized		137	.234		28.50	4.78	.58	33.86	41	
	0500	Steel, 2 rail, on stairs, primed, 1-1/4" diameter		160	.200		8.90	4.09	.50	13.49	18.40	
	0520	1-1/2" diameter		160	.200		9.80	4.09	.50	14.39	19.35	
	0540	Galvanized, 1-1/4" diameter		160	.200		12.35	4.09	.50	16.94	22	
	0560	1-1/2" diameter		160	.200		13.85	4.09	.50	18.44	24	
	0580	Steel, 3 rail, primed, 1-1/4" diameter		137	.234		13.30	4.78	.58	18.66	24.50	
	0600	1-1/2" diameter		137	.234		14.10	4.78	.58	19.46	25.50	
	0620	Galvanized, 1-1/4" diameter		137	.234		18.65	4.78	.58	24.01	30.50	
	0640	1-1/2" diameter		137	.234		20.50	4.78	.58	25.86	32.50	
	0700	Stainless steel, 2 rail, 1-1/4" diam. #4 finish		137	.234		30.50	4.78	.58	35.86	43.50	
	0720	High polish		137	.234		49	4.78	.58	54.36	64	
	0740	Mirror polish		137	.234		61.50	4.78	.58	66.86	77.50	
	0760	Stainless steel, 3 rail, 1-1/2" diam., #4 finish		120	.267		46	5.45	.67	52.12	62	
	0770	High polish		120	.267		75.50	5.45	.67	81.62	95	
	0780	Mirror finish		120	.267		92	5.45	.67	98.12	112	
	0900	Wall rail, alum. pipe, 1-1/4" diam., satin finish		213	.150		7.05	3.07	.38	10.50	14.25	
	0905	Clear anodized		213	.150		8.65	3.07	.38	12.10	15.95	
	0910	Dark anodized		213	.150		10.45	3.07	.38	13.90	17.95	
	0915	1-1/2" diameter, satin finish		213	.150		7.85	3.07	.38	11.30	15.05	
	0920	Clear anodized		213	.150		9.85	3.07	.38	13.30	17.30	
	0925	Dark anodized		213	.150		12.15	3.07	.38	15.60	19.85	
	0930	Steel pipe, 1-1/4" diameter, primed		213	.150		5.40	3.07	.38	8.85	12.40	
	0935	Galvanized		213	.150		7.85	3.07	.38	11.30	15.05	
	0940	1-1/2" diameter		176	.182		5.55	3.72	.45	9.72	13.95	
	0945	Galvanized		213	.150		7.90	3.07	.38	11.35	15.10	
	0955	Stainless steel pipe, 1-1/2" diam., #4 finish		107	.299		24	6.10	.75	30.85	39.50	
	0960	High polish		107	.299		49.50	6.10	.75	56.35	67.50	
	0965	Mirror polish		107	.299		58	6.10	.75	64.85	77	
208	0010	**RAILINGS, ORNAMENTAL** Aluminum, bronze or stainless, minimum	1 Sswk	24	.333	L.F.	95	6.65		101.65	118	208
	0400	Hand-forged wrought iron, minimum	"	12	.667	"	93	13.30		106.30	128	

			CREW	DAILY OUTPUT	LABOR-HOURS	UNIT	1998 BARE COSTS				TOTAL INCL O&P	
		055 500 | Metal Specialties					MAT.	LABOR	EQUIP.	TOTAL		
504	0010	**LAMP POSTS** Only, 7' high, stock units, aluminum	1 Carp	16	.500	Ea.	63.50	9.25		72.75	86	**504**
	0100	Mild steel, plain	"	16	.500	"	38	9.25		47.25	58	

For information about Means Estimating Seminars, see yellow pages 11 and 12 in back of book

Important: See the Reference Section for critical supporting data - Reference Nos., Crews, & Location Factors

Division 6
Wood & Plastics

Estimating Tips
060 Fasteners & Adhesives

- Common to any wood framed structure are the accessory connector items such as screws, nails, adhesives, hangers, connector plates, straps, angles and holdowns. For typical wood framed buildings, such as residential projects, the aggregate total for these items can be significant, especially in areas where seismic loading is a concern. For floor and wall framing, nail quantities can be figured on a "pounds per thousand board feet basis", with 10 to 25 lbs. per MBF the range. Holdowns, hangers and other connectors should be taken off by the piece.

061 Rough Carpentry

- Lumber is a traded commodity and therefore sensitive to supply and demand in the marketplace. Even in "budgetary" estimating of wood framed projects, it is advisable to call local suppliers for the latest market pricing.
- Common quantity units for wood framed projects are "thousand board feet" (MBF). A board foot is a volume of wood, 1″ x 1′ x 1′, or 144 cubic inches. Board foot quantities are generally calculated using nominal material dimensions—dressed sizes are ignored. Board foot per lineal foot of any stick of lumber can be calculated by dividing the nominal cross sectional area by 12. As an example, 2,000 lineal feet of 2 x 12 equates to 4 MBF by dividing the nominal area, 2 x 12, by 12, which equals 2, and multiplying by 2,000 to give 4,000 board feet. This simple rule applies to all nominal dimensioned lumber.
- Waste is an issue of concern at the quantity takeoff for any area of construction. Framing lumber is sold in even foot lengths, i.e., 10′, 12′, 14′, 16′, and depending on spans, wall heights and the grade of lumber, waste is inevitable. A rule of thumb for lumber waste is 5% to 10% depending on material quality and the complexity of the framing.
- Wood in various forms and shapes is used in many projects, even where the main structural framing is steel, concrete or masonry. Plywood as a back-up partition material and 2x boards used as blocking and cant strips around roof edges are two common examples. The estimator should ensure that the costs of all wood materials are included in the final estimate.

062 Finish Carpentry

- It is necessary to consider the grade of workmanship when estimating labor costs for erecting millwork and interior finish. In practice, there are three grades: premium, custom and economy. The Means daily output for base and case moldings is in the range of 200 to 250 L.F. per carpenter per day. This is appropriate for most average custom grade projects. For premium projects an adjustment to productivity of 25% to 50% should be made depending on the complexity of the job.

Reference Numbers

Reference numbers are shown in bold squares at the beginning of some major classifications. These numbers refer to related items in the Reference Section. The reference information may be an estimating procedure, an alternate pricing method or technical information.

Note: Not all subdivisions listed here necessarily appear in this publication.

060 500 | Fasteners & Adhesives	CREW	DAILY OUTPUT	LABOR-HOURS	UNIT	1998 BARE COSTS				TOTAL INCL O&P
					MAT.	LABOR	EQUIP.	TOTAL	
504 0010 **NAILS** Prices of material only, based on 50# box purchase, copper, plain				Lb.	4			4	4.40
0400 Stainless steel, plain					5			5	5.50
0500 Box, 3d to 20d, bright					.70			.70	.77
0520 Galvanized					.90			.90	.99
0600 Common, 3d to 60d, plain					.55			.55	.61
0700 Galvanized					.75			.75	.83
0800 Aluminum					4.40			4.40	4.84
1000 Annular or spiral thread, 4d to 60d, plain					.60			.60	.66
1200 Galvanized					.80			.80	.88
1400 Drywall nails, plain					.70			.70	.77
1600 Galvanized					1			1	1.10
1800 Finish nails, 4d to 10d, plain					.60			.60	.66
2000 Galvanized					.85			.85	.94
2100 Aluminum					4.40			4.40	4.84
2300 Flooring nails, hardened steel, 2d to 10d, plain					1			1	1.10
2400 Galvanized					1.10			1.10	1.21
2500 Gypsum lath nails, 1-1/8", 13 ga. flathead, blued					1.40			1.40	1.54
2600 Masonry nails, hardened steel, 3/4" to 3" long, plain					1			1	1.10
2700 Galvanized					1.15			1.15	1.27
2900 Roofing nails, threaded, galvanized					.90			.90	.99
3100 Aluminum					4.70			4.70	5.15
3300 Compressed lead head, threaded, galvanized					1			1	1.10
3600 Siding nails, plain shank, galvanized					.70			.70	.77
3800 Aluminum					4			4	4.40
5000 Add to prices above for cement coating					.05			.05	.06
5200 Zinc or tin plating					.10			.10	.11
5500 Vinyl coated sinkers, 8d to 16d					.50			.50	.55
506 0010 **NAILS** mat. only, for pneumatic tools, framing, per carton of 5000, 2"				Ea.	36.50			36.50	40
0100 2-3/8"					41.50			41.50	45.50
0200 Per carton of 4000, 3"					37			37	40.50
0300 3-1/4"					39			39	43
0400 Per carton of 5000, 2-3/8", galv.					56.50			56.50	62
0500 Per carton of 4000, 3", galv.					63.50			63.50	70
0600 3-1/4", galv.					60			60	66
0700 Roofing, per carton of 7200, 1"					35			35	38.50
0800 1-1/4"					35			35	38.50
0900 1-1/2"					45			45	49.50
1000 1-3/4"					55			55	60.50
508 0010 **SHEET METAL SCREWS** Steel, standard, #8 x 3/4", plain				C	2.63			2.63	2.89
0100 Galvanized					3.23			3.23	3.55
0300 #10 x 1", plain					3.54			3.54	3.89
0400 Galvanized					4.14			4.14	4.55
1500 Self-drilling, with washers, (pinch point) #8 x 3/4", plain					4.55			4.55	5
1600 Galvanized					6.10			6.10	6.70
1800 #10 x 3/4", plain					6.05			6.05	6.65
1900 Galvanized					6.90			6.90	7.60
3000 Stainless steel w/aluminum or neoprene washers, #14 x 1", plain					15.95			15.95	17.55
3100 #14 x 2", plain					23.50			23.50	26
512 0010 **TIMBER CONNECTORS** Add up cost of each part for total									
0020 cost of connection									
0100 Connector plates, steel, with bolts, straight	2 Carp	75	.213	Ea.	16.20	3.95		20.15	24.50
0110 Tee	"	50	.320		24	5.90		29.90	36.50
0200 Bolts, machine, sq. hd. with nut & washer, 1/2" diameter, 4" long	1 Carp	140	.057		.65	1.06		1.71	2.53
0300 7-1/2" long		130	.062		.90	1.14		2.04	2.93

Important: See the Reference Section for critical supporting data - Reference Nos., Crews, & Location Factors

060 500	Fasteners & Adhesives	CREW	DAILY OUTPUT	LABOR-HOURS	UNIT	MAT.	LABOR	EQUIP.	TOTAL	TOTAL INCL O&P	
								1998 BARE COSTS			
512	0500	3/4" diameter, 7-1/2" long	1 Carp	130	.062	Ea.	1.53	1.14		2.67	3.63	**512**
	0600	15" long		95	.084	↓	2.01	1.56		3.57	4.88	
	0720	Machine bolts, sq. hd. w/nut & wash		150	.053	Lb.	1.79	.99		2.78	3.66	
	0800	Drilling bolt holes in timber, 1/2" diameter		450	.018	Inch		.33		.33	.56	
	0900	1" diameter		350	.023	"		.42		.42	.72	
	1100	Framing anchors, 2 or 3 dimensional, 10 gauge, no nails incl.		175	.046	Ea.	.37	.85		1.22	1.86	
	1250	Holdowns, 3 gauge base, 10 gauge body		8	1		13.35	18.50		31.85	46	
	1300	Joist and beam hangers, 18 ga. galv., for 2" x 4" joist		175	.046		.46	.85		1.31	1.95	
	1400	2" x 6" to 2" x 10" joist		165	.048		.56	.90		1.46	2.16	
	1600	16 ga. galv., 3" x 6" to 3" x 10" joist		160	.050		2.24	.93		3.17	4.05	
	1700	3" x 10" to 3" x 14" joist		160	.050		2.60	.93		3.53	4.44	
	1800	4" x 6" to 4" x 10" joist		155	.052		1.96	.95		2.91	3.80	
	1900	4" x 10" to 4" x 14" joist		155	.052		2.67	.95		3.62	4.58	
	2000	Two-2" x 6" to two-2" x 10" joists		150	.053		2.15	.99		3.14	4.06	
	2100	Two-2" x 10" to two-2" x 14" joists		150	.053		2.15	.99		3.14	4.06	
	2300	3/16" thick, 6" x 8" joist		145	.055		4.71	1.02		5.73	6.95	
	2400	6" x 10" joist		140	.057		5.55	1.06		6.61	7.90	
	2500	6" x 12" joist		135	.059		6.70	1.10		7.80	9.25	
	2700	1/4" thick, 6" x 14" joist		130	.062		8.30	1.14		9.44	11.05	
	2800	Joist anchors, 1/4" x 1-1/4" x 18"		140	.057		3.23	1.06		4.29	5.35	
	2900	Plywood clips, extruded aluminum H clip, for 3/4" panels					.11			.11	.12	
	3000	Galvanized 18 ga. back-up clip					.10			.10	.11	
	3200	Post framing, 16 ga. galv. for 4" x 4" base, 2 piece	1 Carp	130	.062		4.65	1.14		5.79	7.05	
	3300	Cap		130	.062		2.29	1.14		3.43	4.46	
	3500	Rafter anchors, 18 ga. galv., 1-1/2" wide, 5-1/4" long		145	.055		.37	1.02		1.39	2.16	
	3600	10-3/4" long		145	.055		.76	1.02		1.78	2.59	
	3800	Shear plates, 2-5/8" diameter		120	.067		1.38	1.23		2.61	3.63	
	3900	4" diameter		115	.070		3.20	1.29		4.49	5.75	
	4000	Sill anchors, embedded in concrete or block, 18-5/8" long		115	.070		.97	1.29		2.26	3.27	
	4100	Spike grids, 4" x 4", flat or curved		120	.067		.40	1.23		1.63	2.54	
	4400	Split rings, 2-1/2" diameter		120	.067		1.15	1.23		2.38	3.38	
	4500	4" diameter		110	.073		1.77	1.35		3.12	4.26	
	4700	Strap ties, 16 ga., 1-3/8" wide, 12" long		180	.044		.91	.82		1.73	2.41	
	4800	24" long		160	.050		1.42	.93		2.35	3.15	
	5000	Toothed rings, 2-5/8" or 4" diameter		90	.089	↓	.97	1.64		2.61	3.89	
	5200	Truss plates, nailed, 20 gauge, up to 32' span		17	.471	Truss	7	8.70		15.70	22.50	
	5400	Washers, 2" x 2" x 1/8"				Ea.	.22			.22	.24	
	5500	3" x 3" x 3/16"				"	.57			.57	.63	
516	0010	**WOOD SCREWS** #8, 1" long, steel				C	3.30			3.30	3.63	**516**
	0100	Brass					15.90			15.90	17.50	
	0200	#8, 2" long, steel					6.25			6.25	6.85	
	0300	Brass					31			31	34	
	0400	#10, 1" long, steel					4.02			4.02	4.42	
	0500	Brass					21.50			21.50	23.50	
	0600	#10, 2" long, steel					7.15			7.15	7.90	
	0700	Brass					38			38	41.50	
	0800	#10, 3" long, steel					13.10			13.10	14.40	
	1000	#12, 2" long, steel					9.60			9.60	10.55	
	1100	Brass					47.50			47.50	52.50	
	1500	#12, 3" long, steel					15.15			15.15	16.65	
	2000	#12, 4" long, steel				↓	27			27	29.50	

For expanded coverage of these items see *Means Building Construction Cost Data 1998*

061 100 | Wood Framing

			CREW	DAILY OUTPUT	LABOR-HOURS	UNIT	1998 BARE COSTS				TOTAL INCL O&P	
							MAT.	LABOR	EQUIP.	TOTAL		
102	0010	**BLOCKING**										102
	1950	Miscellaneous, to wood construction	R061 -010									
	2000	2" x 4"	1 Carp	250	.032	L.F.	.38	.59		.97	1.43	
	2005	Pneumatic nailed		305	.026		.38	.49		.87	1.25	
	2050	2" x 6"		222	.036		.55	.67		1.22	1.74	
	2055	Pneumatic nailed		271	.030	C	.55	.55		1.10	1.54	
	2100	2" x 8"		200	.040		.78	.74		1.52	2.13	
	2105	Pneumatic nailed		244	.033		.78	.61		1.39	1.90	
	2150	2" x 10"		178	.045		1.09	.83		1.92	2.62	
	2155	Pneumatic nailed		217	.037		1.09	.68		1.77	2.37	
	2200	2" x 12"		151	.053		1.41	.98		2.39	3.23	
	2205	Pneumatic nailed		185	.043		1.41	.80		2.21	2.92	
	2300	To steel construction										
	2320	2" x 4"	1 Carp	208	.038	L.F.	.38	.71		1.09	1.64	
	2340	2" x 6"		180	.044		.55	.82		1.37	2.01	
	2360	2" x 8"		158	.051		.78	.94		1.72	2.47	
	2380	2" x 10"		136	.059		1.09	1.09		2.18	3.06	
	2400	2" x 12"		109	.073		1.41	1.36		2.77	3.88	
104	0010	**BRACING** Let-in, with 1" x 6" boards, studs @ 16" O.C.	1 Carp	1.50	5.333	C.L.F.	44.50	98.50		143	218	104
	0200	Studs @ 24" O.C.	"	2.30	3.478	"	44.50	64.50		109	159	
106	0010	**BRIDGING** Wood, for joists 16" O.C., 1" x 3"	1 Carp	1.30	6.154	C.Pr.	42.50	114		156.50	242	106
	0015	Pneumatic nailed		1.70	4.706		42.50	87		129.50	196	
	0100	2" x 3" bridging		1.30	6.154		40.50	114		154.50	240	
	0105	Pneumatic nailed		1.70	4.706		40.50	87		127.50	194	
	0300	Steel, galvanized, 18 ga., for 2" x 10" joists at 12" O.C.		1.30	6.154		85	114		199	289	
	0400	24" O.C.		1.40	5.714		125	106		231	320	
	0600	For 2" x 14" joists at 16" O.C.		1.30	6.154		168	114		282	380	
	0900	Compression type, 16" O.C., 2" x 8" joists		2	4		115	74		189	254	
	1000	2" x 12" joists		2	4		126	74		200	266	
110	0010	**FRAMING, BEAMS & GIRDERS**										110
	1000	Single, 2" x 6"	F-2	700	.023	L.F.	.55	.37		.92	1.23	
	1005	Pneumatic nailed	F-2A	812	.020		.55	.32		.87	1.15	
	1020	2" x 8"	F-2	650	.025		.78	.40		1.18	1.54	
	1025	Pneumatic nailed	F-2A	754	.021		.78	.34		1.12	1.45	
	1040	2" x 10"	F-2	600	.027		1.09	.43		1.52	1.94	
	1045	Pneumatic nailed	F-2A	696	.023		1.09	.37		1.46	1.84	
	1060	2" x 12"	F-2	550	.029		1.41	.47		1.88	2.36	
	1065	Pneumatic nailed	F-2A	638	.025		1.41	.41		1.82	2.24	
	1080	2" x 14"	F-2	500	.032		1.73	.52		2.25	2.80	
	1085	Pneumatic nailed	F-2A	580	.028		1.73	.45		2.18	2.67	
	1100	3" x 8"	F-2	550	.029		2.20	.47		2.67	3.23	
	1120	3" x 10"		500	.032		2.75	.52		3.27	3.92	
	1140	3" x 12"		450	.036		3.30	.58		3.88	4.61	
	1160	3" x 14"		400	.040		3.85	.65		4.50	5.35	
	1180	4" x 8"	F-3	1,000	.040		2.28	.68	.44	3.40	4.14	
	1200	4" x 10"		950	.042		2.85	.71	.46	4.02	4.86	
	1220	4" x 12"		900	.044		3.42	.75	.49	4.66	5.60	
	1240	4" x 14"		850	.047		3.99	.79	.52	5.30	6.30	
	2000	Double, 2" x 6"	F-2	625	.026		1.09	.41		1.50	1.91	
	2005	Pneumatic nailed	F-2A	725	.022		1.09	.36		1.45	1.81	
	2020	2" x 8"	F-2	575	.028		1.55	.45		2	2.48	
	2025	Pneumatic nailed	F-2A	667	.024		1.55	.39		1.94	2.37	
	2040	2" x 10"	F-2	550	.029		2.18	.47		2.65	3.21	
	2045	Pneumatic nailed	F-2A	638	.025		2.18	.41		2.59	3.09	
	2060	2" x 12"	F-2	525	.030		2.82	.49		3.31	3.95	

Important: See the Reference Section for critical supporting data - Reference Nos., Crews, & Location Factors

			DAILY	LABOR-		1998 BARE COSTS				TOTAL	
061 100	**Wood Framing**	CREW	OUTPUT	HOURS	UNIT	MAT.	LABOR	EQUIP.	TOTAL	INCL O&P	
110	2065	Pneumatic nailed	F-2A	610	.026	L.F.	2.82	.42		3.24	3.84	**110**
	2080	2" x 14"	F-2	475	.034		3.47	.55		4.02	4.74	
	2085	Pneumatic nailed	F-2A	551	.029		3.47	.47		3.94	4.61	
	3000	Triple, 2" x 6"	F-2	550	.029		1.64	.47		2.11	2.62	
	3005	Pneumatic nailed	F-2A	638	.025		1.64	.41		2.05	2.50	
	3020	2" x 8"	F-2	525	.030		2.33	.49		2.82	3.41	
	3025	Pneumatic nailed	F-2A	609	.026		2.33	.43		2.76	3.30	
	3040	2" x 10"	F-2	500	.032		3.27	.52		3.79	4.49	
	3045	Pneumatic nailed	F-2A	580	.028		3.27	.45		3.72	4.36	
	3060	2" x 12"	F-2	475	.034		4.24	.55		4.79	5.60	
	3065	Pneumatic nailed	F-2A	551	.029		4.24	.47		4.71	5.45	
	3080	2" x 14"	F-2	450	.036		5.20	.58		5.78	6.70	
	3085	Pneumatic nailed	F-2A	522	.031		5.20	.50		5.70	6.55	
112	0010	**FRAMING, CEILINGS**										**112**
	6000	Suspended, 2" x 3"	F-2	1,000	.016	L.F.	.27	.26		.53	.74	
	6050	2" x 4"		900	.018		.38	.29		.67	.91	
	6100	2" x 6"		800	.020		.55	.32		.87	1.15	
	6150	2" x 8"		650	.025		.78	.40		1.18	1.54	
114	0010	**FRAMING, JOISTS**										**114**
	2000	Joists, 2" x 4"	F-2	1,250	.013	L.F.	.38	.21		.59	.77	
	2005	Pneumatic nailed	F-2A	1,438	.011		.38	.18		.56	.73	
	2100	2" x 6"	F-2	1,250	.013		.55	.21		.76	.95	
	2105	Pneumatic nailed	F-2A	1,438	.011		.55	.18		.73	.91	
	2150	2" x 8"	F-2	1,100	.015		.78	.24		1.02	1.26	
	2155	Pneumatic nailed	F-2A	1,265	.013		.78	.20		.98	1.21	
	2200	2" x 10"	F-2	900	.018		1.09	.29		1.38	1.69	
	2205	Pneumatic nailed	F-2A	1,035	.015		1.09	.25		1.34	1.63	
	2250	2" x 12"	F-2	875	.018		1.41	.30		1.71	2.06	
	2255	Pneumatic nailed	F-2A	1,006	.016		1.41	.26		1.67	1.99	
	2300	2" x 14"	F-2	770	.021		1.73	.34		2.07	2.49	
	2305	Pneumatic nailed	F-2A	886	.018		1.73	.29		2.02	2.41	
	2350	3" x 6"	F-2	925	.017		1.65	.28		1.93	2.30	
	2400	3" x 10"		780	.021		2.75	.33		3.08	3.60	
	2450	3" x 12"		600	.027		3.30	.43		3.73	4.37	
	2500	4" x 6"		800	.020		1.71	.32		2.03	2.43	
	2550	4" x 10"		600	.027		2.85	.43		3.28	3.88	
	2600	4" x 12"		450	.036		3.42	.58		4	4.74	
	2605	Sister joist, 2" x 6"		800	.020		.55	.32		.87	1.15	
	2606	Pneumatic nailed	F-2A	960	.017		.55	.27		.82	1.06	
	3000	Composite wood joist 9-1/2" deep	F-2	.90	17.778	M.L.F.	1,875	288		2,163	2,575	
	3010	11-1/2" deep		.88	18.182		2,025	294		2,319	2,725	
	3020	14" deep		.82	19.512		2,175	315		2,490	2,950	
	3030	16" deep		.78	20.513		2,875	330		3,205	3,750	
	4000	Open web joist 12" deep		.88	18.182		2,125	294		2,419	2,850	
	4010	14" deep		.82	19.512		2,350	315		2,665	3,125	
	4020	16" deep		.78	20.513		2,450	330		2,780	3,275	
	4030	18" deep		.74	21.622		2,550	350		2,900	3,400	
116	0010	**FRAMING, MISCELLANEOUS** R061 -010										**116**
	2000	Firestops, 2" x 4"	F-2	780	.021	L.F.	.38	.33		.71	.99	
	2005	Pneumatic nailed R061 -030	F-2A	952	.017		.38	.27		.65	.89	
	2100	2" x 6"	F-2	600	.027		.55	.43		.98	1.34	
	2105	Pneumatic nailed	F-2A	732	.022		.55	.35		.90	1.21	
	5000	Nailers, treated, wood construction, 2" x 4"	F-2	800	.020		.53	.32		.85	1.13	

		061 100 | Wood Framing		CREW	DAILY OUTPUT	LABOR-HOURS	UNIT	MAT.	LABOR	EQUIP.	TOTAL	TOTAL INCL O&P	
116	5005	Pneumatic nailed	R061 -010	F-2A	960	.017	L.F.	.53	.27		.80	1.04	**116**
	5100	2" x 6"		F-2	750	.021		.79	.35		1.14	1.46	
	5105	Pneumatic nailed	R061 -030	F-2A	900	.018		.79	.29		1.08	1.36	
	5120	2" x 8"		F-2	700	.023		1.09	.37		1.46	1.83	
	5125	Pneumatic nailed		F-2A	840	.019		1.09	.31		1.40	1.73	
	5200	Steel construction, 2" x 4"		F-2	750	.021		.53	.35		.88	1.17	
	5220	2" x 6"			700	.023		.79	.37		1.16	1.50	
	5240	2" x 8"			·650	.025		1.09	.40		1.49	1.88	
	7000	Rough bucks, treated, for doors or windows, 2" x 6"			400	.040		.79	.65		1.44	1.98	
	7005	Pneumatic nailed		F-2A	480	.033		.79	.54		1.33	1.79	
	7100	2" x 8"		F-2	380	.042		1.09	.68		1.77	2.37	
	7105	Pneumatic nailed		F-2A	456	.035		1.09	.57		1.66	2.17	
	8000	Stair stringers, 2" x 10"		F-2	130	.123		1.09	1.99		3.08	4.61	
	8100	2" x 12"			130	.123		1.41	1.99		3.40	4.96	
	8150	3" x 10"			125	.128		2.75	2.07		4.82	6.60	
	8200	3" x 12"			125	.128		3.30	2.07		5.37	7.20	
118	0010	**FRAMING, COLUMNS**											**118**
	0100	4" x 4"		F-2	390	.041	L.F.	.99	.66		1.65	2.23	
	0150	4" x 6"			275	.058		1.71	.94		2.65	3.49	
	0200	4" x 8"			220	.073		2.28	1.18		3.46	4.52	
	0250	6" x 6"			215	.074		4.94	1.20		6.14	7.50	
	0300	6" x 8"			175	.091		4.98	1.48		6.46	8.05	
	0350	6" x 10"			150	.107		7.05	1.73		8.78	10.70	
120	0010	**FRAMING, ROOFS**											**120**
	2000	Fascia boards, 2" x 8"		F-2	225	.071	L.F.	.78	1.15		1.93	2.83	
	2100	2" x 10"			180	.089		1.09	1.44		2.53	3.66	
	5000	Rafters, to 4 in 12 pitch, 2" x 6", ordinary			1,000	.016		.55	.26		.81	1.04	
	5020	On steep roofs			800	.020		.55	.32		.87	1.15	
	5040	On dormers or complex roofs			590	.027		.55	.44		.99	1.35	
	5060	2" x 8", ordinary			950	.017		.78	.27		1.05	1.33	
	5080	On steep roofs			750	.021		.78	.35		1.13	1.45	
	5100	On dormers or complex roofs			540	.030		.78	.48		1.26	1.68	
	5120	2" x 10", ordinary			630	.025		1.09	.41		1.50	1.90	
	5140	On steep roofs			495	.032		1.09	.52		1.61	2.10	
	5160	On dormers or complex roofs			425	.038		1.09	.61		1.70	2.24	
	5180	2" x 12", ordinary			575	.028		1.41	.45		1.86	2.32	
	5200	On steep roofs			455	.035		1.41	.57		1.98	2.52	
	5220	On dormers or complex roofs			395	.041		1.41	.66		2.07	2.67	
	5300	Hip and valley rafters, 2" x 6", ordinary			760	.021		.55	.34		.89	1.18	
	5320	On steep roofs			585	.027		.55	.44		.99	1.36	
	5340	On dormers or complex roofs			510	.031		.55	.51		1.06	1.47	
	5360	2" x 8", ordinary			720	.022		.78	.36		1.14	1.48	
	5380	On steep roofs			545	.029		.78	.48		1.26	1.67	
	5400	On dormers or complex roofs			470	.034		.78	.55		1.33	1.80	
	5420	2" x 10", ordinary			570	.028		1.09	.45		1.54	1.98	
	5440	On steep roofs			440	.036		1.09	.59		1.68	2.21	
	5460	On dormers or complex roofs			380	.042		1.09	.68		1.77	2.37	
	5480	Hip and valley rafters, 2" x 12", ordinary			525	.030		1.41	.49		1.90	2.39	
	5500	On steep roofs			410	.039		1.41	.63		2.04	2.63	
	5520	On dormers or complex roofs			355	.045		1.41	.73		2.14	2.80	
	5540	Hip and valley jacks, 2" x 6", ordinary			600	.027		.55	.43		.98	1.34	
	5560	On steep roofs			475	.034		.55	.55		1.10	1.53	
	5580	On dormers or complex roofs			410	.039		.55	.63		1.18	1.68	
	5600	2" x 8", ordinary			490	.033		.78	.53		1.31	1.76	
	5620	On steep roofs			385	.042		.78	.67		1.45	2.01	

Important: See the Reference Section for critical supporting data – Reference Nos., Crews, & Location Factors

061 100	Wood Framing	CREW	DAILY OUTPUT	LABOR-HOURS	UNIT	MAT.	LABOR	EQUIP.	TOTAL	TOTAL INCL O&P	
							1998 BARE COSTS				
120 5640	On dormers or complex roofs	F-2	335	.048	L.F.	.78	.77		1.55	2.18	**120**
5660	2" x 10", ordinary		450	.036		1.09	.58		1.67	2.18	
5680	On steep roofs		350	.046		1.09	.74		1.83	2.47	
5700	On dormers or complex roofs		305	.052		1.09	.85		1.94	2.65	
5720	2" x 12", ordinary		375	.043		1.41	.69		2.10	2.73	
5740	On steep roofs		295	.054		1.41	.88		2.29	3.05	
5760	On dormers or complex roofs		255	.063		1.41	1.02		2.43	3.29	
5780	Rafter tie, 1" x 4", #3		800	.020		.37	.32		.69	.96	
5790	2" x 4", #3		800	.020		.38	.32		.70	.97	
5800	Ridge board, #2 or better, 1" x 6"		600	.027		.47	.43		.90	1.26	
5820	1" x 8"		550	.029		.66	.47		1.13	1.53	
5840	1" x 10"		500	.032		.91	.52		1.43	1.89	
5860	2" x 6"		500	.032		.55	.52		1.07	1.49	
5880	2" x 8"		450	.036		.78	.58		1.36	1.84	
5900	2" x 10"		400	.040		1.09	.65		1.74	2.31	
5920	Roof cants, split, 4" x 4"		650	.025		.99	.40		1.39	1.77	
5940	6" x 6"		600	.027		4.94	.43		5.37	6.20	
5960	Roof curbs, untreated, 2" x 6"		520	.031		.55	.50		1.05	1.45	
5980	2" x 12"		400	.040		1.41	.65		2.06	2.66	
6000	Sister rafters, 2" x 6"		800	.020		.55	.32		.87	1.15	
6020	2" x 8"		640	.025		.78	.40		1.18	1.55	
6040	2" x 10"		535	.030		1.09	.48		1.57	2.03	
6060	2" x 12"		455	.035		1.41	.57		1.98	2.52	
122 0010	**FRAMING, SILLS**										**122**
2000	Ledgers, nailed, 2" x 4"	F-2	755	.021	L.F.	.38	.34		.72	1.01	
2050	2" x 6"		600	.027		.55	.43		.98	1.34	
2100	Bolted, not including bolts, 3" x 6"		325	.049		1.65	.80		2.45	3.18	
2150	3" x 12"		233	.069		3.30	1.11		4.41	5.55	
2600	Mud sills, redwood, construction grade, 2" x 4"		895	.018		2.67	.29		2.96	3.43	
2620	2" x 6"		780	.021		4	.33		4.33	4.97	
4000	Sills, 2" x 4"		600	.027		.38	.43		.81	1.16	
4050	2" x 6"		550	.029		.55	.47		1.02	1.41	
4080	2" x 8"		500	.032		.78	.52		1.30	1.75	
4100	2" x 10"		450	.036		1.09	.58		1.67	2.18	
4120	2" x 12"		400	.040		1.41	.65		2.06	2.66	
4200	Treated, 2" x 4"		550	.029		.53	.47		1	1.39	
4220	2" x 6"		500	.032		.79	.52		1.31	1.76	
4240	2" x 8"		450	.036		1.09	.58		1.67	2.18	
4260	2" x 10"		400	.040		1.58	.65		2.23	2.84	
4280	2" x 12"		350	.046		2.04	.74		2.78	3.51	
4400	4" x 4"		450	.036		1.77	.58		2.35	2.93	
4420	4" x 6"		350	.046		2.71	.74		3.45	4.25	
4460	4" x 8"		300	.053		3.64	.86		4.50	5.50	
4480	4" x 10"		260	.062		4.55	1		5.55	6.70	
124 0010	**FRAMING, SLEEPERS**										**124**
0100	On concrete, treated, 1" x 2"	F-2	2,350	.007	L.F.	.17	.11		.28	.37	
0150	1" x 3"		2,000	.008		.25	.13		.38	.50	
0200	2" x 4"		1,500	.011		.53	.17		.70	.88	
0250	2" x 6"		1,300	.012		.79	.20		.99	1.21	
126 0010	**FRAMING, SOFFITS & CANOPIES**										**126**
1000	Canopy or soffit framing , 1" x 4"	F-2	900	.018	L.F.	.39	.29		.68	.91	
1020	1" x 6"		850	.019		.47	.30		.77	1.04	
1040	1" x 8"		750	.021		.66	.35		1.01	1.31	

061 100 | Wood Framing

		CREW	DAILY OUTPUT	LABOR-HOURS	UNIT	MAT.	LABOR	EQUIP.	TOTAL	TOTAL INCL O&P	
126	1100	2" x 4"	F-2	620	.026	L.F.	.38	.42		.80	1.13	**126**
	1120	2" x 6"		560	.029		.55	.46		1.01	1.39	
	1140	2" x 8"		500	.032		.78	.52		1.30	1.75	
	1200	3" x 4"		500	.032		1.10	.52		1.62	2.10	
	1220	3" x 6"		400	.040		1.65	.65		2.30	2.93	
	1240	3" x 10"		300	.053		2.75	.86		3.61	4.51	
127	0010	**FRAMING, TREATED LUMBER**										**127**
	0020	Water-borne salt, C.C.A., A.C.A., wet, .40 P.C.F. retention										
	0100	2" x 4"				M.B.F.	790			790	865	
	0110	2" x 6"					790			790	865	
	0120	2" x 8"					820			820	900	
	0130	2" x 10"					945			945	1,050	
	0140	2" x 12"					1,025			1,025	1,125	
	0200	4" x 4"					1,325			1,325	1,450	
	0210	4" x 6"					1,350			1,350	1,500	
	0220	4" x 8"					1,375			1,375	1,500	
	0250	Add for .60 P.C.F. retention					40%					
	0260	Add for 2.5 P.C.F. retention					200%					
	0270	Add for K.D.A.T.					20%					
128	0010	**FRAMING, WALLS** R061 -010										**128**
	2000	Headers over openings, 2" x 6"	F-2	360	.044	L.F.	.55	.72		1.27	1.83	
	2005	2" x 6", pneumatic nailed R061 -030	F-2A	432	.037		.55	.60		1.15	1.63	
	2050	2" x 8"	F-2	340	.047		.78	.76		1.54	2.16	
	2055	2" x 8", pneumatic nailed	F-2A	408	.039		.78	.63		1.41	1.95	
	2100	2" x 10"	F-2	320	.050		1.09	.81		1.90	2.59	
	2105	2" x 10", pneumatic nailed	F-2A	384	.042		1.09	.67		1.76	2.35	
	2150	2" x 12"	F-2	300	.053		1.41	.86		2.27	3.03	
	2155	2" x 12", pneumatic nailed	F-2A	360	.044		1.41	.72		2.13	2.78	
	2190	4" x 10"	F-2	240	.067		2.85	1.08		3.93	4.99	
	2195	4" x 10", pneumatic nailed	F-2A	288	.056		2.85	.90		3.75	4.68	
	2200	4" x 12"	F-2	190	.084		3.42	1.36		4.78	6.10	
	2205	4" x 12", pneumatic nailed	F-2A	228	.070		3.42	1.14		4.56	5.70	
	2240	6" x 10"	F-2	165	.097		7.05	1.57		8.62	10.45	
	2245	6" x 10", pneumatic nailed	F-2A	198	.081		7.05	1.31		8.36	10	
	2250	6" x 12"	F-2	140	.114		8.45	1.85		10.30	12.45	
	2255	6" x 12", pneumatic nailed	F-2A	168	.095		8.45	1.54		9.99	11.95	
	5000	Plates, untreated, 2" x 3"	F-2	850	.019		.27	.30		.57	.82	
	5005	2" x 3", pneumatic nailed	F-2A	1,020	.016		.27	.25		.52	.73	
	5020	2" x 4"	F-2	800	.020		.38	.32		.70	.97	
	5025	2" x 4", pneumatic nailed	F-2A	960	.017		.38	.27		.65	.88	
	5040	2" x 6"	F-2	750	.021		.55	.35		.90	1.19	
	5045	2" x 6", pneumatic nailed	F-2A	900	.018		.55	.29		.84	1.09	
	5060	Treated, 2" x 3"	F-2	850	.019		.42	.30		.72	.98	
	5065	2" x 3", treated, pneumatic nailed	F-2A	1,020	.016		.42	.25		.67	.89	
	5080	2" x 4"	F-2	800	.020		.53	.32		.85	1.13	
	5085	2" x 4", treated, pneumatic nailed	F-2A	960	.017		.53	.27		.80	1.04	
	5100	2" x 6"	F-2	750	.021		.79	.35		1.14	1.46	
	5103	2" x 6", treated, pneumatic nailed	F-2A	900	.018		.79	.29		1.08	1.36	
	5120	Studs, 8' high wall, 2" x 3"	F-2	1,200	.013		.27	.22		.49	.67	
	5125	2" x 3", pneumatic nailed	F-2A	1,440	.011		.27	.18		.45	.61	
	5140	2" x 4"	F-2	1,100	.015		.38	.24		.62	.82	
	5146	2" x 4", pneumatic nailed	F-2A	1,320	.012		.38	.20		.58	.76	
	5160	2" x 6"	F-2	1,000	.016		.55	.26		.81	1.04	
	5166	2" x 6", pneumatic nailed	F-2A	1,200	.013		.55	.22		.77	.97	
	5180	3" x 4"	F-2	800	.020		1.10	.32		1.42	1.76	

Important: See the Reference Section for critical supporting data - Reference Nos., Crews, & Location Factors

061 100 | Wood Framing

		CREW	DAILY OUTPUT	LABOR-HOURS	UNIT	1998 BARE COSTS				TOTAL INCL O&P
						MAT.	LABOR	EQUIP.	TOTAL	
128 5185	3" x 4", pneumatic nailed	F-2A	960	.017	L.F.	1.10	.27		1.37	1.67 **128**
5200	Installed on second story, 2" x 3"	F-2	1,170	.014		.27	.22		.49	.68
5205	2" x 3", pneumatic nailed	F-2A	1,404	.011		.27	.18		.45	.62
5220	2" x 4"	F-2	1,015	.016		.38	.26		.64	.86
5225	2" x 4", pneumatic nailed	F-2A	1,218	.013		.38	.21		.59	.78
5240	2" x 6"	F-2	890	.018		.55	.29		.84	1.10
5245	2" x 6", pneumatic nailed	F-2A	1,080	.015		.55	.24		.79	1.01
5260	3" x 4"	F-2	800	.020		1.10	.32		1.42	1.76
5265	3" x 4", pneumatic nailed	F-2A	960	.017		1.10	.27		1.37	1.67
5280	Installed on dormer or gable, 2" x 3"	F-2	1,045	.015		.27	.25		.52	.72
5285	2" x 3", pneumatic nailed	F-2A	1,254	.013		.27	.21		.48	.65
5300	2" x 4"	F-2	905	.018		.38	.29		.67	.91
5305	2" x 4", pneumatic nailed	F-2A	1,086	.015		.38	.24		.62	.83
5320	2" x 6"	F-2	800	.020		.55	.32		.87	1.15
5325	2" x 6", pneumatic nailed	F-2A	960	.017		.55	.27		.82	1.06
5340	3" x 4"	F-2	700	.023		1.10	.37		1.47	1.84
5345	3" x 4", pneumatic nailed	F-2A	840	.019		1.10	.31		1.41	1.74
5360	6' high wall, 2" x 3"	F-2	970	.016		.27	.27		.54	.76
5365	2" x 3", pneumatic nailed	F-2A	1,164	.014		.27	.22		.49	.68
5380	2" x 4"	F-2	850	.019		.38	.30		.68	.94
5385	2" x 4", pneumatic nailed	F-2A	1,020	.016		.38	.25		.63	.85
5400	2" x 6"	F-2	740	.022		.55	.35		.90	1.20
5405	2" x 6", pneumatic nailed	F-2A	888	.018		.55	.29		.84	1.10
5420	3" x 4"	F-2	600	.027		1.10	.43		1.53	1.95
5425	3" x 4", pneumatic nailed	F-2A	720	.022		1.10	.36		1.46	1.83
5440	Installed on second story, 2" x 3"	F-2	950	.017		.27	.27		.54	.77
5445	2" x 3", pneumatic nailed	F-2A	1,140	.014		.27	.23		.50	.69
5460	2" x 4"	F-2	810	.020		.38	.32		.70	.97
5465	2" x 4", pneumatic nailed	F-2A	972	.016		.38	.27		.65	.88
5480	2" x 6"	F-2	700	.023		.55	.37		.92	1.23
5485	2" x 6", pneumatic nailed	F-2A	840	.019		.55	.31		.86	1.13
5500	3" x 4"	F-2	550	.029		1.10	.47		1.57	2.02
5505	3" x 4", pneumatic nailed	F-2A	660	.024		1.10	.39		1.49	1.88
5520	Installed on dormer or gable, 2" x 3"	F-2	850	.019		.27	.30		.57	.82
5525	2" x 3", pneumatic nailed	F-2A	1,020	.016		.27	.25		.52	.73
5540	2" x 4"	F-2	720	.022		.38	.36		.74	1.04
5545	2" x 4", pneumatic nailed	F-2A	864	.019		.38	.30		.68	.93
5560	2" x 6"	F-2	620	.026		.55	.42		.97	1.31
5565	2" x 6", pneumatic nailed	F-2A	744	.022		.55	.35		.90	1.20
5580	3" x 4"	F-2	480	.033		1.10	.54		1.64	2.13
5585	3" x 4", pneumatic nailed	F-2A	576	.028		1.10	.45		1.55	1.98
5600	3' high wall, 2" x 3"	F-2	740	.022		.27	.35		.62	.90
5605	2" x 3", pneumatic nailed	F-2A	888	.018		.27	.29		.56	.80
5620	2" x 4"	F-2	640	.025		.38	.40		.78	1.11
5625	2" x 4", pneumatic nailed	F-2A	768	.021		.38	.34		.72	1
5640	2" x 6"	F-2	550	.029		.55	.47		1.02	1.41
5645	2" x 6", pneumatic nailed	F-2A	660	.024		.55	.39		.94	1.27
5660	3" x 4"	F-2	440	.036		1.10	.59		1.69	2.22
5665	3" x 4", pneumatic nailed	F-2A	528	.030		1.10	.49		1.59	2.05
5680	Installed on second story, 2" x 3"	F-2	700	.023		.27	.37		.64	.93
5685	2" x 3", pneumatic nailed	F-2A	840	.019		.27	.31		.58	.83
5700	2" x 4"	F-2	610	.026		.38	.42		.80	1.15
5705	2" x 4", pneumatic nailed	F-2A	732	.022		.38	.35		.73	1.03
5720	2" x 6"	F-2	520	.031		.55	.50		1.05	1.45
5725	2" x 6", pneumatic nailed	F-2A	624	.026	L.F.	.55	.41		.96	1.31
5740	3" x 4"	F-2	430	.037		1.10	.60		1.70	2.24

(R061 -010, R061 -030 reference cross-references appear in the description column.)

WOOD & PLASTICS 6

061 100 | Wood Framing

		CREW	DAILY OUTPUT	LABOR-HOURS	UNIT	1998 BARE COSTS MAT.	LABOR	EQUIP.	TOTAL	TOTAL INCL O&P	
128	5745	3" x 4", pneumatic nailed	F-2A	516	.031	L.F.	1.10	.50		1.60	2.07	**128**
	5760	Installed on dormer or gable, 2" x 3"	F-2	625	.026		.27	.41		.68	1.01	
	5765	2" x 3", pneumatic nailed	F-2A	750	.021		.27	.35		.62	.89	
	5780	2" x 4"	F-2	545	.029		.38	.48		.86	1.23	
	5785	2" x 4", pneumatic nailed	F-2A	654	.024		.38	.40		.78	1.10	
	5800	2" x 6"	F-2	465	.034		.55	.56		1.11	1.55	
	5805	2" x 6", pneumatic nailed	F-2A	558	.029		.55	.46		1.01	1.39	
	5820	3" x 4"	F-2	380	.042		1.10	.68		1.78	2.38	
	5825	3" x 4", pneumatic nailed	F-2A	456	.035		1.10	.57		1.67	2.18	
	8250	For second story & above, add						5%				
	8300	For dormer & gable, add						15%				
130	0010	**FURRING** Wood strips, 1" x 2", on walls, on wood	1 Carp	550	.015	L.F.	.18	.27		.45	.66	**130**
	0015	On wood, pneumatic nailed		710	.011		.18	.21		.39	.56	
	0300	On masonry		495	.016		.18	.30		.48	.71	
	0400	On concrete		260	.031		.18	.57		.75	1.18	
	0600	1" x 3", on walls, on wood		550	.015		.28	.27		.55	.77	
	0605	On wood, pneumatic nailed		710	.011		.28	.21		.49	.67	
	0700	On masonry		495	.016		.28	.30		.58	.82	
	0800	On concrete		260	.031		.28	.57		.85	1.29	
	0850	On ceilings, on wood		350	.023		.28	.42		.70	1.03	
	0855	On wood, pneumatic nailed		450	.018		.28	.33		.61	.87	
	0900	On masonry		320	.025		.28	.46		.74	1.10	
	0950	On concrete		210	.038		.28	.70		.98	1.52	
132	0010	**GROUNDS** For casework, 1" x 2" wood strips, on wood	1 Carp	330	.024	L.F.	.18	.45		.63	.97	**132**
	0100	On masonry		285	.028		.18	.52		.70	1.09	
	0200	On concrete		250	.032		.18	.59		.77	1.21	
	0400	For plaster, 3/4" deep, on wood		450	.018		.18	.33		.51	.76	
	0500	On masonry		225	.036		.18	.66		.84	1.33	
	0600	On concrete		175	.046		.18	.85		1.03	1.65	
	0700	On metal lath		200	.040		.18	.74		.92	1.47	
134	0010	**INSULATION** See division 072										**134**
138	0010	**PARTITIONS** Wood stud with single bottom plate and										**138**
	0020	double top plate, no waste, std. & better lumber										
	0180	2" x 4" studs, 8' high, studs 12" O.C.	F-2	80	.200	L.F.	4.20	3.24		7.44	10.15	
	0185	12" O.C., pneumatic nailed	F-2A	96	.167		4.20	2.70		6.90	9.25	
	0200	16" O.C.	F-2	100	.160		3.44	2.59		6.03	8.20	
	0205	16" O.C., pneumatic nailed	F-2A	120	.133		3.44	2.16		5.60	7.45	
	0300	24" O.C.	F-2	125	.128		2.67	2.07		4.74	6.50	
	0305	24" O.C., pneumatic nailed	F-2A	150	.107		2.67	1.73		4.40	5.90	
	0380	10' high, studs 12" O.C.	F-2	80	.200		4.97	3.24		8.21	11	
	0385	12" O.C., pneumatic nailed	F-2A	96	.167		4.97	2.70		7.67	10.05	
	0400	16" O.C.	F-2	100	.160		4.01	2.59		6.60	8.85	
	0405	16" O.C., pneumatic nailed	F-2A	120	.133		4.01	2.16		6.17	8.10	
	0500	24" O.C.	F-2	125	.128		3.06	2.07		5.13	6.90	
	0505	24" O.C., pneumatic nailed	F-2A	150	.107		3.06	1.73		4.79	6.30	
	0580	12' high, studs 12" O.C.	F-2	65	.246		5.75	3.98		9.73	13.10	
	0585	12" O.C., pneumatic nailed	F-2A	78	.205		5.75	3.32		9.07	12	
	0600	16" O.C.	F-2	80	.200		4.58	3.24		7.82	10.60	
	0605	16" O.C., pneumatic nailed	F-2A	96	.167		4.58	2.70		7.28	9.65	
	0700	24" O.C.	F-2	100	.160		3.44	2.59		6.03	8.20	
	0705	24" O.C., pneumatic nailed	F-2A	120	.133		3.44	2.16		5.60	7.45	
	0780	2" x 6" studs, 8' high, studs 12" O.C.	F-2	70	.229		6	3.70		9.70	12.95	
	0785	12" O.C., pneumatic nailed	F-2A	84	.190		6	3.08		9.08	11.90	

Note: R061-010 (referenced at 5745), R061-030 (referenced at 5765)

Important: See the Reference Section for critical supporting data - Reference Nos., Crews, & Location Factors

		061 100	**Wood Framing**	CREW	DAILY OUTPUT	LABOR-HOURS	UNIT	MAT.	LABOR	EQUIP.	TOTAL	TOTAL INCL O&P	
138	0800		16" O.C.	F-2	90	.178	L.F.	4.92	2.88		7.80	10.30	138
	0805		16" O.C., pneumatic nailed	F-2A	108	.148		4.92	2.40		7.32	9.50	
	0900		24" O.C.	F-2	115	.139		3.83	2.25		6.08	8.05	
	0905		24" O.C., pneumatic nailed	F-2A	138	.116		3.83	1.88		5.71	7.40	
	0980		10' high, studs 12" O.C.	F-2	70	.229		7.10	3.70		10.80	14.15	
	0985		12" O.C., pneumatic nailed	F-2A	84	.190		7.10	3.08		10.18	13.10	
	1000		16" O.C.	F-2	90	.178		5.75	2.88		8.63	11.20	
	1005		16" O.C., pneumatic nailed	F-2A	108	.148		5.75	2.40		8.15	10.40	
	1100		24" O.C.	F-2	115	.139		4.38	2.25		6.63	8.65	
	1105		24" O.C., pneumatic nailed	F-2A	138	.116		4.38	1.88		6.26	8	
	1180		12' high, studs 12" O.C.	F-2	55	.291		8.20	4.71		12.91	17.10	
	1185		12" O.C., pneumatic nailed	F-2A	66	.242		8.20	3.92		12.12	15.75	
	1200		16" O.C.	F-2	70	.229		6.55	3.70		10.25	13.55	
	1205		16" O.C., pneumatic nailed	F-2A	84	.190		6.55	3.08		9.63	12.50	
	1300		24" O.C.	F-2	90	.178		4.92	2.88		7.80	10.30	
	1305		24" O.C., pneumatic nailed	F-2A	108	.148		4.92	2.40		7.32	9.50	
	1400	For horizontal blocking, 2" x 4", add		F-2	600	.027		.38	.43		.81	1.16	
	1500	2" x 6", add			600	.027		.55	.43		.98	1.34	
	1600	For openings, add			250	.064			1.04		1.04	1.77	
	1700	Headers for above openings, material only, add					M.B.F.	585			585	640	
140	0010	**ROUGH HARDWARE** Average % of carpentry material, minimum						.50%					140
	0200	Maximum						1.50%					

		061 150	**Sheathing**	CREW	DAILY OUTPUT	LABOR-HOURS	UNIT	MAT.	LABOR	EQUIP.	TOTAL	TOTAL INCL O&P	
154	0010	**SHEATHING** Plywood on roof, CDX	R061 -020										154
	0030	5/16" thick		F-2	1,600	.010	S.F.	.32	.16		.48	.64	
	0035	Pneumatic nailed		F-2A	1,952	.008		.32	.13		.45	.59	
	0050	3/8" thick		F-2	1,525	.010		.33	.17		.50	.66	
	0055	Pneumatic nailed		F-2A	1,860	.009		.33	.14		.47	.61	
	0100	1/2" thick		F-2	1,400	.011		.43	.18		.61	.79	
	0105	Pneumatic nailed		F-2A	1,708	.009		.43	.15		.58	.73	
	0200	5/8" thick		F-2	1,300	.012		.51	.20		.71	.90	
	0205	Pneumatic nailed		F-2A	1,586	.010		.51	.16		.67	.84	
	0300	3/4" thick		F-2	1,200	.013		.61	.22		.83	1.04	
	0305	Pneumatic nailed		F-2A	1,464	.011		.61	.18		.79	.97	
	0500	Plywood on walls with exterior CDX, 3/8" thick		F-2	1,200	.013		.33	.22		.55	.74	
	0505	Pneumatic nailed		F-2A	1,488	.011		.33	.17		.50	.67	
	0600	1/2" thick		F-2	1,125	.014		.43	.23		.66	.86	
	0605	Pneumatic nailed		F-2A	1,395	.011		.43	.19		.62	.79	
	0700	5/8" thick		F-2	1,050	.015		.51	.25		.76	.98	
	0705	Pneumatic nailed		F-2A	1,302	.012		.51	.20		.71	.90	
	0800	3/4" thick		F-2	975	.016		.61	.27		.88	1.12	
	0805	Pneumatic nailed		F-2A	1,209	.013		.61	.21		.82	1.04	
	1000	For shear wall construction, add							20%				
	1200	For structural 1 exterior plywood, add					S.F.	10%					
	1400	With boards, on roof 1" x 6" boards, laid horizontal		F-2	725	.022		.90	.36		1.26	1.60	
	1500	Laid diagonal			650	.025		.90	.40		1.30	1.67	
	1700	1" x 8" boards, laid horizontal			875	.018		.95	.30		1.25	1.56	
	1800	Laid diagonal			725	.022		.95	.36		1.31	1.66	
	2000	For steep roofs, add							40%				
	2200	For dormers, hips and valleys, add						5%	50%				
	2400	Boards on walls, 1" x 6" boards, laid regular		F-2	650	.025		.90	.40		1.30	1.67	
	2500	Laid diagonal			585	.027		.90	.44		1.34	1.75	
	2700	1" x 8" boards, laid regular			765	.021		.95	.34		1.29	1.63	
	2800	Laid diagonal			650	.025		.95	.40		1.35	1.73	
	2850	Gypsum, weatherproof, 1/2" thick			1,125	.014		.17	.23		.40	.58	

WOOD & PLASTICS　6

061 150 | Sheathing

			DAILY OUTPUT	LABOR-HOURS	UNIT	1998 BARE COSTS				TOTAL INCL O&P	
						MAT.	LABOR	EQUIP.	TOTAL		
154	2900	Sealed, 4/10" thick		1,100	.015	S.F.	.30	.24		.54	.73	154
	3000	Wood fiber, regular, no vapor barrier, 1/2" thick		1,200	.013		.37	.22		.59	.78	
	3100	5/8" thick		1,200	.013		.49	.22		.71	.91	
	3300	No vapor barrier, in colors, 1/2" thick		1,200	.013		.53	.22		.75	.95	
	3400	5/8" thick		1,200	.013		.65	.22		.87	1.08	
	3600	With vapor barrier one side, white, 1/2" thick		1,200	.013		.52	.22		.74	.94	
	3700	Vapor barrier 2 sides, 1/2" thick		1,200	.013		.81	.22		1.03	1.26	
	3800	Asphalt impregnated, 25/32" thick		1,200	.013		.33	.22		.55	.73	
	3850	Intermediate, 1/2" thick		1,200	.013		.27	.22		.49	.67	
	4000	Wafer board on roof, 1/2" thick		1,455	.011		.34	.18		.52	.67	
	4100	5/8" thick		1,330	.012		.40	.19		.59	.77	

Crew F-2. Reference R061-020.

061 160 | Subfloor

			CREW	DAILY OUTPUT	LABOR-HOURS	UNIT	MAT.	LABOR	EQUIP.	TOTAL	TOTAL INCL O&P	
161	0010	**FLOORING, WOOD** See division 095-604										161
164	0010	**SUBFLOOR** Plywood, CDX, 1/2" thick	F-2	1,500	.011	SF Flr.	.43	.17		.60	.77	164
	0015	Pneumatic nailed	F-2A	1,860	.009		.43	.14		.57	.71	
	0100	5/8" thick	F-2	1,350	.012		.51	.19		.70	.89	
	0105	Pneumatic nailed	F-2A	1,674	.010		.51	.15		.66	.82	
	0200	3/4" thick	F-2	1,250	.013		.61	.21		.82	1.02	
	0205	Pneumatic nailed	F-2A	1,550	.010		.61	.17		.78	.96	
	0300	1-1/8" thick, 2-4-1 including underlayment	F-2	1,050	.015		1.18	.25		1.43	1.72	
	0500	With boards, 1" x 10" S4S, laid regular		1,100	.015		.96	.24		1.20	1.46	
	0600	Laid diagonal		900	.018		.97	.29		1.26	1.56	
	0800	1" x 8" S4S, laid regular		1,000	.016		.86	.26		1.12	1.39	
	0900	Laid diagonal		850	.019		.86	.30		1.16	1.46	
	1100	Wood fiber, T&G, 2' x 8' planks, 1" thick		1,000	.016		1.19	.26		1.45	1.75	
	1200	1-3/8" thick		900	.018		1.46	.29		1.75	2.10	
	1500	Wafer board, 5/8" thick		1,330	.012	S.F.	.44	.19		.63	.81	
	1600	3/4" thick		1,230	.013	"	.48	.21		.69	.89	
168	0010	**UNDERLAYMENT** Plywood, underlayment grade, 3/8" thick	F-2	1,500	.011	SF Flr.	.45	.17		.62	.80	168
	0015	Pneumatic nailed	F-2A	1,860	.009		.45	.14		.59	.74	
	0100	1/2" thick	F-2	1,450	.011		.53	.18		.71	.89	
	0105	Pneumatic nailed	F-2A	1,798	.009		.53	.14		.67	.83	
	0200	5/8" thick	F-2	1,400	.011		.63	.18		.81	1.01	
	0205	Pneumatic nailed	F-2A	1,736	.009		.63	.15		.78	.95	
	0300	3/4" thick	F-2	1,300	.012		.75	.20		.95	1.17	
	0305	Pneumatic nailed	F-2A	1,612	.010		.75	.16		.91	1.10	
	0500	Particle board, 3/8" thick	F-2	1,500	.011		.31	.17		.48	.64	
	0505	Pneumatic nailed	F-2A	1,860	.009		.31	.14		.45	.58	
	0600	1/2" thick	F-2	1,450	.011		.33	.18		.51	.67	
	0605	Pneumatic nailed	F-2A	1,798	.009		.33	.14		.47	.61	
	0800	5/8" thick	F-2	1,400	.011		.35	.18		.53	.71	
	0805	Pneumatic nailed	F-2A	1,736	.009		.35	.15		.50	.65	
	0900	3/4" thick	F-2	1,300	.012		.41	.20		.61	.79	
	0905	Pneumatic nailed	F-2A	1,612	.010		.41	.16		.57	.72	
	1100	Hardboard, underlayment grade, 4' x 4', .215" thick	F-2	1,500	.011		.41	.17		.58	.75	

Reference R061-020.

061 250 | Wood Decking

			CREW	DAILY OUTPUT	LABOR-HOURS	UNIT	MAT.	LABOR	EQUIP.	TOTAL	TOTAL INCL O&P	
258	0010	**ROOF DECKS**										258
	0400	Cedar planks, 3" thick	F-2	320	.050	S.F.	5.40	.81		6.21	7.35	
	0500	4" thick		250	.064		7.30	1.04		8.34	9.75	
	0700	Douglas fir, 3" thick		320	.050		2.02	.81		2.83	3.61	

Important: See the Reference Section for critical supporting data - Reference Nos., Crews, & Location Factors

061 250 | Wood Decking

		CREW	DAILY OUTPUT	LABOR-HOURS	UNIT	1998 BARE COSTS				TOTAL INCL O&P	
						MAT.	LABOR	EQUIP.	TOTAL		
258	0800	4" thick	F-2	250	.064	S.F.	2.70	1.04		3.74	4.74	**258**
	1000	Hemlock, 3" thick		320	.050		2.02	.81		2.83	3.61	
	1100	4" thick		250	.064		2.70	1.04		3.74	4.74	
	1300	Western white spruce, 3" thick		320	.050		1.95	.81		2.76	3.54	
	1400	4" thick	↓	250	.064	↓	2.60	1.04		3.64	4.63	

061 300 | Heavy Timber Constr.

		CREW	DAILY OUTPUT	LABOR-HOURS	UNIT	MAT.	LABOR	EQUIP.	TOTAL	TOTAL INCL O&P	
304	0010	**FRAMING, HEAVY** Mill timber, beams, single 6" x 10"	F-2	1.10	14.545	M.B.F.	1,425	235		1,660	1,975	**304**
	0100	Single 8" x 16"		1.20	13.333		1,625	216		1,841	2,175	
	0200	Built from 2" lumber, multiple 2" x 14"		.90	17.778		745	288		1,033	1,300	
	0210	Built from 3" lumber, multiple 3" x 6"		.70	22.857		1,100	370		1,470	1,825	
	0220	Multiple 3" x 8"		.80	20		1,100	325		1,425	1,750	
	0230	Multiple 3" x 10"		.90	17.778		1,100	288		1,388	1,700	
	0240	Multiple 3" x 12"		1	16		1,100	259		1,359	1,650	
	0250	Built from 4" lumber, multiple 4" x 6"		.80	20		855	325		1,180	1,500	
	0260	Multiple 4" x 8"		.90	17.778		855	288		1,143	1,425	
	0270	Multiple 4" x 10"		1	16		855	259		1,114	1,375	
	0280	Multiple 4" x 12"		1.10	14.545		855	235		1,090	1,350	
	0290	Columns, structural grade, 1500f, 4" x 4"		.60	26.667		1,250	430		1,680	2,125	
	0300	6" x 6"		.65	24.615		1,900	400		2,300	2,750	
	0400	8" x 8"		.70	22.857		2,025	370		2,395	2,850	
	0500	10" x 10"		.75	21.333		1,900	345		2,245	2,675	
	0600	12" x 12"		.80	20		1,900	325		2,225	2,650	
	0800	Floor planks, 2" thick, T & G, 2" x 6"		1.05	15.238		700	247		947	1,200	
	0900	2" x 10"		1.10	14.545		760	235		995	1,250	
	1100	3" thick, 3" x 6"		1.05	15.238		1,150	247		1,397	1,700	
	1200	3" x 10"		1.10	14.545		1,150	235		1,385	1,675	
	1400	Girders, structural grade, 12" x 12"		.80	20		1,900	325		2,225	2,625	
	1500	10" x 16"	↓	1	16	↓	1,850	259		2,109	2,475	
	2050	Roof planks, see division 061-258										
	2300	Roof purlins, 4" thick, structural grade	F-2	1.05	15.238	M.B.F.	1,125	247		1,372	1,675	
	2500	Roof trusses, add timber connectors, division 060-512	"	.45	35.556	"	1,100	575		1,675	2,175	

061 500 | Wood-Metal Systems

		CREW	DAILY OUTPUT	LABOR-HOURS	UNIT	MAT.	LABOR	EQUIP.	TOTAL	TOTAL INCL O&P	
508	0010	**STRUCTURAL JOISTS** Fabricated "I" joists with wood flanges,										**508**
	0100	Plywood webs, incl. bridging & blocking, panels 24" O.C.										
	1200	15' to 24' span, 50 psf live load	F-5	2,400	.013	SF Flr.	1.44	.22		1.66	1.95	
	1300	55 psf live load		2,250	.014		1.55	.23		1.78	2.09	
	1400	24' to 30' span, 45 psf live load		2,600	.012		1.80	.20		2	2.32	
	1500	55 psf live load	↓	2,400	.013		1.80	.22		2.02	2.35	
	1600	Tubular steel open webs, 45 psf, 24" O.C., 40' span	F-3	6,250	.006		1.80	.11	.07	1.98	2.24	
	1700	55' span		7,750	.005		1.75	.09	.06	1.90	2.14	
	1800	70' span		9,250	.004		2.27	.07	.05	2.39	2.67	
	1900	85 psf live load, 26' span	↓	2,300	.017	↓	2.11	.29	.19	2.59	3.03	

061 800 | Glued-Laminated Const

		CREW	DAILY OUTPUT	LABOR-HOURS	UNIT	MAT.	LABOR	EQUIP.	TOTAL	TOTAL INCL O&P	
804	0010	**LAMINATED FRAMING** Not including decking										**804**
	0020	30 lb., short term live load, 15 lb. dead load										
	0200	Straight roof beams, 20' clear span, beams 8' O.C.	F-3	2,560	.016	SF Flr.	1.34	.26	.17	1.77	2.11	
	0300	Beams 16' O.C.		3,200	.013		.98	.21	.14	1.33	1.59	
	0500	40' clear span, beams 8' O.C.		3,200	.013		2.58	.21	.14	2.93	3.35	
	0600	Beams 16' O.C.	↓	3,840	.010		2.11	.18	.11	2.40	2.75	
	0800	60' clear span, beams 8' O.C.	F-4	2,880	.014		4.43	.23	.27	4.93	5.55	
	0900	Beams 16' O.C.	"	3,840	.010	↓	3.30	.18	.20	3.68	4.15	

061 800 | Glued-Laminated Const

			CREW	DAILY OUTPUT	LABOR-HOURS	UNIT	1998 BARE COSTS				TOTAL INCL O&P
							MAT.	LABOR	EQUIP.	TOTAL	
804	1100	Tudor arches, 30' to 40' clear span, frames 8' O.C.	F-3	1,680	.024	SF Flr.	5.75	.40	.26	6.41	7.30
	1200	Frames 16' O.C.	"	2,240	.018		4.53	.30	.20	5.03	5.70
	1400	50' to 60' clear span, frames 8' O.C.	F-4	2,200	.018		6.25	.31	.36	6.92	7.75
	1500	Frames 16' O.C.		2,640	.015		5.30	.26	.30	5.86	6.60
	1700	Radial arches, 60' clear span, frames 8' O.C.		1,920	.021		5.80	.35	.41	6.56	7.45
	1800	Frames 16' O.C.		2,880	.014		4.48	.23	.27	4.98	5.65
	2000	100' clear span, frames 8' O.C.		1,600	.025		6.05	.42	.49	6.96	7.90
	2100	Frames 16' O.C.		2,400	.017		5.30	.28	.33	5.91	6.70
	2300	120' clear span, frames 8' O.C.		1,440	.028		8.05	.47	.55	9.07	10.25
	2400	Frames 16' O.C.		1,920	.021		7.30	.35	.41	8.06	9.10
	2600	Bowstring trusses, 20' O.C., 40' clear span	F-3	2,400	.017		3.61	.28	.18	4.07	4.65
	2700	60' clear span	F-4	3,600	.011		3.24	.19	.22	3.65	4.12
	2800	100' clear span		4,000	.010		4.58	.17	.20	4.95	5.55
	2900	120' clear span		3,600	.011		4.94	.19	.22	5.35	6
	3100	For premium appearance, add to S.F. prices					5%				
	3300	For industrial type, deduct					15%				
	3500	For stain and varnish, add					5%				
	3900	For 3/4" laminations, add to straight					25%				
	4100	Add to curved					15%				
	4300	Alternate pricing method: (use nominal footage of									
	4310	components). Straight beams, camber less than 6"	F-3	3.50	11.429	M.B.F.	2,050	193	125	2,368	2,750
	4400	Columns, including hardware		2	20		2,225	340	219	2,784	3,250
	4600	Curved members, radius over 32'		2.50	16		2,275	270	175	2,720	3,150
	4700	Radius 10' to 32'		3	13.333		2,250	225	146	2,621	3,025
	4900	For complicated shapes, add maximum					100%				
	5100	For pressure treating, add to straight					35%				
	5200	Add to curved					45%				
	6000	Laminated veneer members, southern pine or western species									
	6050	1-3/4" wide x 5-1/2" deep	F-2	480	.033	L.F.	3.50	.54		4.04	4.77
	6100	9-1/2" deep		480	.033		3.50	.54		4.04	4.77
	6150	14" deep		450	.036		5.25	.58		5.83	6.80
	6200	18" deep		450	.036		6.50	.58		7.08	8.15
	6300	Parallel strand members, southern pine or western species									
	6350	1-3/4" wide x 9-1/4" deep	F-2	480	.033	L.F.	4.50	.54		5.04	5.85
	6400	11-1/4" deep		450	.036		5.40	.58		5.98	6.95
	6450	14" deep		400	.040		6.50	.65		7.15	8.25
	6500	3-1/2" wide x 9-1/4" deep		480	.033		8.65	.54		9.19	10.40
	6550	11-1/4" deep		450	.036		10.75	.58		11.33	12.80
	6600	14" deep		400	.040		12.75	.65		13.40	15.15
	6650	7" wide x 9-1/4" deep		450	.036		17	.58		17.58	19.70
	6700	11-1/4" deep		420	.038		21	.62		21.62	24
	6750	14" deep		400	.040		25	.65		25.65	28.50
806	0010	**LAMINATED BEAMS** Fb 2400 psi									
	0050	3" x 15"	F-3	480	.083	L.F.	7.50	1.41	.91	9.82	11.65
	0100	3" x 18"		450	.089		9	1.50	.97	11.47	13.50
	0150	5" x 15"		360	.111		11.25	1.88	1.22	14.35	16.95
	0200	5" x 18"		290	.138		14	2.33	1.51	17.84	21
	0250	5" x 22-1/2"		220	.182		18	3.07	1.99	23.06	27
	0300	6-3/4" x 18"		320	.125		14.30	2.11	1.37	17.78	21
	0350	6-3/4" x 25-1/2"		260	.154		18.70	2.60	1.68	22.98	27
	0400	6-3/4" x 33"		210	.190		24	3.22	2.09	29.31	34.50
	0450	8-3/4" x 34-1/2"		160	.250		33	4.22	2.74	39.96	46.50
	0500	For premium appearance, add to S.F. prices					5%				
	0550	For industrial type, deduct					15%				
	0600	For stain and varnish, add					5%				
	0650	For 3/4" laminations, add					25%				

Important: See the Reference Section for critical supporting data - Reference Nos., Crews, & Location Factors

061 800 | Glued-Laminated Const

			CREW	DAILY OUTPUT	LABOR-HOURS	UNIT	1998 BARE COSTS				TOTAL INCL O&P	
							MAT.	LABOR	EQUIP.	TOTAL		
808	0010	**LAMINATED ROOF DECK** Pine or hemlock, 3" thick	F-2	425	.038	S.F.	2.70	.61		3.31	4.01	808
	0100	4" thick		325	.049		3.60	.80		4.40	5.30	
	0300	Cedar, 3" thick		425	.038		3.50	.61		4.11	4.89	
	0400	4" thick		325	.049		4.70	.80		5.50	6.50	
	0600	Fir, 3" thick		425	.038		2.70	.61		3.31	4.01	
	0700	4" thick		325	.049		3.60	.80		4.40	5.30	

061 900 | Wood Trusses

			CREW	DAILY OUTPUT	LABOR-HOURS	UNIT	1998 BARE COSTS				TOTAL INCL O&P	
							MAT.	LABOR	EQUIP.	TOTAL		
908	0010	**ROOF TRUSSES** [R061-100]										908
	0020	For timber connectors, see div. 060-512										
	5000	Common wood, 2" x 4" metal plate connected, 24" O.C., 4/12 slope										
	5010	1' overhang, 12' span	F-5	55	.582	Ea.	27.50	9.40		36.90	46.50	
	5050	20' span	F-6	62	.645		36	10.80	7.05	53.85	65.50	
	5100	24' span		60	.667		42.50	11.15	7.30	60.95	74	
	5150	26' span		57	.702		65	11.75	7.70	84.45	100	
	5200	28' span		53	.755		50.50	12.60	8.25	71.35	86	
	5240	30' span		51	.784		76	13.10	8.60	97.70	115	
	5250	32' span		50	.800		71	13.35	8.75	93.10	111	
	5280	34' span		48	.833		88	13.90	9.15	111.05	131	
	5350	8/12 pitch, 1' overhang, 20' span		57	.702		60	11.75	7.70	79.45	94.50	
	5400	24' span		55	.727		71	12.15	7.95	91.10	107	
	5450	26' span		52	.769		77	12.85	8.40	98.25	116	
	5500	28' span		49	.816		83	13.65	8.95	105.60	124	
	5550	32' span		45	.889		93	14.85	9.75	117.60	138	
	5600	36' span		41	.976		105	16.30	10.70	132	155	
	5650	38' span		40	1		114	16.70	10.95	141.65	166	
	5700	40' span		40	1		130	16.70	10.95	157.65	184	

062 | Finish Carpentry

062 200 | Millwork Moldings

			CREW	DAILY OUTPUT	LABOR-HOURS	UNIT	1998 BARE COSTS				TOTAL INCL O&P	
							MAT.	LABOR	EQUIP.	TOTAL		
204	0010	**MILLWORK** Rule of thumb: Milled material cost [R061-030]										204
	0020	equals three times cost of lumber										
	1000	Typical finish hardwood milled material										
	1020	1" x 12", custom birch				L.F.	2.65			2.65	2.92	
	1040	Cedar					2.30			2.30	2.53	
	1060	Oak					5.95			5.95	6.55	
	1080	Redwood					5.90			5.90	6.50	
	1100	Southern yellow pine					1.28			1.28	1.41	
	1120	Sugar pine					2.72			2.72	2.99	
	1140	Teak					11.90			11.90	13.10	
	1160	Walnut					7.70			7.70	8.45	
	1180	White pine					2.95			2.95	3.25	
208	0010	**MOLDINGS, BASE**										208
	0500	Base, stock pine, 9/16" x 3-1/2"	1 Carp	240	.033	L.F.	1.28	.62		1.90	2.47	
	0501	Oak or birch, 9/16" x 3-1/2"		240	.033		2.16	.62		2.78	3.44	
	0550	9/16" x 4-1/2"		200	.040		1.22	.74		1.96	2.62	
	0561	Base shoe, oak, 3/4" x 1"		240	.033		.76	.62		1.38	1.90	
	0570	Base, prefinished, 2-1/2" x 9/16"	F-2	242	.033		.99	.61		1.60	2.14	

For expanded coverage of these items see *Means Building Construction Cost Data 1998*

WOOD & PLASTICS — **6**

062 200 | Millwork Moldings

			CREW	DAILY OUTPUT	LABOR-HOURS	UNIT	MAT.	LABOR	EQUIP.	TOTAL	TOTAL INCL O&P	
							1998 BARE COSTS					
208	0580	Shoe, prefinished, 3/8″ x 5/8″	1 Carp	266	.030	L.F.	.42	.56		.98	1.41	**208**
	0585	Flooring cant strip, 3/4″ x 1/2″		500	.016		.22	.30		.52	.75	
212	0010	**MOLDINGS, CASINGS**										**212**
	0090	Apron, stock pine, 5/8″ x 2″	1 Carp	250	.032	L.F.	.67	.59		1.26	1.75	
	0110	5/8″ x 3-1/2″		220	.036		1.39	.67		2.06	2.68	
	0300	Band, stock pine, 11/16″ x 1-1/8″		270	.030		.37	.55		.92	1.35	
	0350	11/16″ x 1-3/4″		250	.032		.53	.59		1.12	1.59	
	0700	Casing, stock pine, 11/16″ x 2-1/2″		240	.033		.74	.62		1.36	1.87	
	0701	Oak or birch		240	.033		1.73	.62		2.35	2.96	
	0750	11/16″ x 3-1/2″		215	.037		1.43	.69		2.12	2.75	
	0760	Door & window casing, exterior, 1-1/4″ x 2″		200	.040		1.37	.74		2.11	2.78	
	0770	Finger jointed, 1-1/4″ x 2″		200	.040		1.24	.74		1.98	2.63	
	4600	Mullion casing, stock pine, 5/16″ x 2″		200	.040		.60	.74		1.34	1.93	
	4601	Oak or birch		200	.040		.71	.74		1.45	2.05	
	4700	Teak, custom, nominal 1″ x 1″		215	.037		1.10	.69		1.79	2.39	
	4800	Nominal 1″ x 3″		200	.040		3.25	.74		3.99	4.85	
216	0010	**MOLDINGS, CEILINGS**										**216**
	0600	Bed, stock pine, 9/16″ x 1-3/4″	1 Carp	270	.030	L.F.	.53	.55		1.08	1.52	
	0650	9/16″ x 2″		240	.033		.80	.62		1.42	1.94	
	1200	Cornice molding, stock pine, 9/16″ x 1-3/4″		330	.024		.55	.45		1	1.38	
	1300	9/16″ x 2-1/4″		300	.027		.74	.49		1.23	1.67	
	2400	Cove scotia, stock pine, 9/16″ x 1-3/4″		270	.030		.55	.55		1.10	1.55	
	2401	Oak or birch, 9/16″ x 1-3/4″		270	.030		.80	.55		1.35	1.82	
	2500	11/16″ x 2-3/4″		255	.031		1.18	.58		1.76	2.29	
	2600	Crown, stock pine, 9/16″ x 3-5/8″		250	.032		1.15	.59		1.74	2.28	
	2700	11/16″ x 4-5/8″		220	.036		2.16	.67		2.83	3.53	
220	0010	**MOLDINGS, EXTERIOR**										**220**
	1500	Cornice, boards, pine, 1″ x 2″	1 Carp	330	.024	L.F.	.22	.45		.67	1.01	
	1600	1″ x 4″		250	.032		.44	.59		1.03	1.50	
	1700	1″ x 6″		250	.032		.65	.59		1.24	1.73	
	1800	1″ x 8″		200	.040		.76	.74		1.50	2.11	
	1900	1″ x 10″		180	.044		1.16	.82		1.98	2.68	
	2000	1″ x 12″		180	.044		1.21	.82		2.03	2.74	
	2200	Three piece, built-up, pine, minimum		80	.100		1.40	1.85		3.25	4.71	
	2300	Maximum		65	.123		4.40	2.28		6.68	8.75	
	3000	Corner board, sterling pine, 1″ x 4″		200	.040		.31	.74		1.05	1.61	
	3100	1″ x 6″		200	.040		.47	.74		1.21	1.79	
	3200	2″ x 6″		165	.048		1	.90		1.90	2.64	
	3300	2″ x 8″		165	.048		1.46	.90		2.36	3.15	
	3350	Fascia, sterling pine, 1″ x 6″		250	.032		.47	.59		1.06	1.53	
	3370	1″ x 8″		225	.036		.62	.66		1.28	1.81	
	3372	2″ x 6″		225	.036		1	.66		1.66	2.23	
	3374	2″ x 8″		200	.040		1.46	.74		2.20	2.88	
	3376	2″ x 10″		180	.044		1.62	.82		2.44	3.19	
	3395	Grounds, 1″ x 1″ redwood		300	.027		.18	.49		.67	1.05	
	3400	Trim, exterior, sterling pine, back band		250	.032		.53	.59		1.12	1.59	
	3500	Casing		250	.032		.50	.59		1.09	1.56	
	3600	Crown		250	.032		1.15	.59		1.74	2.28	
	3700	Porch rail with balusters		22	.364		8.45	6.75		15.20	21	
	3800	Screen		395	.020		.30	.37		.67	.97	
	4100	Verge board, sterling pine, 1″ x 4″		200	.040		.31	.74		1.05	1.61	
	4200	1″ x 6″		200	.040		.47	.74		1.21	1.79	
	4300	2″ x 6″		165	.048		1	.90		1.90	2.64	
	4400	2″ x 8″		165	.048		1.33	.90		2.23	3	
	4700	For redwood trim, add					200%					
	5000	Casing/fascia, rough-sawn cedar										

Important: See the Reference Section for critical supporting data - Reference Nos., Crews, & Location Factors

062 200 | Millwork Moldings

		CREW	DAILY OUTPUT	LABOR-HOURS	UNIT	1998 BARE COSTS				TOTAL INCL O&P	
						MAT.	LABOR	EQUIP.	TOTAL		
220	5100	1" x 2"	1 Carp	275	.029	L.F.	.22	.54		.76	1.16	**220**
	5200	1" x 6"		250	.032		.65	.59		1.24	1.72	
	5300	1" x 8"		230	.035		.86	.64		1.50	2.05	
	5400	2" x 4"		220	.036		.86	.67		1.53	2.10	
	5500	2" x 6"		220	.036		1.30	.67		1.97	2.58	
	5600	2" x 8"		200	.040		1.73	.74		2.47	3.17	
	5700	2" x 10"		180	.044		2.16	.82		2.98	3.79	
	5800	2" x 12"		170	.047		2.59	.87		3.46	4.34	
224	0010	**MOLDINGS, TRIM**										**224**
	0200	Astragal, stock pine, 11/16" x 1-3/4"	1 Carp	255	.031	L.F.	.77	.58		1.35	1.84	
	0250	1-5/16" x 2-3/16"		240	.033		2.20	.62		2.82	3.48	
	0800	Chair rail, stock pine, 5/8" x 2-1/2"		270	.030		.79	.55		1.34	1.81	
	0900	5/8" x 3-1/2"		240	.033		1.03	.62		1.65	2.19	
	1000	Closet pole, stock pine, 1-1/8" diameter		200	.040		.80	.74		1.54	2.15	
	1100	Fir, 1-5/8" diameter		200	.040		1.10	.74		1.84	2.48	
	1150	Corner, inside, 5/16" x 1"		225	.036		.27	.66		.93	1.43	
	1160	Outside, 1-1/16" x 1-1/16"		240	.033		.60	.62		1.22	1.72	
	1161	1-5/16" x 1-5/16"		240	.033		.86	.62		1.48	2.01	
	3300	Half round, stock pine, 1/4" x 1/2"		270	.030		.55	.55		1.10	1.55	
	3350	1/2" x 1"		255	.031		.65	.58		1.23	1.70	
	3400	Handrail, fir, single piece, stock, hardware not included										
	3450	1-1/2" x 1-3/4"	1 Carp	80	.100	L.F.	1.05	1.85		2.90	4.32	
	3470	Pine, 1-1/2" x 1-3/4"		80	.100		1.05	1.85		2.90	4.32	
	3500	1-1/2" x 2-1/2"		76	.105		1.43	1.95		3.38	4.91	
	3600	Lattice, stock pine, 1/4" x 1-1/8"		270	.030		.24	.55		.79	1.20	
	3700	1/4" x 1-3/4"		250	.032		.27	.59		.86	1.31	
	3800	Miscellaneous, custom, pine or cedar, 1" x 1"		270	.030		.92	.55		1.47	1.95	
	3900	1" x 3"		240	.033		.80	.62		1.42	1.94	
	4100	Birch or oak, nominal 1" x 1"		240	.033		.44	.62		1.06	1.54	
	4200	Nominal 1" x 3"		215	.037		1.52	.69		2.21	2.85	
	4400	Walnut, nominal 1" x 1"		215	.037		.70	.69		1.39	1.95	
	4500	Nominal 1" x 3"		200	.040		2.10	.74		2.84	3.58	
	4700	Teak, nominal 1" x 1"		215	.037		1.05	.69		1.74	2.33	
	4800	Nominal 1" x 3"		200	.040		3	.74		3.74	4.57	
	4900	Quarter round, stock pine, 1/4" x 1/4"		275	.029		.20	.54		.74	1.14	
	4950	3/4" x 3/4"		255	.031		.41	.58		.99	1.44	
	5600	Wainscot moldings, 1-1/8" x 9/16", 2' high, minimum		76	.105	S.F.	5.60	1.95		7.55	9.50	
	5700	Maximum		65	.123	"	12.35	2.28		14.63	17.50	
228	0010	**MOLDINGS, WINDOW AND DOOR**										**228**
	2800	Door moldings, stock, decorative, 1-1/8" wide, plain	1 Carp	17	.471	Set	27	8.70		35.70	44.50	
	2900	Detailed		17	.471	"	65	8.70		73.70	86.50	
	3150	Door trim set, 1 head and 2 sides, pine, 2 1/2 wide		5.90	1.356	Opng.	12.45	25		37.45	56.50	
	3170	4-1/2" wide		5.30	1.509	"	22.50	28		50.50	73	
	3200	Glass beads, stock pine, 1/4" x 11/16"		285	.028	L.F.	.29	.52		.81	1.21	
	3250	3/8" x 1/2"		275	.029		.35	.54		.89	1.31	
	3270	3/8" x 7/8"		270	.030		.39	.55		.94	1.37	
	4850	Parting bead, stock pine, 3/8" x 3/4"		275	.029		.28	.54		.82	1.23	
	4870	1/2" x 3/4"		255	.031		.35	.58		.93	1.38	
	5000	Stool caps, stock pine, 11/16" x 3-1/2"		200	.040		1.28	.74		2.02	2.68	
	5100	1-1/16" x 3-1/4"		150	.053		3.01	.99		4	5	
	5300	Threshold, oak, 3' long, inside, 5/8" x 3-5/8"		32	.250	Ea.	6.10	4.63		10.73	14.65	
	5400	Outside, 1-1/2" x 7-5/8"		16	.500	"	19	9.25		28.25	37	
	5900	Window trim sets, including casings, header, stops,										
	5910	stool and apron, 2-1/2" wide, minimum	1 Carp	13	.615	Opng.	15	11.40		26.40	36	
	5950	Average		10	.800		26	14.80		40.80	54	
	6000	Maximum		6	1.333		37	24.50		61.50	83	

062 300 | Shelving

		CREW	DAILY OUTPUT	LABOR-HOURS	UNIT	1998 BARE COSTS				TOTAL INCL O&P	
						MAT.	LABOR	EQUIP.	TOTAL		
304	0010	**SHELVING** Pine, clear grade, no edge band, 1" x 8"	1 Carp	115	.070	L.F.	1.77	1.29		3.06	4.15	**304**
	0100	1" x 10"		110	.073		2.38	1.35		3.73	4.93	
	0200	1" x 12"		105	.076		3.26	1.41		4.67	6	
	0400	For lumber edge band, by hand, add					1.49			1.49	1.64	
	0420	By machine, add					.95			.95	1.05	
	0600	Plywood, 3/4" thick with lumber edge, 12" wide	1 Carp	75	.107		1.24	1.97		3.21	4.75	
	0700	24" wide		70	.114		2.35	2.11		4.46	6.20	
	0900	Bookcase, clear grade pine, shelves 12" O.C., 8" deep		70	.114	S.F.	4.24	2.11		6.35	8.30	
	1000	12" deep shelves		65	.123	"	4.64	2.28		6.92	9	
	1200	Adjustable closet rod and shelf, 12" wide, 3' long		20	.400	Ea.	41.50	7.40		48.90	58.50	
	1300	8' long		15	.533	"	58.50	9.85		68.35	81.50	
	1500	Prefinished shelves with supports, stock, 8" wide		75	.107	L.F.	3.65	1.97		5.62	7.40	
	1600	10" wide		70	.114	"	4.07	2.11		6.18	8.10	
	1800	Custom, high quality dadoed pine shelving units, minimum				S.F.					28	
	1900	Maximum				"					40	

062 400 | Plastic Laminate

		CREW	DAILY OUTPUT	LABOR-HOURS	UNIT	1998 BARE COSTS				TOTAL INCL O&P	
						MAT.	LABOR	EQUIP.	TOTAL		
408	0010	**COUNTER TOP** Stock, plastic lam., 24" wide w/backsplash, min.	1 Carp	30	.267	L.F.	4.89	4.93		9.82	13.85	**408**
	0100	Maximum		25	.320		14.25	5.90		20.15	26	
	0300	Custom plastic, 7/8" thick, aluminum molding, no splash		30	.267		15.70	4.93		20.63	26	
	0400	Cove splash		30	.267		20.50	4.93		25.43	31	
	0600	1-1/4" thick, no splash		28	.286		18.50	5.30		23.80	29.50	
	0700	Square splash		28	.286		23	5.30		28.30	34.50	
	0900	Square edge, plastic face, 7/8" thick, no splash		30	.267		19.80	4.93		24.73	30.50	
	1000	With splash		30	.267		26	4.93		30.93	37	
	1200	For stainless channel edge, 7/8" thick, add					2.21			2.21	2.43	
	1300	1-1/4" thick, add					2.58			2.58	2.84	
	1500	For solid color suede finish, add					1.96			1.96	2.16	
	1700	For end splash, add				Ea.	12.35			12.35	13.60	
	1900	For cut outs, standard, add, minimum	1 Carp	32	.250		2.63	4.63		7.26	10.85	
	2000	Maximum		8	1		3.14	18.50		21.64	35	
	2010	Cut out in blacksplash for elec. wall outlet		38	.211			3.89		3.89	6.65	
	2020	Cut out for sink		20	.400			7.40		7.40	12.70	
	2030	Cut out for stove top		18	.444			8.20		8.20	14.10	
	2100	Postformed, including backsplash and front edge		30	.267	L.F.	8.55	4.93		13.48	17.85	
	2110	Mitred, add		12	.667	Ea.		12.35		12.35	21	
	2200	Built-in place, 25" wide, plastic laminate		25	.320	L.F.	10.80	5.90		16.70	22	
	2300	Ceramic tile mosaic		25	.320		24.50	5.90		30.40	37	
	2500	Marble, stock, with splash, 1/2" thick, minimum	1 Bric	17	.471		30	8.90		38.90	48	
	2700	3/4" thick, maximum	"	13	.615		76	11.65		87.65	104	
	2900	Maple, solid, laminated, 1-1/2" thick, no splash	1 Carp	28	.286		31	5.30		36.30	43	
	3000	With square splash		28	.286		35	5.30		40.30	47.50	
	3200	Stainless steel		24	.333	S.F.	73	6.15		79.15	91	
	3400	Recessed cutting block with trim, 16" x 20" x 1"		8	1	Ea.	41	18.50		59.50	77	
	3410	Replace cutting block only		8	1	"	29	18.50		47.50	63	
	3600	Table tops, plastic laminate, square edge, 7/8" thick		45	.178	S.F.	7	3.29		10.29	13.35	
	3700	1-1/8" thick		40	.200	"	7.20	3.70		10.90	14.30	

062 500 | Prefin. Wood Paneling

		CREW	DAILY OUTPUT	LABOR-HOURS	UNIT	1998 BARE COSTS				TOTAL INCL O&P	
						MAT.	LABOR	EQUIP.	TOTAL		
504	0010	**PANELING, PLYWOOD**										**504**
	2400	Plywood, prefinished, 1/4" thick, 4' x 8' sheets										
	2410	with vertical grooves. Birch faced, minimum	F-2	500	.032	S.F.	.72	.52		1.24	1.68	
	2420	Average		420	.038		1.03	.62		1.65	2.19	
	2430	Maximum		350	.046		1.49	.74		2.23	2.91	
	2600	Mahogany, African		400	.040		1.91	.65		2.56	3.21	

Important: See the Reference Section for critical supporting data - Reference Nos., Crews, & Location Factors

062 500 | Prefin. Wood Paneling

		CREW	DAILY OUTPUT	LABOR-HOURS	UNIT	MAT.	LABOR	EQUIP.	TOTAL	TOTAL INCL O&P	
504						1998 BARE COSTS					**504**
2700	Philippine (Lauan)	F-2	500	.032	S.F.	.82	.52		1.34	1.79	
2900	Oak or Cherry, minimum		500	.032		1.60	.52		2.12	2.65	
3000	Maximum		400	.040		2.68	.65		3.33	4.06	
3200	Rosewood		320	.050		3.81	.81		4.62	5.60	
3400	Teak		400	.040		2.68	.65		3.33	4.06	
3600	Chestnut		375	.043		3.97	.69		4.66	5.55	
3800	Pecan		400	.040		1.70	.65		2.35	2.98	
3900	Walnut, minimum		500	.032		2.27	.52		2.79	3.39	
3950	Maximum		400	.040		4.33	.65		4.98	5.85	
4000	Plywood, prefinished, 3/4" thick, stock grades, minimum		320	.050		1.03	.81		1.84	2.52	
4100	Maximum		224	.071		4.48	1.16		5.64	6.90	
4300	Architectural grade, minimum		224	.071		3.30	1.16		4.46	5.60	
4400	Maximum		160	.100		5.05	1.62		6.67	8.30	
4600	Plywood, "A" face, birch, V.C., 1/2" thick, natural		450	.036		1.55	.58		2.13	2.68	
4700	Select		450	.036		1.70	.58		2.28	2.85	
4900	Veneer core, 3/4" thick, natural		320	.050		1.65	.81		2.46	3.21	
5000	Select		320	.050		1.85	.81		2.66	3.43	
5200	Lumber core, 3/4" thick, natural		320	.050		2.47	.81		3.28	4.11	
5500	Plywood, knotty pine, 1/4" thick, A2 grade		450	.036		1.34	.58		1.92	2.45	
5600	A3 grade		450	.036		1.70	.58		2.28	2.85	
5800	3/4" thick, veneer core, A2 grade		320	.050		1.75	.81		2.56	3.32	
5900	A3 grade		320	.050		1.96	.81		2.77	3.55	
6100	Aromatic cedar, 1/4" thick, plywood		400	.040		1.70	.65		2.35	2.98	
6200	1/4" thick, particle board		400	.040		.82	.65		1.47	2.01	

062 550 | Prefin. Hardboard Panel

		CREW	DAILY OUTPUT	LABOR-HOURS	UNIT	MAT.	LABOR	EQUIP.	TOTAL	TOTAL INCL O&P	
554	0010	**PANELING, HARDBOARD**									**554**
0050	Not incl. furring or trim, hardboard, tempered, 1/8" thick	F-2	500	.032	S.F.	.33	.52		.85	1.25	
0100	1/4" thick		500	.032		.37	.52		.89	1.30	
0300	Tempered pegboard, 1/8" thick		500	.032		.33	.52		.85	1.25	
0400	1/4" thick		500	.032		.41	.52		.93	1.34	
0600	Untempered hardboard, natural finish, 1/8" thick		500	.032		.28	.52		.80	1.20	
0700	1/4" thick		500	.032		.31	.52		.83	1.23	
0900	Untempered pegboard, 1/8" thick		500	.032		.32	.52		.84	1.24	
1000	1/4" thick		500	.032		.35	.52		.87	1.28	
1200	Plastic faced hardboard, 1/8" thick		500	.032		.52	.52		1.04	1.46	
1300	1/4" thick		500	.032		.71	.52		1.23	1.67	
1500	Plastic faced pegboard, 1/8" thick		500	.032		.50	.52		1.02	1.44	
1600	1/4" thick		500	.032		.62	.52		1.14	1.57	
1800	Wood grained, plain or grooved, 1/4" thick, minimum		500	.032		.47	.52		.99	1.41	
1900	Maximum		425	.038		.89	.61		1.50	2.02	
2100	Moldings for hardboard, wood or aluminum, minimum		500	.032	L.F.	.32	.52		.84	1.24	
2200	Maximum		425	.038	"	.90	.61		1.51	2.03	

062 600 | Board Paneling

		CREW	DAILY OUTPUT	LABOR-HOURS	UNIT	MAT.	LABOR	EQUIP.	TOTAL	TOTAL INCL O&P	
604	0010	**PANELING, BOARDS**									**604**
6400	Wood board paneling, 3/4" thick, knotty pine	F-2	300	.053	S.F.	1.24	.86		2.10	2.84	
6500	Rough sawn cedar		300	.053		1.60	.86		2.46	3.24	
6700	Redwood, clear, 1" x 4" boards		300	.053		3.75	.86		4.61	5.60	
6900	Aromatic cedar, closet lining, boards		275	.058		2.83	.94		3.77	4.72	

062 700 | Misc. Finish Carpentry

		CREW	DAILY OUTPUT	LABOR-HOURS	UNIT	MAT.	LABOR	EQUIP.	TOTAL	TOTAL INCL O&P	
704	0010	**BEAMS, DECORATIVE** Rough sawn cedar, non-load bearing, 4" x 4"	2 Carp	180	.089	L.F.	2.19	1.64		3.83	5.25	**704**
0100	4" x 6"		170	.094		3.30	1.74		5.04	6.60	

062 700 | Misc. Finish Carpentry

			CREW	DAILY OUTPUT	LABOR-HOURS	UNIT	1998 BARE COSTS				TOTAL INCL O&P	
							MAT.	LABOR	EQUIP.	TOTAL		
704	0200	4″ x 8″	2 Carp	160	.100	L.F.	4.38	1.85		6.23	8	**704**
	0300	4″ x 10″		150	.107		5.45	1.97		7.42	9.40	
	0400	4″ x 12″		140	.114		6.60	2.11		8.71	10.85	
	0500	8″ x 8″		130	.123		8.75	2.28		11.03	13.55	
	0600	Plastic beam, ″hewn finish″, 6″ x 2″		240	.067		2.78	1.23		4.01	5.15	
	0601	6″ x 4″		220	.073		3.24	1.35		4.59	5.85	
	1100	Beam connector plates see div. 060-512										
720	0010	**FIREPLACE MANTEL BEAMS** Rough texture wood, 4″ x 8″	1 Carp	36	.222	L.F.	4.24	4.11		8.35	11.70	**720**
	0100	4″ x 10″		35	.229	″	5.30	4.23		9.53	13.10	
	0300	Laminated hardwood, 2-1/4″ x 10-1/2″ wide, 6′ long		5	1.600	Ea.	95.50	29.50		125	156	
	0400	8′ long		5	1.600	″	133	29.50		162.50	197	
	0600	Brackets for above, rough sawn		12	.667	Pr.	8.75	12.35		21.10	30.50	
	0700	Laminated		12	.667	″	13.25	12.35		25.60	35.50	
725	0010	**FIREPLACE MANTELS** 6″ molding, 6′ x 3′-6″ opening, minimum	1 Carp	5	1.600	Opng.	100	29.50		129.50	161	**725**
	0100	Maximum		5	1.600		150	29.50		179.50	216	
	0300	Prefabricated pine, colonial type, stock, deluxe		2	4		745	74		819	945	
	0400	Economy		3	2.667		250	49.50		299.50	360	
730	0010	**GRILLES** and panels, hardwood, sanded										**730**
	0020	2′ x 4′ to 4′ x 8′, custom designs, unfinished, minimum	1 Carp	38	.211	S.F.	12	3.89		15.89	19.85	
	0050	Average		30	.267		26	4.93		30.93	37	
	0100	Maximum		19	.421		40	7.80		47.80	57.50	
	0300	As above, but prefinished, minimum		38	.211		12	3.89		15.89	19.85	
	0400	Maximum		19	.421		45	7.80		52.80	63	
735	0010	**HARDWARE** Finish, see divisions 064-108 & 087										**735**
	0100	Rough, see division 050										
740	0010	**LOUVERS** Redwood, 2′-0″ diameter, full circle	1 Carp	16	.500	Ea.	105	9.25		114.25	132	**740**
	0100	Half circle		16	.500		98	9.25		107.25	124	
	0200	Octagonal		16	.500		84	9.25		93.25	108	
	0300	Triangular, 5/12 pitch, 5′-0″ at base		16	.500		88.50	9.25		97.75	113	
760	0010	**SHUTTERS, EXTERIOR** Aluminum, louvered, 1′-4″ wide, 3′-0″ long	1 Carp	10	.800	Pr.	29.50	14.80		44.30	58	**760**
	0200	4′-0″ long		10	.800		32.50	14.80		47.30	61.50	
	0300	5′-4″ long		10	.800		37.50	14.80		52.30	67	
	0400	6′-8″ long		9	.889		48	16.45		64.45	80.50	
	1000	Pine, louvered, primed, each 1′-2″ wide, 3′-3″ long		10	.800		41	14.80		55.80	70.50	
	1001	Pine, louvered, primed, each 1′-2″ wide, 3′-3″ long		20	.400	Ea.	24	7.40		31.40	39	
	1100	4′-7″ long		10	.800	Pr.	49	14.80		63.80	79.50	
	1101	4′-7″ long		20	.400	Ea.	34.50	7.40		41.90	50.50	
	1250	Each 1′-4″ wide, 3′-0″ long		10	.800	Pr.	37.50	14.80		52.30	67	
	1251	Each 1′-4″ wide, 3′-0″ long		20	.400	Ea.	25.50	7.40		32.90	40.50	
	1350	5′-3″ long		10	.800	Pr.	53	14.80		67.80	84	
	1351	5′-3″ long		20	.400	Ea.	31	7.40		38.40	46.50	
	1500	Each 1′-6″ wide, 3′-3″ long		10	.800	Pr.	45	14.80		59.80	75	
	1600	4′-7″ long		10	.800	″	61	14.80		75.80	92.50	
	1601	4′-7″ long		20	.400	Ea.	38	7.40		45.40	54.50	
	1620	Hemlock, louvered, 1′-2″ wide, 5′-7″ long		10	.800	Pr.	59.50	14.80		74.30	91	
	1630	Each 1′-4″ wide, 2′-2″ long		10	.800		37	14.80		51.80	66.50	
	1670	4′-3″ long		10	.800		49	14.80		63.80	79	
	1680	5′-3″ long		10	.800		55	14.80		69.80	86	
	1690	5′-11″ long		10	.800		60.50	14.80		75.30	92	
	1700	Door blinds, 6′-9″ long, each 1′-3″ wide		9	.889		70	16.45		86.45	105	
	1710	1′-6″ wide		9	.889		88	16.45		104.45	125	
	2500	Polystyrene, solid raised panel, each 1′-4″ wide, 3′-3″ long		10	.800		51	14.80		65.80	81.50	
	2600	3′-11″ long		10	.800		58.50	14.80		73.30	89.50	
	2700	4′-7″ long		10	.800		61.50	14.80		76.30	93	
	2800	5′-3″ long		10	.800		69	14.80		83.80	102	

Important: See the Reference Section for critical supporting data - Reference Nos., Crews, & Location Factors

062 700	Misc. Finish Carpentry	CREW	DAILY OUTPUT	LABOR-HOURS	UNIT	1998 BARE COSTS				TOTAL INCL O&P	
						MAT.	LABOR	EQUIP.	TOTAL		
760 2900	6'-8" long	1 Carp	9	.889	Pr.	95.50	16.45		111.95	133	**760**
3500	Polystyrene, solid raised panel, each 3'-3" wide, 3'-0" long		10	.800		73	14.80		87.80	106	
3600	3'-11" long		10	.800		82.50	14.80		97.30	116	
3700	4'-7" long		10	.800		92.50	14.80		107.30	128	
3800	5'-3" long		10	.800		102	14.80		116.80	138	
3900	6'-8" long		9	.889		136	16.45		152.45	178	
4500	Polystyrene, louvered, each 1'-2" wide, 3'-3" long		10	.800		39	14.80		53.80	68.50	
4600	4'-7" long		10	.800		49	14.80		63.80	79	
4750	5'-3" long		10	.800		52	14.80		66.80	82.50	
4850	6'-8" long		9	.889		85	16.45		101.45	122	
6000	Vinyl, louvered, each 1'-2" x 4'-7" long		10	.800		51	14.80		65.80	81.50	
6200	Each 1'-4" x 6'-8" long		9	.889		77.50	16.45		93.95	113	
8000	PVC exterior rolling shutters										
8100	including crank control	1 Carp	8	1	Ea.	340	18.50		358.50	405	
8500	Insulative - 6' x 6'8" stock unit	"	8	1	"	485	18.50		503.50	560	
775 0010	**SOFFITS** Wood fiber, no vapor barrier, 15/32" thick	F-2	525	.030	S.F.	.52	.49		1.01	1.41	**775**
0100	5/8" thick		525	.030		.58	.49		1.07	1.48	
0300	As above, 5/8" thick, with factory finish		525	.030		.74	.49		1.23	1.65	
0500	Hardboard, 3/8" thick, slotted		525	.030		.72	.49		1.21	1.63	
1000	Exterior AC plywood, 1/4" thick		420	.038		.52	.62		1.14	1.63	
1100	1/2" thick		420	.038		.71	.62		1.33	1.84	
778 0010	**DOORS AND FRAMES** See division 081 & 082										**778**

063 | Wood Treatment

063 100	Preservative Treatment	CREW	DAILY OUTPUT	LABOR-HOURS	UNIT	1998 BARE COSTS				TOTAL INCL O&P	
						MAT.	LABOR	EQUIP.	TOTAL		
102 0011	**LUMBER TREATMENT**										**102**
0400	Fire retardant, wet				M.B.F.	266			266	292	
0500	KDAT					250			250	275	
0700	Salt treated, water borne, .40 lb. retention					120			120	132	
0800	Oil borne, 8 lb. retention					140			140	154	
1000	Kiln dried lumber, 1" & 2" thick, softwoods					80			80	88	
1100	Hardwoods					85			85	93.50	
1500	For small size 1" stock, add					10			10	11	
1700	For full size rough lumber, add					20%					
104 0010	**PLYWOOD TREATMENT** Fire retardant, 1/4" thick				M.S.F.	200			200	220	**104**
0030	3/8" thick					220			220	242	
0050	1/2" thick					235			235	259	
0070	5/8" thick					250			250	275	
0100	3/4" thick					275			275	305	
0200	For KDAT, add					60			60	66	
0500	Salt treated water borne, .25 lb., wet, 1/4" thick					110			110	121	
0530	3/8" thick					115			115	127	
0550	1/2" thick					120			120	132	
0570	5/8" thick					130			130	143	
0600	3/4" thick					135			135	149	
0800	For KDAT add					60			60	66	

For expanded coverage of these items see *Means Building Construction Cost Data 1998*

063 | Wood Treatment

063 100	Preservative Treatment	CREW	DAILY OUTPUT	LABOR-HOURS	UNIT	1998 BARE COSTS				TOTAL INCL O&P
						MAT.	LABOR	EQUIP.	TOTAL	
104											104
0900	For .40 lb., per C.F. retention, add				M.S.F.	50			50	55
1000	For certification stamp, add				↓	30			30	33

064 | Architectural Woodwork

064 100	Custom Casework	CREW	DAILY OUTPUT	LABOR-HOURS	UNIT	1998 BARE COSTS				TOTAL INCL O&P
						MAT.	LABOR	EQUIP.	TOTAL	
102 0010	**CABINETS** Corner china cabinets, stock pine,										102
0020	80″ high, unfinished, minimum	2 Carp	6.60	2.424	Ea.	229	45		274	330
0100	Maximum	″	4.40	3.636	″	700	67.50		767.50	885
0700	Kitchen base cabinets, hardwood, not incl. counter tops,									
0710	24″ deep, 35″ high, prefinished									
0800	One top drawer, one door below, 12″ wide	2 Carp	24.80	.645	Ea.	111	11.95		122.95	143
0820	15″ wide		24	.667		112	12.35		124.35	144
0840	18″ wide		23.30	.687		127	12.70		139.70	162
0860	21″ wide		22.70	.705		135	13.05		148.05	172
0880	24″ wide		22.30	.717		138	13.25		151.25	174
1000	Four drawers, 12″ wide		24.80	.645		110	11.95		121.95	142
1020	15″ wide		24	.667		142	12.35		154.35	177
1040	18″ wide		23.30	.687		154	12.70		166.70	191
1060	24″ wide		22.30	.717		173	13.25		186.25	213
1200	Two top drawers, two doors below, 27″ wide		22	.727		172	13.45		185.45	212
1220	30″ wide		21.40	.748		188	13.85		201.85	231
1240	33″ wide		20.90	.766		189	14.15		203.15	233
1260	36″ wide		20.30	.788		198	14.60		212.60	243
1280	42″ wide		19.80	.808		216	14.95		230.95	264
1300	48″ wide		18.90	.847		235	15.65		250.65	286
1500	Range or sink base, two doors below, 30″ wide		21.40	.748		144	13.85		157.85	182
1520	33″ wide		20.90	.766		150	14.15		164.15	190
1540	36″ wide		20.30	.788		156	14.60		170.60	197
1560	42″ wide		19.80	.808		171	14.95		185.95	214
1580	48″ wide	↓	18.90	.847		185	15.65		200.65	231
1800	For sink front units, deduct					50			50	55
2000	Corner base cabinets, 36″ wide, standard	2 Carp	18	.889		159	16.45		175.45	202
2100	Lazy Susan with revolving door	″	16.50	.970	↓	197	17.95		214.95	248
4000	Kitchen wall cabinets, hardwood, 12″ deep with two doors									
4050	12″ high, 30″ wide	2 Carp	24.80	.645	Ea.	87	11.95		98.95	117
4100	36″ wide		24	.667		97.50	12.35		109.85	128
4400	15″ high, 30″ wide		24	.667		91	12.35		103.35	121
4420	33″ wide		23.30	.687		91	12.70		103.70	122
4440	36″ wide		22.70	.705		99	13.05		112.05	132
4450	42″ wide		22.70	.705		113	13.05		126.05	147
4700	24″ high, 30″ wide		23.30	.687		115	12.70		127.70	149
4720	36″ wide		22.70	.705		125	13.05		138.05	160
4740	42″ wide		22.30	.717		133	13.25		146.25	169
5000	30″ high, one door, 12″ wide		22	.727		82.50	13.45		95.95	114
5020	15″ wide		21.40	.748		80.50	13.85		94.35	112
5040	18″ wide		20.90	.766		92	14.15		106.15	126
5060	24″ wide		20.30	.788		107	14.60		121.60	143
5300	Custom		19.80	.808		117	14.95		131.95	155
5320	30″ wide	↓	19.30	.829	↓	131	15.35		146.35	171

Important: See the Reference Section for critical supporting data - Reference Nos., Crews, & Location Factors

064 100 | Custom Casework

		CREW	DAILY OUTPUT	LABOR-HOURS	UNIT	1998 BARE COSTS				TOTAL INCL O&P	
						MAT.	LABOR	EQUIP.	TOTAL		
102	5340	36" wide	2 Carp	18.80	.851	Ea.	145	15.75		160.75	186	**102**
	5360	42" wide		18.50	.865		158	16		174	201	
	5380	48" wide		18.40	.870		171	16.10		187.10	216	
	6000	Corner wall, 30" high, 24" wide		18	.889		122	16.45		138.45	162	
	6050	30" wide		17.20	.930		157	17.20		174.20	203	
	6100	36" wide		16.50	.970		182	17.95		199.95	231	
	6500	Revolving Lazy Susan		15.20	1.053		162	19.45		181.45	212	
	7000	Broom cabinet, 84" high, 24" deep, 18" wide		10	1.600		224	29.50		253.50	297	
	7500	Oven cabinets, 84" high, 24" deep, 27" wide		8	2		288	37		325	380	
	7750	Valance board trim		396	.040	L.F.	7.30	.75		8.05	9.35	
	9000	For deluxe models of all cabinets, add					40%					
	9500	For custom built in place, add					25%	10%				
	9550	Rule of thumb, kitchen cabinets not including										
	9560	appliances & counter top, minimum	2 Carp	30	.533	L.F.	75	9.85		84.85	99.50	
	9600	Maximum	"	25	.640	"	220	11.85		231.85	263	
104	0010	**CASEWORK, FRAMES**										**104**
	0050	Base cabinets, counter storage, 36" high, one bay										
	0100	18" wide	1 Carp	2.70	2.963	Ea.	89.50	55		144.50	192	
	0400	Two bay, 36" wide		2.20	3.636		137	67.50		204.50	265	
	1100	Three bay, 54" wide		1.50	5.333		163	98.50		261.50	350	
	2800	Book cases, one bay, 7' high, 18"-wide		2.40	3.333		105	61.50		166.50	222	
	3500	Two bay, 36" wide		1.60	5		152	92.50		244.50	325	
	4100	Three bay, 54" wide		1.20	6.667		252	123		375	490	
	6100	Wall mounted cabinet, one bay, 24" high, 18" wide		3.60	2.222		58	41		99	134	
	6800	Two bay, 36" wide		2.20	3.636		84	67.50		151.50	208	
	7400	Three bay, 54" wide		1.70	4.706		105	87		192	265	
	8400	30" high, one bay, 18" wide		3.60	2.222		63	41		104	140	
	9000	Two bay, 36" wide		2.15	3.721		84	69		153	211	
	9400	Three bay, 54" wide		1.60	5		105	92.50		197.50	275	
	9800	Wardrobe, 7' high, single, 24" wide		2.70	2.963		116	55		171	221	
	9950	Partition, adjustable shelves & drawers, 48" wide		1.40	5.714		221	106		327	425	
106	0010	**CABINET DOORS**										**106**
	2000	Glass panel, hardwood frame										
	2200	12" wide, 18" high	1 Carp	34	.235	Ea.	14	4.35		18.35	23	
	2400	24" high		33	.242		18	4.48		22.48	27.50	
	2600	30" high		32	.250		23	4.63		27.63	33.50	
	2800	36" high		30	.267		28	4.93		32.93	39.50	
	3000	48" high		23	.348		37	6.45		43.45	51.50	
	3200	60" high		17	.471		46	8.70		54.70	65.50	
	3400	72" high		15	.533		56	9.85		65.85	78.50	
	3600	15" wide x 18" high		33	.242		14	4.48		18.48	23	
	3800	24" high		32	.250		18	4.63		22.63	28	
	4000	30" high		30	.267		23	4.93		27.93	34	
	4250	36" high		28	.286		28	5.30		33.30	40	
	4300	48" high		22	.364		37	6.75		43.75	52	
	4350	60" high		16	.500		46	9.25		55.25	66.50	
	4400	72" high		14	.571		56	10.55		66.55	79.50	
	4450	18" wide, 18" high		32	.250		14	4.63		18.63	23.50	
	4500	24" high		30	.267		18	4.93		22.93	28.50	
	4550	30" high		29	.276		23	5.10		28.10	34.50	
	4600	36" high		27	.296		28	5.50		33.50	40.50	
	4650	48" high		21	.381		37	7.05		44.05	52.50	
	4700	60" high		15	.533		46	9.85		55.85	67.50	
	4750	72" high		13	.615		56	11.40		67.40	81	
	5000	Hardwood, raised panel										

WOOD & PLASTICS — 6

064 100	Custom Casework	CREW	DAILY OUTPUT	LABOR-HOURS	UNIT	1998 BARE COSTS				TOTAL INCL O&P
						MAT.	LABOR	EQUIP.	TOTAL	
5100	12" wide, 18" high	1 Carp	16	.500	Ea.	19	9.25		28.25	37
5150	24" high		15.50	.516		25	9.55		34.55	44
5200	30" high		15	.533		31	9.85		40.85	51
5250	36" high		14	.571		37.50	10.55		48.05	59.50
5300	48" high		11	.727		50	13.45		63.45	78
5320	60" high		8	1		62.50	18.50		81	100
5340	72" high		7	1.143		75	21		96	119
5360	15" wide x 18" high		15.50	.516		25	9.55		34.55	44
5380	24" high		15	.533		31	9.85		40.85	51
5400	30" high		14.50	.552		40.50	10.20		50.70	62
5420	36" high		13.50	.593		47	10.95		57.95	70.50
5440	48" high		10.50	.762		62	14.10		76.10	92
5460	60" high		7.50	1.067		78	19.75		97.75	120
5480	72" high		6.50	1.231		93	23		116	141
5500	18" wide, 18" high		15	.533		28	9.85		37.85	48
5550	24" high		14.50	.552		37	10.20		47.20	58
5600	30" high		14	.571		47	10.55		57.55	69.50
5650	36" high		13	.615		56	11.40		67.40	81
5700	48" high		10	.800		75	14.80		89.80	108
5750	60" high		7	1.143		93	21		114	138
5800	72" high		6	1.333		112	24.50		136.50	166
6000	Plastic laminate on particle board									
6100	12" wide, 18" high	1 Carp	25	.320	Ea.	12	5.90		17.90	23.50
6120	24" high		24	.333		17	6.15		23.15	29.50
6140	30" high		23	.348		21	6.45		27.45	34
6160	36" high		21	.381		26	7.05		33.05	40.50
6200	48" high		16	.500		34	9.25		43.25	53.50
6250	60" high		13	.615		43	11.40		54.40	67
6300	72" high		12	.667		52	12.35		64.35	78
6320	15" wide x 18" high		24.50	.327		16	6.05		22.05	28
6340	24" high		23.50	.340		22	6.30		28.30	35
6360	30" high		22.50	.356		27	6.60		33.60	41
6380	36" high		20.50	.390		32	7.20		39.20	47.50
6400	48" high		15.50	.516		43	9.55		52.55	64
6450	60" high		12.50	.640		54	11.85		65.85	80
6480	72" high		11.50	.696		65	12.85		77.85	93.50
6500	18" wide, 18" high		24	.333		19	6.15		25.15	31.50
6550	24" high		23	.348		26	6.45		32.45	39.50
6600	30" high		22	.364		32	6.75		38.75	46.50
6650	36" high		20	.400		39	7.40		46.40	55.50
6700	48" high		15	.533		52	9.85		61.85	74
6750	60" high		12	.667		65	12.35		77.35	92.50
6800	72" high		11	.727		77	13.45		90.45	108
7000	Plywood, with edge band									
7010	12" wide, 18" high	1 Carp	27	.296	Ea.	16.50	5.50		22	27.50
7100	24" high		26	.308		22	5.70		27.70	34
7120	30" high		25	.320		27.50	5.90		33.40	40.50
7140	36" high		23	.348		33	6.45		39.45	47.50
7180	48" high		18	.444		44	8.20		52.20	62.50
7200	60" high		15	.533		55	9.85		64.85	77.50
7250	72" high		14	.571		66	10.55		76.55	90.50
7300	15" wide x 18" high		26.50	.302		21	5.60		26.60	32.50
7350	24" high		25.50	.314		27.50	5.80		33.30	40.50
7400	30" high		24.50	.327		34	6.05		40.05	48
7450	36" high		22.50	.356		41	6.60		47.60	56.50
7500	48" high		17.50	.457		55	8.45		63.45	75

Important: See the Reference Section for critical supporting data - Reference Nos., Crews, & Location Factors

064 100	Custom Casework	CREW	DAILY OUTPUT	LABOR-HOURS	UNIT	1998 BARE COSTS				TOTAL INCL O&P	
						MAT.	LABOR	EQUIP.	TOTAL		
106 7550	60" high	1 Carp	14.50	.552	Ea.	69	10.20		79.20	93.50	**106**
7600	72" high		13.50	.593		82.50	10.95		93.45	110	
7650	18" wide, 18" high		26	.308		25	5.70		30.70	37.50	
7700	24" high		25	.320		33	5.90		38.90	46.50	
7750	30" high		24	.333		41	6.15		47.15	55.50	
108 0010	**CABINET HARDWARE**										**108**
1000	Catches, minimum	1 Carp	235	.034	Ea.	.70	.63		1.33	1.85	
1020	Average		119.40	.067		2.10	1.24		3.34	4.43	
1040	Maximum		80	.100		4	1.85		5.85	7.55	
2000	Door/drawer pulls, handles										
2200	Handles and pulls, projecting, metal, minimum	1 Carp	160	.050	Ea.	1.50	.93		2.43	3.24	
2220	Average		95.24	.084		3	1.55		4.55	5.95	
2240	Maximum		68	.118		7	2.18		9.18	11.45	
2300	Wood, minimum		160	.050		1.50	.93		2.43	3.24	
2320	Average		95.24	.084		2	1.55		3.55	4.86	
2340	Maximum		68	.118		4	2.18		6.18	8.15	
2600	Flush, metal, minimum		160	.050		1.40	.93		2.33	3.13	
2620	Average		95.24	.084		3	1.55		4.55	5.95	
2640	Maximum		68	.118		10	2.18		12.18	14.75	
3000	Drawer tracks/glides, minimum		48	.167	Pr.	5.50	3.08		8.58	11.35	
3020	Average		32	.250		9.30	4.63		13.93	18.20	
3040	Maximum		24	.333		16	6.15		22.15	28	
4000	Cabinet hinges, minimum		160	.050		1.40	.93		2.33	3.13	
4020	Average		95.24	.084		2.45	1.55		4	5.35	
4040	Maximum		68	.118		6	2.18		8.18	10.35	
110 0010	**DRAWERS**										**110**
0100	Solid hardwood front										
1000	4" high, 12" wide	1 Carp	17	.471	Ea.	18.55	8.70		27.25	35.50	
1200	18" wide		16	.500		23.50	9.25		32.75	42	
1400	24" wide		15	.533		31	9.85		40.85	51	
1600	6" high, 12" wide		16	.500		23.50	9.25		32.75	42	
1800	18" wide		15	.533		31	9.85		40.85	51	
2000	24" wide		14	.571		38	10.55		48.55	60	
2200	9" high, 12" wide		15	.533		31	9.85		40.85	51	
2400	18" wide		14	.571		38	10.55		48.55	60	
2600	24" wide		13	.615		46.50	11.40		57.90	70.50	
2800	Plastic laminate on particle board front										
3000	4" high, 12" wide	1 Carp	17	.471	Ea.	18.55	8.70		27.25	35.50	
3200	18" wide		16	.500		21.50	9.25		30.75	40	
3600	24" wide		15	.533		26	9.85		35.85	45.50	
3800	6" high, 12" wide		16	.500		21.50	9.25		30.75	40	
4000	18" wide		15	.533		27	9.85		36.85	46.50	
4500	24" wide		14	.571		32	10.55		42.55	53	
4800	9" high, 12" wide		15	.533		23.50	9.85		33.35	43	
5000	18" wide		14	.571		26	10.55		36.55	46.50	
5200	24" wide		13	.615		41	11.40		52.40	65	
5400	Plywood, flush panel front										
6000	4" high, 12" wide	1 Carp	17	.471	Ea.	19.55	8.70		28.25	36.50	
6200	18" wide	"	16	.500	"	23.50	9.25		32.75	42	
140 0010	**VANITIES**										**140**
8000	Vanity bases, 2 doors, 30" high, 21" deep, 24" wide	2 Carp	20	.800	Ea.	119	14.80		133.80	157	
8050	30" wide		16	1		125	18.50		143.50	170	
8100	36" wide		13.33	1.200		140	22		162	192	
8150	48" wide		11.43	1.400		177	26		203	240	
9000	For deluxe models of all vanities, add to above					40%					

064 100 | Custom Casework

		CREW	DAILY OUTPUT	LABOR-HOURS	UNIT	1998 BARE COSTS				TOTAL INCL O&P	
						MAT.	LABOR	EQUIP.	TOTAL		
140	9500	For custom built in place, add to above				Ea.	25%	10%				**140**

064 300 | Stairwork & Handrails

		CREW	DAILY OUTPUT	LABOR-HOURS	UNIT	MAT.	LABOR	EQUIP.	TOTAL	TOTAL INCL O&P	
306	0011	**STAIRS, PREFABRICATED**									**306**
	0100	Box stairs, prefabricated, 3'-0" wide									
	0110	Oak treads, no handrails, 2' high	2 Carp	5	3.200	Flight	149	59		208	265
	0200	4' high		4	4		320	74		394	475
	0300	6' high		3.50	4.571		520	84.50		604.50	715
	0400	8' high		3	5.333		650	98.50		748.50	885
	0600	With pine treads for carpet, 2' high		5	3.200		118	59		177	231
	0700	4' high		4	4		201	74		275	350
	0800	6' high		3.50	4.571		305	84.50		389.50	480
	0900	8' high		3	5.333		380	98.50		478.50	590
	1100	For 4' wide stairs, add					25%				
	1500	Prefabricated stair rail with balusters, 5 risers	2 Carp	15	1.067	Ea.	175	19.75		194.75	227
	1700	Basement stairs, prefabricated, soft wood,									
	1710	open risers, 3' wide, 8' high	2 Carp	4	4	Flight	375	74		449	540
	1900	Open stairs, prefabricated prefinished poplar, metal stringers,									
	1910	treads 3'-6" wide, no railings									
	2000	3' high	2 Carp	5	3.200	Flight	435	59		494	575
	2100	4' high		4	4		545	74		619	725
	2200	6' high		3.50	4.571		950	84.50		1,034.50	1,200
	2300	8' high		3	5.333		1,525	98.50		1,623.50	1,850
	2500	For prefab. 3 piece wood railings & balusters, add for									
	2600	3' high stairs	2 Carp	15	1.067	Ea.	149	19.75		168.75	198
	2700	4' high stairs		14	1.143		191	21		212	246
	2800	6' high stairs		13	1.231		294	23		317	365
	2900	8' high stairs		12	1.333		365	24.50		389.50	445
	3100	For 3'-6" x 3'-6" platform, add		4	4		170	74		244	315
	3300	Curved stairways, 3'-3" wide, prefabricated, oak, unfinished,									
	3310	incl. curved balustrade system, open one side									
	3400	9' high	2 Carp	.70	22.857	Flight	6,375	425		6,800	7,750
	3500	10' high		.70	22.857		7,200	425		7,625	8,650
	3700	Open two sides, 9' high		.50	32		10,000	590		10,590	12,000
	3800	10' high		.50	32		10,800	590		11,390	12,900
	4000	Residential, wood, oak treads, prefabricated		1.50	10.667		1,150	197		1,347	1,625
	4200	Built in place		.44	36.364		1,325	675		2,000	2,600
	4400	Spiral, oak, 4'-6" diameter, unfinished, prefabricated,									
	4500	incl. railing, 9' high	2 Carp	1.50	10.667	Flight	4,000	197		4,197	4,750
308	0010	**STAIR PARTS** Balusters, turned, 30" high, pine, minimum R064-100	1 Carp	28	.286	Ea.	3.50	5.30		8.80	12.90	**308**
	0100	Maximum		26	.308		9	5.70		14.70	19.65
	0300	30" high birch balusters, minimum		28	.286		6	5.30		11.30	15.65
	0400	Maximum		26	.308		10	5.70		15.70	21
	0600	42" high, pine balusters, minimum		27	.296		6	5.50		11.50	16
	0700	Maximum		25	.320		9.50	5.90		15.40	20.50
	0900	42" high birch balusters, minimum		27	.296		7	5.50		12.50	17.10
	1000	Maximum		25	.320		15	5.90		20.90	26.50
	1050	Baluster, stock pine, 1-1/16" x 1-1/16"		240	.033	L.F.	1.97	.62		2.59	3.23
	1100	1-5/8" x 1-5/8"		220	.036	"	1.82	.67		2.49	3.15
	1200	Newels, 3-1/4" wide, starting, minimum		7	1.143	Ea.	40	21		61	80
	1300	Maximum		6	1.333		120	24.50		144.50	175
	1500	Landing, minimum		5	1.600		70	29.50		99.50	128
	1600	Maximum		4	2		170	37		207	251
	1800	Railings, oak, built-up, minimum		60	.133	L.F.	5.30	2.47		7.77	10.10
	1900	Maximum		55	.145		15	2.69		17.69	21

Important: See the Reference Section for critical supporting data - Reference Nos., Crews, & Location Factors

064 300 | Stairwork & Handrails

			DAILY OUTPUT	LABOR-HOURS	UNIT	1998 BARE COSTS				TOTAL INCL O&P	
						MAT.	LABOR	EQUIP.	TOTAL		
308	2100	Add for sub rail	1 Carp	110	.073	L.F.	4	1.35		5.35	6.70	308
	2300	Risers, beech, 3/4" x 7-1/2" high		64	.125		5.20	2.31		7.51	9.70	
	2400	Fir, 3/4" x 7-1/2" high		64	.125		1.50	2.31		3.81	5.60	
	2600	Oak, 3/4" x 7-1/2" high		64	.125		4.75	2.31		7.06	9.15	
	2800	Pine, 3/4" x 7-1/2" high		66	.121		1.50	2.24		3.74	5.50	
	2850	Skirt board, pine, 1" x 10"		55	.145		1.65	2.69		4.34	6.45	
	2900	1" x 12"		52	.154		2	2.85		4.85	7.10	
	3000	Treads, 1-1/16" x 9-1/2" wide, 3' long, oak		18	.444	Ea.	22	8.20		30.20	38	
	3100	4' long, oak		17	.471		29	8.70		37.70	47	
	3300	1-1/16" x 11-1/2" wide, 3' long, oak		18	.444		23.50	8.20		31.70	39.50	
	3400	6' long, oak		14	.571		56	10.55		66.55	79.50	
	3600	Beech treads, add					40%					
	3800	For mitered return nosings, add				L.F.	8.40			8.40	9.25	
310	0010	**RAILING** Custom design, architectural grade, hardwood, minimum	1 Carp	38	.211	"	12	3.89		15.89	19.85	310
	0100	Maximum		30	.267	L.F.	44	4.93		48.93	57	
	0300	Stock interior railing with spindles 6" O.C., 4' long		40	.200		28	3.70		31.70	37.50	
	0400	8' long		48	.167		26	3.08		29.08	34	

064 400 | Misc. Ornamental Items

			DAILY OUTPUT	LABOR-HOURS	UNIT	MAT.	LABOR	EQUIP.	TOTAL	TOTAL INCL O&P	
402	0011	**COLUMNS**										402
	0050	Aluminum, round colonial, 6" diameter	2 Carp	80	.200	V.L.F.	14	3.70		17.70	22	
	0100	8" diameter		62.25	.257		17	4.76		21.76	27	
	0200	10" diameter		55	.291		22	5.40		27.40	33	
	0250	Fir, stock units, hollow round, 6" diameter		80	.200		13	3.70		16.70	20.50	
	0300	8" diameter		80	.200		14	3.70		17.70	22	
	0350	10" diameter		70	.229		18	4.23		22.23	27	
	0400	Solid turned, to 8' high, 3-1/2" diameter		80	.200		7	3.70		10.70	14.05	
	0500	4-1/2" diameter		75	.213		10	3.95		13.95	17.75	
	0600	5-1/2" diameter		70	.229		13	4.23		17.23	21.50	
	0800	Square columns, built-up, 5" x 5"		65	.246		11	4.55		15.55	19.90	
	0900	Solid, 3-1/2" x 3-1/2"		130	.123		6	2.28		8.28	10.50	
	1600	Hemlock, tapered, T & G, 12" diam, 10' high		100	.160		30	2.96		32.96	38	
	1700	16' high		65	.246		46	4.55		50.55	58.50	
	1900	10' high, 14" diameter		100	.160		63	2.96		65.96	74.50	
	2000	18' high		65	.246		60	4.55		64.55	74	
	2200	18" diameter, 12' high		65	.246		76	4.55		80.55	91.50	
	2300	20' high		50	.320		83	5.90		88.90	102	
	2500	20" diameter, 14' high		40	.400		95	7.40		102.40	118	
	2600	20' high		35	.457		110	8.45		118.45	136	
	2800	For flat pilasters, deduct					33%					
	3000	For splitting into halves, add				Ea.	60			60	66	
	4000	Rough sawn cedar posts, 4" x 4"	2 Carp	250	.064	V.L.F.	2.43	1.18		3.61	4.70	
	4100	4" x 6"		235	.068		3.62	1.26		4.88	6.15	
	4200	6" x 6"		220	.073		5.45	1.35		6.80	8.30	
	4300	8" x 8"		200	.080		5.60	1.48		7.08	8.70	
410	0010	**MILLWORK, HIGH DENSITY POLYMER**										410
	0100	Base, 9/16" x 3-3/16"	1 Carp	230	.035	L.F.	1.20	.64		1.84	2.42	
	0200	Casing, fluted, 5/8" x 3-1/4"		215	.037		1.20	.69		1.89	2.50	
	0300	Chair rail, 9/16" x 2-1/4"		260	.031		.61	.57		1.18	1.65	
	0400	5/8" x 3-1/8"		230	.035		1.15	.64		1.79	2.37	
	0500	Corner, inside, 1/2" x 1-1/8"		220	.036		.60	.67		1.27	1.81	
	0600	Cove, 13/16" x 3-3/4"		260	.031		1.20	.57		1.77	2.30	
	0700	Crown, 3/4" x 3-13/16"		260	.031		1.20	.57		1.77	2.30	
	0800	Half round, 15/16" x 2"		240	.033		.66	.62		1.28	1.79	

WOOD & PLASTICS 6

066 500 | Solid Surfacing Material

		CREW	DAILY OUTPUT	LABOR-HOURS	UNIT	1998 BARE COSTS				TOTAL INCL O&P
						MAT.	LABOR	EQUIP.	TOTAL	
503	**0010**	**COUNTERTOPS**								
	0100	24" wide, no backsplash, solid colors	F-2	28	.571	L.F.	41	9.25		50.25	61
	0200	Standard mineral colors		28	.571		61	9.25		70.25	83
	0300	Premium mineral colors		28	.571		79	9.25		88.25	103
	0400	4" backsplash, solid colors		22	.727		45	11.75		56.75	69.50
	0500	Standard mineral colors		22	.727		66	11.75		77.75	92.50
	0600	Premium mineral colors		22	.727		84	11.75		95.75	113
	0700	For kitchen sink cutout, add	1 Carp	4	2	Ea.		37		37	63.50
	0800	For cooktop cutout, add		4	2			37		37	63.50
	0900	Kitchen sink, drop in, single bowl, 22" x 25", solid colors		5.60	1.429		227	26.50		253.50	296
	1000	Mineral colors		5.60	1.429		320	26.50		346.50	395
	1100	Double bowl, 22" x 33", solid colors		4.80	1.667		262	31		293	340
	1200	Mineral colors		4.80	1.667		365	31		396	455
	1300	Triple bowl, 22" x 44", solid colors		4.80	1.667		360	31		391	450
	1500	Kitchen or bath sink, under counter mounted, minimum		2.80	2.857		180	53		233	289
	1600	Maximum		2.40	3.333		495	61.50		556.50	650
	1700	For warranty, minimum installation charge	F-2	1	16	Job		259		259	445
505	**0010**	**VANITY TOPS**								
	0015	Solid surface, center bowl, 17" x 19"	1 Carp	12	.667	Ea.	163	12.35		175.35	201
	0020	19" x 25"		12	.667		197	12.35		209.35	238
	0030	19" x 31"		12	.667		240	12.35		252.35	285
	0040	19" x 37"		12	.667		279	12.35		291.35	325
	0050	22" x 25"		10	.800		222	14.80		236.80	270
	0060	22" x 31"		10	.800		259	14.80		273.80	310
	0070	22" x 37"		10	.800		300	14.80		314.80	355
	0080	22" x 43"		10	.800		345	14.80		359.80	405
	0090	22" x 49"		10	.800		380	14.80		394.80	445
	0110	22" x 55"		8	1		430	18.50		448.50	505
	0120	22" x 61"		8	1		495	18.50		513.50	570
	0130	22" x 68"		8	1		635	18.50		653.50	725
	0140	22" x 73"		8	1		715	18.50		733.50	815
	0150	22" x 85"		8	1		825	18.50		843.50	935
	0160	Offset bowl, left or right, 22" x 49"		10	.800		455	14.80		469.80	525
	0170	22" x 55"		8	1		520	18.50		538.50	600
	0180	22" x 61"		8	1		590	18.50		608.50	675
	0190	22" x 68"		8	1		690	18.50		708.50	790
	0200	22" x 73"		8	1		850	18.50		868.50	965
	0210	22" x 85"		8	1		995	18.50		1,013.50	1,125
	0220	Double bowl, 22" x 61"		8	1		540	18.50		558.50	625
	0230	Corner top/bowl, 22" x 49"		8	1		420	18.50		438.50	495
	0240	For aggregate colors, add					35%				
	0250	For faucets and fittings see 151-141									

For information about Means Estimating Seminars, see yellow pages 11 and 12 in back of book

Important: See the Reference Section for critical supporting data - Reference Nos., Crews, & Location Factors

Division 7
Thermal & Moisture Protection

Estimating Tips

071 Waterproofing & Dampproofing
- Be sure of the job specifications before pricing this subdivision. The difference in cost between waterproofing and dampproofing can be great. Waterproofing will hold back standing water. Dampproofing prevents the transmission of water vapor. Also included in this section are vapor retarding membranes.

072 Insulation & Fireproofing
- Insulation and fireproofing products are measured by area, thickness, volume or R value. Specifications may only give what the specific R value should be in a certain situation. The estimator may need to choose the type of insulation to meet that R value.

073 Shingles & Roofing Tiles
074 Preformed Roofing & Siding
- Many roofing and siding products are bought and sold by the square. One square is equal to an area that measures 100 square feet.

This simple change in unit of measure could create a large error if the estimator is not observant. Accessories and fasteners necessary for a complete installation must be figured into any calculations for both material and labor.

075 Membrane Roofing
076 Flashing & Sheet Metal
077 Roof Specialties & Accessories
078 Skylights
- The items in these subdivisions compose a roofing system. No one component completes the installation and all must be estimated. Built-up or single ply membrane roofing systems are made up of many products and installation trades. Wood blocking at roof perimeters or penetrations, parapet coverings, reglets, roof drains, gutters, downspouts, sheet metal flashing, skylights, smoke vents or roof hatches all need to be considered along with the roofing material. Several different installation trades will need to work together on the roofing system. Inherent difficulties in the scheduling and coordination of various trades must be accounted for when estimating labor costs.

079 Joint Sealers
- To complete the weather-tight shell the sealants and caulkings must be estimated. Where different materials meet—at expansion joints, at flashing penetrations, and at hundreds of other locations throughout a construction project—they provide another line of defense against water penetration. Often, an entire system is based on the proper location and placement of caulking or sealants. The detail drawings that are included as part of a set of architectural plans, show typical locations for these materials. When caulking or sealants are shown at typical locations, this means the estimator must include them for all the locations where this detail is applicable. Be careful to keep different types of sealants separate, and remember to consider backer rods and primers if necessary.

Reference Numbers
Reference numbers are shown in bold squares at the beginning of some major classifications. These numbers refer to related items in the Reference Section. The reference information may be an estimating procedure, an alternate pricing method or technical information.

Note: Not all subdivisions listed here necessarily appear in this publication.

071 600 | Bitum. Dampproofing

			DAILY OUTPUT	LABOR-HOURS	UNIT	1998 BARE COSTS				TOTAL INCL O&P	
						MAT.	LABOR	EQUIP.	TOTAL		
602	0010	BITUMINOUS ASPHALT COATING For foundation									602
	0030	Brushed on, below grade, 1 coat	1 Rofc	665	.012	S.F.	.06	.19		.25	.43	
	0100	2 coat		500	.016		.10	.26		.36	.59	
	0300	Sprayed on, below grade, 1 coat, 25.6 S.F./gal.		830	.010		.08	.16		.24	.37	
	0400	2 coat, 20.5 S.F./gal.		500	.016		.14	.26		.40	.64	
	0600	Troweled on, asphalt with fibers, 1/16″ thick		500	.016		.15	.26		.41	.65	
	0700	1/8″ thick		400	.020		.26	.32		.58	.89	
	1000	1/2″ thick		350	.023		.85	.37		1.22	1.63	

071 750 | Water Repellent Coat

			DAILY OUTPUT	LABOR-HOURS	UNIT	MAT.	LABOR	EQUIP.	TOTAL	INCL O&P	
754	0010	RUBBER COATING Water base liquid, roller applied	2 Rofc	7,000	.002	S.F.	.55	.04		.59	.68	754
	0200	Silicone or stearate, sprayed on CMU, 1 coat	1 Rofc	4,000	.002		.24	.03		.27	.33	
	0300	2 coats	″	3,000	.003		.48	.04		.52	.61	

071 800 | Cementitious Dampprfng

				DAILY OUTPUT	LABOR-HOURS	UNIT	MAT.	LABOR	EQUIP.	TOTAL	INCL O&P	
802	0010	CEMENT PARGING 2 coats, 1/2″ thick, regular P.C.	R071 -010	D-1	250	.064	S.F.	.15	1.08		1.23	1.98	802
	0100	Waterproofed Portland cement		″	250	.064	″	.16	1.08		1.24	2	

071 920 | Vapor Retarders

			DAILY OUTPUT	LABOR-HOURS	UNIT	MAT.	LABOR	EQUIP.	TOTAL	INCL O&P	
922	0010	BUILDING PAPER Aluminum and kraft laminated, foil 1 side	1 Carp	37	.216	Sq.	3.55	4		7.55	10.75	922
	0100	Foil 2 sides		37	.216		5.70	4		9.70	13.10	
	0300	Asphalt, two ply, 30#, for subfloors		19	.421		11.50	7.80		19.30	26	
	0400	Asphalt felt sheathing paper, 15#		37	.216		2.78	4		6.78	9.90	
	0450	Housewrap, exterior, spun bonded polypropylene										
	0470	Small roll	1 Carp	3,800	.002	S.F.	.10	.04		.14	.18	
	0480	Large roll	″	4,000	.002	″	.09	.04		.13	.16	
	0500	Material only, 3′ x 111.1′ roll				Ea.	33			33	36	
	0520	9′ x 111.1′ roll				″	93			93	102	
	0600	Polyethylene vapor barrier, standard, .002″ thick	1 Carp	37	.216	Sq.	1.02	4		5.02	7.95	
	0700	.004″ thick		37	.216		2.33	4		6.33	9.40	
	0900	.006″ thick		37	.216		2.90	4		6.90	10.05	
	1200	.010″ thick		37	.216		5.35	4		9.35	12.70	
	1500	Red rosin paper, 5 sq rolls, 4 lb per square		37	.216		1.55	4		5.55	8.55	
	1600	5 lbs. per square		37	.216		2	4		6	9.05	
	1800	Reinf. waterproof, .002″ polyethylene backing, 1 side		37	.216		4.92	4		8.92	12.25	
	1900	2 sides		37	.216		6.50	4		10.50	14	
	3000	Building wrap, spunbonded polyethylene	2 Carp	8,000	.002	S.F.	.10	.04		.14	.17	

072 100 | Building Insulation

			DAILY OUTPUT	LABOR-HOURS	UNIT	1998 BARE COSTS				TOTAL INCL O&P	
						MAT.	LABOR	EQUIP.	TOTAL		
101	0010	BLOWN-IN INSULATION Ceilings, with open access										101
	0020	Cellulose, 3-1/2″ thick, R13	G-4	5,000	.005	S.F.	.13	.07	.05	.25	.32	
	0030	5-3/16″ thick, R19		3,800	.006		.19	.09	.07	.35	.44	
	0050	6-1/2″ thick, R22		3,000	.008		.23	.11	.09	.43	.54	
	1000	Fiberglass, 5″ thick, R11		3,800	.006		.15	.09	.07	.31	.40	
	1050	6″ thick, R13		3,000	.008		.15	.11	.09	.35	.45	
	1100	8-1/2″ thick, R19		2,200	.011		.21	.15	.12	.48	.62	
	1300	12″ thick, R26		1,500	.016		.29	.23	.17	.69	.90	

Important: See the Reference Section for critical supporting data - Reference Nos., Crews, & Location Factors

072 100 | Building Insulation

		CREW	DAILY OUTPUT	LABOR-HOURS	UNIT	1998 BARE COSTS MAT.	LABOR	EQUIP.	TOTAL	TOTAL INCL O&P	
101	2000	Mineral wool, 4" thick, R12	G-4	3,500	.007	S.F.	.16	.10	.07	.33	.43	**101**
	2050	6" thick, R17		2,500	.010		.18	.14	.10	.42	.55	
	2100	9" thick, R23		1,750	.014		.27	.19	.15	.61	.79	
	2500	Wall installation, incl. drilling & patching from outside, two 1"										
	2510	diam. holes @ 16" O.C., top & mid-point of wall, add to above										
	2700	For masonry	G-4	415	.058	S.F.	.06	.81	.63	1.50	2.16	
	2800	For wood siding		840	.029		.06	.40	.31	.77	1.10	
	2900	For stucco/plaster		665	.036		.06	.51	.39	.96	1.37	
106	0010	**FLOOR INSULATION, NONRIGID** Including										**106**
	0020	spring type wire fasteners										
	2000	Fiberglass, blankets or batts, paper or foil backing										
	2100	1 side, 3-1/2" thick, R11	1 Carp	700	.011	S.F.	.25	.21		.46	.64	
	2150	6" thick, R19		600	.013		.33	.25		.58	.78	
	2200	8-1/2" thick, R30		550	.015		.60	.27		.87	1.12	
108	0010	**MASONRY INSULATION** Vermiculite or perlite, poured										**108**
	0100	In cores of concrete block, 4" thick wall, .115 CF/SF	D-1	4,800	.003	S.F.	.16	.06		.22	.27	
	0700	Foamed in place, urethane in 2-5/8" cavity	G-2	1,035	.023		.36	.35	.22	.93	1.23	
	0800	For each 1" added thickness, add	"	2,372	.010		.10	.15	.10	.35	.48	
109	0600	**PERIMETER INSULATION,** polystyrene, expanded, 1" thick, R4	1 Carp	680	.012	S.F.	.20	.22		.42	.59	**109**
	0700	2" thick, R8		675	.012	"	.32	.22		.54	.73	
110	0010	**POURED INSULATION** Cellulose fiber, R3.8 per inch		200	.040	C.F.	.40	.74		1.14	1.71	**110**
	0080	Fiberglass wool, R4 per inch		200	.040		.28	.74		1.02	1.58	
	0100	Mineral wool, R3 per inch		200	.040		.30	.74		1.04	1.59	
	0300	Polystyrene, R4 per inch		200	.040		1.72	.74		2.46	3.16	
	0400	Vermiculite or perlite, R2.7 per inch		200	.040		1.42	.74		2.16	2.83	
111	0010	**REFLECTIVE INSULATION,** aluminum foil on reinforced scrim	1 Carp	19	.421	C.S.F.	13.90	7.80		21.70	28.50	**111**
	0100	Reinforced with woven polyolefin		19	.421		16.90	7.80		24.70	32	
	0500	With single bubble air space, R8.8		15	.533		26	9.85		35.85	45.50	
	0600	With double bubble air space, R9.8		15	.533		28	9.85		37.85	48	
116	0010	**WALL INSULATION, RIGID**										**116**
	0040	Fiberglass, 1.5#/CF, unfaced, 1" thick, R4.1	1 Carp	1,000	.008	S.F.	.23	.15		.38	.50	
	0060	1-1/2" thick, R6.2		1,000	.008		.30	.15		.45	.58	
	0080	2" thick, R8.3		1,000	.008		.37	.15		.52	.65	
	0120	3" thick, R12.4		800	.010		.43	.19		.62	.80	
	0370	3#/CF, unfaced, 1" thick, R4.3		1,000	.008		.32	.15		.47	.60	
	0390	1-1/2" thick, R6.5		1,000	.008		.56	.15		.71	.87	
	0400	2" thick, R8.7		890	.009		.76	.17		.93	1.12	
	0420	2-1/2" thick, R10.9		800	.010		.86	.19		1.05	1.27	
	0440	3" thick, R13		800	.010		1.02	.19		1.21	1.44	
	0520	Foil faced, 1" thick, R4.3		1,000	.008		.68	.15		.83	1	
	0540	1-1/2" thick, R6.5		1,000	.008		.91	.15		1.06	1.25	
	0560	2" thick, R8.7		890	.009		1.13	.17		1.30	1.52	
	0580	2-1/2" thick, R10.9		800	.010		1.34	.19		1.53	1.79	
	0600	3" thick, R13		800	.010		1.46	.19		1.65	1.93	
	0670	6#/CF, unfaced, 1" thick, R4.3		1,000	.008		.65	.15		.80	.96	
	0690	1-1/2" thick, R6.5		890	.009		1.01	.17		1.18	1.39	
	0700	2" thick, R8.7		800	.010		1.41	.19		1.60	1.87	
	0721	2-1/2" thick, R10.9		800	.010		1.55	.19		1.74	2.02	
	0741	3" thick, R13		730	.011		1.86	.20		2.06	2.40	
	0821	Foil faced, 1" thick, R4.3		1,000	.008		.92	.15		1.07	1.26	
	0840	1-1/2" thick, R6.5		890	.009		1.32	.17		1.49	1.73	
	0850	2" thick, R8.7		800	.010		1.72	.19		1.91	2.21	
	0880	2-1/2" thick, R10.9		800	.010		2.07	.19		2.26	2.60	
	0900	3" thick, R13		730	.011		2.47	.20		2.67	3.07	
	1500	Foamglass, 1-1/2" thick, R3.9		800	.010		1.40	.19		1.59	1.87	

Note at 0040: R074 -030

For expanded coverage of these items see *Means Building Construction Cost Data 1998*

072 100 | Building Insulation

			CREW	DAILY OUTPUT	LABOR-HOURS	UNIT	MAT.	LABOR	EQUIP.	TOTAL	TOTAL INCL O&P	
									1998 BARE COSTS			
116	1550	3" thick, R9	1 Carp	730	.011	S.F.	2.63	.20		2.83	3.24	116
	1600	Isocyanurate, 4' x 8' sheet, foil faced, both sides [R074 -030]										
	1610	1/2" thick, R3.9	1 Carp	800	.010	S.F.	.27	.19		.46	.62	
	1620	5/8" thick, R4.5		800	.010		.28	.19		.47	.63	
	1630	3/4" thick, R5.4		800	.010		.30	.19		.49	.65	
	1640	1" thick, R7.2		800	.010		.33	.19		.52	.69	
	1650	1-1/2" thick, R10.8		730	.011		.37	.20		.57	.76	
	1660	2" thick, R14.4		730	.011		.47	.20		.67	.86	
	1670	3" thick, R21.6		730	.011		1.21	.20		1.41	1.68	
	1680	4" thick, R28.8		730	.011		1.53	.20		1.73	2.03	
	1700	Perlite, 1" thick, R2.77		800	.010		.28	.19		.47	.63	
	1750	2" thick, R5.55		730	.011		.52	.20		.72	.92	
	1900	Extruded polystyrene, 25 PSI compressive strength, 1" thick, R5		800	.010		.31	.19		.50	.66	
	1940	2" thick R10		730	.011		.62	.20		.82	1.03	
	1960	3" thick R15		730	.011		.91	.20		1.11	1.36	
	2100	Expanded polystyrene, 1" thick, R3.85		800	.010		.19	.19		.38	.52	
	2120	2" thick, R7.69		730	.011		.35	.20		.55	.74	
	2140	3" thick, R11.49		730	.011		.57	.20		.77	.98	
	2350	Sheathing, insulating foil faced fiberboard, 3/8" thick		670	.012		.27	.22		.49	.68	
118	0010	**WALL OR CEILING INSUL., NON-RIGID**										118
	0040	Fiberglass, kraft faced, batts or blankets										
	0060	3-1/2" thick, R11, 11" wide	1 Carp	1,150	.007	S.F.	.18	.13		.31	.42	
	0140	6" thick, R19, 11" wide		1,000	.008		.27	.15		.42	.55	
	0200	9" thick, R30, 15" wide		1,150	.007		.55	.13		.68	.83	
	0240	12" thick, R38, 15" wide		1,000	.008		.65	.15		.80	.96	
	0400	Fiberglass, foil faced, batts or blankets										
	0420	3-1/2" thick, R11, 15" wide	1 Carp	1,600	.005	S.F.	.26	.09		.35	.44	
	0460	6" thick, R19, 15" wide		1,350	.006		.30	.11		.41	.52	
	0500	9" thick, R30, 15" wide		1,150	.007		.54	.13		.67	.81	
	0800	Fiberglass, unfaced, batts or blankets										
	0820	3-1/2" thick, R11, 15" wide	1 Carp	1,350	.006	S.F.	.18	.11		.29	.39	
	0860	6" thick, R19, 15" wide		1,150	.007		.30	.13		.43	.54	
	0900	9" thick, R30, 15" wide		1,000	.008		.50	.15		.65	.80	
	0940	12" thick, R38, 15" wide		1,000	.008		.64	.15		.79	.95	
	1300	Mineral fiber batts, kraft faced										
	1320	3-1/2" thick, R12	1 Carp	1,600	.005	S.F.	.23	.09		.32	.41	
	1340	6" thick, R19		1,600	.005		.35	.09		.44	.55	
	1380	10" thick, R30		1,350	.006		.58	.11		.69	.83	
	1850	Friction fit wire insulation supports, 16" O.C.		960	.008	Ea.	.05	.15		.20	.32	
	1900	For foil backing, add				S.F.	.04			.04	.04	

072 200 | Roof & Deck Insulation

			CREW	DAILY OUTPUT	LABOR-HOURS	UNIT	MAT.	LABOR	EQUIP.	TOTAL	TOTAL INCL O&P	
203	0010	**ROOF DECK INSULATION**										203
	0012	Basalt rock wool, 1" thick, R4.2	1 Rofc	800	.010	S.F.	.43	.16		.59	.77	
	0014	2" thick R8.4		700	.011		.86	.19		1.05	1.30	
	0016	3" thick R12.6		700	.011		1.29	.19		1.48	1.77	
	0018	Tapered for drainage		800	.010	B.F.	.71	.16		.87	1.08	
	0020	Fiberboard low density, 1/2" thick R1.39		1,000	.008	S.F.	.17	.13		.30	.43	
	0030	1" thick R2.78		800	.010		.32	.16		.48	.65	
	0080	1 1/2" thick R4.17		800	.010		.48	.16		.64	.83	
	0100	2" thick R5.56		800	.010		.64	.16		.80	1	
	0110	Fiberboard high density, 1/2" thick R1.3		1,000	.008		.18	.13		.31	.44	
	0120	1" thick R2.5		800	.010		.34	.16		.50	.67	
	0130	1-1/2" thick R3.8		800	.010		.55	.16		.71	.91	

Important: See the Reference Section for critical supporting data - Reference Nos., Crews, & Location Factors

072 200 | Roof & Deck Insulation

		CREW	DAILY OUTPUT	LABOR-HOURS	UNIT	MAT.	LABOR	EQUIP.	TOTAL	TOTAL INCL O&P	
203	0200	Fiberglass, 3/4" thick R2.78	1 Rofc	1,000	.008	S.F.	.39	.13		.52	.67	**203**
0400	15/16" thick R3.70		1,000	.008		.50	.13		.63	.79	
0460	1-1/16" thick R4.17		1,000	.008		.63	.13		.76	.93	
0600	1-5/16" thick R5.26		1,000	.008		.87	.13		1	1.20	
0650	2-1/16" thick R8.33		800	.010		.93	.16		1.09	1.32	
0700	2-7/16" thick R10		800	.010		1.05	.16		1.21	1.45	
1500	Foamglass, 1-3/4" thick R4.5		800	.010		1.36	.16		1.52	1.80	
1530	3" thick R7.89		700	.011		2.73	.19		2.92	3.35	
1600	Tapered for drainage		600	.013	B.F.	.88	.22		1.10	1.37	
1650	Perlite, 1/2" thick R1.32		1,050	.008	S.F.	.27	.12		.39	.53	
1655	3/4" thick R2.08		800	.010		.32	.16		.48	.65	
1660	1" thick R2.78		800	.010		.24	.16		.40	.56	
1670	1-1/2" thick R4.17		800	.010		.35	.16		.51	.69	
1680	2" thick R5.56		700	.011		.47	.19		.66	.87	
1685	2-1/2" thick R6.67		700	.011		.77	.19		.96	1.20	
1690	Tapered for drainage		800	.010	B.F.	.52	.16		.68	.87	
1700	Polyisocyanurate, 2#/CF density, 3/4" thick, R5.1		1,500	.005	S.F.	.30	.09		.39	.49	
1705	1" thick R7.14		1,400	.006		.31	.09		.40	.51	
1715	1-1/2" thick R10.87		1,250	.006		.33	.10		.43	.56	
1725	2" thick R14.29		1,100	.007		.44	.12		.56	.70	
1735	2-1/2" thick R16.67		1,050	.008		.50	.12		.62	.78	
1745	3" thick R21.74		1,000	.008		.63	.13		.76	.93	
1755	3-1/2" thick R25		1,000	.008		.73	.13		.86	1.04	
1765	Tapered for drainage		1,400	.006	B.F.	.37	.09		.46	.58	
1900	Extruded Polystyrene										
1910	15 PSI compressive strength, 1" thick, R5	1 Rofc	1,500	.005	S.F.	.24	.09		.33	.42	
1920	2" thick, R10		1,250	.006		.32	.10		.42	.54	
1930	3" thick R15		1,000	.008		.72	.13		.85	1.04	
1932	4" thick R20		1,000	.008		.97	.13		1.10	1.31	
1934	Tapered for drainage		1,500	.005	B.F.	.34	.09		.43	.53	
1940	25 PSI compressive strength, 1" thick R5		1,500	.005	S.F.	.30	.09		.39	.49	
1942	2" thick R10		1,250	.006		.60	.10		.70	.85	
1944	3" thick R15		1,000	.008		.82	.13		.95	1.14	
1946	4" thick R20		1,000	.008		1.07	.13		1.20	1.42	
1948	Tapered for drainage		1,500	.005	B.F.	.38	.09		.47	.58	
1950	40 psi compressive strength, 1" thick R5		1,500	.005	S.F.	.34	.09		.43	.53	
1952	2" thick R10		1,250	.006		.66	.10		.76	.92	
1954	3" thick R15		1,000	.008		.99	.13		1.12	1.33	
1956	4" thick R20		1,000	.008		1.32	.13		1.45	1.69	
1958	Tapered for drainage		1,400	.006	B.F.	.47	.09		.56	.69	
1960	60 PSI compressive strength, 1" thick R5		1,450	.006	S.F.	.40	.09		.49	.61	
1962	2" thick R10		1,200	.007		.74	.11		.85	1.01	
1964	3" thick R15		975	.008		1.11	.13		1.24	1.47	
1966	4" thick R20		950	.008		1.50	.14		1.64	1.90	
1968	Tapered for drainage		1,400	.006	B.F.	.57	.09		.66	.80	
1970	115 PSI compressive strength, 1" thick R5		1,400	.006	S.F.	.86	.09		.95	1.12	
1972	2" thick R10		1,150	.007		1.72	.11		1.83	2.10	
1974	3" thick R15		950	.008		2.55	.14		2.69	3.05	
1976	4" thick R20		900	.009		3.50	.14		3.64	4.12	
1978	Tapered for drainage		1,400	.006	B.F.	.96	.09		1.05	1.23	
2010	Expanded polystyrene, 1#/CF density, 3/4" thick R2.89		1,500	.005	S.F.	.20	.09		.29	.38	
2020	1" thick R3.85		1,500	.005		.20	.09		.29	.38	
2100	2" thick R7.69		1,250	.006		.32	.10		.42	.54	
2110	3" thick R11.49		1,250	.006		.43	.10		.53	.66	
2120	4" thick R15.38		1,200	.007		.55	.11		.66	.81	
2130	5" thick R19.23		1,150	.007		.69	.11		.80	.97	

THERMAL & MOISTURE PROTECTION 7

072 200 | Roof & Deck Insulation

			CREW	DAILY OUTPUT	LABOR-HOURS	UNIT	MAT.	LABOR	EQUIP.	TOTAL	TOTAL INCL O&P	
203	2140	6" thick R23.26	1 Rofc	1,150	.007	S.F.	.81	.11		.92	1.10	203
	2150	Tapered for drainage	↓	1,500	.005	B.F.	.32	.09		.41	.51	
	2400	Composites with 2" EPS										
	2410	1" fiberboard	1 Rofc	950	.008	S.F.	.72	.14		.86	1.04	
	2420	7/16" oriented strand board		800	.010		.85	.16		1.01	1.24	
	2430	1/2" plywood		800	.010		.91	.16		1.07	1.30	
	2440	1" perlite	↓	800	.010	↓	.76	.16		.92	1.14	
	2450	Composites with 1 1/2" polyisocyanurate										
	2460	1" fiberboard	1 Rofc	800	.010	S.F.	.77	.16		.93	1.15	
	2470	1" perlite	.	850	.009		.79	.15		.94	1.15	
	2480	7/16" oriented strand board	↓	800	.010	↓	.90	.16		1.06	1.29	

072 500 | Fireproofing

			CREW	DAILY OUTPUT	LABOR-HOURS	UNIT	MAT.	LABOR	EQUIP.	TOTAL	TOTAL INCL O&P	
554	0010	**SPRAYED** Mineral fiber or cementitious for fireproofing,										554
	0050	not incl tamping or canvas protection										
	0100	1" thick, on flat plate steel	G-2	3,000	.008	S.F.	.41	.12	.08	.61	.74	
	0200	Flat decking		2,400	.010		.41	.15	.10	.66	.81	
	0400	Beams		1,500	.016		.41	.24	.15	.80	1.03	
	0500	Corrugated or fluted decks		1,250	.019		.62	.29	.18	1.09	1.37	
	0700	Columns, 1-1/8" thick		1,100	.022		.46	.33	.21	1	1.30	
	0800	2-3/16" thick	↓	700	.034	↓	.86	.52	.33	1.71	2.19	
	0850	For tamping, add						10%				
	0900	For canvas protection, add	G-2	5,000	.005	S.F.	.06	.07	.05	.18	.24	

073 | Shingles & Roofing Tiles

073 100 | Shingles

			CREW	DAILY OUTPUT	LABOR-HOURS	UNIT	MAT.	LABOR	EQUIP.	TOTAL	TOTAL INCL O&P	
101	0010	**ALUMINUM** Shingles, mill finish, .019" thick	1 Carp	5	1.600	Sq.	150	29.50		179.50	216	101
	0100	.020" thick	"	5	1.600	↓	143	29.50		172.50	209	
	0300	For colors, add				↓	15.15			15.15	16.65	
	0600	Ridge cap, .024" thick	1 Carp	170	.047	L.F.	1.85	.87		2.72	3.53	
	0700	End wall flashing, .024" thick		170	.047		1.26	.87		2.13	2.88	
	0900	Valley section, .024" thick		170	.047		2.30	.87		3.17	4.02	
	1000	Starter strip, .024" thick		400	.020		1.20	.37		1.57	1.95	
	1200	Side wall flashing, .024" thick	↓	170	.047	↓	1.25	.87		2.12	2.87	
103	0010	**FIBER CEMENT** shingles, 16" x 9.35", 500 lb per square	1 Carp	4	2	Sq.	244	37		281	330	103
	0200	Shakes, 16" x 9.35", 550 lb per square		2.20	3.636	"	221	67.50		288.50	360	
	0300	Hip & ridge, 4.75 x 14"		1	8	C.L.F.	600	148		748	915	
	0400	Hexagonal, 16" x 16"		3	2.667	Sq.	165	49.50		214.50	267	
	0500	Square, 16" x 16"	↓	3	2.667		148	49.50		197.50	248	
	2000	For steep roofs (7/12 pitch or greater), add				↓		50%				
104	0010	**ASPHALT SHINGLES**										104
	0100	Standard strip shingles										
	0150	Inorganic, class A, 210-235 lb/sq	1 Rofc	5.50	1.455	Sq.	26	23.50		49.50	72.50	
	0155	Pneumatic nailed		7	1.143		26	18.50		44.50	63	
	0200	Organic, class C, 235-240 lb/sq		5	1.600		35	26		61	87	
	0205	Pneumatic nailed	↓	6.25	1.280	↓	35	20.50		55.50	77	
	0250	Standard, laminated multi-layered shingles										
	0300	Class A, 240-260 lb/sq	1 Rofc	4.50	1.778	Sq.	35	29		64	92.50	

Important: See the Reference Section for critical supporting data - Reference Nos., Crews, & Location Factors

	073 100	Shingles	CREW	DAILY OUTPUT	LABOR-HOURS	UNIT	MAT.	LABOR	EQUIP.	TOTAL	TOTAL INCL O&P	
104	0305	Pneumatic nailed	1 Rofc	5.63	1.422	Sq.	35	23		58	81.50	**104**
	0350	Class C, 260-300 lb/square, 4 bundles/square		4	2		50	32.50		82.50	116	
	0355	Pneumatic nailed		5	1.600		50	26		76	104	
	0400	Premium, laminated multi-layered shingles										
	0450	Class A, 260-300 lb, 4 bundles/sq	1 Rofc	3.50	2.286	Sq.	47	37		84	121	
	0455	Pneumatic nailed		4.37	1.831		47	29.50		76.50	108	
	0500	Class C, 300-385 lb/square, 5 bundles/square		3	2.667		66	43		109	153	
	0505	Pneumatic nailed		3.75	2.133		66	34.50		100.50	137	
	0800	#15 felt underlayment		64	.125		2.78	2.03		4.81	6.85	
	0825	#30 felt underlayment		58	.138		5.75	2.23		7.98	10.50	
	0850	Self adhering polyethylene and rubberized asphalt underlayment		22	.364		39.50	5.90		45.40	54.50	
	0900	Ridge shingles		330	.024	L.F.	.45	.39		.84	1.23	
	0905	Pneumatic nailed		412.50	.019	"	.45	.31		.76	1.09	
	1000	For steep roofs (7 to 12 pitch or greater), add						50%				
106	0010	**SLATE**, Buckingham, Virginia, black										**106**
	0100	3/16" - 1/4" thick	1 Rots	1.75	4.571	Sq.	550	74.50		624.50	745	
	0200	1/4" thick		1.75	4.571		730	74.50		804.50	945	
	0900	Pennsylvania black, Bangor, #1 clear		1.75	4.571		445	74.50		519.50	630	
	1200	Vermont, unfading, green, mottled green		1.75	4.571		435	74.50		509.50	620	
	1300	Semi-weathering green & gray		1.75	4.571		330	74.50		404.50	505	
	1400	Purple		1.75	4.571		415	74.50		489.50	595	
	1500	Black or gray		1.75	4.571		405	74.50		479.50	585	
	2700	Ridge shingles, slate		200	.040	L.F.	8.55	.65		9.20	10.65	
108	0010	**WOOD** 16" No. 1 red cedar shingles, 5" exposure, on roof	1 Carp	2.50	3.200	Sq.	160	59		219	277	**108**
	0015	Pneumatic nailed		3.25	2.462	"	160	45.50		205.50	254	
	0200	7-1/2" exposure, on walls		2.05	3.902	Sq.	107	72		179	241	
	0205	Pneumatic nailed		2.67	2.996		107	55.50		162.50	212	
	0300	18" No. 1 red cedar perfections, 5-1/2" exposure, on roof		2.75	2.909		202	54		256	315	
	0305	Pneumatic nailed		3.57	2.241		202	41.50		243.50	293	
	0500	7-1/2" exposure, on walls		2.25	3.556		149	66		215	276	
	0505	Pneumatic nailed		2.92	2.740		149	50.50		199.50	250	
	0600	Resquared, and rebutted, 5-1/2" exposure, on roof		3	2.667		118	49.50		167.50	215	
	0605	Pneumatic nailed		3.90	2.051		118	38		156	195	
	0900	7-1/2" exposure, on walls		2.45	3.265		87	60.50		147.50	200	
	0905	Pneumatic nailed		3.18	2.516		87	46.50		133.50	176	
	1000	Add to above for fire retardant shingles, 16" long					25			25	27.50	
	1050	18" long					28.50			28.50	31.50	
	1060	Preformed ridge shingles	1 Carp	400	.020	L.F.	1.80	.37		2.17	2.61	
	1100	Hand-split red cedar shakes, 1/2" thick x 24" long, 10" exp. on roof		2.50	3.200	Sq.	182	59		241	300	
	1105	Pneumatic nailed		3.25	2.462		182	45.50		227.50	278	
	1110	3/4" thick x 24" long, 10" exp. on roof		2.25	3.556		182	66		248	315	
	1115	Pneumatic nailed		2.92	2.740		182	50.50		232.50	287	
	1200	1/2" thick, 18" long, 8-1/2" exp. on roof		2	4		143	74		217	285	
	1205	Pneumatic nailed		2.60	3.077		143	57		200	256	
	1210	3/4" thick x 18" long, 8 1/2" exp. on roof		1.80	4.444		143	82		225	299	
	1215	Pneumatic nailed		2.34	3.419		143	63.50		206.50	266	
	1255	10" exp. on walls		2	4		162	74		236	305	
	1260	10" exposure on walls, pneumatic nailed		2.60	3.077		162	57		219	276	
	1700	Add to above for fire retardant shakes, 24" long					35			35	38.50	
	1800	18" long					33			33	36.50	
	1810	Ridge shakes	1 Carp	350	.023	L.F.	2.52	.42		2.94	3.49	
	2000	White cedar shingles, 16" long, extras, 5" exposure, on roof		2.40	3.333	Sq.	100	61.50		161.50	216	
	2005	Pneumatic nailed		3.12	2.564		100	47.50		147.50	192	
	2050	5" exposure on walls		2	4		100	74		174	237	
	2055	Pneumatic nailed		2.60	3.077		100	57		157	208	

R073 -020 (ref. at 106 0100)
R061 -030 (ref. at 108 0010)

073 100 | Shingles

			CREW	DAILY OUTPUT	LABOR-HOURS	UNIT	1998 BARE COSTS MAT.	LABOR	EQUIP.	TOTAL	TOTAL INCL O&P	
108	2100	7-1/2" exposure, on walls	1 Carp	2	4	Sq.	71.50	74		145.50	206	108
	2105	Pneumatic nailed		2.60	3.077		71.50	57		128.50	176	
	2150	"B" grade, 5" exposure on walls		2	4		85	74		159	221	
	2155	Pneumatic nailed		2.60	3.077		85	57		142	191	
	2300	For 15# organic felt underlayment on roof, 1 layer, add		64	.125		2.78	2.31		5.09	7	
	2400	2 layers, add		32	.250		5.55	4.63		10.18	14.05	
	2600	For steep roofs (7/12 pitch or greater), add to above						50%				
	3000	Ridge shakes or shingle wood	1 Carp	280	.029	L.F.	2.52	.53		3.05	3.68	

R061 -030 (reference at row 2100)

073 200 | Roofing Tile

			CREW	DAILY OUTPUT	LABOR-HOURS	UNIT	1998 BARE COSTS MAT.	LABOR	EQUIP.	TOTAL	TOTAL INCL O&P	
201	0010	**ALUMINUM** Tiles with accessories, .032" thick, mission tile	1 Carp	2.50	3.200	Sq.	345	59		404	475	201
	0200	Spanish tiles	"	3	2.667	"	345	49.50		394.50	460	
202	0010	**CLAY TILE** ASTM C1167, GR 1, severe weathering, acces. incl.										202
	0200	Lanai tile or Classic tile, 158 pc per sq	1 Rots	1.65	4.848	Sq.	490	79		569	690	
	0300	Americana, 158 pc per sq, most colors		1.65	4.848		490	79		569	690	
	0350	Green, gray or brown		1.65	4.848		490	79		569	690	
	0400	Blue		1.65	4.848		490	79		569	690	
	0600	Spanish tile, 171 pc per sq, red		1.80	4.444		305	72.50		377.50	475	
	0800	Blend		1.80	4.444		420	72.50		492.50	595	
	0900	Glazed white		1.80	4.444		500	72.50		572.50	685	
	1100	Mission tile, 192 pc per sq, machine scored finish, red		1.15	6.957		635	113		748	905	
	1700	French tile, 133 pc per sq, smooth finish, red		1.35	5.926		575	96.50		671.50	810	
	1750	Blue or green		1.35	5.926		685	96.50		781.50	935	
	1800	Norman black 317 pc per sq		1	8		805	130		935	1,125	
	2200	Williamsburg tile, 158 pc per sq, aged cedar		1.35	5.926		490	96.50		586.50	720	
	2250	Gray or green		1.35	5.926		490	96.50		586.50	720	
	2350	Ridge shingles, clay tile		200	.040	L.F.	8.65	.65		9.30	10.70	
	3000	For steep roofs (7/12 pitch or greater), add to above				Sq.		50%				
204	0010	**CONCRETE TILE** Including installation of accessories										204
	0020	Corrugated, 13" x 16-1/2", 90 per sq, 950 lb per sq										
	0050	Earthtone colors, nailed to wood deck	1 Rots	1.35	5.926	Sq.	103	96.50		199.50	293	
	0150	Custom blues		1.35	5.926		114	96.50		210.50	305	
	0200	Custom greens		1.35	5.926		114	96.50		210.50	305	
	0250	Premium colors		1.35	5.926		153	96.50		249.50	350	
	0500	Shakes, 13" x 16-1/2", 90 per sq, 950 lb per sq										
	0600	All colors, nailed to wood deck	1 Rots	1.50	5.333	Sq.	185	87		272	365	
	1500	Accessory pieces, ridge & hip, 10" x 16-1/2", 8 lbs. each				Ea.	2.25			2.25	2.48	
	1700	Rake, 6-1/2" x 16-3/4", 9 lbs. each					2.25			2.25	2.48	
	1800	Mansard hip, 10" x 16-1/2", 9.2 lbs. each					2.25			2.25	2.48	
	1900	Hip starter, 10" x 16-1/2", 10.5 lbs. each					9.50			9.50	10.45	
	2000	3 or 4 way apex, 10" each side, 11.5 lbs. each					10.25			10.25	11.30	

074 | Preformed Roofing & Siding

074 100 | Preformed Panels

			CREW	DAILY OUTPUT	LABOR-HOURS	UNIT	1998 BARE COSTS MAT.	LABOR	EQUIP.	TOTAL	TOTAL INCL O&P	
101	0010	**ALUMINUM ROOFING** Corrugated or ribbed, .0155" thick, natural	G-3	1,200	.027	S.F.	.66	.45		1.11	1.49	101
	0300	Painted	"	1,200	.027	"	.94	.45		1.39	1.79	
104	0010	**FIBERGLASS** Corrugated panels, roofing, 8 oz per SF	G-3	1,000	.032	S.F.	2.13	.54		2.67	3.25	104
	0300	Corrugated siding, 6 oz per SF		880	.036		1.84	.61		2.45	3.06	

Important: See the Reference Section for critical supporting data - Reference Nos., Crews, & Location Factors

074 100 | Preformed Panels

			CREW	DAILY OUTPUT	LABOR-HOURS	UNIT	MAT.	LABOR	EQUIP.	TOTAL	TOTAL INCL O&P	
104	0400	8 oz per SF	G-3	880	.036	S.F.	2.13	.61		2.74	3.38	**104**
	0600	12 oz. siding, textured		880	.036		2.65	.61		3.26	3.96	
	0900	Flat panels, 6 oz per SF, clear or colors		880	.036		1.65	.61		2.26	2.86	
	1300	8 oz per SF, clear or colors		880	.036		2.13	.61		2.74	3.38	
107	0010	**STEEL ROOFING** on steel frame, corrugated or ribbed, 30 ga galv	G-3	1,100	.029	S.F.	.75	.49		1.24	1.66	**107**
	0100	28 ga		1,050	.030		.79	.51		1.30	1.74	
	0300	26 ga		1,000	.032		.86	.54		1.40	1.86	
	0400	24 ga		950	.034		1.02	.56		1.58	2.08	
	0600	Colored, 28 ga		1,050	.030		1.05	.51		1.56	2.02	
	0700	26 ga		1,000	.032		1.12	.54		1.66	2.14	
	0710	Flat profile, 1-3/4" standing seams, 10" wide, standard finish, 26 ga		1,000	.032		2.30	.54		2.84	3.44	
	0715	24 ga		950	.034		2.67	.56		3.23	3.90	
	0720	22 ga		900	.036		3.29	.60		3.89	4.63	
	0725	Zinc aluminum alloy finish, 26 ga		1,000	.032		1.75	.54		2.29	2.84	
	0730	24 ga		950	.034		2.09	.56		2.65	3.26	
	0735	22 ga		900	.036		2.40	.60		3	3.65	
	0740	12" wide, standard finish, 26 ga		1,000	.032		2.29	.54		2.83	3.43	
	0745	24 ga		950	.034		2.66	.56		3.22	3.89	
	0750	Zinc aluminum alloy finish, 26 ga		1,000	.032		1.65	.54		2.19	2.73	
	0755	24 ga		950	.034		1.97	.56		2.53	3.13	
	0840	Flat profile, 1" x 3/8" batten, 12" wide, standard finish, 26 ga		1,000	.032		1.79	.54		2.33	2.88	
	0845	24 ga		950	.034		2.10	.56		2.66	3.27	
	0850	22 ga		900	.036		2.52	.60		3.12	3.78	
	0855	Zinc aluminum alloy finish, 26 ga		1,000	.032		1.47	.54		2.01	2.53	
	0860	24 ga		950	.034		1.64	.56		2.20	2.76	
	0865	22 ga		900	.036		1.89	.60		2.49	3.09	
	0870	16-1/2" wide, standard finish, 24 ga		950	.034		2.07	.56		2.63	3.24	
	0875	22 ga		900	.036		2.32	.60		2.92	3.56	
	0880	Zinc aluminum alloy finish, 24 ga		950	.034		1.55	.56		2.11	2.66	
	0885	22 ga		900	.036		1.73	.60		2.33	2.91	
	0890	Flat profile, 2" x 2" batten, 12" wide, standard finish, 26 ga		1,000	.032		2.05	.54		2.59	3.17	
	0895	24 ga		950	.034		2.45	.56		3.01	3.66	
	0900	22 ga		900	.036		3	.60		3.60	4.31	
	0905	Zinc aluminum alloy finish, 26 ga		1,000	.032		1.64	.54		2.18	2.71	
	0910	24 ga		950	.034		1.87	.56		2.43	3.02	
	0915	22 ga		900	.036		2.18	.60		2.78	3.41	
	0920	16-1/2" wide, standard finish, 24 ga		950	.034		2.27	.56		2.83	3.46	
	0925	22 ga		900	.036		2.64	.60		3.24	3.91	
	0930	Zinc aluminum alloy finish, 24 ga		950	.034		1.75	.56		2.31	2.89	
	0935	22 ga		900	.036		1.99	.60		2.59	3.20	

074 600 | Cladding/Siding

			CREW	DAILY OUTPUT	LABOR-HOURS	UNIT	MAT.	LABOR	EQUIP.	TOTAL	TOTAL INCL O&P	
602	0011	**ALUMINUM SIDING** R074 -030	F-2									**602**
	6040	.024 thick smooth white single 8" wide		515	.031	S.F.	1.23	.50		1.73	2.21	
	6060	Double 4" pattern		515	.031		1.17	.50		1.67	2.15	
	6080	Double 5" pattern		550	.029		1.17	.47		1.64	2.10	
	6120	Embossed white, 8" wide		515	.031		1.45	.50		1.95	2.46	
	6140	Double 4" pattern		515	.031		1.32	.50		1.82	2.31	
	6160	Double 5" pattern		550	.029		1.32	.47		1.79	2.26	
	6170	Vertical, embossed white, 12" wide		590	.027		1.32	.44		1.76	2.20	
	6320	.019 thick, insulated, smooth white, 8" wide		515	.031		1.18	.50		1.68	2.16	
	6340	Double 4" pattern		515	.031		1.16	.50		1.66	2.14	
	6360	Double 5" pattern		550	.029		1.16	.47		1.63	2.09	
	6400	Embossed white, 8" wide		515	.031		1.35	.50		1.85	2.35	
	6420	Double 4" pattern		515	.031		1.37	.50		1.87	2.37	
	6440	Double 5" pattern		550	.029		1.37	.47		1.84	2.32	

THERMAL & MOISTURE PROTECTION   **7**

074 600 | Cladding/Siding

			CREW	DAILY OUTPUT	LABOR-HOURS	UNIT	1998 BARE COSTS				TOTAL INCL O&P	
							MAT.	LABOR	EQUIP.	TOTAL		
602	6500	Shake finish 10″ wide white	F-2	550	.029	S.F.	1.45	.47		1.92	2.41	**602**
	6600	Vertical pattern, 12″ wide, white		590	.027		1.37	.44		1.81	2.26	
	6640	For colors add					.07			.07	.08	
	6700	Accessories, white										
	6720	Starter strip 2-1/8″	F-2	610	.026	L.F.	.18	.42		.60	.93	
	6740	Sill trim		450	.036		.28	.58		.86	1.29	
	6760	Inside corner		610	.026		.85	.42		1.27	1.67	
	6780	Outside corner post		610	.026		1.45	.42		1.87	2.33	
	6800	Door & window trim		440	.036		.27	.59		.86	1.31	
	6820	For colors add					.07			.07	.08	
	6900	Soffit & fascia 1′ overhang solid	F-2	110	.145		1.97	2.35		4.32	6.20	
	6920	Vented		110	.145		1.97	2.35		4.32	6.20	
	6940	2′ overhang solid		100	.160		2.88	2.59		5.47	7.60	
	6960	Vented		100	.160		2.88	2.59		5.47	7.60	
606	0010	**STEEL SIDING**, Beveled, vinyl coated, 8″ wide, including fasteners	1 Carp	265	.030	S.F.	1.10	.56		1.66	2.17	**606**
	0050	10″ wide	″	275	.029		1.17	.54		1.71	2.21	
	0080	Galv, corrugated or ribbed, on steel frame, 30 gauge	G-3	800	.040		.75	.67		1.42	1.97	
	0100	28 gauge		795	.040		.79	.68		1.47	2.02	
	0300	26 gauge		790	.041		.86	.68		1.54	2.11	
	0400	24 gauge		785	.041		1.03	.68		1.71	2.29	
	0600	22 gauge		770	.042		1.17	.70		1.87	2.48	
	0700	Colored, corrugated/ribbed, on steel frame, 10 yr fnsh, 28 ga.		800	.040		1.08	.67		1.75	2.33	
	0900	26 gauge		795	.040		1.02	.68		1.70	2.27	
	1000	24 gauge		790	.041		1.18	.68		1.86	2.46	
	1020	20 gauge		785	.041		1.49	.68		2.17	2.80	
607	0010	**VINYL SIDING** Solid PVC panels, 8″ to 10″ wide, plain	1 Carp	255	.031	S.F.	.61	.58		1.19	1.66	**607**
	2000	Smooth, white, single, 8″ wide	F-2	495	.032		.61	.52		1.13	1.57	
	2020	Dutch lap, 10″ wide		550	.029		.59	.47		1.06	1.46	
	2100	Double 4″ pattern, 8″ wide		495	.032		.52	.52		1.04	1.47	
	2120	Double 5″ pattern, 10″ wide		550	.029		.55	.47		1.02	1.42	
	2200	Embossed, white, single, 8″ wide		495	.032		.63	.52		1.15	1.59	
	2220	10″ wide		550	.029		.67	.47		1.14	1.55	
	2300	Double 4″ pattern, 8″ wide		495	.032		.52	.52		1.04	1.47	
	2320	5″ pattern, 10″ wide		550	.029		.57	.47		1.04	1.44	
	2400	Shake finish, 10″ wide, white		550	.029		1.72	.47		2.19	2.70	
	2600	Vertical pattern, double 5″, 10″ wide, white		550	.029		1.27	.47		1.74	2.21	
	2620											
	2700	For colors, add				S.F.	.07			.07	.08	
	2720	1/4″ extruded polystyrene fan folded insulation	F-2	2,000	.008	″	.14	.13		.27	.37	
	3000	Accessories, starter strip		700	.023	L.F.	.22	.37		.59	.87	
	3100	″J″ channel, 1/2″		700	.023		.24	.37		.61	.89	
	3120	5/8″		700	.023		.26	.37		.63	.92	
	3140	3/4″		695	.023		.26	.37		.63	.93	
	3160	1″		690	.023		.28	.38		.66	.95	
	3180	1-1/8″		685	.023		.27	.38		.65	.95	
	3190	1-1/4″		680	.024		.29	.38		.67	.97	
	3200	Under sill trim		500	.032		.26	.52		.78	1.18	
	3300	Outside corner post, 3″ face, pocket 5/8″		700	.023		1.08	.37		1.45	1.82	
	3320	7/8″		690	.023		1.05	.38		1.43	1.79	
	3340	1-1/4″		680	.024		1.06	.38		1.44	1.82	
	3400	Inside corner post, pocket 5/8″		700	.023		.55	.37		.92	1.24	
	3420	7/8″		690	.023		.62	.38		1	1.33	
	3440	1-1/4″		680	.024		.66	.38		1.04	1.38	
	3500	Door & window trim, 2-1/2″ face, pocket 5/8″		510	.031		.54	.51		1.05	1.46	
	3520	7/8″		500	.032		.52	.52		1.04	1.46	

Important: See the Reference Section for critical supporting data - Reference Nos., Crews, & Location Factors

074 600	Cladding/Siding	CREW	DAILY OUTPUT	LABOR-HOURS	UNIT	MAT.	LABOR	EQUIP.	TOTAL	TOTAL INCL O&P	
607						**1998 BARE COSTS**					**607**
3540	1-1/4"	F-2	490	.033	L.F.	.60	.53		1.13	1.56	
3600	Soffit & fascia, 1' overhang, solid		120	.133		1.27	2.16		3.43	5.10	
3620	Vented		120	.133		1.27	2.16		3.43	5.10	
3700	2' overhang, solid		110	.145		1.88	2.35		4.23	6.10	
3720	Vented		110	.145		1.90	2.35		4.25	6.10	
609	**0010** **WOOD SIDING, BOARDS**										**609**
2000	Board & batten, cedar, "B" grade, 1" x 10"	1 Carp	400	.020	S.F.	1.91	.37		2.28	2.73	
2200	Redwood, clear, vertical grain, 1" x 10"		400	.020		3.46	.37		3.83	4.44	
2400	White pine, #2 & better, 1" x 10"		400	.020		.67	.37		1.04	1.37	
2410	Board & batten siding, white pine #2, 1" x 12"		450	.018		.76	.33		1.09	1.40	
3200	Wood, cedar bevel, A grade, 1/2" x 6"		250	.032		2.06	.59		2.65	3.28	
3300	1/2" x 8"		275	.029		1.67	.54		2.21	2.76	
3500	3/4" x 10", clear grade		300	.027		2.17	.49		2.66	3.24	
3600	"B" grade		300	.027		3.22	.49		3.71	4.39	
3800	Cedar, rough sawn, 1" x 4", A grade, natural		240	.033		3.01	.62		3.63	4.37	
3900	Stained		240	.033		3.39	.62		4.01	4.79	
4100	1" x 12", board & batten, #3 & Btr., natural		260	.031		2.08	.57		2.65	3.27	
4200	Stained		260	.031		2.42	.57		2.99	3.64	
4400	1" x 8" channel siding, #3 & Btr., natural		250	.032		2.02	.59		2.61	3.23	
4500	Stained		250	.032		2.30	.59		2.89	3.54	
4700	Redwood, clear, beveled, vertical grain, 1/2" x 4"		200	.040		3.22	.74		3.96	4.81	
4750	1/2" x 6"		225	.036		2.70	.66		3.36	4.10	
4800	1/2" x 8"		250	.032		2.19	.59		2.78	3.42	
5000	3/4" x 10"		300	.027		3.57	.49		4.06	4.78	
5200	Channel siding, 1" x 10", B grade		285	.028		2.30	.52		2.82	3.42	
5250	Redwood, T&G boards, B grade, 1" x 4"	F-2	300	.053		2.74	.86		3.60	4.49	
5270	1" x 8"	"	375	.043		2.36	.69		3.05	3.78	
5400	White pine, rough sawn, 1" x 8", natural	1 Carp	275	.029		.70	.54		1.24	1.69	
5500	Stained	"	275	.029		1.04	.54		1.58	2.06	
5600	Tongue and groove, 1" x 8", horizontal	F-2	375	.043		.64	.69		1.33	1.88	
611	**0010** **WOOD PRODUCT SIDING**										**611**
0030	Lap siding, hardboard, 7/16" x 8", primed										
0050	Wood grain texture finish	F-2	650	.025	S.F.	.95	.40		1.35	1.73	
0100	Panels, 7/16" thick, smooth, textured or grooved, primed		700	.023		.81	.37		1.18	1.52	
0200	Stained		700	.023		.88	.37		1.25	1.60	
0700	Particle board, overlaid, 3/8" thick		750	.021		.63	.35		.98	1.28	
0900	Plywood, medium density overlaid, 3/8" thick		750	.021		1.08	.35		1.43	1.78	
1000	1/2" thick		700	.023		1.26	.37		1.63	2.02	
1100	3/4" thick		650	.025		1.47	.40		1.87	2.30	
1600	Texture 1-11, cedar, 5/8" thick, natural		675	.024		1.50	.38		1.88	2.31	
1700	Factory stained		675	.024		1.74	.38		2.12	2.57	
1900	Texture 1-11, fir, 5/8" thick, natural		675	.024		1.01	.38		1.39	1.77	
2000	Factory stained		675	.024		1.14	.38		1.52	1.91	
2050	Texture 1-11, S.Y.P., 5/8" thick, natural		675	.024		.83	.38		1.21	1.57	
2100	Factory stained		675	.024		.93	.38		1.31	1.68	
2200	Rough sawn cedar, 3/8" thick, natural		675	.024		1.14	.38		1.52	1.91	
2300	Factory stained		675	.024		1.26	.38		1.64	2.05	
2500	Rough sawn fir, 3/8" thick, natural		675	.024		.61	.38		.99	1.33	
2600	Factory stained		675	.024		.68	.38		1.06	1.41	
2800	Redwood, textured siding, 5/8" thick		675	.024		1.89	.38		2.27	2.74	
615	**0010** **FIBER CEMENT SIDING**										**615**
0020	Lap siding, 5/16" thick, 6" wide, smooth texture	2 Carp	415	.039	S.F.	.91	.71		1.62	2.22	
0025	Woodgrain texture		415	.039		.91	.71		1.62	2.22	
0030	7-1/2" wide, smooth texture		425	.038		.90	.70		1.60	2.18	

7 **THERMAL & MOISTURE PROTECTION**

074 600	Cladding/Siding	CREW	DAILY OUTPUT	LABOR-HOURS	UNIT	MAT.	LABOR	EQUIP.	TOTAL	TOTAL INCL O&P	
						1998 BARE COSTS					
615 0035	Woodgrain texture	2 Carp	425	.038	S.F.	.90	.70		1.60	2.18	**615**
0040	8" wide, smooth texture		425	.038		.91	.70		1.61	2.19	
0045	Roughsawn texture		425	.038		.91	.70		1.61	2.19	
0050	9-1/2" wide, smooth texture		440	.036		.87	.67		1.54	2.11	
0055	Woodgrain texture		440	.036		.87	.67		1.54	2.11	
0060	12" wide, smooth texture		455	.035		.85	.65		1.50	2.04	
0065	Woodgrain texture		455	.035		.85	.65		1.50	2.04	
0070	Panel siding, 5/16" thick, smooth texture		750	.021		.76	.39		1.15	1.51	
0075	Stucco texture		750	.021		.76	.39		1.15	1.51	
0080	Grooved woodgrain texture		750	.021		.76	.39		1.15	1.51	
0085	V - grooved woodgrain texture		750	.021		.76	.39		1.15	1.51	
0090	Wood starter strip		400	.040	L.F.	.25	.74		.99	1.55	

075 | Membrane Roofing

075 100	Built-Up Roofing	CREW	DAILY OUTPUT	LABOR-HOURS	UNIT	MAT.	LABOR	EQUIP.	TOTAL	TOTAL INCL O&P	
						1998 BARE COSTS					
101 0010	**ASPHALT** Coated felt, #30, 2 sq per roll, not mopped	1 Rofc	58	.138	Sq.	5.75	2.23		7.98	10.50	**101**
0200	#15, 4 sq per roll, plain or perforated, not mopped		58	.138		2.78	2.23		5.01	7.25	
0250	Perforated		58	.138		2.78	2.23		5.01	7.25	
0300	Roll roofing, smooth, #65		15	.533		12.10	8.65		20.75	29.50	
0500	#90		15	.533		12.65	8.65		21.30	30	
0520	Mineralized		15	.533		15.50	8.65		24.15	33	
0540	D.C. (Double coverage), 19" selvage edge		10	.800		30.50	12.95		43.45	57.50	
0580	Adhesive (lap cement)				Gal.	3.39			3.39	3.73	
102 0010	**BUILT-UP ROOFING**										**102**
0120	Asphalt flood coat with gravel/slag surfacing, not including										
0140	Insulation, flashing or wood nailers										
0200	Asphalt base sheet, 3 plies #15 asphalt felt, mopped	G-1	22	2.545	Sq.	34.50	39	18.15	91.65	130	
0350	On nailable decks		21	2.667		38.50	40.50	19.05	98.05	140	
0500	4 plies #15 asphalt felt, mopped		20	2.800		49	42.50	20	111.50	155	
0550	On nailable decks		19	2.947		44.50	45	21	110.50	156	
0700	Coated glass base sheet, 2 plies glass (type IV), mopped		22	2.545		37.50	39	18.15	94.65	134	
0850	3 plies glass, mopped		20	2.800		44.50	42.50	20	107	151	
0950	On nailable decks		19	2.947		42.50	45	21	108.50	154	
1100	4 plies glass fiber felt (type IV), mopped		20	2.800		53.50	42.50	20	116	161	
1150	On nailable decks		19	2.947		49.50	45	21	115.50	162	
1200	Coated & saturated base sheet, 3 plies #15 asph. felt, mopped		20	2.800		40.50	42.50	20	103	147	
1250	On nailable decks		19	2.947		38.50	45	21	104.50	150	
1300	4 plies #15 asphalt felt, mopped		22	2.545		47	39	18.15	104.15	144	
2000	Asphalt flood coat, smooth surface										
2200	Asphalt base sheet & 3 plies #15 asphalt felt, mopped	G-1	24	2.333	Sq.	35.50	35.50	16.65	87.65	124	
2400	On nailable decks		23	2.435		33	37	17.40	87.40	125	
2600	4 plies #15 asphalt felt, mopped		24	2.333		41.50	35.50	16.65	93.65	130	
2700	On nailable decks		23	2.435		39.50	37	17.40	93.90	132	
2900	Coated glass fiber base sheet, mopped, and 2 plies of										
2910	glass fiber felt (type IV)	G-1	25	2.240	Sq.	32	34	16	82	117	
3100	On nailable decks		24	2.333		30.50	35.50	16.65	82.65	118	
3200	3 plies, mopped		23	2.435		39	37	17.40	93.40	132	
3300	On nailable decks		22	2.545		37	39	18.15	94.15	134	
3800	4 plies glass fiber felt (type IV), mopped		23	2.435		46	37	17.40	100.40	139	

075 100 | Built-Up Roofing

			CREW	DAILY OUTPUT	LABOR-HOURS	UNIT	MAT.	LABOR	EQUIP.	TOTAL	TOTAL INCL O&P	
102	3900	On nailable decks	G-1	22	2.545	Sq.	44	39	18.15	101.15	141	102
	4000	Coated & saturated base sheet, 3 plies #15 asph. felt, mopped		24	2.333		35.50	35.50	16.65	87.65	124	
	4200	On nailable decks		23	2.435		33.50	37	17.40	87.90	125	
	4300	4 plies #15 organic felt, mopped		22	2.545		41.50	39	18.15	98.65	138	
	4500	Coal tar pitch with gravel/slag surfacing										
	4600	4 plies #15 tarred felt, mopped	G-1	21	2.667	Sq.	97.50	40.50	19.05	157.05	204	
	4800	3 plies glass fiber felt (type IV), mopped	"	19	2.947	"	80	45	21	146	195	
	5000	Coated glass fiber base sheet, and 2 plies of										
	5010	glass fiber felt, (type IV), mopped	G-1	19	2.947	Sq.	80	45	21	146	195	
	5300	On nailable decks		18	3.111		70.50	47.50	22	140	191	
	5600	4 plies glass fiber felt (type IV), mopped		21	2.667		110	40.50	19.05	169.55	217	
	5800	On nailable decks		20	2.800		100	42.50	20	162.50	212	
103	0010	**CANTS** 4" x 4", treated timber, cut diagonally	1 Rofc	325	.025	L.F.	.80	.40		1.20	1.62	103
	0100	Foamglass		325	.025		1.87	.40		2.27	2.79	
	0300	Mineral or fiber, trapezoidal, 1"x 4" x 48"		325	.025		.17	.40		.57	.93	
	0400	1-1/2" x 5-5/8" x 48"		325	.025		.29	.40		.69	1.06	
104	0010	**FELT** Glass fibered, #15, no mopping	1 Rofc	58	.138	Sq.	3.65	2.23		5.88	8.20	104
	0300	Base sheet, #45, channel vented		58	.138		15.90	2.23		18.13	21.50	
	0400	#50, coated		58	.138		7.50	2.23		9.73	12.40	
	0500	Cap, mineral surfaced		58	.138		15.50	2.23		17.73	21	
	0600	Flashing membrane, #65		16	.500		36.50	8.10		44.60	55.50	
	0800	Coal tar fibered, #15, no mopping		58	.138		7.80	2.23		10.03	12.75	
	0900	Asphalt felt, #15, 4 sq per roll, no mopping		58	.138		2.78	2.23		5.01	7.25	
	1100	#30, 2 sq per roll		58	.138		5.75	2.23		7.98	10.50	
	1200	Double coated, #33		58	.138		5.90	2.23		8.13	10.65	
	1400	#40, base sheet		58	.138		6.30	2.23		8.53	11.05	
	1450	Coated and saturated		58	.138		6.45	2.23		8.68	11.20	
	1500	Tarred felt, organic, #15, 4 sq rolls		58	.138		7.80	2.23		10.03	12.75	
	1550	#30, 2 sq roll		58	.138		15.60	2.23		17.83	21.50	
	1700	Add for mopping above felts, per ply, asphalt, 24 lb per sq	G-1	192	.292		2.90	4.44	2.08	9.42	13.80	
	1800	Coal tar mopping, 30 lb per sq		186	.301		8.05	4.59	2.15	14.79	19.80	
	1900	Flood coat, with asphalt, 60 lb per sq		60	.933		7.25	14.20	6.65	28.10	42	
	2000	With coal tar, 75 lb per sq		56	1		20	15.25	7.15	42.40	58.50	

075 150 | Cold Applied Roofing

			CREW	DAILY OUTPUT	LABOR-HOURS	UNIT	MAT.	LABOR	EQUIP.	TOTAL	TOTAL INCL O&P	
152	0010	**COLD APPLIED** 3-ply system (components listed below)	G-5	50	.800	Sq.		11.90	3.30	15.20	25.50	152
	0100	Spunbond poly. fabric, 1.35 oz/SY, 36"W, 10.8 Sq/roll				Ea.	122			122	135	
	0200	49" wide, 14.6 Sq./roll					168			168	185	
	0300	2.10 oz./S.Y., 36" wide, 10.8 Sq./roll					184			184	202	
	0400	49" wide, 14.6 Sq./roll					250			250	275	
	0500	Base & finish coat, 3 gal./Sq., 5 gal./can				Gal.	3.21			3.21	3.53	
	0600	Coating, ceramic granules, 1/2 Sq./bag				Ea.	11.10			11.10	12.25	
	0700	Aluminum, 2 gal./Sq.				Gal.	9.35			9.35	10.30	
	0800	Emulsion, fibered or non-fibered, 4 gal./Sq.				"	4.31			4.31	4.74	

075 200 | Prepared Roll Roofing

			CREW	DAILY OUTPUT	LABOR-HOURS	UNIT	MAT.	LABOR	EQUIP.	TOTAL	TOTAL INCL O&P	
204	0010	**ROLL ROOFING**										204
	0100	Asphalt, mineral surface										
	0200	1 ply #15 organic felt, 1 ply mineral surfaced										
	0300	Selvage roofing, lap 19", nailed & mopped	G-1	27	2.074	Sq.	36	31.50	14.80	82.30	115	
	0400	3 plies glass fiber felt (type IV), 1 ply mineral surfaced										
	0500	Selvage roofing, lapped 19", mopped	G-1	25	2.240	Sq.	53	34	16	103	140	
	0600	Coated glass fiber base sheet, 2 plies of glass fiber										
	0700	Felt (type IV), 1 ply mineral surfaced selvage										

075 200 | Prepared Roll Roofing

			DAILY OUTPUT	LABOR-HOURS	UNIT	1998 BARE COSTS				TOTAL INCL O&P	
						MAT.	LABOR	EQUIP.	TOTAL		
204	0800	Roofing, lapped 19", mopped	G-1	25	2.240	Sq.	57	34	16	107	144	204
	0900	On nailable decks	"	24	2.333	"	54	35.50	16.65	106.15	144	
	1000	3 plies glass fiber felt (type III), 1 ply mineral surfaced										
	1100	Selvage roofing, lapped 19", mopped	G-1	25	2.240	Sq.	53	34	16	103	140	

075 300 | Elastomeric Roofing

			DAILY OUTPUT	LABOR-HOURS	UNIT	MAT.	LABOR	EQUIP.	TOTAL	TOTAL INCL O&P	
302	0010	**SINGLE-PLY MEMBRANE**										302
	0800	Chlorosulfonated polyethylene-hypalon (CSPE), 45 mils,										
	0900	0.29 P.S.F., fully adhered	G-5	26	1.538	Sq.	122	23	6.35	151.35	183	
	1100	Loose-laid & ballasted with stone (10 P.S.F.)		51	.784		129	11.65	3.23	143.88	168	
	1200	Partially adhered with fastening strips		35	1.143		124	17	4.71	145.71	173	
	1300	Plates with adhesive attachment		35	1.143		122	17	4.71	143.71	171	
	3500	Ethylene propylene diene monomer (EPDM), 45 mils, 0.28 P.S.F.										
	3600	Loose-laid & ballasted with stone (10 P.S.F.)	G-5	51	.784	Sq.	61.50	11.65	3.23	76.38	93	
	3700	Partially adhered		35	1.143		53.50	17	4.71	75.21	95.50	
	3800	Fully adhered with adhesive		26	1.538		75.50	23	6.35	104.85	132	
	4000	55 mils, 0.35 P.S.F.										
	4100	Loose-laid & ballasted with stone, (10 P.S.F.)	G-5	51	.784	Sq.	71	11.65	3.23	85.88	104	
	4200	Partially adhered		35	1.143		56	17	4.71	77.71	98	
	4300	Fully adhered with adhesive		26	1.538		88.50	23	6.35	117.85	146	
	4500	60 mils, 0.40 P.S.F.										
	4600	Loose-laid & ballasted with stone (10 P.S.F.)	G-5	51	.784	Sq.	75	11.65	3.23	89.88	108	
	4700	Partially adhered		35	1.143		66	17	4.71	87.71	109	
	4800	Fully adhered with adhesive		26	1.538		88.50	23	6.35	117.85	146	
	4810	45 mil, .28 PSF, membrane only					34.50			34.50	38	
	4815	55 mil, .35 PSF, membrane only					42			42	46	
	4820	65 mil, .40 PSF, membrane only					46			46	50.50	
	4850	Seam tape for membrane, 4" x 100' roll				Ea.	67			67	73.50	
	4900	Batten strips, 10' sections					2.45			2.45	2.70	
	4910	Cover tape for batten strips, 6" x 100' roll					126			126	139	
	4930	Plate anchors				M	123			123	135	
	4970	Adhesive for fully adhered systems, 60 S.F./gal.				Gal.	13.60			13.60	14.95	
	7500	Polyisobutylene (PIB), 100 mils, 0.57 P.S.F.										
	7600	Loose-laid & ballasted with stone/gravel (10 P.S.F.)	G-5	51	.784	Sq.	123	11.65	3.23	137.88	161	
	7700	Partially adhered with adhesive		35	1.143		154	17	4.71	175.71	206	
	7800	Hot asphalt attachment		35	1.143		147	17	4.71	168.71	199	
	7900	Fully adhered with contact cement		26	1.538		159	23	6.35	188.35	224	
	8160											
	8200	Polyvinyl chloride (PVC), heat welded seams										
	8700	Reinforced, 48 mils, 0.33 P.S.F.										
	8750	Loose-laid & ballasted with stone/gravel (12 P.S.F.)	G-5	51	.784	Sq.	86.50	11.65	3.23	101.38	121	
	8800	Partially adhered with mechanical fasteners		35	1.143		77.50	17	4.71	99.21	122	
	8850	Fully adhered with adhesive		26	1.538		111	23	6.35	140.35	171	
	8860	Reinforced, 60 mils, .40 P.S.F.										
	8870	Loose-laid & ballasted with stone/gravel (12 P.S.F.)	G-5	51	.784	Sq.	87	11.65	3.23	101.88	121	
	8880	Partially adhered with mechanical fasteners		35	1.143		78	17	4.71	99.71	123	
	8890	Fully adhered with adhesive		26	1.538		111	23	6.35	140.35	171	

075 350 | Modified Bit. Roofing

				DAILY OUTPUT	LABOR-HOURS	UNIT	MAT.	LABOR	EQUIP.	TOTAL	TOTAL INCL O&P	
352	0010	**MODIFIED BITUMEN ROOFING**	R075 -030									352
	0020	Base sheet, #15 glass fiber felt, nailed to deck		1 Rofc	58	.138	Sq.	4.16	2.23		6.39	8.75	
	0030	Spot mopped to deck		G-1	295	.190		5.10	2.89	1.36	9.35	12.50	
	0040	Fully mopped to deck		"	192	.292		6.55	4.44	2.08	13.07	17.80	
	0050	#15 organic felt, nailed to deck		1 Rofc	58	.138		3.29	2.23		5.52	7.80	
	0060	Spot mopped to deck		G-1	295	.190		4.23	2.89	1.36	8.48	11.55	

Important: See the Reference Section for critical supporting data - Reference Nos., Crews, & Location Factors

075 350 | Modified Bit. Roofing

		CREW	DAILY OUTPUT	LABOR-HOURS	UNIT	1998 BARE COSTS MAT.	LABOR	EQUIP.	TOTAL	TOTAL INCL O&P	
352	0070	Fully mopped to deck	G-1	192	.292	Sq.	5.70	4.44	2.08	12.22	16.85	352
	0080	SBS modified, granule surf cap sheet, polyester rein., mopped R075 -030										
	0600	150 mils	G-1	2,000	.028	S.F.	.38	.43	.20	1.01	1.44	
	1100	160 mils		2,000	.028		.44	.43	.20	1.07	1.50	
	1500	Glass fiber reinforced, mopped, 160 mils		2,000	.028		.36	.43	.20	.99	1.42	
	1600	Smooth surface cap sheet, mopped, 145 mils		2,100	.027		.36	.41	.19	.96	1.37	
	1700	Smooth surface flashing, 145 mils		1,260	.044		.36	.68	.32	1.36	2.01	
	1800	150 mils		1,260	.044		.35	.68	.32	1.35	2	
	1900	Granular surface flashing, 150 mils		1,260	.044		.38	.68	.32	1.38	2.03	
	2000	160 mils		1,260	.044		.44	.68	.32	1.44	2.09	
	2100	APP mod., smooth surf. cap sheet, poly. reinf., torched, 160 mils	G-5	2,100	.019		.35	.28	.08	.71	1.01	
	2150	170 mils		2,100	.019		.39	.28	.08	.75	1.05	
	2200	Granule surface cap sheet, poly. reinf., torched, 180 mils		2,000	.020		.38	.30	.08	.76	1.06	
	2250	Smooth surface flashing, torched, 160 mils		1,260	.032		.35	.47	.13	.95	1.41	
	2300	170 mils		1,260	.032		.39	.47	.13	.99	1.45	
	2350	Granule surface flashing, torched, 180 mils		1,260	.032		.38	.47	.13	.98	1.44	
	2400	Fibrated aluminum coating	1 Rofc	3,800	.002		.08	.03		.11	.15	

075 600 | Roof Maint. & Repairs

		CREW	DAILY OUTPUT	LABOR-HOURS	UNIT	1998 BARE COSTS MAT.	LABOR	EQUIP.	TOTAL	TOTAL INCL O&P	
604	0010	**ROOF COATINGS** Asphalt				Gal.	3			3	3.30	604
	0200	Asphalt base, fibered aluminum coating					8.45			8.45	9.25	
	0300	Asphalt primer, 5 gallon					3.54			3.54	3.89	
	0600	Coal tar pitch, 200 lb. barrels				Ton	540			540	590	
	0700	Tar roof cement, 5 gal. lots				Gal.	5.55			5.55	6.10	
	0800	Glass fibered roof & patching cement, 5 gallon				"	3.75			3.75	4.13	
	0900	Reinforcing glass membrane, 450 S.F./roll				Ea.	45.50			45.50	50	
	1000	Neoprene roof coating, 5 gal, 2 gal/sq				Gal.	20.50			20.50	22.50	
	1100	Roof patch & flashing cement, 5 gallon					18.95			18.95	21	
	1200	Roof resaturant, glass fibered, 3 gal/sq					7.30			7.30	8.05	
	1300	Mineral rubber, 3 gal/sq					4.71			4.71	5.20	

076 100 | Sheet Metal Roofing

		CREW	DAILY OUTPUT	LABOR-HOURS	UNIT	1998 BARE COSTS MAT.	LABOR	EQUIP.	TOTAL	TOTAL INCL O&P	
101	0010	**COPPER ROOFING** Batten seam, over 10 sq, 16 oz, 130 lb/sq	1 Shee	1.10	7.273	Sq.	435	147		582	730	101
	0200	18 oz, 145 lb per sq		1	8		485	161		646	810	
	0400	Standing seam, over 10 squares, 16 oz, 125 lb per sq		1.30	6.154		420	124		544	670	
	0600	18 oz, 140 lb per sq		1.20	6.667		470	134		604	740	
	0900	Flat seam, over 10 squares, 16 oz, 115 lb per sq		1.20	6.667		385	134		519	650	
	1200	For abnormal conditions or small areas, add					25%	100%				
	1300	For lead-coated copper, add					25%					
102	0010	**LEAD ROOFING** 5 lb. per SF, batten seam	1 Shee	1.20	6.667	Sq.	410	134		544	675	102
	0100	Flat seam	"	1.30	6.154	"	410	124		534	660	
105	0010	**ZINC** Copper alloy roofing, batten seam, .020" thick	1 Shee	1.20	6.667	Sq.	525	134		659	805	105
	0100	.027" thick		1.15	6.957		635	140		775	935	
	0300	.032" thick		1.10	7.273		715	147		862	1,025	
	0400	.040" thick		1.05	7.619		845	154		999	1,200	

THERMAL & MOISTURE PROTECTION — 7

076 100 | Sheet Metal Roofing

			DAILY OUTPUT	LABOR-HOURS	UNIT	1998 BARE COSTS				TOTAL INCL O&P	
		CREW				MAT.	LABOR	EQUIP.	TOTAL		
105	0600	For standing seam construction, deduct				Sq.	2%					**105**
	0700	For flat seam construction, deduct				↓	3%					

076 200 | Sheet Mtl Flash & Trim

			DAILY OUTPUT	LABOR-HOURS	UNIT	MAT.	LABOR	EQUIP.	TOTAL	TOTAL INCL O&P	
201	0010	**DOWNSPOUTS** Aluminum 2″ x 3″, .020″ thick, embossed	1 Shee	190	.042	L.F.	.70	.85		1.55	2.21	**201**
	0100	Enameled		190	.042		.65	.85		1.50	2.15	
	0300	Enameled, .024″ thick, 2″ x 3″		180	.044		1.01	.90		1.91	2.63	
	0400	3″ x 4″		140	.057		1.35	1.15		2.50	3.44	
	0600	Round, corrugated aluminum, 3″ diameter, .020″ thick		190	.042		.85	.85		1.70	2.38	
	0700	4″ diameter, .025″ thick		140	.057	↓	1.29	1.15		2.44	3.37	
	0900	Wire strainer, round, 2″ diameter		155	.052	Ea.	1.80	1.04		2.84	3.74	
	1000	4″ diameter		155	.052		1.85	1.04		2.89	3.80	
	1200	Rectangular, perforated, 2″ x 3″		145	.055		2.17	1.11		3.28	4.27	
	1300	3″ x 4″		145	.055	↓	3.12	1.11		4.23	5.30	
	1500	Copper, round, 16 oz., stock, 2″ diameter		190	.042	L.F.	4.88	.85		5.73	6.80	
	1600	3″ diameter		190	.042		3.82	.85		4.67	5.65	
	1800	4″ diameter		145	.055		5.20	1.11		6.31	7.60	
	1900	5″ diameter		130	.062		6.70	1.24		7.94	9.50	
	2100	Rectangular, corrugated copper, stock, 2″ x 3″		190	.042		4.02	.85		4.87	5.85	
	2200	3″ x 4″		145	.055		5.40	1.11		6.51	7.80	
	2400	Rectangular, plain copper, stock, 2″ x 3″		190	.042		4.71	.85		5.56	6.65	
	2500	3″ x 4″		145	.055	↓	6.20	1.11		7.31	8.70	
	2700	Wire strainers, rectangular, 2″ x 3″		145	.055	Ea.	2.76	1.11		3.87	4.92	
	2800	3″ x 4″		145	.055		4.37	1.11		5.48	6.70	
	3000	Round, 2″ diameter		145	.055		2.59	1.11		3.70	4.73	
	3100	3″ diameter		145	.055		3.62	1.11		4.73	5.85	
	3300	4″ diameter		145	.055		5.60	1.11		6.71	8.05	
	3400	5″ diameter		115	.070	↓	8.10	1.40		9.50	11.25	
	3600	Lead-coated copper, round, stock, 2″ diameter		190	.042	L.F.	7	.85		7.85	9.15	
	3700	3″ diameter		190	.042		4.84	.85		5.69	6.75	
	3900	4″ diameter		145	.055		6.30	1.11		7.41	8.80	
	4300	Rectangular, corrugated, stock, 2″ x 3″		190	.042		5.05	.85		5.90	7	
	4500	Plain, stock, 2″ x 3″		190	.042		6	.85		6.85	8.05	
	4600	3″ x 4″		145	.055		7.55	1.11		8.66	10.20	
	4800	Steel, galvanized, round, corrugated, 2″ or 3″ diam, 28 ga		190	.042		.71	.85		1.56	2.22	
	4900	4″ diameter, 28 gauge		145	.055		.92	1.11		2.03	2.89	
	5700	Rectangular, corrugated, 28 gauge, 2″ x 3″		190	.042		.53	.85		1.38	2.02	
	5800	3″ x 4″		145	.055		1.48	1.11		2.59	3.51	
	6000	Rectangular, plain, 28 gauge, galvanized, 2″ x 3″		190	.042		.79	.85		1.64	2.31	
	6100	3″ x 4″		145	.055		1.21	1.11		2.32	3.21	
	6300	Epoxy painted, 24 gauge, corrugated, 2″ x 3″		190	.042		1.01	.85		1.86	2.55	
	6400	3″ x 4″		145	.055	↓	1.71	1.11		2.82	3.76	
	6600	Wire strainers, rectangular, 2″ x 3″		145	.055	Ea.	1.62	1.11		2.73	3.66	
	6700	3″ x 4″		145	.055		2.61	1.11		3.72	4.75	
	6900	Round strainers, 2″ or 3″ diameter		145	.055		1.20	1.11		2.31	3.20	
	7000	4″ diameter		145	.055		1.42	1.11		2.53	3.44	
	7200	5″ diameter		145	.055		2.20	1.11		3.31	4.30	
	7300	6″ diameter		115	.070	↓	2.63	1.40		4.03	5.25	
	8200	Vinyl, rectangular, 2″ x 3″		210	.038	L.F.	.72	.77		1.49	2.09	
	8300	Round, 2-1/2″		220	.036	″	.72	.73		1.45	2.03	
202	0010	**DRIP EDGE**, aluminum, .016″ thick, 5″ wide, mill finish	1 Carp	400	.020	L.F.	.19	.37		.56	.84	**202**
	0100	White finish		400	.020		.21	.37		.58	.86	
	0200	8″ wide, mill finish		400	.020		.28	.37		.65	.94	
	0300	Ice belt, 28″ wide, mill finish		100	.080		1.64	1.48		3.12	4.34	
	0310	Vented, mill finish		400	.020		1.39	.37		1.76	2.16	
	0320	Painted finish		400	.020	↓	1.50	.37		1.87	2.28	

Important: See the Reference Section for critical supporting data - Reference Nos., Crews, & Location Factors

076 200	Sheet Mtl Flash & Trim	CREW	DAILY OUTPUT	LABOR-HOURS	UNIT	1998 BARE COSTS				TOTAL INCL O&P	
						MAT.	LABOR	EQUIP.	TOTAL		
202 0400	Galvanized, 5" wide	1 Carp	400	.020	L.F.	.21	.37		.58	.86	**202**
0500	8" wide, mill finish		400	.020		.30	.37		.67	.96	
0510	Rake edge, aluminum, 1-1/2" x 1-1/2"		400	.020		.12	.37		.49	.76	
0520	3-1/2" x 1-1/2"		400	.020		.18	.37		.55	.83	
203 0010	**ELBOWS** Aluminum, 2" x 3", embossed	1 Shee	100	.080	Ea.	.96	1.61		2.57	3.79	**203**
0100	Enameled		100	.080		1.67	1.61		3.28	4.57	
0200	3" x 4", .025" thick, embossed		100	.080		2.95	1.61		4.56	6	
0300	Enameled		100	.080		2.95	1.61		4.56	6	
0400	Round corrugated, 3", embossed, .020" thick		100	.080		1.95	1.61		3.56	4.88	
0500	4", .025" thick		100	.080		2.10	1.61		3.71	5.05	
0600	Copper, 16 oz. round, 2" diameter		100	.080		11	1.61		12.61	14.85	
0700	3" diameter		100	.080		16.05	1.61		17.66	20.50	
0800	4" diameter		100	.080		16.95	1.61		18.56	21.50	
1000	2" x 3" corrugated		100	.080		15.75	1.61		17.36	20	
1100	3" x 4" corrugated		100	.080		16.15	1.61		17.76	20.50	
1300	Vinyl, 2-1/2" diameter, 45° or 75°		100	.080		2	1.61		3.61	4.93	
1400	Tee Y junction		75	.107		8.50	2.15		10.65	13	
204 0010	**FLASHING** Aluminum, mill finish, .013" thick	1 Shee	145	.055	S.F.	.33	1.11		1.44	2.24	**204**
0030	.016" thick		145	.055		.49	1.11		1.60	2.42	
0060	.019" thick		145	.055		.79	1.11		1.90	2.75	
0100	.032" thick		145	.055		.96	1.11		2.07	2.93	
0200	.040" thick		145	.055		1.40	1.11		2.51	3.42	
0300	.050" thick		145	.055		1.93	1.11		3.04	4	
0325	Mill finish 5" x 7" step flashing, .016" thick		1,920	.004	Ea.	.10	.08		.18	.25	
0350	Mill finish 12" x 12" step flashing, .016" thick		1,600	.005	"	.42	.10		.52	.63	
0400	Painted finish, add				S.F.	.25			.25	.28	
0500	Fabric-backed 2 sides, .004" thick	1 Shee	330	.024		.85	.49		1.34	1.77	
0700	.005" thick		330	.024		1.01	.49		1.50	1.94	
0750	Mastic-backed, self adhesive		460	.017		2.20	.35		2.55	3.01	
0800	Mastic-coated 2 sides, .004" thick		330	.024		.81	.49		1.30	1.72	
1000	.005" thick		330	.024		.96	.49		1.45	1.89	
1100	.016" thick		330	.024		1.18	.49		1.67	2.13	
1300	Asphalt flashing cement, 5 gallon				Gal.	3.19			3.19	3.51	
1600	Copper, 16 oz, sheets, under 1000 lbs.	1 Shee	115	.070	S.F.	3.35	1.40		4.75	6.05	
1700	Over 4000 lbs.		155	.052		3.11	1.04		4.15	5.20	
1900	20 oz sheets, under 1000 lbs.		110	.073		4.18	1.47		5.65	7.10	
2000	Over 4000 lbs.		145	.055		3.88	1.11		4.99	6.15	
2200	24 oz sheets, under 1000 lbs.		105	.076		5	1.54		6.54	8.10	
2500	32 oz sheets, under 1000 lbs.		100	.080		6.70	1.61		8.31	10.10	
2700	W shape for valleys, 16 oz, 24" wide		100	.080	L.F.	6.50	1.61		8.11	9.90	
2800	Copper, paperbacked 1 side, 2 oz		330	.024	S.F.	.85	.49		1.34	1.77	
2900	3 oz		330	.024		1.11	.49		1.60	2.05	
3100	Paperbacked 2 sides, 2 oz		330	.024		.90	.49		1.39	1.82	
3150	3 oz		330	.024		1.21	.49		1.70	2.16	
3200	5 oz		330	.024		1.70	.49		2.19	2.70	
3400	Mastic-backed 2 sides, copper, 2 oz		330	.024		1.03	.49		1.52	1.96	
3500	3 oz		330	.024		1.33	.49		1.82	2.29	
3700	5 oz		330	.024		1.92	.49		2.41	2.94	
3800	Fabric-backed 2 sides, copper, 2 oz		330	.024		1.08	.49		1.57	2.02	
4000	3 oz		330	.024		1.32	.49		1.81	2.28	
4100	5 oz		330	.024		1.88	.49		2.37	2.90	
4300	Copper-clad stainless steel, .015" thick, under 500 lbs.		115	.070		3	1.40		4.40	5.65	
4600	.018" thick, under 500 lbs.		100	.080		4.08	1.61		5.69	7.20	
4650	CPE Clad Metal		285	.028	L.F.	1.80	.57		2.37	2.94	
4710	CSPE Bonded		285	.028		1.60	.57		2.17	2.72	

076 200 | Sheet Mtl Flash & Trim	CREW	DAILY OUTPUT	LABOR-HOURS	UNIT	1998 BARE COSTS				TOTAL INCL O&P
					MAT.	LABOR	EQUIP.	TOTAL	
204 4750 EPDM Cured	1 Shee	285	.028	S.F.	1.05	.57		1.62	2.11 **204**
4800 Uncured	↓	285	.028	↓	1.15	.57		1.72	2.23
4900 Fabric, asphalt-saturated cotton, specification grade	1 Rofc	35	.229	S.Y.	1.55	3.70		5.25	8.60
5000 Utility grade		35	.229		1.12	3.70		4.82	8.15
5200 Open-mesh fabric, saturated, 40 oz per S.Y.		35	.229		1.30	3.70		5	8.35
5300 Close-mesh fabric, saturated, 17 oz per S.Y.		35	.229		1.38	3.70		5.08	8.40
5500 Fiberglass, resin-coated		35	.229		1.10	3.70		4.80	8.10
5600 Asphalt-coated, 40 oz per S.Y.	↓	35	.229	↓	7.90	3.70		11.60	15.55
5650 Hypalon Clad Metal	1 Shee	285	.028	S.F.	2.65	.57		3.22	3.88
5800 Lead, 2.5 lb. per SF, up to 12″ wide	1 Rofc	135	.059		3.37	.96		4.33	5.50
5900 Over 12″ wide	″	135	.059		3.37	.96		4.33	5.50
6100 Lead-coated copper, fabric-backed, 2 oz	1 Shee	330	.024		1.40	.49		1.89	2.37
6200 5 oz		330	.024		1.64	.49		2.13	2.63
6400 Mastic-backed 2 sides, 2 oz		330	.024		1.05	.49		1.54	1.98
6500 5 oz		330	.024		1.41	.49		1.90	2.38
6700 Paperbacked 1 side, 2 oz		330	.024		.94	.49		1.43	1.86
6800 3 oz		330	.024		1.10	.49		1.59	2.04
7000 Paperbacked 2 sides, 2 oz		330	.024		.96	.49		1.45	1.89
7100 5 oz		330	.024		1.58	.49		2.07	2.57
7150 Neoprene, 60 mil		285	.028		1.78	.57		2.35	2.92
7160 Self-curing		285	.028		1.37	.57		1.94	2.47
7170 Uncured	↓	285	.028		1.33	.57		1.90	2.42
7300 Polyvinyl chloride, black, .010″ thick	1 Rofc	285	.028		.13	.45		.58	.99
7400 .020″ thick		285	.028		.20	.45		.65	1.07
7600 .030″ thick		285	.028		.30	.45		.75	1.18
7700 .056″ thick		285	.028		.74	.45		1.19	1.66
7900 Black or white for exposed roofs, .060″ thick		285	.028		1.59	.45		2.04	2.60
8000 Mineral Fiber-backed for parking decks, .045″ thick		285	.028		1.08	.45		1.53	2.04
8050 PVC (19 mils) coated galv steel (24 mils), 4′ x 8′ sheets	↓	240	.033	↓	2.08	.54		2.62	3.30
8060 PVC tape, 5″ x 45 mils, for joint covers, 100 L.F./roll				Ea.	79.50			79.50	87
8100 Rubber, butyl, 1/32″ thick	1 Rofc	285	.028	S.F.	.68	.45		1.13	1.60
8200 1/16″ thick		285	.028		1.04	.45		1.49	1.99
8300 Neoprene, cured, 1/16″ thick		285	.028		1.47	.45		1.92	2.47
8400 1/8″ thick	↓	285	.028		2.98	.45		3.43	4.13
8500 Shower pan, bituminous membrane, 7 oz	1 Shee	155	.052		1.07	1.04		2.11	2.94
8550 3 ply copper and fabric, 3 oz		155	.052		1.55	1.04		2.59	3.46
8600 7 oz		155	.052		3.25	1.04		4.29	5.35
8650 Copper, 16 oz		100	.080		3.35	1.61		4.96	6.40
8700 Lead on copper and fabric, 5 oz		155	.052		1.64	1.04		2.68	3.56
8800 7 oz		155	.052		2.60	1.04		3.64	4.62
8900 Stainless steel sheets, 32 ga, .010″ thick		155	.052		2.15	1.04		3.19	4.13
9000 28 ga, .015″ thick		155	.052		2.55	1.04		3.59	4.56
9100 26 ga, .018″ thick		155	.052		3.15	1.04		4.19	5.25
9200 24 ga, .025″ thick	↓	155	.052	↓	4.08	1.04		5.12	6.25
9290 For mechanically keyed flashing, add					40%				
9300 Stainless steel, paperbacked 2 sides, .005″ thick	1 Shee	330	.024	S.F.	1.89	.49		2.38	2.91
9320 Steel sheets, galvanized, 20 gauge		130	.062		.67	1.24		1.91	2.84
9340 30 gauge		160	.050		.28	1.01		1.29	2.01
9400 Terne coated stainless steel, .015″ thick, 28 ga		155	.052		3.97	1.04		5.01	6.15
9500 .018″ thick, 26 ga		155	.052		4.48	1.04		5.52	6.70
9600 Zinc and copper alloy (brass), .020″ thick		155	.052		3.20	1.04		4.24	5.30
9700 .027″ thick		155	.052		4.28	1.04		5.32	6.45
9800 .032″ thick		155	.052		5	1.04		6.04	7.25
9900 .040″ thick	↓	155	.052	↓	6.10	1.04		7.14	8.45
205 0010 **GUTTERS** Aluminum, stock units, 5″ box, .027″ thick, plain	1 Shee	120	.067	L.F.	.87	1.34		2.21	3.22 **205**
0020 Inside corner	↓	25	.320	Ea.	4.70	6.45		11.15	16.05

Important: See the Reference Section for critical supporting data - Reference Nos., Crews, & Location Factors

076 200 | Sheet Mtl Flash & Trim

			CREW	DAILY OUTPUT	LABOR-HOURS	UNIT	MAT.	LABOR	EQUIP.	TOTAL	TOTAL INCL O&P	
205	0030	Outside corner	1 Shee	25	.320	Ea.	4.70	6.45		11.15	16.05	205
	0100	Enameled		120	.067	L.F.	1.05	1.34		2.39	3.42	
	0110	Inside corner		25	.320	Ea.	4.75	6.45		11.20	16.10	
	0120	Outside corner		25	.320	"	4.75	6.45		11.20	16.10	
	0600	5" x 6" combination fascia & gutter, .032" thick, enameled		60	.133	L.F.	3.45	2.69		6.14	8.35	
	3000	Vinyl, O.G., 4" wide	1 Carp	110	.073		.85	1.35		2.20	3.25	
	3100	5" wide		110	.073		1	1.35		2.35	3.41	
	3200	4" half round, stock units		110	.073		.68	1.35		2.03	3.06	
	3250	Joint connectors				Ea.	1.36			1.36	1.50	
	3300	Wood, clear treated cedar, fir or hemlock, 3" x 4"	1 Carp	100	.080	L.F.	6.30	1.48		7.78	9.45	
	3400	4" x 5"	"	100	.080	"	7.30	1.48		8.78	10.55	
206	0010	**GUTTER GUARD** 6" wide strip, aluminum mesh	1 Carp	500	.016	L.F.	.40	.30		.70	.95	206
	0100	Vinyl mesh	"	500	.016	"	.22	.30		.52	.75	
210	0010	**REGLET** Aluminum, .025" thick, in concrete parapet	1 Carp	225	.036	L.F.	.92	.66		1.58	2.14	210
	0100	Copper, 10 oz.		225	.036		1.59	.66		2.25	2.88	
	0300	16 oz.		225	.036		2.12	.66		2.78	3.46	
	0400	Galvanized steel, 24 gauge		225	.036		.78	.66		1.44	1.99	
	0600	Stainless steel, .020" thick		225	.036		1.57	.66		2.23	2.86	
	0700	Zinc and copper alloy, 20 oz.		225	.036		1.75	.66		2.41	3.06	
	0900	Counter flashing for above, 12" wide, .032" aluminum	1 Shee	150	.053		1.19	1.07		2.26	3.13	
	1000	Copper, 10 oz.		150	.053		3.32	1.07		4.39	5.45	
	1200	16 oz.		150	.053		3.70	1.07		4.77	5.90	
	1300	Galvanized steel, .020" thick		150	.053		.62	1.07		1.69	2.50	
	1500	Stainless steel, .020" thick		150	.053		2.76	1.07		3.83	4.86	
	1600	Zinc and copper alloy, 20 oz.		150	.053		3.11	1.07		4.18	5.25	

077 100 | Prefab Roof Specialties

			CREW	DAILY OUTPUT	LABOR-HOURS	UNIT	MAT.	LABOR	EQUIP.	TOTAL	TOTAL INCL O&P	
104	0010	**FASCIA** Aluminum, reverse board and batten,										104
	0100	.032" thick, colored, no furring included	1 Shee	145	.055	S.F.	2.02	1.11		3.13	4.10	
	0200	Residential type, aluminum	1 Carp	200	.040	L.F.	1.21	.74		1.95	2.60	
	0300	Steel, galv and enameled, stock, no furring, long panels	1 Shee	145	.055	S.F.	2.19	1.11		3.30	4.29	
	0600	Short panels	"	115	.070	"	3.31	1.40		4.71	6	
105	0010	**GRAVEL STOP** Aluminum, .050" thick, 4" face height, mill finish	1 Shee	145	.055	L.F.	2.72	1.11		3.83	4.87	105
	0080	Duranodic finish		145	.055		3.69	1.11		4.80	5.95	
	0100	Painted		145	.055		4.26	1.11		5.37	6.55	
	1350	Galv steel, 24 ga., 4" leg, plain, with continuous cleat, 4" face		145	.055		1.50	1.11		2.61	3.53	
	1500	Polyvinyl chloride, 6" face height		135	.059		3.28	1.19		4.47	5.65	
	1800	Stainless steel, 24 ga., 6" face height		135	.059		7.15	1.19		8.34	9.85	

077 200 | Roof Accessories

			CREW	DAILY OUTPUT	LABOR-HOURS	UNIT	MAT.	LABOR	EQUIP.	TOTAL	TOTAL INCL O&P	
206	0010	**ROOF HATCHES** With curb, 1" fiberglass insulation, 2'-6" x 3'-0"										206
	0500	Aluminum curb and cover	G-3	10	3.200	Ea.	405	53.50		458.50	535	
	0520	Galvanized steel curb and aluminum cover		10	3.200		340	53.50		393.50	465	
	0540	Galvanized steel curb and cover		10	3.200		300	53.50		353.50	420	
	0600	2'-6" x 4'-6", aluminum curb and cover		9	3.556		560	59.50		619.50	715	
	0800	Galvanized steel curb and aluminum cover		9	3.556		475	59.50		534.50	625	

077 200	Roof Accessories	CREW	DAILY OUTPUT	LABOR-HOURS	UNIT	1998 BARE COSTS				TOTAL INCL O&P	
						MAT.	LABOR	EQUIP.	TOTAL		
206 0900	Galvanized steel curb and cover	G-3	9	3.556	Ea.	460	59.50		519.50	605	**206**
1200	2'-6" x 8'-0", aluminum curb and cover		6.60	4.848		995	81.50		1,076.50	1,250	
1400	Galvanized steel curb and aluminum cover		6.60	4.848		920	81.50		1,001.50	1,150	
1500	Galvanized steel curb and cover		6.60	4.848		905	81.50		986.50	1,125	
1800	For plexiglass panels, 2'-6" x 3'-0", add to above					345			345	380	
212 0010	**VENTS, ONE-WAY** For insul. decks, 1 per M.S.F., plastic, min.	1 Rofc	40	.200	Ea.	11.50	3.24		14.74	18.70	**212**
0100	Maximum		20	.400		26	6.50		32.50	40.50	
0300	Aluminum		30	.267		11.95	4.32		16.27	21	
0800	Polystyrene baffles, 12" wide for 16" O.C. rafter spacing	1 Carp	90	.089		.50	1.64		2.14	3.37	
0900	For 24" O.C. rafter spacing	"	110	.073		1.05	1.35		2.40	3.46	
215 0010	**PITCH POCKETS**										**215**
0100	Adjustable, 4" to 7", welded corners, 4" deep	1 Rofc	48	.167	Ea.	10.50	2.70		13.20	16.60	
0200	Side extenders, 6"	"	240	.033	"	1.70	.54		2.24	2.88	

078 100	Plastic Skylights	CREW	DAILY OUTPUT	LABOR-HOURS	UNIT	1998 BARE COSTS				TOTAL INCL O&P	
						MAT.	LABOR	EQUIP.	TOTAL		
101 0010	**SKYLIGHT** Plastic domes, flush or curb mounted, ten or										**101**
0100	more units, curb not included, "L" frames										
0300	Nominal size under 10 S.F., double	G-3	130	.246	S.F.	16.45	4.13		20.58	25	
0400	Single		160	.200		13.10	3.35		16.45	20	
0600	10 S.F. to 20 S.F., double		315	.102		13.40	1.70		15.10	17.60	
0700	Single		395	.081		13.10	1.36		14.46	16.70	
0900	20 S.F. to 30 S.F., double		395	.081		13.70	1.36		15.06	17.35	
1000	Single		465	.069		11.60	1.15		12.75	14.70	
1200	30 S.F. to 65 S.F., double		465	.069		14.20	1.15		15.35	17.60	
1300	Single		610	.052		11.90	.88		12.78	14.60	
1500	For insulated 4" curbs, double, add					25%					
1600	Single, add					30%					
1800	For integral insulated 9" curbs, double, add					30%					
1900	Single, add					40%					
2120	Ventilating insulated plexiglass dome with										
2130	curb mounting, 36" x 36"	G-3	12	2.667	Ea.	330	44.50		374.50	440	
2150	52" x 52"		12	2.667		495	44.50		539.50	620	
2160	28" x 52"		10	3.200		385	53.50		438.50	515	
2170	36" x 52"		10	3.200		420	53.50		473.50	550	
2180	For electric opening system, add					248			248	273	
2210	Operating skylight, with thermopane glass, 24" x 48"	G-3	10	3.200		485	53.50		538.50	625	
2220	32" x 48"	"	9	3.556		510	59.50		569.50	660	
2310	Non venting insulated plexiglass dome skylight with										
2320	Flush mount 22" x 46"	G-3	15.23	2.101	Ea.	270	35		305	355	
2330	30" x 30"		16	2		248	33.50		281.50	330	
2340	46" x 46"		13.91	2.301		460	38.50		498.50	570	
2350	Curb mount 22" x 46"		15.23	2.101		237	35		272	320	
2360	30" x 30"		16	2		226	33.50		259.50	305	
2370	46" x 46"		13.91	2.301		425	38.50		463.50	530	
2381	Non-insulated flush mount 22" x 46"		15.23	2.101		182	35		217	260	

Important: See the Reference Section for critical supporting data - Reference Nos., Crews, & Location Factors

078 100 | Plastic Skylights

		CREW	DAILY OUTPUT	LABOR-HOURS	UNIT	1998 BARE COSTS				TOTAL INCL O&P
						MAT.	LABOR	EQUIP.	TOTAL	
101 2382	30" x 30"	G-3	16	2	Ea.	166	33.50		199.50	239
2383	46" x 46"		13.91	2.301		310	38.50		348.50	405
2384	Curb mount 22" x 46"		15.23	2.101		155	35		190	230
2385	30" x 30"		16	2		149	33.50		182.50	221
2400	Sandwich panels, fiberglass, for walls, 1-9/16" thick, to 250 SF		200	.160	S.F.	13.55	2.68		16.23	19.45
2500	250 SF and up		265	.121		12.15	2.03		14.18	16.80
2700	As above, but for roofs, 2-3/4" thick, to 250 SF		295	.108		19.55	1.82		21.37	24.50
2800	250 SF and up		330	.097		16	1.63		17.63	20.50

079 204 | Sealants & Caulkings

		CREW	DAILY OUTPUT	LABOR-HOURS	UNIT	1998 BARE COSTS				TOTAL INCL O&P
						MAT.	LABOR	EQUIP.	TOTAL	
204 0010	**CAULKING AND SEALANTS**									
0020	Acoustical sealant, elastomeric, cartridges				Ea.	1.95			1.95	2.15
0100	Acrylic latex caulk, white									
0200	11 fl. oz cartridge				Ea.	1.97			1.97	2.16
0500	1/4" x 1/2"	1 Bric	260	.031	L.F.	.16	.58		.74	1.17
0600	1/2" x 1/2"		250	.032		.32	.61		.93	1.38
0800	3/4" x 3/4"		230	.035		.72	.66		1.38	1.92
0900	3/4" x 1"		200	.040		.96	.76		1.72	2.34
1000	1" x 1"		180	.044		1.21	.84		2.05	2.76
1400	Butyl based, bulk				Gal.	22			22	24
1500	Cartridges				"	26.50			26.50	29.50
1700	Bulk, in place 1/4" x 1/2", 154 L.F./gal.	1 Bric	230	.035	L.F.	.14	.66		.80	1.28
1800	1/2" x 1/2", 77 L.F./gal.	"	180	.044	"	.29	.84		1.13	1.74
2000	Latex acrylic based, bulk				Gal.	23			23	25.50
2100	Cartridges				"	26			26	28.50
2200	Bulk in place, 1/4" x 1/2", 154 L.F./gal.	1 Bric	230	.035	L.F.	.15	.66		.81	1.28
2300	Polysulfide compounds, 1 component, bulk				Gal.	43.50			43.50	47.50
2400	Cartridges				"	46			46	51
2600	1 or 2 component, in place, 1/4" x 1/4", 308 L.F./gal.	1 Bric	145	.055	L.F.	.14	1.05		1.19	1.92
2700	1/2" x 1/4", 154 L.F./gal.		135	.059		.28	1.12		1.40	2.21
2900	3/4" x 3/8", 68 L.F./gal.		130	.062		.64	1.17		1.81	2.68
3000	1" x 1/2", 38 L.F./gal.		130	.062		1.14	1.17		2.31	3.23
3200	Polyurethane, 1 or 2 component				Gal.	41			41	45
3300	Cartridges				"	46			46	51
3500	Bulk, in place, 1/4" x 1/4"	1 Bric	150	.053	L.F.	.13	1.01		1.14	1.86
3600	1/2" x 1/4"		145	.055		.27	1.05		1.32	2.06
3800	3/4" x 3/8", 68 L.F./gal.		130	.062		.60	1.17		1.77	2.65
3900	1" x 1/2"		110	.073		1.08	1.38		2.46	3.52
4100	Silicone rubber, bulk				Gal.	34			34	37.50
4200	Cartridges				"	40			40	44

7

THERMAL & MOISTURE PROTECTION

For information about Means Estimating Seminars, see yellow pages 11 and 12 in back of book

For expanded coverage of these items see *Means Building Construction Cost Data 1998*

	CREW	DAILY OUTPUT	LABOR-HOURS	UNIT	1998 BARE COSTS				TOTAL INCL O&P
					MAT.	LABOR	EQUIP.	TOTAL	

Division 8
Doors & Windows

Estimating Tips

081 Metal Doors & Frames

- Most metal doors and frames look alike, but there may be significant differences among them. When estimating these items be sure to choose the line item that most closely compares to the specification or door schedule requirements regarding:
  - type of metal
  - metal gauge
  - door core material
  - fire rating
  - finish

082 Wood & Plastic Doors

- Wood and plastic doors vary considerably in price. The primary determinant is the veneer material. Lauan, birch and oak are the most common veneers. Other variables include the following:
  - hollow or solid core
  - fire rating
  - flush or raised panel
  - finish
- If the specifications require compliance with AWI (Architectural Woodwork Institute) standards or acoustical standards, the cost of the door may increase substantially. All wood doors are priced pre-mortised for hinges and predrilled for cylindrical locksets.

083 Special Doors

- There are many varieties of special doors, and they are usually priced per each. Add frames, hardware or operators required for a complete installation.

085 Metal Windows

- Most metal windows are delivered preglazed. However, some metal windows are priced without glass. Refer to 088 Glazing for glass pricing. The grade C indicates commercial grade windows, usually ASTM C 35.

086 Wood & Plastic Windows

- All wood windows are priced preglazed. The two glazing options priced are single pane float glass and insulating glass 1/2" thick. Add the cost of screens and grills if required.

087 Hardware

- Hardware costs add considerably to the cost of a door. The most efficient method to determine the hardware requirements for a project is to review the door schedule. This schedule, in conjunction with the specifications, is all you should need to take off the door hardware.
- Door hinges are priced by the pair, with most doors requiring 1-1/2 pairs per door. The hinge prices do not include installation labor because it is included in door installation. Hinges are classified according to the frequency of use.

088 Glazing

- Different openings require different types of glass. The three most common types are:
  - float
  - tempered
  - insulating
- Most exterior windows are glazed with insulating glass. Entrance doors and window walls, where the glass is less than 18" from the floor, are generally glazed with tempered glass. Interior windows and some residential windows are glazed with float glass.

089 Glazed Curtain Walls

- Glazed curtain walls consist of the metal tube framing and the glazing material. The cost data in this subdivision is presented for the metal tube framing alone or the composite wall. If your estimate requires a detailed takeoff of the framing, be sure to add the glazing cost.

Reference Numbers

Reference numbers are shown in bold squares at the beginning of some major classifications. These numbers refer to related items in the Reference Section. The reference information may be an estimating procedure, an alternate pricing method or technical information.

Note: Not all subdivisions listed here necessarily appear in this publication.

081 100 | Steel Doors & Frames

		CREW	DAILY OUTPUT	LABOR-HOURS	UNIT	MAT.	LABOR	EQUIP.	TOTAL	TOTAL INCL O&P	
103	**0010**	**COMMERCIAL STEEL DOORS**										**103**
	0015	Flush, full panel, hollow core	R081 -010								
	0020	1-3/8" thick, 20 ga., 2'-0"x 6'-8"	F-2	20	.800	Ea.	158	12.95		170.95	195
	0040	2'-8" x 6'-8"		18	.889		162	14.40		176.40	203
	0060	3'-0" x 6'-8"		17	.941		165	15.25		180.25	207
	0100	3'-0" x 7'-0"		17	.941		173	15.25		188.25	216
	0120	For vision lite, add					54.50			54.50	60
	0140	For narrow lite, add					58.50			58.50	64.50
	0160	For bottom louver, add					97			97	107
	0230	For baked enamel finish, add					30%	15%			
	0260	For galvanizing, add					15%				
	0320	Half glass, 20 ga., 2'-0" x 6'-8"	F-2	20	.800	Ea.	202	12.95		214.95	244
	0340	2'-8" x 6'-8"		18	.889		213	14.40		227.40	260
	0360	3'-0" x 6'-8"		17	.941		218	15.25		233.25	266
	0400	3'-0" x 7'-0"		17	.941		221	15.25		236.25	270
	0500	Hollow core, 1-3/4" thick, full panel, 20 ga., 2'-8" x 6'-8"		18	.889		152	14.40		166.40	192
	0520	3'-0" x 6'-8"		17	.941		173	15.25		188.25	217
	0640	3'-0" x 7'-0"		17	.941		181	15.25		196.25	225
	0680	4'-0" x 7'-0"		15	1.067		232	17.25		249.25	285
	0700	4'-0" x 8'-0"		13	1.231		268	19.90		287.90	330
	1000	18 ga., 2'-8" x 6'-8"		17	.941		199	15.25		214.25	245
	1020	3'-0" x 6'-8"		16	1		194	16.20		210.20	242
	1120	3'-0" x 7'-0"		17	.941		191	15.25		206.25	236
	1180	4'-0" x 7'-0"		14	1.143		256	18.50		274.50	315
	1200	4'-0" x 8'-0"		17	.941		305	15.25		320.25	360
	1230	Half glass, 20 ga., 2'-8" x 6'-8"		20	.800		246	12.95		258.95	293
	1240	3'-0" x 6'-8"		18	.889		246	14.40		260.40	296
	1260	3'-0" x 7'-0"		18	.889		258	14.40		272.40	310
	1280	4'-0" x 7'-0"		16	1		284	16.20		300.20	345
	1300	4'-0" x 8'-0"		13	1.231		330	19.90		349.90	400
	1320	18 ga., 2'-8" x 6'-8"		18	.889		271	14.40		285.40	325
	1340	3'-0" x 6'-8"		17	.941		266	15.25		281.25	320
	1360	3'-0" x 7'-0"		17	.941		276	15.25		291.25	330
	1380	4'-0" x 7'-0"		15	1.067		390	17.25		407.25	460
	1400	4'-0" x 8'-0"		14	1.143		450	18.50		468.50	525
	1720	Insulated, 1-3/4" thick, full panel, 18 ga., 3'-0" x 6'-8"		15	1.067		211	17.25		228.25	262
	1740	2'-8" x 7'-0"		16	1		208	16.20		224.20	257
	1760	3'-0" x 7'-0"		15	1.067		218	17.25		235.25	270
	1800	4'-0" x 8'-0"		13	1.231		305	19.90		324.90	370
	1820	Half glass, 18 ga., 3'-0" x 6'-8"		16	1		272	16.20		288.20	325
	1840	2'-8" x 7'-0"		17	.941		269	15.25		284.25	320
	1860	3'-0" x 7'-0"		16	1		278	16.20		294.20	335
	1900	4'-0" x 8'-0"		14	1.143		410	18.50		428.50	480
106	**0010**	**DOOR FRAMES**										**106**
	0020	Steel channels with anchors and bar stops									
	0100	6" channel @ 8.2#/L.F., 3' x 7' door, weighs 200#	E-4	13	2.462	Ea.	183	50.50	6.15	239.65	310
	0200	8" channel @ 11.5#/L.F., 6' x 8' door, weighs 300#		9	3.556		266	72.50	8.90	347.40	445
	0300	8' x 12' door, weighs 450#		6.50	4.923		400	101	12.30	513.30	650
	0800	For frames without bar stops, light sections, deduct					15%				
	0900	Heavy sections, deduct					10%				
110	**0010**	**FIRE DOOR**										**110**
	0015	Steel, flush, "B" label, 90 minute									
	0020	Full panel, 20 ga., 2'-0" x 6'-8"	F-2	20	.800	Ea.	181	12.95		193.95	221
	0040	2'-8" x 6'-8"		18	.889		186	14.40		200.40	230

Important: See the Reference Section for critical supporting data - Reference Nos., Crews, & Location Factors

081 100 | Steel Doors & Frames

		DAILY OUTPUT	LABOR-HOURS	UNIT	1998 BARE COSTS				TOTAL INCL O&P	
		CREW				MAT.	LABOR	EQUIP.	TOTAL	
110	0060	3'-0" x 6'-8"	F-2	17	.941	Ea.	190	15.25		205.25	234	**110**
	0080	3'-0" x 7'-0"		17	.941		197	15.25		212.25	242
	0140	18 ga., 3'-0" x 6'-8"		16	1		210	16.20		226.20	259
	0160	2'-8" x 7'-0"		17	.941		219	15.25		234.25	267
	0180	3'-0" x 7'-0"		16	1		216	16.20		232.20	266
	0200	4'-0" x 7'-0"		15	1.067		276	17.25		293.25	335
	0220	For "A" label, 3 hour, 18 ga., use same price as "B" label									
	0240	For vision lite, add				Ea.	56.50			56.50	62.50
	0520	Flush, "B" label 90 min., composite, 20 ga., 2'-0" x 6'-8"	F-2	18	.889		206	14.40		220.40	252
	0540	2'-8" x 6'-8"		17	.941		212	15.25		227.25	259
	0560	3'-0" x 6'-8"		16	1		222	16.20		238.20	273
	0580	3'-0" x 7'-0"		16	1		228	16.20		244.20	278
	0640	Flush, "A" label 3 hour, composite, 18 ga., 3'-0" x 6'-8"		15	1.067		252	17.25		269.25	310
	0660	2'-8" x 7'-0"		16	1		259	16.20		275.20	310
	0680	3'-0" x 7'-0"		15	1.067		264	17.25		281.25	320
	0700	4'-0" x 7'-0"		14	1.143		320	18.50		338.50	385
114	0011	**RESIDENTIAL DOOR**									**114**
	2510	Bi-passing closet, incl. hardware, no frame or trim incl.									
	2511	Mirrored, metal frame, 6'-8" x 4'-0" wide	F-2	10	1.600	Opng.	155	26		181	216
	2512	5'-0" wide		10	1.600		180	26		206	243
	2513	6'-0" wide		10	1.600		206	26		232	271
	2514	7'-0" wide		9	1.778		243	29		272	315
	2515	8'-0" wide		9	1.778		310	29		339	390
	2611	Mirrored, metal, 8'-0" x 4'-0" wide		10	1.600		196	26		222	261
	2612	5'-0" wide		10	1.600		235	26		261	305
	2613	6'-0" wide		10	1.600		265	26		291	335
	2614	7'-0" wide		9	1.778		310	29		339	390
	2615	8'-0" wide		9	1.778		395	29		424	485
118	0010	**STEEL FRAMES, KNOCK DOWN** R081-020									**118**
	0020	18 ga., up to 5-3/4" deep									
	0025	6'-8" high, 3'-0" wide, single	F-2	16	1	Ea.	69.50	16.20		85.70	104
	0040	6'-0" wide, double		14	1.143		81	18.50		99.50	121
	0100	7'-0" high, 3'-0" wide, single		16	1		72	16.20		88.20	107
	0140	6'-0" wide, double		14	1.143		83	18.50		101.50	123
	1000	18 ga., up to 4-7/8" deep, 7'-0" H, 3'-0" W, single		16	1		78	16.20		94.20	113
	1140	6'-0" wide, double		14	1.143		84	18.50		102.50	124
	2800	16 ga., up to 3-7/8" deep, 7'-0" high, 3'-0" wide, single		16	1		64.50	16.20		80.70	98.50
	2840	6'-0" wide, double		14	1.143		96.50	18.50		115	138
	3600	5-3/4" deep, 7'-0" high, 4'-0" wide, single		15	1.067		82	17.25		99.25	120
	3640	8'-0" wide, double		12	1.333		108	21.50		129.50	155
	3700	8'-0" high, 4'-0" wide, single		15	1.067		88	17.25		105.25	127
	3740	8'-0" wide, double		12	1.333		111	21.50		132.50	159
	4000	6-3/4" deep, 7'-0" high, 4'-0" wide, single		15	1.067		81.50	17.25		98.75	119
	4040	8'-0" wide, double		12	1.333		101	21.50		122.50	148
	4100	8'-0" high, 4'-0" wide, single		15	1.067		95.50	17.25		112.75	135
	4140	8'-0" wide, double		12	1.333		119	21.50		140.50	168
	4400	8-3/4" deep, 7'-0" high, 4'-0" wide, single		15	1.067		91	17.25		108.25	130
	4440	8'-0" wide, double		12	1.333		124	21.50		145.50	173
	4500	8'-0" high, 4'-0" wide, single		15	1.067		105	17.25		122.25	145
	4540	8'-0" wide, double		12	1.333		129	21.50		150.50	179
	4900	For welded frames, add					29.50			29.50	32.50
	5400	16 ga., "B" label, up to 5-3/4" deep, 7'-0" high, 4'-0" wide, single	F-2	15	1.067		90	17.25		107.25	129
	5440	8'-0" wide, double		12	1.333		111	21.50		132.50	160
	5800	6-3/4" deep, 7'-0" high, 4'-0" wide, single		15	1.067		91	17.25		108.25	130
	5840	8'-0" wide, double		12	1.333		105	21.50		126.50	152
	6200	8-3/4" deep, 7'-0" high, 4'-0" wide, single		15	1.067		104	17.25		121.25	144

8

DOORS & WINDOWS

For expanded coverage of these items see *Means Building Construction Cost Data 1998*

081 100 | Steel Doors & Frames

118			CREW	DAILY OUTPUT	LABOR-HOURS	UNIT	1998 BARE COSTS				TOTAL INCL O&P	118
							MAT.	LABOR	EQUIP.	TOTAL		
	6240	8'-0" wide, double	F-2	12	1.333	Ea.	122	21.50		143.50	171	
	6300	For "A" label use same price as "B" label	R081 -020									
	6400	For baked enamel finish, add					30%	15%				
	6500	For galvanizing, add					15%					
	7900	Transom lite frames, fixed, add	F-2	155	.103	S.F.	23.50	1.67		25.17	29	
	8000	Movable, add	"	130	.123	"	27.50	1.99		29.49	34	

082 050 | Wood & Plastic Doors

054			CREW	DAILY OUTPUT	LABOR-HOURS	UNIT	1998 BARE COSTS				TOTAL INCL O&P	054
							MAT.	LABOR	EQUIP.	TOTAL		
	0010	**WOOD FRAMES**										
	0400	Exterior frame, incl. ext. trim, pine, 5/4 x 4-9/16" deep	R082 -120 / F-2	375	.043	L.F.	3.62	.69		4.31	5.15	
	0420	5-3/16" deep		375	.043		4.75	.69		5.44	6.40	
	0440	6-9/16" deep		375	.043		5.30	.69		5.99	7	
	0600	Oak, 5/4 x 4-9/16" deep		350	.046		7.20	.74		7.94	9.15	
	0620	5-3/16" deep		350	.046		7.75	.74		8.49	9.75	
	0640	6-9/16" deep		350	.046		8.85	.74		9.59	11	
	0800	Walnut, 5/4 x 4-9/16" deep		350	.046		9.35	.74		10.09	11.55	
	0820	5-3/16" deep		350	.046		12.10	.74		12.84	14.55	
	0840	6-9/16" deep		350	.046		14.20	.74		14.94	16.85	
	1000	Sills, 8/4 x 8" deep, oak, no horns		100	.160		10.10	2.59		12.69	15.55	
	1020	2" horns		100	.160		11.20	2.59		13.79	16.80	
	1040	3" horns		100	.160		12.20	2.59		14.79	17.90	
	1100	8/4 x 10" deep, oak, no horns		90	.178		13.55	2.88		16.43	19.80	
	1120	2" horns		90	.178		15.10	2.88		17.98	21.50	
	1140	3" horns		90	.178		16.45	2.88		19.33	23	
	2000	Exterior, colonial, frame & trim, 3' opng., in-swing, minimum		22	.727	Ea.	279	11.75		290.75	325	
	2010	Average		21	.762		570	12.35		582.35	645	
	2020	Maximum		20	.800		845	12.95		857.95	950	
	2100	5'-4" opening, in-swing, minimum		17	.941		475	15.25		490.25	545	
	2120	Maximum		15	1.067		885	17.25		902.25	1,000	
	2140	Out-swing, minimum		17	.941		505	15.25		520.25	580	
	2160	Maximum		15	1.067		920	17.25		937.25	1,025	
	2400	6'-0" opening, in-swing, minimum		16	1		505	16.20		521.20	585	
	2420	Maximum		10	1.600		920	26		946	1,050	
	2460	Out-swing, minimum		16	1		625	16.20		641.20	715	
	2480	Maximum		10	1.600		1,150	26		1,176	1,300	
	2600	For two sidelights, add, minimum		30	.533	Opng.	177	8.65		185.65	210	
	2620	Maximum		20	.800	"	940	12.95		952.95	1,050	
	2700	Custom birch frame, 3'-0" opening		16	1	Ea.	169	16.20		185.20	213	
	2750	6'-0" opening		16	1		240	16.20		256.20	292	
	2900	Exterior, modern, plain trim, 3' opng., in-swing, minimum		26	.615		24	9.95		33.95	43.50	
	2920	Average		24	.667		26	10.80		36.80	47	
	2940	Maximum		22	.727		34.50	11.75		46.25	58	
	3000	Interior frame, pine, 11/16" x 3-5/8" deep		375	.043	L.F.	3.57	.69		4.26	5.10	
	3020	4-9/16" deep		375	.043		4.22	.69		4.91	5.80	
	3200	Oak, 11/16" x 3-5/8" deep		350	.046		4.86	.74		5.60	6.60	
	3220	4-9/16" deep		350	.046		4.86	.74		5.60	6.60	
	3240	5-3/16" deep		350	.046		5	.74		5.74	6.75	
	3400	Walnut, 11/16" x 3-5/8" deep		350	.046		8	.74		8.74	10.05	

Important: See the Reference Section for critical supporting data - Reference Nos., Crews, & Location Factors

082 050	Wood & Plastic Doors	CREW	DAILY OUTPUT	LABOR-HOURS	UNIT	1998 BARE COSTS				TOTAL INCL O&P	
						MAT.	LABOR	EQUIP.	TOTAL		
054											**054**
3420	4-9/16" deep	F-2	350	.046	L.F.	8	.74		8.74	10.05	
3440	5-3/16" deep		350	.046		8.30	.74		9.04	10.40	
3600	Pocket door frame		16	1	Ea.	68.50	16.20		84.70	103	
3800	Threshold, oak, 5/8" x 3-5/8" deep		200	.080	L.F.	2.92	1.29		4.21	5.45	
3820	4-5/8" deep		190	.084		3.72	1.36		5.08	6.40	
3840	5-5/8" deep		180	.089		4.47	1.44		5.91	7.40	
4000	For casing see division 062-212										
066											**066**
0010	**WOOD DOORS, DECORATOR**										
3000	Solid wood, 1-3/4" thick stile and rail										
3020	Mahogany, 3'-0" x 7'-0", minimum	F-2	14	1.143	Ea.	340	18.50		358.50	405	
3030	Maximum		10	1.600		570	26		596	675	
3040	3'-6" x 8'-0", minimum		10	1.600		620	26		646	730	
3050	Maximum		8	2		660	32.50		692.50	780	
3100	Pine, 3'-0" x 7'-0", minimum		14	1.143		283	18.50		301.50	340	
3120	3'-6" x 8'-0", minimum		10	1.600		560	26		586	660	
3130	Maximum		8	2		1,525	32.50		1,557.50	1,725	
3200	Red oak, 3'-0" x 7'-0", minimum		14	1.143		640	18.50		658.50	735	
3210	Maximum		10	1.600		1,250	26		1,276	1,425	
3220	3'-6" x 8'-0", minimum		10	1.600		910	26		936	1,050	
3230	Maximum		8	2		1,650	32.50		1,682.50	1,850	
4000	Hand carved door, mahogany										
4020	3'-0" x 7'-0", minimum	F-2	14	1.143	Ea.	630	18.50		648.50	725	
4040	3'-6" x 8'-0", minimum		10	1.600		1,025	26		1,051	1,175	
4050	Maximum		8	2		2,150	32.50		2,182.50	2,425	
4200	Red oak, 3'-0" x 7'-0", minimum		14	1.143		1,250	18.50		1,268.50	1,400	
4210	Maximum		11	1.455		2,925	23.50		2,948.50	3,250	
4220	3'-6" x 8'-0", minimum		10	1.600		2,500	26		2,526	2,800	
4280	For 6'-8" high door, deduct from 7'-0" door					22.50			22.50	25	
4400	For custom finish, add					99.50			99.50	109	
4600	Side light, mahogany, 7'-0" x 1'-6" wide, minimum	F-2	18	.889		250	14.40		264.40	300	
4610	Maximum		14	1.143		680	18.50		698.50	780	
4620	8'-0" x 1'-6" wide, minimum		14	1.143		310	18.50		328.50	370	
4630	Maximum		10	1.600		790	26		816	910	
4640	Side light, oak, 7'-0" x 1'-6" wide, minimum		18	.889		335	14.40		349.40	395	
4650	Maximum		14	1.143		785	18.50		803.50	890	
4660	8'-0" x 1-6" wide, minimum		14	1.143		410	18.50		428.50	480	
4670	Maximum		10	1.600		920	26		946	1,050	
6520	Interior cafe doors, 2'-6" opening, stock, panel pine		16	1		135	16.20		151.20	177	
6540	3'-0" opening		16	1		140	16.20		156.20	182	
6550	Louvered pine										
6560	2'-6" opening	F-2	16	1	Ea.	118	16.20		134.20	157	
8000	3'-0" opening		16	1		125	16.20		141.20	166	
8010	2'-6" opening, hardwood		16	1		169	16.20		185.20	214	
8020	3'-0" opening		16	1		197	16.20		213.20	245	
8800	Pre-hung doors, see division 082-082										
074											**074**
0010	**WOOD DOORS, PANELED**										
0020	Interior, six panel, hollow core, 1-3/8" thick										
0040	Molded hardboard, 2'-0" x 6'-8"	F-2	17	.941	Ea.	40	15.25		55.25	70	
0060	2'-6" x 6'-8"		17	.941		43	15.25		58.25	73.50	
0080	3'-0" x 6'-8"		17	.941		47.50	15.25		62.75	78	
0140	Embossed print, molded hardboard, 2'-0" x 6'-8"		17	.941		43	15.25		58.25	73.50	
0160	2'-6" x 6'-8"		17	.941		43	15.25		58.25	73.50	
0180	3'-0" x 6'-8"		17	.941		47.50	15.25		62.75	78	
0540	Six panel, solid, 1-3/8" thick, pine, 2'-0" x 6'-8"		15	1.067	L.F.	126	17.25		143.25	168	
0560	2'-6" x 6'-8"		14	1.143		130	18.50		148.50	175	

Note: line 3420 references **R082 -120**.

082 050 | Wood & Plastic Doors

		CREW	DAILY OUTPUT	LABOR-HOURS	UNIT	MAT.	LABOR	EQUIP.	TOTAL	TOTAL INCL O&P	
						1998 BARE COSTS					
074	0580	3'-0" x 6'-8"	F-2	13	1.231	Ea.	155	19.90		174.90	204	**074**
	1020	Two panel, bored rail, solid, 1-3/8" thick, pine, 1'-6" x 6'-8"		16	1		203	16.20		219.20	251	
	1040	2'-0" x 6'-8"		15	1.067		299	17.25		316.25	360	
	1060	2'-6" x 6'-8"		14	1.143		345	18.50		363.50	410	
	1340	Two panel, solid, 1-3/8" thick, fir, 2'-0" x 6'-8"		15	1.067		178	17.25		195.25	226	
	1360	2'-6" x 6'-8"		14	1.143		188	18.50		206.50	239	
	1380	3'-0" x 6'-8"		13	1.231		196	19.90		215.90	250	
	1740	Five panel, solid, 1-3/8" thick, fir, 2'-0" x 6'-8"		15	1.067		186	17.25		203.25	235	
	1760	2'-6" x 6'-8"		14	1.143		194	18.50		212.50	245	
	1780	3'-0" x 6'-8"		13	1.231		199	19.90		218.90	253	
082	0010	**PRE-HUNG DOORS**										**082**
	0300	Exterior, wood, combination storm & screen, 6'-9" x 2'-6" wide	F-2	15	1.067	Ea.	201	17.25		218.25	252	
	0320	2'-8" wide		15	1.067		201	17.25		218.25	252	
	0340	3'-0" wide		15	1.067		208	17.25		225.25	259	
	0360	For 7'-0" high door, add					27			27	29.50	
	0370	For aluminum storm doors, see division 083-900										
	1600	Entrance door, flush, birch, solid core										
	1620	4-5/8" solid jamb, 1-3/4" x 6'-8" x 2'-8" wide	F-2	16	1	Ea.	201	16.20		217.20	249	
	1640	3'-0" wide	"	16	1		207	16.20		223.20	256	
	1680	For 7'-0" high door, add					17			17	18.70	
	2000	Entrance door, colonial, 6 panel pine										
	2020	4-5/8" solid jamb, 1-3/4" x 6'-8" x 2'-8" wide	F-2	16	1	Ea.	415	16.20		431.20	485	
	2040	3'-0" wide	"	16	1		440	16.20		456.20	510	
	2060	For 7'-0" high door, add					17			17	18.70	
	2200	For 5-5/8" solid jamb, add					17			17	18.70	
	4000	Interior, passage door, 4-5/8" solid jamb										
	4400	Lauan, flush, solid core, 1-3/8" x 6'-8" x 2'-6" wide	F-2	20	.800	Ea.	160	12.95		172.95	197	
	4420	2'-8" wide		20	.800		161	12.95		173.95	199	
	4440	3'-0" wide		19	.842		165	13.65		178.65	205	
	4600	Hollow core, 1-3/8" x 6'-8" x 2'-6" wide		20	.800		130	12.95		142.95	165	
	4620	2'-8" wide		20	.800		131	12.95		143.95	166	
	4640	3'-0" wide		19	.842		133	13.65		146.65	170	
	4700	For 7'-0" high door, add					10.45			10.45	11.50	
	5000	Birch, flush, solid core, 1-3/8" x 6'-8" x 2'-6" wide	F-2	20	.800		163	12.95		175.95	201	
	5020	2'-8" wide		20	.800		163	12.95		175.95	201	
	5040	3'-0" wide		19	.842		168	13.65		181.65	209	
	5200	Hollow core, 1-3/8" x 6'-8" x 2'-6" wide		20	.800		136	12.95		148.95	172	
	5220	2'-8" wide		20	.800		152	12.95		164.95	189	
	5240	3'-0" wide		19	.842		155	13.65		168.65	195	
	5280	For 7'-0" high door, add					10.45			10.45	11.50	
	5500	Hardboard paneled, 1-3/8" x 6'-8" x 2'-6" wide	F-2	20	.800		163	12.95		175.95	201	
	5520	2'-8" wide		20	.800		166	12.95		178.95	204	
	5540	3'-0" wide		19	.842		168	13.65		181.65	209	
	6000	Pine paneled, 1-3/8" x 6'-8" x 2'-6" wide		20	.800		221	12.95		233.95	266	
	6020	2'-8" wide		20	.800		230	12.95		242.95	275	
	6500	For 5-5/8" solid jamb, add					10.35			10.35	11.40	
	6520	For split jamb, deduct					15.55			15.55	17.10	

083 100 | Sliding Doors

		CREW	DAILY OUTPUT	LABOR-HOURS	UNIT	MAT.	LABOR	EQUIP.	TOTAL	TOTAL INCL O&P	
102	0010 **GLASS, SLIDING**										102
	0012 Vinyl clad, 1" insul. glass, 6'-0" x 6'-10" high	2 Carp	4	4	Opng.	630	74		704	820	
	0030 6'-0" x 8'-0" high		4	4	Ea.	1,200	74		1,274	1,450	
	0100 8'-0" x 6'-10" high		4	4	Opng.	1,375	74		1,449	1,625	
	0500 3 leaf, 9'-0" x 6'-10" high		3	5.333		1,575	98.50		1,673.50	1,900	
	0600 12'-0" x 6'-10" high		3	5.333		2,100	98.50		2,198.50	2,475	
104	0010 **GLASS, SLIDING**										104
	0020 Wood, 5/8" tempered insul. glass, 6' wide, premium	2 Carp	4	4	Ea.	955	74		1,029	1,175	
	0100 Economy		4	4		725	74		799	920	
	0150 8' wide, wood, premium		3	5.333		1,275	98.50		1,373.50	1,575	
	0200 Economy		3	5.333		970	98.50		1,068.50	1,250	
	0250 12' wide, wood, premium		2.50	6.400		1,975	118		2,093	2,375	
	0300 Economy		2.50	6.400		1,500	118		1,618	1,850	
	0350 Aluminum, 5/8" tempered insulated glass, 6' wide										
	0400 Premium	2 Carp	4	4	Ea.	1,275	74		1,349	1,525	
	0450 Economy		4	4		1,050	74		1,124	1,275	
	0500 8' wide, premium		3	5.333		560	98.50		658.50	785	
	0550 Economy		3	5.333		490	98.50		588.50	710	
	0600 12' wide, premium		2.50	6.400		1,975	118		2,093	2,350	
	0650 Economy		2.50	6.400		1,675	118		1,793	2,050	

083 200 | Metal-Clad Doors

		CREW	DAILY OUTPUT	LABOR-HOURS	UNIT	MAT.	LABOR	EQUIP.	TOTAL	TOTAL INCL O&P	
202	0010 **KALAMEIN**										202
	0020 Interior, flush type, 3' x 7'	2 Carp	4.30	3.721	Opng.	165	69		234	300	
204	0010 **TIN CLAD** R083-100										204
	0020 3 ply, 6' x 7', double sliding, manual with hardware	2 Carp	1	16	Opng.	1,300	296		1,596	1,925	
	1000 For electric operator, add	1 Elec	2	4	"	2,300	82.50		2,382.50	2,650	

083 300 | Coiling Doors

		CREW	DAILY OUTPUT	LABOR-HOURS	UNIT	MAT.	LABOR	EQUIP.	TOTAL	TOTAL INCL O&P	
302	0010 **COUNTER DOORS**										302
	0020 Manual, incl. frm and hdwe, galv. stl., 4' roll-up, 6' long	2 Carp	2	8	Opng.	655	148		803	975	
	0300 Galvanized steel, UL label		1.80	8.889		1,025	164		1,189	1,400	
	0600 Stainless steel, 4' high roll-up, 6' long		2	8		1,075	148		1,223	1,450	
	0700 10' long		1.80	8.889		1,625	164		1,789	2,075	
	2000 Aluminum, 4' high, 4' long		2.20	7.273		560	135		695	845	
	2020 6' long		2	8		690	148		838	1,000	
	2040 8' long		1.90	8.421		820	156		976	1,175	
	2060 10' long		1.80	8.889		950	164		1,114	1,325	
	2080 14' long		1.40	11.429		1,250	211		1,461	1,725	
	2100 6' high, 4' long		2	8		720	148		868	1,050	
	2120 6' long		1.60	10		890	185		1,075	1,300	
	2140 10' long		1.40	11.429		1,250	211		1,461	1,725	

083 400 | Coiling Grilles

		CREW	DAILY OUTPUT	LABOR-HOURS	UNIT	MAT.	LABOR	EQUIP.	TOTAL	TOTAL INCL O&P	
404	0010 **COILING GRILLE**										404
	2020 Aluminum, manual operated, mill finish	2 Sswk	82	.195	S.F.	16.25	3.89		20.14	25.50	
	2040 Bronze anodized		82	.195	"	26	3.89		29.89	36	
	2060 Steel, manual operated, 10' x 10' high		1	16	Opng.	1,450	320		1,770	2,225	
	2080 15' x 8' high		.80	20	"	1,700	400		2,100	2,650	
	3000 For safety edge bottom bar, electric, add				L.F.	30			30	33	
	8000 For motor operation, add	2 Sswk	5	3.200	Opng.	730	64		794	930	

083 600 | Sectional Overhead Drs		CREW	DAILY OUTPUT	LABOR-HOURS	UNIT	1998 BARE COSTS				TOTAL INCL O&P
						MAT.	LABOR	EQUIP.	TOTAL	
604 0010	**OVERHEAD, COMMERCIAL** Frames not included									**604**
1000	Stock, sectional, heavy duty, wood, 1-3/4" thick, 8' x 8' high	2 Carp	2	8	Ea.	375	148		523	670
1100	10' x 10' high		1.80	8.889		575	164		739	915
1200	12' x 12' high		1.50	10.667		805	197		1,002	1,225
1300	Chain hoist, 14' x 14' high		1.30	12.308		1,400	228		1,628	1,925
1400	12' x 16' high		1	16		1,225	296		1,521	1,850
1500	20' x 8' high		1.30	12.270		1,150	227		1,377	1,675
1600	20' x 16' high		.65	24.615		2,900	455		3,355	3,950
1800	Center mullion openings, 8' high		4	4		450	74		524	620
1900	20' high		2	8		840	148		988	1,175
2100	For medium duty custom door, deduct					5%	5%			
2150	For medium duty stock doors, deduct					10%	5%			
2300	Fiberglass and aluminum, heavy duty, sectional, 12' x 12' high	2 Carp	1.50	10.667	Ea.	1,175	197		1,372	1,650
2450	Chain hoist, 20' x 20' high		.50	32		4,025	590		4,615	5,450
2600	Steel, 24 ga. sectional, manual, 8' x 8' high		2	8		335	148		483	625
2650	10' x 10' high		1.80	8.889		440	164		604	760
2700	12' x 12' high		1.50	10.667		590	197		787	990
2800	Chain hoist, 20' x 14' high		.70	22.857		1,475	425		1,900	2,350
2850	For 1-1/4" rigid insulation and 26 ga. galv.									
2860	back panel, add				S.F.	1.55			1.55	1.70
2900	For electric trolley operator, 1/3 H.P., to 12' x 12', add	1 Carp	2	4	Ea.	370	74		444	530
2950	Over 12' x 12', 1/2 H.P., add	"	1	8	"	395	148		543	690

083 720 | Special Purpose Doors		CREW	DAILY OUTPUT	LABOR-HOURS	UNIT	MAT.	LABOR	EQUIP.	TOTAL	TOTAL INCL O&P
721 0010	**BULKHEAD CELLAR DOORS**									**721**
0020	Steel, not incl. sides, 44" x 62"	1 Carp	5.50	1.455	Ea.	180	27		207	244
0100	52" x 73"		5.10	1.569		200	29		229	270
0500	With sides and foundation plates, 57" x 45" x 24"		4.70	1.702		185	31.50		216.50	257
0600	42" x 49" x 51"		4.30	1.860		225	34.50		259.50	305
723 0010	**FLOOR, COMMERCIAL**									**723**
0020	Aluminum tile, steel frame, one leaf, 2' x 2' opng.	2 Sswk	3.50	4.571	Opng.	340	91		431	555
0050	3'-6" x 3'-6" opening		3.50	4.571		615	91		706	855
0500	Double leaf, 4' x 4' opening		3	5.333		900	106		1,006	1,200
0550	5' x 5' opening		3	5.333		1,325	106		1,431	1,650
725 0010	**FLOOR, INDUSTRIAL**									**725**
0020	Steel 300 psf L.L., single leaf, 2' x 2', 175#	2 Sswk	6	2.667	Opng.	450	53		503	600
0050	3' x 3' opening, 300#		5.50	2.909		620	58		678	800
0300	Double leaf, 4' x 4' opening, 455#		5	3.200		935	64		999	1,150
0350	5' x 5' opening, 645#		4.50	3.556		1,225	71		1,296	1,500
732 0010	**ROLLING SERVICE DOORS** Steel, manual, 20 ga., incl. hardware									**732**
0050	8' x 8' high	2 Sswk	1.60	10	Ea.	645	200		845	1,100
0100	10' x 10' high		1.40	11.429		865	228		1,093	1,400
0120	8' x 8' high, class A fire door		1.40	11.429		1,175	228		1,403	1,750
0130	12' x 12' high, standard		1.20	13.333		1,325	266		1,591	1,975
0140	12' x 12' high, class A fire door		1	16		2,150	320		2,470	2,975
0160	10' x 20' high, standard		.50	32		1,950	640		2,590	3,400
0180	10' x 20' high, class A fire door		.40	40		3,125	800		3,925	5,000
3000	For 18 ga. doors, add				S.F.	.92			.92	1.01
3300	For enamel finish, add				"	.95			.95	1.05
3600	For safety edge bottom bar, pneumatic, add				L.F.	12.25			12.25	13.50
3700	Electric, add					16			16	17.60
4000	For weatherstripping, extruded rubber, jambs, add					5.60			5.60	6.15
4100	Hood, add					4.50			4.50	4.95

Important: See the Reference Section for critical supporting data - Reference Nos., Crews, & Location Factors

083 720 | Special Purpose Doors

		CREW	DAILY OUTPUT	LABOR-HOURS	UNIT	MAT.	LABOR	EQUIP.	TOTAL	TOTAL INCL O&P	
732	4200	Sill, add				L.F.	1.90			1.90	2.09	732
	4500	Motor operators, to 14' x 14' opening	2 Sswk	5	3.200	Ea.	860	64		924	1,075	
	4700	For fire door, additional fusible link, add				"	20			20	22	
736	0010	**SHOCK ABSORBING DOORS**										736
	0020	Rigid, no frame, 1-1/2" thick, 5' x 7'	2 Sswk	1.90	8.421	Opng.	1,125	168		1,293	1,550	
	0100	8' x 8'		1.80	8.889		1,575	177		1,752	2,100	
	0500	Flexible, no frame, insulated, .16" thick, economy, 5' x 7'		2	8		1,375	160		1,535	1,825	
	0600	Deluxe		1.90	8.421		2,075	168		2,243	2,625	
	1000	8' x 8' opening, economy		2	8		2,150	160		2,310	2,675	
	1100	Deluxe		1.90	8.421		2,750	168		2,918	3,350	

083 750 | Swing Doors

		CREW	DAILY OUTPUT	LABOR-HOURS	UNIT	MAT.	LABOR	EQUIP.	TOTAL	TOTAL INCL O&P	
752	0010	**DOUBLE ACTING**										752
	1000	.063" aluminum, 7'-0" high, 4'-0" wide	2 Carp	4.20	3.810	Pr.	345	70.50		415.50	500	
	1050	6'-8" wide	2 Carp	4	4	Pr.	470	74		544	640	
	2000	Solid core wood, 3/4" thick, metal frame, stainless steel										
	2010	base plate, 7' high opening, 4' wide	2 Carp	4	4	Pr.	705	74		779	900	
	2050	7' wide	"	3.80	4.211	"	1,000	78		1,078	1,225	
754	0010	**GLASS DOOR, SWING**										754
	0020	Including hardware, 1/2" thick, tempered, 3' x 7' opening	2 Glaz	2	8	Opng.	1,375	146		1,521	1,775	
	0100	6' x 7' opening	"	1.40	11.429	"	3,200	209		3,409	3,875	

083 800 | Sound Retardant Doors

		CREW	DAILY OUTPUT	LABOR-HOURS	UNIT	MAT.	LABOR	EQUIP.	TOTAL	TOTAL INCL O&P	
804	0010	**ACOUSTICAL DOORS**										804
	0020	Including framed seals, 3' x 7', wood, 27 STC rating	F-2	1.50	10.667	Ea.	325	173		498	655	
	0100	Steel, 40 STC rating		1.50	10.667		1,250	173		1,423	1,675	
	0200	45 STC rating		1.50	10.667		1,650	173		1,823	2,125	
	0300	48 STC rating		1.50	10.667		2,075	173		2,248	2,600	
	0400	52 STC rating		1.50	10.667		2,500	173		2,673	3,050	

083 900 | Screen & Storm Doors

		CREW	DAILY OUTPUT	LABOR-HOURS	UNIT	MAT.	LABOR	EQUIP.	TOTAL	TOTAL INCL O&P	
904	0010	**STORM DOORS & FRAMES** Aluminum, residential,										904
	0020	combination storm and screen										
	0400	Clear anodic coating, 6'-8" x 2'-6" wide	F-2	15	1.067	Ea.	154	17.25		171.25	199	
	0420	2'-8" wide		14	1.143		171	18.50		189.50	220	
	0440	3'-0" wide		14	1.143		171	18.50		189.50	220	
	0500	For 7' door height, add					5%					
	1000	Mill finish, 6'-8" x 2'-6" wide	F-2	15	1.067	Ea.	212	17.25		229.25	263	
	1020	2'-8" wide		14	1.143		216	18.50		234.50	270	
	1040	3'-0" wide		14	1.143		216	18.50		234.50	270	
	1100	For 7'-0" door, add					5%					
	1500	White painted, 6'-8" x 2'-6" wide	F-2	15	1.067		154	17.25		171.25	199	
	1520	2'-8" wide		14	1.143		171	18.50		189.50	220	
	1540	3'-0" wide		14	1.143		171	18.50		189.50	220	
	1541	Storm door, painted, alum., insul., 6'-8" x 2'-6" wide		14	1.143		213	18.50		231.50	267	
	1545	2'-8" wide		14	1.143		213	18.50		231.50	267	
	1600	For 7'-0" door, add					5%					
	1800	Aluminum screen door, minimum, 6'-8" x 2'-8" wide	F-2	14	1.143		92.50	18.50		111	134	
	1810	3'-0" wide		14	1.143		92.50	18.50		111	134	
	1820	Average, 6'-8" x 2'-8" wide		14	1.143		169	18.50		187.50	218	
	1830	3'-0" wide		14	1.143		175	18.50		193.50	225	

DOORS & WINDOWS　8

083 900	Screen & Storm Doors	CREW	DAILY OUTPUT	LABOR-HOURS	UNIT	1998 BARE COSTS				TOTAL INCL O&P	
						MAT.	LABOR	EQUIP.	TOTAL		
904 1840	Maximum, 6'-8" x 2'-8" wide	F-2	14	1.143	Ea.	420	18.50		438.50	490	**904**
1850	3'-0" wide	↓	14	1.143	↓	238	18.50		256.50	294	
2000	Wood door & screen, see division 082-078										
2020											

084 | **Entrances & Storefronts**

8 DOORS & WINDOWS

084 100	Aluminum	CREW	DAILY OUTPUT	LABOR-HOURS	UNIT	1998 BARE COSTS				TOTAL INCL O&P	
						MAT.	LABOR	EQUIP.	TOTAL		
101 0010	**ALUMINUM FRAMES**										**101**
0020	Entrance, 3' x 7' opening, clear anodized finish	2 Sswk	7	2.286	Opng.	220	45.50		265.50	330	
0040	Bronze finish		7	2.286		270	45.50		315.50	385	
0060	Black finish		7	2.286		305	45.50		350.50	425	
0200	3'-6" x 7'-0", mill finish		7	2.286		240	45.50		285.50	355	
0220	Bronze finish		7	2.286		290	45.50		335.50	410	
0240	Black finish		7	2.286		315	45.50		360.50	435	
0500	6' x 7' opening, clear finish		6	2.667		270	53		323	400	
0520	Bronze finish		6	2.667		305	53		358	440	
0540	Black finish		6	2.667	↓	360	53		413	500	
0600	7'-0" x 7'-0", mill finish		6	2.667	Ea.	251	53		304	380	
0620	Bronze finish		6	2.667		285	53		338	420	
0640	Black finish		6	2.667		335	53		388	475	
2000	Transoms, 3'-0" x 3'-0", mill finish		80	.200		95	3.99		98.99	113	
2020	Bronze finish		80	.200		105	3.99		108.99	124	
2040	Black finish		80	.200		120	3.99		123.99	140	
2200	3'-6" x 3'-0", mill finish		80	.200		110	3.99		113.99	129	
2220	Bronze finish		80	.200		125	3.99		128.99	146	
2240	Black finish		80	.200		137	3.99		140.99	159	
2500	6'-0" x 3'-0", mill finish		65	.246		115	4.91		119.91	137	
2520	Bronze finish		65	.246		143	4.91		147.91	167	
2540	Black finish		65	.246		155	4.91		159.91	181	
2700	7'-0" x 3'-0", mill finish		65	.246		126	4.91		130.91	149	
2720	Bronze finish		65	.246		143	4.91		147.91	167	
2740	Black finish	↓	65	.246	↓	154	4.91		158.91	179	
103 0010	**ALUMINUM DOORS** Commercial entrance										**103**
0800	Narrow stile, no glazing, standard hardware, pair of 2'-6" x 7'-0"	F-2	1.70	9.412	Pr.	740	152		892	1,075	
1000	3'-0" x 7'-0", single		3	5.333	Ea.	445	86.50		531.50	640	
1050	Bronze finish		3	5.333		420	86.50		506.50	610	
1100	Black finish		3	5.333	↓	480	86.50		566.50	680	
1200	Pair of 3'-0" x 7'-0"		1.70	9.412	Pr.	760	152		912	1,100	
1500	3'-6" x 7'-0", single		3	5.333	Ea.	455	86.50		541.50	650	
1550	Bronze finish		3	5.333		450	86.50		536.50	645	
1600	Black finish		3	5.333	↓	510	86.50		596.50	710	
2000	Medium stile, pair of 2'-6" x 7'-0"		1.70	9.412	Pr.	975	152		1,127	1,325	
2100	3'-0" x 7'-0", single		3	5.333	Ea.	550	86.50		636.50	755	
2200	Pair of 3'-0" x 7'-0"		1.70	9.412	Pr.	980	152		1,132	1,325	
2300	3'-6" x 7'-0", single	↓	5.33	3.002	Ea.	565	48.50		613.50	705	
5000	Flush panel doors, pair of 2'-6" x 7'-0"	2 Sswk	2	8	Pr.	805	160		965	1,200	
5050	3'-0" x 7'-0", single		2.50	6.400	Ea.	400	128		528	690	
5100	Pair of 3'-0" x 7'-0"	↓	2	8	Pr.	830	160		990	1,225	

Important: See the Reference Section for critical supporting data - Reference Nos., Crews, & Location Factors

084 100	Aluminum	CREW	DAILY OUTPUT	LABOR-HOURS	UNIT	1998 BARE COSTS MAT.	LABOR	EQUIP.	TOTAL	TOTAL INCL O&P	
103	5150	3'-6" x 7'-0", single	2 Sswk	2.50	6.400	Ea.	495	128		623	795	103
105	0010	**ALUMINUM DOORS & FRAMES** Entrance, narrow stile, including										105
	0015	hardware & closer, clear finish, not incl. glass, 2'-6" x 7'-0" opng.	2 Sswk	2	8	Ea.	460	160		620	820	
	0020	3'-0" x 7'-0" opening		2	8		400	160		560	755	
	0030	3'-6" x 7'-0" opening		2	8		410	160		570	765	
	0100	3'-0" x 10'-0" opening, 3' high transom		1.80	8.889		735	177		912	1,150	
	0200	3'-6" x 10'-0" opening, 3' high transom		1.80	8.889		745	177		922	1,175	
	0280	5'-0" x 7'-0" opening		2	8		680	160		840	1,075	
	0300	6'-0" x 7'-0" opening		1.30	12.308	Pr.	665	246		911	1,225	
	0400	6'-0" x 10'-0" opening, 3' high transom		1.10	14.545		1,125	290		1,415	1,800	
	0420	7'-0" x 7'-0" opening		1	16		765	320		1,085	1,475	
	0500	Wide stile, 2'-6" x 7'-0" opening		2	8	Ea.	675	160		835	1,050	
	0520	3'-0" x 7'-0" opening		2	8		600	160		760	975	
	0540	3'-6" x 7'-0" opening		2	8		630	160		790	1,000	
	0560	5'-0" x 7'-0" opening		2	8		965	160		1,125	1,375	
	0580	6'-0" x 7'-0" opening		1.30	12.308	Pr.	920	246		1,166	1,500	
	0600	7'-0" x 7'-0" opening		1	16	"	1,050	320		1,370	1,775	
	1100	For full vision doors, with 1/2" glass, add				Leaf	55%					
	1200	For non-standard size, add					67%					
	1300	Light bronze finish, add					36%					
	1400	Dark bronze finish, add					18%					
	1500	For black finish, add					36%					
	1600	Concealed panic device, add					885			885	975	
	1700	Electric striker release, add				Opng.	228			228	251	
	1800	Floor check, add				Leaf	675			675	745	
	1900	Concealed closer, add				"	450			450	495	
	2000	Flush 3' x 7' Insulated, 12"x 12" lite, clear finish	2 Sswk	2	8	Ea.	710	160		870	1,100	
109	0010	**BALANCED DOORS**										109
	0020	Hardware & frame, alum. & glass, 3' x 7', econ.	2 Sswk	.90	17.778	Ea.	3,275	355		3,630	4,300	
	0150	Premium	"	.70	22.857	"	4,525	455		4,980	5,875	
111	0010	**STOREFRONT SYSTEMS** Aluminum frame, clear 3/8" plate glass,										111
	0020	incl. 3' x 7' door with hardware (400 sq. ft. max. wall)										
	0500	Wall height to 12' high, commercial grade	2 Glaz	150	.107	S.F.	11	1.95		12.95	15.35	
	0600	Institutional grade		130	.123		14.60	2.25		16.85	19.80	
	0700	Monumental grade		115	.139		21	2.55		23.55	27	
	1000	6' x 7' door with hardware, commercial grade		135	.119		11.25	2.17		13.42	15.95	
	1100	Institutional grade		115	.139		15.40	2.55		17.95	21	
	1200	Monumental grade		100	.160		28.50	2.93		31.43	36.50	
	1500	For bronze anodized finish, add					15%					
	1600	For black anodized finish, add					30%					
	1700	For stainless steel framing, add to monumental					75%					

084 300	Stainless Steel	CREW	DAILY OUTPUT	LABOR-HOURS	UNIT	MAT.	LABOR	EQUIP.	TOTAL	TOTAL INCL O&P	
301	0010	**STAINLESS STEEL AND GLASS** Entrance unit, narrow stiles										301
	0020	3' x 7' opening, including hardware, minimum	2 Sswk	1.60	10	Opng.	3,825	200		4,025	4,600	
	0050	Average		1.40	11.429		4,150	228		4,378	5,000	
	0100	Maximum		1.20	13.333		4,450	266		4,716	5,400	
	1000	For solid bronze entrance units, statuary finish, add					60%					
	1100	Without statuary finish, add					45%					
	2000	Balanced doors, 3' x 7', economy	2 Sswk	.90	17.778	Ea.	5,175	355		5,530	6,400	
	2100	Premium	"	.70	22.857	"	8,925	455		9,380	10,700	

The right margin contains the vertical tab text: **DOORS & WINDOWS 8**

084 600 | Automatic Doors

		CREW	DAILY OUTPUT	LABOR-HOURS	UNIT	1998 BARE COSTS				TOTAL INCL O&P	
						MAT.	LABOR	EQUIP.	TOTAL		
602	0010	**SLIDING ENTRANCE** 12' x 7'-6" opng., 5' x 7' door, 2 way traf.,									602
	0020	mat activated, panic pushout, incl. operator & hardware,									
	0030	not including glass or glazing	2 Glaz	.70	22.857	Opng.	5,100	420		5,520	6,300	
604	0010	**SLIDING PANELS**									604
	0020	Mall fronts, aluminum & glass, 15' x 9' high	2 Glaz	1.30	12.308	Opng.	2,000	225		2,225	2,575	
	0100	24' x 9' high		.70	22.857		2,900	420		3,320	3,900	
	0200	48' x 9' high, with fixed panels		.90	17.778		5,400	325		5,725	6,500	
	0500	For bronze finish, add					17%					

085 | Metal Windows

085 100 | Steel Windows

		CREW	DAILY OUTPUT	LABOR-HOURS	UNIT	1998 BARE COSTS				TOTAL INCL O&P	
						MAT.	LABOR	EQUIP.	TOTAL		
102	0010	**STEEL SASH** Custom units, glazing and trim not included										102
	0100	Casement, 100% vented	2 Sswk	200	.080	S.F.	32.50	1.60		34.10	39	
	0200	50% vented		200	.080		28	1.60		29.60	33.50	
	0300	Fixed		200	.080		18.65	1.60		20.25	23.50	
	1000	Projected, commercial, 40% vented		200	.080		29	1.60		30.60	35	
	1100	Intermediate, 50% vented		200	.080		32	1.60		33.60	38	
	1500	Industrial, horizontally pivoted		200	.080		24	1.60		25.60	29.50	
	1600	Fixed		200	.080		21	1.60		22.60	26	
	2000	Industrial security sash, 50% vented		200	.080		37	1.60		38.60	43.50	
	2100	Fixed		200	.080		31	1.60		32.60	37	
	2500	Picture window		200	.080		13.35	1.60		14.95	17.80	
	3000	Double hung		200	.080		35.50	1.60		37.10	42	
	5000	Mullions for above, open interior face		240	.067	L.F.	6.65	1.33		7.98	9.90	
	5100	With interior cover		240	.067	"	11.20	1.33		12.53	14.90	
	6000	Double glazing for above, add	2 Glaz	200	.080	S.F.	6.65	1.46		8.11	9.80	
	6100	Triple glazing for above, add	"	85	.188	"	8.10	3.44		11.54	14.70	
104	0010	**STEEL WINDOWS** Stock, including frame, trim and insulating glass										104
	1000	Custom units, double hung, 2'-8" x 4'-6" opening	2 Sswk	12	1.333	Ea.	465	26.50		491.50	570	
	1100	2'-4" x 3'-9" opening		12	1.333		385	26.50		411.50	480	
	1500	Commercial projected, 3'-9" x 5'-5" opening		10	1.600		815	32		847	965	
	1600	6'-9" x 4'-1" opening		7	2.286		1,075	45.50		1,120.50	1,275	
	2000	Intermediate projected, 2'-9" x 4'-1" opening		12	1.333		455	26.50		481.50	560	
	2100	4'-1" x 5'-5" opening		10	1.600		925	32		957	1,100	

085 200 | Aluminum Windows

		CREW	DAILY OUTPUT	LABOR-HOURS	UNIT	1998 BARE COSTS				TOTAL INCL O&P	
						MAT.	LABOR	EQUIP.	TOTAL		
202	0010	**ALUMINUM SASH**										202
	0020	Stock, grade C, glaze & trim not incl., casement	2 Sswk	200	.080	S.F.	25.50	1.60		27.10	31	
	0050	Double hung		200	.080		24.50	1.60		26.10	30	
	0100	Fixed casement		200	.080		9.30	1.60		10.90	13.40	
	0150	Picture window		200	.080		10.70	1.60		12.30	14.90	
	0200	Projected window		200	.080		21.50	1.60		23.10	26.50	
	0250	Single hung		200	.080		11.05	1.60		12.65	15.30	
	0300	Sliding		200	.080		14.30	1.60		15.90	18.85	
	1000	Mullions for above, tubular		240	.067	L.F.	3.79	1.33		5.12	6.80	
	3000	Double glazing for above, add	2 Glaz	200	.080	S.F.	8.35	1.46		9.81	11.60	

Important: See the Reference Section for critical supporting data - Reference Nos., Crews, & Location Factors

085 200 | Aluminum Windows

		CREW	DAILY OUTPUT	LABOR-HOURS	UNIT	1998 BARE COSTS				TOTAL INCL O&P	
						MAT.	LABOR	EQUIP.	TOTAL		
202	3100	Triple glazing for above, add	2 Glaz	85	.188	S.F.	9.75	3.44		13.19	16.50	202
204	0010	**ALUMINUM WINDOWS** Incl. frame and glazing, grade C										204
	1000	Stock units, casement, 3'-1" x 3'-2" opening	2 Sswk	10	1.600	Ea.	248	32		280	335	
	1040	Insulating glass	"	10	1.600		248	32		280	335	
	1050	Add for storms					50			50	54.50	
	1600	Projected, with screen, 3'-1" x 3'-2" opening	2 Sswk	10	1.600		177	32		209	258	
	1650	Insulating glass	"	10	1.600		166	32		198	245	
	1700	Add for storms					47			47	51.50	
	2000	4'-5" x 5'-3" opening	2 Sswk	8	2		250	40		290	355	
	2050	Insulating glass	"	8	2		288	40		328	395	
	2100	Add for storms					64			64	70.50	
	2500	Enamel finish windows, 3'-1" x 3'-2"	2 Sswk	10	1.600		159	32		191	238	
	2550	Insulating glass		10	1.600		173	32		205	253	
	2600	4'-5" x 5'-3"		8	2		238	40		278	340	
	2700	Insulating glass		8	2		310	40		350	420	
	3000	Single hung, 2' x 3' opening, enameled, standard glazed		10	1.600		119	32		151	194	
	3100	Insulating glass		10	1.600		144	32		176	222	
	3300	2'-8" x 6'-8" opening, standard glazed		8	2		253	40		293	355	
	3400	Insulating glass		8	2		325	40		365	440	
	3700	3'-4" x 5'-0" opening, standard glazed		9	1.778		163	35.50		198.50	249	
	3800	Insulating glass		9	1.778		229	35.50		264.50	320	
	4000	Sliding aluminum, 3' x 2' opening, standard glazed		10	1.600		134	32		166	210	
	4100	Insulating glass		10	1.600		148	32		180	226	
	4300	5' x 3' opening, standard glazed		9	1.778		170	35.50		205.50	257	
	4400	Insulating glass		9	1.778		238	35.50		273.50	330	
	4600	8' x 4' opening, standard glazed		6	2.667		245	53		298	375	
	4700	Insulating glass		6	2.667		395	53		448	540	
	5000	9' x 5' opening, standard glazed		4	4		370	80		450	565	
	5100	Insulating glass		4	4		595	80		675	805	
	5500	Sliding, with thermal barrier and screen, 6' x 4', 2 track		8	2		505	40		545	635	
	5700	4 track		8	2		615	40		655	755	
	6000	For above units with bronze finish, add					12%					
	6200	For installation in concrete openings, add					5%					

085 500 | Metal Jalousie Windows

		CREW	DAILY OUTPUT	LABOR-HOURS	UNIT	MAT.	LABOR	EQUIP.	TOTAL	TOTAL INCL O&P	
502	0010	**JALOUSIES**									502
	0020	Aluminum incl. glazing & screens, stock, 1'-7" x 3'-2"	2 Sswk	10	1.600	Ea.	108	32		140	181	
	0100	2'-3" x 4'-0"		10	1.600		154	32		186	232	
	0200	3'-1" x 2'-0"		10	1.600		106	32		138	179	
	0300	3'-1" x 5'-3"		10	1.600		219	32		251	305	
	1000	Mullions for above, 2'-0" long		80	.200		8.30	3.99		12.29	16.95	
	1100	5'-3" long		80	.200		14.15	3.99		18.14	23.50	

085 600 | Metal Storm Windows

		CREW	DAILY OUTPUT	LABOR-HOURS	UNIT	MAT.	LABOR	EQUIP.	TOTAL	TOTAL INCL O&P	
601	0010	**STORM WINDOWS** Aluminum, residential										601
	0300	Basement, mill finish, incl. fiberglass screen										
	0320	1'-10" x 1'-0" high	F-2	30	.533	Ea.	24.50	8.65		33.15	42	
	0340	2'-9" x 1'-6" high		30	.533		27	8.65		35.65	44.50	
	0360	3'-4" x 2'-0" high		30	.533		32.50	8.65		41.15	50.50	
	1600	Double-hung, combination, storm & screen										
	1700	Custom, clear anodic coating, 2'-0" x 3'-5" high	F-2	30	.533	Ea.	63.50	8.65		72.15	85	
	1720	2'-6" x 5'-0" high		28	.571		85	9.25		94.25	109	
	1740	4'-0" x 6'-0" high		25	.640		180	10.35		190.35	216	
	1800	White painted, 2'-0" x 3'-5" high		30	.533		75.50	8.65		84.15	98.50	

For expanded coverage of these items see *Means Building Construction Cost Data 1998*

085 600 | Metal Storm Windows

		CREW	DAILY OUTPUT	LABOR-HOURS	UNIT	1998 BARE COSTS				TOTAL INCL O&P
						MAT.	LABOR	EQUIP.	TOTAL	
601	1820	2'-6" x 5'-0" high	F-2	28	.571	Ea.	121	9.25		130.25	150
	1840	4'-0" x 6'-0" high		25	.640		218	10.35		228.35	258
	2000	Average quality, clear anodic coating, 2'-0" x 3'-5" high		30	.533		64.50	8.65		73.15	86
	2020	2'-6" x 5'-0" high		28	.571		81.50	9.25		90.75	105
	2040	4'-0" x 6'-0" high		25	.640		95.50	10.35		105.85	123
	2400	White painted, 2'-0" x 3'-5" high		30	.533		63.50	8.65		72.15	85
	2420	2'-6" x 5'-0" high		28	.571		70	9.25		79.25	93
	2440	4'-0" x 6'-0" high		25	.640		77	10.35		87.35	102
	2600	Mill finish, 2'-0" x 3'-5" high		30	.533		58	8.65		66.65	78.50
	2620	2'-6" x 5'-0" high		28	.571		64.50	9.25		73.75	87
	2640	4'-0" x 6-8" high		25	.640		72.50	10.35		82.85	97.50
	4000	Picture window, storm, 1 lite, white or bronze finish									
	4020	4'-6" x 4'-6" high	F-2	25	.640	Ea.	95	10.35		105.35	123
	4040	5'-8" x 4'-6" high		20	.800		97.50	12.95		110.45	129
	4400	Mill finish, 4'-6" x 4'-6" high		25	.640		83.50	10.35		93.85	110
	4420	5'-8" x 4'-6" high		20	.800		88	12.95		100.95	119
	4600	3 lite, white or bronze finish									
	4620	4'-6" x 4'-6" high	F-2	25	.640	Ea.	111	10.35		121.35	140
	4640	5'-8" x 4'-6" high		20	.800		116	12.95		128.95	149
	4800	Mill finish, 4'-6" x 4'-6" high		25	.640		104	10.35		114.35	132
	4820	5'-8" x 4'-6" high		20	.800		108	12.95		120.95	141
	5000	Sliding glass door, storm 6' x 6'-8", standard	1 Glaz	2	4		300	73		373	450
	5100	Economy	"	2	4		202	73		275	345
	6000	Sliding window, storm, 2 lite, white or bronze finish									
	6020	3'-4" x 2'-7" high	F-2	28	.571	Ea.	60	9.25		69.25	82
	6040	4'-4" x 3'-3" high		25	.640		70	10.35		80.35	95
	6060	5'-4" x 6'-0" high		20	.800		104	12.95		116.95	137
	6400	3 lite, white or bronze finish									
	6420	4'-4" x 3'-3" high	F-2	25	.640	Ea.	93	10.35		103.35	120
	6440	5'-4" x 6'-0" high		20	.800		145	12.95		157.95	181
	6460	6'-0" x 6'-0" high		18	.889		149	14.40		163.40	189
	6800	Mill finish, 4'-4" x 3'-3" high		25	.640		83.50	10.35		93.85	110
	6820	5'-4" x 6'-0" high		20	.800		119	12.95		131.95	153
	6840	6'-0" x 6-0" high		18	.889		127	14.40		141.40	165
	9000	Magnetic interior storm window									
	9100	3/16" plate glass	1 Glaz	107	.075	S.F.	2.04	1.37		3.41	4.52

085 700 | Screens

		CREW	DAILY OUTPUT	LABOR-HOURS	UNIT	1998 BARE COSTS				TOTAL INCL O&P
						MAT.	LABOR	EQUIP.	TOTAL	
701	0010	**SCREENS**									
	0020	For metal sash, aluminum or bronze mesh, flat screen	2 Sswk	1,200	.013	S.F.	2.91	.27		3.18	3.72
	0500	Wicket screen, inside window	"	1,000	.016	"	4.45	.32		4.77	5.55
	0600	Residential, aluminum mesh and frame, 2' x 3'	F-2	32	.500	Ea.	14.85	8.10		22.95	30
	0610	Rescreen		50	.320		2.06	5.20		7.26	11.10
	0620	3' x 5'		32	.500		23.50	8.10		31.60	40
	0630	Rescreen		45	.356		4.90	5.75		10.65	15.25
	0640	4' x 8'		25	.640		44.50	10.35		54.85	66.50
	0650	Rescreen		40	.400		10.05	6.45		16.50	22
	0660	Patio door		25	.640		36	10.35		46.35	57.50
	0680	Rescreening		1,600	.010	S.F.	.31	.16		.47	.62
	0800	Security screen, aluminum frame with stainless steel cloth	2 Sswk	1,200	.013		15.70	.27		15.97	17.75
	0900	Steel grate, painted, on steel frame		1,600	.010		8.45	.20		8.65	9.65
	1000	For solar louvers, add		160	.100		16.25	2		18.25	22

Important: See the Reference Section for critical supporting data - Reference Nos., Crews, & Location Factors

086 100 | Wood Windows

		CREW	DAILY OUTPUT	LABOR-HOURS	UNIT	1998 BARE COSTS				TOTAL INCL O&P
						MAT.	LABOR	EQUIP.	TOTAL	
104	0010	**AWNING WINDOW** Including frame, screen, and exterior trim									**104**
	0100	Average quality, builders model, 34" x 22", standard glazed	1 Carp	10	.800	Ea.	174	14.80		188.80	218
	0200	Insulating glass		10	.800		207	14.80		221.80	254
	0300	40" x 28", standard glazed		9	.889		225	16.45		241.45	276
	0400	Insulating glass		9	.889		251	16.45		267.45	305
	0500	48" x 36", standard glazed		8	1		256	18.50		274.50	315
	0600	Insulating glass		8	1		286	18.50		304.50	345
	1000	34" x 22"		10	.800		188	14.80		202.80	232
	1100	40" x 22"		10	.800		206	14.80		220.80	252
	1200	36" x 28"		9	.889		221	16.45		237.45	271
	1300	36" x 36"		9	.889		296	16.45		312.45	355
	1400	48" x 28"		8	1		262	18.50		280.50	320
	1500	60" x 36"		8	1		288	18.50		306.50	345
	2000	Metal clad, deluxe, insulating glass, 34" x 22"		10	.800		170	14.80		184.80	213
	2100	40" x 22"		10	.800		186	14.80		200.80	230
	2200	36" x 25"		9	.889		179	16.45		195.45	224
	2300	40" x 30"		9	.889		202	16.45		218.45	251
	2400	48" x 28"		8	1		216	18.50		234.50	270
	2500	60" x 36"		8	1		245	18.50		263.50	300
108	0010	**BOW-BAY WINDOW** Including frame, screen and exterior trim,									**108**
	0020	end panels operable									
	1000	Fixed type, builders model, 8' x 5' high, std. glazed, 4 panels	2 Carp	10	1.600	Ea.	755	29.50		784.50	880
	1050	Insulating glass		10	1.600		870	29.50		899.50	1,000
	1100	10'-0" x 5'-0" high, standard glazed		6	2.667		985	49.50		1,034.50	1,150
	1200	Insulating glass, 6 panels		6	2.667		1,025	49.50		1,074.50	1,225
	1300	Vinyl clad, premium, insulating glass, 6'-0" x 4'-0"		10	1.600		1,225	29.50		1,254.50	1,375
	1340	9'-0" x 4'-0"		8	2		1,700	37		1,737	1,950
	1380	10'-0" x 6'-0"		7	2.286		2,075	42.50		2,117.50	2,350
	1420	12'-0" x 6'-0"		6	2.667		2,850	49.50		2,899.50	3,200
	1600	Metal clad, deluxe, insul. glass, 6'-0" x 4'-0" high, 3 panels		10	1.600		760	29.50		789.50	890
	1640	9'-0" x 4'-0" high, 4 panels		8	2		1,075	37		1,112	1,250
	1680	10'-0" x 5'-0" high, 5 panels		7	2.286		1,475	42.50		1,517.50	1,700
	1720	12'-0" x 6'-0" high, 6 panels		6	2.667		2,050	49.50		2,099.50	2,325
	2000	Casement, builders model, bow, 8' x 5' high, std. glazed, 4 panels		10	1.600		1,150	29.50		1,179.50	1,325
	2050	Insulating glass		10	1.600		1,150	29.50		1,179.50	1,325
	2100	12'-0" x 6'-0" high, 6 panels, standard glazed		6	2.667		2,575	49.50		2,624.50	2,900
	2200	Insulating glass		6	2.667		2,625	49.50		2,674.50	2,975
	2280	6'-0" x 4'-0"		11	1.455		900	27		927	1,025
	2300	Vinyl clad, premium, insulating glass, 8'-0" x 5'-0"		10	1.600		1,300	29.50		1,329.50	1,475
	2340	10'-0" x 5'-0"		8	2		1,700	37		1,737	1,950
	2380	10' 0" x 6' 0"		7	2.286		1,950	42.50		1,992.50	2,225
	2420	12'-0" x 6'-0"		6	2.667		2,475	49.50		2,524.50	2,800
	2600	Metal clad, deluxe, insul. glass, 8'-0" x 5'-0" high, 4 panels		10	1.600		1,325	29.50		1,354.50	1,500
	2640	10'-0" x 5'-0" high, 5 panels		8	2		1,525	37		1,562	1,775
	2680	10'-0" x 6'-0" high, 5 panels		7	2.286		1,800	42.50		1,842.50	2,050
	2720	12'-0" x 6'-0" high, 6 panels		6	2.667		2,050	49.50		2,099.50	2,325
	3000	Double hung, bldrs. model, bay, 8' x 4' high, std. glazed		10	1.600		1,125	29.50		1,154.50	1,300
	3050	Insulating glass		10	1.600		1,225	29.50		1,254.50	1,375
	3100	9'-0" x 5'-0" high, standard glazed		6	2.667		1,375	49.50		1,424.50	1,600
	3200	Insulating glass		6	2.667		1,425	49.50		1,474.50	1,650
	3300	Vinyl clad, premium, insulating glass, 7'-0" x 4'-6"		10	1.600		1,250	29.50		1,279.50	1,450
	3340	8'-0" x 4'-6"		8	2		1,350	37		1,387	1,550
	3380	8'-0" x 5'-0"		7	2.286		1,375	42.50		1,417.50	1,575
	3420	9'-0" x 5'-0"		6	2.667		1,600	49.50		1,649.50	1,825
	3600	Metal clad, deluxe, insul. glass, 7'-0" x 4'-0" high		10	1.600		830	29.50		859.50	960

			CREW	DAILY OUTPUT	LABOR-HOURS	UNIT	MAT.	LABOR	EQUIP.	TOTAL	TOTAL INCL O&P	
		**086 100	Wood Windows**					1998 BARE COSTS					
108	3640	8'-0" x 4'-0" high	2 Carp	8	2	Ea.	875	37		912	1,025	108
	3680	8'-0" x 5'-0" high		7	2.286		950	42.50		992.50	1,125	
	3720	9'-0" x 5'-0" high		6	2.667		1,025	49.50		1,074.50	1,200	
	7000	Drip cap, premolded vinyl, 8' long		30	.533		66.50	9.85		76.35	90.50	
	7040	12' long		26	.615		72.50	11.40		83.90	99.50	
120	0010	**CASEMENT WINDOW** Including frame, screen, and exterior trim										120
	0100	Avg. quality, bldrs. model, 2'-0" x 3'-0" H, standard glazed	1 Carp	10	.800	Ea.	187	14.80		201.80	231	
	0150	Insulating glass		10	.800		192	14.80		206.80	237	
	0200	2'-0" x 4'-6" high, standard glazed		9	.889		238	16.45		254.45	290	
	0250	Insulating glass		9	.889		248	16.45		264.45	300	
	0300	2'-3" x 6'-0" high, standard glazed		8	1		261	18.50		279.50	320	
	0350	Insulating glass		8	1		320	18.50		338.50	385	
	8000	Solid vinyl, premium, insulating glass, 2'-0" x 3'-0" high		10	.800		161	14.80		175.80	203	
	8020	2'-0" x 4'-0" high		9	.889		192	16.45		208.45	240	
	8040	2'-0" x 5'-0" high		8	1		218	18.50		236.50	272	
	8100	Metal clad, deluxe, insulating glass, 2'-0" x 3'-0" high		10	.800		166	14.80		180.80	209	
	8120	2'-0" x 4'-0" high		9	.889		192	16.45		208.45	239	
	8140	2'-0" x 5'-0" high		8	1		219	18.50		237.50	273	
	8160	2'-0" x 6'-0" high		8	1		225	18.50		243.50	279	
	8200	For multiple leaf units, deduct for stationary sash										
	8220	2' high				Ea.	38			38	41.50	
	8240	4'-6" high					46			46	51	
	8260	6' high					46			46	50.50	
	8300	For installation, add per leaf						15%				
124	0010	**DOUBLE HUNG** Including frame, screen, and exterior trim										124
	0100	Avg. quality, bldrs. model, 2'-0" x 3'-0" high, standard glazed	1 Carp	10	.800	Ea.	92	14.80		106.80	127	
	0150	Insulating glass		10	.800		130	14.80		144.80	169	
	0200	3'-0" x 4'-0" high, standard glazed		9	.889		123	16.45		139.45	163	
	0250	Insulating glass		9	.889		168	16.45		184.45	213	
	0300	4'-0" x 4'-6" high, standard glazed		8	1		133	18.50		151.50	178	
	0350	Insulating glass		8	1		199	18.50		217.50	251	
	1000	Vinyl clad, premium, insulating glass, 2'-6" x 3'-0"		10	.800		237	14.80		251.80	287	
	1100	3'-0" x 3'-6"		10	.800		287	14.80		301.80	340	
	1200	3'-0" x 4'-0"		9	.889		310	16.45		326.45	370	
	1300	3'-0" x 4'-6"		9	.889		325	16.45		341.45	385	
	1400	3'-0" x 5'-0"		8	1		335	18.50		353.50	400	
	1500	3'-6" x 6'-0"		8	1		560	18.50		578.50	650	
	1540	Solid vinyl, average quality, insulated glass, 2'-0" x 3'-0"		10	.800		120	14.80		134.80	157	
	1542	3'-0" x 4'-0"		9	.889		145	16.45		161.45	187	
	1544	4'-0" x 4'-6"		8	1		153	18.50		171.50	201	
	1560	Premium, insulating glass, 2'-6" x 3'-0"		10	.800		136	14.80		150.80	175	
	1562	3'-0" x 3'-6"		9	.889		157	16.45		173.45	201	
	1564	3'-0" x 4'-0"		9	.889		167	16.45		183.45	212	
	1566	3'-0" x 4'-6"		9	.889		172	16.45		188.45	217	
	1568	3'-0" x 5'-0"		8	1		176	18.50		194.50	226	
	1570	3'-6" x 6'-0"		8	1		194	18.50		212.50	245	
	2000	Metal clad, deluxe, insulating glass, 2'-6" x 3'-0" high		10	.800		156	14.80		170.80	197	
	2100	3'-0" x 3'-6" high		10	.800		186	14.80		200.80	230	
	2200	3'-0" x 4'-0" high		9	.889		199	16.45		215.45	247	
	2300	3'-0" x 4'-6" high		9	.889		211	16.45		227.45	260	
	2400	3'-0" x 5'-0" high		8	1		224	18.50		242.50	279	
	2500	3'-6" x 6'-0" high		8	1		285	18.50		303.50	345	
130	0010	**PALLADIAN WINDOWS**										130
	0020	Aluminum clad, including frame and exterior trim										

Important: See the Reference Section for critical supporting data - Reference Nos., Crews, & Location Factors

086 100	Wood Windows	CREW	DAILY OUTPUT	LABOR-HOURS	UNIT	1998 BARE COSTS				TOTAL INCL O&P	
						MAT.	LABOR	EQUIP.	TOTAL		
130	0040	3'-2" x 2'-0" high	2 Carp	11	1.455	Ea.	895	27		922	1,025	**130**
	0060	3'-2" x 4'-10"		11	1.455		1,025	27		1,052	1,175	
	0080	3'-2" x 6'-4"		10	1.600		1,200	29.50		1,229.50	1,375	
	0100	4'-0" x 4'-0"		10	1.600		985	29.50		1,014.50	1,125	
	0120	4'-0" x 5'-4"	3 Carp	10	2.400		1,150	44.50		1,194.50	1,325	
	0140	4'-0" x 6'-0"		9	2.667		1,200	49.50		1,249.50	1,400	
	0160	4'-0" x 7'-4"		9	2.667		1,300	49.50		1,349.50	1,500	
	0180	5'-5" x 4'-10"		9	2.667		1,275	49.50		1,324.50	1,475	
	0200	5'-5" x 6'-10"		9	2.667		1,525	49.50		1,574.50	1,750	
	0220	5'-5" x 7'-9"		9	2.667		1,650	49.50		1,699.50	1,900	
	0240	6'-0" x 7'-11"		8	3		2,200	55.50		2,255.50	2,525	
	0260	8'-0" x 6'-0"		8	3		1,875	55.50		1,930.50	2,175	
132	0010	**PICTURE WINDOW** Including frame and exterior trim										**132**
	0100	Average quality, bldrs. model, 3'-6" x 4'-0" high, standard glazed	2 Carp	12	1.333	Ea.	169	24.50		193.50	229	
	0150	Insulating glass		12	1.333		201	24.50		225.50	265	
	0200	4'-0" x 4'-6" high, standard glazed		11	1.455		227	27		254	296	
	0250	Insulating glass		11	1.455		244	27		271	315	
	0300	5'-0" x 4'-0" high, standard glazed		11	1.455		267	27		294	340	
	0350	Insulating glass		11	1.455		276	27		303	350	
	0400	6'-0" x 4'-6" high, standard glazed		10	1.600		325	29.50		354.50	405	
	0450	Insulating glass		10	1.600		325	29.50		354.50	410	
	1000	Vinyl clad, premium, insulating glass, 4'-0" x 4'-0"		12	1.333		405	24.50		429.50	495	
	1100	4'-0" x 6'-0"		11	1.455		560	27		587	660	
	1200	5'-0" x 6'-0"		10	1.600		725	29.50		754.50	845	
	1300	6'-0" x 6'-0"		10	1.600		1,025	29.50		1,054.50	1,175	
	2000	Metal clad, deluxe, insulating glass, 4'-0" x 4'-0" high		12	1.333		258	24.50		282.50	325	
	2100	4'-0" x 6'-0" high		11	1.455		380	27		407	460	
	2200	5'-0" x 6'-0" high		10	1.600		475	29.50		504.50	575	
	2300	6'-0" x 6'-0" high		10	1.600		585	29.50		614.50	695	
138	0010	**SOLID VINYL REPLACEMENT WINDOWS** R086 -200										**138**
	0020	White, double hung, up to 83 united inches	2 Carp	8	2	Ea.	185	37		222	268	
	0040	84 to 93		8	2		205	37		242	290	
	0060	94 to 101		6	2.667		233	49.50		282.50	340	
	0080	102 to 111		6	2.667		257	49.50		306.50	370	
	0100	112 to 120		6	2.667		288	49.50		337.50	400	
	0120	For each united inch over 120 , add		800	.020	Inch	2.90	.37		3.27	3.82	
	0140	Casement windows, one operating sash , 42 to 60 united inches		8	2	Ea.	163	37		200	244	
	0160	61 to 70		8	2		186	37		223	269	
	0180	71 to 80		8	2		202	37		239	286	
	0200	81 to 96		8	2		215	37		252	300	
	0220	Two operating sash, 58 to 78 united inches		8	2		296	37		333	390	
	0240	79 to 88		8	2		330	37		367	430	
	0260	89 to 98		8	2		360	37		397	465	
	0280	99 to 108		6	2.667		380	49.50		429.50	505	
	0300	109 to 121		6	2.667		415	49.50		464.50	545	
	0320	Three operating sash, 73 to 108 united inches		8	2		445	37		482	555	
	0340	109 to 118		8	2		475	37		512	590	
	0360	119 to 128		6	2.667		520	49.50		569.50	655	
	0380	129 to 138		6	2.667		565	49.50		614.50	705	
	0400	139 to 156		6	2.667		600	49.50		649.50	745	
	0420	Four operating sash, 89 to 98 united inches		8	2		595	37		632	720	
	0440	99 to 108		8	2		635	37		672	765	
	0460	109 to 118		6	2.667		690	49.50		739.50	845	
	0480	119 to 128		6	2.667		745	49.50		794.50	905	
	0500	129 to 138		6	2.667		800	49.50		849.50	965	

For expanded coverage of these items see *Means Building Construction Cost Data 1998*

086 100 | Wood Windows

		CREW	DAILY OUTPUT	LABOR-HOURS	UNIT	1998 BARE COSTS				TOTAL INCL O&P	
						MAT.	LABOR	EQUIP.	TOTAL		
138	0520	139 to 148	2 Carp	6	2.667	Ea.	860	49.50		909.50	1,025	138
	0540	149 to 168		4	4		915	74		989	1,125	
	0560	Fixed picture window, up to 63 united inches		8	2		124	37		161	200	
	0580	64 to 83		8	2		166	37		203	246	
	0600	84 to 101		8	2		211	37		248	296	
	0620	For each united inch over 101, add		900	.018	Inch	2.30	.33		2.63	3.09	
	0640	Picture window options, low E glazing, up to 101 united in.				Ea.	17.50			17.50	19.25	
	0660	102 to 124					22.50			22.50	25	
	0680	124 and over					34			34	37.50	
	0700	Options, low E glazing, up to 101 united inches					8.75			8.75	9.65	
	0720	102 to 124					11.25			11.25	12.40	
	0740	124 and over					17			17	18.70	
	0760	Muntins, between glazing, square, per lite					2.10			2.10	2.31	
	0780	Diamond shape, per full or partial diamond					2.60			2.60	2.86	
	0800	Celluose fiber insulation, poured into sash balance cavity	1 Carp	36	.222	C.F.	.40	4.11		4.51	7.50	
	0820	Silicone caulking at perimeter	"	800	.010	L.F.	.13	.19		.32	.46	
140	0010	**SLIDING WINDOW** Including frame, screen, and exterior trim										140
	0100	Average quality, bldrs. model, 3'-0" x 3'-0" high, standard glazed	1 Carp	10	.800	Ea.	118	14.80		132.80	156	
	0120	Insulating glass		10	.800		167	14.80		181.80	210	
	0200	4'-0" x 3'-6" high, standard glazed		9	.889		151	16.45		167.45	194	
	0220	Insulating glass		9	.889		234	16.45		250.45	285	
	0300	6'-0" x 5'-0" high, standard glazed		8	1		255	18.50		273.50	310	
	0320	Insulating glass		8	1		385	18.50		403.50	450	
	2000	Metal clad, deluxe, insulating glass, 3'-0" x 3'-0" high		10	.800		241	14.80		255.80	291	
	2050	4'-0" x 3'-6" high		9	.889		305	16.45		321.45	365	
	2100	5'-0" x 4'-0" high		9	.889		365	16.45		381.45	435	
	2150	6'-0" x 5'-0" high		8	1		450	18.50		468.50	525	
142	0010	**WEATHERSTRIPPING** See division 087-300										142
144	0010	**WINDOW GRILLE OR MUNTIN** Snap-in type										144
	0020	Colonial or diamond pattern										
	2000	Wood, awning window, glass size 28" x 16" high	1 Carp	30	.267	Ea.	17.25	4.93		22.18	27.50	
	2060	44" x 24" high		32	.250		17.75	4.63		22.38	27.50	
	2100	Casement, glass size, 20" x 36" high		30	.267		13	4.93		17.93	23	
	2180	20" x 56" high		32	.250		27	4.63		31.63	37.50	
	2200	Double hung, glass size, 16" x 24" high		24	.333	Set	18.05	6.15		24.20	30.50	
	2280	32" x 32" high		34	.235	"	41.50	4.35		45.85	53.50	
	2500	Picture, glass size, 48" x 48" high		30	.267	Ea.	79.50	4.93		84.43	96	
	2580	60" x 68" high		28	.286	"	104	5.30		109.30	123	
	2600	Sliding, glass size, 14" x 36" high		24	.333	Set	26.50	6.15		32.65	39.50	
	2680	36" x 36" high		22	.364	"	44	6.75		50.75	60	
148	0010	**WOOD SASH** Including glazing but not including trim										148
	0050	Custom, 5'-0" x 4'-0", 1" dbl. glazed, 3/16" thick lites	2 Carp	3.20	5	Ea.	180	92.50		272.50	355	
	0100	1/4" thick lites		5	3.200		184	59		243	305	
	0200	1" thick, triple glazed		5	3.200		420	59		479	565	
	0300	7'-0" x 4'-6" high, 1" double glazed, 3/16" thick lites		4.30	3.721		430	69		499	595	
	0400	1/4" thick lites		4.30	3.721		485	69		554	650	
	0500	1" thick, triple glazed		4.30	3.721		555	69		624	730	
	0600	8'-6" x 5'-0" high, 1" double glazed, 3/16" thick lites		3.50	4.571		580	84.50		664.50	785	
	0700	1/4" thick lites		3.50	4.571		635	84.50		719.50	840	
	0800	1" thick, triple glazed		3.50	4.571		640	84.50		724.50	850	
	0900	Window frames only, based on perimeter length				L.F.	3.54			3.54	3.89	
	7000	Sash, single lite, 2'-0" x 2'-0" high	1 Carp	20	.400	Ea.	27.50	7.40		34.90	42.50	
	7050	2'-6" x 2'-0" high		19	.421		34	7.80		41.80	51	
	7100	2'-6" x 2'-6" high		18	.444		39.50	8.20		47.70	57	

Important: See the Reference Section for critical supporting data - Reference Nos., Crews, & Location Factors

086 100	Wood Windows	CREW	DAILY OUTPUT	LABOR-HOURS	UNIT	1998 BARE COSTS MAT.	LABOR	EQUIP.	TOTAL	TOTAL INCL O&P	
148 7150	3'-0" x 2'-0" high	1 Carp	17	.471	Ea.	45.50	8.70		54.20	65	**148**

087 100	Finish Hardware	CREW	DAILY OUTPUT	LABOR-HOURS	UNIT	1998 BARE COSTS MAT.	LABOR	EQUIP.	TOTAL	TOTAL INCL O&P	
101 0010	**AVERAGE** Percentage for hardware, total job cost, minimum									.75%	**101**
0050	Maximum									3.50%	
0500	Total hardware for building, average distribution					85%	15%				
1000	Door hardware, apartment, interior				Door	102			102	112	
1500	Hospital bedroom, minimum					232			232	255	
2000	Maximum				↓	500			500	550	
2100	Pocket door				Ea.	103			103	113	
2250	School, single exterior, incl. lever, not incl. panic device				Door	340			340	375	
2500	Single interior, regular use, no lever included					226			226	249	
2550	Including handicap lever					310			310	340	
2600	Heavy use, incl. lever and closer					395			395	435	
2850	Stairway, single interior				↓	565			565	620	
3100	Double exterior, with panic device				Pr.	795			795	875	
3600	Toilet, public, single interior				Door	124			124	137	
110 0010	**DOORSTOPS** Holder and bumper, floor or wall	1 Carp	32	.250	Ea.	12.40	4.63		17.03	21.50	**110**
1300	Wall bumper, 4" diameter, with rubber pad, aluminum		32	.250	"	6	4.63		10.63	14.55	
1600	Door bumper, floor type, aluminum		32	.250	Ea.	3.53	4.63		8.16	11.85	
1900	Plunger type, door mounted		32	.250	"	21.50	4.63		26.13	31.50	
112 0010	**ENTRANCE LOCKS** Cylinder, grip handle, deadlocking latch	1 Carp	9	.889	Ea.	97	16.45		113.45	135	**112**
0020	Deadbolt		8	1		117	18.50		135.50	161	
0100	Push and pull plate, dead bolt		8	1	↓	112	18.50		130.50	155	
0900	For handicapped lever, add					122			122	135	
116 0010	**HINGES** Full mortise, avg. freq., steel base, 4-1/2" x 4-1/2", USP				Pr.	17.05			17.05	18.75	**116**
0100	5" x 5", USP					27.50			27.50	30.50	
0200	6" x 6", USP					57.50			57.50	63.50	
0400	Brass base, 4-1/2" x 4-1/2", US10					35			35	38.50	
0500	5" x 5", US10					50			50	55	
0600	6" x 6", US10					83			83	91	
0800	Stainless steel base, 4-1/2" x 4-1/2", US32				↓	58			58	64	
0900	For non removable pin, add				Ea.	2			2	2.20	
0910	For floating pin, driven tips, add					3.75			3.75	4.13	
0930	For hospital type tip on pin, add					11			11	12.10	
0940	For steeple type tip on pin, add				↓	7.35			7.35	8.10	
0950	Full mortise, high frequency, steel base, 3-1/2" x 3-1/2", US26D				Pr.	25			25	27.50	
1000	4-1/2" x 4-1/2", USP					42.50			42.50	46.50	
1100	5" x 5", USP					49			49	54	
1200	6" x 6", USP					92.50			92.50	102	
1400	Brass base, 3-1/2" x 3-1/2", US4					37.50			37.50	41	
1430	4-1/2" x 4-1/2", US10					36.50			36.50	40	
1500	5" x 5", US10					92.50			92.50	102	
1600	6" x 6", US10					118			118	129	
1800	Stainless steel base, 4-1/2" x 4-1/2", US32				↓	90			90	99	
1930	For hospital type tip on pin, add				Ea.	14.45			14.45	15.90	
1950	Full mortise, low frequency, steel base, 3-1/2" x 3-1/2", US26D				Pr.	11.75			11.75	12.95	

		087 100 | Finish Hardware	CREW	DAILY OUTPUT	LABOR-HOURS	UNIT	1998 BARE COSTS MAT.	LABOR	EQUIP.	TOTAL	TOTAL INCL O&P	
116	2000	4-1/2" x 4-1/2", USP				Pr.	12.95			12.95	14.25	116
	2100	5" x 5", USP					20.50			20.50	22.50	
	2200	6" x 6", USP					40			40	44	
	2300	4-1/2" x 4-1/2", US3					9.65			9.65	10.60	
	2310	5" x 5", US3					29			29	31.50	
	2400	Brass bass, 4-1/2" x 4-1/2", US10					29			29	32	
	2500	5" x 5", US10					44			44	48	
	2800	Stainless steel base, 4-1/2" x 4-1/2", US32					48.50			48.50	53.50	
118	0010	**KICK PLATE** 6" high, for 3' door, stainless steel	1 Carp	15	.533	Ea.	15.45	9.85		25.30	34	118
	0500	Bronze		15	.533		18.75	9.85		28.60	37.50	
	2000	Aluminum, .050, with 3 beveled edges, 10" x 28"		15	.533		13.75	9.85		23.60	32	
	2010	10" x 30"		15	.533		14.70	9.85		24.55	33	
	2020	10" x 34"		15	.533		16.70	9.85		26.55	35.50	
	2040	10" x 38"		15	.533		18.35	9.85		28.20	37	
120	0010	**LOCKSET** Standard duty, cylindrical, with sectional trim										120
	0020	Non-keyed, passage	1 Carp	12	.667	Ea.	35	12.35		47.35	59.50	
	0100	Privacy		12	.667		42.50	12.35		54.85	68	
	0400	Keyed, single cylinder function		10	.800		62.50	14.80		77.30	94	
	0420	Hotel		8	1		84.50	18.50		103	125	
	0500	Lever handled, keyed, single cylinder function		10	.800		84.50	14.80		99.30	119	
	0600	Bedroom, bathroom and inner office doors		10	.800		107	14.80		121.80	144	
	0900	Apartment, office and corridor doors		10	.800		135	14.80		149.80	174	
	1400	Keyed, single cylinder function		10	.800		147	14.80		161.80	188	
	1700	Residential, interior door, minimum		16	.500		11.85	9.25		21.10	29	
	1720	Maximum		8	1		31.50	18.50		50	66.50	
	1800	Exterior, minimum		14	.571		26.50	10.55		37.05	47	
	1810	Average		8	1		56	18.50		74.50	93	
	1820	Maximum		8	1		111	18.50		129.50	155	
126	0010	**HASP** Steel, 3" assembly	1 Carp	26	.308	Ea.	2.03	5.70		7.73	12	126
	0020	4-1/2"		13	.615		2.60	11.40		14	22.50	
	0040	6"		12.50	.640		3.94	11.85		15.79	25	
127	0010	**PANIC DEVICE** For rim locks, single door, exit only	1 Carp	6	1.333	Ea.	300	24.50		324.50	375	127
	1000	Mortise, bar, exit only		4	2	"	400	37		437	505	
	4000	Double doors, exit only		2	4	Pr.	610	74		684	795	
	4500	Exit & entrance		2	4	"	695	74		769	890	
129	0010	**PUSH-PULL PLATE**										129
	0100	Push plate, .050 thick, 4" x 16", aluminum	1 Carp	12	.667	Ea.	4.58	12.35		16.93	26	
	0500	Bronze		12	.667		10.25	12.35		22.60	32.50	
	1500	Pull handle and push bar, aluminum		11	.727		99	13.45		112.45	132	
	2000	Bronze		10	.800		128	14.80		142.80	167	
	4000	Door pull, designer style, cast aluminum, minimum		12	.667		55.50	12.35		67.85	82	
	5000	Maximum		8	1		259	18.50		277.50	315	
134	0010	**WINDOW HARDWARE**										134
	1000	Handles, surface mounted, aluminum	1 Carp	24	.333	Ea.	1.64	6.15		7.79	12.35	
	1020	Brass		24	.333		1.91	6.15		8.06	12.65	
	1040	Chrome		24	.333		1.74	6.15		7.89	12.45	
	1500	Recessed, aluminum		12	.667		.95	12.35		13.30	22	
	1520	Brass		12	.667		1.06	12.35		13.41	22	
	1540	Chrome		12	.667		1	12.35		13.35	22	
	2000	Latches, aluminum		20	.400		1.37	7.40		8.77	14.20	
	2020	Brass		20	.400		1.64	7.40		9.04	14.50	
	2040	Chrome		20	.400		1.54	7.40		8.94	14.40	

Important: See the Reference Section for critical supporting data - Reference Nos., Crews, & Location Factors

087 200 | Operators

		CREW	DAILY OUTPUT	LABOR-HOURS	UNIT	1998 BARE COSTS MAT.	LABOR	EQUIP.	TOTAL	TOTAL INCL O&P	
202	0010	**AUTOMATIC OPENERS** Swing doors, single	2 Skwk	.80	20	Ea.	2,050	375		2,425	2,900	**202**
	0100	Single operating pair		.50	32	Pr.	3,800	600		4,400	5,200	
	0400	For double simultaneous doors, one way, add		1.20	13.333		315	250		565	775	
	0500	Two way, add		.90	17.778		390	335		725	1,000	
	1000	Sliding doors, 3' wide, including track & hanger, single		.60	26.667	Opng.	3,500	500		4,000	4,700	
	1300	Bi-parting		.50	32		4,225	600		4,825	5,675	
	1450	Activating carpet, single door, one way add		2.20	7.273		635	136		771	935	
	1550	Two way, add		1.30	12.308		910	231		1,141	1,400	
	1750	Handicap opener, button operating	1 Carp	1.50	5.333	Ea.	1,200	98.50		1,298.50	1,500	
206	0010	**DOOR CLOSER** Rack and pinion	1 Carp	6.50	1.231	Ea.	99	23		122	148	**206**
	0020	Adjustable backcheck, 3 way mount, all sizes, regular arm		6	1.333		105	24.50		129.50	158	
	0040	Hold open arm		6	1.333		114	24.50		138.50	168	
	0100	Fusible link		6.50	1.231		88.50	23		111.50	137	
	0200	Non sized, regular arm		6	1.333		99	24.50		123.50	152	
	0240	Hold open arm		6	1.333		123	24.50		147.50	179	
	0400	4 way mount, non sized, regular arm		6	1.333		136	24.50		160.50	193	
	0440	Hold open arm		6	1.333		146	24.50		170.50	204	

087 300 | Weatherstripping/Seals

		CREW	DAILY OUTPUT	LABOR-HOURS	UNIT	1998 BARE COSTS MAT.	LABOR	EQUIP.	TOTAL	TOTAL INCL O&P	
304	0010	**THRESHOLD** 3' long door saddles, aluminum	1 Carp	48	.167	L.F.	3.25	3.08		6.33	8.90	**304**
	0100	Aluminum, 8" wide, 1/2" thick		12	.667	Ea.	28	12.35		40.35	51.50	
	0500	Bronze		60	.133	L.F.	26	2.47		28.47	33	
	0600	Bronze, panic threshold, 5" wide, 1/2" thick		12	.667	Ea.	54.50	12.35		66.85	81	
	0700	Rubber, 1/2" thick, 5-1/2" wide		20	.400		27.50	7.40		34.90	43	
	0800	2-3/4" wide		20	.400		12.75	7.40		20.15	27	
306	0010	**WEATHERSTRIPPING** Window, double hung, 3' x 5', zinc	1 Carp	7.20	1.111	Opng.	10	20.50		30.50	46	**306**
	0100	Bronze		7.20	1.111		18.80	20.50		39.30	55.50	
	0200	Vinyl V strip		7	1.143		3.37	21		24.37	39.50	
	0500	As above but heavy duty, zinc		4.60	1.739		12.50	32		44.50	69	
	0600	Bronze		4.60	1.739		22	32		54	79	
	1000	Doors, wood frame, interlocking, for 3' x 7' door, zinc		3	2.667		11.20	49.50		60.70	97	
	1100	Bronze		3	2.667		17.65	49.50		67.15	104	
	1300	6' x 7' opening, zinc		2	4		12.25	74		86.25	141	
	1400	Bronze		2	4		23	74		97	153	
	1700	Wood frame, spring type, bronze										
	1800	3' x 7' door	1 Carp	7.60	1.053	Opng.	14.70	19.45		34.15	49.50	
	1900	6' x 7' door		7	1.143		15.35	21		36.35	53	
	1920	Felt, 3' x 7' door		14	.571		1.83	10.55		12.38	20	
	1930	6' x 7' door		13	.615		1.99	11.40		13.39	21.50	
	1950	Rubber, 3' x 7' door		7.60	1.053		3.98	19.45		23.43	38	
	1960	6' x 7' door		7	1.143		4.54	21		25.54	41	
	2200	Metal frame, spring type, bronze										
	2300	3' x 7' door	1 Carp	3	2.667	Opng.	24	49.50		73.50	111	
	2400	6' x 7' door	"	2.50	3.200	"	33.50	59		92.50	138	
	2500	For stainless steel, spring type, add					133%					
	2700	Metal frame, extruded sections, 3' x 7' door, aluminum	1 Carp	2	4	Opng.	32.50	74		106.50	163	
	2800	Bronze		2	4		82	74		156	218	
	3100	6' x 7' door, aluminum		1.20	6.667		41.50	123		164.50	257	
	3200	Bronze		1.20	6.667		97	123		220	320	
	3500	Threshold weatherstripping										
	3650	Door sweep, flush mounted, aluminum	1 Carp	25	.320	Ea.	9.35	5.90		15.25	20.50	
	3700	Vinyl		25	.320	"	10.40	5.90		16.30	21.50	
	4000	Astragal for double doors, aluminum		4	2	Opng.	16.30	37		53.30	81.50	
	4100	Bronze		4	2	"	26.50	37		63.50	92.50	
	5000	Garage door bottom weatherstrip, 12' aluminum, clear		14	.571	Ea.	15.45	10.55		26	35	

DOORS & WINDOWS — 8

087 | Hardware

087 300 | Weatherstripping/Seals

			CREW	DAILY OUTPUT	LABOR-HOURS	UNIT	MAT.	LABOR	EQUIP.	TOTAL	TOTAL INCL O&P	
306	5010	Bronze	1 Carp	14	.571	Ea.	58.50	10.55		69.05	82.50	306
	5050	Bottom protection, 12' aluminum, clear		14	.571		18.70	10.55		29.25	38.50	
	5100	Bronze	↓	14	.571	↓	73	10.55		83.55	98	

087 500 | Door/Window Acces.

			CREW	DAILY OUTPUT	LABOR-HOURS	UNIT	MAT.	LABOR	EQUIP.	TOTAL	TOTAL INCL O&P	
504	0010	**DETECTION SYSTEMS** See division 168-120										504
506	0010	**DOOR ACCESSORIES**										506
	1000	Knockers, brass, standard	1 Carp	16	.500	Ea.	33.50	9.25		42.75	53	
	1100	Deluxe		10	.800		103	14.80		117.80	139	
	4000	Security chain, standard		18	.444		5.80	8.20		14	20.50	
	4100	Deluxe	↓	18	.444	↓	34.50	8.20		42.70	52	

088 | Glazing

088 100 | Glass

			CREW	DAILY OUTPUT	LABOR-HOURS	UNIT	MAT.	LABOR	EQUIP.	TOTAL	TOTAL INCL O&P	
118	0010	**FLOAT GLASS** 3/16″ thick, clear, plain	2 Glaz	130	.123	S.F.	3.02	2.25		5.27	7.05	118
	0200	Tempered, clear		130	.123		3.47	2.25		5.72	7.55	
	0300	Tinted		130	.123		4.56	2.25		6.81	8.75	
	0600	1/4″ thick, clear, plain		120	.133		3.63	2.44		6.07	8.05	
	0700	Tinted		120	.133		3.71	2.44		6.15	8.15	
	0800	Tempered, clear		120	.133		3.99	2.44		6.43	8.45	
	0900	Tinted		120	.133		6.05	2.44		8.49	10.70	
	1600	3/8″ thick, clear, plain		75	.213		6.05	3.90		9.95	13.20	
	1700	Tinted		75	.213		7.20	3.90		11.10	14.40	
	1800	Tempered, clear		75	.213		8.30	3.90		12.20	15.65	
	1900	Tinted		75	.213		11.15	3.90		15.05	18.75	
	2200	1/2″ thick, clear, plain		55	.291		11.55	5.30		16.85	21.50	
	2300	Tinted		55	.291		12.30	5.30		17.60	22.50	
	2400	Tempered, clear		55	.291		13.30	5.30		18.60	23.50	
	2500	Tinted		55	.291		16.70	5.30		22	27.50	
	2800	5/8″ thick, clear, plain		45	.356		12.30	6.50		18.80	24.50	
	2900	Tempered, clear	↓	45	.356		14.30	6.50		20.80	26.50	
	8900	For low emissivity coating for 3/16″ and 1/4″ only, add to above				↓	3					
120	0010	**FULL VISION** Window system with 3/4″ glass mullions, 10′ high	H-2	130	.185	S.F.	35.50	3.08		38.58	44	120
	0100	10′ to 20′ high, minimum		110	.218		37.50	3.64		41.14	47	
	0150	Average		100	.240		40.50	4		44.50	51	
	0200	Maximum	↓	80	.300	↓	45.50	5		50.50	58.50	
128	0010	**GLAZING VARIABLES**										128
	0500	For high rise glazing, from exterior, add per S.F. per story				S.F.					.08	
	0600	For glass replacement, add				″		100%				
	0700	For gasket settings, add				L.F.	2.46			2.46	2.71	
	0800	For concrete reglet settings, add				S.F.	20%	25%				
	0900	For sloped glazing, add				″		25%				
	2000	Fabrication, polished edges, 1/4″ thick				Inch	.20			.20	.22	
	2100	1/2″ thick					.40			.40	.44	
	2500	Mitered edges, 1/4″ thick					.61			.61	.67	
	2600	1/2″ thick				↓	1.01			1.01	1.11	

R088 -010

 Important: See the Reference Section for critical supporting data - Reference Nos., Crews, & Location Factors

088 100 | Glass

		CREW	DAILY OUTPUT	LABOR-HOURS	UNIT	MAT.	LABOR	EQUIP.	TOTAL	TOTAL INCL O&P	
132	0010	**INSULATING GLASS** 2 lites 1/8" float, 1/2" thk, under 15 S.F.										132
	0100	Tinted	2 Glaz	95	.168	S.F.	7.75	3.08		10.83	13.65	
	0280	Double glazed, 5/8" thk unit, 3/16" float, 15-30 S.F., clear		90	.178		5.75	3.25		9	11.70	
	0400	1" thk, dbl. glazed, 1/4" float, 30-70 S.F., clear		75	.213		9	3.90		12.90	16.40	
	0500	Tinted		75	.213		8.30	3.90		12.20	15.65	
	2000	Both lites, light & heat reflective		85	.188		13.80	3.44		17.24	21	
	2500	Heat reflective, film inside, 1" thick unit, clear		85	.188		12.65	3.44		16.09	19.65	
	2600	Tinted		85	.188		13.70	3.44		17.14	21	
	3000	Film on weatherside, clear, 1/2" thick unit		95	.168		9	3.08		12.08	15.05	
	3100	5/8" thick unit		90	.178		11.40	3.25		14.65	17.95	
	3200	1" thick unit		85	.188		12.40	3.44		15.84	19.40	
144	0010	**MIRRORS** No frames, wall type, 1/4" plate glass, polished edge										144
	0100	Up to 5 S.F.	2 Glaz	125	.128	S.F.	4.95	2.34		7.29	9.35	
	0200	Over 5 S.F.		160	.100		4.80	1.83		6.63	8.35	
	0500	Door type, 1/4" plate glass, up to 12 S.F.		160	.100		5.15	1.83		6.98	8.70	
	1000	Float glass, up to 10 S.F., 1/8" thick		160	.100		3.03	1.83		4.86	6.40	
	1100	3/16" thick		150	.107		3.52	1.95		5.47	7.15	
	1500	12" x 12" wall tiles, square edge, clear		195	.082		1.29	1.50		2.79	3.92	
	1600	Veined		195	.082		3.29	1.50		4.79	6.10	
	2010	Bathroom, unframed, laminated		160	.100		8.60	1.83		10.43	12.50	
176	0010	**WINDOW GLASS** Clear float, stops, putty bed, 1/8" thick	2 Glaz	480	.033	S.F.	2.50	.61		3.11	3.77	176
	0500	3/16" thick, clear		480	.033		3.05	.61		3.66	4.37	
	0600	Tinted		480	.033		3.45	.61		4.06	4.82	
	0700	Tempered		480	.033		4.20	.61		4.81	5.65	

(R088 -100)

089 200 | Glazed Curtain Wall

		CREW	DAILY OUTPUT	LABOR-HOURS	UNIT	MAT.	LABOR	EQUIP.	TOTAL	TOTAL INCL O&P	
204	0010	**TUBE FRAMING** For window walls and store fronts, aluminum, stock										204
	0050	Plain tube frame, mill finish, 1-3/4" x 1-3/4"	2 Glaz	103	.155	L.F.	5.80	2.84		8.64	11.15	
	0150	1-3/4" x 4"		98	.163		7.30	2.99		10.29	13	
	0200	1-3/4" x 4-1/2"		95	.168		8.10	3.08		11.18	14.05	
	0250	2" x 6"		89	.180		12.25	3.29		15.54	18.95	
	0350	4" x 4"		87	.184		12	3.37		15.37	18.80	
	0400	4-1/2" x 4-1/2"		85	.188		13.40	3.44		16.84	20.50	
	0450	Glass bead		240	.067		1.55	1.22		2.77	3.73	
	1000	Flush tube frame, mill finish, 1/4" glass, 1-3/4" x 4", open header		80	.200		7.15	3.66		10.81	14	
	1050	Open sill		82	.195		6.25	3.57		9.82	12.80	
	1100	Closed back header		83	.193		10.05	3.53		13.58	16.90	
	1150	Closed back sill		85	.188		9.50	3.44		12.94	16.20	
	1200	Vertical mullion, one piece		75	.213		10.60	3.90		14.50	18.20	
	1250	Two piece		73	.219		11.40	4.01		15.41	19.15	
	1300	90° or 180° vertical corner post		75	.213		17.90	3.90		21.80	26	
	1400	1-3/4" x 4-1/2", open header		80	.200		8.75	3.66		12.41	15.75	
	1450	Open sill		82	.195		7.20	3.57		10.77	13.90	
	1500	Closed back header		83	.193		11.05	3.53		14.58	18	
	1550	Closed back sill		85	.188		10.30	3.44		13.74	17.05	
	1600	Vertical mullion, one piece		75	.213		11.50	3.90		15.40	19.15	

(R089 -200)

089 200	Glazed Curtain Wall	CREW	DAILY OUTPUT	LABOR-HOURS	UNIT	1998 BARE COSTS				TOTAL INCL O&P
						MAT.	LABOR	EQUIP.	TOTAL	
204 1650	Two piece	2 Glaz	73	.219	L.F.	12.15	4.01		16.16	20	**204**
1700	90° or 180° vertical corner post		75	.213		12.45	3.90		16.35	20
2000	Flush tube frame, mill fin. for ins. glass, 2" x 4-1/2", open header		75	.213		10.20	3.90		14.10	17.70
2050	Open sill		77	.208		8.95	3.80		12.75	16.15
2100	Closed back header		78	.205		11.10	3.75		14.85	18.50
2150	Closed back sill		80	.200		10.95	3.66		14.61	18.15
2200	Vertical mullion, one piece		70	.229		12.30	4.18		16.48	20.50
2250	Two piece		68	.235		13.15	4.31		17.46	21.50
2300	90° or 180° vertical corner post		70	.229		12.45	4.18		16.63	20.50
5000	Flush tube frame, mill fin., thermal brk., 2-1/4"x 4-1/2", open header		74	.216		11.20	3.96		15.16	18.95
5050	Open sill		75	.213		9.80	3.90		13.70	17.30
5100	Vertical mullion, one piece		69	.232		13.55	4.24		17.79	22
5150	Two piece		67	.239		14.50	4.37		18.87	23
5200	90° or 180° vertical corner post		69	.232		13	4.24		17.24	21.50
6980	Door stop (snap in)		380	.042		2.10	.77		2.87	3.59
7000	For joints, 90°, clip type, add				Ea.	17.40			17.40	19.15
7050	Screw spline joint, add					13.50			13.50	14.85
7100	For joint other than 90°, add					28.50			28.50	31
8000	For bronze anodized aluminum, add					15%				
8050	For stainless steel materials, add					350%				
8100	For monumental grade, add					50%				
8150	For steel stiffener, add	2 Glaz	200	.080	L.F.	6.90	1.46		8.36	10.05
8200	For 2 to 5 stories, add per story				Story		5%			

For information about Means Estimating Seminars, see yellow pages 11 and 12 in back of book

Important: See the Reference Section for critical supporting data - Reference Nos., Crews, & Location Factors

Division 9
Finishes

Estimating Tips
General

- Room Finish Schedule: A complete set of plans should contain a room finish schedule. If one is not available, it would be well worth the time and effort to put one together. A room finish schedule should contain the room number, room name (for clarity), floor materials, base materials, wainscot materials, wainscot height, wall materials (for each wall), ceiling materials and special instructions.

- Surplus Finishes: Review the specifications to determine if there is any requirement to provide certain amounts of extra materials for the owner's maintenance department. In some cases the owner may require a substantial amount of materials, especially when it is a special order item or long lead time item.

092 Lath, Plaster & Gypsum Board

- Lath is estimated by the square yard for both gypsum and metal lath, plus usually 5% allowance for waste. Furring, channels and accessories are measured by the linear foot. An extra foot should be allowed for each accessory miter or stop.

- Plaster is also estimated by the square yard. Deductions for openings vary by preference, from zero deduction to 50% of all openings over 2 feet in width. Some estimators deduct a percentage of the total yardage for openings. The estimator should allow one extra square foot for each linear foot of horizontal interior or exterior angle located below the ceiling level. Also, double the areas of small radius work.

- Each room should be measured, perimeter times maximum wall height. Ceiling areas are equal to length times width.

- Drywall accessories, studs, track, and acoustical caulking are all measured by the linear foot. Drywall taping is figured by the square foot. Gypsum wallboard is estimated by the square foot. No material deductions should be made for door or window openings under 32 S.F. Coreboard can be obtained in a 1" thickness for solid wall and shaft work. Additions should be made to price out the inside or outside corners.

- Different types of partition construction should be listed separately on the quantity sheets. There may be walls with studs of various widths, double studded, and similar or dissimilar surface materials. Shaft work is usually different construction from surrounding partitions requiring separate quantities and pricing of the work.

093 Tile
094 Terrazzo

- Tile and terrazzo areas are taken off on a square foot basis. Trim and base materials are measured by the linear foot. Accent tiles are listed per each. Two basic methods of installation are used. Mud set is approximately 30% more expensive than the thin set. In terrazzo work, be sure to include the linear footage of embedded decorative strips, grounds, machine rubbing and power cleanup.

095 Acoustical Treatment & Wood Flooring

- Acoustical systems fall into several categories. The takeoff of these materials is by the square foot of area with a 5% allowance for waste. Do not forget about scaffolding, if applicable, when estimating these systems.

- Wood flooring is available in strip, parquet, or block configuration. The latter two types are set in adhesives with quantities estimated by the square foot. The laying pattern will influence labor costs and material waste. In addition to the material and labor for laying wood floors, the estimator must make allowances for sanding and finishing these areas unless the flooring is prefinished.

096 Flooring & Carpet

- Most of the various types of flooring are all measured on a square foot basis. Base is measured by the linear foot. If adhesive materials are to be quantified, they are estimated at a specified coverage rate by the gallon depending upon the specified type and the manufacturer's recommendations.

- Sheet flooring is measured by the square yard. Roll widths vary, so consideration should be given to use the most economical width, as waste must be figured into the total quantity. Consider also the installation methods available, direct glue down or stretched.

099 Painting & Wall Coverings

- Painting is one area where bids vary to a greater extent than almost any other section of a project. This arises from the many methods of measuring surfaces to be painted. The estimator should check the plans and specifications carefully to be sure of the required number of coats.

- Protection of adjacent surfaces is not included in painting costs. When considering the method of paint application, an important factor is the amount of protection and masking required. These must be estimated separately and may be the determining factor in choosing the method of application.

- Wall coverings are estimated by the square foot. The area to be covered is measured, length by height of wall above baseboards, to calculate the square footage of each wall. This figure is divided by the number of square feet in the single roll which is being used. Deduct, in full, the areas of openings such as doors and windows. Where a pattern match is required allow 25%-30% waste. One gallon of paste should be sufficient to hang 12 single rolls of light to medium weight paper.

Reference Numbers

Reference numbers are shown in bold squares at the beginning of some major classifications. These numbers refer to related items in the Reference Section. The reference information may be an estimating procedure, an alternate pricing method or technical information.

Note: Not all subdivisions listed here necessarily appear in this publication.

091 300 | Suspension Systems

		CREW	DAILY OUTPUT	LABOR-HOURS	UNIT	1998 BARE COSTS				TOTAL INCL O&P
						MAT.	LABOR	EQUIP.	TOTAL	
304	0010	**SUSPENSION SYSTEMS** For boards and tile									**304**
	0050	Class A suspension system, 15/16" T bar, 2' x 4' grid	1 Carp	800	.010	S.F.	.30	.19		.49	.65
	0300	2' x 2' grid	"	650	.012		.37	.23		.60	.80
	0350	For 9/16" grid, add					.12			.12	.13
	0360	For fire rated grid, add					.07			.07	.08
	0370	For colored grid, add					.14			.14	.15
	0400	Concealed Z bar suspension system, 12" module	1 Carp	520	.015		.33	.28		.61	.85
	0600	1-1/2" carrier channels, 4' O.C., add		470	.017		.16	.31		.47	.72
	0650	1-1/2" x 3-1/2" channels		470	.017		.36	.31		.67	.94
	0700	Carrier channels for ceilings with									
	0900	recessed lighting fixtures, add	1 Carp	460	.017	S.F.	.32	.32		.64	.90
	5000	Wire hangers, #12 wire	"	300	.027	Ea.	.35	.49		.84	1.24

092 050 | Furring & Lathing

		CREW	DAILY OUTPUT	LABOR-HOURS	UNIT	1998 BARE COSTS				TOTAL INCL O&P
						MAT.	LABOR	EQUIP.	TOTAL	
052	0010	**ACCESSORIES, PLASTER** Casing bead, expanded flange, galvanized	1 Lath	2.70	2.963	C.L.F.	35	54		89	128	**052**
	0100	Zinc alloy	"	2.70	2.963		75	54		129	172
	0900	Channels, cold rolled, 16 ga., 3/4" deep, galvanized					17.50			17.50	19.25
	1200	1-1/2" deep, 16 ga., galvanized					25			25	27.50
	1620	Corner bead, expanded bullnose, 3/4" radius, #10, galvanized	1 Lath	2.60	3.077		56	56.50		112.50	154
	1640	Zinc alloy		2.70	2.963		59	54		113	154
	1650	#1, galvanized		2.55	3.137		34	57.50		91.50	132
	1660	Zinc alloy		2.70	2.963		60.50	54		114.50	156
	1670	Expanded wing, 2-3/4" wide, galv. #1		2.65	3.019		49	55.50		104.50	145
	1680	Zinc alloy		2.70	2.963		58.50	54		112.50	154
	1700	Inside corner, (corner rite) 3" x 3", painted		2.60	3.077		25	56.50		81.50	120
	1750	Strip-ex, 4" wide, painted		2.55	3.137		16	57.50		73.50	112
	1800	Expansion joint, 3/4" grounds, limited expansion, galv., 1 piece		2.70	2.963		52	54		106	146
	1900	Zinc alloy		2.70	2.963		67	54		121	163
	2100	Extreme expansion, galvanized, 2 piece		2.60	3.077		76.50	56.50		133	177
	2300	Zinc alloy		2.70	2.963		136	54		190	239
	2500	Joist clips for lath, 2-1/2" flange		1.90	4.211	M	70.50	77		147.50	204
	2600	4-1/2" flange		1.80	4.444	"	82	81.50		163.50	224
	2800	Metal base, galvanized and painted, 2-1/2" high		2.40	3.333	C.L.F.	42.50	61		103.50	147
	2900	Stud clips for gypsum lath, field clip		2.35	3.404	M	68	62.50		130.50	177
	3100	Resilient		2.30	3.478		103	63.50		166.50	217
	3200	Starter/finisher		2.20	3.636		70.50	66.50		137	187
054	0010	**FURRING** Beams & columns, 3/4" galvanized channels,									**054**
	0030	12" O.C.	1 Lath	155	.052	S.F.	.19	.94		1.13	1.76
	0050	16" O.C.		170	.047		.16	.86		1.02	1.58
	0070	24" O.C.		185	.043		.10	.79		.89	1.42
	0100	Ceilings, on steel, 3/4" channels, galvanized, 12" O.C.		210	.038		.17	.70		.87	1.33
	0300	16" O.C.		290	.028		.16	.50		.66	1
	0400	24" O.C.		420	.019		.10	.35		.45	.69
	0600	1-1/2" channels, galvanized, 12" O.C.		190	.042		.25	.77		1.02	1.53
	0700	16" O.C.		260	.031		.22	.56		.78	1.17
	0900	24" O.C.		390	.021		.15	.38		.53	.78
	1000	Walls, 3/4" channels, galvanized, 12" O.C.		235	.034		.17	.62		.79	1.21
	1200	16" O.C.		265	.030		.16	.55		.71	1.08

Important: See the Reference Section for critical supporting data - Reference Nos., Crews, & Location Factors

092 050 | Furring & Lathing

		CREW	DAILY OUTPUT	LABOR-HOURS	UNIT	1998 BARE COSTS				TOTAL INCL O&P	
						MAT.	LABOR	EQUIP.	TOTAL		
054	1300	24" O.C.	1 Lath	350	.023	S.F.	.10	.42		.52	.81	**054**
	1500	1-1/2" channels, galvanized, 12" O.C.		210	.038		.25	.70		.95	1.41	
	1600	16" O.C.		240	.033		.22	.61		.83	1.25	
	1800	24" O.C.		305	.026		.15	.48		.63	.95	
	8000	Suspended ceilings, including carriers										
	8200	1-1/2" carriers, 24" O.C. with:										
	8300	3/4" channels, 16" O.C.	1 Lath	165	.048	S.F.	.48	.89		1.37	1.98	
	8320	24" O.C.		200	.040		.43	.73		1.16	1.67	
	8400	1-1/2" channels, 16" O.C.		155	.052		.54	.94		1.48	2.15	
	8420	24" O.C.		190	.042		.47	.77		1.24	1.78	
	8600	2" carriers, 24" O.C. with:										
	8700	3/4" channels, 16" O.C.	1 Lath	155	.052	S.F.	.16	.94		1.10	1.72	
	8720	24" O.C.		190	.042		.11	.77		.88	1.38	
	8800	1-1/2" channels, 16" O.C.		145	.055		.56	1.01		1.57	2.28	
	8820	24" O.C.		180	.044		.49	.81		1.30	1.87	
056	0010	**GYPSUM LATH** Plain or perforated, nailed, 3/8" thick (R092 -050)	1 Lath	85	.094	S.Y.	3.24	1.72		4.96	6.40	**056**
	0100	1/2" thick, nailed		80	.100		3.87	1.83		5.70	7.25	
	0300	Clipped to steel studs, 3/8" thick		75	.107		3.69	1.95		5.64	7.25	
	0400	1/2" thick		70	.114		3.87	2.09		5.96	7.70	
	0600	Firestop gypsum base, to steel studs, 3/8" thick		70	.114		3.87	2.09		5.96	7.70	
	0700	1/2" thick		65	.123		3.96	2.25		6.21	8.05	
	0900	Foil back, to steel studs, 3/8" thick		75	.107		3.60	1.95		5.55	7.15	
	1000	1/2" thick		70	.114		3.69	2.09		5.78	7.50	
	1500	For ceiling installations, add		216	.037			.68		.68	1.11	
	1600	For columns and beams, add		170	.047			.86		.86	1.41	
058	0011	**METAL LATH** (R092 -060)										**058**
	3600	2.5 lb. diamond painted, on wood framing, on walls	1 Lath	85	.094	S.Y.	1.79	1.72		3.51	4.79	
	3700	On ceilings		75	.107		1.79	1.95		3.74	5.15	
	4200	3.4 lb. diamond painted, wired to steel framing		75	.107		2.05	1.95		4	5.45	
	4300	On ceilings		60	.133		2.05	2.44		4.49	6.25	
	5100	Rib lath, painted, wired to steel, on walls, 2.75 lb.		75	.107		1.69	1.95		3.64	5.05	
	5200	3.4 lb.		70	.114		3.18	2.09		5.27	6.95	
	5700	Suspended ceiling system, incl. 3.4 lb. diamond lath, painted		15	.533		9.50	9.75		19.25	26.50	
	5800	Galvanized		15	.533		9.80	9.75		19.55	27	

092 100 | Gypsum Plaster

		CREW	DAILY OUTPUT	LABOR-HOURS	UNIT	1998 BARE COSTS				TOTAL INCL O&P	
						MAT.	LABOR	EQUIP.	TOTAL		
108	0010	**GYPSUM PLASTER** 80# bag, less than 1 ton (R092 -105)				Bag	13.55			13.55	14.90	**108**
	0300	2 coats, no lath included, on walls	J-1	105	.381	S.Y.	2.88	6.25	.39	9.52	14.05	
	0400	On ceilings		92	.435		2.88	7.15	.45	10.48	15.60	
	0900	3 coats, no lath included, on walls		87	.460		4.05	7.55	.47	12.07	17.60	
	1000	On ceilings		78	.513		4.05	8.45	.53	13.03	19.15	
	1600	For irregular or curved surfaces, add						30%				
	1800	For columns & beams, add						50%				
116	0010	**PERLITE OR VERMICULITE PLASTER** 100 lb. bags Under 200 bags				Bag	13.95			13.95	15.35	**116**
	0300	2 coats, no lath included, on walls	J-1	92	.435	S.Y.	3.51	7.15	.45	11.11	16.30	
	0400	On ceilings		79	.506		3.51	8.30	.52	12.33	18.35	
	0900	3 coats, no lath included, on walls (R092 -115)		74	.541		5.60	8.90	.56	15.06	21.50	
	1000	On ceilings		63	.635		5.60	10.45	.65	16.70	24.50	
	1700	For irregular or curved surfaces, add to above						30%				
	1800	For columns and beams, add to above						50%				
	1900	For soffits, add to ceiling prices						40%				

FINISHES 9

092 150 | Veneer Plaster

		CREW	DAILY OUTPUT	LABOR-HOURS	UNIT	1998 BARE COSTS				TOTAL INCL O&P	
						MAT.	LABOR	EQUIP.	TOTAL		
154	0010	**THIN COAT** Plaster, 1 coat veneer, not incl. lath	J-1	3,600	.011	S.F.	.12	.18	.01	.31	.45	**154**
	1000	In 50 lb. bags				Bag	9.60			9.60	10.55	

092 300 | Aggregate Coatings

		CREW	DAILY OUTPUT	LABOR-HOURS	UNIT	1998 BARE COSTS				TOTAL INCL O&P	
304	0010	**STUCCO**, 3 coats 1" thick, float finish, with mesh, on wood frame	J-2	135	.356	S.Y.	5.60	5.95	.31	11.86	16.45	**304**
	0100	On masonry construction, no mesh incl.	J-1	200	.200		1.93	3.29	.21	5.43	7.85	
	0150	2 coats, 3/4" thick, float finish, no lath incl. [R092-300]	"	110	.364		1.83	5.95	.37	8.15	12.40	
	0300	For trowel finish, add	1 Plas	170	.047			.83		.83	1.38	
	0600	For coloring and special finish, add, minimum	J-1	685	.058		.35	.96	.06	1.37	2.06	
	0700	Maximum		200	.200		1.20	3.29	.21	4.70	7.05	
	1000	Exterior stucco, with bonding agent, 3 coats, on walls, no mesh incl.		200	.200		3.10	3.29	.21	6.60	9.15	
	1200	Ceilings		180	.222		3.10	3.65	.23	6.98	9.75	
	1300	Beams		80	.500		3.10	8.20	.51	11.81	17.75	
	1500	Columns		100	.400		3.10	6.55	.41	10.06	14.85	
	1600	Mesh, painted, nailed to wood, 1.8 lb.	1 Lath	60	.133		3.33	2.44		5.77	7.65	
	1800	3.6 lb.		55	.145		3.69	2.66		6.35	8.40	
	1900	Wired to steel, painted, 1.8 lb.		53	.151		3.33	2.76		6.09	8.20	
	2100	3.6 lb.		50	.160		3.69	2.93		6.62	8.85	

092 600 | Gypsum Board Systems

		CREW	DAILY OUTPUT	LABOR-HOURS	UNIT	1998 BARE COSTS				TOTAL INCL O&P	
602	0010	**BLUEBOARD** For use with thin coat										**602**
	0100	plaster application (see division 092-154)										
	1000	3/8" thick, on walls or ceilings, standard, no finish included	2 Carp	1,900	.008	S.F.	.16	.16		.32	.45	
	1100	With thin coat plaster finish		875	.018		.28	.34		.62	.89	
	1400	On beams, columns, or soffits, standard, no finish included		675	.024		.18	.44		.62	.95	
	1450	With thin coat plaster finish		475	.034		.32	.62		.94	1.42	
	3000	1/2" thick, on walls or ceilings, standard, no finish included		1,900	.008		.16	.16		.32	.45	
	3100	With thin coat plaster finish		875	.018		.28	.34		.62	.89	
	3300	Fire resistant, no finish included		1,900	.008		.16	.16		.32	.45	
	3400	With thin coat plaster finish		875	.018		.28	.34		.62	.89	
	3450	On beams, columns, or soffits, standard, no finish included		675	.024		.18	.44		.62	.95	
	3500	With thin coat plaster finish		475	.034		.32	.62		.94	1.42	
	3700	Fire resistant, no finish included		675	.024		.18	.44		.62	.95	
	3800	With thin coat plaster finish		475	.034		.32	.62		.94	1.42	
	5000	5/8" thick, on walls or ceilings, fire resistant, no finish included		1,900	.008		.18	.16		.34	.47	
	5100	With thin coat plaster finish		875	.018		.30	.34		.64	.91	
	5500	On beams, columns, or soffits, no finish included		675	.024		.21	.44		.65	.98	
	5600	With thin coat plaster finish		475	.034		.35	.62		.97	1.45	
	6000	For high ceilings, over 8' high, add		3,060	.005		.10	.10		.20	.28	
	6500	For over 3 stories high, add per story		6,100	.003		.05	.05		.10	.14	
608	0010	**DRYWALL** Gypsum plasterboard, nailed or screwed to studs										**608**
	0100	unless otherwise noted										
	0150	3/8" thick, on walls, standard, no finish included	2 Carp	2,000	.008	S.F.	.15	.15		.30	.42	
	0200	On ceilings, standard, no finish included		1,800	.009		.15	.16		.31	.45	
	0250	On beams, columns, or soffits, no finish included		675	.024		.17	.44		.61	.94	
	0300	1/2" thick, on walls, standard, no finish included		2,000	.008		.15	.15		.30	.41	
	0350	Taped and finished		965	.017		.24	.31		.55	.79	
	0400	Fire resistant, no finish included		2,000	.008		.21	.15		.36	.48	
	0450	Taped and finished		965	.017		.30	.31		.61	.86	
	0500	Water resistant, no finish included		2,000	.008		.23	.15		.38	.50	
	0550	Taped and finished		965	.017		.32	.31		.63	.88	
	0600	Prefinished, vinyl, clipped to studs		900	.018		.54	.33		.87	1.15	
	1000	On ceilings, standard, no finish included		1,800	.009		.15	.16		.31	.44	
	1050	Taped and finished		765	.021		.24	.39		.63	.92	
	1100	Fire resistant, no finish included		1,800	.009		.21	.16		.37	.51	
	1150	Taped and finished		765	.021		.30	.39		.69	.99	

Important: See the Reference Section for critical supporting data - Reference Nos., Crews, & Location Factors

092 600 | Gypsum Board Systems

		Crew	Daily Output	Labor-Hours	Unit	1998 Bare Costs Mat.	Labor	Equip.	Total	Total Incl O&P	
608	1200	Water resistant, no finish included	2 Carp	1,800	.009	S.F.	.23	.16		.39	.53	**608**
	1250	Taped and finished		765	.021		.32	.39		.71	1.01	
	1500	On beams, columns, or soffits, standard, no finish included		675	.024		.17	.44		.61	.94	
	1550	Taped and finished		475	.034		.28	.62		.90	1.37	
	1600	Fire resistant, no finish included		675	.024		.24	.44		.68	1.02	
	1650	Taped and finished		475	.034		.35	.62		.97	1.45	
	1700	Water resistant, no finish included		675	.024		.26	.44		.70	1.04	
	1750	Taped and finished		475	.034		.37	.62		.99	1.47	
	2000	5/8" thick, on walls, standard, no finish included		2,000	.008		.20	.15		.35	.47	
	2050	Taped and finished		965	.017		.29	.31		.60	.85	
	2100	Fire resistant, no finish included		2,000	.008		.21	.15		.36	.48	
	2150	Taped and finished		965	.017		.30	.31		.61	.86	
	2200	Water resistant, no finish included		2,000	.008		.25	.15		.40	.53	
	2250	Taped and finished		965	.017		.34	.31		.65	.90	
	2300	Prefinished, vinyl, clipped to studs		900	.018		.62	.33		.95	1.24	
	3000	On ceilings, standard, no finish included		1,800	.009		.20	.16		.36	.50	
	3050	Taped and finished		765	.021		.29	.39		.68	.98	
	3100	Fire resistant, no finish included		1,800	.009		.21	.16		.37	.51	
	3150	Taped and finished		765	.021		.30	.39		.69	.99	
	3200	Water resistant, no finish included		1,800	.009		.25	.16		.41	.56	
	3250	Taped and finished		765	.021		.34	.39		.73	1.03	
	3500	On beams, columns, or soffits, no finish included		675	.024		.23	.44		.67	1.01	
	3550	Taped and finished		475	.034		.34	.62		.96	1.44	
	3600	Fire resistant, no finish included		675	.024		.24	.44		.68	1.02	
	3650	Taped and finished		475	.034		.35	.62		.97	1.45	
	3700	Water resistant, no finish included		675	.024		.29	.44		.73	1.07	
	3750	Taped and finished		475	.034		.39	.62		1.01	1.50	
	4000	Fireproofing, beams or columns, 2 layers, 1/2" thick, incl finish		330	.048		.51	.90		1.41	2.10	
	4050	5/8" thick		300	.053		.60	.99		1.59	2.35	
	4100	3 layers, 1/2" thick		225	.071		.90	1.32		2.22	3.24	
	4150	5/8" thick		210	.076		.90	1.41		2.31	3.41	
	5200	For high ceilings, over 8' high, add		3,060	.005		.09	.10		.19	.27	
	5270	For textured spray, add	2 Lath	1,600	.010		.12	.18		.30	.43	
	5300	For over 3 stories high, add per story	2 Carp	6,100	.003		.05	.05		.10	.14	
	5350	For finishing corners, inside or outside, add	"	1,100	.015	L.F.	.06	.27		.33	.53	
	5500	For acoustical sealant, add per bead	1 Carp	500	.016	"	.03	.30		.33	.54	
	5550	Sealant, 1 quart tube				Ea.	4.64			4.64	5.10	
	5600	Sound deadening board, 1/4" gypsum	2 Carp	1,800	.009	S.F.	.15	.16		.31	.45	
	5650	1/2" wood fiber	"	1,800	.009	"	.26	.16		.42	.57	
612	0010	**METAL STUDS, DRYWALL** Partitions, 10' high, with runners (R092-610)										**612**
	2000	Non-load bearing, galvanized, 25 ga. 1-5/8" wide, 16" O.C.	1 Carp	450	.018	S.F.	.19	.33		.52	.77	
	2100	24" O.C.		520	.015		.15	.28		.43	.65	
	2200	2-1/2" wide, 16" O.C.		440	.018		.21	.34		.55	.81	
	2250	24" O.C.		510	.016		.17	.29		.46	.68	
	2300	3-5/8" wide, 16" O.C.		430	.019		.24	.34		.58	.85	
	2350	24" O.C.		500	.016		.20	.30		.50	.73	
	2400	4" wide, 16" O.C.		420	.019		.26	.35		.61	.89	
	2450	24" O.C.		490	.016		.21	.30		.51	.75	
	2500	6" wide, 16" O.C.		410	.020		.33	.36		.69	.98	
	2550	24" O.C.		480	.017		.26	.31		.57	.82	
	2600	20 ga. studs, 1-5/8" wide, 16" O.C.		450	.018		.44	.33		.77	1.04	
	2650	24" O.C.		520	.015		.35	.28		.63	.88	
	2700	2-1/2" wide, 16" O.C.		440	.018		.48	.34		.82	1.11	
	2750	24" O.C.		510	.016		.38	.29		.67	.92	
	2800	3-5/8" wide, 16" O.C.		430	.019		.58	.34		.92	1.23	

092 600 | Gypsum Board Systems

		CREW	DAILY OUTPUT	LABOR-HOURS	UNIT	1998 BARE COSTS				TOTAL INCL O&P
						MAT.	LABOR	EQUIP.	TOTAL	
612	2850	24″ O.C.	1 Carp	500	.016	S.F.	.46	.30		.76	1.02
	2900	4″ wide, 16″ O.C.		420	.019		.61	.35		.96	1.27
	2950	24″ O.C.		490	.016		.49	.30		.79	1.06
	3000	6″ wide, 16″ O.C.		410	.020		.78	.36		1.14	1.48
	3050	24″ O.C.		480	.017		.62	.31		.93	1.22
	4000	LB studs, light ga. structural, galv., 18 ga., 2-1/2″, 16″ O.C.		300	.027		.59	.49		1.08	1.50
	4100	24″ O.C.		360	.022		.47	.41		.88	1.22
	4200	3-5/8″ wide, 16″ O.C.		285	.028		.68	.52		1.20	1.63
	4250	24″ O.C.		345	.023		.54	.43		.97	1.34
	4300	4″ wide, 16″ O.C.		270	.030		.72	.55		1.27	1.73
	4350	24″ O.C.		330	.024		.57	.45		1.02	1.40
	4400	6″ wide, 16″ O.C.		255	.031		.88	.58		1.46	1.95
	4450	24″ O.C.		315	.025		.70	.47		1.17	1.58
	4600	16 ga. studs, 2-1/2″, 16″ O.C.		270	.030		.70	.55		1.25	1.71
	4650	24″ O.C.		330	.024		.56	.45		1.01	1.39
	4700	3-5/8″ wide, 16″ O.C.		255	.031		.91	.58		1.49	1.99
	4750	24″ O.C.		315	.025		.78	.47		1.25	1.66
	4800	4″ wide, 16″ O.C.		240	.033		.98	.62		1.60	2.14
	4850	24″ O.C.		300	.027		.78	.49		1.27	1.71
	4900	6″ wide, 16″ O.C.		190	.042		1.20	.78		1.98	2.65
	4950	24″ O.C.		225	.036		.96	.66		1.62	2.18

Reference: R092-610

092 800 | Drywall Accessories

		CREW	DAILY OUTPUT	LABOR-HOURS	UNIT	1998 BARE COSTS				TOTAL INCL O&P
						MAT.	LABOR	EQUIP.	TOTAL	
804	0010	**ACCESSORIES, DRYWALL** Casing bead, galvanized steel	1 Carp	2.90	2.759	C.L.F.	15.75	51		66.75	105
	0100	Vinyl		3	2.667		23	49.50		72.50	110
	0300	Corner bead, galvanized steel, 1″ x 1″		4	2		9.65	37		46.65	74
	0400	1-1/4″ x 1-1/4″		3.50	2.286		11.05	42.50		53.55	84.50
	0600	Vinyl corner bead		4	2		15.25	37		52.25	80.50
	0700	Door casing, vinyl, for 2″ wall systems		2.50	3.200		26.50	59		85.50	130
	0900	Furring channel, galv. steel, 7/8″ deep, standard		2.60	3.077		17.50	57		74.50	117
	1000	Resilient		2.55	3.137		18.50	58		76.50	120
	1100	J trim, galvanized steel, 1/2″ wide		3	2.667		16.50	49.50		66	103
	1120	5/8″ wide		2.95	2.712		17.35	50		67.35	105
	1160	Screws #6 x 1″ A				M	6.75			6.75	7.45
	1170	#6 x 1-5/8″ A				″	10			10	11
	1500	Z stud, galvanized steel, 1-1/2″ wide	1 Carp	2.60	3.077	C.L.F.	25.50	57		82.50	126

---

093 | Tile

093 100 | Ceramic Tile

		CREW	DAILY OUTPUT	LABOR-HOURS	UNIT	1998 BARE COSTS				TOTAL INCL O&P
						MAT.	LABOR	EQUIP.	TOTAL	
102	0010	**CERAMIC TILE**									
	0050	Base, using 1′ x 4″ high pc. with 1″ x 1″ tiles, mud set	D-7	82	.195	L.F.	3.46	3.22		6.68	9
	0100	Thin set	″	128	.125		3.74	2.06		5.80	7.45
	0300	For 6″ high base, 1″ x 1″ tile face, add					.57			.57	.63
	0400	For 2″ x 2″ tile face, add to above					.30			.30	.33
	0600	Cove base, 4-1/4″ x 4-1/4″ high, mud set	D-7	91	.176		2.98	2.90		5.88	7.95
	0700	Thin set		128	.125		2.71	2.06		4.77	6.30
	0900	6″ x 4-1/4″ high, mud set		100	.160		2.96	2.64		5.60	7.55

Important: See the Reference Section for critical supporting data - Reference Nos., Crews, & Location Factors

093 100 | Ceramic Tile

		CREW	DAILY OUTPUT	LABOR-HOURS	UNIT	1998 BARE COSTS				TOTAL INCL O&P	
						MAT.	LABOR	EQUIP.	TOTAL		
102	1000	Thin set	D-7	137	.117	L.F.	2.81	1.92		4.73	6.20	102
	1200	Sanitary cove base, 6" x 4-1/4" high, mud set		93	.172		3.04	2.84		5.88	7.95
	1300	Thin set		124	.129		2.77	2.13		4.90	6.50
	1500	6" x 6" high, mud set		84	.190		3.69	3.14		6.83	9.15
	1600	Thin set		117	.137		3.35	2.25		5.60	7.35
	1800	Bathroom accessories, average		82	.195	Ea.	9.65	3.22		12.87	15.85
	1900	Bathtub, 5', rec. 4-1/4" x 4-1/4" tile wainscot, adhesive set 6' high		2.90	5.517		244	91		335	415
	2100	7' high wainscot		2.50	6.400		288	105		393	485
	2200	8' high wainscot		2.20	7.273		330	120		450	555
	2400	Bullnose trim, 4-1/4" x 4-1/4", mud set		82	.195	L.F.	3.16	3.22		6.38	8.70
	2500	Thin set		128	.125		2.81	2.06		4.87	6.40
	2700	6" x 4-1/4" bullnose trim, mud set		84	.190		4.01	3.14		7.15	9.50
	2800	Thin set		124	.129		3.74	2.13		5.87	7.55
	3000	Floors, natural clay, random or uniform, thin set, color group 1		183	.087	S.F.	3.26	1.44		4.70	5.90
	3100	Color group 2		183	.087		3.50	1.44		4.94	6.20
	3300	Porcelain type, 1 color, color group 2, 1" x 1"		183	.087		3.77	1.44		5.21	6.50
	3400	2" x 2" or 2" x 1", thin set		190	.084		3.98	1.39		5.37	6.65
	3600	For random blend, 2 colors, add					.69			.69	.76
	3700	4 colors, add					.99			.99	1.09
	4300	Specialty tile, 4-1/4" x 4-1/4" x 1/2", decorator finish	D-7	183	.087		6.40	1.44		7.84	9.35
	4500	Add for epoxy grout, 1/16" joint, 1" x 1" tile		800	.020		.49	.33		.82	1.07
	4600	2" x 2" tile		820	.020		.46	.32		.78	1.03
	4800	Pregrouted sheets, walls, 4-1/4" x 4-1/4", 6" x 4-1/4"									
	4810	and 8-1/2" x 4-1/4", 4 S.F. sheets, silicone grout	D-7	240	.067	S.F.	3.73	1.10		4.83	5.90
	5100	Floors, unglazed, 2 S.F. sheets,									
	5110	urethane adhesive	D-7	180	.089	S.F.	3.72	1.46		5.18	6.45
	5400	Walls, interior, thin set, 4-1/4" x 4-1/4" tile		190	.084		2.03	1.39		3.42	4.48
	5500	6" x 4-1/4" tile		190	.084		2.06	1.39		3.45	4.52
	5700	8-1/2" x 4-1/4" tile		190	.084		2.93	1.39		4.32	5.45
	5800	6" x 6" tile		200	.080		2.35	1.32		3.67	4.71
	5810	8" x 8" tile		225	.071		2.44	1.17		3.61	4.58
	5820	12" x 12" tile		300	.053		2.68	.88		3.56	4.37
	5830	16" x 16" tile		500	.032		2.91	.53		3.44	4.05
	6000	Decorated wall tile, 4-1/4" x 4-1/4", minimum		270	.059		2.70	.98		3.68	4.55
	6100	Maximum		180	.089		14.05	1.46		15.51	17.85
	6600	Crystalline glazed, 4-1/4" x 4-1/4", mud set, plain		100	.160		3.37	2.64		6.01	8
	6700	4-1/4" x 4-1/4", scored tile		100	.160		3.64	2.64		6.28	8.25
	6900	6" x 6" plain		93	.172		4.06	2.84		6.90	9.05
	7000	For epoxy grout, 1/16" joints, 4-1/4" tile, add		800	.020		.29	.33		.62	.85
	7200	For tile set in dry mortar, add		1,735	.009			.15		.15	.25
	7300	For tile set in portland cement mortar, add		290	.055			.91		.91	1.47

093 300 | Quarry Tile

		CREW	DAILY OUTPUT	LABOR-HOURS	UNIT	MAT.	LABOR	EQUIP.	TOTAL	INCL O&P	
304	0010	**QUARRY TILE** Base, cove or sanitary, 2" or 5" high, mud set										304
	0100	1/2" thick	D-7	110	.145	L.F.	3.21	2.40		5.61	7.40
	0300	Bullnose trim, red, mud set, 6" x 6" x 1/2" thick		120	.133		3.43	2.20		5.63	7.35
	0400	4" x 4" x 1/2" thick		110	.145		3.44	2.40		5.84	7.65
	0600	4" x 8" x 1/2" thick, using 8" as edge		130	.123		3.27	2.03		5.30	6.90
	0700	Floors, mud set, 1,000 S.F. lots, red, 4" x 4" x 1/2" thick		120	.133	S.F.	3.37	2.20		5.57	7.25
	0900	6" x 6" x 1/2" thick		140	.114		2.91	1.88		4.79	6.25
	1000	4" x 8" x 1/2" thick		130	.123		3.37	2.03		5.40	7
	1300	For waxed coating, add					.47			.47	.52
	1500	For colors other than green, add					.31			.31	.34
	1600	For abrasive surface, add					.39			.39	.43
	1800	Brown tile, imported, 6" x 6" x 3/4"	D-7	120	.133		3.99	2.20		6.19	7.95

093 300 | Quarry Tile

		CREW	DAILY OUTPUT	LABOR-HOURS	UNIT	MAT.	LABOR	EQUIP.	TOTAL	TOTAL INCL O&P	
						1998 BARE COSTS					
304	1900	8" x 8" x 1"	D-7	110	.145	S.F.	4.52	2.40		6.92	8.85	**304**
	2100	For thin set mortar application, deduct		700	.023			.38		.38	.61	
	2700	Stair tread, 6" x 6" x 3/4", plain		50	.320		3.58	5.25		8.83	12.50	
	2800	Abrasive		47	.340		4.13	5.60		9.73	13.65	
	3000	Wainscot, 6" x 6" x 1/2", thin set, red		105	.152		3.07	2.51		5.58	7.45	
	3100	Colors other than green		105	.152		3.43	2.51		5.94	7.85	
	3300	Window sill, 6" wide, 3/4" thick		90	.178	L.F.	3.58	2.93		6.51	8.70	
	3400	Corners		80	.200	Ea.	3.22	3.30		6.52	8.90	

093 700 | Metal Tile

		CREW	DAILY OUTPUT	LABOR-HOURS	UNIT	MAT.	LABOR	EQUIP.	TOTAL	TOTAL INCL O&P	
						1998 BARE COSTS					
701	0010	METAL TILE 4' x 4' sheet, 24 ga., tile pattern, nailed										**701**
	0200	Stainless steel	2 Carp	512	.031	S.F.	19.50	.58		20.08	22.50	
	0400	Aluminized steel	"	512	.031	"	10.50	.58		11.08	12.55	

094 100 | Portland Cem. Terrazzo

			CREW	DAILY OUTPUT	LABOR-HOURS	UNIT	MAT.	LABOR	EQUIP.	TOTAL	TOTAL INCL O&P	
							1998 BARE COSTS					
104	0010	TERRAZZO, CAST IN PLACE Cove base, 6" high, 16ga. zinc	R094-100	1 Mstz	20	.400	L.F.	2.60	7.35		9.95	14.75	**104**
	0100	Curb, 6" high and 6" wide			6	1.333		4.25	24.50		28.75	44	
	0300	Divider strip for floors, 14 ga., 1-1/4" deep, zinc			375	.021		.85	.39		1.24	1.57	
	0400	Brass			375	.021		1.60	.39		1.99	2.39	
	0600	Heavy top strip 1/4" thick, 1-1/4" deep, zinc			300	.027		1.45	.49		1.94	2.39	
	1200	For thin set floors, 16 ga., 1/2" x 1/2", zinc			350	.023		.75	.42		1.17	1.51	
	1500	Floor, bonded to concrete, 1-3/4" thick, gray cement		J-3	130	.123	S.F.	2.20	2.06	.82	5.08	6.65	
	1600	White cement, mud set			130	.123		2.50	2.06	.82	5.38	7	
	1800	Not bonded, 3" total thickness, gray cement			115	.139		2.75	2.33	.93	6.01	7.80	
	1900	White cement, mud set			115	.139		3	2.33	.93	6.26	8.10	
108	0010	TILE OR TERRAZZO BASE Scratch coat only		1 Mstz	150	.053	S.F.	.29	.98		1.27	1.90	**108**
	0500	Scratch and brown coat only		"	75	.107	"	.53	1.96		2.49	3.75	

094 200 | Precast Terrazzo

		CREW	DAILY OUTPUT	LABOR-HOURS	UNIT	MAT.	LABOR	EQUIP.	TOTAL	TOTAL INCL O&P	
						1998 BARE COSTS					
201	0010	TERRAZZO, PRECAST Base, 6" high, straight	1 Mstz	35	.229	L.F.	8.50	4.19		12.69	16.15	**201**
	0100	Cove		30	.267		10.50	4.89		15.39	19.45	
	0300	8" high base, straight		30	.267		9.50	4.89		14.39	18.35	
	0400	Cove		25	.320		11.55	5.85		17.40	22	
	0600	For white cement, add					.32			.32	.35	
	0700	For 16 ga. zinc toe strip, add					.95			.95	1.05	
	0900	Curbs, 4" x 4" high	1 Mstz	19	.421		22	7.75		29.75	36.50	
	1000	8" x 8" high	"	15	.533		29.50	9.80		39.30	48.50	
	1200	Floor tiles, non-slip, 1" thick, 12" x 12"	D-1	29	.552	S.F.	14.85	9.25		24.10	32	
	1300	1-1/4" thick, 12" x 12"		29	.552		16.10	9.25		25.35	33.50	
	1500	16" x 16"		23	.696		17.50	11.70		29.20	39	
	1600	1-1/2" thick, 16" x 16"		21	.762		16	12.80		28.80	39	
	4800	Wainscot, 12" x 12" x 1" tiles	1 Mstz	12	.667		8.50	12.25		20.75	29	
	4900	16" x 16" x 1-1/2" tiles	"	8	1		11.70	18.35		30.05	42.50	

Important: See the Reference Section for critical supporting data - Reference Nos., Crews, & Location Factors

095 100	Acoustical Ceilings	CREW	DAILY OUTPUT	LABOR-HOURS	UNIT	1998 BARE COSTS				TOTAL INCL O&P	
						MAT.	LABOR	EQUIP.	TOTAL		
102	0010	**CEILING TILE,** Stapled or cemented										102
	0100	12" x 12" or 12" x 24", not including furring									
	0600	Mineral fiber, vinyl coated, 5/8" thick	1 Carp	1,000	.008	S.F.	.51	.15		.66	.81
	0700	3/4" thick		1,000	.008		1.12	.15		1.27	1.48
	0900	Fire rated, 3/4" thick, plain faced		1,000	.008		.76	.15		.91	1.08
	1000	Plastic coated face		1,000	.008		.88	.15		1.03	1.21
	1200	Aluminum faced, 5/8" thick, plain	↓	1,000	.008		.97	.15		1.12	1.31
	3300	For flameproofing, add					.08			.08	.09
	3400	For sculptured 3 dimensional, add					.21			.21	.23
	3900	For ceiling primer, add					.11			.11	.12
	4000	For ceiling cement, add				↓	.28			.28	.31
104	0010	**SUSPENDED ACOUSTIC CEILING TILES,** Not including										104
	0100	suspension system									
	0300	Fiberglass boards, film faced, 2' x 2' or 2' x 4', 5/8" thick	1 Carp	625	.013	S.F.	.45	.24		.69	.91
	0400	3/4" thick		600	.013		.95	.25		1.20	1.46
	0500	3" thick, thermal, R11		450	.018		1.11	.33		1.44	1.78
	0600	Glass cloth faced fiberglass, 3/4" thick		500	.016		1.47	.30		1.77	2.13
	0700	1" thick		485	.016		1.66	.31		1.97	2.35
	0820	1-1/2" thick, nubby face		475	.017		1.97	.31		2.28	2.70
	1110	Mineral fiber tile, lay-in, 2' x 2' or 2' x 4', 5/8" thick, fine texture		625	.013		.42	.24		.66	.87
	1115	Rough textured		625	.013		.80	.24		1.04	1.29
	1125	3/4" thick, fine textured		600	.013		1	.25		1.25	1.52
	1130	Rough textured		600	.013		1.25	.25		1.50	1.80
	1135	Fissured		600	.013		1.31	.25		1.56	1.86
	1150	Tegular, 5/8" thick, fine textured		470	.017		.97	.31		1.28	1.61
	1155	Rough textured		470	.017		1.15	.31		1.46	1.81
	1165	3/4" thick, fine textured		450	.018		1.25	.33		1.58	1.94
	1170	Rough textured		450	.018		1.40	.33		1.73	2.10
	1175	Fissured	↓	450	.018		1.71	.33		2.04	2.44
	1180	For aluminum face, add					4			4	4.40
	1185	For plastic film face, add					.30			.30	.33
	1190	For fire rating, add					.30			.30	.33
	1300	Mirror faced panels, 15/16" thick, 2' x 2'	1 Carp	500	.016		5.20	.30		5.50	6.20
	1900	Eggcrate, acrylic, 1/2" x 1/2" x 1/2" cubes		500	.016		1.21	.30		1.51	1.84
	2100	Polystyrene eggcrate, 3/8" x 3/8" x 1/2" cubes		510	.016		1.01	.29		1.30	1.61
	2200	1/2" x 1/2" x 1/2" cubes		500	.016		1.36	.30		1.66	2.01
	2400	Luminous panels, prismatic, acrylic		400	.020		1.46	.37		1.83	2.24
	2500	Polystyrene		400	.020		.75	.37		1.12	1.46
	2700	Flat white acrylic		400	.020		2.56	.37		2.93	3.45
	2800	Polystyrene		400	.020		1.74	.37		2.11	2.54
	3000	Drop pan, white, acrylic		400	.020		3.74	.37		4.11	4.74
	3100	Polystyrene		400	.020		3.13	.37		3.50	4.07
	3600	Perforated aluminum sheets, .024" thick, corrugated, painted		490	.016		1.49	.30		1.79	2.16
	3700	Plain	↓	500	.016	↓	1.34	.30		1.64	1.98
106	0010	**SUSPENDED CEILINGS, COMPLETE** Including standard										106
	0100	suspension system but not incl. 1-1/2" carrier channels									
	0600	Fiberglass ceiling board, 2' x 4' x 5/8", plain faced,	1 Carp	500	.016	S.F.	.91	.30		1.21	1.51
	0700	Offices, 2' x 4' x 3/4"		380	.021		1.01	.39		1.40	1.78
	1800	Tile, Z bar suspension, 5/8" mineral fiber tile		150	.053		1.31	.99		2.30	3.13
	1900	3/4" mineral fiber tile	↓	150	.053	↓	1.39	.99		2.38	3.22

095 250	Acoustical Space Units										
254	0010	**SOUND ABSORBING PANELS** Perforated steel facing, painted with										254
	0100	fiberglass or mineral filler, no backs, 2-1/4" thick, modular									

095 250 | Acoustical Space Units

			CREW	DAILY OUTPUT	LABOR-HOURS	UNIT	MAT.	LABOR	EQUIP.	TOTAL	TOTAL INCL O&P	
254	0200	space units, ceiling or wall hung, white or colored	1 Carp	100	.080	S.F.	6.90	1.48		8.38	10.15	254
	0300	Fiberboard sound deadening panels, 1/2" thick	"	600	.013	"	.26	.25		.51	.71	
	0500	Fiberglass panels, 4' x 8' x 1" thick, with										
	0600	glass cloth face for walls, cemented	1 Carp	155	.052	S.F.	3.94	.95		4.89	5.95	
	0700	1-1/2" thick, dacron covered, inner aluminum frame,										
	0710	wall mounted	1 Carp	300	.027	S.F.	6.70	.49		7.19	8.25	

095 300 | Acoustical Insulation

			CREW	DAILY OUTPUT	LABOR-HOURS	UNIT	MAT.	LABOR	EQUIP.	TOTAL	TOTAL INCL O&P	
308	0010	SOUND ATTENUATION Blanket, 1" thick	1 Carp	925	.009	S.F.	.20	.16		.36	.49	308
	0500	1-1/2" thick		920	.009		.24	.16		.40	.54	
	1000	2" thick		915	.009		.29	.16		.45	.60	
	1500	3" thick		910	.009		.43	.16		.59	.76	

095 600 | Wood Strip Flooring

			CREW	DAILY OUTPUT	LABOR-HOURS	UNIT	MAT.	LABOR	EQUIP.	TOTAL	TOTAL INCL O&P	
604	0010	WOOD Fir, vertical grain, 1" x 4", not incl. finish, B & better	1 Carp	255	.031	S.F.	2.30	.58		2.88	3.52	604
	0100	C grade & better		255	.031		2.16	.58		2.74	3.37	
	0300	Flat grain, 1" x 4", not incl. finish, B & better		255	.031		2.63	.58		3.21	3.88	
	0400	C & better		255	.031		2.53	.58		3.11	3.77	
	4000	Maple, strip, 25/32" x 2-1/4", not incl. finish, select		170	.047		2.95	.87		3.82	4.74	
	4100	#2 & better		170	.047		2.65	.87		3.52	4.41	
	4300	33/32" x 3-1/4", not incl. finish, #1 grade		170	.047		3.25	.87		4.12	5.05	
	4400	#2 & better		170	.047		2.90	.87		3.77	4.68	
	4600	Oak, white or red, 25/32" x 2-1/4", not incl. finish										
	4700	#1 common	1 Carp	170	.047	S.F.	2.75	.87		3.62	4.52	
	4900	Select quartered, 2-1/4" wide		170	.047		2.96	.87		3.83	4.75	
	5000	Clear		170	.047		3.35	.87		4.22	5.15	
	5200	Parquetry, standard, 5/16" thick, not incl. finish, oak, minimum		160	.050		1.55	.93		2.48	3.29	
	5300	Maximum		100	.080		5.20	1.48		6.68	8.25	
	5500	Teak, minimum		160	.050		4.05	.93		4.98	6.05	
	5600	Maximum		100	.080		7.15	1.48		8.63	10.40	
	5650	13/16" thick, select grade oak, minimum		160	.050		8.10	.93		9.03	10.50	
	5700	Maximum		100	.080		11.85	1.48		13.33	15.60	
	5800	Custom parquetry, including finish, minimum		100	.080		13.20	1.48		14.68	17.05	
	5900	Maximum		50	.160		17.55	2.96		20.51	24.50	
	6100	Prefinished, white oak, prime grade, 2-1/4" wide		170	.047		6.15	.87		7.02	8.25	
	6200	3-1/4" wide		185	.043		7.60	.80		8.40	9.70	
	6400	Ranch plank		145	.055		7.55	1.02		8.57	10.05	
	6500	Hardwood blocks, 9" x 9", 25/32" thick		160	.050		4.75	.93		5.68	6.80	
	6700	Parquetry, 5/16" thick, oak, minimum		160	.050		3.25	.93		4.18	5.15	
	6800	Maximum		100	.080		9.10	1.48		10.58	12.55	
	7000	Walnut or teak, parquetry, minimum		160	.050		4.35	.93		5.28	6.40	
	7100	Maximum		100	.080		7.65	1.48		9.13	10.95	
	7200	Acrylic wood parquet blocks, 12" x 12" x 5/16",										
	7210	irradiated, set in epoxy	1 Carp	160	.050	S.F.	6.35	.93		7.28	8.60	
	7400	Yellow pine, 3/4" x 3-1/8", T & G, C & better, not incl. finish	"	200	.040		2.05	.74		2.79	3.53	
	7500	Refinish wood floor, sand, 2 cts poly, wax, soft wood, min.	1 Clab	400	.020		.63	.27		.90	1.15	
	7600	Hard wood, max		130	.062		.95	.82		1.77	2.46	
	7800	Sanding and finishing, 2 coats polyurethane		295	.027		.64	.36		1	1.32	
	7900	Subfloor and underlayment, see division 061-164 & 168										
	8015	Transition molding, 2 1/4" wide, 5' long	1 Carp	19.20	.417	Ea.	10.15	7.70		17.85	24.50	

095 650 | Wood Block Flooring

			CREW	DAILY OUTPUT	LABOR-HOURS	UNIT	MAT.	LABOR	EQUIP.	TOTAL	TOTAL INCL O&P	
651	0010	WOOD BLOCK FLOORING End grain flooring, coated, 2" thick	1 Carp	295	.027	S.F.	2.65	.50		3.15	3.78	651
	0400	Natural finish, 1" thick, fir		125	.064		2.75	1.18		3.93	5.05	
	0600	1-1/2" thick, pine		125	.064		2.70	1.18		3.88	5	
	0700	2" thick, pine		125	.064		2.75	1.18		3.93	5.05	

Important: See the Reference Section for critical supporting data - Reference Nos., Crews, & Location Factors

096 150 | Marble Flooring

			CREW	DAILY OUTPUT	LABOR-HOURS	UNIT	1998 BARE COSTS				TOTAL INCL O&P	
							MAT.	LABOR	EQUIP.	TOTAL		
151	0010	**MARBLE** Thin gauge tile, 12" x 6", 3/8", White Carara	D-7	60	.267	S.F.	6.70	4.39		11.09	14.50	151
	0100	Travertine		60	.267		7.40	4.39		11.79	15.25	
	0200	12" x 12" x 3/8", thin set, floors		60	.267		6.70	4.39		11.09	14.50	
	0300	On walls		52	.308		6.70	5.05		11.75	15.60	

096 250 | Slate Flooring

| 251 | 0010 | **SLATE TILE** Vermont, 6" x 6" x 1/4" thick, thin set | D-7 | 180 | .089 | S.F. | 3.75 | 1.46 | | 5.21 | 6.50 | 251 |
| 252 | 0010 | **SLATE & STONE FLOORS** See division 025-166 | | | | | | | | | | 252 |

096 350 | Brick Flooring

354	0010	**FLOORING**										354
	0020	Acid proof shales, red, 8" x 3-3/4" x 1-1/4" thick	D-7	.43	37.209	M	675	615		1,290	1,725	
	0050	2-1/4" thick	D-1	.40	40		735	670		1,405	1,950	
	0200	Acid proof clay brick, 8" x 3-3/4" x 2-1/4" thick	"	.40	40		735	670		1,405	1,950	
	0260	Cast ceramic, pressed, 4" x 8" x 1/2", unglazed	D-7	100	.160	S.F.	4.80	2.64		7.44	9.55	
	0270	Glazed		100	.160		6.40	2.64		9.04	11.30	
	0280	Hand molded flooring, 4" x 8" x 3/4", unglazed		95	.168		6.35	2.78		9.13	11.50	
	0290	Glazed		95	.168		7.95	2.78		10.73	13.25	
	0300	8" hexagonal, 3/4" thick, unglazed		85	.188		6.95	3.10		10.05	12.65	
	0310	Glazed		85	.188		12.55	3.10		15.65	18.80	
	0450	Acid proof joints, 1/4" wide	D-1	65	.246		1.10	4.14		5.24	8.20	
	0500	Pavers, 8" x 4", 1" to 1-1/4" thick, red	D-7	95	.168		2.80	2.78		5.58	7.55	
	0510	Ironspot	"	95	.168		3.95	2.78		6.73	8.85	
	0540	1-3/8" to 1-3/4" thick, red	D-1	95	.168		2.70	2.83		5.53	7.75	
	0560	Ironspot		95	.168		3.90	2.83		6.73	9.10	
	0580	2-1/4" thick, red		90	.178		2.75	2.99		5.74	8.10	
	0590	Ironspot		90	.178		4.25	2.99		7.24	9.75	
	0800	For sidewalks and patios with pavers, see division 025-158										
	0870	For epoxy joints, add	D-1	600	.027	S.F.	2.09	.45		2.54	3.06	
	0880	For Furan underlayment, add	"	600	.027		1.73	.45		2.18	2.66	
	0890	For waxed surface, steam cleaned, add	D-5	1,000	.008		.15	.12		.27	.37	

096 600 | Resilient Flooring

601	0010	**RESILIENT FLOORING**										601
	0800	Base, cove, rubber or vinyl, .080" thick	R096 -600									
	1100	Standard colors, 2-1/2" high	1 Tilf	315	.025	L.F.	.39	.46		.85	1.18	
	1150	4" high		315	.025		.49	.46		.95	1.29	
	1200	6" high		315	.025		.73	.46		1.19	1.55	
	1450	1/8" thick, standard colors, 2-1/2" high		315	.025		.40	.46		.86	1.19	
	1500	4" high		315	.025		.55	.46		1.01	1.36	
	1550	6" high		315	.025		.81	.46		1.27	1.64	
	1600	Corners, 2-1/2" high		315	.025	Ea.	1.02	.46		1.48	1.87	
	1630	4" high		315	.025		1.07	.46		1.53	1.93	
	1660	6" high		315	.025		1.38	.46		1.84	2.27	
	1700	Conductive flooring, rubber tile, 1/8" thick		315	.025	S.F.	2.50	.46		2.96	3.50	
	1800	Homogeneous vinyl tile, 1/8" thick		315	.025		3.42	.46		3.88	4.51	
	2200	Cork tile, standard finish, 1/8" thick		315	.025		2.42	.46		2.88	3.41	
	2250	3/16" thick		315	.025		2.84	.46		3.30	3.87	
	2300	5/16" thick		315	.025		4.26	.46		4.72	5.45	
	2350	1/2" thick		315	.025		4.74	.46		5.20	5.95	
	2500	Urethane finish, 1/8" thick		315	.025		3.65	.46		4.11	4.77	
	2550	3/16" thick		315	.025		4.81	.46		5.27	6.05	
	2600	5/16" thick		315	.025		6.60	.46		7.06	8	

096 600 | Resilient Flooring

		CREW	DAILY OUTPUT	LABOR-HOURS	UNIT	MAT.	LABOR	EQUIP.	TOTAL	TOTAL INCL O&P	
601	2650	1/2" thick	1 Tilf	315	.025	S.F.	9.20	.46		9.66	10.85	**601**
	3700	Polyethylene, in rolls, no base incl., landscape surfaces [R096-600]		275	.029		2.19	.53		2.72	3.27	
	3800	Nylon action surface, 1/8" thick		275	.029		2.35	.53		2.88	3.44	
	3900	1/4" thick		275	.029		3.39	.53		3.92	4.59	
	4000	3/8" thick		275	.029		4.25	.53		4.78	5.55	
	5900	Rubber, sheet goods, 36" wide, 1/8" thick		120	.067		3.01	1.22		4.23	5.30	
	5950	3/16" thick		100	.080		4.28	1.46		5.74	7.05	
	6000	1/4" thick		90	.089		4.95	1.62		6.57	8.10	
	6050	Tile, marbleized colors, 12" x 12", 1/8" thick		400	.020		3.03	.36		3.39	3.92	
	6100	3/16" thick		400	.020		4.34	.36		4.70	5.35	
	6300	Special tile, plain colors, 1/8" thick		400	.020		3.83	.36		4.19	4.80	
	6350	3/16" thick		400	.020		5.15	.36		5.51	6.25	
	7000	Vinyl composition tile, 12" x 12", 1/16" thick		500	.016		.68	.29		.97	1.22	
	7050	Embossed		500	.016		.86	.29		1.15	1.42	
	7100	Marbleized		500	.016		.86	.29		1.15	1.42	
	7150	Solid		500	.016		.97	.29		1.26	1.54	
	7200	3/32" thick, embossed		500	.016		.83	.29		1.12	1.39	
	7250	Marbleized		500	.016		.98	.29		1.27	1.55	
	7300	Solid		500	.016		1.43	.29		1.72	2.04	
	7350	1/8" thick, marbleized		500	.016		.98	.29		1.27	1.55	
	7400	Solid		500	.016		1.85	.29		2.14	2.51	
	7450	Conductive		500	.016		3.43	.29		3.72	4.24	
	7500	Vinyl tile, 12" x 12", .050" thick, minimum		500	.016		1.54	.29		1.83	2.16	
	7550	Maximum		500	.016		3.01	.29		3.30	3.78	
	7600	1/8" thick, minimum		500	.016		1.94	.29		2.23	2.60	
	7650	Solid colors		500	.016		4.47	.29		4.76	5.40	
	7700	Marbleized or Travertine pattern		500	.016		3.11	.29		3.40	3.89	
	7750	Florentine pattern		500	.016		3.57	.29		3.86	4.40	
	7800	Maximum		500	.016		7.35	.29		7.64	8.50	
	8000	Vinyl sheet goods, backed, .065" thick, minimum		250	.032		1.25	.58		1.83	2.33	
	8050	Maximum		200	.040		2.30	.73		3.03	3.71	
	8100	.080" thick, minimum		230	.035		1.48	.63		2.11	2.66	
	8150	Maximum		200	.040		2.61	.73		3.34	4.05	
	8200	.125" thick, minimum		230	.035		1.63	.63		2.26	2.82	
	8250	Maximum		200	.040		3.57	.73		4.30	5.10	
	8700	Adhesive cement, 1 gallon does 200 to 300 S.F.				Gal.	13.40			13.40	14.75	
	8800	Asphalt primer, 1 gallon per 300 S.F.					8.40			8.40	9.25	
	8900	Emulsion, 1 gallon per 140 S.F.					8.65			8.65	9.55	
	8950	Latex underlayment, liquid, fortified					26.50			26.50	29	

096 850 | Sheet Carpet

		CREW	DAILY OUTPUT	LABOR-HOURS	UNIT	MAT.	LABOR	EQUIP.	TOTAL	TOTAL INCL O&P	
852	0010	**CARPET** Commercial grades, direct cement										**852**
	0700	Nylon, level loop, 26 oz., light to medium traffic	1 Tilf	75	.107	S.Y.	12.70	1.95		14.65	17.15	
	0720	28 oz., light to medium traffic		75	.107		13.95	1.95		15.90	18.50	
	0900	32 oz., medium traffic		75	.107		18	1.95		19.95	23	
	1100	40 oz., medium to heavy traffic		75	.107		27	1.95		28.95	32.50	
	2920	Nylon plush, 30 oz., medium traffic		57	.140		13.30	2.56		15.86	18.80	
	3000	36 oz., medium traffic		75	.107		16.75	1.95		18.70	21.50	
	3100	42 oz., medium to heavy traffic		70	.114		19.30	2.09		21.39	25	
	3200	46 oz., medium to heavy traffic		70	.114		23	2.09		25.09	29	
	3300	54 oz., heavy traffic		70	.114		26	2.09		28.09	32	
	4500	50 oz., medium to heavy traffic		75	.107		59	1.95		60.95	68	
	4700	Patterned, 32 oz., medium to heavy traffic		70	.114		58.50	2.09		60.59	67.50	
	4900	48 oz., heavy traffic		70	.114		59.50	2.09		61.59	69	
	5000	For less than full roll, add					25%					

Important: See the Reference Section for critical supporting data - Reference Nos., Crews, & Location Factors

096 850 | Sheet Carpet

			CREW	DAILY OUTPUT	LABOR-HOURS	UNIT	1998 BARE COSTS MAT.	LABOR	EQUIP.	TOTAL	TOTAL INCL O&P	
852	5100	For small rooms, less than 12' wide, add						25%				852
	5200	For large open areas (no cuts), deduct						25%				
	5600	For bound carpet baseboard, add	1 Tilf	300	.027	L.F.	1	.49		1.49	1.89	
	5610	For stairs, not incl. price of carpet, add	"	30	.267	Riser		4.87		4.87	7.90	
	5620	For borders and patterns, add to labor						18%				
	8950	For tackless, stretched installation, add padding to above										
	9000	Sponge rubber pad, minimum	1 Tilf	150	.053	S.Y.	2.47	.97		3.44	4.29	
	9100	Maximum		150	.053		6.65	.97		7.62	8.90	
	9200	Felt pad, minimum		150	.053		2.95	.97		3.92	4.83	
	9300	Maximum		150	.053		4.95	.97		5.92	7.05	
	9400	Bonded urethane pad, minimum		150	.053		3.10	.97		4.07	4.99	
	9500	Maximum		150	.053		5.35	.97		6.32	7.50	
	9600	Prime urethane pad, minimum		150	.053		1.75	.97		2.72	3.51	
	9700	Maximum		150	.053		3.20	.97		4.17	5.10	

096 900 | Carpet Tile

			CREW	DAILY OUTPUT	LABOR-HOURS	UNIT	1998 BARE COSTS MAT.	LABOR	EQUIP.	TOTAL	TOTAL INCL O&P	
901	0010	**CARPET TILE**										901
	0100	Tufted nylon, 18" x 18", hard back, 20 oz.	1 Tilf	150	.053	S.Y.	16.10	.97		17.07	19.30	
	0110	26 oz.		150	.053		27.50	.97		28.47	32	
	0200	Cushion back, 20 oz.		150	.053		20.50	.97		21.47	24	
	0210	26 oz.		150	.053		31.50	.97		32.47	36.50	

097 200 | Epoxy-Marble Flooring

			CREW	DAILY OUTPUT	LABOR-HOURS	UNIT	1998 BARE COSTS MAT.	LABOR	EQUIP.	TOTAL	TOTAL INCL O&P	
201	0010	**COMPOSITION FLOORING** Acrylic, 1/4" thick	C-6	520	.092	S.F.	1.10	1.34	.14	2.58	3.64	201
	0600	Epoxy, with colored quartz chips, broadcast, minimum		675	.071		1.95	1.03	.11	3.09	4.02	
	0700	Maximum		490	.098		2.35	1.42	.15	3.92	5.15	
	0900	Trowelled, minimum		560	.086		2.50	1.24	.13	3.87	5	
	1000	Maximum		480	.100		3.65	1.45	.15	5.25	6.65	
	1200	Heavy duty epoxy topping, 1/4" thick,										
	1300	500 to 1,000 S.F.	C-6	420	.114	S.F.	3.75	1.65	.18	5.58	7.15	
	1500	1,000 to 2,000 S.F.		450	.107		2.55	1.54	.17	4.26	5.60	
	1600	Over 10,000 S.F.		480	.100		2.45	1.45	.15	4.05	5.30	
	1800	Epoxy terrazzo, 1/4" thick, chemical resistant, minimum	J-3	200	.080		4.05	1.34	.53	5.92	7.20	
	1900	Maximum	"	150	.107		6.75	1.78	.71	9.24	11.10	

098 150 | Glazed Coatings

			CREW	DAILY OUTPUT	LABOR-HOURS	UNIT	1998 BARE COSTS MAT.	LABOR	EQUIP.	TOTAL	TOTAL INCL O&P	
150	0010	**WALL COATINGS** Acrylic glazed coatings, minimum	1 Pord	525	.015	S.F.	.22	.26		.48	.68	150
	0100	Maximum		305	.026		.47	.45		.92	1.27	

FINISHES 9

098 | Special Coatings

098 150 | Glazed Coatings

150			CREW	DAILY OUTPUT	LABOR-HOURS	UNIT	1998 BARE COSTS				TOTAL INCL O&P	
							MAT.	LABOR	EQUIP.	TOTAL		
150	0300	Epoxy coatings, minimum	1 Pord	525	.015	S.F.	.29	.26		.55	.76	150
	0400	Maximum		170	.047		.89	.81		1.70	2.33	
	0600	Exposed aggregate, troweled on, 1/16" to 1/4", minimum		235	.034		.45	.59		1.04	1.48	
	0700	Maximum (epoxy or polyacrylate)		130	.062		.96	1.06		2.02	2.83	
	0900	1/2" to 5/8" aggregate, minimum		130	.062		.88	1.06		1.94	2.74	
	1000	Maximum		80	.100		1.51	1.72		3.23	4.54	
	1200	1" aggregate size, minimum		90	.089		1.54	1.53		3.07	4.25	
	1300	Maximum		55	.145		2.36	2.50		4.86	6.80	
	1500	Exposed aggregate, sprayed on, 1/8" aggregate, minimum		295	.027		.42	.47		.89	1.24	
	1600	Maximum		145	.055		.77	.95		1.72	2.44	
	1800	High build epoxy, 50 mil, minimum		390	.021		.48	.35		.83	1.12	
	1900	Maximum		95	.084		.84	1.45		2.29	3.34	
	2100	Laminated epoxy with fiberglass, minimum		295	.027		.53	.47		1	1.36	
	2200	Maximum		145	.055		.95	.95		1.90	2.64	
	2400	Sprayed perlite or vermiculite, 1/16" thick, minimum		2,935	.003		.18	.05		.23	.28	
	2500	Maximum		640	.013		.54	.22		.76	.95	
	2700	Vinyl plastic wall coating, minimum		735	.011		.24	.19		.43	.57	
	2800	Maximum		240	.033		.60	.57		1.17	1.62	
	3000	Urethane on smooth surface, 2 coats, minimum		1,135	.007		.17	.12		.29	.39	
	3100	Maximum		665	.012		.41	.21		.62	.80	
	3300	3 coat, minimum		840	.010		.24	.16		.40	.53	
	3400	Maximum		470	.017		.54	.29		.83	1.08	
	3600	Ceramic-like glazed coating, cementitious, minimum		440	.018		.36	.31		.67	.92	
	3700	Maximum		345	.023		.59	.40		.99	1.32	
	3900	Resin base, minimum		640	.013		.24	.22		.46	.62	
	4000	Maximum		330	.024		.40	.42		.82	1.14	

099 | Painting & Wall Coverings

099 100 | Exterior Painting

104			CREW	DAILY OUTPUT	LABOR-HOURS	UNIT	1998 BARE COSTS				TOTAL INCL O&P	
							MAT.	LABOR	EQUIP.	TOTAL		
104	0010	**DOORS AND WINDOWS**	R099 -100									104
	0100	Door frames & trim, only										
	0110	Brushwork, primer	R099 -220	1 Pord / 512	.016	L.F.	.04	.27		.31	.50	
	0120	Finish coat, exterior latex		512	.016		.05	.27		.32	.50	
	0130	Primer & 1 coat, exterior latex		300	.027		.09	.46		.55	.87	
	0135	2 coats, exterior latex, both sides										
	0140	Primer & 2 coats, exterior latex	1 Pord	265	.030	L.F.	.14	.52		.66	1.03	
	0150	Doors, flush, both sides, incl. frame & trim										
	0160	Roll & brush, primer	1 Pord	10	.800	Ea.	3.32	13.75		17.07	26.50	
	0170	Finish coat, exterior latex		10	.800		3.79	13.75		17.54	27	
	0180	Primer & 1 coat, exterior latex		7	1.143		7.10	19.65		26.75	41	
	0190	Primer & 2 coats, exterior latex		5	1.600		10.90	27.50		38.40	58	
	0200	Brushwork, stain, sealer & 2 coats polyurethane		4	2		12.80	34.50		47.30	71.50	
	0210	Doors, French, both sides, 10-15 lite, incl. frame & trim										
	0220	Brushwork, primer	1 Pord	6	1.333	Ea.	1.66	23		24.66	40.50	
	0230	Finish coat, exterior latex		6	1.333		1.90	23		24.90	40.50	
	0240	Primer & 1 coat, exterior latex		3	2.667		3.56	46		49.56	80.50	
	0250	Primer & 2 coats, exterior latex		2	4		5.35	69		74.35	121	
	0260	Brushwork, stain, sealer & 2 coats polyurethane		2.50	3.200		4.58	55		59.58	97	
	0270	Doors, louvered, both sides, incl. frame & trim										

Important: See the Reference Section for critical supporting data - Reference Nos., Crews, & Location Factors

099 100 | Exterior Painting

			DAILY OUTPUT	LABOR-HOURS	UNIT	1998 BARE COSTS				TOTAL INCL O&P	
		CREW				MAT.	LABOR	EQUIP.	TOTAL		
104	0280	Brushwork, primer [R099-100]	1 Pord	7	1.143	Ea.	3.32	19.65		22.97	36.50	104
	0290	Finish coat, exterior latex		7	1.143		3.79	19.65		23.44	37	
	0300	Primer & 1 coat, exterior latex [R099-220]		4	2		7.10	34.50		41.60	65.50	
	0310	Primer & 2 coats, exterior latex		3	2.667		10.70	46		56.70	88.50	
	0320	Brushwork, stain, sealer & 2 coats polyurethane		4.50	1.778		12.80	30.50		43.30	65	
	0330	Doors, panel, both sides, incl. frame & trim										
	0340	Roll & brush, primer	1 Pord	6	1.333	Ea.	3.32	23		26.32	42	
	0350	Finish coat, exterior latex		6	1.333		3.79	23		26.79	42.50	
	0360	Primer & 1 coat, exterior latex		3	2.667		7.10	46		53.10	84.50	
	0370	Primer & 2 coats, exterior latex		2.50	3.200		10.70	55		65.70	104	
	0380	Brushwork, stain, sealer & 2 coats polyurethane		3	2.667		12.80	46		58.80	90.50	
	0400	Windows, per ext. side, based on 15 SF										
	0410	1 to 6 lite										
	0420	Brushwork, primer	1 Pord	13	.615	Ea.	.66	10.60		11.26	18.40	
	0430	Finish coat, exterior latex		13	.615		.75	10.60		11.35	18.50	
	0440	Primer & 1 coat, exterior latex		8	1		1.40	17.20		18.60	30.50	
	0450	Primer & 2 coats, exterior latex		6	1.333		2.11	23		25.11	41	
	0460	Stain, sealer & 1 coat varnish		7	1.143		1.81	19.65		21.46	35	
	0470	7 to 10 lite										
	0480	Brushwork, primer	1 Pord	11	.727	Ea.	.66	12.50		13.16	21.50	
	0490	Finish coat, exterior latex		11	.727		.75	12.50		13.25	22	
	0500	Primer & 1 coat, exterior latex		7	1.143		1.40	19.65		21.05	34.50	
	0510	Primer & 2 coats, exterior latex		5	1.600		2.11	27.50		29.61	48.50	
	0520	Stain, sealer & 1 coat varnish		6	1.333		1.81	23		24.81	40.50	
	0530	12 lite										
	0540	Brushwork, primer	1 Pord	10	.800	Ea.	.66	13.75		14.41	23.50	
	0550	Finish coat, exterior latex		10	.800		.75	13.75		14.50	24	
	0560	Primer & 1 coat, exterior latex		6	1.333		1.40	23		24.40	40	
	0570	Primer & 2 coats, exterior latex		5	1.600		2.11	27.50		29.61	48.50	
	0580	Stain, sealer & 1 coat varnish		6	1.333		1.79	23		24.79	40.50	
	0590	For oil base paint, add				Ea.	10%					
106	0010	**SIDING** Exterior, Alkyd (oil base) [R099-100]	2 Pord	4,550	.004	S.F.	.07	.06		.13	.18	106
	0500	Spray		4,550	.004		.07	.06		.13	.18	
	0800	Paint 2 coats, brushwork		1,300	.012		.09	.21		.30	.45	
	1000	Spray		4,550	.004		.14	.06		.20	.25	
	1200	Stucco, rough, oil base, paint 2 coats, brushwork		1,300	.012		.09	.21		.30	.45	
	1400	Roller		1,625	.010		.10	.17		.27	.39	
	1600	Spray		2,925	.005		.10	.09		.19	.27	
	1800	Texture 1-11 or clapboard, oil base, primer coat, brushwork		1,300	.012		.07	.21		.28	.43	
	2000	Spray		4,550	.004		.07	.06		.13	.18	
	2100	Paint 1 coat, brushwork		1,300	.012		.07	.21		.28	.42	
	2200	Spray		4,550	.004		.07	.06		.13	.17	
	2400	Paint 2 coats, brushwork		810	.020		.14	.34		.48	.72	
	2600	Spray		2,600	.006		.15	.11		.26	.35	
	3000	Stain 1 coat, brushwork		1,520	.011		.04	.18		.22	.34	
	3200	Spray		5,320	.003		.04	.05		.09	.14	
	3400	Stain 2 coats, brushwork		950	.017		.08	.29		.37	.57	
	4000	Spray		3,050	.005		.09	.09		.18	.24	
	4200	Wood shingles, oil base primer coat, brushwork		1,300	.012		.07	.21		.28	.43	
	4400	Spray		3,900	.004		.06	.07		.13	.19	
	4600	Paint 1 coat, brushwork		1,300	.012		.06	.21		.27	.41	
	4800	Spray		3,900	.004		.07	.07		.14	.20	
	5000	Paint 2 coats, brushwork		810	.020		.11	.34		.45	.69	
	5200	Spray		2,275	.007		.11	.12		.23	.32	
	5800	Stain 1 coat, brushwork		1,500	.011		.04	.18		.22	.35	

For expanded coverage of these items see *Means Building Construction Cost Data 1998*

099 100 | Exterior Painting

			CREW	DAILY OUTPUT	LABOR-HOURS	UNIT	1998 BARE COSTS MAT.	LABOR	EQUIP.	TOTAL	TOTAL INCL O&P	
106	6000	Spray	2 Pord	3,900	.004	S.F.	.04	.07		.11	.16	**106**
	6500	Stain 2 coats, brushwork		950	.017		.08	.29		.37	.57	
	7000	Spray		2,660	.006		.11	.10		.21	.29	
	8000	For latex paint, deduct					10%					
108	0010	**FENCES**										**108**
	0100	Chain link or wire metal, water base										
	0110	Roll & brush, first coat	1 Pord	960	.008	S.F.	.05	.14		.19	.29	
	0120	Second coat		1,280	.006		.05	.11		.16	.23	
	0130	Spray, first coat		2,275	.004		.05	.06		.11	.15	
	0140	Second coat		2,600	.003		.05	.05		.10	.14	
	0150	Picket, water base										
	0160	Roll & brush, first coat	1 Pord	865	.009	S.F.	.05	.16		.21	.33	
	0170	Second coat		1,050	.008		.05	.13		.18	.28	
	0180	Spray, first coat		2,275	.004		.05	.06		.11	.16	
	0190	Second coat		2,600	.003		.05	.05		.10	.15	
	0200	Stockade, water base										
	0210	Roll & brush, first coat	1 Pord	1,040	.008	S.F.	.05	.13		.18	.28	
	0220	Second coat		1,200	.007	S.F.	.05	.11		.16	.25	
	0230	Spray, first coat		2,275	.004		.05	.06		.11	.16	
	0240	Second coat		2,600	.003		.05	.05		.10	.15	
112	0010	**MISCELLANEOUS**, for painting metals see Div. 051-255										**112**
	0100	Railing, ext., decorative wood, incl. cap & baluster										
	0110	newels & spindles @ 12" O.C.										
	0120	Brushwork, stain, sand, seal & varnish										
	0130	First coat	1 Pord	90	.089	L.F.	.36	1.53		1.89	2.96	
	0140	Second coat	"	120	.067	"	.36	1.15		1.51	2.32	
	0150	Rough sawn wood, 42" high, 2"x2" verticals, 6" O.C.										
	0160	Brushwork, stain, each coat	1 Pord	90	.089	L.F.	.12	1.53		1.65	2.69	
	0170	Wrought iron, 1" rail, 1/2" sq. verticals										
	0180	Brushwork, zinc chromate, 60" high, bars 6" O.C.										
	0190	Primer	1 Pord	130	.062	L.F.	.23	1.06		1.29	2.02	
	0200	Finish coat		130	.062		.14	1.06		1.20	1.93	
	0210	Additional coat		190	.042		.17	.72		.89	1.39	
	0220	Shutters or blinds, single panel, 2'x4', paint all sides										
	0230	Brushwork, primer	1 Pord	20	.400	Ea.	.48	6.90		7.38	12.05	
	0240	Finish coat, exterior latex		20	.400		.40	6.90		7.30	11.95	
	0250	Primer & 1 coat, exterior latex		13	.615		.77	10.60		11.37	18.55	
	0260	Spray, primer		35	.229		.70	3.93		4.63	7.30	
	0270	Finish coat, exterior latex		35	.229		.85	3.93		4.78	7.50	
	0280	Primer & 1 coat, exterior latex		20	.400		.75	6.90		7.65	12.35	
	0290	For louvered shutters, add				S.F.	10%					
	0300	Stair stringers, exterior, metal										
	0310	Roll & brush, zinc chromate, to 14", each coat	1 Pord	320	.025	L.F.	.05	.43		.48	.77	
	0320	Rough sawn wood, 4" x 12"										
	0330	Roll & brush, exterior latex, each coat	1 Pord	215	.037	L.F.	.06	.64		.70	1.13	
	0340	Trellis/lattice, 2"x2" @ 3" O.C. with 2"x8" supports										
	0350	Spray, latex, per side, each coat	1 Pord	475	.017	S.F.	.06	.29		.35	.54	
	0450	Decking, Ext., sealer, alkyd, brushwork, sealer coat		1,140	.007		.04	.12		.16	.25	
	0460	1st coat		1,140	.007		.04	.12		.16	.25	
	0470	2nd coat		1,300	.006		.03	.11		.14	.22	
	0500	Paint, alkyd, brushwork, primer coat		1,140	.007		.06	.12		.18	.26	
	0510	1st coat		1,140	.007		.06	.12		.18	.26	
	0520	2nd coat		1,300	.006		.04	.11		.15	.23	
	0600	Sand paint, alkyd, brushwork, 1 coat		150	.053		.06	.92		.98	1.60	
116	0010	**SIDING**										**116**
	0100	Aluminum siding										

Important: See the Reference Section for critical supporting data - Reference Nos., Crews, & Location Factors

		099 100	Exterior Painting	CREW	DAILY OUTPUT	LABOR-HOURS	UNIT	1998 BARE COSTS				TOTAL INCL O&P	
								MAT.	LABOR	EQUIP.	TOTAL		
116	0110		Brushwork, primer	2 Pord	2,275	.007	S.F.	.04	.12		.16	.25	116
	0120		Finish coat, exterior latex		2,275	.007	S.F.	.04	.12		.16	.24	
	0130		Primer & 1 coat exterior latex		1,300	.012	S.F.	.09	.21		.30	.44	
	0140		Primer & 2 coats exterior latex		975	.016	"	.12	.28		.40	.60	
	0150		Mineral Fiber shingles										
	0160		Brushwork, primer	2 Pord	1,495	.011	S.F.	.07	.18		.25	.39	
	0170		Finish coat, industrial enamel		1,495	.011	S.F.	.07	.18		.25	.39	
	0180		Primer & 1 coat enamel		810	.020	S.F.	.15	.34		.49	.73	
	0190		Primer & 2 coats enamel		540	.030		.22	.51		.73	1.10	
	0200		Roll, primer		1,625	.010		.08	.17		.25	.37	
	0210		Finish coat, industrial enamel		1,625	.010		.08	.17		.25	.37	
	0220		Primer & 1 coat enamel		975	.016		.16	.28		.44	.65	
	0230		Primer & 2 coats enamel		650	.025		.24	.42		.66	.98	
	0240		Spray, primer		3,900	.004		.06	.07		.13	.19	
	0250		Finish coat, industrial enamel		3,900	.004		.07	.07		.14	.19	
	0260		Primer & 1 coat enamel		2,275	.007		.13	.12		.25	.34	
	0270		Primer & 2 coats enamel		1,625	.010		.20	.17		.37	.50	
	0280		Waterproof sealer, first coat		4,485	.004		.06	.06		.12	.16	
	0290		Second coat		5,235	.003		.06	.05		.11	.15	
	0300		Rough wood incl. shingles, shakes or rough sawn siding										
	0310		Brushwork, primer	2 Pord	1,280	.013	S.F.	.10	.22		.32	.47	
	0320		Finish coat, exterior latex		1,280	.013	S.F.	.06	.22		.28	.43	
	0330		Primer & 1 coat exterior latex		960	.017	S.F.	.16	.29		.45	.65	
	0340		Primer & 2 coats exterior latex		700	.023		.22	.39		.61	.90	
	0350		Roll, primer		2,925	.005		.13	.09		.22	.30	
	0360		Finish coat, exterior latex		2,925	.005		.07	.09		.16	.24	
	0370		Primer & 1 coat exterior latex		1,790	.009		.20	.15		.35	.48	
	0380		Primer & 2 coats exterior latex		1,300	.012		.28	.21		.49	.66	
	0390		Spray, primer		3,900	.004		.11	.07		.18	.24	
	0400		Finish coat, exterior latex		3,900	.004		.06	.07		.13	.18	
	0410		Primer & 1 coat exterior latex		2,600	.006		.17	.11		.28	.36	
	0420		Primer & 2 coats exterior latex		2,080	.008		.22	.13		.35	.47	
	0430		Waterproof sealer, first coat		4,485	.004		.11	.06		.17	.22	
	0440		Second coat		4,485	.004		.06	.06		.12	.16	
	0450		Smooth wood incl. butt, T&G, beveled, drop or B&B siding										
	0460		Brushwork, primer	2 Pord	2,325	.007	S.F.	.07	.12		.19	.28	
	0470		Finish coat, exterior latex		1,280	.013	S.F.	.06	.22		.28	.43	
	0480		Primer & 1 coat exterior latex		800	.020	S.F.	.13	.34		.47	.71	
	0490		Primer & 2 coats exterior latex		630	.025		.19	.44		.63	.94	
	0500		Roll, primer		2,275	.007		.08	.12		.20	.29	
	0510		Finish coat, exterior latex		2,275	.007		.07	.12		.19	.27	
	0520		Primer & 1 coat exterior latex		1,300	.012		.14	.21		.35	.51	
	0530		Primer & 2 coats exterior latex		975	.016		.21	.28		.49	.70	
	0540		Spray, primer		4,550	.004		.06	.06		.12	.17	
	0550		Finish coat, exterior latex		4,550	.004		.06	.06		.12	.16	
	0560		Primer & 1 coat exterior latex		2,600	.006		.12	.11		.23	.31	
	0570		Primer & 2 coats exterior latex		1,950	.008		.18	.14		.32	.43	
	0580		Waterproof sealer, first coat		5,230	.003		.06	.05		.11	.15	
	0590		Second coat		5,980	.003		.06	.05		.11	.14	
	0600		For oil base paint, add				S.F.	10%					
120	0010		**TRIM**										120
	0100		Door frames & trim (see Doors, interior or exterior)										
	0110		Fascia, latex paint, one coat coverage										
	0120		1" x 4", brushwork	1 Pord	640	.013	L.F.	.01	.22		.23	.38	
	0130		Roll		1,280	.006		.02	.11		.13	.20	
	0140		Spray		2,080	.004		.01	.07		.08	.12	

099 100 | Exterior Painting

		CREW	DAILY OUTPUT	LABOR-HOURS	UNIT	1998 BARE COSTS MAT.	LABOR	EQUIP.	TOTAL	TOTAL INCL O&P	
120	0150	1" x 6" to 1" x 10", brushwork	1 Pord	640	.013	L.F.	.05	.22		.27	.42	**120**
	0160	Roll		1,230	.007		.06	.11		.17	.25	
	0170	Spray		2,100	.004		.04	.07		.11	.16	
	0180	1" x 12", brushwork		640	.013		.05	.22		.27	.42	
	0190	Roll		1,050	.008		.06	.13		.19	.28	
	0200	Spray		2,200	.004		.04	.06		.10	.15	
	0210	Gutters & downspouts, metal, zinc chromate paint										
	0220	Brushwork, gutters, 5", first coat	1 Pord	640	.013	L.F.	.05	.22		.27	.42	
	0230	Second coat		960	.008		.05	.14		.19	.29	
	0240	Third coat		1,280	.006		.04	.11		.15	.22	
	0250	Downspouts, 4", first coat		640	.013		.05	.22		.27	.42	
	0260	Second coat		960	.008		.05	.14		.19	.29	
	0270	Third coat		1,280	.006		.04	.11		.15	.22	
	0280	Gutters & downspouts, wood										
	0290	Brushwork, gutters, 5", primer	1 Pord	640	.013	L.F.	.04	.22		.26	.41	
	0300	Finish coat, exterior latex		640	.013		.04	.22		.26	.41	
	0310	Primer & 1 coat exterior latex		400	.020		.09	.34		.43	.67	
	0320	Primer & 2 coats exterior latex		325	.025		.14	.42		.56	.87	
	0330	Downspouts, 4", primer		640	.013		.04	.22		.26	.41	
	0340	Finish coat, exterior latex		640	.013		.04	.22		.26	.41	
	0350	Primer & 1 coat exterior latex		400	.020		.09	.34		.43	.67	
	0360	Primer & 2 coats exterior latex		325	.025		.07	.42		.49	.79	
	0370	Molding, exterior, up to 14" wide										
	0380	Brushwork, primer	1 Pord	640	.013	L.F.	.05	.22		.27	.42	
	0390	Finish coat, exterior latex		640	.013		.05	.22		.27	.42	
	0400	Primer & 1 coat exterior latex		400	.020		.11	.34		.45	.69	
	0410	Primer & 2 coats exterior latex		315	.025		.11	.44		.55	.85	
	0420	Stain & fill		1,050	.008		.05	.13		.18	.27	
	0430	Shellac		1,850	.004		.05	.07		.12	.18	
	0440	Varnish		1,275	.006		.06	.11		.17	.25	
124	0350	**WALLS, MASONRY (CMU)**										**124**
	0360	Concrete masonry units (CMU), smooth surface										
	0370	Brushwork, latex, first coat	1 Pord	640	.013	S.F.	.03	.22		.25	.39	
	0380	Second coat		960	.008		.03	.14		.17	.27	
	0390	Waterproof sealer, first coat		736	.011		.06	.19		.25	.37	
	0400	Second coat		1,104	.007		.04	.12		.16	.25	
	0410	Roll, latex, paint, first coat		1,465	.005		.04	.09		.13	.20	
	0420	Second coat		1,790	.004		.03	.08		.11	.16	
	0430	Waterproof sealer, first coat		1,680	.005		.06	.08		.14	.20	
	0440	Second coat		2,060	.004		.04	.07		.11	.15	
	0450	Spray, latex, paint, first coat		1,950	.004		.03	.07		.10	.15	
	0460	Second coat		2,600	.003		.02	.05		.07	.12	
	0470	Waterproof sealer, first coat		2,245	.004		.04	.06		.10	.14	
	0480	Second coat		2,990	.003		.03	.05		.08	.11	
	0490	Concrete masonry unit (CMU), porous										
	0500	Brushwork, latex, first coat	1 Pord	640	.013	S.F.	.06	.22		.28	.43	
	0510	Second coat		960	.008		.03	.14		.17	.27	
	0520	Waterproof sealer, first coat		736	.011		.05	.19		.24	.36	
	0530	Second coat		1,104	.007		.04	.12		.16	.25	
	0540	Roll latex, first coat		1,465	.005		.05	.09		.14	.21	
	0550	Second coat		1,790	.004		.03	.08		.11	.16	
	0560	Waterproof sealer, first coat		1,400	.006		.06	.10		.16	.22	
	0570	Second coat		2,200	.004		.04	.06		.10	.15	
	0580	Spray latex, first coat		5,600	.001		.03	.02		.05	.07	
	0590	Second coat		6,400	.001		.02	.02		.04	.07	
	0600	Waterproof sealer, first coat		6,000	.001		.06	.02		.08	.10	

Important: See the Reference Section for critical supporting data - Reference Nos., Crews, & Location Factors

			DAILY	LABOR-		1998 BARE COSTS				TOTAL	
099 100	**Exterior Painting**	CREW	OUTPUT	HOURS	UNIT	MAT.	LABOR	EQUIP.	TOTAL	INCL O&P	
124	0610	Second coat	1 Pord	6,400	.001	S.F.	.03	.02		.05	.08	**124**

						1998 BARE COSTS				TOTAL	
099 200	**Interior Painting**	CREW	OUTPUT	HOURS	UNIT	MAT.	LABOR	EQUIP.	TOTAL	INCL O&P	
204	0010	**CABINETS AND CASEWORK**										**204**
	1000	Primer coat, oil base, brushwork	1 Pord	650	.012	S.F.	.04	.21		.25	.40	
	2000	Paint, oil base, brushwork, 1 coat		650	.012		.05	.21		.26	.40	
	2500	2 coats		400	.020		.10	.34		.44	.68	
	3000	Stain, brushwork, wipe off		650	.012		.04	.21		.25	.39	
	4000	Shellac, 1 coat, brushwork		650	.012		.04	.21		.25	.40	
	4500	Varnish, 3 coats, brushwork, sand after 1st coat		325	.025		.15	.42		.57	.88	
	5000	For latex paint, deduct					10%					
214	0010	**DOORS & WINDOWS, LATEX**										**214**
	0100	Doors flush, both sides, incl. frame & trim										
	0110	Roll & brush, primer	1 Pord	10	.800	Ea.	2.90	13.75		16.65	26	
	0120	Finish coat, latex		10	.800		2.98	13.75		16.73	26.50	
	0130	Primer & 1 coat latex		7	1.143		5.85	19.65		25.50	39.50	
	0140	Primer & 2 coats latex		5	1.600		8.70	27.50		36.20	55.50	
	0160	Spray, both sides, primer		20	.400		3.05	6.90		9.95	14.85	
	0170	Finish coat, latex		20	.400		3.12	6.90		10.02	14.95	
	0180	Primer & 1 coat latex		11	.727		6.20	12.50		18.70	28	
	0190	Primer & 2 coats latex		8	1		9.20	17.20		26.40	39	
	0200	Doors, French, both sides, 10-15 lite, incl. frame & trim										
	0210	Roll & brush, primer	1 Pord	6	1.333	Ea.	1.45	23		24.45	40	
	0220	Finish coat, latex		6	1.333		1.49	23		24.49	40	
	0230	Primer & 1 coat latex		3	2.667		2.94	46		48.94	79.50	
	0240	Primer & 2 coats latex		2	4		4.34	69		73.34	120	
	0260	Doors, louvered, both sides, incl. frame & trim										
	0270	Roll & brush, primer	1 Pord	7	1.143	Ea.	2.90	19.65		22.55	36	
	0280	Finish coat, latex		7	1.143		2.98	19.65		22.63	36.50	
	0290	Primer & 1 coat, latex		4	2		5.70	34.50		40.20	64	
	0300	Primer & 2 coats, latex		3	2.667		8.85	46		54.85	86.50	
	0320	Spray, both sides, primer		20	.400		3.05	6.90		9.95	14.85	
	0330	Finish coat, latex		20	.400		3.12	6.90		10.02	14.95	
	0340	Primer & 1 coat, latex		11	.727		6.20	12.50		18.70	28	
	0350	Primer & 2 coats, latex		8	1		9.40	17.20		26.60	39.50	
	0360	Doors, panel, both sides, incl. frame & trim										
	0370	Roll & brush, primer	1 Pord	6	1.333	Ea.	3.05	23		26.05	42	
	0380	Finish coat, latex		6	1.333		2.98	23		25.98	42	
	0390	Primer & 1 coat, latex		3	2.667		5.85	46		51.85	83	
	0400	Primer & 2 coats, latex		2.50	3.200		8.85	55		63.85	102	
	0420	Spray, both sides, primer		10	.800		3.05	13.75		16.80	26.50	
	0430	Finish coat, latex		10	.800		3.12	13.75		16.87	26.50	
	0440	Primer & 1 coat, latex		5	1.600		6.20	27.50		33.70	53	
	0450	Primer & 2 coats, latex		4	2		9.40	34.50		43.90	68	
	0460	Windows, per interior side, base on 15 SF										
	0470	1 to 6 lite										
	0480	Brushwork, primer	1 Pord	13	.615	Ea.	.57	10.60		11.17	18.35	
	0490	Finish coat, enamel		13	.615		.59	10.60		11.19	18.35	
	0500	Primer & 1 coat enamel		8	1		1.16	17.20		18.36	30.50	
	0510	Primer & 2 coats enamel		6	1.333		1.75	23		24.75	40.50	
	0530	7 to 10 lite										
	0540	Brushwork, primer	1 Pord	11	.727	Ea.	.57	12.50		13.07	21.50	
	0550	Finish coat, enamel		11	.727		.59	12.50		13.09	21.50	
	0560	Primer & 1 coat enamel		7	1.143		1.16	19.65		20.81	34.50	
	0570	Primer & 2 coats enamel		5	1.600		1.75	27.50		29.25	48	

FINISHES **9**

099 200 | Interior Painting

			DAILY OUTPUT	LABOR-HOURS	UNIT	1998 BARE COSTS				TOTAL INCL O&P	
			CREW			MAT.	LABOR	EQUIP.	TOTAL		
214	0590	12 lite									**214**
	0600	Brushwork, primer	1 Pord	10	.800	Ea.	.57	13.75		14.32	23.50
	0610	Finish coat, enamel		10	.800		.59	13.75		14.34	23.50
	0620	Primer & 1 coat enamel		6	1.333		1.16	23		24.16	40
	0630	Primer & 2 coats enamel		5	1.600		1.75	27.50		29.25	48
	0650	For oil base paint, add				Ea.	10%				
216	0010	**DOORS AND WINDOWS, ALKYD (OIL BASE)**									**216**
	0500	Flush door & frame, 3' x 7', oil, primer, brushwork	1 Pord	10	.800	Ea.	1.66	13.75		15.41	25
	1000	Paint, 1 coat		10	.800		1.73	13.75		15.48	25
	1200	2 coats		7	1.143		2.94	19.65		22.59	36
	1400	Stain, brushwork, wipe off		18	.444		.81	7.65		8.46	13.70
	1600	Shellac, 1 coat, brushwork		25	.320		.92	5.50		6.42	10.20
	1800	Varnish, 3 coats, brushwork, sand after 1st coat		9	.889		3.19	15.30		18.49	29
	2000	Panel door & frame, 3' x 7', oil, primer, brushwork		6	1.333		1.56	23		24.56	40
	2200	Paint, 1 coat		6	1.333		1.73	23		24.73	40.50
	2400	2 coats		3	2.667		5.05	46		51.05	82
	2600	Stain, brushwork, panel door, 3' x 7', not incl. frame		16	.500		.81	8.60		9.41	15.30
	2800	Shellac, 1 coat, brushwork		22	.364		.92	6.25		7.17	11.45
	3000	Varnish, 3 coats, brushwork, sand after 1st coat		7.50	1.067		3.19	18.35		21.54	34
	3020	French door, incl. 3' x 7', 6 lites, frame & trim									
	3022	Paint, 1 coat, over existing paint	1 Pord	5	1.600	Ea.	3.47	27.50		30.97	50
	3024	2 coats, over existing paint		5	1.600		6.75	27.50		34.25	53.50
	3026	Primer & 1 coat		3.50	2.286		6.60	39.50		46.10	73
	3028	Primer & 2 coats		3	2.667		10.05	46		56.05	87.50
	3032	Varnish or polyurethane, 1 coat		5	1.600		3.61	27.50		31.11	50
	3034	2 coats, sanding between		3	2.667		7.20	46		53.20	84.50
	4400	Windows, including frame and trim, per side									
	4600	Colonial type, 6/6 lites, 2' x 3', oil, primer, brushwork	1 Pord	14	.571	Ea.	.25	9.85		10.10	16.70
	5800	Paint, 1 coat		14	.571		.27	9.85		10.12	16.75
	6000	2 coats		9	.889		.53	15.30		15.83	26
	6200	3' x 5' opening, 6/6 lites, primer coat, brushwork		12	.667		.62	11.45		12.07	19.85
	6400	Paint, 1 coat		12	.667		.68	11.45		12.13	19.90
	6600	2 coats		7	1.143		1.33	19.65		20.98	34.50
	6800	4' x 8' opening, 6/6 lites, primer coat, brushwork		8	1		1.32	17.20		18.52	30.50
	7000	Paint, 1 coat		8	1		1.46	17.20		18.66	30.50
	7200	2 coats		5	1.600		2.84	27.50		30.34	49
	8000	Single lite type, 2' x 3', oil base, primer coat, brushwork		33	.242		.25	4.17		4.42	7.20
	8200	Paint, 1 coat		33	.242		.27	4.17		4.44	7.25
	8400	2 coats		20	.400		.53	6.90		7.43	12.10
	8600	3' x 5' opening, primer coat, brushwork		20	.400		.62	6.90		7.52	12.20
	8800	Paint, 1 coat		20	.400		.68	6.90		7.58	12.25
	9000	2 coats		13	.615		1.33	10.60		11.93	19.15
	9200	4' x 8' opening, primer coat, brushwork		14	.571		1.32	9.85		11.17	17.90
	9400	Paint, 1 coat		14	.571		1.46	9.85		11.31	18.05
	9600	2 coats		8	1		2.84	17.20		20.04	32
218	0010	**FLOORS**									**218**
	0100	Concrete									
	0120	1st coat	1 Pord	975	.008	S.F.	.09	.14		.23	.34
	0130	2nd coat		1,150	.007		.06	.12		.18	.27
	0140	3rd coat		1,300	.006		.05	.11		.16	.24
	0150	Roll, latex, block filler									
	0160	1st coat	1 Pord	2,600	.003	S.F.	.13	.05		.18	.23
	0170	2nd coat		3,250	.002		.07	.04		.11	.15
	0180	3rd coat		3,900	.002		.06	.04		.10	.12
	0190	Spray, latex, block filler									

099 200 | Interior Painting

		CREW	DAILY OUTPUT	LABOR-HOURS	UNIT	1998 BARE COSTS				TOTAL INCL O&P
						MAT.	LABOR	EQUIP.	TOTAL	
218 0200	1st coat	1 Pord	2,600	.003	S.F.	.11	.05		.16	.21
0210	2nd coat		3,250	.002		.06	.04		.10	.13
0220	3rd coat	↓	3,900	.002	↓	.05	.04		.09	.11
220 0010	**MISCELLANEOUS PAINTING**									
2400	Floors, conc./wood, oil base, primer/sealer coat, brushwork	2 Pord	1,950	.008	S.F.	.05	.14		.19	.29
2450	Roller		5,200	.003		.05	.05		.10	.14
2600	Spray		6,000	.003		.05	.05		.10	.13
2650	Paint 1 coat, brushwork		1,950	.008		.04	.14		.18	.28
2800	Roller		5,200	.003		.04	.05		.09	.14
2850	Spray		6,000	.003		.04	.05		.09	.13
3000	Stain, wood floor, brushwork, 1 coat		4,550	.004		.04	.06		.10	.14
3200	Roller		5,200	.003		.04	.05		.09	.13
3250	Spray		6,000	.003		.04	.05		.09	.12
3400	Varnish, wood floor, brushwork		4,550	.004		.05	.06		.11	.16
3450	Roller		5,200	.003		.05	.05		.10	.15
3600	Spray	↓	6,000	.003		.06	.05		.11	.14
3800	Grilles, per side, oil base, primer coat, brushwork	1 Pord	520	.015		.08	.26		.34	.53
3850	Spray		1,140	.007		.09	.12		.21	.29
3880	Paint 1 coat, brushwork		520	.015		.09	.26		.35	.54
3900	Spray		1,140	.007		.10	.12		.22	.31
3920	Paint 2 coats, brushwork		325	.025		.18	.42		.60	.91
3940	Spray	↓	650	.012	↓	.20	.21		.41	.57
4250	Paint 1 coat, brushwork	2 Pord	1,300	.012	L.F.	.09	.21		.30	.45
4500	Louvers, one side, primer, brushwork	1 Pord	524	.015	S.F.	.05	.26		.31	.49
4520	Paint one coat, brushwork		520	.015		.04	.26		.30	.49
4530	Spray		1,140	.007		.05	.12		.17	.25
4540	Paint two coats, brushwork		325	.025		.09	.42		.51	.81
4550	Spray		650	.012		.10	.21		.31	.46
4560	Paint three coats, brushwork		270	.030		.13	.51		.64	.99
4570	Spray	↓	500	.016	↓	.14	.28		.42	.62
5000	Pipe, to 4" diameter, primer or sealer coat, oil base, brushwork	2 Pord	1,250	.013	L.F.	.05	.22		.27	.42
5100	Spray		2,165	.007		.05	.13		.18	.26
5200	Paint 1 coat, brushwork		1,250	.013		.05	.22		.27	.43
5300	Spray		2,165	.007		.05	.13		.18	.26
5350	Paint 2 coats, brushwork		775	.021		.09	.36		.45	.69
5400	Spray		1,240	.013		.10	.22		.32	.48
5450	To 8" diameter, primer or sealer coat, brushwork		620	.026		.10	.44		.54	.85
5500	Spray		1,085	.015		.16	.25		.41	.60
5550	Paint 1 coat, brushwork		620	.026		.14	.44		.58	.89
5600	Spray		1,085	.015		.15	.25		.40	.59
5650	Paint 2 coats, brushwork		305	.042		.18	.71		.89	1.39
5700	Spray		620	.026		.20	.44		.64	.96
5750	To 12" diameter, primer or sealer coat, brushwork		415	.039		.15	.66		.81	1.27
5800	Spray		725	.022		.17	.38		.55	.81
5850	Paint 1 coat, brushwork		415	.039		.14	.66		.80	1.27
6000	Spray		725	.022		.16	.38		.54	.80
6200	Paint 2 coats, brushwork		260	.062		.27	1.06		1.33	2.07
6250	Spray		415	.039		.30	.66		.96	1.44
6300	To 16" diameter, primer or sealer coat, brushwork		310	.052		.20	.89		1.09	1.70
6350	Spray		540	.030		.22	.51		.73	1.09
6400	Paint 1 coat, brushwork		310	.052		.19	.89		1.08	1.69
6450	Spray		540	.030		.21	.51		.72	1.08
6500	Paint 2 coats, brushwork		195	.082		.37	1.41		1.78	2.76
6550	Spray	↓	310	.052	↓	.41	.89		1.30	1.93
6600	Radiators, per side, primer, brushwork	1 Pord	520	.015	S.F.	.05	.26		.31	.49

099 200	Interior Painting	CREW	DAILY OUTPUT	LABOR-HOURS	UNIT	1998 BARE COSTS				TOTAL INCL O&P
						MAT.	LABOR	EQUIP.	TOTAL	
220 6620	Paint one coat, brushwork	1 Pord	520	.015	S.F.	.04	.26		.30	.49 **220**
6640	Paint two coats, brushwork		340	.024		.09	.40		.49	.78
6660	Paint three coats, brushwork		283	.028		.13	.49		.62	.95
7000	Trim, wood, incl. puttying, under 6" wide									
7200	Primer coat, oil base, brushwork	1 Pord	650	.012	L.F.	.02	.21		.23	.37
7250	Paint, 1 coat, brushwork		650	.012		.02	.21		.23	.38
7400	2 coats		400	.020		.04	.34		.38	.62
7450	3 coats		325	.025		.07	.42		.49	.78
7500	Over 6" wide, primer coat, brushwork		650	.012		.04	.21		.25	.40
7550	Paint, 1 coat, brushwork		650	.012		.05	.21		.26	.40
7600	2 coats		400	.020		.09	.34		.43	.67
7650	3 coats		325	.025		.13	.42		.55	.85
8000	Cornice, simple design, primer coat, oil base, brushwork		650	.012	S.F.	.04	.21		.25	.40
8250	Paint, 1 coat		650	.012		.05	.21		.26	.40
8300	2 coats		400	.020		.09	.34		.43	.67
8350	Ornate design, primer coat		350	.023		.04	.39		.43	.71
8400	Paint, 1 coat		350	.023		.05	.39		.44	.71
8450	2 coats		400	.020		.09	.34		.43	.67
8600	Balustrades, primer coat, oil base, brushwork		520	.015		.04	.26		.30	.49
8650	Paint, 1 coat		520	.015		.05	.26		.31	.49
8700	2 coats		325	.025		.09	.42		.51	.81
8900	Trusses and wood frames, primer coat, oil base, brushwork		800	.010		.04	.17		.21	.34
8950	Spray		1,200	.007		.04	.11		.15	.24
9000	Paint 1 coat, brushwork		750	.011		.05	.18		.23	.36
9200	Spray		1,200	.007		.05	.11		.16	.25
9220	Paint 2 coats, brushwork		500	.016		.09	.28		.37	.56
9240	Spray		600	.013		.10	.23		.33	.49
9260	Stain, brushwork, wipe off		600	.013		.04	.23		.27	.42
9280	Varnish, 3 coats, brushwork		275	.029		.15	.50		.65	1.01
9350	For latex paint, deduct					10%				
224 0010	**WALLS AND CEILINGS**									**224**
0100	Concrete, dry wall or plaster, oil base, primer or sealer coat									
0200	Smooth finish, brushwork	1 Pord	1,150	.007	S.F.	.04	.12		.16	.25
0240	Roller		2,040	.004		.04	.07		.11	.15
0300	Sand finish, brushwork		975	.008		.04	.14		.18	.28
0340	Roller		1,150	.007		.04	.12		.16	.25
0380	Spray		2,275	.004		.03	.06		.09	.14
0400	Paint 1 coat, smooth finish, brushwork		1,200	.007		.04	.11		.15	.23
0440	Roller		1,300	.006		.04	.11		.15	.22
0480	Spray		2,275	.004		.03	.06		.09	.13
0500	Sand finish, brushwork		1,050	.008		.03	.13		.16	.26
0540	Roller		1,600	.005		.04	.09		.13	.18
0580	Spray		2,100	.004		.03	.07		.10	.14
0800	Paint 2 coats, smooth finish, brushwork		680	.012		.07	.20		.27	.42
0840	Roller		800	.010		.07	.17		.24	.37
0880	Spray		1,625	.005		.06	.08		.14	.21
0900	Sand finish, brushwork		605	.013		.07	.23		.30	.45
0940	Roller		1,020	.008		.07	.13		.20	.31
0980	Spray		1,700	.005		.06	.08		.14	.21
1200	Paint 3 coats, smooth finish, brushwork		510	.016		.10	.27		.37	.56
1240	Roller		650	.012		.11	.21		.32	.47
1280	Spray		1,625	.005		.09	.08		.17	.24
1300	Sand finish, brushwork		454	.018		.10	.30		.40	.62
1340	Roller		680	.012		.11	.20		.31	.46
1380	Spray		1,133	.007		.09	.12		.21	.30
1600	Glaze coating, 5 coats, spray, clear		900	.009		.60	.15		.75	.92

Important: See the Reference Section for critical supporting data - Reference Nos., Crews, & Location Factors

099 200 | Interior Painting

		CREW	DAILY OUTPUT	LABOR-HOURS	UNIT	1998 BARE COSTS				TOTAL INCL O&P	
						MAT.	LABOR	EQUIP.	TOTAL		
224	1640	Multicolor	1 Pord	900	.009	S.F.	.77	.15		.92	1.11	**224**
	1700	For latex paint, deduct					10%					
	1800	For ceiling installations, add						25%				
	2000	Masonry or concrete block, oil base, primer or sealer coat										
	2100	Smooth finish, brushwork	1 Pord	1,224	.007	S.F.	.04	.11		.15	.24	
	2180	Spray		2,400	.003		.06	.06		.12	.17	
	2200	Sand finish, brushwork		1,089	.007		.07	.13		.20	.28	
	2280	Spray		2,400	.003		.06	.06		.12	.17	
	2400	Paint 1 coat, smooth finish, brushwork		1,100	.007		.07	.13		.20	.28	
	2480	Spray		2,400	.003		.06	.06		.12	.17	
	2500	Sand finish, brushwork		979	.008		.07	.14		.21	.30	
	2580	Spray		2,400	.003		.06	.06		.12	.17	
	2800	Paint 2 coats, smooth finish, brushwork		756	.011		.13	.18		.31	.45	
	2880	Spray		1,360	.006		.12	.10		.22	.31	
	2900	Sand finish, brushwork		672	.012		.13	.20		.33	.49	
	2980	Spray		1,360	.006		.12	.10		.22	.31	
	3200	Paint 3 coats, smooth finish, brushwork		560	.014		.20	.25		.45	.63	
	3280	Spray		1,088	.007		.19	.13		.32	.41	
	3300	Sand finish, brushwork		498	.016		.20	.28		.48	.68	
	3380	Spray		1,088	.007		.19	.13		.32	.41	
	3600	Glaze coating, 5 coats, spray, clear		900	.009		.60	.15		.75	.92	
	3620	Multicolor		900	.009		.77	.15		.92	1.11	
	4000	Block filler, 1 coat, brushwork		425	.019		.11	.32		.43	.66	
	4100	Silicone, water repellent, 2 coats, spray		2,000	.004		.23	.07		.30	.36	
	4120	For latex paint, deduct					10%					
228	0010	**VARNISH** 1 coat + sealer, on wood trim, no sanding included	1 Pord	400	.020	S.F.	.06	.34		.40	.64	**228**
	0100	Hardwood floors, 2 coats, no sanding included, roller	"	1,890	.004	"	.11	.07		.18	.24	

099 700 | Wallpaper

		CREW	DAILY OUTPUT	LABOR-HOURS	UNIT	1998 BARE COSTS				TOTAL INCL O&P	
						MAT.	LABOR	EQUIP.	TOTAL		
701	0010	**WALL COVERING** Including sizing, add 10% waste at takeoff										**701**
	0050	Aluminum foil	1 Pape	275	.029	S.F.	.75	.51		1.26	1.68	
	0100	Copper sheets, .025" thick, vinyl backing		240	.033		3.97	.58		4.55	5.35	
	0300	Phenolic backing		240	.033		5.15	.58		5.73	6.60	
	0600	Cork tiles, light or dark, 12" x 12" x 3/16"		240	.033		2.55	.58		3.13	3.77	
	0700	5/16" thick		235	.034		2.65	.59		3.24	3.91	
	0900	1/4" basketweave		240	.033		4.07	.58		4.65	5.45	
	1000	1/2" natural, non-directional pattern		240	.033		5.85	.58		6.43	7.40	
	1100	3/4" natural, non-directional pattern		240	.033		7.05	.58		7.63	8.70	
	1200	Granular surface, 12" x 36", 1/2" thick		385	.021		.88	.36		1.24	1.58	
	1300	1" thick		370	.022		1.14	.38		1.52	1.88	
	1500	Polyurethane coated, 12" x 12" x 3/16" thick		240	.033		2.75	.58		3.33	4	
	1600	5/16" thick		235	.034		3.90	.59		4.49	5.30	
	1800	Cork wallpaper, paperbacked, natural		480	.017		1.56	.29		1.85	2.21	
	1900	Colors		480	.017		1.93	.29		2.22	2.61	
	2100	Flexible wood veneer, 1/32" thick, plain woods		100	.080		1.57	1.40		2.97	4.06	
	2200	Exotic woods		95	.084		2.38	1.47		3.85	5.05	
	2400	Gypsum-based, fabric-backed, fire										
	2500	resistant for masonry walls, minimum, 21 oz./S.Y.	1 Pape	800	.010	S.F.	.62	.17		.79	.97	
	2600	Average		720	.011		.86	.19		1.05	1.27	
	2700	Maximum, (small quantities)		640	.013		.96	.22		1.18	1.42	
	2750	Acrylic, modified, semi-rigid PVC, .028" thick	2 Carp	330	.048		.79	.90		1.69	2.41	
	2800	.040" thick	"	320	.050		1.03	.93		1.96	2.72	
	3000	Vinyl wall covering, fabric-backed, lightweight, (12-15 oz./S.Y.)	1 Pape	640	.013		.50	.22		.72	.91	
	3300	Medium weight, type 2, (20-24 oz./S.Y.)		480	.017		.63	.29		.92	1.18	
	3400	Heavy weight, type 3, (28 oz./S.Y.)		435	.018		1.13	.32		1.45	1.78	

FINISHES 9

099 700 | Wallpaper

		CREW	DAILY OUTPUT	LABOR-HOURS	UNIT	1998 BARE COSTS				TOTAL INCL O&P	
						MAT.	LABOR	EQUIP.	TOTAL		
701	3600	Adhesive, 5 gal. lots, (18SY/Gal.)				Gal.	7.95			7.95	8.75	**701**
	3700	Wallpaper, average workmanship, solid pattern, low cost paper	1 Pape	640	.013	S.F.	.26	.22		.48	.65	
	3900	basic patterns (matching required), avg. cost paper		535	.015		.46	.26		.72	.95	
	4000	Paper at $50 per double roll, quality workmanship		435	.018		.88	.32		1.20	1.51	
	4100	Linen wall covering, paper backed										
	4150	Flame treatment, minimum				S.F.	.57			.57	.63	
	4180	Maximum					1.12			1.12	1.23	
	4200	Grass cloths with lining paper, minimum	1 Pape	400	.020		.56	.35		.91	1.20	
	4300	Maximum	"	350	.023		1.80	.40		2.20	2.65	

099 860 | Special Wall Surfaces

		CREW	DAILY OUTPUT	LABOR-HOURS	UNIT	MAT.	LABOR	EQUIP.	TOTAL	TOTAL INCL O&P	
861	0010	FIBERGLASS REINFORCED PLASTIC panels, .090" thick, on walls										**861**
	0020	Adhesive mounted, embossed surface	2 Carp	640	.025	S.F.	1.06	.46		1.52	1.96	
	0030	Smooth surface		640	.025		1.20	.46		1.66	2.11	
	0040	Fire rated, embossed surface		640	.025		1.53	.46		1.99	2.47	
	0050	Nylon rivet mounted, on drywall, embossed surface		480	.033		1.18	.62		1.80	2.36	
	0060	Smooth surface		480	.033		1.32	.62		1.94	2.51	
	0070	Fire rated, embossed surface		480	.033		1.65	.62		2.27	2.88	
	0080	On masonry, embossed surface		320	.050		1.18	.93		2.11	2.89	
	0090	Smooth surface		320	.050		1.32	.93		2.25	3.04	
	0100	Fire rated, embossed surface		320	.050		1.65	.93		2.58	3.41	
	0110	Nylon rivet and adhesive mounted, on drywall, embossed surface		240	.067		1.27	1.23		2.50	3.51	
	0120	Smooth surface		240	.067		1.42	1.23		2.65	3.67	
	0130	Fire rated, embossed surface		240	.067		1.75	1.23		2.98	4.04	
	0140	On masonry, embossed surface		190	.084		1.27	1.56		2.83	4.07	
	0150	Smooth surface		190	.084		1.42	1.56		2.98	4.23	
	0160	Fire rated, embossed surface		190	.084		1.75	1.56		3.31	4.60	
	0170	For moldings add	1 Carp	250	.032	L.F.	.21	.59		.80	1.24	
	0180	On ceilings, for lay in grid system, embossed surface		400	.020	S.F.	1.06	.37		1.43	1.80	
	0190	Smooth surface		400	.020		1.20	.37		1.57	1.95	
	0200	Fire rated, embossed surface		400	.020		1.53	.37		1.90	2.31	

099 900 | Surface Preparation

		CREW	DAILY OUTPUT	LABOR-HOURS	UNIT	MAT.	LABOR	EQUIP.	TOTAL	TOTAL INCL O&P	
902	0010	REMOVAL Existing lead paint, by chemicals, per application										**902**
	0020	See also, Div 020-800, Haz. Mat'l Abatement	R020-890									
	0050	Baseboard, to 6" wide	1 Pord	64	.125	L.F.	1.32	2.15		3.47	5.05	
	0070	To 12" wide		32	.250	"	2.60	4.30		6.90	10.05	
	0200	Balustrades, one side		28	.286	S.F.	2.95	4.91		7.86	11.45	
	1400	Cabinets, simple design		32	.250		2.58	4.30		6.88	10.05	
	1420	Ornate design		25	.320		3.31	5.50		8.81	12.85	
	1600	Cornice, simple design		60	.133		1.38	2.29		3.67	5.35	
	1620	Ornate design		20	.400		4.08	6.90		10.98	16	
	2800	Doors, one side, flush		84	.095		1	1.64		2.64	3.84	
	2820	Two panel		80	.100		1.04	1.72		2.76	4.02	
	2840	Four panel		45	.178		1.83	3.06		4.89	7.10	
	2880	For trim, one side, add		64	.125	L.F.	1.32	2.15		3.47	5.05	
	3000	Fence, picket, one side		30	.267	S.F.	2.77	4.59		7.36	10.70	
	3200	Grilles, one side, simple design		30	.267		2.77	4.59		7.36	10.70	
	3220	Ornate design		25	.320		3.31	5.50		8.81	12.85	
	4400	Pipes, to 4" diameter		90	.089	L.F.	.94	1.53		2.47	3.59	
	4420	To 8" diameter		50	.160		1.65	2.75		4.40	6.40	
	4440	To 12" diameter		36	.222		2.29	3.82		6.11	8.90	
	4460	To 16" diameter		20	.400		4.11	6.90		11.01	16	
	4500	For hangers, add		40	.200	Ea.	2.06	3.44		5.50	8	
	4800	Siding		90	.089	S.F.	.94	1.53		2.47	3.59	

			CREW	DAILY OUTPUT	LABOR-HOURS	UNIT	MAT.	LABOR	EQUIP.	TOTAL	TOTAL INCL O&P	
099 900	**Surface Preparation**							1998 BARE COSTS				
902	5000	Trusses, open	R020 -890	1 Pord	55	.145	SF Face	1.50	2.50		4	5.85	902
	6200	Windows, one side only, double hung, 1/1 light, 24" x 48" high			4	2	Ea.	21	34.50		55.50	80.50	
	6220	30" x 60" high			3	2.667		27.50	46		73.50	107	
	6240	36" x 72" high			2.50	3.200		33	55		88	129	
	6280	40" x 80" high			2	4		41.50	69		110.50	161	
	6400	Colonial window, 6/6 light, 24" x 48" high			2	4		41.50	69		110.50	161	
	6420	30" x 60" high			1.50	5.333		55.50	91.50		147	214	
	6440	36" x 72" high			1	8		83	138		221	320	
	6480	40" x 80" high			1	8		83	138		221	320	
	6600	8/8 light, 24" x 48" high			2	4		41.50	69		110.50	161	
	6620	40" x 80" high			1	8		83	138		221	320	
	6800	12/12 light, 24" x 48" high			1	8		83	138		221	320	
	6820	40" x 80" high			.75	10.667		111	183		294	425	
	6840	Window frame & trim items, included in pricing above											
903	0010	**LEAD PAINT ENCAPSULATION**, water based polymer coating,14 mil DFT											903
	0020	Interior, brushwork, trim, under 6"		1 Pord	240	.033	L.F.	1.94	.57		2.51	3.09	
	0030	6" to 12" wide			180	.044		2.58	.76		3.34	4.12	
	0040	Balustrades			300	.027		1.55	.46		2.01	2.47	
	0050	Pipe to 4" diameter			500	.016		.93	.28		1.21	1.48	
	0060	To 8" diameter			375	.021		1.24	.37		1.61	1.97	
	0070	To 12" diameter			250	.032		1.85	.55		2.40	2.96	
	0080	To 16" diameter			170	.047		2.73	.81		3.54	4.35	
	0090	Cabinets, ornate design			200	.040	S.F.	2.32	.69		3.01	3.70	
	0100	Simple design			250	.032	"	1.85	.55		2.40	2.96	
	0110	Doors, 3'x 7', both sides, incl. frame & trim											
	0120	Flush		1 Pord	6	1.333	Ea.	24	23		47	64.50	
	0130	French, 10-15 lite			3	2.667		4.76	46		50.76	82	
	0140	Panel			4	2		28.50	34.50		63	89	
	0150	Louvered			2.75	2.909		26	50		76	113	
	0160	Windows, per interior side, per 15 S.F.											
	0170	1 to 6 lite		1 Pord	14	.571	Ea.	16.45	9.85		26.30	34.50	
	0180	7 to 10 lite			7.50	1.067		18.15	18.35		36.50	50.50	
	0190	12 lite			5.75	1.391		24.50	24		48.50	67	
	0200	Radiators			8	1		58	17.20		75.20	92.50	
	0210	Grilles, vents			275	.029	S.F.	1.69	.50		2.19	2.70	
	0220	Walls, roller, drywall or plaster			1,000	.008		.46	.14		.60	.74	
	0230	With spunbonded reinforcing fabric			720	.011		.52	.19		.71	.89	
	0240	Wood			800	.010		.58	.17		.75	.93	
	0250	Ceilings, roller, drywall or plaster			900	.009		.52	.15		.67	.83	
	0260	Wood			700	.011		.66	.20		.86	1.06	
	0270	Exterior, brushwork, gutters and downspouts			300	.027	L.F.	1.55	.46		2.01	2.47	
	0280	Columns			400	.020	S.F.	1.16	.34		1.50	1.85	
	0290	Spray, siding			600	.013	"	.77	.23		1	1.23	
	0300	Miscellaneous											
	0310	Electrical conduit, brushwork, to 2" diameter		1 Pord	500	.016	L.F.	.93	.28		1.21	1.48	
	0320	Brick, block or concrete, spray			500	.016	S.F.	.93	.28		1.21	1.48	
	0330	Steel, flat surfaces and tanks to 12"			500	.016		.93	.28		1.21	1.48	
	0340	Beams, brushwork			400	.020		1.16	.34		1.50	1.85	
	0350	Trusses			400	.020		1.16	.34		1.50	1.85	
904	0010	**SANDING** And puttying interior trim, compared to											904
	0100	Painting 1 coat, on quality work					L.F.		100%				
	0300	Medium work							50%				
	0400	Industrial grade							25%				
	0500	Surface protection, placement and removal											
	0510	Basic drop cloths		1 Pord	6,400	.001	S.F.		.02		.02	.04	

099 900 | Surface Preparation

			DAILY OUTPUT	LABOR-HOURS	UNIT	1998 BARE COSTS				TOTAL INCL O&P	
						MAT.	LABOR	EQUIP.	TOTAL		
904	0520	Masking with paper	1 Pord	800	.010	S.F.	.02	.17		.19	.31	**904**
	0530	Volume cover up (using plastic sheathing, or building paper)	↓	16,000	.001	↓		.01		.01	.01	
906	0010	**SCRAPE AFTER FIRE DAMAGE**										**906**
	0050	Boards, 1″ x 4″	1 Pord	336	.024	L.F.		.41		.41	.68	
	0060	1″ x 6″		260	.031			.53		.53	.88	
	0070	1″ x 8″		207	.039			.66		.66	1.11	
	0080	1″ x 10″		174	.046			.79		.79	1.32	
	0500	Framing, 2″ x 4″		265	.030			.52		.52	.87	
	0510	2″ x 6″		221	.036			.62		.62	1.04	
	0520	2″ x 8″		190	.042			.72		.72	1.21	
	0530	2″ x 10″		165	.048			.83		.83	1.39	
	0540	2″ x 12″		144	.056			.96		.96	1.60	
	1000	Heavy framing, 3″ x 4″		226	.035			.61		.61	1.02	
	1010	4″ x 4″		210	.038			.66		.66	1.10	
	1020	4″ x 6″		191	.042			.72		.72	1.20	
	1030	4″ x 8″		165	.048			.83		.83	1.39	
	1040	4″ x 10″		144	.056			.96		.96	1.60	
	1060	4″ x 12″		131	.061	↓		1.05		1.05	1.76	
	2900	For sealing, minimum		825	.010	S.F.	.11	.17		.28	.40	
	2920	Maximum	↓	460	.017	″	.23	.30		.53	.75	
910	0010	**SURFACE PREPARATION, EXTERIOR**										**910**
	0015	Doors, per side, not incl. frames or trim										
	0020	Scrape & sand										
	0030	Wood, flush	1 Pord	616	.013	S.F.		.22		.22	.37	
	0040	Wood, detail		496	.016			.28		.28	.46	
	0050	Wood, louvered		280	.029			.49		.49	.82	
	0060	Wood, overhead	↓	616	.013	↓		.22		.22	.37	
	0070	Wire brush										
	0080	Metal, flush	1 Pord	640	.013	S.F.		.22		.22	.36	
	0090	Metal, detail		520	.015			.26		.26	.44	
	0100	Metal, louvered		360	.022			.38		.38	.64	
	0110	Metal or fibr., overhead		640	.013			.22		.22	.36	
	0120	Metal, roll up		560	.014			.25		.25	.41	
	0130	Metal, bulkhead	↓	640	.013	↓		.22		.22	.36	
	0140	Power wash, based on 2500 lb. operating pressure										
	0150	Metal, flush	B-9	2,240	.018	S.F.		.25	.08	.33	.51	
	0160	Metal, detail		2,120	.019			.26	.08	.34	.54	
	0170	Metal, louvered		2,000	.020			.28	.09	.37	.57	
	0180	Metal or fibr., overhead		2,400	.017			.23	.07	.30	.47	
	0190	Metal, roll up		2,400	.017			.23	.07	.30	.47	
	0200	Metal, bulkhead	↓	2,200	.018	↓		.25	.08	.33	.52	
	0400	Windows, per side, not incl. trim										
	0410	Scrape & sand										
	0420	Wood, 1-2 lite	1 Pord	320	.025	S.F.		.43		.43	.72	
	0430	Wood, 3-6 lite		280	.029			.49		.49	.82	
	0440	Wood, 7-10 lite		240	.033			.57		.57	.96	
	0450	Wood, 12 lite		200	.040			.69		.69	1.15	
	0460	Wood, Bay / Bow	↓	320	.025	↓		.43		.43	.72	
	0470	Wire brush										
	0480	Metal, 1-2 lite	1 Pord	480	.017	S.F.		.29		.29	.48	
	0490	Metal, 3-6 lite		400	.020			.34		.34	.57	
	0500	Metal, Bay / Bow	↓	480	.017	↓		.29		.29	.48	
	0510	Power wash, based on 2500 lb. operating pressure										
	0520	1-2 lite	B-9	4,400	.009	S.F.		.13	.04	.17	.25	
	0530	3-6 lite		4,320	.009			.13	.04	.17	.26	
	0540	7-10 lite	↓	4,240	.009	↓		.13	.04	.17	.27	

Important: See the Reference Section for critical supporting data - Reference Nos., Crews, & Location Factors

099 900	Surface Preparation	CREW	DAILY OUTPUT	LABOR-HOURS	UNIT	1998 BARE COSTS				TOTAL INCL O&P	
						MAT.	LABOR	EQUIP.	TOTAL		
910											**910**
0550	12 lite	B-9	4,160	.010	S.F.		.13	.04	.17	.28	
0560	Bay / Bow	↓	4,400	.009	↓		.13	.04	.17	.25	
0600	Siding, scrape and sand, light=10-30%, med.=30-70%										
0610	Heavy=70-100%, % of surface to sand										
0620	For Chemical Washing, see Division 045-108										
0630	For Steam Cleaning, see Division 042-166										
0640	For Sand Blasting, see Division 042-150										
0650	Texture 1-11, light	1 Pord	480	.017	S.F.		.29		.29	.48	
0660	Med.		440	.018			.31		.31	.52	
0670	Heavy		360	.022			.38		.38	.64	
0680	Wood shingles, shakes, light		440	.018			.31		.31	.52	
0690	Med.		360	.022			.38		.38	.64	
0700	Heavy		280	.029			.49		.49	.82	
0710	Clapboard, light		520	.015			.26		.26	.44	
0720	Med.		480	.017			.29		.29	.48	
0730	Heavy	↓	400	.020	↓		.34		.34	.57	
0740	Wire brush										
0750	Aluminum, light	1 Pord	600	.013	S.F.		.23		.23	.38	
0760	Med.		520	.015			.26		.26	.44	
0770	Heavy	↓	440	.018	↓		.31		.31	.52	
0780	Pressure wash, based on 2500 lb.. operating pressure										
0790	Stucco	B-6	3,080	.008	S.F.		.12	.07	.19	.28	
0800	Aluminum or vinyl		3,200	.008			.11	.07	.18	.26	
0810	Siding, masonry, brick & block	↓	2,400	.010	↓		.15	.09	.24	.35	
1300	Miscellaneous, wire brush										
1310	Metal, pedestrian gate	1 Pord	100	.080	S.F.		1.38		1.38	2.30	
920	**SURFACE PREPARATION, INTERIOR**										**920**
0020	Doors										
0030	Scrape & sand										
0040	Wood, flush	1 Pord	616	.013	S.F.		.22		.22	.37	
0050	Wood, detail		496	.016			.28		.28	.46	
0060	Wood, louvered	↓	280	.029	↓		.49		.49	.82	
0070	Wire brush										
0080	Metal, flush	1 Pord	640	.013	S.F.		.22		.22	.36	
0090	Metal, detail		520	.015			.26		.26	.44	
0100	Metal, louvered	↓	360	.022	↓		.38		.38	.64	
0110	Hand wash										
0120	Wood, flush	1 Pord	2,160	.004	S.F.		.06		.06	.11	
0130	Wood, detailed		2,000	.004			.07		.07	.11	
0140	Wood, louvered		1,360	.006			.10		.10	.17	
0150	Metal, flush		2,160	.004			.06		.06	.11	
0160	Metal, detail		2,000	.004			.07		.07	.11	
0170	Metal, louvered	↓	1,360	.006	↓		.10		.10	.17	
0400	Windows, per side, not incl. trim										
0410	Scrape & sand										
0420	Wood, 1-2 lite	1 Pord	360	.022	S.F.		.38		.38	.64	
0430	Wood, 3-6 lite		320	.025			.43		.43	.72	
0440	Wood, 7-10 lite		280	.029			.49		.49	.82	
0450	Wood, 12 lite		240	.033			.57		.57	.96	
0460	Wood, Bay / Bow	↓	360	.022	↓		.38		.38	.64	
0470	Wire brush										
0480	Metal, 1-2 lite	1 Pord	520	.015	S.F.		.26		.26	.44	
0490	Metal, 3-6 lite		440	.018			.31		.31	.52	
0500	Metal, Bay / Bow	↓	520	.015	↓		.26		.26	.44	
0600	Walls, sanding, light=10-30%										
0610	Med.=30-70%, heavy=70-100%, % of surface to sand										

099 900	Surface Preparation	CREW	DAILY OUTPUT	LABOR-HOURS	UNIT	1998 BARE COSTS				TOTAL INCL O&P
						MAT.	LABOR	EQUIP.	TOTAL	
920 0620	For Chemical Washing, see Division 045-108									**920**
0630	For Steam Cleaning, see Division 042-166									
0640	For Sand Blasting, see Division 042-150									
0650	Walls, sand									
0660	Drywall, gypsum, plaster, light	1 Pord	3,077	.003	S.F.		.04		.04	.07
0670	Drywall, gypsum, plaster, med.		2,160	.004			.06		.06	.11
0680	Drywall, gypsum, plaster, heavy		923	.009			.15		.15	.25
0690	Wood, T&G, light		2,400	.003			.06		.06	.10
0700	Wood, T&G, med.		1,600	.005			.09		.09	.14
0710	Wood, T&G, heavy		800	.010			.17		.17	.29
0720	Walls, wash									
0730	Drywall, gypsum, plaster	1 Pord	3,200	.002	S.F.		.04		.04	.07
0740	Wood, T&G		3,200	.002			.04		.04	.07
0750	Masonry, brick & block, smooth		2,800	.003			.05		.05	.08
0760	Masonry, brick & block, coarse		2,000	.004			.07		.07	.11

For information about Means Estimating Seminars, see yellow pages 11 and 12 in back of book

Important: See the Reference Section for critical supporting data - Reference Nos., Crews, & Location Factors

Division 10
Specialties

Estimating Tips
General
- The items in this division are usually priced per square foot or each.
- Many items in Division 10 require some type of support system that is not usually furnished with the item. Examples of these systems include blocking for the attachment of grab bars and support angles for ceiling hung toilet partitions. The required blocking or supports must be added to the estimate in the appropriate division.
- Some items in Division 10, such as lockers, may require assembly before installation. Verify the amount of assembly required. Assembly can often exceed installation time.

101 Visual Display Boards, Compartments & Cubicles
- Toilet partitions are priced by the stall. A stall consists of a side wall, pilaster and door with hardware. Toilet tissue holders and grab bars are extra.

106 Partitions & Storage Shelving
- The required acoustical rating of a folding partition can have a significant impact on costs. Verify the sound transmission coefficient rating of the panel priced to the specification requirements.

Reference Numbers
Reference numbers are shown in bold squares at the beginning of some major classifications. These numbers refer to related items in the Reference Section. The reference information may be an estimating procedure, an alternate pricing method or technical information.

Note: Not all subdivisions listed here necessarily appear in this publication.

101 100 | Chalkboards

		CREW	DAILY OUTPUT	LABOR-HOURS	UNIT	1998 BARE COSTS				TOTAL INCL O&P
						MAT.	LABOR	EQUIP.	TOTAL	
104	0010	**CHALKBOARDS** Porcelain enamel steel									104
	0100	Freestanding, reversible									
	0120	Economy, wood frame, 4′ x 6′									
	0140	Chalkboard both sides				Ea.	460			460	510
	0160	Chalkboard one side, cork other side				″	395			395	435
	0200	Standard, lightweight satin finished aluminum, 4′ x 6′									
	0220	Chalkboard both sides				Ea.	475			475	525
	0240	Chalkboard one side, cork other side				″	420			420	460
	0300	Deluxe, heavy duty extruded aluminum, 4′ x 6′									
	0320	Chalkboard both sides				Ea.	1,025			1,025	1,125
	0340	Chalkboard one side, cork other side				″	1,200			1,200	1,325
	3900	Wall hung									
	4000	Aluminum frame and chalktrough									
	4300	3′ x 5′	2 Carp	15	1.067	Ea.	183	19.75		202.75	235
	4600	4′ x 12′	″	13	1.231	″	405	23		428	485
	4700	Wood frame and chalktrough									
	4800	3′ x 4′	2 Carp	16	1	Ea.	124	18.50		142.50	168
	5300	4′ x 8′	″	13	1.231	″	259	23		282	325
	5400	Liquid chalk, white porcelain enamel, wall hung									
	5420	Deluxe units, aluminum trim and chalktrough									
	5450	4′ x 4′	2 Carp	16	1	Ea.	171	18.50		189.50	221
	5550	4′ x 12′	″	12	1.333	″	415	24.50		439.50	500
	5700	Wood trim and chalktrough									
	5900	4′ x 4′	2 Carp	16	1	Ea.	360	18.50		378.50	430
	6200	4′ x 8′	″	14	1.143	″	515	21		536	600

101 600 | Toilet Compartments

		CREW	DAILY OUTPUT	LABOR-HOURS	UNIT	1998 BARE COSTS				TOTAL INCL O&P
						MAT.	LABOR	EQUIP.	TOTAL	
602	0010	**PARTITIONS, TOILET**									602
	0100	Cubicles, ceiling hung, marble	2 Marb	2	8	Ea.	1,250	150		1,400	1,625
	0200	Painted metal	2 Carp	4	4		385	74		459	545
	0250	Phenolic		4	4		820	74		894	1,025
	0300	Plastic laminate on particle board		4	4		515	74		589	690
	0500	Stainless steel		4	4		940	74		1,014	1,150
	0600	For handicap units, incl. 52″ grab bars, add					244			244	268
	0800	Floor & ceiling anchored, marble	2 Marb	2.50	6.400		1,375	120		1,495	1,700
	1000	Painted metal	2 Carp	5	3.200		390	59		449	525
	1050	Phenolic		5	3.200		820	59		879	1,000
	1100	Plastic laminate on particle board		5	3.200		515	59		574	665
	1300	Stainless steel		5	3.200		1,075	59		1,134	1,300
	1400	For handicap units, incl. 52″ grab bars, add					244			244	268
	1600	Floor mounted, marble	2 Marb	3	5.333		810	100		910	1,050
	1700	Painted metal	2 Carp	7	2.286		350	42.50		392.50	460
	1750	Phenolic		7	2.286		880	42.50		922.50	1,050
	1800	Plastic laminate on particle board		7	2.286		515	42.50		557.50	640
	2000	Stainless steel		7	2.286		1,050	42.50		1,092.50	1,225
	2100	For handicap units, incl. 52″ grab bars, add					244			244	268
	2200	For juvenile units, deduct					35			35	38.50
	2400	Floor mounted, headrail braced, marble	2 Marb	3	5.333		940	100		1,040	1,200
	2500	Painted metal	2 Carp	6	2.667		360	49.50		409.50	480
	2550	Phenolic		6	2.667		880	49.50		929.50	1,050
	2600	Plastic laminate on particle board		6	2.667		500	49.50		549.50	635
	2800	Stainless steel		6	2.667		1,075	49.50		1,124.50	1,250
	2900	For handicap units, incl. 52″ grab bars, add					244			244	268
	3000	Wall hung partitions, painted metal	2 Carp	7	2.286		440	42.50		482.50	560
	3300	Stainless steel	″	7	2.286		1,050	42.50		1,092.50	1,225

Important: See the Reference Section for critical supporting data - Reference Nos., Crews, & Location Factors

101 600 | Toilet Compartments

		CREW	DAILY OUTPUT	LABOR-HOURS	UNIT	MAT.	LABOR	EQUIP.	TOTAL	TOTAL INCL O&P	
602	3400	For handicap units, incl. 52" grab bars, add				Ea.	244			244	268	602
	4000	Screens, entrance, floor mounted, 58" high, 48" wide										
	4100	Marble	2 Marb	9	1.778	Ea.	515	33.50		548.50	620	
	4200	Painted metal	2 Carp	15	1.067		160	19.75		179.75	211	
	4300	Plastic laminate on particle board		15	1.067		325	19.75		344.75	390	
	4500	Stainless steel		15	1.067		580	19.75		599.75	675	
	4600	Urinal screen, 18" wide, ceiling braced, marble	D-1	6	2.667		515	45		560	640	
	4700	Painted metal	2 Carp	8	2		172	37		209	253	
	4800	Plastic laminate on particle board		8	2		258	37		295	350	
	5000	Stainless steel		8	2		410	37		447	515	
	5100	Floor mounted, head rail braced										
	5200	Marble	D-1	6	2.667	Ea.	480	45		525	605	
	5300	Painted metal	2 Carp	8	2		167	37		204	247	
	5400	Plastic laminate on particle board		8	2		233	37		270	320	
	5600	Stainless steel		8	2		485	37		522	595	
	5700	Pilaster, flush, marble	D-1	9	1.778		595	30		625	705	
	5800	Painted metal	2 Carp	10	1.600		216	29.50		245.50	289	
	5900	Plastic laminate on particle board		10	1.600		297	29.50		326.50	375	
	6100	Stainless steel		10	1.600		360	29.50		389.50	445	
	6200	Post braced, marble	D-1	9	1.778		580	30		610	685	
	6300	Painted metal	2 Carp	10	1.600		216	29.50		245.50	289	
	6400	Plastic laminate on particle board		10	1.600		305	29.50		334.50	385	
	6600	Stainless steel		10	1.600		355	29.50		384.50	445	
	6700	Wall hung, bracket supported										
	6800	Painted metal	2 Carp	10	1.600	Ea.	222	29.50		251.50	295	
	6900	Plastic laminate on particle board		10	1.600		110	29.50		139.50	172	
	7100	Stainless steel		10	1.600		325	29.50		354.50	410	
	7400	Flange supported, painted metal		10	1.600		155	29.50		184.50	222	
	7500	Plastic laminate on particle board		10	1.600		177	29.50		206.50	245	
	7700	Stainless steel		10	1.600		380	29.50		409.50	470	
	7800	Wedge type, painted metal		10	1.600		188	29.50		217.50	258	
	8100	Stainless steel		10	1.600		390	29.50		419.50	475	

101 850 | Shower Compartments

		CREW	DAILY OUTPUT	LABOR-HOURS	UNIT	MAT.	LABOR	EQUIP.	TOTAL	TOTAL INCL O&P	
852	0010	**PARTITIONS, SHOWER** Floor mounted, no plumbing										852
	0100	Cabinet, incl. base, no door, painted steel, 1" thick walls	2 Shee	5	3.200	Ea.	565	64.50		629.50	735	
	0300	With door, fiberglass		4.50	3.556		465	71.50		536.50	630	
	0600	Galvanized and painted steel, 1" thick walls		5	3.200		605	64.50		669.50	775	
	0800	Stall, 1" thick wall, no base, enameled steel		5	3.200		580	64.50		644.50	745	
	1500	Circular fiberglass, cabinet 36" diameter,		4	4		475	80.50		555.50	660	
	1700	One piece, 36" diameter, less door		4	4		410	80.50		490.50	585	
	1800	With door		3.50	4.571		660	92		752	880	
	2400	Glass stalls, with doors, no receptors, chrome on brass		3	5.333		985	107		1,092	1,250	
	2700	Anodized aluminum		4	4		680	80.50		760.50	885	
	3200	Receptors, precast terrazzo, 32" x 32"	2 Marb	14	1.143		289	21.50		310.50	355	
	3300	48" x 34"		9.50	1.684		365	31.50		396.50	455	
	3500	Plastic, simulated terrazzo receptor, 32" x 32"		14	1.143		84	21.50		105.50	129	
	3600	32" x 48"		12	1.333		116	25		141	170	
	3800	Precast concrete, colors, 32" x 32"		14	1.143		144	21.50		165.50	195	
	3900	48" x 48"		8	2		177	37.50		214.50	258	
	4100	Shower doors, economy plastic, 24" wide	1 Shee	9	.889		85.50	17.90		103.40	125	
	4200	Tempered glass door, economy		8	1		165	20		185	215	
	4400	Folding, tempered glass, aluminum frame		6	1.333		201	27		228	267	
	4700	Deluxe, tempered glass, chrome on brass frame, minimum		8	1		210	20		230	265	
	4800	Maximum		1	8		605	161		766	940	
	4850	On anodized aluminum frame, minimum		2	4		97.50	80.50		178	243	

101 850 | Shower Compartments

			CREW	DAILY OUTPUT	LABOR-HOURS	UNIT	1998 BARE COSTS MAT.	LABOR	EQUIP.	TOTAL	TOTAL INCL O&P	
852	4900	Maximum	1 Shee	1	8	Ea.	330	161		491	635	852
	5100	Shower enclosure, tempered glass, anodized alum. frame										
	5120	2 panel & door, corner unit, 32" x 32"	1 Shee	2	4	Ea.	330	80.50		410.50	495	
	5140	Neo-angle corner unit, 16" x 24" x 16"	"	2	4		580	80.50		660.50	770	
	5200	Shower surround, 3 wall, polypropylene, 32" x 32"	1 Carp	4	2		183	37		220	265	
	5220	PVC, 32" x 32"		4	2		213	37		250	299	
	5240	Fiberglass		4	2		256	37		293	345	
	5250	2 wall, polypropylene, 32" x 32"		4	2		177	37		214	258	
	5270	PVC		4	2		213	37		250	299	
	5290	Fiberglass		4	2		238	37		275	325	
	5300	Tub doors, tempered glass & frame, minimum	1 Shee	8	1		140	20		160	188	
	5400	Maximum		6	1.333		325	27		352	400	
	5600	Chrome plated, brass frame, minimum		8	1		183	20		203	235	
	5700	Maximum		6	1.333		365	27		392	445	
	5900	Tub/shower enclosure, temp. glass, alum. frame, minimum		2	4		250	80.50		330.50	410	
	6200	Maximum		1.50	5.333		485	107		592	715	
	6500	On chrome-plated brass frame, minimum		2	4		345	80.50		425.50	515	
	6600	Maximum		1.50	5.333		715	107		822	965	
	6800	Tub surround, 3 wall, polypropylene	1 Carp	4	2		140	37		177	218	
	6900	PVC		4	2		213	37		250	299	
	7000	Fiberglass, minimum		4	2		238	37		275	325	
	7100	Maximum		3	2.667		410	49.50		459.50	535	

10 SPECIALTIES

102 100 | Metal Wall Louvers

			CREW	DAILY OUTPUT	LABOR-HOURS	UNIT	1998 BARE COSTS MAT.	LABOR	EQUIP.	TOTAL	TOTAL INCL O&P	
104	0010	LOUVERS Aluminum with screen, residential, 8" x 8"	1 Carp	38	.211	Ea.	6.60	3.89		10.49	13.90	104
	0100	12" x 12"		38	.211		8.30	3.89		12.19	15.80	
	0200	12" x 18"		35	.229		11.45	4.23		15.68	19.85	
	0250	14" x 24"		30	.267		15.55	4.93		20.48	25.50	
	0300	18" x 24"		27	.296		17.10	5.50		22.60	28	
	0500	24" x 30"	"	24	.333		30	6.15		36.15	43.50	
	0700	Triangle, adjustable, small		20	.400		25	7.40		32.40	40	
	0800	Large		15	.533		67.50	9.85		77.35	91	
	2100	Midget, aluminum, 3/4" deep, 1" diameter		85	.094		.54	1.74		2.28	3.57	
	2150	3" diameter		60	.133		.95	2.47		3.42	5.30	
	2200	4" diameter		50	.160		1.36	2.96		4.32	6.55	
	2250	6" diameter		30	.267		2.15	4.93		7.08	10.80	
	2300	Ridge vent strip, mill finish	1 Shee	155	.052	L.F.	1.88	1.04		2.92	3.83	
	2400	Under eaves vent, aluminum, mill finish, 16" x 4"	1 Carp	48	.167	Ea.	1.64	3.08		4.72	7.10	
	2500	16" x 8"		48	.167		1.86	3.08		4.94	7.35	
	7000	Vinyl gable vent, 8" x 8"		38	.211		9.25	3.89		13.14	16.85	
	7020	12" x 12"		38	.211		11.45	3.89		15.34	19.25	
	7080	12" x 18"		35	.229		14.95	4.23		19.18	23.50	
	7200	14" x 24"		30	.267		19.55	4.93		24.48	30	

102 600 | Wall & Corner Guards

			CREW	DAILY OUTPUT	LABOR-HOURS	UNIT	1998 BARE COSTS MAT.	LABOR	EQUIP.	TOTAL	TOTAL INCL O&P	
612	0010	WALLGUARD See division 055										612
	0400	Rub rail, vinyl, adhesive mounted	1 Carp	185	.043	L.F.	4.02	.80		4.82	5.80	

Important: See the Reference Section for critical supporting data - Reference Nos., Crews, & Location Factors

102 600 | Wall & Corner Guards

			CREW	DAILY OUTPUT	LABOR-HOURS	UNIT	1998 BARE COSTS				TOTAL INCL O&P	
							MAT.	LABOR	EQUIP.	TOTAL		
612	0500	Neoprene, aluminum backing, 1-1/2" x 2"	1 Carp	110	.073	L.F.	5.90	1.35		7.25	8.80	**612**
	1000	Trolley rail, PVC, clipped to wall, 5" high		185	.043		4.65	.80		5.45	6.45	
	1050	8" high		180	.044		5.65	.82		6.47	7.60	
	1200	Bed bumper, vinyl acrylic, alum. retainer, 21" long		10	.800	Ea.	32.50	14.80		47.30	61	
	1300	53" long with aligner		9	.889	"	96	16.45		112.45	134	
	1400	Bumper, vinyl cover, alum. retain., cush. mnt., 1-1/2" x 2-3/4"		80	.100	L.F.	9.70	1.85		11.55	13.80	
	1500	2" x 4-1/4"		80	.100		12.95	1.85		14.80	17.40	
	1600	Surface mounted, 1-3/4" x 3-5/8"		80	.100		8.75	1.85		10.60	12.80	
	2000	Crash rail, vinyl cover, alum. retainer, 1" x 4"		110	.073		9.45	1.35		10.80	12.70	
	2100	1" x 8"		90	.089		14.50	1.64		16.14	18.75	
	2150	Vinyl inserts, aluminum plate, 1" x 2-1/2"		110	.073		9.70	1.35		11.05	12.95	
	2200	1" x 5"		90	.089		13.95	1.64		15.59	18.15	
	3000	Handrail/bumper, vinyl cover, alum. retainer										
	3010	bracket mounted, flat rail, 5-1/2"	1 Carp	80	.100	L.F.	15.20	1.85		17.05	19.85	
	3100	6-1/2"		80	.100		16.55	1.85		18.40	21.50	
	3200	Bronze bracket, 1-3/4" diam. rail		80	.100		14.10	1.85		15.95	18.65	

103 050 | Prefabricated Fireplaces

			CREW	DAILY OUTPUT	LABOR-HOURS	UNIT	1998 BARE COSTS				TOTAL INCL O&P	
							MAT.	LABOR	EQUIP.	TOTAL		
054	0010	**FIREPLACE, PREFABRICATED** Free standing or wall hung										**054**
	0100	with hood & screen, minimum	1 Carp	1.30	6.154	Ea.	905	114		1,019	1,200	
	0150	Average		1	8		1,075	148		1,223	1,450	
	0200	Maximum		.90	8.889		2,600	164		2,764	3,150	
	0500	Chimney dbl. wall, all stainless, over 8'-6", 7" diam., add		33	.242	V.L.F.	30.50	4.48		34.98	41	
	0600	10" diameter, add		32	.250		43.50	4.63		48.13	56	
	0700	12" diameter, add		31	.258		57	4.77		61.77	70.50	
	0800	14" diameter, add		30	.267		72	4.93		76.93	87.50	
	1000	Simulated brick chimney top, 4' high, 16" x 16"		10	.800	Ea.	155	14.80		169.80	197	
	1100	24" x 24"		7	1.143	"	289	21		310	355	
	1500	Simulated logs, gas fired, 40,000 BTU, 2' long, minimum		7	1.143	Set	405	21		426	480	
	1600	Maximum		6	1.333		590	24.50		614.50	695	
	1700	Electric, 1,500 BTU, 1'-6" long, minimum		7	1.143		107	21		128	154	
	1800	11,500 BTU, maximum		6	1.333		232	24.50		256.50	298	
	2000	Fireplace, built-in, 36" hearth, radiant		1.30	6.154	Ea.	465	114		579	705	
	2100	Recirculating, small fan		1	8		665	148		813	985	
	2150	Large fan		.90	8.889		1,225	164		1,389	1,625	
	2200	42" hearth, radiant		1.20	6.667		595	123		718	860	
	2300	Recirculating, small fan		.90	8.889		775	164		939	1,125	
	2350	Large fan		.80	10		1,475	185		1,660	1,950	
	2400	48" hearth, radiant		1.10	7.273		1,100	135		1,235	1,450	
	2500	Recirculating, small fan		.80	10		1,375	185		1,560	1,825	
	2550	Large fan		.70	11.429		2,125	211		2,336	2,675	
	3000	See through, including doors		.80	10		1,750	185		1,935	2,250	
	3200	Corner (2 wall)		1	8		865	148		1,013	1,200	

103 460 | Cupolas

			CREW	DAILY OUTPUT	LABOR-HOURS	UNIT	1998 BARE COSTS				TOTAL INCL O&P	
							MAT.	LABOR	EQUIP.	TOTAL		
464	0010	**CUPOLA** Stock units, pine, painted, 18" sq., 28" high, alum. roof	1 Carp	4.10	1.951	Ea.	131	36		167	206	**464**
	0100	Copper roof		3.80	2.105		134	39		173	214	

103 460 | Cupolas

		CREW	DAILY OUTPUT	LABOR-HOURS	UNIT	1998 BARE COSTS				TOTAL INCL O&P	
						MAT.	LABOR	EQUIP.	TOTAL		
464	0300	23" square, 33" high, aluminum roof	1 Carp	3.70	2.162	Ea.	220	40		260	310	**464**
	0400	Copper roof		3.30	2.424		222	45		267	320	
	0600	30" square, 37" high, aluminum roof		3.70	2.162		340	40		380	445	
	0700	Copper roof		3.30	2.424		345	45		390	455	
	0900	Hexagonal, 31" wide, 46" high, copper roof		4	2		480	37		517	595	
	1000	36" wide, 50" high, copper roof		3.50	2.286		590	42.50		632.50	720	
	1200	For deluxe stock units, add to above					25%					
	1400	For custom built units, add to above					50%	50%				
	1600	Fiberglass, 5'-0" base, 63" high minimum	F-3	6	6.667		2,225	113	73	2,411	2,725	
	1650	Maximum		4	10		2,575	169	110	2,854	3,250	
	1700	6'-0" base, 63" high, minimum		5	8		3,450	135	87.50	3,672.50	4,125	
	1750	Maximum		3	13.333		3,550	225	146	3,921	4,450	

103 520 | Ground Set Flagpoles

		CREW	DAILY OUTPUT	LABOR-HOURS	UNIT	1998 BARE COSTS				TOTAL INCL O&P	
						MAT.	LABOR	EQUIP.	TOTAL		
524	0010	FLAGPOLE										**524**
	0050	Not including base or foundation										
	0100	Aluminum, tapered, ground set 20' high	K-1	2	8	Ea.	645	134	87.50	866.50	1,025	
	0200	25' high		1.70	9.412		665	158	103	926	1,100	
	0300	30' high		1.50	10.667		835	179	117	1,131	1,350	
	0500	40' high		1.20	13.333		1,675	224	146	2,045	2,375	

104 100 | Directories

		CREW	DAILY OUTPUT	LABOR-HOURS	UNIT	1998 BARE COSTS				TOTAL INCL O&P	
						MAT.	LABOR	EQUIP.	TOTAL		
104	0010	DIRECTORY BOARDS										**104**
	0050	Plastic, glass covered, 30" x 20"	2 Carp	3	5.333	Ea.	263	98.50		361.50	460	
	0100	36" x 48"		2	8		540	148		688	850	
	0900	Outdoor, weatherproof, black plastic, 36" x 24"		2	8		550	148		698	860	
	1000	36" x 36"		1.50	10.667		635	197		832	1,025	

104 150 | Bulletin Boards

		CREW	DAILY OUTPUT	LABOR-HOURS	UNIT	1998 BARE COSTS				TOTAL INCL O&P	
						MAT.	LABOR	EQUIP.	TOTAL		
151	0010	BULLETIN BOARD Cork sheets, unbacked, no frame, 1/4" thick	2 Carp	290	.055	S.F.	2.44	1.02		3.46	4.43	**151**
	2120	Prefabricated, 1/4" cork, 3' x 5' with aluminum frame		16	1	Ea.	95	18.50		113.50	136	
	2140	Wood frame		16	1	"	124	18.50		142.50	168	
	2300	Glass enclosed cabinets, alum., cork panel, hinged doors										
	2600	4' x 7', 3 door	2 Carp	10	1.600	Ea.	1,175	29.50		1,204.50	1,350	

104 300 | Signs

		CREW	DAILY OUTPUT	LABOR-HOURS	UNIT	1998 BARE COSTS				TOTAL INCL O&P	
						MAT.	LABOR	EQUIP.	TOTAL		
304	0012	SIGNS Plaques										**304**
	3900	Plaques, custom, 20" x 30", for up to 450 letters, cast aluminum	2 Carp	4	4	Ea.	810	74		884	1,025	
	4000	Cast bronze		4	4		1,075	74		1,149	1,300	
	4200	30" x 36", up to 900 letters cast aluminum		3	5.333		1,425	98.50		1,523.50	1,750	
	4300	Cast bronze		3	5.333		1,850	98.50		1,948.50	2,200	
	5100	Exit signs, 24 ga. alum., 14" x 12" surface mounted	1 Carp	30	.267		9.30	4.93		14.23	18.70	
	5200	10" x 7"	"	20	.400		7.10	7.40		14.50	20.50	
	6400	Replacement sign faces, 6" or 8"	1 Clab	50	.160		21	2.14		23.14	26.50	

Important: See the Reference Section for critical supporting data - Reference Nos., Crews, & Location Factors

105 050 | Metal Lockers

			DAILY OUTPUT	LABOR-HOURS	UNIT	1998 BARE COSTS				TOTAL INCL O&P	
						MAT.	LABOR	EQUIP.	TOTAL		
054	0011	**LOCKERS** Steel, baked enamel	CREW								**054**
	0110	Single tier box locker, 12" x 15" x 72"	1 Shee	8	1	Ea.	106	20		126	151
	0120	18" x 15" x 72"		8	1		126	20		146	172
	0130	12" x 18" x 72"		8	1		112	20		132	157
	0140	18" x 18" x 72"		8	1		135	20		155	183
	0410	Double tier, 12" x 15" x 36"		21	.381		94	7.70		101.70	116
	0420	18" x 15" x 36"		21	.381		107	7.70		114.70	131
	0430	12" x 18" x 36"		21	.381		99	7.70		106.70	122
	0440	18" x 18" x 36"		21	.381		110	7.70		117.70	134
	0500	Two person, 18" x 15" x 72"		8	1		189	20		209	242
	0510	18" x 18" x 72"		8	1		216	20		236	271
	0520	Duplex, 15" x 15" x 72"		8	1		183	20		203	235
	0530	15" x 21" x 72"		8	1		208	20		228	262
	1100	Wire meshed wardrobe, floor. mtd., open front varsity type		7.50	1.067		144	21.50		165.50	196
	3250	Rack w/ 24 wire mesh baskets		1.50	5.333	Set	257	107		364	465
	3260	30 baskets		1.25	6.400		221	129		350	460
	3270	36 baskets		.95	8.421		310	170		480	625
	3280	42 baskets		.80	10		350	202		552	725
	3600	For hanger rods, add				Ea.	1.50			1.50	1.65

105 200 | Fire Protect. Specialties

			DAILY OUTPUT	LABOR-HOURS	UNIT	MAT.	LABOR	EQUIP.	TOTAL	TOTAL INCL O&P	
220	0010	**FIRE EQUIPMENT CABINETS** Not equipped, 20 ga. steel box,									**220**
	0040	recessed, D.S. glass in door, box size given									
	1000	Portable extinguisher, single, 8" x 12" x 27", alum. door & frame	Q-12	8	2	Ea.	108	38		146	181
	1100	Steel door and frame	"	8	2	"	66	38		104	136
	3000	Hose rack assy., 1-1/2" valve & 100' hose, 24" x 40" x 5-1/2"									
	3100	Aluminum door and frame	Q-12	6	2.667	Ea.	252	50.50		302.50	360
	3200	Steel door and frame	"	6	2.667	"	144	50.50		194.50	242

			DAILY OUTPUT	LABOR-HOURS	UNIT	MAT.	LABOR	EQUIP.	TOTAL	TOTAL INCL O&P	
225	0010	**FIRE EXTINGUISHERS**									**225**
	0120	CO2, portable with swivel horn, 5 lb.				Ea.	101			101	111
	0140	With hose and "H" horn, 10 lb.				"	150			150	165
	1000	Dry chemical, pressurized									
	1040	Standard type, portable, painted, 2-1/2 lb.				Ea.	25			25	27.50
	1080	10 lb.					65			65	71.50
	1100	20 lb.					90			90	99
	1120	30 lb.					145			145	160
	2000	ABC all purpose type, portable, 2-1/2 lb.					25			25	27.50
	2080	9-1/2 lb.					60			60	66

105 380 | Canopies

			DAILY OUTPUT	LABOR-HOURS	UNIT	MAT.	LABOR	EQUIP.	TOTAL	TOTAL INCL O&P	
384	0010	**CANOPIES** Wall hung, .032", aluminum, prefinished, 8' x 10'	K-2	1.30	18.462	Ea.	1,150	350	134	1,634	2,075	**384**
	0300	8' x 20'	"	1.10	21.818	"	2,175	415	159	2,749	3,350
	2300	Aluminum entrance canopies, flat soffit, .032"									
	2500	3'-6" x 4'-0", clear anodized	2 Carp	4	4	Ea.	510	74		584	685
	4700	Canvas awnings, including canvas, frame & lettering									
	5000	Minimum	2 Carp	100	.160	S.F.	35.50	2.96		38.46	44
	5300	Average		90	.178		42.50	3.29		45.79	52.50
	5500	Maximum		80	.200		86.50	3.70		90.20	101

105 510 | Mail Chutes

			DAILY OUTPUT	LABOR-HOURS	UNIT	MAT.	LABOR	EQUIP.	TOTAL	TOTAL INCL O&P	
511	0010	**MAIL CHUTES** Aluminum & glass, 14-1/4" wide, 4-5/8" deep	2 Shee	4	4	Floor	610	80.50		690.50	810	**511**
	0100	8-5/8" deep		3.80	4.211	"	680	85		765	890
	0600	Lobby collection boxes, aluminum		5	3.200	Ea.	1,675	64.50		1,739.50	1,950
	0700	Bronze or stainless		4.50	3.556	"	2,075	71.50		2,146.50	2,400

105 | Lockers, Protective Covers & Postal Specialties

105 520 | Mail Boxes

		CREW	DAILY OUTPUT	LABOR-HOURS	UNIT	1998 BARE COSTS				TOTAL INCL O&P
						MAT.	LABOR	EQUIP.	TOTAL	
521	**0010** **MAIL BOXES** Horiz., key lock, 5"H x 6"W x 15"D, alum., rear load	1 Carp	34	.235	Ea.	34	4.35		38.35	44.50	**521**
0100	Front loading		34	.235		39	4.35		43.35	50.50
0200	Double, 5"H x 12"W x 15"D, rear loading		26	.308		61	5.70		66.70	77
0300	Front loading		26	.308		69.50	5.70		75.20	86.50
0500	Quadruple, 10"H x 12"W x 15"D, rear loading		20	.400		116	7.40		123.40	140
0600	Front loading		20	.400		128	7.40		135.40	154
1600	Vault type, horizontal, for apartments, 4" x 5"		34	.235		35.50	4.35		39.85	46.50
1700	Alphabetical directories, 120 names		10	.800		140	14.80		154.80	179

106 | Partitions & Storage Shelving

106 150 | Demountable Partitions

		CREW	DAILY OUTPUT	LABOR-HOURS	UNIT	1998 BARE COSTS				TOTAL INCL O&P
						MAT.	LABOR	EQUIP.	TOTAL	
152	**0010** **PARTITIONS, MOVABLE OFFICE** Demountable, add for doors										**152**
0100	Do not deduct door openings from total L.F.									
0900	Demountable gypsum system on 2" to 2-1/2"									
1000	steel studs, 9' high, 3" to 3-3/4" thick									
1200	Vinyl clad gypsum	2 Carp	48	.333	L.F.	22.50	6.15		28.65	35
1300	Fabric clad gypsum		44	.364		69	6.75		75.75	87.50
1500	Steel clad gypsum		40	.400		73.50	7.40		80.90	93.50
1600	1.75 system, aluminum framing, vinyl clad hardboard,									
1800	paper honeycomb core panel, 1-3/4" to 2-1/2" thick									
1900	9' high	2 Carp	48	.333	L.F.	47	6.15		53.15	62
2100	7' high		60	.267		43	4.93		47.93	55.50
2200	5' high		80	.200		37	3.70		40.70	47
2250	Unitized gypsum system									
2300	Unitized panel, 9' high, 2" to 2-1/2" thick									
2350	Vinyl clad gypsum	2 Carp	48	.333	L.F.	63	6.15		69.15	80
2400	Fabric clad gypsum	"	44	.364	"	114	6.75		120.75	137
2500	Unitized mineral fiber system									
2510	Unitized panel, 9' high, 2-1/4" thick, aluminum frame									
2550	Vinyl clad mineral fiber	2 Carp	48	.333	L.F.	80.50	6.15		86.65	99
2600	Fabric clad mineral fiber	"	44	.364	"	102	6.75		108.75	125
2800	Movable steel walls, modular system									
2900	Unitized panels, 9' high, 48" wide									
3100	Baked enamel, pre-finished	2 Carp	60	.267	L.F.	73.50	4.93		78.43	89.50
3200	Fabric clad steel	"	56	.286	"	112	5.30		117.30	133
5500	For acoustical partitions, add, minimum				S.F.	1.30			1.30	1.43
5550	Maximum				"	4.75			4.75	5.20
5700	For doors, see div. 081 & 082									
5800	For door hardware, see div. 087									
6100	In-plant modular office system, w/prehung steel door									
6200	3" thick honeycomb core panels									
6250	12' x 12', 2 wall	2 Clab	3.80	4.211	Ea.	3,125	56.50		3,181.50	3,525
6300	4 wall		1.90	8.421		4,800	113		4,913	5,500
6350	16' x 16', 2 wall		3.60	4.444		4,650	59.50		4,709.50	5,225
6400	4 wall		1.80	8.889		6,700	119		6,819	7,550

106 300 | Portable Partitions

		CREW	DAILY OUTPUT	LABOR-HOURS	UNIT	1998 BARE COSTS				TOTAL INCL O&P
						MAT.	LABOR	EQUIP.	TOTAL	
304	**0010** **PARTITIONS, PORTABLE** Divider panels, free standing, fiber core										**304**
0020	Fabric face straight									

Important: See the Reference Section for critical supporting data - Reference Nos., Crews, & Location Factors

106 300 | Portable Partitions

			DAILY OUTPUT	LABOR-HOURS	UNIT	MAT.	LABOR	EQUIP.	TOTAL	TOTAL INCL O&P	
304	1500	6"-0" high	2 Carp	125	.128	L.F.	86	2.37		88.37	98.50	304
	1600	6'-0" long, 5'-0" high		162	.099		70	1.83		71.83	80	
	3100	Curved, 3'-0" long, 5'-0" high		90	.178		131	3.29		134.29	150	
	3150	6'-0" high		75	.213		142	3.95		145.95	164	
	3200	Economical panels, fabric face, 4'-0" long, 5'-0" high		132	.121		77	2.24		79.24	88.50	
	3250	6'-0" high		112	.143		88	2.64		90.64	102	
	3300	5'-0" long, 5'-0" high		150	.107		70	1.97		71.97	80.50	
	3350	6'-0" high		125	.128		78	2.37		80.37	90	
	3380	3'-0" curved, 5'-0" high		90	.178		202	3.29		205.29	228	
	3390	6'-0" high		75	.213		226	3.95		229.95	256	
	3450	Acoustical panels, 60 to 90 NRC, 3'-0" long, 5'-0" high		90	.178		117	3.29		120.29	134	
	3550	6'-0" high		75	.213		143	3.95		146.95	164	
	3600	5'-0" long, 5'-0" high		150	.107		93	1.97		94.97	105	
	3650	6'-0" high		125	.128		103	2.37		105.37	118	
	3700	6'-0" long, 5'-0" high		162	.099		85	1.83		86.83	96.50	
	3750	6'-0" high		138	.116		104	2.14		106.14	118	
	3800	Economy acoustical panels, 40 NRC, 4'-0" long, 5'-0" high		132	.121		89.50	2.24		91.74	102	
	3850	6'-0" high		112	.143		104	2.64		106.64	119	
	3900	5'-0" long, 6'-0" high		125	.128		95.50	2.37		97.87	109	
	3950	6'-0" long, 5'-0" high		162	.099		79	1.83		80.83	90	

106 520 | Panel Partitions

			DAILY OUTPUT	LABOR-HOURS	UNIT	MAT.	LABOR	EQUIP.	TOTAL	TOTAL INCL O&P	
522	0010	**PARTITIONS, FOLDING LEAF** Acoustic, wood									522
	0100	Vinyl faced, to 18' high, 6 psf, minimum	2 Carp	60	.267	S.F.	34.50	4.93		39.43	46.50	
	0150	Average		45	.356		41	6.60		47.60	56.50	
	0200	Maximum		30	.533		53	9.85		62.85	75.50	
	0400	Formica or hardwood finish, minimum		60	.267		35.50	4.93		40.43	47.50	
	0500	Maximum		30	.533		38	9.85		47.85	58.50	
	0600	Wood, low acoustical type, 4.5 psf, to 14' high		50	.320		25.50	5.90		31.40	38.50	

106 550 | Accordion Partitions

			DAILY OUTPUT	LABOR-HOURS	UNIT	MAT.	LABOR	EQUIP.	TOTAL	TOTAL INCL O&P	
552	0010	**PARTITIONS, FOLDING ACCORDION**									552
	0100	Vinyl covered, over 150 S.F., frame not included										
	0300	Residential, 1.25 lb. per S.F., 8' maximum height	2 Carp	300	.053	S.F.	11.60	.99		12.59	14.45	
	0400	Commercial, 1.75 lb. per S.F., 8' maximum height		225	.071		11.60	1.32		12.92	15	
	0900	Acoustical, 3 lb. per S.F., 17' maximum height		100	.160		17.50	2.96		20.46	24.50	
	1200	5 lb. per S.F., 20' maximum height		95	.168		26	3.12		29.12	34	
	1500	Vinyl clad wood or steel, electric operation, 5.0 psf		160	.100		29	1.85		30.85	35	
	1900	Wood, non-acoustic, birch or mahogany, to 10' high		300	.053		15.55	.99		16.54	18.80	

106 750 | Storage & Shelving

			DAILY OUTPUT	LABOR-HOURS	UNIT	MAT.	LABOR	EQUIP.	TOTAL	TOTAL INCL O&P	
754	0010	**SHELVING** Metal, industrial, cross-braced, 3' wide, 12" deep	1 Sswk	175	.046	SF Shlf	6.05	.91		6.96	8.45	754
	0100	24" deep		330	.024		4.30	.48		4.78	5.70	
	2200	Wide span, 1600 lb. capacity per shelf, 6' wide, 24" deep		380	.021		9.75	.42		10.17	11.60	
	2400	36" deep		440	.018		8.10	.36		8.46	9.60	
	3000	Residential, adjustable metal, 12" deep										
	3100	23" to 37" wide	1 Carp	16	.500	Ea.	7.35	9.25		16.60	24	
	3200	47" to 61" wide		16	.500		10.40	9.25		19.65	27.50	
	3300	71" to 88" wide		16	.500		15.70	9.25		24.95	33	
	3400	95" to 109"		16	.500		15.70	9.25		24.95	33	
	3500	For closet rod, add					22%	2%				
	3600	For 14" shelves, add					12%	5%				

107 550	Telephone Enclosures	CREW	DAILY OUTPUT	LABOR-HOURS	UNIT	1998 BARE COSTS				TOTAL INCL O&P	
						MAT.	LABOR	EQUIP.	TOTAL		
551	0010	**TELEPHONE ENCLOSURE**										**551**
	0300	Shelf type, wall hung, minimum	2 Carp	5	3.200	Ea.	705	59		764	880
	0400	Maximum		5	3.200		2,425	59		2,484	2,775
	0600	Booth type, painted steel, indoor or outdoor, minimum		1.50	10.667		2,975	197		3,172	3,625
	0700	Maximum (stainless steel)		1.50	10.667		9,850	197		10,047	11,100
	1900	Outdoor, drive-up type, wall mounted		4	4		775	74		849	980
	2000	Post mounted, stainless steel posts		3	5.333		1,200	98.50		1,298.50	1,500

108 200	Bath Accessories	CREW	DAILY OUTPUT	LABOR-HOURS	UNIT	1998 BARE COSTS				TOTAL INCL O&P	
						MAT.	LABOR	EQUIP.	TOTAL		
204	0010	**BATH ACCESSORIES**										**204**
	0200	Curtain rod, stainless steel, 5' long, 1" diameter	1 Carp	13	.615	Ea.	27.50	11.40		38.90	49.50
	0300	1-1/4" diameter	"	13	.615	"	29.50	11.40		40.90	52
	0500	Dispenser units, combined soap & towel dispensers,									
	0510	mirror and shelf, flush mounted	1 Carp	10	.800	Ea.	262	14.80		276.80	315
	0600	Towel dispenser and waste receptacle,									
	0610	18 gallon capacity	1 Carp	10	.800	Ea.	274	14.80		288.80	325
	0800	Grab bar, straight, 1-1/4" diameter, stainless steel, 18" long		24	.333		35.50	6.15		41.65	49.50
	1100	36" long		20	.400		44.50	7.40		51.90	61.50
	3000	Mirror, with stainless steel 3/4" square frame, 18" x 24"		20	.400		63	7.40		70.40	81.50
	3100	36" x 24"		15	.533		163	9.85		172.85	196
	3300	72" x 24"		6	1.333		291	24.50		315.50	365
	3500	With 5" stainless steel shelf, 18" x 24"		20	.400		90.50	7.40		97.90	112
	3600	36" x 24"		15	.533		272	9.85		281.85	315
	3800	72" x 24"		6	1.333		620	24.50		644.50	725
	4200	Napkin/tampon dispenser, recessed		15	.533		360	9.85		369.85	410
	4300	Robe hook, single, regular		36	.222		11	4.11		15.11	19.15
	4400	Heavy duty, concealed mounting		36	.222		13.60	4.11		17.71	22
	4600	Soap dispenser, chrome, surface mounted, liquid		20	.400		38.50	7.40		45.90	55
	5000	Recessed stainless steel, liquid		10	.800		71.50	14.80		86.30	105
	5600	Shelf, stainless steel, 5" wide, 18 ga., 24" long		24	.333		35	6.15		41.15	49.50
	6100	Toilet tissue dispenser, surface mounted, SS, single roll		30	.267		9.60	4.93		14.53	19
	6200	Double roll		24	.333		14.65	6.15		20.80	26.50
	6400	Towel bar, stainless steel, 18" long		23	.348		27.50	6.45		33.95	41
	6500	30" long		21	.381		31.50	7.05		38.55	46.50
	6700	Towel dispenser, stainless steel, surface mounted		16	.500		41	9.25		50.25	61
	6800	Flush mounted, recessed		10	.800		133	14.80		147.80	172
	7400	Tumbler holder, tumbler only		30	.267		19.25	4.93		24.18	29.50
	7500	Soap, tumbler & toothbrush		30	.267		19.30	4.93		24.23	29.50
	7700	Wall urn ash receiver, surface mount, 11" long		12	.667		97	12.35		109.35	127
	8000	Waste receptacles, stainless steel, with top, 13 gallon		10	.800		180	14.80		194.80	223
208	0010	**MEDICINE CABINETS** With mirror, st. st. frame, 16" x 22", unlighted	1 Carp	14	.571	Ea.	62	10.55		72.55	86	**208**
	0100	Wood frame		14	.571		90	10.55		100.55	117
	0300	Sliding mirror doors, 20" x 16" x 4-3/4", unlighted		7	1.143		80.50	21		101.50	125
	0400	24" x 19" x 8-1/2", lighted		5	1.600		135	29.50		164.50	200
	0600	Triple door, 30" x 32", unlighted, plywood body		7	1.143		203	21		224	259
	0700	Steel body		7	1.143		265	21		286	330

Important: See the Reference Section for critical supporting data - Reference Nos., Crews, & Location Factors

108 200	Bath Accessories	CREW	DAILY OUTPUT	LABOR-HOURS	UNIT	1998 BARE COSTS				TOTAL INCL O&P	
						MAT.	LABOR	EQUIP.	TOTAL		
208 0900	Oak door, wood body, beveled mirror, single door	1 Carp	7	1.143	Ea.	116	21		137	163	**208**
1000	Double door		6	1.333		295	24.50		319.50	370	
1200	Hotel cabinets, stainless, with lower shelf, unlighted		10	.800		162	14.80		176.80	204	
1300	Lighted		5	1.600		242	29.50		271.50	315	

For information about Means Estimating Seminars, see yellow pages 11 and 12 in back of book

		CREW	DAILY OUTPUT	LABOR-HOURS	UNIT	1998 BARE COSTS				TOTAL INCL O&P
						MAT.	LABOR	EQUIP.	TOTAL	

Division 11
Equipment

Estimating Tips
General
- The items in this division are usually priced per square foot or each. Many of these items are purchased by the owner for installation by the contractor. Check the specifications for responsibilities, and include time for receiving, storage, installation and mechanical and electrical hook-ups in the appropriate divisions.

- Many items in Division 11 require some type of support system that is not usually furnished with the item. Examples of these systems include blocking for the attachment of casework and support angles for ceiling hung projection screens. The required blocking or supports must be added to the estimate in the appropriate division.
- Some items in Division 11 may require assembly or electrical hook-ups. Verify the amount of assembly required or the need for a hard electrical connection and add the appropriate costs.

Reference Numbers
Reference numbers are shown in bold squares at the beginning of some major classifications. These numbers refer to related items in the Reference Section. The reference information may be an estimating procedure, an alternate pricing method or technical information.

Note: Not all subdivisions listed here necessarily appear in this publication.

110 | Equipment

110 100 | Maintenance Equipment

			CREW	DAILY OUTPUT	LABOR-HOURS	UNIT	1998 BARE COSTS				TOTAL INCL O&P	
							MAT.	LABOR	EQUIP.	TOTAL		
121	0010	**VACUUM CLEANING**										121
	0020	Central, 3 inlet, residential	1 Skwk	.90	8.889	Total	955	167		1,122	1,325	
	0400	5 inlet system, residential		.50	16		1,025	300		1,325	1,650	
	0600	7 inlet system, commercial		.40	20		1,100	375		1,475	1,850	
	0800	9 inlet system		.30	26.667		1,200	500		1,700	2,175	
	4010	Rule of thumb: First 1200 S.F., installed									1,044	
	4020	For each additional S.F., add				S.F.					.17	

111 | Mercantile, Commercial & Detention Equipment

111 320 | Projection Screens

			CREW	DAILY OUTPUT	LABOR-HOURS	UNIT	1998 BARE COSTS				TOTAL INCL O&P	
							MAT.	LABOR	EQUIP.	TOTAL		
321	0010	**PROJECTION SCREENS** Wall or ceiling hung, matte white										321
	0100	Manually operated, economy	2 Carp	500	.032	S.F.	5.30	.59		5.89	6.85	
	0300	Intermediate		450	.036		5.70	.66		6.36	7.45	
	0400	Deluxe		400	.040		8.70	.74		9.44	10.85	

111 600 | Loading Dock Equipment

			CREW	DAILY OUTPUT	LABOR-HOURS	UNIT	1998 BARE COSTS				TOTAL INCL O&P	
							MAT.	LABOR	EQUIP.	TOTAL		
603	0010	**DOCK BUMPERS** Bolts not included										603
	0020	2" x 6" to 4" x 8", average	1 Carp	.30	26.667	M.B.F.	865	495		1,360	1,800	

114 | Food Service, Residential, Darkroom, Athletic Equipment

114 000 | Food Service Equipment

			CREW	DAILY OUTPUT	LABOR-HOURS	UNIT	1998 BARE COSTS				TOTAL INCL O&P	
							MAT.	LABOR	EQUIP.	TOTAL		
002	0010	**RESIDENTIAL APPLIANCES**										002
	0020	Cooking range, 30" free standing, 1 oven, minimum	2 Clab	10	1.600	Ea.	300	21.50		321.50	365	
	0050	Maximum		4	4		1,050	53.50		1,103.50	1,250	
	0150	2 oven, minimum		10	1.600		670	21.50		691.50	770	
	0200	Maximum		10	1.600		1,500	21.50		1,521.50	1,675	
	0350	Built-in, 30" wide, 1 oven, minimum	1 Clab	20	.400		415	5.35		420.35	465	
	0400	Maximum	2 Carp	2	8		1,050	148		1,198	1,400	
	0500	2 oven, conventional, minimum		4	4		800	74		874	1,000	
	0550	1 conventional, 1 microwave, maximum		2	8		1,500	148		1,648	1,900	
	0700	Free-standing, 1 oven, 21" wide range, minimum	2 Clab	10	1.600		300	21.50		321.50	365	
	0750	21 wide, maximum	"	4	4		410	53.50		463.50	540	
	0900	Counter top cook tops, 4 burner, standard, minimum	1 Elec	6	1.333		181	27.50		208.50	244	
	0950	Maximum		3	2.667		455	55		510	595	
	1050	As above, but with grille and griddle attachment, minimum		6	1.333		330	27.50		357.50	405	
	1100	Maximum		3	2.667		470	55		525	610	
	1250	Microwave oven, minimum		4	2		97	41		138	175	
	1300	Maximum		2	4		1,525	82.50		1,607.50	1,800	
	1500	Combination range, refrigerator and sink, 30" wide, minimum	L-1	2	5		630	104		734	865	
	1550	Maximum		1	10		1,250	208		1,458	1,725	
	1570	60" wide, average		1.40	7.143		2,050	149		2,199	2,500	

Important: See the Reference Section for critical supporting data - Reference Nos., Crews, & Location Factors

114 000	Food Service Equipment	CREW	DAILY OUTPUT	LABOR-HOURS	UNIT	1998 BARE COSTS MAT.	LABOR	EQUIP.	TOTAL	TOTAL INCL O&P	
1590	72" wide, average	L-1	1.20	8.333	Ea.	2,300	173		2,473	2,825	002
1640	Combination range, refrigerator, sink, microwave										
1660	oven and ice maker	L-1	.80	12.500	Ea.	3,425	260		3,685	4,175	
1750	Compactor, residential size, 4 to 1 compaction, minimum	1 Carp	5	1.600		300	29.50		329.50	380	
1800	Maximum	"	3	2.667		415	49.50		464.50	545	
2000	Deep freeze, 15 to 23 C.F., minimum	2 Clab	10	1.600		395	21.50		416.50	470	
2050	Maximum		5	3.200		640	43		683	780	
2200	30 C.F., minimum		8	2		710	27		737	825	
2250	Maximum	↓	3	5.333		825	71.50		896.50	1,025	
2450	Dehumidifier, portable, automatic, 15 pint					180			180	198	
2550	40 pint					226			226	248	
2750	Dishwasher, built-in, 2 cycles, minimum	L-1	4	2.500		218	52		270	325	
2800	Maximum		2	5		430	104		534	645	
2950	4 or more cycles, minimum		4	2.500		247	52		299	360	
2960	Average		4	2.500		330	52		382	445	
3000	Maximum	↓	2	5		645	104		749	880	
3200	Dryer, automatic, minimum	L-2	3	5.333		268	86.50		354.50	445	
3250	Maximum	"	2	8		615	129		744	895	
3300	Garbage disposer, sink type, minimum	L-1	10	1		36.50	21		57.50	75	
3350	Maximum	"	10	1		154	21		175	204	
3550	Heater, electric, built-in, 1250 watt, ceiling type, minimum	1 Elec	4	2		45	41		86	117	
3600	Maximum		3	2.667		100	55		155	200	
3700	Wall type, minimum		4	2		40	41		81	112	
3750	Maximum		3	2.667		90	55		145	189	
3900	1500 watt wall type, with blower		4	2		90	41		131	167	
3950	3000 watt	↓	3	2.667		130	55		185	233	
4150	Hood for range, 2 speed, vented, 30" wide, minimum	L-3	5	2		37	38		75	105	
4200	Maximum		3	3.333		243	63		306	375	
4300	42" wide, minimum		5	2		162	38		200	242	
4330	Custom		5	2		203	38		241	287	
4350	Maximum	↓	3	3.333		247	63		310	380	
4500	For ventless hood, 2 speed, add					13			13	14.30	
4650	For vented 1 speed, deduct from maximum					25			25	27.50	
4850	Humidifier, portable, 8 gallons per day					90			90	99	
5000	15 gallons per day					145			145	160	
5200	Icemaker, automatic, 20 lb. per day	1 Plum	7	1.143		490	24		514	580	
5350	51 lb. per day	"	2	4		840	83.50		923.50	1,075	
5380	Oven, built in, standard	1 Elec	4	2		345	41		386	450	
5390	Deluxe	"	2	4		695	82.50		777.50	900	
5500	Refrigerator, no frost, 10 C.F. to 12 C.F. minimum	2 Clab	10	1.600		415	21.50		436.50	490	
5600	Maximum		6	2.667		620	35.50		655.50	745	
5750	14 C.F. to 16 C.F., minimum		9	1.778		470	24		494	560	
5800	Maximum		5	3.200		675	43		718	820	
5950	18 C.F. to 20 C.F., minimum		8	2		545	27		572	645	
6000	Maximum		4	4		1,175	53.50		1,228.50	1,400	
6150	21 C.F. to 29 C.F., minimum		7	2.286		745	30.50		775.50	875	
6200	Maximum	↓	3	5.333		2,275	71.50		2,346.50	2,625	
6400	Sump pump cellar drainer, 1/3 H.P., minimum	1 Plum	3	2.667		70	55.50		125.50	169	
6450	Maximum	"	2	4	↓	310	83.50		393.50	480	
6460	Sump pump, see also division 152-480										
6650	Washing machine, automatic, minimum	1 Plum	3	2.667	Ea.	330	55.50		385.50	455	
6700	Maximum	"	1	8		890	167		1,057	1,250	
6900	Water heater, electric, glass lined, 30 gallon, minimum	L-1	5	2		140	41.50		181.50	223	
6950	Maximum		3	3.333		325	69.50		394.50	475	
7100	80 gallon, minimum		2	5		245	104		349	440	
7150	Maximum	↓	1	10	↓	550	208		758	950	

114 000 | Food Service Equipment

		CREW	DAILY OUTPUT	LABOR-HOURS	UNIT	1998 BARE COSTS				TOTAL INCL O&P	
						MAT.	LABOR	EQUIP.	TOTAL		
002	7180	Water heater, gas, glass lined, 30 gallon, minimum	2 Plum	5	3.200	Ea.	185	66.50		251.50	315	**002**
	7220	Maximum		3	5.333		465	111		576	695	
	7260	50 gallon, minimum		2.50	6.400		215	133		348	460	
	7300	Maximum		1.50	10.667		485	222		707	905	
	7310	Water heater, see also division 153-110										
	7350	Water softener, automatic, to 30 grains per gallon	2 Plum	5	3.200	Ea.	450	66.50		516.50	605	
	7400	To 100 grains per gallon	"	4	4		550	83.50		633.50	745	
	7450	Vent kits for dryers	1 Carp	10	.800		14	14.80		28.80	41	
004	0011	**KITCHEN EQUIPMENT**										**004**
	1050	Butter pat dispenser	1 Clab	13	.615	Ea.	415	8.25		423.25	475	
	1100	Bread dispenser, counter top	"	13	.615		415	8.25		423.25	475	
	1650	Cabinet, heated, 1 compartment, reach-in	R-18	5.60	4.643		1,875	76.50		1,951.50	2,200	
	1655	Pass-thru roll-in		5.60	4.643		2,625	76.50		2,701.50	3,000	
	1660	2 compartment, reach-in		4.80	5.417		4,325	89.50		4,414.50	4,900	
	1670	Mobile					1,625			1,625	1,800	
	1700	Choppers, 5 pounds	R-18	7	3.714		1,550	61		1,611	1,825	
	1720	16 pounds		5	5.200		1,825	85.50		1,910.50	2,175	
	1740	35 to 40 pounds		4	6.500		3,525	107		3,632	4,075	
	1840	Coffee brewer, 5 burners	1 Plum	3	2.667		810	55.50		865.50	980	
	1860	Single, 3 gallon	"	3	2.667		3,800	55.50		3,855.50	4,275	
	1900	Cup and glass dispenser, drop in	1 Clab	4	2		735	27		762	855	
	1920	Disposable cup, drop in		16	.500		208	6.70		214.70	241	
	2650	Dish dispenser, drop in, 12"		11	.727		885	9.75		894.75	990	
	2660	Mobile		10	.800		1,450	10.70		1,460.70	1,625	
	2950	Dishwasher hood, canopy type	L-3A	10	1.200	L.F.	3,125	24.50		3,149.50	3,475	
	2960	Pant leg type	"	2.50	4.800	Ea.	4,150	98		4,248	4,750	
	2970	Exhaust hood, sst, gutter on all sides, 4' x 4' x 2'	1 Carp	1.80	4.444		2,300	82		2,382	2,675	
	2980	4' x 4' x 7'	"	1.60	5		3,600	92.50		3,692.50	4,100	
	3600	Well, hot food, built-in, rectangular, 12" x 20"	R-30	10	2.600		2,075	42.50		2,117.50	2,375	
	3610	Circular, 7 qt		10	2.600		4,150	42.50		4,192.50	4,650	
	3620	Refrigerated, 2 compartments		10	2.600		1,450	42.50		1,492.50	1,675	
	3630	3 compartments		9	2.889		1,450	47		1,497	1,675	
	3640	4 compartments		8	3.250		2,075	53		2,128	2,400	
	3850	40 quarts	L-7	5.40	4.815		6,250	76		6,326	7,000	
	4040	80 quarts		3.90	6.667		9,350	105		9,455	10,500	
	4080	130 quarts		2.20	11.818		14,200	187		14,387	15,900	
	4100	Floor type, 20 quarts		15	1.733		3,175	27.50		3,202.50	3,525	
	4120	60 quarts		14	1.857		8,300	29.50		8,329.50	9,175	
	4140	80 quarts		12	2.167		10,400	34		10,434	11,500	
	4160	140 quarts		8.60	3.023		17,200	48		17,248	19,100	
	4600	Freezer, pre-fab, 8' x 8' w/refrigeration	2 Carp	.45	35.556		6,625	660		7,285	8,400	
	4620	8' x 12'		.35	45.714		7,875	845		8,720	10,100	
	4640	8' x 16'		.25	64		9,300	1,175		10,475	12,200	
	4660	8' x 20'		.17	94.118		12,000	1,750		13,750	16,200	
	4680	Reach-in, 1 compartment	Q-1	4	4		1,775	75		1,850	2,075	
	4700	2 compartment	"	3	5.333		3,125	100		3,225	3,625	
	5700	Hot chocolate dispenser	1 Plum	4	2		570	41.50		611.50	700	
	5900	250 pounds per day	Q-1	1.20	13.333		1,675	250		1,925	2,250	
	6060	With bin		1.20	13.333		3,275	250		3,525	4,025	
	6090	1000 pounds per day, with bin		1	16		6,775	300		7,075	7,975	
	6100	Ice flakers, 300 pounds per day		1.60	10		3,375	188		3,563	4,025	
	6120	600 pounds per day		.95	16.842		4,375	315		4,690	5,325	
	6130	1000 pounds per day		.75	21.333		5,725	400		6,125	6,975	
	6140	2000 pounds per day		.65	24.615		10,300	460		10,760	12,100	
	6160	Ice storage bin, 500 pound capacity	Q-5	1	16		1,300	300		1,600	1,925	
	6180	1000 pound	"	.56	28.571		2,125	540		2,665	3,250	

Important: See the Reference Section for critical supporting data - Reference Nos., Crews, & Location Factors

			CREW	DAILY OUTPUT	LABOR-HOURS	UNIT	1998 BARE COSTS				TOTAL INCL O&P	
114 000		**Food Service Equipment**					MAT.	LABOR	EQUIP.	TOTAL		
004	6200	Iced tea brewer	1 Plum	3.44	2.326	Ea.	520	48.50		568.50	650	**004**
	6300	Juice dispenser, concentrate	R-18	4.50	5.778		730	95		825	960	
	6690	Milk dispenser, bulk, 2 flavor	R-30	8	3.250		830	53		883	1,000	
	6695	3 flavor	"	8	3.250		1,250	53		1,303	1,475	
	6700	Peelers, small	R-18	8	3.250		1,775	53.50		1,828.50	2,050	
	6720	Large	"	6	4.333		3,025	71.50		3,096.50	3,450	
	6750	Pot sink, 3 compartment	1 Plum	7.25	1.103	L.F.	380	23		403	460	
	6760	Pot washer, small		1.60	5	Ea.	11,400	104		11,504	12,800	
	6770	Large		1.20	6.667		28,600	139		28,739	31,700	
	6800	Pulper/extractor, close coupled, 5 HP		1.90	4.211		2,850	88		2,938	3,300	
	6850	Mobile rack w/pan slide					885			885	970	
	8320	Refrigerator, reach-in, 1 compartment	R-18	7.80	3.333		1,825	55		1,880	2,125	
	8330	2 compartment		6.20	4.194		2,400	69		2,469	2,775	
	8340	3 compartment		5.60	4.643		3,375	76.50		3,451.50	3,850	
	8350	Pre-fab, with refrigeration, 8' x 8'	2 Carp	.45	35.556		4,875	660		5,535	6,500	
	8360	8' x 12'		.35	45.714		6,150	845		6,995	8,200	
	8370	8' x 16'		.25	64		7,200	1,175		8,375	9,950	
	8380	8' x 20'		.17	94.118		8,625	1,750		10,375	12,500	
	8390	Pass-thru/roll-in, 1 compartment	R-18	7.80	3.333		2,600	55		2,655	2,950	
	8400	2 compartment		6.24	4.167		3,625	68.50		3,693.50	4,100	
	8410	3 compartment		5.60	4.643		4,950	76.50		5,026.50	5,575	
	8420	Walk-in, alum, door & floor only, no refrig, 6' x 6' x 7'-6"	2 Carp	1.40	11.429		6,350	211		6,561	7,325	
	8430	10' x 6' x 7'-6"		.55	29.091		8,725	540		9,265	10,500	
	8440	12' x 14' x 7'-6"		.25	64		12,300	1,175		13,475	15,600	
	8450	12' x 20' x 7'-6"		.17	94.118		14,600	1,750		16,350	19,100	
	8460	Refrigerated cabinets, mobile					2,500			2,500	2,750	
	8470	Refrigerator/freezer, reach-in, 1 compartment	R-18	5.60	4.643		4,025	76.50		4,101.50	4,550	
	8480	2 compartment		4.80	5.417		5,550	89.50		5,639.50	6,250	
	8580	Slicer with table		9	2.889		3,025	47.50		3,072.50	3,400	
	8600	Stainless steel shelving, louvered 4-tier, 20" x 3'	1 Clab	6	1.333		900	17.85		917.85	1,025	
	8605	20" x 4'		6	1.333		1,050	17.85		1,067.85	1,200	
	8610	20" x 6'		6	1.333		1,450	17.85		1,467.85	1,625	
	8615	24" x 3'		6	1.333		935	17.85		952.85	1,050	
	8620	24" x 4'		6	1.333		1,125	17.85		1,142.85	1,250	
	8625	24" x 6'		6	1.333		1,550	17.85		1,567.85	1,725	
	8630	Flat 4-tier, 20" x 3'		6	1.333		830	17.85		847.85	940	
	8635	20" x 4'		6	1.333		1,000	17.85		1,017.85	1,150	
	8640	20" x 5'		6	1.333		1,150	17.85		1,167.85	1,300	
	8645	24" x 3'		6	1.333		885	17.85		902.85	1,000	
	8650	24" x 4'		6	1.333		1,050	17.85		1,067.85	1,175	
	8655	24" x 6'		6	1.333		2,100	17.85		2,117.85	2,350	
	8700	Galvanized shelving, louvered 4-tier, 20" x 3'		6	1.333		360	17.85		377.85	425	
	8705	20" x 4'		6	1.333		410	17.85		427.85	480	
	8710	20" x 6'		6	1.333		555	17.85		572.85	645	
	8715	24" x 3'		6	1.333		390	17.85		407.85	460	
	8720	24" x 4'		6	1.333		435	17.85		452.85	510	
	8725	24" x 6'		6	1.333		620	17.85		637.85	715	
	8730	Flat 4-tier, 20" x 3'		6	1.333		335	17.85		352.85	395	
	8735	20" x 4'		6	1.333		380	17.85		397.85	450	
	8740	20" x 6'		6	1.333		515	17.85		532.85	595	
	8745	24" x 3'		6	1.333		360	17.85		377.85	430	
	8750	24" x 4'		6	1.333		410	17.85		427.85	480	
	8755	24" x 6'		6	1.333		555	17.85		572.85	640	
	8760	Stainless steel dunnage rack, 24" x 3'		8	1		495	13.40		508.40	570	
	8765	24" x 4'		8	1		635	13.40		648.40	725	
	8770	Galvanized dunnage rack, 24" x 3'		8	1		105	13.40		118.40	138	

EQUIPMENT 11

			DAILY	LABOR-		1998 BARE COSTS				TOTAL	
114 000	**Food Service Equipment**	CREW	OUTPUT	HOURS	UNIT	MAT.	LABOR	EQUIP.	TOTAL	INCL O&P	
004	8775	24" x 4'	1 Clab	8	1	Ea.	116	13.40		129.40	151	004
	8800	Serving counter, straight	1 Carp	40	.200	L.F.	435	3.70		438.70	485	
	8820	Curved section	"	30	.267	"	560	4.93		564.93	625	
	8830	Soft serve ice cream machine, medium	R-18	11	2.364	Ea.	5,575	39		5,614	6,200	
	8840	Large	"	9	2.889		8,375	47.50		8,422.50	9,275	
	9160	Pop-up, 2 slot					565			565	620	
	9170	Trash compactor, small, up to 125 lb. compacted weight	L-4	4	4		12,400	65		12,465	13,800	
	9175	Large, up to 175 lb. compacted weight	"	3	5.333		15,200	87		15,287	16,800	
	9180	Tray and silver dispenser, mobile	1 Clab	16	.500		570	6.70		576.70	640	

			DAILY	LABOR-		1998 BARE COSTS				TOTAL	
114 580	**Disappearing Stairs**	CREW	OUTPUT	HOURS	UNIT	MAT.	LABOR	EQUIP.	TOTAL	INCL O&P	
581	0010	**DISAPPEARING STAIRWAY** No trim included										581
	0020	One piece, yellow pine, 8'-0" ceiling	2 Carp	4	4	Ea.	850	74		924	1,050	
	0030	9'-0" ceiling		4	4		875	74		949	1,075	
	0040	10'-0" ceiling		3	5.333		920	98.50		1,018.50	1,175	
	0050	11'-0" ceiling		3	5.333		1,075	98.50		1,173.50	1,375	
	0060	12'-0" ceiling		3	5.333		1,125	98.50		1,223.50	1,425	
	0100	Custom grade, pine, 8'-6" ceiling, minimum	1 Carp	4	2		100	37		137	174	
	0150	Average		3.50	2.286		105	42.50		147.50	189	
	0200	Maximum		3	2.667		185	49.50		234.50	289	
	0500	Heavy duty, pivoted, from 7'7" to 12'10" floor to floor		3	2.667		315	49.50		364.50	435	
	0600	16'-0" ceiling		2	4		1,025	74		1,099	1,275	
	0800	Economy folding, pine, 8'-6" ceiling		4	2		83	37		120	155	
	0900	9'-6" ceiling		4	2		93.50	37		130.50	167	
	1000	Fire escape, galvanized steel, 8'-0" to 10'-4" ceiling	2 Carp	1	16		1,125	296		1,421	1,725	
	1010	10'-6" to 13'-6" ceiling		1	16		1,400	296		1,696	2,050	
	1100	Automatic electric, aluminum, floor to floor height, 8' to 9'		1	16		5,500	296		5,796	6,575	

For information about Means Estimating Seminars, see yellow pages 11 and 12 in back of book

Important: See the Reference Section for critical supporting data - Reference Nos., Crews, & Location Factors

Division 12
Furnishings

Estimating Tips

General

- The items in this division are usually priced per square foot or each. Most of these items are purchased by the owner and placed by the supplier. Do not assume the items in Division 12 will be purchased and installed by the supplier. Check the specifications for responsibilities and include receiving, storage, installation and mechanical and electrical hook-ups in the appropriate divisions.

- Some items in this division require some type of support system that is not usually furnished with the item. Examples of these systems include blocking for the attachment of casework and heavy drapery rods. The required blocking must be added to the estimate in the appropriate division.

Reference Numbers

Reference numbers are shown in bold squares at the beginning of some major classifications. These numbers refer to related items in the Reference Section. The reference information may be an estimating procedure, an alternate pricing method or technical information.

Note: Not all subdivisions listed here necessarily appear in this publication.

125 100 | Blinds

		CREW	DAILY OUTPUT	LABOR-HOURS	UNIT	1998 BARE COSTS				TOTAL INCL O&P	
						MAT.	LABOR	EQUIP.	TOTAL		
103	0010	**BLINDS, INTERIOR**									103
	0020	Solid colors	1 Carp	590	.014	S.F.	2.72	.25		2.97	3.42
	0090	Horizontal, 1" aluminum slats, custom, minimum		590	.014		2.73	.25		2.98	3.43
	0100	Maximum		440	.018		7.35	.34		7.69	8.70
	0450	Stock, minimum		590	.014		2.55	.25		2.80	3.23
	0500	Maximum		440	.018		4.73	.34		5.07	5.80
	3000	Wood folding panels with movable louvers, 7" x 20" each		17	.471	Pr.	23	8.70		31.70	40
	3300	8" x 28" each		17	.471		32	8.70		40.70	50
	3450	9" x 36" each		17	.471		41	8.70		49.70	60
	3600	10" x 40" each		17	.471		45.50	8.70		54.20	65
	4000	Fixed louver type, stock units, 8" x 20" each		17	.471		26	8.70		34.70	44
	4150	10" x 28" each		17	.471		36.50	8.70		45.20	55
	4300	12" x 36" each		17	.471		51.50	8.70		60.20	71.50
	4450	18" x 40" each		17	.471		72	8.70		80.70	94
	5000	Insert panel type, stock, 7" x 20" each		17	.471		13.70	8.70		22.40	30
	5150	8" x 28" each		17	.471		25	8.70		33.70	42.50
	5300	9" x 36" each		17	.471		32	8.70		40.70	50
	5450	10" x 40" each		17	.471		34	8.70		42.70	52.50
	5600	Raised panel type, stock, 10" x 24" each		17	.471		36.50	8.70		45.20	55
	5650	12" x 26" each		17	.471		38.50	8.70		47.20	57.50
	5700	14" x 30" each		17	.471		45.50	8.70		54.20	65
	5750	16" x 36" each		17	.471		58	8.70		66.70	79
	6000	For custom built pine, add					22%				
	6500	For custom built hardwood blinds, add					42%				
201	0011	**SHADES** Basswood roll-up, stain finish, 3/8" slats	1 Carp	300	.027	S.F.	13.65	.49		14.14	15.90	201
	5011	Insulative shades		125	.064		7.30	1.18		8.48	10.10
	6011	Solar screening, fiberglass		85	.094		3.61	1.74		5.35	6.95
	8011	Interior insulative shutter									
	8111	Stock unit, 15" x 60"	1 Carp	17	.471	Pr.	7.30	8.70		16	23

125 300 | Drape/Curtain Hardware

		CREW	DAILY OUTPUT	LABOR-HOURS	UNIT	1998 BARE COSTS				TOTAL INCL O&P	
301	0010	**DRAPERY HARDWARE**										301
	0030	Standard traverse, per foot, minimum	1 Carp	59	.136	L.F.	1.30	2.51		3.81	5.75
	0100	Maximum		51	.157	"	6.75	2.90		9.65	12.40
	0200	Decorative traverse, 28"-48", minimum		22	.364	Ea.	12	6.75		18.75	25
	0220	Maximum		21	.381		28	7.05		35.05	43
	0300	48"-84", minimum		20	.400		16	7.40		23.40	30.50
	0320	Maximum		19	.421		45.50	7.80		53.30	63.50
	0400	66"-120", minimum		18	.444		18	8.20		26.20	34
	0420	Maximum		17	.471		57.50	8.70		66.20	78.50
	0500	84"-156", minimum		16	.500		20	9.25		29.25	38
	0520	Maximum		15	.533		65.50	9.85		75.35	89.50
	0600	130"-240", minimum		14	.571		24	10.55		34.55	44.50
	0620	Maximum		13	.615		99	11.40		110.40	129
	0700	Slide rings, each, minimum					.60			.60	.66
	0720	Maximum					1.25			1.25	1.38
	3000	Ripplefold, snap-a-pleat system, 3' or less, minimum	1 Carp	15	.533		34.50	9.85		44.35	55
	3020	Maximum	"	14	.571		47	10.55		57.55	70
	3200	Each additional foot, add, minimum				L.F.	2			2	2.20
	3220	Maximum				"	3.90			3.90	4.29
	4000	Traverse rods, adjustable, 28" to 48"	1 Carp	22	.364	Ea.	16.85	6.75		23.60	30
	4020	48" to 84"		20	.400		20	7.40		27.40	34.50
	4040	66" to 120"		18	.444		22.50	8.20		30.70	39
	4060	84" to 156"		16	.500		25	9.25		34.25	43.50
	4080	100" to 180"		14	.571		31	10.55		41.55	52

12 FURNISHINGS

Important: See the Reference Section for critical supporting data - Reference Nos., Crews, & Location Factors

125 300	Drape/Curtain Hardware	CREW	DAILY OUTPUT	LABOR-HOURS	UNIT	1998 BARE COSTS				TOTAL INCL O&P	
						MAT.	LABOR	EQUIP.	TOTAL		
301											**301**
4100	228" to 312"	1 Carp	13	.615	Ea.	45.50	11.40		56.90	69.50	
4500	Curtain rod, 28" to 48", single		22	.364		3.90	6.75		10.65	15.85	
4510	Double		22	.364		6.70	6.75		13.45	18.90	
4520	48" to 86", single		20	.400		6.70	7.40		14.10	20	
4530	Double		20	.400		11.15	7.40		18.55	25	
4540	66" to 120", single		18	.444		11.15	8.20		19.35	26.50	
4550	Double		18	.444		17.55	8.20		25.75	33.50	
4600	Valance, pinch pleated fabric, 12" deep, up to 54" long, minimum					28			28	31	
4610	Maximum					70			70	77	
4620	Up to 77" long, minimum					43			43	47.50	
4630	Maximum					113			113	124	
5000	Stationary rods, first 2 feet					8.15			8.15	8.95	

For information about Means Estimating Seminars, see yellow pages 11 and 12 in back of book

For expanded coverage of these items see *Means Building Construction Cost Data 1998*

Division Notes

	CREW	DAILY OUTPUT	LABOR-HOURS	UNIT	1998 BARE COSTS				TOTAL INCL O&P
					MAT.	LABOR	EQUIP.	TOTAL	

Division 13
Special Construction

Estimating Tips
General

- The items and systems in this division are usually estimated, purchased, supplied and installed as a unit by one or more subcontractors. The estimator must ensure that all parties are operating from the same set of specifications and assumptions and that all necessary items are estimated and will be provided. Many times the complex items and systems are covered but the more common ones such as excavation or a crane are overlooked for the very reason that everyone assumes nobody could miss them. The estimator should be the central focus and be able to ensure that all systems are complete.
- Another area where problems can develop in this division is at the interface between systems. The estimator must ensure, for instance, that anchor bolts, nuts and washers are estimated and included for the air-supported structures and pre-engineered buildings to be bolted to their foundations.

Utility supply is a common area where essential items or pieces of equipment can be missed or overlooked due to the fact that each subcontractor may feel it is the others' responsibility. The estimator should also be aware of certain items which may be supplied as part of a package but installed by others, and ensure that the installing contractor's estimate includes the cost of installation. Conversely, the estimator must also ensure that items are not costed by two different subcontractors, resulting in an inflated overall estimate.

131 Pre-Engineered Structures, Aquatic Facilities & Ice Rinks

- The foundations and floor slab, as well as rough mechanical and electrical, should be estimated, as this work is required for the assembly and erection of the structure. Generally, as noted in the book, the pre-engineered building comes as a shell and additional features must be included by the estimator. Here again, the estimator must have a clear understanding of the scope of each portion of the work and all the necessary interfaces.

132 Tanks, Tank Covers, Filtration Equipment

- The prices in this subdivision for above and below ground storage tanks do not include foundations or hold-down slabs. The estimator should refer to Divisions 2 and 3 for foundation system pricing. In addition to the foundations, required tank accessories such as tank gauges, leak detection devices, and additional manholes and piping must be added to the tank prices.

Reference Numbers

Reference numbers are shown in bold squares at the beginning of some major classifications. These numbers refer to related items in the Reference Section. The reference information may be an estimating procedure, an alternate pricing method or technical information.

Note: Not all subdivisions listed here necessarily appear in this publication.

130 | Special Construction

130 520 | Saunas

			DAILY OUTPUT	LABOR-HOURS	UNIT	1998 BARE COSTS				TOTAL INCL O&P	
						MAT.	LABOR	EQUIP.	TOTAL		
521	0010	**SAUNA** Prefabricated, incl. heater & controls, 7' high, 6' x 4', C/C	L-7	2.20	11.818	Ea.	2,925	187		3,112	3,550	521
	0050	6' x 4', C/P		2	13		2,525	205		2,730	3,125	
	0400	6' x 5', C/C		2	13		3,200	205		3,405	3,875	
	0450	6' x 5', C/P		2	13		2,875	205		3,080	3,500	
	0600	6' x 6', C/C		1.80	14.444		3,425	228		3,653	4,175	
	0650	6' x 6', C/P		1.80	14.444		3,000	228		3,228	3,700	
	0800	6' x 9', C/C		1.60	16.250		4,675	257		4,932	5,600	
	0850	6' x 9', C/P		1.60	16.250		4,675	257		4,932	5,600	
	1000	8' x 12', C/C		1.10	23.636		5,975	375		6,350	7,200	
	1050	8' x 12', C/P		1.10	23.636		5,275	375		5,650	6,425	
	1400	8' x 10', C/C		1.20	21.667		5,400	340		5,740	6,500	
	1450	8' x 10', C/P		1.20	21.667		4,725	340		5,065	5,775	
	1600	10' x 12', C/C		1	26		7,100	410		7,510	8,500	
	1650	10' x 12', C/P		1	26		6,375	410		6,785	7,725	
	1700	Door only, cedar, 2'x6', with tempered insulated glass window	2 Carp	3.40	4.706		279	87		366	455	
	1800	Prehung, incl. jambs, pulls & hardware	"	12	1.333		330	24.50		354.50	405	
	2500	Heaters only (incl. above), wall mounted, to 200 C.F.					500			500	550	
	2750	To 300 C.F.					530			530	585	
	3000	Floor standing, to 720 C.F., 10,000 watts, w/controls	1 Elec	3	2.667		1,075	55		1,130	1,300	
	3250	To 1,000 C.F., 16,000 watts	"	3	2.667		1,575	55		1,630	1,825	

130 540 | Steam Baths

			DAILY OUTPUT	LABOR-HOURS	UNIT	MAT.	LABOR	EQUIP.	TOTAL	TOTAL INCL O&P	
541	0010	**STEAM BATH** Heater, timer & head, single, to 140 C.F.	1 Plum	1.20	6.667	Ea.	600	139		739	890	541
	0500	To 300 C.F.	"	1.10	7.273		760	152		912	1,075	
	2700	Conversion unit for residential tub, including door					2,225			2,225	2,425	

131 | Pre-Eng. Structures, Aquatic Facilities and Ice Rinks

131 230 | Greenhouses

			DAILY OUTPUT	LABOR-HOURS	UNIT	1998 BARE COSTS				TOTAL INCL O&P	
						MAT.	LABOR	EQUIP.	TOTAL		
231	0010	**GREENHOUSE** Shell only, stock units, not incl. 2' stub walls,										231
	0020	foundation, floors, heat or compartments										
	0300	Residential type, free standing, 8'-6" long x 7'-6" wide	2 Carp	59	.271	SF Flr.	35	5		40	47	
	0400	10'-6" wide		85	.188		27	3.48		30.48	35.50	
	0600	13'-6" wide		108	.148		24	2.74		26.74	31	
	0700	17'-0" wide		160	.100		27	1.85		28.85	32.50	
	0900	Lean-to type, 3'-10" wide		34	.471		31	8.70		39.70	49	
	1000	6'-10" wide		58	.276		24	5.10		29.10	35.50	
	1100	Wall mounted, to existing window, 3' x 3'	1 Carp	4	2	Ea.	335	37		372	435	
	1120	4' x 5'	"	3	2.667	"	500	49.50		549.50	635	
	1200	Deluxe quality, free standing, 7'-6" wide	2 Carp	55	.291	SF Flr.	69	5.40		74.40	85	
	1220	10'-6" wide		81	.198		64	3.65		67.65	77	
	1240	13'-6" wide		104	.154		60	2.85		62.85	71	
	1260	17'-0" wide		150	.107		51	1.97		52.97	59.50	
	1400	Lean-to type, 3'-10" wide		31	.516		80	9.55		89.55	104	
	1420	6'-10" wide		55	.291		75	5.40		80.40	91.50	
	1440	8'-0" wide		97	.165		70	3.05		73.05	82.50	
	1500	Commercial, custom, truss frame, incl. equip., plumbing, elec.,										
	1510	benches and controls, under 2,000 S.F., minimum				SF Flr.					30	
	1550	Maximum									38	

Important: See the Reference Section for critical supporting data - Reference Nos., Crews, & Location Factors

131 230	Greenhouses	CREW	DAILY OUTPUT	LABOR-HOURS	UNIT	1998 BARE COSTS MAT.	LABOR	EQUIP.	TOTAL	TOTAL INCL O&P	
231 1700	Over 5,000 S.F., minimum				SF Flr.					23	**231**
1750	Maximum				↓					30	

131 520	Swimming Pools	CREW	DAILY OUTPUT	LABOR-HOURS	UNIT	MAT.	LABOR	EQUIP.	TOTAL	INCL O&P	
521 0010	**SWIMMING POOL ENCLOSURE** Translucent, free standing,										**521**
0020	not including foundations, heat or light										
0200	Economy, minimum	2 Carp	200	.080	SF Hor.	10	1.48		11.48	13.55	
0300	Maximum		100	.160		21	2.96		23.96	28	
0400	Deluxe, minimum		100	.160		23	2.96		25.96	30.50	
0600	Maximum	↓	70	.229	↓	250	4.23		254.23	282	
523 0010	**SWIMMING POOL EQUIPMENT** Diving stand, stainless steel, 3 meter	2 Carp	.40	40	Ea.	4,950	740		5,690	6,725	**523**
0600	Diving boards, 16' long, aluminum		2.70	5.926		1,875	110		1,985	2,250	
0700	Fiberglass	↓	2.70	5.926	↓	1,375	110		1,485	1,725	
0900	Filter system, sand or diatomite type, incl. pump, 6,000 gal./hr.	2 Plum	1.80	8.889	Total	1,000	185		1,185	1,400	
1020	Add for chlorination system, 800 S.F. pool	"	3	5.333	Ea.	270	111		381	480	
1200	Ladders, heavy duty, stainless steel, 2 tread	2 Carp	7	2.286		210	42.50		252.50	305	
1500	4 tread	"	6	2.667		281	49.50		330.50	395	
2100	Lights, underwater, 12 volt, with transformer, 300 watt	1 Elec	.40	20		173	410		583	865	
2200	110 volt, 500 watt, standard	"	.40	20	S.F.	150	410		560	840	
3000	Pool covers, reinforced vinyl, material only					.15			.15	.17	
3100	Vinyl water tube, material only, minimum					.80			.80	.88	
3200	Maximum				↓	.26			.26	.29	
3300	Slides, tubular, fiberglass, aluminum handrails & ladder, 5'-0", straight	2 Carp	1.60	10	Ea.	4,125	185		4,310	4,875	
3320	8'-0", curved	"	3	5.333	"	9,450	98.50		9,548.50	10,600	
525 0010	**SWIMMING POOLS** Residential in-ground, vinyl lined, concrete sides										**525**
0020	Sides including equipment, sand bottom	B-52	300	.187	SF Surf	10.05	2.85	1.40	14.30	17.50	
0100	Metal or polystyrene sides	B-14	410	.117		8.40	1.70	.52	10.62	12.70	
0200	Add for vermiculite bottom					.65			.65	.71	
0500	Gunite bottom and sides, white plaster finish										
0600	12' x 30' pool	B-52	145	.386	SF Surf	16.65	5.90	2.89	25.44	31.50	
0720	16' x 32' pool		155	.361		15	5.50	2.70	23.20	29	
0750	20' x 40' pool	↓	250	.224	↓	13.40	3.42	1.68	18.50	22.50	
0810	Concrete bottom and sides, tile finish										
0820	12' x 30' pool	B-52	80	.700	SF Surf	16.80	10.70	5.25	32.75	42.50	
0830	16' x 32' pool		95	.589		13.90	9	4.41	27.31	35.50	
0840	20' x 40' pool	↓	130	.431	↓	11.05	6.60	3.22	20.87	27	
1600	For water heating system, see division 155-150										
1700	Filtration and deck equipment only, as % of total				Total				20%	20%	
1800	Deck equipment, rule of thumb, 20' x 40' pool				SF Pool					1.30	
3000	Painting pools, preparation + 3 coats, 20' x 40' pool, epoxy	2 Pord	.33	48.485	Total	575	835		1,410	2,025	
3100	Rubber base paint, 18 gallons	"	.33	48.485	"	435	835		1,270	1,875	

Note: Row 0100 references **R131 -520**.

SPECIAL CONSTRUCTION 13

132 150	Undergrnd Storage Tanks	CREW	DAILY OUTPUT	LABOR-HOURS	UNIT	1998 BARE COSTS MAT.	LABOR	EQUIP.	TOTAL	TOTAL INCL O&P	
151 0010	**TANKS**										**151**
0210	Fiberglass, underground, single wall, U.L. listed, not including										
0220	manway or hold-down strap										
0230	1,000 gallon capacity	Q-5	2.46	6.504	Ea.	1,550	123		1,673	1,925	

		132 150	Undergrnd Storage Tanks	CREW	DAILY OUTPUT	LABOR-HOURS	UNIT	1998 BARE COSTS MAT.	LABOR	EQUIP.	TOTAL	TOTAL INCL O&P	
151	0250		4,000 gallon capacity	Q-7	3.55	9.014	Ea.	3,225	170		3,395	3,825	151
	0270		8,000 gallon capacity		2.29	13.974		4,325	264		4,589	5,225	
	0280		10,000 gallon capacity		2	16		4,950	300		5,250	5,950	
	0500		For manway, fittings and hold-downs, add					20%	15%				
	1150		For hold-downs 500-4000 gal, add	Q-7	16	2	Set	157	38		195	235	
	1160		For hold-downs 5000-15000 gal, add		8	4		315	75.50		390.50	470	
	1170		For hold-downs 20,000 gal, add		5.33	6.004		470	113		583	705	
	1180		For hold-downs 25,000 gal, add		4	8		625	151		776	940	
	1190		For hold-downs 30,000 gal, add		2.60	12.308		940	233		1,173	1,400	
	2210		Fiberglass, underground, single wall, U.L. listed, including										
	2220		hold-down straps, no manways										
	2230		1,000 gallon capacity	Q-5	1.88	8.511	Ea.	1,725	161		1,886	2,150	
	2250		4,000 gallon capacity	Q-7	2.90	11.034		3,375	209		3,584	4,075	
	2270		8,000 gallon capacity		1.78	17.978		4,650	340		4,990	5,675	
	2280		10,000 gallon capacity		1.60	20		5,275	380		5,655	6,425	
	5000		Steel underground, sti-P3, set in place, not incl. hold-down bars.										
	5500		Excavation, pad, pumps and piping not included										
	5510		Single wall, 500 gallon capacity, 7 gauge shell	Q-5	2.70	5.926	Ea.	800	112		912	1,075	
	5520		1,000 gallon capacity, 7 gauge shell	"	2.50	6.400		1,575	121		1,696	1,925	
	5530		2,000 gallon capacity, 1/4" thick shell	Q-7	4.60	6.957		2,375	131		2,506	2,825	
	5535		2,500 gallon capacity, 7 gauge shell	Q-5	3	5.333		2,775	101		2,876	3,225	
	5540		5,000 gallon capacity, 1/4" thick shell	Q-7	3.20	10		4,400	189		4,589	5,175	
	5560		10,000 gallon capacity, 1/4" thick shell		2	16		5,725	300		6,025	6,800	
	5610		25,000 gallon capacity, 3/8" thick shell		1.30	24.615		17,600	465		18,065	20,200	
	5630		40,000 gallon capacity, 3/8" thick shell		.90	35.556		30,700	670		31,370	34,800	
	5640		50,000 gallon capacity, 3/8" thick shell		.80	40		38,300	755		39,055	43,500	
	6200		Steel, underground, 360°, double wall, U.L. listed,										
	6210		with sti-P3 corrosion protection,										
	6220		(dielectric coating, cathodic protection, electrical										
	6230		isolation) 30 year warranty,										
	6240		not incl. manholes or hold-downs.										
	6260		1,000 gallon capactiy	Q-5	2.25	7.111	Ea.	2,725	134		2,859	3,225	
	6280		3,000 gallon capacity	Q-7	3.90	8.205		4,625	155		4,780	5,325	
	6300		5,000 gallon capacity		2.91	10.997		6,825	208		7,033	7,850	
	6310		6,000 gallon capacity		2.42	13.223		8,200	250		8,450	9,425	
	6330		10,000 gallon capacity		1.82	17.582		9,650	330		9,980	11,200	
	6400		For hold-downs 500-2000 gal, add		16	2	Set	284	38		322	375	
	6410		For hold-downs 3000-6000 gal, add		12	2.667		475	50.50		525.50	605	
	6420		For hold-downs 8000-12,000 gal, add		11	2.909		520	55		575	660	

For information about Means Estimating Seminars, see yellow pages 11 and 12 in back of book

Important: See the Reference Section for critical supporting data - Reference Nos., Crews, & Location Factors

Division 14
Conveying Systems

Estimating Tips
General
- Many products in Division 14 will require some type of support or blocking for installation not included with the item itself. Examples are supports for conveyors or tube systems, attachment points for lifts, and footings for hoists or cranes. Add these supports in the appropriate division.

141 Dumbwaiters
142 Elevators
- Dumbwaiters and elevators are estimated and purchased in a method similar to buying a car. The manufacturer has a base unit with standard features. Added to this base unit price will be whatever options the owner or specifications require. Increased load capacity, additional vertical travel, additional stops, higher speed, and cab finish options are items to be considered. When developing an estimate for dumbwaiters and elevators, remember that some items needed by the installers may have to be included as part of the general contract.

Examples are:
- shaftway
- rail support brackets
- machine room
- electrical supply
- sill angles
- electrical connections
- pits
- roof penthouses
- pit ladders

Check the job specifications and drawings before pricing.

143 Escalators & Moving Walks
- Escalators and moving walks are specialty items installed by specialty contractors. There are numerous options associated with these items. For specific options contact a manufacturer or contractor. In a method similar to estimating dumbwaiters and elevators, you should verify the extent of general contract work and add items as necessary.

144 Lifts
145 Material Handling Systems
146 Hoists & Cranes
- Products such as correspondence lifts, conveyors, chutes, pneumatic tube systems, material handling cranes and hoists as well as other items specified in this subdivision may require trained installers. The general contractor might not have any choice as to who will perform the installation or when it will be performed. Long lead times are often required for these products, making early decisions in scheduling necessary.

Reference Numbers
Reference numbers are shown in bold squares at the beginning of some major classifications. These numbers refer to related items in the Reference Section. The reference information may be an estimating procedure, an alternate pricing method or technical information.

Note: Not all subdivisions listed here necessarily appear in this publication.

141 100	Manual Dumbwaiters	CREW	DAILY OUTPUT	LABOR-HOURS	UNIT	1998 BARE COSTS				TOTAL INCL O&P	
						MAT.	LABOR	EQUIP.	TOTAL		
101	0010 **DUMBWAITERS** 2 stop, hand, minimum	2 Elev	.75	21.333	Ea.	2,175	455		2,630	3,125	101
	0100 Maximum		.50	32	"	4,925	685		5,610	6,525	
	0300 For each additional stop, add	↓	.75	21.333	Stop	785	455		1,240	1,625	

141 200	Electric Dumbwaiters	CREW	DAILY OUTPUT	LABOR-HOURS	UNIT	1998 BARE COSTS				TOTAL INCL O&P	
						MAT.	LABOR	EQUIP.	TOTAL		
201	0010 **DUMBWAITERS** 2 stop, electric, minimum	2 Elev	.13	123	Ea.	5,475	2,625		8,100	10,400	201
	0100 Maximum		.11	145	"	16,400	3,100		19,500	23,100	
	0600 For each additional stop, add	↓	.54	29.630	Stop	2,400	635		3,035	3,700	

142 010	Elevators	CREW	DAILY OUTPUT	LABOR-HOURS	UNIT	1998 BARE COSTS				TOTAL INCL O&P	
						MAT.	LABOR	EQUIP.	TOTAL		
011	0012 **ELEVATORS OR LIFTS**										011
	7000 Residential, cab type, 1 floor, 2 stop, minimum	2 Elev	.20	80	Ea.	7,650	1,700		9,350	11,300	
	7100 Maximum		.10	160		12,900	3,425		16,325	19,900	
	7200 2 floor, 3 stop, minimum		.12	133		11,400	2,850		14,250	17,200	
	7300 Maximum		.06	266		18,600	5,700		24,300	29,800	
	7700 Stair climber (chair lift), single seat, minimum		1	16		3,700	340		4,040	4,650	
	7800 Maximum	↓	.20	80	↓	5,100	1,700		6,800	8,425	

For information about Means Estimating Seminars, see yellow pages 11 and 12 in back of book

Important: See the Reference Section for critical supporting data - Reference Nos., Crews, & Location Factors

Division 15
Mechanical

Estimating Tips

151 Pipe & Fittings

This subdivision is primarily basic pipe and related materials. The pipe may be used by any of the mechanical disciplines, i.e., plumbing, fire protection, heating, and air conditioning.

- The piping section lists the add to labor for elevated pipe installation. These adds apply to all elevated pipe, fittings, valves, insulation, etc., that are placed above 10' high. CAUTION: the correct percentage may vary for the same pipe. For example, the percentage add for the basic pipe installation should be based on the maximum height that the craftsman must install for that particular section. If the pipe is to be located 14' above the floor but it is suspended on threaded rod from beams, the bottom flange of which is 18' high (4' rods), then the height is actually 18' and the add is 20%. The pipe coverer, however, does not have to go above the 14' and so his add should be 10%.

- Most pipe is priced first as straight pipe with a joint (coupling, weld, etc.) every 10' and a hanger usually every 10'. There are exceptions with hanger spacing such as: for cast iron pipe (5') and plastic pipe (3 per 10'). Following each type of pipe there are several lines listing sizes and the amount to be subtracted to delete couplings and hangers. This is for pipe that is to be buried or supported together on trapeze hangers. The reason that the couplings are deleted is that these runs are usually long and frequently longer lengths of pipe are used. By deleting the couplings the estimator is expected to look up and add back the correct reduced number of couplings.

- When preparing an estimate it may be necessary to approximate the fittings. Fittings usually run between 25% and 50% of the cost of the pipe. The lower percentage is for simpler runs, and the higher number is for complex areas like mechanical rooms.

152 Plumbing Fixtures

- Plumbing fixture costs usually require two lines, the fixture itself and its "rough-in, supply and waste".

153 Plumbing Appliances

- Remember that gas and oil fired units need venting.

154 Fire Protection

- Include all valves needed for fire protection system.

155 Heating

- When estimating the cost of an HVAC system, check to see who is responsible for providing and installing the temperature control system. It is possible to overlook controls, assuming that they would be included in the electrical estimate.

- When looking up a boiler be careful on specified capacity. Some manufacturers rate their products on output while others use input.

- Include HVAC insulation for pipe, boiler and duct (wrap and liner).

156 HVAC Piping Specialties

- Be careful when looking up mechanical items to get the correct pressure rating and connection type (thread, weld, flange).

157 Air Conditioning & Ventilation

- Combination heating and cooling units are sized by the air conditioning requirements. (See Reference No. R157-020 for preliminary sizing guide.)

- A ton of air conditioning is nominally 400 CFM.

- Rectangular duct is taken off by the linear foot for each size, but its cost is usually estimated by the pound. Remember that SMACNA standards now base duct on internal pressure.

- Prefabricated duct is estimated and purchased like pipe: straight sections and fittings.

- Note that cranes or other lifting equipment are not included on any lines in Division 15. For example, if a crane is required to lift a heavy piece of pipe into place high above a gym floor, or to put a rooftop unit on the roof of a four-story building, etc., it must be added. Due to the potential for extreme variation—from nothing additional required, to a major crane or helicopter—we feel that including a nominal amount for "lifting contingency" would be useless and detract from the accuracy of the estimate. When using equipment rental from Means do not forget to include the cost of the operator(s).

Reference Numbers

Reference numbers are shown in bold squares at the beginning of some major classifications. These numbers refer to related items in the Reference Section. The reference information may be an estimating procedure, an alternate pricing method or technical information.

Note: Not all subdivisions listed here necessarily appear in this publication.

151 100	Miscellaneous Fittings	CREW	DAILY OUTPUT	LABOR-HOURS	UNIT	1998 BARE COSTS				TOTAL INCL O&P
						MAT.	LABOR	EQUIP.	TOTAL	
105	**0010 BACKFLOW PREVENTER** Includes valves									**105**
	0020 and four test cocks, corrosion resistant, automatic operation									
	4100 Threaded, valves are ball									
	4120 3/4" pipe size	1 Plum	16	.500	Ea.	665	10.45		675.45	745
	4140 1" pipe size		14	.571		705	11.90		716.90	800
	4150 1-1/4" pipe size		12	.667		985	13.90		998.90	1,100
	4160 1-1/2" pipe size		10	.800		960	16.70		976.70	1,075
	4180 2" pipe size		7	1.143		1,025	24		1,049	1,175
	5000 Flanged, bronze, valves are OS&Y									
	5060 2-1/2" pipe size	Q-1	5	3.200	Ea.	2,650	60		2,710	3,025
	5080 3" pipe size		4.50	3.556		3,275	67		3,342	3,700
	5100 4" pipe size		3	5.333		4,675	100		4,775	5,300
	5120 6" pipe size	Q-2	3	8		9,900	145		10,045	11,100
115	**0010 CLEANOUT TEE**									**115**
	0100 Cast iron, B&S, with countersunk plug									
	0220 3" pipe size	1 Plum	3.60	2.222	Ea.	49	46.50		95.50	131
	0240 4" pipe size	"	3.30	2.424		62.50	50.50		113	153
	0500 For round smooth access cover, same price									
	4000 Plastic, tees and adapters. Add plugs									
	4010 ABS, DWV									
	4020 Cleanout tee, 1-1/2" pipe size	1 Plum	15	.533	Ea.	.99	11.10		12.09	19.55
125	**0010 DRAINS**									**125**
	0140 Cornice, C.I., 45° or 90° outlet									
	0200 3" and 4" pipe size	Q-1	12	1.333	Ea.	55	25		80	102
	0260 For galvanized body, add					10.90			10.90	12
	0280 For polished bronze dome, add					9.35			9.35	10.30
	2000 Floor, medium duty, C.I., deep flange, 7" top									
	2040 2" and 3" pipe size	Q-1	12	1.333	Ea.	35.50	25		60.50	80.50
	2080 For galvanized body, add					14.15			14.15	15.55
	2120 For polished bronze top, add					23			23	25
	2500 Heavy duty, cleanout & trap w/bucket, C.I., 15" top									
	2540 2", 3", and 4" pipe size	Q-1	6	2.667	Ea.	835	50		885	1,000
	2560 For galvanized body, add					250			250	275
	2580 For polished bronze top, add					291			291	320
	3860 Roof, flat metal deck, C.I. body, 12" C.I. dome									
	3890 3" pipe size	Q-1	14	1.143	Ea.	67.50	21.50		89	110
141	**0010 FAUCETS/FITTINGS**									**141**
	0150 Bath, faucets, diverter spout combination, sweat	1 Plum	8	1	Ea.	63.50	21		84.50	105
	0200 For integral stops, IPS unions, add					31			31	34
	0500 Drain, central lift, 1-1/2" IPS male	1 Plum	20	.400		36	8.35		44.35	53.50
	0600 Trip lever, 1-1/2" IPS male		20	.400		37	8.35		45.35	54.50
	1000 Kitchen sink faucets, top mount, cast spout		10	.800		44	16.70		60.70	76
	1100 For spray, add		24	.333		12	6.95		18.95	24.50
	2000 Laundry faucets, shelf type, IPS or copper unions		12	.667		37	13.90		50.90	63.50
	2020									
	2100 Lavatory faucet, centerset, without drain	1 Plum	10	.800	Ea.	32	16.70		48.70	62.50
	2200 For pop-up drain, add		16	.500		13	10.45		23.45	31.50
	2800 Self-closing, center set		10	.800		92.50	16.70		109.20	130
	3000 Service sink faucet, cast spout, pail hook, hose end		14	.571		66.50	11.90		78.40	93
	4000 Shower by-pass valve with union		18	.444		46	9.25		55.25	66
	4200 Shower thermostatic mixing valve, concealed		8	1		222	21		243	279
	4300 For inlet strainer, check, and stops, add					66.50			66.50	73
	5000 Sillcock, compact, brass, IPS or copper to hose	1 Plum	24	.333		4.55	6.95		11.50	16.50

Important: See the Reference Section for critical supporting data - Reference Nos., Crews, & Location Factors

151 100	**Miscellaneous Fittings**	CREW	DAILY OUTPUT	LABOR-HOURS	UNIT	1998 BARE COSTS				TOTAL INCL O&P
						MAT.	LABOR	EQUIP.	TOTAL	
165	0010 **SHOCK ABSORBERS**										165
0490	Copper									
0500	3/4″ male I.P.S. For 1 to 11 fixtures	1 Plum	12	.667	Ea.	15.60	13.90		29.50	40
0600	1″ male I.P.S., For 12 to 32 fixtures		8	1		40	21		61	78.50
0700	1-1/4″ male I.P.S. For 33 to 60 fixtures	▼	8	1	▼	46	21		67	85
170	0010 **SUPPORTS/CARRIERS** For plumbing fixtures										170
0500	Drinking fountain, wall mounted									
0600	Plate type with studs, top back plate	1 Plum	7	1.143	Ea.	15.20	24		39.20	56
0700	Top front and back plate	″	7	1.143	″	19.05	24		43.05	60.50
3000	Lavatory, concealed arm									
3050	Floor mounted, single									
3100	High back fixture	1 Plum	6	1.333	Ea.	151	28		179	212
3200	Flat slab fixture		6	1.333		175	28		203	239
3220	Paraplegic	▼	6	1.333	▼	81	28		109	135
3250	Floor mounted, back to back									
3300	High back fixtures	1 Plum	5	1.600	Ea.	215	33.50		248.50	293
3400	Flat slab fixtures		5	1.600		264	33.50		297.50	345
3430	Paraplegic	▼	5	1.600	▼	103	33.50		136.50	169
4600	Sink, floor mounted									
4650	Exposed arm system									
4700	Single heavy fixture	1 Plum	5	1.600	Ea.	125	33.50		158.50	193
5400	Wall mounted, exposed arms, single heavy fixture		5	1.600		68	33.50		101.50	131
6000	Urinal, floor mounted, 2″ or 3″ coupling, blowout type		6	1.333		149	28		177	210
6100	With fixture or hanger bolts, blowout or washout		6	1.333		106	28		134	163
6300	Wall mounted, plate type system	▼	6	1.333	▼	61.50	28		89.50	114
6980	Water closet, siphon jet									
7000	Horizontal, adjustable, caulk									
7040	Single, 4″ pipe size	1 Plum	6	1.333	Ea.	114	28		142	172
7050	4″ pipe size, paraplegic		6	1.333		114	28		142	172
7100	Double, 4″ pipe size		5	1.600		224	33.50		257.50	300
7110	4″ pipe size, paraplegic	▼	5	1.600	▼	224	33.50		257.50	300
8201	Water closet, residential, vert. centerline, floor mount									
8300	4″ copper sweat, 4″ vent	1 Plum	6	1.333	Ea.	165	28		193	228
9000	Water cooler (electric), floor mounted									
9100	Plate type with bearing plate, single	1 Plum	6	1.333	Ea.	62.50	28		90.50	115
181	0010 **TRAPS**										181
0030	Cast iron, service weight									
0050	Long P trap, 2″ pipe size									
1100	12″ long	Q-1	16	1	Ea.	21.50	18.80		40.30	54.50
3000	P trap, B&S, 2″ pipe size		16	1		14	18.80		32.80	46.50
3040	3″ pipe size	▼	14	1.143		21	21.50		42.50	58.50
3800	Drum trap, 4″ x 5″, 1-1/2″ tapping	Q-2	17	1.412	▼	13.50	25.50		39	57.50
4700	Copper, drainage, drum trap									
4840	3″ x 6″ swivel, 1-1/2″ pipe size	1 Plum	16	.500	Ea.	50	10.45		60.45	72.50
5100	P trap, standard pattern									
5200	1-1/4″ pipe size	1 Plum	18	.444	Ea.	36	9.25		45.25	55
5240	1-1/2″ pipe size		17	.471		36	9.80		45.80	56
5260	2″ pipe size		15	.533		42	11.10		53.10	64.50
5280	3″ pipe size	▼	11	.727		101	15.15		116.15	136
6710	ABS DWV P trap, solvent weld joint									
6720	1-1/2″ pipe size	1 Plum	18	.444	Ea.	7.45	9.25		16.70	23.50
6722	2″ pipe size		17	.471		12	9.80		21.80	29.50
6724	3″ pipe size		15	.533	▼	21	11.10		32.10	41.50
6726	4″ pipe size		14	.571		42	11.90		53.90	66
6760	PP DWV, dilution trap, 1-1/2″ pipe size	▼	16	.500	▼	88	10.45		98.45	114

For expanded coverage of these items see *Means Mechanical or Plumbing Cost Data 1998*

151 100	Miscellaneous Fittings	CREW	DAILY OUTPUT	LABOR-HOURS	UNIT	1998 BARE COSTS MAT.	LABOR	EQUIP.	TOTAL	TOTAL INCL O&P	
181	6770	P trap, 1-1/2" pipe size	1 Plum	17	.471	Ea.	23.50	9.80		33.30	42.50	**181**
	6780	2" pipe size		16	.500		32.50	10.45		42.95	53	
	6790	3" pipe size		14	.571		74.50	11.90		86.40	102	
	6800	4" pipe size		13	.615		95	12.85		107.85	126	
	6860	PVC DWV hub x hub, basin trap, 1-1/4" pipe size		18	.444		11.15	9.25		20.40	27.50	
	6870	Sink P trap, 1-1/2" pipe size		18	.444		10.35	9.25		19.60	27	
	6880	Tubular S trap, 1-1/2" pipe size		17	.471		13.95	9.80		23.75	31.50	
	6890	PVC sch. 40 DWV, drum trap										
	6900	1-1/2" pipe size	1 Plum	16	.500	Ea.	14.70	10.45		25.15	33.50	
	6910	P trap, 1-1/2" pipe size		18	.444		2.15	9.25		11.40	17.70	
	6920	2" pipe size		17	.471		2.93	9.80		12.73	19.45	
	6930	3" pipe size		15	.533		14.55	11.10		25.65	34.50	
	6940	4" pipe size		14	.571		35	11.90		46.90	58.50	
	6950	P trap w/clean out, 1-1/2" pipe size		18	.444		5.20	9.25		14.45	21	
	6960	2" pipe size		17	.471		8.85	9.80		18.65	26	
195	0010	**VENT FLASHING**										**195**
	1000	Aluminum with lead ring										
	1040	2" pipe	1 Plum	18	.444	Ea.	6.60	9.25		15.85	22.50	
	1060	4" pipe	"	16	.500	"	8.80	10.45		19.25	27	
	1350	Copper with neoprene ring										
	1430	1-1/2" pipe	1 Plum	20	.400	Ea.	20.50	8.35		28.85	36.50	
	1440	2" pipe		18	.444		21.50	9.25		30.75	39	
	1460	4" pipe		16	.500		28	10.45		38.45	48.50	

151 200	Piping	CREW	DAILY OUTPUT	LABOR-HOURS	UNIT	1998 BARE COSTS MAT.	LABOR	EQUIP.	TOTAL	TOTAL INCL O&P	
200	0010	**PIPING** See also divisions 026 & 027 for site work										**200**
	1000	Add to labor for elevated installation										
	1080	10' to 15' high						10%				
	1100	15' to 20' high						20%				
	1120	20' to 25' high						25%				
	1140	25' to 30' high						35%				
	1160	30' to 35' high						40%				
	1180	35' to 40' high						50%				
	1200	Over 40' high						55%				

151 300	Cast Iron Pipe	CREW	DAILY OUTPUT	LABOR-HOURS	UNIT	1998 BARE COSTS MAT.	LABOR	EQUIP.	TOTAL	TOTAL INCL O&P	
301	0010	**PIPE, CAST IRON** Soil, on hangers 5' O.C.										**301**
	0020	Single hub, service wt., lead & oakum joints 10' O.C.										
	2120	2" diameter	Q-1	63	.254	L.F.	4.30	4.77		9.07	12.60	
	2140	3" diameter		60	.267		5.90	5		10.90	14.80	
	2160	4" diameter		55	.291		8.30	5.45		13.75	18.15	
	2180	5" diameter	Q-2	76	.316		10.15	5.70		15.85	20.50	
	2200	6" diameter	"	73	.329		12.35	5.95		18.30	23.50	
	2320	For service weight, double hub, add					65%					
	2340	For extra heavy, single hub, add					48%	4%				
	2360	For extra heavy, double hub, add					29%	4%				
	2400	Lead for caulking				Lb.	1.04			1.04	1.14	
	2420	Oakum for caulking				"	3.17			3.17	3.49	
	4000	No hub, couplings 10' O.C.										
	4100	1-1/2" diameter	Q-1	71	.225	L.F.	5.10	4.23		9.33	12.60	
	4120	2" diameter		67	.239		5.10	4.48		9.58	13.05	
	4140	3" diameter		64	.250		6.75	4.69		11.44	15.20	
	4160	4" diameter		58	.276		8.55	5.20		13.75	18	
	4180	5" diameter	Q-2	83	.289		12.20	5.25		17.45	22	

Important: See the Reference Section for critical supporting data - Reference Nos., Crews, & Location Factors

151 300 | Cast Iron Pipe

		CREW	DAILY OUTPUT	LABOR-HOURS	UNIT	MAT.	LABOR	EQUIP.	TOTAL	TOTAL INCL O&P	
301	4200	6″ diameter	Q-2	79	.304	L.F.	15.55	5.50		21.05	26	**301**
320	0010	**PIPE, CAST IRON, FITTINGS** Soil										**320**
	0040	Hub and spigot, service weight, lead & oakum joints										
	0080	1/4 bend, 2″	Q-1	16	1	Ea.	5.85	18.80		24.65	37.50	
	0120	3″		14	1.143		10.10	21.50		31.60	46.50	
	0140	4″	↓	13	1.231		15.75	23		38.75	56	
	0160	5″	Q-2	18	1.333		22	24		46	64	
	0180	6″	″	17	1.412		27.50	25.50		53	73	
	0500	Sanitary tee, 2″	Q-1	10	1.600		9.25	30		39.25	60	
	0540	3″		9	1.778		15.75	33.50		49.25	73	
	0620	4″	↓	8	2		21	37.50		58.50	85	
	0700	5″	Q-2	12	2		41.50	36		77.50	106	
	0800	6″	″	11	2.182	↓	125	39.50		164.50	203	
	5990	No hub										
	6000	Cplg. & labor required at joints not incl. in fitting										
	6010	price. Add 1 coupling per joint for installed price										
	6020	1/4 Bend, 1-1/2″				Ea.	4.25			4.25	4.68	
	6060	2″					4.63			4.63	5.10	
	6080	3″					6.45			6.45	7.10	
	6120	4″					9.25			9.25	10.20	
	6140	5″					21			21	23	
	6160	6″					23.50			23.50	26	
	6184	1/4 Bend, long sweep, 1-1/2″					9.95			9.95	10.95	
	6186	2″					9.95			9.95	10.95	
	6188	3″					11.85			11.85	13.05	
	6189	4″					18.90			18.90	21	
	6190	5″					35			35	38.50	
	6191	6″					42.50			42.50	47	
	6192	8″					131			131	144	
	6193	10″					236			236	260	
	6380	Sanitary Tee, tapped, 1-1/2″					8.40			8.40	9.25	
	6382	2″ x 1-1/2″					8.40			8.40	9.25	
	6384	2″					9.30			9.30	10.25	
	6386	3″ x 2″					11.90			11.90	13.10	
	6388	3″					22			22	24	
	6390	4″ x 1-1/2″					11			11	12.10	
	6392	4″ x 2″					11.90			11.90	13.10	
	6394	6″ x 1-1/2″					25			25	27.50	
	6396	6″ x 2″					25			25	27.50	
	6459	Sanitary Tee, 1-1/2″					5.80			5.80	6.40	
	6460	2″					6.45			6.45	7.10	
	6470	3″					7.80			7.80	8.60	
	6472	4″					12.10			12.10	13.30	
	6474	5″					34.50			34.50	38	
	6476	6″				↓	35			35	38.50	
	8000	Coupling, standard (by CISPI Mfrs.)										
	8020	1-1/2″	Q-1	48	.333	Ea.	3.63	6.25		9.88	14.35	
	8040	2″		44	.364		3.63	6.85		10.48	15.30	
	8080	3″		38	.421		4.32	7.90		12.22	17.85	
	8120	4″	↓	33	.485		5.15	9.10		14.25	21	
	8160	5″	Q-2	44	.545		12.50	9.85		22.35	30	
	8180	6″	″	40	.600	↓	13.10	10.85		23.95	32.50	

151 400 | Copper Pipe & Tubing

			DAILY	LABOR-			1998 BARE COSTS				TOTAL	
			CREW	OUTPUT	HOURS	UNIT	MAT.	LABOR	EQUIP.	TOTAL	INCL O&P	
401	0010	**PIPE, COPPER** Solder joints										**401**
	0020	Type K tubing, couplings & clevis hangers 10' O.C.										
	1180	3/4" diameter	1 Plum	74	.108	L.F.	2.79	2.25		5.04	6.80	
	1200	1" diameter		66	.121		3.57	2.53		6.10	8.10	
	1240	1-1/2" diameter		50	.160		6.35	3.34		9.69	12.55	
	1260	2" diameter		40	.200		8.60	4.17		12.77	16.35	
	1280	2-1/2" diameter	Q-1	60	.267		12.70	5		17.70	22.50	
	1300	3" diameter	"	54	.296		17.55	5.55		23.10	28.50	
	2000	Type L tubing, couplings & hangers 10' O.C.										
	2140	1/2" diameter	1 Plum	81	.099	L.F.	1.51	2.06		3.57	5.05	
	2180	3/4" diameter		76	.105		2.07	2.19		4.26	5.90	
	2200	1" diameter		68	.118		2.27	2.45		4.72	6.55	
	2220	1-1/4" diameter		58	.138		3.83	2.88		6.71	9	
	2240	1-1/2" diameter		52	.154		4.76	3.21		7.97	10.55	
	2260	2" diameter		42	.190		7.25	3.97		11.22	14.60	
	2280	2-1/2" diameter	Q-1	62	.258		10.85	4.85		15.70	20	
	2300	3" diameter		56	.286		14.70	5.35		20.05	25	
	2320	3-1/2" diameter		43	.372		19	7		26	32.50	
	2340	4" diameter		39	.410		23.50	7.70		31.20	39	
	3000	Type M tubing, couplings & hangers 10' O.C.										
	3140	1/2" diameter	1 Plum	84	.095	L.F.	1.22	1.99		3.21	4.63	
	3180	3/4" diameter		78	.103		1.65	2.14		3.79	5.35	
	3200	1" diameter		70	.114		2.27	2.38		4.65	6.45	
	3220	1-1/4" diameter		60	.133		3.19	2.78		5.97	8.10	
	3240	1-1/2" diameter		54	.148		4.26	3.09		7.35	9.80	
	3260	2" diameter		44	.182		6.75	3.79		10.54	13.75	
	3280	2-1/2" diameter	Q-1	64	.250		9.60	4.69		14.29	18.35	
	3300	3" diameter		58	.276		12.40	5.20		17.60	22	
	3320	3-1/2" diameter		45	.356		16.05	6.70		22.75	28.50	
	3340	4" diameter		40	.400		21	7.50		28.50	36	
	4000	Type DWV tubing, couplings & hangers 10' O.C.										
	4100	1-1/4" diameter	1 Plum	60	.133	L.F.	3.13	2.78		5.91	8.05	
	4120	1-1/2" diameter		54	.148		3.83	3.09		6.92	9.30	
	4140	2" diameter		44	.182		5.05	3.79		8.84	11.85	
	4160	3" diameter	Q-1	58	.276		8.70	5.20		13.90	18.20	
	4180	4" diameter	"	40	.400		15.55	7.50		23.05	29.50	
430	0010	**PIPE, COPPER, FITTINGS** Wrought unless otherwise noted										**430**
	0040	Solder joints, copper x copper										
	0100	1/2"	1 Plum	20	.400	Ea.	.50	8.35		8.85	14.35	
	0120	3/4"		19	.421		1.07	8.80		9.87	15.75	
	0130	1"		16	.500		2.47	10.45		12.92	19.95	
	0140	1-1/4"		15	.533		3.77	11.10		14.87	22.50	
	0150	1-1/2"		13	.615		5.85	12.85		18.70	28	
	0160	2"		11	.727		10.70	15.15		25.85	37	
	0170	2-1/2"	Q-1	13	1.231		20.50	23		43.50	61	
	0180	3"		11	1.455		28.50	27.50		56	76.50	
	0190	3-1/2"		10	1.600		84	30		114	143	
	0200	4"		9	1.778		65.50	33.50		99	128	
	0450	Tee, 1/4"	1 Plum	14	.571		2.99	11.90		14.89	23	
	0480	1/2"		13	.615		.82	12.85		13.67	22.50	
	0500	3/4"		12	.667		1.98	13.90		15.88	25	
	0510	1"		10	.800		5.75	16.70		22.45	34	
	0520	1-1/4"		9	.889		8.70	18.55		27.25	40	
	0530	1-1/2"		8	1		12.10	21		33.10	48	
	0540	2"		7	1.143		18.85	24		42.85	60	
	0550	2-1/2"	Q-1	8	2		38	37.50		75.50	104	

Important: See the Reference Section for critical supporting data - Reference Nos., Crews, & Location Factors

151 400 | Copper Pipe & Tubing

		CREW	DAILY OUTPUT	LABOR-HOURS	UNIT	MAT.	LABOR	EQUIP.	TOTAL	TOTAL INCL O&P	
430	0560	3"	Q-1	7	2.286	Ea.	58	43		101	135	**430**
	0580	4"	↓	5	3.200		126	60		186	238	
	0612	Tee, reducing on the outlet, 1/4"	1 Plum	15	.533		4.84	11.10		15.94	24	
	0613	3/8"		15	.533		4.44	11.10		15.54	23.50	
	0614	1/2"		14	.571		3.88	11.90		15.78	24	
	0615	5/8"		13	.615		6.35	12.85		19.20	28.50	
	0616	3/4"		12	.667		1.81	13.90		15.71	25	
	0617	1"		11	.727		5.75	15.15		20.90	31.50	
	0618	1-1/4"		10	.800		8.50	16.70		25.20	37	
	0619	1-1/2"		9	.889		9	18.55		27.55	40.50	
	0620	2"	↓	8	1		14.25	21		35.25	50	
	0621	2-1/2"	Q-1	9	1.778		43.50	33.50		77	104	
	0622	3"		8	2		48	37.50		85.50	115	
	0623	4"		6	2.667		85	50		135	177	
	0624	5"	↓	5	3.200		390	60		450	530	
	0625	6"	Q-2	7	3.429		535	62		597	695	
	0626	8"	"	6	4		2,175	72.50		2,247.50	2,525	
	0630	Tee, reducing on the run, 1/4"	1 Plum	15	.533		5.70	11.10		16.80	24.50	
	0631	3/8"		15	.533		6.25	11.10		17.35	25.50	
	0632	1/2"		14	.571		5.15	11.90		17.05	25.50	
	0633	5/8"		13	.615		7.70	12.85		20.55	30	
	0634	3/4"		12	.667		2.20	13.90		16.10	25.50	
	0635	1"		11	.727		6.85	15.15		22	32.50	
	0636	1-1/4"		10	.800		10.90	16.70		27.60	39.50	
	0637	1-1/2"		9	.889		18.05	18.55		36.60	50.50	
	0638	2"	↓	8	1		24	21		45	61	
	0639	2-1/2"	Q-1	9	1.778		54.50	33.50		88	116	
	0640	3"		8	2		80	37.50		117.50	150	
	0641	4"		6	2.667		177	50		227	278	
	0642	5"	↓	5	3.200		390	60		450	530	
	0643	6"	Q-2	7	3.429		535	62		597	695	
	0644	8"	"	6	4		2,175	72.50		2,247.50	2,525	
	2000	DWV, solder joints, copper x copper										
	2030	90° Elbow, 1-1/4"	1 Plum	13	.615	Ea.	4.38	12.85		17.23	26.50	
	2050	1-1/2"		12	.667		5.85	13.90		19.75	29.50	
	2070	2"	↓	10	.800		8.55	16.70		25.25	37	
	2090	3"	Q-1	10	1.600		20.50	30		50.50	72.50	
	2100	4"	"	9	1.778		99.50	33.50		133	166	
	2250	Tee, Sanitary, 1-1/4"	1 Plum	9	.889		8.45	18.55		27	40	
	2290	2"	"	7	1.143		12.50	24		36.50	53.50	
	2310	3"	Q-1	7	2.286	↓	37.50	43		80.50	112	

151 550 | Plastic Pipe

		CREW	DAILY OUTPUT	LABOR-HOURS	UNIT	MAT.	LABOR	EQUIP.	TOTAL	TOTAL INCL O&P	
551	0011	**PIPE, PLASTIC**										**551**
	1800	PVC, couplings 10' O.C., hangers 3 per 10'										
	1820	Schedule 40										
	1860	1/2" diameter	1 Plum	54	.148	L.F.	1.63	3.09		4.72	6.90	
	1870	3/4" diameter		51	.157		1.66	3.27		4.93	7.20	
	1880	1" diameter		46	.174		1.83	3.63		5.46	8	
	1890	1-1/4" diameter		42	.190		1.97	3.97		5.94	8.75	
	1900	1-1/2" diameter	↓	36	.222		2.08	4.63		6.71	10	
	1910	2" diameter	Q-1	59	.271		2.56	5.10		7.66	11.25	
	1920	2-1/2" diameter		56	.286		3.23	5.35		8.58	12.45	
	1930	3" diameter		53	.302		3.77	5.65		9.42	13.55	
	1940	4" diameter	↓	48	.333	↓	4.93	6.25		11.18	15.75	

MECHANICAL 15

151 550 | Plastic Pipe

		CREW	DAILY OUTPUT	LABOR-HOURS	UNIT	1998 BARE COSTS				TOTAL INCL O&P
						MAT.	LABOR	EQUIP.	TOTAL	
551										**551**
4100	DWV type, schedule 40, couplings 10' O.C., hangers 3 per 10'									
4120	ABS									
4140	1-1/4" diameter	1 Plum	42	.190	L.F.	1.99	3.97		5.96	8.80
4150	1-1/2" diameter	"	36	.222		1.92	4.63		6.55	9.80
4160	2" diameter	Q-1	59	.271		2.11	5.10		7.21	10.75
4170	3" diameter		53	.302		3.35	5.65		9	13.10
4180	4" diameter		48	.333		5.20	6.25		11.45	16.05
5360	CPVC, couplings 10' O.C., hangers 3 per 10'									
5380	Schedule 40									
5460	1/2" diameter	1 Plum	54	.148	L.F.	2.34	3.09		5.43	7.65
5470	3/4" diameter		51	.157		2.61	3.27		5.88	8.25
5480	1" diameter		46	.174		3.19	3.63		6.82	9.50
5490	1-1/4" diameter		42	.190		3.86	3.97		7.83	10.85
5500	1-1/2" diameter		36	.222		4.39	4.63		9.02	12.55
5510	2" diameter	Q-1	59	.271		5.25	5.10		10.35	14.25
9000	Perforated									
9040	4" diameter				L.F.	.40			.40	.44
558	0010 **PIPE, PLASTIC, FITTINGS**									**558**
2700	PVC (white), schedule 40, socket joints									
2760	90° elbow, 1/2"	1 Plum	22	.364	Ea.	.23	7.60		7.83	12.80
2770	3/4"		21	.381		.25	7.95		8.20	13.45
2780	1"		18	.444		.45	9.25		9.70	15.85
2790	1-1/4"		17	.471		.80	9.80		10.60	17.15
2800	1-1/2"		16	.500		.85	10.45		11.30	18.20
2810	2"	Q-1	28	.571		1.34	10.75		12.09	19.20
2820	2-1/2"		22	.727		4.06	13.65		17.71	27
2830	3"		17	.941		4.86	17.70		22.56	35
2840	4"		14	1.143		8.70	21.50		30.20	45
3180	Tee, 1/2"	1 Plum	14	.571		.28	11.90		12.18	20
3190	3/4"		13	.615		.32	12.85		13.17	22
3200	1"		12	.667		.59	13.90		14.49	23.50
3210	1-1/4"		11	.727		.94	15.15		16.09	26
3220	1-1/2"		10	.800		1.13	16.70		17.83	28.50
3230	2"	Q-1	17	.941		1.65	17.70		19.35	31.50
3240	2-1/2"		14	1.143		5.45	21.50		26.95	41.50
3250	3"		11	1.455		7.15	27.50		34.65	53
3260	4"		9	1.778		12.90	33.50		46.40	69.50
3380	Coupling, 1/2"	1 Plum	22	.364		.18	7.60		7.78	12.75
3390	3/4"		21	.381		.24	7.95		8.19	13.40
3400	1"		18	.444		.41	9.25		9.66	15.80
3410	1-1/4"		17	.471		.57	9.80		10.37	16.90
3420	1-1/2"		16	.500		.61	10.45		11.06	17.90
3430	2"	Q-1	28	.571		.95	10.75		11.70	18.80
3440	2-1/2"		20	.800		2.08	15		17.08	27.50
3450	3"		19	.842		3.27	15.80		19.07	29.50
3460	4"		16	1		4.72	18.80		23.52	36
4500	DWV, ABS, non pressure, socket joints									
4540	1/4 Bend, 1-1/4"	1 Plum	17	.471	Ea.	2.71	9.80		12.51	19.25
4560	1-1/2"	"	16	.500		1.17	10.45		11.62	18.55
4570	2"	Q-1	28	.571		1.67	10.75		12.42	19.60
4580	3"		17	.941		5.15	17.70		22.85	35
4590	4"		14	1.143		10.05	21.50		31.55	46.50
4600	6"		8	2		61.50	37.50		99	130
4650	1/8 Bend, same as 1/4 Bend									
4800	Tee, sanitary									

Important: See the Reference Section for critical supporting data - Reference Nos., Crews, & Location Factors

151 550 | Plastic Pipe

		CREW	DAILY OUTPUT	LABOR-HOURS	UNIT	MAT.	LABOR	EQUIP.	TOTAL	TOTAL INCL O&P
4820	1-1/4″	1 Plum	11	.727	Ea.	3.41	15.15		18.56	29
4830	1-1/2″	″	10	.800		1.99	16.70		18.69	29.50
4840	2″	Q-1	17	.941		2.99	17.70		20.69	33
4850	3″		11	1.455		7.60	27.50		35.10	53.50
4860	4″		9	1.778		14.65	33.50		48.15	71.50
4862	Tee, sanitary, reducing, 2″ x 1-1/2″		17	.941		2.99	17.70		20.69	33
4864	3″ x 2″		11	1.455		7.60	27.50		35.10	53.50
4868	4″ x 3″		10	1.600		23	30		53	75.50
4870	Combination Y and 1/8 bend									
4872	1-1/2″	1 Plum	10	.800	Ea.	6.40	16.70		23.10	34.50
4874	2″	Q-1	17	.941		7.75	17.70		25.45	38
4876	3″		11	1.455		13.70	27.50		41.20	60
4878	4″		9	1.778		27	33.50		60.50	85
4880	3″ x 1-1/2″		11	1.455		15.95	27.50		43.45	62.50
4882	4″ x 3″		10	1.600		24	30		54	76.50
4900	Wye, 1-1/4″	1 Plum	11	.727		3.94	15.15		19.09	29.50
4902	1-1/2″	″	10	.800		4.01	16.70		20.71	32
4904	2″	Q-1	17	.941		4.11	17.70		21.81	34
4906	3″		11	1.455		5.15	27.50		32.65	50.50
4908	4″		9	1.778		16.95	33.50		50.45	74
4910	6″		5	3.200		85	60		145	193
4918	3″ x 1-1/2″		11	1.455		9.35	27.50		36.85	55.50
4920	4″ x 3″		10	1.600		16.20	30		46.20	68
4922	6″ x 4″		6	2.667		70.50	50		120.50	161
4930	Double Wye, 1-1/2″	1 Plum	8	1		8.50	21		29.50	44
4932	2″	Q-1	12	1.333		10.95	25		35.95	53.50
4934	3″		8	2		28.50	37.50		66	93
4936	4″		6	2.667		57.50	50		107.50	146
4940	2″ x 1-1/2″		11	1.455		10.95	27.50		38.45	57
4942	3″ x 2″		8	2		21	37.50		58.50	85
4944	4″ x 3″		7	2.286		45.50	43		88.50	121
4946	6″ x 4″		5	3.200		95	60		155	204
4950	Reducer bushing, 2″ x 1-1/2″		30	.533		.94	10		10.94	17.65
4952	3″ x 1-1/2″		24	.667		5.05	12.50		17.55	26
4954	4″ x 2″		20	.800		11.65	15		26.65	38
4956	6″ x 4″		17	.941		33.50	17.70		51.20	66.50
4960	Couplings, 1-1/2″	1 Plum	16	.500		.89	10.45		11.34	18.25
4962	2″	Q-1	28	.571		.94	10.75		11.69	18.80
4963	3″		22	.727		2.55	13.65		16.20	25.50
4964	4″		17	.941		4.21	17.70		21.91	34
4966	6″		12	1.333		25.50	25		50.50	69.50
4970	2″ x 1-1/2″		30	.533		3.97	10		13.97	21
4972	3″ x 1-1/2″		24	.667		12.60	12.50		25.10	34.50
4974	4″ x 3″		19	.842		22	15.80		37.80	50
4978	Closet flange, 4″	1 Plum	32	.250		6.10	5.20		11.30	15.35
4980	4″ x 3″	″	34	.235		5.70	4.91		10.61	14.45
5500	CPVC, Schedule 80, threaded joints									
5540	90° Elbow, 1/4″	1 Plum	20	.400	Ea.	5.35	8.35		13.70	19.65
5560	1/2″		18	.444		2.09	9.25		11.34	17.65
5570	3/4″		17	.471		2.67	9.80		12.47	19.20
5580	1″		15	.533		4.24	11.10		15.34	23
5590	1-1/4″		14	.571		9.20	11.90		21.10	30
5600	1-1/2″		13	.615		10.25	12.85		23.10	33
5610	2″	Q-1	22	.727		12.40	13.65		26.05	36
5620	2-1/2″		18	.889		28.50	16.70		45.20	59
5630	3″		14	1.143		32	21.50		53.50	71

151 550 | Plastic Pipe

			CREW	DAILY OUTPUT	LABOR-HOURS	UNIT	1998 BARE COSTS				TOTAL INCL O&P	
							MAT.	LABOR	EQUIP.	TOTAL		
558	6000	Coupling, 1/4"	1 Plum	20	.400	Ea.	5.70	8.35		14.05	20	558
	6020	1/2"		18	.444		2.21	9.25		11.46	17.80	
	6030	3/4"		17	.471		3.09	9.80		12.89	19.65	
	6040	1"		15	.533		4.16	11.10		15.26	23	
	6050	1-1/4"		14	.571		6.25	11.90		18.15	26.50	
	6060	1-1/2"		13	.615		7.85	12.85		20.70	30	
	6070	2"	Q-1	22	.727		9.15	13.65		22.80	32.50	
	6080	2-1/2"		20	.800		20.50	15		35.50	47.50	
	6090	3"		19	.842		22	15.80		37.80	50.50	

151 700 | Steel Pipe

			CREW	DAILY OUTPUT	LABOR-HOURS	UNIT	MAT.	LABOR	EQUIP.	TOTAL	TOTAL INCL O&P	
701	0010	**PIPE, STEEL**										701
	0050	Schedule 40, threaded, with couplings, and clevis type										
	0060	hangers sized for covering, 10' O.C.										
	0540	Black, 1/4" diameter	1 Plum	66	.121	L.F.	1.50	2.53		4.03	5.85	
	0570	3/4" diameter		61	.131		1.66	2.73		4.39	6.35	
	0580	1" diameter		53	.151		2.17	3.15		5.32	7.60	
	0590	1-1/4" diameter	Q-1	89	.180		2.79	3.38		6.17	8.65	
	0600	1-1/2" diameter		80	.200		3.05	3.76		6.81	9.55	
	0610	2" diameter		64	.250		4.12	4.69		8.81	12.35	
	2000	Welded, sch. 40, on yoke & roll hangers, sized for covering,										
	2010	10' O.C. (no hangers incl. for 14" diam. and up)										
	2080	2-1/2" diameter	Q-15	47	.340	L.F.	5	6.40	1.03	12.43	17.25	
	2090	3" diameter		43	.372		5.95	7	1.13	14.08	19.35	
	2100	3-1/2" diameter		39	.410		7.40	7.70	1.24	16.34	22.50	
	2110	4" diameter		37	.432		8.45	8.10	1.31	17.86	24	
	2120	5" diameter		32	.500		13.80	9.40	1.52	24.72	32.50	
	2130	6" diameter	Q-16	36	.667		17.40	13	1.35	31.75	42	
716	0010	**PIPE, STEEL, FITTINGS** Threaded										716
	0020	Cast Iron										
	0040	Standard weight, black										
	0060	90° Elbow, straight										
	0100	3/4"	1 Plum	14	.571	Ea.	1.82	11.90		13.72	22	
	0110	1"	"	13	.615		2.24	12.85		15.09	24	
	0130	1-1/2"	Q-1	20	.800		4.37	15		19.37	30	
	0140	2"	"	18	.889		6.80	16.70		23.50	35	
	0500	Tee, straight										
	0530	1/2"	1 Plum	9	.889	Ea.	2.69	18.55		21.24	33.50	
	0540	3/4"		9	.889		3.17	18.55		21.72	34	
	0550	1"		8	1		2.90	21		23.90	37.50	
	0570	1-1/2"	Q-1	13	1.231		6.95	23		29.95	46	
	0580	2"	"	11	1.455		9.55	27.50		37.05	55.50	
	6000	For galvanized elbows, tees, and couplings add						20%				

151 800 | Grooved-Joint Pipe

			CREW	DAILY OUTPUT	LABOR-HOURS	UNIT	MAT.	LABOR	EQUIP.	TOTAL	TOTAL INCL O&P	
801	0010	**PIPE, GROOVED-JOINT STEEL FITTINGS & VALVES**										801
	0020	Pipe includes coupling & clevis type hanger 10' O.C.										
	1000	Schedule 40, black										
	1040	3/4" diameter	1 Plum	71	.113	L.F.	1.80	2.35		4.15	5.85	
	1050	1" diameter		63	.127		1.97	2.65		4.62	6.55	
	1060	1-1/4" diameter		58	.138		2.48	2.88		5.36	7.50	
	1070	1-1/2" diameter		51	.157		2.80	3.27		6.07	8.50	
	1080	2" diameter		40	.200		3.36	4.17		7.53	10.60	

Important: See the Reference Section for critical supporting data - Reference Nos., Crews, & Location Factors

151 800 | Grooved-Joint Pipe

			CREW	DAILY OUTPUT	LABOR-HOURS	UNIT	MAT.	LABOR	EQUIP.	TOTAL	TOTAL INCL O&P	
801	1090	2-1/2" diameter	Q-1	57	.281	L.F.	4.83	5.25		10.08	14.05	**801**
	1100	3" diameter		50	.320		5.95	6		11.95	16.50	
	1110	4" diameter		45	.356		8.35	6.70		15.05	20	
	4000	Elbow, 90° or 45°, painted										
	4040	1" diameter	1 Plum	50	.160	Ea.	10.65	3.34		13.99	17.30	
	4050	1-1/4" diameter		40	.200		10.65	4.17		14.82	18.65	
	4060	1-1/2" diameter		33	.242		10.65	5.05		15.70	20	
	4070	2" diameter		25	.320		10.65	6.65		17.30	23	
	4080	2-1/2" diameter	Q-1	40	.400		10.65	7.50		18.15	24	
	4100	4" diameter	"	25	.640		20.50	12		32.50	43	
	4250	For galvanized elbows, add					26%					
	4690	Tee, painted										
	4700	3/4" diameter	1 Plum	38	.211	Ea.	16.45	4.39		20.84	25.50	
	4740	1" diameter		33	.242		16.45	5.05		21.50	26.50	
	4750	1-1/4" diameter		27	.296		16.45	6.20		22.65	28.50	
	4760	1-1/2" diameter		22	.364		16.45	7.60		24.05	30.50	
	4770	2" diameter		17	.471		16.45	9.80		26.25	34.50	
	4780	2-1/2" diameter	Q-1	27	.593		16.45	11.15		27.60	36.50	
	4800	4" diameter	"	17	.941		35	17.70		52.70	68	
	4900	For galvanized tees, add					24%					

151 950 | Valves

			CREW	DAILY OUTPUT	LABOR-HOURS	UNIT	MAT.	LABOR	EQUIP.	TOTAL	TOTAL INCL O&P	
955	0010	**VALVES, BRONZE**										**955**
	1380	Ball, 150 psi, threaded										
	1450	1/2" size	1 Plum	22	.364	Ea.	6.10	7.60		13.70	19.25	
	1460	3/4" size		20	.400		10.05	8.35		18.40	25	
	1470	1" size		19	.421		12.65	8.80		21.45	28.50	
	1480	1-1/4" size		15	.533		21.50	11.10		32.60	42	
	1490	1-1/2" size		13	.615		27	12.85		39.85	51.50	
	1500	2" size		11	.727		34	15.15		49.15	62.50	
	1600	Butterfly, 175 psi, full port, solder or threaded ends										
	1610	Stainless steel disc and stem										
	1620	1/4" size	1 Plum	24	.333	Ea.	5.55	6.95		12.50	17.60	
	1630	3/8" size		24	.333		5.55	6.95		12.50	17.60	
	1640	1/2" size		22	.364		6.25	7.60		13.85	19.45	
	1650	3/4" size		20	.400		9.55	8.35		17.90	24.50	
	1660	1" size		19	.421		11.65	8.80		20.45	27.50	
	1670	1-1/4" size		15	.533		18.75	11.10		29.85	39	
	1680	1-1/2" size		13	.615		24.50	12.85		37.35	48	
	1690	2" size		11	.727		30.50	15.15		45.65	58.50	
	1750	Check, swing, class 150, regrinding disc, threaded										
	1850	1/2" size	1 Plum	24	.333	Ea.	25	6.95		31.95	39	
	1860	3/4" size		20	.400		36	8.35		44.35	53.50	
	1870	1" size		19	.421		53	8.80		61.80	73	
	1880	1-1/4" size		15	.533		74.50	11.10		85.60	100	
	1900	2" size		11	.727		129	15.15		144.15	167	
	2850	Gate, N.R.S., soldered, 125 psi										
	2920	1/2" size	1 Plum	24	.333	Ea.	13.40	6.95		20.35	26	
	2940	3/4" size		20	.400		15.60	8.35		23.95	31	
	2950	1" size		19	.421		22	8.80		30.80	38.50	
	2960	1-1/4" size		15	.533		29.50	11.10		40.60	51	
	2970	1-1/2" size		13	.615		35.50	12.85		48.35	60.50	
	2980	2" size		11	.727		49.50	15.15		64.65	79.50	
	4250	Threaded, class 150										
	4340	3/4" size	1 Plum	20	.400	Ea.	21	8.35		29.35	37	
	4350	1" size		19	.421		27.50	8.80		36.30	44.50	

MECHANICAL 15

151 950 | Valves

			CREW	DAILY OUTPUT	LABOR-HOURS	UNIT	MAT.	LABOR	EQUIP.	TOTAL	TOTAL INCL O&P	
955	4360	1-1/4" size	1 Plum	15	.533	Ea.	37.50	11.10		48.60	59.50	955
	4370	1-1/2" size	↓	13	.615	↓	45.50	12.85		58.35	72	
	5600	Relief, pressure & temperature, self-closing, ASME, threaded										
	5650	1" size	1 Plum	24	.333	Ea.	86.50	6.95		93.45	107	
	5660	1-1/4" size	"	20	.400	"	173	8.35		181.35	204	
	6400	Pressure, water, ASME, threaded										
	6440	3/4" size	1 Plum	28	.286	Ea.	39.50	5.95		45.45	53.50	
	6450	1" size		24	.333		73	6.95		79.95	91.50	
	6460	1-1/4" size		20	.400		125	8.35		133.35	152	
	6470	1-1/2" size		18	.444		173	9.25		182.25	205	
	6480	2" size	↓	16	.500	↓	238	10.45		248.45	279	
	6900	Reducing, water pressure										
	6940	1/2" size	1 Plum	24	.333	Ea.	69.50	6.95		76.45	88	
	6950	3/4" size		20	.400		81.50	8.35		89.85	104	
	6960	1" size	↓	19	.421	↓	126	8.80		134.80	154	
	8350	Tempering, water, sweat connections										
	8400	1/2" size	1 Plum	24	.333	Ea.	36	6.95		42.95	51	
	8440	3/4" size	"	20	.400	"	44	8.35		52.35	62.50	
960	0010	**VALVES, IRON BODY**										960
	1650	Gate, 125 lb., N.R.S.										
	2150	Flanged										
	2240	2-1/2" size	Q-1	5	3.200	Ea.	244	60		304	370	
	2260	3" size	↓	4.50	3.556		274	67		341	410	
	2280	4" size	↓	3	5.333	↓	390	100		490	595	
	3550	OS&Y, 125 lb., flanged										
	3660	3" size	Q-1	4.50	3.556	Ea.	157	67		224	283	
	3680	4" size	"	3	5.333	↓	222	100		322	410	
	3700	6" size	Q-2	3	8	↓	370	145		515	645	
	5450	Swing check, 125 lb., threaded										
	5500	2" size	1 Plum	11	.727	Ea.	270	15.15		285.15	320	
	5950	Flanged										
	6040	2-1/2" size	Q-1	5	3.200	Ea.	152	60		212	267	
	6050	3" size	↓	4.50	3.556		227	67		294	360	
	6060	4" size	↓	3	5.333		258	100		358	450	
	6070	6" size	Q-2	3	8	↓	440	145		585	725	

152 100 | Fixtures

			CREW	DAILY OUTPUT	LABOR-HOURS	UNIT	MAT.	LABOR	EQUIP.	TOTAL	TOTAL INCL O&P	
104	0010	**BATHS**										104
	0100	Tubs, recessed porcelain enamel on cast iron, with trim										
	0180	48" x 42"	Q-1	4	4	Ea.	1,150	75		1,225	1,400	
	0300	Mat bottom, 4' long		5.50	2.909		975	54.50		1,029.50	1,175	
	0380	5' long		4.40	3.636		375	68.50		443.50	525	
	0480	Above floor drain, 5' long		4	4		625	75		700	810	
	0560	Corner 48" x 44"		4.40	3.636		1,325	68.50		1,393.50	1,575	
	2000	Enameled formed steel, 4'-6" long		5.80	2.759		264	52		316	375	
	2200	5' long	↓	5.50	2.909	↓	246	54.50		300.50	360	
	4600	Module tub & showerwall surround, molded fiberglass										

The note beside row 0010/0100: R151 -420

152 100	Fixtures		CREW	DAILY OUTPUT	LABOR-HOURS	UNIT	1998 BARE COSTS				TOTAL INCL O&P	
							MAT.	LABOR	EQUIP.	TOTAL		
104	4610	5' long x 34" wide x 76" high	Q-1	4	4	Ea.	680	75		755	875	**104**
	4750	Handicap with 1-1/2" OD grab bar, antiskid bottom	R151 -420									
	4760	60" x 32-3/4" x 72" high	Q-1	4	4	Ea.	675	75		750	865	
	4770	60" x 30" x 71" high with molded seat		3.50	4.571		825	86		911	1,050	
	5300	Whirlpool, porcelain enamel on cast iron, 72" x 36"		1	16		2,800	300		3,100	3,575	
	6000	Whirlpool, bath with vented overflow, molded fiberglass										
	6100	66" x 48" x 24"	Q-1	1	16	Ea.	2,425	300		2,725	3,175	
	6400	72" x 36" x 24"		1	16		2,325	300		2,625	3,050	
	6500	60" x 30" x 21"		1	16		1,875	300		2,175	2,550	
	7000	Redwood hot tub system										
	7050	4' diameter x 4' deep	Q-1	1	16	Ea.	925	300		1,225	1,525	
	7150	6' diameter x 4' deep		.80	20		1,525	375		1,900	2,300	
	7200	8' diameter x 4' deep		.80	20		2,125	375		2,500	2,975	
	9600	Rough-in, supply, waste and vent, for all above tubs, add		2.07	7.729		136	145		281	390	
128	0010	**INTERCEPTORS**										**128**
	0150	Grease, cast iron, 4 GPM, 8 lb. fat capacity	1 Plum	4	2	Ea.	252	41.50		293.50	345	
	0200	7 GPM, 14 lb. fat capacity		4	2		350	41.50		391.50	455	
	1040	15 GPM, 30 lb. fat capacity		4	2		610	41.50		651.50	740	
	1060	20 GPM, 40 lb. fat capacity		3	2.667		745	55.50		800.50	910	
	3000	Hair, cast iron, 1-1/4" and 1-1/2" pipe connection		8	1		94.50	21		115.50	139	
	3100	For chrome-plated cast iron, add					58			58	64	
	4000	Oil, fabricated steel, 10 GPM, 2" pipe size	1 Plum	4	2		535	41.50		576.50	655	
	4100	15 GPM, 2" or 3" pipe size		4	2		730	41.50		771.50	875	
	4120	20 GPM, 2" or 3" pipe size		3	2.667		885	55.50		940.50	1,050	
	6000	Solids, precious metals recovery, C.I., 1-1/4" to 2" pipe		4	2		143	41.50		184.50	226	
	6100	Dental Lab., large, C.I., 1-1/2" to 2" pipe		3	2.667		500	55.50		555.50	640	
136	0010	**LAVATORIES** With trim, white unless noted otherwise										**136**
	0500	Vanity top, porcelain enamel on cast iron										
	0600	20" x 18"	Q-1	6.40	2.500	Ea.	175	47		222	271	
	0640	33" x 19" oval	"	6.40	2.500	"	355	47		402	470	
	0860	For color, add					25%					
	1000	Cultured marble, 19" x 17", single bowl	Q-1	6.40	2.500	Ea.	97	47		144	184	
	1120	25" x 22", single bowl		6.40	2.500		133	47		180	225	
	1900	Stainless steel, self-rimming, 25" x 22", single bowl, ledge		6.40	2.500		147	47		194	240	
	1960	17" x 22", single bowl		6.40	2.500		143	47		190	236	
	2600	Steel, enameled, 20" x 17", single bowl		5.80	2.759		87	52		139	182	
	2900	Vitreous china, 20" x 16", single bowl		5.40	2.963		179	55.50		234.50	289	
	3200	22" x 13", single bowl		5.40	2.963		189	55.50		244.50	300	
	3580	Rough-in, supply, waste and vent for all above lavatories		2.30	6.957		93.50	131		224.50	320	
	4000	Wall hung										
	4040	Porcelain enamel on cast iron, 16" x 14", single bowl	Q-1	8	2	Ea.	370	37.50		407.50	465	
	4180	20" x 18", single bowl	"	8	2	"	204	37.50		241.50	286	
	4580	For color, add					30%					
	6000	Vitreous china, 18" x 15", single bowl with backsplash	Q-1	7	2.286	Ea.	196	43		239	287	
	6060	19" x 17", single bowl		7	2.286		137	43		180	222	
	6960	Rough-in, supply, waste and vent for above lavatories		1.66	9.639		180	181		361	500	
140	0010	**LAUNDRY SINKS** With trim										**140**
	0020	Porcelain enamel on cast iron, black iron frame										
	0050	24" x 20", single compartment	Q-1	6	2.667	Ea.	295	50		345	410	
	3000	Plastic, on wall hanger or legs										
	3100	20" x 24", single compartment	Q-1	6.50	2.462	Ea.	105	46		151	192	
	3200	36" x 23", double compartment		5.50	2.909		123	54.50		177.50	227	
	9600	Rough-in, supply, waste and vent, for all laundry sinks		2.14	7.477		106	140		246	350	
148	0010	**SHOWERS**										**148**
	1500	Stall, with door and trim										

MECHANICAL 15

152 100		Fixtures	CREW	DAILY OUTPUT	LABOR-HOURS	UNIT	1998 BARE COSTS				TOTAL INCL O&P
							MAT.	LABOR	EQUIP.	TOTAL	
148	1520	32" square	Q-1	2	8	Ea.	350	150		500	635
	1530	36" square		2	8		400	150		550	690
	1540	Terrazzo receptor, 32" square		2	8		745	150		895	1,075
	1560	36" square		1.80	8.889		800	167		967	1,150
	1580	36" corner angle		1.80	8.889		730	167		897	1,075
	3000	Fiberglass, one piece, with 3 walls, 32" x 32" square		2.40	6.667		380	125		505	625
	3100	36" x 36" square		2.40	6.667		435	125		560	680
	3200	Handicap, 1-1/2" O.D. grab bars, nonskid floor									
	3210	48" x 34-1/2" x 72" corner seat	Q-1	2.40	6.667	Ea.	650	125		775	920
	3250	64" x 65-3/4" x 81-1/2" fold. seat, whlchr.		1.80	8.889		3,275	167		3,442	3,875
	4000	Polypropylene, with molded-stone floor, 30" x 30"		2	8		345	150		495	625
	4200	Rough-in, supply, waste and vent for above showers		2.05	7.805		102	147		249	355
	5000	Built-in, head, arm, 4 GPM valve	1 Plum	4	2		78	41.50		119.50	155
	5200	Head, arm, by-pass, integral stops, handles		3.60	2.222		230	46.50		276.50	330
	5800	Mixing valve, built-in		6	1.333		98.50	28		126.50	154
152	0010	**SINKS** With faucets and drain									
	2000	Kitchen, counter top style, P.E. on C.I., 24" x 21" single bowl	Q-1	5.60	2.857	Ea.	214	53.50		267.50	325
	2100	30" x 21" single bowl		5.60	2.857		282	53.50		335.50	400
	2200	32" x 21" double bowl		4.80	3.333		315	62.50		377.50	450
	3000	Stainless steel, self rimming, 19" x 18" single bowl		5.60	2.857		249	53.50		302.50	365
	3100	25" x 22" single bowl		5.60	2.857		274	53.50		327.50	390
	3300	43" x 22" double bowl		4.80	3.333		455	62.50		517.50	605
	4000	Steel, enameled, with ledge, 24" x 21" single bowl		5.60	2.857		102	53.50		155.50	201
	4100	32" x 21" double bowl		4.80	3.333		119	62.50		181.50	234
	4960	For color sinks except stainless steel, add					10%				
	4980	For rough-in, supply, waste and vent, counter top sinks	Q-1	2.14	7.477		106	140		246	350
	5000	Kitchen, raised deck, P.E. on C.I.									
	5100	32" x 21", dual level, double bowl	Q-1	2.60	6.154	Ea.	310	116		426	530
	5790	For rough-in, supply, waste & vent, sinks		1.85	8.649		106	162		268	385
	6650	Service, floor, corner, P.E. on C.I., 28" x 28"		4.40	3.636		555	68.50		623.50	725
	6750	Vinyl coated rim guard, add					66			66	72.50
	6790	For rough-in, supply, waste & vent, floor service sinks	Q-1	1.64	9.756		234	183		417	560
160	0010	**SINK WASTE TREATMENT** System for commercial kitchens									
	0100	includes clock timer, & fittings									
	0200	System less chemical, wall mounted cabinet	1 Plum	16	.500	Ea.	270	10.45		280.45	315
	2000	Chemical, 1 gallon, add					29			29	32
	2100	6 gallons, add					159			159	175
	2200	15 gallons, add					360			360	395
	2300	30 gallons, add					670			670	735
	2400	55 gallons, add					1,125			1,125	1,250
168	0010	**URINALS**									
	3000	Wall hung, vitreous china, with hanger & self-closing valve									
	3100	Siphon jet type	Q-1	3	5.333	Ea.	325	100		425	520
	3120	Blowout type		3	5.333		375	100		475	580
	3300	Rough-in, supply, waste & vent		2.83	5.654		90.50	106		196.50	276
	5000	Stall type, vitreous china, includes valve		2.50	6.400		580	120		700	835
	5100	3" seam cover, add		12	1.333		132	25		157	187
	6980	Rough-in, supply, waste and vent		1.99	8.040		133	151		284	395
176	0010	**WASH FOUNTAINS** Rigging not included									
	1900	Group, foot control									
	2000	Precast terrazzo, circular, 36" diam., 5 or 6 persons	Q-2	3	8	Ea.	1,800	145		1,945	2,250
	2100	54" diameter for 8 or 10 persons		2.50	9.600		2,250	174		2,424	2,775
	2400	Semi-circular, 36" diam. for 3 persons		3	8		1,650	145		1,795	2,075
	2500	54" diam. for 4 or 5 persons		2.50	9.600		2,025	174		2,199	2,525

Important: See the Reference Section for critical supporting data - Reference Nos., Crews, & Location Factors

152 100 | Fixtures

		CREW	DAILY OUTPUT	LABOR-HOURS	UNIT	1998 BARE COSTS MAT.	LABOR	EQUIP.	TOTAL	TOTAL INCL O&P	
176	5610	Group, infrared control, barrier free										**176**
	5614	Precast terrazzo									
	5620	Semi-circular 36″ diam. for 3 persons	Q-2	3	8	Ea.	2,525	145		2,670	3,050
	5630	46″ diam. for 4 persons		2.80	8.571		2,775	155		2,930	3,300
	5640	Circular, 54″ diam. for 8 persons, button control		2.50	9.600		4,200	174		4,374	4,900
180	0010	**WATER CLOSETS**										**180**
	0150	Tank type, vitreous china, incl. seat, supply pipe w/stop									
	0200	Wall hung, one piece	Q-1	5.30	3.019	Ea.	650	56.50		706.50	810
	0960	For rough-in, supply, waste, vent and carrier		2.73	5.861		166	110		276	365
	1000	Floor mounted, one piece		5.30	3.019		435	56.50		491.50	575
	1020	One piece, low profile		5.30	3.019		795	56.50		851.50	970
	1100	Two piece, close coupled, water saver		5.30	3.019		145	56.50		201.50	254
	1960	For color, add					30%				
	1980	For rough-in, supply, waste and vent	Q-1	3.05	5.246	Ea.	121	98.50		219.50	296
	3000	Bowl only, with flush valve, seat									
	3100	Wall hung	Q-1	5.80	2.759	Ea.	375	52		427	500
	3200	For rough-in, supply, waste and vent, single WC		2.56	6.250		177	117		294	390
	3300	Floor mounted		5.80	2.759		310	52		362	425
	3400	For rough-in, supply, waste and vent, single WC		2.84	5.634		132	106		238	320
184	0010	**WATER FILTERS** Purification and treatment										**184**
	1000	Cartridge style, dirt and rust type	1 Plum	12	.667	Ea.	25.50	13.90		39.40	51
	1200	Replacement cartridge		32	.250		.80	5.20		6	9.55
	1600	Taste and odor type		12	.667		25	13.90		38.90	50
	1700	Replacement cartridge		32	.250		1.86	5.20		7.06	10.70
	8000	Commercial, fully automatic or push button automatic									
	8200	Iron removal, 660 GPH, 1″ pipe size	Q-1	1.50	10.667	Ea.	1,275	200		1,475	1,725
	8240	1500 GPH, 1-1/4″ pipe size	″	1	16	″	2,025	300		2,325	2,725

152 400 | Pumps

		CREW	DAILY OUTPUT	LABOR-HOURS	UNIT	1998 BARE COSTS MAT.	LABOR	EQUIP.	TOTAL	TOTAL INCL O&P	
410	0010	**PUMPS, CIRCULATING** Heated or chilled water application										**410**
	0600	Bronze, sweat connections, 1/40 HP, in line									
	0640	3/4″ size	Q-1	16	1	Ea.	107	18.80		125.80	149
	1000	Flange connection, 3/4″ to 1-1/2″ size									
	1040	1/12 HP	Q-1	6	2.667	Ea.	293	50		343	405
	1060	1/8 HP		6	2.667		505	50		555	640
	1100	1/3 HP		6	2.667		550	50		600	690
	1140	2″ size, 1/6 HP		5	3.200		720	60		780	895
	2000	Cast iron, flange connection									
	2040	3/4″ to 1-1/2″ size, in line, 1/12 HP	Q-1	6	2.667	Ea.	191	50		241	293
	2060	1/8 HP		6	2.667		320	50		370	435
	2100	1/3 HP		6	2.667		355	50		405	475
	2101	Pumps, circulating, 3/4″ to 1-1/2″ size, 1/3 HP		6	2.667		355	50		405	475
	2140	2″ size, 1/6 HP		5	3.200		390	60		450	530
	2180	2-1/2″ size, 1/4 HP		5	3.200		510	60		570	660
	2220	3″ size, 1/4 HP		4	4		520	75		595	695
	2260	1/3 HP		4	4		700	75		775	895
	2300	1/2 HP		4	4		730	75		805	925
	2340	3/4 HP		4	4		840	75		915	1,050
	2380	1 HP		4	4		1,200	75		1,275	1,450
	4000	Close coupled, end suction, bronze impeller									
	4040	1-1/2″ size, 1-1/2 HP, to 40 GPM	Q-1	3	5.333	Ea.	1,000	100		1,100	1,275
	4090	2″ size, 2 HP, to 50 GPM		3	5.333		1,025	100		1,125	1,300
	4100	2″ size, 3 HP, to 90 GPM		2.30	6.957		1,125	131		1,256	1,475
	4190	2-1/2″ size, 3 HP, to 150 GPM		2	8		1,225	150		1,375	1,600
	4300	3″ size, 5 HP, to 225 GPM		1.80	8.889		1,325	167		1,492	1,725

152 400 | Pumps

		CREW	DAILY OUTPUT	LABOR-HOURS	UNIT	1998 BARE COSTS MAT.	LABOR	EQUIP.	TOTAL	TOTAL INCL O&P	
410	4410	3" size, 10 HP, to 350 GPM	Q-1	1.60	10	Ea.	1,900	188		2,088	2,400	**410**
	4420	4" size, 7-1/2 HP, to 350 GPM	↓	1.60	10	↓	1,850	188		2,038	2,325	
	5000	Base mounted, bronze impeller, coupling guard										
	5190	2-1/2" size, 3 HP, to 150 GPM	Q-1	1.80	8.889	Ea.	1,650	167		1,817	2,100	
	5300	3" size, 5 HP, to 225 GPM		1.60	10		1,750	188		1,938	2,225	
	5410	4" size, 5 HP, to 350 GPM	↓	1.50	10.667	↓	1,900	200		2,100	2,425	
450	0010	**PUMPS, PRESSURE BOOSTER SYSTEM**										**450**
	0200	Pump system, with diaphragm tank, control, press. switch										
	0300	1 HP pump	Q-1	1.30	12.308	Ea.	2,775	231		3,006	3,425	
	0400	1-1/2 HP pump		1.25	12.800		2,800	240		3,040	3,500	
	0420	2 HP pump		1.20	13.333		2,875	250		3,125	3,600	
	0440	3 HP pump	↓	1.10	14.545		2,925	273		3,198	3,675	
	0460	5 HP pump	Q-2	1.50	16		3,225	289		3,514	4,025	
	0480	7-1/2 HP pump	↓	1.42	16.901		3,600	305		3,905	4,450	
	0500	10 HP pump	↓	1.34	17.910	↓	3,775	325		4,100	4,675	
	1000	Pump/ energy storage system, diaphragm tank, 3 HP pump										
	1100	motor, PRV, switch, gauge, control center, flow switch										
	1200	125 lb. working pressure	Q-2	.70	34.286	Ea.	8,100	620		8,720	9,925	
	1300	250 lb. working pressure	"	.64	37.500	"	8,850	680		9,530	10,900	
460	0010	**PUMPS, PEDESTAL SUMP** With float control										**460**
	0400	Molded PVC base, 21 GPM at 15' head, 1/3 HP	1 Plum	5	1.600	Ea.	86.50	33.50		120	151	
	0800	Iron base, 21 GPM at 15' head, 1/3 HP		5	1.600		110	33.50		143.50	177	
	1200	Solid brass, 21 GPM at 15' head, 1/3 HP	↓	5	1.600	↓	179	33.50		212.50	·253	
465	0010	**PUMPS, SEWAGE EJECTOR** With operating and level controls										**465**
	0100	Simplex system incl. tank, cover, pump 15' head										
	0500	37 gal PE tank, 12 GPM, 1/2 HP, 2" discharge	Q-1	3.20	5	Ea.	370	94		464	560	
	0510	3" discharge		3.10	5.161		395	97		492	595	
	0600	45 gal. coated stl tank, 12 GPM, 1/2 HP, 2" discharge		3	5.333		630	100		730	860	
	0610	3" discharge		2.90	5.517		660	104		764	895	
	0700	70 gal. PE tank, 12 GPM, 1/2 HP, 2" discharge		2.60	6.154		720	116		836	980	
	0710	3" discharge		2.40	6.667		750	125		875	1,025	
	0730	87 GPM, 0.7 HP, 2" discharge		2.50	6.400		905	120		1,025	1,200	
	0740	3" discharge		2.30	6.957		965	131		1,096	1,300	
	0760	134 GPM, 1 HP, 2" discharge		2.20	7.273		990	137		1,127	1,325	
	0770	3" discharge	↓	2	8	↓	1,050	150		1,200	1,400	
	1040	Duplex system incl. tank, covers, pumps										
	1060	110 gal. fiberglass tank, 24 GPM, 1/2 HP, 2" discharge	Q-1	1.60	10	Ea.	1,400	188		1,588	1,850	
	1080	3" discharge		1.40	11.429		1,475	215		1,690	1,975	
	1100	174 GPM, .7 HP, 2" discharge		1.50	10.667		1,850	200		2,050	2,350	
	1120	3" discharge		1.30	12.308		1,925	231		2,156	2,500	
	1140	268 GPM, 1 HP, 2" discharge		1.20	13.333		1,975	250		2,225	2,600	
	1160	3" discharge	↓	1	16		2,075	300		2,375	2,775	
	1260	135 gal. coated stl. tank, 24 GPM, 1/2 HP, 2" discharge	Q-2	1.70	14.118		1,350	255		1,605	1,900	
	2000	3" discharge		1.60	15		1,400	271		1,671	1,975	
	2640	174 GPM, .7 HP, 2" discharge		1.60	15		1,725	271		1,996	2,350	
	2660	3" discharge		1.50	16		1,825	289		2,114	2,500	
	2700	268 GPM, 1 HP, 2" discharge		1.30	18.462		1,900	335		2,235	2,625	
	3040	3" discharge	↓	1.10	21.818	↓	2,000	395		2,395	2,850	

Important: See the Reference Section for critical supporting data - Reference Nos., Crews, & Location Factors

153 100 | Water Appliances

		CREW	DAILY OUTPUT	LABOR-HOURS	UNIT	1998 BARE COSTS				TOTAL INCL O&P	
						MAT.	LABOR	EQUIP.	TOTAL		
105	**0010**	**WATER COOLER**									**105**
	0100	Wall mounted, non-recessed									
	0180	8.2 GPH	Q-1	4	4	Ea.	470	75		545	640
	0600	For hot and cold water, add					120			120	132
	1000	Dual height, 8.2 GPH	Q-1	3.80	4.211		645	79		724	840
	2600	Wheelchair type, 8 GPH	"	4	4	↓	1,100	75		1,175	1,350
	4600	Floor mounted, flush-to-wall									
	4680	8.2 GPH	1 Plum	3	2.667	Ea.	475	55.50		530.50	615
	9800	For supply, waste & vent, all coolers	"	2.21	3.620	"	71	75.50		146.50	203
110	**0010**	**WATER HEATERS**									**110**
	1000	Residential, electric, glass lined tank, 5 yr, 10 gal., single element	1 Plum	2.30	3.478	Ea.	190	72.50		262.50	330
	1060	30 gallon, double element		2.20	3.636		264	76		340	415
	1100	52 gallon, double element		2	4		320	83.50		403.50	490
	2000	Gas fired, glass lined tank, 5 yr, vent not incl., 20 gallon		2.10	3.810		280	79.50		359.50	440
	2040	30 gallon		2	4		287	83.50		370.50	455
	2100	75 gallon		1.50	5.333		630	111		741	880
	3000	Oil fired, glass lined tank, 5 yr, vent not included, 30 gallon		2	4		735	83.50		818.50	945
	3040	50 gallon	↓	1.80	4.444	↓	1,050	92.50		1,142.50	1,300
	4000	Commercial, 100° rise. NOTE: for each size tank, a range of									
	4010	heaters between the ones shown are available									
	4020	Electric									
	4100	5 gal., 3 kW, 12 GPH	1 Plum	2	4	Ea.	1,625	83.50		1,708.50	1,950
	4120	10 gal., 6 kW, 25 GPH		2	4		1,800	83.50		1,883.50	2,150
	4140	50 gal., 9 kW, 37 GPH		1.80	4.444		2,475	92.50		2,567.50	2,875
	4160	50 gal., 36 kW, 148 GPH		1.80	4.444		3,800	92.50		3,892.50	4,325
	4200	80 gal., 36 kW, 148 GPH		1.50	5.333		4,250	111		4,361	4,850
	4220	100 gal., 36 kW, 148 GPH		1.20	6.667		4,425	139		4,564	5,075
	4240	120 gal., 36 kW, 148 GPH		1.20	6.667		4,600	139		4,739	5,300
	4280	150 gal., 120 kW, 490 GPH	↓	1	8		16,700	167		16,867	18,600
	4300	200 gal., 15 kW, 61 GPH	Q-1	1.70	9.412	↓	11,900	177		12,077	13,400
	6000	Gas fired, flush jacket, std. controls, vent not incl.									
	6040	75 MBH input, 73 GPH	1 Plum	1.40	5.714	Ea.	1,100	119		1,219	1,425
	6060	98 MBH input, 95 GPH		1.40	5.714		1,900	119		2,019	2,300
	6080	120 MBH input, 114 GPH		1.20	6.667		2,100	139		2,239	2,525
	6120	135 MBH input, 130 GPH		1	8		1,775	167		1,942	2,225
	6140	155 MBH input, 150 GPH		.80	10		1,825	209		2,034	2,350
	6180	200 MBH input, 192 GPH	↓	.60	13.333		2,450	278		2,728	3,150
	6220	295 MBH input, 278 GPH	Q-1	.80	20		3,725	375		4,100	4,725
	6240	365 MBH input, 374 GPH		.80	20		4,175	375		4,550	5,200
	6260	500 MBH input, 480 GPH		.70	22.857		5,700	430		6,130	6,975
	6280	600 MBH input, 576 GPH		.60	26.667		6,575	500		7,075	8,050
	6320	1000 MBH input, 960 GPH		.50	32		9,450	600		10,050	11,400
	6340	1200 MBH input, 1150 GPH	↓	.40	40	↓	11,800	750		12,550	14,200
	8000	Oil fired, flush jacket, std. controls, vent not incl.									
	8060	103 MBH gross output, 116 GPH	1 Plum	1.10	7.273	Ea.	1,700	152		1,852	2,125
	8080	122 MBH gross output, 141 GPH		1	8		1,750	167		1,917	2,200
	8120	168 MBH gross output, 192 GPH		.60	13.333		2,250	278		2,528	2,925
	8140	195 MBH gross output, 224 GPH	↓	.50	16		2,375	335		2,710	3,175
	8180	262 MBH gross output, 315 GPH	Q-1	.70	22.857		3,075	430		3,505	4,075
	8200	315 MBH gross output, 409 GPH		.70	22.857		4,050	430		4,480	5,150
	8220	420 MBH gross output, 504 GPH		.60	26.667		4,650	500		5,150	5,950
	8240	525 MBH gross output, 630 GPH	↓	.50	32		5,950	600		6,550	7,550
120	**0010**	**WATER HEATER PACKAGED SYSTEMS**									**120**
	1000	Car wash package, continuous duty, high recovery, gas fired									

MECHANICAL 15

For expanded coverage of these items see *Means Mechanical or Plumbing Cost Data 1998*

		153 100	Water Appliances	CREW	DAILY OUTPUT	LABOR-HOURS	UNIT	1998 BARE COSTS MAT.	LABOR	EQUIP.	TOTAL	TOTAL INCL O&P	
120	1040		100° rise, 180 MBH input, 174 GPH	1 Plum	3	2.667	Ea.	2,875	55.50		2,930.50	3,275	120
	1060		280 MBH input, 270 GPH		2.50	3.200		3,075	66.50		3,141.50	3,475	
	1100		480 MBH input, 464 GPH		2	4		3,725	83.50		3,808.50	4,250	
	1120		605 MBH input, 584 GPH		1.50	5.333		4,250	111		4,361	4,850	
	1140		700 MBH input, 676 GPH		1	8		4,425	167		4,592	5,150	
	3000		Combination dishwasher & general purpose, 2 temp. gas fired										
	3040		124 GPH @ 140° rise; 434 GPH @ 40° rise	1 Plum	2	4	Ea.	2,925	83.50		3,008.50	3,375	
	3060		193 GPH @ 140° rise; 677 GPH @ 40° rise	"	1.60	5	"	3,350	104		3,454	3,875	
	5000		Coin laundry units, gas fired, 100° rise										
	5020		Single heater,										
	5040		280 MBH input, 270 GPH	1 Plum	1.60	5	Ea.	3,350	104		3,454	3,875	
	5060		400 MBH input, 386 GPH	"	1.30	6.154	"	4,425	128		4,553	5,075	
160	0010		WATER SUPPLY METERS										160
	2000		Domestic/commercial, bronze										
	2020		Threaded										
	2080		3/4" diameter, to 30 GPM	1 Plum	14	.571	Ea.	115	11.90		126.90	147	
	2100		1" diameter, to 50 GPM	"	12	.667	"	157	13.90		170.90	195	
	2300		Threaded/flanged										
	2340		1-1/2" diameter, to 100 GPM	1 Plum	8	1	Ea.	485	21		506	570	
	2360		2" diameter, to 160 GPM	"	6	1.333	"	645	28		673	755	
	2600		Flanged, compound										
	2601		Water supply meters, bronze, flanged, compound										
	2640		3" diameter, 320 GPM	Q-1	3	5.333	Ea.	2,900	100		3,000	3,375	
	2660		4" diameter, to 500 GPM	"	1.50	10.667	"	4,525	200		4,725	5,300	

		154 100	Fire Systems	CREW	DAILY OUTPUT	LABOR-HOURS	UNIT	1998 BARE COSTS MAT.	LABOR	EQUIP.	TOTAL	TOTAL INCL O&P	
101	0010		AUTOMATIC FIRE SUPPRESSION SYSTEMS										101
	0040		For detectors and control stations, see division 168-120										
	0100		Control panel, single zone with batteries (2 zones det., 1 suppr.)	1 Elec	1	8	Ea.	1,375	165		1,540	1,800	
	0150		Multizone (4) with batteries (8 zones det., 4 suppr.)	"	.50	16		2,450	330		2,780	3,250	
	1000		Dispersion nozzle, CO2, 3" x 5"	1 Plum	18	.444		50	9.25		59.25	70.50	
	1100		FM200, 1-1/2"	"	14	.571		50	11.90		61.90	75	
	2000		Extinguisher, CO2 system, high pressure, 75 lb. cylinder	Q-1	6	2.667		950	50		1,000	1,125	
	2400		FM200 system, filled, with mounting bracket										
	2460		26 lb. container	Q-1	8	2	Ea.	1,600	37.50		1,637.50	1,800	
	2500		63 lb. container		6	2.667		2,800	50		2,850	3,150	
	2520		101 lb. container		5	3.200		3,750	60		3,810	4,225	
	2540		196 lb. container		4	4		5,800	75		5,875	6,500	
	3000		Electro/mechanical release	L-1	4	2.500		115	52		167	213	
	3400		Manual pull station	1 Plum	6	1.333		45	28		73	95.50	
	4000		Pneumatic damper release	"	8	1		175	21		196	228	
	6000		Average FM200 system, minimum				C.F.	1.25			1.25	1.38	
	6020		Maximum				"	2.50			2.50	2.75	
135	0010		FIRE HOSE AND EQUIPMENT										135
	0200		Adapters, rough brass, straight hose threads										
	2200		Hose, less couplings										
	2260		Synthetic jacket, lined, 300 lb. test, 1-1/2" diameter	Q-12	2,600	.006	L.F.	1.92	.12		2.04	2.30	

Important: See the Reference Section for critical supporting data - Reference Nos., Crews, & Location Factors

			DAILY	LABOR-		1998 BARE COSTS				TOTAL	
154 100	**Fire Systems**	CREW	OUTPUT	HOURS	UNIT	MAT.	LABOR	EQUIP.	TOTAL	INCL O&P	
135 2280	2-1/2" diameter	Q-12	2,200	.007	L.F.	3.10	.14		3.24	3.64	**135**
2360	High strength, 500 lb. test, 1-1/2" diameter		2,600	.006		2.27	.12		2.39	2.69	
2380	2-1/2" diameter	↓	2,200	.007	↓	3.91	.14		4.05	4.53	
2600	Hose rack, swinging, for 1-1/2" diameter hose,										
2620	Enameled steel, 50' & 75' lengths of hose	Q-12	20	.800	Ea.	29.50	15.20		44.70	57.50	
2640	100' and 125' lengths of hose		20	.800		29.50	15.20		44.70	57.50	
2680	Chrome plated, 50' and 75' lengths of hose		20	.800		47	15.20		62.20	76.50	
2700	100' and 125' lengths of hose	↓	20	.800	↓	47	15.20		62.20	76.50	
2990	Hose reel, swinging, for 1-1/2" polyester neoprene lined hose										
3000	50' long	Q-12	14	1.143	Ea.	79	21.50		100.50	123	
3020	100' long		14	1.143		108	21.50		129.50	154	
3060	For 2-1/2" cotton rubber hose, 75' long		14	1.143		128	21.50		149.50	177	
3100	150' long	↓	14	1.143	↓	146	21.50		167.50	196	
3750	Hydrants, wall, w/caps, single, flush, polished brass										
3800	2-1/2" x 2-1/2"	Q-12	5	3.200	Ea.	125	61		186	238	
3840	2-1/2" x 3"		5	3.200		166	61		227	284	
3860	3" x 3"	↓	4.80	3.333	↓	204	63.50		267.50	330	
5600	Nozzles, brass										
5620	Adjustable fog, 3/4" booster line				Ea.	72.50			72.50	80	
5630	1" booster line					75.50			75.50	83	
5640	1-1/2" leader line					80			80	88	
5660	2-1/2" direct connection					159			159	175	
5680	2-1/2" playpipe nozzle				↓	128			128	141	
5780	For chrome plated, add					8%					
5850	Electrical fire, adjustable fog, no shock										
5900	1-1/2"				Ea.	243			243	267	
5920	2-1/2"				↓	330			330	365	
5980	For polished chrome, add					6%					
6200	Heavy duty, comb. adj. fog and str. stream, with handle										
6210	1" booster line				Ea.	233			233	256	
6240	1-1/2"					230			230	253	
6260	2-1/2", for playpipe					263			263	289	
6280	2-1/2" direct connection					330			330	360	
6300	2-1/2" playpipe combination				↓	470			470	515	
7140	Standpipe connections, wall, w/plugs & chains										
7160	Single, flush, brass, 2-1/2" x 2-1/2"	Q-12	5	3.200	Ea.	93	61		154	203	
7180	2-1/2" x 3"	"	5	3.200	"	96.50	61		157.50	207	
7240	For polished chrome, add					15%					
7280	Double, flush, polished brass										
7300	2-1/2" x 2-1/2" x 4"	Q-12	5	3.200	Ea.	315	61		376	445	
8800	Sidewalk siamese unit, polished brass, two way										
8820	2-1/2" x 2-1/2" x 4"	Q-12	2.50	6.400	Ea.	272	122		394	500	
160 0010	**FIRE VALVES**										**160**
0020	Angle, combination pressure adjustable/restricting, rough brass										
0030	1-1/2"	1 Spri	12	.667	Ea.	49.50	14.05		63.55	78	
0040	2-1/2"	"	7	1.143	"	105	24		129	156	
170 0010	**SPRINKLER SYSTEM COMPONENTS**										**170**
0800	Air compressor for dry pipe system, automatic, complete										
0820	280 gal. system capacity, 3/4 HP	1 Spri	1.30	6.154	Ea.	700	130		830	985	
1100	Alarm, electric pressure switch (circuit closer)	"	26	.308	"	110	6.50		116.50	131	
2600	Sprinkler heads, not including supply piping										
2640	Dry, pendent, 1/2" orifice, 3/4" or 1" NPT										
2700	11" to 12-3/4" length	1 Spri	14	.571	Ea.	30.50	12.05		42.55	53.50	
3700	Standard spray, pendent or upright, brass, 135° to 286°F										
3730	1/2" NPT, 7/16" orifice	1 Spri	16	.500	Ea.	5.05	10.55		15.60	23	
5600	Concealed, complete with cover plate										

154 100 | Fire Systems

		CREW	DAILY OUTPUT	LABOR-HOURS	UNIT	1998 BARE COSTS				TOTAL INCL O&P	
						MAT.	LABOR	EQUIP.	TOTAL		
170	5620	1/2" NPT, 1/2" orifice, 135°F to 212°F	1 Spri	9	.889	Ea.	9.35	18.75		28.10	41.50	**170**
	6200	Valves									
	6210	Alarm, includes									
	6220	retard chamber, trim, gauges, alarm line strainer									
	6260	3" size	Q-12	3	5.333	Ea.	620	101		721	855
	6280	4" size	"	2	8	"	630	152		782	945

155 100 | Boilers

		CREW	DAILY OUTPUT	LABOR-HOURS	UNIT	1998 BARE COSTS				TOTAL INCL O&P	
						MAT.	LABOR	EQUIP.	TOTAL		
105	0010	**BOILERS, GENERAL** Prices do not include flue piping, elec. wiring,										**105**
	0020	gas or oil piping, boiler base, pad, or tankless unless noted									
110	0010	**BOILERS, ELECTRIC, ASME** Standard controls and trim R155-050	Q-19	1.20	20	Ea.	7,900	390		8,290	9,350	**110**
	1000	Steam, 6 KW, 20.5 MBH									
	1160	60 KW, 205 MBH		1	24		9,575	465		10,040	11,300
	2000	Hot water, 12 KW, 41 MBH		1.30	18.462		3,325	360		3,685	4,250
	2040	24 KW, 82 MBH		1.20	20		3,550	390		3,940	4,575
	2070	36 KW, 123 MBH		1.20	20		3,875	390		4,265	4,900
115	0010	**BOILERS, GAS FIRED** Natural or propane, standard controls										**115**
	1000	Cast iron, with insulated jacket									
	2000	Steam, gross output, 81 MBH	Q-7	1.40	22.857	Ea.	1,050	430		1,480	1,900
	2080	203 MBH		.90	35.556		1,800	670		2,470	3,100
	2180	400 MBH		.60	53.333		2,975	1,000		3,975	4,925
	2280	1275 MBH		.36	88.889		7,225	1,675		8,900	10,700
	2380	3060 MBH		.23	139		14,500	2,625		17,125	20,400
	2400	3570 MBH		.18	177		16,200	3,350		19,550	23,400
	3000	Hot water, gross output, 80 MBH		1.46	21.918		1,050	415		1,465	1,850
	3020	100 MBH		1.35	23.704		1,200	450		1,650	2,075
	3080	203 MBH		1	32		1,800	605		2,405	2,975
	3180	400 MBH		.70	45.714		2,975	865		3,840	4,675
	3280	1,275 MBH		.46	69.565		7,000	1,325		8,325	9,875
	3340	2,312 MBH		.36	88.889		11,900	1,675		13,575	15,900
	3400	3,808 MBH		.26	123		17,000	2,325		19,325	22,600
	4000	Steel, insulating jacket									
	6000	Hot water, including burner & one zone valve, gross output									
	6010	51.2 MBH	Q-6	2	12	Ea.	1,575	218		1,793	2,075
	6020	72 MBH		2	12		1,750	218		1,968	2,275
	6040	89 MBH		1.90	12.632		1,775	230		2,005	2,350
	6060	105 MBH		1.80	13.333		2,000	243		2,243	2,600
	6080	132 MBH		1.70	14.118		2,275	257		2,532	2,925
	6100	155 MBH		1.50	16		2,650	291		2,941	3,375
	6110	186 MBH		1.40	17.143		3,200	310		3,510	4,025
	6140	292 MBH		1.20	20		4,575	365		4,940	5,625
	7000	For tankless water heater on smaller gas units, add						10%			
	7990	Special feature gas fired boilers									
	8000	Pulse combustion, standard controls / trim, 44,000 BTU	Q-5	1.50	10.667	Ea.	1,975	202		2,177	2,500
	8050	88,000 BTU		1.40	11.429		2,350	216		2,566	2,925
	8080	134,000 BTU		1.20	13.333		3,125	252		3,377	3,875
120	0010	**BOILERS, OIL FIRED** Standard controls, flame retention burner										**120**
	1000	Cast iron, with insulated flush jacket									

155 100 | Boilers

		CREW	DAILY OUTPUT	LABOR-HOURS	UNIT	MAT.	LABOR	EQUIP.	TOTAL	TOTAL INCL O&P	
120							**1998 BARE COSTS**				**120**
2000	Steam, gross output, 109 MBH	Q-7	1.20	26.667	Ea.	1,175	505		1,680	2,125	
2060	207 MBH		.90	35.556		1,625	670		2,295	2,900	
2180	1,084 MBH		.42	76.190		6,225	1,450		7,675	9,225	
3000	Hot water, same price as steam										
5000	Steel, insulated jacket, burner										
7000	Hot water, gross output, 103 MBH	Q-6	1.90	12.632	Ea.	2,075	230		2,305	2,650	
7020	122 MBH		1.80	13.333		2,100	243		2,343	2,700	
7060	168 MBH		1.50	16		2,625	291		2,916	3,375	
7080	225 MBH		1.40	17.143		3,125	310		3,435	3,950	
135	**BOILERS, PACKAGED SCOTCH MARINE** Steam or hot water										**135**
0010											
1000	Packaged fire tube, #2 oil, gross output										
1020	3348 MBH, 100 HP	Q-7	.12	266	Ea.	37,900	5,050		42,950	50,000	
1120	To fire #6, add	"	.83	38.554	"	6,350	730		7,080	8,200	
2000	Packaged water tube, #2 oil, gross output										
2040	1200 MBH	Q-7	.50	64	Ea.	14,500	1,200		15,700	17,900	
2060	1600 MBH		.40	80		15,100	1,500		16,600	19,200	
2080	2400 MBH		.30	106		19,400	2,025		21,425	24,700	
2100	3200 MBH		.25	128		20,600	2,425		23,025	26,700	
2120	4800 MBH		.20	160		25,800	3,025		28,825	33,400	
2140	For gas fired, add		.40	80		2,250	1,500		3,750	4,975	
150	**SWIMMING POOL HEATERS** Not including wiring, external										**150**
0010											
0020	piping, base or pad,										
0060	Gas fired, input, 115 MBH	Q-6	3	8	Ea.	1,500	146		1,646	1,900	
0100	135 MBH		2	12		1,675	218		1,893	2,200	
0160	155 MBH		1.50	16		1,775	291		2,066	2,425	
0220	235 MBH		.70	34.286		2,825	625		3,450	4,125	
0240	285 MBH		.60	40		3,625	730		4,355	5,200	
0280	500 MBH		.40	60		5,600	1,100		6,700	7,950	
0370	1,200 MBH		.21	114		11,100	2,075		13,175	15,800	
0400	1,800 MBH		.14	171		15,600	3,125		18,725	22,300	
0420	3,000 MBH		.11	218		25,300	3,975		29,275	34,400	
0440	3,750 MBH		.09	266		31,400	4,850		36,250	42,700	
2000	Electric, 12 KW, 4,800 gallon pool	Q-19	3	8		1,475	156		1,631	1,875	
2020	15 KW, 7,200 gallon pool		2.80	8.571		1,475	167		1,642	1,900	
2040	24 KW, 9,600 gallon pool		2.40	10		1,975	195		2,170	2,500	
2100	55 KW, 24,000 gallon pool		1.20	20		2,825	390		3,215	3,750	

155 400 | Warm Air Systems

		CREW	DAILY OUTPUT	LABOR-HOURS	UNIT	MAT.	LABOR	EQUIP.	TOTAL	TOTAL INCL O&P	
401	**DUCT FURNACES** Includes burner, controls, stainless steel										**401**
0010											
0020	heat exchanger. Gas fired, electric ignition										
0030	Indoor installation										
0080	100 MBH output	Q-5	5	3.200	Ea.	980	60.50		1,040.50	1,175	
0100	120 MBH output		4	4		1,125	75.50		1,200.50	1,350	
0130	200 MBH output		2.70	5.926		1,550	112		1,662	1,875	
0140	240 MBH output		2.30	6.957		1,600	131		1,731	2,000	
0160	280 MBH output		2	8		1,800	151		1,951	2,225	
0180	320 MBH output		1.60	10		1,950	189		2,139	2,475	
0300	For powered venter and adapter, add					238			238	262	
420	**FURNACES** Hot air heating, blowers, standard controls										**420**
0010											
0020	not including gas, oil or flue piping										
3000	Gas, AGA certified, direct drive models										
3040	60 MBH input	Q-9	3.80	4.211	Ea.	580	76.50		656.50	770	
3100	100 MBH input		3.20	5		655	90.50		745.50	875	
3130	150 MBH input		2.80	5.714		775	104		879	1,025	

155 400 | Warm Air Systems

		CREW	DAILY OUTPUT	LABOR-HOURS	UNIT	1998 BARE COSTS MAT.	LABOR	EQUIP.	TOTAL	TOTAL INCL O&P	
420	3140	200 MBH input	Q-9	2.60	6.154	Ea.	1,850	112		1,962	2,225	**420**
	4000	For starter plenum, add	↓	16	1	↓	60	18.15		78.15	96.50	
	6000	Oil, UL listed, atomizing gun type burner										
	6020	56 MBH output	Q-9	3.60	4.444	Ea.	745	80.50		825.50	955	
	6040	95 MBH output		3.40	4.706		790	85.50		875.50	1,025	
	6080	151 MBH output		3	5.333		1,225	96.50		1,321.50	1,500	
	6100	200 MBH input	↓	2.60	6.154	↓	2,050	112		2,162	2,475	
440	0010	**HEATING & VENTILATING UNITS** Classroom										**440**
	0020	Includes filter, heating/cooling coils, standard controls										
	0080	750 CFM, 2 tons cooling	Q-6	2	12	Ea.	2,900	218		3,118	3,525	
	0100	1000 CFM, 2-1/2 tons cooling		1.60	15		3,400	273		3,673	4,200	
	0120	1250 CFM, 3 tons cooling		1.40	17.143		3,525	310		3,835	4,400	
	0140	1500 CFM, 4 tons cooling	↓	.80	30		3,775	545		4,320	5,050	
	0500	For electric heat, add					35%					
	1000	For no cooling, deduct				↓	25%	10%				
451	0010	**INFRA-RED UNIT**										**451**
	0020	Gas fired, unvented, electric ignition, 100% shutoff. Piping and										
	0030	wiring not included										
	0060	Input, 15 MBH	Q-5	7	2.286	Ea.	340	43		383	440	
	0120	45 MBH		5	3.200		380	60.50		440.50	520	
	0160	60 MBH		4	4		445	75.50		520.50	615	
	0180	75 MBH		3	5.333		485	101		586	700	
	0220	105 MBH	↓	2	8	↓	680	151		831	1,000	
471	0010	**SOLAR ENERGY**										**471**
	0020	System/Package prices, not including connecting										
	0030	pipe, insulation, or special heating/plumbing fixtures										
	0500	Hot water, standard package, low temperature										
	0540	1 collector, circulator, fittings, 65 gal. tank	Q-1	.50	32	Ea.	1,875	600		2,475	3,075	
	0580	2 collectors, circulator, fittings, 120 gal. tank	"	.40	40	"	2,725	750		3,475	4,250	
	0701	Solar energy, medium temperature package										
	0720	1 collector, circulator, fittings, 80 gal. tank	Q-1	.50	32	Ea.	2,025	600		2,625	3,225	
	0740	2 collectors, circulator, fittings, 120 gal. tank	"	.40	40	↓	2,925	750		3,675	4,475	
	0980	For each additional 120 gal. tank, add					720			720	795	
480	0010	**SPACE HEATERS** Cabinet, grilles, fan, controls, burner,										**480**
	0020	thermostat, no piping. For flue see division 155-680										
	1000	Gas fired, floor mounted										
	1100	60 MBH output	Q-5	10	1.600	Ea.	535	30		565	635	
	1140	100 MBH output		8	2		580	38		618	705	
	1180	180 MBH output	↓	6	2.667	↓	795	50.50		845.50	960	
	1500	Rooftop mounted, gravity vent, stainless steel exchanger										
	1520	75 MBH output	Q-6	4	6	Ea.	3,275	109		3,384	3,775	
	1560	120 MBH output		3.30	7.273		3,575	132		3,707	4,150	
	1600	190 MBH output		2.60	9.231		4,150	168		4,318	4,850	
	1620	225 MBH output		2.30	10.435		4,225	190		4,415	4,950	
	1640	300 MBH output		1.90	12.632		5,025	230		5,255	5,900	
	1660	375 MBH output		1.40	17.143		5,700	310		6,010	6,800	
	1700	600 MBH output		1	24		7,000	435		7,435	8,425	
	1720	750 MBH output		.80	30		12,000	545		12,545	14,100	
	1760	1200 MBH output	↓	.30	80		16,300	1,450		17,750	20,400	
	1800	For power vent, add					10%					
	2000	Suspension mounted, propeller fan, 20 MBH output	Q-5	8.50	1.882		400	35.50		435.50	500	
	2100	130 MBH output		5	3.200		730	60.50		790.50	905	
	2240	320 MBH output	↓	2	8		1,500	151		1,651	1,900	
	2500	For powered venter and adapter, add					254			254	279	
	5000	Wall furnace, 17.5 MBH output	Q-5	6	2.667	↓	440	50.50		490.50	565	

Important: See the Reference Section for critical supporting data - Reference Nos., Crews, & Location Factors

		155 400	**Warm Air Systems**	CREW	DAILY OUTPUT	LABOR-HOURS	UNIT	MAT.	LABOR	EQUIP.	TOTAL	TOTAL INCL O&P	
480	5020		24 MBH output	Q-5	5	3.200	Ea.	455	60.50		515.50	600	**480**
	5040		35 MBH output	↓	4	4	↓	610	75.50		685.50	795	

155 600 | Heating System Access.

				CREW	DAILY OUTPUT	LABOR-HOURS	UNIT	MAT.	LABOR	EQUIP.	TOTAL	TOTAL INCL O&P	
630	0010	**HYDRONIC HEATING** Terminal units, not incl. main supply pipe											**630**
	1000	Radiation											
	1150	Fin tube, wall hung, 14" slope top cover, with damper											
	1200	1-1/4" copper tube, 4-1/4" alum. fin		Q-5	38	.421	L.F.	34.50	7.95		42.45	51	
	1255	2" steel tube, 4-1/4" steel fin			32	.500		33.50	9.45		42.95	52.50	
	1310	Baseboard, pkgd, 1/2" copper tube, alum. fin, 7" high			60	.267		8.45	5.05		13.50	17.60	
	1320	3/4" copper tube, alum. fin, 7" high			58	.276		8.85	5.20		14.05	18.35	
	1340	1" copper tube, alum. fin, 8-7/8" high			56	.286		17.20	5.40		22.60	28	
	1360	1-1/4" copper tube, alum. fin, 8-7/8" high		↓	54	.296	↓	24.50	5.60		30.10	36.50	
	3950	Unit heaters, propeller, 230/460V 2 psi steam, 60°F entering air											
	4000	Horizontal, 12 MBH		Q-5	12	1.333	Ea.	610	25		635	715	
	4060	43.9 MBH			8	2		815	38		853	965	
	4240	286.9 MBH			2	8		1,700	151		1,851	2,100	
	4250	326.0 MBH			1.90	8.421		1,800	159		1,959	2,250	
	4260	364.0 MBH			1.80	8.889		2,075	168		2,243	2,575	
	4270	404.0 MBH		↓	1.60	10		2,450	189		2,639	3,025	
	4300	For vertical diffuser, add						130			130	143	
	4310	Vertical flow, 40.0 MBH		Q-5	11	1.455		990	27.50		1,017.50	1,150	
	4326	131.0 MBH		"	4	4		1,150	75.50		1,225.50	1,400	
	4354	420 MBH		Q-6	1.80	13.333		2,300	243		2,543	2,925	
	4358	500 MBH			1.40	17.143		2,600	310		2,910	3,400	
	4362	570 MBH			1.40	17.143		2,725	310		3,035	3,525	
	4366	620 MBH			1.30	18.462		3,125	335		3,460	4,000	
	4370	960 MBH		↓	1.10	21.818	↓	5,475	395		5,870	6,675	
640	0010	**HUMIDIFIERS**											**640**
	0520	Steam, room or duct, filter, regulators, auto. controls, 220 V											
	0580	33 lb. per hour		Q-5	4	4	Ea.	2,250	75.50		2,325.50	2,600	
	0620	100 lb. per hour		"	3	5.333	"	3,300	101		3,401	3,825	
651	0010	**INSULATION**											**651**
	3000	Ductwork											
	3020	Blanket type, fiberglass, flexible											
	3030	Fire resistant liner, black coating one side											
	3050	1/2" thick, 2 lb. density		Q-14	380	.042	S.F.	.28	.74		1.02	1.61	
	3060	1" thick, 1-1/2 lb. density		"	350	.046	"	.37	.80		1.17	1.82	
	3140	FRK vapor barrier wrap, .75 lb. density											
	3160	1" thick		Q-14	350	.046	S.F.	.26	.80		1.06	1.70	
	3170	1-1/2" thick		"	320	.050	"	.23	.88		1.11	1.79	
	3210	Vinyl jacket, same as FRK											
	3490	Board type, fiberglass, 3 lb. density											
	3500	Fire resistant, black pigmented, 1 side											
	3520	1" thick		Q-14	150	.107	S.F.	1.10	1.88		2.98	4.50	
	3540	1-1/2" thick			130	.123		1.36	2.17		3.53	5.30	
	3560	2" thick		↓	120	.133	↓	1.64	2.35		3.99	5.90	
	3600	FRK vapor barrier											
	3620	1" thick		Q-14	150	.107	S.F.	1.21	1.88		3.09	4.62	
	3630	1-1/2" thick		"	130	.123	"	1.56	2.17		3.73	5.50	
	4000	Pipe covering (price copper tube one size less than IPS)											
	6600	Fiberglass, with all service jacket											
	6840	1" wall, 1/2" iron pipe size		Q-14	240	.067	L.F.	.85	1.17		2.02	2.99	
	6860	3/4" iron pipe size			230	.070		.97	1.22		2.19	3.21	
	6870	1" iron pipe size			220	.073		.99	1.28		2.27	3.33	
	6900	2" iron pipe size		↓	200	.080	↓	1.32	1.41		2.73	3.92	

MECHANICAL 15

155 600 | Heating System Access.

		CREW	DAILY OUTPUT	LABOR-HOURS	UNIT	1998 BARE COSTS				TOTAL INCL O&P	
						MAT.	LABOR	EQUIP.	TOTAL		
651	6910	2-1/2" iron pipe size	Q-14	190	.084	L.F.	1.45	1.48		2.93	4.20	**651**
	6920	3" iron pipe size		180	.089		1.64	1.56		3.20	4.54	
	6940	4" iron pipe size		150	.107		2.11	1.88		3.99	5.60	
	6960	6" iron pipe size		120	.133		2.58	2.35		4.93	6.95	
	7660	For fittings, add 3 L.F. for each fitting										
	7680	plus 4 L.F. for each flange of the fitting										
	7700	For equipment or congested areas, add						20%				
	7879	Rubber tubing, flexible closed cell foam										
	8300	3/4" wall, 1/4" iron pipe size	1 Asbe	90	.089	L.F.	.86	1.74		2.60	3.99	
	8330	1/2" iron pipe size		89	.090		1.12	1.76		2.88	4.31	
	8340	3/4" iron pipe size		89	.090		1.38	1.76		3.14	4.60	
	8350	1" iron pipe size		88	.091		1.56	1.78		3.34	4.83	
	8360	1-1/4" iron pipe size		87	.092		2.11	1.80		3.91	5.45	
	8380	2" iron pipe size		86	.093		2.80	1.82		4.62	6.25	
	8390	2-1/2" iron pipe size		86	.093		3.73	1.82		5.55	7.30	
	8400	3" iron pipe size		85	.094		4.26	1.84		6.10	7.90	
	8420	4" iron pipe size		80	.100		5.40	1.95		7.35	9.35	
	8440	6" iron pipe size		80	.100		7.85	1.95		9.80	12.05	
	8444	1" wall, 1/2" iron pipe size		86	.093		2.18	1.82		4	5.60	
	8445	3/4" iron pipe size		84	.095		2.64	1.86		4.50	6.15	
	8446	1" iron pipe size		84	.095		3.07	1.86		4.93	6.65	
	8447	1-1/4" iron pipe size		82	.098		3.47	1.91		5.38	7.15	
	8448	1-1/2" iron pipe size		82	.098		4.03	1.91		5.94	7.75	
	8449	2" iron pipe size		80	.100		5.40	1.95		7.35	9.35	
	8450	2-1/2" iron pipe size		80	.100		7	1.95		8.95	11.10	
	8456	Rubber insulation tape, 1/8" x 2" x 30'				Ea.	13.45			13.45	14.80	
671	0010	**TANKS**										**671**
	1505	Fiberglass and steel single / double wall storage, see Div 132-150										
	1510	Tank leak detection systems, see Div 132-190										
	2000	Steel, liquid expansion, ASME, painted, 15 gallon capacity	Q-5	17	.941	Ea.	310	17.80		327.80	370	
	2040	30 gallon capacity		12	1.333		350	25		375	425	
	2120	100 gallon capacity		6	2.667		635	50.50		685.50	780	
	3000	Steel ASME expansion, rubber diaphragm, 19 gal. cap. accept.		12	1.333		1,225	25		1,250	1,375	
	3020	31 gallon capacity		8	2		1,350	38		1,388	1,550	
	3080	119 gallon capacity		4	4		2,125	75.50		2,200.50	2,475	
680	0010	**VENT CHIMNEY** Prefab metal, U.L. listed										**680**
	0020	Gas, double wall, galvanized steel										
	0080	3" diameter	Q-9	72	.222	V.L.F.	3.38	4.03		7.41	10.50	
	0100	4" diameter		68	.235		4.14	4.27		8.41	11.75	
	0120	5" diameter		64	.250		4.91	4.53		9.44	13.05	
	0140	6" diameter		60	.267		5.70	4.83		10.53	14.50	
	0160	7" diameter		56	.286		8.40	5.20		13.60	18	
	0180	8" diameter		52	.308		9.35	5.60		14.95	19.75	
	0200	10" diameter		48	.333		19.75	6.05		25.80	31.50	
	7800	All fuel, double wall, stainless steel, 6" diameter		60	.267		28	4.83		32.83	38.50	
	7802	7" diameter		56	.286		36	5.20		41.20	48.50	
	7806	10" diameter		48	.333		61	6.05		67.05	77.50	

488

Important: See the Reference Section for critical supporting data - Reference Nos., Crews, & Location Factors

156 200 | Heat/Cool Piping Misc.

		CREW	DAILY OUTPUT	LABOR-HOURS	UNIT	1998 BARE COSTS				TOTAL INCL O&P
						MAT.	LABOR	EQUIP.	TOTAL	
240	**0010**	**HEATING CONTROL VALVES**									**240**
	0050	Hot water, nonelectric, thermostatic								
	0100	Radiator supply, 1/2" diameter	1 Stpi	24	.333	Ea.	26.50	7		33.50	41
	0120	3/4" diameter		20	.400		27	8.40		35.40	44
	0140	1" diameter		19	.421		33.50	8.85		42.35	51.50
	0160	1-1/4" diameter		15	.533		47.50	11.20		58.70	70.50
	0500	For low pressure steam, add					25%				
	1000	Manual, radiator supply									
	1010	1/2" pipe size, angle union	1 Stpi	24	.333	Ea.	20.50	7		27.50	34
	1020	3/4" pipe size, angle union		20	.400		25.50	8.40		33.90	42
	1030	1" pipe size, angle union		19	.421		33	8.85		41.85	50.50
	1140	Balance and stop valve 1/2" size		22	.364		20.50	7.65		28.15	35
	1150	3/4" size		20	.400		13.45	8.40		21.85	28.50
	1160	1" size		19	.421		16.90	8.85		25.75	33.50
	1170	1-1/4" size		15	.533		21	11.20		32.20	41.50
	8000	System balancing and shut-off									
	8020	Butterfly, quarter turn, calibrated, threaded or solder									
	8040	Bronze, -30° F to +350° F, pressure to 175 psi									
	8060	1/2" size	1 Stpi	22	.364	Ea.	8.90	7.65		16.55	22.50
	8070	3/4" size		20	.400		8.15	8.40		16.55	23
	8080	1" size		19	.421		10.65	8.85		19.50	26.50
	8090	1-1/4" size		15	.533		17.45	11.20		28.65	38
	8110	2" size		11	.727		26	15.25		41.25	54

156 600 | Strainers

		CREW	DAILY OUTPUT	LABOR-HOURS	UNIT	1998 BARE COSTS				TOTAL INCL O&P
						MAT.	LABOR	EQUIP.	TOTAL	
612	**0010**	**STRAINERS, Y TYPE** Iron body									**612**
	0100	1/2" pipe size	1 Stpi	20	.400	Ea.	8.75	8.40		17.15	23.50
	0120	3/4" pipe size		18	.444		10.50	9.35		19.85	27
	0140	1" pipe size		16	.500		14.10	10.50		24.60	33
	0150	1-1/4" pipe size		15	.533		19.35	11.20		30.55	40
	0180	2" pipe size		8	1		36	21		57	74.50
	1000	Flanged, 125 lb., 1-1/2" pipe size		11	.727		113	15.25		128.25	151
	1020	2" pipe size		8	1		84.50	21		105.50	128
	1030	2-1/2" pipe size	Q-5	5	3.200		95.50	60.50		156	205
	1040	3" pipe size		4.50	3.556		112	67		179	234
	1060	4" pipe size		3	5.333		191	101		292	375

157 100 | A.C. & Vent. Units

		CREW	DAILY OUTPUT	LABOR-HOURS	UNIT	1998 BARE COSTS				TOTAL INCL O&P
						MAT.	LABOR	EQUIP.	TOTAL	
110	**0010**	**ABSORPTION COLD GENERATORS** Water chiller	R157 -040								**110**
	0020	Steam or hot water, water cooled									
	0050	100 ton	Q-8	.07	484	Ea.	86,000	9,675	735	96,410	111,500
	0100	148 ton	"	.06	533	"	108,000	10,600	810	119,410	137,000
	3000	Gas fired, air cooled									
	3270	10 ton	Q-5	.40	40	Ea.	11,100	755		11,855	13,500
125	**0010**	**CENTRAL STATION AIR-HANDLING UNIT** Chilled water									**125**
	1000	Modular, capacity at 700 FPM face velocity									

157 100 | A.C. & Vent. Units

			CREW	DAILY OUTPUT	LABOR-HOURS	UNIT	1998 BARE COSTS				TOTAL INCL O&P	
							MAT.	LABOR	EQUIP.	TOTAL		
125	1100	1300 CFM	Q-5	1.20	13.333	Ea.	2,300	252		2,552	2,950	125
	1200	1900 CFM		1.10	14.545		2,550	275		2,825	3,250	
	1300	3200 CFM		.80	20		3,025	380		3,405	3,950	
	1400	5400 CFM	Q-6	.80	30		4,550	545		5,095	5,900	
130	0010	**COMPUTER ROOM UNITS**										130
	1000	Air cooled, includes remote condenser but not										
	1020	interconnecting tubing or refrigerant										
	1080	3 ton	Q-5	.50	32	Ea.	6,375	605		6,980	8,025	
	1120	5 ton		.45	35.556		8,575	670		9,245	10,600	
	1160	6 ton		.30	53.333		15,800	1,000		16,800	19,000	
	1200	8 ton		.27	59.259		17,800	1,125		18,925	21,500	
	1240	10 ton		.25	64		18,500	1,200		19,700	22,400	
	1280	15 ton		.22	72.727		20,700	1,375		22,075	25,100	
	1290	18 ton		.20	80		23,600	1,500		25,100	28,400	
	1320	20 ton	Q-6	.29	82.759		24,700	1,500		26,200	29,700	
150	0010	**FAN COIL AIR CONDITIONING** Cabinet mounted, filters, controls										150
	0100	Chilled water, 1/2 ton cooling	Q-5	8	2	Ea.	720	38		758	855	
	0120	1 ton cooling		6	2.667		815	50.50		865.50	980	
	0180	3 ton cooling		4	4		1,625	75.50		1,700.50	1,925	
	0940	Direct expansion, for use w/air cooled condensing, 1.5 ton cooling		5	3.200		435	60.50		495.50	575	
	1000	5 ton cooling		3	5.333		990	101		1,091	1,250	
	1510	For condensing unit add see division 157-230.										
160	0010	**HEAT PUMPS** (Not including interconnecting tubing)										160
	1000	Air to air, split system, not including curbs, pads, or ductwork										
	1060	5 ton cooling, 27 MBH heat @ 0°F	Q-5	.50	32	Ea.	4,175	605		4,780	5,600	
	1080	7.5 ton cooling, 33 MBH heat @ 0°F	"	.30	53.333		6,425	1,000		7,425	8,750	
	1100	10 ton cooling, 50 MBH heat @ 0°F	Q-6	.38	63.158		8,350	1,150		9,500	11,100	
	1120	15 ton cooling, 64 MBH heat @ 0°F		.26	92.308		11,500	1,675		13,175	15,400	
	1130	20 ton cooling, 85 MBH heat @ 0°F		.20	120		16,800	2,175		18,975	22,100	
	1160	30 ton cooling, 163 MBH heat @ 0°F	Q-7	.17	188		23,300	3,550		26,850	31,500	
	1200	50 ton cooling, 220 MBH heat @ 0°F	"	.10	320		46,500	6,050		52,550	61,000	
	2000	Water source to air, single package										
	2220	5 ton cooling, 29 MBH heat @ 75°F	Q-5	.90	17.778	Ea.	2,100	335		2,435	2,850	
	2240	7.5 ton cooling, 35 MBH heat @ 75°F		.60	26.667		2,375	505		2,880	3,425	
	2260	10 ton cooling, 50 MBH heat @ 75°F		.53	30.189		7,450	570		8,020	9,150	
	2280	15 ton cooling, 64 MBH heat @ 75°F	Q-6	.47	51.064		8,000	930		8,930	10,400	
	2300	20 ton cooling, 100 MBH heat @ 75°F		.41	58.537		8,250	1,075		9,325	10,900	
	2320	30 ton cooling, (twin 15 ton units)		.23	102		16,000	1,850		17,850	20,700	
	2340	40 ton cooling, (twin 20 ton units)		.20	117		16,500	2,125		18,625	21,600	
	2360	50 ton cooling, (twin 15 + 20 ton unit)		.15	160		24,300	2,900		27,200	31,500	
	3960	For supplementary heat coil, add					10%					
170	0010	**PACKAGED TERMINAL AIR CONDITIONER** Cabinet, wall sleeve,										170
	0100	louver, electric heat, thermostat, manual changeover, 208 V										
	0200	6,000 BTUH cooling, 8800 BTU heat	Q-5	6	2.667	Ea.	875	50.50		925.50	1,050	
	0220	9,000 BTUH cooling, 13,900 BTU heat		5	3.200		895	60.50		955.50	1,075	
	0240	12,000 BTUH cooling, 13,900 BTU heat		4	4		925	75.50		1,000.50	1,150	
	0260	15,000 BTUH cooling, 13,900 BTU heat		3	5.333		1,000	101		1,101	1,275	
	0320	30,000 BTUH cooling, 10 KW heat		1.40	11.429		1,550	216		1,766	2,050	
	0340	36,000 BTUH cooling, 10 KW heat		1.25	12.800		1,625	242		1,867	2,200	
	0360	42,000 BTUH cooling, 10 KW heat		1	16		2,050	300		2,350	2,775	
	0380	48,000 BTUH cooling, 10 KW heat		.90	17.778		2,225	335		2,560	3,000	
180	0010	**ROOF TOP AIR CONDITIONERS** Standard controls, curb, economizer										180
	1000	Single zone, electric cool, gas heat										

Important: See the Reference Section for critical supporting data - Reference Nos., Crews, & Location Factors

157 100 | A.C. & Vent. Units

		CREW	DAILY OUTPUT	LABOR-HOURS	UNIT	1998 BARE COSTS				TOTAL INCL O&P
						MAT.	LABOR	EQUIP.	TOTAL	
180 1140	5 ton cooling, 112 MBH heating	Q-5	.56	28.571	Ea.	4,600	540		5,140	5,950 **180**
1160	10 ton cooling, 200 MBH heating	Q-6	.46	52.174		9,300	950		10,250	11,800
1180	15 ton cooling, 270 MBH heating	"	.31	77.419		12,300	1,400		13,700	15,800
1220	30 ton cooling, 540 MBH heating	Q-7	.22	145		28,600	2,750		31,350	36,000
1260	50 ton cooling, 810 MBH heating		.13	246		45,100	4,650		49,750	57,500
1270	80 ton cooling, 1000 MBH heating		.09	355		85,000	6,725		91,725	104,500
185 0010	**SELF-CONTAINED SINGLE PACKAGE**									**185**
0100	Air cooled, for free blow or duct, including remote condenser									
0220	5 ton cooling	Q-6	1.20	20	Ea.	5,300	365		5,665	6,425
0221	Self-contained, air cooled, 5 ton cooling		1.20	20			365		365	605
0230	7.5 ton cooling		.90	26.667		7,200	485		7,685	8,725
0240	10 ton cooling	Q-7	1	32		9,025	605		9,630	11,000
0250	15 ton cooling		.95	33.684		12,800	635		13,435	15,100
0260	20 ton cooling		.90	35.556		14,700	670		15,370	17,300
0270	25 ton cooling		.85	37.647		17,700	710		18,410	20,700
1000	Water cooled for free blow or duct, not including tower									
1160	20 ton cooling	Q-7	.80	40	Ea.	12,800	755		13,555	15,300
1170	25 ton cooling		.75	42.667		15,100	805		15,905	17,900
1180	30 ton cooling		.70	45.714		16,900	865		17,765	20,000
1200	40 ton cooling		.40	80		21,400	1,500		22,900	26,100
190 0010	**WATER CHILLERS**, With standard controls									**190**
0020	Centrifugal liquid chiller, water cooled									
0030	not including water tower									
0200	Packaged unit, water cooled, not incl. tower									
0240	200 ton	Q-8	.09	347	Ea.	82,500	6,950	530	89,980	102,500
0490	Reciprocating, packaged w/integral air cooled condenser, 15 ton cool	Q-7	.37	86.486		12,000	1,625		13,625	15,900
0500	20 ton cooling		.35	91.429		15,200	1,725		16,925	19,700
0510	25 ton cooling		.32	100		17,300	1,900		19,200	22,100
0980	Water cooled, multiple compress., semi-hermetic, tower not incl.									
1000	15 ton cooling	Q-6	.30	80	Ea.	15,600	1,450		17,050	19,600
1040	25 ton cooling	Q-7	.32	100		18,400	1,900		20,300	23,300
1060	30 ton cooling		.27	118		19,600	2,250		21,850	25,200
1100	50 ton cooling		.21	152		24,600	2,875		27,475	31,800
1130	70 ton cooling		.16	200		33,400	3,775		37,175	43,100
1150	90 ton cooling	Q-8	.13	246		46,700	4,900	375	51,975	60,000
1160	100 ton cooling	"	.12	266		47,700	5,325	405	53,430	62,000
4000	Packaged chiller, with remote air cooled condensers incl.									
4020	15 ton cooling	Q-7	.35	91.429	Ea.	13,500	1,725		15,225	17,800
4040	25 ton cooling		.30	106		18,700	2,025		20,725	23,900
4050	30 ton cooling		.27	118		21,200	2,250		23,450	27,000
4070	50 ton cooling		.18	177		30,800	3,350		34,150	39,400
4080	60 ton cooling		.14	228		33,500	4,325		37,825	44,100
195 0010	**WINDOW UNIT AIR CONDITIONERS**									**195**
4000	Portable/window, 15 amp 125V grounded receptacle required									
4060	5000 BTUH	1 Carp	8	1	Ea.	287	18.50		305.50	345
4340	6000 BTUH		8	1		365	18.50		383.50	430
4480	8000 BTUH		6	1.333		465	24.50		489.50	555
4500	10,000 BTUH		6	1.333		480	24.50		504.50	570
4520	12,000 BTUH	L-2	8	2		560	32.50		592.50	670

157 200 | System Components

		CREW	DAILY OUTPUT	LABOR-HOURS	UNIT	1998 BARE COSTS				TOTAL INCL O&P
						MAT.	LABOR	EQUIP.	TOTAL	
225 0010	**CONDENSERS** Ratings are for 30°F TD, R-22									**225**
0080	Air cooled, belt drive, propeller fan									
0220	45 ton	Q-6	.54	44.444	Ea.	8,850	810		9,660	11,100
0240	50 ton		.48	50		8,875	910		9,785	11,300

For expanded coverage of these items see *Means Mechanical or Plumbing Cost Data 1998*

157 200 | System Components

			CREW	DAILY OUTPUT	LABOR-HOURS	UNIT	1998 BARE COSTS				TOTAL INCL O&P	
							MAT.	LABOR	EQUIP.	TOTAL		
225	0260	54 ton	Q-6	.40	60	Ea.	9,450	1,100		10,550	12,200	**225**
240	0010	**COOLING TOWERS** Packaged units										**240**
	0070	Galvanized steel										
	0080	Draw thru, single flow										
	0100	Belt drive, 60 tons	Q-6	90	.267	TonAC	71.50	4.85		76.35	87	
	0150	95 tons		100	.240		63	4.37		67.37	77	
	0200	110 tons		109	.220		62	4.01		66.01	74.50	
	1000	For higher capacities, use multiples										
	5000	Fiberglass										
	5010	Draw thru										
	5100	60 tons	Q-6	1.50	16	Ea.	2,775	291		3,066	3,525	
	5120	125 tons		.99	24.242		5,675	440		6,115	6,975	
	5140	300 tons		.43	55.814		13,300	1,025		14,325	16,300	
	5160	600 tons		.22	109		24,200	1,975		26,175	29,900	
	5180	1000 tons		.15	160		41,500	2,900		44,400	50,500	
	6000	Stainless steel										
	6010	Draw thru										
	6100	60 tons	Q-6	1.50	16	Ea.	7,000	291		7,291	8,175	
	6120	110 tons		.99	24.242		11,200	440		11,640	13,000	
	6140	300 tons		.43	55.814		31,000	1,025		32,025	35,800	
	6160	600 tons		.22	109		48,000	1,975		49,975	56,500	
	6180	1000 tons		.15	160		78,000	2,900		80,900	91,000	
250	0010	**DUCTWORK**	R157 -050									**250**
	0020	Fabricated rectangular, includes fittings, joints, supports,										
	0030	allowance for flexible connections, no insulation										
	0031	NOTE: Fabrication and installation are combined										
	0040	as LABOR cost.										
	0050	Add to labor for elevated installation										
	0051	of fabricated ductwork										
	0052	10' to 15' high						6%				
	0053	15' to 20' high						12%				
	0054	20' to 25' high						15%				
	0100	Aluminum, alloy 3003-H14, under 100 lb.	Q-10	75	.320	Lb.	3.09	6		9.09	13.60	
	0140	1,000 to 2,000 lb.		120	.200		1.37	3.76		5.13	7.85	
	0150	2,000 to 5,000 lb.		130	.185		1.26	3.47		4.73	7.25	
	0500	Galvanized steel, under 200 lb.		235	.102		1.01	1.92		2.93	4.36	
	0520	200 to 500 lb.		245	.098		.65	1.84		2.49	3.83	
	0540	500 to 1,000 lb.		255	.094		.55	1.77		2.32	3.60	
	0560	1,000 to 2,000 lb.		265	.091		.47	1.70		2.17	3.40	
	0570	2,000 to 5,000 lb.		275	.087		.44	1.64		2.08	3.26	
	1300	Flexible, coated fiberglass fabric on corr. resist. metal helix										
	1400	pressure to 12" (WG) UL-181										
	1500	Non-insulated, 3" diameter	Q-9	400	.040	L.F.	.78	.73		1.51	2.09	
	1520	4" diameter		360	.044		.78	.81		1.59	2.22	
	1540	5" diameter		320	.050		.91	.91		1.82	2.53	
	1560	6" diameter		280	.057		1.02	1.04		2.06	2.87	
	1561	Ductwork, flexible, non-insulated, 6" diameter		280	.057		1.02	1.04		2.06	2.87	
	1580	7" diameter		240	.067		1.34	1.21		2.55	3.51	
	1600	8" diameter		200	.080		1.39	1.45		2.84	3.98	
	1620	9" diameter		180	.089		1.57	1.61		3.18	4.46	
	1640	10" diameter		160	.100		1.78	1.81		3.59	5.05	
	1900	Insulated, 1" thick with 3/4 lb., PE jacket, 3" diameter		380	.042		1.34	.76		2.10	2.76	
	1920	5" diameter		300	.053		1.57	.97		2.54	3.37	
	1940	6" diameter		260	.062		1.73	1.12		2.85	3.79	

Important: See the Reference Section for critical supporting data - Reference Nos., Crews, & Location Factors

157 200 | System Components

			CREW	DAILY OUTPUT	LABOR-HOURS	UNIT	MAT.	LABOR	EQUIP.	TOTAL	TOTAL INCL O&P	
250	1960	7" diameter	Q-9	220	.073	L.F.	2.11	1.32		3.43	4.55	**250**
	1980	8" diameter		180	.089		2.16	1.61		3.77	5.10	
	2000	9" diameter		160	.100		2.60	1.81		4.41	5.95	
	2020	10" diameter		140	.114		2.81	2.07		4.88	6.60	
	3490	Rigid fiberglass duct board, foil reinf. kraft facing										
	3500	Rectangular, 1" thick, alum. faced, (FRK), std. weight	Q-10	350	.069	SF Surf	.81	1.29		2.10	3.07	
290	0010	**FANS**										**290**
	0020	Air conditioning and process air handling										
	2500	Ceiling fan, right angle, extra quiet, 0.10" S.P.										
	2520	95 CFM	Q-20	20	1	Ea.	136	18.60		154.60	182	
	2540	210 CFM	"	19	1.053	"	149	19.60		168.60	197	
	5000	Utility set, centrifugal, V belt drive, motor										
	5020	1/4" S.P., 1200 CFM, 1/4 HP	Q-20	6	3.333	Ea.	2,075	62		2,137	2,375	
	5040	1520 CFM, 1/3 HP		5	4		2,100	74.50		2,174.50	2,450	
	5060	1850 CFM, 1/2 HP		4	5		2,100	93		2,193	2,475	
	5080	2180 CFM, 3/4 HP		3	6.667		2,125	124		2,249	2,525	
	6000	Propeller exhaust, wall shutter, 1/4" S.P.										
	6020	Direct drive, two speed										
	6100	375 CFM, 1/10 HP	Q-20	10	2	Ea.	198	37		235	280	
	6140	1000 CFM, 1/8 HP		8	2.500		305	46.50		351.50	415	
	6160	1890 CFM, 1/4 HP		7	2.857		274	53		327	390	
	6650	Residential, bath exhaust, grille, back draft damper										
	6660	50 CFM	Q-20	24	.833	Ea.	13.95	15.50		29.45	41.50	
	6670	110 CFM	"	22	.909	"	49.50	16.95		66.45	83	
	7000	Roof exhauster, centrifugal, aluminum housing, 12" galvanized										
	7020	curb, bird screen, back draft damper, 1/4" S.P.										
	7100	Direct drive, 320 CFM, 11" sq. damper	Q-20	7	2.857	Ea.	251	53		304	365	
	7120	600 CFM, 11" sq. damper		6	3.333		276	62		338	410	
	7140	815 CFM, 13" sq. damper		5	4		395	74.50		469.50	560	
	7160	1450 CFM, 13" sq. damper		4.20	4.762		405	88.50		493.50	595	
	7180	2050 CFM, 16" sq. damper		4	5		455	93		548	655	
	7200	V-belt drive, 1650 CFM, 12" sq. damper		6	3.333		720	62		782	900	
	7220	2750 CFM, 21" sq. damper		5	4		920	74.50		994.50	1,150	
	7230	3500 CFM, 21" sq. damper		4.50	4.444		1,075	83		1,158	1,325	
	7240	4910 CFM, 23" sq. damper		4	5		1,225	93		1,318	1,500	

157 400 | Accessories

			CREW	DAILY OUTPUT	LABOR-HOURS	UNIT	MAT.	LABOR	EQUIP.	TOTAL	TOTAL INCL O&P	
401	0010	**AIR FILTERS**										**401**
	0050	Activated charcoal type, full flow				MCFM	600			600	660	
	2000	Electronic air cleaner, duct mounted										
	2150	400 - 1000 CFM	1 Shee	2.30	3.478	Ea.	485	70		555	655	
	2200	1000 - 1400 CFM		2.20	3.636		625	73.50		698.50	815	
	2250	1400 - 2000 CFM		2.10	3.810		690	77		767	890	
	2950	Mechanical media filtration units										
	3000	High efficiency type, with frame, non-supported				MCFM	40			40	44	
	3100	Supported type					55			55	60.50	
	4000	Medium efficiency, extended surface					5			5	5.50	
	4500	Permanent washable					60			60	66	
	5000	Renewable disposable roll					120			120	132	
	5500	Throwaway glass or paper media type				Ea.	3.89			3.89	4.28	
420	0010	**CONTROL COMPONENTS**										**420**
	5000	Thermostats										
	5030	Manual	1 Shee	8	1	Ea.	23.50	20		43.50	59.50	
	5040	1 set back, electric, timed	"	8	1	"	75.50	20		95.50	117	

157 400	Accessories	CREW	DAILY OUTPUT	LABOR-HOURS	UNIT	1998 BARE COSTS				TOTAL INCL O&P
						MAT.	LABOR	EQUIP.	TOTAL	
425 0010	**CONTROL SYSTEMS, PNEUMATIC** (Sub's quote incl. mat. & labor)									
0011	and nominal 50 Ft. of tubing.									
0100	Heating and Ventilating, split system									
0200	Mixed air control, economizer cycle, panel readout, tubing									
0220	Up to 10 tons				Ea.					4,472
0240	For 10 to 20 tons, add									8%
0260	For over 20 tons, add									17%
0300	Heating coil, hot water, 3 way valve,									
0320	Freezestat, limit control on discharge, readout				Ea.					3,200
0500	Cooling coil, chilled water, room									
0520	Thermostat, 3 way valve				Ea.					1,355
0600	Cooling tower, fan cycle, damper control,									
0620	Control system including water readout in/out at panel	Q-19	.67	35.821	Ea.	3,325	695		4,020	4,800
1000	Unit ventilator, day/night operation,									
1100	freezestat, ASHRAE, cycle 2				Ea.					3,310
3000	Boiler room combustion air, damper to 5 SF, controls									2,675
3500	Fan coil, heating and cooling valves, 4 pipe control system	Q-19	3	8		770	156		926	1,100
3600	Heat exchanger system controls	"	.86	27.907		1,450	545		1,995	2,500
3900	Multizone control (one per zone), includes thermostat, damper									
3910	motor and reset of discharge temperature	Q-5	.51	31.373	Ea.	1,450	595		2,045	2,575
4000	Pneumatic thermostat, including controlling room radiator valve									715
4060	Pump control system	Q-19	3	8		770	156		926	1,100
4500	Air supply for pneumatic control system									
4600	Tank mounted duplex compressor, starter, alternator,									
4620	piping, dryer, PRV station and filter									
4640	3/4 HP				Ea.					8,975
4650	1 HP									9,750
4680	3 HP									13,510
4690	5 HP									22,645
4800	Main air supply, includes 3/8" copper main and labor				C.L.F.					656
4810	If poly tubing used, deduct									30%
8600	VAV boxes, incl. thermostat, damper motor, reheat coil & tubing				Ea.					1,390
8610	If no reheat coil, deduct				"					180
430 0010	**CONTROL SYSTEMS, ELECTRONIC**									
0020	For electronic costs, add to division 157-425				Ea.					15%
1000	Electronic hydronic controller	1 Plum	8	1	"	31	21		52	68.50
450 0010	**DIFFUSERS** Aluminum, opposed blade damper unless noted									
0100	Ceiling, linear, also for sidewall									
0120	2" wide	1 Shee	32	.250	L.F.	23.50	5.05		28.55	34.50
0160	4" wide		26	.308	"	31.50	6.20		37.70	45
0500	Perforated, 24" x 24" lay-in panel size, 6" x 6"		16	.500	Ea.	64	10.05		74.05	87
0520	8" x 8"		15	.533		66	10.75		76.75	90.50
0530	9" x 9"		14	.571		66	11.50		77.50	92
0590	16" x 16"		11	.727		85	14.65		99.65	119
1000	Rectangular, 1 to 4 way blow, 6" x 6"		16	.500		37.50	10.05		47.55	58
1010	8" x 8"		15	.533		40	10.75		50.75	62
1014	9" x 9"		15	.533		43.50	10.75		54.25	66
1016	10" x 10"		15	.533		45	10.75		55.75	67.50
1020	12" x 6"		15	.533		48	10.75		58.75	70.50
1040	12" x 9"		14	.571		56.50	11.50		68	82
1060	12" x 12"		12	.667		66	13.45		79.45	95
1070	14" x 6"		13	.615		60	12.40		72.40	87
1074	14" x 14"		12	.667		73	13.45		86.45	103
1150	18" x 18"		9	.889		114	17.90		131.90	156

Important: See the Reference Section for critical supporting data - Reference Nos., Crews, & Location Factors

157 400 | Accessories

		CREW	DAILY OUTPUT	LABOR-HOURS	UNIT	1998 BARE COSTS				TOTAL INCL O&P	
						MAT.	LABOR	EQUIP.	TOTAL		
450	1170	24" x 12"	1 Shee	10	.800	Ea.	112	16.10		128.10	151	**450**
	1180	24" x 24"		7	1.143		190	23		213	248	
	1500	Round, butterfly damper, 6" diameter		18	.444		14.55	8.95		23.50	31	
	1520	8" diameter		16	.500		15.40	10.05		25.45	34	
	2000	T bar mounting, 24" x 24" lay-in frame, 6" x 6"		16	.500		60.50	10.05		70.55	83.50	
	2020	9" x 9"		14	.571		67.50	11.50		79	94	
	2040	12" x 12"		12	.667		88.50	13.45		101.95	120	
	2060	15" x 15"		11	.727		92	14.65		106.65	126	
	2080	18" x 18"		10	.800		121	16.10		137.10	161	
	6000	For steel diffusers instead of aluminum, deduct					10%					
460	0010	**GRILLES**										**460**
	0020	Aluminum										
	1000	Air return, 6" x 6"	1 Shee	26	.308	Ea.	9.90	6.20		16.10	21.50	
	1020	10" x 6"		24	.333		12.10	6.70		18.80	24.50	
	1080	16" x 8"		22	.364		17.60	7.35		24.95	32	
	1100	12" x 12"		22	.364		17.60	7.35		24.95	32	
	1120	24" x 12"		18	.444		31.50	8.95		40.45	49.50	
	1180	16" x 16"		22	.364		25.50	7.35		32.85	40.50	
470	0010	**REGISTERS**										**470**
	0980	Air supply										
	3000	Baseboard, hand adj. damper, enameled steel										
	3012	8" x 6"	1 Shee	26	.308	Ea.	7.70	6.20		13.90	19	
	3020	10" x 6"		24	.333		9.10	6.70		15.80	21.50	
	3060	12" x 6"		23	.348		9.10	7		16.10	22	
	4000	Floor, toe operated damper, enameled steel										
	4020	4" x 6"	1 Shee	32	.250	Ea.	13.80	5.05		18.85	23.50	
	4040	4" x 12"	"	26	.308	"	16.90	6.20		23.10	29	
480	0010	**DUCT ACCESSORIES**										**480**
	0050	Air extractors, 12" x 4"	1 Shee	24	.333	Ea.	12.25	6.70		18.95	25	
	0100	8" x 6"		22	.364		12.25	7.35		19.60	26	
	0200	20" x 8"		16	.500		27.50	10.05		37.55	47.50	
	3000	Fire damper, curtain type, 1-1/2 hr rated, vertical, 6" x 6"		24	.333		16.65	6.70		23.35	29.50	
	3020	8" x 6"		22	.364		16.65	7.35		24	31	
	3240	16" x 14"		18	.444		26.50	8.95		35.45	44	
	5990	Multi-blade dampers, opposed blade, 8" x 6"		24	.333		18	6.70		24.70	31	
	5994	8" x 8"		22	.364		18.80	7.35		26.15	33	
	5996	10" x 10"		21	.381		21	7.70		28.70	36.50	
	6000	12" x 12"		21	.381		24.50	7.70		32.20	40	
	6020	12" x 18"		18	.444		32.50	8.95		41.45	50.50	
	6030	14" x 10"		20	.400		23.50	8.05		31.55	39.50	
	6031	14" x 14"		17	.471		29	9.50		38.50	48	
	6033	16" x 12"		17	.471		29	9.50		38.50	48	
	6035	16" x 16"		16	.500		35.50	10.05		45.55	56	
	6037	18" x 14"		16	.500		35.50	10.05		45.55	56	
	6038	18" x 18"		15	.533		42	10.75		52.75	64	
	6070	20" x 16"		14	.571		43	11.50		54.50	66.50	
	6072	20" x 20"		13	.615		51	12.40		63.40	77	
	6074	22" x 18"		14	.571		51	11.50		62.50	75.50	
	6076	24" x 16"		11	.727		50	14.65		64.65	80	
	6078	24" x 20"		8	1		59.50	20		79.50	99.50	
	6080	24" x 24"		8	1		69	20		89	110	
	6110	26" x 26"		6	1.333		82.50	27		109.50	136	
	6133	30" x 30"	Q-9	6.60	2.424		109	44		153	195	
	6135	32" x 32"	"	6.40	2.500		124	45.50		169.50	213	
	7500	Variable volume modulating motorized damper, incl. elect. mtr.										

157 400	Accessories	CREW	DAILY OUTPUT	LABOR-HOURS	UNIT	1998 BARE COSTS				TOTAL INCL O&P
						MAT.	LABOR	EQUIP.	TOTAL	
480 7504	8" x 6"	1 Shee	15	.533	Ea.	119	10.75		129.75	149 **480**
7506	10" x 6"		14	.571		119	11.50		130.50	151
7510	10" x 10"		13	.615		123	12.40		135.40	156
7520	12" x 12"		12	.667		126	13.45		139.45	162
7522	12" x 16"		11	.727		132	14.65		146.65	170
7524	16" x 10"		12	.667		127	13.45		140.45	163
7526	16" x 14"		10	.800		133	16.10		149.10	174
7528	16" x 18"		9	.889		140	17.90		157.90	185
7542	18" x 18"		8	1		146	20		166	195
7544	20" x 14"		8	1		149	20		169	198
7546	20" x 18"		7	1.143		158	23		181	213
7560	24" x 12"		8	1		151	20		171	200
7562	24" x 18"		7	1.143		168	23		191	224
7568	28" x 10"		7	1.143		250	23		273	315
7590	30" x 14"		5	1.600		266	32		298	350
7610	30" x 24"		3.80	2.105		320	42.50		362.50	425
7700	For thermostat, add		8	1		30	20		50	67
8000	Multi-blade dampers, parallel blade									
8100	8" x 8"	1 Shee	24	.333	Ea.	51	6.70		57.70	67.50
8160	18" x 12"	"	18	.444	"	55	8.95		63.95	75.50
9000	Silencers, noise control for air flow, duct				MCFM	41			41	45
490 0010	**VENTILATORS** Base, damper & bird screen, CFM in 5 MPH wind									**490**
4200	Stationary mushroom, aluminum, 16" orifice diameter	Q-9	10	1.600	Ea.	246	29		275	320
4240	38" orifice diameter		8	2		760	36.50		796.50	895
4250	42" orifice diameter		7.60	2.105		1,000	38		1,038	1,175
4260	50" orifice diameter		7	2.286		1,200	41.50		1,241.50	1,400

For information about Means Estimating Seminars, see yellow pages 11 and 12 in back of book

Important: See the Reference Section for critical supporting data - Reference Nos., Crews, & Location Factors

Division 16 Electrical

Estimating Tips

160 Raceways

- Conduit should be taken off in three main categories: power distribution, branch power, and branch lighting, so the estimator can concentrate on systems and components, therefore making it easier to ensure all items have been accounted for.
- For cost modifications for elevated conduit installation, add the percentages to labor according to the height of installation and only the quantities exceeding the different height levels, not to the total conduit quantities.

161 Conductors & Grounding

- Remember that aluminum wiring of equal ampacity is larger in diameter than copper and may require larger conduit.
- If more than three wires at a time are being pulled, deduct percentages from the labor hours of that grouping of wires.
- When taking off grounding system, identify separately the type and size of wire and list each unique type of ground connection.

162 Boxes & Wiring Devices

- The estimator should take the weights of materials into consideration when completing a takeoff. Topics to consider include: How will the materials be supported? What methods of support are available? How high will the support structure have to reach? Will the final support structure be able to withstand the total burden? Is the support material included or separate from the fixture, equipment and material specified?

163 Motors, Starters, Boards & Switches

- Supports and concrete pads may be shown on drawings for the larger equipment, or the support system may be just a piece of plywood for the back of a panelboard. In either case, it must be included in the costs.

164 Transformers & Bus Ducts

- Do not overlook the costs for equipment used in the installation. If scaffolding or highlifts are available in the field, contractors may use them in lieu of the proposed ladders and rolling staging.

166 Lighting

- Fixtures should be taken off room by room, using the fixture schedule, specifications, and the ceiling plan. For large concentrations of lighting fixtures in the same area deduct the percentages from labor hours.

168 Special Systems

- When estimating material costs for special systems, it is always prudent to obtain manufacturers' quotations for equipment prices and special installation requirements which will affect the total costs.

Reference Numbers

Reference numbers are shown in bold squares at the beginning of some major classifications. These numbers refer to related items in the Reference Section. The reference information may be an estimating procedure, an alternate pricing method or technical information.

Note: Not all subdivisions listed here necessarily appear in this publication.

160 100 | Cable Trays

		CREW	DAILY OUTPUT	LABOR-HOURS	UNIT	1998 BARE COSTS				TOTAL INCL O&P
						MAT.	LABOR	EQUIP.	TOTAL	
105	0010	**CABLE TRAY LADDER TYPE** w/ ftngs & supports, 4" dp., to 15' elev.									**105**
	0160	Galvanized steel tray								
	0170	4" rung spacing, 6" wide	1 Elec	49	.163	L.F.	8.15	3.36		11.51	14.45
	0800	6" rung spacing, 6" wide		50	.160		7.40	3.30		10.70	13.55
	0850	9" wide		47	.170		8.15	3.51		11.66	14.70
	0860	12" wide		44	.182		8.55	3.75		12.30	15.55
	0880	24" wide		40	.200		10.90	4.12		15.02	18.75
130	0010	**CABLE TRAY, COVERS AND DIVIDER**S To 15' high									**130**
	0100	Covers, ventilated galv. steel, straight, 6" wide tray size	1 Elec	260	.031	L.F.	3.15	.63		3.78	4.51
	0200	9" wide tray size		230	.035		3.90	.72		4.62	5.45
	0300	12" wide tray size		200	.040		4.65	.82		5.47	6.45
	0400	18" wide tray size		150	.053		6.35	1.10		7.45	8.80
	0500	24" wide tray size		110	.073		7.80	1.50		9.30	11.05
	3250	Covers, aluminum, straight, 6" wide tray size		260	.031		3.05	.63		3.68	4.39
	3270	9" wide tray size		230	.035		3.70	.72		4.42	5.25
	3290	12" wide tray size		200	.040		4.40	.82		5.22	6.20
	3330	24" wide tray size		130	.062		7.35	1.27		8.62	10.15
150	0010	**WIREWAY** to 15' high									**150**
	0100	Screw cover, with fittings and supports, 2-1/2" x 2-1/2"	1 Elec	45	.178	L.F.	7.80	3.66		11.46	14.60
	0200	4" x 4"		40	.200		8.65	4.12		12.77	16.25
	0400	6" x 6"		30	.267		14.65	5.50		20.15	25
	0600	8" x 8"		20	.400		19.60	8.25		27.85	35

160 200 | Conduits

		CREW	DAILY OUTPUT	LABOR-HOURS	UNIT	1998 BARE COSTS				TOTAL INCL O&P
						MAT.	LABOR	EQUIP.	TOTAL	
205	0010	**CONDUIT** To 15' high, includes 2 terminations, 2 elbows and									**205**
	0020	11 beam clamps per 100 L.F. R160 -205									
	1750	Rigid galvanized steel, 1/2" diameter	1 Elec	90	.089	L.F.	1.40	1.83		3.23	4.54
	1770	3/4" diameter		80	.100		1.65	2.06		3.71	5.20
	1800	1" diameter		65	.123		2.30	2.54		4.84	6.70
	1850	1-1/2" diameter		55	.145		3.55	3		6.55	8.80
	1870	2" diameter		45	.178		4.70	3.66		8.36	11.15
	2500	Steel, intermediate conduit (IMC), 1/2" diameter		100	.080		1.10	1.65		2.75	3.91
	2530	3/4" diameter		90	.089		1.30	1.83		3.13	4.43
	2550	1" diameter		70	.114		1.75	2.35		4.10	5.80
	2600	1-1/2" diameter		60	.133		2.80	2.75		5.55	7.55
	2630	2" diameter		50	.160		3.40	3.30		6.70	9.15
	2650	2-1/2" diameter		40	.200		6.05	4.12		10.17	13.40
	2670	3" diameter		30	.267		7.95	5.50		13.45	17.75
	2700	3-1/2" diameter		27	.296		10.75	6.10		16.85	22
	2730	4" diameter		25	.320		12.75	6.60		19.35	25
	5000	Electric metallic tubing (EMT), 1/2" diameter		170	.047		.40	.97		1.37	2.03
	5020	3/4" diameter		130	.062		.59	1.27		1.86	2.71
	5040	1" diameter		115	.070		.88	1.43		2.31	3.30
	5080	1-1/2" diameter		90	.089		1.58	1.83		3.41	4.74
	5100	2" diameter		80	.100		2.12	2.06		4.18	5.70
	5120	2-1/2" diameter		60	.133		5.05	2.75		7.80	10.05
	5140	3" diameter		50	.160		6.15	3.30		9.45	12.20
	5160	3-1/2" diameter		45	.178		7.90	3.66		11.56	14.70
	5180	4" diameter		40	.200		9.05	4.12		13.17	16.70
	9100	PVC, #40, 1/2" diameter		190	.042		.56	.87		1.43	2.04
	9110	3/4" diameter		145	.055		.65	1.14		1.79	2.57
	9120	1" diameter		125	.064		.90	1.32		2.22	3.15
	9140	1-1/2" diameter		100	.080		1.40	1.65		3.05	4.24
	9150	2" diameter		90	.089		1.85	1.83		3.68	5.05

Important: See the Reference Section for critical supporting data - Reference Nos., Crews, & Location Factors

160 200 | Conduits

			CREW	DAILY OUTPUT	LABOR-HOURS	UNIT	1998 BARE COSTS				TOTAL INCL O&P	
							MAT.	LABOR	EQUIP.	TOTAL		
205	9160	2-1/2" diameter	1 Elec	65	.123	L.F.	3	2.54		5.54	7.45	**205**
	9170	3" diameter		55	.145		3.50	3		6.50	8.75	
	9880	Heat bender, to 6" diameter				Ea.	960			960	1,050	
	9910	15' to 20' high, add						10%				
	9920	20' to 25' high, add						20%				
	9980	Allow. for cond. ftngs., 5% min.-20% max.										
230	0010	**CONDUIT IN CONCRETE SLAB** Including terminations,										**230**
	0020	fittings and supports										
	3230	PVC, schedule 40, 1/2" diameter	1 Elec	270	.030	L.F.	.35	.61		.96	1.39	
	3250	3/4" diameter		230	.035		.38	.72		1.10	1.59	
	3270	1" diameter		200	.040		.54	.82		1.36	1.94	
	3330	1-1/2" diameter		140	.057		.88	1.18		2.06	2.90	
	3350	2" diameter		120	.067		1.14	1.37		2.51	3.50	
	4350	Rigid galvanized steel, 1/2" diameter		200	.040		1.19	.82		2.01	2.66	
	4400	3/4" diameter		170	.047		1.45	.97		2.42	3.18	
	4450	1" diameter		130	.062		2.07	1.27		3.34	4.35	
	4600	1-1/2" diameter		100	.080		3.16	1.65		4.81	6.20	
	4800	2" diameter		90	.089		4.26	1.83		6.09	7.70	
240	0010	**CONDUIT IN TRENCH** Includes terminations and fittings										**240**
	0200	Rigid galvanized steel, 2" diameter	1 Elec	150	.053	L.F.	4.10	1.10		5.20	6.30	
	0600	3" diameter		80	.100		9.25	2.06		11.31	13.55	
	0800	3-1/2" diameter		70	.114		11.80	2.35		14.15	16.85	
250	0010	**CONDUIT FITTINGS FOR RIGID GALVANIZED STEEL**										**250**
	2280	LB, LR or LL fittings & covers, 1/2" diameter	1 Elec	16	.500	Ea.	7.40	10.30		17.70	25	
	2290	3/4" diameter		13	.615		8.95	12.70		21.65	30.50	
	2300	1" diameter		11	.727		13.20	15		28.20	39	
	2350	1-1/2" diameter		6	1.333		24.50	27.50		52	72	
	2370	2" diameter		5	1.600		41	33		74	99	
	2380	2-1/2" diameter		4	2		82.50	41		123.50	159	
	2390	3" diameter		3.50	2.286		107	47		154	195	
	2400	3-1/2" diameter		3	2.667		172	55		227	279	
	2410	4" diameter		2.50	3.200		199	66		265	325	
	5280	Service entrance cap, 1/2" diameter		16	.500		5.80	10.30		16.10	23.50	
	5300	3/4" diameter		13	.615		6.70	12.70		19.40	28	
	5320	1" diameter		10	.800		7.90	16.50		24.40	35.50	
	5360	1-1/2" diameter		6.50	1.231		12.25	25.50		37.75	55	
	5380	2" diameter		5.50	1.455		32.50	30		62.50	84.50	
	5400	2-1/2" diameter		4	2		100	41		141	178	
	5420	3" diameter		3.40	2.353		153	48.50		201.50	248	
	5440	3-1/2" diameter		3	2.667		187	55		242	296	
	5460	4" diameter		2.70	2.963		285	61		346	415	
272	0010	**ELECTRICAL NONMETALLIC TUBING (ENT)**										**272**
	0050	Flexible, 1/2" diameter	1 Elec	270	.030	L.F.	.27	.61		.88	1.30	
	0100	3/4" diameter		230	.035		.37	.72		1.09	1.58	
	0200	1" diameter		145	.055		.59	1.14		1.73	2.51	
	0300	Connectors, to outlet box, 1/2" diameter		230	.035	Ea.	.35	.72		1.07	1.56	
	0310	3/4" diameter		210	.038		.64	.78		1.42	1.98	
	0320	1" diameter		200	.040		.83	.82		1.65	2.26	
	0400	Couplings, to conduit, 1/2" diameter		145	.055		.33	1.14		1.47	2.22	
	0410	3/4" diameter		130	.062		.55	1.27		1.82	2.68	
	0420	1" diameter		125	.064		.61	1.32		1.93	2.83	
275	0010	**MOTOR CONNECTIONS**										**275**
	0020	Flexible conduit and fittings, 115 volt, 1 phase, up to 1 HP motor	1 Elec	8	1	Ea.	3.85	20.50		24.35	37.50	
	0050	2 HP motor		6.50	1.231		3.85	25.50		29.35	45.50	
	0100	3 HP motor		5.50	1.455		6.75	30		36.75	56.50	

R160 -205

For expanded coverage of these items see *Means Electrical Cost Data 1998*

160 200 | Conduits

		CREW	DAILY OUTPUT	LABOR-HOURS	UNIT	1998 BARE COSTS				TOTAL INCL O&P	
						MAT.	LABOR	EQUIP.	TOTAL		
275	0120	230 volt, 10 HP motor, 3 phase	1 Elec	4.20	1.905	Ea.	7.80	39		46.80	72.50	**275**
	0150	15 HP motor		3.30	2.424		13.20	50		63.20	96
	0200	25 HP motor		2.70	2.963		14.20	61		75.20	116
	1500	460 volt, 5 HP motor, 3 phase		8	1		4.05	20.50		24.55	38
	1520	10 HP motor		8	1		4.05	20.50		24.55	38
	1530	25 HP motor		6	1.333		6.60	27.50		34.10	52.50
290	0010	**WIREMOLD RACEWAY**										**290**
	0100	No. 500	1 Elec	100	.080	L.F.	.61	1.65		2.26	3.37
	0110	No. 700		100	.080		.69	1.65		2.34	3.46
	0200	No. 1000		90	.089		1.25	1.83		3.08	4.38
	0400	No. 1500, small pancake		90	.089		1.28	1.83		3.11	4.41
	0600	No. 2000, base & cover, blank		90	.089		1.26	1.83		3.09	4.39
	0800	No. 3000, base & cover, blank		75	.107		2.44	2.20		4.64	6.25
	1000	No. 4000, base & cover, blank		65	.123		4.18	2.54		6.72	8.75
	1200	No. 6000, base & cover, blank		50	.160		6.75	3.30		10.05	12.85
	2400	Fittings, elbows, No. 500		40	.200	Ea.	1.11	4.12		5.23	7.95
	2800	Elbow cover, No. 2000		40	.200		2.30	4.12		6.42	9.30
	2880	Tee, No. 500		42	.190		2.15	3.92		6.07	8.75
	2900	No. 2000		27	.296		6.85	6.10		12.95	17.55
	3000	Switch box, No. 500		16	.500		8.10	10.30		18.40	26
	3400	Telephone outlet, No. 1500		16	.500		8.25	10.30		18.55	26
	3600	Junction box, No. 1500		16	.500		5.75	10.30		16.05	23
	3800	Plugmold wired sections, No. 2000									
	4000	1 circuit, 6 outlets, 3 ft. long	1 Elec	8	1	Ea.	21	20.50		41.50	56.50
	4100	2 circuits, 8 outlets, 6 ft. long		5.30	1.509		35	31		66	89.50
	4200	Tele-power poles, aluminum, 4 outlets		3.70	2.162		139	44.50		183.50	226

161 100 | Conductors

		CREW	DAILY OUTPUT	LABOR-HOURS	UNIT	1998 BARE COSTS				TOTAL INCL O&P	
						MAT.	LABOR	EQUIP.	TOTAL		
105	0010	**ARMORED CABLE**										**105**
	0050	600 volt, copper (BX), #14, 2 conductor, solid	1 Elec	2.40	3.333	C.L.F.	42	68.50		110.50	158
	0100	3 conductor, solid		2.20	3.636		54.50	75		129.50	183
	0120	4 conductor, solid		2	4		73.50	82.50		156	216
	0150	#12, 2 conductor, solid		2.30	3.478		43.50	71.50		115	165
	0200	3 conductor, solid		2	4		63.50	82.50		146	205
	0220	4 conductor, solid		1.80	4.444		89	91.50		180.50	248
	0250	#10, 2 conductor, solid		2	4		76.50	82.50		159	219
	0300	3 conductor, solid		1.60	5		100	103		203	279
	0320	4 conductor, solid		1.40	5.714		151	118		269	360
	0350	#8, 3 conductor, solid		1.30	6.154		164	127		291	385
	0370	4 conductor, stranded		1.10	7.273		256	150		406	525
	9010	600 volt, copper (MC) steel clad, #14, 2 wire		2.40	3.333		41	68.50		109.50	157
	9020	3 wire		2.20	3.636		53	75		128	182
	9030	4 wire		2	4		73.50	82.50		156	216
	9040	#12, 2 wire		2.30	3.478		44.50	71.50		116	166
	9050	3 wire		2	4		65.50	82.50		148	207
	9060	4 wire		1.80	4.444		85.50	91.50		177	244
	9070	#10, 2 wire		2	4		76	82.50		158.50	219
	9080	3 wire		1.60	5		114	103		217	295

Important: See the Reference Section for critical supporting data - Reference Nos., Crews, & Location Factors

161 100 | Conductors

		CREW	DAILY OUTPUT	LABOR-HOURS	UNIT	1998 BARE COSTS				TOTAL INCL O&P	
						MAT.	LABOR	EQUIP.	TOTAL		
105	9090	4 wire	1 Elec	1.40	5.714	C.L.F.	176	118		294	385	**105**
	9100	#8, 2 wire, stranded		1.80	4.444		145	91.50		236.50	310	
	9110	3 wire, stranded		1.30	6.154		198	127		325	425	
	9120	4 wire, stranded		1.10	7.273		269	150		419	540	
137	0010	**FIBER OPTICS**										**137**
	0020	Fiber optics cable only. Added costs depend on the type of fiber										
	0030	special connectors, optical modems, and networking parts.										
	0040	Specialized tools & techniques cause installation costs to vary.										
	0070	Cable, minimum, bulk simplex	1 Elec	8	1	C.L.F.	24.50	20.50		45	60.50	
	0080	Cable, maximum, bulk plenum quad	"	2.29	3.493	"	122	72		194	252	
	0150	Fiber optic jumper				Ea.	55.50			55.50	61	
	0200	Fiber optic pigtail					30			30	33	
	0300	Fiber optic connector	1 Elec	24	.333		18.50	6.85		25.35	32	
	0350	Fiber optic finger splice		32	.250		32	5.15		37.15	43.50	
	0400	Transceiver (low cost bi-directional)		8	1		200	20.50		220.50	254	
	0450	Multi-channel rack enclosure (10 modules)		2	4		295	82.50		377.50	460	
	0500	Fiber optic patch panel (12 ports)		6	1.333		180	27.50		207.50	243	
145	0010	**NON-METALLIC SHEATHED CABLE** 600 volt										**145**
	0100	Copper with ground wire, (Romex)										
	0150	#14, 2 conductor	1 Elec	2.70	2.963	C.L.F.	17.50	61		78.50	119	
	0200	3 conductor		2.40	3.333		29.50	68.50		98	145	
	0220	4 conductor		2.20	3.636		61	75		136	190	
	0250	#12, 2 conductor		2.50	3.200		26	66		92	137	
	0300	3 conductor		2.20	3.636		42	75		117	169	
	0320	4 conductor		2	4		90	82.50		172.50	234	
	0350	#10, 2 conductor		2.20	3.636		41.50	75		116.50	169	
	0400	3 conductor		1.80	4.444		71	91.50		162.50	228	
	0420	4 conductor		1.60	5		118	103		221	299	
	0450	#8, 3 conductor		1.50	5.333		133	110		243	325	
	0480	4 conductor		1.40	5.714		218	118		336	435	
	0500	#6, 3 conductor		1.40	5.714		206	118		324	420	
	0550	SE type SER aluminum cable, 3 RHW and										
	0600	1 bare neutral, 3 #8 & 1 #8	1 Elec	1.60	5	C.L.F.	76	103		179	253	
	0650	3 #6 & 1 #6		1.40	5.714		86	118		204	288	
	0700	3 #4 & 1 #6		1.20	6.667		100	137		237	335	
	0750	3 #2 & 1 #4		1.10	7.273		142	150		292	400	
	0800	3 #1/0 & 1 #2		1	8		215	165		380	505	
	0850	3 #2/0 & 1 #1		.90	8.889		253	183		436	580	
	0900	3 #4/0 & 1 #2/0		.80	10		310	206		516	680	
	1450	UF underground feeder cable, copper with ground, #14, 2 conductor		4	2		17.90	41		58.90	87	
	1500	#12, 2 conductor		3.50	2.286		25.50	47		72.50	105	
	1550	#10, 2 conductor		3	2.667		40	55		95	134	
	1600	#14, 3 conductor		3.50	2.286		26	47		73	106	
	1650	#12, 3 conductor		3	2.667		37	55		92	131	
	1700	#10, 3 conductor		2.50	3.200		57	66		123	171	
	2400	SEU service entrance cable, copper 2 conductors, #8 + #8 neut.		1.50	5.333		107	110		217	298	
	2600	#6 + #8 neutral		1.30	6.154		147	127		274	370	
	2800	#6 + #6 neutral		1.30	6.154		160	127		287	385	
	3000	#4 + #6 neutral		1.10	7.273		217	150		367	485	
	3200	#4 + #4 neutral		1.10	7.273		243	150		393	510	
	3400	#3 + #5 neutral		1.05	7.619		268	157		425	550	
	3600	#3 + #3 neutral		1.05	7.619		305	157		462	590	
	3800	#2 + #4 neutral		1	8		345	165		510	650	
	4000	#1 + #1 neutral		.95	8.421		590	173		763	935	
	4200	1/0 + 1/0 neutral		.90	8.889		650	183		833	1,025	

161 100 | Conductors

		CREW	DAILY OUTPUT	LABOR-HOURS	UNIT	MAT.	LABOR	EQUIP.	TOTAL	TOTAL INCL O&P
4400	2/0 + 2/0 neutral	1 Elec	.85	9.412	C.L.F.	780	194		974	1,175
4600	3/0 + 3/0 neutral		.80	10		980	206		1,186	1,400
4800	Aluminum 2 conductors, #8 + #8 neutral		1.60	5		74	103		177	251
5000	#6 + #6 neutral		1.40	5.714		75	118		193	276
5100	#4 + #6 neutral		1.25	6.400		97	132		229	325
5200	#4 + #4 neutral		1.20	6.667		97	137		234	330
5300	#2 + #4 neutral		1.15	6.957		120	143		263	365
5400	#2 + #2 neutral		1.10	7.273		129	150		279	385
5450	1/0 + #2 neutral		·1.05	7.619		185	157		342	460
5500	1/0 + 1/0 neutral		1	8		195	165		360	485
5550	2/0 + #1 neutral		.95	8.421		207	173		380	510
5600	2/0 + 2/0 neutral		.90	8.889		224	183		407	545
5800	3/0 + 1/0 neutral		.85	9.412		262	194		456	605
6000	3/0 + 3/0 neutral		.85	9.412		286	194		480	630
6200	4/0 + 2/0 neutral		.80	10		278	206		484	640
6400	4/0 + 4/0 neutral		.80	10		315	206		521	680
6500	Service entrance cap for copper SEU									
6600	100 amp	1 Elec	12	.667	Ea.	6.40	13.75		20.15	29.50
6700	150 amp		10	.800		12.05	16.50		28.55	40.50
6800	200 amp		8	1		17.55	20.50		38.05	53

155 | SPECIAL WIRES & FITTINGS

		CREW	DAILY OUTPUT	LABOR-HOURS	UNIT	MAT.	LABOR	EQUIP.	TOTAL	TOTAL INCL O&P
0010	**SPECIAL WIRES & FITTINGS**									
0100	Fixture TFFN 600 volt 90°C stranded, #18	1 Elec	13	.615	C.L.F.	6	12.70		18.70	27
0150	#16		13	.615		8	12.70		20.70	29.50
0500	Thermostat, no jacket, twisted, #18-2 conductor		8	1		8.55	20.50		29.05	43
0550	#18-3 conductor		7	1.143		11.70	23.50		35.20	51.50
0600	#18-4 conductor		6.50	1.231		16.50	25.50		42	59.50
0650	#18-5 conductor		6	1.333		19.80	27.50		47.30	67
0700	#18-6 conductor		5.50	1.455		25	30		55	76
0750	#18-7 conductor		5	1.600		27	33		60	83.50
0800	#18-8 conductor		4.80	1.667		31.50	34.50		66	90.50
0900	TV antenna lead-in, 300 ohm, #20-2 conductor		7	1.143		13.60	23.50		37.10	53.50
0950	Coaxial, feeder outlet		7	1.143		14.40	23.50		37.90	54.50
1000	Coaxial, main riser		6	1.333		21	27.50		48.50	68
1100	Sound, shielded with drain, #22-2 conductor		8	1		14.30	20.50		34.80	49.50
1150	#22-3 conductor		7.50	1.067		18.10	22		40.10	56
1200	#22-4 conductor		6.50	1.231		22	25.50		47.50	65.50
1250	Nonshielded, #22-2 conductor		10	.800		10.10	16.50		26.60	38
1300	#22-3 conductor		9	.889		16.50	18.30		34.80	48
1350	#22-4 conductor		8	1		21.50	20.50		42	57
1400	Microphone cable		8	1		51	20.50		71.50	89.50
2200	Telephone twisted, PVC insulation, #22-2 conductor		10	.800		8	16.50		24.50	36
2250	#22-3 conductor		9	.889		10.80	18.30		29.10	42
2300	#22-4 conductor		8	1		13.65	20.50		34.15	48.50
2350	#19-2 conductor		9	.889		10.10	18.30		28.40	41

160 | UNDERCARPET

		CREW	DAILY OUTPUT	LABOR-HOURS	UNIT	MAT.	LABOR	EQUIP.	TOTAL	TOTAL INCL O&P
0010	**UNDERCARPET**									
0020	Power System									
0100	Cable flat, 3 conductor, #12, w/attached bottom shield	1 Elec	982	.008	L.F.	3.90	.17		4.07	4.56
0200	Shield, top, steel		1,768	.005	"	4.20	.09		4.29	4.77
0250	Splice, 3 conductor		48	.167	Ea.	12.85	3.43		16.28	19.75
0300	Top shield		96	.083		1.15	1.72		2.87	4.08
0350	Tap		40	.200		16.45	4.12		20.57	25
0400	Insulating patch, splice, tap, & end		48	.167		37.50	3.43		40.93	47
0450	Fold		230	.035			.72		.72	1.17
0500	Top shield, tap & fold		96	.083		1.15	1.72		2.87	4.08
0700	Transition, block assembly		77	.104		32	2.14		34.14	38.50
0750	Receptacle frame & base		32	.250		33	5.15		38.15	45

Important: See the Reference Section for critical supporting data - Reference Nos., Crews, & Location Factors

161 100	Conductors	CREW	DAILY OUTPUT	LABOR-HOURS	UNIT	1998 BARE COSTS				TOTAL INCL O&P
						MAT.	LABOR	EQUIP.	TOTAL	
160 0800	Cover receptacle	1 Elec	120	.067	Ea.	2.80	1.37		4.17	5.35
0850	Cover blank		160	.050		3.30	1.03		4.33	5.30
0860	Receptacle, direct connected, single		25	.320		70	6.60		76.60	88
0870	Dual		16	.500		115	10.30		125.30	144
0880	Combination Hi & Lo, tension		21	.381		85	7.85		92.85	106
0900	Box, floor with cover		20	.400		71	8.25		79.25	91.50
0920	Floor service w/barrier		4	2		200	41		241	288
1000	Wall, surface, with cover		20	.400		46.50	8.25		54.75	64.50
1100	Wall, flush, with cover		20	.400		33	8.25		41.25	50
2500	Telephone System									
2510	Transition fitting wall box, surface	1 Elec	24	.333	Ea.	25.50	6.85		32.35	39.50
2520	Flush		24	.333		25.50	6.85		32.35	39.50
2530	Flush, for PC board		24	.333		24	6.85		30.85	38
2540	Floor service box		4	2		200	41		241	288
2550	Cover, surface					11.20			11.20	12.30
2560	Flush					11.20			11.20	12.30
2570	Flush for PC board					11.20			11.20	12.30
2700	Floor fitting w/duplex jack & cover	1 Elec	21	.381		34.50	7.85		42.35	51
2720	Low profile		53	.151		12	3.11		15.11	18.30
2740	Miniature w/duplex jack		53	.151		18.70	3.11		21.81	25.50
2760	25 pair kit		21	.381		37	7.85		44.85	53.50
2780	Low profile		53	.151		12.30	3.11		15.41	18.65
2800	Call director kit for 5 cable		19	.421		58	8.65		66.65	78
2820	4 pair kit		19	.421		69	8.65		77.65	90
2840	3 pair kit		19	.421		73	8.65		81.65	94.50
2860	Comb. 25 pair & 3 cond power		21	.381		58	7.85		65.85	77
2880	5 cond power		21	.381		66	7.85		73.85	85.50
2900	PC board, 8-3 pair		161	.050		50	1.02		51.02	56.50
2920	6-4 pair		161	.050		50	1.02		51.02	56.50
2940	3 pair adapter		161	.050		50	1.02		51.02	56.50
2950	Plug		77	.104		2.10	2.14		4.24	5.80
2960	Couplers		321	.025		5.95	.51		6.46	7.40
3000	Bottom shield for 25 pr. cable		4,420	.002	L.F.	.60	.04		.64	.72
3020	4 pair		4,420	.002		.28	.04		.32	.37
3040	Top shield for 25 pr. cable		4,420	.002		.60	.04		.64	.72
3100	Cable assembly, double-end, 50', 25 pr.		11.80	.678	Ea.	150	13.95		163.95	188
3110	3 pair		23.60	.339		52	7		59	68.50
3120	4 pair		23.60	.339		58	7		65	75.50
3140	Bulk 3 pair		1,473	.005	L.F.	.70	.11		.81	.95
3160	4 pair		1,473	.005	"	.80	.11		.91	1.06
3500	Data System									
3520	Cable 25 conductor w/conn. 40', 75 ohm	1 Elec	14.50	.552	Ea.	54	11.35		65.35	78
3530	Single lead		22	.364		147	7.50		154.50	174
3540	Dual lead		22	.364		186	7.50		193.50	217
3560	Shields same for 25 cond. as 25 pair tele.									
3570	Single & dual, none req'd.									
3590	BNC coax connectors, Plug	1 Elec	40	.200	Ea.	6.25	4.12		10.37	13.65
3600	TNC coax connectors, Plug	"	40	.200	"	6.25	4.12		10.37	13.65
3700	Cable-bulk									
3710	Single lead	1 Elec	1,473	.005	L.F.	2.40	.11		2.51	2.82
3720	Dual lead	"	1,473	.005	"	3.10	.11		3.21	3.59
3730	Hand tool crimp				Ea.	270			270	297
3740	Hand tool notch				"	14.50			14.50	15.95
3750	Boxes & floor fitting same as telephone									
165 0010	**WIRE**									
0020	600 volt type THW, copper, solid, #14	1 Elec	13	.615	C.L.F.	4.40	12.70		17.10	25.50

161 100 | Conductors

			DAILY OUTPUT	LABOR-HOURS	UNIT	1998 BARE COSTS				TOTAL INCL O&P	
						MAT.	LABOR	EQUIP.	TOTAL		
165	0030	#12	1 Elec	11	.727	C.L.F.	6.45	15		21.45	31.50	165
	0040	#10		10	.800		9.95	16.50		26.45	38	
	0050	Stranded, #14		13	.615		5.65	12.70		18.35	26.50	
	0051	Wire, 600 volt, stranded										
	0100	#12	1 Elec	11	.727	C.L.F.	7.40	15		22.40	32.50	
	0120	#10		10	.800		11.50	16.50		28	39.50	
	0140	#8		8	1		18.80	20.50		39.30	54	
	0400	250 MCM(kcmil)		2	4		245	82.50		327.50	405	
	0420	300 MCM		1.90	4.211		290	86.50		376.50	460	
	0450	350 MCM		1.80	4.444		335	91.50		426.50	520	
	0480	400 MCM		1.70	4.706		390	97		487	590	
	0490	500 MCM		1.60	5		475	103		578	695	
	0530	Aluminum, stranded, #8		9	.889		12.30	18.30		30.60	43.50	
	0540	#6		8	1		15.40	20.50		35.90	50.50	
	0560	#4		6.50	1.231		19.40	25.50		44.90	63	
	0580	#2		5.30	1.509		26	31		57	79.50	
	0600	#1		4.50	1.778		38.50	36.50		75	103	
	0620	1/0		4	2		44	41		85	116	
	0640	2/0		3.60	2.222		52	46		98	132	
	0680	3/0		3.30	2.424		64.50	50		114.50	153	
	0700	4/0		3.10	2.581		73	53		126	168	
	0720	250 MCM(kcmil)		2.90	2.759		88	57		145	190	
	0740	300 MCM		2.70	2.963		120	61		181	232	
	0760	350 MCM		2.50	3.200		124	66		190	244	
	0780	400 MCM		2.30	3.478		142	71.50		213.50	273	
	0800	500 MCM		2	4		160	82.50		242.50	310	
	0850	600 MCM		1.90	4.211		202	86.50		288.50	365	
	0880	700 MCM		1.70	4.706		227	97		324	410	
	0900	750 MCM		1.60	5		230	103		333	420	
	0920	Type THWN-THHN, copper, solid, #14		13	.615		4.40	12.70		17.10	25.50	
	0940	#12		11	.727		6.45	15		21.45	31.50	
	0960	#10		10	.800		9.95	16.50		26.45	38	
	1000	Stranded, #14		13	.615		5.65	12.70		18.35	26.50	
	1010	2 wire, #14, install wire only		655	.012	L.F.	.11	.25		.36	.53	
	1020	3 wire, #14, install wire only		440	.018	"	.17	.37		.54	.80	
	1200	#12		11	.727	C.L.F.	7.40	15		22.40	32.50	
	1250	#10		10	.800		11.50	16.50		28	39.50	
	1300	#8		8	1		18.80	20.50		39.30	54	
	1350	#6		6.50	1.231		28	25.50		53.50	72.50	
	1400	#4		5.30	1.509		44	31		75	99.50	

161 800 | Grounding

			DAILY OUTPUT	LABOR-HOURS	UNIT	MAT.	LABOR	EQUIP.	TOTAL	TOTAL INCL O&P	
810	0010	**GROUNDING**									810
	0030	Rod, copper clad, 8' long, 1/2" diameter	1 Elec	5.50	1.455	Ea.	15.20	30		45.20	66	
	0050	3/4" diameter		5.30	1.509		29.50	31		60.50	83	
	0080	10' long, 1/2" diameter		4.80	1.667		17.50	34.50		52	75.50	
	0100	3/4" diameter		4.40	1.818		29.50	37.50		67	94	
	0130	15' long, 3/4" diameter		4	2		79	41		120	155	
	0260	Wire ground bare armored, #8-1 conductor		2	4	C.L.F.	57	82.50		139.50	198	
	0270	#6-1 conductor		1.80	4.444		72	91.50		163.50	229	
	0390	Bare copper wire, #8 stranded		11	.727		17	15		32	43	
	0400	#6		10	.800		24.50	16.50		41	54	
	0600	#2		5	1.600		60	33		93	120	
	0800	3/0		3.30	2.424		146	50		196	243	
	1000	4/0		2.85	2.807		183	58		241	296	
	1200	250 MCM(kcmil)		2.40	3.333		215	68.50		283.50	350	

Important: See the Reference Section for critical supporting data - Reference Nos., Crews, & Location Factors

161 800 | Grounding

		CREW	DAILY OUTPUT	LABOR-HOURS	UNIT	1998 BARE COSTS				TOTAL INCL O&P
						MAT.	LABOR	EQUIP.	TOTAL	
810	1800	Water pipe ground clamps, heavy duty									810
	2000	Bronze, 1/2" to 1" diameter	1 Elec	8	1	Ea.	11.10	20.50		31.60	46
	2100	1-1/4" to 2" diameter		8	1		14.15	20.50		34.65	49
	2200	2-1/2" to 3" diameter		6	1.333		38	27.50		65.50	86.50
	2800	Brazed connections, #6 wire		12	.667		9.95	13.75		23.70	33.50
	3000	#2 wire		10	.800		13.30	16.50		29.80	41.50
	3100	3/0 wire		8	1		20	20.50		40.50	55.50
	3200	4/0 wire		7	1.143		23	23.50		46.50	64
	3400	250 MCM wire		5	1.600		27	33		60	83.50

162 100 | Boxes

		CREW	DAILY OUTPUT	LABOR-HOURS	UNIT	1998 BARE COSTS				TOTAL INCL O&P
						MAT.	LABOR	EQUIP.	TOTAL	
110	0010	**OUTLET BOXES**									110
	0020	Pressed steel, octagon, 4"	1 Elec	20	.400	Ea.	1.44	8.25		9.69	15.10
	0060	Covers, blank		64	.125		.59	2.58		3.17	4.86
	0100	Extension		40	.200		2.29	4.12		6.41	9.25
	0150	Square, 4"		20	.400		1.90	8.25		10.15	15.60
	0200	Extension		40	.200		2.41	4.12		6.53	9.40
	0250	Covers, blank		64	.125		.68	2.58		3.26	4.96
	0300	Plaster rings		64	.125		1.14	2.58		3.72	5.45
	0550	Handy box		27	.296		1.64	6.10		7.74	11.80
	0560	Covers, device		64	.125		.55	2.58		3.13	4.82
	0650	Switchbox		27	.296		2.11	6.10		8.21	12.30
	0700	Masonry, 1 gang, 2-1/2" deep		27	.296		5.05	6.10		11.15	15.55
	0710	3-1/2" deep		27	.296		4.41	6.10		10.51	14.85
	0750	2 gang, 2-1/2" deep		20	.400		7.75	8.25		16	22
	0760	3-1/2" deep		20	.400		7.40	8.25		15.65	21.50
	1100	Concrete, floor, 1 gang		5.30	1.509		55	31		86	112
	1400	Cast, 1 gang, FS (2" deep), 1/2" hub		12	.667		10.85	13.75		24.60	34.50
	1410	3/4" hub		12	.667		11.45	13.75		25.20	35
	1420	FD (2-11/16" deep), 1/2" hub		12	.667		13.10	13.75		26.85	37
	1430	3/4" hub		12	.667		14.25	13.75		28	38
	1450	2 gang, FS, 1/2" hub		10	.800		19.45	16.50		35.95	48.50
	1460	3/4" hub		10	.800		20	16.50		36.50	49
	1470	FD, 1/2" hub		10	.800		23.50	16.50		40	53
	1480	3/4" hub		10	.800		23.50	16.50		40	52.50
	1500	3 gang, FS, 3/4" hub		9	.889		30.50	18.30		48.80	63.50
	1510	Switch cover, 1 gang, FS		64	.125		2.89	2.58		5.47	7.40
	1520	2 gang		53	.151		4.76	3.11		7.87	10.35
	1530	Duplex receptacle cover, 1 gang, FS		64	.125		2.89	2.58		5.47	7.40
	1540	2 gang, FS		53	.151		4.76	3.11		7.87	10.35
	1550	Weatherproof switch cover		64	.125		5.50	2.58		8.08	10.25
	1600	Weatherproof receptacle cover		64	.125		5.50	2.58		8.08	10.25
	1750	FSC, 1 gang, 1/2" hub		11	.727		12.05	15		27.05	38
	1760	3/4" hub		11	.727		13.15	15		28.15	39
	1770	2 gang, 1/2" hub		9	.889		21.50	18.30		39.80	53.50
	1780	3/4" hub		9	.889		23	18.30		41.30	55
	1790	FDC, 1 gang, 1/2" hub		11	.727		14.35	15		29.35	40.50

ELECTRICAL 16

162 100 | Boxes

			DAILY OUTPUT	LABOR-HOURS	UNIT	1998 BARE COSTS				TOTAL INCL O&P	
						MAT.	LABOR	EQUIP.	TOTAL		
110	1800	3/4" hub	1 Elec	11	.727	Ea.	15.45	15		30.45	41.50	110
	1810	2 gang, 1/2" hub		9	.889		26	18.30		44.30	58.50
	1820	3/4" hub		9	.889		25	18.30		43.30	57.50
120	0010	**OUTLET BOXES, PLASTIC**										120
	0050	4" diameter, round with 2 mounting nails	1 Elec	25	.320	Ea.	1.54	6.60		8.14	12.50
	0100	Bar hanger mounted		25	.320		2.38	6.60		8.98	13.40
	0200	4", square with 2 mounting nails		25	.320		1.92	6.60		8.52	12.90
	0300	Plaster ring		64	.125		.82	2.58		3.40	5.10
	0400	Switch box with 2 mounting nails, 1 gang		30	.267		.87	5.50		6.37	9.95
	0500	2 gang		25	.320		1.87	6.60		8.47	12.85
	0600	3 gang		20	.400		2.97	8.25		11.22	16.75
130	0010	**PULL BOXES & CABINETS**										130
	0100	Sheet metal, pull box, NEMA 1, type SC, 6" W x 6" H x 4" D	1 Elec	8	1	Ea.	8.35	20.50		28.85	42.50
	0200	8" W x 8" H x 4" D		8	1		11.45	20.50		31.95	46
	0300	10" W x 12" H x 6" D		5.30	1.509		20	31		51	73
	0800	12" W x 16" H x 6" D		4.70	1.702		28	35		63	88
	1000	20" W x 20" H x 6" D		3.60	2.222		56	46		102	137
	7000	Cabinets, current transformer									
	7050	Single door, 24" H x 24" W x 10" D	1 Elec	1.60	5	Ea.	116	103		219	297
	7100	30" H x 24" W x 10" D		1.30	6.154		133	127		260	355
	7150	36" H x 24" W x 10" D		1.10	7.273		149	150		299	410
	7200	30" H x 30" W x 10" D		1	8		160	165		325	445
	7250	36" H x 30" W x 10" D		.90	8.889		217	183		400	540
	7300	36" H x 36" W x 10" D		.80	10		241	206		447	600
	7500	Double door, 48" H x 36" W x 10" D		.60	13.333		360	275		635	845
	7550	24" H x 24" W x 12" D		1	8		187	165		352	475

162 300 | Wiring Devices

			CREW	DAILY OUTPUT	LABOR-HOURS	UNIT	MAT.	LABOR	EQUIP.	TOTAL	TOTAL INCL O&P	
310	0010	**LOW VOLTAGE SWITCHING**										310
	3600	Relays, 120 V or 277 V standard	1 Elec	12	.667	Ea.	26	13.75		39.75	51
	3800	Flush switch, standard		40	.200		9.05	4.12		13.17	16.70
	4000	Interchangeable		40	.200		11.80	4.12		15.92	19.75
	4100	Surface switch, standard		40	.200		6.60	4.12		10.72	14.05
	4200	Transformer 115 V to 25 V		12	.667		93	13.75		106.75	125
	4400	Master control, 12 circuit, manual		4	2		94	41		135	171
	4500	25 circuit, motorized		4	2		102	41		143	180
	4600	Rectifier, silicon		12	.667		30.50	13.75		44.25	56
	4800	Switchplates, 1 gang, 1, 2 or 3 switch, plastic		80	.100		3	2.06		5.06	6.65
	5000	Stainless steel		80	.100		8.10	2.06		10.16	12.25
	5400	2 gang, 3 switch, stainless steel		53	.151		15.65	3.11		18.76	22.50
320	0010	**WIRING DEVICES**										320
	0200	Toggle switch, quiet type, single pole, 15 amp	1 Elec	40	.200	Ea.	4.10	4.12		8.22	11.25
	1650	Dimmer switch, 120 volt, incandescent, 600 watt, 1 pole		16	.500		10.80	10.30		21.10	29
	2460	Receptacle, duplex, 120 volt, grounded, 15 amp		40	.200		2.25	4.12		6.37	9.25
	2470	20 amp		27	.296		5.95	6.10		12.05	16.55
	2490	Dryer, 30 amp		15	.533		11.70	11		22.70	31
	2500	Range, 50 amp		11	.727		16	15		31	42
	2600	Wall plates, stainless steel, 1 gang		80	.100		1.70	2.06		3.76	5.25
	2800	2 gang		53	.151		3.90	3.11		7.01	9.40
	3200	Lampholder, keyless		26	.308		5.20	6.35		11.55	16.05
	3400	Pullchain with receptacle		22	.364		8.25	7.50		15.75	21.50

Important: See the Reference Section for critical supporting data - Reference Nos., Crews, & Location Factors

163 100 | Starters & Controls

			DAILY OUTPUT	LABOR-HOURS	UNIT	1998 BARE COSTS				TOTAL INCL O&P	
		CREW				MAT.	LABOR	EQUIP.	TOTAL		
110	0010	**MOTOR CONTROL CENTER** Consists of starters & structures									110
	0050	Starters, class 1, type B, comb. MCP, FVNR, with									
	0100	control transformer, 10 HP, size 1, 12" high	1 Elec	2.70	2.963	Ea.	925	61		986	1,125
	0200	25 HP, size 2, 18" high		2	4		1,050	82.50		1,132.50	1,275
	0300	50 HP, size 3, 24" high		1	8		1,625	165		1,790	2,075
	0350	75 HP, size 4, 24" high		.80	10		2,400	206		2,606	2,975
	0400	100 HP, size 4, 30" high		.70	11.429		2,725	235		2,960	3,375
	0500	200 HP, size 5, 48" high		.50	16		4,850	330		5,180	5,875
130	0010	**MOTOR STARTERS & CONTROLS**									130
	0050	Magnetic, FVNR, with enclosure and heaters, 480 volt									
	0080	2 HP, size 00	1 Elec	3.50	2.286	Ea.	153	47		200	245
	0100	5 HP, size 0		2.30	3.478		185	71.50		256.50	320
	0200	10 HP, size 1		1.60	5		218	103		321	410
	0300	25 HP, size 2	2 Elec	2.20	7.273		410	150		560	695
	0400	50 HP, size 3	"	1.80	8.889		670	183		853	1,025
	1400	Combination, with fused switch, 5 HP, size 0	1 Elec	1.80	4.444		460	91.50		551.50	655
	1600	10 HP, size 1		1.30	6.154		480	127		607	735
	1800	25 HP, size 2		1	8		745	165		910	1,100

163 200 | Boards

			DAILY OUTPUT	LABOR-HOURS	UNIT	1998 BARE COSTS				TOTAL INCL O&P	
		CREW				MAT.	LABOR	EQUIP.	TOTAL		
205	0010	**CIRCUIT BREAKERS** (in enclosure)									205
	0100	Enclosed (NEMA 1), 600 volt, 3 pole, 30 amp	1 Elec	3.20	2.500	Ea.	300	51.50		351.50	415
	0200	60 amp		2.80	2.857		300	59		359	425
	0400	100 amp		2.30	3.478		335	71.50		406.50	485
	0600	225 amp		1.50	5.333		880	110		990	1,150
	0700	400 amp		.80	10		1,500	206		1,706	1,975
230	0010	**LOAD CENTERS** (residential type)									230
	0100	3 wire, 120/240V, 1 phase, including 1 pole plug-in breakers									
	0200	100 amp main lugs, indoor, 8 circuits	1 Elec	1.40	5.714	Ea.	107	118		225	310
	0300	12 circuits		1.20	6.667		146	137		283	385
	0400	Rainproof, 8 circuits		1.40	5.714		129	118		247	335
	0500	12 circuits		1.20	6.667		176	137		313	420
	0600	200 amp main lugs, indoor, 16 circuits		.90	8.889		249	183		432	575
	0700	20 circuits		.75	10.667		310	220		530	705
	0800	24 circuits		.65	12.308		370	254		624	825
	1200	Rainproof, 16 circuits		.90	8.889		297	183		480	625
	1300	20 circuits		.75	10.667		345	220		565	735
	1400	24 circuits		.65	12.308		440	254		694	900
	1500	30 circuits		.60	13.333		530	275		805	1,025
	1600	42 circuits		.40	20		820	410		1,230	1,575
	1800	400 amp main lugs, indoor, 42 circuits		.36	22.222		935	460		1,395	1,775
	1900	Rainproof, 42 circuit		.36	22.222		1,100	460		1,560	1,975
	2200	Plug in breakers, 20 amp, 1 pole, 4 wire, 120/208 volts									
	2210	125 amp main lugs, indoor, 12 circuits	1 Elec	1.20	6.667	Ea.	229	137		366	475
	2300	18 circuits		.80	10		335	206		541	705
	2400	Rainproof, 12 circuits		1.20	6.667		269	137		406	520
	2500	18 circuits		.80	10		375	206		581	750
	2600	200 amp main lugs, indoor, 24 circuits		.65	12.308		455	254		709	915
	2700	30 circuits		.60	13.333		510	275		785	1,025
	2800	36 circuits		.50	16		645	330		975	1,250
	2900	42 circuits		.40	20		700	410		1,110	1,450
	3000	Rainproof, 24 circuits		.65	12.308		505	254		759	970
	3100	30 circuits		.60	13.333		560	275		835	1,075
	3200	36 circuits		.50	16		800	330		1,130	1,425
	3300	42 circuits		.40	20		860	410		1,270	1,625

ELECTRICAL 16

163 200 | **Boards**		CREW	DAILY OUTPUT	LABOR-HOURS	UNIT	1998 BARE COSTS				TOTAL INCL O&P
						MAT.	LABOR	EQUIP.	TOTAL	
240	**0010**	**METER CENTERS AND SOCKETS**									**240**
0100	Sockets, single position, 4 terminal, 100 amp	1 Elec	3.20	2.500	Ea.	29	51.50		80.50	117
0200	150 amp		2.30	3.478		33.50	71.50		105	154
0300	200 amp		1.90	4.211		40	86.50		126.50	186
0400	20 amp		3.20	2.500		29	51.50		80.50	117
0500	Double position, 4 terminal, 100 amp		2.80	2.857		122	59		181	231
0600	150 amp		2.10	3.810		126	78.50		204.50	267
0700	200 amp		1.70	4.706		274	97		371	460
2590	Basic meter device									
2600	1P 3W 120/240V 4 jaw 125A sockets, 3 meter	1 Elec	.50	16	Ea.	370	330		700	950
2610	4 meter		.45	17.778		445	365		810	1,100
2620	5 meter		.40	20		560	410		970	1,300
2630	6 meter		.30	26.667		640	550		1,190	1,600
2640	7 meter		.28	28.571		820	590		1,410	1,875
2650	8 meter		.26	30.769		890	635		1,525	2,000
2660	10 meter		.24	33.333		1,125	685		1,810	2,350
2680	Rainproof 1P 3W 120/240V 4 jaw 125A sockets									
2690	3 meter	1 Elec	.50	16	Ea.	370	330		700	950
2700	4 meter		.45	17.778		445	365		810	1,100
2710	6 meter		.30	26.667		640	550		1,190	1,600
2720	7 meter		.28	28.571		820	590		1,410	1,875
2730	8 meter		.26	30.769		890	635		1,525	2,000
2750	1P 3W 120/240V 4 jaw sockets									
2760	with 125A circuit breaker, 3 meter	1 Elec	.50	16	Ea.	520	330		850	1,125
2770	4 meter		.45	17.778		645	365		1,010	1,300
2780	5 meter		.40	20		810	410		1,220	1,575
2790	6 meter		.30	26.667		940	550		1,490	1,925
2800	7 meter		.28	28.571		1,175	590		1,765	2,250
2810	8 meter		.26	30.769		1,300	635		1,935	2,450
2820	10 meter		.24	33.333		1,625	685		2,310	2,900
3250	1P 3W 120/240V 4 jaw sockets									
3260	with 200A circuit breaker, 3 meter	1 Elec	.50	16	Ea.	1,050	330		1,380	1,700
3270	4 meter		.45	17.778		1,400	365		1,765	2,150
3290	6 meter		.30	26.667		2,100	550		2,650	3,200
3300	7 meter		.28	28.571		2,450	590		3,040	3,675
3310	8 meter		.28	28.571		2,825	590		3,415	4,075
245	**0010**	**PANELBOARDS** (Commercial use)									**245**
0050	NQOD, w/20 amp 1 pole bolt-on circuit breakers									
0600	4 wire, 120/208 volts, 100 amp main lugs, 12 circuits	1 Elec	1	8	Ea.	440	165		605	755
0650	16 circuits		.75	10.667		505	220		725	915
0700	20 circuits		.65	12.308		590	254		844	1,075
0750	24 circuits		.60	13.333		640	275		915	1,150
0800	30 circuits		.53	15.094		740	310		1,050	1,325
0850	225 amp main lugs, 32 circuits		.45	17.778		830	365		1,195	1,525
0900	34 circuits		.42	19.048		850	390		1,240	1,575
0950	36 circuits		.40	20		870	410		1,280	1,625
1000	42 circuits		.34	23.529		975	485		1,460	1,875
1200	NEHB, w/20 amp, 1 pole bolt-on circuit breakers									
1250	4 wire, 277/480 volts, 100 amp main lugs, 12 circuits	1 Elec	.88	9.091	Ea.	840	187		1,027	1,225
1300	20 circuits		.60	13.333		1,250	275		1,525	1,825
1350	225 amp main lugs, 24 circuits		.45	17.778		1,425	365		1,790	2,175
1400	30 circuits		.40	20		1,725	410		2,135	2,575
1450	36 circuits		.36	22.222		2,000	460		2,460	2,950
1500	42 circuits		.30	26.667		2,300	550		2,850	3,425
1510	225 amp main lugs, NEMA 7, 12 circuits		.45	17.778		3,050	365		3,415	3,950
1590	24 circuits		.15	53.333		3,600	1,100		4,700	5,775

Important: See the Reference Section for critical supporting data - Reference Nos., Crews, & Location Factors

163 200 | Boards

		CREW	DAILY OUTPUT	LABOR-HOURS	UNIT	1998 BARE COSTS				TOTAL INCL O&P
						MAT.	LABOR	EQUIP.	TOTAL	
250	0010	**PANELBOARD & LOAD CENTER CIRCUIT BREAKERS**								
	0050	Bolt-on, 10,000 amp IC, 120 volt, 1 pole								
	0100	15 to 50 amp	1 Elec	10	.800	Ea.	10	16.50		26.50	38
	0200	60 amp		8	1		10	20.50		30.50	44.50
	0300	70 amp		8	1		19	20.50		39.50	54.50
	0350	240 volt, 2 pole									
	0400	15 to 50 amp	1 Elec	8	1	Ea.	22	20.50		42.50	57.50
	0500	60 amp		7.50	1.067		22	22		44	60
	0600	80 to 100 amp		5	1.600		85	33		118	148
	0700	3 pole, 15 to 60 amp		6.20	1.290		69.50	26.50		96	120
	0800	70 amp		5	1.600		87.50	33		120.50	151
	0900	80 to 100 amp		3.60	2.222		99.50	46		145.50	184
	1000	22,000 amp I.C., 240 volt, 2 pole, 70 - 225 amp		2.70	2.963		430	61		491	570
	1100	3 pole, 70 - 225 amp		2.30	3.478		475	71.50		546.50	640
	2000	Plug-in panel or load center, 120/240 volt, to 60 amp, 1 pole		12	.667		7.75	13.75		21.50	31
	2010	2 pole		9	.889		17.50	18.30		35.80	49.50
	2020	3 pole		7.50	1.067		60.50	22		82.50	103
	2030	100 amp, 2 pole		6	1.333		81	27.50		108.50	134
	2040	3 pole		4.50	1.778		90	36.50		126.50	159
	2050	150 amp, 2 pole		3	2.667		113	55		168	214

163 300 | Switches

		CREW	DAILY OUTPUT	LABOR-HOURS	UNIT	MAT.	LABOR	EQUIP.	TOTAL	TOTAL INCL O&P
310	0010	**CONTACTORS, AC** Enclosed (NEMA 1)								
	0050	Lighting, 600 volt 3 pole, electrically held								
	0100	20 amp	1 Elec	4	2	Ea.	165	41		206	250
	0200	30 amp		3.60	2.222		174	46		220	266
	0300	60 amp		3	2.667		350	55		405	475
	0400	100 amp		2.50	3.200		575	66		641	745
	0500	200 amp		1.40	5.714		1,350	118		1,468	1,700
	0600	300 amp		.80	10		2,900	206		3,106	3,500
	0800	600 volt 3 pole, mechanically held, 30 amp		3.60	2.222		270	46		316	370
	0900	60 amp		3	2.667		540	55		595	685
	1000	75 amp		2.80	2.857		760	59		819	930
	1100	100 amp		2.50	3.200		760	66		826	945
	1200	150 amp		2	4		2,100	82.50		2,182.50	2,425
	1300	200 amp		1.40	5.714		2,100	118		2,218	2,500
	1500	Magnetic with auxiliary contact, size 00, 9 amp		4	2		141	41		182	224
	1600	Size 0, 18 amp		4	2		169	41		210	254
	1700	Size 1, 27 amp		3.60	2.222		192	46		238	286
	1800	Size 2, 45 amp		3	2.667		355	55		410	480
	1900	Size 3, 90 amp		2.50	3.200		575	66		641	740
	2000	Size 4, 135 amp		2.30	3.478		1,325	71.50		1,396.50	1,575
	2100	Size 5, 270 amp		.90	8.889		2,775	183		2,958	3,350
	2310	Size 8, 1215 amp		.40	20		16,900	410		17,310	19,300
	2500	Magnetic, 240 volt, 1-2 pole, .75 HP motor		4	2		114	41		155	193
	2520	2 HP motor		3.60	2.222		128	46		174	215
	2540	5 HP motor		2.50	3.200		310	66		376	450
	2560	10 HP motor		1.40	5.714		510	118		628	755
	2600	240 volt or less, 3 pole, .75 HP motor		4	2		114	41		155	193
	2620	5 HP motor		3.60	2.222		141	46		187	231
	2640	10 HP motor		3.60	2.222		164	46		210	256
	2660	15 HP motor		2.50	3.200		330	66		396	470
	2700	25 HP motor		2.50	3.200		330	66		396	470
	2720	30 HP motor		1.40	5.714		545	118		663	795
	2740	40 HP motor		1.40	5.714		545	118		663	795
	2760	50 HP motor		.80	10		545	206		751	935

ELECTRICAL 16

For expanded coverage of these items see *Means Electrical Cost Data 1998*

163 300 | Switches

		CREW	DAILY OUTPUT	LABOR-HOURS	UNIT	1998 BARE COSTS				TOTAL INCL O&P	
						MAT.	LABOR	EQUIP.	TOTAL		
310	2800	75 HP motor	1 Elec	.80	10	Ea.	1,275	206		1,481	1,750	**310**
	2820	100 HP motor		.50	16		1,275	330		1,605	1,975
	2860	150 HP motor		.50	16		2,725	330		3,055	3,550
	2880	200 HP motor		.50	16		2,725	330		3,055	3,550
	3000	600 volt, 3 pole, 5 HP motor		4	2		141	41		182	224
	3020	10 HP motor		3.60	2.222		164	46		210	256
	3040	25 HP motor		3	2.667		330	55		385	450
	3100	50 HP motor		2.50	3.200		545	66		611	710
	3160	100 HP motor		1.40	5.714		1,275	118		1,393	1,625
	3220	200 HP motor		.80	10		2,725	206		2,931	3,325
320	0010	**CONTROL STATIONS**										**320**
	0050	NEMA 1, heavy duty, stop/start	1 Elec	8	1	Ea.	105	20.50		125.50	149
	0100	Stop/start, pilot light		6.20	1.290		141	26.50		167.50	200
	0200	Hand/off/automatic		6.20	1.290		77.50	26.50		104	129
360	0010	**SAFETY SWITCHES**										**360**
	0100	General duty 240 volt, 3 pole, fused, 30 amp	1 Elec	3.20	2.500	Ea.	50.50	51.50		102	140
	0200	60 amp		2.30	3.478		85	71.50		156.50	211
	0300	100 amp		1.90	4.211		148	86.50		234.50	305
	0400	200 amp		1.30	6.154		320	127		447	555
	0500	400 amp		.90	8.889		825	183		1,008	1,200

16 ELECTRICAL

164 100 | Transformers

		CREW	DAILY OUTPUT	LABOR-HOURS	UNIT	1998 BARE COSTS				TOTAL INCL O&P	
						MAT.	LABOR	EQUIP.	TOTAL		
110	0010	**BUCK-BOOST TRANSFORMER**										**110**
	0100	Single phase, 120/240 volt primary, 12/24 volt secondary									
	0200	0.10 KVA	1 Elec	8	1	Ea.	49	20.50		69.50	87.50
	0400	0.25 KVA		5.70	1.404		64.50	29		93.50	118
	0600	0.50 KVA		4	2		88	41		129	164
	0800	0.75 KVA		3.10	2.581		113	53		166	211
	1200	1.5 KVA		1.80	4.444		172	91.50		263.50	340
	1600	3.0 KVA		1.40	5.714		280	118		398	505
	2000	3 phase, 240/208 volt, 15 KVA		1.20	6.667		385	137		522	650
	2200	30 KVA		.80	10		495	206		701	880
	2400	45 KVA		.70	11.429		665	235		900	1,125
	2600	75 KVA		.60	13.333		935	275		1,210	1,475
	3000	150 KVA		.45	17.778		1,525	365		1,890	2,275
	3200	225 KVA		.40	20		2,150	410		2,560	3,050
120	0010	**DRY TYPE TRANSFORMER**										**120**
	0050	Single phase, 240/480 volt primary, 120/240 volt secondary									
	0100	1 KVA	1 Elec	2	4	Ea.	131	82.50		213.50	280
	0300	2 KVA		1.60	5		197	103		300	385
	0500	3 KVA		1.40	5.714		244	118		362	460
	0700	5 KVA		1.20	6.667		335	137		472	595
	1100	10 KVA		.80	10		600	206		806	995
	1300	15 KVA	2 Elec	1.20	13.333		815	275		1,090	1,350
	1700	37.5 KVA		.80	20		1,450	410		1,860	2,275
	1900	50 KVA		.70	22.857		1,725	470		2,195	2,675
	2110	100 KVA	R-3	.90	22.222		2,950	460	150	3,560	4,175
	2310	Ventilated, 3 KVA	1 Elec	1	8		425	165		590	740

Important: See the Reference Section for critical supporting data - Reference Nos., Crews, & Location Factors

164 100 | Transformers

			DAILY	LABOR-		1998 BARE COSTS				TOTAL	
			CREW	OUTPUT	HOURS	UNIT	MAT.	LABOR	EQUIP.	TOTAL	INCL O&P
120	2900	9 KVA	1 Elec	.70	11.429	Ea.	660	235		895	1,125	**120**
	3100	15 KVA	2 Elec	1.10	14.545		860	300		1,160	1,425
	3300	30 KVA		.90	17.778		1,025	365		1,390	1,725
	3700	75 KVA		.70	22.857		1,875	470		2,345	2,825
	3900	112.5 KVA	R-3	.90	22.222		2,475	460	150	3,085	3,650
	4100	150 KVA		.85	23.529		3,225	485	158	3,868	4,525
	4500	300 KVA		.55	36.364		5,550	750	245	6,545	7,600
	4700	500 KVA		.45	44.444		9,175	915	299	10,389	11,900
	5380	3 phase, 5 KV primary 277/480 volt secondary									
	5400	High voltage, 112.5 KVA	R-3	.85	23.529	Ea.	7,150	485	158	7,793	8,825
	5440	500 KVA		.35	57.143		15,100	1,175	385	16,660	19,000
	5480	2000 KVA		.25	80		32,600	1,650	540	34,790	39,200
	5590	15 KV primary 277/480 volt secondary									
	5600	High voltage, 112.5 KVA	R-3	.85	23.529	Ea.	10,900	485	158	11,543	13,000
	5640	500 KVA		.35	57.143		20,900	1,175	385	22,460	25,400
	5680	2000 KVA		.25	80		38,000	1,650	540	40,190	45,100
170	0010	**TRANSFORMER, SILICON FILLED** Pad mounted										**170**
	0020	5 KV or 15 KV primary, 277/480 volt secondary, 3 phase									
	0050	225 KVA	R-3	.55	36.364	Ea.	9,600	750	245	10,595	12,100
	0100	300 KVA		.45	44.444		10,800	915	299	12,014	13,700
	0200	500 KVA		.40	50		12,600	1,025	335	13,960	16,000
	0250	750 KVA		.38	52.632		15,600	1,075	355	17,030	19,400

164 300 | Computer Power Supplies

			CREW	OUTPUT	HOURS	UNIT	MAT.	LABOR	EQUIP.	TOTAL	INCL O&P	
301	0010	**VOLTAGE MONITOR SYSTEMS** (test equipment)										**301**
	0100	AC voltage monitor system, 120/240 V, one-channel				Ea.	3,000			3,000	3,300
	0110	Modem adapter					375			375	415
	0120	Add-on detector only					1,575			1,575	1,725
	0150	AC voltage remote monitor sys., 3 channel, 120, 230, or 480 V					5,450			5,450	6,000
	0160	With internal modem					5,750			5,750	6,325
	0170	Combination temperature and humidity probe					845			845	930
	0180	Add-on detector only					3,950			3,950	4,350
	0190	With internal modem					4,300			4,300	4,725
305	0010	**AUTOMATIC VOLTAGE REGULATORS**										**305**
	0100	Computer grade, solid state, variable trans. volt. regulator									
	0110	Single-phase, 120 V, 8.6 KVA	2 Elec	1.33	12.030	Ea.	3,350	248		3,598	4,075
	0120	17.3 KVA		1.14	14.035		5,300	289		5,589	6,300
	0130	208/240 V, 7.5/8.6 KVA		1.33	12.030		3,350	248		3,598	4,075
	0140	13.5/15.6 KVA		1.33	12.030		4,600	248		4,848	5,450
	0150	27.0/31.2 KVA		1.14	14.035		8,250	289		8,539	9,550
	0210	Two-phase, single control, 208/240 V, 15.0/17.3 KVA		1.14	14.035		4,825	289		5,114	5,800
	0220	Individual phase control, 15.0/17.3 KVA		1.14	14.035		5,400	289		5,689	6,425
	0230	30.0/34.6 KVA	3 Elec	1.33	18.045		8,600	370		8,970	10,100
	0310	Three-phase single control, 208/240 V, 26/30 KVA	2 Elec	1	16		7,050	330		7,380	8,300
	0320	380/480 V, 24/30 KVA	"	1	16		7,950	330		8,280	9,300
	0330	43/54 KVA	3 Elec	1.33	18.045		8,900	370		9,270	10,400
	0340	Individual phase control, 208 V, 26 KVA	"	1.33	18.045		8,900	370		9,270	10,400
	0350	52 KVA	R-3	.91	21.978		9,600	455	148	10,203	11,400
	0360	340/480 V, 24/30 KVA	"	.91	21.978		10,600	455	148	11,203	12,600
	0370	43/54 KVA	2 Elec	1	16		4,700	330		5,030	5,725
	0380	48/60 KVA	3 Elec	1.33	18.045		7,900	370		8,270	9,300
	0390	86/108 KVA	R-3	.91	21.978		11,000	455	148	11,603	13,000
	0500	Standard grade, solid state, variable transformer volt. regulator									
	0510	Single-phase, 115 V, 2.3 KVA	1 Elec	2	4	Ea.	1,625	82.50		1,707.50	1,925
	0520	4.2 KVA		2.29	3.493		2,600	72		2,672	2,975

ELECTRICAL 16

164 300 | Computer Power Supplies

		CREW	DAILY OUTPUT	LABOR-HOURS	UNIT	1998 BARE COSTS				TOTAL INCL O&P
						MAT.	LABOR	EQUIP.	TOTAL	
305	0530	6.6 KVA	1 Elec	1.14	7.018	Ea.	3,200	145		3,345	3,725
	0540	13.0 KVA	↓	1.14	7.018		3,600	145		3,745	4,175
	0550	16.6 KVA	2 Elec	1.23	13.008		4,400	268		4,668	5,300
	0610	230 V, 8.3 KVA		1.33	12.030		4,300	248		4,548	5,125
	0620	21.4 KVA		1.23	13.008		4,500	268		4,768	5,400
	0630	29.9 KVA		1.23	13.008		4,800	268		5,068	5,725
	0710	460 V, 9.2 KVA		1.33	12.030		4,700	248		4,948	5,575
	0720	20.7 KVA	↓	1.23	13.008		5,075	268		5,343	6,025
	0810	Three-phase, 230 V, 13.1 KVA	3 Elec	1.41	17.021		6,600	350		6,950	7,825
	0820	19.1 KVA		1.41	17.021		7,450	350		7,800	8,775
	0830	25.1 KVA	↓	1.60	15		8,500	310		8,810	9,850
	0840	57.8 KVA	R-3	.95	21.053		11,900	435	142	12,477	14,000
	0850	74.9 KVA	"	.91	21.978		15,300	455	148	15,903	17,700
	0910	460 V, 14.3 KVA	3 Elec	1.41	17.021		6,800	350		7,150	8,050
	0920	19.1 KVA		1.41	17.021		7,900	350		8,250	9,275
	0930	27.9 KVA	↓	1.50	16		8,500	330		8,830	9,900
	0940	59.8 KVA	R-3	1	20		11,200	415	135	11,750	13,100
	0950	79.7 KVA		.95	21.053		14,800	435	142	15,377	17,200
	0960	118 KVA	↓	.95	21.053	↓	17,400	435	142	17,977	20,000
	1000	Laboratory grade, precision, electronic voltage regulator									
	1110	Single-phase, 115 V, 0.5 KVA	1 Elec	2.29	3.493	Ea.	1,350	72		1,422	1,600
	1120	1.0 KVA		2	4		1,450	82.50		1,532.50	1,700
	1130	3.0 KVA	↓	.80	10		2,000	206		2,206	2,525
	1140	6.0 KVA	2 Elec	1.46	10.959		3,725	226		3,951	4,475
	1150	10.0 KVA	3 Elec	1	24		4,825	495		5,320	6,125
	1160	15.0 KVA	"	1.50	16		5,525	330		5,855	6,625
	1210	230 V, 3.0 KVA	1 Elec	.80	10		2,275	206		2,481	2,825
	1220	6.0 KVA	2 Elec	1.46	10.959		3,800	226		4,026	4,550
	1230	10.0 KVA	3 Elec	1.71	14.035		5,050	289		5,339	6,025
	1240	15.0 KVA	"	1.60	15	↓	5,700	310		6,010	6,775
310	0010	**ISOLATION TRANSFORMER**									
	0100	Computer grade									
	0110	Single-phase, 120/240 V, 0.5 KVA	1 Elec	4	2	Ea.	360	41		401	465
	0120	1.0 KVA		2.67	2.996		515	61.50		576.50	665
	0130	2.5 KVA		2	4		730	82.50		812.50	940
	0140	5 KVA	↓	1.14	7.018	↓	830	145		975	1,150
315	0010	**TRANSIENT VOLTAGE SUPPRESSOR TRANSFORMER**									
	0110	Single-phase, 120 V, 1.8 KVA	1 Elec	4	2	Ea.	430	41		471	545
	0120	3.6 KVA		4	2		650	41		691	785
	0130	7.2 KVA		3.20	2.500		1,100	51.50		1,151.50	1,275
	0150	240 V, 3.6 KVA		4	2		580	41		621	710
	0160	7.2 KVA		4	2		840	41		881	995
	0170	14.4 KVA		3.20	2.500		1,175	51.50		1,226.50	1,350
	0210	Plug-in unit, 120 V, 1.8 KVA	↓	8	1	↓	470	20.50		490.50	550
325	0010	**COMPUTER REGULATOR TRANSFORMER**									
	0100	Ferro-resonant, constant voltage, variable transformer									
	0110	Single-phase, 240 V, 0.5 KVA	1 Elec	2.67	2.996	Ea.	440	61.50		501.50	585
	0120	1.0 KVA		2	4		605	82.50		687.50	800
	0130	2.0 KVA		1	8		1,025	165		1,190	1,400
	0210	Plug-in unit 120 V, 0.14 KVA		8	1		254	20.50		274.50	315
	0220	0.25 KVA		8	1		297	20.50		317.50	360
	0230	0.5 KVA		8	1		440	20.50		460.50	520
	0240	1.0 KVA		5.33	1.501		605	31		636	715
	0250	2.0 KVA	↓	4	2	↓	1,025	41		1,066	1,200
330	0010	**POWER CONDITIONER TRANSFORMER**									
	0100	Electronic solid state, buck-boost, transformer, w/tap switch									

Important: See the Reference Section for critical supporting data - Reference Nos., Crews, & Location Factors

164 300	Computer Power Supplies	CREW	DAILY OUTPUT	LABOR-HOURS	UNIT	1998 BARE COSTS				TOTAL INCL O&P	
						MAT.	LABOR	EQUIP.	TOTAL		
330											**330**
0110	Single-phase, 115 V, 3.0 KVA, + or - 3% accuracy	2 Elec	1.60	10	Ea.	2,375	206		2,581	2,950	
0120	208, 220, 230, or 240 V, 5.0 KVA, + or - 1.5% accuracy	3 Elec	1.60	15		3,075	310		3,385	3,875	
0130	5.0 KVA, + or - 6% accuracy	2 Elec	1.14	14.035		2,750	289		3,039	3,500	
0140	7.5 KVA, + or - 1.5% accuracy	3 Elec	1.50	16		3,900	330		4,230	4,850	
0150	7.5 KVA, + or - 6% accuracy		1.60	15		3,250	310		3,560	4,075	
0160	10.0 KVA, + or - 1.5% accuracy		1.33	18.045		5,200	370		5,570	6,325	
0170	10.0 KVA, + or - 6% accuracy		1.41	17.021		4,425	350		4,775	5,450	
335	**UNINTERRUPTIBLE POWER SUPPLY/CONDITIONER TRANSFORMERS**										**335**
0010											
0100	Volt. regulating, isolating trans., w/invert. & 10 min. battery pack										
0110	Single-phase, 120 V, 0.35 KVA	1 Elec	2.29	3.493	Ea.	890	72		962	1,100	
0120	0.5 KVA		2	4		1,000	82.50		1,082.50	1,225	
0130	For additional 55 min. battery, add to .35 KVA		2.29	3.493		480	72		552	650	
0140	Add to 0.5 KVA		1.14	7.018		605	145		750	900	
0150	Single-phase, 120 V, 0.75 KVA		.80	10		1,200	206		1,406	1,650	
0160	1.0 KVA		.80	10		1,900	206		2,106	2,400	
0170	1.5 KVA	2 Elec	1.14	14.035		2,875	289		3,164	3,625	
0180	2 KVA	"	.89	17.978		3,150	370		3,520	4,075	
0190	3 KVA	R-3	.63	31.746		3,725	655	214	4,594	5,375	
0200	5 KVA		.42	47.619		5,350	980	320	6,650	7,825	
0210	7.5 KVA		.33	60.606		10,000	1,250	410	11,660	13,500	
0220	10 KVA		.28	71.429		10,500	1,475	480	12,455	14,600	
0230	15 KVA		.22	90.909		13,800	1,875	610	16,285	18,900	
0500	For options & accessories add to above, minimum									10%	
0520	Maximum									35%	
0600	For complex & special design systems to meet specific										
0610	requirements, obtain quote from vendor.										

165 100	Power Systems	CREW	DAILY OUTPUT	LABOR-HOURS	UNIT	1998 BARE COSTS				TOTAL INCL O&P	
						MAT.	LABOR	EQUIP.	TOTAL		
110	**AUTOMATIC TRANSFER SWITCHES**										**110**
0010											
0100	Switches, enclosed 480 volt, 3 pole, 30 amp	1 Elec	2.30	3.478	Ea.	1,500	71.50		1,571.50	1,775	
0200	60 amp	"	1.90	4.211	"	2,025	86.50		2,111.50	2,375	
120	**GENERATOR SET**										**120**
0010											
0020	Gas or gasoline operated, includes battery,										
0050	charger, muffler & transfer switch										
0200	3 phase 4 wire, 277/480 volt, 7.5 KW	R-3	.83	24.096	Ea.	6,000	495	162	6,657	7,600	
0300	10 KW		.71	28.169		8,200	580	190	8,970	10,200	
0400	20 KW		.63	31.746		10,000	655	214	10,869	12,300	
0500	35 KW		.55	36.364		12,000	750	245	12,995	14,700	
2000	Diesel engine, including battery, charger,										
2010	muffler, transfer switch & fuel tank, 30 KW	R-3	.55	36.364	Ea.	16,000	750	245	16,995	19,100	
2100	50 KW		.42	47.619		19,700	980	320	21,000	23,600	
2200	75 KW		.35	57.143		25,700	1,175	385	27,260	30,700	
2300	100 KW		.31	64.516		28,500	1,325	435	30,260	34,100	

ELECTRICAL 16

166 100	Lighting	CREW	DAILY OUTPUT	LABOR-HOURS	UNIT	1998 BARE COSTS				TOTAL INCL O&P
						MAT.	LABOR	EQUIP.	TOTAL	
110 0010	**EXIT AND EMERGENCY LIGHTING**									**110**
0080	Exit light ceiling or wall mount, incandescent, single face	1 Elec	8	1	Ea.	35	20.50		55.50	72
0100	Double face		6.70	1.194		44	24.50		68.50	88.50
0120	Explosion proof		3.80	2.105		410	43.50		453.50	520
0150	Fluorescent, single face		8	1		115	20.50		135.50	161
0160	Double face		6.70	1.194		130	24.50		154.50	183
0300	Emergency light units, battery operated									
0350	Twin sealed beam light, 25 watt, 6 volt each									
0500	Lead battery operated	1 Elec	4	2	Ea.	215	41		256	305
0700	Nickel cadmium battery operated		4	2		480	41		521	600
0780	Additional remote mount, sealed beam, 25W 6V		26.70	.300		19	6.15		25.15	31
0790	Twin sealed beam light, 25W 6V each		26.70	.300		38	6.15		44.15	52
115 0010	**EXTERIOR FIXTURES** With lamps									**115**
0400	Quartz, 500 watt	1 Elec	5.30	1.509	Ea.	51	31		82	107
0800	Wall pack, mercury vapor, 175 watt		4	2		160	41		201	244
1000	250 watt		4	2		210	41		251	299
1100	Low pressure sodium, 35 watt		4	2		190	41		231	277
1150	55 watt		4	2		255	41		296	350
6420	Wood pole, 4-1/2″ x 5-1/8″, 8′ high		6	1.333		220	27.50		247.50	287
6440	12′ high		5.70	1.404		280	29		309	360
6460	20′ high		4	2		395	41		436	505
6500	Bollard light, lamp & ballast, 42″ high with polycarbonate lens									
6700	Mercury vapor, 175 watt	1 Elec	3	2.667	Ea.	495	55		550	630
7200	Incandescent, 150 watt	″	3	2.667	″	385	55		440	510
7380	Landscape recessed uplight, incl. housing, ballast, transformer									
7390	& reflector									
7420	Incandescent, 250 watt	1 Elec	5	1.600	Ea.	400	33		433	495
7440	Quartz, 250 watt	″	5	1.600	″	380	33		413	475
127 0010	**MODULAR FLEXIBLE WIRING SYSTEM**									**127**
0050	Standard selector cable, 3 conductor, 15′ long	1 Elec	40	.200	Ea.	38.50	4.12		42.62	49.50
0100	Conversion Module		16	.500		14	10.30		24.30	32.50
0150	Cable set, 3 conductor, 15′ long		24	.333		15	6.85		21.85	28
0200	Switching assembly, 3 conductor, 10′ long		32	.250		29.50	5.15		34.65	41
130 0010	**INTERIOR LIGHTING FIXTURES** Including lamps, mounting									**130**
0030	hardware and connections									
0100	Fluorescent, C.W. lamps, troffer, recess mounted in grid, RS									
0130	grid ceiling mount									
0200	Acrylic lens, 1′W x 4′L, two 40 watt	1 Elec	5.70	1.404	Ea.	46	29		75	98
0300	2′W x 2′L, two U40 watt		5.70	1.404		50	29		79	103
0600	2′W x 4′L, four 40 watt		4.70	1.702		56	35		91	119
0910	Acrylic lens, 1′W x 4′L, two 32 watt		5.70	1.404		55	29		84	108
0930	2′W x 2′L, two U32 watt		5.70	1.404		75	29		104	130
0940	2′W x 4′L, two 32 watt		5.30	1.509		67	31		98	125
0950	2′W x 4′L, three 32 watt		5	1.600		72	33		105	133
0960	2′W x 4′L, four 32 watt		4.70	1.702		75	35		110	140
1000	Surface mounted, RS									
2100	Strip fixture									
2200	4′ long, one 40 watt RS	1 Elec	8.50	.941	Ea.	26	19.40		45.40	60
2300	4′ long, two 40 watt RS		8	1		28	20.50		48.50	64.50
2600	8′ long, one 75 watt, SL		6.70	1.194		39	24.50		63.50	83
2700	8′ long, two 75 watt, SL		6.20	1.290		47	26.50		73.50	95
3580	Mercury vapor, integral ballast, ceiling, recess mounted,									
3590	prismatic glass lens, floating door									
3600	2′W x 2′L, 250 watt DX lamp	1 Elec	3.20	2.500	Ea.	235	51.50		286.50	345
3700	2′W x 2′L, 400 watt DX lamp		2.90	2.759		255	57		312	375

Important: See the Reference Section for critical supporting data - Reference Nos., Crews, & Location Factors

166 100	Lighting	CREW	DAILY OUTPUT	LABOR-HOURS	UNIT	1998 BARE COSTS				TOTAL INCL O&P	
						MAT.	LABOR	EQUIP.	TOTAL		
130 3800	Surface mtd., prismatic lens, 2'W x 2'L, 250 watt DX lamp	1 Elec	2.70	2.963	Ea.	240	61		301	365	**130**
3900	2'W x 2'L, 400 watt DX lamp	↓	2.40	3.333	↓	270	68.50		338.50	410	
4000	High bay, aluminum reflector										
4030	Single unit, 400 watt DX lamp	1 Elec	2.30	3.478	Ea.	245	71.50		316.50	385	
4100	Single unit, 1000 watt DX lamp		2	4		390	82.50		472.50	565	
4200	Twin unit, two 400 watt DX lamps		1.60	5		460	103		563	675	
4210	Low bay, aluminum reflector, 250W DX lamp	↓	3.20	2.500	↓	280	51.50		331.50	395	
4220	Metal halide, integral ballast, ceiling, recess mounted										
4230	prismatic glass lens, floating door										
4240	2'W x 2'L, 250 watt	1 Elec	3.20	2.500	Ea.	265	51.50		316.50	375	
4250	2'W x 2'L, 400 watt		2.90	2.759		310	57		367	435	
4260	Surface mounted, 2'W x 2'L, 250 watt		2.70	2.963		245	61		306	370	
4270	400 watt	↓	2.40	3.333	↓	290	68.50		358.50	430	
4280	High bay, aluminum reflector,										
4290	Single unit, 400 watt	1 Elec	2.30	3.478	Ea.	340	71.50		411.50	490	
4300	Single unit, 1000 watt		2	4		485	82.50		567.50	670	
4310	Twin unit, 400 watt		1.60	5		680	103		783	915	
4320	Low bay, aluminum reflector, 250W DX lamp	↓	3.20	2.500	↓	325	51.50		376.50	445	
4450	Incandescent, high hat can, round alzak reflector, prewired										
4470	100 watt	1 Elec	8	1	Ea.	52	20.50		72.50	90.50	
4480	150 watt	"	8	1	"	76	20.50		96.50	117	
5200	Ceiling, surface mounted, opal glass drum										
5300	8", one 60 watt lamp	1 Elec	10	.800	Ea.	31	16.50		47.50	61	
5400	10", two 60 watt lamps		8	1		33	20.50		53.50	70	
5500	12", four 60 watt lamps		6.70	1.194		50	24.50		74.50	95	
6900	Mirror light, fluorescent, RS, acrylic enclosure, two 40 watt		8	1		80	20.50		100.50	122	
6910	One 40 watt		8	1		62	20.50		82.50	102	
6920	One 20 watt	↓	12	.667	↓	49	13.75		62.75	76.50	
140 0010	**LAMPS**										**140**
0080	Fluorescent, rapid start, cool white, 2' long, 20 watt	1 Elec	1	8	C	249	165		414	545	
0100	4' long, 40 watt		.90	8.889		215	183		398	535	
1000	Metal halide, mogul base, 175 watt		.30	26.667		4,625	550		5,175	5,975	
1100	250 watt		.30	26.667		5,225	550		5,775	6,650	
1350	Sodium high pressure, 70 watt		.30	26.667		4,700	550		5,250	6,075	
1370	150 watt	↓	.30	26.667	↓	5,075	550		5,625	6,475	
150 0010	**TRACK LIGHTING**										**150**
0080	Track, 1 circuit, 4' section	1 Elec	6.70	1.194	Ea.	33	24.50		57.50	76.50	
0100	8' section		5.30	1.509		55	31		86	112	
0200	12' section		4.40	1.818		87	37.50		124.50	157	
0300	3 circuits, 4' section		6.70	1.194		44	24.50		68.50	88.50	
0400	8' section		5.30	1.509		68	31		99	126	
0500	12' section		4.40	1.818		136	37.50		173.50	212	
1000	Feed kit, surface mounting		16	.500		8	10.30		18.30	25.50	
1100	End cover		24	.333		3	6.85		9.85	14.55	
1200	Feed kit, stem mounting, 1 circuit		16	.500		22.50	10.30		32.80	42	
1300	3 circuit		16	.500		22.50	10.30		32.80	42	
2000	Electrical joiner, for continuous runs, 1 circuit		32	.250		10.50	5.15		15.65	20	
2100	3 circuit		32	.250		24	5.15		29.15	35	
2200	Fixtures, spotlight, 75W PAR halogen		16	.500		82.50	10.30		92.80	108	
2210	50W MR16 halogen		16	.500		98	10.30		108.30	125	
3000	Wall washer, 250 watt tungsten halogen		16	.500		84	10.30		94.30	109	
3100	Low voltage, 25/50 watt, 1 circuit		16	.500		98	10.30		108.30	125	
3120	3 circuit	↓	16	.500	↓	100	10.30		110.30	127	

ELECTRICAL 16

167 100 | Electric Utilities	CREW	DAILY OUTPUT	LABOR-HOURS	UNIT	1998 BARE COSTS				TOTAL INCL O&P	
					MAT.	LABOR	EQUIP.	TOTAL		
110 0010 **ELECTRIC & TELEPHONE SITE WORK** Not including excavation										**110**
0200 backfill and cast in place concrete										
4200 Underground duct, banks ready for concrete fill, min. of 7.5"										
4400 between conduits, ctr. to ctr.(for wire & cable see div. 161)										
4601 PVC, type EB, 2 @ 2" diameter	2 Elec	240	.067	L.F.	1.05	1.37		2.42	3.40	
4800 4 @ 2" diameter		120	.133		2.10	2.75		4.85	6.80	
5600 4 @ 4" diameter		80	.200		4.50	4.12		8.62	11.70	
6200 Rigid galvanized steel, 2 @ 2" diameter		180	.089		8.25	1.83		10.08	12.05	
6400 4 @ 2" diameter		90	.178		16.50	3.66		20.16	24	
7400 4 @ 4" diameter		34	.471		53	9.70		62.70	74	

168 100 | Special Systems	CREW	DAILY OUTPUT	LABOR-HOURS	UNIT	1998 BARE COSTS				TOTAL INCL O&P	
					MAT.	LABOR	EQUIP.	TOTAL		
120 0010 **DETECTION SYSTEMS**, not including wires & conduits										**120**
0100 Burglar alarm, battery operated, mechanical trigger	1 Elec	4	2	Ea.	249	41		290	340	
0200 Electrical trigger		4	2		297	41		338	395	
0400 For outside key control, add		8	1		70.50	20.50		91	111	
0600 For remote signaling circuitry, add		8	1		112	20.50		132.50	157	
0800 Card reader, flush type, standard		2.70	2.963		835	61		896	1,025	
1000 Multi-code		2.70	2.963		1,075	61		1,136	1,275	
1200 Door switches, hinge switch		5.30	1.509		52.50	31		83.50	109	
1400 Magnetic switch		5.30	1.509		62	31		93	119	
2800 Ultrasonic motion detector, 12 volt		2.30	3.478		206	71.50		277.50	345	
3000 Infrared photoelectric detector		2.30	3.478		170	71.50		241.50	305	
3200 Passive infrared detector		2.30	3.478		254	71.50		325.50	395	
3420 Switchmats, 30" x 5'		5.30	1.509		76	31		107	135	
3440 30" x 25'		4	2		182	41		223	268	
3460 Police connect panel		4	2		219	41		260	310	
3480 Telephone dialer		5.30	1.509		345	31		376	430	
3500 Alarm bell		4	2		69.50	41		110.50	144	
3520 Siren		4	2		131	41		172	212	
5200 Smoke detector, ceiling type		6.20	1.290		75	26.50		101.50	126	
5600 Strobe and horn		5.30	1.509		95	31		126	156	
5800 Fire alarm horn		6.70	1.194		36.50	24.50		61	80	
6600 Drill switch		8	1		86.50	20.50		107	129	
6800 Master box		2.70	2.963		3,100	61		3,161	3,500	
7800 Remote annunciator, 8 zone lamp		1.80	4.444		175	91.50		266.50	345	
8000 12 zone lamp		1.30	6.154		300	127		427	535	
8200 16 zone lamp		1.10	7.273		300	150		450	575	
8400 Standpipe or sprinkler alarm, alarm device		8	1		125	20.50		145.50	172	
8600 Actuating device		8	1		290	20.50		310.50	355	
130 0010 **ELECTRIC HEATING**										**130**
0400 Cable heating, radiant heat plaster, no controls, in South	1 Elec	130	.062	S.F.	7.25	1.27		8.52	10.05	
0600 In North		90	.089		7.25	1.83		9.08	11	
0800 Cable on 1/2" board not incl. controls, tract housing		90	.089		6.10	1.83		7.93	9.70	
1000 Custom housing		80	.100		7.10	2.06		9.16	11.15	
1100 Rule of thumb: Baseboard units, including control		4.40	1.818	KW	75	37.50		112.50	144	
1300 Baseboard heaters, 2' long, 375 watt		8	1	Ea.	36	20.50		56.50	73	
1600 4' long, 750 watt		6.70	1.194		48	24.50		72.50	93	

Important: See the Reference Section for critical supporting data - Reference Nos., Crews, & Location Factors

		168 100	Special Systems	CREW	DAILY OUTPUT	LABOR-HOURS	UNIT	1998 BARE COSTS				TOTAL INCL O&P	
								MAT.	LABOR	EQUIP.	TOTAL		
130	2000	6' long, 1125 watt	1 Elec	5	1.600	Ea.	65	33		98	126	**130**
	2800	10' long, 1875 watt	↓	3.30	2.424	↓	134	50		184	229	
	2950	Wall heaters with fan, 120 to 277 volt										
	2970	surface mounted, residential, 750 watt	1 Elec	7	1.143	Ea.	88	23.50		111.50	136	
	2990	1250 watt		6	1.333		97	27.50		124.50	152	
	3000	1500 watt		5	1.600		118	33		151	184	
	3010	2000 watt		5	1.600		123	33		156	189	
	3050	2500 watt		4	2		223	41		264	315	
	3070	4000 watt		3.50	2.286		223	47		270	320	
	5000	Radiant heating ceiling panels, 2' x 4', 500 watt		16	.500		160	10.30		170.30	193	
	5050	750 watt		16	.500		176	10.30		186.30	211	
	5300	Infra-red quartz heaters, 120 volts, 1000 watts		6.70	1.194		157	24.50		181.50	213	
	5350	1500 watt		5	1.600		119	33		152	185	
	5400	240 volts, 1500 watt		5	1.600		119	33		152	185	
	5450	2000 watt	↓	4	2	↓	119	41		160	199	
140	0010	**LIGHTNING PROTECTION**										**140**
	0200	Air terminals & base, copper										
	0400	3/8" diameter x 10" (to 75' high)	1 Elec	8	1	Ea.	21.50	20.50		42	57	
	0500	1/2" diameter x 12" (over 75' high)		8	1		26	20.50		46.50	62	
	1000	Aluminum, 1/2" diameter x 12" (to 75' high)		8	1		17	20.50		37.50	52	
	1100	5/8" diameter x 12" (over 75' high)		8	1	↓	22	20.50		42.50	57.50	
	2000	Cable, copper, 220 lb. per thousand ft. (to 75' high)		320	.025	L.F.	1.07	.52		1.59	2.02	
	2100	375 lb. per thousand ft. (over 75' high)		230	.035		1.69	.72		2.41	3.03	
	2500	Aluminum, 101 lb. per thousand ft. (to 75' high)		280	.029		.39	.59		.98	1.39	
	2600	199 lb. per thousand ft. (over 75' high)		240	.033	↓	.74	.69		1.43	1.93	
	3000	Arrester, 175 volt AC to ground		8	1	Ea.	36	20.50		56.50	73	
	3100	650 volt AC to ground	↓	6.70	1.194	"	70	24.50		94.50	117	
155	0010	**SOUND SYSTEM**										**155**
	2000	Intercom, 25 station capacity, master station	1 Elec	1	8	Ea.	1,300	165		1,465	1,725	
	2020	11 station capacity		2	4		645	82.50		727.50	845	
	2200	Remote station		8	1		105	20.50		125.50	150	
	2400	Intercom outlets		8	1		61.50	20.50		82	101	
	2600	Handset		4	2		190	41		231	277	
	2800	Emergency call system, 12 zones, annunciator		1.30	6.154		615	127		742	880	
	3000	Bell		5.30	1.509		64	31		95	122	
	3200	Light or relay		8	1		32	20.50		52.50	68.50	
	3400	Transformer		4	2		140	41		181	222	
	3600	House telephone, talking station		1.60	5		300	103		403	500	
	3800	Press to talk, release to listen	↓	5.30	1.509		70	31		101	128	
	4000	System-on button					42			42	46	
	4200	Door release	1 Elec	4	2		74	41		115	149	
	4400	Combination speaker and microphone		8	1		128	20.50		148.50	175	
	4600	Termination box		3.20	2.500		40	51.50		91.50	129	
	4800	Amplifier or power supply		5.30	1.509	↓	460	31		491	555	
	5000	Vestibule door unit		16	.500	Name	84.50	10.30		94.80	110	
	5200	Strip cabinet		27	.296	Ea.	160	6.10		166.10	186	
	5400	Directory		16	.500		75	10.30		85.30	99.50	
	6000	Master door, button buzzer type, 100 unit		.27	29.630		765	610		1,375	1,850	
	6020	200 unit	↓	.15	53.333	↓	1,450	1,100		2,550	3,375	
160	0010	**T.V. SYSTEMS**										**160**
	0100	Master TV antenna system										
	0200	VHF reception & distribution, 12 outlets	1 Elec	6	1.333	Outlet	152	27.50		179.50	212	
	0800	VHF & UHF reception & distribution, 12 outlets		6	1.333	"	151	27.50		178.50	211	
	5000	T.V. Antenna only, minimum		6	1.333	Ea.	33	27.50		60.50	81.50	
	5100	Maximum	↓	4	2	"	140	41		181	222	

168 100	Special Systems	CREW	DAILY OUTPUT	LABOR-HOURS	UNIT	1998 BARE COSTS				TOTAL INCL O&P
						MAT.	LABOR	EQUIP.	TOTAL	
170 0010	**RESIDENTIAL WIRING**									170
0020	20' avg. runs and #14/2 wiring incl. unless otherwise noted									
1000	Service & panel, includes 24' SE-AL cable, service eye, meter,									
1010	Socket, panel board, main bkr., ground rod, 15 or 20 amp									
1020	1-pole circuit breakers, and misc. hardware									
1100	100 amp, with 10 branch breakers	1 Elec	1.19	6.723	Ea.	370	138		508	635
1110	With PVC conduit and wire		.92	8.696		395	179		574	730
1120	With RGS conduit and wire	.	.73	10.959		505	226		731	930
1200	200 amp, with 18 branch breakers		.90	8.889		740	183		923	1,125
1220	With PVC conduit and wire		.73	10.959		795	226		1,021	1,250
1230	With RGS conduit and wire		.62	12.903		1,050	266		1,316	1,600
1800	Lightning surge suppressor for above services, add		32	.250		39.50	5.15		44.65	52
2000	Switch devices									
2100	Single pole, 15 amp, Ivory, with a 1-gang box, cover plate,									
2110	Type NM (Romex) cable	1 Elec	17.10	.468	Ea.	7.60	9.65		17.25	24
2120	Type MC (BX) cable		14.30	.559		15.90	11.50		27.40	36.50
2130	EMT & wire		5.71	1.401		18.60	29		47.60	67.50
2150	3-way, #14/3, type NM cable		14.55	.550		11.35	11.35		22.70	31
2170	Type MC cable		12.31	.650		19.75	13.40		33.15	43.50
2180	EMT & wire		5	1.600		21	33		54	77
2250	S.P., 20 amp, #12/2, type NM cable		13.33	.600		13.25	12.35		25.60	34.50
2270	Type MC cable		11.43	.700		20	14.40		34.40	45.50
2280	EMT & wire		4.85	1.649		24.50	34		58.50	82.50
2300	S.P. rotary dimmer, 600W, type NM cable		14.55	.550		18.60	11.35		29.95	39
2320	Type MC cable		12.31	.650		27	13.40		40.40	51.50
2330	EMT & wire		5	1.600		30.50	33		63.50	87.50
2350	3-way rotary dimmer, type NM cable		13.33	.600		21.50	12.35		33.85	43.50
2370	Type MC cable		11.43	.700		29.50	14.40		43.90	56
2380	EMT & wire		4.85	1.649		33	34		67	92
2400	Interval timer wall switch, 20 amp, 1-30 min., #12/2									
2410	Type NM cable	1 Elec	14.55	.550	Ea.	29	11.35		40.35	50.50
2420	Type MC cable		12.31	.650		33	13.40		46.40	58.50
2430	EMT & wire		5	1.600		40.50	33		73.50	98.50
2500	Decorator style									
2510	S.P., 15 amp, type NM cable	1 Elec	17.10	.468	Ea.	10.45	9.65		20.10	27.50
2520	Type MC cable		14.30	.559		18.80	11.50		30.30	39.50
2530	EMT & wire		5.71	1.401		21.50	29		50.50	70.50
2550	3-way, #14/3, type NM cable		14.55	.550		14.20	11.35		25.55	34
2570	Type MC cable		12.31	.650		22.50	13.40		35.90	47
2580	EMT & wire		5	1.600		23.50	33		56.50	80
2650	S.P., 20 amp, #12/2, type NM cable		13.33	.600		16.10	12.35		28.45	38
2670	Type MC cable		11.43	.700		23	14.40		37.40	49
2680	EMT & wire		4.85	1.649		27.50	34		61.50	86
2700	S.P., slide dimmer, type NM cable		17.10	.468		26	9.65		35.65	44.50
2720	Type MC cable		14.30	.559		34.50	11.50		46	57
2730	EMT & wire		5.71	1.401		38	29		67	89
2750	S.P., touch dimmer, type NM cable		17.10	.468		25.50	9.65		35.15	44
2770	Type MC cable		14.30	.559		34	11.50		45.50	56.50
2780	EMT & wire		5.71	1.401		37.50	29		66.50	88
2800	3-way touch dimmer, type NM cable		13.33	.600		39.50	12.35		51.85	63.50
2820	Type MC cable		11.43	.700		48	14.40		62.40	76
2830	EMT & wire		4.85	1.649		51.50	34		85.50	112
3000	Combination devices									
3100	S.P. switch/15 amp recpt., Ivory, 1-gang box, plate									
3110	Type NM cable	1 Elec	11.43	.700	Ea.	14.85	14.40		29.25	40
3120	Type MC cable		10	.800		23	16.50		39.50	52.50

Important: See the Reference Section for critical supporting data - Reference Nos., Crews, & Location Factors

168 100	Special Systems	CREW	DAILY OUTPUT	LABOR-HOURS	UNIT	1998 BARE COSTS				TOTAL INCL O&P
						MAT.	LABOR	EQUIP.	TOTAL	
3130	EMT & wire	1 Elec	4.40	1.818	Ea.	26.50	37.50		64	91
3150	S.P. switch/pilot light, type NM cable		11.43	.700		15.45	14.40		29.85	40.50
3170	Type MC cable		10	.800		23.50	16.50		40	53
3180	EMT & wire		4.43	1.806		27.50	37		64.50	91
3200	2-S.P. switches, 2-#14/2, type NM cables		10	.800		17.70	16.50		34.20	46.50
3220	Type MC cable		8.89	.900		31	18.55		49.55	64.50
3230	EMT & wire		4.10	1.951		28.50	40		68.50	97.50
3250	3-way switch/15 amp recpt., #14/3, type NM cable		10	.800		21.50	16.50		38	50.50
3270	Type MC cable		8.89	.900		30	18.55		48.55	63
3280	EMT & wire		4.10	1.951		31	40		71	100
3300	2-3 way switches, 2-#14/3, type NM cables		8.89	.900		29	18.55		47.55	62.50
3320	Type MC cable		8	1		42.50	20.50		63	80.50
3330	EMT & wire		4	2		35.50	41		76.50	107
3350	S.P. switch/20 amp recpt., #12/2, type NM cable		10	.800		25.50	16.50		42	55
3370	Type MC cable		8.89	.900		29.50	18.55		48.05	63
3380	EMT & wire		4.10	1.951		37	40		77	107
4000	Receptacle devices									
4010	Duplex outlet, 15 amp recpt., Ivory, 1-gang box, plate									
4015	Type NM cable	1 Elec	12.31	.650	Ea.	7.20	13.40		20.60	30
4020	Type MC cable		12.31	.650		15.50	13.40		28.90	39
4030	EMT & wire		5.33	1.501		18.20	31		49.20	70.50
4050	With #12/2, type NM cable		12.31	.650		8.90	13.40		22.30	32
4070	Type MC cable		10.67	.750		15.80	15.45		31.25	43
4080	EMT & wire		4.71	1.699		20.50	35		55.50	79.50
4100	20 amp recpt., #12/2, type NM cable		12.31	.650		12.60	13.40		26	36
4120	Type MC cable		10.67	.750		19.50	15.45		34.95	47
4130	EMT & wire		4.71	1.699		24	35		59	83.50
4500	Weather-proof cover for above receptacles, add		32	.250		4.40	5.15		9.55	13.30
4550	Air conditioner outlet, 20 amp-240 volt recpt.									
4560	30' of #12/2, 2 pole circuit breaker									
4570	Type NM cable	1 Elec	10	.800	Ea.	35.50	16.50		52	66
4580	Type MC cable		9	.889		44	18.30		62.30	78.50
4590	EMT & wire		4	2		46	41		87	119
4600	Decorator style, type NM cable		10	.800		37	16.50		53.50	68
4620	Type MC cable		9	.889		46	18.30		64.30	80.50
4630	EMT & wire		4	2		48	41		89	120
4650	Dryer outlet, 30 amp-240 volt recpt., 20' of #10/3									
4660	2 pole circuit breaker									
4670	Type NM cable	1 Elec	6.41	1.248	Ea.	55	25.50		80.50	102
4680	Type MC cable		5.71	1.401		53.50	29		82.50	106
4690	EMT & wire		3.48	2.299		56	47.50		103.50	139
4700	Range outlet, 50 amp-240 volt recpt., 30' of #8/3									
4710	Type NM cable	1 Elec	4.21	1.900	Ea.	84.50	39		123.50	157
4720	Type MC cable		4	2		97	41		138	175
4730	EMT & wire		2.96	2.703		76	55.50		131.50	175
4750	Central vacuum outlet, Type NM cable		6.40	1.250		45	26		71	91.50
4770	Type MC cable		5.71	1.401		54.50	29		83.50	107
4780	EMT & wire		3.48	2.299		55	47.50		102.50	138
4800	30 amp-110 volt locking recpt., #10/2 circ. bkr.									
4810	Type NM cable	1 Elec	6.20	1.290	Ea.	51	26.50		77.50	99.50
4820	Type MC cable		5.40	1.481		66	30.50		96.50	123
4830	EMT & wire		3.20	2.500		63.50	51.50		115	155
4900	Low voltage outlets									
4910	Telephone recpt., 20' of 4/C phone wire	1 Elec	26	.308	Ea.	7.40	6.35		13.75	18.50
4920	TV recpt., 20' of RG59U coax wire, F type connector	"	16	.500	"	11.05	10.30		21.35	29
4950	Door bell chime, transformer, 2 buttons, 60' of bellwire									

			DAILY	LABOR-		1998 BARE COSTS				TOTAL	
168 100 | Special Systems		CREW	OUTPUT	HOURS	UNIT	MAT.	LABOR	EQUIP.	TOTAL	INCL O&P	
170	4970	Economy model	1 Elec	11.50	.696	Ea.	51.50	14.35		65.85	80	170
	4980	Custom model	"	11.50	.696	"	86.50	14.35		100.85	119	
	6000	Lighting outlets										
	6050	Wire only (for fixture), type NM cable	1 Elec	32	.250	Ea.	4.39	5.15		9.54	13.30	
	6070	Type MC cable		24	.333		10.20	6.85		17.05	22.50	
	6080	EMT & wire		10	.800		12.05	16.50		28.55	40.50	
	6100	Box (4"), and wire (for fixture), type NM cable		25	.320		8.30	6.60		14.90	19.90	
	6120	Type MC cable		20	.400		14.10	8.25		22.35	29	
	6130	EMT & wire	↓	11	.727	↓	15.95	15		30.95	42	
	6200	Fixtures (use with lines 6050 or 6100 above)										
	6210	Canopy style, economy grade	1 Elec	40	.200	Ea.	25.50	4.12		29.62	35	
	6220	Custom grade		40	.200		46	4.12		50.12	57.50	
	6250	Dining room chandelier, economy grade		19	.421		76	8.65		84.65	97.50	
	6260	Custom grade		19	.421		225	8.65		233.65	262	
	6310	Kitchen fixture (fluorescent), economy grade		30	.267		51.50	5.50		57	65.50	
	6320	Custom grade		25	.320		161	6.60		167.60	188	
	6350	Outdoor, wall mounted, economy grade		30	.267		27	5.50		32.50	38.50	
	6360	Custom grade		30	.267		101	5.50		106.50	120	
	6410	Outdoor PAR floodlights, 1 lamp, 150 watt		20	.400		20	8.25		28.25	35.50	
	6420	2 lamp, 150 watt each		20	.400		33.50	8.25		41.75	50.50	
	6430	For infrared security sensor, add		32	.250		87	5.15		92.15	104	
	6450	Outdoor, quartz-halogen, 300 watt flood		20	.400		38	8.25		46.25	55.50	
	6600	Recessed downlight, round, pre-wired, 50 or 75 watt trim		30	.267		35	5.50		40.50	47.50	
	6610	With shower light trim		30	.267		42	5.50		47.50	55	
	6620	With wall washer trim	↓	28	.286		52.50	5.90		58.40	67.50	
	6640	For direct contact with insulation, add					1.60			1.60	1.76	
	6700	Porcelain lamp holder	1 Elec	40	.200		3.50	4.12		7.62	10.60	
	6710	With pull switch	·	40	.200		3.75	4.12		7.87	10.90	
	6750	Fluorescent strip, 1-20 watt tube, wrap around diffuser, 24"		24	.333		51.50	6.85		58.35	68	
	6760	1-40 watt tube, 48"		24	.333		65	6.85		71.85	83	
	6770	2-40 watt tubes, 48"		20	.400		78	8.25		86.25	99.50	
	6780	With residential ballast		20	.400		89.50	8.25		97.75	112	
	6800	Bathroom heat lamp, 1-250 watt		28	.286		30.50	5.90		36.40	43	
	6810	2-250 watt lamps	↓	28	.286	↓	51	5.90		56.90	65.50	
	6820	For timer switch, see line 2400										
	6900	Outdoor post lamp, incl. post, fixture, 35' of #14/2										
	6910	Type NMC cable	1 Elec	3.50	2.286	Ea.	161	47		208	254	
	6920	Photo-eye, add		27	.296		29	6.10		35.10	42	
	6950	Clock dial time switch, 24 hr., w/enclosure, type NM cable		11.43	.700		51.50	14.40		65.90	80	
	6970	Type MC cable		11	.727		60	15		75	90	
	6980	EMT & wire	↓	4.85	1.649	↓	62.50	34		96.50	124	
	7000	Alarm systems										
	7050	Smoke detectors, box, #14/3, type NM cable	1 Elec	14.55	.550	Ea.	29.50	11.35		40.85	51	
	7070	Type MC cable		12.31	.650		35	13.40		48.40	60.50	
	7080	EMT & wire	↓	5	1.600		36	33		69	94	
	7090	For relay output to security system, add				↓	11.75			11.75	12.95	
	8000	Residential equipment										
	8050	Disposal hook-up, incl. switch, outlet box, 3' of flex										
	8060	20 amp-1 pole circ. bkr., and 25' of #12/2										
	8070	Type NM cable	1 Elec	10	.800	Ea.	21	16.50		37.50	50	
	8080	Type MC cable		8	1		28.50	20.50		49	65	
	8090	EMT & wire	↓	5	1.600	↓	33.50	33		66.50	91	
	8100	Trash compactor or dishwasher hook-up, incl. outlet box,										
	8110	3' of flex, 15 amp-1 pole circ. bkr., and 25' of #14/2										
	8130	Type MC cable	1 Elec	8	1	Ea.	24	20.50		44.50	60	
	8140	EMT & wire	"	5	1.600	"	27.50	33		60.50	84.50	

Important: See the Reference Section for critical supporting data - Reference Nos., Crews, & Location Factors

168 100	Special Systems	CREW	DAILY OUTPUT	LABOR-HOURS	UNIT	1998 BARE COSTS				TOTAL INCL O&P
						MAT.	LABOR	EQUIP.	TOTAL	
8150	Hot water sink dispensor hook-up, use line 8100									
8200	Vent/exhaust fan hook-up, type NM cable	1 Elec	32	.250	Ea.	4.39	5.15		9.54	13.30
8220	Type MC cable		24	.333		10.20	6.85		17.05	22.50
8230	EMT & wire		10	.800		12.05	16.50		28.55	40.50
8250	Bathroom vent fan, 50 CFM (use with above hook-up)									
8260	Economy model	1 Elec	15	.533	Ea.	21.50	11		32.50	41.50
8280	Custom model	"	12	.667	"	111	13.75		124.75	145
8300	Bathroom or kitchen vent fan, 110 CFM									
8310	Economy model	1 Elec	15	.533	Ea.	53	11		64	76.50
8320	Low noise model	"	15	.533	"	72	11		83	97
8350	Paddle fan, variable speed (w/o lights)									
8360	Economy model (AC motor)	1 Elec	10	.800	Ea.	94.50	16.50		111	131
8370	Custom model (AC motor)		10	.800		165	16.50		181.50	209
8390	Remote speed switch for above, add		12	.667		21	13.75		34.75	45.50
8500	Whole house exhaust fan, ceiling mount, 36", variable speed									
8510	Remote switch, incl. shutters, 20 amp-1 pole circ. bkr.									
8520	30' of #12/2, type NM cable	1 Elec	4	2	Ea.	495	41		536	615
8530	Type MC cable		3.50	2.286		500	47		547	625
8540	EMT & wire		3	2.667		510	55		565	650
8600	Whirlpool tub hook-up, incl. timer switch, outlet box									
8610	3' of flex, 20 amp-1 pole GFI circ. bkr.									
8620	30' of #12/2, type NM cable	1 Elec	5	1.600	Ea.	76.50	33		109.50	138
8630	Type MC cable		4.20	1.905		81	39		120	153
8640	EMT & wire		3.40	2.353		85.50	48.50		134	174
8650	Hot water heater hook-up, incl. 1-2 pole circ. bkr., box;									
8660	3' of flex, 20' of #10/2, type NM cable	1 Elec	5	1.600	Ea.	21	33		54	77
8670	Type MC cable		4.20	1.905		31.50	39		70.50	98.50
8680	EMT & wire		3.40	2.353		29	48.50		77.50	112
9000	Heating/air conditioning									
9050	Furnace/boiler hook-up, incl. firestat, local on-off switch									
9060	Emergency switch, and 40' of type NM cable	1 Elec	4	2	Ea.	42	41		83	114
9070	Type MC cable		3.50	2.286		55.50	47		102.50	138
9080	EMT & wire		1.50	5.333		60.50	110		170.50	247
9100	Air conditioner hook-up, incl. local 60 amp disc. switch									
9110	3' Sealtite, 40 amp, 2 pole circuit breaker									
9130	40' of #8/2, type NM cable	1 Elec	3.50	2.286	Ea.	124	47		171	214
9140	Type MC cable		3	2.667		155	55		210	260
9150	EMT & wire		1.30	6.154		131	127		258	350
9200	Heat pump hook-up, 1-40 & 1-100 amp 2 pole circ. bkr.									
9210	Local disconnect switch, 3' Sealtite									
9220	40' of #8/2 & 30' of #3/2									
9230	Type NM cable	1 Elec	1.30	6.154	Ea.	350	127		477	590
9240	Type MC cable		1.08	7.407		400	153		553	690
9250	EMT & wire		.94	8.511		330	175		505	645
9500	Thermostat hook-up, using low voltage wire									
9520	Heating only	1 Elec	24	.333	Ea.	5.10	6.85		11.95	16.85
9530	Heating/cooling	"	20	.400	"	6.30	8.25		14.55	20.50

For information about Means Estimating Seminars, see yellow pages 11 and 12 in back of book

For expanded coverage of these items see *Means Electrical Cost Data 1998*

		CREW	DAILY OUTPUT	LABOR-HOURS	UNIT	1998 BARE COSTS				TOTAL INCL O&P
						MAT.	LABOR	EQUIP.	TOTAL	

Reference
Section

All the reference information is in one section making it easy to find what you need to know . . . and easy to use the book on a daily basis. This section is visually identified by a vertical gray bar on the edge of pages.

In the reference number information that follows, you'll see the background that relates to the "reference numbers" that appeared in the Unit Price Sections. You'll find reference tables, explanations and estimating information that support how we arrived at the unit price data. Also included are alternate pricing methods, technical data and estimating procedures along with information on design and economy in construction.

Also in this Reference Section, we've included Crew Listings, a full listing of all the crews, equipment and their costs, Historical Cost Indexes for cost comparisons over time, Location Factors for adjusting costs to the region you are in, an explanation of all abbreviations used in the book, and an index.

Table of Contents

R010-005   Tips for Accurate Estimating

1. Use pre-printed or columnar forms for orderly sequence of dimensions and locations and for recording telephone quotations.

2. Use only the front side of each paper or form except for certain pre-printed summary forms.

3. Be consistent in listing dimensions: For example, length x width x height. This helps in rechecking to ensure that, the total length of partitions is appropriate for the building area.

4. Use printed (rather than measured) dimensions where given.

5. Add up multiple printed dimensions for a single entry where possible.

6. Measure all other dimensions carefully.

7. Use each set of dimensions to calculate multiple related quantities.

8. Convert foot and inch measurements to decimal feet when listing. Memorize decimal equivalents to .01 parts of a foot (1/8″ equals approximately .01′).

9. Do not "round off" quantities until the final summary.

10. Mark drawings with different colors as items are taken off.

11. Keep similar items together, different items separate.

12. Identify location and drawing numbers to aid in future checking for completeness.

13. Measure or list everything on the drawings or mentioned in the specifications.

14. It may be necessary to list items not called for to make the job complete.

15. Be alert for: Notes on plans such as N.T.S. (not to scale); changes in scale throughout the drawings; reduced size drawings; discrepancies between the specifications and the drawings.

16. Develop a consistent pattern of performing an estimate. For example:
    a. Start the quantity takeoff at the lower floor and move to the next higher floor.
    b. Proceed from the main section of the building to the wings.
    c. Proceed from south to north or vice versa, clockwise or counterclockwise.
    d. Take off floor plan quantities first, elevations next, then detail drawings.

17. List all gross dimensions that can be either used again for different quantities, or used as a rough check of other quantities for verification (exterior perimeter, gross floor area, individual floor areas, etc.).

18. Utilize design symmetry or repetition (repetitive floors, repetitive wings, symmetrical design around a center line, similar room layouts, etc.). Note: Extreme caution is needed here so as not to omit or duplicate an area.

19. Do not convert units until the final total is obtained. For instance, when estimating concrete work, keep all units to the nearest cubic foot, then summarize and convert to cubic yards.

20. When figuring alternatives, it is best to total all items involved in the basic system, then total all items involved in the alternates. Therefore you work with positive numbers in all cases. When adds and deducts are used, it is often confusing whether to add or subtract a portion of an item; especially on a complicated or involved alternate.

R010-010   Architectural Fees

Tabulated below are typical percentage fees by project size, for good professional architectural service. Fees may vary from those listed depending upon degree of design difficulty and economic conditions in any particular area.

Rates can be interpolated horizontally and vertically. Various portions of the same project requiring different rates should be adjusted proportionately.

For alterations, add 50% to the fee for the first $500,000 of project cost and add 25% to the fee for project cost over $500,000.

Architectural fees tabulated below include Engineering Fees.

Building Types	Total Project Size in Thousands of Dollars						
	100	250	500	1,000	5,000	10,000	50,000
Factories, garages, warehouses, repetitive housing	9.0%	8.0%	7.0%	6.2%	5.3%	4.9%	4.5%
Apartments, banks, schools, libraries, offices, municipal buildings	11.7	10.8	8.5	7.3	6.4	6.0	5.6
Churches, hospitals, homes, laboratories, museums, research	14.0	12.8	11.9	10.9	8.5	7.8	7.2
Memorials, monumental work, decorative furnishings	—	16.0	14.5	13.1	10.0	9.0	8.3

R010-040   Builder's Risk Insurance

Builder's Risk Insurance is insurance on a building during construction. Premiums are paid by the owner or the contractor. Blasting, collapse and underground insurance would raise total insurance costs above those listed. Floater policy for materials delivered to the job runs $.75 to $1.25 per $100 value. Contractor equipment insurance runs $.50 to $1.50 per $100 value. Insurance for miscellaneous tools to $1,500 value runs from $3.00 to $7.50 per $100 value.

Tabulated below are New England Builder's Risk insurance rates in dollars per $100 value for $1,000 deductible. For $25,000 deductible, rates can be reduced 13% to 34%. On contracts over $1,000,000, rates may be lower than those tabulated. Policies are written annually for the total completed value in place. For "all risk" insurance (excluding flood, earthquake and certain other perils) add $.025 to total rates below.

Coverage	Frame Construction (Class 1)		Average	Brick Construction (Class 4)		Average	Fire Resistive (Class 6)		Average
	Range			Range			Range		
Fire Insurance	$.300 to $.420		$.394	$.132 to $.189		$.174	$.052 to $.080		$.070
Extended Coverage	.115 to .150		.144	.080 to .105		.101	.081 to .105		.100
Vandalism	.012 to .016		.015	.008 to .011		.011	.008 to .011		.010
Total Annual Rate	$.427 to $.586		$.553	$.220 to .305		$.286	$.141 to $.196		$.180

R010-050 General Contractor's Overhead

The table below shows a contractor's overhead as a percentage of direct cost in two ways. The figures on the right are for the overhead, markup based on both material and labor. The figures on the left are based on the entire overhead applied only to the labor. This figure would be used if the owner supplied the materials or if a contract is for labor only. Note: Some of these markups are included in the labor rates shown on Reference Table R010-070.

Items of General Contractor's Indirect Costs	% of Direct Costs	
	As a Markup of Labor Only	As a Markup of Both Material and Labor
Field Supervision	6.0%	3.2%
Main Office Expense (see details below)	14.7	7.7
Tools and Minor Equipment	1.0	0.5
Workers' Compensation & Employers' Liability. See R010-060	18.5	9.7
Field Office, Sheds, Photos, Etc.	1.5	0.8
Performance and Payment Bond, 0.5% to 0.9%. See R010-080	0.7	0.4
Unemployment Tax See R010-100 (Combined Federal and State)	7.0	3.7
Social Security and Medicare. See R010-100	7.7	4.0
Sales Tax — add if applicable 42/80 x % as markup of total direct costs including both material and labor. See R010-090		
Sub Total	57.1%	30.0%
*Builder's Risk Insurance ranges from .141% to .586%. See R010-040	0.3	0.3
*Public Liability Insurance	1.5	1.5
Grand Total	58.9%	31.8%

*Paid by Owner or Contractor

Main Office Expense

A General Contractor's main office expense consists of many items not detailed in the front portion of the book. The percentage of main office expense declines with increased annual volume of the contractor. Typical main office expense ranges from 2% to 20% with the median about 7.2% of total volume. This equals about 7.7% of direct costs. The following are approximate percentages of total overhead for different items usually included in a General Contractor's main office overhead. With different accounting procedures, these percentages may vary.

Item	Typical Range			Average
Managers', clerical and estimators' salaries	40 %	to	55 %	48%
Profit sharing, pension and bonus plans	2	to	20	12
Insurance	5	to	8	6
Estimating and project management (not including salaries)	5	to	9	7
Legal, accounting and data processing	0.5	to	5	3
Automobile and light truck expense	2	to	8	5
Depreciation of overhead capital expenditures	2	to	6	4
Maintenance of office equipment	0.1	to	1.5	1
Office rental	3	to	5	4
Utilities including phone and light	1	to	3	2
Miscellaneous	5	to	15	8
Total				100%

R010-060 Workers' Compensation Insurance Rates by Trade

The table below tabulates the national averages for Workers' Compensation insurance rates by trade and type of building. The average "Insurance Rate" is multiplied by the "% of Building Cost" for each trade. This produces the "Workers' Compensation Cost" by % of total labor cost, to be added for each trade by building type to determine the weighted average Workers' Compensation rate for the building types analyzed.

Trade	Insurance Rate (% Labor Cost)		% of Building Cost			Workers' Compensation		
	Range	Average	Office Bldgs.	Schools & Apts.	Mfg.	Office Bldgs.	Schools & Apts.	Mfg.
Excavation, Grading, etc.	3.8 % to 26.6%	11.6%	4.8%	4.9%	4.5%	0.56%	0.57%	0.52%
Piles & Foundations	8.1 to 80.1	30.5	7.1	5.2	8.7	2.17	1.59	2.65
Concrete	7.0 to 40.7	19.2	5.0	14.8	3.7	0.96	2.84	0.71
Masonry	4.5 to 48.6	17.9	6.9	7.5	1.9	1.24	1.34	0.34
Structural Steel	8.1 to 132.9	42.6	10.7	3.9	17.6	4.56	1.66	7.50
Miscellaneous & Ornamental Metals	5.2 to 44.3	14.3	2.8	4.0	3.6	0.40	0.57	0.51
Carpentry & Millwork	7.0 to 49.9	19.9	3.7	4.0	0.5	0.74	0.80	0.10
Metal or Composition Siding	8.1 to 37.1	18.3	2.3	0.3	4.3	0.42	0.05	0.79
Roofing	8.1 to 88.8	35.2	2.3	2.6	3.1	0.81	0.92	1.09
Doors & Hardware	3.4 to 28.8	11.8	0.9	1.4	0.4	0.11	0.17	0.05
Sash & Glazing	4.6 to 30.0	14.9	3.5	4.0	1.0	0.52	0.60	0.15
Lath & Plaster	5.3 to 37.4	15.6	3.3	6.9	0.8	0.51	1.08	0.12
Tile, Marble & Floors	3.4 to 29.5	10.3	2.6	3.0	0.5	0.27	0.31	0.05
Acoustical Ceilings	4.2 to 26.7	12.3	2.4	0.2	0.3	0.30	0.02	0.04
Painting	4.6 to 41.1	15.6	1.5	1.6	1.6	0.23	0.25	0.25
Interior Partitions	7.0 to 49.9	19.9	3.9	4.3	4.4	0.78	0.86	0.88
Miscellaneous Items	2.8 to 109.2	19.6	5.2	3.7	9.7	1.02	0.72	1.90
Elevators	2.0 to 20.0	8.9	2.1	1.1	2.2	0.19	0.10	0.20
Sprinklers	2.7 to 20.6	9.3	0.5	—	2.0	0.05	—	0.19
Plumbing	2.8 to 16.1	9.2	4.9	7.2	5.2	0.45	0.66	0.48
Heat., Vent., Air Conditioning	3.9 to 29.2	12.8	13.5	11.0	12.9	1.73	1.41	1.65
Electrical	2.7 to 11.6	7.1	10.1	8.4	11.1	0.72	0.60	0.79
Total	2.0 % to 132.9%	—	100.0%	100.0%	100.0%	18.74%	17.12%	20.96%
Overall Weighted Average 18.94%								

Workers' Compensation Insurance Rates by States

The table below lists the weighted average Workers' Compensation base rate for each state with a factor comparing this with the national average of 18.5%.

State	Weighted Average	Factor	State	Weighted Average	Factor	State	Weighted Average	Factor
Alabama	24.8%	134	Kentucky	19.8%	107	North Dakota	16.7%	90
Alaska	12.6	68	Louisiana	27.2	147	Ohio	16.1	87
Arizona	16.7	90	Maine	21.8	118	Oklahoma	21.9	118
Arkansas	14.5	78	Maryland	11.9	64	Oregon	19.2	104
California	18.6	101	Massachusetts	26.5	143	Pennsylvania	22.8	123
Colorado	26.4	143	Michigan	22.6	122	Rhode Island	22.3	121
Connecticut	22.7	123	Minnesota	37.6	203	South Carolina	13.9	75
Delaware	12.3	66	Mississippi	20.5	111	South Dakota	18.8	102
District of Columbia	26.5	143	Missouri	15.8	85	Tennessee	16.0	86
Florida	27.9	151	Montana	37.3	202	Texas	24.7	134
Georgia	24.9	135	Nebraska	15.7	85	Utah	12.9	70
Hawaii	16.7	90	Nevada	17.8	96	Vermont	16.5	89
Idaho	13.2	71	New Hampshire	27.2	147	Virginia	12.0	65
Illinois	27.5	149	New Jersey	11.0	59	Washington	12.0	65
Indiana	6.6	36	New Mexico	22.6	122	West Virginia	14.0	76
Iowa	14.6	79	New York	17.0	92	Wisconsin	13.5	73
Kansas	12.3	66	North Carolina	12.7	69	Wyoming	8.9	48
Weighted Average for U.S. is 18.9% of payroll = 100%								

Rates in the following table are the base or manual costs per $100 of payroll for Workers' Compensation in each state. Rates are usually applied to straight time wages only and not to premium time wages and bonuses.

The weighted average skilled worker rate for 35 trades is 18.5%. For bidding purposes, apply the full value of Workers' Compensation directly to total labor costs, or if labor is 38%, materials 42% and overhead and profit 20% of total cost, carry 38/80 x 18.5% = 8.8% of cost (before overhead and profit) into overhead. Rates vary not only from state to state but also with the experience rating of the contractor.

Rates are the most current available at the time of publication.

R010-060 Workers' Compensation Insurance Rates by Trade and State (cont.)

State	Carpentry — 3 stories or less	Carpentry — interior cab. work	Carpentry — general	Concrete Work — NOC	Concrete Work — flat (flr., sdwk.)	Electrical Wiring — inside	Excavation — earth NOC	Excavation — rock	Glaziers	Insulation Work	Lathing	Masonry	Painting & Decorating	Pile Driving	Plastering	Plumbing	Roofing	Sheet Metal Work (HVAC)	Steel Erection — door & sash	Steel Erection — inter., ornam.	Steel Erection — structure	Steel Erection — NOC	Tile Work — (interior ceramic)	Waterproofing	Wrecking
	5651	5437	5403	5213	5221	5190	6217	6217	5462	5479	5443	5022	5474	6003	5480	5183	5551	5538	5102	5102	5040	5057	5348	9014	5701
AL	29.72	14.08	29.32	24.41	13.28	8.21	15.78	15.78	18.46	19.50	11.49	26.63	19.78	54.72	25.41	10.36	43.33	23.96	15.14	15.14	41.12	50.13	13.77	6.07	41.12
AK	9.82	6.54	11.30	8.47	6.80	5.41	8.26	8.26	14.08	11.07	6.57	8.78	12.69	34.98	9.10	4.79	15.78	7.67	13.35	13.35	23.66	23.66	6.89	5.65	23.66
AZ	19.59	8.13	24.24	16.75	10.46	7.75	9.83	9.83	12.55	23.53	8.09	15.15	13.28	22.26	12.52	8.87	19.74	9.46	13.23	13.23	45.71	31.76	6.67	5.95	45.71
AR	18.19	11.07	15.52	13.78	8.42	6.79	10.33	10.33	13.58	13.16	8.51	13.05	13.10	20.51	10.53	8.26	28.30	11.65	10.65	10.65	21.83	27.91	7.41	5.12	21.83
CA	28.28	9.07	28.28	13.18	13.18	9.93	7.88	7.88	16.37	23.34	10.90	14.55	19.45	21.18	18.12	11.59	39.79	15.33	13.55	13.55	23.93	21.91	7.74	19.45	21.91
CO	32.22	13.54	20.95	20.96	15.13	9.20	15.99	15.99	15.98	21.31	14.68	31.06	21.85	41.53	37.40	14.50	59.10	12.54	11.83	11.83	73.46	43.49	15.04	12.26	73.46
CT	19.99	15.66	25.33	24.46	15.81	7.68	10.83	10.83	24.15	35.23	19.15	27.08	17.75	24.26	16.79	14.44	34.63	14.28	13.12	13.12	45.44	50.30	14.75	4.80	45.44
DE	13.83	13.83	11.56	9.28	6.29	5.82	8.54	8.54	11.18	11.56	10.51	9.85	13.41	11.64	10.51	5.41	22.89	10.29	10.21	10.21	26.69	10.21	7.54	9.85	25.69
DC	15.87	12.68	21.60	24.27	13.97	10.28	18.06	18.06	25.12	23.67	11.75	25.81	13.97	49.63	14.44	16.13	28.69	15.25	44.32	44.32	55.08	56.98	17.73	5.31	55.08
FL	37.10	21.45	29.77	31.32	16.35	11.51	14.88	14.88	20.14	25.74	25.02	28.18	28.47	55.84	26.52	13.50	52.41	18.56	17.82	17.82	40.27	48.25	11.58	7.69	40.27
GA	31.04	16.02	32.71	20.67	15.12	9.82	19.55	19.55	21.97	19.15	26.66	23.03	21.20	40.88	19.50	11.14	40.72	18.65	13.74	13.74	36.03	53.09	13.08	10.16	36.03
HI	17.38	11.62	28.20	13.74	12.27	7.10	7.73	7.73	20.55	21.19	10.94	17.87	11.33	20.15	15.61	6.06	33.24	8.02	11.11	11.11	31.71	21.70	9.78	11.54	31.71
ID	14.63	7.15	17.11	11.14	10.02	5.88	7.38	7.38	10.19	13.27	8.36	10.33	10.08	20.19	9.45	6.06	26.23	9.00	7.69	7.69	28.85	27.53	5.62	8.28	28.85
IL	18.94	12.54	23.02	36.76	13.56	9.82	10.43	10.43	27.14	25.52	17.50	23.22	14.93	52.40	15.82	14.48	37.32	17.03	25.56	25.56	78.98	70.79	15.71	7.31	78.98
IN	8.05	3.37	6.97	7.04	3.36	2.73	3.84	3.84	4.59	7.36	4.20	4.47	4.61	11.67	5.29	2.80	10.80	4.20	5.21	5.21	16.06	12.80	3.36	4.05	16.06
IA	10.26	6.79	14.54	12.94	6.78	4.20	5.47	5.47	13.54	18.19	5.82	12.97	7.52	19.90	13.44	6.97	21.56	7.44	12.92	12.92	53.87	36.20	7.13	5.05	17.98
KS	11.81	6.71	11.00	13.98	7.68	4.88	4.97	4.97	7.51	14.41	9.05	13.25	8.45	19.60	11.48	6.50	31.34	6.66	6.07	6.07	23.95	30.42	4.70	4.70	23.95
KY	16.47	15.26	24.34	22.12	9.53	8.95	13.67	13.67	15.96	14.68	10.62	22.00	19.83	30.44	13.94	10.38	32.83	15.92	13.21	13.21	45.21	33.73	10.60	8.85	45.21
LA	29.63	22.95	26.01	20.86	14.80	10.48	24.73	24.73	23.88	20.82	14.00	21.27	27.63	54.39	21.42	13.78	49.11	21.35	15.51	15.51	66.41	36.94	12.82	10.26	66.41
ME	13.43	10.42	43.59	24.29	11.12	10.02	14.75	14.75	13.35	15.39	14.03	17.26	16.07	31.44	17.95	11.12	32.07	17.70	15.08	15.08	37.84	61.81	11.66	9.37	37.84
MD	10.55	5.95	10.55	11.35	5.05	5.15	9.25	9.25	13.20	13.05	6.45	11.35	6.75	27.45	6.65	5.55	22.80	7.00	9.15	9.15	26.80	18.50	6.35	3.50	26.80
MA	15.25	11.38	22.82	39.38	16.79	6.14	10.72	10.72	16.78	27.76	14.02	29.03	14.95	29.58	13.75	8.49	67.48	13.20	16.07	16.07	80.76	79.51	16.46	6.62	62.87
MI	14.80	10.56	18.34	32.85	12.74	5.68	12.49	12.49	11.74	22.33	11.51	19.97	17.86	64.20	23.74	8.60	44.62	11.92	13.62	13.62	53.03	39.73	10.26	10.42	53.03
MN	27.66	28.76	49.94	30.20	22.59	8.28	19.73	19.73	30.01	37.20	23.93	29.89	19.91	80.08	23.93	15.88	88.83	13.91	27.21	27.21	132.92	31.83	29.54	8.71	132.92
MS	24.53	13.32	22.38	16.58	11.16	8.64	10.38	10.38	15.53	16.90	14.52	19.10	20.18	62.11	15.99	8.72	28.96	17.82	14.94	14.94	29.82	34.43	12.21	8.26	29.82
MO	18.72	7.52	11.84	13.28	10.66	5.79	11.05	11.05	8.97	19.99	9.67	13.57	11.77	16.88	14.22	7.10	25.74	9.76	13.82	13.82	47.98	31.96	5.83	5.71	47.98
MT	30.63	15.19	47.53	32.78	24.74	11.01	26.63	26.63	25.04	38.54	19.54	48.55	41.06	53.68	27.00	15.03	85.06	16.93	33.95	33.95	86.98	55.25	13.30	12.10	86.98
NE	16.19	9.23	15.94	17.31	12.73	5.43	10.19	10.19	12.36	16.79	9.34	18.07	11.58	22.01	14.62	9.06	33.91	13.86	10.75	10.75	25.03	26.96	7.88	5.47	25.03
NV	15.42	15.42	15.42	12.86	12.86	8.82	10.62	10.62	12.65	17.78	20.58	14.66	27.12	10.39	20.58	11.30	25.86	29.24	12.56	12.56	27.12	27.12	10.61	10.39	27.12
NH	22.49	9.49	19.59	40.72	12.91	7.83	14.19	14.19	17.36	42.34	10.91	27.18	18.02	38.80	14.43	9.21	87.03	16.74	19.82	19.82	65.07	48.37	13.24	9.53	65.07
NJ	10.40	8.85	10.40	10.83	8.37	4.23	7.95	7.95	7.36	12.78	7.88	10.77	11.49	11.32	7.88	5.48	29.41	6.20	8.03	8.03	22.52	12.96	5.74	5.37	29.92
NM	25.46	13.01	19.56	20.47	14.42	6.58	11.55	11.55	14.48	25.34	12.80	28.10	13.28	32.91	15.58	12.09	46.15	12.97	19.70	19.70	74.42	29.13	9.16	9.86	74.42
NY	15.39	6.82	14.50	24.92	15.16	6.89	9.69	9.69	16.92	15.82	13.00	18.61	12.55	27.49	13.03	9.03	32.78	14.13	14.12	14.12	27.11	24.09	10.56	7.48	32.53
NC	13.73	9.27	15.40	13.18	6.71	7.60	7.76	7.76	10.18	12.28	6.79	10.43	9.07	15.77	11.81	6.59	23.73	9.56	10.94	10.94	37.63	12.46	6.40	3.69	37.63
ND	13.89	13.89	13.89	12.17	12.17	6.16	9.91	9.91	9.85	9.38	13.16	12.20	11.42	29.82	13.16	8.05	31.37	8.05	13.89	13.89	29.82	29.82	9.63	31.37	29.82
OH	13.27	13.27	13.27	13.45	13.45	6.02	13.45	13.45	18.28	16.31	16.31	15.62	18.28	13.45	16.31	7.96	28.62	24.12	13.45	13.45	13.45	13.45	13.08	13.45	13.45
OK	26.16	11.79	19.33	15.51	11.50	7.41	18.45	18.45	13.55	18.00	12.05	16.80	17.10	36.26	18.29	8.72	42.26	12.45	11.17	11.17	69.45	46.76	11.42	8.20	69.45
OR	29.51	9.28	20.03	16.50	10.61	6.63	12.94	12.94	14.95	19.68	12.27	12.43	19.83	22.05	13.85	8.38	38.11	11.85	11.01	11.01	61.69	25.54	9.92	12.10	61.69
PA	15.80	15.80	19.85	28.18	11.84	7.80	12.74	12.74	15.69	19.85	18.94	18.96	22.05	32.12	18.94	11.57	43.30	12.58	24.71	24.71	59.05	24.71	11.83	18.96	81.34
RI	21.32	11.02	18.77	22.11	15.55	5.86	12.16	12.16	17.02	20.95	11.89	20.66	19.86	41.76	14.19	7.25	29.48	10.52	13.18	13.18	78.79	49.67	14.34	9.90	78.79
SC	20.63	13.45	18.64	10.77	8.42	7.13	8.76	8.76	14.61	12.03	7.43	10.83	10.94	19.82	17.30	7.52	26.81	11.23	9.33	9.33	17.82	23.27	6.07	4.47	17.82
SD	14.18	12.00	20.31	21.01	9.04	5.56	14.16	14.16	15.19	17.21	11.16	15.13	12.87	31.19	19.40	9.78	35.44	12.07	11.45	11.45	45.19	39.81	9.57	5.43	45.19
TN	19.19	8.38	14.68	16.00	9.42	5.81	11.19	11.19	13.21	19.26	8.45	14.09	14.13	26.27	12.09	7.49	25.88	13.52	12.81	12.81	39.01	23.63	6.94	6.40	39.01
TX	28.29	18.95	28.29	25.59	19.49	11.60	18.16	18.16	14.17	25.84	13.47	23.12	18.29	43.91	20.99	12.88	47.24	22.92	14.29	14.29	50.47	31.01	9.94	11.38	54.81
UT	11.17	11.17	11.17	17.68	7.99	7.05	6.38	6.38	9.55	10.95	12.32	13.75	15.29	17.24	10.60	6.41	26.27	6.30	8.99	8.99	23.51	23.51	6.06	5.83	26.53
VT	9.31	6.60	17.05	30.92	11.53	5.21	9.76	9.76	13.18	14.90	8.31	13.57	10.36	21.53	11.08	9.95	28.65	10.26	9.81	9.81	30.85	55.10	7.40	8.29	30.85
VA	10.75	8.07	9.65	14.28	6.00	4.31	7.24	7.24	11.39	10.80	6.21	9.90	8.05	33.02	9.29	5.70	17.42	7.69	9.86	9.86	23.21	25.38	4.87	2.80	23.21
WA	8.78	8.78	8.78	8.53	8.53	3.68	8.82	8.82	8.71	9.93	8.78	11.54	10.23	20.99	11.78	5.41	23.39	3.94	16.21	16.21	16.21	16.21	9.81	10.23	16.21
WV	15.43	15.43	15.43	20.87	20.87	5.01	9.14	9.14	7.33	7.33	16.88	16.02	16.88	10.56	16.88	7.40	12.04	7.33	14.98	14.98	13.27	14.98	16.02	4.87	13.27
WI	11.08	9.43	17.02	9.23	8.35	5.49	6.61	6.61	9.92	13.77	10.12	13.73	10.88	15.46	14.40	6.09	29.22	8.00	11.03	11.03	37.40	18.86	9.39	4.08	37.40
WY	8.13	8.13	8.13	8.13	8.13	8.13	8.13	8.13	8.13	8.13	8.13	8.13	8.13	8.13	8.13	8.13	8.13	8.13	8.13	8.13	8.13	8.13	8.13	8.13	8.13
AVG.	18.32	11.75	19.88	19.18	11.84	7.13	11.63	11.63	14.85	18.65	12.25	17.87	15.60	30.47	15.59	9.18	35.21	12.73	14.28	14.28	42.56	33.17	10.30	8.52	42.28

R010-060 Workers' Compensation (cont.) (Canada in Canadian dollars)

Province		Alberta	British Columbia	Manitoba	Ontario	New Brunswick	Newfndld. & Labrador	Northwest Territories	Nova Scotia	Prince Edward Island	Quebec	Saskatchewan	Yukon
Carpentry—3 stories or less	Rate	4.17	6.38	8.73	7.71	3.53	5.11	6.75	5.77	5.98	11.84	6.96	2.35
	Code	25401	70600	40102	723	422	403	4-41	4013	401	40010	B12-02	4-042
Carpentry—interior cab. work	Rate	2.54	4.12	8.73	7.71	3.92	5.11	6.75	5.77	5.98	11.84	5.85	2.35
	Code	42133	60412	40102	723	427	403	4-41	4013	401	40010	B11-25	4-042
CARPENTRY—general	Rate	4.17	6.38	8.73	7.71	3.53	5.11	6.75	5.77	5.98	11.84	6.96	2.35
	Code	25401	70600	40102	723	422	403	4-41	4013	401	40010	B12-02	4-042
CONCRETE WORK—NOC	Rate	5.96	6.38	11.50	12.65	3.53	5.11	6.75	7.39	5.98	17.38	10.95	3.25
	Code	42104	70604	40110	745	422	403	4-41	4222	401	40080	B14-04	2-032
CONCRETE WORK—flat (flr. sidewalk)	Rate	5.96	6.38	11.50	12.65	3.53	5.11	6.75	7.39	5.98	17.38	10.95	3.25
	Code	42104	70604	40110	745	422	403	4-41	4222	401	40080	B14-04	2-032
ELECTRICAL Wiring—inside	Rate	2.58	3.59	5.47	3.63	1.20	4.14	4.50	3.25	3.34	6.96	5.85	2.35
	Code	42124	71100	40203	704	426	400	4-46	4261	402	40150	B11-05	4-041
EXCAVATION—earth NOC	Rate	3.01	4.21	8.73	6.89	2.26	5.11	4.50	4.27	5.98	9.73	7.86	3.25
	Code	40604	72607	40706	711	421	403	4-43	4214	401	40020	R11-06	2-016
EXCAVATION—rock	Rate	3.01	4.21	8.73	6.89	2.26	5.11	4.50	4.27	5.98	9.73	7.86	3.25
	Code	40604	72607	40706	711	421	403	4-43	4214	401	40020	R11-06	2-016
GLAZIERS	Rate	2.70	2.68	6.45	7.89	3.92	3.90	6.75	7.75	3.34	12.33	10.95	2.35
	Code	42121	60236	40109	751	423	402	4-41	4233	402	40110	B13-04	4-042
INSULATION WORK	Rate	3.63	5.92	8.73	7.89	3.92	3.90	6.75	7.75	5.98	15.00	6.96	3.25
	Code	42184	70504	40102	751	423	402	4-41	4234	401	40170	B12-07	2-035
LATHING	Rate	6.77	5.92	8.73	7.71	3.92	3.90	6.75	6.14	3.34	15.00	10.95	3.25
	Code	42135	70500	40102	723	427	402	4-41	4271	402	40170	B13-02	2-036
MASONRY	Rate	5.96	6.38	8.73	9.95	3.92	5.11	6.75	7.75	5.98	17.38	10.95	3.25
	Code	42102	70602	40102	741	423	403	4-41	4231	401	40080	B15-01	2-032
PAINTING & DECORATING	Rate	4.21	5.92	12.20	9.93	3.92	3.90	6.75	6.14	3.34	15.00	6.96	3.25
	Code	42111	70501	40105	719	427	402	4-41	4275	402	40170	B12-01	2-036
PILE DRIVING	Rate	5.96	17.50	8.73	8.97	3.53	9.15	4.50	7.39	5.98	9.73	6.96	3.25
	Code	42159	72502	40706	732	422	404	4-43	4221	401	40020	B12-10	2-030
PLASTERING	Rate	6.77	5.92	12.20	9.93	3.92	3.90	6.75	6.14	3.34	15.00	10.95	3.25
	Code	42135	70502	40108	719	427	402	4-41	4271	402	40170	B13-02	2-036
PLUMBING	Rate	2.58	4.06	5.47	5.06	1.87	3.70	4.50	3.79	3.34	8.79	5.85	2.35
	Code	42122	70712	40204	707	424	401	4-46	4241	402	40130	B11-01	4-039
ROOFING	Rate	9.97	6.38	12.25	9.95	6.13	5.11	6.75	7.75	5.98	19.61	10.95	3.25
	Code	42118	70600	40403	728	430	403	4-41	4235	401	40120	B15-02	2-031
SHEET METAL WORK (HVAC)	Rate	2.58	4.06	8.26	5.06	1.87	3.70	4.50	7.75	3.34	8.79	5.85	2.35
	Code	42117	70714	40402	707	424	401	4-46	4236	402	40130	B11-07	4-040
STEEL ERECTION—door & sash	Rate	3.63	17.50	8.73	16.58	3.53	9.15	6.75	7.39	5.98	23.08	10.95	3.25
	Code	42106	72509	40502	748	422	404	4-41	4223	401	40100	B15-03	2-012
STEEL ERECTION—inter., ornam.	Rate	3.63	17.50	8.73	16.58	3.53	5.11	6.75	7.39	5.98	23.08	10.95	3.25
	Code	42106	72509	40502	748	422	403	4-41	4223	401	40100	B15-03	2-012
STEEL ERECTION—structure	Rate	3.63	17.50	8.73	16.58	3.53	9.15	6.75	10.08	5.98	23.08	10.95	3.25
	Code	42106	72509	40502	748	422	404	4-41	4227	401	40100	B15-04	2-012
STEEL ERECTION—NOC	Rate	3.63	17.50	8.73	16.58	3.53	9.15	6.75	10.08	5.98	23.08	10.95	3.25
	Code	42106	72509	40502	748	422	404	4-41	4227	401	40100	B15-03	2-012
TILE WORK—inter. (ceramic)	Rate	3.07	5.92	5.47	9.93	3.92	3.90	6.75	6.14	3.34	15.00	10.95	3.25
	Code	42113	70506	40103	719	427	402	4-41	4276	402	40170	B13-01	2-034
WATERPROOFING	Rate	4.21	6.38	8.73	7.71	3.92	5.11	6.75	7.75	3.34	19.61	5.85	3.25
	Code	42139	70620	40102	723	423	403	4-41	4239	402	40120	B11-17	2-030
WRECKING	Rate	12.86	6.38	9.84	16.58	2.26	5.11	4.50	4.27	5.98	38.63	10.95	3.25
	Code	42108	70600	40106	748	421	403	4-43	4211	401	40050	B14-07	2-030

R010-070 Contractor's Overhead & Profit

Below are the **average** installing contractor's percentage mark-ups applied to base labor rates to arrive at typical billing rates.

Column A: Labor rates are based on average open shop wages for 7 major U.S. regions. Base rates including fringe benefits are listed hourly and daily. These figures are the sum of the wage rate and employer-paid fringe benefits such as vacation pay, and employer-paid health costs.

Column B: Workers' Compensation rates are the national average of state rates established for each trade.

Column C: Column C lists average fixed overhead figures for all trades. Included are Federal and State Unemployment costs set at 7.0%; Social Security Taxes (FICA) set at 7.65%; Builder's Risk Insurance costs set at 0.34%; and Public Liability costs set at 1.55%. All the percentages except those for Social Security Taxes vary from state to state as well as from company to company.

Columns D and E: Percentages in Columns D and E are based on the presumption that the installing contractor has annual billing of $1,000,000 and up. Overhead percentages may increase with smaller annual billing. The overhead percentages for any given contractor may vary greatly and depend on a number of factors, such as the contractor's annual volume, engineering and logistical support costs, and staff requirements. The figures for overhead and profit will also vary depending on the type of job, the job location, and the prevailing economic conditions. All factors should be examined very carefully for each job.

Column F: Column F lists the total of Columns B, C, D, and E.

Column G: Column G is Column A (hourly base labor rate) multiplied by the percentage in Column F (O&P percentage).

Column H: Column H is the total of Column A (hourly base labor rate) plus Column G (Total O&P).

Column I: Column I is Column H multiplied by eight hours.

		A		B	C	D	E	F		G	H		I
		Base Rate Incl. Fringes		Workers' Comp. Ins.	Average Fixed Overhead	Overhead	Profit	Total Overhead & Profit			Rate with O & P		
Abbr.	Trade	Hourly	Daily					%	Amount		Hourly	Daily	
Skwk	Skilled Workers Average (35 trades)	$18.75	$150.00	18.5%	16.5%	27.0%	10.0%	72.0%	$13.50		$32.25	$258.00	
	Helpers Average (5 trades)	13.85	110.80	19.8		25.0		71.3	9.85		23.70	189.60	
	Foreman Average, Inside ($.50 over trade)	19.25	154.00	18.5		27.0		72.0	13.85		33.10	264.80	
	Foreman Average, Outside ($2.00 over trade)	20.75	166.00	18.5		27.0		72.0	14.95		35.70	285.60	
Clab	Common Building Laborers	13.40	107.20	19.9		25.0		71.4	9.55		22.95	183.60	
Asbe	Asbestos/Insulation Workers/Pipe Coverers	19.55	156.40	18.7		30.0		75.2	14.70		34.25	274.00	
Boil	Boilermakers	21.10	168.80	16.9		30.0		73.4	15.50		36.60	292.80	
Bric	Bricklayers	18.95	151.60	17.9		25.0		69.4	13.15		32.10	256.80	
Brhe	Bricklayer Helpers	14.65	117.20	17.9		25.0		69.4	10.15		24.80	198.40	
Carp	Carpenters	18.50	148.00	19.9		25.0		71.4	13.20		31.70	253.60	
Cefi	Cement Finishers	17.80	142.40	11.8		25.0		63.3	11.25		29.05	232.40	
Elec	Electricians	20.60	164.80	7.1		30.0		63.6	13.10		33.70	269.60	
Elev	Elevator Constructors	21.35	170.80	8.9		30.0		65.4	13.95		35.30	282.40	
Eqhv	Equipment Operators, Crane or Shovel	19.75	158.00	11.6		28.0		66.1	13.05		32.80	262.40	
Eqmd	Equipment Operators, Medium Equipment	19.00	152.00	11.6		28.0		66.1	12.55		31.55	252.40	
Eqlt	Equipment Operators, Light Equipment	18.15	145.20	11.6		28.0		66.1	12.00		30.15	241.20	
Eqol	Equipment Operators, Oilers	16.20	129.60	11.6		28.0		66.1	10.70		26.90	215.20	
Eqmm	Equipment Operators, Master Mechanics	20.20	161.60	11.6		28.0		66.1	13.35		33.55	268.40	
Glaz	Glaziers	18.30	146.40	14.9		25.0		66.4	12.15		30.45	243.60	
Lath	Lathers	18.30	146.40	12.3		25.0		63.8	11.70		30.00	240.00	
Marb	Marble Setters	18.75	150.00	17.9		25.0		69.4	13.00		31.75	254.00	
Mill	Millwrights	19.50	156.00	12.0		25.0		63.5	12.40		31.90	255.20	
Mstz	Mosaic and Terrazzo Workers	18.35	146.80	10.3		25.0		61.8	11.35		29.70	237.60	
Pord	Painters, Ordinary	17.20	137.60	15.6		25.0		67.1	11.55		28.75	230.00	
Psst	Painters, Structural Steel	18.00	144.00	53.3		25.0		104.8	18.85		36.85	294.80	
Pape	Paper Hangers	17.45	139.60	15.6		25.0		67.1	11.70		29.15	233.20	
Pile	Pile Drivers	18.65	149.20	30.5		30.0		87.0	16.25		34.90	279.20	
Plas	Plasterers	17.55	140.40	15.6		25.0		67.1	11.80		29.35	234.80	
Plah	Plasterer Helpers	14.75	118.00	15.6		25.0		67.1	9.90		24.65	197.20	
Plum	Plumbers	20.85	166.80	9.2		30.0		65.7	13.70		34.55	276.40	
Rodm	Rodmen (Reinforcing)	19.75	158.00	33.2		28.0		87.7	17.30		37.05	296.40	
Rofc	Roofers, Composition	16.20	129.60	35.2		25.0		86.7	14.05		30.25	242.00	
Rots	Roofers, Tile and Slate	16.30	130.40	35.2		25.0		86.7	14.15		30.45	243.60	
Rohe	Roofer Helpers (Composition)	11.85	94.80	35.2		25.0		86.7	10.25		22.10	176.80	
Shee	Sheet Metal Workers	20.15	161.20	12.7		30.0		69.2	13.95		34.10	272.80	
Spri	Sprinkler Installers	21.10	168.80	9.3		30.0		65.8	13.90		35.00	280.00	
Stpi	Steamfitters or Pipefitters	21.00	168.00	9.2		30.0		65.7	13.80		34.80	278.40	
Ston	Stone Masons	18.45	147.60	17.9		25.0		69.4	12.80		31.25	250.00	
Sswk	Structural Steel Workers	19.95	159.60	42.6		28.0		97.1	19.35		39.30	314.40	
Tilf	Tile Layers (Floor)	18.25	146.00	10.3		25.0		61.8	11.30		29.55	236.40	
Tilh	Tile Layer Helpers	14.70	117.60	10.3		25.0		61.8	9.10		23.80	190.40	
Trlt	Truck Drivers, Light	15.10	120.80	15.9		25.0		67.4	10.20		25.30	202.40	
Trhv	Truck Drivers, Heavy	15.40	123.20	15.9		25.0		67.4	10.40		25.80	206.40	
Sswl	Welders, Structural Steel	19.95	159.60	42.6		28.0		97.1	19.35		39.30	314.40	
Wrck	*Wrecking	13.85	110.80	42.3		25.0		93.8	13.00		26.85	214.80	

*Not included in Averages.

R010-080   Performance Bond

This table shows the cost of a Performance Bond for a construction job scheduled to be completed in 12 months. Add 1% of the premium cost per month for jobs requiring more than 12 months to complete. The rates are "standard" rates offered to contractors that the bonding company considers financially sound and capable of doing the work. Preferred rates are offered by some bonding companies based upon financial strength of the contractor. Actual rates vary from contractor to contractor and from bonding company to bonding company. Contractors should prequalify through a bonding agency before submitting a bid on a contract that requires a bond.

Contract Amount	Building Construction Class B Projects			Highways & Bridges					
				Class A New Construction			Class A-1 Highway Resurfacing		
First $ 100,000 bid	$25.00 per M			$15.00 per M			$9.40 per M		
Next 400,000 bid	$ 2,500	plus	$15.00 per M	$ 1,500	plus	$10.00 per M	$ 940	plus	$7.20 per M
Next 2,000,000 bid	8,500	plus	10.00 per M	5,500	plus	7.00 per M	3,820	plus	5.00 per M
Next 2,500,000 bid	28,500	plus	7.50 per M	19,500	plus	5.50 per M	15,820	plus	4.50 per M
Next 2,500,000 bid	47,250	plus	7.00 per M	33,250	plus	5.00 per M	28,320	plus	4.50 per M
Over 7,500,000 bid	64,750	plus	6.00 per M	45,750	plus	4.50 per M	39,570	plus	4.00 per M

R010-090   Sales Tax by State

State sales tax on materials is tabulated below (5 states have no sales tax). Many states allow local jurisdictions, such as a county or city, to levy additional sales tax.

Some projects may be sales tax exempt, particularly those constructed with public funds.

State	Tax (%)	State	Tax (%)	State	Tax (%)	State	Tax (%)
Alabama	4	Illinois	6.25	Montana	0	Rhode Island	7
Alaska	0	Indiana	5	Nebraska	5	South Carolina	5
Arizona	5	Iowa	5	Nevada	6.75	South Dakota	4
Arkansas	4.5	Kansas	4.9	New Hampshire	0	Tennessee	6
California	7.25	Kentucky	6	New Jersey	6	Texas	6.25
Colorado	3	Louisiana	4	New Mexico	5	Utah	6.125
Connecticut	6	Maine	6	New York	4	Vermont	5
Delaware	0	Maryland	5	North Carolina	4	Virginia	4.5
District of Columbia	5.75	Massachusetts	5	North Dakota	5	Washington	6.5
Florida	6	Michigan	6	Ohio	5	West Virginia	6
Georgia	6	Minnesota	6.5	Oklahoma	4.5	Wisconsin	5
Hawaii	4	Mississippi	7	Oregon	0	Wyoming	4
Idaho	5	Missouri	4.225	Pennsylvania	6	Average	4.78 %

R010-100   Unemployment Taxes and Social Security Taxes

Mass. State Unemployment tax ranges from 2.6% to 8.5% plus an experience rating assessment the following year, on the first $10,800 of wages. Federal Unemployment tax is 6.2% of the first $7,000 of wages. This is reduced by a credit for payment to the state. The minimum Federal Unemployment tax is .8% after all credits.

Combined rates in Mass. thus vary from 3.4% to 9.3% of the first $10,800 of wages. Combined average U.S. rate is about 7.0% of the first $7,000. Contractors with permanent workers will pay less since the average annual wages for skilled workers is $27.30 x 2,000 hours or about $54,600 per year. The average combined rate for U.S. would thus be 7.0% x $7,000 ÷ $54,600 = 0.9% of total wages for permanent employees.

Rates vary not only from state to state but also with the experience rating of the contractor.

Social Security (FICA) for 1998 is estimated at time of publication to be 7.65% of wages up to $65,400.

R011-010   Repair and Remodeling

Cost figures are based on new construction utilizing the most cost-effective combination of labor, equipment and material with the work scheduled in proper sequence to allow the various trades to accomplish their work in an efficient manner.

The costs for repair and remodeling work must be modified due to the following factors that may be present in any given repair and remodeling project.

1. Equipment usage curtailment due to the physical limitations of the project, with only hand-operated equipment being used.

2. Increased requirement for shoring and bracing to hold up the building while structural changes are being made and to allow for temporary storage of construction materials on above-grade floors.

3. Material handling becomes more costly due to having to move within the confines of an enclosed building. For multi-story construction, low capacity elevators and stairwells may be the only access to the upper floors.

4. Large amount of cutting and patching and attempting to match the existing construction is required. It is often more economical to remove entire walls rather than create many new door and window openings. This sort of trade-off has to be carefully analyzed.

5. Cost of protection of completed work is increased since the usual sequence of construction usually cannot be accomplished.

6. Economies of scale usually associated with new construction may not be present. If small quantities of components must be custom fabricated due to job requirements, unit costs will naturally increase. Also, if only small work areas are available at a given time, job scheduling between trades becomes difficult and subcontractor quotations may reflect the excessive start-up and shut-down phases of the job.

7. Work may have to be done on other than normal shifts and may have to be done around an existing production facility which has to stay in production during the course of the repair and remodeling.

8. Dust and noise protection of adjoining non-construction areas can involve substantial special protection and alter usual construction methods.

9. Job may be delayed due to unexpected conditions discovered during demolition or removal. These delays ultimately increase construction costs.

10. Piping and ductwork runs may not be as simple as for new construction. Wiring may have to be snaked through walls and floors.

11. Matching "existing construction" may be impossible because materials may no longer be manufactured. Substitutions may be expensive.

12. Weather protection of existing structure requires additional temporary structures to protect building at openings.

13. On small projects, because of local conditions, it may be necessary to pay a tradesman for a minimum of four hours for a task that is completed in one hour.

All of the above areas can contribute to increased costs for a repair and remodeling project. Each of the above factors should be considered in the planning, bidding and construction stage in order to minimize the increased costs associated with repair and remodeling jobs.

R015-100 Steel Tubular Scaffolding

On new construction, tubular scaffolding is efficient up to 60' high or five stories. Above this it is usually better to use a hung scaffolding if construction permits. Swing scaffolding operations may interfere with tenants. In this case, the tubular is more practical at all heights.

In repairing or cleaning the front of an existing building the cost of tubular scaffolding per S.F. of building front increases as the height increases above the first tier. The first tier cost is relatively high due to leveling and alignment.

The minimum efficient crew for erection is three workers. For heights over 50', a crew of four is more efficient. Use two or more on top and two at the bottom for handing up or hoisting. Four workers can erect and dismantle about nine frames per hour up to five stories. From five to eight stories they will average six frames per hour. With 7' horizontal spacing this will run about 400 S.F. and 265 S.F. of wall surface, respectively. Time for placing planks must be added to the above. On heights above 50', five planks can be placed per labor-hour.

The cost per 1,000 S.F. of building front in the table below was developed by pricing the materials required for a typical tubular scaffolding system eleven frames long and two frames high. Planks were figured five wide for standing plus two wide for materials.

Frames are 5' wide and usually spaced 7' O.C. horizontally. Sidewalk frames are 6' wide. Rental rates will be lower for jobs over three months duration.

For jobs under twenty-five frames, add 50% to rental cost. These figures do not include accessories which are listed separately below. Large quantities for long periods can reduce rental rates by 20%.

Item	Unit	Purchase, Each		Monthly Rent, Each		Per 1,000 S. F. of Building Front	
		Regular	Heavy Duty	Regular	Heavy Duty	No. of Frames	Rental per Mo.
5' Wide Standard Frames, 3' High	Ea.	$ 58.00	—	$ 3.75	—	—	—
*5'-0" High		67.00	—	3.75	—	—	—
*6'-4" High		84.00	—	3.75	—	24	$ 90.00
2' & 4' Wide Shoring Frames, 5' High		—	$ 85.00	—	$ 5.00	—	—
6'-0" High		—	101.00	—	5.00	—	—
6' Wide Canopy (Sidewalk) Frame, 7'-6" High		145.00	—	7.00	—	—	—
Side Arm Bracket, 20"		28.00	—	1.50	—	12	18.00
Guardrail Post		15.00	—	1.00	—	12	12.00
Guardrail, 7' section		7.25	—	.75	—	11	8.25
Cross Braces		15.00	16.00	.60	1.00	44	26.40
Leveling Jacks & Plates		24.00	44.00	1.50	3.00	24	36.00
8" Casters		33.00	—	6.00	—	—	—
16' Laminated Wood Plank, 2" x 10"		43.00	—	5.00	—	35	175.00
8' Laminated Wood Plank, 2" x 10"		22.00	—	2.50	—	7	17.50
Shoring Stringers, steel, 10' to 12' long	L.F.	—	7.00	—	2.00	—	—
Aluminum, 12' to 16' long		—	16.00	—	2.00	—	—
Aluminum joists with nailers, 10' to 22' long		—	12.50	—	2.00	—	—
24' Powered Swing Staging (with accessories)	Ea.	—	—	650.00	—	—	—
						Total	$383.15
						2 Use/Mo.	$191.58

*Most commonly used

Scaffolding is often used as falsework over 15' high during construction of cast-in-place concrete beams and slabs. Two foot wide scaffolding is generally used for heavy beam construction. The span between frames depends upon the load to be carried with a maximum span of 5'.

Heavy duty shoring frames with a capacity of 10,000#/leg can be spaced up to 10' O. C. depending upon form support design and loading.

Scaffolding used as horizontal shoring requires less than half the material required with conventional shoring.

On new construction, erection is done by carpenters.

Rolling towers supporting horizontal shores can reduce labor and speed the job. For maintenance work, catwalks with spans up to 70' can be supported by the rolling towers.

R020-820   Asbestos Removal Process

Asbestos removal is accomplished by a specialty contractor who understands the federal and state regulations regarding the handling and disposal of the material. The process of asbestos removal is divided into many individual steps. An accurate estimate can be calculated only after all the steps have been priced.

The steps are generally as follows:

1. Obtain an asbestos abatement plan from an industrial hygienist.
2. Monitor the air quality in and around the removal area and along the path of travel between the removal area and transport area. This establishes the background contamination.
3. Construct a two part decontamination chamber at entrance to removal area.
4. Install a HEPA filter to create a negative pressure in the removal area.
5. Install wall, floor and ceiling protection as required by the plan, usually 2 layers of fireproof 6 mil polyethylene.
6. Industrial hygienist visually inspects work area to verify compliance with plan.
7. Provide temporary supports for conduit and piping affected by the removal process.
8. Proceed with asbestos removal and bagging process. Monitor air quality as described in Step #2. Discontinue operations when contaminate levels exceed applicable standards.
9. Document the legal disposal of materials in accordance with EPA standards.
10. Thoroughly clean removal area including all ledges, crevices and surfaces.
11. Post abatement inspection by industrial hygienist to verify plan compliance.
12. Provide a certificate from a licensed industrial hygienist attesting that contaminate levels are within acceptable standards before returning area to regular use.

R022-220   Compacting Backfill

Compaction of fill in embankments, around structures, in trenches, and under slabs is important to control settlement. Factors affecting compaction are:

1. Soil gradation
2. Moisture content
3. Equipment used
4. Depth of fill per lift
5. Density required

The costs for testing and soil analyses are listed in Division 014-108. Also, see Division 022 for further backfill, borrow, and compaction costs.

Example:

Compact granular fill around a building foundation using a 21" wide x 24" vibratory plate in 8" lifts. Operator moves at 50 FPM working a 50 minute hour to develop 95% Modified Proctor Density with 4 passes.

Production Rate:

$$\frac{1.75' \text{ plate width x 50 F.P.M. x 50 min./hr. x .67' lift}}{27 \text{ C.F. per C.Y.}} = 108.5 \text{ C.Y./hr.}$$

Production Rate for 4 Passes:

$$\frac{108.5 \text{ C.Y.}}{4 \text{ passes}} = 27.125 \text{ C.Y./hr. x 8 hrs.} = 217 \text{ C.Y./day}$$

	Compacting 217 C.Y. with 21" Wide Vibratory Plate	L.H./ Day	Hourly Cost	Daily Cost	CY Cost
1	Laborer	8	$13.40	$107.20	$.50
1	Vibratory Plate Compactor			49.60	.23
	Total for 217 C.Y./day			$156.80	$.73

R022-240   Excavating

The selection of equipment used for structural excavation and bulk excavation or for grading is determined by the following factors.
1. Quantity of material.
2. Type of material.
3. Depth or height of cut.
4. Length of haul.
5. Condition of haul road.
6. Accessibility of site.
7. Moisture content and dewatering requirements.
8. Availability of excavating and hauling equipment.

Some additional costs must be allowed for hand trimming the sides and bottom of concrete pours and other excavation below the general excavation.

Number of B.C.Y. per truck = 1.5 C.Y. bucket × 8 passes = 12 loose C.Y.

$$= 12 \times \frac{100}{118} = 10.2 \text{ B.C.Y. per truck}$$

Truck Haul Cycle:

Load truck 8 passes	=	4 minutes
Haul distance 1 mile	=	9 minutes
Dump time	=	2 minutes
Return 1 mile	=	7 minutes
Spot under machine	=	1 minute
		23 minute cycle

When planning excavation and fill, the following should also be considered.
1. Swell factor.
2. Compaction factor.
3. Moisture content.
4. Density requirements.

A typical example for scheduling and estimating the cost of excavation of a 15′ deep basement on a dry site when the material must be hauled off the site, is outlined below.

Assumptions:
1. Swell factor, 18%.
2. No mobilization or demobilization.
3. Allowance included for idle time and moving on job.
4. No dewatering, sheeting, or bracing.
5. No truck spotter or hand trimming.

Fleet Haul Production per day in B.C.Y.

$$4 \text{ trucks} \times \frac{50 \text{ min. hour}}{23 \text{ min. haul cycle}} \times 8 \text{ hrs.} \times 10.2 \text{ B.C.Y.}$$

$$= 4 \times 2.2 \times 8 \times 10.2 = 718 \text{ B.C.Y./day}$$

Excavating Cost with a 1-1/2 C.Y. Hydraulic Excavator 15′ Deep, 2 Mile Round Trip Haul		L.H./Day	Hourly Cost	Daily Cost	Subtotal	Unit Price
1	Equipment Operator	8	$19.75	$ 158.00		
1	Oiler	8	16.20	129.60		
4	Truck Drivers	32	15.40	492.80	$ 780.40	$1.08
1	Hydraulic Excavator			706.55		
4	Dump Trucks			1,726.00	2,432.55	3.38
Total for 720 BCY				$3,212.95	$3,212.95	$4.46

Description		1-1/2 C.Y. Hyd. Backhoe 15′ Deep		1-1/2 C.Y. Power Shovel 7′ Bank		1-1/2 C.Y. Dragline 7′ Deep		2-1/2 C.Y. Trackloader Stockpile
Operator (and Oiler, if required)		$ 287.60		$ 287.60		$ 287.60		$ 287.60
Truck Drivers	3 Ea.	369.60	4 Ea.	492.80	3 Ea.	369.60	4 Ea.	492.80
Equipment Rental		706.55		913.90		932.95		838.00
20 C.Y. Trailer Dump Trucks	3 Ea.	1,294.50	4 Ea.	1,726.00	3 Ea.	1,294.50	4 Ea.	1,726.00
Total Cost per Day		$2,658.25		$3,420.30		$2,884.65		$3,344.40
Daily Production, C.Y. Bank Measure		720.00		960.00		640.00		1000.00
Cost per C.Y.		$ 3.69		$ 3.56		$ 4.51		$ 3.34

Add the mobilization and demobilization costs to the total excavation costs. When equipment is rented for more than three days, there is often no mobilization charge by the equipment dealer. On larger jobs outside of urban areas, scrapers can move earth economically provided a dump site or fill area and adequate haul roads are available. Excavation within sheeting bracing or cofferdam bracing is usually done with a clamshell and production is low, since the clamshell may have to be guided by hand between the bracing. When excavating or filling an area enclosed with a wellpoint system, add 10% to 15% to the cost to allow for restricted access. When estimating earth excavation quantities for structures, allow work space outside the building footprint for construction of the foundation, and a slope of 1:1 unless sheeting is used.

R022-250 Excavating Equipment

The table below lists THEORETICAL hourly production in C.Y./hr. bank measure for some typical excavation equipment. Figures assume 50 minute hours, 83% job efficiency, 100% operator efficiency, 90° swing and properly sized hauling units, which must be modified for adverse digging and loading conditions. Actual production costs in the front of the book average about 50% of the theoretical values listed here.

Equipment	Soil Type	B.C.Y. Weight	% Swell	1 C.Y.	1-1/2 C.Y	2 C.Y.	2-1/2 C.Y.	3 C.Y.	3-1/2 C.Y.	4 C.Y.
Hydraulic Excavator	Moist loam, sandy clay	3400 lb.	40%	85	125	175	220	275	330	380
"Backhoe"	Sand and gravel	3100	18	80	120	160	205	260	310	365
15' Deep Cut	Common earth	2800	30	70	105	150	190	240	280	330
	Clay, hard, dense	3000	33	65	100	130	170	210	255	300
	Moist loam, sandy clay	3400	40	170	245	295	335	385	435	475
				(6.0)	(7.0)	(7.8)	(8.4)	(8.8)	(9.1)	(9.4)
	Sand and gravel	3100	18	165	225	275	325	375	420	460
Power Shovel				(6.0)	(7.0)	(7.8)	(8.4)	(8.8)	(9.1)	(9.4)
Optimum Cut (Ft.)	Common earth	2800	30	145	200	250	295	335	375	425
				(7.8)	(9.2)	(10.2)	(11.2)	(12.1)	(13.0)	(13.8)
	Clay, hard, dense	3000	33	120	175	220	255	300	335	375
				(9.0)	(10.7)	(12.2)	(13.3)	(14.2)	(15.1)	(16.0)
	Moist loam, sandy clay	3400	40	130	180	220	250	290	325	385
				(6.6)	(7.4)	(8.0)	(8.5)	(9.0)	(9.5)	(10.0)
	Sand and gravel	3100	18	130	175	210	245	280	315	375
Drag Line				(6.6)	(7.4)	(8.0)	(8.5)	(9.0)	(9.5)	(10.0)
Optimum Cut (Ft.)	Common earth	2800	30	110	160	190	220	250	280	310
				(8.0)	(9.0)	(9.9)	(10.5)	(11.0)	(11.5)	(12.0)
	Clay, hard, dense	3000	33	90	130	160	190	225	250	280
				(9.3)	(10.7)	(11.8)	(12.3)	(12.8)	(13.3)	(12.0)

Equipment	Soil Type	B.C.Y. Weight	% Swell	Wheel Loaders				Track Loaders		
				3 C.Y.	4 C.Y.	6 C.Y.	8 C.Y.	2-1/4 C.Y.	3 C.Y.	4 C.Y.
	Moist loam, sandy clay	3400	40	260	340	510	690	135	180	250
	Sand and gravel	3100	18	245	320	480	650	130	170	235
Loading Tractors	Common earth	2800	30	230	300	460	620	120	155	220
	Clay, hard, dense	3000	33	200	270	415	560	110	145	200
	Rock, well-blasted	4000	50	180	245	380	520	100	130	180

R025-110   Bituminous Paving

City	Bituminous Asphalt per Ton*	Pavement (3") 6.13 S.Y./ton				Sidewalks (2") 9.2 S.Y./ton			
		Cost per S.Y.			Per Ton	Cost per S.Y.			Per Ton
		Material*	Installation	Total	Total	Material*	Installation	Total	Total
Atlanta	$24.10	$3.93	$.61	$4.54	$27.84	$2.62	$ .85	$3.47	$31.92
Baltimore	27.50	4.49	.64	5.13	31.48	2.99	.96	3.95	36.34
Boston	31.55	5.15	.83	5.98	36.64	3.43	1.57	5.00	46.00
Buffalo	31.00	5.06	.81	5.87	35.97	3.37	1.49	4.86	44.71
Chicago	29.85	4.87	.82	5.69	34.91	3.24	1.57	4.81	44.25
Cincinnati	29.50	4.81	.73	5.54	33.94	3.21	1.26	4.47	41.12
Cleveland	27.80	4.54	.77	5.31	32.58	3.02	1.44	4.46	41.03
Columbus	24.65	4.02	.72	4.74	29.04	2.68	1.23	3.91	35.97
Dallas	25.15	4.10	.59	4.69	28.74	2.73	.78	3.51	32.29
Denver	24.25	3.96	.63	4.59	28.12	2.64	.92	3.56	32.75
Detroit	28.50	4.65	.79	5.44	33.32	3.10	1.46	4.56	41.95
Houston	27.75	4.53	.62	5.15	31.58	3.02	.89	3.91	35.97
Indianapolis	24.00	3.92	.72	4.64	28.42	2.61	1.21	3.82	35.14
Kansas City	23.15	3.78	.70	4.48	27.44	2.52	1.17	3.69	33.95
Los Angeles	29.75	4.85	.81	5.66	34.69	3.23	1.52	4.75	43.70
Memphis	28.10	4.58	.59	5.17	31.72	3.05	.81	3.86	35.51
Milwaukee	25.90	4.23	.79	5.02	30.78	2.82	1.45	4.27	39.28
Minneapolis	26.90	4.39	.77	5.16	31.65	2.92	1.43	4.35	40.02
Nashville	27.70	4.52	.62	5.14	31.53	3.01	.91	3.92	36.06
New Orleans	38.75	6.32	.59	6.91	42.38	4.21	.83	5.04	46.37
New York City	41.50	6.77	.94	7.71	47.24	4.51	1.92	6.43	59.16
Philadelphia	24.15	3.94	.79	4.73	28.96	2.63	1.49	4.12	37.90
Phoenix	23.00	3.75	.65	4.40	26.98	2.50	1.01	3.51	32.29
Pittsburgh	31.20	5.09	.74	5.83	35.73	3.39	1.30	4.69	43.15
St. Louis	23.85	3.89	.79	4.68	28.68	2.59	1.47	4.06	37.35
San Antonio	25.00	4.08	.58	4.66	28.57	2.72	.75	3.47	31.92
San Diego	26.95	4.40	.81	5.21	31.93	2.93	1.52	4.45	40.94
San Francisco	29.35	4.79	.83	5.62	34.43	3.19	1.57	4.76	43.79
Seattle	31.60	5.15	.77	5.92	36.29	3.43	1.43	4.86	44.71
Washington, D.C.	35.00	5.71	.64	6.35	38.95	3.80	.96	4.76	43.79
Average	$28.25	$4.61	$.72	$5.33	$32.68	$3.07	$1.24	$4.31	$39.65

Assumed density is 145 lb. per C.F.

*Includes delivery within 20 miles

Table below shows quantities and bare costs for 1000 S.Y. of Bituminous Paving.

Item	Roads and Parking Areas, 3" Thick (025-104-0460)			Sidewalks, 2" Thick (025-128-0010)		
	Quantities		Cost	Quantities		Cost
Bituminous asphalt	163 tons @ $28.25 per ton		$4,604.75	109 tons @ $28.25 per ton		$3,079.25
Installation using	Crew B-25B @ $3,456.80 /4900SY/ day x 1000		705.47	Crew B-37 @ $845.60 /720 SY/day x 1000		1,174.44
Total per 1000 S.Y.			$5,310.22			$4,253.69
Total per S.Y.			$ 5.33			$ 4.25
Total per Ton			$ 32.58			$ 39.02

R027-110   Concrete Pipe

Prices given are for inside 20 mile delivery zone. Add $1.70 per ton of pipe for each additional 10 miles. Minimum truckload is 10 tons. The non-reinforced pipe listed in the front of the book is designation ASTM C14-59 extra strength. The reinforced pipe listed is ASTM C76-65T class 3, no gaskets. The installation cost given includes shaping bottom of the trench, placing the pipe, and backfilling and tamping to the top of the pipe only.

R029-310   Seeding

The type of grass is determined by light, shade and moisture content of soil plus intended use. Fertilizer should be disked 4″ before seeding. For steep slopes disk five tons of mulch and lay two tons of hay or straw on surface per acre after seeding. Surface mulch can be staked, lightly disked or tar emulsion sprayed. Material for mulch can be wood chips, peat moss, partially rotted hay or straw, wood fibers and sprayed emulsions. Hemp seed blankets with fertilizer are also available. For spring seeding, watering is necessary. Late fall seeding may have to be reseeded in the spring. Hydraulic seeding, power mulching, and aerial seeding can be used on large areas.

R029-545   Cost of Trees: based on Pin Oak (Quercus palustris)

Tree Diameter	Normal Height	Catalog List Price of Tree	Guying Material	Equipment Charge	Installation Labor	Total
2 to 3 inch	14 feet	$ 123	$ 13.50	$ 56.90	$ 48.28	$ 241.26
3 to 4 inch	16 feet	226	15.00	94.83	80.47	416.56
4 to 5 inch	18 feet	335	63.00	113.80	96.56	608.16
6 to 7 inch	22 feet	799	81.00	142.25	120.70	1,143.15
8 to 9 inch	26 feet	1620	100.00	189.67	160.93	2,070.60

Installation Time & Cost for Planting Trees, Bare Costs													
Ball Size Diam. X Depth	Soil in Ball	Weight of Ball	Hole Diam. Req'd	Hole Excavation	Amount of Soil Displ.	Topsoil Handled	Time Required in Labor-Hours						Cost	
							Dig & Lace	Handle Ball	Dig Hole	Plant & Prune	Water & Guy	Total L.H.	Crew	Total per Tree
Inches	C.F.	Lbs.	Feet	C.F.	C.F.	C.F.								
12 x 12	.70	56.00	2.00	4.00	3.00	11.00	.25	.17	.33	.25	.07	1.10	1 Clab	$ 14.74
18 x 16	2.00	160.00	2.50	8.00	6.00	21.00	.50	.33	.47	.35	.08	1.70	2 Clab	22.78
24 x 18	4.00	320.00	3.00	13.00	9.00	38.00	1.00	.67	1.08	.82	.20	3.80	3 Clab	50.92
30 x 21	7.50	600.00	4.00	27.00	19.50	76.00	.82	.71	.79	1.22	.26	3.80		90.78
36 x 24	12.50	980.00	4.50	38.00	25.50	114.00	1.08	.95	1.11	1.32	.30	4.76		113.72
42 x 27	19.00	1,520.00	5.50	64.00	45.00	185.00	1.90	1.27	1.87	1.43	.34	6.80	B-6	162.45
48 x 30	28.00	2,040.00	6.00	85.00	57.00	254.00	2.41	1.60	2.06	1.55	.39	8.00	@	191.12
54 x 33	38.50	3,060.00	7.00	127.00	88.50	370.00	2.86	1.90	2.39	1.76	.45	9.40	$23.89	224.57
60 x 36	52.00	4,160.00	7.50	159.00	107.00	474.00	3.26	2.17	2.73	2.00	.51	10.70	per	255.62
66 x 39	68.00	5,440.00	8.00	196.00	128.00	596.00	3.61	2.41	3.07	2.26	.58	11.90	Labor-	284.29
72 x 42	87.00	7,160.00	9.00	267.00	180.00	785.00	3.90	2.60	3.71	2.78	.70	13.70	hour	327.29

R033-090   Placing Ready Mixed Concrete

For ground pours allow for 5% waste when figuring quantities.

Prices in the front of the book assume normal deliveries. If deliveries are made before 8 A.M. or after 5 P.M. or on Saturday afternoons add $21 per C.Y. Large volume discounts are not included in prices in front of book.

For the lower floors without truck access, concrete may be wheeled in rubber tired buggies, conveyer handled, crane handled or pumped. Pumping is economical if there is top steel. Conveyers are more efficient for thick slabs. Concrete pump with an operator can be rented from $585 per day for up to 25 C.Y. to $1,375 per day for a 400 C.Y. pour. Figures include travel time if done at straight time. Pumping lightweight concrete costs an extra $2.20 per C.Y.

At higher floors the rubber tired buggies may be hoisted by a hoisting tower then wheeled to location. Placement by a conveyer is limited to three floors and is best for high volume pours. Pumped concrete is best when building has no crane access. Concrete may be pumped directly as high as thirty-six stories using special pumping techniques. Normal maximum height is about fifteen stories.

Best pumping aggregate is screened and graded bank gravel rather than crushed stone.

Pumping downward is more difficult than pumping upwards. Horizontal distance from pump to pour may increase preparation time prior to pour. Placing by cranes, either mobile, climbing or tower types continues as the most efficient method for high rise concrete buildings.

	Cost per C.Y. for Wheeled Concrete, Dumped Only (Add to appropriate placing cost)							
	10 C.F. Walking Cart				18 C.F. Riding Cart			
Item	Hourly Cost	Wheeled up to 50 ft.	Wheeled up to 150 ft.	Wheeled up to 250 ft.	Hourly Cost	Wheeled up to 50 ft.	Wheeled up to 150 ft.	Wheeled up to 250 ft.
Laborer	$13.40	$3.35	$4.48	$5.96	$13.40	$1.34	$1.74	$2.28
.125 Labor foreman	1.93	.48	.64	.86	1.93	.19	.25	.33
Concrete cart	6.33	1.58	2.11	2.81	10.60	1.06	1.38	1.80
Total Cost/C.Y.		$5.41	$7.23	$9.63		$2.59	$3.37	$4.41
Hourly production		4 C.Y.				10 C.Y.		

R034-010   Precast Concrete Wall Panels

Panels are either solid or insulated with plain, colored or textured finishes. Transportation is an important cost factor. Prices below are based on delivery within 50 miles of a plant including fabricators' overhead and profit. Engineering data is available from fabricators to assist with construction details. Usual minimum job size for economical use of panels is about 5000 S.F. Small jobs can double the prices below. For large, highly repetitive jobs, deduct up to 15% from the prices below.

Panel Cost and Maximum Size   Base price for panels based on 50 S.F. or larger panels is as follows:

Thickness	Cost per S.F.	Maximum Size	Thickness	Cost per S.F.	Maximum Size
3″	$ 7.60	50 S.F.	6″	$10.50	300 S.F.
4″	8.30	150 S.F.	7″	10.70	300 S.F.
5″	10.30	200 S.F.	8″	10.95	300 S.F.

The above prices are for gray smooth form or board finish one side. Add $1.55 per S.F. for white facing; $8.05 per C.F. for solid white for full thickness. For fluted surface (not broken) add $.45 to $1.00 per S.F. For broken rib finish add $.80 to $1.50 to fluted surface. Add $.35 to $1.25 per S.F. for exposed local aggregate and $1.00 to $3.80 per S.F. for special facing aggregates.

Sandblasting runs $.55 to $1.00 per S.F., and bush hammering runs $1.70 to $3.35 per S.F. Loose hardware (not included above) runs $1.00 to $2.00 per S.F.

2″ thick panels cost about the same as 3″ thick panels and maximum panel size is less. For building panels faced with granite, marble or stone, add the material prices from Division 044 to the plain panel price above. There is a growing trend toward aggregate facings and broken rib finish rather than plain gray concrete panels.

Composite Panels   Add to above panel prices for core insulation, mesh and shear ties per S.F.

Type	1″ Thick	1-1/2″ Thick	2″ Thick
Fiberglass	$1.65	$2.30	$2.95
Polystyrene (E.P.S. Board)	1.10	1.55	2.00

Erection   Table shows cost ranges for erection including Subcontractor's O & P using a six man crew, crane, operator & oiler.

Panel Size L x H	Area per Panel	Low Rise Hyd. 55 Ton Crane @ $4,260 Daily Cost Daily Production Range Pieces	Area	Erection Cost Per Piece	Per S.F.	High Rise Daily Production Range Pieces	Area	55 Ton Crane $4,260 Daily Cost Piece Cost	S.F. Cost	90 Ton Crane $4,500 Daily Cost Piece Cost	S.F. Cost	150 Ton Crane $4,875 Daily Cost Piece Cost	S.F. Cost
4′ x 4′	16 S.F.	10	160 S.F.	$370	$23.15	9	144 S.F.	$411	$25.69	$437	$27.31	$ 511	$31.94
		20	320	185	11.55	18	288	206	12.85	218	13.65	256	15.97
4′ x 8′	32	10	320	370	11.55	9	288	411	12.85	437	13.65	511	15.97
		19	608	195	6.10	17	544	218	6.80	231	7.23	271	8.46
8′ x 8′	64	9	576	411	6.40	8	512	463	7.23	492	7.68	575	8.98
		18	1152	206	3.20	16	1024	231	3.61	246	3.84	288	4.49
10′ x 10′	100	8	800	463	4.65	7	700	529	5.29	562	5.62	657	6.57
		15	1500	247	2.45	14	1400	264	2.64	281	2.81	329	3.29
15′ x 10′	150	7	1050	529	3.50	6	900	—	—	655	4.37	767	5.11
		12	1800	308	2.05	11	1650	—	—	358	2.38	418	2.79
20 x 10′	200	6	1200	617	3.10	5	1000	—	—	787	3.93	920	4.60
		8	1600	463	2.30	7	1400	—	—	562	2.81	657	3.29
30′ x 10′	300	5	1500	740	2.45	4	1200	—	—	983	3.28	1150	3.83
		7	2100	529	1.75	7	2100	—	—	562	1.87	657	2.19

Total Cost in Place for Low Rise Construction Including Subcontractor's O & P

Description	Gray 4′ x 8′ x 4″	20′ x 10′ x 6″	White Face 4′ x 8′ x 4″	20′ x 10′ x 6″	Exposed Aggregate 4′ x 8′ x 4″	20′ x 10′ x 6″
Panel, steel form, broomed finish	$ 8.30	$10.50	$ 9.65	$11.75	$ 9.00	$11.25
Caulking, grout, etc. (runs about $2.45 per L.F.)	.95	.36	.90	.36	.90	.36
Erect, plumb, align (from above)	11.55	3.10	11.55	3.10	11.55	3.10
Total in place per S.F.	$20.80	$13.96	$22.10	$15.21	$21.45	$14.71

No allowance has been made for supporting steel framework. On one story buildings, panels may rest on grade beams and require only wind bracing and fasteners. On multi-story buildings panels can span from column to column and floor to floor. Plastic designed steel framed structures may have large deflections which slow down erection and raise costs.

Large panels are more economical than small panels on a S.F. basis. When figuring areas include all protrusions, returns, etc. Overhangs can triple erection costs. Panels over 45′ have been produced. Larger flat units should be prestressed. Vacuum lifting of smooth finish panels eliminates inserts and can speed erection.

R034-020 Tilt Up Concrete Panels

The advantage of tilt up construction is in the low cost of forms and the placing of concrete and reinforcing. Panels up to 75' high and 5-1/2" thick have been tilted using strongbacks. Tilt up has been used for one to five story buildings and is well suited for warehouses, stores, offices, schools and residences.

The panels are cast in forms on the floor slab. Most jobs use 5-1/2" thick solid reinforced concrete panels. Sandwich panels with a layer of insulating materials are also used. Where dampness is a factor, lightweight aggregate is used at an added cost of $.42 per square foot. Optimum panel size is 300 to 500 S.F.

Slabs are usually poured with 3000 psi concrete which permits tilting seven days after pouring. Slabs may be stacked on top of each other and are separated from each other by either two coats of bond breaker or a film of polyethylene. Use of high early strength cement allows tilting two days after a pour. Tilting up is done with a roller outrigger crane with a capacity of at least 1-1/2 times the weight of the panel at the required reach. Exterior precast columns can be set at the same time as the panels; interior precast columns can be set first and the panels clipped directly to them. The use of cast-in-place concrete columns is diminishing due to shrinkage problems. Structural steel columns are sometimes used if crane rails are planned. Panels can be clipped to the columns or lowered between the flanges. Steel channels with anchors may be used as edge forms for the slab. When the panels are lifted the channels form an integral steel column to take structural loads. Roof loads can be carried directly by the panels for wall heights to 14'. For soft ground requiring mats for cranes, add 100% to costs below.

Below are typical costs per S.F., for panels of 300 to 500 S.F. 20' high, not including contractor's overhead and profit.

Item	5-1/2" Thick			7-1/2" Thick		
	Material	Installation	Total	Material	Installation	Total
Prepare pouring surface	$ .02	$ .05	$ .07	$ .02	$ .05	$ .07
Erect, strip side forms	.07	.20	.27	.12	.27	.39
Place concrete, 3000 psi	1.00	.09	1.09	1.36	.31	1.67
Steel trowel finish & curing	$ .02	.23	.25	.02	.23	.25
Reinforcing, inserts & misc. items	1.03	.38	1.41	1.43	.51	1.94
Panel erection and aligning	—	.65	.65	—	.65	.65
Total per S.F. of Wall	$ 2.14	$ 1.60	$ 3.74	$ 2.95	$ 2.02	$ 4.97
Site precast concrete columns, add	1.07	.88	1.95	1.03	.88	1.91
Total per S.F. of Wall	$ 3.21	$ 2.48	$ 5.69	$ 3.98	$ 2.90	$ 6.88
Panels only, per C.Y.	$115.56	$86.40	$201.96	$127.44	$87.26	$214.70

Requirements of local building codes may be a limiting factor and should be checked. Building floor slabs should be poured first and should be a minimum of 5" thick with 100% compaction of soil or 6" thick with less than 100% compaction.

Setting times as fast as nine minutes per panel have been observed, but a safer expectation would be four panels per hour with a crane and a four person setting crew. If crane erects from inside building, some provision must be made to get crane out after walls are erected. Good yarding procedure is important to minimize delays. Equalizing three point lifting beams and self-releasing pick-up hooks speed erection. If panels must be carried to their final location, setting time per panel will be increased and erection costs may fall in the erection cost range of architectural precast wall panels. Placing panels into slots formed in continuous footers will speed erection.

Reinforcing should be with #5 bars with vertical bars on the bottom. If surface is to be sandblasted, stainless steel chairs should be used to prevent rust staining.

Use of a broom finish is popular since the unavoidable surface blemishes are concealed. Many West Coast jobs have used exposed aggregate panels which cost an additional $.50 to $3.60 per S.F. depending on the aggregate and construction method employed. Eastern prices for exposed aggregate tend to be considerably higher. Panel connections run $1.60 to $6.00 per L.F. Precast columns run from three to five times the C.Y. price of the panels only.

R035-010 Lightweight Concrete

Vermiculite or Perlite come in bags of 4 C.F. under various trade names. Weight is about 8 lbs. per C.F. For insulating roof fill use 1:6 mix. For structural deck use 1:4 mix over gypsum boards, steeltex, steel centering, etc. supported by closely spaced joists or bulb trees. For structural slabs use 1:3:2 vermiculite sand concrete over steeltex, metal lath, steel centering, etc. on joists spaced 2′-0″ O.C. for maximum L.L. of 80#/S.F. Use same mix for slab base fill over steel flooring or regular reinforced concrete slab when tile, terrazzo or other finish is to be laid over.

For slabs on grade use 1:3:2 mix when tile, etc. finish is to be laid over. If radiant heating units are installed use a 1:6 mix for a base. After coils are in place, cover with a regular granolithic finish (mix 1:3:2) to a minimum depth of 1-1/2″ over top of units.

Reinforce all slabs with 6 x 6 or 10 x 10 welded wire mesh.

Vermiculite concrete can be purchased ready mixed but the following breakdown is included for field mix. Prices given below are for a one story building and assume 50 C.Y. or more. For less than 50 C.Y. add 10%. For over one story add $3.25 per C.Y. Screed finish cost is included below. Ready mix 1:6 costs $84.10 per C.Y. delivered in truckload lots.

See R33-070 for prices of ready mix lightweight concrete.

Quantities per C.Y., Field Mix			(Line 0110) 1:6 Mix Insulating Roof Fill		1:3:2 Mix Lightweight Structural Concrete	
Portland cement @	$ 7.00	per bag	5.0 bags	$ 35.00	6.2 bags	$ 43.40
Vermiculite or Perlite @	9.45	per bag	7.5 bags	70.88		
treated type @	10.75	per bag			4.7 bags	50.53
Sand @	16.65	per C.Y.			12.5 C.F.	7.71
Plant and Water				6.37		6.37
Labor, machine mix, hoist and place, Crew C-8 @	$27.04	per L.H.	1.12 L.H.	30.28	.91 L.H.	24.61
Total in place, per C.Y.				$142.53		$132.62

R041-100   Cement Mortar (material only)

Type N - 1:1:6 mix by volume. Use everywhere above grade except as noted below.

- 1:3 mix using conventional masonry cement which saves handling two separate bagged materials.

Type M - 1:1/4:3 mix by volume, or 1 part cement, 1/4 (10% by wt.) lime, 3 parts sand. Use for heavy loads and where earthquakes or hurricanes may occur. Also for reinforced brick, sewers, manholes and everywhere below grade.

Cost and Mix Proportions of Various Types of Mortar

Components	Type Mortar and Mix Proportions by Volume										
	M		S		N		O		K	PM	PL
	1:1:6	1:1/4:3	1/2:1:4	1:1/2:4	1:3	1:1:6	1:3	1:2:9	1:3:12	1:1:6	1:1/2:4
Portland cement @ $7.00 per bag	$ 7.00	$ 7.00	$ 3.50	$ 7.00	—	$ 7.00	—	$ 7.00	$ 7.00	$ 7.00	$ 7.00
Masonry cement @ $5.40 per bag	5.40	—	5.40	—	$5.40	—	$5.40	—	—	5.40	—
Lime @ $5.15 per 50 lb. bag	—	1.29	—	2.58	—	5.15	—	10.30	15.45	—	2.58
Masonry sand @ $18.50 per C.Y.*	4.11	2.06	2.74	2.74	2.06	4.11	2.06	6.17	8.22	4.11	2.74
Mixing machine incl. fuel**	1.51	.76	1.01	1.01	.76	1.51	0.76	2.27	3.02	1.51	1.01
Total for Materials	$18.02	$11.11	$12.65	$13.33	$8.22	$17.77	$8.22	$25.74	$33.69	$18.02	$13.33
Total C.F.	6	3	4	4	3	6	3	9	12	6	4
Approximate Cost per C.F.	$ 3.00	$ 3.70	$ 3.16	$ 3.33	$2.74	$ 2.96	$2.74	$ 2.86	$ 2.81	$ 3.00	$ 3.33

*Includes 10 mile haul
**Based on a daily rental, 10 C.F., 25 H.P. mixer, mix 200 C.F./Day

Mix Proportions by Volume, Compressive Strength and Cost of Mortar

Where Used	Mortar Type	Allowable Proportions by Volume				Compressive Strength @ 28 days	Cost per Cubic Foot
		Portland Cement	Masonry Cement	Hydrated Lime	Masonry Sand		
Plain Masonry	M	1	1	—	6		$3.00
		1	—	1/4	3	2500 psi	3.70
	S	1/2	1	—	4		3.16
		1	—	1/4 to 1/2	4	1800 psi	3.33
	N	—	1	—	3		2.74
		1	—	1/2 to 1-1/4	6	750 psi	2.96
	O	—	1	—	3		2.74
		1	—	1-1/4 to 2-1/2	9	350 psi	2.86
	K	1	—	2-1/2 to 4	12	75 psi	2.81
Reinforced Masonry	PM	1	1	—	6	2500 psi	3.00
	PL	1	—	1/4 to 1/2	4	2500 psi	3.33

Note: The total aggregate should be between 2.25 to 3 times the sum of the cement and lime used.

The labor cost to mix the mortar is included in the labor cost on brickwork.

Machine mixing is usually specified on jobs of any size. There is a large price saving over hand mixing and mortar is more uniform.

There are two types of mortar color used. Prices in Section 041-012 are for the inert additive type with about 100 lbs. per M brick as the typical quantity required. These colors are also available in smaller batch size bags (1 lb. to 15 lb.) which can be placed directly into the mixer without measuring. The other type is premixed and replaces the masonry cement with ranges in price from $5 to $15 per 70 lb. bag. Dark green color has the highest cost.

R041-200   Miscellaneous Mortar (material only)

Quantities	Glass Block Mortar		Gypsum Cement Mortar	
White Portland cement at $18.00 per bag	7 bags	$126.00		
Gypsum cement at $10.90 per 80 lb. bag			11.25 bags	$122.63
Lime at $5.15 per 50 lb. bag	280 lbs.	28.84		
Sand at $18.50 per C.Y.*	1 C.Y.	18.50	1 C.Y.	18.50
Mixing machine and fuel		6.80		6.80
Total per C.Y.		$180.14		$147.93
Approximate Total per C.F.		$ 6.67		$ 5.48

* Includes 10 mile haul

R041-500 Masonry Reinforcing

Horizontal joint reinforcing helps prevent wall cracks where wall movement may occur. Reinforcing strips come in 10′ and 12′ lengths. Field labor runs between 2.7 to 5.3 hours per 1000 L.F. for wall thicknesses up to 12″.

Table below is cost per 1000 L.F., material only, all galvanized (mill std.). For hot dip galvanizing, add 65%.

Galvanized Reinforcing Size		Type	Wall Thickness						
			4″	6″	8″	10″	12″	14″	16″
9 ga. sides, 9 ga. ties		Regular	$120	$120	$125	$130	$135	$150	$155
3/16″ sides, 9 ga. ties		truss	150	160	165	170	175	190	195
3/16″ sides, 3/16″ ties			200	205	210	220	230	250	275
9 ga. sides, 9 ga. ties		Cavity	120	125	130	140	150	175	180
3/16″ sides, 9 ga. ties		truss	150	160	170	180	185	200	210
9 ga. sides, 9 ga. ties		Ladder	90	95	100	105	110	130	170
9 ga. sides, 9 ga. ties		Cavity	100	120	120	130	150	160	190
3/16″ sides, 9 ga. ties		Ladder	150	155	160	175	180	200	210

R042-050 Brick Chimneys

Quantities	16″ x 16″		20″ x 20″		20″ x 24″		20″ x 32″	
Brick at $300 per M	28 brick	$ 8.40	37 brick	$11.10	42 brick	$12.60	51 brick	$15.30
Type M mortar at $3.70 per C.F.	.5 C.F.	1.85	.6 C.F.	2.22	1.0 C.F.	3.70	1.3 C.F.	4.81
Flue tile (square)(per foot)	8″ x 8″	2.75	12″ x 12″	5.00	2 @ 8″ x 12″	8.00	2 @ 12″ x 12″	10.00
Install tile & brick, crew D-1	.055 day	14.78	.073 day	19.62	.083 day	22.31	.10 day	26.88
Total per L.F. high		$27.78		$37.94		$46.61		$56.99

Material costs include 3% waste for brick and 25% waste for mortar.

Labor costs are bare costs and do not include contractor's O&P.

Labor for chimney brick using D-1 crew is 31 hours per thousand brick or about $500 per thousand brick. An 8″ x 12″ flue takes 33 brick and two 8″ x 8″ flues take 37 brick.

R042-100 Economy in Bricklaying

Have adequate supervision. Be sure bricklayers are always supplied with materials so there is no waiting. Place best bricklayers at corners and openings.

Use only screened sand for mortar. Otherwise, labor time will be wasted picking out pebbles. Use seamless metal tubs for mortar as they do not leak or catch the trowel. Locate stack and mortar for easy wheeling.

Have brick delivered for stacking. This makes for faster handling, reduces chipping and breakage, and requires less storage space. Many dealers will deliver select common in 2′ x 3′ x 4′ pallets or face brick packaged. This affords quick handling with a crane or forklift and easy tonging in units of ten, which reduces waste.

Use wider bricks for one wythe wall construction. Keep scaffolding away from wall to allow mortar to fall clear and not stain wall.

On large jobs develop specialized crews for each type of masonry unit.

Consider designing for prefabricated panel construction on high rise projects.

Avoid excessive corners or openings. Each opening adds about 50% to labor cost for area of opening.

Bolting stone panels and using window frames as stops reduces labor costs and speeds up erection.

R042-120 Common and Face Brick Prices

Prices are based on truckload lot purchases for Common Brick and carload lots for Face Brick. Prices are per M, (thousand), brick.

City	Material			Installation				Total			
	Brick per M Delivered		Mortar 3/8" Joint	Common in 8" Wall		Face Brick, 4" Veneer		Common in 8" Wall		Face Brick, 4" Veneer	
	Common	Face		Bare Costs	Incl. O & P	Bare Costs	Incl. O & P	Bare Costs	Incl. O & P	Bare Costs	Incl. O & P
Atlanta	$180	$325	$37.40	$254	$ 430	$305	$ 516	$477	$ 675	$ 670	$ 918
Baltimore	225	250	for 8" Wall	307	519	368	623	576	815	656	940
Boston	295	395	and	536	909	644	1,090	878	1,284	1,081	1,572
Buffalo	295	380	$30.80	457	775	549	930	799	1,150	971	1,394
Chicago	250	295	for 4" Wall	482	817	579	980	777	1,141	913	1,348
Cincinnati	180	290		373	633	448	759	596	878	778	1,122
Cleveland	225	330		428	724	513	869	697	1,021	884	1,277
Columbus	195	250		366	619	439	743	604	881	727	1,060
Dallas	190	320		240	407	288	488	473	663	649	884
Denver	190	335		298	505	358	606	531	762	734	1,020
Detroit	240	320		466	789	559	947	750	1,102	919	1,343
Houston	240	275		280	474	336	568	564	787	650	914
Indianapolis	245	340		373	632	448	758	663	951	829	1,177
Kansas City	210	410		350	592	420	711	603	871	873	1,209
Los Angeles	240	285		473	802	568	962	758	1,115	892	1,319
Memphis	190	340		251	426	302	511	484	682	683	930
Milwaukee	270	365		436	739	523	886	751	1,086	930	1,334
Minneapolis	245	320		434	736	521	883	724	1,054	881	1,279
Nashville	175	215		267	453	321	543	485	692	573	821
New Orleans	200	380		236	400	283	480	480	668	706	944
New York City	290	335		636	1,078	764	1,293	972	1,448	1,139	1,707
Philadelphia	235	285		470	796	564	955	749	1,104	888	1,312
Phoenix	320	410		281	476	337	572	648	880	791	1,070
Pittsburgh	210	230		410	695	492	834	664	974	760	1,128
St. Louis	225	255		438	743	526	891	708	1,039	820	1,214
San Antonio	265	305		229	389	275	466	540	730	620	846
San Diego	275	300		412	698	494	837	732	1,050	834	1,211
San Francisco	325	385		548	929	658	1,115	921	1,338	1,085	1,585
Seattle	335	425		433	734	520	880	815	1,154	988	1,396
Washington, D.C.	240	255		300	508	360	610	585	821	654	933
Average	$235	$305	▼	$380	$ 650	$460	$ 777	$670	$ 960	$ 820	$1,174

Common building brick manufactured according to ASTM C62 and facing brick manufactured according to ASTM C216 are the two standard bricks available for general building use. Building brick is made in three grades; SW, where high resistance to damage caused by cyclic freezing is required; MW, where moderate resistance to cyclic freezing is needed; and NW, where little resistance to cyclic freezing is needed. Facing brick is made in only the two grades SW and MW. Additionally, facing brick is available in three types; FBS, for general use; FBX, for general use where a higher degree of precision and lower permissible variation in size than FBS is needed; and FBA, for general use to produce characteristic architectural effects resulting from non-uniformity in size and texture of the units.

In figuring above installation costs, a D-8 Crew (with a daily output of 1.5 M) was used for the 4" veneer. A D-8 Crew (with a daily output of 1.8 M) was used for the 8" solid wall.

In figuring the total cost including overhead and profit, an allowance of 10% was added to the sum of the cost of the brick and mortar. Also, 3% breakage was included for both the bare costs and the costs with overhead and profit. If bricks are delivered palletized with 280 to 300 per pallet, or packaged, allow only 1-1/2% for breakage. Then add $10 per M to the cost of brick and deduct two hours helper time. The net result is a savings of $30 to $40 per M in place. Packaged or palletized delivery is practical when a job is big enough to have a crane or other equipment available to handle a package of brick. This is so on all industrial work but not always true on small commercial buildings.

There are many types of red face brick. The prices above are for the most usual type used in commercial, apartment house or industrial construction. If it is possible to obtain the price of the actual brick to be used, it should be done and substituted in the table. The use of buff and gray face is increasing, and there is a continuing trend to the Norman, Roman, Jumbo and SCR brick.

See R042-500 for brick quantities per S.F. and mortar quantities per M brick. (Average prices for the various sizes are listed in Division 4)

Common red clay brick for backup is not used that often. Concrete block is the most usual backup material with occasional use of sand lime or cement brick. Sand lime cost about $15 per M less than red clay and cement brick are about $5 per M less than red clay. These figures may be substituted in the common brick breakdown for the cost of these items in place, as labor is about the same. Occasionally common brick is being used in solid walls for strength and as a fire stop.

Brick panels built on the ground floor and then crane erected to the upper floors have proven to be economical. This allows the work to be done under cover and without scaffolding.

R042-180   Brick in Place

Table below is for common bond with 3/8" concave joints and
includes 3% waste for brick and 25% waste for mortar.
Crew costs are bare costs.

Item	8" Common Brick Wall 8" x 2-2/3" x 4"		Select Common Face 8" x 2-2/3" x 4"		Red Face Brick 8" x 2-2/3" x 4"	
1030 brick delivered	$250 per M	$247.20	$305 per M	$309.00	$330 per M	$329.60
Type N mortar @ $2.99 per C.F.	12.5 C.F.	37.00	10.3 C.F.	30.49	10.3 C.F.	30.49
Installation using indicated crew	Crew D-8 @ .556 Days	383.20	Crew D-8 @ .667 Days	459.70	Crew D-8 @ .667 Days	459.7
Total per M in place		$667.40		$799.19		$819.79
Total per S.F. of wall	13.5 bricks/S.F.	$ 9.01	6.75 bricks/S.F.	$ 5.39	6.75 bricks/S.F.	$ 5.53

R042-185   Reinforced Brick in Walls

Table below is for common bond with 3/8" concave joints and includes 3%
waste. Standard 8" x 2-2/3" x 4" bricks.

Item	8" to 9" Thick Wall		16" to 17" Thick Wall	
1030 brick (select common)	$305 per M	$309.00	$280 per M	$309.00
Type PL mortar @ $3.33 per C.F.	12.5 C.F.	41.63	13.9 C.F.	46.29
Reinforcing bars delivered	50 lb. @ .26 per lb.	12.88	50 lb. @ .26 per lb.	12.88
Installation incl. reinforcing using indicated crew	Crew D-8 @ .517 Days	356.32	Crew D-8 @ .513 Days	353.56
Total per M in place		$719.83		$721.72
Total per S.F. of wall	8" wall, 13.5 brick/S.F.	$ 9.72	16" wall, 27.0 brick/S.F.	$ 19.49

R042-200   Concrete Block

8″ x 16″ block, sand aggregate blocks with 3/8″ joints for partitions.

| City | Material | | | | City | Material | | | |
| | Per Block, Delivered | | 113 Block, Delivered | | | Per Block, Delivered | | 113 Block, Delivered | |
	4″ Thick	8″ Thick	4″ Thick	8″ Thick		4″ Thick	8″ Thick	4″ Thick	8″ Thick
Atlanta	$.70	$ .97	$79.10	$109.61	Memphis	.65	.90	$73.45	$101.70
Baltimore	.63	.86	71.19	97.18	Milwaukee	.62	.90	70.06	101.70
Boston	.59	.85	66.67	96.05	Minneapolis	.72	.91	81.36	102.83
Buffalo	.66	.98	74.58	110.74	Nashville	.76	1.00	85.88	113.00
Chicago	.59	.87	66.67	98.31	New Orleans	.79	1.10	89.27	124.30
Cincinnati	.59	.85	66.67	96.05	New York City	.73	1.00	82.49	113.00
Cleveland	.66	.92	74.58	103.96	Philadelphia	.68	.80	76.84	90.40
Columbus	.59	.80	66.67	90.40	Phoenix	.60	.87	67.80	98.31
Dallas	.71	1.10	80.23	124.30	Pittsburgh	.70	.94	79.10	106.22
Denver	.76	1.20	85.88	135.60	St. Louis	.70	1.08	79.10	122.04
Detroit	.73	.98	82.49	110.74	San Antonio	.75	1.10	84.75	124.30
Houston	.69	.95	77.97	107.35	San Diego	.61	.88	68.93	99.44
Indianapolis	.70	1.02	79.10	115.26	San Francisco	.80	1.32	90.40	149.16
Kansas City	.78	1.20	88.14	135.60	Seattle	.85	1.30	96.05	146.90
Los Angeles	.62	.90	70.06	101.70	Washington D.C.	.65	1.00	73.45	113.00
					Average	$.69	$ .99	$77.63	$111.31

The cost of sand aggregate block is based on the delivered price of 113 blocks, which is the equivalent of a 100 S.F. wall area. Mortar for a 100 S.F. wall area will average $10.50 for 4″ thick block and $21.10 for 8″ thick block.

Cost for 100 S.F. of 8″ x 16″ Concrete Block Partitions to Four Stories High, Tooled Joint One Side						
8″ x 16″ Sand Aggregate	4″ Thick Block		8″ Thick Block		12″ Thick Block	
113 block delivered	$.69 Ea.	$ 76.84	$.99 Ea.	$110.74	$1.52 Ea.	$168.37
Type N mortar at $2.96 per C.F.	3.5 C.F.	10.47	7.0 C.F.	20.93	10.4 C.F.	31.10
Installation Crew (Bare Costs)	D-2 @ 9.32 L.H.	160.58	D-2 @ 10.68 L.H.	184.02	D-6 @ 14.70 L.H.	296.35
Total per 100 S.F.		$247.89		$315.69		$495.82
Add for filling cores solid	6.7 C.F.	$ 77.76	25.8 C.F.	$170.90	42.2 C.F.	$231.88

Cost for 100 S.F. of 8″ x 16″ Concrete Block Backup with Trowel Cut Joints						
8″ x 16″ Sand Aggregate	4″ Thick Block		8″ Thick Block		12″ Thick Block	
113 block delivered	$.69 Ea.	$ 76.84	$.99 Ea.	$110.74	$1.52 Ea.	$168.37
Type N mortar at $2.96 per C.F.	5.8 C.F.	17.34	9.3 C.F.	27.81	12.7 C.F.	37.97
Installation Crew (Bare Costs)	D-2 @ 9.20 L.H.	158.52	D-2 @ 10.10 L.H.	174.02	D-6 @ 13.20 L.H.	266.11
Total per 100 S.F.		$252.70		$312.57		$472.45

Special block: corner, jamb and head block are same price as ordinary block of same size. Tabulated on the next page are national average prices per block. Labor on specials is about the same as equal sized regular block. Bond beam and 16″ high lintel blocks cost 30% more than regular units of equal size. Lintel blocks are 8″ long and 8″ or 16″ high. Costs in individual cities may be factored from the table above.

Use of motorized mortar spreader box will speed construction of continuous walls.

R042-200   Concrete Block (cont.)

Size in Inches	Type	Net Area Strength in PSI, Material Cost per Block									Lightweight	
		2000	2500	3000	3500	4000	4500	5000	5500	6000	Std.	Prem.
2 x 8 x 16	Solid	.80	.85	.65	.65	.75	.78	.82	.85	.90	.70	.82
3 x 8 x 16	Solid	.97	.98	.85	.88	1.05	1.10	1.16	1.25	1.20	1.15	1.72
4 x 8 x 16	Hollow	.69	.74	.70	.70	.85	.90	.86	.93	.95	.80	.85
	75% Solid	.96	.95	1.00	.99	.98	1.10	1.20	1.25	1.30	1.05	1.50
	Solid	1.00	1.05	.95	.99	1.05	1.15	1.20	1.26	1.35	1.19	1.20
6 x 8 x 16	Hollow	.81	.85	.88	.88	1.00	1.05	1.10	1.18	1.25	1.05	1.08
	75% Solid	1.15	1.18	1.30	1.32	1.35	1.40	1.40	1.52	1.60	1.45	1.72
	Solid	1.28	1.40	1.35	1.38	1.55	1.65	1.70	1.80	1.80	1.55	1.90
8 x 8 x 16	Hollow	.99	1.00	1.00	1.07	1.25	1.37	1.45	1.55	1.62	1.40	1.10
	75% Solid	1.32	1.42	1.55	1.62	1.57	1.65	1.65	1.72	1.85	1.80	2.30
	Solid	1.60	1.70	1.75	1.80	1.90	2.00	2.00	2.12	2.15	2.10	2.45
	Bond Beam	1.22	—	—	—	—	—	—	—	—	1.75	2.22
	Sash	1.19	—	—	—	—	—	—	—	—	1.50	1.80
	Interlocking	1.24	—	1.50	—	1.52	—	—	—	—	—	—
10 x 8 x 16	Hollow	2.00	1.65	1.52	1.55	1.80	1.90	2.05	2.10	2.25	1.99	1.45
	75% Solid	2.00	2.05	2.16	2.25	2.30	2.40	2.45	2.50	2.60	2.50	3.00
	Solid	2.42	2.55	1.85	1.86	2.82	2.80	2.90	2.95	3.10	3.00	3.65
12 x 8 x 16	Hollow	1.52	1.65	1.50	1.55	1.80	1.85	2.05	2.05	2.15	2.06	1.50
	75% Solid	2.05	2.15	2.32	2.40	2.60	2.60	2.65	2.75	2.90	2.65	3.15
	Solid	2.45	2.60	2.15	2.20	2.82	2.95	3.05	3.10	3.20	3.20	3.80
	Bond Beam	2.08	—	—	—	—	—	—	—	—	2.50	3.10
	Sash	1.65	—	—	—	—	—	—	—	—	2.20	2.80
	Fire Rated	1.70	—	—	—	—	—	—	—	—	2.80	2.60
	Interlocking	1.55	—	1.70	—	2.00	—	—	—	—	—	—
16 x 8 x 16	Interlocking	2.40	—	2.50	—	2.80	—	—	—	—	—	—

R042-500   Brick, Block & Mortar Quantities

Type Brick	Nominal Size (incl. mortar) L H W	Modular Coursing	Number of Brick per S.F.	C.F. of Mortar per M Bricks, Waste Included 3/8" Joint	1/2" Joint	Bond Type	Description	Factor
Standard	8 x 2-2/3 x 4	3C=8"	6.75	10.3	12.9	Common	full header every fifth course	+20%
Economy	8 x 4 x 4	1C=4"	4.50	11.4	14.6		full header every sixth course	+16.7%
Engineer	8 x 3-1/5 x 4	5C=16"	5.63	10.6	13.6	English	full header every second course	+50%
Fire	9 x 2-1/2 x 4-1/2	2C=5"	6.40	550 # Fireclay	—	Flemish	alternate headers every course	+33.3%
Jumbo	12 x 4 x 6 or 8	1C=4"	3.00	23.8	30.8		every sixth course	+5.6%
Norman	12 x 2-2/3 x 4	3C=8"	4.50	14.0	17.9	Header = W x H exposed		+100%
Norwegian	12 x 3-1/5 x 4	5C=16"	3.75	14.6	18.6	Rowlock = H x W exposed		+100%
Roman	12 x 2 x 4	2C=4"	6.00	13.4	17.0	Rowlock stretcher = L x W exposed		+33.3%
SCR	12 x 2-2/3 x 6	3C=8"	4.50	21.8	28.0	Soldier = H x L exposed		—
Utility	12 x 4 x 4	1C=4"	3.00	15.4	19.6	Sailor = W x L exposed		-33.3%

Running Bond — Number of Brick per S.F. of Wall - Single Wythe with 3/8" Joints. For Other Bonds Standard Size Add to S.F. Quantities in Table to Left.

Concrete Blocks Nominal Size	Approximate Weight per S.F. Standard	Lightweight	Blocks per 100 S.F.	Mortar per M Block Partitions	Back up
2" x 8" x 16"	20 PSF	15 PSF	113	16 C.F.	36 C.F.
4"	30	20		31	51
6"	42	30		46	66
8"	55	38		62	82
10"	70	47		77	97
12"	85	55		92	112

R042-550   Brick Veneer in Place

Table below is for running bond with 3/8″ concave joints and includes 3%
waste for brick and 25% waste for mortar.

Item	Buff Face Brick 8″ x 2-2/3″ x 4″		Red Norman Face Brick 12″ x 2-2/3″ x 4″		Buff Roman Face Brick 12″ x 2″ x 4″	
1030 brick delivered	$370 per M	$381.10	$660 per M	$ 679.80	$675 per M	$ 695.25
Type N mortar @ $2.96 per C.F.	10.3 C.F.	30.49	14.0 C.F.	41.44	13.4 C.F.	39.66
Installation using indicated crew	Crew D-8 @ .667 days	459.70	Crew D-8 @ .690 days	475.55	Crew D-8 @ .667 days	459.70
Total per M in place		$871.29		$1,196.79		$1,194.61
Total per S.F. of wall	6.75 brick/S.F.	$ 5.88	4.50 brick/S.F.	$ 5.39	6.00 brick/S.F.	$ 7.17

R042-700   Glass Block

Cost of Blocks Each, Zone 1 Contractor Prices								
Nominal Size (Including Mortar)	Truckload or Carload				Less Than Truckload or Carload			
	Regular	Thinline	Essex	Solar Reflective	Regular	Thinline	Essex	Solar Reflective
6″ x 6″	$ 4.10	$ 3.50	—	$ 7.00	$ 4.00	$ 3.70	—	$ 8.00
8″ x 8″	5.00	3.60	4.50	11.00	5.00	4.20	5.00	12.00
Solid 8″ x 8″	20.00	—	—	—	23.50	—	—	—
4″ x 8″	4.10	3.50	—	—	4.20	4.00	—	—
12″ x 12″	13.50	—	—	—	15.50	—	—	—

	Per 100 S.F.		Per 1000 Block				
Size	No. of Block	Mortar 1/4″ Joint	Asphalt Emulsion	Caulk	Expansion Joint	Panel Anchors	Wall Mesh
6″ x 6″	410 ea.	5.0 C.F.	.17 gal.	1.5 gal.	80 L.F.	20 ea.	500 L.F.
8″ x 8″	230	3.6	.33	2.8	140	36	670
12″ x 12″	102	2.3	.67	6.0	312	80	1000
Approximate quantity per 100 S.F.			.07 gal.	0.6 gal.	32 L.F.	9 ea.	51, 68, 102 L.F.

Accessories required for installation — Wall ties: run galvanized double steel mesh full length of joint at $.30 per L.F.— fiberglass expansion joints at sides and top at $.35 per L.F.— silicone caulking one gallon at $50.00 does 95 L.F. asphalt emulsion one gallon does 600 L.F. at $4.00 per gallon in one gallon lots. If blocks are not set in wall chase use 2′-0″ long wall anchors at $1.50 each at 2′-0″ O.C.

	Cost per 100 S.F. (Truckload or Carload Price)					
Quantities	6″ x 6″ Block		8″ x 8″ Block		12″ x 12″ Block	
Block, delivered	410 @ $3.50	$1,435.00	230 @ $4.50	$1,035.00	102 @ $12.00	$1,224.00
Mortar @ $6.10 per C.F.	5.0 C.F.	30.50	3.6 C.F.	21.96	2.3 C.F.	14.03
Ties, expansion joints, etc.		91.29		93.12		101.62
Crew D-8@ $689.20 per day	.690 days	475.55	.465 days	320.48	.417 days	287.40
Total per 100 S.F.		$2,032.34		$1,470.56		$1,627.05

R044-500   Ashlar Veneer

Quantities			Low Price Stone		High Price Stone	
4″ Stone, random ashlar,	2.5	tons average	$195 per ton	$ 487.50	$350 per ton	$ 875.00
Type N mortar @	$2.96	C.F.	20 C.F.	59.20	20 C.F.	59.20
50 - 1″ x 1/4″ x 6″ galv. stone anchors @	$1.05	each		52.50		52.50
Using crew D-8 @	$689.20	per day	.714 days	492.00	.833 days	574.10
Total per 100 S.F.				$1,091.20		$1,560.80

Stone coverage varies from 30 S.F. to 50 S.F. per ton. Mortar quantities include 1″ backing bed.

R045-100　Cleaning Face Brick

On smooth brick a person can clean 70 S.F. an hour; on rough brick 50 S.F. per hour. Use one gallon muriatic acid to 20 gallons of water for 1000 S.F. Do not use acid solution until wall is at least seven days old, but a mild soap solution may be used after two days. Commercial cleaners cost from $9 to $12 per gallon.

Time has been allowed for clean-up in brick prices.

R051-220　Steel Estimating Quantities

One estimate on erection is that a crane can handle 35 to 60 pieces per day. Say average is 45. With usual sizes of beams, girders, and columns, this would amount to about 20 tons. Allowing $1,188 per day rental for the crane including fuel and operator, this would amount to about $59 per ton for crane charge. The type of connection greatly affects the speed of erection. Moment connections for continuous design slow down production and increase erection costs.

Short open web bar joists can be set at the rate of 75 to 80 per day, with 50 per day being the average for setting long span joists.

After main members are calculated, add the following for usual allowances: base plates 2% to 3%; column splices 4% to 5%; and miscellaneous details 4% to 5%, for a total of 10% to 13% in addition to main members.

Ratio of column to beam tonnage varies depending on type of steels used, typical spans, story heights and live loads.

It is more economical to keep the column size constant and to vary the strength of the column by using high strength steels. This also saves floor space. Buildings have recently gone as high as ten stories with 8″ high strength columns. For light columns under W8X31 lb. sections, concrete filled steel columns are economical.

High strength steels may be used in columns and beams to save floor space and to meet head room requirements. High strength steels in some sizes sometimes require long lead times.

Round, square and rectangular columns, both plain and concrete filled, are readily available and save floor area, but are higher in cost per pound than rolled columns. For high unbraced columns, tube columns may be less expensive.

Below are average minimum figures for the weights of the structural steel frame for different types of buildings using A36 steel, rolled shapes and simple joints. For economy in domes, rise to span ratio = .13. Open web joist framing systems will reduce weights by 10% to 40%. Composite design can reduce steel weight by up to 25% but additional concrete floor slab thickness may be required. Continuous design can reduce the weights up to 20%. There are many building codes with different live load requirements and different structural requirements, such as hurricane and earthquake loadings which can alter the figures.

*See R131-310 for Domes and Thin Shell Structures.

Structural Steel Weights per S.F. of Floor Area									
Type of Building	No. of Stories	Avg. Spans	L.L. #/S.F.	Lbs. Per S.F.	Type of Building	No. of Stories	Avg. Spans	L.L. #/S.F.	Lbs. Per S.F.
Steel Frame Mfg.	1	20′x20′	40	8	Apartments	2-8	20′x20′	40	8
		30′x30′		13		9-25			14
		40′x40′		18	Office	to 10	Various	80	10
Parking garage	4	Various	80	8.5		20			18
Domes (Schwedler)*	1	200′	30	10		30			26
		300′		15		over 50			35

R051-250　Subpurlins

Table is based on subpurlins 32-5/8″ O.C. with simple spans, 40 psf, L.L., material only, 10,000 lb. to 30,000 lb. lots.

Type	Bulb Tees, Painted			Truss Tees, Painted							
	Wt. per L.F.	Cost per L.F.	Max. Span	Size	Wt. per L.F.	Cost per L.F.	Max. Span	Size	Wt. per L.F.	Cost per L.F.	Max. Span
112	1.48 #	$ .78	5′-6″	2″	1.10 #	$ .72	5′-9″	2-1/2″	1.39 #	$ .82	7′-0″
158	1.68	.80	6′-4″		1.27	.77	6′-0″		1.85	1.12	8′-9″
168	1.87	.85	7′-8″		1.33	.78	6′-0″	3″	1.14	.72	7′-3″
178	2.15	.94	8′-9″		1.78	1.05	7′-6″		1.88	1.12	9′-0″
218	3.19	1.20	10′-2″	2-1/2″	1.12	.72	6′-9″	3-1/2″	1.17	.73	7′-9″
228	3.87	1.54	12′-0″		1.34	.78	6′-9″		1.90	1.14	10′-9″

R054-210 Lightgauge Joists

Material price, delivered in truckload lots is $.50 per pound for galvanized. Weights run from 1.4# per L.F. to 6.0# per L.F. depending on gauge, depth and flange width. Connections are made with self tapping screws.

Prices in front of book include allowance for usual clip angles, welding plates, anchor bolts, etc. See Reference Number R092-610.

R061-010   Thirty City Lumber Prices

Prices for boards are for #2 or better or sterling, whichever is in best supply. Dimension lumber is "Standard or Better" either Southern Yellow Pine (S.Y.P.), Spruce-Pine-Fir (S.P.F.), Hem-Fir (H.F.) or Douglas Fir (D.F.). The species of lumber used in a geographic area is listed by city. Plyform is 3/4" BB oil sealed fir or S.Y.P. whichever prevails locally, 3/4" CDX is S.Y.P. or Fir.

These are prices at the time of publication and should be checked against the current market price. Relative differences between cities will stay approximately constant.

City	Species	Contractor Purchases per M.B.F. — S4S — Dimensions 2"x4"	2"x6"	2"x8"	2"x10"	2"x12"	4"x4"	Boards 1"x6"	1"x12"	Contractor Purchases per M.S.F. 3/4" Ext. Plyform	3/4" Thick CDX T&G
Atlanta	S.P.F.	$508	$498	$575	$624	$739	$660	$ 886	$1,177	$ 792	$638
Baltimore	S.P.F.	639	621	629	768	790	782	428	520	1,028	928
Boston	S.P.F.	578	493	540	629	693	986	459	1,156	1,011	710
Buffalo	H.F.	512	490	500	583	601	995	598	1,411	1,089	638
Chicago	S.P.F.	536	529	553	638	768	848	902	1,264	995	638
Cincinnati	S.P.F.	686	580	757	733	808	938	510	1,241	1,048	716
Cleveland	S.P.F.	799	647	698	820	850	927	850	1,097	1,030	749
Columbus	S.P.F.	599	558	564	633	702	527	620	1,768	1,100	640
Dallas	S.P.F.	446	438	480	527	587	557	676	956	1,017	825
Denver	H.F.	569	569	551	661	672	871	1,027	1,189	1,066	831
Detroit	S.P.F.	554	502	626	774	790	518	1,207	1,796	978	674
Houston	S.Y.P.	532	547	547	614	694	729	580	825	1,012	647
Indianapolis	S.Y.F.	666	629	626	643	723	663	1,921	2,252	1,012	760
Kansas City	D.F.	560	578	587	638	638	740	968	1,360	1,040	715
Los Angeles	D.F.	450	450	510	510	536	590	782	1,232	1,034	599
Memphis	S.P.F.	540	506	536	591	595	672	849	1,228	875	780
Milwaukee	S.P.F.	660	578	578	701	761	663	723	1,088	770	687
Minneapolis	S.P.F.	523	472	530	593	663	478	904	1,158	1,140	560
Nashville	S.P.F.	450	550	587	615	702	602	1,345	1,770	921	682
New Orleans	S.Y.P.	574	565	597	612	683	659	525	1,613	820	691
New York City	S.P.F.	818	680	583	783	918	961	1,188	1,382	1,300	870
Philadelphia	H.F.	544	530	558	632	664	927	990	1,385	1,260	680
Phoenix	D.F.	475	548	543	608	603	786	1,100	1,318	1,050	928
Pittsburgh	H.F.	720	637	698	734	769	717	1,122	1,351	1,120	624
St. Louis	S.P.F.	480	463	548	578	693	663	723	997	781	627
San Antonio	S.Y.P.	468	497	497	591	638	561	506	1,020	825	583
San Diego	D.F.	519	531	587	608	625	667	989	1,101	1,065	630
San Francisco	D.F.	527	603	603	672	663	999	935	1,268	1,156	704
Seattle	S.P.F.	639	569	673	769	832	836	684	1,445	900	650
Washington, DC	S.P.F.	689	541	637	737	786	816	982	1,351	1,109	755
Average		$575	$547	$583	$654	$706	$745	$ 866	$1,291	$1,011	$705

To convert square feet of surface to board feet, 4% waste included.

S4S Size	Multiply S.F. by	T & G Size	Multiply S.F. by	Flooring Size	Multiply S.F. by
				25/32" x 2-1/4"	1.37
1 x 4	1.18	1 x 4	1.27	25/32" x 3-1/4"	1.29
1 x 6	1.13	1 x 6	1.18	15/32" x 1-1/2"	1.54
1 x 8	1.11	1 x 8	1.14	1" x 3"	1.28
1 x 10	1.09	2 x 6	2.36	1" x 4"	1.24

R061-020 Plywood

There are two types of plywood used in construction: interior, which is moisture resistant but not waterproofed, and exterior, which is waterproofed.

The grade of the exterior surface of the plywood sheets is designated by the first letter: A, for smooth surface with patches allowed; B, for solid surface with patches and plugs allowed; C, which may be surface plugged or may have knot holes up to 1″ wide; and D, which is used only for interior type plywood and may have knot holes up to 2-1/2″ wide. "Structural Grade" is specifically designed for engineered applications such as box beams. All CC & DD grades have roof and floor spans marked on them.

Underlayment grade plywood runs from 1/4″ to 1-1/4″ thick. Thicknesses 5/8″ and over have optional tongue and groove joints which eliminates the need for blocking the edges. Underlayment 19/32″ and over may be referred to as Sturd-i-Floor.

The price of plywood can fluctuate widely due to geographic and economic conditions. When one or two local prices are known, the relative prices for other types and sizes may be found by direct factoring of the prices in the table below.

Typical uses for various plywood grades are as follows:

AA-AD Interior — cupboards, shelving, paneling, furniture

BB Plyform — concrete form plywood

CDX — wall and roof sheathing

Structural — box beams, girders, stressed skin panels

AA-AC Exterior — fences, signs, siding, soffits, etc.

Underlayment — base for resilient floor coverings

Overlaid HDO — high density for concrete forms & highway signs

Overlaid MDO — medium density for painting, siding, soffits & signs

303 Siding — exterior siding, textured, striated, embossed, etc.

Grade	Type	4′x8′	Type	4′x8′	4′x10′
			National Average Price in Lots of 10 MSF, per MSF-January 1998		
Sanded Grade	1/4″ Interior AD	$ 475	1/4″ Exterior AC	$ 506	$ 536
	3/8″	553	3/8″	567	590
	1/2″	661	1/2″	675	726
	5/8″	776	5/8″	790	882
	3/4″	844	3/4″	857	998
	1″	956	1″	1,112	1,150
	1-1/4″	1,146	Exterior AA, add	150	155
	Interior AA, add	150	Exterior AB, add	145	140
			CD Structural 1	**Underlayment**	
Unsanded Grade 4′ x 8′ Sheets	5/16″ CDX	$ 324			
	3/8″	334	5/16″, 4′x8′ sheets $378	3/8″ 4′x8′ sheets	$ 453
	1/2″	427	3/8″ 475	1/2″	526
	5/8″	513	1/2″ 545	5/8″ T&G	627
	3/4″	607	5/8″ 648	3/4″ T&G	810
	3/4″ T&G	707	3/4″ 724	1-1/8″ 2-4-1 T&G	1,269
Form Plywood	5/8″ Exterior, oiled BB, plyform	$ 914	5/8″ HDO (overlay 2 sides)		$1,701
	3/4″ Exterior, oiled BB, plyform	1,012	3/4″ HDO (overlay 2 sides)		2,214
Overlaid 4′x8′ Sheets	Overlay 2 Sides MDO 3/8″ thick	$1,057	Overlay 1 Side MDO 3/8″ thick		$ 918
	1/2″	1,236	1/2″		1,058
	5/8″	1,339	5/8″		1,188
	3/4″	1,442	3/4″		1,393
303 Siding	Fir, rough sawn, natural finish, 3/8″ thick	$ 682	Texture 1-11	5/8″ thick, Fir	$1,037
	Redwood	1,890		Redwood	1,890
	Cedar	1,890		Cedar	1,537
	Southern Yellow Pine	693		Southern Yellow	832
Waferboard/O.S.B.	1/4″ sheathing	$ 167	19/32″ T&G		$ 296
	7/16″ sheathing	190	23/32″ T&G		330

For 2 MSF to 10 MSF, add 10%. For less than 2 MSF, add 15%.

R061-030 Lumber Product Material Prices

The price of forest products fluctuates widely from location to location and from season to season depending upon economic conditions. The table below indicates National Average material prices in effect Jan. 1, 1998. The table shows relative differences between various sizes, grades and species. These percentage differentials remain fairly constant even though lumber prices in general may change significantly during the year.

Availability of certain items depends upon geographic location and must be checked prior to firm price bidding.

National Average Contractor Price, Quantity Purchase

Dimension Lumber, S4S, #2 & Better, KD / **Heavy Timbers, Fir**

	Species	2"x4"	2"x6"	2"x8"	2"x10"	2"x12"	Heavy Timbers, Fir	
Framing	Douglas Fir	$ 512	$ 543	$ 566	$ 606	$ 602	3" x 4" thru 3" x 12"	$1,100
Lumber	Spruce	597	547	593	672	741	4" x 4" thru 4" x 12"	1,253
per MBF	Southern Yellow Pine	560	559	566	615	684	6" x 6" thru 6" x 12"	1,890
	Hem-Fir	458	557	579	653	677	8" x 8" thru 8" x 12"	2,030
							10" x 10" and 10" x 12"	1,890

S4S "D" Quality or Clear, KD / **S4S # 2 & Better or Sterling, KD**

	Species	1"x4"	1"x6"	1"x8"	1"x10"	1"x12"	Species	1"x4"	1"x6"	1"x8"	1"x10"	1"x12"
Boards per MBF	Sugar Pine	$1,638	$2,044	$2,184	$2,254	$2,716	Sugar Pine	$ 728	$ 882	$ 882	$1,078	$1,302
*See also	Idaho Pine	1,428	1,764	1,834	1,834	2,261	Idaho Pine	1,106	1,106	1,134	1,162	1,274
Cedar Siding	Engleman Spruce	1,400	1,841	1,834	1,890	2,310	Engleman Spruce	658	896	798	1,036	1,232
	So. Yellow Pine	1,036	1,190	1,246	1,092	1,281	So. Yellow Pine	518	840	959	875	1,015
	Ponderosa Pine	1,428	1,764	1,834	1,834	2,261	Ponderosa Pine	623	896	784	1,008	1,232
	Redwood, CVG	3,940	3,800	3,750	3,770	5,940						

Flooring

Flooring per MSF	1" x 4" Vertical grain, Fir "B" & better	$2,300	2-1/4" x 25/32" Maple, select	$2,970
	2-1/4" x 25/32", Oak, clear	3,200	#2 & better	2,700
	Select	3,050	2-1/4" x 33/32" Maple, #2 & better	3,132
	#1 common	2,700	3-1/4" x 33/32" Maple, #2 & better	2,916
	Oak, prefinished, standard & better	3,750	Parquet, unfinished, 5/16", minimum	1,539
	Standard	3,200	Maximum	4,800

Siding

Siding per MBF	Clapboard, Cedar, beveled		*Rough sawn, Cedar, T&G, "A" grade 1" x 4"	$2,484
	1/2" x 6" thru 1/2" x 8", clear	$1,404	1" x 6"	3,267
	"A" grade	1,377	"STK" grade, 1" x 6"	1,647
	"B" grade	972	Board, "STK" grade, 1" x 8"	1,593
	3/4" x 10" "clear"	2,700	Board & Batten 1" x 12"	1,593
	"A" grade	2,430	Cedar channel siding 1" x 8", #3 & better	$1,544
	Redwood, beveled		Factory stained	1,728
	1/2" x 6" thru 1/2" x 8", vertical grain, clear	1,907	White Pine siding, T&G, rough sawn	$ 517
	3/4" x 10" vertical grain, clear	2,889	Factory stained	567

Shingles

Shingles per CSF	Red Cedar		White Cedar shingles	
	5X—16" long #1 regular	$ 157	16" long, extra grade (East Coast)	$ 100
	#2	177	Clear, 1st grade (East Coast)	85
	18" long perfections #1	202		
	#2	133	Fire retardant Red Cedar	
	Resquared & Rebutted #1	118	5X — 16" long	$ 257
			18" long perfections	275
	Handsplit shakes, resawn		Handsplit & resawn	
	24" long, 1/2" to 3/4"	182	24" long, 1/2" to 3/4"	275
	18" long, 1/2" to 3/4"	162	18" long, 1/2" to 3/4"	290

R061-100   Wood Roof Trusses

Loading figures represent live load. An additional load of 10 psf on the
top chord and 10 psf on the bottom chord is included in the truss
design. Spacing is 24″ O.C.

Span in Feet	Cost per Truss for Different Live Loads and Roof Pitches					
	Flat	4 in 12 Pitch		5 in 12 Pitch		8 in 12 Pitch
	40 psf	30 psf	40 psf	30 psf	40 psf	30 psf
20	$ 70.00	$39.00	$41.00	$42.00	$46.00	$ 54.00
22	77.00	43.00	45.00	46.00	50.00	58.00
24	84.00	47.00	49.00	50.00	54.00	62.00
26	91.00	50.00	52.00	53.00	57.00	65.00
28	98.00	52.00	54.00	55.00	59.00	67.00
30	105.00	63.00	65.00	66.00	70.00	78.00
32	112.00	65.00	67.00	68.00	72.00	80.00
34	119.00	76.00	78.00	79.00	83.00	91.00
36	126.00	80.00	82.00	83.00	87.00	95.00
38	133.00	83.00	85.00	86.00	90.00	98.00
40	140.00	90.00	92.00	93.00	97.00	105.00

R064-100   Wood Stair, Residential

Item	Quantity	Unit Cost	Bare Costs	Costs Incl. Subs O & P
One Flight with 8′-6″ Story Height, 3′-6″ Wide Oak Treads Open One Side, Built in Place				
Treads 10-1/2″ x 1-1/16″ thick	11 Ea.	$ 26.00	$ 286.00	$ 314.60
Landing tread nosing, rabbeted	1 Ea.	15.75	15.75	17.40
Risers 3/4″ thick	12 Ea.	11.11	133.32	146.60
Single end starting step (range $90 to $190)	1 Ea.	193.00	193.00	212.40
Balusters (range $3.50 to $12.00)	22 Ea.	7.75	170.50	187.60
Newels, starting & landing (range $36 to $110)	2 Ea.	73.00	146.00	160.60
Rail starter (range $29 to $105)	1 Ea.	67.00	67.00	73.80
Handrail (range $3.50 to $9.80)	26 L.F.	7.00	182.00	200.20
Cove trim	50 L.F.	.55	27.50	30.20
Rough stringers three - 2 x 12′s, 14′ long	84 B.F.	.71	59.64	65.60
Carpenter's installation: Bare Cost	36 Hrs.	$ 18.50	$ 666.00	
Incl. Subs O & P		$ 31.70		$1,141.20
	Total per Flight		$1,946.71	$2,550.20

Add for rail return on second floor and for varnishing or other finish. Adjoining walls or landings must be figured separately.

R071-010   1/2″ Pargeting (rough dampproofing plaster)

1:2-1/2 Mix, 4.5 C.F. Covers 100 S.F., 2 Coats, Waste Included	Regular Portland Cement		Waterproofed Portland Cement	
1.7 lbs. integral waterproofing admixture			$.64 per lb.	$ 1.09
1.7 bags Portland cement	$6.85 per bag	$ 11.65	$6.85 per bag	11.65
4.25 C.F. sand @ $20.00 per C.Y.		3.15		3.15
Labor mix, apply, crew D-1 @ $268.80 per day	.40 days	107.52	.40 days	107.52
Total Bare Cost per 100 S.F.		$122.32		$123.41

R073-020   Roof Slate

16″, 18″ and 20″ are standard lengths and slate usually comes in random widths. For standard 3/16″ thickness use 1-1/2″ copper nails. Allow for 3% breakage.

Quantities per Square	Unfading Vermont Colored	Weathering Sea Green	Buckingham, Virginia Black
Slate delivered (incl. punching)	$418.50	$314.83	$531.50
# 30 Felt, 2.5 lbs. copper nails	18.26	18.26	18.26
Slate roofer 4.6 hrs. @ $16.30 per hr.	74.98	74.98	74.98
Total Bare Cost per Square	$511.74	$408.07	$624.74

R074-030   Metal and Fiberglass Sandwich Panels

Aluminum facing panels may be field erected with corrugated sheets or perforated acoustical type on the inside face and either corrugated, V-beam or ribbed exterior sheets. The most usual gauges are .032″ for the exterior and .024″ for the interior sheets. The V-beam type is available in .032″, .040″ and .050″. The insulation is generally 1″ thick. Individual sheets can be as large as 30′ x 2′.

The cost of the particular sandwich can be figured by adding the appropriate siding costs (Div. 074-602) to the insulation costs (Div. 072-116). Then add 15% per S.F. for multi-story construction and 25% per S.F. additional for jobs less than 2000 S.F. These figures do not include any supporting framework or flashings; add 2% to 4% for the usual flashings.

R075-020   Built-Up Roofing

Asphalt is available in kegs of 100 lbs. each; coal tar pitch in 560 lb. kegs. Prepared roofing felts are available in a wide range of sizes, weights & characteristics. However, the most commonly used are #15 (432 S.F. per roll, 13 lbs. per square) and #30 (216 S.F. per roll, 27 lbs. per square).

Inter-ply bitumen varies from 24 lbs. per sq. (asphalt) to 30 lbs. per sq. (coal tar) per ply, ± 25%. Flood coat bitumen also varies from 60 lbs. per sq. (asphalt) to 75 lbs. per sq. (coal tar), ± 25%. Expendable equipment (mops, brooms, screeds, etc.) runs about 16% of the bitumen cost. For new, inexperienced crews this factor may be much higher.

Rigid insulation board is typically applied in two layers. The first is mechanically attached to nailable decks or spot or solid mopped to non nailable decks; the second layer is then spot or solid mopped to the first layer. Membrane application follows the insulation, except in protected membrane roofs, where the membrane goes down first and the insulation on top, followed with ballast (stone or concrete pavers). Insulation and related labor costs are NOT included in Div. 075-102; they can be found in Div. 072-203.

Prices shown are bare costs only.

4-Ply Felt on Insulated Deck	Asphalt/Glass Fiber Felt		Coal Tar/Organic Felt	
4 ply #15 felt (1/4 roll per ply)	$3.65 per ply	$ 14.60	$7.80 per ply	$ 31.20
156 lbs. hot asphalt/195 lbs. hot tar	$242 per ton	18.88	$538 per ton	52.46
500 lbs. roofing stone (1/4″ to 1/2″)	$17.00 per ton	4.25	$17.00 per ton	4.25
Installation, crew G-1, @ $1,253.60 per day	.050 days	62.68	.048 days	60.17
Total Bare Cost per Square in-place		$100.41		$148.08

R075-030   Modified Bitumen Roofing

The cost of modified bitumen roofing is highly dependent on the type of installation that is planned. Installation is based on the type of modifier used in the bitumen. The two most popular modifiers are atactic polypropylene (APP) and styrene butadiene styrene (SBS). The modifiers are added to heated bitumen during the manufacturing process to change its characteristics. A polyethylene, polyester or fiberglass reinforcing sheet is then sandwiched between layers of this bitumen. When completed, the result is a pre-assembled, built-up roof that has increased elasticity and weatherablility. Some manufacturers include a surfacing material such as ceramic or mineral granules, metal particles or sand.

The preferred method of adhering SBS-modified bitumen roofing to the substrate is with hot-mopped asphalt (much the same as built-up roofing). This installation method requires a tar kettle/pot to heat the asphalt, as well as the labor, tools and equipment necessary to distribute and spread the hot asphalt.

The alternative method for applying APP and SBS modified bitumen is as follows. A skilled installer uses a torch to melt a small pool of bitumen off the membrane. This pool must form across the entire roll for proper adhesion. The installer must unroll the roofing at a pace slow enough to melt the bitumen, but fast enough to prevent damage to the rest of the membrane.

Modified bitumen roofing provides the advantages of both built-up and single-ply roofing. Labor costs are reduced over those of built-up roofing because only a single ply is necessary. The elasticity of single-ply roofing is attained with the reinforcing sheet and polymer modifiers. Modifieds have some self-healing characteristics and because of their multi-layer construction, they offer the reliability and safety of built-up roofing.

R081-010   Hollow Metal Doors

Table below lists material prices only, not including hardware or labor.

Door Thickness and Size		Full Flush Doors, 18 Ga.				Flush Fire Doors			
		Hollow Metal		Composite Core		Hollow Metal "B"		Hollow Metal Class "A"	
		Plain	Glazed	Plain	Glazed	20 Ga.	18 Ga.	16 Ga.	18 Ga.
1-3/4"	3'-0" x 6'-8"	$198	$255	$225	$280	$197	$220	$268	$231
	3'-0" x 7'-0"	209	264	231	286	202	226	285	238
	3'-6" x 7'-0"	237	291	278	333	237	278	315	291
	3'-0" x 8'-0"	268	322	299	358	246	291	335	303
	4'-0" x 8'-0"	315	367	344	401	317	349	388	360
1-3/8"	2'-0" x 6'-8"	148	202	—	—	171	—	—	—
	2'-6" x 6'-8"	160	214	—	—	183	—	—	—
	3'-0" x 7'-0"	169	223	—	—	193	—	—	—

*Indicates 20 gauge doors.

R081-020   Hollow Metal Frames

Table below lists material prices only.

Frame for Opening Size	16 Ga. Frames		16 Ga. UL Frames		16 Ga. Drywall Frames		16 Ga. UL Drywall Frames	
	6-3/4" Deep	8-3/4" Deep	6-3/4" Deep	8-3/4" Deep	7-1/8" Deep	8-1/4" Deep	7-1/8" Deep	8-1/4" Deep
2'-0" x 6'-8"	$ 74	$78	$ 86	$ 91	$ 80	$ 88	$ 98	$104
2'-6" x 6'-8"	74	78	86	91	80	88	98	104
3'-0" x 7'-0"	76	80	88	94	84	90	101	107
3'-6" x 7'-0"	80	83	91	96	86	92	102	111
4'-0" x 7'-0"	87	92	102	107	94	102	111	117
6'-0" x 7'-0"	87	92	102	107	94	102	111	117
8'-0" x 8'-0"	107	97	119	112	111	103	129	139

For welded frames, add $26.00.
For galvanized frames, add $22.00.

R082-120   Wood Doors

Table below lists price per door only, not including frame, hardware or labor. For pre-hung exterior door units up to 3′ x 7′, add $162 per door for wood frame and hardware for types not listed under pre-hung. Pricing is for ten or more doors. Doors are factory trimmed for butts and locksets.

Door Thickness and Size		Flush Type Doors (Interior)					Architectural (1-3/4″ Ext., 1-3/8″ Int.)					
		Hollow Core		Solid Particle Core			Pine		Fir		Pre-hung	
		Lauan	Birch	Lauan	Birch	Oak	Panel	Glazed	Panel	Glazed	Pine Panel	Flush Birch S. C.
1-3/4″	2′-6″ x 6′-8″	$47	$48	$70	$70 *	$ 88	$348	$375	$217	$215	$472	$190
	3′-0″ x 6′-8″	48	55	75	82 *	91	362	397	226	221	495	205
	3′-0″ x 7′-0″	54	60	85	89 *	99	408	430	233	234	538	220
	3′-6″ x 7′-0″	69	69	90	93 *	18	—	—	—	—	—	—
	4′-0″ x 7′-0″	72	73	93	96 *	115	—	—	—	—	—	—
1-3/8″	2′-0″ x 6′-8″	30	41	68	80 *	84	126	—	111	—	242	173
	2′-6″ x 6′-8″	35	46	70	84 *	88	135	—	120	—	250	179
	3′-0″ x 6′-8″	39	49	76	89 *	94	158	—	141	—	276	185
	3′-0″ x 7′-0″	48	57	87	94 *	98	194	—	175	—	—	—

*Add to the above for the following birch face door types:

Solid wood stave core, add $31

3/4 hour label door, add $61

1 hour label door, add $72

1-1/2 hour label door, add $82

8′ high door, add 31%

R083-100   Tin Clad Fire Door

6′ x 7′ Opening, A Label	Double Sliding Door		Double Swing Door	
Doors, 3 Ply	($12.00 to $13.00/S.F.)	$ 510.00	($12.00 to $13.00/S.F.)	$ 510.00
Frame, 9″ channel (15 lb. per L.F.), installed	325 lb.	478.00	325 lb.	478.00
Track, hangers, hardware	13.75 L.F.	693.00	Total	326.00
2 Carpenters @ $18.50 per hour each	16 hrs. total	296.00	16 hrs. total	296.00
Complete in place		$1,977.00		$1,610.00

R086-200   Replacement Windows

Replacement windows are typically measured per United Inch.

United Inches are calculated by rounding the width and height of the window opening up to the nearest inch, then adding the two figures.

The labor cost for replacement windows includes removal of sash, existing sash balance or weights, parting bead where necessary and installation of new window.

Debris hauling and dump fees are not included.

R088-010   Glazing Labor

Glass sizes are estimated by the "united inch" (height + width). Table below shows the number of lights glazed in an eight hour period by the crew size indicated, for glass up to 1/4" thick. Square or nearly square lights are more economical on a S.F. basis. Long slender lights will have a high S.F. installation cost. For insulated glass reduce production by 33%. For 1/2" plate glass reduce production by 50%. Production time for glazing with two glaziers per day averages: 1/4" float glass 120 S.F.; 1/2" float glass 55 S.F.; 1/2" insulated glass 95 S.F.; 3/4" insulated glass 75 S.F.

Glazing Method	United Inches per Light							
	40"	60"	80"	100"	135"	165"	200"	240"
Number of Men in Crew	1	1	1	1	2	3	3	4
Industrial sash, putty	60	45	24	15	18	—	—	—
With stops, putty bed	50	36	21	12	16	8	4	3
Wood stops, rubber	40	27	15	9	11	6	3	2
Metal stops, rubber	30	24	14	9	9	6	3	2
Structural glass	10	7	4	3	—	—	—	—
Corrugated glass	12	9	7	4	4	4	3	—
Storefronts	16	15	13	11	7	6	4	4
Skylights, putty glass	60	36	21	12	16	—	—	—
Thiokol set	15	15	11	9	9	6	3	2
Vinyl set, snap on	18	18	13	12	12	7	5	4
Maximum area per light	2.8 S.F.	6.3 S.F.	11.1 S.F.	17.4 S.F.	31.6 S.F.	47 S.F.	69 S.F.	100 S.F.
Daily Crew Cost	$148.00	$148.00	$148.00	$148.00	$296.00	$444.00	$444.00	$592.00

R088-100   Commercial Window Glass

Description	1/2" Insulated Glass		3/8" Float Glass	
Glass per S.F.	$ 5.50	R Value	$ 5.99	R Value
Allow for breakage 10%	.55	1.68	.60	.93
Glazing compound 1/2 lb. per S.F. @ $1.00 per lb.	.55	Average	.55	Average
Labor, 2 glaziers @ $18.50 per hour each (75 S.F./day)	3.95		3.95	
Total in place, per S.F.	$10.55		$11.09	

R089-200   Window Walls

The table below shows the S.F. costs for 1-3/4" x 4-1/2" clear anodized tubular aluminum framing, flush glazed with fixed 1/4" clear polished float glass for jobs over 250 S.F. and includes overhead & profit.

Description	Total Window Wall Height								
	3'-0"			6'-0"			10'-0"		
Mullion Spacing	3'	5'	7'	3'	5'	7'	3'	5'	7'
Mullions Only	$24.00	$21.00	$20.00	$18.00	$16.00	—	$17.00	—	—
1 Intermediate horizontal member	—	27.00	26.00	21.00	19.00	$18.00	18.00	$16.00	—
2	—	—	—	24.00	22.00	21.00	20.00	18.00	$17.00
3	—	—	—	26.00	24.00	23.00	21.00	19.00	18.00

Add to the above erected costs for the following: 1" insulating glass, plus 2" x 4-1/2" clear tube frame, add $9.00 per S.F.

Bronze anodized aluminum tubing, add 15% to 20%. For screw applied square stops, add 15%. For operating sash, add cost of each sash to total job.

R092-050   Gypsum Lath

Item	Nailed to Wood Studs			Clipped to Steel Studs		
	Quantities and Bare Cost		Incl. O & P	Quantities and Bare Cost		Incl. O & P
105 S.Y. 3/8″ gypsum lath	$3.24 per S.Y.	$340.20	$374.22	$3.24 per S.Y.	$340.20	$374.22
Fasteners, nails & clips	6 lb. @ $1.50/lb.	9.00	9.90	600 @ $.06 ea.	36.00	39.60
Lather @ $18.30 and $30.00 per hr.	9.4 hours	172.02	282.00	10.7 hours	195.81	321.00
Total per 100 S.Y. in place		$521.22	$666.12		$572.01	$734.82

Regular lath comes in 16″ x 32″ and 16″ x 48″ sheets. Firestop gypsum base comes in 4′ x 8′ sheets. For nailing use 1-1/8″ No. 13 ga. flathead blued nails.

R092-060   Metal Lath

Painted Diamond Expanded	Nailed to Wood Studs			Screwed to Steel Studs		
	Quantity and Bare Cost		Incl. O & P	Quantity and Bare Cost		Incl. O & P
105 S.Y. 3.4 lb. lath	$2.30 per S.Y.	$241.50	$265.65	$2.30 per S.Y.	$241.50	$265.65
Fasteners, corner beads, etc.		14.00	15.40		12.30	13.53
Lather @ $18.30 and $30.00 per hr.	10.0 hrs.	183.00	300.00	10.7 hrs.	195.81	321.00
Total per 100 S.Y. in place		$438.50	$581.05		$449.61	$600.18

Material prices of other types of lath in Div. 092 may be substituted in the table above to arrive at their costs in place. For nailing on studs use a 1″ roofing nail with 7/16″ head. For nailing on ceiling use 1-1/2″ No. 11 ga. barbed roofers nail with 7/16″ head for holding power.

R092-105   Gypsum Plaster

Quantities for 100 S.Y.	2 Coat, 5/8″ Thick		3 Coat, 3/4″ Thick		
	Base	Finish	Scratch	Brown	Finish
	1:3 Mix	2:1 Mix	1:2 Mix	1:3 Mix	2:1 Mix
Gypsum plaster	1300 lb.		1350 lb.	650 lb.	
Sand	1.75 C.Y.		1.16 C.Y.	.84 C.Y.	
Finish hydrated lime		340 lb.			340 lb.
Gauging plaster		170 lb.			170 lb.

Total, in Place for 100 S.Y. on Walls				2 Coat, 5/8″ Thick			3 Coat, 3/4″ Thick		
				Quantities	Bare Cost	Incl. O & P	Quantities	Bare Cost	Incl. O & P
Gypsum plaster @	$12.40	per 80	lb. bag	1300 lb.	$201.50	$ 221.65	2000 lb.	$ 310.00	$ 341.00
Finish hydrated lime @	$ 5.15	per 50	lb. bag	340 lb.	35.02	38.52	340 lb.	35.02	38.52
Gauging plaster @	$15.35	per 100	lb. bag	170 lb.	26.10	28.71	170 lb.	26.10	28.71
Sand @	$17.00	per C.Y.		1.7 C.Y.	28.90	31.79	2.0 C.Y.	34.00	37.40
J-1 crew @	$16.43	& $27.47	per L.H.	34.8 L.H.	571.76	955.96	42.0 L.H.	690.06	1,153.74
Cleaning, staging, handling, patching				3.6 L.H.	59.15	98.89	4.0 L.H.	65.72	109.88
Total per 100 S.Y. in place					$922.43	$1,375.52		$1,160.90	$1,709.25

R092-115 Vermiculite or Perlite Plaster

Proportions: Over lath, scratch coat and brown coat,
100# gypsum plaster to 2 C.F. aggregate; over masonry,
100# gypsum plaster to 3 C.F. aggregate.

Quantities for 100 S.Y.				2 Coat, 5/8" Thick			3 Coat, 3/4" Thick		
				Quantities	Bare Cost	Incl. O & P	Quantities	Bare Cost	Incl. O & P
Gypsum plaster @	$12.40	per 80	lb. bag	1250 lb.	$ 193.75	$ 213.13	2250 lb.	$ 348.75	$ 383.63
Vermiculite or perlite @	$12.85	per	bag	7.8 bags	100.23	110.25	11.3 bags	145.21	159.73
Finish hydrated lime @	$ 5.15	per 50	lb. bag	340 lb.	35.02	38.52	340 lb.	35.02	38.52
Gauging plaster @	$15.35	per 100	lb. bag	170 lb.	26.10	28.71	170 lb.	26.10	28.71
J-1 crew @	$17.46	& $28.60	per L.H.	40.0 L.H.	698.40	1,144.10	50.0 L.H.	873.00	1,430.13
Cleaning, staging, handling, patching				3.6 L.H.	62.86	102.97	4.0 L.H.	69.84	114.41
Total per 100 S.Y. in place					$1,116.36	$1,637.68		$1,497.92	$2,155.13

R092-300 Stucco

Quantities for 100 S.Y. 3 Coats, 1" Thick		On Wood Frame			On Masonry		
		Quantities	Bare Cost	Incl. O & P	Quantities	Bare Cost	Incl. O & P
Portland cement @	$ 7.00 per bag	29 bags	$ 203.00	$ 223.30	21 bags	$147.00	$161.70
Sand @	$17.00 per C.Y.	2.6 C.Y.	44.20	48.62	2 C.Y.	34.00	37.40
Hydrated lime @	$ 5.15 per 50 lb. bag	180 lb.	18.54	20.39	120 lb.	12.36	13.60
Painted stucco mesh @	$ 3.56 per S.Y., 3.6#	105 S.Y.	373.80	411.18	—	—	—
Furring nails and jute fiber			17.84	19.62	—	—	—
Mix and install with crew indicated		J-2 @ .74 days	625.15	1,024.23	J-1 @ 0.5 days	349.20	572.05
Cleaning, staging, handling, patching		.10 days	80.36	134.17	.10 days	80.36	134.17
Total per 100 S.Y. in place			$1,362.89	$1,881.51		$622.92	$918.92

R092-610 Studs, Joists and Track

Material prices per 1000 L.F., for galvanized studs, joists and tracks. Panhead, framing screws, 7/16″ long are $9.30 per thousand; 1-5/8″ long are $14.50 per thousand.

Non-load bearing, 20 ga. stud and track are primarily used for curtain wall. (C.W.)

| | Non-Load Bearing | | | | Load Bearing 1-5/8″ Flange—Light Gauge Structural | | | | | |
| | 25 Ga. | | 20 Ga. (C.W.) | | 18 Ga. | | 16 Ga. | | 14 Ga. | |
Size	Stud	Track	Stud	Track	Stud	Track	Stud	Track	Stud	Track
1-5/8″	$180	$175	$300	$290						
2-1/2″	200	195	340	330	$ 500	$490	$ 600	$ 590	$ 700	$ 690
3-5/8″	240	235	400	390	620	610	720	710	820	810
4″	280	275	500	490	640	630	740	730	840	830
6″	400	395	600	590	800	790	900	890	1,000	900
8″					1,000	990	1,100	1,090	1,200	1,100

| | Non-Load Bearing | | | | Load Bearing—Extra Wide Flange — 2″ Flange | | | | | | | |
| | 25 Ga. | | 22 Ga. | | 18 Ga. | | 16 Ga. | | 14 Ga. | | 12 Ga. | |
Size	C-H Stud	J-Track	C-H Stud	J-Track	Joist	Track	Joist	Track	Joist	Track	Joist	Track
2-1/2″	$600	$470	$ 950	$630	—	$ 670		$ 950		$1,100	—	—
4″	720	670	1,230	940	$ 860	850	$1,160	1,150	$1,450	1,440	$2,000	$1,900
6″					960	950	1,260	1,160	1,560	1,550	2,100	2,000
8″					1,060	1,050	1,360	1,260	1,660	1,650	2,200	2,100
10″							1,460	1,360	1,760	1,750	2,300	2,200
12″							1,500	1,460	1,860	1,850	2,400	2,300

R094-100 Terrazzo Floor

5000 S.F. 5/8″ Terrazzo Topping Quantities for 100 S.F.	Bonded to Concrete 1-1/8″ Bed, 1:4 Mix			Not Bonded 2-1/8″ Bed and 1/4″ Sand		
	Quantities	Bare Cost	Incl. O & P	Quantities	Bare Cost	Incl. O & P
Portland cement @ $7.00 per bag	6 bags	$ 42.00	$ 46.20	8 bags	$ 56.00	$ 61.60
Sand @ $17.00 per C.Y.	10 C.F.	6.30	6.93	20 C.F.	12.59	13.85
Divider strips for 4′ square panels, zinc, 14 ga.	55 L.F.	47.85	52.64	55 L.F.	47.85	52.64
Terrazzo fill, domestic aggregates	600 lb.	162.00	178.20	600 lb.	162.00	178.20
15 lb. tarred felt		—	—		3.00	3.30
Mesh 2 x 2 #14 galvanized		—	—		13.38	14.72
Crew J-3 @ $23.37 & $32.33 per day	12.30 L.H.	287.45	397.66	13.90 L.H.	324.84	449.39
Total per 100 S.F. in place	Bonded	$545.60	$681.63	Not Bonded	$619.66	$773.70

2′ x 2′ panels require 1 L.F. divider strip per S.F.; 6′ x 6′ panels need .33 L.F. per S.F.

R096-600 Resilient Flooring and Base

Description	12″ x 12″ x 3/32″ V.C. Tile			4″ x 1/8″ Vinyl Base		
	Quantities	Bare Cost	Incl. O & P	Quantities	Bare Cost	Incl. O & P
Vinyl composition, tile, standard line	100 S.F.	$106.00	$116.60	100 L.F.	$ 60.00	$ 66.00
Vinyl cement @ $14.25 per gallon	0.8 gallon	11.40	12.54	0.5 gallon	7.13	7.84
Tile layer @ $18.25 and $29.55 per L.H.	1.5 L.H.	27.38	44.33	2.7 L.H.	49.28	79.79
Total per 100 S.F. in place		$144.78	$173.47	100 L.F.	$116.41	$153.63

R099-100 Paint

Material prices per gallon in 5 gallon lots, up to 25 gallons. For 100 gallons, deduct 10%.

Exterior, Alkyd (oil base)	
Flat	$20.30
Gloss	19.10
Primer	18.70

Exterior, Latex (water base)	
Acrylic stain	16.25
Gloss enamel	21.20
Flat	16.65
Primer	17.50
Semi-gloss	20.50

Interior, Alkyd (oil base)	
Enamel undercoater	17.50
Flat	15.25
Gloss	21.00
Primer sealer	16.45
Semi-gloss	19.40

Interior, Latex (water base)	
Enamel undercoater	15.50
Flat	15.20
Floor and deck	16.00
Gloss	21.00
Primer sealer	15.25
Semi-gloss	16.65

Masonry, Exterior	
Alkali resistant primer	$18.50
Block filler, epoxy	19.50
Block filler, latex	9.40
Latex, flat or semi-gloss	16.25

Masonry, Interior	
Alkali resistant primer	14.85
Block filler, epoxy	22.50
Block filler, latex	9.50
Floor, alkyd	16.60
Floor, latex	18.75
Latex, flat acrylic	10.75
Latex, flat emulsion	10.50
Latex, sealer	11.75
Latex, semi-gloss	18.00

Varnish and Stain	
Alkyd clear	20.25
Polyurethane, clear	18.95
Primer sealer	13.75
Semi-transparent stain	15.50
Solid color stain	15.00

Metal Coatings	
Galvanized	23.50
High heat	31.25
Machinery enamel, alkyd	20.00
Normal heat	21.30

Rust inhibitor ferrous metal	$20.80
Zinc chromate	19.00

Heavy Duty Coatings	
Acrylic urethane	49.00
Chlorinated rubber	28.80
Coal tar epoxy	21.50
Metal pretreatment (polyvinyl butyral)	22.50
Polyamide epoxy finish	30.00
Polyamide epoxy primer	30.00
Silicone alkyd	30.75
2 component solvent based acrylic epoxy	31.50
2 component solvent based polyester epoxy	39.45
Vinyl	25.60
Zinc rich primer	46.00

Special Coatings/Miscellaneous	
Aluminum	21.75
Dry fall out, flat	11.50
Fire retardant, intumescent	34.85
Linseed oil	9.75
Shellac	17.60
Swimming pool, epoxy or urethane base	34.75
Swimming pool, rubber base	24.00
Texture paint	10.80
Turpentine	10.00
Water repellent 5% silicone	17.00

R099-220   Painting

Item	Coat	One Gallon Covers			In 8 Hours a Laborer Covers			Labor-Hours per 100 S.F.		
		Brush	Roller	Spray	Brush	Roller	Spray	Brush	Roller	Spray
Paint wood siding	prime	250 S.F.	225 S.F.	290 S.F.	1150 S.F.	1300 S.F.	2275 S.F.	.695	.615	.351
	others	270	250	290	1300	1625	2600	.615	.492	.307
Paint exterior trim	prime	400	—	—	650	—	—	1.230	—	—
	1st	475	—	—	800	—	—	1.000	—	—
	2nd	520	—	—	975	—	—	.820	—	—
Paint shingle siding	prime	270	255	300	650	975	1950	1.230	.820	.410
	others	360	340	380	800	1150	2275	1.000	.695	.351
Stain shingle siding	1st	180	170	200	750	1125	2250	1.068	.711	.355
	2nd	270	250	290	900	1325	2600	.888	.603	.307
Paint brick masonry	prime	180	135	160	750	800	1800	1.066	1.000	.444
	1st	270	225	290	815	975	2275	.981	.820	.351
	2nd	340	305	360	815	1150	2925	.981	.695	.273
Paint interior plaster or drywall	prime	400	380	495	1150	2000	3250	.695	.400	.246
	others	450	425	495	1300	2300	4000	.615	.347	.200
Paint interior doors and windows	prime	400	—	—	650	—	—	1.230	—	—
	1st	425	—	—	800	—	—	1.000	—	—
	2nd	450	—	—	975	—	—	.820	—	—

R099-700   Wall Covering

Quantities for 100 S.F.		Medium Price Paper			Expensive Paper		
		Quantities	Bare Cost	Incl. O & P	Quantities	Bare Cost	Incl. O & P
Paper @	$26.10 and $52.20 per double roll	1.6 dbl. rolls	$41.76	$45.94	1.6 dbl. rolls	$ 83.52	$ 91.87
Wall sizing @	$15.00 per gallon	.25 gallon	3.75	4.13	.25 gallon	3.75	4.13
Vinyl wall paste @	$ 7.80 per gallon	0.4 gallon	3.12	3.43	0.4 gallon	3.12	3.43
Apply sizing @	$17.45 and $29.15 per hour	0.3 hour	5.24	8.75	0.3 hour	5.24	8.75
Apply paper @	$17.45 and $29.15 per hour	1.2 hours	20.94	34.98	1.5 hours	26.18	43.73
Total including waste allowance per 100 S.F.			$74.81	$97.23		$121.81	$151.91

This is equivalent to about $67.50 and $135.00 per double roll
complete in place. Most wallpapers now come in double rolls
only. To remove old paper allow 1.3 hours per 100 S.F.

R131-210 Pre-engineered Steel Buildings

These buildings are manufactured by many companies and normally erected by franchised dealers throughout the U.S. The four basic types are: Rigid Frames, Truss type, Post and Beam and the Sloped Beam type. Most popular roof slope is low pitch of 1″ in 12″. The minimum economical area of these buildings is about 3000 S.F. of floor area. Bay sizes are usually 20′ to 24′ but can go as high as 30′ with heavier girts and purlins. Eave heights are usually 12′ to 24′ with 18′ to 20′ most typical.

Material prices shown below are bare costs for the building shell only and do not include floors, foundations, interior finishes or utilities. Typical erection cost, including erector's overhead and profit, with siding and roofing depends on the building shape and runs $2.21 to $3.97 per S.F of floor for one in twelve roof slope and $2.28 to $5.56 for four in twelve roof slope. Site, weather, labor source, shape and size of the project will determine the actual erection cost of the building.

Table below is based on 30 psf roof load, 20 psf wind load and no unusual structural requirements. Costs assume at least three bays of 24′ each. Material costs include the structural frame, 26 ga. colored steel roofing, 26 ga. colored steel siding, fasteners, closures and flashing but no allowance for doors, windows, gutters or skylights. Very large projects would generally cost less than the prices listed below. Typical budget figures for above material delivered to the job runs $1385 to $1745 per ton. Fasteners and flashings (included below) run $.50 to $.75 per S.F.

Material Costs per S.F. of Floor Area Above the Foundations						
Type of Building	Total Width in Feet	Eave Height				
		10 Ft.	14 Ft.	16 Ft.	20 Ft.	24 Ft.
Rigid Frame Clear Span	30-40	$4.15	$4.21	$4.55	$4.95	$6.00
	50-100	3.86	3.85	4.15	4.39	5.00
	110	—	—	—	—	—
	120	—	—	—	4.54	—
	130	—	—	—	—	—
Tapered Beam Clear Span	30	4.54	5.10	5.55	6.25	—
	40	4.20	4.55	4.75	5.30	—
	50-80	4.21	4.29	4.47	5.30	—
Post & Beam 1 Post at Center	80	—	3.53	3.77	4.05	4.43
	100	—	3.42	3.57	4.03	4.24
	120	—	3.34	3.47	3.70	4.11
Post & Beam 2 Posts @ 1/3 Points	120	—	—	—	—	—
	150	—	3.12	3.08	3.26	4.12
	180	—	—	—	—	—
Post & Beam 3 Posts @ 1/4 points	160	—	2.88	2.99	3.26	3.96
	200	—	2.90	3.06	3.26	3.56
	240	—	—	—	—	—

Typical accessory items are listed in the front of the book. All normal interior work, floors, foundations, utilities and site work should be figured the same as usual.

Costs in the table below include allowance for erection, normal doors, windows, gutters and erector's overhead and profit. Figures do not include foundations, floors, interior finishes, electrical, mechanical or installed equipment.

Total Cost per S.F. Above the Foundations, 16′ Eave Height				
Project Size:	Basic Building	Add to Basic Building Price		
Rigid Frame 30′ to 60′ Spans 1 in 12 Roof Slope	Using 26 ga. Galvanized Roof & Siding	R13 Field Insulation	Exterior Finish	S.F. of Skin
	S.F. Floor Area	S.F. Floor Area	Sandwich wall	$6.96
4,000 S.F.	$8.50	$1.61	Corrugated fiberglass	7.12
10,000 S.F.	7.85	1.37	Corr. fiberglass-insulated	4.56
20,000 S.F.	7.81	1.31	10 year paint	.26

R131-520   Swimming Pools

Pool prices given per square foot of surface area include pool structure, filter and chlorination equipment where required, pumps, related piping, diving boards, ladders, maintenance kit, skimmer and vacuum system. Decks and electrical service to equipment are not included.

Residential in-ground pool construction can be divided into two categories: vinyl lined and gunite. Vinyl lined pool walls are constructed of different materials including wood, concrete, plastic or metal. The bottom is often graded with sand over which the vinyl liner is installed. Costs are generally in the $15 to $22 per S.F. range. Vermiculite or soil cement bottoms may be substituted for an added cost of $1.00 per S.F. surface.

Gunite pool construction is used both in residential and municipal installations. These structures are steel reinforced for strength and finished with a white cement limestone plaster. Residential costs run from $18 to $36 per S.F. surface. Municipal costs vary from $35 to $55 because plumbing codes require more expensive materials, chlorination equipment and higher filtration rates.

Municipal pools greater than 1,800 S.F. require gutter systems to control waves. This gutter may be formed into the concrete wall. Often a vinyl, stainless steel gutter or gutter and wall system is specified, which will raise the pool cost an additional $51 to $180 per L.F. of gutter installed and up to $290 per L.F. if a gutter and wall system is installed.

Competition pools usually require tile bottoms and sides with contrasting lane striping. Add $12 per S.F. of wall or bottom to be tiled.

R151-420   Plumbing Fixture Installation Time

Item	Rough-In	Set	Total Hours	Item	Rough-In	Set	Total Hours
Bathtub	5	5	10	Shower head only	2	1	3
Bathtub and shower, cast iron	6	6	12	Shower drain	3	1	4
Fire hose reel and cabinet	4	2	6	Shower stall, slate		15	15
Floor drain to 4 inch diameter	3	1	4	Slop sink	5	3	8
Grease trap, single, cast iron	5	3	8	Test 6 fixtures			14
Kitchen gas range		4	4	Urinal, wall	6	2	8
Kitchen sink, single	4	4	8	Urinal, pedestal or floor	6	4	10
Kitchen sink, double	6	6	12	Water closet and tank	4	3	7
Laundry tubs	4	2	6	Water closet and tank, wall hung	5	3	8
Lavatory wall hung	5	3	8	Water heater, 45 gals. gas, automatic	5	2	7
Lavatory pedestal	5	3	8	Water heaters, 65 gals. gas, automatic	5	2	7
Shower and stall	6	4	10	Water heaters, electric, plumbing only	4	2	6

Fixture prices in front of book are based on the cost per fixture set in place. The rough-in cost, which must be added for each fixture, includes carrier, if required, some supply, waste and vent pipe connecting fittings and stops. The lengths of rough-in pipe are nominal runs which would connect to the larger runs and stacks. The supply runs and DWV runs and stacks must be accounted for in separate entries. In the eastern half of the United States it is common for the plumber to carry these to a point 5' outside the building.

R155-050 Factor for Determining Heat Loss for Various Types of Buildings

General: While the most accurate estimates of heating requirements would naturally be based on detailed information about the building being considered, it is possible to arrive at a reasonable approximation using the following procedure:

1. Calculate the cubic volume of the room or building.
2. Select the appropriate factor from Table 1 below. Note that the factors apply only to inside temperatures listed in the first column and to 0°F outside temperature.
3. If the building has bad north and west exposures, multiply the heat loss factor by 1.1.
4. If the outside design temperature is other than 0°F, multiply the factor from Table 1 by the factor from Table 2.
5. Multiply the cubic volume by the factor selected from Table 1. This will give the estimated BTUH heat loss which must be made up to maintain inside temperature.

Table 1 — Building Type	Conditions	Qualifications	Loss Factor*
Factories & Industrial Plants General Office Areas at 70°F	One Story	Skylight in Roof	6.2
		No Skylight in Roof	5.7
	Multiple Story	Two Story	4.6
		Three Story	4.3
		Four Story	4.1
		Five Story	3.9
		Six Story	3.6
	All Walls Exposed	Flat Roof	6.9
		Heated Space Above	5.2
	One Long Warm Common Wall	Flat Roof	6.3
		Heated Space Above	4.7
	Warm Common Walls on Both Long Sides	Flat Roof	5.8
		Heated Space Above	4.1
Warehouses at 60°F	All Walls Exposed	Skylights in Roof	5.5
		No Skylight in Roof	5.1
		Heated Space Above	4.0
	One Long Warm Common Wall	Skylight in Roof	5.0
		No Skylight in Roof	4.9
		Heated Space Above	3.4
	Warm Common Walls on Both Long Sides	Skylight in Roof	4.7
		No Skylight in Roof	4.4
		Heated Space Above	3.0

*Note: This table tends to be conservative particularly for new buildings designed for minimum energy consumption.

Table 2 — Outside Design Temperature Correction Factor (for Degrees Fahrenheit)									
Outside Design Temperature	50	40	30	20	10	0	-10	-20	-30
Correction Factor	.29	.43	.57	.72	.86	1.00	1.14	1.28	1.43

R157-020 Air Conditioning Requirements

BTU's per hour per S.F. of floor area and S.F. per ton of air conditioning.

Type of Building	BTU per S.F.	S.F. per Ton	Type of Building	BTU per S.F.	S.F. per Ton	Type of Building	BTU per S.F.	S.F. per Ton
Apartments, Individual	26	450	Dormitory, Rooms	40	300	Libraries	50	240
Corridors	22	550	Corridors	30	400	Low Rise Office, Exterior	38	320
Auditoriums & Theaters	40	300/18*	Dress Shops	43	280	Interior	33	360
Banks	50	240	Drug Stores	80	150	Medical Centers	28	425
Barber Shops	48	250	Factories	40	300	Motels	28	425
Bars & Taverns	133	90	High Rise Office—Ext. Rms.	46	263	Office (small suite)	43	280
Beauty Parlors	66	180	Interior Rooms	37	325	Post Office, Individual Office	42	285
Bowling Alleys	68	175	Hospitals, Core	43	280	Central Area	46	260
Churches	36	330/20*	Perimeter	46	260	Residences	20	600
Cocktail Lounges	68	175	Hotel, Guest Rooms	44	275	Restaurants	60	200
Computer Rooms	141	85	Corridors	30	400	Schools & Colleges	46	260
Dental Offices	52	230	Public Spaces	55	220	Shoe Stores	55	220
Dept. Stores, Basement	34	350	Industrial Plants, Offices	38	320	Shop'g. Ctrs., Supermarkets	34	350
Main Floor	40	300	General Offices	34	350	Retail Stores	48	250
Upper Floor	30	400	Plant Areas	40	300	Specialty	60	200

*Persons per ton
12,000 BTU = 1 ton of air conditioning

R157-050 Ductwork

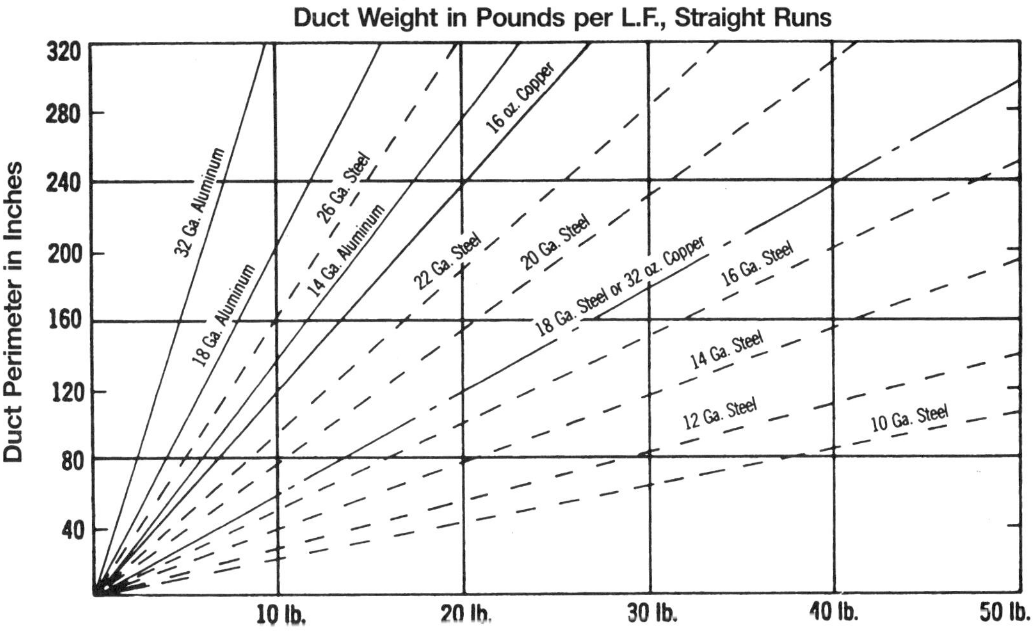

Add to the above for fittings; 90° elbow is 3 L.F.; 45° elbow is 2.5 L.F.; offset is 4 L.F.; transition offset is 6 L.F.; square-to-round transition is 4 L.F.; 90° reducing elbow is 5 L.F. For bracing and waste, add 20% to aluminum and copper, 15% to steel.

R160-205　Conductors in Conduit

Table below lists maximum number of conductors for various sized conduit using THW, TW or THWN insulations.

Copper Wire Size	½″			¾″			1″			1-¼″			1-½″			2″			2-½″			3″		3-½″		4″	
	TW	THW	THWN	TW	THW	THWN	TW	THW	THWN	TW	THW	THWN	TW	THW	THWN	TW	THW	THWN	TW	THW	THWN	THW	THWN	THW	THWN	THW	THWN
#14	9	6	13	15	10	24	25	16	39	44	29	69	60	40	94	99	65	154	142	93		143		192			
#12	7	4	10	12	8	18	19	13	29	35	24	51	47	32	70	78	53	114	111	76	164	117		157			
#10	5	4	6	9	6	11	15	11	18	26	19	32	36	26	44	60	43	73	85	61	104	95	160	127		163	
#8	2	1	3	4	3	5	7	5	9	12	10	16	17	13	22	28	22	36	40	32	51	49	79	66	106	85	136
#6		1	1		2	4		4	6		7	11		10	15		16	26		23	37	36	57	48	76	62	98
#4		1	1		1	2		3	4		5	7		7	9		12	16		17	22	27	35	36	47	47	60
#3		1	1		1	1		2	3		4	6		6	8		10	13		15	19	23	29	31	39	40	51
#2		1	1		1	1		2	3		4	5		5	7		9	11		13	16	20	25	27	33	34	43
#1					1	1		1	1		3	3		4	5		6	8		9	12	14	18	19	25	25	32
1/0					1	1		1	1		2	3		3	4		5	7		8	10	12	15	16	21	21	27
2/0					1	1		1	1		1	2		3	3		5	6		7	8	10	13	14	17	18	22
3/0						1			1		1	1		2	3		4	5		6	7	9	11	12	14	15	18
4/0						1		1	1		1	1		1	2		3	4		5	6	7	9	10	12	13	15
250　MCM								1	1		1	1		1	1		2	3		4	4	6	7	8	10	10	12
300								1	1		1	1		1	1		2	3		3	4	5	6	7	8	9	11
350									1		1	1		1	1		1	2		3	3	4	5	6	7	8	9
400											1	1		1	1		1	1		2	3	4	5	5	6	7	8
500											1	1		1	1		1	1		1	2	3	4	4	5	6	7
600												1		1	1		1	1		1	1	3	3	4	4	5	5
700														1	1		1	1		1	1	2	3	3	3	4	5
750														1	1		1	1		1	1	2	2	3	3	4	4

Crew A-1

Crew No.	Hr.	Daily	Hr.	Daily	Bare Costs	Incl. O&P
1 Building Laborer	$13.40	$107.20	$22.95	$183.60	$13.40	$22.95
1 Gas Eng. Power Tool		67.00		73.70	8.38	9.21
8 L.H., Daily Totals		$174.20		$257.30	$21.78	$32.16

Crew A-2

Crew No.	Hr.	Daily	Hr.	Daily	Bare Costs	Incl. O&P
2 Laborers	$13.40	$214.40	$22.95	$367.20	$13.97	$23.73
1 Truck Driver (light)	15.10	120.80	25.30	202.40		
1 Light Truck, 1.5 Ton		167.40		184.15	6.98	7.67
24 L.H., Daily Totals		$502.60		$753.75	$20.95	$31.40

Crew A-2A

Crew No.	Hr.	Daily	Hr.	Daily	Bare Costs	Incl. O&P
2 Laborers	$13.40	$214.40	$22.95	$367.20	$13.97	$23.73
1 Truck Driver (light)	15.10	120.80	25.30	202.40		
1 Light Truck, 1.5 ton		167.40		184.15		
1 Concrete Saw		107.35		118.10	11.45	12.59
24 L.H., Daily Totals		$609.95		$871.85	$25.42	$36.32

Crew A-3

Crew No.	Hr.	Daily	Hr.	Daily	Bare Costs	Incl. O&P
1 Truck Driver (heavy)	$15.40	$123.20	$25.80	$206.40	$15.40	$25.80
1 Dump Truck, 12 Ton		355.20		390.70	44.40	48.84
8 L.H., Daily Totals		$478.40		$597.10	$59.80	$74.64

Crew A-4

Crew No.	Hr.	Daily	Hr.	Daily	Bare Costs	Incl. O&P
2 Carpenters	$18.50	$296.00	$31.70	$507.20	$18.07	$30.72
1 Painter, Ordinary	17.20	137.60	28.75	230.00		
24 L.H., Daily Totals		$433.60		$737.20	$18.07	$30.72

Crew A-5

Crew No.	Hr.	Daily	Hr.	Daily	Bare Costs	Incl. O&P
2 Laborers	$13.40	$214.40	$22.95	$367.20	$13.59	$23.21
.25 Truck Driver (light)	15.10	30.20	25.30	50.60		
.25 Light Truck, 1.5 Ton		41.85		46.05	2.33	2.56
18 L.H., Daily Totals		$286.45		$463.85	$15.92	$25.77

Crew A-6

Crew No.	Hr.	Daily	Hr.	Daily	Bare Costs	Incl. O&P
1 Chief Of Party	$18.15	$145.20	$30.15	$241.20	$17.18	$28.52
1 Instrument Man	16.20	129.60	26.90	215.20		
16 L.H., Daily Totals		$274.80		$456.40	$17.18	$28.52

Crew A-7

Crew No.	Hr.	Daily	Hr.	Daily	Bare Costs	Incl. O&P
1 Chief Of Party	$18.15	$145.20	$30.15	$241.20	$16.07	$26.92
1 Instrument Man	16.20	129.60	26.90	215.20		
1 Rodman/Chainman	13.85	110.80	23.70	189.60		
24 L.H., Daily Totals		$385.60		$646.00	$16.07	$26.92

Crew A-8

Crew No.	Hr.	Daily	Hr.	Daily	Bare Costs	Incl. O&P
1 Chief Of Party	$18.15	$145.20	$30.15	$241.20	$15.51	$26.11
1 Instrument Man	16.20	129.60	26.90	215.20		
2 Rodmen/Chainmen	13.85	221.60	23.70	379.20		
32 L.H., Daily Totals		$496.40		$835.60	$15.51	$26.11

Crew A-9

Crew No.	Hr.	Daily	Hr.	Daily	Bare Costs	Incl. O&P
1 Asbestos Foreman	$20.05	$160.40	$35.15	$281.20	$19.61	$34.36
7 Asbestos Workers	19.55	1094.80	34.25	1918.00		
64 L.H., Daily Totals		$1255.20		$2199.20	$19.61	$34.36

Crew A-10

Crew No.	Hr.	Daily	Hr.	Daily	Bare Costs	Incl. O&P
1 Asbestos Foreman	$20.05	$160.40	$35.15	$281.20	$19.61	$34.36
7 Asbestos Workers	19.55	1094.80	34.25	1918.00		
64 L.H., Daily Totals		$1255.20		$2199.20	$19.61	$34.36

Crew A-10A

Crew No.	Hr.	Daily	Hr.	Daily	Bare Costs	Incl. O&P
1 Asbestos Foreman	$20.05	$160.40	$35.15	$281.20	$19.72	$34.55
2 Asbestos Workers	19.55	312.80	34.25	548.00		
24 L.H., Daily Totals		$473.20		$829.20	$19.72	$34.55

Crew A-10B

Crew No.	Hr.	Daily	Hr.	Daily	Bare Costs	Incl. O&P
1 Asbestos Foreman	$20.05	$160.40	$35.15	$281.20	$19.67	$34.48
3 Asbestos Workers	19.55	469.20	34.25	822.00		
32 L.H., Daily Totals		$629.60		$1103.20	$19.67	$34.48

Crew A-10C

Crew No.	Hr.	Daily	Hr.	Daily	Bare Costs	Incl. O&P
3 Asbestos Workers	$19.55	$469.20	$34.25	$822.00	$19.55	$34.25
1 Flatbed truck		167.40		184.15	6.98	7.67
24 L.H., Daily Totals		$636.60		$1006.15	$26.53	$41.92

Crew A-10D

Crew No.	Hr.	Daily	Hr.	Daily	Bare Costs	Incl. O&P
2 Asbestos Workers	$19.55	$312.80	$34.25	$548.00	$18.76	$32.05
1 Equip. Oper. (crane)	19.75	158.00	32.80	262.40		
1 Equip. Oper. Oiler	16.20	129.60	26.90	215.20		
1 Hydraulic crane, 33 ton		689.30		758.25	21.54	23.69
32 L.H., Daily Totals		$1289.70		$1783.85	$40.30	$55.74

Crew A-11

Crew No.	Hr.	Daily	Hr.	Daily	Bare Costs	Incl. O&P
1 Asbestos Foreman	$20.05	$160.40	$35.15	$281.20	$19.61	$34.36
7 Asbestos Workers	19.55	1094.80	34.25	1918.00		
2 Chipping Hammers		25.90		28.50	.40	.45
64 L.H., Daily Totals		$1281.10		$2227.70	$20.01	$34.81

Crew A-12

Crew No.	Hr.	Daily	Hr.	Daily	Bare Costs	Incl. O&P
1 Asbestos Foreman	$20.05	$160.40	$35.15	$281.20	$19.61	$34.36
7 Asbestos Workers	19.55	1094.80	34.25	1918.00		
1 Large Prod. Vac. Loader		512.10		563.30	8.00	8.80
64 L.H., Daily Totals		$1767.30		$2762.50	$27.61	$43.16

Crew A-13

Crew No.	Hr.	Daily	Hr.	Daily	Bare Costs	Incl. O&P
1 Equip. Oper. (light)	$18.15	$145.20	$30.15	$241.20	$18.15	$30.15
1 Large Prod. Vac. Loader		512.10		563.30	64.01	70.41
8 L.H., Daily Totals		$657.30		$804.50	$82.16	$100.56

Crew B-1

Crew No.	Hr.	Daily	Hr.	Daily	Bare Costs	Incl. O&P
1 Labor Foreman (outside)	$15.40	$123.20	$26.40	$211.20	$14.07	$24.10
2 Laborers	13.40	214.40	22.95	367.20		
24 L.H., Daily Totals		$337.60		$578.40	$14.07	$24.10

Crew B-2

Crew No.	Hr.	Daily	Hr.	Daily	Bare Costs	Incl. O&P
1 Labor Foreman (outside)	$15.40	$123.20	$26.40	$211.20	$13.80	$23.64
4 Laborers	13.40	428.80	22.95	734.40		
40 L.H., Daily Totals		$552.00		$945.60	$13.80	$23.64

Crew B-3

Crew No.	Hr.	Daily	Hr.	Daily	Bare Costs	Incl. O&P
1 Labor Foreman (outside)	$15.40	$123.20	$26.40	$211.20	$15.33	$25.91
2 Laborers	13.40	214.40	22.95	367.20		
1 Equip. Oper. (med.)	19.00	152.00	31.55	252.40		
2 Truck Drivers (heavy)	15.40	246.40	25.80	412.80		
1 F.E. Loader, T.M., 2.5 C.Y.		838.00		921.80		
2 Dump Trucks, 16 Ton		863.00		949.30	35.44	38.98
48 L.H., Daily Totals		$2437.00		$3114.70	$50.77	$64.89

Crew B-3A

Crew No.	Bare Costs Hr.	Daily	Incl. Subs O & P Hr.	Daily	Cost Per Labor-Hour Bare Costs	Incl. O&P
4 Laborers	$13.40	$428.80	$22.95	$734.40	$14.52	$24.67
1 Equip. Oper. (med.)	19.00	152.00	31.55	252.40		
1 Hyd. Excavator, 1.5 C.Y.		706.55		777.20	17.66	19.43
40 L.H., Daily Totals		$1287.35		$1764.00	$32.18	$44.10

Crew B-3B

Crew No.	Bare Costs Hr.	Daily	Incl. Subs O & P Hr.	Daily	Cost Per Labor-Hour Bare Costs	Incl. O&P
2 Laborers	$13.40	$214.40	$22.95	$367.20	$15.30	$25.81
1 Equip. Oper. (med.)	19.00	152.00	31.55	252.40		
1 Truck Drivers (heavy)	15.40	123.20	25.80	206.40		
1 Backhoe Loader, 80 H.P.		287.20		315.90		
1 Dump Trucks, 16 Ton		431.50		474.65	22.46	24.71
32 L.H., Daily Totals		$1208.30		$1616.55	$37.76	$50.52

Crew B-3C

Crew No.	Bare Costs Hr.	Daily	Incl. Subs O & P Hr.	Daily	Cost Per Labor-Hour Bare Costs	Incl. O&P
3 Laborers	$13.40	$321.60	$22.95	$550.80	$14.80	$25.10
1 Equip. Oper. (med.)	19.00	152.00	31.55	252.40		
1 F.E. Loader, 3.75 C.Y.		562.00		618.20	17.56	19.32
32 L.H., Daily Totals		$1035.60		$1421.40	$32.36	$44.42

Crew B-4

Crew No.	Bare Costs Hr.	Daily	Incl. Subs O & P Hr.	Daily	Cost Per Labor-Hour Bare Costs	Incl. O&P
1 Labor Foreman (outside)	$15.40	$123.20	$26.40	$211.20	$14.07	$24.00
4 Laborers	13.40	428.80	22.95	734.40		
1 Truck Driver (heavy)	15.40	123.20	25.80	206.40		
1 Tractor, 4 x 2, 195 H.P.		316.50		348.15		
1 Platform Trailer		146.30		160.95	9.64	10.61
48 L.H., Daily Totals		$1138.00		$1661.10	$23.71	$34.61

Crew B-5

Crew No.	Bare Costs Hr.	Daily	Incl. Subs O & P Hr.	Daily	Cost Per Labor-Hour Bare Costs	Incl. O&P
1 Labor Foreman (outside)	$15.40	$123.20	$26.40	$211.20	$14.92	$25.36
3 Laborers	13.40	321.60	22.95	550.80		
1 Equip. Oper. (med.)	19.00	152.00	31.55	252.40		
1 Air Compr., 250 C.F.M.		120.40		132.45		
2 Air Tools & Accessories		36.90		40.60		
2-50 Ft. Air Hoses, 1.5" Dia.		16.40		18.05		
1 F.E. Loader, T.M., 2.5 C.Y.		838.00		921.80	25.29	27.82
40 L.H., Daily Totals		$1608.50		$2127.30	$40.21	$53.18

Crew B-5A

Crew No.	Bare Costs Hr.	Daily	Incl. Subs O & P Hr.	Daily	Cost Per Labor-Hour Bare Costs	Incl. O&P
1 Foreman	$15.40	$123.20	$26.40	$211.20	$15.23	$25.75
6 Laborers	13.40	643.20	22.95	1101.60		
2 Equip. Oper. (med.)	19.00	304.00	31.55	504.80		
1 Equip. Oper. (light)	18.15	145.20	30.15	241.20		
2 Truck Drivers (heavy)	15.40	246.40	25.80	412.80		
1 Air Compr. 365 C.F.M.		171.15		188.25		
2 Pavement Breakers		36.90		40.60		
8 Air Hoses w/Coup.,1"		38.40		42.25		
2 Dump Trucks, 12 Ton		710.40		781.45	9.97	10.96
96 L.H., Daily Totals		$2418.85		$3524.15	$25.20	$36.71

Crew B-5B

Crew No.	Bare Costs Hr.	Daily	Incl. Subs O & P Hr.	Daily	Cost Per Labor-Hour Bare Costs	Incl. O&P
1 Powderman	$18.75	$150.00	$32.25	$258.00	$17.16	$28.79
2 Equip. Oper. (med.)	19.00	304.00	31.55	504.80		
3 Truck Drivers (heavy)	15.40	369.60	25.80	619.20		
1 F.E. Ldr. 2-1/2 CY		436.10		479.70		
3 Dump Trucks, 16 Ton		1294.50		1423.95		
1 Air Compr. 365 C.F.M.		171.15		188.25	39.62	43.58
48 L.H., Daily Totals		$2725.35		$3473.90	$56.78	$72.37

Crew B-5C

Crew No.	Bare Costs Hr.	Daily	Incl. Subs O & P Hr.	Daily	Cost Per Labor-Hour Bare Costs	Incl. O&P
3 Laborers	$13.40	$321.60	$22.95	$550.80	$15.74	$26.46
1 Equip. Oper. (medium)	19.00	152.00	31.55	252.40		
2 Truck Drivers (heavy)	15.40	246.40	25.80	412.80		
1 Equip. Oper. (crane)	19.75	158.00	32.80	262.40		
1 Equip. Oper. Oiler	16.20	129.60	26.90	215.20		
2 Dump Truck, 16 Ton		863.00		949.30		
1 F.E. Lder, 3.75 C.Y.		1132.00		1245.20		
1 Hyd. Crane (20T) W/B		563.05		619.35	39.97	43.97
64 L.H., Daily Totals		$3565.65		$4507.45	$55.71	$70.43

Crew B-6

Crew No.	Bare Costs Hr.	Daily	Incl. Subs O & P Hr.	Daily	Cost Per Labor-Hour Bare Costs	Incl. O&P
2 Laborers	$13.40	$214.40	$22.95	$367.20	$14.98	$25.35
1 Equip. Oper. (light)	18.15	145.20	30.15	241.20		
1 Backhoe Loader, 48 H.P.		213.80		235.20	8.91	9.80
24 L.H., Daily Totals		$573.40		$843.60	$23.89	$35.15

Crew B-7

Crew No.	Bare Costs Hr.	Daily	Incl. Subs O & P Hr.	Daily	Cost Per Labor-Hour Bare Costs	Incl. O&P
1 Labor Foreman (outside)	$15.40	$123.20	$26.40	$211.20	$14.67	$24.96
4 Laborers	13.40	428.80	22.95	734.40		
1 Equip. Oper. (med.)	19.00	152.00	31.55	252.40		
1 Chipping Machine		208.20		229.00		
1 F.E. Loader, T.M., 2.5 C.Y.		838.00		921.80		
2 Chain Saws		105.60		116.15	24.00	26.40
48 L.H., Daily Totals		$1855.80		$2464.95	$38.67	$51.36

Crew B-7A

Crew No.	Bare Costs Hr.	Daily	Incl. Subs O & P Hr.	Daily	Cost Per Labor-Hour Bare Costs	Incl. O&P
2 Laborers	$13.40	$214.40	$22.95	$367.20	$14.98	$25.35
1 Equip. Oper. (light)	18.15	145.20	30.15	241.20		
1 Rake w/Tractor		217.20		238.90		
2 Chain Saws		54.30		59.75	11.31	12.44
24 L.H., Daily Totals		$631.10		$907.05	$26.29	$37.79

Crew B-8

Crew No.	Bare Costs Hr.	Daily	Incl. Subs O & P Hr.	Daily	Cost Per Labor-Hour Bare Costs	Incl. O&P
1 Labor Foreman (outside)	$15.40	$123.20	$26.40	$211.20	$15.86	$26.71
2 Laborers	13.40	214.40	22.95	367.20		
2 Equip. Oper. (med.)	19.00	304.00	31.55	504.80		
2 Truck Drivers (heavy)	15.40	246.40	25.80	412.80		
1 Hyd. Crane, 25 Ton		538.85		592.75		
1 F.E. Loader, T.M., 2.5 C.Y.		838.00		921.80		
2 Dump Trucks, 16 Ton		863.00		949.30	40.00	44.00
56 L.H., Daily Totals		$3127.85		$3959.85	$55.86	$70.71

Crew B-9

Crew No.	Bare Costs Hr.	Daily	Incl. Subs O & P Hr.	Daily	Cost Per Labor-Hour Bare Costs	Incl. O&P
1 Labor Foreman (outside)	$15.40	$123.20	$26.40	$211.20	$13.80	$23.64
4 Laborers	13.40	428.80	22.95	734.40		
1 Air Compr., 250 C.F.M.		120.40		132.45		
2 Air Tools & Accessories		36.90		40.60		
2-50 Ft. Air Hoses, 1.5" Dia.		16.40		18.05	4.34	4.78
40 L.H., Daily Totals		$725.70		$1136.70	$18.14	$28.42

Crew B-9A

Crew No.	Bare Costs Hr.	Daily	Incl. Subs O & P Hr.	Daily	Cost Per Labor-Hour Bare Costs	Incl. O&P
2 Laborers	$13.40	$214.40	$22.95	$367.20	$14.07	$23.90
1 Truck Driver (heavy)	15.40	123.20	25.80	206.40		
1 Water Tanker		203.20		223.50		
1 Tractor		316.50		348.15		
2-50 Ft. Disch. Hoses		10.80		11.90	22.10	24.31
24 L.H., Daily Totals		$868.10		$1157.15	$36.17	$48.21

Crew B-9B

Crew No.	Bare Costs Hr.	Bare Costs Daily	Incl. Subs O&P Hr.	Incl. Subs O&P Daily	Cost Per LH Bare Costs	Cost Per LH Incl. O&P
2 Laborers	$13.40	$214.40	$22.95	$367.20	$14.07	$23.90
1 Truck Driver (heavy)	15.40	123.20	25.80	206.40		
2-50 Ft. Disch. HoseS		10.80		11.90		
1 Water Tanker		203.20		223.50		
1 Tractor		316.50		348.15		
1 Pressure Washer		55.40		60.95	24.41	26.85
24 L.H., Daily Totals		$923.50		$1218.10	$38.48	$50.75

Crew B-9C

Crew No.	Bare Costs Hr.	Bare Costs Daily	Incl. Subs O&P Hr.	Incl. Subs O&P Daily	Cost Per LH Bare Costs	Cost Per LH Incl. O&P
1 Labor Foreman (outside)	$15.40	$123.20	$26.40	$211.20	$13.80	$23.64
4 Laborers	13.40	428.80	22.95	734.40		
1 Air Compr., 250 C.F.M.		120.40		132.45		
2-50 Ft. Air Hoses, 1.5" Dia.		16.40		18.05		
2 Breaker, Pavement, 60 lb.		36.90		40.60	4.34	4.78
40 L.H., Daily Totals		$725.70		$1136.70	$18.14	$28.42

Crew B-10

Crew No.	Bare Costs Hr.	Bare Costs Daily	Incl. Subs O&P Hr.	Incl. Subs O&P Daily	Cost Per LH Bare Costs	Cost Per LH Incl. O&P
1 Equip. Oper. (med.)	$19.00	$152.00	$31.55	$252.40	$19.00	$31.55
8 L.H., Daily Totals		$152.00		$252.40	$19.00	$31.55

Crew B-10A

Crew No.	Bare Costs Hr.	Bare Costs Daily	Incl. Subs O&P Hr.	Incl. Subs O&P Daily	Cost Per LH Bare Costs	Cost Per LH Incl. O&P
1 Equip. Oper. (med.)	$19.00	$152.00	$31.55	$252.40	$19.00	$31.55
1 Roll. Compact., 2K Lbs.		90.65		99.70	11.33	12.46
8 L.H., Daily Totals		$242.65		$352.10	$30.33	$44.01

Crew B-10B

Crew No.	Bare Costs Hr.	Bare Costs Daily	Incl. Subs O&P Hr.	Incl. Subs O&P Daily	Cost Per LH Bare Costs	Cost Per LH Incl. O&P
1 Equip. Oper. (med.)	$19.00	$152.00	$31.55	$252.40	$19.00	$31.55
1 Dozer, 200 H.P.		832.00		915.20	104.00	114.40
8 L.H., Daily Totals		$984.00		$1167.60	$123.00	$145.95

Crew B-10C

Crew No.	Bare Costs Hr.	Bare Costs Daily	Incl. Subs O&P Hr.	Incl. Subs O&P Daily	Cost Per LH Bare Costs	Cost Per LH Incl. O&P
1 Equip. Oper. (med.)	$19.00	$152.00	$31.55	$252.40	$19.00	$31.55
1 Dozer, 200 H.P.		832.00		915.20		
1 Vibratory Roller, Towed		110.00		121.00	117.75	129.53
8 L.H., Daily Totals		$1094.00		$1288.60	$136.75	$161.08

Crew B-10D

Crew No.	Bare Costs Hr.	Bare Costs Daily	Incl. Subs O&P Hr.	Incl. Subs O&P Daily	Cost Per LH Bare Costs	Cost Per LH Incl. O&P
1 Equip. Oper. (med.)	$19.00	$152.00	$31.55	$252.40	$19.00	$31.55
1 Dozer, 200 H.P		832.00		915.20		
1 Sheepsft. Roller, Towed		120.80		132.90	119.10	131.01
8 L.H., Daily Totals		$1104.80		$1300.50	$138.10	$162.56

Crew B-10E

Crew No.	Bare Costs Hr.	Bare Costs Daily	Incl. Subs O&P Hr.	Incl. Subs O&P Daily	Cost Per LH Bare Costs	Cost Per LH Incl. O&P
1 Equip. Oper. (med.)	$19.00	$152.00	$31.55	$252.40	$19.00	$31.55
1 Tandem Roller, 5 Ton		148.40		163.25	18.55	20.41
8 L.H., Daily Totals		$300.40		$415.65	$37.55	$51.96

Crew B-10F

Crew No.	Bare Costs Hr.	Bare Costs Daily	Incl. Subs O&P Hr.	Incl. Subs O&P Daily	Cost Per LH Bare Costs	Cost Per LH Incl. O&P
1 Equip. Oper. (med.)	$19.00	$152.00	$31.55	$252.40	$19.00	$31.55
1 Tandem Roller, 10 Ton		238.00		261.80	29.75	32.73
8 L.H., Daily Totals		$390.00		$514.20	$48.75	$64.28

Crew B-10G

Crew No.	Bare Costs Hr.	Bare Costs Daily	Incl. Subs O&P Hr.	Incl. Subs O&P Daily	Cost Per LH Bare Costs	Cost Per LH Incl. O&P
1 Equip. Oper. (med.)	$19.00	$152.00	$31.55	$252.40	$19.00	$31.55
1 Sheepsft. Roll., 130 H.P.		542.80		597.10	67.85	74.64
8 L.H., Daily Totals		$694.80		$849.50	$86.85	$106.19

Crew B-10H

Crew No.	Bare Costs Hr.	Bare Costs Daily	Incl. Subs O&P Hr.	Incl. Subs O&P Daily	Cost Per LH Bare Costs	Cost Per LH Incl. O&P
1 Equip. Oper. (med.)	$19.00	$152.00	$31.55	$252.40	$19.00	$31.55
1 Diaphr. Water Pump, 2"		28.00		30.80		
1-20 Ft. Suction Hose, 2"		6.50		7.15		
2-50 Ft. Disch. Hoses, 2"		8.80		9.70	5.41	5.95
8 L.H., Daily Totals		$195.30		$300.05	$24.41	$37.50

Crew B-10I

Crew No.	Bare Costs Hr.	Bare Costs Daily	Incl. Subs O&P Hr.	Incl. Subs O&P Daily	Cost Per LH Bare Costs	Cost Per LH Incl. O&P
1 Equip. Oper. (med.)	$19.00	$152.00	$31.55	$252.40	$19.00	$31.55
1 Diaphr. Water Pump, 4"		62.00		68.20		
1-20 Ft. Suction Hose, 4"		12.50		13.75		
2-50 Ft. Disch. Hoses, 4"		17.00		18.70	11.44	12.58
8 L.H., Daily Totals		$243.50		$353.05	$30.44	$44.13

Crew B-10J

Crew No.	Bare Costs Hr.	Bare Costs Daily	Incl. Subs O&P Hr.	Incl. Subs O&P Daily	Cost Per LH Bare Costs	Cost Per LH Incl. O&P
1 Equip. Oper. (med.)	$19.00	$152.00	$31.55	$252.40	$19.00	$31.55
1 Centr. Water Pump, 3"		36.60		40.25		
1-20 Ft. Suction Hose, 3"		9.50		10.45		
2-50 Ft. Disch. Hoses, 3"		10.80		11.90	7.11	7.82
8 L.H., Daily Totals		$208.90		$315.00	$26.11	$39.37

Crew B-10K

Crew No.	Bare Costs Hr.	Bare Costs Daily	Incl. Subs O&P Hr.	Incl. Subs O&P Daily	Cost Per LH Bare Costs	Cost Per LH Incl. O&P
1 Equip. Oper. (med.)	$19.00	$152.00	$31.55	$252.40	$19.00	$31.55
1 Centr. Water Pump, 6"		155.20		170.70		
1-20 Ft. Suction Hose, 6"		20.50		22.55		
2-50 Ft. Disch. Hoses, 6"		37.00		40.70	26.59	29.25
8 L.H., Daily Totals		$364.70		$486.35	$45.59	$60.80

Crew B-10L

Crew No.	Bare Costs Hr.	Bare Costs Daily	Incl. Subs O&P Hr.	Incl. Subs O&P Daily	Cost Per LH Bare Costs	Cost Per LH Incl. O&P
1 Equip. Oper. (med.)	$19.00	$152.00	$31.55	$252.40	$19.00	$31.55
1 Dozer, 75 H.P.		305.20		335.70	38.15	41.97
8 L.H., Daily Totals		$457.20		$588.10	$57.15	$73.52

Crew B-10M

Crew No.	Bare Costs Hr.	Bare Costs Daily	Incl. Subs O&P Hr.	Incl. Subs O&P Daily	Cost Per LH Bare Costs	Cost Per LH Incl. O&P
1 Equip. Oper. (med.)	$19.00	$152.00	$31.55	$252.40	$19.00	$31.55
1 Dozer, 300 H.P.		1122.00		1234.20	140.25	154.28
8 L.H., Daily Totals		$1274.00		$1486.60	$159.25	$185.83

Crew B-10N

Crew No.	Bare Costs Hr.	Bare Costs Daily	Incl. Subs O&P Hr.	Incl. Subs O&P Daily	Cost Per LH Bare Costs	Cost Per LH Incl. O&P
1 Equip. Oper. (med.)	$19.00	$152.00	$31.55	$252.40	$19.00	$31.55
1 F.E. Loader, T.M., 1.5 C.Y		357.20		392.90	44.65	49.12
8 L.H., Daily Totals		$509.20		$645.30	$63.65	$80.67

Crew B-10O

Crew No.	Bare Costs Hr.	Bare Costs Daily	Incl. Subs O&P Hr.	Incl. Subs O&P Daily	Cost Per LH Bare Costs	Cost Per LH Incl. O&P
1 Equip. Oper. (med.)	$19.00	$152.00	$31.55	$252.40	$19.00	$31.55
1 F.E. Loader, T.M., 2.25 C.Y.		496.00		545.60	62.00	68.20
8 L.H., Daily Totals		$648.00		$798.00	$81.00	$99.75

Crew B-10P

Crew No.	Bare Costs Hr.	Bare Costs Daily	Incl. Subs O&P Hr.	Incl. Subs O&P Daily	Cost Per LH Bare Costs	Cost Per LH Incl. O&P
1 Equip. Oper. (med.)	$19.00	$152.00	$31.55	$252.40	$19.00	$31.55
1 F.E. Loader, T.M., 2.5 C.Y.		838.00		921.80	104.75	115.23
8 L.H., Daily Totals		$990.00		$1174.20	$123.75	$146.78

Crew B-10Q

Crew No.	Bare Costs Hr.	Bare Costs Daily	Incl. Subs O&P Hr.	Incl. Subs O&P Daily	Cost Per LH Bare Costs	Cost Per LH Incl. O&P
1 Equip. Oper. (med.)	$19.00	$152.00	$31.55	$252.40	$19.00	$31.55
1 F.E. Loader, T.M., 5 C.Y.		1132.00		1245.20	141.50	155.65
8 L.H., Daily Totals		$1284.00		$1497.60	$160.50	$187.20

Crew B-10R

	Bare Costs Hr.	Daily	Incl. Subs O & P Hr.	Daily	Cost Per Labor-Hour Bare Costs	Incl. O&P
1 Equip. Oper. (med.)	$19.00	$152.00	$31.55	$252.40	$19.00	$31.55
1 F.E. Loader, W.M., 1 C.Y.		236.50		260.15	29.56	32.52
8 L.H., Daily Totals		$388.50		$512.55	$48.56	$64.07

Crew B-10S

	Bare Costs Hr.	Daily	Incl. Subs O & P Hr.	Daily	Cost Per Labor-Hour Bare Costs	Incl. O&P
1 Equip. Oper. (med.)	$19.00	$152.00	$31.55	$252.40	$19.00	$31.55
1 F.E. Loader, W.M., 1.5 C.Y.		312.30		343.55	39.04	42.94
8 L.H., Daily Totals		$464.30		$595.95	$58.04	$74.49

Crew B-10T

	Bare Costs Hr.	Daily	Incl. Subs O & P Hr.	Daily	Cost Per Labor-Hour Bare Costs	Incl. O&P
1 Equip. Oper. (med.)	$19.00	$152.00	$31.55	$252.40	$19.00	$31.55
1 F.E. Ldr, W.M.,2.5CY		436.10		479.70	54.51	59.96
8 L.H., Daily Totals		$588.10		$732.10	$73.51	$91.51

Crew B-10U

	Bare Costs Hr.	Daily	Incl. Subs O & P Hr.	Daily	Cost Per Labor-Hour Bare Costs	Incl. O&P
1 Equip. Oper. (med.)	$19.00	$152.00	$31.55	$252.40	$19.00	$31.55
1 F.E. Loader, W.M., 5.5 C.Y.		897.50		987.25	112.19	123.41
8 L.H., Daily Totals		$1049.50		$1239.65	$131.19	$154.96

Crew B-10V

	Bare Costs Hr.	Daily	Incl. Subs O & P Hr.	Daily	Cost Per Labor-Hour Bare Costs	Incl. O&P
1 Equip. Oper. (med.)	$19.00	$152.00	$31.55	$252.40	$19.00	$31.55
1 Dozer, 700 H.P.		2742.00		3016.20	342.75	377.03
8 L.H., Daily Totals		$2894.00		$3268.60	$361.75	$408.58

Crew B-10W

	Bare Costs Hr.	Daily	Incl. Subs O & P Hr.	Daily	Cost Per Labor-Hour Bare Costs	Incl. O&P
1 Equip. Oper. (med.)	$19.00	$152.00	$31.55	$252.40	$19.00	$31.55
1 Dozer, 105 H.P.		414.00		455.40	51.75	56.93
8 L.H., Daily Totals		$566.00		$707.80	$70.75	$88.48

Crew B-10X

	Bare Costs Hr.	Daily	Incl. Subs O & P Hr.	Daily	Cost Per Labor-Hour Bare Costs	Incl. O&P
1 Equip. Oper. (med.)	$19.00	$152.00	$31.55	$252.40	$19.00	$31.55
1 Dozer, 410 H.P.		1329.00		1461.90	166.13	182.74
8 L.H., Daily Totals		$1481.00		$1714.30	$185.13	$214.29

Crew B-10Y

	Bare Costs Hr.	Daily	Incl. Subs O & P Hr.	Daily	Cost Per Labor-Hour Bare Costs	Incl. O&P
1 Equip. Oper. (med.)	$19.00	$152.00	$31.55	$252.40	$19.00	$31.55
1 Vibratory Drum Roller		341.20		375.30	42.65	46.92
8 L.H., Daily Totals		$493.20		$627.70	$61.65	$78.47

Crew B-11

	Bare Costs Hr.	Daily	Incl. Subs O & P Hr.	Daily	Cost Per Labor-Hour Bare Costs	Incl. O&P
1 Equipment Oper. (med.)	$19.00	$152.00	$31.55	$252.40	$16.20	$27.25
1 Laborer	13.40	107.20	22.95	183.60		
16 L.H., Daily Totals		$259.20		$436.00	$16.20	$27.25

Crew B-11A

	Bare Costs Hr.	Daily	Incl. Subs O & P Hr.	Daily	Cost Per Labor-Hour Bare Costs	Incl. O&P
1 Equipment Oper. (med.)	$19.00	$152.00	$31.55	$252.40	$16.20	$27.25
1 Laborer	13.40	107.20	22.95	183.60		
1 Dozer, 200 H.P.		832.00		915.20	52.00	57.20
16 L.H., Daily Totals		$1091.20		$1351.20	$68.20	$84.45

Crew B-11B

	Bare Costs Hr.	Daily	Incl. Subs O & P Hr.	Daily	Cost Per Labor-Hour Bare Costs	Incl. O&P
1 Equipment Oper. (med.)	$19.00	$152.00	$31.55	$252.40	$16.20	$27.25
1 Laborer	13.40	107.20	22.95	183.60		
1 Dozer, 200 H.P.		832.00		915.20		
1 Air Powered Tamper		14.80		16.30		
1 Air Compr. 365 C.F.M.		171.15		188.25		
2-50 Ft. Air Hoses, 1.5" Dia.		16.40		18.05	64.65	71.11
16 L.H., Daily Totals		$1293.55		$1573.80	$80.85	$98.36

Crew B-11C

	Bare Costs Hr.	Daily	Incl. Subs O & P Hr.	Daily	Cost Per Labor-Hour Bare Costs	Incl. O&P
1 Equipment Oper. (med.)	$19.00	$152.00	$31.55	$252.40	$16.20	$27.25
1 Laborer	13.40	107.20	22.95	183.60		
1 Backhoe Loader, 48 H.P.		213.80		235.20	13.36	14.70
16 L.H., Daily Totals		$473.00		$671.20	$29.56	$41.95

Crew B-11K

	Bare Costs Hr.	Daily	Incl. Subs O & P Hr.	Daily	Cost Per Labor-Hour Bare Costs	Incl. O&P
1 Equipment Oper. (med.)	$19.00	$152.00	$31.55	$252.40	$16.20	$27.25
1 Laborer	13.40	107.20	22.95	183.60		
1 Trencher, 8' D., 16" W.		526.80		579.50	32.93	36.22
16 L.H., Daily Totals		$786.00		$1015.50	$49.13	$63.47

Crew B-11L

	Bare Costs Hr.	Daily	Incl. Subs O & P Hr.	Daily	Cost Per Labor-Hour Bare Costs	Incl. O&P
1 Equipment Oper. (med.)	$19.00	$152.00	$31.55	$252.40	$16.20	$27.25
1 Laborer	13.40	107.20	22.95	183.60		
1 Grader, 30,000 Lbs.		558.40		614.25	34.90	38.39
16 L.H., Daily Totals		$817.60		$1050.25	$51.10	$65.64

Crew B-11M

	Bare Costs Hr.	Daily	Incl. Subs O & P Hr.	Daily	Cost Per Labor-Hour Bare Costs	Incl. O&P
1 Equipment Oper. (med.)	$19.00	$152.00	$31.55	$252.40	$16.20	$27.25
1 Laborer	13.40	107.20	22.95	183.60		
1 Backhoe Loader, 80 H.P.		287.20		315.90	17.95	19.75
16 L.H., Daily Totals		$546.40		$751.90	$34.15	$47.00

Crew B-12

	Bare Costs Hr.	Daily	Incl. Subs O & P Hr.	Daily	Cost Per Labor-Hour Bare Costs	Incl. O&P
1 Equip. Oper. (crane)	$19.75	$158.00	$32.80	$262.40	$19.75	$32.80
8 L.H., Daily Totals		$158.00		$262.40	$19.75	$32.80

Crew B-12A

	Bare Costs Hr.	Daily	Incl. Subs O & P Hr.	Daily	Cost Per Labor-Hour Bare Costs	Incl. O&P
1 Equip. Oper. (crane)	$19.75	$158.00	$32.80	$262.40	$19.75	$32.80
1 Hyd. Excavator, 1 C.Y.		532.40		585.65	66.55	73.21
8 L.H., Daily Totals		$690.40		$848.05	$86.30	$106.01

Crew B-12B

	Bare Costs Hr.	Daily	Incl. Subs O & P Hr.	Daily	Cost Per Labor-Hour Bare Costs	Incl. O&P
1 Equip. Oper. (crane)	$19.75	$158.00	$32.80	$262.40	$19.75	$32.80
1 Hyd. Excavator, 1.5 C.Y.		706.55		777.20	88.32	97.15
8 L.H., Daily Totals		$864.55		$1039.60	$108.07	$129.95

Crew B-12C

	Bare Costs Hr.	Daily	Incl. Subs O & P Hr.	Daily	Cost Per Labor-Hour Bare Costs	Incl. O&P
1 Equip. Oper. (crane)	$19.75	$158.00	$32.80	$262.40	$19.75	$32.80
1 Hyd. Excavator, 2 C.Y.		1013.00		1114.30	126.63	139.29
8 L.H., Daily Totals		$1171.00		$1376.70	$146.38	$172.09

Crew B-12D

	Bare Costs Hr.	Daily	Incl. Subs O & P Hr.	Daily	Cost Per Labor-Hour Bare Costs	Incl. O&P
1 Equip. Oper. (crane)	$19.75	$158.00	$32.80	$262.40	$19.75	$32.80
1 Hyd. Excavator, 3.5 C.Y.		2117.00		2328.70	264.63	291.09
8 L.H., Daily Totals		$2275.00		$2591.10	$284.38	$323.89

Crew B-12E

	Bare Costs Hr.	Daily	Incl. Subs O & P Hr.	Daily	Cost Per Labor-Hour Bare Costs	Incl. O&P
1 Equip. Oper. (crane)	$19.75	$158.00	$32.80	$262.40	$19.75	$32.80
1 Hyd. Excavator, .5 C.Y.		333.80		367.20	41.73	45.90
8 L.H., Daily Totals		$491.80		$629.60	$61.48	$78.70

Crew B-12F

	Bare Costs Hr.	Daily	Incl. Subs O & P Hr.	Daily	Cost Per Labor-Hour Bare Costs	Incl. O&P
1 Equip. Oper. (crane)	$19.75	$158.00	$32.80	$262.40	$19.75	$32.80
1 Hyd. Excavator, .75 C.Y.		435.05		478.55	54.38	59.82
8 L.H., Daily Totals		$593.05		$740.95	$74.13	$92.62

Crew B-12G

Crew No.	Bare Costs Hr.	Bare Costs Daily	Incl. Subs O & P Hr.	Incl. Subs O & P Daily	Cost Per Labor-Hour Bare Costs	Cost Per Labor-Hour Incl. O&P
1 Equip. Oper. (crane)	$19.75	$158.00	$32.80	$262.40	$19.75	$32.80
1 Power Shovel, .5 C.Y.		447.60		492.35		
1 Clamshell Bucket, .5 C.Y		48.65		53.50	62.03	68.23
8 L.H., Daily Totals		$654.25		$808.25	$81.78	$101.03

Crew B-12H

Crew No.	Bare Costs Hr.	Bare Costs Daily	Incl. Subs O & P Hr.	Incl. Subs O & P Daily	Cost Per Labor-Hour Bare Costs	Cost Per Labor-Hour Incl. O&P
1 Equip. Oper. (crane)	$19.75	$158.00	$32.80	$262.40	$19.75	$32.80
1 Power Shovel, 1 C.Y.		496.90		546.60		
1 Clamshell Bucket, 1 C.Y.		64.80		71.30	70.21	77.23
8 L.H., Daily Totals		$719.70		$880.30	$89.96	$110.03

Crew B-12I

Crew No.	Bare Costs Hr.	Bare Costs Daily	Incl. Subs O & P Hr.	Incl. Subs O & P Daily	Cost Per Labor-Hour Bare Costs	Cost Per Labor-Hour Incl. O&P
1 Equip. Oper. (crane)	$19.75	$158.00	$32.80	$262.40	$19.75	$32.80
1 Power Shovel, .75 C.Y.		462.90		509.20		
1 Dragline Bucket, .75 C.Y.		32.40		35.65	61.91	68.10
8 L.H., Daily Totals		$653.30		$807.25	$81.66	$100.90

Crew B-12J

Crew No.	Bare Costs Hr.	Bare Costs Daily	Incl. Subs O & P Hr.	Incl. Subs O & P Daily	Cost Per Labor-Hour Bare Costs	Cost Per Labor-Hour Incl. O&P
1 Equip. Oper. (crane)	$19.75	$158.00	$32.80	$262.40	$19.75	$32.80
1 Gradall, 3 Ton, .5 C.Y.		625.60		688.15	78.20	86.02
8 L.H., Daily Totals		$783.60		$950.55	$97.95	$118.82

Crew B-12K

Crew No.	Bare Costs Hr.	Bare Costs Daily	Incl. Subs O & P Hr.	Incl. Subs O & P Daily	Cost Per Labor-Hour Bare Costs	Cost Per Labor-Hour Incl. O&P
1 Equip. Oper. (crane)	$19.75	$158.00	$32.80	$262.40	$19.75	$32.80
1 Gradall, 3 Ton, 1 C.Y.		789.20		868.10	98.65	108.52
8 L.H., Daily Totals		$947.20		$1130.50	$118.40	$141.32

Crew B-12L

Crew No.	Bare Costs Hr.	Bare Costs Daily	Incl. Subs O & P Hr.	Incl. Subs O & P Daily	Cost Per Labor-Hour Bare Costs	Cost Per Labor-Hour Incl. O&P
1 Equip. Oper. (crane)	$19.75	$158.00	$32.80	$262.40	$19.75	$32.80
1 Power Shovel, .5 C.Y.		447.60		492.35		
1 F.E. Attachment, .5 C.Y.		53.80		59.20	62.68	68.94
8 L.H., Daily Totals		$659.40		$813.95	$82.43	$101.74

Crew B-12M

Crew No.	Bare Costs Hr.	Bare Costs Daily	Incl. Subs O & P Hr.	Incl. Subs O & P Daily	Cost Per Labor-Hour Bare Costs	Cost Per Labor-Hour Incl. O&P
1 Equip. Oper. (crane)	$19.75	$158.00	$32.80	$262.40	$19.75	$32.80
1 Power Shovel, .75 C.Y.		462.90		509.20		
1 F.E. Attachment, .75 C.Y.		99.80		109.80	70.34	77.37
8 L.H., Daily Totals		$720.70		$881.40	$90.09	$110.17

Crew B-12N

Crew No.	Bare Costs Hr.	Bare Costs Daily	Incl. Subs O & P Hr.	Incl. Subs O & P Daily	Cost Per Labor-Hour Bare Costs	Cost Per Labor-Hour Incl. O&P
1 Equip. Oper. (crane)	$19.75	$158.00	$32.80	$262.40	$19.75	$32.80
1 Power Shovel, 1 C.Y.		496.90		546.60		
1 F.E. Attachment, 1 C.Y.		127.20		139.90	78.01	85.81
8 L.H., Daily Totals		$782.10		$948.90	$97.76	$118.61

Crew B-12O

Crew No.	Bare Costs Hr.	Bare Costs Daily	Incl. Subs O & P Hr.	Incl. Subs O & P Daily	Cost Per Labor-Hour Bare Costs	Cost Per Labor-Hour Incl. O&P
1 Equip. Oper. (crane)	$19.75	$158.00	$32.80	$262.40	$19.75	$32.80
1 Power Shovel, 1.5 C.Y.		757.20		832.90		
1 F.E. Attachment, 1.5 C.Y.		156.70		172.35	114.24	125.66
8 L.H., Daily Totals		$1071.90		$1267.65	$133.99	$158.46

Crew B-12P

Crew No.	Bare Costs Hr.	Bare Costs Daily	Incl. Subs O & P Hr.	Incl. Subs O & P Daily	Cost Per Labor-Hour Bare Costs	Cost Per Labor-Hour Incl. O&P
1 Equip. Oper. (crane)	$19.75	$158.00	$32.80	$262.40	$19.75	$32.80
1 Crawler Crane, 40 Ton		757.20		832.90		
1 Dragline Bucket, 1.5 C.Y.		47.00		51.70	100.53	110.58
8 L.H., Daily Totals		$962.20		$1147.00	$120.28	$143.38

Crew B-12Q

Crew No.	Bare Costs Hr.	Bare Costs Daily	Incl. Subs O & P Hr.	Incl. Subs O & P Daily	Cost Per Labor-Hour Bare Costs	Cost Per Labor-Hour Incl. O&P
1 Equip. Oper. (crane)	$19.75	$158.00	$32.80	$262.40	$19.75	$32.80
1 Hyd. Excavator, 5/8 C.Y.		395.65		435.20	49.46	54.40
8 L.H., Daily Totals		$553.65		$697.60	$69.21	$87.20

Crew B-12R

Crew No.	Bare Costs Hr.	Bare Costs Daily	Incl. Subs O & P Hr.	Incl. Subs O & P Daily	Cost Per Labor-Hour Bare Costs	Cost Per Labor-Hour Incl. O&P
1 Equip. Oper. (crane)	$19.75	$158.00	$32.80	$262.40	$19.75	$32.80
1 Hyd. Excavator, 1.5 C.Y.		706.55		777.20	88.32	97.15
8 L.H., Daily Totals		$864.55		$1039.60	$108.07	$129.95

Crew B-12S

Crew No.	Bare Costs Hr.	Bare Costs Daily	Incl. Subs O & P Hr.	Incl. Subs O & P Daily	Cost Per Labor-Hour Bare Costs	Cost Per Labor-Hour Incl. O&P
1 Equip. Oper. (crane)	$19.75	$158.00	$32.80	$262.40	$19.75	$32.80
1 Hyd. Excavator, 2.5 C.Y.		1681.00		1849.10	210.13	231.14
8 L.H., Daily Totals		$1839.00		$2111.50	$229.88	$263.94

Crew B-12T

Crew No.	Bare Costs Hr.	Bare Costs Daily	Incl. Subs O & P Hr.	Incl. Subs O & P Daily	Cost Per Labor-Hour Bare Costs	Cost Per Labor-Hour Incl. O&P
1 Equip. Oper. (crane)	$19.75	$158.00	$32.80	$262.40	$19.75	$32.80
1 Crawler Crane, 75 Ton		959.10		1055.00		
1 F.E. Attachment, 3 C.Y.		288.40		317.25	155.94	171.53
8 L.H., Daily Totals		$1405.50		$1634.65	$175.69	$204.33

Crew B-12V

Crew No.	Bare Costs Hr.	Bare Costs Daily	Incl. Subs O & P Hr.	Incl. Subs O & P Daily	Cost Per Labor-Hour Bare Costs	Cost Per Labor-Hour Incl. O&P
1 Equip. Oper. (crane)	$19.75	$158.00	$32.80	$262.40	$19.75	$32.80
1 Crawler Crane, 75 Ton		959.10		1055.00		
1 Dragline Bucket, 3 C.Y.		80.40		88.45	129.94	142.93
8 L.H., Daily Totals		$1197.50		$1405.85	$149.69	$175.73

Crew B-13

Crew No.	Bare Costs Hr.	Bare Costs Daily	Incl. Subs O & P Hr.	Incl. Subs O & P Daily	Cost Per Labor-Hour Bare Costs	Cost Per Labor-Hour Incl. O&P
1 Labor Foreman (outside)	$15.40	$123.20	$26.40	$211.20	$14.79	$25.17
4 Laborers	13.40	428.80	22.95	734.40		
1 Equip. Oper. (crane)	19.75	158.00	32.80	262.40		
1 Hyd. Crane, 55 Ton		785.20		863.70	16.36	17.99
48 L.H., Daily Totals		$1495.20		$2071.70	$31.15	$43.16

Crew B-13A

Crew No.	Bare Costs Hr.	Bare Costs Daily	Incl. Subs O & P Hr.	Incl. Subs O & P Daily	Cost Per Labor-Hour Bare Costs	Cost Per Labor-Hour Incl. O&P
1 Foreman	$15.40	$123.20	$26.40	$211.20	$15.86	$26.71
2 Laborers	13.40	214.40	22.95	367.20		
2 Equipment Operator	19.00	304.00	31.55	504.80		
2 Truck Drivers (heavy)	15.40	246.40	25.80	412.80		
1 Crane, 75 Ton		959.10		1055.00		
1 F.E. Lder, 3.75 C.Y.		1132.00		1245.20		
2 Dump Trucks, 12 Ton		710.40		781.45	50.03	55.03
56 L.H., Daily Totals		$3689.50		$4577.65	$65.89	$81.74

Crew B-13B

Crew No.	Bare Costs Hr.	Bare Costs Daily	Incl. Subs O & P Hr.	Incl. Subs O & P Daily	Cost Per Labor-Hour Bare Costs	Cost Per Labor-Hour Incl. O&P
1 Labor Foreman (outside)	$15.40	$123.20	$26.40	$211.20	$14.99	$25.41
4 Laborers	13.40	428.80	22.95	734.40		
1 Equip. Oper. (crane)	19.75	158.00	32.80	262.40		
1 Equip. Oper. Oiler	16.20	129.60	26.90	215.20		
1 Hyd. Crane, 55 Ton		785.20		863.70	14.02	15.42
56 L.H., Daily Totals		$1624.80		$2286.90	$29.01	$40.83

Crew B-13C

Crew No.	Bare Costs Hr.	Bare Costs Daily	Incl. Subs O & P Hr.	Incl. Subs O & P Daily	Cost Per Labor-Hour Bare Costs	Cost Per Labor-Hour Incl. O&P
1 Labor Foreman (outside)	$15.40	$123.20	$26.40	$211.20	$14.99	$25.41
4 Laborers	13.40	428.80	22.95	734.40		
1 Equip. Oper. (crane)	19.75	158.00	32.80	262.40		
1 Equip. Oper. Oiler	16.20	129.60	26.90	215.20		
1 Crawler Crane, 100 Ton		1162.00		1278.20	20.75	22.83
56 L.H., Daily Totals		$2001.60		$2701.40	$35.74	$48.24

Crew B-14

	Bare Costs Hr.	Bare Costs Daily	Incl. Subs O & P Hr.	Incl. Subs O & P Daily	Cost Per Labor-Hour Bare Costs	Cost Per Labor-Hour Incl. O&P
1 Labor Foreman (outside)	$15.40	$123.20	$26.40	$211.20	$14.53	$24.73
4 Laborers	13.40	428.80	22.95	734.40		
1 Equip. Oper. (light)	18.15	145.20	30.15	241.20		
1 Backhoe Loader, 48 H.P.		213.80		235.20	4.45	4.90
48 L.H., Daily Totals		$911.00		$1422.00	$18.98	$29.63

Crew B-15

	Bare Costs Hr.	Bare Costs Daily	Incl. Subs O & P Hr.	Incl. Subs O & P Daily	Cost Per Labor-Hour Bare Costs	Cost Per Labor-Hour Incl. O&P
1 Equipment Oper. (med)	$19.00	$152.00	$31.55	$252.40	$16.14	$27.04
.5 Laborer	13.40	53.60	22.95	91.80		
2 Truck Drivers (heavy)	15.40	246.40	25.80	412.80		
2 Dump Trucks, 16 Ton		863.00		949.30		
1 Dozer, 200 H.P.		832.00		915.20	60.54	66.59
28 L.H., Daily Totals		$2147.00		$2621.50	$76.68	$93.63

Crew B-16

	Bare Costs Hr.	Bare Costs Daily	Incl. Subs O & P Hr.	Incl. Subs O & P Daily	Cost Per Labor-Hour Bare Costs	Cost Per Labor-Hour Incl. O&P
1 Labor Foreman (outside)	$15.40	$123.20	$26.40	$211.20	$14.40	$24.52
2 Laborers	13.40	214.40	22.95	367.20		
1 Truck Driver (heavy)	15.40	123.20	25.80	206.40		
1 Dump Truck, 16 Ton		431.50		474.65	13.48	14.83
32 L.H., Daily Totals		$892.30		$1259.45	$27.88	$39.35

Crew B-17

	Bare Costs Hr.	Bare Costs Daily	Incl. Subs O & P Hr.	Incl. Subs O & P Daily	Cost Per Labor-Hour Bare Costs	Cost Per Labor-Hour Incl. O&P
2 Laborers	$13.40	$214.40	$22.95	$367.20	$15.09	$25.46
1 Equip. Oper. (light)	18.15	145.20	30.15	241.20		
1 Truck Driver (heavy)	15.40	123.20	25.80	206.40		
1 Backhoe Loader, 48 H.P.		213.80		235.20		
1 Dump Truck, 12 Ton		355.20		390.70	17.78	19.56
32 L.H., Daily Totals		$1051.80		$1440.70	$32.87	$45.02

Crew B-18

	Bare Costs Hr.	Bare Costs Daily	Incl. Subs O & P Hr.	Incl. Subs O & P Daily	Cost Per Labor-Hour Bare Costs	Cost Per Labor-Hour Incl. O&P
1 Labor Foreman (outside)	$15.40	$123.20	$26.40	$211.20	$14.07	$24.10
2 Laborers	13.40	214.40	22.95	367.20		
1 Vibrating Compactor		49.60		54.55	2.07	2.27
24 L.H., Daily Totals		$387.20		$632.95	$16.14	$26.37

Crew B-19

	Bare Costs Hr.	Bare Costs Daily	Incl. Subs O & P Hr.	Incl. Subs O & P Daily	Cost Per Labor-Hour Bare Costs	Cost Per Labor-Hour Incl. O&P
1 Pile Driver Foreman	$20.65	$165.20	$38.60	$308.80	$18.34	$33.42
4 Pile Drivers	18.65	596.80	34.90	1116.80		
1 Equip. Oper. (crane)	19.75	158.00	32.80	262.40		
1 Building Laborer	13.40	107.20	22.95	183.60		
1 Crane, 40 Ton & Access.		757.20		832.90		
60 L.F. Leads, 15K Ft. Lbs.		105.00		115.50		
1 Hammer, 15K Ft. Lbs.		312.80		344.10		
1 Air Compr., 600 C.F.M.		268.15		294.95		
2-50 Ft. Air Hoses, 3" Dia.		33.50		36.85	26.37	29.01
56 L.H., Daily Totals		$2503.85		$3495.90	$44.71	$62.43

Crew B-19A

	Bare Costs Hr.	Bare Costs Daily	Incl. Subs O & P Hr.	Incl. Subs O & P Daily	Cost Per Labor-Hour Bare Costs	Cost Per Labor-Hour Incl. O&P
1 Pile Driver Foreman	$20.65	$165.20	$38.60	$308.80	$18.87	$33.84
4 Pile Drivers	18.65	596.80	34.90	1116.80		
2 Equip. Oper. (crane)	19.75	316.00	32.80	524.80		
1 Equip. Oper. Oiler	16.20	129.60	26.90	215.20		
1 Crawler Crane, 75 Ton		959.10		1055.00		
60 Leads, 25K ft.lbs		126.00		138.60		
1 Air Compressor, 750 CFM		282.50		310.75		
4-50 Ft. Air Hose, 3" Dia.		67.00		73.70	22.42	24.66
64 L.H., Daily Totals		$2642.20		$3743.65	$41.29	$58.50

Crew B-20

	Bare Costs Hr.	Bare Costs Daily	Incl. Subs O & P Hr.	Incl. Subs O & P Daily	Cost Per Labor-Hour Bare Costs	Cost Per Labor-Hour Incl. O&P
1 Labor Foreman (out)	$15.40	$123.20	$26.40	$211.20	$14.07	$24.10
2 Laborer	13.40	214.40	22.95	367.20		
24 L.H., Daily Totals		$337.60		$578.40	$14.07	$24.10

Crew B-20A

	Bare Costs Hr.	Bare Costs Daily	Incl. Subs O & P Hr.	Incl. Subs O & P Daily	Cost Per Labor-Hour Bare Costs	Cost Per Labor-Hour Incl. O&P
1 Labor Foreman	$15.40	$123.20	$26.40	$211.20	$16.59	$27.89
1 Laborer	13.40	107.20	22.95	183.60		
1 Plumber	20.85	166.80	34.55	276.40		
1 Plumber Apprentice	16.70	133.60	27.65	221.20		
32 L.H., Daily Totals		$530.80		$892.40	$16.59	$27.89

Crew B-21

	Bare Costs Hr.	Bare Costs Daily	Incl. Subs O & P Hr.	Incl. Subs O & P Daily	Cost Per Labor-Hour Bare Costs	Cost Per Labor-Hour Incl. O&P
1 Labor Foreman (out)	$15.40	$123.20	$26.40	$211.20	$14.88	$25.34
2 Laborer	13.40	214.40	22.95	367.20		
.5 Equip. Oper. (crane)	19.75	79.00	32.80	131.20		
.5 S.P. Crane, 5 Ton		134.52		148.00	4.80	5.28
28 L.H., Daily Totals		$551.12		$857.60	$19.68	$30.62

Crew B-21A

	Bare Costs Hr.	Bare Costs Daily	Incl. Subs O & P Hr.	Incl. Subs O & P Daily	Cost Per Labor-Hour Bare Costs	Cost Per Labor-Hour Incl. O&P
1 Labor Foreman	$15.40	$123.20	$26.40	$211.20	$17.22	$28.87
1 Laborer	13.40	107.20	22.95	183.60		
1 Plumber	20.85	166.80	34.55	276.40		
1 Plumber Apprentice	16.70	133.60	27.65	221.20		
1 Equip. Oper. (crane)	19.75	158.00	32.80	262.40		
1 S.P. Crane, 12 Ton		467.50		514.25	11.69	12.86
40 L.H., Daily Totals		$1156.30		$1669.05	$28.91	$41.73

Crew B-22

	Bare Costs Hr.	Bare Costs Daily	Incl. Subs O & P Hr.	Incl. Subs O & P Daily	Cost Per Labor-Hour Bare Costs	Cost Per Labor-Hour Incl. O&P
1 Labor Foreman (out)	$15.40	$123.20	$26.40	$211.20	$15.20	$25.84
2 Laborer	13.40	214.40	22.95	367.20		
.75 Equip. Oper. (crane)	19.75	118.50	32.80	196.80		
.75 S.P. Crane, 5 Ton		201.79		221.95	6.73	7.40
30 L.H., Daily Totals		$657.89		$997.15	$21.93	$33.24

Crew B-23

	Bare Costs Hr.	Bare Costs Daily	Incl. Subs O & P Hr.	Incl. Subs O & P Daily	Cost Per Labor-Hour Bare Costs	Cost Per Labor-Hour Incl. O&P
1 Labor Foreman (outside)	$15.40	$123.20	$26.40	$211.20	$13.80	$23.64
4 Laborers	13.40	428.80	22.95	734.40		
1 Drill Rig, Wells		1725.00		1897.50		
1 Light Truck, 3 Ton		174.80		192.30	47.50	52.24
40 L.H., Daily Totals		$2451.80		$3035.40	$61.30	$75.88

Crew B-23A

	Bare Costs Hr.	Bare Costs Daily	Incl. Subs O & P Hr.	Incl. Subs O & P Daily	Cost Per Labor-Hour Bare Costs	Cost Per Labor-Hour Incl. O&P
1 Labor Foreman (outside)	$15.40	$123.20	$26.40	$211.20	$15.93	$26.97
1 Laborers	13.40	107.20	22.95	183.60		
1 Equip. Operator, medium	19.00	152.00	31.55	252.40		
1 Drill Rig, Wells		1725.00		1897.50		
1 Pick-up Truck		125.50		138.05	77.10	84.81
24 L.H., Daily Totals		$2232.90		$2682.75	$93.03	$111.78

Crew B-23B

	Bare Costs Hr.	Bare Costs Daily	Incl. Subs O & P Hr.	Incl. Subs O & P Daily	Cost Per Labor-Hour Bare Costs	Cost Per Labor-Hour Incl. O&P
1 Labor Foreman (outside)	$15.40	$123.20	$26.40	$211.20	$15.93	$26.97
1 Laborers	13.40	107.20	22.95	183.60		
1 Equip. Operator, medium	19.00	152.00	31.55	252.40		
1 Drill Rig, Wells		1725.00		1897.50		
1 Pick-up Truck		125.50		138.05		
1 Pump, Cntfgl, 6"		155.20		170.70	83.57	91.93
24 L.H., Daily Totals		$2388.10		$2853.45	$99.50	$118.90

Crew B-24

Crew No.	Bare Costs Hr.	Bare Costs Daily	Incl. Subs O & P Hr.	Incl. Subs O & P Daily	Cost Per Labor-Hour Bare Costs	Cost Per Labor-Hour Incl. O&P
1 Cement Finisher	$17.80	$142.40	$29.05	$232.40	$16.57	$27.90
1 Laborer	13.40	107.20	22.95	183.60		
1 Carpenter	18.50	148.00	31.70	253.60		
24 L.H., Daily Totals		$397.60		$669.60	$16.57	$27.90

Crew B-25

Crew No.	Bare Costs Hr.	Bare Costs Daily	Incl. Subs O & P Hr.	Incl. Subs O & P Daily	Cost Per Labor-Hour Bare Costs	Cost Per Labor-Hour Incl. O&P
1 Labor Foreman	$15.40	$123.20	$26.40	$211.20	$15.11	$25.61
7 Laborers	13.40	750.40	22.95	1285.20		
3 Equip. Oper. (med.)	19.00	456.00	31.55	757.20		
1 Asphalt Paver, 130 H.P		1262.00		1388.20		
1 Tandem Roller, 10 Ton		238.00		261.80		
1 Roller, Pneumatic Wheel		237.20		260.90	19.74	21.72
88 L.H., Daily Totals		$3066.80		$4164.50	$34.85	$47.33

Crew B-25B

Crew No.	Bare Costs Hr.	Bare Costs Daily	Incl. Subs O & P Hr.	Incl. Subs O & P Daily	Cost Per Labor-Hour Bare Costs	Cost Per Labor-Hour Incl. O&P
1 Labor Foreman	$15.40	$123.20	$26.40	$211.20	$15.43	$26.10
7 Laborers	13.40	750.40	22.95	1285.20		
4 Equip. Oper. (medium)	19.00	608.00	31.55	1009.60		
1 Asphalt Paver, 130 H.P.		1262.00		1388.20		
2 Rollers, Steel Wheel		476.00		523.60		
1 Roller, Pneumatic Wheel		237.20		260.90	20.58	22.63
96 L.H., Daily Totals		$3456.80		$4678.70	$36.01	$48.73

Crew B-26

Crew No.	Bare Costs Hr.	Bare Costs Daily	Incl. Subs O & P Hr.	Incl. Subs O & P Daily	Cost Per Labor-Hour Bare Costs	Cost Per Labor-Hour Incl. O&P
1 Labor Foreman (outside)	$15.40	$123.20	$26.40	$211.20	$15.58	$26.66
6 Laborers	13.40	643.20	22.95	1101.60		
2 Equip. Oper. (med.)	19.00	304.00	31.55	504.80		
1 Rodman (reinf.)	19.75	158.00	37.05	296.40		
1 Cement Finisher	17.80	142.40	29.05	232.40		
1 Grader, 30,000 Lbs.		558.40		614.25		
1 Paving Mach. & Equip.		1313.00		1444.30	21.27	23.39
88 L.H., Daily Totals		$3242.20		$4404.95	$36.85	$50.05

Crew B-27

Crew No.	Bare Costs Hr.	Bare Costs Daily	Incl. Subs O & P Hr.	Incl. Subs O & P Daily	Cost Per Labor-Hour Bare Costs	Cost Per Labor-Hour Incl. O&P
1 Labor Foreman (outside)	$15.40	$123.20	$26.40	$211.20	$13.90	$23.81
3 Laborers	13.40	321.60	22.95	550.80		
1 Berm Machine		69.80		76.80	2.18	2.40
32 L.H., Daily Totals		$514.60		$838.80	$16.08	$26.21

Crew B-28

Crew No.	Bare Costs Hr.	Bare Costs Daily	Incl. Subs O & P Hr.	Incl. Subs O & P Daily	Cost Per Labor-Hour Bare Costs	Cost Per Labor-Hour Incl. O&P
2 Carpenters	$18.50	$296.00	$31.70	$507.20	$16.80	$28.78
1 Laborer	13.40	107.20	22.95	183.60		
24 L.H., Daily Totals		$403.20		$690.80	$16.80	$28.78

Crew B-29

Crew No.	Bare Costs Hr.	Bare Costs Daily	Incl. Subs O & P Hr.	Incl. Subs O & P Daily	Cost Per Labor-Hour Bare Costs	Cost Per Labor-Hour Incl. O&P
1 Labor Foreman (outside)	$15.40	$123.20	$26.40	$211.20	$14.79	$25.17
4 Laborers	13.40	428.80	22.95	734.40		
1 Equip. Oper. (crane)	19.75	158.00	32.80	262.40		
1 Gradall, 3 Ton, 1/2 C.Y.		625.60		688.15	13.03	14.34
48 L.H., Daily Totals		$1335.60		$1896.15	$27.82	$39.51

Crew B-30

Crew No.	Bare Costs Hr.	Bare Costs Daily	Incl. Subs O & P Hr.	Incl. Subs O & P Daily	Cost Per Labor-Hour Bare Costs	Cost Per Labor-Hour Incl. O&P
1 Equip. Oper. (med.)	$19.00	$152.00	$31.55	$252.40	$16.60	$27.72
2 Truck Drivers (heavy)	15.40	246.40	25.80	412.80		
1 Hyd. Excavator, 1.5 C.Y.		706.55		777.20		
2 Dump Trucks, 16 Ton		863.00		949.30	65.40	71.94
24 L.H., Daily Totals		$1967.95		$2391.70	$82.00	$99.66

Crew B-31

Crew No.	Bare Costs Hr.	Bare Costs Daily	Incl. Subs O & P Hr.	Incl. Subs O & P Daily	Cost Per Labor-Hour Bare Costs	Cost Per Labor-Hour Incl. O&P
1 Labor Foreman (outside)	$15.40	$123.20	$26.40	$211.20	$13.80	$23.64
4 Laborers	13.40	428.80	22.95	734.40		
1 Air Compr., 250 C.F.M.		120.40		132.45		
1 Sheeting Driver		11.20		12.30		
2-50 Ft. Air Hoses, 1.5" Dia.		16.40		18.05	3.70	4.07
40 L.H., Daily Totals		$700.00		$1108.40	$17.50	$27.71

Crew B-32

Crew No.	Bare Costs Hr.	Bare Costs Daily	Incl. Subs O & P Hr.	Incl. Subs O & P Daily	Cost Per Labor-Hour Bare Costs	Cost Per Labor-Hour Incl. O&P
1 Laborer	$13.40	$107.20	$22.95	$183.60	$17.60	$29.40
3 Equip. Oper. (med.)	19.00	456.00	31.55	757.20		
1 Grader, 30,000 Lbs.		558.40		614.25		
1 Tandem Roller, 10 Ton		238.00		261.80		
1 Dozer, 200 H.P.		832.00		915.20	50.89	55.98
32 L.H., Daily Totals		$2191.60		$2732.05	$68.49	$85.38

Crew B-32A

Crew No.	Bare Costs Hr.	Bare Costs Daily	Incl. Subs O & P Hr.	Incl. Subs O & P Daily	Cost Per Labor-Hour Bare Costs	Cost Per Labor-Hour Incl. O&P
1 Laborer	$13.40	$107.20	$22.95	$183.60	$17.13	$28.68
2 Equip. Oper. (medium)	19.00	304.00	31.55	504.80		
1 Grader, 30,000 Lbs.		558.40		614.25		
1 Roller, Vibratory, 29,000 Lbs.		408.80		449.70	40.30	44.33
24 L.H., Daily Totals		$1378.40		$1752.35	$57.43	$73.01

Crew B-32B

Crew No.	Bare Costs Hr.	Bare Costs Daily	Incl. Subs O & P Hr.	Incl. Subs O & P Daily	Cost Per Labor-Hour Bare Costs	Cost Per Labor-Hour Incl. O&P
1 Laborer	$13.40	$107.20	$22.95	$183.60	$17.13	$28.68
2 Equip. Oper. (medium)	19.00	304.00	31.55	504.80		
1 Dozer, 200 H.P.		832.00		915.20		
1 Roller, Vibratory, 29,000 Lbs.		408.80		449.70	51.70	56.87
24 L.H., Daily Totals		$1652.00		$2053.30	$68.83	$85.55

Crew B-32C

Crew No.	Bare Costs Hr.	Bare Costs Daily	Incl. Subs O & P Hr.	Incl. Subs O & P Daily	Cost Per Labor-Hour Bare Costs	Cost Per Labor-Hour Incl. O&P
1 Labor Foreman	$15.40	$123.20	$26.40	$211.20	$16.53	$27.82
2 Laborers	13.40	214.40	22.95	367.20		
3 Equip. Oper. (medium)	19.00	456.00	31.55	757.20		
1 Grader, 30,000 Lbs.		558.40		614.25		
1 Roller, Steel Wheel		238.00		261.80		
1 Dozer, 200 H.P.		832.00		915.20	33.93	37.32
48 L.H., Daily Totals		$2422.00		$3126.85	$50.46	$65.14

Crew B-33

Crew No.	Bare Costs Hr.	Bare Costs Daily	Incl. Subs O & P Hr.	Incl. Subs O & P Daily	Cost Per Labor-Hour Bare Costs	Cost Per Labor-Hour Incl. O&P
1 Equip. Oper. (med.)	$19.00	$152.00	$31.55	$252.40	$19.00	$31.55
.25 Equip. Oper. (med.)	19.00	38.00	31.55	63.10		
10 L.H., Daily Totals		$190.00		$315.50	$19.00	$31.55

Crew B-33A

Crew No.	Bare Costs Hr.	Bare Costs Daily	Incl. Subs O & P Hr.	Incl. Subs O & P Daily	Cost Per Labor-Hour Bare Costs	Cost Per Labor-Hour Incl. O&P
1 Equip. Oper. (med.)	$19.00	$152.00	$31.55	$252.40	$19.00	$31.55
.25 Equip. Oper. (med.)	19.00	38.00	31.55	63.10		
1 Scraper, Towed, 7 C.Y.		74.80		82.30		
1 Dozer, 300 H.P.		1122.00		1234.20		
.25 Dozer, 300 H.P.		280.50		308.55	147.73	162.50
10 L.H., Daily Totals		$1667.30		$1940.55	$166.73	$194.05

Crew B-33B

Crew No.	Bare Costs Hr.	Bare Costs Daily	Incl. Subs O & P Hr.	Incl. Subs O & P Daily	Cost Per Labor-Hour Bare Costs	Cost Per Labor-Hour Incl. O&P
1 Equip. Oper. (med.)	$19.00	$152.00	$31.55	$252.40	$19.00	$31.55
.25 Equip. Oper. (med.)	19.00	38.00	31.55	63.10		
1 Scraper, Towed, 10 C.Y.		196.10		215.70		
1 Dozer, 300 H.P.		1122.00		1234.20		
.25 Dozer, 300 H.P.		280.50		308.55	159.86	175.85
10 L.H., Daily Totals		$1788.60		$2073.95	$178.86	$207.40

Crew B-33C

Crew No.	Bare Costs Hr.	Bare Costs Daily	Incl. Subs O & P Hr.	Incl. Subs O & P Daily	Cost Per Labor-Hour Bare Costs	Cost Per Labor-Hour Incl. O&P
1 Equip. Oper. (med.)	$19.00	$152.00	$31.55	$252.40	$19.00	$31.55
.25 Equip. Oper. (med.)	19.00	38.00	31.55	63.10		
1 Scraper, Towed, 12 C.Y.		196.10		215.70		
1 Dozer, 300 H.P.		1122.00		1234.20		
.25 Dozer, 300 H.P.		280.50		308.55	159.86	175.85
10 L.H., Daily Totals		$1788.60		$2073.95	$178.86	$207.40

Crew B-33D

Crew No.	Bare Costs Hr.	Bare Costs Daily	Incl. Subs O & P Hr.	Incl. Subs O & P Daily	Cost Per Labor-Hour Bare Costs	Cost Per Labor-Hour Incl. O&P
1 Equip. Oper. (med.)	$19.00	$152.00	$31.55	$252.40	$19.00	$31.55
.25 Equip. Oper. (med.)	19.00	38.00	31.55	63.10		
1 S.P. Scraper, 14 C.Y.		1573.00		1730.30		
.25 Dozer, 300 H.P.		280.50		308.55	185.35	203.89
10 L.H., Daily Totals		$2043.50		$2354.35	$204.35	$235.44

Crew B-33E

Crew No.	Bare Costs Hr.	Bare Costs Daily	Incl. Subs O & P Hr.	Incl. Subs O & P Daily	Cost Per Labor-Hour Bare Costs	Cost Per Labor-Hour Incl. O&P
1 Equip. Oper. (med.)	$19.00	$152.00	$31.55	$252.40	$19.00	$31.55
.25 Equip. Oper. (med.)	19.00	38.00	31.55	63.10		
1 S.P. Scraper, 24 C.Y.		1854.00		2039.40		
.25 Dozer, 300 H.P.		280.50		308.55	213.45	234.80
10 L.H., Daily Totals		$2324.50		$2663.45	$232.45	$266.35

Crew B-33F

Crew No.	Bare Costs Hr.	Bare Costs Daily	Incl. Subs O & P Hr.	Incl. Subs O & P Daily	Cost Per Labor-Hour Bare Costs	Cost Per Labor-Hour Incl. O&P
1 Equip. Oper. (med.)	$19.00	$152.00	$31.55	$252.40	$19.00	$31.55
.25 Equip. Oper. (med.)	19.00	38.00	31.55	63.10		
1 Elev. Scraper, 11 C.Y.		667.15		733.85		
.25 Dozer, 300 H.P.		280.50		308.55	94.77	104.24
10 L.H., Daily Totals		$1137.65		$1357.90	$113.77	$135.79

Crew B-33G

Crew No.	Bare Costs Hr.	Bare Costs Daily	Incl. Subs O & P Hr.	Incl. Subs O & P Daily	Cost Per Labor-Hour Bare Costs	Cost Per Labor-Hour Incl. O&P
1 Equip. Oper. (med.)	$19.00	$152.00	$31.55	$252.40	$19.00	$31.55
.25 Equip. Oper. (med.)	19.00	38.00	31.55	63.10		
1 Elev. Scraper, 20 C.Y.		956.90		1052.60		
.25 Dozer, 300 H.P.		280.50		308.55	123.74	136.11
10 L.H., Daily Totals		$1427.40		$1676.65	$142.74	$167.66

Crew B-34A

Crew No.	Bare Costs Hr.	Bare Costs Daily	Incl. Subs O & P Hr.	Incl. Subs O & P Daily	Cost Per Labor-Hour Bare Costs	Cost Per Labor-Hour Incl. O&P
1 Truck Driver (heavy)	$15.40	$123.20	$25.80	$206.40	$15.40	$25.80
1 Dump Truck, 12 Ton		355.20		390.70	44.40	48.84
8 L.H., Daily Totals		$478.40		$597.10	$59.80	$74.64

Crew B-34B

Crew No.	Bare Costs Hr.	Bare Costs Daily	Incl. Subs O & P Hr.	Incl. Subs O & P Daily	Cost Per Labor-Hour Bare Costs	Cost Per Labor-Hour Incl. O&P
1 Truck Driver (heavy)	$15.40	$123.20	$25.80	$206.40	$15.40	$25.80
1 Dump Truck, 16 Ton		431.50		474.65	53.94	59.33
8 L.H., Daily Totals		$554.70		$681.05	$69.34	$85.13

Crew B-34C

Crew No.	Bare Costs Hr.	Bare Costs Daily	Incl. Subs O & P Hr.	Incl. Subs O & P Daily	Cost Per Labor-Hour Bare Costs	Cost Per Labor-Hour Incl. O&P
1 Truck Driver (heavy)	$15.40	$123.20	$25.80	$206.40	$15.40	$25.80
1 Truck Tractor, 40 Ton		414.10		455.50		
1 Dump Trailer, 16.5 C.Y.		128.80		141.70	67.86	74.65
8 L.H., Daily Totals		$666.10		$803.60	$83.26	$100.45

Crew B-34D

Crew No.	Bare Costs Hr.	Bare Costs Daily	Incl. Subs O & P Hr.	Incl. Subs O & P Daily	Cost Per Labor-Hour Bare Costs	Cost Per Labor-Hour Incl. O&P
1 Truck Driver (heavy)	$15.40	$123.20	$25.80	$206.40	$15.40	$25.80
1 Truck Tractor, 40 Ton		414.10		455.50		
1 Dump Trailer, 20 C.Y.		131.30		144.45	68.18	74.99
8 L.H., Daily Totals		$668.60		$806.35	$83.58	$100.79

Crew B-34E

Crew No.	Bare Costs Hr.	Bare Costs Daily	Incl. Subs O & P Hr.	Incl. Subs O & P Daily	Cost Per Labor-Hour Bare Costs	Cost Per Labor-Hour Incl. O&P
1 Truck Driver (heavy)	$15.40	$123.20	$25.80	$206.40	$15.40	$25.80
1 Truck, Off Hwy., 25 Ton		679.20		747.10	84.90	93.39
8 L.H., Daily Totals		$802.40		$953.50	$100.30	$119.19

Crew B-34F

Crew No.	Bare Costs Hr.	Bare Costs Daily	Incl. Subs O & P Hr.	Incl. Subs O & P Daily	Cost Per Labor-Hour Bare Costs	Cost Per Labor-Hour Incl. O&P
1 Truck Driver (heavy)	$15.40	$123.20	$25.80	$206.40	$15.40	$25.80
1 Truck, Off Hwy., 22 C.Y.		1011.00		1112.10	126.38	139.01
8 L.H., Daily Totals		$1134.20		$1318.50	$141.78	$164.81

Crew B-34G

Crew No.	Bare Costs Hr.	Bare Costs Daily	Incl. Subs O & P Hr.	Incl. Subs O & P Daily	Cost Per Labor-Hour Bare Costs	Cost Per Labor-Hour Incl. O&P
1 Truck Driver (heavy)	$15.40	$123.20	$25.80	$206.40	$15.40	$25.80
1 Truck, Off Hwy., 34 C.Y.		1361.00		1497.10	170.13	187.14
8 L.H., Daily Totals		$1484.20		$1703.50	$185.53	$212.94

Crew B-34H

Crew No.	Bare Costs Hr.	Bare Costs Daily	Incl. Subs O & P Hr.	Incl. Subs O & P Daily	Cost Per Labor-Hour Bare Costs	Cost Per Labor-Hour Incl. O&P
1 Truck Driver (heavy)	$15.40	$123.20	$25.80	$206.40	$15.40	$25.80
1 Truck, Off Hwy., 42 C.Y.		1551.00		1706.10	193.88	213.26
8 L.H., Daily Totals		$1674.20		$1912.50	$209.28	$239.06

Crew B-34J

Crew No.	Bare Costs Hr.	Bare Costs Daily	Incl. Subs O & P Hr.	Incl. Subs O & P Daily	Cost Per Labor-Hour Bare Costs	Cost Per Labor-Hour Incl. O&P
1 Truck Driver (heavy)	$15.40	$123.20	$25.80	$206.40	$15.40	$25.80
1 Truck, Off Hwy., 60 C.Y.		2171.00		2388.10	271.38	298.51
8 L.H., Daily Totals		$2294.20		$2594.50	$286.78	$324.31

Crew B-34K

Crew No.	Bare Costs Hr.	Bare Costs Daily	Incl. Subs O & P Hr.	Incl. Subs O & P Daily	Cost Per Labor-Hour Bare Costs	Cost Per Labor-Hour Incl. O&P
1 Truck Driver (heavy)	$15.40	$123.20	$25.80	$206.40	$15.40	$25.80
1 Truck Tractor, 240 H.P.		517.85		569.65		
1 Low Bed Trailer		378.20		416.00	112.01	123.21
8 L.H., Daily Totals		$1019.25		$1192.05	$127.41	$149.01

Crew B-35

Crew No.	Bare Costs Hr.	Bare Costs Daily	Incl. Subs O & P Hr.	Incl. Subs O & P Daily	Cost Per Labor-Hour Bare Costs	Cost Per Labor-Hour Incl. O&P
1 Laborer Foreman (out)	$15.40	$123.20	$26.40	$211.20	$17.63	$29.79
1 Skilled Worker	18.75	150.00	32.25	258.00		
1 Welder (plumber)	20.85	166.80	34.55	276.40		
1 Laborer	13.40	107.20	22.95	183.60		
1 Equip. Oper. (crane)	19.75	158.00	32.80	262.40		
1 Electric Welding Mach.		48.50		53.35		
1 Hyd. Excavator, .75 C.Y.		435.05		478.55	12.09	13.30
40 L.H., Daily Totals		$1188.75		$1723.50	$29.72	$43.09

Crew B-35A

Crew No.	Bare Costs Hr.	Bare Costs Daily	Incl. Subs O & P Hr.	Incl. Subs O & P Daily	Cost Per Labor-Hour Bare Costs	Cost Per Labor-Hour Incl. O&P
1 Laborer Foreman (out)	$15.40	$123.20	$26.40	$211.20	$16.82	$28.40
2 Laborers	13.40	214.40	22.95	367.20		
1 Skilled Worker	18.75	150.00	32.25	258.00		
1 Welder (plumber)	20.85	166.80	34.55	276.40		
1 Equip. Oper. (crane)	19.75	158.00	32.80	262.40		
1 Equip. Oper. Oiler	16.20	129.60	26.90	215.20		
1 Welder, 300 amp		80.15		88.15		
1 Crane, 75 Ton		959.10		1055.00	18.56	20.41
56 L.H., Daily Totals		$1981.25		$2733.55	$35.38	$48.81

Crew B-36

Crew No.	Bare Costs Hr.	Bare Costs Daily	Incl. Subs O & P Hr.	Incl. Subs O & P Daily	Cost Per Labor-Hour Bare Costs	Cost Per Labor-Hour Incl. O&P
1 Labor Foreman (outside)	$15.40	$123.20	$26.40	$211.20	$16.04	$27.08
2 Laborers	13.40	214.40	22.95	367.20		
2 Equip. Oper. (med.)	19.00	304.00	31.55	504.80		
1 Dozer, 200 H.P.		832.00		915.20		
1 Aggregate Spreader		68.80		75.70		
1 Tandem Roller, 10 Ton		238.00		261.80	28.47	31.32
40 L.H., Daily Totals		$1780.40		$2335.90	$44.51	$58.40

Crew No.	Bare Costs		Incl. Subs O & P		Cost Per Labor-Hour	

Crew B-36A	Hr.	Daily	Hr.	Daily	Bare Costs	Incl. O&P
1 Labor Foreman	$15.40	$123.20	$26.40	$211.20	$16.89	$28.36
2 Laborers	13.40	214.40	22.95	367.20		
4 Equip. Oper. (med.)	19.00	608.00	31.55	1009.60		
1 Dozer, 200 H.P.		832.00		915.20		
1 Aggregate Spreader		68.80		75.70		
1 Roller, Steel Wheel		238.00		261.80		
1 Roller, Pneumatic Wheel		237.20		260.90	24.57	27.03
56 L.H., Daily Totals		$2321.60		$3101.60	$41.46	$55.39

Crew B-36B	Hr.	Daily	Hr.	Daily	Bare Costs	Incl. O&P
1 Labor Foreman	$15.40	$123.20	$26.40	$211.20	$16.70	$28.04
2 Laborers	13.40	214.40	22.95	367.20		
4 Equip. Oper. (medium)	19.00	608.00	31.55	1009.60		
1 Truck Driver, Heavy	15.40	123.20	25.80	206.40		
1 Grader, 30,000 Lbs.		558.40		614.25		
1 F.E. Loader, crl, 1.5 C.Y.		384.15		422.55		
1 Dozer, 300 H.P.		1122.00		1234.20		
1 Roller, vibratory		408.80		449.70		
1 Truck, Tractor, 240 H.P.		517.85		569.65		
1 Water Tanker, 5000 Gal.		203.20		223.50	49.91	54.90
64 L.H., Daily Totals		$4263.20		$5308.25	$66.61	$82.94

Crew B-37	Hr.	Daily	Hr.	Daily	Bare Costs	Incl. O&P
1 Labor Foreman (outside)	$15.40	$123.20	$26.40	$211.20	$14.53	$24.73
4 Laborers	13.40	428.80	22.95	734.40		
1 Equip. Oper. (light)	18.15	145.20	30.15	241.20		
1 Tandem Roller, 5 Ton		148.40		163.25	3.09	3.40
48 L.H., Daily Totals		$845.60		$1350.05	$17.62	$28.13

Crew B-38	Hr.	Daily	Hr.	Daily	Bare Costs	Incl. O&P
2 Laborers	$13.40	$214.40	$22.95	$367.20	$14.98	$25.35
1 Equip. Oper. (light)	18.15	145.20	30.15	241.20		
1 Backhoe Loader, 48 H.P.		213.80		235.20		
1 Demol.Hammer,(1200 I		174.60		192.05	16.18	17.80
24 L.H., Daily Totals		$748.00		$1035.65	$31.16	$43.15

Crew B-39	Hr.	Daily	Hr.	Daily	Bare Costs	Incl. O&P
1 Labor Foreman (outside)	$15.40	$123.20	$26.40	$211.20	$13.73	$23.53
5 Laborers	13.40	536.00	22.95	918.00		
1 Air Compr., 250 C.F.M.		120.40		132.45		
2 Air Tools & Accessories		36.90		40.60		
2-50 Ft. Air Hoses, 1.5" Dia.		16.40		18.05	3.62	3.98
48 L.H., Daily Totals		$832.90		$1320.30	$17.35	$27.51

Crew B-40	Hr.	Daily	Hr.	Daily	Bare Costs	Incl. O&P
1 Pile Driver Foreman	$20.65	$165.20	$38.60	$308.80	$18.34	$33.42
4 Pile Drivers	18.65	596.80	34.90	1116.80		
1 Building Laborer	13.40	107.20	22.95	183.60		
1 Equip. Oper. (crane)	19.75	158.00	32.80	262.40		
1 Crane, 40 Ton		757.20		832.90		
1 Vibratory Hammer & Gen.		1184.00		1302.40	34.66	38.13
56 L.H., Daily Totals		$2968.40		$4006.90	$53.00	$71.55

Crew B-41	Hr.	Daily	Hr.	Daily	Bare Costs	Incl. O&P
1 Labor Foreman (outside)	$15.40	$123.20	$26.40	$211.20	$14.18	$24.20
4 Laborers	13.40	428.80	22.95	734.40		
.25 Equip. Oper. (crane)	19.75	39.50	32.80	65.60		
.25 Equip. Oper. Oiler	16.20	32.40	26.90	53.80		
.25 Crawler Crane, 40 Ton		189.30		208.25	4.30	4.73
44 L.H., Daily Totals		$813.20		$1273.25	$18.48	$28.93

Crew B-42	Hr.	Daily	Hr.	Daily	Bare Costs	Incl. O&P
1 Labor Foreman (outside)	$15.40	$123.20	$26.40	$211.20	$15.66	$26.51
4 Laborers	13.40	428.80	22.95	734.40		
1 Equip. Oper. (crane)	19.75	158.00	32.80	262.40		
1 Welder	20.85	166.80	34.55	276.40		
1 Hyd. Crane, 25 Ton		538.85		592.75		
1 Gas Welding Machine		80.15		88.15		
1 Horz. Boring Csg. Mch.		476.80		524.50	19.57	21.52
56 L.H., Daily Totals		$1972.60		$2689.80	$35.23	$48.03

Crew B-43	Hr.	Daily	Hr.	Daily	Bare Costs	Incl. O&P
1 Labor Foreman (outside)	$15.40	$123.20	$26.40	$211.20	$13.80	$23.64
4 Laborers	13.40	428.80	22.95	734.40		
1 Drill Rig & Augers		1725.00		1897.50	43.13	47.44
40 L.H., Daily Totals		$2277.00		$2843.10	$56.93	$71.08

Crew B-44	Hr.	Daily	Hr.	Daily	Bare Costs	Incl. O&P
1 Pile Driver Foreman	$20.65	$165.20	$38.60	$308.80	$17.73	$32.11
4 Pile Drivers	18.65	596.80	34.90	1116.80		
1 Equip. Oper. (crane)	19.75	158.00	32.80	262.40		
2 Laborer	13.40	214.40	22.95	367.20		
1 Crane, 40 Ton, & Access.		757.20		832.90		
45 L.F. Leads, 15K Ft. Lbs.		78.75		86.65	13.06	14.37
64 L.H., Daily Totals		$1970.35		$2974.75	$30.79	$46.48

Crew B-45	Hr.	Daily	Hr.	Daily	Bare Costs	Incl. O&P
1 Building Laborer	$13.40	$107.20	$22.95	$183.60	$14.40	$24.38
1 Truck Driver (heavy)	15.40	123.20	25.80	206.40		
1 Dist. Tank Truck, 3K Gal.		353.05		388.35	22.07	24.27
16 L.H., Daily Totals		$583.45		$778.35	$36.47	$48.65

Crew B-46	Hr.	Daily	Hr.	Daily	Bare Costs	Incl. O&P
1 Pile Driver Foreman	$20.65	$165.20	$38.60	$308.80	$16.36	$29.54
2 Pile Drivers	18.65	298.40	34.90	558.40		
3 Laborers	13.40	321.60	22.95	550.80		
1 Chain Saw, 36" Long		52.80		58.10	1.10	1.21
48 L.H., Daily Totals		$838.00		$1476.10	$17.46	$30.75

Crew B-47	Hr.	Daily	Hr.	Daily	Bare Costs	Incl. O&P
1 Blast Foreman	$15.40	$123.20	$26.40	$211.20	$14.40	$24.68
1 Driller	13.40	107.20	22.95	183.60		
1 Crawler Type Drill, 4"		302.10		332.30		
1 Air Compr., 600 C.F.M.		268.15		294.95		
2-50 Ft. Air Hoses, 3" Dia.		33.50		36.85	37.73	41.51
16 L.H., Daily Totals		$834.15		$1058.90	$52.13	$66.19

Crew B-47A	Hr.	Daily	Hr.	Daily	Bare Costs	Incl. O&P
1 Drilling Foreman	$15.40	$123.20	$26.40	$211.20	$17.12	$28.70
1 Equip. Oper. (heavy)	19.75	158.00	32.80	262.40		
1 Oiler	16.20	129.60	26.90	215.20		
1 Quarry Drill		468.40		515.25	19.52	21.47
24 L.H., Daily Totals		$879.20		$1204.05	$36.64	$50.17

Crew B-47C	Hr.	Daily	Hr.	Daily	Bare Costs	Incl. O&P
1 Laborer	$13.40	$107.20	$22.95	$183.60	$15.77	$26.55
1 Equip. Oper. (light)	18.15	145.20	30.15	241.20		
1 Air Compressor, 750 CFM		282.50		310.75		
2-50' Air Hose, 3"		33.50		36.85		
1 Air Track Drill, 4"		302.10		332.30	38.63	42.49
16 L.H., Daily Totals		$870.50		$1104.70	$54.40	$69.04

Crew No.	Bare Costs		Incl. Subs O & P		Cost Per Labor-Hour	

Crew B-47E	Hr.	Daily	Hr.	Daily	Bare Costs	Incl. O&P
1 Laborer Forman	$15.40	$123.20	$26.40	$211.20	$13.90	$23.81
3 Laborers	13.40	321.60	22.95	550.80		
1 Truck, Flatbed, 3 ton		174.80		192.30	5.46	6.01
32 L.H., Daily Totals		$619.60		$954.30	$19.36	$29.82

Crew B-48	Hr.	Daily	Hr.	Daily	Bare Costs	Incl. O&P
1 Labor Foreman (outside)	$15.40	$123.20	$26.40	$211.20	$14.79	$25.17
4 Laborers	13.40	428.80	22.95	734.40		
1 Equip. Oper. (crane)	19.75	158.00	32.80	262.40		
1 Centr. Water Pump, 6"		155.20		170.70		
1-20 Ft. Suction Hose, 6"		20.50		22.55		
1-50 Ft. Disch. Hose, 6"		18.50		20.35		
1 Drill Rig & Augers		1725.00		1897.50	39.98	43.98
48 L.H., Daily Totals		$2629.20		$3319.10	$54.77	$69.15

Crew B-49	Hr.	Daily	Hr.	Daily	Bare Costs	Incl. O&P
1 Labor Foreman (outside)	$15.40	$123.20	$26.40	$211.20	$15.49	$27.08
5 Laborers	13.40	536.00	22.95	918.00		
1 Equip. Oper. (crane)	19.75	158.00	32.80	262.40		
2 Pile Drivers	18.65	298.40	34.90	558.40		
1 Hyd. Crane, 25 Ton		538.85		592.75		
1 Centr. Water Pump, 6"		155.20		170.70		
1-20 Ft. Suction Hose, 6"		20.50		22.55		
1-50 Ft. Disch. Hose, 6"		18.50		20.35		
1 Drill Rig & Augers		1725.00		1897.50	34.14	37.55
72 L.H., Daily Totals		$3573.65		$4653.85	$49.63	$64.63

Crew B-50	Hr.	Daily	Hr.	Daily	Bare Costs	Incl. O&P
1 Pile Driver Foremen	$20.65	$165.20	$38.60	$308.80	$16.87	$30.43
6 Pile Drivers	18.65	895.20	34.90	1675.20		
1 Equip. Oper. (crane)	19.75	158.00	32.80	262.40		
5 Laborers	13.40	536.00	22.95	918.00		
1 Crane, 40 Ton		757.20		832.90		
60 L.F. Leads, 15K Ft. Lbs.		105.00		115.50		
1 Hammer, 15K Ft. Lbs.		312.80		344.10		
1 Air Compr., 600 C.F.M.		268.15		294.95		
2-50 Ft. Air Hoses, 3" Dia.		33.50		36.85		
1 Chain Saw, 36" Long		52.80		58.10	14.71	16.18
104 L.H., Daily Totals		$3283.85		$4846.80	$31.58	$46.61

Crew B-51	Hr.	Daily	Hr.	Daily	Bare Costs	Incl. O&P
1 Labor Foreman (outside)	$15.40	$123.20	$26.40	$211.20	$14.02	$23.92
4 Laborers	13.40	428.80	22.95	734.40		
1 Truck Driver (light)	15.10	120.80	25.30	202.40		
1 Light Truck, 1.5 Ton		167.40		184.15	3.49	3.84
48 L.H., Daily Totals		$840.20		$1332.15	$17.51	$27.76

Crew B-52	Hr.	Daily	Hr.	Daily	Bare Costs	Incl. O&P
1 Labor Foreman	$15.40	$123.20	$26.40	$211.20	$15.27	$26.31
1 Carpenter	18.50	148.00	31.70	253.60		
4 Laborers	13.40	428.80	22.95	734.40		
.5 Rodman (reinf.)	19.75	79.00	37.05	148.20		
.5 Equip. Oper. (med.)	19.00	76.00	31.55	126.20		
.5 F.E. Ldr., T.M., 2.5 C.Y.		419.00		460.90	7.48	8.23
56 L.H., Daily Totals		$1274.00		$1934.50	$22.75	$34.54

Crew B-53	Hr.	Daily	Hr.	Daily	Bare Costs	Incl. O&P
1 Building Laborer	$13.40	$107.20	$22.95	$183.60	$13.40	$22.95
1 Trencher, Chain, 12 H.P.		91.80		101.00	11.48	12.62
8 L.H., Daily Totals		$199.00		$284.60	$24.88	$35.57

Crew B-54	Hr.	Daily	Hr.	Daily	Bare Costs	Incl. O&P
1 Equip. Oper. (light)	$18.15	$145.20	$30.15	$241.20	$18.15	$30.15
1 Trencher, Chain, 40 H.P.		209.35		230.30	26.17	28.79
8 L.H., Daily Totals		$354.55		$471.50	$44.32	$58.94

Crew B-54A	Hr.	Daily	Hr.	Daily	Bare Costs	Incl. O&P
.17 IABOR FOREMAN	$15.40	$20.94	$26.40	$35.90	$18.48	$30.80
1 Equipment Operator (med.)	19.00	152.00	31.55	252.40		
1 Wheel Trencher, 67HP		396.50		436.15	42.36	46.60
9.36 L.H., Daily Totals		$569.44		$724.45	$60.84	$77.40

Crew B-54B	Hr.	Daily	Hr.	Daily	Bare Costs	Incl. O&P
.25 Labor Foreman	$15.40	$30.80	$26.40	$52.80	$18.28	$30.52
1 Equipment Operator (med.)	19.00	152.00	31.55	252.40		
1 Wheel Trencher,150HP		561.50		617.65	56.15	61.77
10 L.H., Daily Totals		$744.30		$922.85	$74.43	$92.29

Crew B-55	Hr.	Daily	Hr.	Daily	Bare Costs	Incl. O&P
1 Laborers	$13.40	$107.20	$22.95	$183.60	$14.25	$24.13
1 Truck Driver (light)	15.10	120.80	25.30	202.40		
1 Auger, 4" to 36" Dia		445.60		490.15		
1 Flatbed 3 Ton Truck		174.80		192.30	38.78	42.65
16 L.H., Daily Totals		$848.40		$1068.45	$53.03	$66.78

Crew B-56	Hr.	Daily	Hr.	Daily	Bare Costs	Incl. O&P
2 Laborer	$13.40	$214.40	$22.95	$367.20	$13.40	$22.95
1 Crawler Type Drill, 4"		302.10		332.30		
1 Air Compr., 600 C.F.M.		268.15		294.95		
1-50 Ft. Air Hose, 3" Dia.		16.75		18.45	36.69	40.36
16 L.H., Daily Totals		$801.40		$1012.90	$50.09	$63.31

Crew B-57	Hr.	Daily	Hr.	Daily	Bare Costs	Incl. O&P
1 Labor Foreman (outside)	$15.40	$123.20	$26.40	$211.20	$15.07	$25.61
3 Laborers	13.40	321.60	22.95	550.80		
1 Equip. Oper. (crane)	19.75	158.00	32.80	262.40		
1 Barge, 400 Ton		439.05		482.95		
1 Power Shovel, 1 C.Y.		496.90		546.60		
1 Clamshell Bucket, 1 C.Y.		64.80		71.30		
1 Centr. Water Pump, 6"		155.20		170.70		
1-20 Ft. Suction Hose, 6"		20.50		22.55		
20-50 Ft. Disch. Hoses, 6"		370.00		407.00	38.66	42.53
40 L.H., Daily Totals		$2149.25		$2725.50	$53.73	$68.14

Crew B-58	Hr.	Daily	Hr.	Daily	Bare Costs	Incl. O&P
2 Laborers	$13.40	$214.40	$22.95	$367.20	$14.98	$25.35
1 Equip. Oper. (light)	18.15	145.20	30.15	241.20		
1 Backhoe Loader, 48 H.P.		213.80		235.20		
1 Small Helicopter		3448.00		3792.80	152.58	167.83
24 L.H., Daily Totals		$4021.40		$4636.40	$167.56	$193.18

Crew B-59	Hr.	Daily	Hr.	Daily	Bare Costs	Incl. O&P
1 Truck Driver (heavy)	$15.40	$123.20	$25.80	$206.40	$15.40	$25.80
1 Truck, 30 Ton		316.50		348.15		
1 Water tank, 5000 Gal.		203.20		223.50	64.96	71.46
8 L.H., Daily Totals		$642.90		$778.05	$80.36	$97.26

Crew B-60

	Bare Costs Hr.	Bare Costs Daily	Incl. Subs O & P Hr.	Incl. Subs O & P Daily	Cost Per Labor-Hour Bare Costs	Cost Per Labor-Hour Incl. O&P
1 Labor Foreman (outside)	$15.40	$123.20	$26.40	$211.20	$15.58	$26.37
3 Laborers	13.40	321.60	22.95	550.80		
1 Equip. Oper. (crane)	19.75	158.00	32.80	262.40		
1 Equip. Oper. (light)	18.15	145.20	30.15	241.20		
1 Crawler Crane, 40 Ton		757.20		832.90		
45 L.F. Leads, 15K Ft. Lbs.		78.75		86.65		
1 Backhoe Loader, 48 H.P.		213.80		235.20	21.87	24.06
48 L.H., Daily Totals		$1797.75		$2420.35	$37.45	$50.43

Crew B-61

	Bare Costs Hr.	Bare Costs Daily	Incl. Subs O & P Hr.	Incl. Subs O & P Daily	Cost Per Labor-Hour Bare Costs	Cost Per Labor-Hour Incl. O&P
1 Labor Foreman (outside)	$15.40	$123.20	$26.40	$211.20	$13.80	$23.64
4 Laborers	13.40	428.80	22.95	734.40		
1 Cement Mixer, 2 C.Y.		257.60		283.35		
1 Air Compr., 160 C.F.M.		94.80		104.30	8.81	9.69
40 L.H., Daily Totals		$904.40		$1333.25	$22.61	$33.33

Crew B-62

	Bare Costs Hr.	Bare Costs Daily	Incl. Subs O & P Hr.	Incl. Subs O & P Daily	Cost Per Labor-Hour Bare Costs	Cost Per Labor-Hour Incl. O&P
2 Laborers	$13.40	$214.40	$22.95	$367.20	$14.98	$25.35
1 Equip. Oper. (light)	18.15	145.20	30.15	241.20		
1 Loader, Skid Steer		105.20		115.70	4.38	4.82
24 L.H., Daily Totals		$464.80		$724.10	$19.36	$30.17

Crew B-63

	Bare Costs Hr.	Bare Costs Daily	Incl. Subs O & P Hr.	Incl. Subs O & P Daily	Cost Per Labor-Hour Bare Costs	Cost Per Labor-Hour Incl. O&P
5 Laborers	$13.40	$536.00	$22.95	$918.00	$13.40	$22.95
1 Loader, Skid Steer		105.20		115.70	2.63	2.89
40 L.H., Daily Totals		$641.20		$1033.70	$16.03	$25.84

Crew B-64

	Bare Costs Hr.	Bare Costs Daily	Incl. Subs O & P Hr.	Incl. Subs O & P Daily	Cost Per Labor-Hour Bare Costs	Cost Per Labor-Hour Incl. O&P
1 Laborer	$13.40	$107.20	$22.95	$183.60	$14.25	$24.13
1 Truck Driver (light)	15.10	120.80	25.30	202.40		
1 Power Mulcher (small)		112.00		123.20		
1 Light Truck, 1.5 Ton		167.40		184.15	17.46	19.21
16 L.H., Daily Totals		$507.40		$693.35	$31.71	$43.34

Crew B-65

	Bare Costs Hr.	Bare Costs Daily	Incl. Subs O & P Hr.	Incl. Subs O & P Daily	Cost Per Labor-Hour Bare Costs	Cost Per Labor-Hour Incl. O&P
1 Laborer	$13.40	$107.20	$22.95	$183.60	$14.25	$24.13
1 Truck Driver (light)	15.10	120.80	25.30	202.40		
1 Power Mulcher (large)		274.60		302.05		
1 Light Truck, 1.5 Ton		167.40		184.15	27.63	30.39
16 L.H., Daily Totals		$670.00		$872.20	$41.88	$54.52

Crew B-66

	Bare Costs Hr.	Bare Costs Daily	Incl. Subs O & P Hr.	Incl. Subs O & P Daily	Cost Per Labor-Hour Bare Costs	Cost Per Labor-Hour Incl. O&P
1 Equip. Oper. (light)	$18.15	$145.20	$30.15	$241.20	$18.15	$30.15
1 Backhoe Ldr. w/Attchmt.		196.80		216.50	24.60	27.06
8 L.H., Daily Totals		$342.00		$457.70	$42.75	$57.21

Crew B-67

	Bare Costs Hr.	Bare Costs Daily	Incl. Subs O & P Hr.	Incl. Subs O & P Daily	Cost Per Labor-Hour Bare Costs	Cost Per Labor-Hour Incl. O&P
1 Millwright	$19.50	$156.00	$31.90	$255.20	$18.83	$31.02
1 Equip. Oper. (light)	18.15	145.20	30.15	241.20		
1 Forklift		175.60		193.15	10.98	12.07
16 L.H., Daily Totals		$476.80		$689.55	$29.81	$43.09

Crew B-68

	Bare Costs Hr.	Bare Costs Daily	Incl. Subs O & P Hr.	Incl. Subs O & P Daily	Cost Per Labor-Hour Bare Costs	Cost Per Labor-Hour Incl. O&P
2 Millwrights	$19.50	$312.00	$31.90	$510.40	$19.05	$31.32
1 Equip. Oper. (light)	18.15	145.20	30.15	241.20		
1 Forklift		175.60		193.15	7.32	8.05
24 L.H., Daily Totals		$632.80		$944.75	$26.37	$39.37

Crew B-69

	Bare Costs Hr.	Bare Costs Daily	Incl. Subs O & P Hr.	Incl. Subs O & P Daily	Cost Per Labor-Hour Bare Costs	Cost Per Labor-Hour Incl. O&P
1 Labor Foreman (outside)	$15.40	$123.20	$26.40	$211.20	$15.26	$25.83
3 Laborers	13.40	321.60	22.95	550.80		
1 Equip. Oper. (crane)	19.75	158.00	32.80	262.40		
1 Equip. Oper. Oiler	16.20	129.60	26.90	215.20		
1 Truck Crane, 80 Ton		1090.00		1199.00	22.71	24.98
48 L.H., Daily Totals		$1822.40		$2438.60	$37.97	$50.81

Crew B-69A

	Bare Costs Hr.	Bare Costs Daily	Incl. Subs O & P Hr.	Incl. Subs O & P Daily	Cost Per Labor-Hour Bare Costs	Cost Per Labor-Hour Incl. O&P
1 Labor Foreman	$15.40	$123.20	$26.40	$211.20	$15.40	$25.98
3 Laborers	13.40	321.60	22.95	550.80		
1 Equip. Oper. (medium)	19.00	152.00	31.55	252.40		
1 Concrete Finisher	17.80	142.40	29.05	232.40		
1 Curb Paver		451.20		496.30	9.40	10.34
48 L.H., Daily Totals		$1190.40		$1743.10	$24.80	$36.32

Crew B-69B

	Bare Costs Hr.	Bare Costs Daily	Incl. Subs O & P Hr.	Incl. Subs O & P Daily	Cost Per Labor-Hour Bare Costs	Cost Per Labor-Hour Incl. O&P
1 Labor Foreman	$15.40	$123.20	$26.40	$211.20	$15.40	$25.98
3 Laborers	13.40	321.60	22.95	550.80		
1 Equip. Oper. (medium)	19.00	152.00	31.55	252.40		
1 Cement Finisher	17.80	142.40	29.05	232.40		
1 Curb/Gutter Paver		901.20		991.30	18.78	20.65
48 L.H., Daily Totals		$1640.40		$2238.10	$34.18	$46.63

Crew B-70

	Bare Costs Hr.	Bare Costs Daily	Incl. Subs O & P Hr.	Incl. Subs O & P Daily	Cost Per Labor-Hour Bare Costs	Cost Per Labor-Hour Incl. O&P
1 Labor Foreman (outside)	$15.40	$123.20	$26.40	$211.20	$16.09	$27.13
3 Laborers	13.40	321.60	22.95	550.80		
3 Equip. Oper. (med.)	19.00	456.00	31.55	757.20		
1 Motor Grader, 30,000 Lb.		558.40		614.25		
1 Grader Attach., Ripper		60.60		66.65		
1 Road Sweeper, S.P.		191.00		210.10		
1 F.E. Loader, 1-3/4 C.Y.		312.30		343.55	20.04	22.05
56 L.H., Daily Totals		$2023.10		$2753.75	$36.13	$49.18

Crew B-71

	Bare Costs Hr.	Bare Costs Daily	Incl. Subs O & P Hr.	Incl. Subs O & P Daily	Cost Per Labor-Hour Bare Costs	Cost Per Labor-Hour Incl. O&P
1 Labor Foreman (outside)	$15.40	$123.20	$26.40	$211.20	$16.09	$27.13
3 Laborers	13.40	321.60	22.95	550.80		
3 Equip. Oper. (med.)	19.00	456.00	31.55	757.20		
1 Pvmt. Profiler, 450 H.P.		3370.00		3707.00		
1 Road Sweeper, S.P.		191.00		210.10		
1 F.E. Loader, 1-3/4 C.Y.		312.30		343.55	69.17	76.08
56 L.H., Daily Totals		$4774.10		$5779.85	$85.26	$103.21

Crew B-72

	Bare Costs Hr.	Bare Costs Daily	Incl. Subs O & P Hr.	Incl. Subs O & P Daily	Cost Per Labor-Hour Bare Costs	Cost Per Labor-Hour Incl. O&P
1 Labor Foreman (outside)	$15.40	$123.20	$26.40	$211.20	$16.45	$27.68
3 Laborers	13.40	321.60	22.95	550.80		
4 Equip. Oper. (med.)	19.00	608.00	31.55	1009.60		
1 Pvmt. Profiler, 450 H.P.		3370.00		3707.00		
1 Hammermill, 250 H.P.		1144.00		1258.40		
1 Windrow Loader		905.45		996.00		
1 Mix Paver 165 H.P.		1392.00		1531.20		
1 Roller, Pneu. Tire, 12 T.		237.20		260.90	110.14	121.15
64 L.H., Daily Totals		$8101.45		$9525.10	$126.59	$148.83

Crew No.	Bare Costs		Incl. Subs O & P		Cost Per Labor-Hour	
Crew B-73	Hr.	Daily	Hr.	Daily	Bare Costs	Incl. O&P
1 Labor Foreman (outside)	$15.40	$123.20	$26.40	$211.20	$17.15	$28.76
2 Laborers	13.40	214.40	22.95	367.20		
5 Equip. Oper. (med.)	19.00	760.00	31.55	1262.00		
1 Road Mixer, 310 H.P.		986.40		1085.05		
1 Roller, Tandem, 12 Ton		238.00		261.80		
1 Hammermill, 250 H.P.		1144.00		1258.40		
1 Motor Grader, 30,000 Lb.		558.40		614.25		
.5 F.E. Loader, 1-3/4 C.Y.		156.15		171.75		
.5 Truck, 30 Ton		158.25		174.05		
.5 Water Tank 5000 Gal.		101.60		111.75	52.23	57.45
64 L.H., Daily Totals		$4440.40		$5517.45	$69.38	$86.21
Crew B-74	Hr.	Daily	Hr.	Daily	Bare Costs	Incl. O&P
1 Labor Foreman (outside)	$15.40	$123.20	$26.40	$211.20	$16.95	$28.39
1 Laborer	13.40	107.20	22.95	183.60		
4 Equip. Oper. (med.)	19.00	608.00	31.55	1009.60		
2 Truck Drivers (heavy)	15.40	246.40	25.80	412.80		
1 Motor Grader, 30,000 Lb.		558.40		614.25		
1 Grader Attach., Ripper		60.60		66.65		
2 Stabilizers, 310 H.P.		1366.10		1502.70		
1 Flatbed Truck, 3 Ton		174.80		192.30		
1 Chem. Spreader, Towed		96.80		106.50		
1 Vibr. Roller, 29,000 Lb.		408.80		449.70		
1 Water Tank 5000 Gal.		203.20		223.50		
1 Truck, 30 Ton		316.50		348.15	49.77	54.75
64 L.H., Daily Totals		$4270.00		$5320.95	$66.72	$83.14
Crew B-75	Hr.	Daily	Hr.	Daily	Bare Costs	Incl. O&P
1 Labor Foreman (outside)	$15.40	$123.20	$26.40	$211.20	$17.17	$28.76
1 Laborer	13.40	107.20	22.95	183.60		
4 Equip. Oper. (med.)	19.00	608.00	31.55	1009.60		
1 Truck Driver (heavy)	15.40	123.20	25.80	206.40		
1 Motor Grader, 30,000 Lb.		558.40		614.25		
1 Grader Attach., Ripper		60.60		66.65		
2 Stabilizers, 310 H.P.		1366.10		1502.70		
1 Dist. Truck, 3000 Gal.		353.05		388.35		
1 Vibr. Roller, 29,000 Lb.		408.80		449.70	49.05	53.96
56 L.H., Daily Totals		$3708.55		$4632.45	$66.22	$82.72
Crew B-76	Hr.	Daily	Hr.	Daily	Bare Costs	Incl. O&P
1 Dock Builder Foreman	$20.65	$165.20	$38.60	$308.80	$18.84	$33.96
5 Dock Builders	18.65	746.00	34.90	1396.00		
2 Equip. Oper. (crane)	19.75	316.00	32.80	524.80		
1 Equip. Oper. Oiler	16.20	129.60	26.90	215.20		
1 Crawler Crane, 50 Ton		876.40		964.05		
1 Barge, 400 Ton		439.05		482.95		
1 Hammer, 15K Ft. Lbs.		312.80		344.10		
60 L.F. Leads, 15K Ft. Lbs.		105.00		115.50		
1 Air Compr., 600 C.F.M.		268.15		294.95		
2-50 Ft. Air Hoses, 3" Dia.		33.50		36.85	28.26	31.09
72 L.H., Daily Totals		$3391.70		$4683.20	$47.10	$65.05
Crew B-77	Hr.	Daily	Hr.	Daily	Bare Costs	Incl. O&P
1 Labor Foreman	$15.40	$123.20	$26.40	$211.20	$14.14	$24.11
3 Laborers	13.40	321.60	22.95	550.80		
1 Truck Driver (light)	15.10	120.80	25.30	202.40		
1 Crack Cleaner, 25 H.P.		72.40		79.65		
1 Crack Filler, Trailer Mtd.		143.50		157.85		
1 Flatbed Truck, 3 Ton		174.80		192.30	9.77	10.74
40 L.H., Daily Totals		$956.30		$1394.20	$23.91	$34.85

Crew No.	Bare Costs		Incl. Subs O & P		Cost Per Labor-Hour	
Crew B-78	Hr.	Daily	Hr.	Daily	Bare Costs	Incl. O&P
1 Labor Foreman	$15.40	$123.20	$26.40	$211.20	$13.80	$23.64
4 Laborers	13.40	428.80	22.95	734.40		
1 Paint Striper, S.P.		202.80		223.10		
1 Flatbed Truck, 3 Ton		174.80		192.30		
1 Pickup Truck, 3/4 Ton		125.50		138.05	12.58	13.84
40 L.H., Daily Totals		$1055.10		$1499.05	$26.38	$37.48
Crew B-79	Hr.	Daily	Hr.	Daily	Bare Costs	Incl. O&P
1 Labor Foreman	$15.40	$123.20	$26.40	$211.20	$13.90	$23.81
3 Laborers	13.40	321.60	22.95	550.80		
1 Thermo. Striper, T.M.		236.70		260.35		
1 Flatbed Truck, 3 Ton		174.80		192.30		
2 Pickup Trucks, 3/4 Ton		251.00		276.10	20.70	22.77
32 L.H., Daily Totals		$1107.30		$1490.75	$34.60	$46.58
Crew B-80	Hr.	Daily	Hr.	Daily	Bare Costs	Incl. O&P
1 Labor Foreman	$15.40	$123.20	$26.40	$211.20	$14.07	$24.10
2 Laborer	13.40	214.40	22.95	367.20		
1 Flatbed Truck, 3 Ton		174.80		192.30		
1 Post Driver, T.M.		292.00		321.20	19.45	21.40
24 L.H., Daily Totals		$804.40		$1091.90	$33.52	$45.50
Crew B-80A	Hr.	Daily	Hr.	Daily	Bare Costs	Incl. O&P
3 Laborer	$13.40	$321.60	$22.95	$550.80	$13.40	$22.95
1 Flatbed Truck, 3 Ton		174.80		192.30	7.28	8.01
24 L.H., Daily Totals		$496.40		$743.10	$20.68	$30.96
Crew B-80B	Hr.	Daily	Hr.	Daily	Bare Costs	Incl. O&P
3 Laborer	$13.40	$321.60	$22.95	$550.80	$14.59	$24.75
1 Equip. Oper. (light)	18.15	145.20	30.15	241.20		
1 Crane, flatbed mnt		267.20		293.90	8.35	9.19
32 L.H., Daily Totals		$734.00		$1085.90	$22.94	$33.94
Crew B-81	Hr.	Daily	Hr.	Daily	Bare Costs	Incl. O&P
1 Laborer	$13.40	$107.20	$22.95	$183.60	$14.40	$24.38
1 Truck Driver (heavy)	15.40	123.20	25.80	206.40		
1 Hydromulcher, T.M.		272.80		300.10		
1 Tractor Truck, 4x2		316.50		348.15	36.83	40.51
16 L.H., Daily Totals		$819.70		$1038.25	$51.23	$64.89
Crew B-82	Hr.	Daily	Hr.	Daily	Bare Costs	Incl. O&P
1 Laborer	$13.40	$107.20	$22.95	$183.60	$15.77	$26.55
1 Equip. Oper. (light)	18.15	145.20	30.15	241.20		
1 Horiz. Borer, 6 H.P.		46.40		51.05	2.90	3.19
16 L.H., Daily Totals		$298.80		$475.85	$18.67	$29.74
Crew B-83	Hr.	Daily	Hr.	Daily	Bare Costs	Incl. O&P
1 Tugboat Captain	$19.00	$152.00	$31.55	$252.40	$16.20	$27.25
1 Tugboat Hand	13.40	107.20	22.95	183.60		
1 Tugboat, 250 H.P.		433.30		476.65	27.08	29.79
16 L.H., Daily Totals		$692.50		$912.65	$43.28	$57.04
Crew B-84	Hr.	Daily	Hr.	Daily	Bare Costs	Incl. O&P
1 Equip. Oper. (med.)	$19.00	$152.00	$31.55	$252.40	$19.00	$31.55
1 Rotary Mower/Tractor		206.50		227.15	25.81	28.39
8 L.H., Daily Totals		$358.50		$479.55	$44.81	$59.94

Crew B-85

Crew No.	Bare Costs Hr.	Daily	Incl. Subs O & P Hr.	Daily	Cost Per Labor-Hour Bare Costs	Incl. O&P
3 Laborers	$13.40	$321.60	$22.95	$550.80	$14.92	$25.24
1 Equip. Oper. (med.)	19.00	152.00	31.55	252.40		
1 Truck Driver (heavy)	15.40	123.20	25.80	206.40		
1 Aerial Lift Truck		565.75		622.35		
1 Brush Chipper, 130 H.P.		208.20		229.00		
1 Pruning Saw, Rotary		22.10		24.30	19.90	21.89
40 L.H., Daily Totals		$1392.85		$1885.25	$34.82	$47.13

Crew B-86

Crew No.	Bare Costs Hr.	Daily	Incl. Subs O & P Hr.	Daily	Cost Per Labor-Hour Bare Costs	Incl. O&P
1 Equip. Oper. (med.)	$19.00	$152.00	$31.55	$252.40	$19.00	$31.55
1 Stump Chipper, S.P.		160.00		176.00	20.00	22.00
8 L.H., Daily Totals		$312.00		$428.40	$39.00	$53.55

Crew B-86A

Crew No.	Bare Costs Hr.	Daily	Incl. Subs O & P Hr.	Daily	Cost Per Labor-Hour Bare Costs	Incl. O&P
1 Equip. Oper. (medium)	$19.00	$152.00	$31.55	$252.40	$19.00	$31.55
1 Grader, 30,000 Lbs.		558.40		614.25	69.80	76.78
8 L.H., Daily Totals		$710.40		$866.65	$88.80	$108.33

Crew B-86B

Crew No.	Bare Costs Hr.	Daily	Incl. Subs O & P Hr.	Daily	Cost Per Labor-Hour Bare Costs	Incl. O&P
1 Equip. Oper. (medium)	$19.00	$152.00	$31.55	$252.40	$19.00	$31.55
1 Dozer, 200 H.P.		832.00		915.20	104.00	114.40
8 L.H., Daily Totals		$984.00		$1167.60	$123.00	$145.95

Crew B-87

Crew No.	Bare Costs Hr.	Daily	Incl. Subs O & P Hr.	Daily	Cost Per Labor-Hour Bare Costs	Incl. O&P
1 Laborer	$13.40	$107.20	$22.95	$183.60	$17.88	$29.83
4 Equip. Oper. (med.)	19.00	608.00	31.55	1009.60		
2 Feller Bunchers, 50 H.P.		755.20		830.70		
1 Log Chipper, 22" Tree		2030.00		2233.00		
1 Dozer, 105 H.P.		305.20		335.70		
1 Chainsaw, Gas, 36" Long		52.80		58.10	78.58	86.44
40 L.H., Daily Totals		$3858.40		$4650.70	$96.46	$116.27

Crew B-88

Crew No.	Bare Costs Hr.	Daily	Incl. Subs O & P Hr.	Daily	Cost Per Labor-Hour Bare Costs	Incl. O&P
1 Laborer	$13.40	$107.20	$22.95	$183.60	$18.20	$30.32
6 Equip. Oper. (med.)	19.00	912.00	31.55	1514.40		
2 Feller Bunchers, 50 H.P.		755.20		830.70		
1 Log Chipper, 22" Tree		2030.00		2233.00		
2 Log Skidders, 50 H.P.		738.40		812.25		
1 Dozer, 105 H.P.		305.20		335.70		
1 Chainsaw, Gas, 36" Long		52.80		58.10	69.31	76.25
56 L.H., Daily Totals		$4900.80		$5967.75	$87.51	$106.57

Crew B-89

Crew No.	Bare Costs Hr.	Daily	Incl. Subs O & P Hr.	Daily	Cost Per Labor-Hour Bare Costs	Incl. O&P
1 Skilled Worker	$18.75	$150.00	$32.25	$258.00	$16.08	$27.60
1 Building Laborer	13.40	107.20	22.95	183.60		
1 Cutting Machine		44.20		48.60	2.76	3.04
16 L.H., Daily Totals		$301.40		$490.20	$18.84	$30.64

Crew B-89A

Crew No.	Bare Costs Hr.	Daily	Incl. Subs O & P Hr.	Daily	Cost Per Labor-Hour Bare Costs	Incl. O&P
1 Skilled Worker	$18.75	$150.00	$32.25	$258.00	$16.08	$27.60
1 Laborer	13.40	107.20	22.95	183.60		
1 Core Drill (large)		59.20		65.10	3.70	4.07
16 L.H., Daily Totals		$316.40		$506.70	$19.78	$31.67

Crew B-90

Crew No.	Bare Costs Hr.	Daily	Incl. Subs O & P Hr.	Daily	Cost Per Labor-Hour Bare Costs	Incl. O&P
1 Labor Foreman (outside)	$15.40	$123.20	$26.40	$211.20	$15.34	$25.89
3 Laborers	13.40	321.60	22.95	550.80		
2 Equip. Oper. (light)	18.15	290.40	30.15	482.40		
2 Truck Drivers (heavy)	15.40	246.40	25.80	412.80		
1 Road Mixer, 310 H.P.		986.40		1085.05		
1 Dist. Truck, 2000 Gal.		328.55		361.40	20.55	22.60
64 L.H., Daily Totals		$2296.55		$3103.65	$35.89	$48.49

Crew B-90A

Crew No.	Bare Costs Hr.	Daily	Incl. Subs O & P Hr.	Daily	Cost Per Labor-Hour Bare Costs	Incl. O&P
1 Labor Foreman	$15.40	$123.20	$26.40	$211.20	$16.89	$28.36
2 Laborers	13.40	214.40	22.95	367.20		
4 Equip. Oper. (medium)	19.00	608.00	31.55	1009.60		
2 Graders, 30,000 Lbs.		1116.80		1228.50		
1 Roller, Steel Wheel		238.00		261.80		
1 Roller, Pneumatic Wheel		237.20		260.90	28.43	31.27
56 L.H., Daily Totals		$2537.60		$3339.20	$45.32	$59.63

Crew B-90B

Crew No.	Bare Costs Hr.	Daily	Incl. Subs O & P Hr.	Daily	Cost Per Labor-Hour Bare Costs	Incl. O&P
1 Labor Foreman	$15.40	$123.20	$26.40	$211.20	$16.53	$27.82
2 Laborers	13.40	214.40	22.95	367.20		
3 Equip. Oper. (medium)	19.00	456.00	31.55	757.20		
1 Roller, Steel Wheel		238.00		261.80		
1 Roller, Pneumatic Wheel		237.20		260.90		
1 Road Mixer, 310 H.P.		986.40		1085.05	30.45	33.50
48 L.H., Daily Totals		$2255.20		$2943.35	$46.98	$61.32

Crew B-91

Crew No.	Bare Costs Hr.	Daily	Incl. Subs O & P Hr.	Daily	Cost Per Labor-Hour Bare Costs	Incl. O&P
1 Labor Foreman (outside)	$15.40	$123.20	$26.40	$211.20	$16.70	$28.04
2 Laborers	13.40	214.40	22.95	367.20		
4 Equip. Oper. (med.)	19.00	608.00	31.55	1009.60		
1 Truck Driver (heavy)	15.40	123.20	25.80	206.40		
1 Dist. Truck, 3000 Gal.		353.05		388.35		
1 Aggreg. Spreader, S.P.		615.20		676.70		
1 Roller, Pneu. Tire, 12 Ton		237.20		260.90		
1 Roller, Steel, 10 Ton		238.00		261.80	22.55	24.81
64 L.H., Daily Totals		$2512.25		$3382.15	$39.25	$52.85

Crew B-92

Crew No.	Bare Costs Hr.	Daily	Incl. Subs O & P Hr.	Daily	Cost Per Labor-Hour Bare Costs	Incl. O&P
1 Labor Foreman (outside)	$15.40	$123.20	$26.40	$211.20	$13.90	$23.81
3 Laborers	13.40	321.60	22.95	550.80		
1 Crack Cleaner, 25 H.P.		72.40		79.65		
1 Air Compressor		74.00		81.40		
1 Tar Kettle, T.M.		21.80		24.00		
1 Flatbed Truck, 3 Ton		174.80		192.30	10.72	11.79
32 L.H., Daily Totals		$787.80		$1139.35	$24.62	$35.60

Crew B-93

Crew No.	Bare Costs Hr.	Daily	Incl. Subs O & P Hr.	Daily	Cost Per Labor-Hour Bare Costs	Incl. O&P
1 Equip. Oper. (med.)	$19.00	$152.00	$31.55	$252.40	$19.00	$31.55
1 Feller Buncher, 50 H.P.		377.60		415.35	47.20	51.92
8 L.H., Daily Totals		$529.60		$667.75	$66.20	$83.47

Crew B-94A

Crew No.	Bare Costs Hr.	Daily	Incl. Subs O & P Hr.	Daily	Cost Per Labor-Hour Bare Costs	Incl. O&P
1 Laborer	$13.40	$107.20	$22.95	$183.60	$13.40	$22.95
1 Diaph. Water Pump, 2"		28.00		30.80		
1-20 Ft. Suction Hose, 2"		6.50		7.15		
2-50 Ft. Disch. Hoses, 2"		8.80		9.70	5.41	5.95
8 L.H., Daily Totals		$150.50		$231.25	$18.81	$28.90

Crew No.	Bare Costs		Incl. Subs O & P		Cost Per Labor-Hour	
Crew B-94B	Hr.	Daily	Hr.	Daily	Bare Costs	Incl. O&P
1 Laborer	$13.40	$107.20	$22.95	$183.60	$13.40	$22.95
1 Diaph. Water Pump, 4"		62.00		68.20		
1-20 Ft. Suction Hose, 4"		12.50		13.75		
2-50 Ft. Disch. Hoses, 4"		17.00		18.70	11.44	12.58
8 L.H., Daily Totals		$198.70		$284.25	$24.84	$35.53
Crew B-94C	Hr.	Daily	Hr.	Daily	Bare Costs	Incl. O&P
1 Laborer	$13.40	$107.20	$22.95	$183.60	$13.40	$22.95
1 Centr. Water Pump, 3"		36.60		40.25		
1-20 Ft. Suction Hose, 3"		9.50		10.45		
2-50 Ft. Disch. Hoses, 3"		10.80		11.90	7.11	7.82
8 L.H., Daily Totals		$164.10		$246.20	$20.51	$30.77
Crew B-94D	Hr.	Daily	Hr.	Daily	Bare Costs	Incl. O&P
1 Laborer	$13.40	$107.20	$22.95	$183.60	$13.40	$22.95
1 Centr. Water Pump, 6"		155.20		170.70		
1-20 Ft. Suction Hose, 6"		20.50		22.55		
2-50 Ft. Disch. Hoses, 6"		37.00		40.70	26.59	29.25
8 L.H., Daily Totals		$319.90		$417.55	$39.99	$52.20
Crew B-95	Hr.	Daily	Hr.	Daily	Bare Costs	Incl. O&P
1 Equip. Oper. (crane)	$19.75	$158.00	$32.80	$262.40	$16.58	$27.88
1 Laborer	13.40	107.20	22.95	183.60		
16 L.H., Daily Totals		$265.20		$446.00	$16.58	$27.88
Crew B-95A	Hr.	Daily	Hr.	Daily	Bare Costs	Incl. O&P
1 Equip. Oper. (crane)	$19.75	$158.00	$32.80	$262.40	$16.58	$27.88
1 Laborer	13.40	107.20	22.95	183.60		
1 Hyd. Excavator, 5/8 C.Y.		395.65		435.20	24.73	27.20
16 L.H., Daily Totals		$660.85		$881.20	$41.31	$55.08
Crew B-95B	Hr.	Daily	Hr.	Daily	Bare Costs	Incl. O&P
1 Equip. Oper. (crane)	$19.75	$158.00	$32.80	$262.40	$16.58	$27.88
1 Laborer	13.40	107.20	22.95	183.60		
1 Hyd. Excavator, 1.5 C.Y.		706.55		777.20	44.16	48.58
16 L.H., Daily Totals		$971.75		$1223.20	$60.74	$76.46
Crew B-95C	Hr.	Daily	Hr.	Daily	Bare Costs	Incl. O&P
1 Equip. Oper. (crane)	$19.75	$158.00	$32.80	$262.40	$16.58	$27.88
1 Laborer	13.40	107.20	22.95	183.60		
1 Hyd. Excavator, 2.5 C.Y.		1681.00		1849.10	105.06	115.57
16 L.H., Daily Totals		$1946.20		$2295.10	$121.64	$143.45
Crew C-1	Hr.	Daily	Hr.	Daily	Bare Costs	Incl. O&P
2 Carpenters	$18.50	$296.00	$31.70	$507.20	$16.06	$27.51
1 Carpenter Helper	13.85	110.80	23.70	189.60		
1 Laborer	13.40	107.20	22.95	183.60		
32 L.H., Daily Totals		$514.00		$880.40	$16.06	$27.51
Crew C-2	Hr.	Daily	Hr.	Daily	Bare Costs	Incl. O&P
1 Carpenter Foreman (out)	$20.50	$164.00	$35.15	$281.20	$16.43	$28.15
2 Carpenters	18.50	296.00	31.70	507.20		
2 Carpenter Helpers	13.85	221.60	23.70	379.20		
1 Laborer	13.40	107.20	22.95	183.60		
48 L.H., Daily Totals		$788.80		$1351.20	$16.43	$28.15

Crew No.	Bare Costs		Incl. Subs O & P		Cost Per Labor-Hour	
Crew C-2A	Hr.	Daily	Hr.	Daily	Bare Costs	Incl. O&P
1 Carpenter Foreman	$20.50	$164.00	$35.15	$281.20	$17.87	$30.38
3 Carpenters	18.50	444.00	31.70	760.80		
1 Cement Finisher	17.80	142.40	29.05	232.40		
1 Laborer	13.40	107.20	22.95	183.60		
48 L.H., Daily Totals		$857.60		$1458.00	$17.87	$30.38
Crew C-3	Hr.	Daily	Hr.	Daily	Bare Costs	Incl. O&P
1 Rodman Foreman	$21.75	$174.00	$40.80	$326.40	$17.42	$31.37
3 Rodmen (reinf.)	19.75	474.00	37.05	889.20		
1 Equip. Oper. (light)	18.15	145.20	30.15	241.20		
3 Laborers	13.40	321.60	22.95	550.80		
3 Stressing Equipment		40.35		44.40		
.5 Grouting Equipment		114.40		125.85	2.42	2.66
64 L.H., Daily Totals		$1269.55		$2177.85	$19.84	$34.03
Crew C-4	Hr.	Daily	Hr.	Daily	Bare Costs	Incl. O&P
1 Rodman Foreman	$21.75	$174.00	$40.80	$326.40	$18.66	$34.46
2 Rodmen (reinf.)	19.75	316.00	37.05	592.80		
1 Building Laborer	13.40	107.20	22.95	183.60		
3 Stressing Equipment		40.35		44.40	1.26	1.39
32 L.H., Daily Totals		$637.55		$1147.20	$19.92	$35.85
Crew C-5	Hr.	Daily	Hr.	Daily	Bare Costs	Incl. O&P
1 Rodman Foreman	$21.75	$174.00	$40.80	$326.40	$17.97	$32.27
2 Rodmen (reinf.)	19.75	316.00	37.05	592.80		
1 Equip. Oper. (crane)	19.75	158.00	32.80	262.40		
2 Building Laborers	13.40	214.40	22.95	367.20		
1 Hyd. Crane, 25 Ton		538.85		592.75	11.23	12.35
48 L.H., Daily Totals		$1401.25		$2141.55	$29.20	$44.62
Crew C-6	Hr.	Daily	Hr.	Daily	Bare Costs	Incl. O&P
1 Labor Foreman (outside)	$15.40	$123.20	$26.40	$211.20	$14.47	$24.54
4 Laborers	13.40	428.80	22.95	734.40		
1 Cement Finisher	17.80	142.40	29.05	232.40		
2 Gas Engine Vibrators		74.40		81.85	1.55	1.71
48 L.H., Daily Totals		$768.80		$1259.85	$16.02	$26.25
Crew C-7	Hr.	Daily	Hr.	Daily	Bare Costs	Incl. O&P
1 Labor Foreman (outside)	$15.40	$123.20	$26.40	$211.20	$15.04	$25.41
5 Laborers	13.40	536.00	22.95	918.00		
1 Cement Finisher	17.80	142.40	29.05	232.40		
1 Equip. Oper. (med.)	19.00	152.00	31.55	252.40		
1 Equip. Oper. (oiler)	16.20	129.60	26.90	215.20		
2 Gas Engine Vibrators		74.40		81.85		
1 Concrete Bucket, 1 C.Y.		26.35		29.00		
1 Hyd. Crane, 55 Ton		785.20		863.70	12.30	13.54
72 L.H., Daily Totals		$1969.15		$2803.75	$27.34	$38.95
Crew C-8	Hr.	Daily	Hr.	Daily	Bare Costs	Incl. O&P
1 Labor Foreman (outside)	$15.40	$123.20	$26.40	$211.20	$15.74	$26.41
3 Laborers	13.40	321.60	22.95	550.80		
2 Cement Finishers	17.80	284.80	29.05	464.80		
1 Equip. Oper. (med.)	19.00	152.00	31.55	252.40		
1 Concrete Pump (small)		632.80		696.10	11.30	12.43
56 L.H., Daily Totals		$1514.40		$2175.30	$27.04	$38.84
Crew C-8A	Hr.	Daily	Hr.	Daily	Bare Costs	Incl. O&P
1 Labor Foreman (outside)	$15.40	$123.20	$26.40	$211.20	$15.20	$25.56
3 Laborers	13.40	321.60	22.95	550.80		
2 Cement Finishers	17.80	284.80	29.05	464.80		
48 L.H., Daily Totals		$729.60		$1226.80	$15.20	$25.56

Crew C-8B

Crew No.	Bare Costs Hr.	Daily	Incl. Subs O & P Hr.	Daily	Cost Per Labor-Hour Bare Costs	Incl. O&P
1 Labor Foreman (outside)	$15.40	$123.20	$26.40	$211.20	$14.92	$25.36
3 Laborers	13.40	321.60	22.95	550.80		
1 Equipment Operator	19.00	152.00	31.55	252.40		
1 Vibrating Screed		47.40		52.15		
1 Vibratory Roller		408.80		449.70		
1 Dozer, 200 HP		832.00		915.20	32.21	35.43
40 L.H., Daily Totals		$1885.00		$2431.45	$47.13	$60.79

Crew C-8C

Crew No.	Bare Costs Hr.	Daily	Incl. Subs O & P Hr.	Daily	Cost Per Labor-Hour Bare Costs	Incl. O&P
1 Labor Forman	$15.40	$123.20	$26.40	$211.20	$15.40	$25.98
3 Laborers	13.40	321.60	22.95	550.80		
1 Cement Finisher	17.80	142.40	29.05	232.40		
1 Equipment Operator (med.)	19.00	152.00	31.55	252.40		
1 Shotcr.Rig(12CY/HR)		376.00		413.60	7.83	8.62
48 L.H., Daily Totals		$1115.20		$1660.40	$23.23	$34.60

Crew C-8D

Crew No.	Bare Costs Hr.	Daily	Incl. Subs O & P Hr.	Daily	Cost Per Labor-Hour Bare Costs	Incl. O&P
1 Labor Foreman	$15.40	$123.20	$26.40	$211.20	$16.19	$27.14
1 Laborers	13.40	107.20	22.95	183.60		
1 Cement Finisher	17.80	142.40	29.05	232.40		
1 Equipment Operator (light)	18.15	145.20	30.15	241.20		
1 Compressor, 250 CFM		120.40		132.45		
2 Hoses, 1", 50'		9.60		10.55	4.06	4.47
32 L.H., Daily Totals		$648.00		$1011.40	$20.25	$31.61

Crew C-8E

Crew No.	Bare Costs Hr.	Daily	Incl. Subs O & P Hr.	Daily	Cost Per Labor-Hour Bare Costs	Incl. O&P
1 Labor Foreman	$15.40	$123.20	$26.40	$211.20	$16.19	$27.14
1 Laborers	13.40	107.20	22.95	183.60		
1 Cement Finisher	17.80	142.40	29.05	232.40		
1 Equipment Operator (light)	18.15	145.20	30.15	241.20		
1 Compressor, 250 CFM		120.40		132.45		
2 Hoses, 1", 50'		9.60		10.55		
1 Concrete Pump (small)		632.80		696.10	23.84	26.22
32 L.H., Daily Totals		$1280.80		$1707.50	$40.03	$53.36

Crew C-10

Crew No.	Bare Costs Hr.	Daily	Incl. Subs O & P Hr.	Daily	Cost Per Labor-Hour Bare Costs	Incl. O&P
1 Laborer	$13.40	$107.20	$22.95	$183.60	$16.33	$27.02
2 Cement Finishers	17.80	284.80	29.05	464.80		
24 L.H., Daily Totals		$392.00		$648.40	$16.33	$27.02

Crew C-11

Crew No.	Bare Costs Hr.	Daily	Incl. Subs O & P Hr.	Daily	Cost Per Labor-Hour Bare Costs	Incl. O&P
1 Skilled Worker Foreman	$20.75	$166.00	$35.70	$285.60	$19.18	$32.82
5 Skilled Worker	18.75	750.00	32.25	1290.00		
1 Equip. Oper. (crane)	19.75	158.00	32.80	262.40		
1 Truck Crane, 150 Ton		1555.00		1710.50	27.77	30.54
56 L.H., Daily Totals		$2629.00		$3548.50	$46.95	$63.36

Crew C-12

Crew No.	Bare Costs Hr.	Daily	Incl. Subs O & P Hr.	Daily	Cost Per Labor-Hour Bare Costs	Incl. O&P
1 Carpenter Foreman (out)	$20.50	$164.00	$35.15	$281.20	$18.19	$31.00
3 Carpenters	18.50	444.00	31.70	760.80		
1 Laborer	13.40	107.20	22.95	183.60		
1 Equip. Oper. (crane)	19.75	158.00	32.80	262.40		
1 Hyd. Crane, 12 Ton		438.10		481.90	9.13	10.04
48 L.H., Daily Totals		$1311.30		$1969.90	$27.32	$41.04

Crew C-13

Crew No.	Bare Costs Hr.	Daily	Incl. Subs O & P Hr.	Daily	Cost Per Labor-Hour Bare Costs	Incl. O&P
2 Struc. Steel Worker	$19.95	$319.20	$39.30	$628.80	$19.47	$36.77
1 Carpenter	18.50	148.00	31.70	253.60		
1 Gas Welding Machine		80.15		88.15	3.34	3.67
24 L.H., Daily Totals		$547.35		$970.55	$22.81	$40.44

Crew C-14

Crew No.	Bare Costs Hr.	Daily	Incl. Subs O & P Hr.	Daily	Cost Per Labor-Hour Bare Costs	Incl. O&P
1 Carpenter Foreman (out)	$20.50	$164.00	$35.15	$281.20	$16.46	$28.34
3 Carpenters	18.50	444.00	31.70	760.80		
2 Carpenter Helpers	13.85	221.60	23.70	379.20		
4 Laborers	13.40	428.80	22.95	734.40		
2 Rodmen (reinf.)	19.75	316.00	37.05	592.80		
2 Rodman Helpers	13.85	221.60	23.70	379.20		
2 Cement Finishers	17.80	284.80	29.05	464.80		
1 Equip. Oper. (crane)	19.75	158.00	32.80	262.40		
1 Crane, 80 Ton, & Tools		1090.00		1199.00	8.01	8.82
136 L.H., Daily Totals		$3328.80		$5053.80	$24.47	$37.16

Crew C-14A

Crew No.	Bare Costs Hr.	Daily	Incl. Subs O & P Hr.	Daily	Cost Per Labor-Hour Bare Costs	Incl. O&P
1 Carpenter Foreman (out)	$20.50	$164.00	$35.15	$281.20	$18.36	$31.88
16 Carpenters	18.50	2368.00	31.70	4057.60		
4 Rodmen (reinf.)	19.75	632.00	37.05	1185.60		
2 Laborers	13.40	214.40	22.95	367.20		
1 Cement Finisher	17.80	142.40	29.05	232.40		
1 Equip. Oper. (med)	19.00	152.00	31.55	252.40		
1 Gas Engine Vibrator		37.20		40.90		
1 Concrete Pump (small)		632.80		696.10	3.35	3.69
200 L.H., Daily Totals		$4342.80		$7113.40	$21.71	$35.57

Crew C-14B

Crew No.	Bare Costs Hr.	Daily	Incl. Subs O & P Hr.	Daily	Cost Per Labor-Hour Bare Costs	Incl. O&P
1 Carpenter Foreman (out)	$20.50	$164.00	$35.15	$281.20	$18.34	$31.77
16 Carpenters	18.50	2368.00	31.70	4057.60		
4 Rodmen (reinf.)	19.75	632.00	37.05	1185.60		
2 Laborers	13.40	214.40	22.95	367.20		
2 Cement Finishers	17.80	284.80	29.05	464.80		
1 Equip. Oper. (med)	19.00	152.00	31.55	252.40		
1 Gas Engine Vibrator		37.20		40.90		
1 Concrete Pump (small)		632.80		696.10	3.22	3.54
208 L.H., Daily Totals		$4485.20		$7345.80	$21.56	$35.31

Crew C-14C

Crew No.	Bare Costs Hr.	Daily	Incl. Subs O & P Hr.	Daily	Cost Per Labor-Hour Bare Costs	Incl. O&P
1 Carpenter Foreman (out)	$20.50	$164.00	$35.15	$281.20	$17.31	$30.02
6 Carpenters	18.50	888.00	31.70	1521.60		
2 Rodmem (reinf)	19.75	316.00	37.05	592.80		
4 Laborers	13.40	428.80	22.95	734.40		
1 Cement Finisher	17.80	142.40	29.05	232.40		
1 Gas Engine Vibrator		37.20		40.90	.33	.37
112 L.H., Daily Totals		$1976.40		$3403.30	$17.64	$30.39

Crew C-14D

Crew No.	Bare Costs Hr.	Daily	Incl. Subs O & P Hr.	Daily	Cost Per Labor-Hour Bare Costs	Incl. O&P
1 Carpenter Foreman (out)	$20.50	$164.00	$35.15	$281.20	$18.26	$31.45
18 Carpenters	18.50	2664.00	31.70	4564.80		
2 Rodmen (reinf.)	19.75	316.00	37.05	592.80		
2 Laborers	13.40	214.40	22.95	367.20		
1 Cement Finisher	17.80	142.40	29.05	232.40		
1 Equip. Oper. (med.)	19.00	152.00	31.55	252.40		
1 Gas Engine Vibrator		37.20		40.90		
1 Concrete Pump (small)		632.80		696.10	3.35	3.69
200 L.H., Daily Totals		$4322.80		$7027.80	$21.61	$35.14

Crew C-14E

Crew No.	Bare Costs Hr.	Daily	Incl. Subs O & P Hr.	Daily	Cost Per Labor-Hour Bare Costs	Incl. O&P
1 Carpenter Foreman (out)	$20.50	$164.00	$35.15	$281.20	$17.68	$31.33
2 Carpenters	18.50	296.00	31.70	507.20		
4 Rodmen (reinf.)	19.75	632.00	37.05	1185.60		
3 Laborers	13.40	321.60	22.95	550.80		
1 Cement Finisher	17.80	142.40	29.05	232.40		
1 Gas Engine Vibrator		37.20		40.90	.42	.47
88 L.H., Daily Totals		$1593.20		$2798.10	$18.10	$31.80

Crew C-14F

Crew No.	Bare Costs Hr.	Bare Costs Daily	Incl. Subs O & P Hr.	Incl. Subs O & P Daily	Cost Per Labor-Hour Bare Costs	Cost Per Labor-Hour Incl. O&P
1 Laborer Foreman (out)	$15.40	$123.20	$26.40	$211.20	$16.56	$27.40
2 Laborers	13.40	214.40	22.95	367.20		
6 Cement Finishers	17.80	854.40	29.05	1394.40		
1 Gas Engine Vibrator		37.20		40.90	.52	.57
72 L.H., Daily Totals		$1229.20		$2013.70	$17.08	$27.97

Crew C-14G

Crew No.	Bare Costs Hr.	Bare Costs Daily	Incl. Subs O & P Hr.	Incl. Subs O & P Daily	Cost Per Labor-Hour Bare Costs	Cost Per Labor-Hour Incl. O&P
1 Laborer Foreman	$15.40	$123.20	$26.40	$211.20	$16.20	$26.93
2 Laborers	13.40	214.40	22.95	367.20		
4 Cement Finishers	17.80	569.60	29.05	929.60		
1 Gas Engine Vibrator		37.20		40.90	.66	.73
56 L.H., Daily Totals		$944.40		$1548.90	$16.86	$27.66

Crew C-14H

Crew No.	Bare Costs Hr.	Bare Costs Daily	Incl. Subs O & P Hr.	Incl. Subs O & P Daily	Cost Per Labor-Hour Bare Costs	Cost Per Labor-Hour Incl. O&P
1 Carpenter Foreman (out)	$20.50	$164.00	$35.15	$281.20	$18.07	$31.27
2 Carpenters	18.50	296.00	31.70	507.20		
1 Rodman (reinf.)	19.75	158.00	37.05	296.40		
1 Laborer	13.40	107.20	22.95	183.60		
1 Cement Finisher	17.80	142.40	29.05	232.40		
1 Gas Engine Vibrator		37.20		40.90	.78	.85
48 L.H., Daily Totals		$904.80		$1541.70	$18.85	$32.12

Crew C-15

Crew No.	Bare Costs Hr.	Bare Costs Daily	Incl. Subs O & P Hr.	Incl. Subs O & P Daily	Cost Per Labor-Hour Bare Costs	Cost Per Labor-Hour Incl. O&P
1 Carpenter Foreman (out)	$20.50	$164.00	$35.15	$281.20	$17.01	$29.17
2 Carpenters	18.50	296.00	31.70	507.20		
3 Laborers	13.40	321.60	22.95	550.80		
2 Cement Finishers	17.80	284.80	29.05	464.80		
1 Rodman (reinf.)	19.75	158.00	37.05	296.40		
72 L.H., Daily Totals		$1224.40		$2100.40	$17.01	$29.17

Crew C-16

Crew No.	Bare Costs Hr.	Bare Costs Daily	Incl. Subs O & P Hr.	Incl. Subs O & P Daily	Cost Per Labor-Hour Bare Costs	Cost Per Labor-Hour Incl. O&P
1 Labor Foreman (outside)	$15.40	$123.20	$26.40	$211.20	$16.63	$28.78
3 Laborers	13.40	321.60	22.95	550.80		
2 Cement Finishers	17.80	284.80	29.05	464.80		
1 Equip. Oper. (med.)	19.00	152.00	31.55	252.40		
2 Rodmen (reinf.)	19.75	316.00	37.05	592.80		
1 Concrete Pump (small)		632.80		696.10	8.79	9.67
72 L.H., Daily Totals		$1830.40		$2768.10	$25.42	$38.45

Crew C-17

Crew No.	Bare Costs Hr.	Bare Costs Daily	Incl. Subs O & P Hr.	Incl. Subs O & P Daily	Cost Per Labor-Hour Bare Costs	Cost Per Labor-Hour Incl. O&P
2 Skilled Worker Foremen	$20.75	$332.00	$35.70	$571.20	$19.15	$32.94
8 Skilled Workers	18.75	1200.00	32.25	2064.00		
80 L.H., Daily Totals		$1532.00		$2635.20	$19.15	$32.94

Crew C-17A

Crew No.	Bare Costs Hr.	Bare Costs Daily	Incl. Subs O & P Hr.	Incl. Subs O & P Daily	Cost Per Labor-Hour Bare Costs	Cost Per Labor-Hour Incl. O&P
2 Skilled Worker Foremen	$20.75	$332.00	$35.70	$571.20	$19.16	$32.94
8 Skilled Workers	18.75	1200.00	32.25	2064.00		
.125 Equip. Oper. (crane)	19.75	19.75	32.80	32.80		
.125 Crane, 80 Ton, & Tools		136.25		149.90	1.68	1.85
81 L.H., Daily Totals		$1688.00		$2817.90	$20.84	$34.79

Crew C-17B

Crew No.	Bare Costs Hr.	Bare Costs Daily	Incl. Subs O & P Hr.	Incl. Subs O & P Daily	Cost Per Labor-Hour Bare Costs	Cost Per Labor-Hour Incl. O&P
2 Skilled Worker Foremen	$20.75	$332.00	$35.70	$571.20	$19.16	$32.94
8 Skilled Workers	18.75	1200.00	32.25	2064.00		
.25 Equip. Oper. (crane)	19.75	39.50	32.80	65.60		
.25 Crane, 80 Ton, & Tools		272.50		299.75		
.25 Hand Held Power Tools		3.08		3.40		
.25 Walk Behind Power Tools		12.60		13.85	3.51	3.87
82 L.H., Daily Totals		$1859.68		$3017.80	$22.67	$36.81

Crew C-17C

Crew No.	Bare Costs Hr.	Bare Costs Daily	Incl. Subs O & P Hr.	Incl. Subs O & P Daily	Cost Per Labor-Hour Bare Costs	Cost Per Labor-Hour Incl. O&P
2 Skilled Worker Foremen	$20.75	$332.00	$35.70	$571.20	$19.17	$32.93
8 Skilled Workers	18.75	1200.00	32.25	2064.00		
.375 Equip. Oper. (crane)	19.75	59.25	32.80	98.40		
.375 Crane, 80 Ton & Tools		408.75		449.65	4.92	5.42
83 L.H., Daily Totals		$2000.00		$3183.25	$24.09	$38.35

Crew C-17D

Crew No.	Bare Costs Hr.	Bare Costs Daily	Incl. Subs O & P Hr.	Incl. Subs O & P Daily	Cost Per Labor-Hour Bare Costs	Cost Per Labor-Hour Incl. O&P
2 Skilled Worker Foremen	$20.75	$332.00	$35.70	$571.20	$19.18	$32.93
8 Skilled Workers	18.75	1200.00	32.25	2064.00		
.5 Equip. Oper. (crane)	19.75	79.00	32.80	131.20		
.5 Crane, 80 Ton & Tools		545.00		599.50	6.49	7.14
84 L.H., Daily Totals		$2156.00		$3365.90	$25.67	$40.07

Crew C-17E

Crew No.	Bare Costs Hr.	Bare Costs Daily	Incl. Subs O & P Hr.	Incl. Subs O & P Daily	Cost Per Labor-Hour Bare Costs	Cost Per Labor-Hour Incl. O&P
2 Skilled Worker Foremen	$20.75	$332.00	$35.70	$571.20	$19.15	$32.94
8 Skilled Workers	18.75	1200.00	32.25	2064.00		
1 Hyd. Jack with Rods		60.50		66.55	.76	.83
80 L.H., Daily Totals		$1592.50		$2701.75	$19.91	$33.77

Crew C-18

Crew No.	Bare Costs Hr.	Bare Costs Daily	Incl. Subs O & P Hr.	Incl. Subs O & P Daily	Cost Per Labor-Hour Bare Costs	Cost Per Labor-Hour Incl. O&P
.125 Labor Foreman (out)	$15.40	$15.40	$26.40	$26.40	$13.62	$23.33
1 Laborer	13.40	107.20	22.95	183.60		
1 Concrete Cart, 10 C.F.		50.60		55.65	5.62	6.18
9 L.H., Daily Totals		$173.20		$265.65	$19.24	$29.51

Crew C-19

Crew No.	Bare Costs Hr.	Bare Costs Daily	Incl. Subs O & P Hr.	Incl. Subs O & P Daily	Cost Per Labor-Hour Bare Costs	Cost Per Labor-Hour Incl. O&P
.125 Labor Foreman (out)	$15.40	$15.40	$26.40	$26.40	$13.62	$23.33
1 Laborer	13.40	107.20	22.95	183.60		
1 Concrete Cart, 18 C.F.		84.80		93.30	9.42	10.36
9 L.H., Daily Totals		$207.40		$303.30	$23.04	$33.69

Crew C-20

Crew No.	Bare Costs Hr.	Bare Costs Daily	Incl. Subs O & P Hr.	Incl. Subs O & P Daily	Cost Per Labor-Hour Bare Costs	Cost Per Labor-Hour Incl. O&P
1 Labor Foreman (outside)	$15.40	$123.20	$26.40	$211.20	$14.90	$25.22
5 Laborers	13.40	536.00	22.95	918.00		
1 Cement Finisher	17.80	142.40	29.05	232.40		
1 Equip. Oper. (med.)	19.00	152.00	31.55	252.40		
2 Gas Engine Vibrators		74.40		81.85		
1 Concrete Pump (small)		632.80		696.10	11.05	12.16
64 L.H., Daily Totals		$1660.80		$2391.95	$25.95	$37.38

Crew C-21

Crew No.	Bare Costs Hr.	Bare Costs Daily	Incl. Subs O & P Hr.	Incl. Subs O & P Daily	Cost Per Labor-Hour Bare Costs	Cost Per Labor-Hour Incl. O&P
1 Labor Foreman (outside)	$15.40	$123.20	$26.40	$211.20	$14.90	$25.22
5 Laborers	13.40	536.00	22.95	918.00		
1 Cement Finisher	17.80	142.40	29.05	232.40		
1 Equip. Oper. (med.)	19.00	152.00	31.55	252.40		
2 Gas Engine Vibrators		74.40		81.85		
1 Concrete Conveyer		180.00		198.00	3.98	4.37
64 L.H., Daily Totals		$1208.00		$1893.85	$18.88	$29.59

Crew C-22

Crew No.	Bare Costs Hr.	Bare Costs Daily	Incl. Subs O & P Hr.	Incl. Subs O & P Daily	Cost Per Labor-Hour Bare Costs	Cost Per Labor-Hour Incl. O&P
1 Rodman Foreman	$21.75	$174.00	$40.80	$326.40	$20.05	$37.42
4 Rodmen (reinf.)	19.75	632.00	37.05	1185.60		
.125 Equip. Oper. (crane)	19.75	19.75	32.80	32.80		
.125 Equip. Oper. Oiler	16.20	16.20	26.90	26.90		
.125 Hyd. Crane, 25 Ton		67.36		74.10	1.60	1.76
42 L.H., Daily Totals		$909.31		$1645.80	$21.65	$39.18

Crew C-23

Crew No.	Bare Costs Hr.	Bare Costs Daily	Incl. Subs O & P Hr.	Incl. Subs O & P Daily	Cost Per Labor-Hour Bare Costs	Cost Per Labor-Hour Incl. O&P
2 Skilled Worker Foremen	$20.75	$332.00	$35.70	$571.20	$18.99	$32.46
6 Skilled Workers	18.75	900.00	32.25	1548.00		
1 Equip. Oper. (crane)	19.75	158.00	32.80	262.40		
1 Equip. Oper. Oiler	16.20	129.60	26.90	215.20		
1 Crane, 90 Ton		1030.00		1133.00	12.88	14.16
80 L.H., Daily Totals		$2549.60		$3729.80	$31.87	$46.62

Crew C-24

Crew No.	Bare Costs Hr.	Bare Costs Daily	Incl. Subs O & P Hr.	Incl. Subs O & P Daily	Cost Per Labor-Hour Bare Costs	Cost Per Labor-Hour Incl. O&P
2 Skilled Worker Foremen	$20.75	$332.00	$35.70	$571.20	$18.99	$32.46
6 Skilled Workers	18.75	900.00	32.25	1548.00		
1 Equip. Oper. (crane)	19.75	158.00	32.80	262.40		
1 Equip. Oper. Oiler	16.20	129.60	26.90	215.20		
1 Truck Crane, 150 Ton		1555.00		1710.50	19.44	21.38
80 L.H., Daily Totals		$3074.60		$4307.30	$38.43	$53.84

Crew C-25

Crew No.	Bare Costs Hr.	Bare Costs Daily	Incl. Subs O & P Hr.	Incl. Subs O & P Daily	Cost Per Labor-Hour Bare Costs	Cost Per Labor-Hour Incl. O&P
2 Rodmen (reinf.)	$19.75	$316.00	$37.05	$592.80	$15.80	$29.58
2 Rodmen Helpers	11.85	189.60	22.10	353.60		
32 L.H., Daily Totals		$505.60		$946.40	$15.80	$29.58

Crew D-1

Crew No.	Bare Costs Hr.	Bare Costs Daily	Incl. Subs O & P Hr.	Incl. Subs O & P Daily	Cost Per Labor-Hour Bare Costs	Cost Per Labor-Hour Incl. O&P
1 Bricklayer	$18.95	$151.60	$32.10	$256.80	$16.80	$28.45
1 Bricklayer Helper	14.65	117.20	24.80	198.40		
16 L.H., Daily Totals		$268.80		$455.20	$16.80	$28.45

Crew D-2

Crew No.	Bare Costs Hr.	Bare Costs Daily	Incl. Subs O & P Hr.	Incl. Subs O & P Daily	Cost Per Labor-Hour Bare Costs	Cost Per Labor-Hour Incl. O&P
3 Bricklayers	$18.95	$454.80	$32.10	$770.40	$17.23	$29.18
2 Bricklayer Helpers	14.65	234.40	24.80	396.80		
40 L.H., Daily Totals		$689.20		$1167.20	$17.23	$29.18

Crew D-3

Crew No.	Bare Costs Hr.	Bare Costs Daily	Incl. Subs O & P Hr.	Incl. Subs O & P Daily	Cost Per Labor-Hour Bare Costs	Cost Per Labor-Hour Incl. O&P
3 Bricklayers	$18.95	$454.80	$32.10	$770.40	$17.29	$29.30
2 Bricklayer Helpers	14.65	234.40	24.80	396.80		
.25 Carpenter	18.50	37.00	31.70	63.40		
42 L.H., Daily Totals		$726.20		$1230.60	$17.29	$29.30

Crew D-4

Crew No.	Bare Costs Hr.	Bare Costs Daily	Incl. Subs O & P Hr.	Incl. Subs O & P Daily	Cost Per Labor-Hour Bare Costs	Cost Per Labor-Hour Incl. O&P
1 Bricklayer	$18.95	$151.60	$32.10	$256.80	$15.26	$25.89
3 Bricklayer Helpers	14.65	351.60	24.80	595.20		
1 Building Laborer	13.40	107.20	22.95	183.60		
1 Grout Pump		120.40		132.45		
1 Hoses & Hopper		30.45		33.50		
1 Accessories		10.50		11.55	4.03	4.44
40 L.H., Daily Totals		$771.75		$1213.10	$19.29	$30.33

Crew D-5

Crew No.	Bare Costs Hr.	Bare Costs Daily	Incl. Subs O & P Hr.	Incl. Subs O & P Daily	Cost Per Labor-Hour Bare Costs	Cost Per Labor-Hour Incl. O&P
1 Block Mason Helper	$14.65	$117.20	$24.80	$198.40	$14.65	$24.80
8 L.H., Daily Totals		$117.20		$198.40	$14.65	$24.80

Crew D-6

Crew No.	Bare Costs Hr.	Bare Costs Daily	Incl. Subs O & P Hr.	Incl. Subs O & P Daily	Cost Per Labor-Hour Bare Costs	Cost Per Labor-Hour Incl. O&P
3 Bricklayers	$18.95	$454.80	$32.10	$770.40	$16.80	$28.45
3 Bricklayer Helpers	14.65	351.60	24.80	595.20		
48 L.H., Daily Totals		$806.40		$1365.60	$16.80	$28.45

Crew D-7

Crew No.	Bare Costs Hr.	Bare Costs Daily	Incl. Subs O & P Hr.	Incl. Subs O & P Daily	Cost Per Labor-Hour Bare Costs	Cost Per Labor-Hour Incl. O&P
1 Tile Layer	$18.25	$146.00	$29.55	$236.40	$16.48	$26.67
1 Tile Layer Helper	14.70	117.60	23.80	190.40		
16 L.H., Daily Totals		$263.60		$426.80	$16.48	$26.67

Crew D-8

Crew No.	Bare Costs Hr.	Bare Costs Daily	Incl. Subs O & P Hr.	Incl. Subs O & P Daily	Cost Per Labor-Hour Bare Costs	Cost Per Labor-Hour Incl. O&P
3 Bricklayers	$18.95	$454.80	$32.10	$770.40	$17.23	$29.18
2 Bricklayer Helpers	14.65	234.40	24.80	396.80		
40 L.H., Daily Totals		$689.20		$1167.20	$17.23	$29.18

Crew D-9

Crew No.	Bare Costs Hr.	Bare Costs Daily	Incl. Subs O & P Hr.	Incl. Subs O & P Daily	Cost Per Labor-Hour Bare Costs	Cost Per Labor-Hour Incl. O&P
3 Block Masons	$18.95	$454.80	$32.10	$770.40	$16.80	$28.45
3 Block Mason Helpers	14.65	351.60	24.80	595.20		
48 L.H., Daily Totals		$806.40		$1365.60	$16.80	$28.45

Crew D-10

Crew No.	Bare Costs Hr.	Bare Costs Daily	Incl. Subs O & P Hr.	Incl. Subs O & P Daily	Cost Per Labor-Hour Bare Costs	Cost Per Labor-Hour Incl. O&P
1 Stone Mason Foreman	$20.45	$163.60	$34.65	$277.20	$17.59	$29.66
1 Stone Mason	18.45	147.60	31.25	250.00		
2 Stone Mason Helpers	14.65	234.40	24.80	396.80		
1 Equip. Oper. (crane)	19.75	158.00	32.80	262.40		
1 Truck Crane, 12.5 Ton		467.50		514.25	11.69	12.86
40 L.H., Daily Totals		$1171.10		$1700.65	$29.28	$42.52

Crew D-11

Crew No.	Bare Costs Hr.	Bare Costs Daily	Incl. Subs O & P Hr.	Incl. Subs O & P Daily	Cost Per Labor-Hour Bare Costs	Cost Per Labor-Hour Incl. O&P
2 Stone Masons	$18.45	$295.20	$31.25	$500.00	$17.18	$29.10
1 Stone Mason Helper	14.65	117.20	24.80	198.40		
24 L.H., Daily Totals		$412.40		$698.40	$17.18	$29.10

Crew D-12

Crew No.	Bare Costs Hr.	Bare Costs Daily	Incl. Subs O & P Hr.	Incl. Subs O & P Daily	Cost Per Labor-Hour Bare Costs	Cost Per Labor-Hour Incl. O&P
2 Stone Masons	$18.45	$295.20	$31.25	$500.00	$16.55	$28.02
2 Stone Mason Helpers	14.65	234.40	24.80	396.80		
32 L.H., Daily Totals		$529.60		$896.80	$16.55	$28.02

Crew D-13

Crew No.	Bare Costs Hr.	Bare Costs Daily	Incl. Subs O & P Hr.	Incl. Subs O & P Daily	Cost Per Labor-Hour Bare Costs	Cost Per Labor-Hour Incl. O&P
1 Stone Mason Foreman	$20.45	$163.60	$34.65	$277.20	$17.73	$29.93
2 Stone Masons	18.45	295.20	31.25	500.00		
2 Stone Mason Helpers	14.65	234.40	24.80	396.80		
1 Equip. Oper. (crane)	19.75	158.00	32.80	262.40		
1 Truck Crane, 12.5 Ton		467.50		514.25	9.74	10.71
48 L.H., Daily Totals		$1318.70		$1950.65	$27.47	$40.64

Crew E-1

Crew No.	Bare Costs Hr.	Bare Costs Daily	Incl. Subs O & P Hr.	Incl. Subs O & P Daily	Cost Per Labor-Hour Bare Costs	Cost Per Labor-Hour Incl. O&P
2 Struc. Steel Workers	$19.95	$319.20	$39.30	$628.80	$19.95	$39.30
1 Gas Welding Machine		80.15		88.15	5.01	5.51
16 L.H., Daily Totals		$399.35		$716.95	$24.96	$44.81

Crew E-2

Crew No.	Bare Costs Hr.	Bare Costs Daily	Incl. Subs O & P Hr.	Incl. Subs O & P Daily	Cost Per Labor-Hour Bare Costs	Cost Per Labor-Hour Incl. O&P
1 Struc. Steel Foreman	$21.95	$175.60	$43.25	$346.00	$20.25	$38.88
4 Struc. Steel Workers	19.95	638.40	39.30	1257.60		
1 Equip. Oper. (crane)	19.75	158.00	32.80	262.40		
1 Crane, 90 Ton		1030.00		1133.00	21.46	23.60
48 L.H., Daily Totals		$2002.00		$2999.00	$41.71	$62.48

Crew E-3

Crew No.	Bare Costs Hr.	Bare Costs Daily	Incl. Subs O & P Hr.	Incl. Subs O & P Daily	Cost Per Labor-Hour Bare Costs	Cost Per Labor-Hour Incl. O&P
1 Struc. Steel Foreman	$21.95	$175.60	$43.25	$346.00	$20.62	$40.62
2 Struc. Steel Worker	19.95	319.20	39.30	628.80		
1 Gas Welding Machine		80.15		88.15	3.34	3.67
24 L.H., Daily Totals		$574.95		$1062.95	$23.96	$44.29

Crew E-4

Crew No.	Bare Costs Hr.	Bare Costs Daily	Incl. Subs O & P Hr.	Incl. Subs O & P Daily	Cost Per Labor-Hour Bare Costs	Cost Per Labor-Hour Incl. O&P
1 Struc. Steel Foreman	$21.95	$175.60	$43.25	$346.00	$20.45	$40.29
3 Struc. Steel Workers	19.95	478.80	39.30	943.20		
1 Gas Welding Machine		80.15		88.15	2.50	2.76
32 L.H., Daily Totals		$734.55		$1377.35	$22.95	$43.05

Crew No.	Bare Costs Hr.	Bare Costs Daily	Incl. Subs O & P Hr.	Incl. Subs O & P Daily	Cost Per Labor-Hour Bare Costs	Cost Per Labor-Hour Incl. O&P
Crew E-5	Hr.	Daily	Hr.	Daily	Bare Costs	Incl. O&P
1 Struc. Steel Foremen	$21.95	$175.60	$43.25	$346.00	$20.15	$39.02
7 Struc. Steel Workers	19.95	1117.20	39.30	2200.80		
1 Equip. Oper. (crane)	19.75	158.00	32.80	262.40		
1 Crane, 90 Ton		1030.00		1133.00		
1 Gas Welding Machine		80.15		88.15	15.42	16.96
72 L.H., Daily Totals		$2560.95		$4030.35	$35.57	$55.98
Crew E-6	Hr.	Daily	Hr.	Daily	Bare Costs	Incl. O&P
1 Struc. Steel Foreman	$21.95	$175.60	$43.25	$346.00	$19.95	$38.52
12 Struc. Steel Workers	19.95	1915.20	39.30	3772.80		
1 Equip. Oper. (crane)	19.75	158.00	32.80	262.40		
1 Equip. Oper. (light)	18.15	145.20	30.15	241.20		
1 Crane, 90 Ton		1030.00		1133.00		
1 Gas Welding Machine		80.15		88.15		
1 Air Compr., 160 C.F.M.		94.80		104.30		
2 Impact Wrenches		55.40		60.95	10.50	11.55
120 L.H., Daily Totals		$3654.35		$6008.80	$30.45	$50.07
Crew E-7	Hr.	Daily	Hr.	Daily	Bare Costs	Incl. O&P
1 Struc. Steel Foreman	$21.95	$175.60	$43.25	$346.00	$20.15	$39.02
7 Struc. Steel Workers	19.95	1117.20	39.30	2200.80		
1 Equip. Oper. (crane)	19.75	158.00	32.80	262.40		
1 Crane, 90 Ton		1030.00		1133.00		
2 Gas Welding Machines		160.30		176.35	16.53	18.19
72 L.H., Daily Totals		$2641.10		$4118.55	$36.68	$57.21
Crew E-8	Hr.	Daily	Hr.	Daily	Bare Costs	Incl. O&P
1 Struc. Steel Foreman	$21.95	$175.60	$43.25	$346.00	$20.11	$39.07
9 Struc. Steel Workers	19.95	1436.40	39.30	2829.60		
1 Equip. Oper. (crane)	19.75	158.00	32.80	262.40		
1 Crane, 90 Ton		1030.00		1133.00		
4 Gas Welding Machines		320.60		352.65	15.35	16.88
88 L.H., Daily Totals		$3120.60		$4923.65	$35.46	$55.95
Crew E-9	Hr.	Daily	Hr.	Daily	Bare Costs	Incl. O&P
2 Struc. Steel Foremen	$21.95	$351.20	$43.25	$692.00	$19.97	$38.29
5 Struc. Steel Workers	19.95	798.00	39.30	1572.00		
1 Welder Foreman	21.95	175.60	43.25	346.00		
5 Welders	19.95	798.00	39.30	1572.00		
1 Equip. Oper. (crane)	19.75	158.00	32.80	262.40		
1 Equip. Oper. Oiler	16.20	129.60	26.90	215.20		
1 Equip. Oper. (light)	18.15	145.20	30.15	241.20		
1 Crane, 90 Ton		1030.00		1133.00		
5 Gas Welding Machines		400.75		440.85	11.18	12.30
128 L.H., Daily Totals		$3986.35		$6474.65	$31.15	$50.59
Crew E-10	Hr.	Daily	Hr.	Daily	Bare Costs	Incl. O&P
1 Struc. Steel Foreman	$21.95	$175.60	$43.25	$346.00	$20.62	$40.62
2 Struc. Steel Workers	19.95	319.20	39.30	628.80		
4 Gas Welding Machines		320.60		352.65		
1 Truck, 3 Ton		174.80		192.30	20.64	22.71
24 L.H., Daily Totals		$990.20		$1519.75	$41.26	$63.33
Crew E-11	Hr.	Daily	Hr.	Daily	Bare Costs	Incl. O&P
2 Painters, Struc. Steel	$18.00	$288.00	$36.85	$589.60	$16.47	$32.22
1 Building Laborer	13.40	107.20	22.95	183.60		
1 Air Compressor 250 C.F.M.		120.40		132.45		
1 Sand Blaster		30.45		33.50		
1 Sand Blasting Accessories		10.50		11.55	6.72	7.40
24 L.H., Daily Totals		$556.55		$950.70	$23.19	$39.62

Crew No.	Bare Costs Hr.	Bare Costs Daily	Incl. Subs O & P Hr.	Incl. Subs O & P Daily	Cost Per Labor-Hour Bare Costs	Cost Per Labor-Hour Incl. O&P
Crew E-12	Hr.	Daily	Hr.	Daily	Bare Costs	Incl. O&P
1 Welder Foreman	$21.95	$175.60	$43.25	$346.00	$20.05	$36.70
1 Equip. Oper. (light)	18.15	145.20	30.15	241.20		
1 Gas Welding Machine		80.15		88.15	5.01	5.51
16 L.H., Daily Totals		$400.95		$675.35	$25.06	$42.21
Crew E-13	Hr.	Daily	Hr.	Daily	Bare Costs	Incl. O&P
1 Welder Foreman	$21.95	$175.60	$43.25	$346.00	$20.68	$38.88
.5 Equip. Oper. (light)	18.15	72.60	30.15	120.60		
1 Gas Welding Machine		80.15		88.15	6.68	7.35
12 L.H., Daily Totals		$328.35		$554.75	$27.36	$46.23
Crew E-14	Hr.	Daily	Hr.	Daily	Bare Costs	Incl. O&P
1 Struc. Steel Worker	$19.95	$159.60	$39.30	$314.40	$19.95	$39.30
1 Gas Welding Machine		80.15		88.15	10.02	11.02
8 L.H., Daily Totals		$239.75		$402.55	$29.97	$50.32
Crew E-16	Hr.	Daily	Hr.	Daily	Bare Costs	Incl. O&P
1 Welder Foreman	$21.95	$175.60	$43.25	$346.00	$20.95	$41.28
1 Welder	19.95	159.60	39.30	314.40		
1 Gas Welding Machine		80.15		88.15	5.01	5.51
16 L.H., Daily Totals		$415.35		$748.55	$25.96	$46.79
Crew E-17	Hr.	Daily	Hr.	Daily	Bare Costs	Incl. O&P
1 Structural Steel Foreman	$21.95	$175.60	$43.25	$346.00	$20.95	$41.28
1 Structural Steel Worker	19.95	159.60	39.30	314.40		
1 Power Tool		2.00		2.20	.13	.14
16 L.H., Daily Totals		$337.20		$662.60	$21.08	$41.42
Crew E-18	Hr.	Daily	Hr.	Daily	Bare Costs	Incl. O&P
1 Structural Steel Foreman	$21.95	$175.60	$43.25	$346.00	$20.16	$38.54
3 Structural Steel Workers	19.95	478.80	39.30	943.20		
1 Equipment Operator (med.)	19.00	152.00	31.55	252.40		
1 Crane, 20 Ton		529.30		582.25	13.23	14.56
40 L.H., Daily Totals		$1335.70		$2123.85	$33.39	$53.10
Crew E-19	Hr.	Daily	Hr.	Daily	Bare Costs	Incl. O&P
1 Structural Steel Worker	$19.95	$159.60	$39.30	$314.40	$20.02	$37.57
1 Structural Steel Foreman	21.95	175.60	43.25	346.00		
1 Equip. Oper. (light)	18.15	145.20	30.15	241.20		
1 Power Tool		2.00		2.20		
1 Crane, 20 Ton		529.30		582.25	22.14	24.35
24 L.H., Daily Totals		$1011.70		$1486.05	$42.16	$61.92
Crew E-20	Hr.	Daily	Hr.	Daily	Bare Costs	Incl. O&P
1 Structural Steel Foreman	$21.95	$175.60	$43.25	$346.00	$19.71	$37.43
5 Structural Steel Workers	19.95	798.00	39.30	1572.00		
1 Equip. Oper. (crane)	19.75	158.00	32.80	262.40		
1 Oiler	16.20	129.60	26.90	215.20		
1 Power Tool		2.00		2.20		
1 Crane, 40 Ton		687.90		756.70	10.78	11.86
64 L.H., Daily Totals		$1951.10		$3154.50	$30.49	$49.29
Crew E-24	Hr.	Daily	Hr.	Daily	Bare Costs	Incl. O&P
3 Structural Steel Worker	$19.95	$478.80	$39.30	$943.20	$19.71	$37.36
1 Equipment Operator (medium)	19.00	152.00	31.55	252.40		
1-25 ton crane		538.85		592.75	16.84	18.52
32 L.H., Daily Totals		$1169.65		$1788.35	$36.55	$55.88

Crew F-2

Crew No.	Bare Costs Hr.	Bare Costs Daily	Incl. Subs O & P Hr.	Incl. Subs O & P Daily	Cost Per Labor-Hour Bare Costs	Cost Per Labor-Hour Incl. O&P
1 Carpenters	$18.50	$148.00	$31.70	$253.60	$16.18	$27.70
1 Carpenter Helper	13.85	110.80	23.70	189.60		
16 L.H., Daily Totals		$258.80		$443.20	$16.18	$27.70

Crew F-2A

Crew No.	Bare Costs Hr.	Bare Costs Daily	Incl. Subs O & P Hr.	Incl. Subs O & P Daily	Cost Per Labor-Hour Bare Costs	Cost Per Labor-Hour Incl. O&P
1 Carpenters	$18.50	$148.00	$31.70	$253.60	$16.18	$27.70
1 Carpenter Helper	13.85	110.80	23.70	189.60		
16 L.H., Daily Totals		$258.80		$443.20	$16.18	$27.70

Crew F-3

Crew No.	Bare Costs Hr.	Bare Costs Daily	Incl. Subs O & P Hr.	Incl. Subs O & P Daily	Cost Per Labor-Hour Bare Costs	Cost Per Labor-Hour Incl. O&P
2 Carpenters	$18.50	$296.00	$31.70	$507.20	$16.89	$28.72
2 Carpenter Helpers	13.85	221.60	23.70	379.20		
1 Equip. Oper. (crane)	19.75	158.00	32.80	262.40		
1 Hyd. Crane, 12 Ton		438.10		481.90	10.95	12.05
40 L.H., Daily Totals		$1113.70		$1630.70	$27.84	$40.77

Crew F-4

Crew No.	Bare Costs Hr.	Bare Costs Daily	Incl. Subs O & P Hr.	Incl. Subs O & P Daily	Cost Per Labor-Hour Bare Costs	Cost Per Labor-Hour Incl. O&P
2 Carpenters	$18.50	$296.00	$31.70	$507.20	$16.89	$28.72
2 Carpenter Helpers	13.85	221.60	23.70	379.20		
1 Equip. Oper. (crane)	19.75	158.00	32.80	262.40		
1 Hyd. Crane, 55 Ton		785.20		863.70	19.63	21.59
40 L.H., Daily Totals		$1460.80		$2012.50	$36.52	$50.31

Crew F-5

Crew No.	Bare Costs Hr.	Bare Costs Daily	Incl. Subs O & P Hr.	Incl. Subs O & P Daily	Cost Per Labor-Hour Bare Costs	Cost Per Labor-Hour Incl. O&P
2 Carpenters	$18.50	$296.00	$31.70	$507.20	$16.18	$27.70
2 Carpenter Helpers	13.85	221.60	23.70	379.20		
32 L.H., Daily Totals		$517.60		$886.40	$16.18	$27.70

Crew F-6

Crew No.	Bare Costs Hr.	Bare Costs Daily	Incl. Subs O & P Hr.	Incl. Subs O & P Daily	Cost Per Labor-Hour Bare Costs	Cost Per Labor-Hour Incl. O&P
2 Carpenters	$18.50	$296.00	$31.70	$507.20	$16.71	$28.42
2 Building Laborers	13.40	214.40	22.95	367.20		
1 Equip. Oper. (crane)	19.75	158.00	32.80	262.40		
1 Hyd. Crane, 12 Ton		438.10		481.90	10.95	12.05
40 L.H., Daily Totals		$1106.50		$1618.70	$27.66	$40.47

Crew F-7

Crew No.	Bare Costs Hr.	Bare Costs Daily	Incl. Subs O & P Hr.	Incl. Subs O & P Daily	Cost Per Labor-Hour Bare Costs	Cost Per Labor-Hour Incl. O&P
2 Carpenters	$18.50	$296.00	$31.70	$507.20	$15.95	$27.33
2 Building Laborers	13.40	214.40	22.95	367.20		
32 L.H., Daily Totals		$510.40		$874.40	$15.95	$27.33

Crew G-1

Crew No.	Bare Costs Hr.	Bare Costs Daily	Incl. Subs O & P Hr.	Incl. Subs O & P Daily	Cost Per Labor-Hour Bare Costs	Cost Per Labor-Hour Incl. O&P
1 Roofer Foreman	$18.20	$145.60	$34.00	$272.00	$15.24	$28.46
4 Roofers, Composition	16.20	518.40	30.25	968.00		
2 Roofer Helpers	11.85	189.60	22.10	353.60		
1 Application Equipment		164.80		181.30		
1 Tar Kettle/Pot		43.85		48.25		
1 Crew Truck		191.35		210.50	7.14	7.86
56 L.H., Daily Totals		$1253.60		$2033.65	$22.38	$36.32

Crew G-2

Crew No.	Bare Costs Hr.	Bare Costs Daily	Incl. Subs O & P Hr.	Incl. Subs O & P Daily	Cost Per Labor-Hour Bare Costs	Cost Per Labor-Hour Incl. O&P
1 Plasterer	$17.55	$140.40	$29.35	$234.80	$15.23	$25.65
1 Plasterer Helper	14.75	118.00	24.65	197.20		
1 Building Laborer	13.40	107.20	22.95	183.60		
1 Grouting Equipment		228.80		251.70	9.53	10.49
24 L.H., Daily Totals		$594.40		$867.30	$24.76	$36.14

Crew G-3

Crew No.	Bare Costs Hr.	Bare Costs Daily	Incl. Subs O & P Hr.	Incl. Subs O & P Daily	Cost Per Labor-Hour Bare Costs	Cost Per Labor-Hour Incl. O&P
2 Sheet Metal Workers	$20.15	$322.40	$34.10	$545.60	$16.77	$28.52
2 Building Laborers	13.40	214.40	22.95	367.20		
32 L.H., Daily Totals		$536.80		$912.80	$16.77	$28.52

Crew G-4

Crew No.	Bare Costs Hr.	Bare Costs Daily	Incl. Subs O & P Hr.	Incl. Subs O & P Daily	Cost Per Labor-Hour Bare Costs	Cost Per Labor-Hour Incl. O&P
1 Labor Foreman (outside)	$15.40	$123.20	$26.40	$211.20	$14.07	$24.10
2 Building Laborers	13.40	214.40	22.95	367.20		
1 Light Truck, 1.5 Ton		167.40		184.15		
1 Air Compr., 160 C.F.M.		94.80		104.30	10.93	12.02
24 L.H., Daily Totals		$599.80		$866.85	$25.00	$36.12

Crew G-5

Crew No.	Bare Costs Hr.	Bare Costs Daily	Incl. Subs O & P Hr.	Incl. Subs O & P Daily	Cost Per Labor-Hour Bare Costs	Cost Per Labor-Hour Incl. O&P
1 Roofer Foreman	$18.20	$145.60	$34.00	$272.00	$14.86	$27.74
2 Roofers, Composition	16.20	259.20	30.25	484.00		
2 Roofer Helpers	11.85	189.60	22.10	353.60		
1 Application Equipment		164.80		181.30	4.12	4.53
40 L.H., Daily Totals		$759.20		$1290.90	$18.98	$32.27

Crew G-6A

Crew No.	Bare Costs Hr.	Bare Costs Daily	Incl. Subs O & P Hr.	Incl. Subs O & P Daily	Cost Per Labor-Hour Bare Costs	Cost Per Labor-Hour Incl. O&P
2 Roofers Composition	$16.20	$259.20	$30.25	$484.00	$16.20	$30.25
1 Small Compressor		20.95		23.05		
2 Pneumatic Nailers		37.90		41.70	3.68	4.05
16 L.H., Daily Totals		$318.05		$548.75	$19.88	$34.30

Crew G-7

Crew No.	Bare Costs Hr.	Bare Costs Daily	Incl. Subs O & P Hr.	Incl. Subs O & P Daily	Cost Per Labor-Hour Bare Costs	Cost Per Labor-Hour Incl. O&P
1 Carpenter	$18.50	$148.00	$31.70	$253.60	$18.50	$31.70
1 Small Compressor		20.95		23.05		
1 Pneumatic Nailer		18.95		20.85	4.99	5.49
8 L.H., Daily Totals		$187.90		$297.50	$23.49	$37.19

Crew H-1

Crew No.	Bare Costs Hr.	Bare Costs Daily	Incl. Subs O & P Hr.	Incl. Subs O & P Daily	Cost Per Labor-Hour Bare Costs	Cost Per Labor-Hour Incl. O&P
2 Glaziers	$18.30	$292.80	$30.45	$487.20	$19.13	$34.88
2 Struc. Steel Workers	19.95	319.20	39.30	628.80		
32 L.H., Daily Totals		$612.00		$1116.00	$19.13	$34.88

Crew H-2

Crew No.	Bare Costs Hr.	Bare Costs Daily	Incl. Subs O & P Hr.	Incl. Subs O & P Daily	Cost Per Labor-Hour Bare Costs	Cost Per Labor-Hour Incl. O&P
2 Glaziers	$18.30	$292.80	$30.45	$487.20	$16.67	$27.95
1 Building Laborer	13.40	107.20	22.95	183.60		
24 L.H., Daily Totals		$400.00		$670.80	$16.67	$27.95

Crew H-3

Crew No.	Bare Costs Hr.	Bare Costs Daily	Incl. Subs O & P Hr.	Incl. Subs O & P Daily	Cost Per Labor-Hour Bare Costs	Cost Per Labor-Hour Incl. O&P
1 Glazier	$18.30	$146.40	$30.45	$243.60	$16.08	$27.08
1 Helper	13.85	110.80	23.70	189.60		
16 L.H., Daily Totals		$257.20		$433.20	$16.08	$27.08

Crew J-1

Crew No.	Bare Costs Hr.	Bare Costs Daily	Incl. Subs O & P Hr.	Incl. Subs O & P Daily	Cost Per Labor-Hour Bare Costs	Cost Per Labor-Hour Incl. O&P
3 Plasterers	$17.55	$421.20	$29.35	$704.40	$16.43	$27.47
2 Plasterer Helpers	14.75	236.00	24.65	394.40		
1 Mixing Machine, 6 C.F.		41.20		45.30	1.03	1.13
40 L.H., Daily Totals		$698.40		$1144.10	$17.46	$28.60

Crew J-2

Crew No.	Bare Costs Hr.	Bare Costs Daily	Incl. Subs O & P Hr.	Incl. Subs O & P Daily	Cost Per Labor-Hour Bare Costs	Cost Per Labor-Hour Incl. O&P
3 Plasterers	$17.55	$421.20	$29.35	$704.40	$16.74	$27.89
2 Plasterer Helpers	14.75	236.00	24.65	394.40		
1 Lather	18.30	146.40	30.00	240.00		
1 Mixing Machine, 6 C.F.		41.20		45.30	.86	.94
48 L.H., Daily Totals		$844.80		$1384.10	$17.60	$28.83

Crew J-3

Crew No.	Bare Costs Hr.	Bare Costs Daily	Incl. Subs O & P Hr.	Incl. Subs O & P Daily	Cost Per Labor-Hour Bare Costs	Cost Per Labor-Hour Incl. O&P
1 Terrazzo Worker	$18.35	$146.80	$29.70	$237.60	$16.73	$27.08
1 Terrazzo Helper	15.10	120.80	24.45	195.60		
1 Terrazzo Grinder, Electric		44.00		48.40		
1 Terrazzo Mixer		62.40		68.65	6.65	7.32
16 L.H., Daily Totals		$374.00		$550.25	$23.38	$34.40

Crew J-4

Crew No.	Bare Costs		Incl. Subs O & P		Cost Per Labor-Hour	
	Hr.	Daily	Hr.	Daily	Bare Costs	Incl. O&P
1 Tile Layer	$18.25	$146.00	$29.55	$236.40	$16.48	$26.67
1 Tile Layer Helper	14.70	117.60	23.80	190.40		
16 L.H., Daily Totals		$263.60		$426.80	$16.48	$26.67

Crew K-1

Crew No.	Hr.	Daily	Hr.	Daily	Bare Costs	Incl. O&P
1 Carpenter	$18.50	$148.00	$31.70	$253.60	$16.80	$28.50
1 Truck Driver (light)	15.10	120.80	25.30	202.40		
1 Truck w/Power Equip.		174.80		192.30	10.93	12.02
16 L.H., Daily Totals		$443.60		$648.30	$27.73	$40.52

Crew K-2

Crew No.	Hr.	Daily	Hr.	Daily	Bare Costs	Incl. O&P
1 Struc. Steel Foreman	$21.95	$175.60	$43.25	$346.00	$19.00	$35.95
1 Struc. Steel Worker	19.95	159.60	39.30	314.40		
1 Truck Driver (light)	15.10	120.80	25.30	202.40		
1 Truck w/Power Equip.		174.80		192.30	7.28	8.01
24 L.H., Daily Totals		$630.80		$1055.10	$26.28	$43.96

Crew L-1

Crew No.	Hr.	Daily	Hr.	Daily	Bare Costs	Incl. O&P
.25 Electrician	$20.60	$41.20	$33.70	$67.40	$20.80	$34.38
1 Plumber	20.85	166.80	34.55	276.40		
10 L.H., Daily Totals		$208.00		$343.80	$20.80	$34.38

Crew L-2

Crew No.	Hr.	Daily	Hr.	Daily	Bare Costs	Incl. O&P
1 Carpenter	$18.50	$148.00	$31.70	$253.60	$16.18	$27.70
1 Carpenter Helper	13.85	110.80	23.70	189.60		
16 L.H., Daily Totals		$258.80		$443.20	$16.18	$27.70

Crew L-3

Crew No.	Hr.	Daily	Hr.	Daily	Bare Costs	Incl. O&P
1 Carpenter	$18.50	$148.00	$31.70	$253.60	$18.92	$32.10
.25 Electrician	20.60	41.20	33.70	67.40		
10 L.H., Daily Totals		$189.20		$321.00	$18.92	$32.10

Crew L-3A

Crew No.	Hr.	Daily	Hr.	Daily	Bare Costs	Incl. O&P
1 Carpenter Foreman (outside)	$20.50	$164.00	$35.15	$281.20	$20.38	$34.80
.5 Sheet Metal Worker	20.15	80.60	34.10	136.40		
12 L.H., Daily Totals		$244.60		$417.60	$20.38	$34.80

Crew L-4

Crew No.	Hr.	Daily	Hr.	Daily	Bare Costs	Incl. O&P
1 Skilled Workers	$18.75	$150.00	$32.25	$258.00	$16.30	$27.98
1 Helper	13.85	110.80	23.70	189.60		
16 L.H., Daily Totals		$260.80		$447.60	$16.30	$27.98

Crew L-5

Crew No.	Hr.	Daily	Hr.	Daily	Bare Costs	Incl. O&P
1 Struc. Steel Foreman	$21.95	$175.60	$43.25	$346.00	$20.21	$38.94
5 Struc. Steel Workers	19.95	798.00	39.30	1572.00		
1 Equip. Oper. (crane)	19.75	158.00	32.80	262.40		
1 Hyd. Crane, 25 Ton		538.85		592.75	9.62	10.58
56 L.H., Daily Totals		$1670.45		$2773.15	$29.83	$49.52

Crew L-5A

Crew No.	Hr.	Daily	Hr.	Daily	Bare Costs	Incl. O&P
1 Structural Steel Foreman	$21.95	$175.60	$43.25	$346.00	$20.40	$38.66
2 Structural Steel Worker	19.95	319.20	39.30	628.80		
1 Equip. Oper. (crane)	19.75	158.00	32.80	262.40		
1 Crane,SP, 25 Ton		563.05		619.35	17.60	19.35
32 L.H., Daily Totals		$1215.85		$1856.55	$38.00	$58.01

Crew L-6

Crew No.	Bare Costs		Incl. Subs O & P		Cost Per Labor-Hour	
	Hr.	Daily	Hr.	Daily	Bare Costs	Incl. O&P
1 Plumber	$20.85	$166.80	$34.55	$276.40	$20.77	$34.27
.5 Electrician	20.60	82.40	33.70	134.80		
12 L.H., Daily Totals		$249.20		$411.20	$20.77	$34.27

Crew L-7

Crew No.	Hr.	Daily	Hr.	Daily	Bare Costs	Incl. O&P
1 Carpenters	$18.50	$148.00	$31.70	$253.60	$15.80	$26.93
2 Carpenter Helpers	13.85	221.60	23.70	379.20		
.25 Electrician	20.60	41.20	33.70	67.40		
26 L.H., Daily Totals		$410.80		$700.20	$15.80	$26.93

Crew L-8

Crew No.	Hr.	Daily	Hr.	Daily	Bare Costs	Incl. O&P
1 Carpenters	$18.50	$148.00	$31.70	$253.60	$17.11	$29.07
1 Carpenter Helper	13.85	110.80	23.70	189.60		
.5 Plumber	20.85	83.40	34.55	138.20		
20 L.H., Daily Totals		$342.20		$581.40	$17.11	$29.07

Crew L-9

Crew No.	Hr.	Daily	Hr.	Daily	Bare Costs	Incl. O&P
1 Skilled Worker Foreman	$20.75	$166.00	$35.70	$285.60	$17.22	$29.38
1 Skilled Worker	18.75	150.00	32.25	258.00		
2 Helpers	13.85	221.60	23.70	379.20		
.5 Electrician	20.60	82.40	33.70	134.80		
36 L.H., Daily Totals		$620.00		$1057.60	$17.22	$29.38

Crew L-10

Crew No.	Hr.	Daily	Hr.	Daily	Bare Costs	Incl. O&P
1 Structural Steel Foreman	$21.95	$175.60	$43.25	$346.00	$20.55	$38.45
1 Structural Steel Worker	19.95	159.60	39.30	314.40		
1 Equip. Oper. (crane)	19.75	158.00	32.80	262.40		
1 Hyd. Crane, 12 Ton		438.10		481.90	18.25	20.08
24 L.H., Daily Totals		$931.30		$1404.70	$38.80	$58.53

Crew M-1

Crew No.	Hr.	Daily	Hr.	Daily	Bare Costs	Incl. O&P
3 Elevator Constructors	$21.35	$512.40	$35.30	$847.20	$20.29	$33.55
1 Elevator Apprentice	17.10	136.80	28.30	226.40		
5 Hand Tools		84.50		92.95	2.64	2.90
32 L.H., Daily Totals		$733.70		$1166.55	$22.93	$36.45

Crew Q-1

Crew No.	Hr.	Daily	Hr.	Daily	Bare Costs	Incl. O&P
1 Plumber	$20.85	$166.80	$34.55	$276.40	$18.78	$31.10
1 Plumber Apprentice	16.70	133.60	27.65	221.20		
16 L.H., Daily Totals		$300.40		$497.60	$18.78	$31.10

Crew Q-1C

Crew No.	Hr.	Daily	Hr.	Daily	Bare Costs	Incl. O&P
1 Plumber	$20.85	$166.80	$34.55	$276.40	$18.85	$31.25
1 Plumber Apprentice	16.70	133.60	27.65	221.20		
1 Equip. Oper. (medium)	19.00	152.00	31.55	252.40		
1 Trencher, Chain		526.80		579.50	21.95	24.15
24 L.H., Daily Totals		$979.20		$1329.50	$40.80	$55.40

Crew Q-2

Crew No.	Hr.	Daily	Hr.	Daily	Bare Costs	Incl. O&P
1 Plumber	$20.85	$166.80	$34.55	$276.40	$18.08	$29.95
2 Plumber Apprentices	16.70	267.20	27.65	442.40		
24 L.H., Daily Totals		$434.00		$718.80	$18.08	$29.95

Crew Q-3

Crew No.	Hr.	Daily	Hr.	Daily	Bare Costs	Incl. O&P
2 Plumbers	$20.85	$333.60	$34.55	$552.80	$18.78	$31.10
2 Plumber Apprentices	16.70	267.20	27.65	442.40		
32 L.H., Daily Totals		$600.80		$995.20	$18.78	$31.10

Crew Q-4

Crew No.	Bare Costs Hr.	Daily	Incl. Subs O & P Hr.	Daily	Cost Per Labor-Hour Bare Costs	Incl. O&P
2 Plumbers	$20.85	$333.60	$34.55	$552.80	$19.81	$32.83
1 Welder (plumber)	20.85	166.80	34.55	276.40		
1 Plumber Apprentice	16.70	133.60	27.65	221.20		
1 Electric Welding Mach.		48.50		53.35	1.52	1.67
32 L.H., Daily Totals		$682.50		$1103.75	$21.33	$34.50

Crew Q-5

Crew No.	Bare Costs Hr.	Daily	Incl. Subs O & P Hr.	Daily	Cost Per Labor-Hour Bare Costs	Incl. O&P
1 Steamfitter	$21.00	$168.00	$34.80	$278.40	$18.90	$31.33
1 Steamfitter Apprentice	16.80	134.40	27.85	222.80		
16 L.H., Daily Totals		$302.40		$501.20	$18.90	$31.33

Crew Q-6

Crew No.	Bare Costs Hr.	Daily	Incl. Subs O & P Hr.	Daily	Cost Per Labor-Hour Bare Costs	Incl. O&P
1 Steamfitters	$21.00	$168.00	$34.80	$278.40	$18.20	$30.17
2 Steamfitter Apprentices	16.80	268.80	27.85	445.60		
24 L.H., Daily Totals		$436.80		$724.00	$18.20	$30.17

Crew Q-7

Crew No.	Bare Costs Hr.	Daily	Incl. Subs O & P Hr.	Daily	Cost Per Labor-Hour Bare Costs	Incl. O&P
2 Steamfitters	$21.00	$336.00	$34.80	$556.80	$18.90	$31.33
2 Steamfitter Apprentices	16.80	268.80	27.85	445.60		
32 L.H., Daily Totals		$604.80		$1002.40	$18.90	$31.33

Crew Q-8

Crew No.	Bare Costs Hr.	Daily	Incl. Subs O & P Hr.	Daily	Cost Per Labor-Hour Bare Costs	Incl. O&P
2 Steamfitters	$21.00	$336.00	$34.80	$556.80	$19.95	$33.06
1 Welder (steamfitter)	21.00	168.00	34.80	278.40		
1 Steamfitter Apprentice	16.80	134.40	27.85	222.80		
1 Electric Welding Mach.		48.50		53.35	1.52	1.67
32 L.H., Daily Totals		$686.90		$1111.35	$21.47	$34.73

Crew Q-9

Crew No.	Bare Costs Hr.	Daily	Incl. Subs O & P Hr.	Daily	Cost Per Labor-Hour Bare Costs	Incl. O&P
1 Sheet Metal Worker	$20.15	$161.20	$34.10	$272.80	$18.13	$30.67
1 Sheet Metal Apprentice	16.10	128.80	27.25	218.00		
16 L.H., Daily Totals		$290.00		$490.80	$18.13	$30.67

Crew Q-10

Crew No.	Bare Costs Hr.	Daily	Incl. Subs O & P Hr.	Daily	Cost Per Labor-Hour Bare Costs	Incl. O&P
2 Sheet Metal Workers	$20.15	$322.40	$34.10	$545.60	$18.80	$31.82
1 Sheet Metal Apprentice	16.10	128.80	27.25	218.00		
24 L.H., Daily Totals		$451.20		$763.60	$18.80	$31.82

Crew Q-11

Crew No.	Bare Costs Hr.	Daily	Incl. Subs O & P Hr.	Daily	Cost Per Labor-Hour Bare Costs	Incl. O&P
2 Sheet Metal Workers	$20.15	$322.40	$34.10	$545.60	$18.13	$30.67
2 Sheet Metal Apprentices	16.10	257.60	27.25	436.00		
32 L.H., Daily Totals		$580.00		$981.60	$18.13	$30.67

Crew Q-12

Crew No.	Bare Costs Hr.	Daily	Incl. Subs O & P Hr.	Daily	Cost Per Labor-Hour Bare Costs	Incl. O&P
1 Sprinkler Installer	$21.10	$168.80	$35.00	$280.00	$19.00	$31.50
1 Sprinkler Apprentice	16.90	135.20	28.00	224.00		
16 L.H., Daily Totals		$304.00		$504.00	$19.00	$31.50

Crew Q-13

Crew No.	Bare Costs Hr.	Daily	Incl. Subs O & P Hr.	Daily	Cost Per Labor-Hour Bare Costs	Incl. O&P
2 Sprinkler Installers	$21.10	$337.60	$35.00	$560.00	$19.00	$31.50
2 Sprinkler Apprentices	16.90	270.40	28.00	448.00		
32 L.H., Daily Totals		$608.00		$1008.00	$19.00	$31.50

Crew Q-14

Crew No.	Bare Costs Hr.	Daily	Incl. Subs O & P Hr.	Daily	Cost Per Labor-Hour Bare Costs	Incl. O&P
1 Asbestos Worker	$19.55	$156.40	$34.25	$274.00	$17.60	$30.83
1 Asbestos Apprentice	15.65	125.20	27.40	219.20		
16 L.H., Daily Totals		$281.60		$493.20	$17.60	$30.83

Crew Q-15

Crew No.	Bare Costs Hr.	Daily	Incl. Subs O & P Hr.	Daily	Cost Per Labor-Hour Bare Costs	Incl. O&P
1 Plumber	$20.85	$166.80	$34.55	$276.40	$18.78	$31.10
1 Plumber Apprentice	16.70	133.60	27.65	221.20		
1 Electric Welding Mach.		48.50		53.35	3.03	3.33
16 L.H., Daily Totals		$348.90		$550.95	$21.81	$34.43

Crew Q-16

Crew No.	Bare Costs Hr.	Daily	Incl. Subs O & P Hr.	Daily	Cost Per Labor-Hour Bare Costs	Incl. O&P
2 Plumbers	$20.85	$333.60	$34.55	$552.80	$19.47	$32.25
1 Plumber Apprentice	16.70	133.60	27.65	221.20		
1 Electric Welding Mach.		48.50		53.35	2.02	2.22
24 L.H., Daily Totals		$515.70		$827.35	$21.49	$34.47

Crew Q-17

Crew No.	Bare Costs Hr.	Daily	Incl. Subs O & P Hr.	Daily	Cost Per Labor-Hour Bare Costs	Incl. O&P
1 Steamfitter	$21.00	$168.00	$34.80	$278.40	$18.90	$31.33
1 Steamfitter Apprentice	16.80	134.40	27.85	222.80		
1 Electric Welding Mach.		48.50		53.35	3.03	3.33
16 L.H., Daily Totals		$350.90		$554.55	$21.93	$34.66

Crew Q-17A

Crew No.	Bare Costs Hr.	Daily	Incl. Subs O & P Hr.	Daily	Cost Per Labor-Hour Bare Costs	Incl. O&P
1 Steamfitter	$21.00	$168.00	$34.80	$278.40	$19.18	$31.82
1 Steamfitter Apprentice	16.80	134.40	27.85	222.80		
1 Equip. Oper. (crane)	19.75	158.00	32.80	262.40		
1 Truck Crane, 12 Ton		438.10		481.90		
1 Electric Welding Mach.		48.50		53.35	20.28	22.30
24 L.H., Daily Totals		$947.00		$1298.85	$39.46	$54.12

Crew Q-18

Crew No.	Bare Costs Hr.	Daily	Incl. Subs O & P Hr.	Daily	Cost Per Labor-Hour Bare Costs	Incl. O&P
2 Steamfitters	$21.00	$336.00	$34.80	$556.80	$19.60	$32.48
1 Steamfitter Apprentice	16.80	134.40	27.85	222.80		
1 Electric Welding Mach.		48.50		53.35	2.02	2.22
24 L.H., Daily Totals		$518.90		$832.95	$21.62	$34.70

Crew Q-19

Crew No.	Bare Costs Hr.	Daily	Incl. Subs O & P Hr.	Daily	Cost Per Labor-Hour Bare Costs	Incl. O&P
1 Steamfitter	$21.00	$168.00	$34.80	$278.40	$19.47	$32.12
1 Steamfitter Apprentice	16.80	134.40	27.85	222.80		
1 Electrician	20.60	164.80	33.70	269.60		
24 L.H., Daily Totals		$467.20		$770.80	$19.47	$32.12

Crew Q-20

Crew No.	Bare Costs Hr.	Daily	Incl. Subs O & P Hr.	Daily	Cost Per Labor-Hour Bare Costs	Incl. O&P
1 Sheet Metal Worker	$20.15	$161.20	$34.10	$272.80	$18.62	$31.28
1 Sheet Metal Apprentice	16.10	128.80	27.25	218.00		
.5 Electrician	20.60	82.40	33.70	134.80		
20 L.H., Daily Totals		$372.40		$625.60	$18.62	$31.28

Crew Q-21

Crew No.	Bare Costs Hr.	Daily	Incl. Subs O & P Hr.	Daily	Cost Per Labor-Hour Bare Costs	Incl. O&P
2 Steamfitters	$21.00	$336.00	$34.80	$556.80	$19.85	$32.79
1 Steamfitter Apprentice	16.80	134.40	27.85	222.80		
1 Electrician	20.60	164.80	33.70	269.60		
32 L.H., Daily Totals		$635.20		$1049.20	$19.85	$32.79

Crew Q-22

Crew No.	Bare Costs Hr.	Daily	Incl. Subs O & P Hr.	Daily	Cost Per Labor-Hour Bare Costs	Incl. O&P
1 Plumber	$20.85	$166.80	$34.55	$276.40	$18.78	$31.10
1 Plumber Apprentice	16.70	133.60	27.65	221.20		
1 Truck Crane, 12 Ton		438.10		481.90	27.38	30.12
16 L.H., Daily Totals		$738.50		$979.50	$46.16	$61.22

Crew Q-22A

Crew No.	Bare Costs Hr.	Bare Costs Daily	Incl. Subs O & P Hr.	Incl. Subs O & P Daily	Cost Per Labor-Hour Bare Costs	Cost Per Labor-Hour Incl. O&P
1 Plumber	$20.85	$166.80	$34.55	$276.40	$17.68	$29.49
1 Plumber Apprentice	16.70	133.60	27.65	221.20		
1 Laborer	13.40	107.20	22.95	183.60		
1 Equip. Oper. (crane)	19.75	158.00	32.80	262.40		
1 Truck Crane, 12 Ton		438.10		481.90	13.69	15.06
32 L.H., Daily Totals		$1003.70		$1425.50	$31.37	$44.55

Crew Q-23

Crew No.	Bare Costs Hr.	Bare Costs Daily	Incl. Subs O & P Hr.	Incl. Subs O & P Daily	Cost Per Labor-Hour Bare Costs	Cost Per Labor-Hour Incl. O&P
1 Plumber Foreman	$22.85	$182.80	$37.85	$302.80	$20.90	$34.65
1 Plumber	20.85	166.80	34.55	276.40		
1 Equip. Oper. (medium)	19.00	152.00	31.55	252.40		
1 Power Tools		2.00		2.20		
1 Crane, 20 Ton		529.30		582.25	22.14	24.35
24 L.H., Daily Totals		$1032.90		$1416.05	$43.04	$59.00

Crew R-1

Crew No.	Bare Costs Hr.	Bare Costs Daily	Incl. Subs O & P Hr.	Incl. Subs O & P Daily	Cost Per Labor-Hour Bare Costs	Cost Per Labor-Hour Incl. O&P
1 Electrician Foreman	$21.10	$168.80	$34.50	$276.00	$18.43	$30.50
3 Electricians	20.60	494.40	33.70	808.80		
2 Helpers	13.85	221.60	23.70	379.20		
48 L.H., Daily Totals		$884.80		$1464.00	$18.43	$30.50

Crew R-2

Crew No.	Bare Costs Hr.	Bare Costs Daily	Incl. Subs O & P Hr.	Incl. Subs O & P Daily	Cost Per Labor-Hour Bare Costs	Cost Per Labor-Hour Incl. O&P
1 Electrician Foreman	$21.10	$168.80	$34.50	$276.00	$18.62	$30.83
3 Electricians	20.60	494.40	33.70	808.80		
2 Helpers	13.85	221.60	23.70	379.20		
1 Equip. Oper. (crane)	19.75	158.00	32.80	262.40		
1 S.P. Crane, 5 Ton		269.05		295.95	4.80	5.28
56 L.H., Daily Totals		$1311.85		$2022.35	$23.42	$36.11

Crew R-3

Crew No.	Bare Costs Hr.	Bare Costs Daily	Incl. Subs O & P Hr.	Incl. Subs O & P Daily	Cost Per Labor-Hour Bare Costs	Cost Per Labor-Hour Incl. O&P
1 Electrician Foreman	$21.10	$168.80	$34.50	$276.00	$20.63	$33.84
1 Electrician	20.60	164.80	33.70	269.60		
.5 Equip. Oper. (crane)	19.75	79.00	32.80	131.20		
.5 S.P. Crane, 5 Ton		134.52		148.00	6.73	7.40
20 L.H., Daily Totals		$547.12		$824.80	$27.36	$41.24

Crew R-4

Crew No.	Bare Costs Hr.	Bare Costs Daily	Incl. Subs O & P Hr.	Incl. Subs O & P Daily	Cost Per Labor-Hour Bare Costs	Cost Per Labor-Hour Incl. O&P
1 Struc. Steel Foreman	$21.95	$175.60	$43.25	$346.00	$20.48	$38.97
3 Struc. Steel Workers	19.95	478.80	39.30	943.20		
1 Electrician	20.60	164.80	33.70	269.60		
1 Gas Welding Machine		80.15		88.15	2.00	2.20
40 L.H., Daily Totals		$899.35		$1646.95	$22.48	$41.17

Crew R-5

Crew No.	Bare Costs Hr.	Bare Costs Daily	Incl. Subs O & P Hr.	Incl. Subs O & P Daily	Cost Per Labor-Hour Bare Costs	Cost Per Labor-Hour Incl. O&P
1 Electrician Foreman	$21.10	$168.80	$34.50	$276.00	$18.19	$30.14
4 Electrician Linemen	20.60	659.20	33.70	1078.40		
2 Electrician Operators	20.60	329.60	33.70	539.20		
4 Electrician Groundmen	13.85	443.20	23.70	758.40		
1 Crew Truck		191.35		210.50		
1 Tool Van		205.70		226.25		
1 Pick-up Truck		125.50		138.05		
.2 Crane, 55 Ton		157.04		172.75		
.2 Crane, 12 Ton		87.62		96.40		
.2 Auger, Truck Mtd.		345.00		379.50		
1 Tractor w/Winch		291.00		320.10	15.95	17.54
88 L.H., Daily Totals		$3004.01		$4195.55	$34.14	$47.68

Crew R-6

Crew No.	Bare Costs Hr.	Bare Costs Daily	Incl. Subs O & P Hr.	Incl. Subs O & P Daily	Cost Per Labor-Hour Bare Costs	Cost Per Labor-Hour Incl. O&P
1 Electrician Foreman	$21.10	$168.80	$34.50	$276.00	$18.19	$30.14
4 Electrician Linemen	20.60	659.20	33.70	1078.40		
2 Electrician Operators	20.60	329.60	33.70	539.20		
4 Electrician Groundmen	13.85	443.20	23.70	758.40		
1 Crew Truck		191.35		210.50		
1 Tool Van		205.70		226.25		
1 Pick-up Truck		125.50		138.05		
.2 Crane, 55 Ton		157.04		172.75		
.2 Crane, 12 Ton		87.62		96.40		
.2 Auger, Truck Mtd.		345.00		379.50		
1 Tractor w/Winch		291.00		320.10		
3 Cable Trailers		417.45		459.20		
.5 Tensioning Rig		135.20		148.70		
.5 Cable Pulling Rig		808.50		889.35	31.41	34.55
88 L.H., Daily Totals		$4365.16		$5692.80	$49.60	$64.69

Crew R-7

Crew No.	Bare Costs Hr.	Bare Costs Daily	Incl. Subs O & P Hr.	Incl. Subs O & P Daily	Cost Per Labor-Hour Bare Costs	Cost Per Labor-Hour Incl. O&P
1 Electrician Foreman	$21.10	$168.80	$34.50	$276.00	$15.06	$25.50
5 Electrician Groundmen	13.85	554.00	23.70	948.00		
1 Crew Truck		191.35		210.50	3.99	4.39
48 L.H., Daily Totals		$914.15		$1434.50	$19.05	$29.89

Crew R-8

Crew No.	Bare Costs Hr.	Bare Costs Daily	Incl. Subs O & P Hr.	Incl. Subs O & P Daily	Cost Per Labor-Hour Bare Costs	Cost Per Labor-Hour Incl. O&P
1 Electrician Foreman	$21.10	$168.80	$34.50	$276.00	$18.43	$30.50
3 Electrician Linemen	20.60	494.40	33.70	808.80		
2 Electrician Groundmen	13.85	221.60	23.70	379.20		
1 Pick-up Truck		125.50		138.05		
1 Crew Truck		191.35		210.50	6.60	7.26
48 L.H., Daily Totals		$1201.65		$1812.55	$25.03	$37.76

Crew R-9

Crew No.	Bare Costs Hr.	Bare Costs Daily	Incl. Subs O & P Hr.	Incl. Subs O & P Daily	Cost Per Labor-Hour Bare Costs	Cost Per Labor-Hour Incl. O&P
1 Electrician Foreman	$21.10	$168.80	$34.50	$276.00	$17.29	$28.80
1 Electrician Lineman	20.60	164.80	33.70	269.60		
2 Electrician Operators	20.60	329.60	33.70	539.20		
4 Electrician Groundmen	13.85	443.20	23.70	758.40		
1 Pick-up Truck		125.50		138.05		
1 Crew Truck		191.35		210.50	4.95	5.45
64 L.H., Daily Totals		$1423.25		$2191.75	$22.24	$34.25

Crew R-10

Crew No.	Bare Costs Hr.	Bare Costs Daily	Incl. Subs O & P Hr.	Incl. Subs O & P Daily	Cost Per Labor-Hour Bare Costs	Cost Per Labor-Hour Incl. O&P
1 Electrician Foreman	$21.10	$168.80	$34.50	$276.00	$19.56	$32.17
4 Electrician Linemen	20.60	659.20	33.70	1078.40		
1 Electrician Groundman	13.85	110.80	23.70	189.60		
1 Crew Truck		191.35		210.50		
3 Tram Cars		512.70		563.95	14.67	16.13
48 L.H., Daily Totals		$1642.85		$2318.45	$34.23	$48.30

Crew R-11

Crew No.	Bare Costs Hr.	Bare Costs Daily	Incl. Subs O & P Hr.	Incl. Subs O & P Daily	Cost Per Labor-Hour Bare Costs	Cost Per Labor-Hour Incl. O&P
1 Electrician Foreman	$21.10	$168.80	$34.50	$276.00	$18.68	$30.85
4 Electricians	20.60	659.20	33.70	1078.40		
1 Helper	13.85	110.80	23.70	189.60		
1 Common Laborer	13.40	107.20	22.95	183.60		
1 Crew Truck		191.35		210.50		
1 Crane, 12 Ton		438.10		481.90	11.24	12.36
56 L.H., Daily Totals		$1675.45		$2420.00	$29.92	$43.21

Crew No.	Bare Costs		Incl. Subs O & P		Cost Per Labor-Hour	
Crew R-12	Hr.	Daily	Hr.	Daily	Bare Costs	Incl. O&P
1 Carpenter Foreman	$19.00	$152.00	$32.55	$260.40	$16.87	$29.27
4 Carpenters	18.50	592.00	31.70	1014.40		
4 Common Laborers	13.40	428.80	22.95	734.40		
1 Equip. Oper. (med.)	19.00	152.00	31.55	252.40		
1 Steel Worker	19.95	159.60	39.30	314.40		
1 Dozer, 200 H.P.		832.00		915.20		
1 Pick-up Truck		125.50		138.05	10.88	11.97
88 L.H., Daily Totals		$2441.90		$3629.25	$27.75	$41.24
Crew R-15	Hr.	Daily	Hr.	Daily	Bare Costs	Incl. O&P
1 Electrician Foreman	$21.10	$168.80	$34.50	$276.00	$20.27	$33.24
4 Electricians	20.60	659.20	33.70	1078.40		
1 Equipment Operator	18.15	145.20	30.15	241.20		
1 Aerial Lift Truck		269.85		296.85	5.62	6.18
48 L.H., Daily Totals		$1243.05		$1892.45	$25.89	$39.42
Crew R-18	Hr.	Daily	Hr.	Daily	Bare Costs	Incl. O&P
.25 Electrician Foreman	$21.10	$42.20	$34.50	$69.00	$16.48	$27.61
1 Electrician	20.60	164.80	33.70	269.60		
2 Helpers	13.85	221.60	23.70	379.20		
26 L.H., Daily Totals		$428.60		$717.80	$16.48	$27.61
Crew R-19	Hr.	Daily	Hr.	Daily	Bare Costs	Incl. O&P
.5 Electrician Foreman	$21.10	$84.40	$34.50	$138.00	$20.70	$33.86
2 Electricians	20.60	329.60	33.70	539.20		
20 L.H., Daily Totals		$414.00		$677.20	$20.70	$33.86
Crew R-21	Hr.	Daily	Hr.	Daily	Bare Costs	Incl. O&P
1 Electrician Foreman	$21.10	$168.80	$34.50	$276.00	$20.68	$33.84
3 Electricians	20.60	494.40	33.70	808.80		
.1 Equip. Oper. (med.)	19.00	15.20	31.55	25.24		
.1 Hyd. Crane (20T)		56.30		61.95	1.72	1.89
32. L.H., Daily Totals		$734.70		$1171.99	$22.40	$35.73
Crew R-22	Hr.	Daily	Hr.	Daily	Bare Costs	Incl. O&P
.66 Electrician Foreman	$21.10	$111.41	$34.50	$182.16	$17.77	$29.52
2 Helpers	13.85	221.60	23.70	379.20		
2 Electricians	20.60	329.60	33.70	539.20		
37.28 L.H., Daily Totals		$662.61		$1100.56	$17.77	$29.52

The following tables are the revised Historical Cost Indexes based on a 30-city national average with a base of 100 on January 1, 1993.

The indexes may be used to:

1. Estimate and compare construction costs for different years in the same city.
2. Estimate and compare construction costs in different cities for the same year.
3. Estimate and compare construction costs in different cities for different years.
4. Compare construction trends in any city with the national average.

EXAMPLES

1. Estimate and compare construction costs for different years in the same city.

   A. To estimate the construction cost of a building in Lexington, KY in 1970, knowing that it cost $900,000 in 1998.

   $$\text{Index Lexington, KY in 1970} = 26.9$$
   $$\text{Index Lexington, KY in 1998} = 98.2$$

   $$\frac{\text{Index 1970}}{\text{Index 1998}} \times \text{Cost 1998} = \text{Cost 1970}$$

   $$\frac{26.9}{98.2} \times \$900,000 = \$246,500$$

   $$\text{Construction Cost in Lexington, KY in 1970} = \$246,500$$

   B. To estimate the current construction cost of a building in Boston, MA that was built in 1978 for $900,000.

   $$\text{Index Boston, MA in 1978} = 54.0$$
   $$\text{Index Boston, MA in 1998} = 133.1$$

   $$\frac{\text{Index 1998}}{\text{Index 1978}} \times \text{Cost 1978} = \text{Cost 1998}$$

   $$\frac{133.1}{54.0} \times \$900,000 = \$2,218,500$$

   $$\text{Construction Cost in Boston in 1998} = \$2,218,500$$

2. Estimate and compare construction costs in different cities for the same year.

   To compare the construction cost of a building in Topeka, KS in 1998 with the known cost of $800,000 in Baltimore, MD in 1998

   $$\text{Index Topeka, KS in 1998} = 97.9$$
   $$\text{Index Baltimore, MD in 1998} = 104.8$$

   $$\frac{\text{Index Topeka}}{\text{Index Baltimore}} \times \text{Cost Baltimore} = \text{Cost Topeka}$$

   $$\frac{97.9}{104.8} \times \$800,000 = \$747,500$$

   $$\text{Construction Cost in Topeka in 1998} = \$747,500$$

3. Estimate and compare construction costs in different cities for different years.

   To compare the construction cost of a building in Detroit, MI in 1998 with the known construction cost of $4,000,000 for the same building in San Francisco, CA in 1978.

   $$\text{Index Detroit, MI in 1998} = 119.9$$
   $$\text{Index San Francisco, CA in 1978} = 62.6$$

   $$\frac{\text{Index Detroit 1998}}{\text{Index San Francisco 1978}} \times \text{Cost San Francisco 1978} = \text{Cost Detroit 1998}$$

   $$\frac{119.9}{62.6} \times \$4,000,000 = \$7,661,500$$

   $$\text{Construction Cost in Detroit in 1998} = \$7,661,500$$

4. Compare construction trends in any city with the national average.

   To compare the construction cost in Las Vegas, NV from 1974 to 1998 with the increase in the National Average during the same time period.

   $$\text{Index Las Vegas, NV for 1974} = 40.0 \quad \text{For 1998} = 116.7$$
   $$\text{Index 30 City Average for 1974} = 39.4 \quad \text{For 1998} = 114.4$$

   A. National Average increase From 1974 to 1998 $= \dfrac{\text{Index 30 City 1998}}{\text{Index 30 City 1974}}$

   $$= \frac{114.4}{39.4}$$

   National Average increase From 1974 to 1998 $= 2.90 \quad$ or 290%

   B. Increase for Las Vegas, NV From 1974 to 1998 $= \dfrac{\text{Index Las Vegas, NV 1998}}{\text{Index Las Vegas, NV 1974}}$

   $$= \frac{116.7}{40.0}$$

   Las Vegas increase 1974 — 1998 $= 2.92 \quad$ or 292%

   Conclusion: Construction costs in Las Vegas are higher than National average costs and increased at a greater rate from 1974 to 1998 than the National Average.

Historical Cost Indexes

Year	National 30 City Average	Alabama — Birmingham	Alabama — Huntsville	Alabama — Mobile	Alabama — Montgomery	Alabama — Tuscaloosa	Alaska — Anchorage	Arizona — Phoenix	Arizona — Tuscon	Arkansas — Fort Smith	Arkansas — Little Rock	California — Anaheim	California — Bakersfield	California — Fresno	California — Los Angeles	California — Oxnard
Jan 1998	114.4E	97.0E	94.7E	95.5E	91.0E	90.5E	144.5E	102.3E	101.2E	91.2E	92.0E	125.7E	120.9E	124.3E	126.4E	125.3E
1997	111.5	94.6	92.4	93.3	88.8	88.3	142.0	101.8	100.7	89.3	90.1	124.0	119.1	122.3	124.6	123.5
1996	108.9	90.9	91.4	92.3	87.5	86.5	140.4	98.3	97.4	86.5	86.7	121.7	117.2	120.2	122.4	121.5
1995	105.6	87.8	88.0	88.8	84.5	83.5	138.0	96.1	95.5	85.0	85.6	120.1	115.7	117.5	120.9	120.0
1994	103.0	85.9	86.1	86.8	82.6	81.7	134.0	93.7	93.1	82.4	83.9	118.1	113.6	114.5	119.0	117.5
1993	100.0	82.8	82.7	86.1	82.1	78.7	132.0	90.9	90.9	80.9	82.4	115.0	111.3	112.9	115.6	115.4
1992	97.9	81.7	81.5	84.9	80.9	77.6	128.6	88.8	89.4	79.7	81.2	113.5	108.1	110.3	113.7	113.7
1991	95.7	80.5	78.9	83.9	79.8	76.6	127.4	88.1	88.5	78.3	80.0	111.0	105.8	108.2	110.9	111.4
1990	93.2	79.4	77.6	82.7	78.4	75.2	125.8	86.4	87.0	77.1	77.9	107.7	102.7	103.3	107.5	107.4
1989	91.0	77.5	76.1	81.1	76.8	73.7	123.4	85.3	85.6	75.6	76.4	105.6	101.0	101.7	105.3	105.4
1988	88.5	75.7	74.7	79.7	75.4	72.3	121.4	84.4	84.3	74.2	75.0	102.7	97.6	99.2	102.6	102.3
1987	85.7	74.0	73.8	77.9	74.5	71.7	119.3	79.7	80.6	72.2	72.8	100.9	96.1	98.0	100.0	101.1
1986	83.7	72.8	71.7	76.7	72.3	69.4	117.3	78.9	78.8	70.7	72.3	99.2	94.2	93.8	97.8	100.4
1985	81.8	71.1	70.5	72.7	70.9	67.9	116.0	78.1	77.7	69.6	71.0	95.4	92.2	92.6	94.6	96.6
1984	80.6	69.7	69.0	75.2	69.4	66.3	113.8	79.4	79.2	69.1	70.3	93.4	89.8	90.6	92.1	94.0
1983	78.2	67.4	68.8	72.9	69.4	65.5	104.5	77.5	77.9	66.8	68.7	90.9	88.8	88.3	89.7	91.8
1982	72.1	63.5	63.1	67.3	64.4	61.8	96.1	72.3	71.7	62.2	64.6	83.1	83.0	82.5	80.9	83.4
1981	66.1	60.0	58.1	62.2	61.3	58.1	91.5	69.0	67.5	57.7	60.2	75.7	76.5	75.2	74.8	77.1
1980	60.7	55.2	54.1	56.8	56.8	54.0	91.4	63.7	62.4	53.1	55.8	68.7	69.4	68.7	67.4	69.9
1979	54.9	49.9	48.9	52.5	50.8	48.7	82.5	55.9	56.5	47.8	49.5	62.9	61.9	62.5	61.7	62.9
1978	51.3	46.7	45.5	48.8	46.9	45.0	75.3	51.9	52.3	44.8	45.4	57.4	57.1	57.7	57.0	58.6
1977	47.9	43.1	43.2	45.1	42.4	40.5	70.6	48.7	48.5	41.6	42.2	53.2	52.8	53.2	53.7	53.4
1976	45.3	40.8	40.7	42.3	40.3	38.6	64.0	46.5	46.4	39.2	39.4	49.6	49.3	50.4	50.2	48.2
1975	43.7	40.0	40.9	41.8	40.1	37.8	57.3	44.5	45.2	39.0	38.7	47.6	46.6	47.7	48.3	47.0
1974	39.4	34.8	35.6	36.6	36.0	34.4	55.9	40.1	40.5	34.6	32.4	44.1	43.7	44.2	42.2	44.1
1970	27.8	24.1	25.1	25.8	25.4	24.2	43.0	27.2	28.5	24.3	22.3	31.0	30.8	31.1	29.0	31.0
1965	21.5	19.6	19.3	19.6	19.5	18.7	34.9	21.8	22.0	18.7	18.5	23.9	23.7	24.0	22.7	23.9
1960	19.5	17.9	17.5	17.8	17.7	16.9	31.7	19.9	20.0	17.0	16.8	21.7	21.5	21.8	20.6	21.7
1955	16.3	14.8	14.7	14.9	14.9	14.2	26.6	16.7	16.7	14.3	14.5	18.2	18.1	18.3	17.3	18.2
1950	13.5	12.2	12.1	12.3	12.3	11.7	21.9	13.8	13.8	11.8	11.6	15.1	15.0	15.1	14.3	15.1
1945	8.6	7.8	7.7	7.8	7.8	7.5	14.0	8.8	8.8	7.5	7.4	9.6	9.5	9.6	9.1	9.6
1940	6.6	6.0	6.0	6.1	6.0	5.8	10.8	6.8	6.8	5.8	5.7	7.4	7.4	7.4	7.0	7.4

Year	National 30 City Average	California — Riverside	California — Sacramento	California — San Diego	California — San Francisco	California — Santa Barbara	California — Stockton	California — Vallejo	Colorado — Colorado Springs	Colorado — Denver	Colorado — Pueblo	Connecticut — Bridge-Port	Connecticut — Bristol	Connecticut — Hartford	Connecticut — New Britain	Connecticut — New Haven
Jan 1998	114.4E	124.3E	126.7E	121.9E	141.5E	124.5E	125.4E	132.8E	103.3E	106.8E	104.1E	120.0E	120.0E	120.4E	120.6E	120.6E
1997	111.5	122.6	124.7	120.3	139.2	122.4	123.3	130.6	101.1	104.4	102.0	119.2	119.5	119.9	119.7	120.0
1996	108.9	120.6	122.4	118.4	136.8	120.5	121.4	128.1	98.5	101.4	99.7	117.4	117.6	117.9	117.8	118.1
1995	105.6	119.2	119.5	115.4	133.8	119.0	119.0	122.5	96.1	98.9	96.8	116.0	116.5	116.9	116.3	116.5
1994	103.0	116.6	114.8	113.4	131.4	116.4	115.5	119.7	94.2	95.9	94.6	114.3	115.0	115.3	114.7	114.8
1993	100.0	114.3	112.5	111.3	129.6	114.3	115.2	117.5	92.1	93.8	92.6	108.7	107.8	108.4	106.3	106.1
1992	97.9	111.8	110.8	109.5	127.9	112.2	112.7	115.2	90.6	91.5	90.9	106.7	106.0	106.6	104.5	104.4
1991	95.7	109.4	108.5	107.7	125.8	110.0	111.0	113.4	88.9	90.4	89.4	97.6	97.3	98.0	97.3	97.9
1990	93.2	107.0	104.9	105.6	121.8	106.4	105.4	111.4	86.9	88.8	88.8	96.3	95.9	96.6	95.9	96.5
1989	91.0	105.2	103.1	103.6	119.2	104.6	103.7	109.8	85.6	88.7	87.4	94.5	94.2	94.9	94.2	94.6
1988	88.5	102.6	100.5	101.0	115.2	101.4	101.2	107.2	83.8	87.0	85.6	92.8	92.6	93.3	92.6	93.0
1987	85.7	100.3	98.7	99.0	111.0	99.3	99.6	103.2	81.6	84.8	83.8	92.4	93.0	93.3	93.0	92.6
1986	83.7	98.6	94.4	96.9	108.0	97.0	96.1	99.1	81.9	83.1	82.7	88.5	88.7	89.4	88.6	88.8
1985	81.8	94.8	92.2	94.2	106.2	93.2	93.7	97.0	79.6	81.0	80.4	86.1	86.2	87.2	86.1	86.3
1984	80.6	91.8	93.8	92.0	102.8	90.6	91.9	95.6	79.8	84.1	80.5	83.5	83.4	84.4	83.0	83.2
1983	78.2	89.4	92.0	90.3	101.0	88.8	91.0	93.0	77.3	80.5	77.4	79.6	79.9	81.1	79.5	79.3
1982	72.1	82.1	84.9	83.8	91.9	81.8	84.9	85.7	71.2	73.4	75.0	71.9	71.2	73.2	71.4	72.0
1981	66.1	75.5	77.7	74.5	82.8	75.9	77.4	78.5	65.5	66.3	65.4	67.0	66.3	67.3	66.3	67.0
1980	60.7	68.7	71.3	68.1	75.2	71.1	71.2	71.9	60.7	60.9	59.5	61.5	60.7	61.9	60.7	61.4
1979	54.9	62.2	64.0	61.6	67.5	65.0	64.2	65.2	54.9	56.0	54.0	56.0	55.3	56.4	55.4	55.8
1978	51.3	57.7	60.0	58.3	62.6	59.2	59.5	59.9	51.1	52.0	50.6	52.5	50.9	51.9	51.8	52.0
1977	47.9	53.3	55.7	53.4	58.1	53.0	54.8	54.2	47.2	48.3	47.0	49.3	48.6	48.6	48.3	49.0
1976	45.3	49.7	51.5	50.1	53.9	49.0	51.1	50.6	44.5	45.6	44.4	47.1	46.6	46.8	46.6	47.8
1975	43.7	47.3	49.1	47.7	49.8	46.1	47.6	46.5	43.1	42.7	42.5	45.2	45.5	46.0	45.2	46.0
1974	39.4	44.0	45.7	43.1	44.7	44.5	45.2	45.1	39.2	37.3	38.8	41.7	40.1	41.8	40.1	41.8
1970	27.8	31.0	32.1	30.7	31.6	31.3	31.8	31.8	27.6	26.1	27.3	29.2	28.2	29.6	28.2	29.3
1965	21.5	23.9	24.8	24.0	23.7	24.1	24.5	24.5	21.3	20.9	21.0	22.4	21.7	22.6	21.7	23.2
1960	19.5	21.7	22.5	21.7	21.5	21.9	22.3	22.2	19.3	19.0	19.1	20.0	19.8	20.0	19.8	20.0
1955	16.3	18.2	18.9	18.2	18.0	18.4	18.7	18.6	16.2	15.9	16.0	16.8	16.6	16.8	16.6	16.8
1950	13.5	15.0	15.6	15.0	14.9	15.2	15.4	15.4	13.4	13.2	13.2	13.9	13.7	13.8	13.7	13.9
1945	8.6	9.6	10.0	9.6	9.5	9.7	9.9	9.8	8.5	8.4	8.4	8.8	8.7	8.8	8.7	8.9
1940	6.6	7.4	7.7	7.4	7.3	7.5	7.6	7.6	6.6	6.5	6.5	6.8	6.8	6.8	6.8	6.8

Historical Cost Indexes

Year	National 30 City Average	Connecticut Norwalk	Connecticut Stamford	Connecticut Waterbury	Delaware Wilmington	D.C. Washington	Florida Fort Lauderdale	Florida Jacksonville	Florida Miami	Florida Orlando	Florida Tallahassee	Florida Tampa	Georgia Albany	Georgia Atlanta	Georgia Columbus	Georgia Macon
Jan 1998	114.4E	119.9E	121.5E	121.0E	113.5E	108.9E	100.8E	96.9E	99.8E	99.4E	91.7E	97.1E	92.3E	101.4E	91.0E	92.9E
1997	111.5	118.9	120.9	120.2	110.7	106.4	98.6	95.4	98.2	97.2	89.9	95.5	89.9	98.5	88.7	91.6
1996	108.9	117.0	119.0	118.4	108.6	105.4	95.9	92.6	95.9	95.1	88.0	93.6	86.6	94.3	83.9	88.3
1995	105.6	115.5	117.9	117.3	106.1	102.3	94.0	90.9	93.7	93.5	86.5	92.2	85.0	92.0	82.6	86.8
1994	103.0	113.9	116.4	115.8	105.0	99.6	92.2	88.9	91.8	91.5	84.5	90.2	82.3	89.6	80.6	83.7
1993	100.0	108.8	110.6	104.8	101.5	96.3	87.4	86.1	87.1	88.5	82.1	87.7	79.5	85.7	77.8	80.9
1992	97.9	107.2	109.0	103.1	100.3	94.7	85.7	84.0	85.3	87.1	80.8	86.2	78.2	84.3	76.5	79.6
1991	95.7	100.6	103.2	96.5	94.5	92.9	85.1	82.8	85.2	85.5	79.7	86.3	76.3	82.6	75.4	78.4
1990	93.2	96.3	98.9	95.1	92.5	90.4	83.9	81.1	84.0	82.9	78.4	85.0	75.0	80.4	74.0	76.9
1989	91.0	94.4	97.0	93.4	89.5	87.5	82.4	79.7	82.5	81.6	76.9	83.5	73.4	78.6	72.4	75.3
1988	88.5	92.3	94.9	91.8	87.7	85.0	80.9	78.0	81.0	80.1	75.4	81.9	71.8	76.9	70.8	73.6
1987	85.7	92.0	92.7	92.5	85.1	82.1	78.8	76.0	77.9	76.6	74.4	79.4	72.2	73.6	70.3	72.3
1986	83.7	88.2	89.7	87.8	83.8	80.8	78.9	75.0	79.9	76.2	72.3	79.1	68.5	72.0	67.5	69.8
1985	81.8	85.3	86.8	85.6	81.1	78.8	76.7	73.4	78.3	73.9	70.6	77.3	66.9	70.3	66.1	68.1
1984	80.6	82.6	83.7	82.5	79.7	79.1	73.8	72.6	75.6	73.0	69.6	76.5	65.9	68.6	65.2	67.5
1983	78.2	78.5	79.8	78.7	76.3	76.0	71.5	70.4	72.9	71.1	67.6	73.6	65.6	68.6	64.7	65.3
1982	72.1	71.7	72.0	71.8	69.1	69.4	65.4	65.2	65.7	66.0	62.8	67.7	60.1	61.9	59.3	60.0
1981	66.1	66.2	66.2	67.3	63.4	64.9	60.3	61.0	60.7	62.0	58.6	62.1	56.2	58.3	55.7	56.1
1980	60.7	60.7	60.9	62.3	58.5	59.6	55.3	55.8	56.5	56.7	53.5	57.2	51.7	54.0	51.1	51.3
1979	54.9	55.2	55.6	57.0	53.6	53.9	50.0	51.7	50.8	51.8	48.9	51.8	46.5	48.3	46.1	46.4
1978	51.3	51.5	51.6	52.9	50.0	51.5	47.0	47.7	47.6	48.0	45.4	48.4	42.7	45.0	42.5	42.4
1977	47.9	48.1	48.3	49.0	47.1	48.4	43.7	43.3	45.7	44.9	42.0	45.9	41.6	42.1	37.7	40.8
1976	45.3	46.3	45.9	47.4	43.3	45.7	42.2	42.1	43.9	42.8	40.1	43.0	39.9	39.6	35.8	37.9
1975	43.7	44.7	45.0	46.3	42.9	43.7	42.1	40.3	43.2	41.5	38.1	41.3	37.5	38.4	36.2	36.5
1974	39.4	39.9	40.0	41.0	38.4	38.5	36.5	34.2	39.6	37.2	34.9	35.8	33.8	34.6	32.2	33.2
1970	27.8	28.1	28.2	28.9	27.0	26.3	25.7	22.8	27.0	26.2	24.5	24.2	23.8	25.2	22.8	23.4
1965	21.5	21.6	21.7	22.2	20.9	21.8	19.8	17.4	19.3	20.2	18.9	18.6	18.3	19.8	17.6	18.0
1960	19.5	19.7	19.7	20.2	18.9	19.4	18.0	15.8	17.6	18.3	17.2	16.9	16.7	17.1	16.0	16.4
1955	16.3	16.5	16.5	17.0	15.9	16.3	15.1	13.2	14.7	15.4	14.4	14.1	14.0	14.4	13.4	13.7
1950	13.5	13.6	13.7	14.0	13.1	13.4	12.5	11.0	12.2	12.7	11.9	11.7	11.5	11.9	11.0	11.3
1945	8.6	8.7	8.7	8.9	8.4	8.6	7.9	7.0	7.8	8.1	7.6	7.5	7.4	7.6	7.0	7.2
1940	6.6	6.7	6.7	6.9	6.4	6.6	6.1	5.4	6.0	6.2	5.8	5.7	5.7	5.8	5.5	5.6

Year	National 30 City Average	Georgia Savannah	Hawaii Honolulu	Idaho Boise	Idaho Pocatello	Illinois Chicago	Illinois Decatur	Illinois Joliet	Illinois Peoria	Illinois Rockford	Illinois Springfield	Indiana Anderson	Indiana Evansville	Indiana Fort Wayne	Indiana Gary	Indiana Indianapolis
Jan 1998	114.4E	94.2E	140.6E	106.7E	106.8E	125.8E	110.5E	120.5E	114.3E	116.0E	111.7E	104.3E	106.1E	103.4E	112.0E	108.6E
1997	111.5	92.0	139.8	104.6	104.7	121.3	107.8	117.5	111.5	113.1	108.9	101.9	104.4	101.8	110.3	105.2
1996	108.9	88.6	134.5	102.2	102.1	118.8	106.6	116.2	109.3	111.5	106.5	100.0	102.1	99.9	107.5	102.7
1995	105.6	87.4	130.3	99.5	98.2	114.2	98.5	110.5	102.3	103.6	98.1	96.4	97.2	95.0	100.7	100.1
1994	103.0	85.3	124.0	94.8	95.0	111.3	97.3	108.9	100.9	102.2	97.0	93.6	95.8	93.6	99.1	97.1
1993	100.0	82.0	122.0	92.2	92.1	107.6	95.6	106.8	98.9	99.6	95.2	91.2	94.3	91.5	96.7	93.9
1992	97.9	80.8	120.0	91.0	91.0	104.3	94.4	104.2	97.3	98.2	94.0	89.5	92.9	89.9	95.0	91.5
1991	95.7	79.5	106.1	89.5	89.4	100.9	92.3	100.0	95.9	95.8	91.5	87.8	91.4	88.3	93.3	89.1
1990	93.2	77.9	104.7	88.2	88.1	98.4	90.9	98.4	93.7	94.0	90.1	84.6	89.3	83.4	88.4	87.1
1989	91.0	76.0	102.8	86.6	86.5	93.7	89.4	92.8	91.6	92.1	88.6	82.3	87.8	81.7	86.5	85.1
1988	88.5	74.5	101.1	83.9	83.7	90.6	87.5	89.8	88.5	87.9	86.8	80.8	85.5	80.0	84.6	83.1
1987	85.7	72.0	99.1	81.4	81.5	86.6	84.4	86.7	86.3	85.2	85.0	78.8	82.4	78.1	81.7	80.4
1986	83.7	70.8	97.5	80.6	80.3	84.4	83.5	85.3	85.1	84.5	83.4	77.0	80.9	76.5	79.6	78.6
1985	81.8	68.9	94.7	78.0	78.0	82.4	81.9	83.4	83.7	83.0	81.5	75.2	79.8	75.0	77.8	77.1
1984	80.6	67.9	90.7	76.6	76.8	80.2	80.4	82.2	83.6	80.6	80.1	73.5	77.4	73.5	77.3	75.9
1983	78.2	66.3	87.2	76.0	75.8	79.0	78.8	80.6	81.5	78.9	78.8	70.9	75.0	71.2	75.2	74.0
1982	72.1	61.1	79.3	71.0	70.1	75.0	72.7	74.4	75.3	72.6	72.7	66.4	69.8	67.1	70.5	68.2
1981	66.1	57.1	73.4	65.4	64.8	68.6	67.2	68.8	70.3	67.3	67.0	61.5	64.3	61.5	65.3	62.4
1980	60.7	52.2	68.9	60.3	59.5	62.8	62.3	63.4	64.5	61.6	61.1	56.5	59.0	56.7	59.8	57.9
1979	54.9	47.2	63.0	54.4	53.7	56.5	56.0	57.0	58.4	54.8	55.6	50.8	53.3	51.2	54.0	51.9
1978	51.3	43.6	58.9	50.0	49.9	52.9	51.9	52.7	53.8	50.4	51.5	46.7	49.1	46.9	49.4	48.7
1977	47.9	40.2	52.0	45.1	45.2	49.0	47.7	49.1	49.4	46.8	47.5	43.2	45.7	43.0	46.7	45.4
1976	45.3	38.3	45.7	41.9	41.9	47.3	45.0	46.0	46.5	44.1	44.1	40.6	42.8	41.4	43.9	43.1
1975	43.7	36.9	44.6	40.8	40.5	45.7	43.1	44.5	44.7	42.7	42.4	39.5	41.7	39.9	41.9	40.6
1974	39.4	31.0	40.6	37.9	37.7	41.7	39.7	40.6	41.2	39.0	39.2	36.0	36.5	36.3	38.4	36.6
1970	27.8	21.0	30.4	26.7	26.6	29.1	28.0	28.6	29.0	27.5	27.6	25.4	26.4	25.5	27.1	26.2
1965	21.5	16.4	21.8	20.6	20.5	22.7	21.5	22.1	22.4	21.2	21.3	19.5	20.4	19.7	20.8	20.7
1960	19.5	14.9	19.8	18.7	18.6	20.2	19.6	20.0	20.3	19.2	19.3	17.7	18.7	17.9	18.9	18.4
1955	16.3	12.5	16.6	15.7	15.6	16.9	16.4	16.8	17.0	16.1	16.2	14.9	15.7	15.0	15.9	15.5
1950	13.5	10.3	13.7	13.0	12.9	14.0	13.6	13.9	14.0	13.3	13.4	12.3	12.9	12.4	13.1	12.8
1945	8.6	6.6	8.8	8.3	8.2	8.9	8.6	8.9	9.0	8.5	8.6	7.8	8.3	7.9	8.4	8.1
1940	6.6	5.1	6.8	6.4	6.3	6.9	6.7	6.8	6.9	6.5	6.6	6.0	6.4	6.1	6.5	6.3

Historical Cost Indexes

Year	National 30 City Average	Indiana Muncie	Indiana South Bend	Indiana Terre Haute	Iowa Cedar Rapids	Iowa Daven-port	Iowa Des Moines	Iowa Sioux City	Iowa Water-loo	Kansas Topeka	Kansas Wichita	Kentucky Lexing-ton	Kentucky Louis-ville	Louisiana Baton Rouge	Louisiana Lake Charles	Louisiana New Orleans
Jan 1998	114.4E	103.2E	103.2E	107.0E	103.4E	107.5E	104.5E	95.9E	95.6E	97.9E	97.8E	98.2E	102.2E	95.2E	97.5E	98.0E
1997	111.5	101.3	101.2	104.6	101.2	105.6	102.5	93.8	93.5	96.3	96.2	96.1	99.9	93.3	96.8	96.2
1996	108.9	99.4	99.1	102.1	99.1	102.0	99.1	90.9	90.7	93.6	93.8	94.4	98.3	91.5	94.9	94.3
1995	105.6	95.6	94.6	97.2	95.4	96.6	94.4	88.2	88.9	91.1	91.0	91.6	94.6	89.6	92.7	91.6
1994	103.0	93.2	93.0	95.8	93.0	92.3	92.4	85.9	86.8	89.4	88.1	89.6	92.2	87.7	89.9	89.5
1993	100.0	91.0	90.7	94.4	91.1	90.5	90.7	84.1	85.2	87.4	86.2	87.3	89.4	86.4	88.5	87.8
1992	97.9	89.4	89.3	93.1	89.8	89.3	89.4	82.9	83.8	86.2	85.0	85.8	88.0	85.2	87.3	86.6
1991	95.7	87.7	87.5	91.7	88.5	88.0	88.4	81.8	81.6	84.4	83.9	84.1	84.4	83.2	85.3	85.8
1990	93.2	83.9	85.1	89.1	87.1	86.5	86.9	80.4	80.3	83.2	82.6	82.8	82.6	82.0	84.1	84.5
1989	91.0	81.9	83.5	86.7	85.3	84.9	85.3	79.0	78.8	81.6	81.2	81.3	80.1	80.6	82.7	83.4
1988	88.5	80.4	82.1	84.9	83.4	83.1	83.6	76.8	77.2	80.1	79.5	79.4	78.2	79.0	81.1	82.0
1987	85.7	78.5	79.6	83.0	82.4	81.1	81.2	75.4	75.6	78.9	78.4	77.3	76.3	77.1	78.0	81.3
1986	83.7	76.9	77.5	81.1	79.5	80.0	79.8	73.8	73.7	76.9	76.2	76.5	75.3	76.1	78.1	79.2
1985	81.8	75.2	76.0	79.2	75.9	77.6	77.1	72.1	72.3	75.0	74.7	75.5	74.5	74.9	78.5	78.2
1984	80.6	73.6	75.1	78.0	80.3	77.8	76.6	74.4	73.7	75.1	74.6	75.6	73.7	77.3	78.3	77.5
1983	78.2	71.2	74.5	76.1	79.6	77.0	75.5	74.6	73.0	73.2	72.4	74.9	74.1	75.1	76.6	74.7
1982	72.1	66.0	68.3	69.9	73.5	71.7	71.1	70.1	67.7	68.4	66.1	69.4	69.2	69.3	70.5	69.2
1981	66.1	61.0	63.1	65.1	68.6	66.7	67.1	65.5	63.5	63.0	62.2	64.7	65.1	64.0	64.8	62.7
1980	60.7	56.1	58.3	59.7	62.7	59.6	61.8	59.3	57.7	58.9	58.0	59.3	59.8	59.1	60.0	57.2
1979	54.9	50.7	52.6	54.0	56.6	54.1	55.6	53.6	52.0	53.5	52.9	53.2	54.5	53.5	54.5	52.6
1978	51.3	46.8	48.2	49.9	52.3	50.0	51.1	49.2	48.1	50.0	48.9	49.2	50.0	49.2	50.2	48.5
1977	47.9	43.4	44.7	46.4	48.1	45.7	47.6	45.3	44.2	46.2	45.8	45.8	46.1	45.2	45.2	46.0
1976	45.3	40.9	42.2	43.9	45.8	43.9	45.1	42.3	42.0	43.7	43.3	43.5	43.7	41.7	42.3	43.5
1975	43.7	39.2	40.3	41.9	43.1	41.0	43.1	41.0	40.0	42.0	43.1	42.7	42.5	40.4	40.6	41.5
1974	39.4	36.0	37.2	38.5	40.1	38.2	38.1	37.8	36.9	38.1	37.4	38.1	37.9	35.3	37.9	37.4
1970	27.8	25.3	26.2	27.0	28.2	26.9	27.6	26.6	25.9	27.0	25.5	26.9	25.9	25.0	26.7	27.2
1965	21.5	19.5	20.2	20.9	21.7	20.8	21.7	20.5	20.0	20.8	19.6	20.7	20.3	19.4	20.6	20.4
1960	19.5	17.8	18.3	18.9	19.7	18.8	19.5	18.6	18.2	18.9	17.8	18.8	18.4	17.6	18.7	18.5
1955	16.3	14.9	15.4	15.9	16.6	15.8	16.3	15.6	15.2	15.8	15.0	15.8	15.4	14.8	15.7	15.6
1950	13.5	12.3	12.7	13.1	13.7	13.1	13.5	12.9	12.6	13.1	12.4	13.0	12.8	12.2	13.0	12.8
1945	8.6	7.8	8.1	8.4	8.7	8.3	8.6	8.2	8.0	8.3	7.9	8.3	8.1	7.8	8.3	8.2
1940	6.6	6.0	6.3	6.5	6.7	6.4	6.6	6.3	6.2	6.4	6.1	6.4	6.3	6.0	6.4	6.3

Year	National 30 City Average	Louisiana Shreve-port	Maine Lewis-ton	Maine Portland	Maryland Balti-more	Massachusetts Boston	Massachusetts Brockton	Massachusetts Fall River	Massachusetts Law-rence	Massachusetts Lowell	Massachusetts New Bedford	Massachusetts Pitts-field	Massachusetts Spring-field	Massachusetts Wor-cester	Michigan Ann Arbor	Michigan Dear-born
Jan 1998	114.4E	92.6E	103.7E	103.4E	104.8E	133.1E	125.4E	125.6E	126.0E	126.4E	125.5E	115.4E	118.3E	123.7E	115.1E	120.1E
1997	111.5	90.1	101.7	101.4	102.2	132.1	124.4	124.7	125.1	125.6	124.7	114.6	117.4	122.9	113.5	117.9
1996	108.9	88.5	99.5	99.2	99.6	128.7	121.6	122.2	121.7	122.1	122.2	112.7	115.0	120.2	112.9	116.2
1995	105.6	86.7	96.8	96.5	96.1	128.6	119.6	117.7	120.5	119.6	117.2	110.7	112.7	114.8	106.1	110.5
1994	103.0	85.0	95.2	94.9	94.4	124.9	114.2	113.7	116.8	116.9	111.6	109.2	111.1	112.9	105.0	109.0
1993	100.0	83.6	93.0	93.0	93.1	121.1	111.7	111.5	114.9	114.3	109.4	107.1	108.9	110.3	102.8	105.8
1992	97.9	82.3	91.7	91.7	90.9	118.0	110.0	109.8	110.9	110.2	107.7	105.7	107.2	108.6	101.5	103.2
1991	95.7	81.6	89.8	89.9	89.1	115.2	105.9	104.7	107.3	105.5	105.1	100.6	103.2	105.7	95.1	98.3
1990	93.2	80.3	88.5	88.6	85.6	110.9	103.7	102.9	105.2	102.8	102.6	98.7	101.4	103.2	93.2	96.5
1989	91.0	78.8	86.7	86.7	83.5	107.2	100.4	99.7	102.9	99.7	99.8	94.7	96.4	99.1	89.6	94.6
1988	88.5	77.0	84.0	84.1	81.2	102.5	96.7	96.4	97.7	96.6	96.4	91.9	92.7	94.4	85.8	91.4
1987	85.7	74.5	81.3	81.4	78.6	97.3	94.5	94.8	94.9	94.8	95.0	89.4	90.5	92.2	87.4	88.7
1986	83.7	74.1	79.0	79.0	75.7	95.1	91.2	90.8	92.5	90.8	90.7	87.1	87.8	89.8	81.7	85.3
1985	81.8	73.6	76.7	77.0	72.7	92.8	88.8	88.7	89.4	88.2	88.6	85.0	85.6	86.7	80.1	82.7
1984	80.6	73.3	75.0	75.1	72.4	88.1	85.5	85.8	86.8	84.1	84.8	82.8	83.5	83.6	78.9	81.1
1983	78.2	72.2	73.1	72.9	70.6	84.0	82.5	82.2	82.7	80.9	81.3	79.5	80.6	81.3	77.7	79.4
1982	72.1	67.0	67.4	67.4	64.7	76.9	75.7	75.0	74.5	73.6	74.2	72.3	73.7	73.1	73.7	75.8
1981	66.1	62.5	62.2	63.1	59.0	67.8	68.7	69.2	68.4	67.1	68.7	66.3	67.0	66.7	68.6	70.1
1980	60.7	58.7	57.3	58.5	53.6	64.0	63.7	64.1	63.4	62.7	63.1	61.8	62.0	62.3	62.9	64.0
1979	54.9	53.1	52.3	52.6	48.2	57.9	58.0	57.5	57.3	57.0	57.4	56.2	55.9	56.0	57.0	57.8
1978	51.3	49.2	48.6	48.8	45.9	54.0	53.4	53.3	53.5	53.1	53.2	52.6	52.3	52.2	52.5	54.1
1977	47.9	44.8	46.2	45.2	44.4	49.3	50.4	50.1	50.5	49.5	49.9	49.2	50.0	49.4	48.6	48.6
1976	45.3	42.0	43.2	42.5	41.5	47.7	46.9	47.8	47.2	46.9	47.3	46.3	47.6	47.1	46.5	46.4
1975	43.7	40.5	41.9	42.1	39.8	46.6	45.8	45.7	46.2	45.7	46.1	45.5	45.8	46.0	44.1	44.9
1974	39.4	37.3	37.5	35.6	37.4	41.7	41.3	41.3	41.3	40.9	41.1	40.4	41.0	40.6	40.4	40.9
1970	27.8	26.4	26.5	25.8	25.1	29.2	29.1	29.2	29.1	28.9	29.0	28.6	28.5	28.7	28.5	28.9
1965	21.5	20.3	20.4	19.4	20.2	23.0	22.5	22.5	22.5	22.2	22.4	22.0	22.4	22.1	22.0	22.3
1960	19.5	18.5	18.6	17.6	17.5	20.5	20.4	20.4	20.4	20.2	20.3	20.0	20.1	20.1	20.0	20.2
1955	16.3	15.5	15.6	14.7	14.7	17.2	17.1	17.1	17.1	16.9	17.1	16.8	16.9	16.8	16.7	16.9
1950	13.5	12.8	12.9	12.2	12.1	14.2	14.1	14.1	14.1	14.0	14.1	13.8	14.0	13.9	13.8	14.0
1945	8.6	8.1	8.2	7.8	7.7	9.1	9.0	9.0	9.0	8.9	9.0	8.9	8.9	8.9	8.8	8.9
1940	6.6	6.3	6.3	6.0	6.0	7.0	6.9	7.0	7.0	6.9	6.9	6.8	6.9	6.8	6.8	6.9

Historical Cost Indexes

| Year | National 30 City Average | Michigan | | | | | | Minnesota | | | Mississippi | | Missouri | | | |
		Detroit	Flint	Grand Rapids	Kala-mazoo	Lansing	Sagi-naw	Duluth	Minne-apolis	Roches-ter	Biloxi	Jackson	Kansas City	St. Joseph	St. Louis	Spring-field
Jan 1998	114.4E	119.9E	112.3E	100.3E	105.5E	110.4E	109.0E	118.5E	125.3E	116.3E	92.8E	90.2E	108.6E	104.8E	116.4E	99.6E
1997	111.5	117.6	110.8	98.9	104.2	108.7	107.5	115.9	121.9	114.1	90.9	88.3	106.4	102.4	113.2	97.3
1996	108.9	116.0	109.7	93.9	103.5	107.8	106.9	115.0	120.4	113.5	87.1	85.4	103.2	99.9	110.1	94.6
1995	105.6	110.1	104.1	91.1	96.4	98.3	102.3	100.3	111.9	102.5	84.2	83.5	99.7	96.1	106.3	89.5
1994	103.0	108.5	102.9	89.7	95.0	97.1	101.1	99.1	109.3	101.1	82.4	81.7	97.3	92.8	103.2	88.2
1993	100.0	105.4	100.7	87.7	92.7	95.3	98.8	99.8	106.7	99.9	80.0	79.4	94.5	90.9	99.9	86.2
1992	97.9	102.9	99.4	86.4	91.4	94.0	97.4	98.5	105.3	98.6	78.8	78.2	92.9	89.5	98.6	84.9
1991	95.7	97.6	93.6	85.3	90.0	92.1	90.3	96.0	102.9	97.3	77.6	76.9	90.9	88.1	96.4	83.9
1990	93.2	96.0	91.7	84.0	87.4	90.2	88.9	94.3	100.5	95.6	76.4	75.6	89.5	86.8	94.0	82.6
1989	91.0	94.0	90.2	82.4	85.8	88.6	87.4	92.5	97.4	93.7	75.0	74.1	87.3	85.1	91.8	80.9
1988	88.5	90.8	87.8	80.7	84.1	86.3	85.6	90.5	94.6	91.9	73.5	72.9	84.9	82.9	89.0	79.2
1987	85.7	87.8	85.4	79.7	82.6	85.2	84.5	88.5	92.0	90.6	72.4	71.3	81.7	82.7	85.5	78.2
1986	83.7	84.3	82.8	77.5	80.9	82.7	81.5	87.2	89.7	88.4	70.6	69.9	79.9	79.4	83.5	75.6
1985	81.8	81.6	80.7	75.8	79.4	80.0	80.5	85.2	87.9	86.5	69.4	68.5	78.2	77.3	80.8	73.2
1984	80.6	79.4	79.5	74.3	78.7	79.0	79.9	84.9	86.9	86.0	68.6	67.5	77.7	76.7	77.7	72.7
1983	78.2	77.0	77.4	74.1	76.2	77.1	78.0	81.6	81.4	82.6	68.0	67.3	76.2	76.8	76.2	71.6
1982	72.1	74.8	73.5	69.9	72.3	73.2	73.7	74.6	75.2	75.6	63.1	62.6	70.2	70.3	69.2	66.5
1981	66.1	68.6	68.2	64.6	67.1	67.5	68.5	68.8	69.6	69.8	58.7	59.1	63.5	65.7	64.5	61.4
1980	60.7	62.5	63.1	59.8	61.7	60.3	63.3	63.4	64.1	64.8	54.3	54.3	59.1	60.4	59.7	56.7
1979	54.9	56.2	56.4	53.7	55.5	55.8	56.8	59.2	58.5	58.7	49.3	49.3	54.4	54.6	54.9	51.3
1978	51.3	52.6	52.0	49.1	51.7	51.4	53.2	53.1	54.1	53.3	45.0	45.8	50.1	51.4	51.1	47.9
1977	47.9	48.7	48.4	44.9	48.4	47.4	49.1	49.5	50.0	48.9	41.2	42.3	46.8	48.8	48.2	44.5
1976	45.3	47.0	46.4	42.7	46.2	45.0	46.5	45.8	47.0	45.1	39.2	39.3	43.5	47.5	44.9	41.7
1975	43.7	45.8	44.9	41.8	43.7	44.1	44.8	45.2	46.0	45.3	37.9	37.7	42.7	45.4	44.8	41.1
1974	39.4	41.8	40.3	37.8	39.8	39.4	40.6	41.0	42.4	40.9	34.5	30.8	37.7	40.0	39.7	36.7
1970	27.8	29.7	28.5	26.7	28.1	27.8	28.7	28.9	29.5	28.9	24.4	21.3	25.3	28.3	28.4	26.0
1965	21.5	22.1	21.9	20.6	21.7	21.5	22.2	22.3	23.4	22.3	18.8	16.4	20.6	21.8	21.8	20.0
1960	19.5	20.1	20.0	18.7	19.7	19.5	20.1	20.3	20.5	20.2	17.1	14.9	19.0	19.8	19.5	18.2
1955	16.3	16.8	16.7	15.7	16.5	16.3	16.9	17.0	17.2	17.0	14.3	12.5	16.0	16.6	16.3	15.2
1950	13.5	13.9	13.8	13.0	13.6	13.5	13.9	14.0	14.2	14.0	11.8	10.3	13.2	13.7	13.5	12.6
1945	8.6	8.9	8.8	8.3	8.7	8.6	8.9	8.9	9.0	8.9	7.6	6.6	8.4	8.8	8.6	8.0
1940	6.6	6.8	6.8	6.4	6.7	6.6	6.9	6.9	7.0	6.9	5.8	5.1	6.5	6.7	6.7	6.2

| Year | National 30 City Average | Montana | | Nebraska | | Nevada | | New Hamphire | | New Jersey | | | | | NM | NY |
		Billings	Great Falls	Lincoln	Omaha	Las Vegas	Reno	Man-chester	Nashua	Camden	Jersey City	Newark	Pater-son	Trenton	Albu-querque	Albany
Jan 1998	114.4E	109.5E	109.7E	95.4E	101.8E	116.7E	111.9E	110.3E	110.3E	123.6E	126.8E	128.2E	128.1E	127.0E	103.9E	112.0E
1997	111.5	107.4	107.6	93.3	99.5	114.6	109.8	108.6	108.6	121.9	125.1	126.4	126.4	125.2	100.8	110.0
1996	108.9	108.0	107.5	91.4	97.4	111.8	108.9	106.5	106.4	119.1	122.5	122.5	123.8	121.8	98.6	108.3
1995	105.6	104.7	104.9	85.6	93.4	108.5	105.0	100.9	100.8	107.0	112.2	111.9	112.1	111.2	96.3	103.6
1994	103.0	100.2	100.7	84.3	91.0	105.3	102.2	97.8	97.7	105.4	110.6	110.1	110.6	108.2	93.4	102.4
1993	100.0	97.9	97.8	82.3	88.7	102.8	99.9	95.4	95.4	103.7	109.0	108.2	109.0	106.4	90.1	99.8
1992	97.9	95.5	96.4	81.1	87.4	99.4	98.4	90.4	90.4	102.0	107.5	107.0	107.8	103.9	87.5	98.5
1991	95.7	94.2	95.1	80.1	86.4	97.5	95.8	87.8	87.8	95.0	98.5	95.6	100.0	96.6	86.2	96.1
1990	93.2	92.9	93.8	78.8	85.0	96.3	94.5	86.3	86.3	93.0	93.5	93.8	97.2	94.5	84.9	93.2
1989	91.0	91.3	92.2	77.3	83.6	94.8	92.3	84.8	84.8	90.4	91.4	91.8	95.5	91.9	83.3	88.4
1988	88.5	89.5	90.4	75.8	82.0	93.0	90.5	83.2	83.2	87.7	89.5	89.7	92.5	89.5	81.7	86.8
1987	85.7	85.4	87.3	75.3	79.8	91.5	88.3	82.4	82.5	86.1	88.3	88.7	90.3	87.9	78.1	84.5
1986	83.7	86.2	87.3	72.6	78.8	90.1	87.2	79.7	79.7	84.6	86.1	86.7	87.8	86.1	78.9	81.7
1985	81.8	83.9	84.3	71.5	77.5	87.6	85.0	78.1	78.1	81.3	83.3	83.5	84.5	82.8	76.7	79.5
1984	80.6	83.2	82.9	70.9	77.6	85.9	83.3	76.4	75.5	78.2	80.3	80.1	81.3	78.5	75.9	77.3
1983	78.2	80.3	80.0	72.1	77.3	83.0	81.9	73.0	72.2	74.6	76.5	76.7	76.5	74.2	73.1	74.2
1982	72.1	74.3	74.1	67.8	72.5	76.9	75.8	67.0	66.7	67.9	69.8	70.4	70.1	69.1	67.1	69.2
1981	66.1	69.5	70.3	64.1	68.2	70.2	68.5	61.5	61.1	62.5	65.0	65.1	64.8	63.4	63.3	64.4
1980	60.7	63.9	64.6	58.5	63.5	64.6	63.0	56.6	56.0	58.6	60.6	60.1	60.0	58.9	59.0	59.5
1979	54.9	57.5	58.9	52.9	56.2	58.4	56.7	51.2	50.8	53.5	55.6	55.3	54.7	54.0	54.3	54.2
1978	51.3	53.4	54.0	49.3	52.2	53.9	51.9	47.5	47.1	50.1	51.7	51.4	51.7	50.8	49.8	50.6
1977	47.9	48.9	49.1	44.6	48.0	49.7	48.1	43.9	43.6	47.2	47.4	47.2	47.7	47.0	45.1	47.7
1976	45.3	45.4	46.2	42.1	45.2	46.3	45.1	40.8	41.1	44.6	44.8	45.4	45.3	45.2	41.9	45.1
1975	43.7	43.1	43.8	40.9	43.2	42.8	41.9	41.3	40.8	42.3	43.4	44.2	43.9	43.8	40.3	43.9
1974	39.4	40.4	41.0	37.3	37.7	40.0	39.7	36.3	36.3	38.4	39.3	40.5	39.4	39.4	35.5	40.4
1970	27.8	28.5	28.9	26.4	26.8	29.4	28.0	26.2	25.6	27.2	27.8	29.0	27.8	27.4	26.4	28.3
1965	21.5	22.0	22.3	20.3	20.6	22.4	21.6	20.6	19.7	20.9	21.4	23.8	21.4	21.3	20.6	22.3
1960	19.5	20.0	20.3	18.5	18.7	20.2	19.6	18.0	17.9	19.0	19.4	19.4	19.4	19.2	18.5	19.3
1955	16.3	16.7	17.0	15.5	15.7	16.9	16.4	15.1	15.0	16.0	16.3	16.3	16.3	16.1	15.6	16.2
1950	13.5	13.9	14.0	12.8	13.0	14.0	13.6	12.5	12.4	13.2	13.5	13.5	13.5	13.3	12.9	13.4
1945	8.6	8.8	9.0	8.1	8.3	8.9	8.7	8.0	7.9	8.4	8.6	8.6	8.6	8.5	8.2	8.5
1940	6.6	6.8	6.9	6.3	6.4	6.9	6.7	6.2	6.1	6.5	6.6	6.6	6.6	6.6	6.3	6.6

Historical Cost Indexes

Year	National 30 City Average	New York Bing-hamton	Buffalo	New York	Roches-ter	Schen-ectady	Syracuse	Utica	Yonkers	North Carolina Charlotte	Durham	Greens-boro	Raleigh	Winston-Salem	N. Dakota Fargo	Ohio Akron
Jan 1998	114.4E	109.3E	118.5E	152.4E	117.4E	113.1E	112.7E	108.9E	140.5E	88.7E	89.6E	89.7E	89.9E	89.5E	95.7E	112.5E
1997	111.5	107.2	115.7	150.3	115.2	111.2	110.6	107.0	138.8	86.8	87.7	87.8	87.9	87.6	93.5	110.6
1996	108.9	105.5	114.0	148.0	113.3	109.5	108.2	105.2	137.2	85.0	85.9	86.0	86.1	85.8	91.7	107.9
1995	105.6	99.3	110.1	140.7	106.6	104.6	104.0	97.7	129.4	81.8	82.6	82.6	82.7	82.6	88.4	103.4
1994	103.0	98.1	107.2	137.1	105.2	103.5	102.3	96.5	128.1	80.3	81.1	81.0	81.2	81.1	87.0	102.2
1993	100.0	95.7	102.2	133.3	102.0	100.8	99.6	94.1	126.3	78.2	78.9	78.9	78.9	78.8	85.8	100.3
1992	97.9	93.7	100.1	128.2	99.3	99.3	98.1	90.5	123.9	77.1	77.7	77.8	77.8	77.6	83.1	98.0
1991	95.7	89.5	96.8	124.4	96.0	96.7	95.5	88.7	121.5	76.0	76.6	76.8	76.7	76.6	82.6	96.8
1990	93.2	87.0	94.1	118.1	94.6	93.9	91.0	85.6	111.4	74.8	75.4	75.5	75.5	75.3	81.3	94.6
1989	91.0	85.4	91.8	114.8	92.6	89.6	88.4	84.1	102.1	73.3	74.0	74.0	74.0	73.8	79.7	92.9
1988	88.5	83.6	89.4	106.5	87.9	87.7	86.5	82.4	99.9	71.5	72.2	72.2	72.2	72.0	78.2	91.1
1987	85.7	82.4	85.8	102.6	85.8	85.3	85.1	82.1	98.9	70.6	71.2	71.2	71.1	71.0	77.4	90.1
1986	83.7	80.2	84.4	99.8	83.7	82.4	83.3	79.2	96.6	68.4	69.1	69.1	69.1	68.9	75.4	87.7
1985	81.8	77.5	83.2	94.9	81.6	80.1	81.0	77.0	92.8	66.9	67.6	67.7	67.6	67.5	73.4	86.8
1984	80.6	75.4	81.3	91.1	80.6	78.5	79.0	76.1	89.5	66.3	66.6	66.8	66.2	66.0	72.2	82.8
1983	78.2	73.2	78.2	86.5	77.1	75.3	76.5	74.5	85.4	64.6	65.1	65.7	64.9	64.6	70.5	79.4
1982	72.1	67.5	70.3	78.3	71.3	69.8	70.2	68.4	77.8	58.7	60.0	60.5	59.5	59.4	66.1	73.2
1981	66.1	62.6	65.2	71.6	65.6	64.7	64.6	63.5	71.0	55.3	56.6	56.7	55.9	56.3	61.7	67.7
1980	60.7	58.0	60.6	66.0	60.5	60.3	61.6	58.5	65.9	51.1	52.2	52.5	51.7	50.8	57.4	62.3
1979	54.9	52.8	56.0	60.2	55.0	54.9	56.2	53.2	60.0	45.9	46.9	47.1	46.4	45.9	52.2	56.3
1978	51.3	50.0	52.6	56.4	51.7	50.8	52.2	49.1	54.7	41.9	43.0	42.9	42.4	41.9	47.8	52.1
1977	47.9	46.5	49.0	52.9	48.3	48.5	49.1	46.0	51.8	40.2	41.5	40.7	40.0	39.6	44.2	48.5
1976	45.3	44.1	46.2	50.6	45.6	45.3	46.6	43.8	48.5	37.5	37.9	38.1	37.6	37.0	41.3	45.8
1975	43.7	42.5	45.1	49.5	44.8	43.7	44.8	42.7	47.1	36.1	37.0	37.0	37.3	36.1	39.3	44.7
1974	39.4	38.2	41.8	45.4	41.5	39.3	40.3	38.1	42.5	30.3	33.5	33.5	33.1	32.6	36.5	40.1
1970	27.8	27.0	28.9	32.8	29.2	27.8	28.5	26.9	30.0	20.9	23.7	23.6	23.4	23.0	25.8	28.3
1965	21.5	20.8	22.2	25.5	22.8	21.4	21.9	20.7	23.1	16.0	18.2	18.2	18.0	17.7	19.9	21.8
1960	19.5	18.9	19.9	21.5	19.7	19.5	19.9	18.8	21.0	14.4	16.6	16.6	16.4	16.1	18.1	19.9
1955	16.3	15.8	16.7	18.1	16.5	16.3	16.7	15.8	17.6	12.1	13.9	13.9	13.7	13.5	15.2	16.6
1950	13.5	13.1	13.8	14.9	13.6	13.5	13.8	13.0	14.5	10.0	11.5	11.5	11.4	11.2	12.5	13.7
1945	8.6	8.3	8.8	9.5	8.7	8.6	8.8	8.3	9.3	6.4	7.3	7.3	7.2	7.1	8.0	8.8
1940	6.6	6.4	6.8	7.4	6.7	6.7	6.8	6.4	7.2	4.9	5.6	5.6	5.6	5.5	6.2	6.8

Year	National 30 City Average	Ohio Canton	Cincin-nati	Cleve-land	Colum-bus	Dayton	Lorain	Spring-field	Toledo	Youngs-town	Oklahoma Lawton	Oklaho-ma City	Tulsa	Oregon Eugene	Port-land	PA Allen-town
Jan 1998	114.4E	108.6E	105.5E	114.8E	107.3E	104.5E	109.4E	104.3E	110.5E	109.1E	94.9E	95.3E	95.0E	119.9E	121.2E	115.7E
1997	111.5	106.4	102.9	112.9	104.7	102.6	107.4	102.2	108.5	107.2	92.9	93.2	93.0	118.1	119.6	113.9
1996	108.9	103.2	100.2	110.1	101.1	98.8	103.8	98.4	105.1	104.1	90.9	91.3	91.2	114.4	116.1	112.0
1995	105.6	98.8	97.1	106.4	99.1	94.6	97.1	92.0	100.6	100.1	85.3	88.0	89.0	112.2	114.3	108.3
1994	103.0	97.7	95.0	104.8	95.4	93.0	96.0	90.1	99.3	98.9	84.0	86.5	87.6	107.6	109.2	106.4
1993	100.0	95.9	92.3	101.9	93.9	90.6	93.9	87.7	98.3	97.2	81.3	83.9	84.8	107.2	108.8	103.6
1992	97.9	94.5	90.6	98.7	92.6	89.2	92.2	86.4	97.1	96.0	80.1	82.7	83.4	101.1	102.6	101.6
1991	95.7	93.4	88.6	97.2	90.6	87.7	91.1	84.9	94.2	92.7	78.3	80.7	81.4	99.5	101.1	98.9
1990	93.2	92.1	86.7	95.1	88.1	85.9	89.4	83.4	92.9	91.8	77.1	79.9	80.0	98.0	99.4	95.0
1989	91.0	90.3	84.6	93.6	85.8	82.6	87.9	81.0	91.4	88.4	75.8	78.5	78.4	95.2	97.0	92.4
1988	88.5	89.1	82.9	92.1	83.9	81.0	85.2	79.5	89.6	86.9	74.3	77.0	76.8	93.5	95.3	89.8
1987	85.7	87.7	80.9	89.6	81.5	79.5	86.9	78.0	86.1	85.4	72.6	74.4	74.7	91.4	91.9	86.9
1986	83.7	84.9	78.8	87.3	79.7	78.0	81.6	76.1	84.6	83.8	71.6	74.1	73.9	90.0	91.7	84.7
1985	81.8	83.7	78.2	86.2	77.7	76.3	80.7	74.3	84.7	82.7	71.2	73.9	73.8	88.7	90.1	81.9
1984	80.6	81.3	77.4	82.5	76.1	75.9	79.5	73.9	83.5	80.3	70.6	72.9	72.8	89.5	89.3	78.4
1983	78.2	76.9	74.9	77.9	74.5	71.6	77.2	71.9	80.4	77.3	69.1	71.5	72.2	89.0	89.2	75.0
1982	72.1	70.8	70.1	71.6	68.6	65.2	70.8	66.5	74.6	71.7	63.5	65.2	66.2	82.3	83.6	68.2
1981	66.1	65.8	65.3	66.0	63.9	60.6	65.7	61.3	69.4	67.2	57.0	60.3	61.5	74.2	74.5	62.7
1980	60.7	60.6	59.8	61.0	58.6	56.6	60.8	56.2	64.0	61.5	52.2	55.5	57.2	68.9	68.4	58.7
1979	54.9	55.2	54.3	54.8	52.7	51.2	55.2	51.0	57.9	56.0	47.7	49.8	51.4	61.9	60.6	54.1
1978	51.3	51.1	50.4	51.3	49.3	47.2	50.3	46.9	54.0	52.2	44.5	46.3	48.0	57.2	56.0	50.4
1977	47.9	47.7	46.9	48.2	45.3	44.3	46.7	42.9	50.2	48.7	40.9	43.0	44.8	51.5	51.7	46.7
1976	45.3	45.4	44.8	45.9	43.7	42.3	44.3	40.7	47.4	45.8	39.0	40.6	41.7	48.7	48.2	45.1
1975	43.7	44.0	43.7	44.6	42.2	40.5	42.7	40.0	44.9	44.5	38.6	38.6	39.2	43.0	40.8	42.3
1974	39.4	39.4	39.9	41.0	38.4	37.6	38.9	35.9	41.9	40.3	34.1	33.3	34.1	40.8	38.6	38.6
1970	27.8	27.8	28.2	29.6	26.7	27.0	27.5	25.4	29.1	29.6	24.1	23.4	24.1	30.4	30.0	27.3
1965	21.5	21.5	20.7	21.1	20.2	20.0	21.1	19.6	21.3	21.0	18.5	17.8	20.6	23.4	22.6	21.0
1960	19.5	19.5	19.3	19.6	18.7	18.1	19.2	17.8	19.4	19.1	16.9	16.1	18.1	21.2	21.1	19.1
1955	16.3	16.3	16.2	16.5	15.7	15.2	16.1	14.9	16.2	16.0	14.1	13.5	15.1	17.8	17.7	16.0
1950	13.5	13.5	13.3	13.6	13.0	12.5	13.3	12.3	13.4	13.2	11.7	11.2	12.5	14.7	14.6	13.2
1945	8.6	8.6	8.5	8.7	8.3	8.0	8.5	7.8	8.6	8.4	7.4	7.1	8.0	9.4	9.3	8.4
1940	6.6	6.7	6.5	6.7	6.4	6.2	6.5	6.1	6.6	6.5	5.7	5.5	6.1	7.2	7.2	6.5

Historical Cost Indexes

Year	National 30 City Average	Pennsylvania						RI	South Carolina		South Dakota		Tennessee			
		Erie	Harrisburg	Philadelphia	Pittsburgh	Reading	Scranton	Providence	Charleston	Columbia	Rapid City	Sioux Falls	Chattanooga	Knoxville	Memphis	Nashville
Jan 1998	114.4E	110.5E	111.2E	126.0E	117.6E	112.7E	114.4E	120.9E	88.5E	87.7E	91.4E	94.3E	95.7E	92.7E	97.9E	97.3E
1997	111.5	108.5	108.8	123.3	113.9	110.8	112.3	118.9	86.5	85.6	89.3	92.1	93.6	90.8	96.1	94.9
1996	108.9	106.8	105.7	120.3	110.8	108.7	109.9	117.2	84.9	84.1	87.2	90.0	91.6	88.8	94.0	91.9
1995	105.6	99.8	100.6	117.1	106.3	103.6	103.8	111.1	82.7	82.2	84.0	84.7	89.2	86.0	91.2	87.6
1994	103.0	98.2	99.4	115.2	103.7	102.2	102.6	109.6	81.2	80.6	82.7	83.1	87.4	84.1	89.0	84.8
1993	100.0	94.9	96.6	107.4	99.0	98.6	99.8	108.2	78.4	77.9	81.2	81.9	85.1	82.3	86.8	81.9
1992	97.9	92.1	94.8	105.3	96.5	96.7	97.8	106.9	77.1	76.8	80.0	80.7	83.6	77.7	85.4	80.7
1991	95.7	90.5	92.5	101.7	93.2	94.1	94.8	96.1	75.0	75.8	78.7	79.7	81.9	76.6	83.0	79.1
1990	93.2	88.5	89.5	98.5	91.2	90.8	91.3	94.1	73.7	74.5	77.2	78.4	79.9	75.1	81.3	77.1
1989	91.0	86.6	86.5	94.2	89.4	87.9	88.8	92.4	72.1	72.8	75.7	77.0	78.5	73.7	81.2	74.8
1988	88.5	84.9	84.0	89.8	87.7	85.3	86.4	89.1	70.5	71.3	74.2	75.5	77.0	72.1	79.7	73.1
1987	85.7	82.6	81.8	86.8	85.6	82.9	83.3	86.3	70.2	70.2	73.6	74.9	74.4	70.2	77.5	71.0
1986	83.7	81.5	79.6	84.9	83.6	79.7	81.3	85.2	67.4	68.1	71.0	72.3	74.2	69.2	75.3	68.1
1985	81.8	79.6	77.2	82.2	81.5	77.7	80.0	83.0	65.9	66.9	69.7	71.2	72.5	67.7	74.3	66.7
1984	80.6	78.3	75.4	79.0	78.9	75.9	78.8	80.3	64.3	65.5	68.7	70.4	71.4	66.4	73.3	66.3
1983	78.2	76.3	72.2	74.4	75.5	74.0	75.0	76.3	65.3	65.2	67.7	69.7	69.3	64.1	72.8	65.1
1982	72.1	71.0	66.3	69.0	70.9	67.7	68.4	70.0	59.9	60.0	64.4	67.6	64.5	59.8	66.5	62.2
1981	66.1	64.6	61.7	63.4	66.3	62.5	62.3	64.3	56.3	55.9	59.7	63.6	59.9	55.9	60.3	57.3
1980	60.7	59.4	56.9	58.7	61.3	57.7	58.4	59.2	50.6	51.1	54.8	57.1	55.0	51.6	55.9	53.1
1979	54.9	54.0	52.6	54.2	55.6	53.8	53.5	53.3	45.7	44.4	49.7	51.5	49.8	46.4	51.0	47.0
1978	51.3	50.2	49.6	51.3	51.4	49.7	50.0	50.2	42.1	40.7	46.0	47.4	45.9	42.8	47.4	43.9
1977	47.9	46.6	45.2	48.1	48.3	46.5	46.9	47.0	39.2	39.4	42.5	44.3	42.5	40.2	45.0	40.7
1976	45.3	44.7	43.7	45.8	45.6	44.2	44.4	44.2	36.6	37.9	40.1	42.0	39.7	38.3	42.0	38.3
1975	43.7	43.4	42.2	44.5	44.5	43.6	41.9	42.7	35.0	36.4	37.7	39.6	39.1	37.3	40.7	37.0
1974	39.4	39.2	36.7	39.6	39.9	38.4	38.2	39.7	32.3	32.4	35.1	36.5	35.2	31.5	33.9	32.1
1970	27.8	27.5	25.8	27.5	28.7	27.1	27.0	27.3	22.8	22.9	24.8	25.8	24.9	22.0	23.0	22.8
1965	21.5	21.1	20.1	21.7	22.4	20.9	20.8	21.9	17.6	17.6	19.1	19.9	19.2	17.1	18.3	17.4
1960	19.5	19.1	18.6	19.4	19.7	19.0	18.9	19.1	16.0	16.0	17.3	18.1	17.4	15.5	16.6	15.8
1955	16.3	16.1	15.6	16.3	16.5	15.9	15.9	16.0	13.4	13.4	14.5	15.1	14.6	13.0	13.9	13.3
1950	13.5	13.3	12.9	13.5	13.6	13.2	13.1	13.2	11.1	11.1	12.0	12.5	12.1	10.7	11.5	10.9
1945	8.6	8.5	8.2	8.6	8.7	8.4	8.3	8.4	7.1	7.1	7.7	8.0	7.7	6.8	7.3	7.0
1940	6.6	6.5	6.3	6.6	6.7	6.4	6.5	6.5	5.4	5.5	5.9	6.2	6.0	5.3	5.6	5.4

Year	National 30 City Average	Texas														Utah
		Abilene	Amarillo	Austin	Beaumont	Corpus Christi	Dallas	El Paso	Fort Worth	Houston	Lubbock	Odessa	San Antonio	Waco	Wichita Falls	Ogden
Jan 1998	114.4E	90.4E	93.4E	93.8E	98.6E	92.4E	98.5E	88.9E	95.1E	102.0E	93.9E	90.7E	95.4E	93.2E	93.6E	98.1E
1997	111.5	88.4	91.3	92.8	96.8	90.3	96.1	87.0	93.3	100.1	91.9	88.8	93.4	91.4	91.5	96.1
1996	108.9	86.8	89.6	90.9	95.3	88.5	94.1	86.7	91.5	97.9	90.2	87.1	92.3	89.7	89.9	94.1
1995	105.6	85.2	87.4	89.3	93.7	87.4	91.4	85.2	89.5	95.9	88.4	85.6	88.9	86.4	86.8	92.2
1994	103.0	83.3	85.5	87.0	91.8	84.6	89.6	82.2	87.4	93.4	87.0	83.9	87.0	84.9	85.4	89.4
1993	100.0	81.4	83.4	84.9	90.0	82.8	87.8	80.2	85.5	91.1	84.8	81.9	85.0	83.0	83.5	87.1
1992	97.9	80.3	82.3	83.8	88.9	81.6	86.2	79.0	84.1	89.8	83.6	80.7	83.9	81.8	82.4	85.1
1991	95.7	79.2	81.1	82.8	87.8	80.6	85.9	78.0	83.3	87.9	82.7	79.8	83.3	80.6	81.5	84.1
1990	93.2	78.0	80.1	81.3	86.5	79.3	84.5	76.7	82.1	85.4	81.5	78.6	80.7	79.6	80.3	83.4
1989	91.0	76.6	78.7	79.7	85.1	77.8	82.5	75.2	80.6	84.0	80.0	77.0	78.7	78.1	78.7	81.9
1988	88.5	75.0	77.2	78.3	83.9	76.3	81.1	73.6	79.8	82.5	78.4	75.5	77.1	76.5	77.4	80.4
1987	85.7	75.9	76.4	76.0	81.3	75.6	79.9	72.2	77.7	81.3	77.3	74.6	75.7	75.2	77.4	79.1
1986	83.7	72.5	74.1	75.3	80.5	73.5	78.9	70.7	76.7	80.3	75.4	72.4	74.3	72.8	74.5	77.3
1985	81.8	71.1	72.5	74.5	79.3	72.3	77.6	69.4	75.1	79.6	74.0	71.2	73.9	71.7	73.3	75.2
1984	80.6	70.5	71.6	72.8	78.7	71.6	79.8	68.3	77.1	80.7	72.7	69.8	73.1	71.3	72.4	74.9
1983	78.2	68.3	70.6	71.0	75.0	70.2	77.8	66.3	74.9	79.8	71.0	69.0	71.8	69.4	70.1	76.1
1982	72.1	62.7	65.5	64.9	65.7	64.7	70.7	62.6	68.2	72.1	65.6	62.5	66.0	64.3	64.8	69.7
1981	66.1	57.9	60.1	59.3	62.4	59.6	63.1	58.6	61.0	65.1	61.0	58.3	60.5	59.2	60.0	65.3
1980	60.7	53.4	55.2	54.5	57.6	54.5	57.9	53.1	57.0	59.4	55.6	57.2	55.0	54.9	55.4	62.2
1979	54.9	48.6	49.8	49.0	52.3	48.7	51.5	48.5	51.4	53.3	50.3	51.1	49.2	49.7	49.4	53.3
1978	51.3	44.7	45.8	46.2	48.0	44.9	48.2	45.3	47.5	50.2	46.5	43.9	46.0	46.4	47.1	49.6
1977	47.9	42.9	42.5	43.1	43.9	42.2	44.9	41.7	44.7	47.6	42.9	41.5	42.5	42.1	41.1	45.4
1976	45.3	40.4	39.7	40.3	40.7	39.6	41.2	39.0	41.5	43.8	40.3	38.6	40.8	39.7	37.7	43.2
1975	43.7	37.6	39.0	39.0	39.6	38.1	40.7	38.0	40.4	41.2	38.9	37.9	39.0	38.6	38.0	40.0
1974	39.4	34.7	35.2	35.3	36.5	34.7	35.8	32.3	36.6	36.1	35.6	34.9	33.8	35.2	34.7	37.9
1970	27.8	24.5	24.9	24.9	25.7	24.5	25.5	23.7	25.9	25.4	25.1	24.6	23.3	24.8	24.5	26.8
1965	21.5	18.9	19.2	19.2	19.9	18.9	19.9	19.0	19.9	20.0	19.4	19.0	18.5	19.2	18.9	20.6
1960	19.5	17.1	17.4	17.4	18.1	17.1	18.2	17.0	18.1	18.2	17.6	17.3	16.8	17.4	17.2	18.8
1955	16.3	14.4	14.6	14.6	15.1	14.4	15.3	14.3	15.2	15.2	14.8	14.5	14.1	14.6	14.4	15.7
1950	13.5	11.9	12.1	12.1	12.5	11.9	12.6	11.8	12.5	12.6	12.2	12.0	11.6	12.1	11.9	13.0
1945	8.6	7.6	7.7	7.7	8.0	7.6	8.0	7.5	8.0	8.0	7.8	7.6	7.4	7.7	7.6	8.3
1940	6.6	5.9	5.9	5.9	6.1	5.8	6.2	5.8	6.2	6.2	6.0	5.9	5.7	5.9	5.8	6.4

Year	National 30 City Average	Utah Salt Lake City	Vermont Burlington	Vermont Rutland	Virginia Alexandria	Virginia Newport News	Virginia Norfolk	Virginia Richmond	Virginia Roanoke	Washington Seattle	Washington Spokane	Washington Tacoma	West Virginia Charleston	West Virginia Huntington	Wisconsin Green Bay	Wisconsin Kenosha
Jan 1998	114.4E	99.2E	98.7E	98.0E	103.6E	93.5E	93.7E	95.0E	88.8E	120.0E	113.3E	119.0E	107.2E	109.6E	109.1E	112.1E
1997	111.5	97.2	96.6	96.3	101.2	91.6	91.7	92.9	86.9	118.1	111.7	117.2	105.3	107.7	105.6	109.1
1996	108.9	94.9	95.1	94.8	99.7	90.2	90.4	91.6	85.5	115.2	109.2	114.3	103.1	104.8	103.8	106.4
1995	105.6	93.1	91.1	90.8	96.3	86.0	86.4	87.8	82.8	113.7	107.4	112.8	95.8	97.2	97.6	97.9
1994	103.0	90.2	89.5	89.3	93.9	84.6	84.8	86.3	81.4	109.9	104.0	108.3	94.3	95.3	96.3	96.2
1993	100.0	87.9	87.6	87.6	91.6	82.9	83.0	84.3	79.5	107.3	103.9	106.7	92.6	93.5	94.0	94.3
1992	97.9	86.0	86.1	86.1	90.1	81.0	81.6	82.0	78.3	105.1	101.4	103.7	91.4	92.3	92.0	92.1
1991	95.7	84.9	84.2	84.2	88.2	77.6	77.9	79.8	77.3	102.2	100.0	102.2	89.7	88.6	88.6	89.8
1990	93.2	84.3	83.0	82.9	86.1	76.3	76.7	77.6	76.1	100.1	98.5	100.5	86.1	86.8	86.7	87.8
1989	91.0	82.8	81.4	81.3	83.4	74.9	75.2	76.0	74.2	96.1	96.8	98.4	84.5	84.8	84.3	85.6
1988	88.5	81.3	79.7	79.7	81.0	73.4	73.7	74.1	72.3	94.2	95.0	96.6	82.8	83.1	82.0	83.0
1987	85.7	79.8	79.0	79.0	77.9	71.0	71.7	72.7	70.2	91.9	92.4	92.9	81.1	81.3	80.2	81.7
1986	83.7	78.1	76.4	76.4	76.8	70.3	70.5	71.0	67.8	90.5	91.9	92.8	79.9	80.2	77.8	79.3
1985	81.8	75.9	74.8	74.9	75.1	68.7	68.8	69.5	67.2	88.3	89.0	91.2	77.7	77.7	76.7	77.4
1984	80.6	75.2	73.7	73.7	75.0	67.9	68.3	68.8	66.2	89.2	88.1	90.0	75.4	75.8	75.0	75.1
1983	78.2	74.5	70.9	71.5	72.9	65.9	66.5	68.9	66.0	87.6	86.9	88.8	73.8	74.4	73.4	73.2
1982	72.1	68.1	64.9	65.4	67.0	60.6	60.9	63.7	61.0	80.7	80.6	81.7	67.6	68.4	67.5	69.1
1981	66.1	61.9	59.3	59.7	61.6	56.0	56.8	58.2	55.6	75.7	72.9	73.8	62.6	63.3	63.6	63.5
1980	60.7	57.0	55.3	58.3	57.3	52.5	52.4	54.3	51.3	67.9	66.3	66.7	57.7	58.3	58.6	58.3
1979	54.9	52.5	49.8	51.2	51.7	47.6	47.6	49.1	46.8	61.1	60.1	60.0	51.8	52.5	52.7	53.2
1978	51.3	49.0	46.1	47.2	48.9	44.0	44.9	46.0	43.6	56.7	55.4	55.6	48.2	48.5	48.7	49.5
1977	47.9	45.0	44.5	46.3	44.6	41.2	41.5	42.5	41.2	51.8	51.1	50.7	44.7	43.2	44.8	45.5
1976	45.3	42.8	42.3	45.0	42.5	38.4	38.7	39.9	38.6	48.5	47.5	47.1	42.3	40.3	42.6	42.9
1975	43.7	40.1	41.8	43.9	41.7	37.2	36.9	37.1	37.1	44.9	44.4	44.5	41.0	40.0	40.9	40.5
1974	39.4	36.1	35.7	38.0	37.2	33.8	31.4	31.1	33.6	39.8	40.1	42.0	37.0	36.5	37.3	37.5
1970	27.8	26.1	25.4	26.8	26.2	23.9	21.5	22.0	23.7	28.8	29.3	29.6	26.1	25.8	26.4	26.5
1965	21.5	20.0	19.8	20.6	20.2	18.4	17.1	17.2	18.3	22.4	22.5	22.8	20.1	19.9	20.3	20.4
1960	19.5	18.4	18.0	18.8	18.4	16.7	15.4	15.6	16.6	20.4	20.8	20.8	18.3	18.1	18.4	18.6
1955	16.3	15.4	15.1	15.7	15.4	14.0	12.9	13.1	13.9	17.1	17.4	17.4	15.4	15.2	15.5	15.6
1950	13.5	12.7	12.4	13.0	12.7	11.6	10.7	10.8	11.5	14.1	14.4	14.4	12.7	12.5	12.8	12.9
1945	8.6	8.1	7.9	8.3	8.1	7.4	6.8	6.9	7.3	9.0	9.2	9.2	8.1	8.0	8.1	8.2
1940	6.6	6.3	6.1	6.4	6.2	5.7	5.3	5.3	5.7	7.0	7.1	7.1	6.2	6.2	6.3	6.3

Year	National 30 City Average	Wisconsin Madison	Wisconsin Milwaukee	Wisconsin Racine	Wyoming Cheyenne	Canada Calgary	Canada Edmonton	Canada Hamilton	Canada London	Canada Montreal	Canada Ottawa	Canada Quebec	Canada Toronto	Canada Vancouver	Canada Winnipeg
Jan 1998	114.4E	108.4E	113.5E	112.2E	95.2E	112.8E	112.7E	126.5E	124.1E	116.7E	124.9E	117.3E	128.0E	124.3E	113.4E
1997	111.5	106.1	110.1	109.3	93.2	110.7	110.6	124.4	121.9	114.6	122.8	115.4	125.7	121.9	111.4
1996	108.9	104.4	107.1	106.5	91.1	109.1	109.0	122.6	120.2	112.9	121.0	113.6	123.9	119.0	109.7
1995	105.6	96.5	103.9	97.8	87.6	107.4	107.4	119.9	117.5	110.8	118.2	111.5	121.6	116.2	107.6
1994	103.0	94.5	100.6	96.1	85.4	106.7	106.6	116.5	114.2	109.5	115.0	110.2	117.8	115.1	105.5
1993	100.0	91.3	96.7	93.8	82.9	104.7	104.5	113.7	111.9	106.9	112.0	107.0	114.9	109.2	102.8
1992	97.9	89.2	93.9	91.7	81.7	103.4	103.2	112.4	110.7	104.2	110.8	103.6	113.7	108.0	101.6
1991	95.7	86.2	91.6	89.3	80.3	102.1	102.0	108.2	106.7	101.8	106.9	100.4	109.0	106.8	98.6
1990	93.2	84.3	88.9	87.3	79.1	98.0	97.1	103.8	101.4	99.0	102.7	96.8	104.6	103.2	95.1
1989	91.0	81.8	86.4	85.0	77.8	95.6	94.7	98.2	96.5	94.9	98.5	92.5	98.5	97.6	92.9
1988	88.5	79.9	84.1	82.4	76.3	93.9	93.0	95.4	93.2	90.6	94.0	88.3	94.8	95.9	90.0
1987	85.7	78.1	81.1	81.1	76.7	90.2	89.3	89.7	90.1	87.2	89.6	85.0	90.4	94.2	87.1
1986	83.7	76.1	78.9	78.6	73.4	91.2	90.1	88.9	87.9	84.8	87.7	82.4	89.3	93.2	85.3
1985	81.8	74.3	77.4	77.0	72.3	90.2	89.1	85.8	84.9	82.4	83.9	79.9	86.5	89.8	83.4
1984	80.6	72.5	76.3	74.7	73.8	89.6	88.1	85.1	84.1	81.6	82.9	80.0	85.2	89.2	82.5
1983	78.2	71.9	74.0	73.0	73.8	84.6	82.8	80.1	79.1	76.4	77.6	75.8	79.9	83.3	77.9
1982	72.1	65.9	70.5	68.6	68.3	76.2	74.1	73.7	71.8	70.2	72.3	69.6	71.8	76.4	70.7
1981	66.1	61.4	64.9	63.3	62.7	70.6	68.1	68.8	67.1	64.6	65.8	64.3	67.5	70.3	65.4
1980	60.7	56.8	58.8	58.1	56.9	64.9	63.3	63.7	61.9	59.2	60.9	59.3	60.9	65.0	61.7
1979	54.9	51.4	53.0	52.8	51.2	58.9	58.5	58.1	56.6	54.4	55.5	53.9	56.0	59.2	56.1
1978	51.3	47.3	49.9	49.2	47.5	54.9	54.4	54.0	52.6	50.4	51.7	49.7	51.9	55.1	51.7
1977	47.9	44.0	46.9	45.7	44.8	49.9	49.9	48.8	48.2	43.7	48.0	43.0	48.2	50.2	47.8
1976	45.3	41.9	44.2	43.5	42.2	45.6	45.4	44.5	43.3	41.7	43.8	40.6	44.1	45.9	42.1
1975	43.7	40.7	43.3	40.7	40.6	42.2	41.6	42.9	41.6	39.7	41.5	39.0	42.2	42.4	39.2
1974	39.4	37.6	40.9	37.6	36.7	40.9	40.5	40.3	39.4	34.7	39.0	36.8	36.4	37.1	33.9
1970	27.8	26.5	29.4	26.5	26.0	28.9	28.6	28.5	27.8	25.6	27.6	26.0	25.6	26.0	23.1
1965	21.5	20.6	21.8	20.4	20.0	22.3	22.0	22.0	21.4	18.7	21.2	20.1	19.4	20.5	17.5
1960	19.5	18.1	19.0	18.6	18.2	20.2	20.0	20.0	19.5	17.0	19.3	18.2	17.6	18.6	15.8
1955	16.3	15.2	15.9	15.6	15.2	17.0	16.8	16.7	16.3	14.3	16.2	15.3	14.8	15.5	13.3
1950	13.5	12.5	13.2	12.9	12.6	14.0	13.9	13.8	13.5	11.8	13.4	12.6	12.2	12.8	10.9
1945	8.6	8.0	8.4	8.2	8.0	9.0	8.8	8.8	8.6	7.5	8.5	8.0	7.8	8.2	7.0
1940	6.6	6.2	6.5	6.3	6.2	6.9	6.8	6.8	6.6	5.8	6.6	6.2	6.0	6.3	5.4

Costs shown in *Means cost data publications* are based on National Averages for materials and installation. To adjust these costs to a specific location, simply multiply the base cost by the factor for that city. The data is arranged alphabetically by state and postal zip code numbers. For a city not listed, use the factor for a nearby city with similar economic characteristics.

STATE/ZIP	CITY	Residential	Commercial
ALABAMA			
350-352	Birmingham	.84	.85
354	Tuscaloosa	.81	.79
355	Jasper	.76	.77
356	Decatur	.82	.83
357-358	Huntsville	.82	.83
359	Gadsden	.81	.82
360-361	Montgomery	.82	.80
362	Anniston	.74	.75
363	Dothan	.81	.79
364	Evergreen	.82	.80
365-366	Mobile	.83	.84
367	Selma	.81	.79
368	Phenix City	.81	.79
369	Butler	.81	.79
ALASKA			
995-996	Anchorage	1.27	1.26
997	Fairbanks	1.27	1.26
998	Juneau	1.26	1.25
999	Ketchikan	1.31	1.30
ARIZONA			
850,853	Phoenix	.93	.90
852	Mesa/Tempe	.88	.86
855	Globe	.91	.88
856-857	Tucson	.91	.88
859	Show Low	.92	.88
860	Flagstaff	.95	.91
863	Prescott	.92	.88
864	Kingman	.91	.88
865	Chambers	.91	.88
ARKANSAS			
716	Pine Bluff	.80	.80
717	Camden	.72	.72
718	Texarkana	.77	.76
719	Hot Springs	.71	.71
720-722	Little Rock	.81	.81
723	West Memphis	.81	.81
724	Jonesboro	.81	.81
725	Batesville	.78	.78
726	Harrison	.79	.79
727	Fayetteville	.71	.68
728	Russellville	.80	.77
729	Fort Smith	.83	.80
CALIFORNIA			
900-902	Los Angeles	1.11	1.11
903-905	Inglewood	1.09	1.09
906-908	Long Beach	1.10	1.10
910-912	Pasadena	1.08	1.08
913-916	Van Nuys	1.10	1.10
917-918	Alhambra	1.10	1.10
919-921	San Diego	1.11	1.07
922	Palm Springs	1.12	1.08
923-924	San Bernardino	1.11	1.07
925	Riverside	1.13	1.09
926-927	Santa Ana	1.11	1.08
928	Anaheim	1.13	1.11
930	Oxnard	1.15	1.10
931	Santa Barbara	1.12	1.09
932-933	Bakersfield	1.11	1.06
934	San Luis Obispo	1.22	1.10
935	Mojave	1.11	1.07
936-938	Fresno	1.13	1.09
939	Salinas	1.12	1.12
940-941	San Francisco	1.21	1.24
942,956-958	Sacramento	1.12	1.11
943	Palo Alto	1.14	1.17
944	San Mateo	1.15	1.18
945	Vallejo	1.14	1.17
946	Oakland	1.16	1.19
947	Berkeley	1.29	1.33
948	Richmond	1.15	1.17
949	San Rafael	1.25	1.19
950	Santa Cruz	1.17	1.15
951	San Jose	1.22	1.20
952	Stockton	1.14	1.10
953	Modesto	1.14	1.10

STATE/ZIP	CITY	Residential	Commercial
CALIFORNIA (CONT'D)			
954	Santa Rosa	1.15	1.19
955	Eureka	1.12	1.11
959	Marysville	1.11	1.10
960	Redding	1.10	1.09
961	Susanville	1.10	1.09
COLORADO			
800-802	Denver	.98	.94
803	Boulder	.90	.86
804	Golden	.95	.91
805	Fort Collins	.98	.92
806	Greeley	.92	.86
807	Fort Morgan	.96	.90
808-809	Colorado Springs	.93	.91
810	Pueblo	.93	.91
811	Alamosa	.91	.89
812	Salida	.91	.89
813	Durango	.89	.87
814	Montrose	.87	.85
815	Grand Junction	.91	.87
816	Glenwood Springs	.96	.91
CONNECTICUT			
060	New Britain	1.05	1.06
061	Hartford	1.05	1.06
062	Willimantic	1.05	1.06
063	New London	1.06	1.05
064	Meriden	1.04	1.05
065	New Haven	1.05	1.06
066	Bridgeport	1.02	1.05
067	Waterbury	1.06	1.06
068	Norwalk	1.01	1.05
069	Stamford	1.03	1.07
D.C.			
200-205	Washington	.94	.95
DELAWARE			
197	Newark	.99	1.00
198	Wilmington	.98	.99
199	Dover	.99	1.00
FLORIDA			
320,322	Jacksonville	.86	.85
321	Daytona Beach	.90	.89
323	Tallahassee	.78	.80
324	Panama City	.72	.74
325	Pensacola	.87	.85
326	Gainesville	.87	.84
327-328,347	Orlando	.89	.87
329	Melbourne	.90	.89
330-332,340	Miami	.86	.88
333	Fort Lauderdale	.86	.88
334,349	West Palm Beach	.87	.84
335-336,346	Tampa	.83	.85
337	St. Petersburg	.84	.86
338	Lakeland	.82	.84
339	Fort Myers	.83	.83
342	Sarasota	.82	.84
GEORGIA			
300-303,399	Atlanta	.84	.89
304	Statesboro	.67	.69
305	Gainesville	.71	.76
306	Athens	.77	.81
307	Dalton	.68	.68
308-309	Augusta	.79	.81
310-312	Macon	.82	.82
313-314	Savannah	.82	.83
315	Waycross	.76	.76
316	Valdosta	.78	.78
317	Albany	.79	.81
318-319	Columbus	.80	.80
HAWAII			
967	Hilo	1.27	1.23
968	Honolulu	1.27	1.23

STATE/ZIP	CITY	Residential	Commercial
STATES & POSS.			
969	Guam	.85	.82
IDAHO			
832	Pocatello	.95	.94
833	Twin Falls	.82	.81
834	Idaho Falls	.86	.85
835	Lewiston	1.11	1.02
836-837	Boise	.95	.94
838	Coeur d'Alene	1.00	.92
ILLINOIS			
600-603	North Suburban	1.08	1.07
604	Joliet	1.07	1.06
605	South Suburban	1.08	1.07
606	Chicago	1.12	1.11
609	Kankakee	1.00	1.00
610-611	Rockford	1.03	1.02
612	Rock Island	1.05	.96
613	La Salle	1.07	.99
614	Galesburg	1.04	.97
615-616	Peoria	1.07	1.00
617	Bloomington	1.03	.99
618-619	Champaign	1.03	1.00
620-622	East St. Louis	.99	.99
623	Quincy	.97	.94
624	Effingham	.99	.96
625	Decatur	1.00	.97
626-627	Springfield	1.01	.98
628	Centralia	.98	.98
629	Carbondale	.96	.96
INDIANA			
460	Anderson	.93	.92
461-462	Indianapolis	.97	.95
463-464	Gary	1.00	.98
465-466	South Bend	.92	.90
467-468	Fort Wayne	.90	.91
469	Kokomo	.91	.90
470	Lawrenceburg	.91	.88
471	New Albany	.92	.88
472	Columbus	.92	.89
473	Muncie	.91	.90
474	Bloomington	.93	.90
475	Washington	.93	.93
476-477	Evansville	.93	.93
478	Terre Haute	.95	.94
479	Lafayette	.91	.91
IOWA			
500-503,509	Des Moines	.96	.92
504	Mason City	.88	.82
505	Fort Dodge	.85	.80
506-507	Waterloo	.90	.84
508	Creston	.91	.86
510-511	Sioux City	.90	.84
512	Sibley	.81	.79
513	Spencer	.83	.81
514	Carroll	.87	.83
515	Council Bluffs	.94	.88
516	Shenandoah	.83	.77
520	Dubuque	.97	.87
521	Decorah	.92	.82
522-524	Cedar Rapids	.99	.91
525	Ottumwa	.95	.86
526	Burlington	.85	.80
527-528	Davenport	.97	.94
KANSAS			
660-662	Kansas City	.95	.93
664-666	Topeka	.87	.86
667	Fort Scott	.88	.86
668	Emporia	.83	.83
669	Belleville	.90	.84
670-672	Wichita	.89	.86
673	Independence	.84	.81
674	Salina	.87	.83
675	Hutchinson	.82	.78
676	Hays	.88	.84
677	Colby	.88	.84
678	Dodge City	.88	.85
679	Liberal	.81	.78
KENTUCKY			
400-402	Louisville	.93	.90
403-405	Lexington	.89	.86

STATE/ZIP	CITY	Residential	Commercial
KENTUCKY (CONT'D)			
406	Frankfort	.95	.89
407-409	Corbin	.80	.75
410	Covington	.96	.93
411-412	Ashland	.95	.96
413-414	Campton	.79	.75
415-416	Pikeville	.84	.84
417-418	Hazard	.79	.75
420	Paducah	.95	.90
421-422	Bowling Green	.94	.89
423	Owensboro	.92	.90
424	Henderson	.93	.91
425-426	Somerset	.77	.74
427	Elizabethtown	.91	.88
LOUISIANA			
700-701	New Orleans	.87	.86
703	Thibodaux	.86	.86
704	Hammond	.85	.84
705	Lafayette	.86	.83
706	Lake Charles	.85	.85
707-708	Baton Rouge	.85	.84
710-711	Shreveport	.81	.81
712	Monroe	.80	.80
713-714	Alexandria	.79	.79
MAINE			
039	Kittery	.80	.82
040-041	Portland	.89	.91
042	Lewiston	.90	.91
043	Augusta	.81	.81
044	Bangor	.93	.93
045	Bath	.80	.80
046	Machias	.85	.85
047	Houlton	.82	.82
048	Rockland	.85	.85
049	Waterville	.81	.80
MARYLAND			
206	Waldorf	.88	.88
207-208	College Park	.90	.90
209	Silver Spring	.89	.89
210-212	Baltimore	.92	.92
214	Annapolis	.90	.91
215	Cumberland	.87	.88
216	Easton	.70	.70
217	Hagerstown	.90	.89
218	Salisbury	.79	.79
219	Elkton	.85	.86
MASSACHUSETTS			
010-011	Springfield	1.06	1.04
012	Pittsfield	1.01	1.01
013	Greenfield	1.03	1.01
014	Fitchburg	1.11	1.06
015-016	Worcester	1.12	1.08
017	Framingham	1.09	1.10
018	Lowell	1.11	1.11
019	Lawrence	1.10	1.10
020-022	Boston	1.16	1.17
023-024	Brockton	1.08	1.10
025	Buzzards Bay	1.05	1.07
026	Hyannis	1.07	1.08
027	New Bedford	1.09	1.10
MICHIGAN			
480,483	Royal Oak	1.00	.99
481	Ann Arbor	1.02	1.01
482	Detroit	1.06	1.05
484-485	Flint	.98	.99
486	Saginaw	.95	.96
487	Bay City	.95	.96
488-489	Lansing	1.00	.97
490	Battle Creek	.99	.93
491	Kalamazoo	.99	.93
492	Jackson	.97	.94
493,495	Grand Rapids	.91	.88
494	Muskegan	.95	.92
496	Traverse City	.91	.88
497	Gaylord	.91	.92
498-499	Iron mountain	.97	.94
MINNESOTA			
550-551	Saint Paul	1.11	1.08
553-554	Minneapolis	1.14	1.10

STATE/ZIP	CITY	Residential	Commercial
MINNESOTA (CONT'D)			
556-558	Duluth	1.03	1.04
559	Rochester	1.06	1.02
560	Mankato	.99	.98
561	Windom	.89	.88
562	Willmar	.89	.88
563	St. Cloud	1.09	1.01
564	Brainerd	1.06	.99
565	Detroit Lakes	.89	.96
566	Bemidji	.91	.98
567	Thief River Falls	.88	.94
MISSISSIPPI			
386	Clarksdale	.71	.68
387	Greenville	.83	.79
388	Tupelo	.72	.73
389	Greenwood	.73	.70
390-392	Jackson	.83	.79
393	Meridian	.78	.77
394	Laurel	.74	.70
395	Biloxi	.85	.81
396	Mccomb	.70	.68
397	Columbus	.72	.73
MISSOURI			
630-631	St. Louis	.99	1.02
633	Bowling Green	.93	.96
634	Hannibal	1.02	.95
635	Kirksville	.86	.90
636	Flat River	.96	.99
637	Cape Girardeau	.95	.98
638	Sikeston	.91	.93
639	Poplar Bluff	.91	.92
640-641	Kansas City	.98	.95
644-645	St. Joseph	.88	.92
646	Chillicothe	.81	.84
647	Harrisonville	.95	.93
648	Joplin	.85	.87
650-651	Jefferson City	.98	.92
652	Columbia	.96	.89
653	Sedalia	.96	.90
654-655	Rolla	.89	.83
656-658	Springfield	.85	.87
MONTANA			
590-591	Billings	.98	.96
592	Wolf Point	.97	.95
593	Miles City	.98	.96
594	Great Falls	.97	.96
595	Havre	.95	.94
596	Helena	.97	.96
597	Butte	.95	.94
598	Missoula	.95	.94
599	Kalispell	.94	.93
NEBRASKA			
680-681	Omaha	.90	.89
683-685	Lincoln	.89	.84
686	Columbus	.76	.75
687	Norfolk	.86	.85
688	Grand Island	.88	.84
689	Hastings	.85	.81
690	Mccook	.76	.72
691	North Platte	.85	.81
692	Valentine	.80	.76
693	Alliance	.77	.73
NEVADA			
889-891	Las Vegas	1.04	1.03
893	Ely	.96	.97
894-895	Reno	.94	.98
897	Carson City	.95	.98
898	Elko	.93	.96
NEW HAMPSHIRE			
030	Nashua	.96	.97
031	Manchester	.96	.97
032-033	Concord	.95	.96
034	Keene	.82	.83
035	Littleton	.84	.85
036	Charleston	.80	.81
037	Claremont	.80	.81
038	Portsmouth	.95	.94

STATE/ZIP	CITY	Residential	Commercial
NEW JERSEY			
070-071	Newark	1.15	1.13
072	Elizabeth	1.11	1.09
073	Jersey City	1.12	1.12
074-075	Paterson	1.13	1.13
076	Hackensack	1.10	1.10
077	Long Branch	1.11	1.09
078	Dover	1.13	1.11
079	Summit	1.10	1.08
080,083	Vineland	1.11	1.07
081	Camden	1.12	1.09
082,084	Atlantic City	1.11	1.08
085-086	Trenton	1.14	1.12
087	Point Pleasant	1.12	1.10
088-089	New Brunswick	1.12	1.10
NEW MEXICO			
870-872	Albuquerque	.89	.91
873	Gallup	.90	.92
874	Farmington	.90	.92
875	Santa Fe	.89	.91
877	Las Vegas	.89	.91
878	Socorro	.89	.91
879	Truth/Consequences	.88	.88
880	Las Cruces	.85	.85
881	Clovis	.91	.91
882	Roswell	.91	.91
883	Carrizozo	.92	.92
884	Tucumcari	.91	.91
NEW YORK			
100-102	New York	1.34	1.34
103	Staten Island	1.29	1.29
104	Bronx	1.28	1.28
105	Mount Vernon	1.21	1.21
106	White Plains	1.20	1.20
107	Yonkers	1.24	1.24
108	New Rochelle	1.22	1.22
109	Suffern	1.15	1.15
110	Queens	1.28	1.28
111	Long Island City	1.29	1.29
112	Brooklyn	1.28	1.28
113	Flushing	1.30	1.30
114	Jamaica	1.28	1.28
115,117,118	Hicksville	1.26	1.26
116	Far Rockaway	1.29	1.29
119	Riverhead	1.27	1.27
120-122	Albany	.98	.98
123	Schenectady	.99	.99
124	Kingston	1.13	1.11
125-126	Poughkeepsie	1.15	1.13
127	Monticello	1.12	1.10
128	Glens Falls	.96	.94
129	Plattsburgh	.97	.95
130-132	Syracuse	1.01	.99
133-135	Utica	.93	.96
136	Watertown	.95	.98
137-139	Binghamton	.96	.96
140-142	Buffalo	1.08	1.04
143	Niagara Falls	1.08	1.04
144-146	Rochester	1.02	1.03
147	Jamestown	.96	.93
148-149	Elmira	.97	.95
NORTH CAROLINA			
270,272-274	Greensboro	.78	.79
271	Winston-Salem	.77	.78
275-276	Raleigh	.79	.79
277	Durham	.78	.79
278	Rocky Mount	.67	.67
279	Elizabeth City	.67	.67
280	Gastonia	.77	.77
281-282	Charlotte	.77	.78
283	Fayetteville	.78	.78
284	Wilmington	.75	.77
285	Kinston	.66	.66
286	Hickory	.64	.65
287-288	Asheville	.76	.78
289	Murphy	.66	.67
NORTH DAKOTA			
580-581	Fargo	.79	.84
582	Grand Forks	.79	.84
583	Devils Lake	.79	.84
584	Jamestown	.79	.84
585	Bismarck	.81	.85

STATE/ZIP	CITY	Residential	Commercial
NORTH DAKOTA (CONT'D)			
586	Dickinson	.80	.84
587	Minot	.81	.85
588	Williston	.80	.83
OHIO			
430-432	Columbus	.96	.94
433	Marion	.91	.92
434-436	Toledo	.98	.97
437-438	Zanesville	.91	.90
439	Steubenville	.96	.96
440	Lorain	1.02	.96
441	Cleveland	1.08	1.01
442-443	Akron	1.00	.99
444-445	Youngstown	.99	.96
446-447	Canton	.96	.95
448-449	Mansfield	.95	.93
450	Hamilton	.98	.93
451-452	Cincinnati	.98	.92
453-454	Dayton	.93	.92
455	Springfield	.94	.92
456	Chillicothe	1.00	.94
457	Athens	.89	.88
458	Lima	.93	.92
OKLAHOMA			
730-731	Oklahoma City	.81	.83
734	Ardmore	.83	.82
735	Lawton	.84	.83
736	Clinton	.80	.82
737	Enid	.83	.82
738	Woodward	.82	.81
739	Guymon	.70	.70
740-741	Tulsa	.86	.83
743	Miami	.85	.82
744	Muskogee	.77	.75
745	Mcalester	.76	.78
746	Ponca City	.83	.82
747	Durant	.79	.81
748	Shawnee	.79	.80
749	Poteau	.86	.82
OREGON			
970-972	Portland	1.09	1.07
973	Salem	1.07	1.06
974	Eugene	1.06	1.05
975	Medford	1.06	1.05
976	Klamath Falls	1.06	1.05
977	Bend	1.07	1.05
978	Pendleton	1.04	1.02
979	Vale	.99	.97
PENNSYLVANIA			
150-152	Pittsburgh	1.05	1.03
153	Washington	1.03	1.01
154	Uniontown	1.02	1.00
155	Bedford	1.04	.97
156	Greensburg	1.03	1.01
157	Indiana	1.05	.99
158	Dubois	1.04	.97
159	Johnstown	1.05	.98
160	Butler	1.01	.98
161	New Castle	1.01	.98
162	Kittanning	1.03	1.00
163	Oil City	.90	.95
164-165	Erie	.98	.97
166	Altoona	1.05	.97
167	Bradford	.99	.97
168	State College	.98	.98
169	Wellsboro	.95	.96
170-171	Harrisburg	.99	.98
172	Chambersburg	.97	.96
173-174	York	.99	.97
175-176	Lancaster	.98	.96
177	Williamsport	.94	.93
178	Sunbury	.97	.96
179	Pottsville	.97	.96
180	Lehigh Valley	1.03	1.03
181	Allentown	1.03	1.02
182	Hazleton	.98	.97
183	Stroudsburg	.98	.97
184-185	Scranton	.97	1.01
186-187	Wilkes-Barre	.94	.97
188	Montrose	.95	.98
189	Doylestown	.94	1.06

STATE/ZIP	CITY	Residential	Commercial
PENNSYLVANIA (CONT'D)			
190-191	Philadelphia	1.12	1.10
193	Westchester	1.06	1.05
194	Norristown	1.09	1.07
195-196	Reading	.98	.99
RHODE ISLAND			
028	Newport	1.04	1.06
029	Providence	1.04	1.06
SOUTH CAROLINA			
290-292	Columbia	.74	.77
293	Spartanburg	.73	.76
294	Charleston	.76	.78
295	Florence	.73	.75
296	Greenville	.73	.76
297	Rock Hill	.66	.69
298	Aiken	.67	.70
299	Beaufort	.70	.72
SOUTH DAKOTA			
570-571	Sioux Falls	.89	.83
572	Watertown	.87	.81
573	Mitchell	.86	.80
574	Aberdeen	.88	.82
575	Pierre	.87	.81
576	Mobridge	.88	.81
577	Rapid City	.86	.80
TENNESSEE			
370-372	Nashville	.85	.85
373-374	Chattanooga	.85	.84
375, 380-381	Memphis	.86	.86
376	Johnson City	.82	.81
377-379	Knoxville	.81	.81
382	Mckenzie	.71	.71
383	Jackson	.70	.77
384	Columbia	.78	.78
385	Cookeville	.70	.70
TEXAS			
750	Mckinney	.91	.84
751	Waxahackie	.84	.84
752-753	Dallas	.91	.86
754	Greenville	.81	.75
755	Texarkana	.91	.80
756	Longview	.86	.76
757	Tyler	.92	.81
758	Palestine	.76	.76
759	Lufkin	.79	.79
760-761	Fort Worth	.84	.83
762	Denton	.90	.81
763	Wichita Falls	.82	.82
764	Eastland	.77	.76
765	Temple	.79	.79
766-767	Waco	.83	.82
768	Brownwood	.75	.74
769	San Angelo	.81	.77
770-772	Houston	.89	.90
773	Huntsville	.76	.76
774	Wharton	.78	.79
775	Galveston	.87	.88
776-777	Beaumont	.85	.87
778	Bryan	.82	.83
779	Victoria	.82	.82
780	Laredo	.79	.80
781-782	San Antonio	.83	.84
783-784	Corpus Christi	.82	.81
785	Mc Allen	.81	.79
786-787	Austin	.80	.83
788	Del Rio	.70	.70
789	Giddings	.76	.75
790-791	Amarillo	.82	.82
792	Childress	.77	.80
793-794	Lubbock	.80	.82
795-796	Abilene	.79	.79
797	Midland	.81	.82
798-799,885	El Paso	.79	.78
UTAH			
840-841	Salt Lake City	.88	.87
842,844	Ogden	.88	.86
843	Logan	.89	.87
845	Price	.83	.82
846-847	Provo	.89	.88

STATE/ZIP	CITY	Residential	Commercial
VERMONT			
050	White River Jct.	.75	.75
051	Bellows Falls	.76	.75
052	Bennington	.72	.71
053	Brattleboro	.76	.76
054	Burlington	.85	.86
056	Montpelier	.84	.85
057	Rutland	.87	.86
058	St. Johnsbury	.77	.78
059	Guildhall	.76	.77
VIRGINIA			
220-221	Fairfax	.89	.90
222	Arlington	.89	.90
223	Alexandria	.90	.91
224-225	Fredericksburg	.85	.86
226	Winchester	.80	.81
227	Culpeper	.79	.80
228	Harrisonburg	.76	.76
229	Charlottesville	.84	.82
230-232	Richmond	.85	.83
233-235	Norfolk	.82	.82
236	Newport News	.83	.82
237	Portsmouth	.81	.81
238	Petersburg	.85	.83
239	Farmville	.77	.75
240-241	Roanoke	.79	.78
242	Bristol	.80	.76
243	Pulaski	.72	.71
244	Staunton	.74	.72
245	Lynchburg	.82	.78
246	Grundy	.71	.71
WASHINGTON			
980-981,987	Seattle	1.00	1.05
982	Everett	.98	1.04
983-984	Tacoma	1.07	1.05
985	Olympia	1.07	1.04
986	Vancouver	1.10	1.04
988	Wenatchee	.97	1.01
989	Yakima	1.04	1.02
990-992	Spokane	1.01	1.00
993	Richland	1.02	1.01
994	Clarkston	1.00	.99
WEST VIRGINIA			
247-248	Bluefield	.86	.86
249	Lewisburg	.90	.90
250-253	Charleston	.94	.94
254	Martinsburg	.78	.79
255-257	Huntington	.94	.96
258-259	Beckley	.92	.92
260	Wheeling	.93	.95
261	Parkersburg	.93	.94
262	Buckhannon	.96	.94
263-264	Clarksburg	.97	.94
265	Morgantown	.97	.94
266	Gassaway	.94	.94
267	Romney	.92	.92
268	Petersburg	.95	.92
WISCONSIN			
530,532	Milwaukee	1.01	1.00
531	Kenosha	1.00	.98
534	Racine	1.04	.99
535	Beloit	.99	.97
537	Madison	.97	.95
538	Lancaster	.92	.90
539	Portage	.96	.94
540	New Richmond	1.00	.93
541-543	Green Bay	.99	.96
544	Wausau	.95	.91
545	Rhinelander	.96	.92
546	La Crosse	.96	.93
547	Eau Claire	1.02	.94
548	Superior	1.00	.94
549	Oshkosh	.95	.92
WYOMING			
820	Cheyenne	.88	.84
821	Yellowstone Nat. Pk.	.84	.81
822	Wheatland	.85	.80
823	Rawlins	.85	.80
824	Worland	.81	.79
825	Riverton	.85	.81
826	Casper	.88	.84

STATE/ZIP	CITY	Residential	Commercial
WYOMING (CONT'D)			
827	Newcastle	.82	.78
828	Sheridan	.86	.84
829-831	Rock Springs	.86	.81
CANADIAN FACTORS (reflect Canadian currency)			
ALBERTA			
	Calgary	1.02	.99
	Edmonton	1.02	.99
BRITISH COLUMBIA			
	Vancouver	1.08	1.09
	Victoria	1.08	1.09
MANITOBA			
	Winnipeg	1.01	1.00
NEW BRUNSWICK			
	Moncton	.96	.94
	Saint John	.99	.97
NEWFOUNDLAND			
	St. John's	.97	.96
NOVA SCOTIA			
	Halifax	.99	.98
ONTARIO			
	Barrie	1.12	1.10
	Brantford	1.14	1.12
	Cornwall	1.12	1.10
	Hamilton	1.15	1.11
	Kingston	1.12	1.10
	Kitchener	1.07	1.05
	London	1.11	1.09
	North Bay	1.10	1.09
	Oshawa	1.12	1.10
	Ottawa	1.12	1.10
	Owen Sound	1.11	1.09
	Peterborough	1.12	1.10
	Sarnia	1.12	1.10
	St. Catharines	1.07	1.05
	Sudbury	1.07	1.05
	Thunder Bay	1.08	1.06
	Toronto	1.14	1.13
	Windsor	1.09	1.07
PRINCE EDWARD ISLAND			
	Charlottetown	.95	.93
QUEBEC			
	Chicoutimi	1.03	1.02
	Montreal	1.10	1.03
	Quebec	1.11	1.03
SASKATCHEWAN			
	Regina	.92	.92
	Saskatoon	.92	.92

A	Area Square Feet; Ampere
ABS	Acrylonitrile Butadiene Stryrene; Asbestos Bonded Steel
A.C.	Alternating Current; Air-Conditioning; Asbestos Cement; Plywood Grade A & C
A.C.I.	American Concrete Institute
AD	Plywood, Grade A & D
Addit.	Additional
Adj.	Adjustable
af	Audio-frequency
A.G.A.	American Gas Association
Agg.	Aggregate
A.H.	Ampere Hours
A hr.	Ampere-hour
A.H.U.	Air Handling Unit
A.I.A.	American Institute of Architects
AIC	Ampere Interrupting Capacity
Allow.	Allowance
alt.	Altitude
Alum.	Aluminum
a.m.	Ante Meridiem
Amp.	Ampere
Anod.	Anodized
Approx.	Approximate
Apt.	Apartment
Asb.	Asbestos
A.S.B.C.	American Standard Building Code
Asbe.	Asbestos Worker
A.S.H.R.A.E.	American Society of Heating, Refrig. & AC Engineers
A.S.M.E.	American Society of Mechanical Engineers
A.S.T.M.	American Society for Testing and Materials
Attchmt.	Attachment
Avg.	Average
A.W.G.	American Wire Gauge
Bbl.	Barrel
B. & B.	Grade B and Better; Balled & Burlapped
B. & S.	Bell and Spigot
B. & W.	Black and White
b.c.c.	Body-centered Cubic
B.C.Y.	Bank Cubic Yards
BE	Bevel End
B.F.	Board Feet
Bg. cem.	Bag of Cement
BHP	Boiler Horsepower; Brake Horsepower
B.I.	Black Iron
Bit.; Bitum.	Bituminous
Bk.	Backed
Bkrs.	Breakers
Bldg.	Building
Blk.	Block
Bm.	Beam
Boil.	Boilermaker
B.P.M.	Blows per Minute
BR	Bedroom
Brg.	Bearing
Brhe.	Bricklayer Helper
Bric.	Bricklayer
Brk.	Brick
Brng.	Bearing
Brs.	Brass
Brz.	Bronze
Bsn.	Basin
Btr.	Better
BTU	British Thermal Unit
BTUH	BTU per Hour
B.U.R.	Built-up Roofing
BX	Interlocked Armored Cable
c	Conductivity, Copper Sweat
C	Hundred; Centigrade
C/C	Center to Center, Cedar on Cedar
Cab.	Cabinet
Cair.	Air Tool Laborer
Calc	Calculated
Cap.	Capacity
Carp.	Carpenter
C.B.	Circuit Breaker
C.C.A.	Chromate Copper Arsenate
C.C.F.	Hundred Cubic Feet
cd	Candela
cd/sf	Candela per Square Foot
CD	Grade of Plywood Face & Back
CDX	Plywood, Grade C & D, exterior glue
Cefi.	Cement Finisher
Cem.	Cement
CF	Hundred Feet
C.F.	Cubic Feet
CFM	Cubic Feet per Minute
c.g.	Center of Gravity
CHW	Chilled Water; Commercial Hot Water
C.I.	Cast Iron
C.I.P.	Cast in Place
Circ.	Circuit
C.L.	Carload Lot
Clab.	Common Laborer
C.L.F.	Hundred Linear Feet
CLF	Current Limiting Fuse
CLP.	Cross Linked Polyethylene
cm	Centimeter
CMP	Corr. Metal Pipe
C.M.U.	Concrete Masonry Unit
CN	Change Notice
Col.	Column
CO₂	Carbon Dioxide
Comb.	Combination
Compr.	Compressor
Conc.	Concrete
Cont.	Continuous; Continued
Corr.	Corrugated
Cos	Cosine
Cot	Cotangent
Cov.	Cover
C/P	Cedar on Paneling
CPA	Control Point Adjustment
Cplg.	Coupling
C.P.M.	Critical Path Method
CPVC	Chlorinated Polyvinyl Chloride
C.Pr.	Hundred Pair
CRC	Cold Rolled Channel
Creos.	Creosote
Crpt.	Carpet & Linoleum Layer
CRT	Cathode-ray Tube
CS	Carbon Steel, Constant Shear Bar Joist
Csc	Cosecant
C.S.F.	Hundred Square Feet
CSI	Construction Specifications Institute
C.T.	Current Transformer
CTS	Copper Tube Size
Cu	Copper, Cubic
Cu. Ft.	Cubic Foot
cw	Continuous Wave
C.W.	Cool White; Cold Water
Cwt.	100 Pounds
C.W.X.	Cool White Deluxe
C.Y.	Cubic Yard (27 cubic feet)
C.Y./Hr.	Cubic Yard per Hour
Cyl.	Cylinder
d	Penny (nail size)
D	Deep; Depth; Discharge
Dis.;Disch.	Discharge
Db.	Decibel
Dbl.	Double
DC	Direct Current
Demob.	Demobilization
d.f.u.	Drainage Fixture Units
D.H.	Double Hung
DHW	Domestic Hot Water
Diag.	Diagonal
Diam.	Diameter
Distrib.	Distribution
Dk.	Deck
D.L.	Dead Load; Diesel
DLH	Deep Long Span Bar Joist
Do.	Ditto
Dp.	Depth
D.P.S.T.	Double Pole, Single Throw
Dr.	Driver
Drink.	Drinking
D.S.	Double Strength
D.S.A.	Double Strength A Grade
D.S.B.	Double Strength B Grade
Dty.	Duty
DWV	Drain Waste Vent
DX	Deluxe White, Direct Expansion
dyn	Dyne
e	Eccentricity
E	Equipment Only; East
Ea.	Each
E.B.	Encased Burial
Econ.	Economy
EDP	Electronic Data Processing
EIFS	Exterior Insulation Finish System
E.D.R.	Equiv. Direct Radiation
Eq.	Equation
Elec.	Electrician; Electrical
Elev.	Elevator; Elevating
EMT	Electrical Metallic Conduit; Thin Wall Conduit
Eng.	Engine, Engineered
EPDM	Ethylene Propylene Diene Monomer
EPS	Expanded Polystyrene
Eqhv.	Equip. Oper., Heavy
Eqlt.	Equip. Oper., Light
Eqmd.	Equip. Oper., Medium
Eqmm.	Equip. Oper., Master Mechanic
Eqol.	Equip. Oper., Oilers
Equip.	Equipment
ERW	Electric Resistance Welded
E.S.	Energy Saver
Est.	Estimated
esu	Electrostatic Units
E.W.	Each Way
EWT	Entering Water Temperature
Excav.	Excavation
Exp.	Expansion, Exposure
Ext.	Exterior
Extru.	Extrusion
f.	Fiber stress
F	Fahrenheit; Female; Fill
Fab.	Fabricated
FBGS	Fiberglass
F.C.	Footcandles
f.c.c.	Face-centered Cubic
f'c.	Compressive Stress in Concrete; Extreme Compressive Stress
F.E.	Front End
FEP	Fluorinated Ethylene Propylene (Teflon)
F.G.	Flat Grain
F.H.A.	Federal Housing Administration
Fig.	Figure
Fin.	Finished
Fixt.	Fixture
Fl. Oz.	Fluid Ounces
Flr.	Floor
F.M.	Frequency Modulation; Factory Mutual
Fmg.	Framing
Fndtn.	Foundation
Fori.	Foreman, Inside
Foro.	Foreman, Outside

Abbrev.	Definition	Abbrev.	Definition	Abbrev.	Definition
Fount.	Fountain	J.I.C.	Joint Industrial Council	MD	Medium Duty
FPM	Feet per Minute	K	Thousand; Thousand Pounds; Heavy Wall Copper Tubing	M.D.O.	Medium Density Overlaid
FPT	Female Pipe Thread	K.A.H.	Thousand Amp. Hours	Med.	Medium
Fr.	Frame	KCMIL	Thousand Circular Mils	MF	Thousand Feet
F.R.	Fire Rating	KD	Knock Down	M.F.B.M.	Thousand Feet Board Measure
FRK	Foil Reinforced Kraft	K.D.A.T.	Kiln Dried After Treatment	Mfg.	Manufacturing
FRP	Fiberglass Reinforced Plastic	kg	Kilogram	Mfrs.	Manufacturers
FS	Forged Steel	kG	Kilogauss	mg	Milligram
FSC	Cast Body; Cast Switch Box	kgf	Kilogram Force	MGD	Million Gallons per Day
Ft.	Foot; Feet	kHz	Kilohertz	MGPH	Thousand Gallons per Hour
Ftng.	Fitting	Kip.	1000 Pounds	MH, M.H.	Manhole; Metal Halide; Man-Hour
Ftg.	Footing	KJ	Kiljoule	MHz	Megahertz
Ft. Lb.	Foot Pound	K.L.	Effective Length Factor	Mi.	Mile
Furn.	Furniture	K.L.F.	Kips per Linear Foot	MI	Malleable Iron; Mineral Insulated
FVNR	Full Voltage Non-Reversing	Km	Kilometer	mm	Millimeter
FXM	Female by Male	K.S.F.	Kips per Square Foot	Mill.	Millwright
Fy.	Minimum Yield Stress of Steel	K.S.I.	Kips per Square Inch	Min., min.	Minimum, minute
g	Gram	K.V.	Kilovolt	Misc.	Miscellaneous
G	Gauss	K.V.A.	Kilovolt Ampere	ml	Milliliter
Ga.	Gauge	K.V.A.R.	Kilovar (Reactance)	M.L.F.	Thousand Linear Feet
Gal.	Gallon	KW	Kilowatt	Mo.	Month
Gal./Min.	Gallon per Minute	KWh	Kilowatt-hour	Mobil.	Mobilization
Galv.	Galvanized	L	Labor Only; Length; Long; Medium Wall Copper Tubing	Mog.	Mogul Base
Gen.	General	Lab.	Labor	MPH	Miles per Hour
G.F.I.	Ground Fault Interrupter	lat	Latitude	MPT	Male Pipe Thread
Glaz.	Glazier	Lath.	Lather	MRT	Mile Round Trip
GPD	Gallons per Day	Lav.	Lavatory	ms	Millisecond
GPH	Gallons per Hour	lb.; #	Pound	M.S.F.	Thousand Square Feet
GPM	Gallons per Minute	L.B.	Load Bearing; L Conduit Body	Mstz.	Mosaic & Terrazzo Worker
GR	Grade	L. & E.	Labor & Equipment	M.S.Y.	Thousand Square Yards
Gran.	Granular	lb./hr.	Pounds per Hour	Mtd.	Mounted
Grnd.	Ground	lb./L.F.	Pounds per Linear Foot	Mthe.	Mosaic & Terrazzo Helper
H	High; High Strength Bar Joist; Henry	lbf/sq.in.	Pound-force per Square Inch	Mtng.	Mounting
H.C.	High Capacity	L.C.L.	Less than Carload Lot	Mult.	Multi; Multiply
H.D.	Heavy Duty; High Density	Ld.	Load	M.V.A.	Million Volt Amperes
H.D.O.	High Density Overlaid	LE	Lead Equivalent	M.V.A.R.	Million Volt Amperes Reactance
Hdr.	Header	LED	Light Emitting Diode	MV	Megavolt
Hdwe.	Hardware	L.F.	Linear Foot	MW	Megawatt
Help.	Helpers Average	Lg.	Long; Length; Large	MXM	Male by Male
HEPA	High Efficiency Particulate Air Filter	L & H	Light and Heat	MYD	Thousand Yards
Hg	Mercury	LH	Long Span Bar Joist, Labor Hours	N	Natural; North
HIC	High Interrupting Capacity	L.L.	Live Load	nA	Nanoampere
HM	Hollow Metal	L.L.D.	Lamp Lumen Depreciation	NA	Not Available; Not Applicable
H.O.	High Output	L-O-L	Lateralolet	N.B.C.	National Building Code
Horiz.	Horizontal	lm	Lumen	NC	Normally Closed
H.P.	Horsepower; High Pressure	lm/sf	Lumen per Square Foot	N.E.M.A.	National Electrical Manufacturers Assoc.
H.P.F.	High Power Factor	lm/W	Lumen per Watt	NEHB	Bolted Circuit Breaker to 600V.
Hr.	Hour	L.O.A.	Length Over All	N.L.B.	Non-Load-Bearing
Hrs./Day	Hours per Day	log	Logarithm	NM	Non-Metallic Cable
HSC	High Short Circuit	L.P.	Liquefied Petroleum; Low Pressure	nm	Nanometer
Ht.	Height	L.P.F.	Low Power Factor	No.	Number
Htg.	Heating	LR	Long Radius	NO	Normally Open
Htrs.	Heaters	L.S.	Lump Sum	N.O.C.	Not Otherwise Classified
HVAC	Heating, Ventilation & Air-Conditioning	Lt.	Light	Nose.	Nosing
Hvy.	Heavy	Lt. Ga.	Light Gauge	N.P.T.	National Pipe Thread
HW	Hot Water	L.T.L.	Less than Truckload Lot	NQOD	Combination Plug-on/Bolt on Circuit Breaker to 240V.
Hyd.;Hydr.	Hydraulic	Lt. Wt.	Lightweight	N.R.C.	Noise Reduction Coefficient
Hz.	Hertz (cycles)	L.V.	Low Voltage	N.R.S.	Non Rising Stem
I.	Moment of Inertia	M	Thousand; Material; Male; Light Wall Copper Tubing	ns	Nanosecond
I.C.	Interrupting Capacity	m/hr; M.H.	Man-hour	nW	Nanowatt
ID	Inside Diameter	mA	Milliampere	OB	Opposing Blade
I.D.	Inside Dimension; Identification	Mach.	Machine	OC	On Center
I.F.	Inside Frosted	Mag. Str.	Magnetic Starter	OD	Outside Diameter
I.M.C.	Intermediate Metal Conduit	Maint.	Maintenance	O.D.	Outside Dimension
In.	Inch	Marb.	Marble Setter	ODS	Overhead Distribution System
Incan.	Incandescent	Mat; Mat'l.	Material	O.G.	Ogee
Incl.	Included; Including	Max.	Maximum	O.H.	Overhead
Int.	Interior	MBF	Thousand Board Feet	O & P	Overhead and Profit
Inst.	Installation	MBH	Thousand BTU's per hr.	Oper.	Operator
Insul.	Insulation/Insulated	MC	Metal Clad Cable	Opng.	Opening
I.P.	Iron Pipe	M.C.F.	Thousand Cubic Feet	Orna.	Ornamental
I.P.S.	Iron Pipe Size	M.C.F.M.	Thousand Cubic Feet per Minute	OSB	Oriented Strand Board
I.P.T.	Iron Pipe Threaded	M.C.M.	Thousand Circular Mils	O. S. & Y.	Outside Screw and Yoke
I.W.	Indirect Waste	M.C.P.	Motor Circuit Protector	Ovhd.	Overhead
J	Joule			OWG	Oil, Water or Gas
				Oz.	Ounce

Abbreviations

P.	Pole; Applied Load; Projection
p.	Page
Pape.	Paperhanger
P.A.P.R.	Powered Air Purifying Respirator
PAR	Weatherproof Reflector
Pc., Pcs.	Piece, Pieces
P.C.	Portland Cement; Power Connector
P.C.F.	Pounds per Cubic Foot
P.C.M.	Phase Contract Microscopy
P.E.	Professional Engineer; Porcelain Enamel; Polyethylene; Plain End
Perf.	Perforated
Ph.	Phase
P.I.	Pressure Injected
Pile.	Pile Driver
Pkg.	Package
Pl.	Plate
Plah.	Plasterer Helper
Plas.	Plasterer
Pluh.	Plumbers Helper
Plum.	Plumber
Ply.	Plywood
p.m.	Post Meridiem
Pntd.	Painted
Pord.	Painter, Ordinary
pp	Pages
PP; PPL	Polypropylene
P.P.M.	Parts per Million
Pr.	Pair
P.E.S.B.	Pre-engineered Steel Building
Prefab.	Prefabricated
Prefin.	Prefinished
Prop.	Propelled
PSF; psf	Pounds per Square Foot
PSI; psi	Pounds per Square Inch
PSIG	Pounds per Square Inch Gauge
PSP	Plastic Sewer Pipe
Pspr.	Painter, Spray
Psst.	Painter, Structural Steel
P.T.	Potential Transformer
P. & T.	Pressure & Temperature
Ptd.	Painted
Ptns.	Partitions
Pu	Ultimate Load
PVC	Polyvinyl Chloride
Pvmt.	Pavement
Pwr.	Power
Q	Quantity Heat Flow
Quan.; Qty.	Quantity
Q.C.	Quick Coupling
r	Radius of Gyration
R	Resistance
R.C.P.	Reinforced Concrete Pipe
Rect.	Rectangle
Reg.	Regular
Reinf.	Reinforced
Req'd.	Required
Res.	Resistant
Resi.	Residential
Rgh.	Rough
R.H.W.	Rubber, Heat & Water Resistant; Residential Hot Water
rms	Root Mean Square
Rnd.	Round
Rodm.	Rodman
Rofc.	Roofer, Composition
Rofp.	Roofer, Precast
Rohe.	Roofer Helpers (Composition)
Rots.	Roofer, Tile & Slate
R.O.W.	Right of Way
RPM	Revolutions per Minute
R.R.	Direct Burial Feeder Conduit
R.S.	Rapid Start
Rsr	Riser
RT	Round Trip
S.	Suction; Single Entrance; South
SCFM	Standard Cubic Feet per Minute
Scaf.	Scaffold
Sch.; Sched.	Schedule
S.C.R.	Modular Brick
S.D.	Sound Deadening
S.D.R.	Standard Dimension Ratio
S.E.	Surfaced Edge
Sel.	Select
S.E.R.; S.E.U.	Service Entrance Cable
S.F.	Square Foot
S.F.C.A.	Square Foot Contact Area
S.F.G.	Square Foot of Ground
S.F. Hor.	Square Foot Horizontal
S.F.R.	Square Feet of Radiation
S.F. Shlf.	Square Foot of Shelf
S4S	Surface 4 Sides
Shee.	Sheet Metal Worker
Sin.	Sine
Skwk.	Skilled Worker
SL	Saran Lined
S.L.	Slimline
Sldr.	Solder
SLH	Super Long Span Bar Joist
S.N.	Solid Neutral
S-O-L	Socketolet
S.P.	Static Pressure; Single Pole; Self-Propelled
Spri.	Sprinkler Installer
Sq.	Square; 100 Square Feet
S.P.D.T.	Single Pole, Double Throw
SPF	Spruce Pine Fir
S.P.S.T.	Single Pole, Single Throw
SPT	Standard Pipe Thread
Sq. Hd.	Square Head
Sq. In.	Square Inch
S.S.	Single Strength; Stainless Steel
S.S.B.	Single Strength B Grade
Sswk.	Structural Steel Worker
Sswl.	Structural Steel Welder
St.; Stl.	Steel
S.T.C.	Sound Transmission Coefficient
Std.	Standard
STK	Select Tight Knot
STP	Standard Temperature & Pressure
Stpi.	Steamfitter, Pipefitter
Str.	Strength; Starter; Straight
Strd.	Stranded
Struct.	Structural
Sty.	Story
Subj.	Subject
Subs.	Subcontractors
Surf.	Surface
Sw.	Switch
Swbd.	Switchboard
S.Y.	Square Yard
Syn.	Synthetic
S.Y.P.	Southern Yellow Pine
Sys.	System
t.	Thickness
T	Temperature; Ton
Tan	Tangent
T.C.	Terra Cotta
T & C	Threaded and Coupled
T.D.	Temperature Difference
T.E.M.	Transmission Electron Microscopy
TFE	Tetrafluoroethylene (Teflon)
T. & G.	Tongue & Groove; Tar & Gravel
Th.; Thk.	Thick
Thn.	Thin
Thrded	Threaded
Tilf.	Tile Layer, Floor
Tilh.	Tile Layer, Helper
THHN	Nylon Jacketed Wire
THW.	Insulated Strand Wire
THWN;	Nylon Jacketed Wire
T.L.	Truckload
T.M.	Track Mounted
Tot.	Total
T-O-L	Threadolet
T.S.	Trigger Start
Tr.	Trade
Transf.	Transformer
Trhv.	Truck Driver, Heavy
Trlr	Trailer
Trlt.	Truck Driver, Light
TV	Television
T.W.	Thermoplastic Water Resistant Wire
UCI	Uniform Construction Index
UF	Underground Feeder
UGND	Underground Feeder
U.H.F.	Ultra High Frequency
U.L.	Underwriters Laboratory
Unfin.	Unfinished
URD	Underground Residential Distribution
US	United States
USP	United States Primed
UTP	Unshielded Twisted Pair
V	Volt
V.A.	Volt Amperes
V.C.T.	Vinyl Composition Tile
VAV	Variable Air Volume
VC	Veneer Core
Vent.	Ventilation
Vert.	Vertical
V.F.	Vinyl Faced
V.G.	Vertical Grain
V.H.F.	Very High Frequency
VHO	Very High Output
Vib.	Vibrating
V.L.F.	Vertical Linear Foot
Vol.	Volume
VRP	Vinyl Reinforced Polyester
W	Wire; Watt; Wide; West
w/	With
W.C.	Water Column; Water Closet
W.F.	Wide Flange
W.G.	Water Gauge
Wldg.	Welding
W. Mile	Wire Mile
W-O-L	Weldolet
W.R.	Water Resistant
Wrck.	Wrecker
W.S.P.	Water, Steam, Petroleum
WT., Wt.	Weight
WWF	Welded Wire Fabric
XFER	Transfer
XFMR	Transformer
XHD	Extra Heavy Duty
XHHW;	Cross Linked Polyethylene Wire Insulation
XLP	Cross-linked Polyethylene
XLPE	Cross-Linked Polyethylene Wire Insulation
Y	Wye
yd	Yard
yr	Year
Δ	Delta
%	Percent
~	Approximately
Ø	Phase
@	At
#	Pound; Number
<	Less Than
>	Greater Than

629

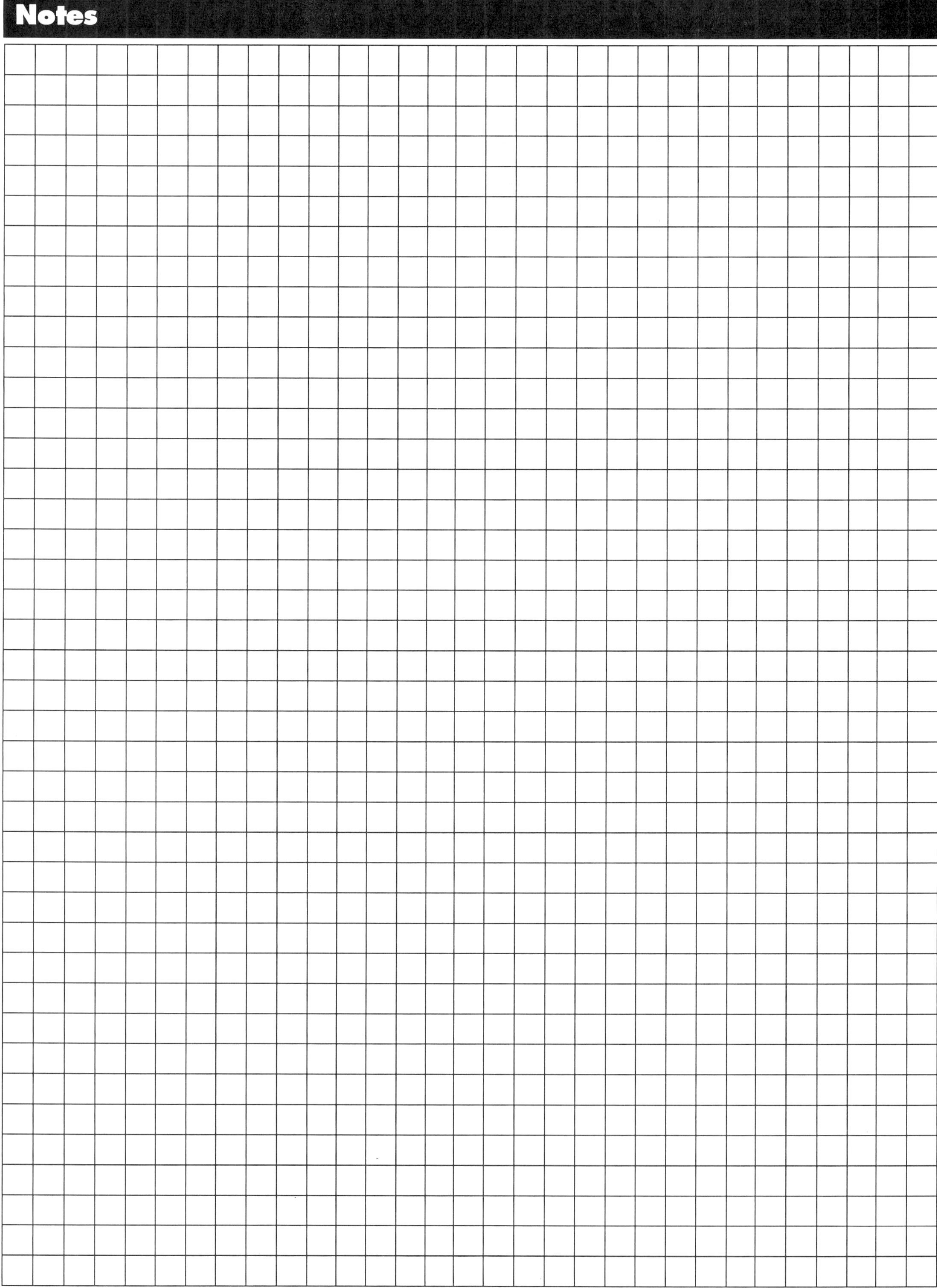

R.S. Means Co., Inc., the leading provider of construction cost data in North America, supplies comprehensive construction cost guides, related technical publications and educational services. Means is a member of the *Construction Market Data Group*, a unique group of companies that serve the construction industry in the U.S., Canada and Mexico.

The *Construction Market Data Group* has a broad mission: To study all the information dynamics that affect construction — products, projects, estimating factors, and costs — and report on them for the entire North American region. No other company offers this scope of coverage.

CMDG has a record of innovative leadership in meeting the evolving needs of its customers. Design/build professionals, product marketers, building owners and facilities departments, developers, government agencies, bankers, insurers, lawyers, as well as homeowners and the media are all served. Other *CMDG* members include:

Architects' First Source for Products - The most extensive and widely circulated North American product selection resource for specifiers of construction materials and products. 800-395-1988 • www.afsonl.com

Construction Market Data - The source for project information and market intelligence throughout the United States. CMD information is used for sales leads, market analysis, organizing the bidding process, identifying key contacts, and other functions that support sales and marketing throughout the construction industry. 800-417-7291 National Services; 800-992-7575 Local Services • www.cmdonl.com

CMD/Canada - Serves the Canadian construction marketplace with the most reliable and comprehensive variety of products available in Canada. Services include:

Daily Commercial News - Serves Ontario with project news and information.
Journal of Commerce - Provides construction news for British Columbia.
CMD Bulletins - Contain detailed project information delivered weekly by post.
CanaData - Publishes forecasts and customized statistics. CanaData is the leader of conferences for manufacturers, planners and industry analysts.
Buildcore - Links architects and designers to Canadian manufacturers and their products.
416-494-4990

Manufacturers' Survey Associates - Provides detailed product and specifier information to building product manufacturers. MSA, in conjunction with Autodesk®, jointly produce DesignBlocks™, an electronic library of product information on CD-ROM. 800-999-5502 • www.msaonl.com

PACRIM/Hospitality Profiles - Publishers of specific hospitality and leisure time construction data in the Americas and Asia. 800-259-1654 • www.expo.net/hospitality

BIMSA/Mexico - (An Affiliate Company) Distributes building project information and construction costs throughout Mexico. 011-525-357-1222 • www.bimsa.com.mx

Information the North American Construction Industry Builds On

4126 Pleasantdale Road, Atlanta, GA 30340 • www.cmdg.com

Means Project Cost Report

By filling out and returning the Project Description, you can receive a discount of $20.00 off any one of the Means products advertised in the following pages. The cost information required includes all items marked (✔) except those where no costs occurred. The sum of all major items should equal the Total Project Cost.

$20.00 Discount per product for each report you submit.

DISCOUNT PRODUCTS AVAILABLE—FOR U.S. CUSTOMERS ONLY—STRICTLY CONFIDENTIAL

Project Description (No remodeling projects, please.)

✔ Type Building _______________________
✔ Location _______________________
 Capacity _______________________
✔ Frame _______________________
✔ Exterior _______________________
✔ Basement: full ☐ partial ☐ none ☐ crawl ☐
✔ Height in Stories _______________________
✔ Total Floor Area _______________________
 Ground Floor Area _______________________
✔ Volume in C.F. _______________________
 % Air Conditioned ___________ Tons _______
 Comments _______________________

Owner _______________________
Architect _______________________
General Contractor _______________________
✔ Bid Date _______________________
 Typical Bay Size _______________________
✔ Labor Force: ______ % Union ______ % Non-Union
✔ Project Description (Circle one number in each line)
 1. Economy 2. Average 3. Custom 4. Luxury
 1. Square 2. Rectangular 3. Irregular 4. Very Irregular

	✔ **Total Project Cost**		$		
A	✔ **General Conditions**		$		
B	✔ **Site Work**		$		
BS	**Site Clearing & Improvement**				
BE	*Excavation*	(		*C.Y.)*	
BF	*Caissons & Piling*	(		*L.F.)*	
BU	*Site Utilities*				
BP	*Roads & Walks Exterior Paving*	(		*S.Y.)*	
C	✔ **Concrete**		$		
C	*Cast in Place*	(		*C.Y.)*	
CP	*Precast*	(		*S.F.)*	
D	✔ **Masonry**		$		
DB	*Brick*	(		*M)*	
DC	*Block*	(		*M)*	
DT	*Tile*	(		*S.F.)*	
DS	*Stone*	(		*S.F.)*	
E	✔ **Metals**		$		
ES	*Structural Steel*	(		*Tons)*	
EM	*Misc. & Ornamental Metals*				
F	✔ **Wood & Plastics**		$		
FR	*Rough Carpentry*	(		*MBF)*	
FF	*Finish Carpentry*				
FM	*Architectural Millwork*				
G	✔ **Thermal & Moisture Protection**		$		
GW	*Waterproofing-Dampproofing*	(		*S.F.)*	
GN	*Insulation*	(		*S.F.)*	
GR	*Roofing & Flashing*	(		*S.F.)*	
GM	*Metal Siding/Curtain Wall*	(		*S.F.)*	
H	✔ **Doors and Windows**		$		
HD	*Doors*	(		*Ea.)*	
HW	*Windows*	(		*S.F.)*	
HH	*Finish Hardware*				
HG	*Glass & Glazing*	(		*S.F.)*	
HS	*Storefronts*	(		*S.F.)*	

J	✔ **Finishes**		$		
JL	*Lath & Plaster*	(		*S.Y.)*	
JD	*Drywall*	(		*S.F.)*	
JM	*Tile & Marble*	(		*S.F.)*	
JT	*Terrazzo*	(		*S.F.)*	
JA	*Acoustical Treatment*	(		*S.F.)*	
JC	*Carpet*	(		*S.Y.)*	
JF	*Hard Surface Flooring*	(		*S.F.)*	
JP	*Painting & Wall Covering*	(		*S.F.)*	
K	✔ **Specialties**		$		
KB	*Bathroom Partitions & Access.*	(		*S.F.)*	
KF	*Other Partitions*	(		*S.F.)*	
KL	*Lockers*	(		*Ea.)*	
L	✔ **Equipment**		$		
LK	*Kitchen*				
LS	*School*				
LO	*Other*				
M	✔ **Furnishings**		$		
MW	*Window Treatment*				
MS	*Seating*	(		*Ea.)*	
N	✔ **Special Construction**		$		
NA	*Acoustical*	(		*S.F.)*	
NB	*Prefab. Bldgs.*	(		*S.F.)*	
NO	*Other*				
P	✔ **Conveying Systems**		$		
PE	*Elevators*	(		*Ea.)*	
PS	*Escalators*	(		*Ea.)*	
PM	*Material Handling*				
Q	✔ **Mechanical**		$		
QP	*Plumbing*	*(No. of fixtures*		*)*	
QS	*Fire Protection (Sprinklers)*				
QF	*Fire Protection (Hose Standpipes)*				
QB	*Heating, Ventilating & A.C.*				
QH	*Heating & Ventilating*	*(BTU Output*		*)*	
QA	*Air Conditioning*	(		*Tons)*	
R	✔ **Electrical**		$		
RL	*Lighting*	(		*S.F.)*	
RP	*Power Service*				
RD	*Power Distribution*				
RA	*Alarms*				
RG	*Special Systems*				
S	✔ **Mech./Elec. Combined**		$		

Product Name _______________________
Product Number _______________________
Your Name _______________________
Title _______________________
Company _______________________
☐ Company
☐ Home Street Address _______________________
City, State, Zip _______________________
☐ Please send _______ forms.

Please specify the Means product you wish to receive. Complete the address information as requested and return this form with your check (product cost less $20.00) to address below.

R.S. Means Company, Inc.,
Square Foot Costs Department
100 Construction Plaza, P.O. Box 800
Kingston, MA 02364-9988

R.S. Means Company, Inc. . . a tradition of excellence in Construction Cost Information and Services since 1942.

Table of Contents

Book Selection Guide

The following table provides definitive information on the content of each cost data publication. The number of lines of data provided in each unit price or assemblies division, as well as the number of reference tables and crews is listed for each book. The presence of other elements such as an historical cost index, city cost indexes, square foot models or cross referenced index is also indicated. You can use the table to help select the Means' book that has the quantity and type of information you most need in your work.

Unit Cost Divisions	Building Construction Costs	Mechanical	Electrical	Repair & Remodel.	Square Foot	Site Work Landsc.	Assemblies	Interior	Concrete Masonry	Open Shop	Heavy Construc.	Resi-dential	Light Commercial	Facil. Construc.	Plumbing	Western Construction Costs
1	1033	521	519	702		945		436	969	1032	1045	372	556	1480	592	1030
2	3366	1384	451	2158		5281		1134	1594	3315	5169	1131	1206	4925	1636	3346
3	1502	91	68	747		1321		188	1932	1478	1449	262	206	1342	43	1480
4	874	27	0	666		805		633	1227	856	709	317	402	1134	0	851
5	1284	170	165	461		787		457	653	1247	1087	267	293	1347	93	1264
6	1370	96	83	1275		475		1287	328	1339	593	1521	1429	1367	59	1727
7	1378	60	9	1304		484		539	458	1377	364	824	1014	1370	70	1377
8	1837	60	0	1861		339		1746	717	1812	9	1094	1151	1967	0	1837
9	1639	50	0	1484		186		1734	322	1637	114	1311	1380	1833	50	1629
10	973	56	30	546		206		804	195	973	0	244	446	971	249	973
11	1019	261	200	586		53		896	34	1002	16	115	234	1016	231	1001
12	353	0	0	47		227		1533	30	344	0	73	68	1536	0	344
13	881	413	201	154		464		356	105	879	325	120	133	966	239	930
14	365	43	0	262		36		330	0	365	40	9	18	363	17	363
15	2724	15477	703	2302		2022		1772	80	2643	2327	986	1613	13058	11712	2600
16	1547	798	10589	1121		1098		1346	62	1562	1034	704	1275	10170	705	1489
17	459	369	368	1		0		0	0	460	0	0	0	459	369	459
Totals	**22604**	**19876**	**13386**	**15677**		**14729**		**15191**	**8706**	**22321**	**14281**	**9350**	**11424**	**45304**	**16065**	**22700**

Assembly Divisions	Building Construction Costs	Mechanical	Electrical	Repair & Remodel.	Square Foot	Site Work Landsc.	Assemblies	Interior	Concrete Masonry	Open Shop	Heavy Construc.	Resi-dential	Light Commercial	Facil. Construc.	Plumbing	Western Construction Costs
1		0	0	202	131	738	775	0	689		752	844	131	89	0	
2		0	0	40	33	35	48	0	49		0	589	32	0	0	
3		0	0	445	1114	0	3064	228	1245		0	1549	806	174	0	
4		0	0	713	1353	0	3175	255	1273		0	2047	1171	26	0	
5		0	0	289	231	0	465	0	0		0	1082	227	28	0	
6		0	0	1006	857	0	1308	1736	152		0	1084	749	333	0	
7		0	0	42	89	0	179	159	0		0	723	33	71	0	
8		2255	149	998	1699	0	2725	925	0		0	1694	1197	1114	2091	
9		0	1347	355	357	0	1284	307	0		0	242	358	319	0	
10		0	0	0	0	0	0	0	0		0	0	0	0	0	
11		0	0	365	465	0	724	174	0		0	0	466	222	0	
12		501	160	539	84	2392	699	0	698		541	0	84	119	850	
Totals		**2756**	**1656**	**4994**	**6413**	**3165**	**14446**	**3784**	**4106**		**1293**	**9854**	**5254**	**2495**	**2941**	

Reference Section	Building Construction Costs	Mechanical	Electrical	Repair & Remodel.	Square Foot	Site Work Landsc.	Assemblies	Interior	Concrete Masonry	Open Shop	Heavy Construc.	Resi-dential	Light Commercial	Facil. Construc.	Plumbing	Western Construction Costs
Tables	156	46	85	70	4	87	223	60	84	156	56	51	74	88	49	158
Models					102							32	43			
Crews	397	397	397	397		397		397	397	397	397	397	397	397	397	397
City Cost Indexes	yes	yes	yes	yes	yes	yes	yes	yes	yes	yes	yes	yes	yes	yes	yes	yes
Historical Cost Indexes	yes	yes	yes	yes	yes	yes	yes	yes	yes	yes	yes	no	yes	yes	yes	yes
Index	yes	yes	yes	yes	no	yes	yes	yes	yes	yes	yes	yes	yes	yes	yes	yes

Annual Cost Guides

Means Building Construction Cost Data 1998

Available in Both Softbound and Looseleaf Editions

The "Bible" of the industry comes in the standard softcover edition or the looseleaf edition.

Many customers enjoy the convenience and flexibility of the looseleaf binder, which increases the usefulness of *Means Building Construction Cost Data 1998* by making it easy to add and remove pages. You can insert your own cost information pages, so everything is in one place. Copying pages for faxing is easier also. Whichever edition you prefer, softbound or the convenient looseleaf edition, you'll get *The DI Quarterly/Change Notice* newsletter at no extra cost.

$84.95 per copy, Softbound
Catalog No. 60018

$109.95 per copy, Looseleaf
Catalog No. 61018

Means Building Construction Cost Data 1998

Offers you unchallenged unit price reliability in an easy-to-use arrangement. Whether used for complete, finished estimates or for periodic checks, it supplies more cost facts better and faster than any comparable source. Over 21,000 unit prices for 1998. The City Cost Indexes now cover over 930 areas, for indexing to any project location in North America. Order and get *The DI Quarterly/Change Notice* newsletter sent to you FREE. You'll have year-long access to the Means Estimating **Hotline** FREE with your subscription. Expert assistance when using Means data is just a phone call away.

$84.95 per copy
Over 650 pages, illustrated, available Oct. 1997
Catalog No. 60018

Means Building Construction Cost Data 1998

Metric Version

The Federal Government has stated that all federal construction projects must now use metric documentation. The *Metric Version* of *Means Building Construction Cost Data 1998* is presented in metric measurements covering all construction areas. Don't miss out on these billion dollar opportunities. Make the switch to metric today.

$89.95 per copy
Over 650 pages, illustrated, available Nov. 1997
Catalog No. 63018

Annual Cost Guides

Means Mechanical Cost Data 1998

• HVAC • Controls

Total unit and systems price guidance for mechanical construction . . . materials, parts, fittings, and complete labor cost information. Includes prices for piping, heating, air conditioning, ventilation, and all related construction.

Plus new 1998 unit costs for:

- Over 3000 installed HVAC/controls assemblies
- "On Site" Location Factors for close to 1,000 cities and towns in the U.S. and Canada
- Crews, labor and equipment

$87.95 per copy
Over 600 pages, illustrated, available Oct. 1997
Catalog No. 60028

Means Plumbing Cost Data 1998

Comprehensive unit prices and assemblies for plumbing, irrigation systems, commercial and residential fire protection, point-of-use water heaters, and the latest approved materials. This publication and its companion, *Means Mechanical Cost Data*, provide full-range cost estimating coverage for all the mechanical trades.

$87.95 per copy
Over 500 pages, illustrated, available Oct. 1997
Catalog No. 60218

Means Electrical Cost Data 1998

Pricing information for every part of electrical cost planning: More than 17,000 unit and systems costs with design tables; clear specifications and drawings; engineering guides and illustrated estimating procedures; complete labor-hour and materials costs for better scheduling and procurement; the latest electrical products and construction methods.

- (NEW FOR 1998) Underground Marking Tape and Polyethylene Pull Rope costs
- A Variety of Special Electrical Systems including Cathodic Protection
- Costs for maintenance, demolition, HVAC/ mechanical, specialties, equipment, and more

$87.95 per copy
Over 450 pages, illustrated, available Oct. 1997
Catalog No. 60038

Means Electrical Change Order Cost Data 1998

You are provided with electrical unit prices exclusively for pricing change orders—based on the recent, direct experience of contractors and suppliers. Analyze and check your own change order estimates against the experience others have had doing the same work. It also covers productivity analysis and change order cost justifications. With useful information for calculating the effects of change orders and dealing with their administration.

$89.95 per copy
Over 440 pages, available Oct. 1997
Catalog No. 60238

Means Facilities Maintenance & Repair Cost Data 1998

Published in a looseleaf format, *Means Facilities Maintenance & Repair Cost Data* gives you a complete system to manage and plan your facility repair and maintenance costs and budget efficiently. Guidelines for auditing a facility and developing an annual maintenance plan. Budgeting is included, along with reference tables on cost and management and information on frequency and productivity of maintenance operations.

The only nationally recognized source of maintenance and repair costs. Developed in cooperation with the Army Corps of Engineers.

$199.95 per copy
Over 600 pages, illustrated, available Dec. 1997
Catalog No. 60308

Means Square Foot Costs 1998

It's Accurate and Easy To Use!

- **NEW 1998 price information,** based on nationwide figures from suppliers, estimators, labor experts and contractors.

- **NEW** "How-to-Use" Sections, with **better, clearer examples** of commercial, residential, industrial, and institutional structures.

- **NEW**, more realistic graphics, offering true-to-life illustrations of building projects.

- **NEW**, more extensive information on using square foot cost data, including **sample estimates** and **alternate pricing methods.**

$97.95 per copy
Over 460 pages, illustrated, available Nov. 1997
Catalog No. 60058

Annual Cost Guides

Means Repair & Remodeling Cost Data 1998
Commercial/Residential

You can use this valuable tool to estimate commercial and residential renovation and remodeling.

Includes: New 1998 costs for hundreds of unique methods, materials and conditions that only come up in repair and remodeling. PLUS:

- 1998 unit costs for over 15,000 construction components
- 1998 installed costs for over 4,000 assemblies.
- 1998 costs for 300+ construction crews
- Over 930 "On Site" localization factors for the U.S. and Canada.

$79.95 per copy
Over 600 pages, illustrated, available Oct. 1997
Catalog No. 60048

Means Facilities Construction Cost Data 1998

For the maintenance and construction of commercial, industrial, municipal, and institutional properties. Costs are shown for new and remodeling construction and are broken down into materials, labor, equipment, overhead, and profit. Special emphasis is given to sections on mechanical, electrical, furnishings, site work, building maintenance, finish work, and demolition. More than 40,000 unit costs plus assemblies and reference sections are included.

$209.95 per copy
Over 1100 pages, illustrated, available Nov. 1997
Catalog No. 60208

Means Residential Cost Data 1998

Now contains square foot costs for 30 basic home models with the look of today—plus hundreds of custom additions and modifications you can quote right off the page. With costs for the 100 residential systems you're most likely to use in the year ahead. Complete with blank estimating forms, sample estimates and step-by-step instructions.

$74.95 per copy
Over 550 pages, illustrated, available Dec. 1997
Catalog No. 60178

Means Light Commercial Cost Data 1998

Specifically addresses the light commercial market, which is an increasingly specialized niche in the industry. Aids you, the owner/designer/contractor, in preparing all types of estimates, from budgets to detailed bids. Includes new advances in methods and materials. Assemblies section allows you to evaluate alternatives in the early stages of design/planning.

Over 10,000 unit costs for 1998 ensure you have the prices you need... when you need them.

$76.95 per copy
Over 600 pages, illustrated, available Nov. 1997
Catalog No. 60188

Means Assemblies Cost Data 1998

Means Assemblies Cost Data 1998 takes the guesswork out of preliminary or conceptual estimates. Now you don't have to try to calculate the assembled cost by working up individual components costs. We've done all the work for you.

Presents detailed illustrations, descriptions, specifications and costs for every conceivable building assembly—240 types in all—arranged in the easy-to-use UniFormat system. Each illustrated "assembled" cost includes a complete grouping of materials and associated installation costs including the installing contractor's overhead and profit.

$139.95 per copy
Over 570 pages, illustrated, available Oct. 1997
Catalog No. 60068

Means Site Work & Landscape Cost Data 1998

Means Site Work & Landscape Cost Data 1998 is organized to assist you in all your estimating needs. Hundreds of fact-filled pages help you make accurate cost estimates efficiently.

New for 1998!

- New or expanded demolition features— ceilings, doors, electrical, flooring, HVAC, millwork, plumbing, roofing, walls and windows
- State-of-the-art segmental retaining walls
- Flywheel trenching costs and details
- Expanded Wells section
- Thousands of landscape materials, flowers, shrubs and trees

$89.95 per copy
Over 570 pages, illustrated, available Nov. 1997
Catalog No. 60288

Annual Cost Guides

Means Open Shop Building Construction Cost Data 1998

The latest costs for accurate budgeting and estimating of new commercial and residential construction... renovation work... change orders... cost engineering. *Means Open Shop BCCD* will assist you to...
· Develop benchmark prices for change orders
· Plug gaps in preliminary estimates, budgets
· Estimate complex projects
· Substantiate invoices on contracts
· Price ADA-related renovations

$86.95 per copy
Over 650 pages, illustrated, available Nov. 1997
Catalog No. 60158

Means Heavy Construction Cost Data 1998

A comprehensive guide to heavy construction costs. Includes costs for highly specialized projects such as tunnels, dams, highways, airports, and waterways. Information on different labor rates, equipment, and material costs is included. Has unit price costs, systems costs, and numerous reference tables for costs and design. Valuable not only to contractors and civil engineers, but also to government agencies and city/town engineers.

$89.95 per copy
Over 450 pages, illustrated, available Dec. 1997
Catalog No. 60168

Means Building Construction Cost Data 1998

Western Edition

This regional edition provides more precise cost information for western North America. Labor rates are based on union rates from 13 western states and western Canada. Included are western practices and materials not found in our national edition: tilt-up concrete walls, glu-lam structural systems, specialized timber construction, seismic restraints, landscape and irrigation systems.

$89.95 per copy
Over 600 pages, illustrated, available Nov. 1997
Catalog No. 60228

Means Heavy Construction Cost Data 1998

Metric Version

Make sure you have the Means industry standard metric costs for the federal, state, municipal and private marketplace. With thousands of up-to-date metric unit prices in tables by CSI standard divisions. Supplies you with assemblies costs using the metric standard for reliable cost projections in the design stage of your project. Helps you determine sizes, material amounts, and has tips for handling metric estimates.

$89.95 per copy
Over 450 pages, illustrated, available Dec. 1997
Catalog No. 63168

Means Construction Cost Indexes 1998

Who knows what 1998 holds? What materials and labor costs will change unexpectedly? By how much?
· Breakdowns for 305 major cities.
· National averages for 30 key cities.
· Expanded five major city indexes.
· Historical construction cost indexes.

$198.00 per year/$49.50 individual quarters
Catalog No. 60148

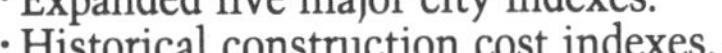

Means Interior Cost Data 1998

Provides you with prices and guidance needed to make accurate interior work estimates. Contains costs on materials, equipment, hardware, custom installations, furnishings, labor costs . . . every cost factor for new and remodel commercial and industrial interior construction, plus more than 50 reference tables. For contractors, facility managers, owners.

Newly expanded information on office furnishings.

$83.95 per copy
Over 550 pages, illustrated, available Nov. 1997
Catalog No. 60098

Means Concrete & Masonry Cost Data 1998

Provides you with cost facts for virtually all concrete/masonry estimating needs, from complicated form work to various sizes and face finishes of brick and block, all in great detail. The comprehensive unit cost section contains more than 15,000 selected entries. The assemblies cost section is illustrated with isometric drawings. A detailed reference section supplements the cost data.

$77.95 per copy
Over 520 pages, illustrated, available Dec. 1997
Catalog No. 60118

Means Labor Rates for the Construction Industry 1998

Complete information for estimating labor costs, making comparisons and negotiating wage rates by trade for over 300 cities (United States and Canada), 46 construction trades listed by local union number in each city, historical wage rates included for comparison. No similar book is available through the trade.

New: Each city chart now lists the county and is alphabetically arranged with handy visual flip tabs for quick reference.

$189.95 per copy
Over 330 pages, available Dec. 1997
Catalog No. 60128

Reference Books

Means Environmental Remediation Estimating Methods

By Richard R. Rast **NEW TITLE**

The first-ever guide to estimating any size environmental remediation project... anywhere in the country.

Field-tested guidelines for estimating 50 standard remediation technologies. This resource will help you: prepare preliminary budgets, develop detailed estimates, compare costs, select solutions, estimate liability, review quotes, and negotiate settlements.

A valuable support tool for the *Environmental Restoration* cost data books.

$99.95 per copy
Over 600 pages, illustrated, Hardcover
Catalog No. 64777

Cyberplaces: The Internet Guide for Architects, Engineers & Contractors

By Paul Doherty **NEW TITLE**

Internet applications for business and project management. Includes book, CD-ROM and Web site.

The CD-ROM offers built-in links to Web sites, tours of captured sites, FREE browser software and a document workshop, plus a test for Continuing Education Credits. **The Web Site** keeps the book current with updates on new technologies, interactive workshops, links to new tools and sites, and reports from top firms.

See Page 14 for full-page description

$59.95 per copy
Over 700 pages, illustrated, Softcover
Catalog No. 67317

Value Engineering: Practical Applications

...For Design, Construction, Maintenance & Operations **NEW TITLE**

By Alphonse Dell'Isola, P.E., the recognized leading authority on Value Engineering

A tool for immediate application—for engineers, architects, facility managers, owners and contractors. Includes: Making the Case for VE—The Management Briefing, Integrating VE into Planning and Budgeting, Conducting Life Cycle Costing, Integrating VE into the Design Process, and Using VE Methodology in Design Review and Consultant Selection, Case Studies (corporate, commercial, hospital, industrial and civil), and an expertly organized VE Workbook.

$79.95 per copy
Over 450 pages, illustrated, Hardcover
Catalog No. 67319

HVAC: Design Criteria, Options, Selection

Expanded Second Edition

By William H. Rowe III, AIA, PE

Now including Indoor Air Quality, CFC Removal, Energy Efficient Systems and Special Systems by Building Type.

The book that helps you solve a wide range of HVAC system design and selection problems effectively and economically. Gives you explanations of the latest ASHRAE standards.

$84.95 per copy
Over 600 pages, illustrated, Hardcover
Catalog No. 67306

The ADA in Practice

(Revised, expanded edition of the award-winning New ADA: Compliance & Costs)*

By Deborah S. Kearney, PhD

Helps you meet and budget for the requirements of the Americans with Disabilities Act. Shows you how to do the job right, by understanding what the law actually requires and allows. Includes an objective "authoritative buyers guide" for 70 ADA-compliant products with specs and purchasing information. With sample evaluation forms and illustrations of ADA-compliant products.

*Winner of the "Distinguished Author of the Year" award from the International Facility Managers Association.

$72.95 per copy
Over 600 pages, illustrated, Softcover
Catalog No. 67147A

Means ADA Compliance Pricing Guide

Gives you detailed cost estimates for budgeting modification projects, including estimates for each of 260 alternates. You get the 75 most commonly needed modifications for ADA compliance, each an assembly estimate with detailed cost breakdown including materials, labor hours and contractor's overhead. With 3,000 additional ADA compliance-related unit cost line items and costs easily adjusted to 900 cities and towns.

$72.95 per copy
Over 350 pages, illustrated, Softcover
Catalog No. 67310

Cost Planning & Estimating for Facilities Maintenance

In this unique book, a team of facilities management authorities shares their expertise at:
· Evaluating and budgeting operations
· Maintaining & repairing key building components
· Applying *Means Facilities Maintenance & Repair Cost Data to your estimating*

With the special maintenance requirements of 10 different building types.

$82.95 per copy
Over 475 pages, Hardcover
Catalog No. 67314

Means Facilities Maintenance Standards

Unique features of this one-of-a-kind working guide for facilities maintenance

A working encyclopedia that points the way to solutions to every kind of maintenance and repair dilemma. With a labor-hours section to provide productivity figures for over 180 maintenance tasks. Included are ready-to-use forms, checklists, worksheets and comparisons, as well as analysis of materials systems and remedies for deterioration and wear.

$159.95 per copy, 600 pages, 205 tables, checklists and diagrams, Hardcover
Catalog No. 67246

The Facilities Manager's Reference

By Harvey H. Kaiser, PhD

The tasks and tools the facility manager needs to accomplish the organization's objectives, and develop individual and staff skills. Includes Facilities and Property Management, Administrative Control, Planning and Operations, Support Services, and a complete building audit with forms and instructions, widely used by facilities managers nationwide.

$86.95 per copy, over 250 pages, with prototype forms and graphics, Hardcover
Catalog No. 67264

Understanding Building Automation Systems

· Direct Digital Control   · Energy Management
· Security/Access Control   · Life Safety
    · Lighting

By Reinhold A. Carlson, PE & Robert Di Giandomenico

The authors, leading authorities on the design and installation of these systems, describe the major building systems in both an overview with estimating and selection criteria, and in system configuration-level detail.

$79.95 per copy
Over 200 pages, illustrated, Hardcover
Catalog No. 67284

Facilities Planning & Relocation

By David D. Owen

An A–Z working guide—complete with the checklists, schematic diagrams and questionnaires to ensure the success of every office relocation. Complete with a step-by-step manual, 100-page technical reference section, over 50 reproducible forms in hard copy and on computer diskettes.
Winner of the International Facility Managers Assoc. "Distinguished Author of the Year" award.

$109.95 per copy, Textbook-384 pages, Forms Binder-137 pages, illustrated, Hardcover
Catalog No. 67301

Maintenance Management Audit

At a time when many companies and institutions are reorganizing or downsizing, this annual audit program is essential for every organization in need of a proper assessment of its maintenance operation. The forms presented in this easy-to-use workbook allow managers to identify and correct problems, enhance productivity, and impact the bottom line.

New reduced price

Now $32.48 per copy, limited quantity
125 pages, spiral bound, illustrated, Hardcover
Catalog No. 67299

Facilities Maintenance Management

By Gregory H. Magee, PE

Now you can get successful management methods and techniques for all aspects of facilities maintenance. This comprehensive reference explains and demonstrates successful management techniques for all aspects of maintenance, repair and improvements for buildings, machinery, equipment and grounds. Plus, guidance for outsourcing and managing internal staffs.

$86.95 per copy
Over 280 pages with illustrations, Hardcover
Catalog No. 67249

Planning and Managing Interior Projects

By Carol E. Farren

Expert, up-to-date guidance for managing interior installation projects. Includes: project phases; winning client support; space planning and design; budgeting, bidding and purchasing, and more.
For interior designers, architects, facilities professionals.

$76.95 per copy
Over 330 pages, illustrated, Hardcover
Catalog No. 67245

Reference Books

Basics For Builders:
Plan Reading & Material Takeoff
By Wayne J. DelPico

For Residential and Light Commercial Construction

A valuable tool for understanding plans and specs, and accurately calculating material quantities. Step-by-step instructions and takeoff procedures based on a full set of working drawings.

$35.95 per copy
Over 420 pages, Softcover
Catalog No. 67307

Means Illustrated Construction
Dictionary Unabridged Edition

Written in contractor's language, the information is adaptable for report writing, specifications or just intelligent discussion. Contains over 17,000 construction terms, words, phrases, acronyms and abbreviations, slang, regional terminology, and hundreds of illustrations.

$99.95 per copy
Over 700 pages, Hardcover
Catalog No. 67292

Superintending for Contractors:
How to Bring Jobs in On-time, On-Budget
By Paul J. Cook

This book examines the complex role of the superintendent/field project manager, and provides guidelines for the efficient organization of this job. Includes administration of contracts, change orders, purchase orders, and more.

$35.95 per copy
Over 220 pages, illustrated, Softcover
Catalog No. 67233

Means Forms for Contractors
For general and specialty contractors

Means editors have created and collected the most needed forms as requested by contractors of various-sized firms and specialties. This book covers all project phases. Includes sample correspondence and personnel administration.
Full-size forms on durable paper for photocopying or reprinting.

$79.95 per copy
Over 400 pages, three-ring binder, more than 80 forms
Catalog No. 67288

Bidding for Contractors:
How to Make Bids that Make Money
By Paul J. Cook

The author shares the benefit of his more than 30 years of experience in construction project management, providing contractors with the tools they need to develop competitive bids.

New reduced price

Now $17.98 per copy, limited quantity
Over 225 pages with graphics, Softcover
Catalog No. 67180

Means Heavy Construction
Handbook

Informed guidance for planning, estimating and performing today's heavy construction projects. Provides expert advice on every aspect of heavy construction work including hazardous waste remediation and estimating. To assist planning, estimating, performing, or overseeing work.

$74.95 per copy
Over 430 pages, illustrated, Hardcover
Catalog No. 67148

Estimating for Contractors:
How to Make Estimates that Win Jobs
By Paul J. Cook

Estimating for Contractors is a reference that will be used over and over, whether to check a specific estimating procedure, or to take a complete course in estimating.

$35.95 per copy
Over 225 pages, illustrated, Softcover
Catalog No. 67160

HVAC Systems Evaluation
By Harold R. Colen, PE

You get direct comparisons of how each type of system works, with the relative costs of installation, operation, maintenance and applications by type of building. With requirements for hooking up electrical power to HVAC components. Contains experienced advice for repairing operational problems in existing HVAC systems, ductwork, fans, cooling coils, and much more!

$84.95 per copy
Over 500 pages, illustrated, Hardcover
Catalog No. 67281

Quantity Takeoff for Contractors:
How to Get Accurate Material Counts
By Paul J. Cook

Contractors who are new to material takeoffs or want to be sure they are using the best techniques will find helpful information in this book, organized by CSI MasterFormat division.

New reduced price

Now $17.48 per copy, limited quantity
Over 250 pages, illustrated, Softcover
Catalog No. 67262

Roofing: Design Criteria,
Options, Selection

By R.D. Herbert, III

This book is required reading for those who specify, install or have to maintain roofing systems. It covers all types of roofing technology and systems. You'll get the facts needed to intelligently evaluate and select both traditional and new roofing systems.

$62.95 per copy
Over 225 pages with illustrations, Hardcover
Catalog No. 67253

Means Estimating Handbook

This comprehensive reference covers a full spectrum of technical data for estimating, with information on sizing, productivity, equipment requirements, codes, design standards and engineering factors.
Means Estimating Handbook will help you: evaluate architectural plans and specifications, prepare accurate quantity takeoffs, prepare estimates from conceptual to detailed, and evaluate change orders.

$99.95 per copy
Over 900 pages, Hardcover
Catalog No. 67276

Means Repair and Remodeling Estimating Third Edition

By Edward B. Wetherill & R.S. Means

Focuses on the unique problems of estimating renovations of existing structures. It helps you determine the true costs of remodeling through careful evaluation of architectural details and a site visit.
New section on disaster restoration costs.

$69.95 per copy
Over 450 pages, illustrated, Hardcover
Catalog No. 67265A

Successful Estimating Methods:

From Concept to Bid

By John D. Bledsoe, PhD, PE

A highly practical, all-in-one guide to the tips and practices of today's successful estimator. Presents techniques for all types of estimates, *and* advanced topics such as life cycle cost analysis, value engineering, and automated estimating.
Estimate spreadsheets available at Means Web site.

$64.95 per copy
Over 300 pages, illustrated, Hardcover
Catalog No. 67287

Means Productivity Standards for Construction

Expanded Edition *(Formerly* Man-Hour Standards)

Here is the working encyclopedia of labor productivity information for construction professionals, with labor requirements for thousands of construction functions in CSI MasterFormat.
Completely updated, with over 3,000 new work items.

$159.95 per copy
Over 800 pages, Hardcover
Catalog No. 67236A

Means Electrical Estimating Methods Second Edition

Expanded version includes sample estimates and cost information in keeping with the latest version of the CSI MasterFormat. Contains new coverage of Fiber Optic and Uninterruptible Power Supply electrical systems, broken down by components and explained in detail. A practical companion to *Means Electrical Cost Data.*

$62.95 per copy
Over 325 pages, Hardcover
Catalog No. 67230A

Means Mechanical Estimating

Second Edition

This guide assists you in making a review of plans, specs and bid packages with suggestions for takeoff procedures, listings, substitutions and pre-bid scheduling. Includes suggestions for budgeting labor and equipment usage. Compares materials and construction methods to allow you to select the best options for your job.

$64.95 per copy
Over 350 pages, illustrated, Hardcover
Catalog No. 67294

Means Scheduling Manual

Third Edition

By F. William Horsley

Fast, convenient expertise for keeping your scheduling skills right in step with today's cost-conscious times. Covers bar charts, PERT, precedence and CPM scheduling methods. Now updated to include computer applications.

$62.95 per copy
Over 200 pages, spiral-bound, Softcover
Catalog No. 67291

Means Graphic Construction Standards

Means Graphic Construction Standards bridges the gap between design and actual construction methods. With illustrations of unit assemblies, systems and components, you can see quickly which construction methods work best to meet design, budget and time objectives.

$124.95 per copy
Over 540 pages, illustrated, Hardcover
Catalog No. 67210

Means Square Foot Estimating Methods Second Edition

By Billy J. Cox and F. William Horsley

Proven techniques for conceptual and design-stage cost planning. Steps you through the square foot cost process, demonstrating faster, better ways to relate the design to the budget. Now updated to the latest version of UniFormat.

$69.95 per copy
Over 300 pages, illustrated, Hardcover
Catalog No. 67145A

The Building Professional's Guide to Contract Documents

By Waller S. Poage, AIA, CSI, CCS

Includes latest changes to AIA prototype contracts. Directions for preparing specifications and technical requirements . . . description writing, proprietary and performance specs, standards. Guidance on owner, designer and contractor liability.

$64.95 per copy
Over 400 pages, illustrated, Hardcover
Catalog No. 67261

Reference Books

Unit Price Estimating Methods
2nd Edition

$59.95 per copy
Catalog No. 67303

Legal Reference for Design & Construction
By Charles R. Heuer, Esq., AIA

$109.95, Now $54.98 per copy, limited quantity
Catalog No. 67266

Plumbing Estimating
By Joseph J. Galeno & Sheldon T. Greene

$59.95 per copy
Catalog No. 67283

Structural Steel Estimating
By S. Paul Bunea, PhD

$79.95 per copy
Catalog No. 67241

Landscape Estimating
2nd Edition

By Sylvia H. Fee

$64.95 per copy
Catalog No. 67295

Business Management for Contractors
How to Make Profits in Today's Market

By Paul J. Cook

$35.95, Now $17.98 per copy, limited quantity
Catalog No. 67250

Understanding Legal Aspects of Design/Build
By Timothy R. Twomey, Esq., AIA

$79.95 per copy
Catalog No. 67259

Construction Paperwork
An Efficient Management System

By J. Edward Grimes

$52.95 per copy
Catalog No. 67268

Contractor's Business Handbook
By Michael S. Milliner

$42.95, Now $21.48 per copy, limited quantity
Catalog No. 67255

Basics for Builders: How to Survive and Prosper in Construction
By Thomas N. Frisby

$34.95 per copy
Catalog No. 67273

Successful Interior Projects Through Effective Contract Documents
By Joel Downey & Patricia K. Gilbert

$69.95 per copy
Catalog No. 67313

Project Planning & Control for Construction
By David R. Pierce, Jr.

$64.95 per copy
Catalog No. 67247

Illustrated Construction Dictionary, Condensed

$59.95 per copy
Catalog No. 67282

Managing Construction Purchasing
By John G. McConville, CCC, CPE

$62.95, Now $31.48 per copy, limited quantity
Catalog No. 67302

Hazardous Material & Hazardous Waste
By Francis J. Hopcroft, PE, David L. Vitale, M. Ed., & Donald L. Anglehart, Esq.

$89.95, Now $44.98 per copy, limited quantity
Catalog No. 67258

Fundamentals of the Construction Process
By Kweku K. Bentil, AIC

$69.95, Now $34.98 per copy, limited quantity
Catalog No. 67260

Construction Delays
By Theodore J. Trauner, Jr., PE, PP

$69.95 per copy
Catalog No. 67278

Basics for Builders: Framing & Rough Carpentry
By Scot Simpson

$24.95 per copy
Catalog No. 67298

Interior Home Improvement Costs
5th Edition

$19.95 per copy
Catalog No. 67308A

Exterior Home Improvement Costs
5th Edition

$19.95 per copy
Catalog No. 67309A

Concrete Repair and Maintenance Illustrated
By Peter H. Emmons

$64.95 per copy
Catalog No. 67146

How to Estimate with Metric Units

$49.95, Now $24.98 per copy, limited quantity
Catalog No. 67304

How to Develop Facility Assessment Programs

This two-day program concentrates on the management process required for planning, conducting, and documenting the physical condition and functional adequacy of buildings and other facilities. Regular facilities condition inspections are one of the facility department's most important duties. However, gathering reliable data hinges on the design of the overall program used to identify and gauge deferred maintenance requirements. Knowing where to look . . . and reporting results effectively are the keys. This seminar is designed to give the facility executive essential steps for conducting facilities inspection programs.

Inspection Program Requirements • Where is the deficiency? • What is the nature of the problem? • How can it be remedied? • How much will it cost in labor, equipment and materials? • When should it be accomplished? • Who is best suited to do the work?

Note: Because of its management focus, this course will not address trade practices and procedures.

Repair and Remodeling Estimating

Repair and remodeling work is becoming increasingly competitive as more professionals enter the market. Recycling existing buildings can pose difficult estimating problems. Labor costs, energy use concerns, building codes, and the limitations of working with an existing structure place enormous importance on the development of accurate estimates. Using the exclusive techniques associated with Means' widely acclaimed **Repair & Remodeling Cost Data**, this seminar sorts out and discusses solutions to the problems of building alteration estimating. Attendees will receive two intensive days of eye-opening methods for handling virtually every kind of repair and remodeling situation . . . from demolition and removal to final restoration.

Mechanical and Electrical Estimating

This seminar is tailored to fit the needs of those seeking to develop or improve their skills and to have a better understanding of how mechanical and electrical estimates are prepared during the conceptual, planning, budgeting and bidding stages. Learn how to avoid costly omissions and overlaps between these two interrelated specialties by preparing complete and thorough cost estimates for both trades. Featured are order of magnitude, assemblies, and unit price estimating. In combination with the use of **Means Mechanical Cost Data**, **Means Plumbing Cost Data** and **Means Electrical Cost Data**, this seminar will ensure more accurate and complete Mechanical/Electrical estimates for both unit price and preliminary estimating procedures.

Unit Price Estimating

This seminar shows how today's advanced estimating techniques and cost information sources can be used to develop more reliable unit price estimates for projects of any size. It demonstrates how to organize data, use plans efficiently, and avoid embarrassing errors by using better methods of checking.

You'll get down-to-earth help and easy-to-apply guidance for:
- making maximum use of construction cost information sources
- organizing estimating procedures in order to save time and reduce mistakes
- sorting out and identifying unusual job requirements to improve estimating accuracy.

Square Foot Cost Estimating

Learn how to make better preliminary estimates with a limited amount of budget and design information. You will benefit from examples of a wide range of systems estimates with specifications limited to building use requirements, budget, building codes, and type of building. And yet, with minimal information, you will obtain a remarkable degree of accuracy.

Workshop sessions will provide you with model square foot estimating problems and other skill-building exercises. The exclusive Means building assemblies square foot cost approach shows how to make very reliable estimates using "bare bones" budget and design information.

Scheduling and Project Management

This seminar helps you successfully establish project priorities, develop realistic schedules, and apply today's advanced management techniques to your construction projects. Hands-on exercises familiarize participants with network approaches such as the Critical Path Method. Special emphasis is placed on cost control, including use of computer-based systems. Through this seminar you'll perfect your scheduling and management skills, ensuring completion of your projects *on time* and *within budget*. Includes hands-on application of **Means Scheduling Manual** and **Means Building Construction Cost Data.**

Facilities Maintenance and Repair Estimating

With our Facilities Maintenance and Repair Estimating seminar, you'll learn how to plan, budget, and estimate the cost of ongoing and preventive maintenance and repair for all your buildings and grounds. Based on R.S. Means' groundbreaking cost estimating book, this two-day seminar will show you how to decide to either contract out or retain maintenance and repair work in-house. In addition, you'll learn how to prepare budgets and schedules that help cut down on unplanned and costly emergency repair projects. Facilities Maintenance and Repair Estimating crystallizes what facilities professionals have learned over the years, but never had time to organize or document. This program covers a variety of maintenance and repair projects, from underground storage tank removal, roof repair and maintenance, exterior wall renovations, and energy source conversions, to service upgrades and estimating energy-saving alternatives.

Managing Facilities Construction and Maintenance

In you're involved in new facility construction, renovation or maintenance projects and are concerned about getting quality work done on time and on or below budget, in Means' seminar **Managing Facilities Construction and Maintenance** you'll learn management techniques needed to effectively plan, organize, control and get the most out of your limited facilities resources.

Learn how to develop budgets, reduce expenditures and check productivity. With the knowledge gained in this course, you'll be better prepared to successfully sell accurate project budgets, timing and manpower needs to senior management . . . plus understand how to evaluate the impact of today's facility decisions on tomorrow's budget.

Call 1-800-448-8182 for more information

Seminars

1998 Means Seminar Schedule

Location	Dates
Las Vegas, NV	March 9-12
Washington, DC	April 20-23
Denver, CO	May 4-7
Los Angeles, CA	May 18-21
San Francisco, CA	June 1-4
New Orleans, LA	June 15-18
Washington, DC	September 14-17
Orlando, FL	November 16-19

Registration Information

How to Register Register by phone today! Means toll-free number for making reservations is: **1-800-448-8182**

Individual Seminar Registration Fee $845 To register by mail, complete the registration form and return with your full fee (or minimum deposit of $300/person for one 2-day seminar, or minimum deposit of $475/person for two 2-day seminars) to: Seminar Division, R.S. Means Company, Inc., 100 Construction Plaza, P.O. Box 800, Kingston, MA 02364-0800.

Federal Government Pricing All federal government employees save 25% off regular seminar price. Other promotional discounts cannot be combined with Federal Government discount.

Team Discount Program Two to four seminar registrations: $745 per person—Five or more seminar registrations: $695 per person—Ten or more seminar registrations: Call for pricing.

Consecutive Seminar Offer One individual signing up for two separate courses at the same location during the designated time period pays only $1,345. You get the second course for only $500 (**a 40% discount**). Payment must be received at least ten days prior to seminar dates to confirm attendance.

Refunds Cancellations will be accepted up to ten days prior to the seminar start. There are no refunds for cancellations postmarked later than ten working days prior to the first day of the seminar. A $150 processing fee will be charged for all cancellations. Written notice or telegram is required for all cancellations. Substitutions can be made at any time before the session starts. **No-shows are subject to the full seminar fee.**

AIA Continuing Education R.S. Means is registered with the AIA Continuing Education System (AIA/CES) and is committed to developing quality learning activities in accordance with the CES criteria. R.S. Means seminars meet the AIA/CES criteria for Quality Level 2. AIA members will receive (28) learning units (LUs) for each two day R.S. Means Course.

Daily Course Schedule The first day of each seminar session begins at 8:30 A.M. and ends at 4:30 P.M. The second day is 8:00 A.M.–4:00 P.M. Participants are urged to bring a hand-held calculator since many actual problems will be worked out in each session.

Continental Breakfast Your registration includes the cost of a continental breakfast, a morning coffee break, and an afternoon cola break. These informal segments will allow you to discuss topics of mutual interest with other members of the seminar. (You are free to make your own lunch and dinner arrangements.)

Hotel/Transportation Arrangements R.S. Means has arranged to hold a block of rooms at each hotel hosting a seminar. To take advantage of special group rates when making your reservation be sure to mention that you are attending the Means Seminar. You are of course free to stay at the lodging place of your choice. (**Hotel reservations and transportation arrangements should be made directly by seminar attendees.**)

Important Class sizes are limited, so please register as soon as possible.

Registration Form   Call 1-800-448-8182, ext. 701 to register or FAX 1-617-585-7466

Please register the following people for the Means Construction Seminars as shown here. Full payment or deposit is enclosed, and we understand that we must make our own hotel reservations if overnight stays are necessary.

☐ Full payment of $ _______________ enclosed.

☐ Deposit of $ _______________ enclosed.

Balance Due is $ _______________

U.S. Funds

Name of Registrant(s)
(To appear on certificate of completion)

____

____

____

Firm Name _______________________________________

Address _______________________________________

City/State/Zip _______________________________________

Telephone No. _______________ Fax No. _______________

E-Mail Address _______________________________________

Charge our registration(s) to: ☐ MasterCard ☐ VISA ☐ American Express ☐ Discover

Account No. _______________ Exp. Date _______________

Cardholder's Signature _______________________________________

Seminar Name	City	Dates

Please mail check to: R.S. MEANS COMPANY, INC., 100 Construction Plaza, P.O. Box 800, Kingston, MA 02364-0800 USA

Consulting Services Group

Proven Solutions for Managing the Costs of Construction and Facility Operations

Business and government leaders go to great lengths today to control costs and increase return on their construction-related activities. But they seldom have an opportunity to achieve optimum success. No matter what the activity . . . planning and design, new construction, facilities management, or introducing new building technologies . . . it comes down to the same problem: To control costs one must accurately predict them. R.S. Means Consulting Services Group has a proven record of success helping businesses meet this challenge while greatly increasing opportunity to maximize return on construction investments. With its extensive, highly specialized experience understanding both construction costs and relational database technology, Means Consulting Services Group offers an unparalleled opportunity for businesses to realize dramatic improvement in their construction/facilities cost control and valuation programs.

Database Management Solutions

- **Custom Database Development**—Means expertise in construction cost engineering and database management can be put to work creating customized cost databases and applications.
- **Data Licensing & Integration**—To enhance applications dealing with construction, any segment of Means vast database can be licensed for use, and harnessed in a format compatible with a previously developed proprietary system.
- **Cost Modeling**—Pre-built custom cost models provide organizations that expend countless hours estimating repetitive work with a systematic time-saving estimating solution.
- **Database Auditing & Maintenance**—For clients with in-house data, Means can help organize it, and by linking it with Means database, fill in any gaps that exist and provide necessary updates to maintain current and relevant proprietary cost data.

Cost Planning Solutions

- **Estimating Service**—Means expertise is available to perform construction cost estimates, as well as to develop baseline schedules and establish management plans for projects of all sizes and types. Conceptual, budget and detailed estimates are available.
- **Benchmarking**—Means can run baseline estimates on existing project estimates. Gauging estimating accuracy and identifying inefficiencies improves the success ratio, precision and productivity of estimates.
- **Project Feasibility Studies**—The Consulting Services Group can assist in the review and clarification of the most sound and practical construction approach in terms of time, cost and use.
- **Professional Review/Expert Witness**—Means is available to provide opinions of value and to supply expert interpretations or testimony in the resolution of construction cost claims, litigation and mediation.

Training and Development Solutions

- **Core Curriculum**—Means educational programs, delivered on-site, are designed to sharpen professional skills and to maximize effective use of cost estimating and management tools. On-site training cuts down on travel expenses and time away from the office. The broad curriculum covers such topics as repair, new construction, and conceptual estimating; scheduling and project management; facilities management; metrication; and delivery order contracting (DOC) methods.
- **Custom Curriculum**—Means can custom-tailor courses to meet the specific needs and requirements of clients. The goal is to simultaneously boost skills and broaden cost estimating and management knowledge while focusing on applications that bring immediate benefits to unique operations, challenges, or markets.
- **Development Programs**—In addition to custom curricula, Means can work with a client's Human Resources Department or with individual operating units to create programs consistent with long-term employee development objectives.

1998 Order Form

Qty.	Book No.	COST ESTIMATING BOOKS	Unit Price	Total
	60068	Assemblies Cost Data 1998	$139.95	
	60018	Building Construction Cost Data 1998	84.95	
	61018	Building Const. Cost Data–Looseleaf Ed. 1998	109.95	
	63018	Building Const. Cost Data–Metric Version 1998	89.95	
	60228	Building Const. Cost Data–Western Ed. 1998	89.95	
	60118	Concrete & Masonry Cost Data 1998	77.95	
	50140	Construction Cost Indexes 1998	198.00	
	60148A	Construction Cost Index–January 1998	49.50	
	60148B	Construction Cost Index–April 1998	49.50	
	60148C	Construction Cost Index–July 1998	49.50	
	60148D	Construction Cost Index–October 1998	49.50	
	60318	Contr. Pricing Guide: Framing/Carpentry	34.95	
	60338	Contr. Pricing Guide: Resid. Detailed	36.95	
	60328	Contr. Pricing Guide: Resid. Sq. Ft.	39.95	
	64028	ECHOS Assemblies Cost Book 1998	149.95	
	64018	ECHOS Unit Cost Book 1998	99.95	
	60238	Electrical Change Order Cost Data 1998	89.95	
	60038	Electrical Cost Data 1998	87.95	
	60208	Facilities Construction Cost Data 1998	209.95	
	60308	Facilities Maintenance & Repair Cost Data 1998	199.95	
	60168	Heavy Construction Cost Data 1998	89.95	
	63168	Heavy Const. Cost Data–Metric Version 1998	89.95	
	60098	Interior Cost Data 1998	83.95	
	60128	Labor Rates for the Const. Industry 1998	189.95	
	60188	Light Commercial Cost Data 1998	76.95	
	60028	Mechanical Cost Data 1998	87.95	
	60158	Open Shop Building Const. Cost Data 1998	86.95	
	60218	Plumbing Cost Data 1998	87.95	
	60048	Repair and Remodeling Cost Data 1998	79.95	
	60178	Residential Cost Data 1998	74.95	
	60288	Site Work & Landscape Cost Data 1998	89.95	
	60058	Square Foot Costs 1998	97.95	
		REFERENCE BOOKS		
	67147A	ADA in Practice	72.95	
	67310	ADA Pricing Guide	72.95	
	67298	Basics for Builders: Framing & Rough Carpentry	24.95	
	67273	Basics for Builders: How to Survive and Prosper	34.95	
	67307	Basics for Builders: Plan Reading & Takeoff	35.95	
	67180	Bidding for Contractors	17.98	
	67261	Building Profess. Guide to Contract Documents	64.95	
	67250	Business Management for Contractors	17.98	
	67146	Concrete Repair & Maintenance Illustrated	64.95	
	67278	Construction Delays	69.95	
	67268	Construction Paperwork	52.95	
	67255	Contractor's Business Handbook	21.48	
	67314	Cost Planning & Est. for Facil. Maint.	82.95	
	67317	Cyberplaces: The Internet Guide for A/E/C	59.95	
	67230A	Electrical Estimating Methods–2nd Ed.	62.95	
	64777	Environmental Remediation Est. Methods	99.95	
	67160	Estimating for Contractors	35.95	
	67276	Estimating Handbook	99.95	
	67249	Facilities Maintenance Management	86.95	

Qty.	Book No.	REFERENCE BOOKS (Con't)	Unit Price	Total
	67246	Facilities Maintenance Standards	159.95	
	67264	Facilities Manager's Reference	86.95	
	67301	Facilities Planning & Relocation	109.95	
	67288	Forms for Contractors	79.95	
	67260	Fundamentals of the Construction Process	34.98	
	67210	Graphic Construction Standards	124.95	
	67258	Hazardous Material & Hazardous Waste	44.98	
	67148	Heavy Construction Handbook	74.95	
	67308A	Home Improvement Costs–Interior Projects	19.95	
	67309A	Home Improvement Costs–Exterior Projects	19.95	
	67304	How to Estimate with Metric Units	24.98	
	67306	HVAC: Design Criteria, Options, Select.–2nd Ed.	84.95	
	67281	HVAC Systems Evaluation	84.95	
	67282	Illustrated Construction Dictionary, Condensed	59.95	
	67292	Illustrated Construction Dictionary, Unabridged	99.95	
	67295	Landscape Estimating–2nd Ed.	64.95	
	67266	Legal Reference for Design & Construction	54.98	
	67299	Maintenance Management Audit	32.48	
	67302	Managing Construction Purchasing	31.48	
	67294	Mechanical Estimating–2nd Ed.	64.95	
	67245	Planning and Managing Interior Projects	76.95	
	67283	Plumbing Estimating Methods	59.95	
	67236A	Productivity Standards for Constr.–3rd Ed.	159.95	
	67247	Project Planning & Control	64.95	
	67262	Quantity Takeoff for Contractors	17.48	
	67265A	Repair & Remodeling Estimating–3rd Ed.	69.95	
	67253	Roofing: Design Criteria, Options, Selection	62.95	
	67291	Scheduling Manual–3rd Ed.	62.95	
	67145A	Square Foot Estimating Methods–2nd Ed.	69.95	
	67241	Structural Steel Estimating	79.95	
	67287	Successful Estimating Methods	64.95	
	67313	Successful Interior Projects	69.95	
	67233	Superintending for Contractors	35.95	
	67284	Understanding Building Automation Systems	79.95	
	67259	Understanding Legal Aspects of Design/Build	79.95	
	67303	Unit Price Estimating Methods–2nd Ed.	59.95	
	67319	Value Engineering: Practical Applications	79.95	

MA residents add 5% state sales tax	
Shipping & Handling**	
Total (U.S. Funds)*	

Prices are subject to change and are for U.S. delivery only. *Canadian customers may call for current prices. **Shipping & handling charges: Add 6.5% of total order for check and credit card payments. Add 9% of total order for invoiced orders.

Send Order To: ADDV-1001

Name (Please Print) _______________________

Company _______________________

☐ **Company**
☐ **Home**   Address _______________________

____

City/State/Zip _______________________

Phone # _________________ P.O. # _______________________

(Must accompany all orders being billed)

Mail To: **R.S. Means Company, Inc.**, P.O. Box 800, Kingston, MA 02364-0800

MeansData™

CONSTRUCTION COSTS FOR SOFTWARE APPLICATIONS
Your construction estimating software is only as good as your cost data.

Software Integration

A proven construction cost database is a mandatory part of any estimating package. We have linked MeansData™ directly into the industry's leading software applications. The following list of software providers can offer you MeansData™ as an added feature for their estimating systems. Visit them on-line at *http://www.rsmeans.com/demo/* for more information and free demos. Or call their numbers listed below.

AEC DATA SYSTEMS
1-800-659-9001

ARMSTRONG & ASSOCIATES
Reserve Study Systems
1-808-263-7732

BSD
Building System Design
1-800-875-0047

CDCI
Construction Data Controls, Inc.
1-800-285-3929

CMS
Computerized Micro Solutions
1-800-255-7407

CONAC GROUP
1-604-273-3463

CONSTRUCTIVE COMPUTING, Inc.
1-800-456-2113

DATAQUIRE
1-401-253-8969

EAGLE POINT
1-800-678-6565

ESTIMATING SYSTEMS, Inc.
1-800-967-8572

G2, Inc.
1-800-657-6312

GEAC COMMERCIAL SYSTEMS, Inc.
1-800-851-1115

GRANTLUN CORPORATION
1-602-897-7750

IQ BENECO
1-801-565-1122

MC^2
Management Computer Controls
1-800-225-5622

PRISM COMPUTER CORPORATION
Facility Management Software
1-800-774-7622

PDA CONSTRUCTION
Handheld Cost Estimating
1-909-653-5878

SANDERS SOFTWARE, Inc.
1-404-934-8423

STN, Inc.
Workline Maintenance Systems
1-800-321-1969

TIMBERLINE SOFTWARE CORP.
1-800-628-6583

TMA SYSTEMS, Inc.
Facility Management Software
1-918-494-2890

US COST, Inc.
1-800-955-1385

VERTIGRAPH, Inc.
1-800-989-4243

WINESTIMATOR, Inc.
1-800-950-2374

DemoSource™ | **One-stop shopping for the latest cost estimating software for just $19.95.** This evaluation tool includes product literature and demo diskettes for ten or more estimating systems, all of which link to MeansData™. **Call 1-800-334-3509 to order.**

FOR MORE INFORMATION ON ELECTRONIC PRODUCTS CALL
1-800-448-8182 OR FAX 1-800-632-6732.

MeansData™ is a registered trademark of R.S. Means Co., Inc., *A **Construction Market Data Group** Company.*